Contemporary
Linear Algebra

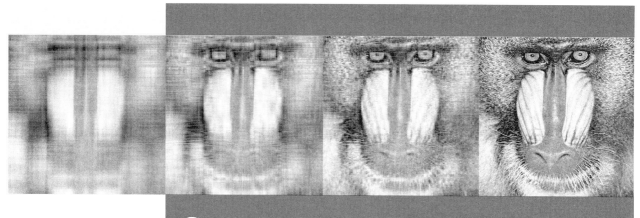

Contemporary
Linear Algebra

Howard Anton
Drexel University

Robert C. Busby
Drexel University

John Wiley & Sons, Inc.

ACQUISITIONS EDITOR	Laurie Rosatone
MARKETING MANAGER	Julie Z. Lindstrom
SENIOR PRODUCTION EDITOR	Ken Santor
PHOTO EDITOR	Sara Wight
COVER DESIGN	Madelyn Lesure
ILLUSTRATION STUDIO	ATI Illustration Services

This book was set in Times Roman by Techsetters, Inc. and printed and bound by Von Hoffman Corporation. The cover was printed by Phoenix Color Corporation.

This book is printed on acid-free paper. ∞

ISBN 0-471-16362-7

Printed in the United States of America

10 9 8 7 6 5 4 3 2 1

Howard Anton obtained his B.A. from Lehigh University, his M.A. from the University of Illinois, and his Ph.D. from the Polytechnic University of Brooklyn, all in mathematics. In the early 1960s he worked for Burroughs Corporation and Avco Corporation at Cape Canaveral, Florida (now the Kennedy Space Center), on mathematical problems related to the manned space program. In 1968 he joined the Mathematics Department at Drexel University, where he taught full time until 1983. Since then he has been an adjunct professor at Drexel and has devoted the majority of his time to textbook writing and projects for mathematical associations. He was president of the Eastern Pennsylvania and Delaware section of the Mathematical Association of America (MAA), served on the Board of Governors of the MAA, and guided the creation of its Student Chapters. He has published numerous research papers in functional analysis, approximation theory, and topology, as well as various pedagogical papers. He is best known for his textbooks in mathematics, which have been widely used for more than thirty years. There are currently more than 125 versions of his books used throughout the world, including translations into Spanish, Arabic, Portuguese, Italian, Indonesian, French, Japanese, Chinese, Hebrew, and German. In 1994 he was awarded the Textbook Excellence Award by the Textbook Authors Association. For relaxation, Dr. Anton enjoys traveling, photography, and art.

Robert C. Busby obtained his B.S. in physics from Drexel University and his M.A. and Ph.D. in mathematics from the University of Pennsylvania. He taught at Oakland University in Rochester, Michigan, and since 1969 has taught full time at Drexel University, where he currently holds the position of Professor in the Department of Mathematics and Computer Science. He has regularly taught courses in calculus, linear algebra, probability and statistics, and modern analysis. Dr. Busby is the author of numerous research articles in functional analysis, representation theory, and operator algebras, and he has coauthored an undergraduate text in discrete mathematical structures and a workbook on the use of *Maple* in calculus. His current professional interests include aspects of signal processing and the use of computer technology in undergraduate education. Professor Busby also enjoys contemporary jazz and computer graphic design. He and his wife, Patricia, have two sons, Robert and Scott.

To:

My wife, Pat

My children, Brian, David, and Lauren

My parents, Shirley and Benjamin

My benefactor, Stephen Girard (1750–1831),
    whose philanthropy changed my life

Howard Anton

To Patricia, my wife, partner, and best friend

Robert C. Busby

This book is for sophomore- and junior-level students of mathematics and such client disciplines as computer science, engineering, chemistry, biology, economics, actuarial science, and business. The challenge in writing a contemporary linear algebra text is to focus on those aspects of the subject that *are most likely to have practical value to the student,* but at the same time *not compromise the intrinsic mathematical form of the subject.* Linear algebra is special in that its key ideas are embodied in the theorems and their proofs as well as in its applications and the problem-solving techniques. Thus, a contemporary linear algebra text must foster mathematical thinking, problem-solving abilities, and exposure to real-world applications. This is our goal.

Anybody who has taught this course is well aware of its pedagogical challenges: the number of important concepts is great, making it difficult to cover them all in one semester, some of the ideas are intrinsically difficult and require thoughtful and skillful exposition to make them clear, and finally, many students struggle so much with the axiomatic formulation of general vector spaces that it impedes their mastery of basic concepts. As you look through the following list of features and peruse the book, you will see that we have gone to great lengths to address these problems. We are hopeful that this book will live up to the promise of the word "contemporary" in its title and that you will find it refreshing and exciting.

The following is a list of key features:

Linear Algebra Curriculum Study Group This text meets the guidelines of the Linear Algebra Curriculum Study Group (LACSG).

Primary Focus Is on R^n We believe that a working knowledge of vectors in R^n and some experience with viewing functions as vectors is an adequate level of generalization for the majority of students taking this course. Accordingly, this text focuses primarily on R^n. The material on axiomatic vector spaces has been placed toward the latter part of the text, thereby allowing this generalization to evolve naturally (as it did historically) and avoiding the "wall" of abstraction that the majority of students find so troublesome. However, that material has been structured so it too can be covered earlier for those who prefer that approach.

Geometry In order to make a successful transition from two and three dimensions to R^n, the student must have a good understanding of the geometry of lines and planes, particularly vector and parametric representations of these objects. There is a strong geometric emphasis throughout the text, thereby providing students a good geometric foundation on which they can build.

Why Vectors Precede Linear Systems There are many reasonable ways to order the introductory topics in a linear algebra text—starting with linear systems, matrices, or vectors are all reasonable possibilities. We have chosen to start with vectors and then proceed to linear systems, since this allows us to interpret parametric solutions of linear systems as geometric objects, such as points, lines, and planes, rather than amorphous sets. This is in keeping with our philosophy of providing a strong geometric foundation for ideas.

Pace Anyone who has taught linear algebra is well aware that it is difficult to cover all of the important topics in one semester, so the pace of a linear algebra text is very important. Linear algebra courses typically range from 28 lectures to 40+ lectures, so we have designed the text to accommodate this range (see the Syllabus Guide below).

Early Introduction of Fundamental Concepts All of the major concepts are introduced early and revisited later in more depth. For example, by the end of Chapter 1 the student will have encountered notions of R^n, orthogonality, linear combinations, spanning, subspaces, linear independence, and dimension at a basic level. By the end of Chapter 4 (roughly the 15th lecture) he or she will have a first view of eigenvalues. This "spiral" approach to concept development ensures that all key topics can be covered in the course and makes it easier for the student to deal with these ideas when they are studied in more depth at a later time.

Conceptual Understanding It is in the nature of the material that certain concepts, which seem innocuous or without purpose at the time they are introduced, assume great importance as the course unfolds. Accordingly, we have employed an element, called *Looking Ahead,* whose purpose is to provide the student some insight into the future role of the material currently being studied (see, p. 9, for example).

Applications A wide range of applications appears throughout the text. They have been chosen and written to give the student a sense of the broad range of applicability of linear algebra without letting the number or complexity of applications interfere with the clarity or logical presentation of the mathematics. In keeping with the word "contemporary" in the title of this text, you will find applications to such modern topics as global positioning and Internet search procedures.

Mathematical Rigor Theorems and proofs are presented precisely, but in a style that is appropriate for beginning students.

Numerical Matters This is *not* intended to be a book on numerical methods of linear algebra. It is our goal to alert the student to numerical issues, but not to discuss them in detail.

Mastering Points of View A successful course in linear algebra should impart to the student the ability to view mathematical objects and structures in a variety of ways —for example, by the end of the course the student should be able to think of a matrix from the viewpoint of a linear system, a coordinate change, or a linear transformation. Considerable effort has been devoted to teaching the student how to change points of view.

Student Involvement A common complaint of instructors is that students don't read the text. In an attempt to address this issue we have created an element called *Concept Problem* in which the student is asked to answer a question about the material he or she has just read. To keep the students involved with the text, instructors may want to devote a few minutes of class time to discussing these questions.

Exercises Each exercise set is divided into four categories: *basic exercises, Discussion and Discovery, Working with Proofs*, and *Technology Exercises*.

- The *basic exercises* progress in level of difficulty from drill to those with more substance. There is a rough match between even and odd exercises, and answers are provided in the text to the odd exercises only. Detailed solutions of most of the odd-numbered exercises appear in the *Student Solutions Manual,* which is available as a supplement to the text.

- The *Discussion and Discovery* exercises tend to be more open ended than the basic exercises. They include true/false problems, problems requiring conjecture, and occasionally some problems requiring writing or Internet research.

- As the name implies, *Working with Proofs* contains exercises in which the student is asked for precise proofs. When we ask the student to "show" something

in the basic exercises, it is our intent that he or she provide a reasonable argument, not necessarily a formal proof.

- The goal of the *Technology Exercises* is to teach the student how to use modern technological tools such as MATLAB, *Mathematica, Maple*, or handheld calculators to solve linear algebra problems. These exercises are stated generically (no reference to syntax). Platform-specific syntax and techniques are discussed in supplements to the text. Certain of the Technology Exercises are marked with a red icon (▬) to indicate that the exercise teaches a basic technique needed in other Technology Exercises.

Data Sets Instructors giving a technology-intensive course may want to have students use technology to solve most numerical exercises—not just those in the Technology Exercises themselves. To relieve the student of the burden of typing in matrices and vectors, we have provided the data to numerical exercises on the companion Web site. These sets are provided in formats suitable for MATLAB, *Mathematica*, and *Maple*.

Scripts It is our goal to have the student use technology utilities as a problem-solving tool without getting involved in program creation. Thus, where a particular utility would require some programming to perform a linear algebra operation, or where that utility provides its results in a form that differs from the text convention, we have provided a "script" to simplify the work for the student. The scripts can be obtained from the Web site for this text.

Historical Perspective An element called *Linear Algebra in History* appears in various places throughout the text. Its purpose is to bring some vitality to the subject by providing vignettes about the people who contributed to the development of linear algebra and about the origin of various ideas and terminology.

Web site This text is supported by a Web site

   http://www.wiley.com/college/anton

that contains a variety of useful materials, including the data sets and scripts mentioned above.

Supplements

SUPPLEMENTS FOR THE STUDENT

Student Solutions Manual This manual contains detailed solutions to most of the odd-numbered exercises in the text. (ISBN 0-471-17059-3)

Technology Resource Manuals Each manual is geared to the technology exercises in the text. These manuals, which discuss platform-specific commands, syntax, and problem-solving techniques, are available for four platforms:

- *Maple Technology Resource Manual*
  (ISBN 0-471-26938-7)
- *Mathematica Technology Resource Manual*
  (ISBN 0-471-26939-5)
- MATLAB *Technology Resource Manual*
  (ISBN 0-471-26940-9)
- *TI-Calculator Technology Resource Manual*
  (ISBN 0-471-26942-5)

Companion Web site The Web site

http://www.wiley.com/college/anton

accompanies this book. Its content will grow dynamically over time and should be checked regularly. Among the materials for the student that are currently on the site are:

- Data sets (on each technology platform) for many of the numerical exercises in the text—These can be used to eliminate tedious data entry for those who want to use technology to solve the text exercises.
- Technology Toolbox—These are programs (or "scripts") that simplify the solution of various kinds of linear algebra problems. They supplement the commands that are available on the various technology platforms.

SUPPLEMENTS FOR THE INSTRUCTOR

SUPPLEMENTS FOR THE INSTRUCTOR CAN BE OBTAINED BY SENDING A REQUEST ON YOUR INSTITUTIONAL LETTERHEAD TO MATHEMATICS MARKETING MANAGER, HIGHER EDUCATION, JOHN WILEY & SONS, INC., 111 RIVER STREET, HOBOKEN, NJ 07030, OR BY CONTACTING YOUR LOCAL WILEY REPRESENTATIVE.

Instructor Solutions Manual This manual contains detailed solutions to most computational exercises. (ISBN 0-471-26941-7)

Technology Resource Manuals As described above for the student.

eGrade An online assessment system that contains a large bank of skill-building problems and solutions. Instructors can now automate the process of assigning, delivering, grading, and routing all kinds of homework, quizzes, and tests while providing students with immediate scoring and feedback on their work. Wiley eGrade "does the math"… and much more. For more information, visit www.wiley.com/college/egrade

Companion Web site In addition to the materials for the student described above, the Web site

http://www.wiley.com/college/anton

contains PowerPoint files for most of the text illustrations. These can be used either for classroom slide demonstrations or for making overhead transparencies.

Acknowledgments

We express our appreciation to the many talented people whose knowledge and skills have contributed to this text:

REVIEWERS

G. Donald Allen, *Texas A&M University*
William C. Bauldry, *Appalachian State University*
Ernest R. Bishop, *Acadia University*
Christine Black, *Seattle University*
Martin Bohner, *University of Missouri–Rolla*
David Carlson, *San Diego State University*
Daniel Cunningham, *SUNY College at Buffalo*
Dennis Garity, *Oregon State University*
William Goldman, *University of Maryland*
Dr. David L. Gross, *University of Connecticut*

Laxmi N. Gupta, *Rochester Institute of Technology*
James Guyker, *SUNY College at Buffalo*
Dean Hickerson, *University of California, Davis*
Alan Krautstengel, *Case Western Reserve University*
Reginald Laursen, *Luther College*
Jeffrey Leader, *Rose–Hulman Institute of Technology*
Joseph Liang, *University of South Florida*
Benjamin Lotto, *Vassar College*
Tim Loughlin, *New York Institute of Technology*
Roy C. Mathias, *College of William & Mary*

Laura R. Perez, *Washtenaw Community College*
Kenneth Pothoven, *University of South Florida*
John Rossi, *Virginia Polytechnic Institute and State University*
David Ryeburn, *Simon Fraser University*
Mohammed Saleem, *San Jose State University*
Herman Senter, *Clemson University*
Derar Serhan, *Arizona State University*
Ann Sitomer, *Portland Community College*
James Snodgrass III, *Xavier University*
Thomas Timchek, *Nassau Community College*
Paul M. Weichsel, *University of Illinois–Urbana Champaign*
Philip B. Yasskin, *Texas A&M University*

EXERCISE SOLVERS

Loren Argabright, *Drexel University*
Elka Block, *Twin Prime Editorial*
Blaise DeSesa, *Drexel University*
Paul Lorczak
Frank Purcell, *Twin Prime Editorial*
Marie Vanisko, *Carroll College*

COMPOSITION, ILLUSTRATION, COPYEDITING, AND PROJECT MANAGEMENT

Techsetters, Inc., Composition and project management
iMedia, Interior design
ATI Illustration Services, Illustration
Lilian Brady, Copyediting and proofreading
Anne Scanlan-Rohrer, Editorial development

TECHNICAL ADVICE AND RESEARCH

Professors Gene Golub of Stanford University and Claude Brezinski of Université des Sciences et Technologies de Lille for their assistance in locating a picture of André-Louis Cholesky.

Dr. Joseph J. Atick of Visionics, Inc. for his assistance with "eigenheads" and face recognition.

Dr. Jignesh S. Panchal of Lucent Technologies for advice on image compression.

Professor Dan Kalman of American University for suggesting various applications in the text.

THE WILEY TEAM

Bruce Spatz, Publisher
Laurie Rosatone, Associate publisher
Lucille Buonocore, Associate production director
Julie Lindstrom, Marketing manager
Maddy Lesure, Design director
Harold Nolan, Senior designer
Ken Santor, Senior production editor
Sara Wight, Photo researcher
Elyse Rieder, Photo researcher
Jennifer Battista, Assistant editor
Stacy French, Editorial assistant

SPECIAL CONTRIBUTIONS

We would like to express a special debt of gratitude to the following people:

David Ryeburn and Dean Hickerson, who critiqued every aspect of our work and contributed greatly to its accuracy and quality.

Lilian Brady whose unerring eye for typography and accuracy has improved this book immensely.

Carol Sawyer, Rena Lam, Peter Welding, and John Rogosich of Techsetters, Inc., whose care and commitment to quality helped us to create this beautiful book on an extremely tight schedule.

Bruce Spatz, who, in addition to his role as Publisher, filled in as our interim editor and worked hard to make this book a reality.

Barbara Holland, our former editor, who helped create the concept for this book and launched it with enthusiasm and dedication.

CONTENTS

CHAPTER 1   Vectors   1

1.1   Vectors and Matrices in Engineering and Mathematics; n-Space   1
1.2   Dot Product and Orthogonality   15
1.3   Vector Equations of Lines and Planes   29

CHAPTER 2   Systems of Linear Equations   39

2.1   Introduction to Systems of Linear Equations   39
2.2   Solving Linear Systems by Row Reduction   48
2.3   Applications of Linear Systems   63

CHAPTER 3   Matrices and Matrix Algebra   79

3.1   Operations on Matrices   79
3.2   Inverses; Algebraic Properties of Matrices   94
3.3   Elementary Matrices; A Method for Finding A^{-1} 109
3.4   Subspaces and Linear Independence   123
3.5   The Geometry of Linear Systems   135
3.6   Matrices with Special Forms   143
3.7   Matrix Factorizations; LU-Decomposition   154
3.8   Partitioned Matrices and Parallel Processing   166

CHAPTER 4   Determinants   175

4.1   Determinants; Cofactor Expansion   175
4.2   Properties of Determinants   184
4.3   Cramer's Rule; Formula for A^{-1}; Applications of Determinants   196
4.4   A First Look at Eigenvalues and Eigenvectors   210

CHAPTER 5   Matrix Models   225

5.1   Dynamical Systems and Markov Chains   225
5.2   Leontief Input-Output Models   235
5.3   Gauss–Seidel and Jacobi Iteration; Sparse Linear Systems   241
5.4   The Power Method; Application to Internet Search Engines   249

CHAPTER 6   Linear Transformations   265

6.1   Matrices as Transformations   265
6.2   Geometry of Linear Operators   280
6.3   Kernel and Range   296
6.4   Composition and Invertibility of Linear Transformations   305
6.5   Computer Graphics   318

CHAPTER 7   Dimension and Structure   329

7.1   Basis and Dimension   329
7.2   Properties of Bases   335
7.3   The Fundamental Spaces of a Matrix   342
7.4   The Dimension Theorem and Its Implications   352
7.5   The Rank Theorem and Its Implications   360
7.6   The Pivot Theorem and Its Implications   370
7.7   The Projection Theorem and Its Implications   379
7.8   Best Approximation and Least Squares   393
7.9   Orthonormal Bases and the Gram–Schmidt Process   406
7.10 QR-Decomposition; Householder Transformations   417
7.11   Coordinates with Respect to a Basis   430

CHAPTER 8   Diagonalization   443

8.1   Matrix Representations of Linear Transformations   443
8.2   Similarity and Diagonalizability   456
8.3   Orthogonal Diagonalizability; Functions of a Matrix   468
8.4   Quadratic Forms   481
8.5   Application of Quadratic Forms to Optimization   495
8.6   Singular Value Decomposition   502
8.7   The Pseudoinverse   518
8.8   Complex Eigenvalues and Eigenvectors   525
8.9   Hermitian, Unitary, and Normal Matrices   535
8.10   Systems of Differential Equations   542

CHAPTER 9   General Vector Spaces   555

9.1   Vector Space Axioms   555
9.2   Inner Product Spaces; Fourier Series   569
9.3   General Linear Transformations; Isomorphism   582

APPENDIX A   How to Read Theorems   A1

APPENDIX B   Complex Numbers   A3

ANSWERS TO ODD-NUMBERED EXERCISES   A9

PHOTO CREDITS   C1

INDEX   I-1

GUIDE FOR THE INSTRUCTOR

Number of Lectures

The Syllabus Guide below provides for a 29-lecture core and a 35-lecture core. The 29-lecture core is for schools with time constraints, as with abbreviated summer courses. Both core programs can be supplemented by starred topics as time permits. The omission of starred topics does not affect the readability or continuity of the core topics.

Pace

The core program is based on covering one section per lecture, but whether you can do this in every instance will depend on your teaching style and the capabilities of your particular students. For longer sections we recommend that you just highlight the main points in class and leave the details for the students to read. Since the reviews of this text have praised the clarity of the exposition, you should find this workable. If, in certain cases, you want to devote more than one lecture to

a core topic, you can do so by adjusting the number of starred topics that you cover.

By the end of Lecture 15 the following concepts will have been covered in a basic form: linear combination, spanning, subspace, dimension, eigenvalues, and eigenvectors. Thus, even with a relatively slow pace you will have no trouble touching on all of the main ideas in the course.

Organization

It is our feeling that the most effective way to teach abstract vector spaces is to place that material at the end (Chapter 9), at which point it occurs as a "natural generalization" of the earlier material, and the student has developed the "linear algebra maturity" to understand its purpose. However, we recognize that not everybody shares that philosophy, so we have designed that chapter so it can be moved forward, if desired.

35-Lecture Course		CONTENTS	29-Lecture Course
SYLLABUS GUIDE			
Chapter 1		**Vectors**	**Chapter 1**
1	1.1	Vectors and Matrices in Engineering and Mathematics; n-Space	1
2	1.2	Dot Product and Orthogonality	2
3	1.3	Vector Equations of Lines and Planes	3
Chapter 2		**Systems of Linear Equations**	**Chapter 2**
4	2.1	Introduction to Systems of Linear Equations	4
5	2.2	Solving Linear Systems by Row Reduction	5
*	2.3	Applications of Linear Systems	*
Chapter 3		**Matrices and Matrix Algebra**	**Chapter 3**
6	3.1	Operations on Matrices	6
7	3.2	Inverses; Algebraic Properties of Matrices	7
8	3.3	Elementary Matrices; A Method for Finding A^{-1}	8
9	3.4	Subspaces and Linear Independence	9
10	3.5	The Geometry of Linear Systems	10
11	3.6	Matrices with Special Forms	11
*	3.7	Matrix Factorizations; LU-Decomposition	*
12	3.8	Partitioned Matrices and Parallel Processing	12
Chapter 4		**Determinants**	**Chapter 4**
13	4.1	Determinants; Cofactor Expansion	13
14	4.2	Properties of Determinants	*

*	4.3	Cramer's Rule; Formula for A^{-1}; Applications of Determinants	*
15	4.4	A First Look at Eigenvalues and Eigenvectors	14

Chapter 5		**Matrix Models**	**Chapter 5**
*	5.1	Dynamical Systems and Markov Chains	*
*	5.2	Leontief Input-Output Models	*
*	5.3	Gauss–Seidel and Jacobi Iteration; Sparse Linear Systems	*
*	5.4	The Power Method; Application to Internet Search Engines	*

Chapter 6		**Linear Transformations**	**Chapter 6**
16	6.1	Matrices as Transformations	15
17	6.2	Geometry of Linear Operators	16
18	6.3	Kernel and Range	*
19	6.4	Composition and Invertibility of Linear Transformations	*
*	6.5	Computer Graphics	*

Chapter 7		**Dimension and Structure**	**Chapter 7**
20	7.1	Basis and Dimension	17
21	7.2	Properties of Bases	18
22	7.3	The Fundamental Spaces of a Matrix	19
23	7.4	The Dimension Theorem and Its Implications	20
24	7.5	The Rank Theorem and Its Implications	21
25	7.6	The Pivot Theorem and Its Implications	22
26	7.7	The Projection Theorem and Its Implications	*
27	7.8	Best Approximation and Least Squares	*
28	7.9	Orthonormal Bases and the Gram–Schmidt Process	*
*	7.10	QR-Decomposition; Householder Transformations	*
29	7.11	Coordinates with Respect to a Basis	23

Chapter 8		**Diagonalization**	**Chapter 8**
30	8.1	Matrix Representations of Linear Transformations	24
31	8.2	Similarity and Diagonalizability	25
32	8.3	Orthogonal Diagonalizability; Functions of a Matrix	26
*	8.4	Quadratic Forms	*
*	8.5	Application of Quadratic Forms to Optimization	*
*	8.6	Singular Value Decomposition	*
*	8.7	The Pseudoinverse	*
*	8.8	Complex Eigenvalues and Eigenvectors	*
*	8.9	Hermitian, Unitary, and Normal Matrices	*
*	8.10	Systems of Differential Equations	*

Chapter 9		**General Vector Spaces**	**Chapter 9**
33	9.1	Vector Space Axioms	27
34	9.2	Inner Product Spaces; Fourier Series	28
35	9.3	General Linear Transformations; Isomorphism	29

	Appendices	
	Appendix A How to Read Theorems	*
	Appendix B Complex Numbers	*

To assist you in planning your course, we have provided below a list of topics that occur in each section. These topics are identified in the text by headings in the margin.

CHAPTER 1   VECTORS

1.1 Vectors and Matrices in Engineering and Mathematics; n-Space   Scalars and Vectors • Equivalent Vectors • Vector Addition • Vector Subtraction • Scalar Multiplication • Vectors in Coordinate Systems • Components of a Vector Whose Initial Point Is Not at the Origin • Vectors in R^n • Equality of Vectors • Sums of Three or More Vectors • Parallel and Collinear Vectors • Linear Combinations • Application to Computer Color Models • Alternative Notations for Vectors • Matrices

1.2 Dot Product and Orthogonality   Norm of a Vector • Unit Vectors • The Standard Unit Vectors • Distance Between Points in R^n • Dot Products • Algebraic Properties of the Dot Product • Angle Between Vectors in R^2 and R^3 • Orthogonality • Orthonormal Sets • Euclidean Geometry in R^n

1.3 Vector Equations of Lines and Planes   Vector and Parametric Equations of Lines • Lines Through Two Points • Point-Normal Equations of Planes • Vector and Parametric Equations of Planes • Lines and Planes in R^n • Comments on Terminology

CHAPTER 2   SYSTEMS OF LINEAR EQUATIONS

2.1 Introduction to Systems of Linear Equations   Linear Systems • Linear Systems with Two and Three Unknowns • Augmented Matrices and Elementary Row Operations

2.2 Solving Linear Systems by Row Reduction   Considerations in Solving Linear Systems • Echelon Forms • General Solutions as Linear Combinations of Column Vectors • Gauss–Jordan and Gaussian Elimination • Some Facts About Echelon Forms • Back Substitution • Homogeneous Linear Systems • The Dimension Theorem for Homogeneous Linear Systems • Stability, Roundoff Error, and Partial Pivoting

2.3 Applications of Linear Systems   Global Positioning • Network Analysis • Electrical Circuits • Balancing Chemical Equations • Polynomial Interpolation

CHAPTER 3   MATRICES AND MATRIX ALGEBRA

3.1 Operations on Matrices   Matrix Notation and Terminology • Operations on Matrices • Row and Column Vectors • The Product $A\mathbf{x}$ • The Product AB • Finding Specific Entries in a Matrix Product • Finding Specific Rows and Columns of a Matrix Product • Matrix Products as Linear Combinations • Transpose of a Matrix • Trace • Inner and Outer Matrix Products

3.2 Inverses; Algebraic Properties of Matrices   Properties of Matrix Addition and Scalar Multiplication • Properties of Matrix Multiplication • Zero Matrices • Identity Matrices • Inverse of a Matrix • Properties of Inverses • Powers of a Matrix • Matrix Polynomials • Properties of the Transpose • Properties of the Trace • Transpose and Dot Product

3.3 Elementary Matrices; A Method for Finding A^{-1}   Elementary Matrices • Characterizations of Invertibility • Row Equivalence • An Algorithm for Inverting Matrices • Solving Linear Systems by Matrix Inversion • Solving Multiple Linear Systems with a Common Coefficient Matrix • Consistency of Linear Systems

3.4 Subspaces and Linear Independence   Subspaces of R^n • Solution Space of a Linear System • Linear Independence • Linear Independence and Homogeneous Linear Systems • Translated Subspaces • A Unifying Theorem

3.5 The Geometry of Linear Systems   The Relationship Between $A\mathbf{x} = \mathbf{b}$ and $A\mathbf{x} = \mathbf{0}$ • Consistency of a Linear System from the Vector Point of View • Hyperplanes • Geometric Interpretations of Solution Spaces

3.6 Matrices with Special Forms   Diagonal Matrices • Triangular Matrices • Linear Systems with Triangular Coefficient Matrices • Properties of Triangular Matrices • Symmetric and Skew-Symmetric Matrices • Invertibility of Symmetric Matrices • Matrices of the Form $A^T A$ and $A A^T$ • Fixed Points of a Matrix • A Technique for Inverting $I - A$ When A Is Nilpotent • Inverting $I - A$ by Power Series

3.7 Matrix Factorizations; LU-Decomposition   Solving Linear Systems by Factorization • Finding LU-Decompositions • The Relationship Between Gaussian Elimination and LU-Decomposition • Matrix Inversion by LU-Decomposition • LDU-Decompositions • Using Permutation Matrices to Deal with Row Interchanges • Flops and the Cost of Solving a Linear System • Cost Estimates for Solving Large Linear Systems • Considerations in Choosing an Algorithm for Solving a Linear System

3.8 Partitioned Matrices and Parallel Processing   General Partitioning • Block Diagonal Matrices • Block Upper Triangular Matrices

CHAPTER 4 DETERMINANTS

4.1 Determinants; Cofactor Expansion   Determinants of 2×2 and 3×3 Matrices • Elementary Products • General Determinants • Evaluation Difficulties for Higher-Order Determinants • Determinants of Matrices with Rows or Columns That Have All Zeros • Determinants of Triangular Matrices • Minors and Cofactors • Cofactor Expansions

4.2 Properties of Determinants   Determinant of A^T • Effect of Elementary Row Operations on a Determinant • Simplifying Cofactor Expansions • Determinants by Gaussian Elimination • A Determinant Test for Invertibility • Determinant of a Product of Matrices • Determinant Evaluation by LU-Decomposition • Determinant of the Inverse of a Matrix • Determinant of $A + B$ • A Unifying Theorem

4.3 Cramer's Rule; Formula for A^{-1}; Applications of Determinants   Adjoint of a Matrix • A Formula for the Inverse of a Matrix • How the Inverse Formula Is Actually Used • Cramer's Rule • Geometric Interpretation of Determinants • Polynomial Interpolation and the Vandermonde Determinant • Cross Products

4.4 A First Look at Eigenvalues and Eigenvectors   Fixed Points • Eigenvalues and Eigenvectors • Eigenvalues of Triangular Matrices • Eigenvalues of Powers of a Matrix • A Unifying Theorem • Complex Eigenvalues • Algebraic Multiplicity • Eigenvalue Analysis of 2×2 Matrices • Eigenvalue Analysis of 2×2 Symmetric Matrices • Expressions for Determinant and Trace in Terms of Eigenvalues • Eigenvalues by Numerical Methods

CHAPTER 5 MATRIX MODELS

5.1 Dynamical Systems and Markov Chains   Dynamical Systems • Markov Chains • Markov Chains as Powers of the Transition Matrix • Long-Term Behavior of a Markov Chain

5.2 Leontief Input-Output Models   Inputs and Outputs in an Economy • The Leontief Model of an Open Economy • Productive Open Economies

5.3 Gauss–Seidel and Jacobi Iteration; Sparse Linear Systems   Iterative Methods • Jacobi Iteration • Gauss–Seidel Iteration • Convergence • Speeding Up Convergence

5.4 The Power Method; Application to Internet Search Engines   The Power Method • The Power Method with Euclidean Scaling • The Power Method with Maximum Entry Scaling • Rate of Convergence • Stopping Procedures • An Application of the Power Method to Internet Searches • Variations of the Power Method

CHAPTER 6 LINEAR TRANSFORMATIONS

6.1 Matrices as Transformations   A Review of Functions • Matrix Transformations • Linear Transformations • Some Properties of Linear Transformations • All Linear Transformations from R^n to R^m Are Matrix Transformations • Rotations About the Origin • Reflections About Lines Through the Origin • Orthogonal Projections onto Lines Through the Origin • Transformations of the Unit Square • Power Sequences

6.2 Geometry of Linear Operators   Norm-Preserving Linear Operators • Orthogonal Operators Preserve Angles and Orthogonality • Orthogonal Matrices • All Orthogonal Linear Operators on R^2 are Rotations or Reflections • Contractions and Dilations of R^2 • Vertical and Horizontal Compressions and Expansions of R^2 • Shears • Linear Operators on R^3 • Reflections About Coordinate Planes • Rotations in R^3 • General Rotations

6.3 Kernel and Range   Kernel of a Linear Transformation • Kernel of a Matrix Transformation • Range of a Linear Transformation • Range of a Matrix Transformation • Existence and Uniqueness Issues • One-to-One and Onto from the Viewpoint of Linear Systems • A Unifying Theorem

6.4 Composition and Invertibility of Linear Transformations   Compositions of Linear Transformations • Compositions of Three or More Linear Transformations • Factoring Linear Operators into Compositions • Inverse of a Linear Transformation • Invertible Linear Operators • Geometric Properties of Invertible Linear Operators on R^2 • Image of the Unit Square Under an Invertible Linear Operator

6.5 Computer Graphics   Wireframes • Matrix Representations of Wireframes • Transforming Wireframes • Translation Using Homogeneous Coordinates • Three-Dimensional Graphics

CHAPTER 7 DIMENSION AND STRUCTURE

7.1 Basis and Dimension   Bases for Subspaces • Dimension of a Solution Space • Dimension of a Hyperplane

7.2 Properties of Bases   Properties of Bases • Subspaces of Subspaces • Sometimes Spanning Implies Linear Independence and Conversely • A Unifying Theorem

7.3 The Fundamental Spaces of a Matrix   The Fundamental Spaces of a Matrix • Orthogonal Complements • Properties of Orthogonal Complements • Finding Bases by Row Reduction • Determining Whether a Vector Is in a Given Subspace

7.4 The Dimension Theorem and Its Implications   The Dimension Theorem for Matrices • Extending a Linearly Independent Set to a Basis • Some Consequences of the Dimension Theorem for Matrices • The Dimension Theorem for Subspaces • A Unifying Theorem • More on Hyperplanes • Rank 1 Matrices • Symmetric Rank 1 Matrices

7.5 The Rank Theorem and Its Implications   The Rank Theorem • Relationship Between Consistency and Rank • Overdetermined and Underdetermined Linear Systems • Matrices of the form A^TA and AA^T • Some Unifying Theorems • Applications of Rank

7.6 The Pivot Theorem and Its Implications   Basis Problems Revisited • Bases for the Fundamental Spaces of a Matrix • A Column-Row Factorization • Column-Row Expansion

7.7 The Projection Theorem and Its Implications   Orthogonal Projections onto Lines Through the Origin in R^2 • Orthogonal Projections onto Lines Through the Origin in R^n • Projection Operators on R^n • Orthogonal Projections onto General Subspaces • When Does a Matrix Represent an Orthogonal Projection? • Strang Diagrams • Full Column Rank and Consistency of a Linear System • The Double Perp Theorem • Orthogonal Projections onto $W^\perp$

7.8 Best Approximation and Least Squares   Minimum Distance Problems • Least Squares Solutions of Linear Systems • Finding Least Squares Solutions of Linear Systems • Orthogonality Property of Least Squares Error Vectors • Strang Diagrams for Least Squares Problems • Fitting a Curve to Experimental Data • Least Squares Fits by Higher-Degree Polynomials • Theory Versus Practice

7.9 Orthonormal Bases and the Gram–Schmidt Process   Orthogonal and Orthonormal Bases • Orthogonal Projections Using Orthogonal Bases • Trace and Orthogonal Projections • Linear Combinations of Orthonormal Basis Vectors • Finding Orthogonal and Orthonormal Bases • A Property of the Gram–Schmidt Process • Extending Orthonormal Sets to Orthonormal Bases

7.10 QR-Decomposition; Householder Transformations QR-Decomposition • The Role of QR-Decomposition in Least Squares Problems • Other Numerical Issues • Householder Reflections • QR-Decomposition Using Householder Reflections • Householder Reflections in Applications

7.11 Coordinates with Respect to a Basis   Nonrectangular Coordinate Systems in R^2 and R^3 • Coordinates with Respect to an Orthonormal Basis • Computing with Coordinates with Respect to an Orthononormal Basis • Change of Basis for R^n • Invertibility of Transition Matrices • A Good Technique for finding Transition Matrices • Coordinate Maps • Transition Between Orthonormal Bases • Application to Rotation of Coordinate Axes • New Ways to Think About Matrices

CHAPTER 8   DIAGONALIZATION

8.1 Matrix Representations of Linear Transformations   Matrix of a Linear Operator with Respect to a Basis • Changing Bases • Matrix of a Linear Transformation with Respect to a Pair of Bases • Effect of Changing Bases on Matrices of Linear Transformations • Representing Linear Operators with Two Bases

8.2 Similarity and Diagonalizability   Similar Matrices • Similarity Invariants • Eigenvectors and Eigenvalues of Similar Matrices • Diagonalization • A Method for Diagonalizing a Matrix • Linear Independence of Eigenvectors • Relationship Between Algebraic and Geometric Multiplicity • A Unifying Theorem on Diagonalizability

8.3 Orthogonal Diagonalizability; Functions of a Matrix   Orthogonal Similarity • A Method for Orthogonally Diagonalizing a Symmetric Matrix • Spectral Decomposition • Powers of a Diagonalizable Matrix • Cayley-Hamilton Theorem • Exponential of a Matrix • Diagonalization and Linear Systems • The Nondiagonalizable Case

8.4 Quadratic Forms   Definition of a Quadratic Form • Change of Variable in a Quadratic Form • Quadratic Forms in Geometry • Identifying Conic Sections • Positive Definite Quadratic Forms • Classifying Conic Sections Using Eigenvalues • Identifying Positive Definite Matrices • Cholesky Factorization

8.5 Application of Quadratic Forms to Optimization   Relative Extrema of Functions of Two Variables • Constrained Extremum Problems • Constrained Extrema and Level Curves

8.6 Singular Value Decomposition   Singular Value Decomposition of Square Matrices • Singular Value Decomposition of Symmetric Matrices • Polar Decomposition • Singular Value Decomposition of Nonsquare Matrices • Singular Value Decomposition and the Fundamental Spaces of a Matrix • Reduced Singular Value Decomposition • Data Compression and Image Processing • Singular Value Decomposition from the Transformation Point of View

8.7 The Pseudoinverse   The Pseudoinverse • Properties of the Pseudoinverse • The Pseudoinverse and Least Squares • Condition Number and Numerical Considerations

8.8 Complex Eigenvalues and Eigenvectors   Vectors in C^n • Algebraic Properties of the Complex Conjugate • The Complex Euclidean Inner Product • Vector Space Concepts in C^n • Complex Eigenvalues of Real Matrices Acting on Vectors in C^n • A Proof That Real Symmetric Matrices Have Real Eigenvalues • A Geometric Interpretation of Complex Eigenvalues of Real Matrices

8.9 Hermitian, Unitary, and Normal Matrices   Hermitian and Unitary Matrices • Unitary Diagonalizability • Skew-Hermitian Matrices • Normal Matrices • A Comparison of Eigenvalues

8.10 Systems of Differential Equations   Terminology • Linear Systems of Differential Equations • Fundamental Solutions • Solutions Using Eigenvalues and Eigenvectors • Exponential Form of a Solution • The Case Where A is Not Diagonalizable

CHAPTER 9   GENERAL VECTOR SPACES

9.1 Vector Space Axioms   Vector Space Axioms • Function Spaces • Matrix Spaces • Unusual Vector Spaces • Subspaces • Linear Independence, Spanning, Basis • Wroński's Test for Linear Independence of Functions • Dimension • The Lagrange Interpolating Polynomials • Lagrange Interpolation from a Vector Point of View

9.2 Inner Product Spaces; Fourier Series   Inner Product Axioms • The Effect of Weighting on Geometry • Algebraic Properties of Inner Products • Orthonormal Bases • Best Approximation • Fourier Series • General Inner Products on R^n

9.3 General Linear Transformations; Isomorphism   General Linear Transformations • Kernel and Range • Properties of the Kernel and Range • Isomorphism • Inner Product Space Isomorphisms

APPENDIX A   HOW TO READ THEOREMS

Contrapositive Form of a Theorem • Converse of a Theorem • Theorems Involving Three or More Implications

APPENDIX B   COMPLEX NUMBERS

Complex Numbers • The Complex Plane • Polar Form of a Complex Number • Geometric Interpretation of Multiplication and Division of Complex Numbers • DeMoivre's Formula • Euler's Formula

Linear algebra is a compilation of diverse but interrelated ideas that provide a way of analyzing and solving problems in many applied fields. As with most mathematics courses, the subject involves theorems, proofs, formulas, and computations of various kinds. However, if all you do is learn to use the formulas and mechanically perform the computations, you will have missed the most important part of the subject—understanding how the different ideas discussed in the course interrelate with one another—and this can only be achieved by reading the theorems and working through the proofs. This is important because the key to solving a problem using linear algebra often rests with looking at the problem from the right point of view. Keep in mind that every problem in this text has already been solved by somebody, so your ability to solve those problems gives you nothing unique. However, if you master the ideas and their interrelationship, then you will have the tools to go beyond what other people have done, limited only by your talents and creativity.

Before starting your studies, you may find it helpful to leaf through the text to get a feeling for its parts:

- At the beginning of each section you will find an introduction that gives you an overview of what you will be reading about in the section.

- Each section ends with a set of exercises that is divided into four groups. The first group consists of exercises that are intended for hand calculation, though there is no reason why you cannot use a calculator or computer program where convenient and appropriate. These exercises tend to become more challenging as they progress. Answers to most odd-numbered exercises are in the back of the text and worked-out solutions to many of them are available in a supplement to the text.

- The second group of exercises, *Discussion and Discovery,* consists of problems that call for more creative thinking than those in the first group.

- The third group of exercises, *Working with Proofs,* consists of problems in which you are asked to give mathematical proofs. By comparison, if you are asked to "show" something in the first group of exercises, a reasonable logical argument will suffice; here we are looking for *precise* proofs.

- The fourth group of exercises, *Technology Exercises,* are specifically designed to be solved using some kind of technology tool: typically, a computer algebra system or a handheld calculator with linear algebra capabilities. Syntax and techniques for using specific types of technology tools are discussed in supplements to this text. Certain Technology Exercises are marked with a red icon (▬) to indicate that they teach basic techniques that are needed in other Technology Exercises.

- Near the end of the text you will find two appendices: Appendix A provides some suggestions on how to read theorems, and Appendix B reviews some results about complex numbers that will be needed toward the later part of the course. We suggest that you read Appendix A as soon as you can.

- Theorems, definitions, and figures are referenced using a triple number system. Thus, for example, Figure 7.3.4 is the fourth figure in Section 7.3. Illustrations in the exercises are identified by the exercise number with which they are associated. Thus, if there is a figure associated with Exercise 7 in a certain section, it would be labeled Figure Ex-7.

- Additional materials relating to this text can be found on the Web site

    http://www.wiley.com/college/anton

Best of luck with your studies.

Howard Anton

Robert C. Busby

CHAPTER 1

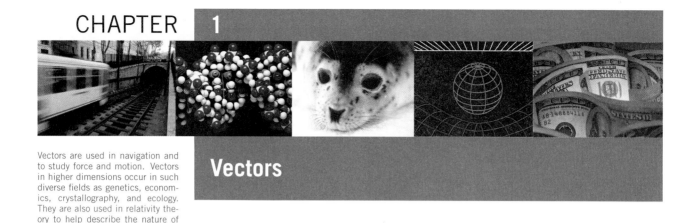

Vectors

Vectors are used in navigation and to study force and motion. Vectors in higher dimensions occur in such diverse fields as genetics, economics, crystallography, and ecology. They are also used in relativity theory to help describe the nature of gravity, space, and matter.

Section 1.1 Vectors and Matrices in Engineering and Mathematics; *n*-Space

Linear algebra is concerned with two basic kinds of quantities: "vectors" and "matrices." The term "vector" has various meanings in engineering, science, and mathematics, some of which will be discussed in this section. We will begin by reviewing the geometric notion of a vector as it is used in basic physics and engineering, next we will discuss vectors in two-dimensional and three-dimensional coordinate systems, and then we will consider how the notion of a vector can be extended to higher-dimensional spaces. Finally, we will talk a little about matrices, explaining how they arise and how they are related to vectors.

SCALARS AND VECTORS

Engineers and physicists distinguish between two types of physical quantities—*scalars*, which are quantities that can be described by a numerical value alone, and *vectors*, which require both a numerical value and a direction for their complete description. For example, temperature, length, and speed are scalars because they are completely described by a number that tells "how much"—say a temperature of 20°C, a length of 5 cm, or a speed of 10 m/s. In contrast, velocity, force, and displacement are vectors because they involve a direction as well as a numerical value:

- **Velocity**—Knowing that a ship has a speed of 10 knots (nautical miles per hour) tells how fast it is going but not which way it is moving. To plot a navigational course, a sailor needs to know the direction as well as the speed of the boat, say 10 knots at a compass heading of 45° north of east (Figure 1.1.1*a*). Speed and direction together form a vector quantity called *velocity*.

- **Force**—When a force is applied to an object, the resulting effect depends on the magnitude of the force and the direction in which it is applied. For example, although the three 10-lb forces illustrated in Figure 1.1.1*b* have the same magnitude, they have different effects on the block because of the differences in their direction. The magnitude and direction of a force together form a vector quantity called a *force vector*.

- **Displacement**—If a particle moves along a path from a point *A* to a point *B* in a plane (2-space) or in three-dimensional space (3-space), then the straight-line distance from *A* to *B* together with the direction from *A* to *B* form a vector quantity that is called the *displacement* from *A* to *B* (Figure 1.1.1*c*). The displacement describes the change in position of the particle without regard to the particular path that the particle traverses.

1

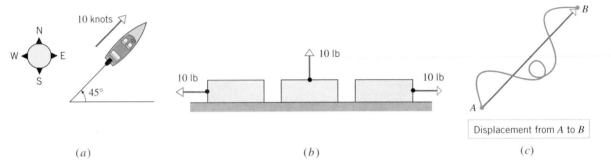

(a)   (b)   (c)

Figure 1.1.1

Vectors in two dimensions (2-space) or three dimensions (3-space) can be represented geometrically by *arrows*—the length of the arrow is proportional to the **magnitude** (or numerical part) of the vector, and the direction of the arrow indicates the direction of the vector. The *tail* of the arrow is called the **initial point** and the *tip* is called the **terminal point** (Figure 1.1.2). In this text we will denote vectors in lowercase boldface type such as **a**, **k**, **v**, **w**, and **x**, and scalars in lowercase italic type such as a, k, v, w, and x. If a vector **v** has initial point A and terminal point B, then we will denote the vector as

$$\mathbf{v} = \overrightarrow{AB}$$

when we want to indicate the initial and terminal points explicitly (Figure 1.1.3).

EQUIVALENT VECTORS

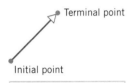

Terminal point

Initial point

The length of the arrow measures the magnitude of the vector and the arrowhead indicates the direction.

Figure 1.1.2

Two types of vectors occur in applications: bound vectors and free vectors. A **bound vector** is one whose physical effect depends on the location of the initial point as well as the magnitude and direction, and a **free vector** is one whose physical effect depends on the magnitude and direction alone. For example, Figure 1.1.4 shows two 10-lb upward forces applied to a block. Although the forces have the same magnitude and direction, the differences in their points of application (the initial points of the vectors) cause differences in the behavior of the block. Thus, these forces need to be treated as bound vectors. In contrast, velocity and displacement are generally treated as free vectors. *In this text we will focus exclusively on free vectors*, leaving the study of bound vectors for courses in engineering and physics.

Because free vectors are not changed when they are translated, we will consider two vectors **v** and **w** to be **equal** (also called **equivalent**) if they are represented by parallel arrows with the same length and direction (Figure 1.1.5). To indicate that **v** and **w** are equivalent vectors we will write **v** = **w**.

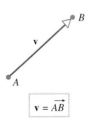

Figure 1.1.3

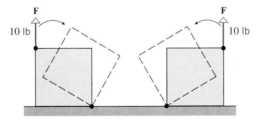

Figure 1.1.4

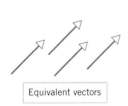

Equivalent vectors

Figure 1.1.5

The vector whose initial and terminal points coincide has length zero, so we call this the **zero vector** and denote it by **0**. The zero vector has no natural direction, so we will agree that it can be assigned any direction that is convenient for the problem at hand.

VECTOR ADDITION

There are a number of important algebraic operations on vectors, all of which have their origin in laws of physics.

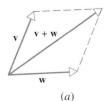

(*a*)

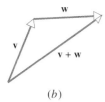

(*b*)

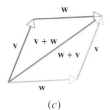

(*c*)

Figure 1.1.6

Parallelogram Rule for Vector Addition   If **v** and **w** are vectors in 2-space or 3-space that are positioned so their initial points coincide, then the two vectors form adjacent sides of a parallelogram, and the *sum* **v** + **w** is the vector represented by the arrow from the common initial point of **v** and **w** to the opposite vertex of the parallelogram (Figure 1.1.6*a*).

Here is another way to form the sum of two vectors.

Triangle Rule for Vector Addition   If **v** and **w** are vectors in 2-space or 3-space that are positioned so the initial point of **w** is at the terminal point of **v**, then the *sum* **v** + **w** is represented by the arrow from the initial point of **v** to the terminal point of **w** (Figure 1.1.6*b*).

In Figure 1.1.6*c* we have constructed the sums **v** + **w** and **w** + **v** by the triangle rule. This construction makes it evident that

$$\mathbf{v} + \mathbf{w} = \mathbf{w} + \mathbf{v} \tag{1}$$

and that the sum obtained by the triangle rule is the same as the sum obtained by the parallelogram rule.

Vector addition can also be viewed as a process of translating points.

Vector Addition Viewed as Translation   If **v**, **w**, and **v** + **w** are positioned so their initial points coincide, then the terminal point of **v** + **w** can be viewed in two ways:

1. The terminal point of **v** + **w** is the point that results when the terminal point of **v** is translated in the direction of **w** by a distance equal to the length of **w** (Figure 1.1.7*a*).

2. The terminal point of **v** + **w** is the point that results when the terminal point of **w** is translated in the direction of **v** by a distance equal to the length of **v** (Figure 1.1.7*b*).

Accordingly, we say that **v** + **w** is the *translation of* **v** *by* **w** or, alternatively, the *translation of* **w** *by* **v**.

Linear Algebra in History

The idea that a directed line segment (an arrow) could be used to represent the magnitude and direction of a velocity, force, or displacement developed gradually over a long period of time. The Greek logician Aristotle, for example, knew that the combined effect of two forces was given by the parallelogram law, and the Italian astronomer Galileo stated the law explicitly in his work on mechanics. Applications of vectors in geometry appeared in a book entitled *Der barycentrische Calcul*, published in 1827 by the German mathematician August Ferdinand Möbius. In 1837 Möbius published a work on statics in which he used the idea of resolving a vector into components. During the same time period the Italian mathematician Giusto Bellavitis proposed an "algebra" of directed line segments in which line segments with the same length and direction are considered to be equal. Bellavitus published his results in 1832.

Aristotle
(384 B.C.–322 B.C.)

Galileo Galilei
(1564–1642)

August Ferdinand Möbius
(1790–1868)

Giusto Bellavitis
(1803–1880)

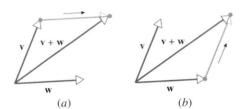

Figure 1.1.7   (*a*)   (*b*)

EXAMPLE 1
Vector Addition
in Physics and
Engineering

The parallelogram rule for vector addition correctly describes the additive behavior of forces, velocities, and displacements in engineering and physics. For example, the effect of applying the forces $\mathbf{F}_1$ and $\mathbf{F}_2$ to the block in Figure 1.1.8*a* is the same as applying the single force $\mathbf{F}_1 + \mathbf{F}_2$ to the block. Similarly, if the engine of the boat in Figure 1.1.8*b* imparts the velocity $\mathbf{v}_1$ and the wind imparts a velocity $\mathbf{v}_2$, then the combined effect of the engine and wind is to impart the velocity $\mathbf{v}_1 + \mathbf{v}_2$ to the boat. Also, if a particle undergoes a displacement $\overrightarrow{AB}$ from A to B, followed by a displacement $\overrightarrow{BC}$ from B to C (Figure 1.1.8*c*), then the successive displacements are the same as the single displacement $\overrightarrow{AC} = \overrightarrow{AB} + \overrightarrow{BC}$.   ■

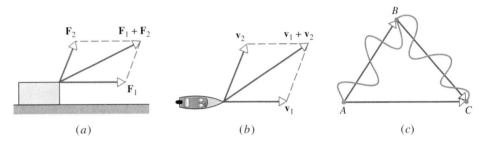

Figure 1.1.8   (*a*)   (*b*)   (*c*)

VECTOR SUBTRACTION   In ordinary arithmetic we can write $a - b = a + (-b)$, which expresses subtraction in terms of addition. There is an analogous idea in vector arithmetic.

> **Vector Subtraction**   The ***negative*** of a vector $\mathbf{v}$, denoted by $-\mathbf{v}$, is the vector that has the same length as $\mathbf{v}$ but is oppositely directed (Figure 1.1.9*a*), and the ***difference*** of $\mathbf{v}$ from $\mathbf{w}$, denoted by $\mathbf{w} - \mathbf{v}$, is taken to be the sum
>
> $$\mathbf{w} - \mathbf{v} = \mathbf{w} + (-\mathbf{v}) \tag{2}$$

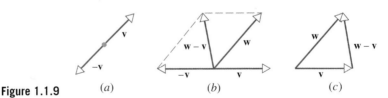

Figure 1.1.9   (*a*)   (*b*)   (*c*)

The difference of $\mathbf{v}$ from $\mathbf{w}$ can be obtained geometrically by the parallelogram method shown in Figure 1.1.9*b*, or more directly by positioning $\mathbf{w}$ and $\mathbf{v}$ so their initial points coincide and drawing the vector from the terminal point of $\mathbf{v}$ to the terminal point of $\mathbf{w}$ (Figure 1.1.9*c*).

SCALAR MULTIPLICATION   Sometimes there is a need to change the length of a vector or change its length and reverse its direction. This is accomplished by a type of multiplication in which vectors are multiplied by scalars. As an example, the product $2\mathbf{v}$ denotes the vector that has the same direction as $\mathbf{v}$ but twice the length, and the product $-2\mathbf{v}$ denotes the vector that is oppositely directed to $\mathbf{v}$ and has twice the length. Here is the general result.

> **Scalar Multiplication**   If $\mathbf{v}$ is a nonzero vector and k is a nonzero scalar, then the ***scalar multiple*** of $\mathbf{v}$ by k, denoted by $k\mathbf{v}$, is the vector whose length is $|k|$ times the length of $\mathbf{v}$ and whose direction is the same as $\mathbf{v}$ if $k > 0$ and opposite to $\mathbf{v}$ if $k < 0$. If $k = 0$ or $\mathbf{v} = \mathbf{0}$, then we take $k\mathbf{v} = \mathbf{0}$.

Figure 1.1.10 shows the geometric relationship between a vector **v** and some of its scalar multiples. In particular, observe that $(-1)\mathbf{v}$ has the same length as **v** but is oppositely directed; therefore,

$$(-1)\mathbf{v} = -\mathbf{v} \tag{3}$$

VECTORS IN COORDINATE SYSTEMS

Although arrows are useful for describing vectors geometrically, it is desirable to have some way of representing vectors algebraically. We will do this by considering vectors in rectangular coordinate systems, and we will begin by briefly reviewing some of the basic ideas about coordinate systems in 2-space and 3-space.

Recall that a ***rectangular coordinate system in 2-space*** consists of two perpendicular coordinate axes, which are usually called the ***x-axis*** and ***y-axis***. The point of intersection of the axes is called the ***origin*** of the coordinate system. In this text we will assume that the same scale of measurement is used on both axes and that the positive y-axis is 90° counterclockwise from the positive x-axis (Figure 1.1.11a).

Once a rectangular coordinate system has been introduced, the construction shown in Figure 1.1.11b produces a one-to-one correspondence between points in the plane and ordered pairs of real numbers; that is, each point P is associated with a unique ordered pair (a, b) of real numbers, and each ordered pair of real numbers (a, b) is associated with a unique point P. The numbers in the ordered pair are called the ***coordinates*** of P, and the point is denoted by $P(a, b)$ when it is important to emphasize its coordinates.

A ***rectangular coordinate system in 3-space*** consists of three mutually perpendicular coordinate axes, which are usually called the ***x-axis***, the ***y-axis***, and the ***z-axis***. The point of intersection of the axes is called the ***origin*** of the coordinate system. As in 2-space, we will assume in this text that equal scales of measurement are used on the coordinate axes.

Coordinate systems in 3-space can be ***left-handed*** or ***right-handed***. To distinguish one from the other, assume that you are standing at the origin with your head in the positive z-direction and your arms extending along the x- and y-axes. The coordinate system is left-handed or right-handed in accordance with which of your arms is in the positive x-direction (Figure 1.1.12a). *In this text we will work exclusively with right-handed coordinate systems.* Note that in a right-handed system an ordinary screw pointing in the positive z-direction will advance if the positive x-axis is turned toward the positive y-axis through the 90° angle between them (Figure 1.1.12b). Once a rectangular coordinate system is introduced in 3-space, then the construction shown in Figure 1.1.12c produces a one-to-one correspondence between points and ordered triples of real numbers; that is, each point P in 3-space is associated with a unique ordered triple (a, b, c) of real numbers, and each ordered triple of real numbers (a, b, c) is associated with a unique point P. The numbers in the ordered triple are called the ***coordinates*** of P, and the point is denoted by $P(a, b, c)$ when it is important to emphasize the associated coordinates.

If a vector **v** in 2-space or 3-space is positioned with its initial point at the origin of a rectangular coordinate system, then the vector is completely determined by the coordinates of its terminal point, and we call these coordinates the ***components*** of **v** relative to the coordinate system. We will write $\mathbf{v} = (v_1, v_2)$ for the vector **v** in 2-space with components (v_1, v_2) and $\mathbf{v} = (v_1, v_2, v_3)$ for the vector **v** in 3-space with components (v_1, v_2, v_3) (Figure 1.1.13).

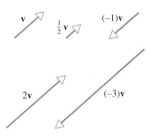

Figure 1.1.10

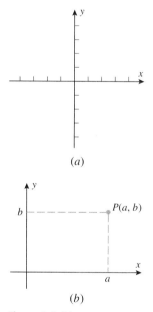

(a)

(b)

Figure 1.1.11

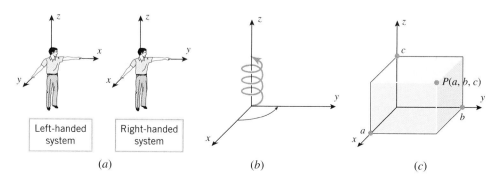

Figure 1.1.12   (a)   (b)   (c)

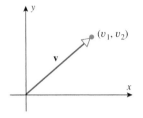

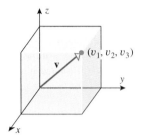

Figure 1.1.13

Note that the component forms of the zero vector in 2-space and 3-space are $\mathbf{0} = (0, 0)$ and $\mathbf{0} = (0, 0, 0)$, respectively.

It should be evident geometrically that two vectors in 2-space or 3-space are equivalent if and only if they have the same terminal point when their initial points are at the origin. Algebraically, this means that two vectors are equivalent if and only if their corresponding components are equal. Thus, the vectors $\mathbf{v} = (v_1, v_2)$ and $\mathbf{w} = (w_1, w_2)$ are equivalent if and only if $v_1 = w_1$ and $v_2 = w_2$; and the vectors $\mathbf{v} = (v_1, v_2, v_3)$ and $\mathbf{w} = (w_1, w_2, w_3)$ are equivalent if and only if $v_1 = w_1$, $v_2 = w_2$, and $v_3 = w_3$.

Algebraically, vectors in 2-space can now be viewed as ordered pairs of real numbers and vectors in 3-space as ordered triples of real numbers. Thus, we will denote the set of all vectors in 2-space by R^2 and the set of all vectors in 3-space by R^3 (the "R" standing for the word "real").

REMARK   It may already have occurred to you that ordered pairs and triples are used to represent both points and vectors in 2-space and 3-space. Thus, in the absence of additional information, there is no way to tell whether the ordered pair (v_1, v_2) represents the *point* with coordinates v_1 and v_2 or the *vector* with components v_1 and v_2 (Figure 1.1.14). The appropriate interpretation depends on the geometric viewpoint that you want to emphasize.

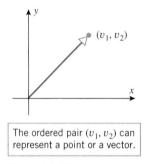

The ordered pair (v_1, v_2) can represent a point or a vector.

Figure 1.1.14

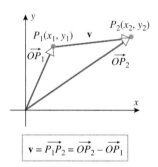

$\mathbf{v} = \overrightarrow{P_1 P_2} = \overrightarrow{OP_2} - \overrightarrow{OP_1}$

Figure 1.1.15

COMPONENTS OF A VECTOR WHOSE INITIAL POINT IS NOT AT THE ORIGIN

Sometimes we will need to find the components of a vector $\mathbf{v}$ in R^2 or R^3 that does not have its initial point at the origin. For this purpose, suppose that $\mathbf{v}$ is a vector in R^2 with initial point $P_1(x_1, y_1)$ and terminal point $P_2(x_2, y_2)$. As suggested by Figure 1.1.15, we can express $\mathbf{v}$ in terms of the vectors $\overrightarrow{OP_1}$ and $\overrightarrow{OP_2}$ as

$$\mathbf{v} = \overrightarrow{P_1 P_2} = \overrightarrow{OP_2} - \overrightarrow{OP_1} = (x_2 - x_1, y_2 - y_1)$$

That is, *the components of* $\mathbf{v}$ *are obtained by subtracting the coordinates of the initial point from the corresponding coordinates of the terminal point.* The same result holds in 3-space, so we have the following theorem.

Theorem 1.1.1

(a) *The vector in 2-space that has initial point $P_1(x_1, y_1)$ and terminal point $P_2(x_2, y_2)$ is*

$$\overrightarrow{P_1 P_2} = (x_2 - x_1, y_2 - y_1) \tag{4}$$

(b) *The vector in 3-space that has initial point $P_1(x_1, y_1, z_1)$ and terminal point $P_2(x_2, y_2, z_2)$ is*

$$\overrightarrow{P_1 P_2} = (x_2 - x_1, y_2 - y_1, z_2 - z_1) \tag{5}$$

EXAMPLE 2
Components of a Vector Whose Initial Point Is Not at the Origin

The component form of the vector that has its initial point at $P_1(2, -1, 4)$ and its terminal point at $P_2(7, 5, -8)$ is

$$\overrightarrow{P_1 P_2} = (7 - 2, 5 - (-1), -8 - 4) = (5, 6, -12)$$

This means that if the vector $\overrightarrow{P_1 P_2}$ is translated so that its initial point is at the origin, then its terminal point will fall at the point $(5, 6, -12)$. ∎

VECTORS IN R^n   The idea of using ordered pairs and triples of real numbers to represent points and vectors in 2-space and 3-space was well known in the eighteenth and nineteenth centuries, but in the late nineteenth and early twentieth centuries mathematicians and physicists began to recognize the physical importance of *higher-dimensional spaces*. One of the most important examples is due to Albert Einstein, who attached a time component t to three space components (x, y, z) to obtain a quadruple (x, y, z, t), which he regarded to be a point in a four-dimensional *space-time* universe. Although we cannot see four-dimensional space in the way that we see two- and three-dimensional space, it is nevertheless possible to extend familiar geometric ideas to four dimensions by working with algebraic properties of quadruples. Indeed, by developing an appropriate geometry of the four-dimensional space-time universe, Einstein developed his general relativity theory, which explained for the first time how gravity works. To explore the concept of higher-dimensional spaces we make the following definition.

Linear Algebra in History

The German-born physicist Albert Einstein immigrated to the United States in 1935, where he settled at Princeton University. Einstein spent the last three decades of his life working unsuccessfully at producing a *unified field theory* that would establish an underlying link between the forces of gravity and electromagnetism. Recently, physicists have made progress on the problem using a framework known as *string theory*. In this theory the smallest, indivisible components of the Universe are not particles but loops that behave like vibrating strings. Whereas Einstein's space-time universe was four-dimensional, strings reside in an 11-dimensional world that is the focus of much current research.

—*Based on an article in Time Magazine, September 30, 1999.*

**Albert Einstein
(1879–1955)**

Definition 1.1.2   If n is a positive integer, then an ***ordered n-tuple*** is a sequence of n real numbers $(v_1, v_2, \ldots, v_n)$. The set of all ordered n-tuples is called ***n-space*** and is denoted by R^n.

We will denote n-tuples using the vector notation $\mathbf{v} = (v_1, v_2, \ldots, v_n)$, and we will write $\mathbf{0} = (0, 0, \ldots, 0)$ for the n-tuple whose components are all zero. We will call this the ***zero vector*** or sometimes the ***origin*** of R^n.

REMARK   You can think of the numbers in an n-tuple $(v_1, v_2, \ldots, v_n)$ as either the coordinates of a *generalized point* or the components of a *generalized vector*, depending on the geometric image you want to bring to mind—the choice makes no difference mathematically, since it is the algebraic properties of n-tuples that are of concern.

An ordered 1-tuple ($n = 1$) is a single real number, so R^1 can be viewed algebraically as the set of real numbers or geometrically as a line. Ordered 2-tuples ($n = 2$) are ordered pairs of real numbers, so we can view R^2 geometrically as a plane. Ordered 3-tuples are ordered triples of real numbers, so we can view R^3 geometrically as the space around us. We will sometimes refer to R^1, R^2, and R^3 as ***visible space*** and R^4, $R^5, \ldots$ as ***higher-dimensional spaces***. Here are some physical examples that lead to higher-dimensional spaces.

EXAMPLE 3
Some Examples of Vectors in Higher-Dimensional Spaces

- **Experimental Data**—A scientist performs an experiment and makes n numerical measurements each time the experiment is performed. The result of each experiment can be regarded as a vector $\mathbf{y} = (y_1, y_2, \ldots, y_n)$ in R^n in which $y_1, y_2, \ldots, y_n$ are the measured values.

- **Storage and Warehousing**—A national trucking company has 15 depots for storing and servicing its trucks. At each point in time the distribution of trucks in the service depots can be described by a 15-tuple $\mathbf{x} = (x_1, x_2, \ldots, x_{15})$ in which x_1 is the number of trucks in the first depot, x_2 is the number in the second depot, and so forth.

- **Electrical Circuits**—A certain kind of processing chip is designed to receive four input voltages and produces three output voltages in response. The input voltages can be regarded as vectors in R^4 and the output voltages as vectors in R^3. Thus, the chip can be viewed as a device that transforms each input vector $\mathbf{v} = (v_1, v_2, v_3, v_4)$ in R^4 into some output vector $\mathbf{w} = (w_1, w_2, w_3)$ in R^3.

- **Graphical Images**—One way in which color images are created on computer screens is by assigning each pixel (an addressable point on the screen) three numbers that describe the ***hue***, ***saturation***, and ***brightness*** of the pixel. Thus, a complete color image can be viewed as a set of 5-tuples of the form $\mathbf{v} = (x, y, h, s, b)$ in which x and y are the screen coordinates of a pixel and h, s, and b are its hue, saturation, and brightness.

- **Economics**—One approach to economic analysis is to divide an economy into sectors (manufacturing, services, utilities, and so forth) and to measure the output of each sector by a dollar value. Thus, in an economy with 10 sectors the economic output of the entire economy can be represented by a 10-tuple $\mathbf{s} = (s_1, s_2, \ldots, s_{10})$ in which the numbers $s_1, s_2, \ldots, s_{10}$ are the outputs of the individual sectors.

- **Mechanical Systems**—Suppose that six particles move along the same coordinate line so that at time t their coordinates are $x_1, x_2, \ldots, x_6$ and their velocities are $v_1, v_2, \ldots, v_6$, respectively. This information can be represented by the vector

$$\mathbf{v} = (x_1, x_2, x_3, x_4, x_5, x_6, v_1, v_2, v_3, v_4, v_5, v_6, t)$$

in R^{13}. This vector is called the **state** of the particle system at time t. ∎

CONCEPT PROBLEM Try to think of some other physical examples in which n-tuples might arise.

EQUALITY OF VECTORS We observed earlier that two vectors in R^2 or R^3 are equivalent if and only if their corresponding components are equal. Thus, we make the following definition.

Definition 1.1.3 Vectors $\mathbf{v} = (v_1, v_2, \ldots, v_n)$ and $\mathbf{w} = (w_1, w_2, \ldots, w_n)$ in R^n are said to be **equivalent** (also called **equal**) if

$$v_1 = w_1, \quad v_2 = w_2, \ldots, \quad v_n = w_n$$

We indicate this by writing $\mathbf{v} = \mathbf{w}$.

Linear Algebra in History

The idea of representing vectors as n-tuples began to crystallize around 1814 when the Swiss accountant (and amateur mathematician) Jean Robert Argand (1768–1822) proposed the idea of representing a complex number $a + bi$ as an ordered pair (a, b) of real numbers. Subsequently, the Irish mathematician William Hamilton developed his theory of *quaternions*, which was the first important example of a four-dimensional space. Hamilton presented his ideas in a paper given to the Irish Academy in 1833. The concept of an n-dimensional space became firmly established in 1844 when the German mathematician Hermann Grassmann published a book entitled *Ausdehnungslehre* in which he developed many of the fundamental ideas that appear in this text.

Sir William Rowan
Hamilton
(1805–1865)

Hermann Günther
Grassmann
(1809–1877)

Thus, for example,

$$(a, b, c, d) = (1, -4, 2, 7)$$

if and only if $a = 1$, $b = -4$, $c = 2$, and $d = 7$.

Our next objective is to define the operations of addition, subtraction, and scalar multiplication for vectors in R^n. To motivate the ideas, we will consider how these operations can be performed on vectors in R^2 using components. By studying Figure 1.1.16 you should be able to deduce that if $\mathbf{v} = (v_1, v_2)$ and $\mathbf{w} = (w_1, w_2)$, then

$$\mathbf{v} + \mathbf{w} = (v_1 + w_1, v_2 + w_2) \tag{6}$$

$$k\mathbf{v} = (kv_1, kv_2) \tag{7}$$

Stated in words, *vectors are added by adding their corresponding components*, and *a vector is multiplied by a scalar by multiplying each component by the scalar*. In particular, it follows from (7) that

$$-\mathbf{v} = (-1)\mathbf{v} = (-v_1, -v_2) \tag{8}$$

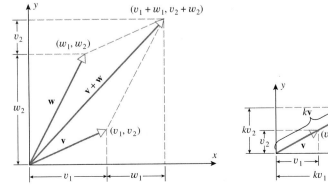

Figure 1.1.16

and hence that

$$\mathbf{w} - \mathbf{v} = \mathbf{w} + (-\mathbf{v}) = (w_1 - v_1, w_2 - v_2) \tag{9}$$

That is, *vectors are subtracted by subtracting corresponding components.*

Motivated by Formulas (6)–(9), we make the following definition.

Definition 1.1.4   If $\mathbf{v} = (v_1, v_2, \dots, v_n)$ and $\mathbf{w} = (w_1, w_2, \dots, w_n)$ are vectors in R^n, and if k is any scalar, then we define

$$\mathbf{v} + \mathbf{w} = (v_1 + w_1, v_2 + w_2, \dots, v_n + w_n) \tag{10}$$

$$k\mathbf{v} = (kv_1, kv_2, \dots, kv_n) \tag{11}$$

$$-\mathbf{v} = (-v_1, -v_2, \dots, -v_n) \tag{12}$$

$$\mathbf{w} - \mathbf{v} = \mathbf{w} + (-\mathbf{v}) = (w_1 - v_1, w_2 - v_2, \dots, w_n - v_n) \tag{13}$$

EXAMPLE 4
Algebraic
Operations
Using
Components

If $\mathbf{v} = (1, -3, 2)$ and $\mathbf{w} = (4, 2, 1)$, then

$$\mathbf{v} + \mathbf{w} = (5, -1, 3), \qquad 2\mathbf{v} = (2, -6, 4)$$
$$-\mathbf{w} = (-4, -2, -1), \qquad \mathbf{v} - \mathbf{w} = \mathbf{v} + (-\mathbf{w}) = (-3, -5, 1) \qquad \blacksquare$$

The following theorem summarizes the most important properties of vector operations.

Theorem 1.1.5   *If* $\mathbf{u}$, $\mathbf{v}$, *and* $\mathbf{w}$ *are vectors in* R^n, *and if* k *and* l *are scalars, then:*

(a) $\mathbf{u} + \mathbf{v} = \mathbf{v} + \mathbf{u}$

(b) $(\mathbf{u} + \mathbf{v}) + \mathbf{w} = \mathbf{u} + (\mathbf{v} + \mathbf{w})$

(c) $\mathbf{u} + \mathbf{0} = \mathbf{0} + \mathbf{u} = \mathbf{u}$

(d) $\mathbf{u} + (-\mathbf{u}) = \mathbf{0}$

(e) $(k + l)\mathbf{u} = k\mathbf{u} + l\mathbf{u}$

(f) $k(\mathbf{u} + \mathbf{v}) = k\mathbf{u} + k\mathbf{v}$

(g) $k(l\mathbf{u}) = (kl)\mathbf{u}$

(h) $1\mathbf{u} = \mathbf{u}$

We will prove part (b) and leave some of the other proofs as exercises.

Proof (b)   Let $\mathbf{u} = (u_1, u_2, \dots, u_n)$, $\mathbf{v} = (v_1, v_2, \dots, v_n)$, and $\mathbf{w} = (w_1, w_2, \dots, w_n)$. Then

$$
\begin{aligned}
(\mathbf{u} + \mathbf{v}) + \mathbf{w} &= [(u_1, u_2, \dots, u_n) + (v_1, v_2, \dots, v_n)] + (w_1, w_2, \dots, w_n) \\
&= (u_1 + v_1, u_2 + v_2, \dots, u_n + v_n) + (w_1, w_2, \dots, w_n) &&\text{[Vector addition]} \\
&= \big((u_1 + v_1) + w_1, (u_2 + v_2) + w_2, \dots, (u_n + v_n) + w_n\big) &&\text{[Vector addition]} \\
&= \big(u_1 + (v_1 + w_1), u_2 + (v_2 + w_2), \dots, u_n + (v_n + w_n)\big) &&\text{[Regroup]} \\
&= (u_1, u_2, \dots, u_n) + (v_1 + w_1, v_2 + w_2, \dots, v_n + w_n) &&\text{[Vector addition]} \\
&= \mathbf{u} + (\mathbf{v} + \mathbf{w}) &&\blacksquare
\end{aligned}
$$

The following additional properties of vectors in R^n can be deduced easily by expressing the vectors in terms of components (verify).

Theorem 1.1.6   *If* $\mathbf{v}$ *is a vector in* R^n *and* k *is a scalar, then:*

(a) $0\mathbf{v} = \mathbf{0}$

(b) $k\mathbf{0} = \mathbf{0}$

(c) $(-1)\mathbf{v} = -\mathbf{v}$

LOOKING AHEAD   Theorem 1.1.5 is one of the most fundamental theorems in linear algebra in that all algebraic properties of vectors can be derived from the eight properties stated in the theorem. For example, even though Theorem 1.1.6 is easy to prove by using components, it can also be derived from the properties in Theorem 1.1.5 without breaking the vectors into components (Exercise P3). Later we will use Theorem 1.1.5 as a starting point for extending the concept of a vector beyond R^n.

SUMS OF THREE OR MORE VECTORS

Part (*b*) of Theorem 1.1.5, called the ***associative law for vector addition***, implies that the expression $\mathbf{u} + \mathbf{v} + \mathbf{w}$ is unambiguous, since the same sum results no matter how the parentheses are inserted. This is illustrated geometrically in Figure 1.1.17*a* for vectors in R^2 and R^3. That figure also shows that the vector $\mathbf{u} + \mathbf{v} + \mathbf{w}$ can be obtained by placing $\mathbf{u}$, $\mathbf{v}$, and $\mathbf{w}$ tip to tail in succession and then drawing the vector from the initial point of $\mathbf{u}$ to the terminal point of $\mathbf{w}$. This result generalizes to sums with four or more vectors in R^2 and R^3 (Figure 1.1.17*b*). The tip-to-tail method makes it evident that if $\mathbf{u}$, $\mathbf{v}$, and $\mathbf{w}$ are vectors in R^3 that are positioned with a common initial point, then $\mathbf{u} + \mathbf{v} + \mathbf{w}$ is a diagonal of the parallelepiped that has the three vectors as adjacent edges (Figure 1.1.17*c*).

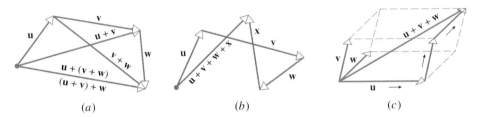

Figure 1.1.17   (*a*)   (*b*)   (*c*)

PARALLEL AND COLLINEAR VECTORS

Suppose that $\mathbf{v}$ and $\mathbf{w}$ are vectors in R^2 or R^3 that are positioned with a common initial point. If one of the vectors is a scalar multiple of the other, then the vectors lie on a common line, so it is reasonable to say that they are *collinear* (Figure 1.1.18*a*). However, if we translate one of the vectors as indicated in Figure 1.1.18*b*, then the vectors are *parallel* but no longer collinear. This creates a linguistic problem because translating a vector does not change it. The only way to resolve this problem is to agree that the terms *parallel* and *collinear* mean the same thing when applied to vectors. Accordingly, we make the following definition.

Definition 1.1.7 Two vectors in R^n are said to be ***parallel*** or, alternatively, ***collinear*** if at least one of the vectors is a scalar multiple of the other. If one of the vectors is a positive scalar multiple of the other, then the vectors are said to have the ***same direction***, and if one of them is a negative scalar multiple of the other, then the vectors are said to have ***opposite directions***.

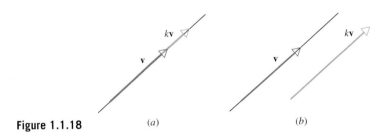

Figure 1.1.18   (*a*)   (*b*)

REMARK The vector $\mathbf{0}$ is parallel to *every* vector $\mathbf{v}$ in R^n, since it can be expressed as the scalar multiple $\mathbf{0} = 0\mathbf{v}$.

LINEAR COMBINATIONS

Frequently, addition, subtraction, and scalar multiplication are used in combination to form new vectors. For example, if $\mathbf{v}_1$, $\mathbf{v}_2$, and $\mathbf{v}_3$ are given vectors, then the vectors

$$\mathbf{w} = 2\mathbf{v}_1 + 3\mathbf{v}_2 + \mathbf{v}_3 \quad \text{and} \quad \mathbf{w} = 7\mathbf{v}_1 - 6\mathbf{v}_2 + 8\mathbf{v}_3$$

are formed in this way. In general, we make the following definition.

Definition 1.1.8 A vector **w** in R^n is said to be a ***linear combination*** of the vectors $\mathbf{v}_1, \mathbf{v}_2, \ldots, \mathbf{v}_k$ in R^n if **w** can be expressed in the form

$$\mathbf{w} = c_1\mathbf{v}_1 + c_2\mathbf{v}_2 + \cdots + c_k\mathbf{v}_k \tag{14}$$

The scalars $c_1, c_2, \ldots, c_k$ are called the ***coefficients*** in the linear combination. In the case where $k = 1$, Formula (14) becomes $\mathbf{w} = c_1\mathbf{v}_1$, so to say that **w** is a linear combination of $\mathbf{v}_1$ is the same as saying that **w** is a scalar multiple of $\mathbf{v}_1$.

APPLICATION TO COMPUTER COLOR MODELS

Colors on computer monitors are commonly based on what is called the **RGB** *color model*. Colors in this system are created by adding together percentages of the primary colors red (R), green (G), and blue (B). One way to do this is to identify the primary colors with the vectors

$$\mathbf{r} = (1, 0, 0) \quad \text{(pure red)}, \qquad \mathbf{g} = (0, 1, 0) \quad \text{(pure green)}, \qquad \mathbf{b} = (0, 0, 1) \quad \text{(pure blue)}$$

in R^3 and to create all other colors by forming linear combinations of **r**, **g**, and **b** using coefficients between 0 and 1, inclusive; these coefficients represent the percentage of each pure color in the mix. The set of all such color vectors is called **RGB** *space* or the **RGB** *color cube* (Figure 1.1.19). Thus, each color vector **c** in this cube is expressible as a linear combination of the form

$$\mathbf{c} = c_1\mathbf{r} + c_2\mathbf{g} + c_3\mathbf{b} = c_1(1, 0, 0) + c_2(0, 1, 0) + c_3(0, 0, 1) = (c_1, c_2, c_3)$$

where $0 \leq c_i \leq 1$. As indicated in the figure, the corners of the cube represent the pure primary colors together with the colors, black, white, magenta, cyan, and yellow. The vectors along the diagonal running from black to white correspond to shades of gray.

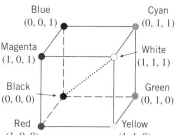

Figure 1.1.19

ALTERNATIVE NOTATIONS FOR VECTORS

Up to now we have been writing vectors in R^n using the notation

$$\mathbf{v} = (v_1, v_2, \ldots, v_n) \tag{15}$$

We call this the ***comma-delimited*** form. However, a vector in R^n is essentially just a list of n numbers (the components) in a definite order, so any notation that displays the components of the vector in their correct order is a valid alternative to the comma-delimited notation. For example, the vector in (15) might be written as

$$\mathbf{v} = [v_1 \quad v_2 \quad \cdots \quad v_n] \tag{16}$$

which is called ***row-vector*** form, or as

$$\mathbf{v} = \begin{bmatrix} v_1 \\ v_2 \\ \vdots \\ v_n \end{bmatrix} \tag{17}$$

which is called ***column-vector*** form. The choice of notation is often a matter of taste or convenience, but sometimes the nature of the problem under consideration will suggest a particular notation. All three notations will be used in this text.

MATRICES   Numerical information is often organized into tables called *matrices* (plural of *matrix*). For example, here is a matrix description of the number of hours a student spent on homework in four subjects over a certain one-week period:

	Monday	Tuesday	Wednesday	Thursday	Friday	Saturday	Sunday
Math	2	1	2	0	3	0	1
English	2	0	1	3	1	0	1
Chemistry	1	3	0	0	1	0	1
Physics	1	2	4	1	0	0	2

Linear Algebra in History

The theory of graphs originated with the Swiss mathematician Leonhard Euler, who developed the ideas to solve a problem that was posed to him in the mid 1700s by the citizens of the Prussian city of Königsberg (now Kaliningrad in Russia). The city is cut by the Pregel River, which encloses an island, as shown in the accompanying old lithograph.

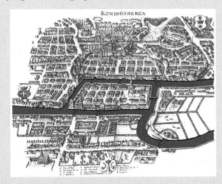

The problem was to determine whether it was possible to start at any point on the shore of the river, or on the island, and walk over all of the bridges, once and only once, returning to the starting spot. In 1736 Euler showed that the walk was impossible by analyzing the graph.

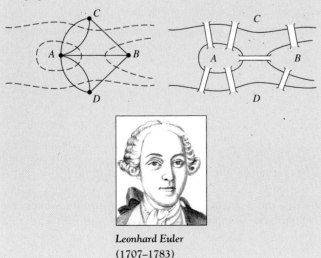

Leonhard Euler
(1707–1783)

If we suppress the headings, then the numerical data that remain form a matrix with four rows and seven columns:

$$\begin{bmatrix} 2 & 1 & 2 & 0 & 3 & 0 & 1 \\ 2 & 0 & 1 & 3 & 1 & 0 & 1 \\ 1 & 3 & 0 & 0 & 1 & 0 & 1 \\ 1 & 2 & 4 & 1 & 0 & 0 & 2 \end{bmatrix} \tag{18}$$

To formalize this idea, we define a ***matrix*** to be a rectangular array of numbers, called the ***entries*** of the matrix. If a matrix has m rows and n columns, then it is said to have ***size*** $m \times n$, where the number of rows is always written first. Thus, for example, the matrix in (18) has size 4×7. A matrix with one row is called a ***row vector***, and a matrix with one column is called a ***column vector*** [see (16) and (17), for example]. You can also think of a matrix as a list of row vectors or column vectors. For example, the matrix in (18) can be viewed as a list of four row vectors in R^7 or as a list of seven column vectors in R^4.

In addition to describing tabular information, matrices are useful for describing *connections* between objects, say connections between cities by airline routes, connections between people in social structures, or connections between elements in an electrical circuit. The idea is to represent the objects being connected as points, called ***vertices***, and to indicate connections between vertices by line segments or arcs, called ***edges***. The vertices and edges form what is called a ***connectivity graph*** or, more simply, a ***graph***. For example, Figure 1.1.20a shows a graph that describes airline routes between four cities; the cities that have a direct airline route between them are connected by an edge. The arrows on the edges distinguish between ***two-way connections*** and ***one-way connections***; for example, the double arrow on the edge joining cities 1 and 3 indicates that there is a route from city 1 to city 3 and one from city 3 to city 1, whereas the single arrow on the edge joining cities 1 and 4 indicates that there is a route from city 1 to city 4 but not one from city 4 to city 1. A graph marked with one-way and two-way connections is called a ***directed graph***.

A directed graph can be described by an $n \times n$ matrix, called an ***adjacency matrix***, in which the 1's and 0's are used to describe connections. Specifically, if the vertices are labeled from 1 to n, then the entry in row i and column j of the matrix is a 1 if there is a connection from vertex i to vertex j and a 0 if there is not. For example, Figure 1.1.20*b* is the adjacency matrix for the directed graph in part (*a*) of the figure (verify).

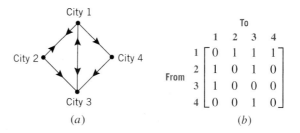

Figure 1.1.20   (*a*)   (*b*)

Exercise Set 1.1

In Exercises 1 and 2, draw the vectors with their initial points at the origin.

1. (a) $\mathbf{v}_1 = (3, 6)$ (b) $\mathbf{v}_2 = (-4, -8)$
    (c) $\mathbf{v}_3 = (3, 3, 0)$ (d) $\mathbf{v}_4 = (0, 0, -3)$

2. (a) $\mathbf{v}_1 = (-1, 2)$ (b) $\mathbf{v}_2 = (3, 4)$
    (c) $\mathbf{v}_3 = (1, 2, 3)$ (d) $\mathbf{v}_4 = (-1, 6, 1)$

In Exercises 3 and 4, draw the vectors with their initial points at the origin, given that $\mathbf{u} = (1, 1)$ and $\mathbf{v} = (-1, 1)$.

3. (a) $2\mathbf{u}$ (b) $\mathbf{u} + \mathbf{v}$ (c) $2\mathbf{u} + 2\mathbf{v}$
    (d) $\mathbf{u} - \mathbf{v}$ (e) $\mathbf{u} + 2\mathbf{v}$

4. (a) $-\mathbf{u} + \mathbf{v}$ (b) $3\mathbf{u} + 2\mathbf{v}$ (c) $2\mathbf{u} + 5\mathbf{v}$
    (d) $-2\mathbf{u} - \mathbf{v}$ (e) $2\mathbf{u} - 3\mathbf{v}$

In Exercises 5 and 6, find the components of the vector, and sketch an equivalent vector with its initial point at the origin.

5. (a) 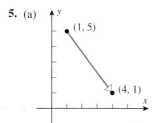   (b)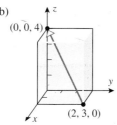

6. (a)   (b)

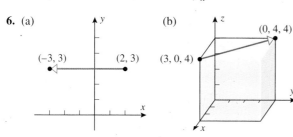

In Exercises 7 and 8, find the components of the vector $\overrightarrow{P_1 P_2}$.

7. (a) $P_1(3, 5)$, $P_2(2, 8)$ (b) $P_1(5, -2, 1)$, $P_2(2, 4, 2)$

8. (a) $P_1(-6, 2)$, $P_2(-4, -1)$ (b) $P_1(0, 0, 0)$, $P_2(-1, 6, 1)$

9. (a) Find the terminal point of the vector that is equivalent to $\mathbf{u} = (1, 2)$ and whose initial point is $A(1, 1)$.
    (b) Find the initial point of the vector that is equivalent to $\mathbf{u} = (1, 1, 3)$ and whose terminal point is $B(-1, -1, 2)$.

10. (a) Find the initial point of the vector that is equivalent to $\mathbf{u} = (1, 2)$ and whose terminal point is $A(2, 0)$.
    (b) Find the terminal point of the vector that is equivalent to $\mathbf{u} = (1, 1, 3)$ and whose initial point is $B(0, 2, 0)$.

11. Let $\mathbf{u} = (-3, 1, 2, 4, 4)$, $\mathbf{v} = (4, 0, -8, 1, 2)$, and $\mathbf{w} = (6, -1, -4, 3, -5)$. Find the components of
    (a) $\mathbf{v} - \mathbf{w}$ (b) $6\mathbf{u} + 2\mathbf{v}$
    (c) $(2\mathbf{u} - 7\mathbf{w}) - (8\mathbf{v} + \mathbf{u})$

12. Let $\mathbf{u} = (1, 2, -3, 5, 0)$, $\mathbf{v} = (0, 4, -1, 1, 2)$, and $\mathbf{w} = (7, 1, -4, -2, 3)$. Find the components of
    (a) $\mathbf{v} + \mathbf{w}$ (b) $3(2\mathbf{u} - \mathbf{v})$
    (c) $(3\mathbf{u} - \mathbf{v}) - (2\mathbf{u} + 4\mathbf{w})$

13. Let $\mathbf{u}$, $\mathbf{v}$, and $\mathbf{w}$ be the vectors in Exercise 11. Find the components of the vector $\mathbf{x}$ that satisfies the equation $2\mathbf{u} - \mathbf{v} + \mathbf{x} = 7\mathbf{x} + \mathbf{w}$.

14. Let $\mathbf{u}$, $\mathbf{v}$, and $\mathbf{w}$ be the vectors in Exercise 12. Find the components of the vector $\mathbf{x}$ that satisfies the equation $3\mathbf{u} + \mathbf{v} - 2\mathbf{w} = 3\mathbf{x} + 2\mathbf{w}$.

15. Which of the following vectors in R^6 are parallel to $\mathbf{u} = (-2, 1, 0, 3, 5, 1)$?
    (a) $(4, 2, 0, 6, 10, 2)$ (b) $(4, -2, 0, -6, -10, -2)$
    (c) $(0, 0, 0, 0, 0, 0)$

16. For what value(s) of t, if any, is the given vector parallel to $\mathbf{u} = (4, -1)$?
    (a) $(8t, -2)$ (b) $(8t, 2t)$
    (c) $(1, t^2)$

17. In each part, sketch the vector $\mathbf{u} + \mathbf{v} + \mathbf{w}$, and express it in component form.

(a)

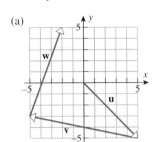

(b)

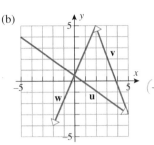

18. In each part of Exercise 17, sketch the vector $\mathbf{u} - \mathbf{v} + \mathbf{w}$, and express it in component form.

19. Let $\mathbf{u} = (1, -1, 3, 5)$ and $\mathbf{v} = (2, 1, 0, -3)$. Find scalars a and b so that $a\mathbf{u} + b\mathbf{v} = (1, -4, 9, 18)$.

20. Let $\mathbf{u} = (2, 1, 0, 1, -1)$ and $\mathbf{v} = (-2, 3, 1, 0, 2)$. Find scalars a and b so that $a\mathbf{u} + b\mathbf{v} = (-8, 8, 3, -1, 7)$.

21. Draw three parallelograms that have points $A = (0, 0)$, $B = (-1, 3)$, and $C = (1, 2)$ as vertices.

22. Verify that one of the parallelograms in Exercise 21 has the terminal point of $\overrightarrow{AB} + \overrightarrow{AC}$ as the fourth vertex, and then express the fourth vertex in each of the other parallelograms in terms of $\overrightarrow{AB}$ and $\overrightarrow{AC}$.

A particle is said to be in **static equilibrium** if the sum of all forces applied to it is zero. In Exercises 23 and 24, find the components of the force $\mathbf{F}$ that must be applied to a particle at the origin to produce static equilibrium. The force $\mathbf{F}$ is applied in addition to the forces shown, and no other force is present.

23.

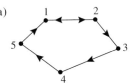

24.

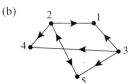

In Exercises 25 and 26, construct an adjacency matrix for the given directed graph.

25. (a)   (b)

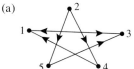

26. (a)   (b)

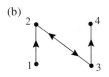

In Exercises 27 and 28, construct a directed graph whose adjacency matrix is equal to the given matrix.

27. $\begin{bmatrix} 0 & 1 & 0 & 0 & 0 \\ 0 & 0 & 0 & 1 & 0 \\ 1 & 0 & 0 & 0 & 0 \\ 0 & 0 & 0 & 0 & 1 \\ 0 & 0 & 1 & 0 & 0 \end{bmatrix}$

28. $\begin{bmatrix} 0 & 1 & 0 & 1 & 0 & 0 \\ 0 & 0 & 1 & 0 & 0 & 0 \\ 0 & 0 & 0 & 1 & 0 & 1 \\ 0 & 0 & 0 & 0 & 0 & 0 \\ 0 & 0 & 0 & 1 & 0 & 0 \\ 1 & 1 & 0 & 0 & 0 & 0 \end{bmatrix}$

Discussion and Discovery

D1. Give some physical examples of quantities that might be described by vectors in R^4.

D2. Is time a vector or a scalar? Write a paragraph to explain your answer.

D3. If the sum of three vectors in R^3 is zero, must they lie in the same plane? Explain.

D4. A monk walks from a monastery gate to the top of a mountain to pray and returns to the monastery gate the next day. What is the monk's displacement? What is the relationship between the monk's displacement going from the monastery gate to the top of the mountain and the displacement going from the top of the mountain back to the gate?

D5. Consider the regular hexagon shown in the accompanying figure.
 (a) What is the sum of the six radial vectors that run from the center to the vertices?

 (b) How is the sum affected if each radial vector is multiplied by $\frac{1}{2}$?
 (c) What is the sum of the five radial vectors that remain if $\mathbf{a}$ is removed?
 (d) Discuss some variations and generalizations of the result in part (c).

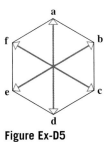

Figure Ex-D5

D6. What is the sum of all radial vectors of a regular n-sided polygon? (See Exercise D5.)

D7. Consider a clock with vectors drawn from the center to each hour as shown in the accompanying figure.
  (a) What is the sum of the 12 vectors that result if the vector terminating at 12 is doubled in length and the other vectors are left alone?
  (b) What is the sum of the 12 vectors that result if the vectors terminating at 3 and 9 are each tripled and the others are left alone?
  (c) What is the sum of the 9 vectors that remain if the vectors terminating at 5, 11, and 8 are removed?

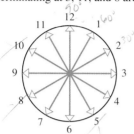

Figure Ex-D7

D8. Draw a picture that shows four nonzero vectors in the plane, one of which is the sum of the other three.

D9. Indicate whether the statement is true (T) or false (F). Justify your answer.
  (a) If $\mathbf{x} + \mathbf{y} = \mathbf{x} + \mathbf{z}$, then $\mathbf{y} = \mathbf{z}$.
  (b) If $\mathbf{u} + \mathbf{v} = \mathbf{0}$, then $a\mathbf{u} + b\mathbf{v} = \mathbf{0}$ for all a and b.
  (c) Parallel vectors with the same length are equal.
  (d) If $a\mathbf{x} = \mathbf{0}$, then either $a = 0$ or $\mathbf{x} = \mathbf{0}$.
  (e) If $a\mathbf{u} + b\mathbf{v} = \mathbf{0}$, then $\mathbf{u}$ and $\mathbf{v}$ are parallel vectors.
  (f) The vectors $\mathbf{u} = (\sqrt{2}, \sqrt{3})$ and $\mathbf{v} = \left(\frac{1}{\sqrt{2}}, \frac{1}{2}\sqrt{3}\right)$ are equivalent.

Working with Proofs

P1. Prove part (*e*) of Theorem 1.1.5.
P2. Prove part (*f*) of Theorem 1.1.5.

P3. Prove Theorem 1.1.6 without using components.

Technology Exercises

T1. (***Numbers and numerical operations***) Read how to enter integers, fractions, decimals, and irrational numbers such as π and $\sqrt{2}$. Check your understanding of the procedures by converting π, $\sqrt{2}$, and $1/3$ to decimal form with various numbers of decimal places in the display. Read about the procedures for performing the operations of addition, subtraction, multiplication, division, raising numbers to powers, and extraction of roots. Experiment with numbers of your own choosing until you feel you have mastered the techniques.

T2. (***Drawing vectors***) Read how to draw line segments in two- or three-dimensional space, and draw some line segments with initial and terminal points of your choice. If your utility allows you to create arrowheads, then you can make your line segments look like geometric vectors.

T3. (***Operations on vectors***) Read how to enter vectors and how to calculate their sums, differences, and scalar multiples. Check your understanding of these operations by performing the calculations in Example 4.

T4. Use your technology utility to compute the components of $\mathbf{u} = (7.1, -3) - 5(\sqrt{2}, 6) + 3(0, \pi)$ to five decimal places.

Section 1.2 Dot Product and Orthogonality

In this section we will be concerned with the concepts of length, angle, distance, and perpendicularity in R^n. We will begin by discussing these concepts geometrically in R^2 and R^3, and then we will extend them algebraically to R^n using components.

NORM OF A VECTOR   The length of a vector $\mathbf{v}$ in R^2 or R^3 is commonly denoted by the symbol $\|\mathbf{v}\|$. It follows from the theorem of Pythagoras that the length of a vector $\mathbf{v} = (v_1, v_2)$ in R^2 is given by the formula

$$\|\mathbf{v}\| = \sqrt{v_1^2 + v_2^2} \tag{1}$$

(Figure 1.2.1a). A companion formula for the length of a vector $\mathbf{v} = (v_1, v_2, v_3)$ in R^3 can be obtained using two applications of the theorem of Pythagoras (Figure 1.2.1b):

$$\|\mathbf{v}\|^2 = (OR)^2 + (RP)^2 = (OQ)^2 + (QR)^2 + (RP)^2 = v_1^2 + v_2^2 + v_3^2$$

Thus,

$$\|\mathbf{v}\| = \sqrt{v_1^2 + v_2^2 + v_3^2} \tag{2}$$

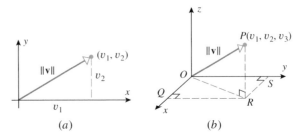

Figure 1.2.1 (a) (b)

Motivated by Formulas (1) and (2), we make the following general definition for the length of a vector in R^n.

Definition 1.2.1   If $\mathbf{v} = (v_1, v_2, \ldots, v_n)$ is a vector in R^n, then the **length** of $\mathbf{v}$, also called the **norm** of $\mathbf{v}$ or the **magnitude** of $\mathbf{v}$, is denoted by $\|\mathbf{v}\|$ and is defined by the formula

$$\|\mathbf{v}\| = \sqrt{v_1^2 + v_2^2 + v_3^2 + \cdots + v_n^2} \tag{3}$$

EXAMPLE 1
Calculating
Norms

From (3), the norm of the vector $\mathbf{v} = (-3, 2, 1)$ in R^3 is

$$\|\mathbf{v}\| = \sqrt{(-3)^2 + 2^2 + 1^2} = \sqrt{14}$$

and the norm of the vector $\mathbf{v} = (2, -1, 3, -5)$ in R^4 is

$$\|\mathbf{v}\| = \sqrt{2^2 + (-1)^2 + 3^2 + (-5)^2} = \sqrt{39}$$ ∎

Since lengths in R^2 and R^3 are nonnegative numbers, and since $\mathbf{0}$ is the only vector that has length zero, it follows that $\|\mathbf{v}\| \geq 0$ and that $\|\mathbf{v}\| = 0$ if and only if $\mathbf{v} = \mathbf{0}$. Also, multiplying $\mathbf{v}$ by a scalar k multiplies the length of $\mathbf{v}$ by $|k|$, so $\|k\mathbf{v}\| = |k|\|\mathbf{v}\|$. We will leave it for you to prove that these three properties also hold in R^n.

Theorem 1.2.2   *If $\mathbf{v}$ is a vector in R^n, and if k is any scalar, then:*
  (a) $\|\mathbf{v}\| \geq 0$
  (b) $\|\mathbf{v}\| = 0$ *if and only if* $\mathbf{v} = \mathbf{0}$
  (c) $\|k\mathbf{v}\| = |k|\|\mathbf{v}\|$

UNIT VECTORS

A vector of length 1 is called a **unit vector**. If $\mathbf{v}$ is a nonzero vector in R^n, then a unit vector $\mathbf{u}$ that has the same direction as $\mathbf{v}$ is given by the formula

$$\mathbf{u} = \frac{1}{\|\mathbf{v}\|}\mathbf{v} \tag{4}$$

In words, Formula (4) states that *a unit vector with the same direction as a vector* $\mathbf{v}$ *can be obtained by multiplying* $\mathbf{v}$ *by the reciprocal of its length.* This process is called **normalizing** $\mathbf{v}$. The vector $\mathbf{u}$ has the same direction as $\mathbf{v}$ since $1/\|\mathbf{v}\|$ is a positive scalar; and it has length 1 since part (c) of Theorem 1.2.2 with $k = 1/\|\mathbf{v}\|$ yields

$$\|\mathbf{u}\| = \|k\mathbf{v}\| = |k|\|\mathbf{v}\| = k\|\mathbf{v}\| = \frac{1}{\|\mathbf{v}\|}\|\mathbf{v}\| = 1$$

Sometimes you will see Formula (4) expressed as

$$\mathbf{u} = \frac{\mathbf{v}}{\|\mathbf{v}\|}$$

This is just a more compact way of writing the scalar product in (4).

EXAMPLE 2
Normalizing a
Vector

Find the unit vector **u** that has the same direction as $\mathbf{v} = (2, 2, -1)$.

Solution The vector **v** has length

$$\|\mathbf{v}\| = \sqrt{2^2 + 2^2 + (-1)^2} = 3$$

Thus, from (4)

$$\mathbf{u} = \tfrac{1}{3}(2, 2, -1) = \left(\tfrac{2}{3}, \tfrac{2}{3}, -\tfrac{1}{3}\right)$$

As a check, you may want to confirm that **u** is in fact a unit vector. ∎

CONCEPT PROBLEM Unit vectors are often used to specify directions in 2-space or 3-space. Find a unit vector that describes the direction that a bug would travel if it walked from the origin of an xy-coordinate system into the first quadrant along a line that makes an angle of $30°$ with the positive x-axis. Also, find a unit vector that describes the direction that the bug would travel if it walked into the third quadrant along the line.

**THE STANDARD UNIT
VECTORS**

When a rectangular coordinate system is introduced in R^2 or R^3, the unit vectors in the positive directions of the coordinate axes are called the ***standard unit vectors***. In R^2 these vectors are denoted by

$$\mathbf{i} = (1, 0) \quad \text{and} \quad \mathbf{j} = (0, 1) \tag{5}$$

and in R^3 they are denoted by

$$\mathbf{i} = (1, 0, 0), \quad \mathbf{j} = (0, 1, 0), \quad \text{and} \quad \mathbf{k} = (0, 0, 1) \tag{6}$$

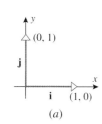

(Figure 1.2.2).

Observe that every vector $\mathbf{v} = (v_1, v_2)$ in R^2 can be expressed in terms of the standard unit vectors as

$$\mathbf{v} = (v_1, v_2) = v_1(1, 0) + v_2(0, 1) = v_1\mathbf{i} + v_2\mathbf{j}$$

and every vector $\mathbf{v} = (v_1, v_2, v_3)$ in R^3 can be expressed in terms of the standard unit vectors as

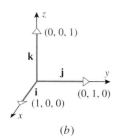

$$\mathbf{v} = (v_1, v_2, v_3) = v_1(1, 0, 0) + v_2(0, 1, 0) + v_3(0, 0, 1) = v_1\mathbf{i} + v_2\mathbf{j} + v_3\mathbf{k}$$

For example,

$$(2, -3, 4) = 2\mathbf{i} - 3\mathbf{j} + 4\mathbf{k}$$

Figure 1.2.2

REMARK The $\mathbf{i}, \mathbf{j}, \mathbf{k}$ notation for vectors in R^2 and R^3 is common in engineering and physics, but it will be used only occasionally in this text.

More generally, we define the ***standard unit vectors in R^n*** to be

$$\mathbf{e}_1 = (1, 0, 0, \ldots, 0), \quad \mathbf{e}_2 = (0, 1, 0, \ldots, 0), \ldots, \quad \mathbf{e}_n = (0, 0, 0, \ldots, 1) \tag{7}$$

We leave it for you to verify that every vector $\mathbf{v} = (v_1, v_2, \ldots, v_n)$ in R^n can be expressed in terms of the standard unit vectors as

$$\mathbf{v} = (v_1, v_2, \ldots, v_n) = v_1\mathbf{e}_1 + v_2\mathbf{e}_2 + \cdots + v_n\mathbf{e}_n \tag{8}$$

**DISTANCE BETWEEN
POINTS IN R^n**

If P_1 and P_2 are points in R^2 or R^3, then the length of the vector $\overrightarrow{P_1 P_2}$ is equal to the distance d between the two points (Figure 1.2.3). Specifically, if $P_1(x_1, y_1)$ and $P_2(x_2, y_2)$ are points in

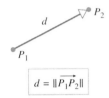

$d = \|\overrightarrow{P_1 P_2}\|$

Figure 1.2.3

R^2, then Theorem 1.1.1(a) implies that

$$d = \|\overrightarrow{P_1 P_2}\| = \sqrt{(x_2 - x_1)^2 + (y_2 - y_1)^2} \tag{9}$$

This is the familiar distance formula from analytic geometry. Similarly, the distance between the points $P_1(x_1, y_1, z_1)$ and $P_2(x_2, y_2, z_2)$ in 3-space is

$$d(\mathbf{u}, \mathbf{v}) = \|\overrightarrow{P_1 P_2}\| = \sqrt{(x_2 - x_1)^2 + (y_2 - y_1)^2 + (z_2 - z_1)^2} \tag{10}$$

Motivated by Formulas (9) and (10), we make the following definition.

Definition 1.2.3 If $\mathbf{u} = (u_1, u_2, \ldots, u_n)$ and $\mathbf{v} = (v_1, v_2, \ldots, v_n)$ are points in R^n, then we denote the ***distance*** between $\mathbf{u}$ and $\mathbf{v}$ by $d(\mathbf{u}, \mathbf{v})$ and define it to be

$$d(\mathbf{u}, \mathbf{v}) = \|\mathbf{u} - \mathbf{v}\| = \sqrt{(u_1 - v_1)^2 + (u_2 - v_2)^2 + \cdots + (u_n - v_n)^2} \tag{11}$$

For example, if

$$\mathbf{u} = (1, 3, -2, 7) \quad \text{and} \quad \mathbf{v} = (0, 7, 2, 2)$$

then the distance between $\mathbf{u}$ and $\mathbf{v}$ is

$$d(\mathbf{u}, \mathbf{v}) = \sqrt{(1 - 0)^2 + (3 - 7)^2 + (-2 - 2)^2 + (7 - 2)^2} = \sqrt{58}$$

We leave it for you to use Formula (11) to show that distances in R^n have the following properties.

Theorem 1.2.4 *If $\mathbf{u}$ and $\mathbf{v}$ are points in R^n, then:*

(a) $d(\mathbf{u}, \mathbf{v}) \geq 0$

(b) $d(\mathbf{u}, \mathbf{v}) = 0$ *if and only if* $\mathbf{u} = \mathbf{v}$

(c) $d(\mathbf{u}, \mathbf{v}) = d(\mathbf{v}, \mathbf{u})$

This theorem states that distances in R^n behave like distances in visible space; that is, distances are nonnegative numbers, the distance between distinct points is nonzero, and the distance is the same whether you measure from $\mathbf{u}$ to $\mathbf{v}$ or from $\mathbf{v}$ to $\mathbf{u}$.

DOT PRODUCTS   We will now define a new kind of multiplication that will be useful for finding angles between vectors and determining whether two vectors are perpendicular.

Definition 1.2.5 If $\mathbf{u} = (u_1, u_2, \ldots, u_n)$ and $\mathbf{v} = (v_1, v_2, \ldots, v_n)$ are vectors in R^n, then the ***dot product*** of $\mathbf{u}$ and $\mathbf{v}$, also called the ***Euclidean inner product*** of $\mathbf{u}$ and $\mathbf{v}$, is denoted by $\mathbf{u} \cdot \mathbf{v}$ and is defined by the formula

$$\mathbf{u} \cdot \mathbf{v} = u_1 v_1 + u_2 v_2 + \cdots + u_n v_n \tag{12}$$

In words, *the dot product is calculated by multiplying corresponding components of the vectors and adding the resulting products.* For example, the dot product of the vectors $\mathbf{u} = (-1, 3, 5, 7)$ and $\mathbf{v} = (5, -4, 7, 0)$ in R^4 is

$$\mathbf{u} \cdot \mathbf{v} = (-1)(5) + (3)(-4) + (5)(7) + (7)(0) = 18$$

REMARK Note the distinction between scalar multiplication and dot products—in scalar multiplication one factor is a scalar, the other is a vector, and the result is a vector; and in a dot product both factors are vectors and the result is a scalar.

EXAMPLE 3
An Application
of Dot Products
to ISBNs

Most books published in the last 25 years have been assigned a unique 10-digit number called an ***International Standard Book Number*** or ISBN. The first nine digits of this number are split into three groups—the first group representing the country or group of countries in which the book originates, the second identifying the publisher, and the third assigned to the book title

Linear Algebra in History

The dot product notation was first introduced by the American physicist and mathematician J. Willard Gibbs in a pamphlet distributed to his students at Yale University in the 1880s. The product was originally written on the baseline, rather than centered as today, and was referred to as the *direct product*. Gibbs's pamphlet was eventually incorporated into a book entitled *Vector Analysis* that was published in 1901 and coauthored by Gibbs and one of his students. Gibbs made major contributions to the fields of thermodynamics and electromagnetic theory and is generally regarded as the greatest American physicist of the nineteenth century.

Josiah Willard Gibbs
(1839–1903)

itself. The tenth and final digit, called a ***check digit***, is computed from the first nine digits and is used to ensure that an electronic transmission of the ISBN, say over the Internet, occurs without error.

To explain how this is done, regard the first nine digits of the ISBN as a vector **b** in R^9, and let **a** be the vector

$$\mathbf{a} = (1, 2, 3, 4, 5, 6, 7, 8, 9)$$

Then the check digit c is computed using the following procedure:

1. Form the dot product $\mathbf{a} \cdot \mathbf{b}$.
2. Divide $\mathbf{a} \cdot \mathbf{b}$ by 11, thereby producing a remainder c that is an integer between 0 and 10, inclusive. The check digit is taken to be c, with the proviso that $c = 10$ is written as X to avoid double digits.

For example, the ISBN of the brief edition of *Calculus*, sixth edition, by Howard Anton is

0-471-15307-9

which has a check digit of 9. This is consistent with the first nine digits of the ISBN, since

$$\mathbf{a} \cdot \mathbf{b} = (1, 2, 3, 4, 5, 6, 7, 8, 9) \cdot (0, 4, 7, 1, 1, 5, 3, 0, 7) = 152$$

Dividing 152 by 11 produces a quotient of 13 and a remainder of 9, so the check digit is $c = 9$. If an electronic order is placed for a book with a certain ISBN, then the warehouse can use the above procedure to verify that the check digit is consistent with the first nine digits, thereby reducing the possibility of a costly shipping error.   ∎

ALGEBRAIC PROPERTIES OF THE DOT PRODUCT

In the special case where $\mathbf{u} = \mathbf{v}$ in Definition 1.2.5, we obtain the relationship

$$\mathbf{v} \cdot \mathbf{v} = v_1^2 + v_2^2 + \cdots + v_n^2 = \|\mathbf{v}\|^2 \tag{13}$$

This yields the following formula for expressing the length of a vector in terms of a dot product:

$$\|\mathbf{v}\| = \sqrt{\mathbf{v} \cdot \mathbf{v}} \tag{14}$$

Dot products have many of the same algebraic properties as products of real numbers.

Theorem 1.2.6   *If* **u**, **v**, *and* **w** *are vectors in* R^n, *and if k is a scalar, then:*

(*a*) $\mathbf{u} \cdot \mathbf{v} = \mathbf{v} \cdot \mathbf{u}$	**[Symmetry property]**
(*b*) $\mathbf{u} \cdot (\mathbf{v} + \mathbf{w}) = \mathbf{u} \cdot \mathbf{v} + \mathbf{u} \cdot \mathbf{w}$	**[Distributive property]**
(*c*) $k(\mathbf{u} \cdot \mathbf{v}) = (k\mathbf{u}) \cdot \mathbf{v}$	**[Homogeneity property]**
(*d*) $\mathbf{v} \cdot \mathbf{v} \geq 0$ *and* $\mathbf{v} \cdot \mathbf{v} = 0$ *if and only if* $\mathbf{v} = \mathbf{0}$	**[Positivity property]**

We will prove parts (*c*) and (*d*) and leave the other proofs as exercises.

Proof (*c*) Let $\mathbf{u} = (u_1, u_2, \ldots, u_n)$ and $\mathbf{v} = (v_1, v_2, \ldots, v_n)$. Then

$$k(\mathbf{u} \cdot \mathbf{v}) = k(u_1 v_1 + u_2 v_2 + \cdots + u_n v_n) = (ku_1)v_1 + (ku_2)v_2 + \cdots + (ku_n)v_n = (k\mathbf{u}) \cdot \mathbf{v}$$

Proof (*d*) The result follows from parts (*a*) and (*b*) of Theorem 1.2.2 and the fact that

$$\mathbf{v} \cdot \mathbf{v} = v_1 v_1 + v_2 v_2 + \cdots + v_n v_n = v_1^2 + v_2^2 + \cdots + v_n^2 = \|\mathbf{v}\|^2 \qquad \blacksquare$$

The following theorem gives some more properties of dot products. The results in this theorem can be proved either by expressing the vectors in terms of components or by using the algebraic properties already established in Theorem 1.2.6.

Theorem 1.2.7   *If* **u**, **v**, *and* **w** *are vectors in* R^n, *and if* k *is a scalar, then:*

(a) $\mathbf{0} \cdot \mathbf{v} = \mathbf{v} \cdot \mathbf{0} = 0$

(b) $(\mathbf{u} + \mathbf{v}) \cdot \mathbf{w} = \mathbf{u} \cdot \mathbf{w} + \mathbf{v} \cdot \mathbf{w}$

(c) $\mathbf{u} \cdot (\mathbf{v} - \mathbf{w}) = \mathbf{u} \cdot \mathbf{v} - \mathbf{u} \cdot \mathbf{w}$

(d) $(\mathbf{u} - \mathbf{v}) \cdot \mathbf{w} = \mathbf{u} \cdot \mathbf{w} - \mathbf{v} \cdot \mathbf{w}$

(e) $k(\mathbf{u} \cdot \mathbf{v}) = \mathbf{u} \cdot (k\mathbf{v})$

We will show how Theorem 1.2.6 can be used to prove part (*b*) without breaking the vectors down into components. Some of the other proofs are left as exercises.

Proof (b)

$$(\mathbf{u} + \mathbf{v}) \cdot \mathbf{w} = \mathbf{w} \cdot (\mathbf{u} + \mathbf{v}) \qquad \text{[By symmetry]}$$
$$= \mathbf{w} \cdot \mathbf{u} + \mathbf{w} \cdot \mathbf{v} \qquad \text{[By distributivity]}$$
$$= \mathbf{u} \cdot \mathbf{w} + \mathbf{v} \cdot \mathbf{w} \qquad \text{[By symmetry]} \qquad \blacksquare$$

Formulas (13) and (14) together with Theorems 1.2.6 and 1.2.7 make it possible to manipulate expressions involving dot products using familiar algebraic techniques.

EXAMPLE 4
Calculating with Dot Products

$$(\mathbf{u} - 2\mathbf{v}) \cdot (3\mathbf{u} + 4\mathbf{v}) = \mathbf{u} \cdot (3\mathbf{u} + 4\mathbf{v}) - 2\mathbf{v} \cdot (3\mathbf{u} + 4\mathbf{v})$$
$$= 3(\mathbf{u} \cdot \mathbf{u}) + 4(\mathbf{u} \cdot \mathbf{v}) - 6(\mathbf{v} \cdot \mathbf{u}) - 8(\mathbf{v} \cdot \mathbf{v})$$
$$= 3\|\mathbf{u}\|^2 - 2(\mathbf{u} \cdot \mathbf{v}) - 8\|\mathbf{v}\|^2 \qquad \blacksquare$$

ANGLE BETWEEN VECTORS IN R^2 AND R^3

To see how dot products can be used to calculate angles between vectors in R^2 and R^3, let **u** and **v** be nonzero vectors in R^2 or R^3, and define the ***angle*** between **u** and **v** to be the smallest nonnegative angle θ through which one of the vectors can be rotated in the plane of the vectors until it coincides with the other (Figure 1.2.4). Algebraically, the radian measure of θ is in the interval $0 \leq \theta \leq \pi$, and in R^2 the angle θ is generated by a counterclockwise rotation.

The following theorem provides an effective way to calculate the angle between vectors in both R^2 and R^3.

Theorem 1.2.8   *If* **u** *and* **v** *are nonzero vectors in* R^2 *or* R^3, *and if* θ *is the angle between these vectors, then*

$$\cos\theta = \frac{\mathbf{u} \cdot \mathbf{v}}{\|\mathbf{u}\|\|\mathbf{v}\|} \quad \text{or equivalently,} \quad \theta = \cos^{-1}\left(\frac{\mathbf{u} \cdot \mathbf{v}}{\|\mathbf{u}\|\|\mathbf{v}\|}\right) \qquad (15\text{--}16)$$

Proof Suppose that the vectors **u**, **v**, and **v** − **u** are positioned to form the sides of a triangle, as shown in Figure 1.2.5. It follows from the law of cosines that

$$\|\mathbf{v} - \mathbf{u}\|^2 = \|\mathbf{u}\|^2 + \|\mathbf{v}\|^2 - 2\|\mathbf{u}\|\|\mathbf{v}\|\cos\theta \qquad (17)$$

Using Formula (13) and the properties of the dot product in Theorems 1.2.6 and 1.2.7, we can rewrite the left side of this equation as

$$\|\mathbf{v} - \mathbf{u}\|^2 = (\mathbf{v} - \mathbf{u}) \cdot (\mathbf{v} - \mathbf{u})$$
$$= (\mathbf{v} - \mathbf{u}) \cdot \mathbf{v} - (\mathbf{v} - \mathbf{u}) \cdot \mathbf{u}$$
$$= \mathbf{v} \cdot \mathbf{v} - \mathbf{u} \cdot \mathbf{v} - \mathbf{v} \cdot \mathbf{u} + \mathbf{u} \cdot \mathbf{u}$$
$$= \|\mathbf{v}\|^2 - 2\mathbf{u} \cdot \mathbf{v} + \|\mathbf{u}\|^2$$

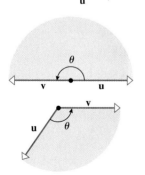

Figure 1.2.4

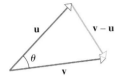

Figure 1.2.5

Substituting the last expression in (17) yields

$$\|\mathbf{v}\|^2 - 2\mathbf{u} \cdot \mathbf{v} + \|\mathbf{u}\|^2 = \|\mathbf{u}\|^2 + \|\mathbf{v}\|^2 - 2\|\mathbf{u}\|\|\mathbf{v}\| \cos \theta$$

which we can simplify and rewrite as

$$\mathbf{u} \cdot \mathbf{v} = \|\mathbf{u}\|\|\mathbf{v}\| \cos \theta$$

Finally, dividing both sides of this equation by $\|\mathbf{u}\|\|\mathbf{v}\|$ yields (15). ■

If $\mathbf{u}$ and $\mathbf{v}$ are nonzero vectors in R^2 or R^3 and $\mathbf{u} \cdot \mathbf{v} = 0$, then it follows from Formula (16) that $\theta = \cos^{-1} 0 = \pi/2$. Conversely, if $\theta = \pi/2$, then $\cos \theta = 0$ and $\mathbf{u} \cdot \mathbf{v} = 0$. Thus, *two nonzero vectors in R^2 or R^3 are perpendicular if and only if their dot product is zero.*

CONCEPT PROBLEM What can you say about the angle between the nonzero vectors $\mathbf{u}$ and $\mathbf{v}$ in R^2 or R^3 if $\mathbf{u} \cdot \mathbf{v} > 0$? What if $\mathbf{u} \cdot \mathbf{v} < 0$?

EXAMPLE 5
An Application
of the Angle
Formula

Find the angle θ between a diagonal of a cube and one of its edges.

Solution Assume that the cube has side a, and introduce a coordinate system as shown in Figure 1.2.6. In this coordinate system the vector

$$\mathbf{d} = (a, a, a)$$

is a diagonal of the cube, and the vectors $\mathbf{v}_1 = (a, 0, 0)$, $\mathbf{v}_2 = (0, a, 0)$, and $\mathbf{v}_3 = (0, 0, a)$ run along the edges. By symmetry, the diagonal makes the same angle with each edge, so it is sufficient to find the angle between $\mathbf{d}$ and $\mathbf{v}_1$. From Formula (15), the cosine of this angle is

$$\cos \theta = \frac{\mathbf{v}_1 \cdot \mathbf{d}}{\|\mathbf{v}_1\|\|\mathbf{d}\|} = \frac{a^2}{a(\sqrt{3a^2})} = \frac{1}{\sqrt{3}}$$

Thus, with the help of a calculating utility,

$$\theta = \cos^{-1}\left(\frac{1}{\sqrt{3}}\right) \approx 54.7°$$

■

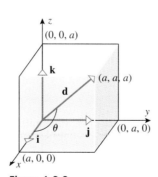

Figure 1.2.6

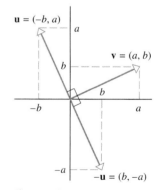

Figure 1.2.7

EXAMPLE 6
Finding a
Vector in R^2
That Is
Perpendicular
to a Given
Vector

Find a nonzero vector in R^2 that is perpendicular to the nonzero vector $\mathbf{v} = (a, b)$.

Solution We are looking for a nonzero vector $\mathbf{u}$ for which $\mathbf{u} \cdot \mathbf{v} = 0$. By experimentation, $\mathbf{u} = (-b, a)$ is such a vector, since

$$\mathbf{u} \cdot \mathbf{v} = (-b, a) \cdot (a, b) = -ba + ab = 0$$

The vector $-\mathbf{u} = (b, -a)$ is also perpendicular to $\mathbf{v}$, as is any scalar multiple of $\mathbf{u}$ (Figure 1.2.7).

■

ORTHOGONALITY   To generalize the notion of perpendicularity to R^n we make the following definition.

> **Definition 1.2.9**   Two vectors $\mathbf{u}$ and $\mathbf{v}$ in R^n are said to be ***orthogonal*** if $\mathbf{u} \cdot \mathbf{v} = 0$, and a nonempty set of vectors in R^n is said to be an ***orthogonal set*** if each pair of distinct vectors in the set is orthogonal.

REMARK Note that we do not require $\mathbf{u}$ and $\mathbf{v}$ to be nonzero in this definition; thus, two vectors in R^2 and R^3 are orthogonal if and only if they are either nonzero and perpendicular or if one or both of them are zero.

EXAMPLE 7
An Orthogonal
Set of Vectors
in R^4

Show that the vectors

$$\mathbf{v}_1 = (1, 2, 2, 4), \quad \mathbf{v}_2 = (-2, 1, -4, 2), \quad \mathbf{v}_3 = (-4, 2, 2, -1)$$

form an orthogonal set in R^4.

Solution Because of the symmetry property of the dot product, we need only confirm that

$$\mathbf{v}_1 \cdot \mathbf{v}_2 = 0, \quad \mathbf{v}_1 \cdot \mathbf{v}_3 = 0, \quad \mathbf{v}_2 \cdot \mathbf{v}_3 = 0$$

We leave the computations to you.   ■

If S is a nonempty set of vectors in R^n, and if $\mathbf{v}$ is orthogonal to every vector in S, then we say that $\mathbf{v}$ is ***orthogonal to the set S***. For example, the vector $\mathbf{k} = (0, 0, 1)$ in R^3 is orthogonal to the xy-plane (Figure 1.2.2b).

EXAMPLE 8
The Zero Vector
Is Orthogonal
to R^n

Part (a) of Theorem 1.2.7 states that if $\mathbf{0}$ is the zero vector in R^n, then $\mathbf{0} \cdot \mathbf{v} = 0$ for every vector $\mathbf{v}$ in R^n. Thus, $\mathbf{0}$ is orthogonal to R^n. Moreover, $\mathbf{0}$ is the only vector in R^n that is orthogonal to R^n, since if $\mathbf{v}$ is a vector in R^n that is orthogonal to R^n, then, in particular, it would be true that $\mathbf{v} \cdot \mathbf{v} = 0$; this implies that $\mathbf{v} = \mathbf{0}$ by part (d) of Theorem 1.2.6.   ■

REMARK Although the result in Example 8 may seem obvious, it will prove to be useful later in the text, since it provides a way of using the dot product to show that a vector $\mathbf{w}$ in R^n is the zero vector—just show that $\mathbf{w} \cdot \mathbf{v} = 0$ for every vector $\mathbf{v}$ in R^n.

ORTHONORMAL SETS   Orthogonal sets of unit vectors have special importance, and there is some terminology associated with them.

> **Definition 1.2.10**   Two vectors $\mathbf{u}$ and $\mathbf{v}$ in R^n are said to be ***orthonormal*** if they are orthogonal and have length 1, and a set of vectors is said to be an ***orthonormal set*** if every vector in the set has length 1 and each pair of distinct vectors is orthogonal.

EXAMPLE 9
The Standard
Unit Vectors in
R^n Are
Orthonormal

The standard unit vectors in R^2 and R^3 form orthonormal sets, since these vectors have length 1 and run along the coordinate axes of rectangular coordinate systems (Figure 1.2.2). More generally, the standard unit vectors

$$\mathbf{e}_1 = (1, 0, 0, \ldots, 0), \quad \mathbf{e}_2 = (0, 1, 0, \ldots, 0), \ldots, \quad \mathbf{e}_n = (0, 0, 0, \ldots, 1)$$

in R^n form an orthonormal set, since

$$\mathbf{e}_i \cdot \mathbf{e}_j = 0 \ \text{ if } i \neq j \quad \text{and} \quad \|\mathbf{e}_1\| = \|\mathbf{e}_2\| = \cdots = \|\mathbf{e}_n\| = 1$$

(verify).   ■

In the following example we form an orthonormal set of three vectors in R^4.

EXAMPLE 10
An Orthonormal
Set in R^4

The vectors

$$\mathbf{q}_1 = \left(\tfrac{1}{5}, \tfrac{2}{5}, \tfrac{2}{5}, \tfrac{4}{5}\right), \quad \mathbf{q}_2 = \left(-\tfrac{2}{5}, \tfrac{1}{5}, -\tfrac{4}{5}, \tfrac{2}{5}\right), \quad \mathbf{q}_3 = \left(-\tfrac{4}{5}, \tfrac{2}{5}, \tfrac{2}{5}, -\tfrac{1}{5}\right)$$

form an orthonormal set in R^4, since

$$\|\mathbf{q}_1\| = \|\mathbf{q}_2\| = \|\mathbf{q}_3\| = 1$$

and

$$\mathbf{q}_1 \cdot \mathbf{q}_2 = 0, \quad \mathbf{q}_1 \cdot \mathbf{q}_3 = 0, \quad \mathbf{q}_2 \cdot \mathbf{q}_3 = 0$$

(verify). ∎

EUCLIDEAN GEOMETRY IN R^n

Formulas (3) and (11) are sometimes called the ***Euclidean norm*** and ***Euclidean distance*** because they produce theorems in R^n that reduce to theorems in Euclidean geometry when applied in R^2 and R^3. Here are just three examples:

1. In a right triangle, the square of the hypotenuse is equal to the sum of the squares of the other two sides (theorem of Pythagoras).
2. The sum of the lengths of two sides of a triangle is at least as large as the length of the third side.
3. The shortest distance between two points is along a straight line.

To extend these theorems to R^n, we need to state them in vector form. For example, a right triangle in R^2 or R^3 can be constructed by placing orthogonal vectors $\mathbf{u}$ and $\mathbf{v}$ tip to tail and using the vector $\mathbf{u} + \mathbf{v}$ as the hypotenuse (Figure 1.2.8). In vector notation, the theorem of Pythagoras now takes the form

$$\|\mathbf{u} + \mathbf{v}\|^2 = \|\mathbf{u}\|^2 + \|\mathbf{v}\|^2$$

Figure 1.2.8

The following theorem is the extension of this result to R^n.

> **Theorem 1.2.11** (***Theorem of Pythagoras***) *If $\mathbf{u}$ and $\mathbf{v}$ are orthogonal vectors in R^n, then*
>
> $$\|\mathbf{u} + \mathbf{v}\|^2 = \|\mathbf{u}\|^2 + \|\mathbf{v}\|^2 \tag{18}$$

Proof

$$\|\mathbf{u} + \mathbf{v}\|^2 = (\mathbf{u} + \mathbf{v}) \cdot (\mathbf{u} + \mathbf{v})$$
$$= \|\mathbf{u}\|^2 + 2(\mathbf{u} \cdot \mathbf{v}) + \|\mathbf{v}\|^2$$
$$= \|\mathbf{u}\|^2 + \|\mathbf{v}\|^2 \quad ∎$$

We have seen that the angle between nonzero vectors in R^2 and R^3 is given by the formula

$$\theta = \cos^{-1}\left(\frac{\mathbf{u} \cdot \mathbf{v}}{\|\mathbf{u}\|\|\mathbf{v}\|}\right) \tag{19}$$

Since this formula involves only the dot product and norms of the vectors $\mathbf{u}$ and $\mathbf{v}$, and since the notions of dot product and norm are applicable to vectors in R^n, it seems reasonable to use Formula (19) as the *definition* of the angle between nonzero vectors $\mathbf{u}$ and $\mathbf{v}$ in R^n. However, this plan would only work if it were true that

$$-1 \leq \frac{\mathbf{u} \cdot \mathbf{v}}{\|\mathbf{u}\|\|\mathbf{v}\|} \leq 1 \tag{20}$$

for all nonzero vectors in R^n. The following theorem gives a result, called the ***Cauchy–Schwarz inequality***, which will show that (20) does in fact hold for all nonzero vectors in R^n.

Linear Algebra in History

The Cauchy–Schwarz inequality is named in honor of the French mathematician Augustin Cauchy and the German mathematician Hermann Schwarz. Variations of this inequality occur in many different settings and under various names. Depending on the context in which the inequality occurs, you may find it called Cauchy's inequality, the Schwarz inequality, or sometimes even the Bunyakovsky inequality, in recognition of the Russian mathematician who published his version of the inequality in 1859, about 25 years before Schwarz.

Augustin Louis Cauchy
(1789–1857)

Hermann Amandus Schwarz
(1843–1921)

Viktor Yakovlevich Bunyakovsky
(1804–1889)

Theorem 1.2.12 (*Cauchy–Schwarz Inequality in R^n*) *If* **u** *and* **v** *are vectors in* R^n, *then*

$$(\mathbf{u} \cdot \mathbf{v})^2 \leq \|\mathbf{u}\|^2 \|\mathbf{v}\|^2 \tag{21}$$

or equivalently (*by taking square roots*),

$$|\mathbf{u} \cdot \mathbf{v}| \leq \|\mathbf{u}\| \|\mathbf{v}\| \tag{22}$$

Proof Observe first that if $\mathbf{u} = \mathbf{0}$ or $\mathbf{v} = \mathbf{0}$, then both sides of (21) are zero (verify), so equality holds in this case. Now consider the case where **u** and **v** are nonzero. As suggested by Figure 1.2.9, the vector **v** can be written as the sum of some scalar multiple of **u**, say $k\mathbf{u}$, and a vector **w** that is orthogonal to **u**. The appropriate scalar k can be computed by setting $\mathbf{w} = \mathbf{v} - k\mathbf{u}$ and using the orthogonality condition $\mathbf{u} \cdot \mathbf{w} = 0$ to write

$$0 = \mathbf{u} \cdot \mathbf{w} = \mathbf{u} \cdot (\mathbf{v} - k\mathbf{u}) = (\mathbf{u} \cdot \mathbf{v}) - k(\mathbf{u} \cdot \mathbf{u})$$

from which it follows that

$$k = \frac{\mathbf{u} \cdot \mathbf{v}}{\mathbf{u} \cdot \mathbf{u}} \tag{23}$$

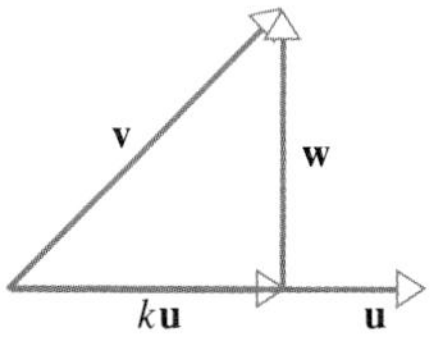

Now apply the theorem of Pythagoras to the vectors in Figure 1.2.9 to obtain

$$\|\mathbf{v}\|^2 = \|k\mathbf{u}\|^2 + \|\mathbf{w}\|^2 = k^2\|\mathbf{u}\|^2 + \|\mathbf{w}\|^2 \tag{24}$$

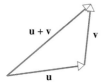

Figure 1.2.9

Substituting (23) for k and multiplying both sides of the resulting equation by $\|\mathbf{u}\|^2$ yields (verify)

$$\|\mathbf{u}\|^2 \|\mathbf{v}\|^2 = (\mathbf{u} \cdot \mathbf{v})^2 + \|\mathbf{u}\|^2 \|\mathbf{w}\|^2 \tag{25}$$

Since $\|\mathbf{u}\|^2 \|\mathbf{w}\|^2 \geq 0$, it follows from (25) that

$$(\mathbf{u} \cdot \mathbf{v})^2 \leq \|\mathbf{u}\|^2 \|\mathbf{v}\|^2$$

This establishes (21) and hence (22). ∎

REMARK The Cauchy–Schwarz inequality now allows us to use Formula (19) as a definition of the angle between nonzero vectors in R^n.

There is a theorem in plane geometry, called the ***triangle inequality***, which states that the sum of the lengths of two sides of a triangle is at least as large as the third side. The following theorem is a generalization of that result to R^n (Figure 1.2.10).

Theorem 1.2.13 (*Triangle Inequality for Vectors*) *If* **u**, **v**, *and* **w** *are vectors in* R^n, *then*

$$\|\mathbf{u} + \mathbf{v}\| \leq \|\mathbf{u}\| + \|\mathbf{v}\| \tag{26}$$

Figure 1.2.10

Proof

$$\begin{aligned}
\|\mathbf{u} + \mathbf{v}\|^2 &= (\mathbf{u} + \mathbf{v}) \cdot (\mathbf{u} + \mathbf{v}) \\
&= \|\mathbf{u}\|^2 + 2(\mathbf{u} \cdot \mathbf{v}) + \|\mathbf{v}\|^2 \\
&\leq \|\mathbf{u}\|^2 + 2|\mathbf{u} \cdot \mathbf{v}| + \|\mathbf{v}\|^2 \qquad \text{[Property of absolute value]} \\
&\leq \|\mathbf{u}\|^2 + 2\|\mathbf{u}\| \|\mathbf{v}\| + \|\mathbf{v}\|^2 \qquad \text{[Cauchy–Schwarz inequality]} \\
&= (\|\mathbf{u}\| + \|\mathbf{v}\|)^2
\end{aligned}$$

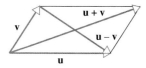

Figure 1.2.11

Formula (26) now follows by taking square roots. ∎

There is a theorem in plane geometry which states that for any parallelogram the sum of the squares of the lengths of the diagonals is equal to the sum of the squares of the lengths of the four sides. The following theorem is a generalization of that result to R^n (Figure 1.2.11).

Theorem 1.2.14 (*Parallelogram Equation for Vectors*) *If* **u** *and* **v** *are vectors in* R^n, *then*

$$\|\mathbf{u} + \mathbf{v}\|^2 + \|\mathbf{u} - \mathbf{v}\|^2 = 2\left(\|\mathbf{u}\|^2 + \|\mathbf{v}\|^2\right) \tag{27}$$

Proof

$$\begin{aligned}
\|\mathbf{u} + \mathbf{v}\|^2 + \|\mathbf{u} - \mathbf{v}\|^2 &= (\mathbf{u} + \mathbf{v}) \cdot (\mathbf{u} + \mathbf{v}) + (\mathbf{u} - \mathbf{v}) \cdot (\mathbf{u} - \mathbf{v}) \\
&= 2(\mathbf{u} \cdot \mathbf{u}) + 2(\mathbf{v} \cdot \mathbf{v}) \\
&= 2\left(\|\mathbf{u}\|^2 + \|\mathbf{v}\|^2\right)
\end{aligned}$$ ∎

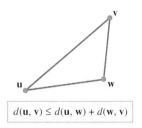

$$d(\mathbf{u}, \mathbf{v}) \le d(\mathbf{u}, \mathbf{w}) + d(\mathbf{w}, \mathbf{v})$$

Figure 1.2.12

Finally, let **u** and **v** be any two points in R^2 or R^3. To say that the shortest distance from **u** to **v** is along a straight line implies that if we choose a third point **w** in R^2 or R^3, then

$$d(\mathbf{u}, \mathbf{v}) \le d(\mathbf{u}, \mathbf{w}) + d(\mathbf{w}, \mathbf{v})$$

(Figure 1.2.12). This is called the ***triangle inequality for distances***. The following theorem is the extension to R^n.

Theorem 1.2.15 (*Triangle Inequality for Distances*) *If* **u**, **v**, *and* **w** *are points in* R^n, *then*

$$d(\mathbf{u}, \mathbf{v}) \le d(\mathbf{u}, \mathbf{w}) + d(\mathbf{w}, \mathbf{v}) \tag{28}$$

Proof

$$\begin{aligned}
d(\mathbf{u}, \mathbf{v}) &= \|\mathbf{u} - \mathbf{v}\| \\
&= \|(\mathbf{u} - \mathbf{w}) + (\mathbf{w} - \mathbf{v})\| && \text{[Add and subtract \textbf{w}.]} \\
&\le \|\mathbf{u} - \mathbf{w}\| + \|\mathbf{w} - \mathbf{v}\| && \text{[Triangle inequality for vectors]} \\
&= d(\mathbf{u}, \mathbf{w}) + d(\mathbf{w}, \mathbf{v}) && \text{[Definition of distance]}
\end{aligned}$$ ∎

LOOKING AHEAD   The notions of length, angle, and distance in R^n can all be expressed in terms of the dot product (which, you may recall, is also called the *Euclidean inner product*):

$$\|\mathbf{v}\| = \sqrt{\mathbf{v} \cdot \mathbf{v}} \tag{29}$$

$$\theta = \cos^{-1}\left(\frac{\mathbf{u} \cdot \mathbf{v}}{\|\mathbf{u}\|\|\mathbf{v}\|}\right) = \cos^{-1}\left(\frac{\mathbf{u} \cdot \mathbf{v}}{\sqrt{\mathbf{u} \cdot \mathbf{u}}\sqrt{\mathbf{v} \cdot \mathbf{v}}}\right) \tag{30}$$

$$d(\mathbf{u}, \mathbf{v}) = \|\mathbf{u} - \mathbf{v}\| = \sqrt{(\mathbf{u} - \mathbf{v}) \cdot (\mathbf{u} - \mathbf{v})} \tag{31}$$

Thus, it is the algebraic properties of the Euclidean inner product that ultimately determine the geometric properties of vectors in R^n. However, the most important algebraic properties of the Euclidean inner product can all be derived from the four properties in Theorem 1.2.6, so this theorem is really the foundation on which the geometry of R^n rests. Because R^n with the Euclidean inner product has so many of the familiar properties of Euclidean geometry, it is often called ***Euclidean n-space*** or ***n-dimensional Euclidean space***.

Exercise Set 1.2

In Exercises 1 and 2, find the norm of **v**, a unit vector that has the same direction as **v**, and a unit vector that is oppositely directed to **v**.

1. (a) $\mathbf{v} = (4, -3)$ (b) $\mathbf{v} = (2, 2, 2)$
    (c) $\mathbf{v} = (1, 0, 2, 1, 3)$

2. (a) $\mathbf{v} = (-5, 12)$ (b) $\mathbf{v} = (1, -1, 2)$
    (c) $\mathbf{v} = (-2, 3, 3, -1)$

In Exercises 3 and 4, evaluate the given expression with $\mathbf{u} = (2, -2, 3)$, $\mathbf{v} = (1, -3, 4)$, and $\mathbf{w} = (3, 6, -4)$.

3. (a) $\|\mathbf{u} + \mathbf{v}\|$ (b) $\|\mathbf{u}\| + \|\mathbf{v}\|$
   (c) $\|-2\mathbf{u} + 2\mathbf{v}\|$ (d) $\|3\mathbf{u} - 5\mathbf{v} + \mathbf{w}\|$

4. (a) $\|\mathbf{u} + \mathbf{v} + \mathbf{w}\|$ (b) $\|\mathbf{u} - \mathbf{v}\|$
   (c) $\|3\mathbf{v}\| - 3\|\mathbf{v}\|$ (d) $\|\mathbf{u}\| - \|\mathbf{v}\|$

In Exercises 5 and 6, evaluate the given expression with $\mathbf{u} = (-2, -1, 4, 5)$, $\mathbf{v} = (3, 1, -5, 7)$, and $\mathbf{w} = (-6, 2, 1, 1)$.

5. (a) $\|3\mathbf{u} - 5\mathbf{v} + \mathbf{w}\|$ (b) $\|3\mathbf{u}\| - 5\|\mathbf{v}\| + \|\mathbf{w}\|$
   (c) $\|-\|\mathbf{u}\|\mathbf{v}\|$

6. (a) $\|\mathbf{u}\| - 2\|\mathbf{v}\| - 3\|\mathbf{w}\|$ (b) $\|\mathbf{u}\| + \|-2\mathbf{v}\| + \|-3\mathbf{w}\|$
   (c) $\big\|\|\mathbf{u} - \mathbf{v}\|\mathbf{w}\big\|$

7. Let $\mathbf{v} = (-2, 3, 0, 6)$. Find all scalars k such that $\|k\mathbf{v}\| = 5$.

8. Let $\mathbf{v} = (1, 1, 2, -3, 1)$. Find all scalars k such that $\|k\mathbf{v}\| = 4$.

In Exercises 9 and 10, find $\mathbf{u} \cdot \mathbf{v}$, $\mathbf{u} \cdot \mathbf{u}$, and $\mathbf{v} \cdot \mathbf{v}$.

9. (a) $\mathbf{u} = (3, 1, 4)$, $\mathbf{v} = (2, 2, -4)$
   (b) $\mathbf{u} = (1, 1, 4, 6)$, $\mathbf{v} = (2, -2, 3, -2)$

10. (a) $\mathbf{u} = (1, 1, -2, 3)$, $\mathbf{v} = (-1, 0, 5, 1)$
    (b) $\mathbf{u} = (2, -1, 1, 0, -2)$, $\mathbf{v} = (1, 2, 2, 2, 1)$

In Exercises 11 and 12, find the Euclidean distance between $\mathbf{u}$ and $\mathbf{v}$.

11. (a) $\mathbf{u} = (3, 3, 3)$, $\mathbf{v} = (1, 0, 4)$
    (b) $\mathbf{u} = (0, -2, -1, 1)$, $\mathbf{v} = (-3, 2, 4, 4)$
    (c) $\mathbf{u} = (3, -3, -2, 0, -3, 13, 5)$,
        $\mathbf{v} = (-4, 1, -1, 5, 0, -11, 4)$

12. (a) $\mathbf{u} = (1, 2, -3, 0)$, $\mathbf{v} = (5, 1, 2, -2)$
    (b) $\mathbf{u} = (2, -1, -4, 1, 0, 6, -3, 1)$,
        $\mathbf{v} = (-2, -1, 0, 3, 7, 2, -5, 1)$
    (c) $\mathbf{u} = (0, 1, 1, 1, 2)$, $\mathbf{v} = (2, 1, 0, -1, 3)$

13. Find the cosine of the angle between the vectors in each part of Exercise 11, and then state whether the angle is acute, obtuse, or a right angle.

14. Find the cosine of the angle between the vectors in each part of Exercise 12, and then state whether the angle is acute, obtuse, or a right angle.

15. A vector $\mathbf{a}$ in the xy-plane has a length of 9 units and points in a direction that is $120°$ counterclockwise from the positive x-axis, and a vector $\mathbf{b}$ in that plane has a length of 5 units and points in the positive y-direction. Find $\mathbf{a} \cdot \mathbf{b}$.

16. A vector $\mathbf{a}$ in the xy-plane points in a direction that is $47°$ counterclockwise from the positive x-axis, and a vector $\mathbf{b}$ in that plane points in a direction that is $43°$ clockwise from the positive x-axis. What can you say about the value of $\mathbf{a} \cdot \mathbf{b}$?

17. Solve the equation $5\mathbf{x} - 2\mathbf{v} = 2(\mathbf{w} - 5\mathbf{x})$ for $\mathbf{x}$, given that $\mathbf{v} = (1, 2, -4, 0)$ and $\mathbf{w} = (-3, 5, 1, 1)$.

18. Solve the equation $5\mathbf{x} - \|\mathbf{v}\|\mathbf{v} = \|\mathbf{w}\|(\mathbf{w} - 5\mathbf{x})$ for $\mathbf{x}$ with $\mathbf{v}$ and $\mathbf{w}$ being the vectors in Exercise 17.

In Exercises 19 and 20, determine whether the expression makes sense mathematically. If not, explain why.

19. (a) $\mathbf{u} \cdot (\mathbf{v} \cdot \mathbf{w})$ (b) $\mathbf{u} \cdot (\mathbf{v} + \mathbf{w})$
    (c) $\|\mathbf{u} \cdot \mathbf{v}\|$ (d) $(\mathbf{u} \cdot \mathbf{v}) - \|\mathbf{u}\|$

20. (a) $\|\mathbf{u}\| \cdot \|\mathbf{v}\|$ (b) $(\mathbf{u} \cdot \mathbf{v}) - \mathbf{w}$
    (c) $(\mathbf{u} \cdot \mathbf{v}) - k$ (d) $k \cdot \mathbf{u}$

In Exercises 21 and 22, verify that the Cauchy–Schwarz inequality holds.

21. (a) $\mathbf{u} = (3, 2)$, $\mathbf{v} = (4, -1)$
    (b) $\mathbf{u} = (-3, 1, 0)$, $\mathbf{v} = (2, -1, 3)$
    (c) $\mathbf{u} = (0, 2, 2, 1)$, $\mathbf{v} = (1, 1, 1, 1)$

22. (a) $\mathbf{u} = (4, 1, 1)$, $\mathbf{v} = (1, 2, 3)$
    (b) $\mathbf{u} = (1, 2, 1, 2, 3)$, $\mathbf{v} = (0, 1, 1, 5, -2)$
    (c) $\mathbf{u} = (1, 3, 5, 2, 0, 1)$, $\mathbf{v} = (0, 2, 4, 1, 3, 5)$

In Exercises 23 and 24, show that the vectors form an orthonormal set.

23. $\mathbf{v}_1 = \left(\frac{1}{2}, \frac{1}{2}, \frac{1}{2}, \frac{1}{2}\right)$, $\mathbf{v}_2 = \left(\frac{1}{2}, -\frac{5}{6}, \frac{1}{6}, \frac{1}{6}\right)$,
    $\mathbf{v}_3 = \left(\frac{1}{2}, \frac{1}{6}, \frac{1}{6}, -\frac{5}{6}\right)$, $\mathbf{v}_4 = \left(\frac{1}{2}, \frac{1}{6}, -\frac{5}{6}, \frac{1}{6}\right)$

24. $\mathbf{v}_1 = \left(-\frac{1}{\sqrt{2}}, \frac{1}{\sqrt{6}}, \frac{1}{\sqrt{3}}\right)$, $\mathbf{v}_2 = \left(0, -\frac{2}{\sqrt{6}}, \frac{1}{\sqrt{3}}\right)$,
    $\mathbf{v}_3 = \left(\frac{1}{\sqrt{2}}, \frac{1}{\sqrt{6}}, \frac{1}{\sqrt{3}}\right)$

25. Find two unit vectors that are orthogonal to the nonzero vector $\mathbf{u} = (a, b)$.

26. For what values of k, if any, are $\mathbf{u}$ and $\mathbf{v}$ orthogonal?
    (a) $\mathbf{u} = (2, k, k)$, $\mathbf{v} = (1, 7, k)$
    (b) $\mathbf{u} = (k, k, 1)$, $\mathbf{v} = (k, 5, 6)$

27. For which values of k, if any, are $\mathbf{u}$ and $\mathbf{v}$ orthogonal?
    (a) $\mathbf{u} = (k, 1, 3)$, $\mathbf{v} = (1, 7, k)$
    (b) $\mathbf{u} = (-2, k, k)$, $\mathbf{v} = (k, 5, k)$

28. Use vectors to find the cosines of the interior angles of the triangle with vertices $A(0, -1)$, $B(1, -2)$, and $C(4, 1)$.

29. Use vectors to show that $A(3, 0, 2)$, $B(4, 3, 0)$, and $C(8, 1, -1)$ are vertices of a right triangle. At which vertex is the right angle?

30. In each part determine whether the given number is a valid ISBN by computing its check digit.
    (a) 1-56592-170-7   (b) 0-471-05333-5

31. In each part determine whether the given number is a valid ISBN by computing its check digit.
    (a) 0-471-06368-1   (b) 0-13-947752-3

It will be convenient to have a more compact way of writing expressions such as $x_1y_1 + x_2y_2 + \cdots + x_ny_n$ and $x_1^2 + x_2^2 + \cdots + x_n^2$ that arise in working with vectors in R^n. For this purpose we will use **sigma notation** (also called **summation notation**), which uses the Greek letter Σ (capital sigma) to indicate that a sum is to be formed. To illustrate how the notation works, consider the sum

$$1^2 + 2^2 + 3^2 + 4^2 + 5^2$$

in which each term is of the form k^2, where k is an integer between 1 and 5, inclusive. This sum can be written in sigma notation as

$$\sum_{k=1}^{5} k^2$$

This directs us to form the sum of the terms that result by substituting successive integers for k, starting with $k = 1$ and ending with $k = 5$. In general, if $f(k)$ is a function of k, and if m and n are integers with $m \leq n$, then

$$\sum_{k=m}^{n} f(k) = f(m) + f(m+1) + \cdots + f(n)$$

This is the sum of the terms that result by substituting successive integers for k, starting with $k = m$ and ending with $k = n$. The number m is called the **lower limit of summation**, the number n the **upper limit of summation**, and the letter k the **index of summation**. It is not essential to use k as the index of summation; any letter can be used, though we will generally use i, j, or k. Thus,

$$\sum_{k=1}^{n} a_k = \sum_{i=1}^{n} a_i = \sum_{j=1}^{n} a_j = a_1 + a_2 + \cdots + a_n$$

If $\mathbf{u} = (u_1, u_2, \ldots, u_n)$ and $\mathbf{v} = (v_1, v_2, \ldots, v_n)$ are vectors in R^n, then the norm of $\mathbf{u}$ and the dot product of $\mathbf{u}$ and $\mathbf{v}$ can

be expressed in sigma notation as

$$\|\mathbf{u}\| = \sqrt{u_1^2 + u_2^2 + \cdots + u_n^2} = \sqrt{\sum_{k=1}^{n} u_k^2}$$

$$\mathbf{u} \cdot \mathbf{v} = u_1v_1 + u_2v_2 + \cdots + u_nv_n = \sum_{k=1}^{n} u_kv_k$$

32. (**Sigma notation**) In each part, write the sum in sigma notation.
   (a) $a_1b_1 + a_2b_2 + a_3b_3 + a_4b_4$
   (b) $c_1^2 + c_2^2 + c_3^2 + c_4^2 + c_5^2$
   (c) $b_3 + b_4 + \cdots + b_n$

33. (**Sigma notation**) Write Formula (11) in sigma notation.

34. (**Sigma notation**) In each part, evaluate the sum for
   $c_1 = 3, c_2 = -1, c_3 = 5, c_4 = -6, c_5 = 4$
   $d_1 = 6, d_2 = 0, d_3 = 7, d_4 = -2, d_5 = -3$
   (a) $\displaystyle\sum_{k=1}^{4} c_k + \sum_{k=2}^{5} d_k$ (b) $\displaystyle\sum_{j=1}^{5}(2c_j - d_j)$
   (c) $\displaystyle\sum_{k=1}^{5}(-1)^k c_k$

35. (**Sigma notation**) In each part, confirm the statement by writing out the sums on the two sides.
   (a) $\displaystyle\sum_{k=1}^{n}(a_k + b_k) = \sum_{k=1}^{n} a_k + \sum_{k=1}^{n} b_k$
   (b) $\displaystyle\sum_{k=1}^{n}(a_k - b_k) = \sum_{k=1}^{n} a_k - \sum_{k=1}^{n} b_k$
   (c) $\displaystyle\sum_{k=1}^{n} ca_k = c\sum_{k=1}^{n} a_k$

Discussion and Discovery

D1. Write a paragraph or two that explains some of the similarities and differences between visible space and higher-dimensional spaces. Include an explanation of why R^n is referred to as *Euclidean space*.

D2. What can you say about k and $\mathbf{v}$ if $\|k\mathbf{v}\| = k\|\mathbf{v}\|$?

D3. (a) The set of all vectors in R^2 that are orthogonal to a nonzero vector is what kind of geometric object?
   (b) The set of all vectors in R^3 that are orthogonal to a nonzero vector is what kind of geometric object?
   (c) The set of all vectors in R^2 that are orthogonal to two noncollinear vectors is what kind of geometric object?
   (d) The set of all vectors in R^3 that are orthogonal to two noncollinear vectors is what kind of geometric object?

D4. Show that $\mathbf{v}_1 = \left(\frac{2}{3}, \frac{1}{3}, \frac{2}{3}\right)$ and $\mathbf{v}_2 = \left(\frac{1}{3}, \frac{2}{3}, -\frac{2}{3}\right)$ are orthonormal vectors, and find a third vector $\mathbf{v}_3$ for which $\{\mathbf{v}_1, \mathbf{v}_2, \mathbf{v}_3\}$ is an orthonormal set.

D5. Something is wrong with one of the following expressions. Which one is it, and what is wrong?
   $$\mathbf{u} \cdot (\mathbf{v} + \mathbf{w}), \quad \mathbf{u} \cdot \mathbf{v} + \mathbf{u} \cdot \mathbf{w}, \quad (\mathbf{u} \cdot \mathbf{v}) + \mathbf{w}$$

D6. Let $\mathbf{x} = (x, y)$ and $\mathbf{x}_0 = (x_0, y_0)$. Write down an equality or inequality involving norms that describes
   (a) the circle of radius 1 centered at $\mathbf{x}_0$;
   (b) the set of points inside the circle in part (a);
   (c) the set of points outside the circle in part (a).

D7. If $\mathbf{u}$ and $\mathbf{v}$ are orthogonal vectors in R^n such that $\|\mathbf{u}\| = 1$ and $\|\mathbf{v}\| = 1$, then $d(\mathbf{u}, \mathbf{v}) = $ _____. Draw a picture to illustrate your result in R^2.

D8. In each part, find $\|\mathbf{u}\|$ for $n = 5$, 10, and 100.
(a) $\mathbf{u} = (1, \sqrt{2}, \sqrt{3}, \ldots, \sqrt{n}\,)$
(b) $\mathbf{u} = (1, 2, 3, \ldots, n)$
[*Hint:* There exist formulas for the sum of the first n positive integers and the sum of the squares of the first n positive integers. If you don't know those formulas, look them up.]

D9. Indicate whether the statement is true (T) or false (F). Justify your answer.
(a) If $\|\mathbf{u} + \mathbf{v}\|^2 = \|\mathbf{u}\|^2 + \|\mathbf{v}\|^2$, then $\mathbf{u}$ and $\mathbf{v}$ are orthogonal.
(b) If $\mathbf{u}$ is orthogonal to $\mathbf{v}$ and $\mathbf{w}$, then $\mathbf{u}$ is orthogonal to $\mathbf{v} + \mathbf{w}$.
(c) If $\mathbf{u}$ is orthogonal to $\mathbf{v} + \mathbf{w}$, then $\mathbf{u}$ is orthogonal to $\mathbf{v}$ and $\mathbf{w}$.
(d) If $\mathbf{a} \cdot \mathbf{b} = \mathbf{a} \cdot \mathbf{c}$ and $\mathbf{a} \neq \mathbf{0}$, then $\mathbf{b} = \mathbf{c}$.

(e) If $\|\mathbf{u} + \mathbf{v}\| = 0$, then $\mathbf{u} = -\mathbf{v}$.
(f) Every orthonormal set of vectors in R^n is also an orthogonal set.

D10. Indicate whether the statement is true (T) or false (F). Justify your answer.
(a) If $k\mathbf{u} = \mathbf{0}$, then either $k = 0$ or $\mathbf{u} = \mathbf{0}$.
(b) If two vectors $\mathbf{u}$ and $\mathbf{v}$ in R^2 are orthogonal to a nonzero vector $\mathbf{w}$ in R^2, then $\mathbf{u}$ and $\mathbf{v}$ are scalar multiples of one another.
(c) There is a vector $\mathbf{u}$ in R^3 such that $\|\mathbf{u} - (1, 1, 1)\| \leq 3$ and $\|\mathbf{u} - (-1, -1, -1)\| \leq 3$.
(d) If $\mathbf{u}$ is a vector in R^3 that is orthogonal to the vectors $(1, 0, 0)$, $(0, 1, 0)$, and $(0, 0, 1)$, then $\mathbf{u} = 0$.
(e) If $\mathbf{u} \cdot \mathbf{v} = 0$ and $\mathbf{v} \cdot \mathbf{w} = 0$, then $\mathbf{u} \cdot \mathbf{w} = 0$.
(f) $\|\mathbf{u} + \mathbf{v}\| = \|\mathbf{u}\| + \|\mathbf{v}\|$.

Working with Proofs

P1. Prove that if $\mathbf{u}_1, \mathbf{u}_2, \ldots, \mathbf{u}_n$ are pairwise orthogonal vectors in R^n, then
$$\|\mathbf{u}_1 + \mathbf{u}_2 + \cdots + \mathbf{u}_n\|^2 = \|\mathbf{u}_1\|^2 + \|\mathbf{u}_2\|^2 + \cdots + \|\mathbf{u}_n\|^2$$
This generalizes Theorem 1.2.11 and hence is called the ***generalized theorem of Pythagoras***.

P2. (a) Use the Cauchy–Schwarz inequality to prove that if a_1 and a_2 are nonnegative numbers, then
$$\sqrt{a_1 a_2} \leq \frac{a_1 + a_2}{2}$$
The expression on the left side is called the ***geometric mean*** of a_1 and a_2, and the expression on the right side is the familiar ***arithmetic mean*** of a and b, so this relationship states that the geometric mean of two numbers cannot exceed the arithmetic mean. [*Hint:* Consider the vectors $\mathbf{u} = (\sqrt{a_1}, \sqrt{a_2}\,)$ and $\mathbf{v} = (\sqrt{a_2}, \sqrt{a_1}\,)$.]
(b) Generalize the result in part (a) for n nonnegative numbers.

P3. Use the Cauchy–Schwarz inequality to prove that
$$(a_1 b_1 + a_2 b_2 + \cdots + a_n b_n)^2$$
$$\leq (a_1^2 + a_2^2 + \cdots + a_n^2)(b_1^2 + b_2^2 + \cdots + b_n^2)$$

P4. (a) Prove the identity $\mathbf{u} \cdot \mathbf{v} = \frac{1}{4}\|\mathbf{u} + \mathbf{v}\|^2 - \frac{1}{4}\|\mathbf{u} - \mathbf{v}\|^2$ for vectors in R^n by expressing the two sides in terms of dot products.
(b) Find $\mathbf{u} \cdot \mathbf{v}$ given that $\|\mathbf{u} + \mathbf{v}\| = 1$ and $\|\mathbf{u} - \mathbf{v}\| = 5$.

P5. Recall that two nonvertical lines in the plane are perpendicular if and only if the product of their slopes is -1. Prove this using dot products by first showing that if a nonzero vector $\mathbf{u} = (a, b)$ is parallel to a line of slope m, then $b/a = m$.

P6. Prove Theorem 1.2.4 using Formula (11).

P7. (a) Prove part (a) of Theorem 1.2.6.
(b) Prove part (b) of Theorem 1.2.6.

P8. (a) Use Theorem 1.2.6 to prove part (e) of Theorem 1.2.7 without breaking the vectors into components.
(b) Use Theorem 1.2.6 and the fact that $\mathbf{0} = (0)\mathbf{0}$ to prove part (a) of Theorem 1.2.7 without breaking the vectors into components.

P9. As shown in the accompanying figure, let a triangle AXB be inscribed in a circle so that one side coincides with a diameter. Express the vectors $\overrightarrow{AX}$ and $\overrightarrow{BX}$ in terms of the vectors $\mathbf{a}$ and $\mathbf{x}$, and then use a dot product to prove that the angle at X is a right angle.

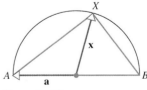

Figure Ex-P9

Technology Exercises

T1. (***Dot product and norm***) Some linear algebra programs provide commands for calculating dot products and norms, and others only provide a command for the dot product. In the latter case, norms can be computed from the formula $\|\mathbf{v}\| = \sqrt{\mathbf{v} \cdot \mathbf{v}}$. Determine how to compute dot products and norms with your technology utility and perform the calculations in Examples 1, 2, and 4.

T2. (*Sigma notation*) Determine how to evaluate expressions involving sigma notation and compute

(a) $\displaystyle\sum_{k=1}^{10} k^3$

(b) $\displaystyle\sum_{k=1}^{20} k^2 \cos(k\pi)$

T3. (a) Find the sine and cosine of the angle between the vectors $\mathbf{u} = (1, -2, 4, 1)$ and $\mathbf{v} = (7, 4, -3, 2)$.

(b) Find the angle between the vectors in part (a).

T4. Use the method of Example 5 to estimate, to the nearest degree, the angles that a diagonal of a box with dimensions 10 cm × 15 cm × 25 cm makes with edges of the box.

T5. (*Sigma notation*) Let $\mathbf{u}$ be the vector in R^{100} whose ith component is i, and let $\mathbf{v}$ be the vector in R^{100} whose ith component is $1/(i+1)$. Evaluate the dot product $\mathbf{u} \cdot \mathbf{v}$ by first writing it in sigma notation.

Section 1.3   Vector Equations of Lines and Planes

In this section we will derive vector equations of lines and planes in R^2 and R^3, and we will use these equations as a foundation for defining lines and planes in higher-dimensional spaces.

VECTOR AND PARAMETRIC EQUATIONS OF LINES

Recall that the ***general equation of a line*** in R^2 has the form

$$Ax + By = C \quad (A \text{ and } B \text{ not both zero}) \tag{1}$$

In the special case where the line passes through the origin, this equation simplifies to

$$Ax + By = 0 \quad (A \text{ and } B \text{ not both zero}) \tag{2}$$

These equations, though useful for many purposes, are only applicable in R^2, so our first objective in this section will be to obtain equations of lines that are applicable in both R^2 and R^3.

A line in R^2 or R^3 can be uniquely determined by specifying a point $\mathbf{x}_0$ on the line and a nonzero vector $\mathbf{v}$ that is parallel to the line (Figure 1.3.1*a*). Thus, if $\mathbf{x}$ is any point on the line through $\mathbf{x}_0$ that is parallel to $\mathbf{v}$, then the vector $\mathbf{x} - \mathbf{x}_0$ is parallel to $\mathbf{v}$ (Figure 1.3.1*b*), so

$$\mathbf{x} - \mathbf{x}_0 = t\mathbf{v}$$

for some scalar t. This can be rewritten as

$$\mathbf{x} = \mathbf{x}_0 + t\mathbf{v} \tag{3}$$

As the variable t, called a ***parameter***, varies from $-\infty$ to $+\infty$, the point $\mathbf{x}$ traces out the line, so the line can be represented by the equation

$$\mathbf{x} = \mathbf{x}_0 + t\mathbf{v} \quad (-\infty < t < +\infty) \tag{4}$$

We call this a ***vector equation of the line*** through $\mathbf{x}_0$ that is parallel to $\mathbf{v}$. In the special case where $\mathbf{x}_0 = \mathbf{0}$, the line passes through the origin, and (4) simplifies to

$$\mathbf{x} = t\mathbf{v} \quad (-\infty < t < +\infty) \tag{5}$$

Note that the line in (4) is the ***translation*** by $\mathbf{x}_0$ of the line in (5) (Figure 1.3.1*c*).

A vector equation of a line can be split into a set of scalar equations by equating corresponding components; these are called ***parametric equations*** of the line. For example, if we let $\mathbf{x} = (x, y)$ be a general point on the line through $\mathbf{x}_0 = (x_0, y_0)$ that is parallel to $\mathbf{v} = (a, b)$, then (4) can be expressed in component form as

$$(x, y) = (x_0, y_0) + t(a, b) \quad (-\infty < t < +\infty)$$

Equating corresponding components yields the parametric equations

$$x = x_0 + at, \ y = y_0 + bt \quad (-\infty < t < +\infty) \tag{6}$$

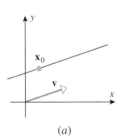

(a)

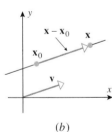

(b)

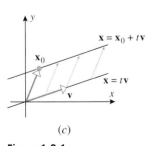

(c)

Figure 1.3.1

Similarly, if we let $\mathbf{x} = (x, y, z)$ be a general point on the line through $\mathbf{x}_0 = (x_0, y_0, z_0)$ that is parallel to $\mathbf{v} = (a, b, c)$, then (4) can be expressed in component form as

$$(x, y, z) = (x_0, y_0, z_0) + t(a, b, c) \quad (-\infty < t < +\infty)$$

Equating corresponding components yields the parametric equations

$$x = x_0 + at, \quad y = y_0 + bt, \quad z = z_0 + ct \quad (-\infty < t < +\infty) \tag{7}$$

REMARK For simplicity, we will often omit the explicit reference to the fact that $-\infty < t < +\infty$ when writing vector or parametric equations of lines.

EXAMPLE 1
Vector
Equations of
Lines

(a) Find a vector equation and parametric equations of the line in R^2 that passes through the origin and is parallel to the vector $\mathbf{v} = (-2, 3)$.

(b) Find a vector equation and parametric equations of the line in R^3 that passes through the point $P_0(1, 2, -3)$ and is parallel to the vector $\mathbf{v} = (4, -5, 1)$.

(c) Use the vector equation obtained in part (b) to find two points on the line that are different from P_0.

Solution (a) It follows from (5) that a vector equation of the line is $\mathbf{x} = t\mathbf{v}$. If we let $\mathbf{x} = (x, y)$, then this equation can be expressed in component form as

$$(x, y) = t(-2, 3)$$

Equating corresponding components on the two sides of this equation yields the parametric equations

$$x = -2t, \quad y = 3t$$

Solution (b) It follows from (4) that a vector equation of the line is $\mathbf{x} = \mathbf{x}_0 + t\mathbf{v}$. If we let $\mathbf{x} = (x, y, z)$, and if we take $\mathbf{x}_0 = (1, 2, -3)$, then this equation can be expressed in component form as

$$(x, y, z) = (1, 2, -3) + t(4, -5, 1) \tag{8}$$

Equating corresponding components on the two sides of this equation yields the parametric equations

$$x = 1 + 4t, \quad y = 2 - 5t, \quad z = -3 + t$$

Solution (c) Specific points on a line represented by a vector equation or by parametric equations can be obtained by substituting numerical values for the parameter t. For example, if we take $t = 0$ in (8), we obtain the point $(x, y, z) = (1, 2, -3)$, which is the given point P_0. Other values of t will yield other points; for example, $t = 1$ yields the point $(5, -3, -2)$ and $t = -1$ yields the point $(-3, 7, -4)$. ∎

LINES THROUGH TWO POINTS

If $\mathbf{x}_0$ and $\mathbf{x}_1$ are distinct points in R^2 or R^3, then the line determined by these points is parallel to the vector $\mathbf{v} = \mathbf{x}_1 - \mathbf{x}_0$ (Figure 1.3.2), so it follows from (4) that the line can be expressed in vector form as

$$\mathbf{x} = \mathbf{x}_0 + t(\mathbf{x}_1 - \mathbf{x}_0) \quad (-\infty < t < +\infty) \tag{9}$$

or, equivalently, as

$$\mathbf{x} = (1 - t)\mathbf{x}_0 + t\mathbf{x}_1 \quad (-\infty < t < +\infty) \tag{10}$$

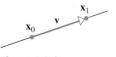

Figure 1.3.2

Equations (9) and (10) are called *two-point vector equations* of the line through $\mathbf{x}_0$ and $\mathbf{x}_1$.

REMARK If the parameter t in (9) or (10) is restricted to the interval $0 \le t \le 1$, then these equations represent the *line segment* from $\mathbf{x}_0$ to $\mathbf{x}_1$ rather than the entire line (see Exercises 41–45).

EXAMPLE 2
Vector and Parametric Equations of a Line Through Two Points

Find vector and parametric equations of the line in R^2 that passes through the points $P(0, 7)$ and $Q(5, 0)$.

Solution If we let $\mathbf{x} = (x, y)$, then it follows from (10) with $\mathbf{x}_0 = (0, 7)$ and $\mathbf{x}_1 = (5, 0)$ that a two-point vector equation of the line is

$$(x, y) = (1 - t)(0, 7) + t(5, 0) \tag{11}$$

Equating corresponding components yields the parametric equations

$$x = 5t, \;\; y = 7 - 7t \tag{12}$$

(verify). ∎

REMARK Had we taken $\mathbf{x}_0 = (5, 0)$ and $\mathbf{x}_1 = (0, 7)$ in the last example, then the resulting vector equation would have been

$$(x, y) = (1 - t)(5, 0) + t(0, 7) \tag{13}$$

and the corresponding parametric equations would have been

$$x = 5 - 5t, \;\; y = 7t \tag{14}$$

(verify). Although (13) and (14) look different from (11) and (12), they all represent the same geometric line. This can be seen by eliminating the parameter t in the parametric equations and finding a direct relationship between x and y. For example, if we solve the first equation in (12) for t in terms of x and substitute in the second equation, we obtain

$$7x + 5y = 35$$

(verify). The same equation results if we solve the second equation in (14) for t in terms of y and substitute in the first equation (verify), so (12) and (14) represent the same geometric line (Figure 1.3.3).

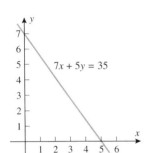

Figure 1.3.3

POINT-NORMAL EQUATIONS OF PLANES

A plane in R^3 can be uniquely determined by specifying a point $\mathbf{x}_0$ in the plane and a *nonzero* vector $\mathbf{n}$ that is perpendicular to the plane (Figure 1.3.4a). The vector $\mathbf{n}$ is said to be ***normal*** to the plane. If $\mathbf{x}$ is any point in this plane, then the vector $\mathbf{x} - \mathbf{x}_0$ is orthogonal to $\mathbf{n}$ (Figure 1.3.4b), so

$$\mathbf{n} \cdot (\mathbf{x} - \mathbf{x}_0) = 0 \tag{15}$$

Conversely, any point $\mathbf{x}$ that satisfies this equation lies in the plane, so (15) is an equation of the plane through $\mathbf{x}_0$ with normal $\mathbf{n}$. If we now let $\mathbf{x} = (x, y, z)$ be any point on the plane through $\mathbf{x}_0 = (x_0, y_0, z_0)$ with normal $\mathbf{n} = (A, B, C)$, then (15) can be expressed in component form as

$$(A, B, C) \cdot (x - x_0, y - y_0, z - z_0) = 0$$

or, equivalently, as

$$A(x - x_0) + B(y - y_0) + C(z - z_0) = 0 \tag{16}$$

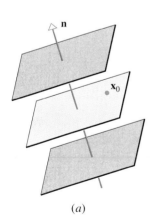

(a)

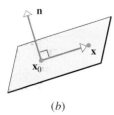

(b)

Figure 1.3.4

where A, B, and C are not all zero. We call this a ***point-normal equation*** of the plane through $\mathbf{x}_0 = (x_0, y_0, z_0)$ with normal $\mathbf{n} = (A, B, C)$. When convenient, the terms on the left side of (16) can be multiplied out and the equation rewritten in the form

$$Ax + By + Cz = D \quad (A, B, \text{ and } C \text{ not all zero}) \tag{17}$$

We call this the ***general equation*** of a plane.

In the special case where $\mathbf{x}_0 = (0, 0, 0)$ (i.e., the plane passes through the origin), Equations (15) and (17) simplify to

$$\mathbf{n} \cdot \mathbf{x} = 0 \tag{18}$$

and

$$Ax + By + Cz = 0 \quad (A, B, \text{ and } C \text{ not all zero}) \tag{19}$$

respectively.

EXAMPLE 3
Finding a
Point-Normal
Equation of a
Plane

Find a point-normal equation and a general equation of the plane that passes through the point $(3, -1, 7)$ and has normal $\mathbf{n} = (4, 2, -5)$.

Solution From (16), a point-normal equation of the plane is

$$4(x - 3) + 2(y + 1) - 5(z - 7) = 0$$

Multiplying out and taking the constant to the right side yields the general equation

$$4x + 2y - 5z = -25 \qquad \blacksquare$$

VECTOR AND PARAMETRIC EQUATIONS OF PLANES

Although point-normal equations of planes are useful, there are many applications in which it is preferable to have vector or parametric equations of a plane. To derive such equations we start with the observation that a plane W is uniquely determined by specifying a point $\mathbf{x}_0$ in W and two nonzero vectors $\mathbf{v}_1$ and $\mathbf{v}_2$ that are parallel to W and are not scalar multiples of one another (Figure 1.3.5*a*). If $\mathbf{x}$ is any point in the plane W, and if $\mathbf{v}_1$ and $\mathbf{v}_2$ are positioned with their initial points at $\mathbf{x}_0$, then by forming suitable scalar multiples of $\mathbf{v}_1$ and $\mathbf{v}_2$, we can create a parallelogram with adjacent sides $t_1\mathbf{v}_1$ and $t_2\mathbf{v}_2$ in which $\mathbf{x} - \mathbf{x}_0$ is the diagonal given by the sum

$$\mathbf{x} - \mathbf{x}_0 = t_1\mathbf{v}_1 + t_2\mathbf{v_2}$$

(Figure 1.3.5*b*) or, equivalently,

$$\mathbf{x} = \mathbf{x}_0 + t_1\mathbf{v}_1 + t_2\mathbf{v}_2$$

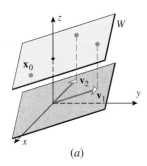

As the variables t_1 and t_2, called *parameters*, vary independently from $-\infty$ to $+\infty$, the point $\mathbf{x}$ in this formula varies over the entire plane W, so the plane through $\mathbf{x}_0$ that is parallel to $\mathbf{v}_1$ and $\mathbf{v}_2$ can be represented by the equation

$$\mathbf{x} = \mathbf{x}_0 + t_1\mathbf{v}_1 + t_2\mathbf{v}_2 \quad (-\infty < t_1 < +\infty, -\infty < t_2 < +\infty) \tag{20}$$

We call this a *vector equation of the plane* through $\mathbf{x}_0$ that is parallel to $\mathbf{v}_1$ and $\mathbf{v}_2$. In the special case where $\mathbf{x}_0 = \mathbf{0}$, the plane passes through the origin and (20) simplifies to

$$\mathbf{x} = t_1\mathbf{v}_1 + t_2\mathbf{v}_2 \quad (-\infty < t_1 < +\infty, -\infty < t_2 < +\infty) \tag{21}$$

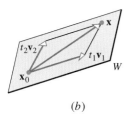

(*b*)

Figure 1.3.5

Note that the plane in (20) is the *translation* by $\mathbf{x}_0$ of the plane in (21).

As with a line, a vector equation of a plane can be split into a set of scalar equations by equating corresponding components; these are called *parametric equations* of the plane. For example, if we let $\mathbf{x} = (x, y, z)$ be a general point in the plane through $\mathbf{x}_0 = (x_0, y_0, z_0)$ that is parallel to the vectors $\mathbf{v}_1 = (a_1, b_1, c_1)$ and $\mathbf{v}_2 = (a_2, b_2, c_2)$, then (20) can be expressed in component form as

$$(x, y, z) = (x_0, y_0, z_0) + t_1(a_1, b_1, c_1) + t_2(a_2, b_2, c_2)$$

Equating corresponding components yields the parametric equations

$$\begin{aligned}
x &= x_0 + a_1t_1 + a_2t_2 \\
y &= y_0 + b_1t_1 + b_2t_2 \quad (-\infty < t_1 < +\infty, -\infty < t_2 < +\infty) \\
z &= z_0 + c_1t_1 + c_2t_2
\end{aligned} \tag{22}$$

for this plane.

EXAMPLE 4

Vector and Parametric Equations of Planes

(a) Find vector and parametric equations of the plane that passes through the origin of R^3 and is parallel to the vectors $\mathbf{v}_1 = (1, -2, 3)$ and $\mathbf{v}_2 = (4, 0, 5)$.

(b) Find three points in the plane obtained in part (a).

Solution (a) It follows from (21) that a vector equation of the plane is $\mathbf{x} = t_1\mathbf{v}_1 + t_2\mathbf{v}_2$. If we let $\mathbf{x} = (x, y, z)$, then this equation can be expressed in component form as

$$(x, y, z) = t_1(1, -2, 3) + t_2(4, 0, 5) \tag{23}$$

Equating corresponding components on the two sides of this equation yields the parametric equations

$$x = t_1 + 4t_2, \quad y = -2t_1, \quad z = 3t_1 + 5t_2$$

Solution (b) Points in the plane can be obtained by assigning values to the parameters t_1 and t_2 in (23). For example,

$t_1 = 0$ and $t_2 = 0$ produces the point $(0, 0, 0)$

$t_1 = -2$ and $t_2 = 1$ produces the point $(2, 4, -1)$

$t_1 = \frac{1}{2}$ and $t_2 = \frac{1}{2}$ produces the point $\left(\frac{5}{2}, -1, 4\right)$ ∎

EXAMPLE 5

A Plane Through Three Points

A plane is uniquely determined by three noncollinear points. If $\mathbf{x}_0$, $\mathbf{x}_1$, and $\mathbf{x}_2$ are three such points, then the vectors $\mathbf{v}_1 = \mathbf{x}_1 - \mathbf{x}_0$ and $\mathbf{v}_2 = \mathbf{x}_2 - \mathbf{x}_0$ are parallel to the plane (draw a picture), so it follows from (20) that a vector equation of the plane is

$$\mathbf{x} = \mathbf{x}_0 + t_1(\mathbf{x}_1 - \mathbf{x}_0) + t_2(\mathbf{x}_2 - \mathbf{x}_0) \tag{24}$$

Use this result to find vector and parametric equations of the plane that passes through the points $P(2, -4, 5)$, $Q(-1, 4, -3)$, and $R(1, 10, -7)$.

Solution If we let $\mathbf{x} = (x, y, z)$, and if we take $\mathbf{x}_0$, $\mathbf{x}_1$, and $\mathbf{x}_2$ to be the points P, Q, and R, respectively, then

$$\mathbf{x}_1 - \mathbf{x}_0 = \overrightarrow{PQ} = (-3, 8, -8) \quad \text{and} \quad \mathbf{x}_2 - \mathbf{x}_0 = \overrightarrow{PR} = (-1, 14, -12) \tag{25}$$

so (24) can be expressed in component form as

$$(x, y, z) = (2, -4, 5) + t_1(-3, 8, -8) + t_2(-1, 14, -12)$$

Equating corresponding components on the two sides of this equation yields the parametric equations

$$x = 2 - 3t_1 - t_2, \quad y = -4 + 8t_1 + 14t_2, \quad z = 5 - 8t_1 - 12t_2 \quad\quad ∎$$

CONCEPT PROBLEM How can you tell from (25) that the points P, Q, and R are not collinear?

EXAMPLE 6

Finding a Vector Equation from Parametric Equations

Find a vector equation of the plane whose parametric equations are

$$x = 4 + 5t_1 - t_2, \quad y = 2 - t_1 + 8t_2, \quad z = t_1 + t_2$$

Solution First we rewrite the three equations as the single vector equation

$$(x, y, z) = (4 + 5t_1 - t_2, 2 - t_1 + 8t_2, t_1 + t_2) \tag{26}$$

Each component on the right side is the sum of a constant (possibly zero), a scalar multiple of t_1, and a scalar multiple of t_2. We now isolate the terms of each type by splitting (26) apart:

$$(x, y, z) = (4, 2, 0) + (5t_1, -t_1, t_1) + (-t_2, 8t_2, t_2)$$

We can rewrite this equation as

$$(x, y, z) = (4, 2, 0) + t_1(5, -1, 1) + t_2(-1, 8, 1)$$

which is a vector equation of the plane that passes through the point $(4, 2, 0)$ and is parallel to the vectors $\mathbf{v}_1 = (5, -1, 1)$ and $\mathbf{v}_2 = (-1, 8, 1)$. ∎

EXAMPLE 7
Finding Parametric Equations from a General Equation

Find parametric equations of the plane $x - y + 2z = 5$.

Solution We will solve for x in terms of y and z, then make y and z into parameters, and then express x in terms of these parameters. Solving for x in terms of y and z yields

$$x = 5 + y - 2z$$

Now setting $y = t_1$ and $z = t_2$ yields the parametric equations

$$x = 5 + t_1 - 2t_2, \quad y = t_1, \quad z = t_2$$

Different parametric equations can be obtained by solving for y in terms of x and z and taking x and z as the parameters or by solving for z in terms of x and y and taking x and y as the parameters. However, they all produce the same plane as the parameters vary independently from $-\infty$ to $+\infty$. ∎

LINES AND PLANES IN R^n

The concepts of line and plane can be extended to R^n. Although we cannot actually see these objects when n is greater than three, lines and planes in R^n will prove to be very useful. Motivated by Formulas (4) and (20), we make the following definitions.

> **Definition 1.3.1**
> (a) If $\mathbf{x}_0$ is a vector in R^n, and if $\mathbf{v}$ is a nonzero vector in R^n, then we define the ***line through $\mathbf{x}_0$ that is parallel to $\mathbf{v}$*** to be the set of all vectors $\mathbf{x}$ in R^n that are expressible in the form
>
> $$\mathbf{x} = \mathbf{x}_0 + t\mathbf{v} \quad (-\infty < t < +\infty) \tag{27}$$
>
> (b) If $\mathbf{x}_0$ is a vector in R^n, and if $\mathbf{v}_1$ and $\mathbf{v}_2$ are nonzero vectors in R^n that are not scalar multiples of one another, then we define ***the plane through $\mathbf{x}_0$ that is parallel to $\mathbf{v}_1$ and $\mathbf{v}_2$*** to be the set of all vectors $\mathbf{x}$ in R^n that are expressible in the form
>
> $$\mathbf{x} = \mathbf{x}_0 + t_1\mathbf{v}_1 + t_2\mathbf{v}_2 \quad (-\infty < t_1 < +\infty, -\infty < t_2 < +\infty) \tag{28}$$

REMARK If $\mathbf{x}_0 = \mathbf{0}$, then the line in (27) and the plane in (28) are said to ***pass through the origin***. In this case Equations (27) and (28) simplify to $\mathbf{x} = t\mathbf{v}$ and $\mathbf{x} = t_1\mathbf{v}_1 + t_2\mathbf{v}_2$, the first of which expresses $\mathbf{x}$ as a linear combination (scalar multiple) of $\mathbf{v}$, and the second of which expresses $\mathbf{x}$ as a linear combination of $\mathbf{v}_1$ and $\mathbf{v}_2$. Thus, *a line through the origin of R^n can be viewed as the set of all linear combinations of a single nonzero vector, and a plane through the origin of R^n as the set of all linear combinations of two nonzero vectors that are not scalar multiples of one another.*

EXAMPLE 8
Parametric Equations of Lines and Planes in R^4

(a) Find vector and parametric equations of the line through the origin of R^4 that is parallel to the vector $\mathbf{v} = (5, -3, 6, 1)$.

(b) Find vector and parametric equations of the plane in R^4 that passes through the point $\mathbf{x}_0 = (2, -1, 0, 3)$ and is parallel to the vectors $\mathbf{v}_1 = (1, 5, 2, -4)$ and $\mathbf{v}_2 = (0, 7, -8, 6)$.

Solution (a) If we let $\mathbf{x} = (x_1, x_2, x_3, x_4)$, then the vector equation $\mathbf{x} = t\mathbf{v}$ can be expressed as

$$(x_1, x_2, x_3, x_4) = t(5, -3, 6, 1)$$

Equating corresponding components yields the parametric equations

$$x_1 = 5t, \ x_2 = -3t, \ x_3 = 6t, \ x_4 = t$$

Solution (b) The vector equation $\mathbf{x} = \mathbf{x}_0 + t_1\mathbf{v}_1 + t_2\mathbf{v}_2$ can be expressed as

$$(x_1, x_2, x_3, x_4) = (2, -1, 0, 3) + t_1(1, 5, 2, -4) + t_2(0, 7, -8, 6)$$

which yields the parametric equations

$$
\begin{aligned}
x_1 &= 2 + t_1 \\
x_2 &= -1 + 5t_1 + 7t_2 \\
x_3 &= 2t_1 - 8t_2 \\
x_4 &= 3 - 4t_1 + 6t_2
\end{aligned}
$$

■

COMMENTS ON TERMINOLOGY

It is evident what we mean when we say that a *point* lies on a line L in R^2 or R^3 or that a point lies in a plane W in R^3, but it is not so clear what we mean when we say that a *vector* lies on L or in W, since vectors can be translated. For example, it is certainly reasonable to say that the vector $\mathbf{v}$ in Figure 1.3.6*a* lies on the line L since it is collinear with L, yet if we translate $\mathbf{v}$ (which does not change $\mathbf{v}$), then the vector and the line no longer coincide (Figure 1.3.6*b*). To complicate life still further, the vector $\mathbf{v}$ in Figure 1.3.6*c* cannot be translated to coincide with the line L, yet the coordinates of its terminal point satisfy the equation of the line when the initial point of the vector is at the origin.

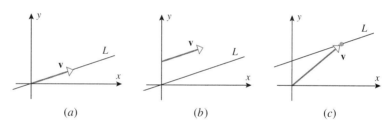

Figure 1.3.6   (*a*)   (*b*)   (*c*)

To resolve these linguistic ambiguities in R^2 and R^3 we will agree to say that *a vector* $\mathbf{v}$ *lies on a line* L in R^2 or R^3 if the terminal point of the vector lies on the line when the vector is positioned with its initial point at the origin. Thus, the vector $\mathbf{v}$ lies on the line L in all three cases shown in Figure 1.3.6. Similarly, we will say that *a vector* $\mathbf{v}$ *lies in a plane* W *in* R^3 if the terminal point of the vector lies in the plane when the initial point of the vector is at the origin.

Exercise Set 1.3

In Exercises 1 and 2, find parametric equations of the lines L_1, L_2, L_3, and L_4 that pass through the indicated vertices of the square and cube.

1. (a) (b)

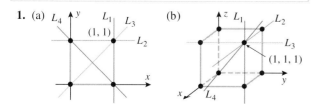

2. (a) (b)

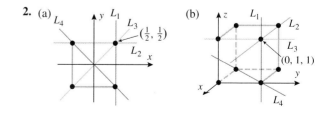

In Exercises 3 and 4, sketch the line whose vector equation is given.

3. (a) $(x, y) = t(2, 3)$
   (b) $(x, y) = (1, 1) + t(1, -1)$

4. (a) $(x, y) = (2, 0) + t(1, 1)$
   (b) $(x, y) = t(-1, -1)$

In Exercises 5 and 6, find vector and parametric equations of the line determined by the given points.

5. (a) $(0, 0)$ and $(3, 5)$ (b) $(1, 1, 1)$ and $(0, 0, 0)$
   (c) $(1, -1, 1)$ and $(2, 1, 1)$

6. (a) $(1, 2)$ and $(-5, 6)$
   (b) $(1, 2, 3)$ and $(-1, -2, -3)$
   (c) $(1, 2, -4)$ and $(3, -1, 1)$

In Exercises 7 and 8, find parametric and vector equations of the line that is parallel to **u** and passes through the point P. Use your vector equation to find two points on the line that are different from P_0.

7. (a) $\mathbf{u} = (1, 2)$; $P_0(1, 1)$
   (b) $\mathbf{u} = (1, -1, 1)$; $P_0(2, 0, 3)$
   (c) $\mathbf{u} = (3, 2, -3)$; $P_0(0, 0, 0)$

8. (a) $\mathbf{u} = (-2, 4)$; $P_0(0, 1)$
   (b) $\mathbf{u} = (5, -2, 1)$; $P_0(1, 6, 2)$
   (c) $\mathbf{u} = (4, 0, -1)$; $P_0(4, 0, -1)$

In Exercises 9 and 10, find a point-normal equation of the plane that passes through the point P and has normal **n**.

9. $\mathbf{n} = (3, 2, 1)$; $P(-1, -1, -1)$

10. $\mathbf{n} = (1, 1, 4)$; $P(3, 5, -2)$

In Exercises 11 and 12, find a vector equation and parametric equations of the plane that passes through the given points. Also, find three points in the plane that are different from those given.

11. $(1, 1, 4)$, $(2, -3, 1)$, and $(3, 5, -2)$

12. $(3, 2, 1)$, $(-1, -1, -1)$, and $(6, 0, 2)$

13. (a) Find a vector equation of the line whose parametric equations are
$$x = 2 + 4t, \ y = -1 + t, \ z = t$$
   (b) Find a vector equation of the plane whose parametric equations are
$$x = 1 + 2t_1 + t_2, \ y = -2 - t_1 + 5t_2, \ z = 4t_1 - t_2$$
   (c) Find parametric equations of the plane
   $3x + 4y - 2z = 4$.

14. (a) Find a vector equation of the line whose parametric equations are
$$x = t, \ y = -3 + 5t, \ z = 1 + t$$

(b) Find a vector equation of the plane whose parametric equations are
$$x = t_1 + t_2, \ y = 4 + 3t_1 - t_2, \ z = 4t_1$$

(c) Find parametric equations of the plane
   $3x - 5y + z = 32$.

In Exercises 15 and 16, find the general equation and a vector equation of the plane that passes through the points.

15. $P(1, 2, 4)$, $Q(1, -1, 6)$, $R(1, 4, 8)$

16. $P(2, 2, 1)$, $Q(0, 3, 4)$, $R(1, -1, -3)$

In Exercises 17 and 18, describe the object in R^4 that is represented by the vector equation and find parametric equations for it.

17. (a) $(x_1, x_2, x_3, x_4) = t(1, -2, 5, 7)$
   (b) $(x_1, x_2, x_3, x_4) = (4, 5, -6, 1) + t(1, 1, 1, 1)$
   (c) $(x_1, x_2, x_3, x_4) = (-1, 0, 4, 2) + t_1(-3, 5, -7, 4) + t_2(6, 3, -1, 2)$

18. (a) $(x_1, x_2, x_3, x_4) = t(-3, 5, -7, 4)$
   (b) $(x_1, x_2, x_3, x_4) = (5, 6, -5, 2) + t(3, 0, 1, 4)$
   (c) $(x_1, x_2, x_3, x_4) = t_1(-4, 7, -1, 5) + t_2(2, 1, -3, 0)$

19. (a) The parametric equations $x_1 = 3t$, $x_2 = 4t$, $x_3 = 7t$, $x_4 = t$, $x_5 = 9t$ represent a _____ passing through _____ and parallel to the vector _____.
   (b) The parametric equations
$$x_1 = 3 - 2t_1 + 5t_2$$
$$x_2 = 4 - 3t_1 + 6t_2$$
$$x_3 = -2 - 2t_1 + 7t_2$$
$$x_4 = 1 - 2t_1 - t_2$$
represent a _____ passing through _____ and parallel to _____.

20. (a) The parametric equations $x_1 = 1 + 2t$, $x_2 = -5 + 3t$, $x_3 = 6t$, $x_4 = -2 + t$, $x_5 = 4 + 9t$ represent a _____ passing through _____ and parallel to the vector _____.
   (b) The parametric equations
$$x_1 = 3t_1 + 5t_2$$
$$x_2 = 4t_1 + 6t_2$$
$$x_3 = -t_1 + 5t_2$$
$$x_4 = t_1 + t_2$$
represent a _____ passing through _____ and parallel to _____.

21. Find parametric equations of the plane that is parallel to the plane $3x + 2y - z = 1$ and passes through the point $P(1, 1, 1)$.

22. Find parametric equations of the plane through the origin that is parallel to the plane $x = t_1 + t_2$, $y = 4 + 3t_1 - t_2$, $z = 4t_1$.

23. Which of the following planes, if any, are parallel to the plane $3x + y - 2z = 5$?
   (a) $x + y - z = 3$
   (b) $3x + y - 2z = 0$
   (c) $x + \frac{1}{3}y - \frac{2}{3}z = 5$

24. Which of the following planes, if any, are parallel to the plane $x + 2y - 3z = 2$?
   (a) $x + 2y - 3z = 3$
   (b) $\frac{1}{4}x + \frac{1}{2}y - \frac{3}{4}z = 0$
   (c) $x + 2y + 3z = 2$

25. Find parametric equations of the line that is perpendicular to the plane $x + y + z = 0$ and passes through the point $P(2, 0, 1)$.

26. Find parametric equations of the line that is perpendicular to the plane $x + 2y + 3z = 0$ and passes through the origin.

27. Find a vector equation of the plane that passes through the origin and contains the points $(5, 4, 3)$ and $(1, -1, -2)$.

28. Find a vector equation of the plane that is perpendicular to the x axis and contains the point $P(1, 1, 3)$.

29. Find parametric equations of the plane that passes through the point $P(-2, 1, 7)$ and is perpendicular to the line whose parametric equations are
$$x = 4 + 2t, \ y = -2 + 3t, \ z = -5t$$

30. Find parametric equations of the plane that passes through the origin and contains the line whose parametric equations are
$$x = 2t, \ y = 1 + t, \ z = 2 - t$$

31. Determine whether the line and plane are parallel.
   (a) $x = -5 - 4t, \ y = 1 - t, \ z = 3 + 2t$;
   $x + 2y + 3z - 9 = 0$
   (b) $x = 3t, \ y = 1 + 2t, \ z = 2 - t$; $4x + y + 2z = 1$

32. Determine whether the line and plane are perpendicular.
   (a) $x = -2 - 4t, \ y = 3 - 2t, \ z = 1 + 2t$; $2x + y - z = 5$
   (b) $x = 2 + t, \ y = 1 - t, \ z = 5 + 3t$; $6x + 6y - 7 = 0$

33. Determine whether the planes are perpendicular.
   (a) $3x - y + z - 4 = 0, \ x + 2z = -1$
   (b) $x - 2y + 3z = 4, \ -2x + 5y + 4z = -1$

34. Determine whether the planes are perpendicular.
   (a) $4x + 3y - z + 1 = 0, \ 2x - 2y + 2z = -3$
   (b) $2x - 3y - z = 1, \ x + 3y - 2z = 12$

35. Show that the line $x = 0, \ y = t, \ z = t$
   (a) lies in the plane $6x + 4y - 4z = 0$;
   (b) is parallel to and below the plane $5x - 3y + 3z = 1$;
   (c) is parallel to and below the plane $6x + 2y - 2z = -3$.

36. Find an equation for the plane whose points are equidistant from $(-1, -4, -2)$ and $(0, -2, 2)$. [*Hint:* Choose an arbitrary point (x, y, z) in the plane, and use the distance formula.]

In Exercises 37 and 38, find parametric equations for the line of intersection, if any, of the planes.

37. (a) $7x - 2y + 3z = -2$ and $-3x + y + 2z + 5 = 0$
   (b) $2x + 3y - 5z = 0$ and $4x + 6y - 10z = 8$

38. (a) $-3x + 2y + z = -5$ and $7x + 3y - 2z = -2$
   (b) $5x - 7y + 2z = 0$ and $y = 0$

In Exercises 39 and 40, find the point of intersection, if any, of the line and plane.

39. (a) $x = 9 - 5t, \ y = -1 - t, \ z = 3 + t$;
   $2x - 3y + 4z + 7 = 0$
   (b) $x = t, \ y = t, \ z = t$; $x + y - 2z = 3$

40. (a) $x = t, \ y = t, \ z = t$; $x + y - 2z = 0$
   (b) $x = 3 - 4t, \ y = -2 - t, \ z = 5 + t$; $3x - 4y + 5z = 0$

As noted in the remark following Formula (10), the equation
$$\mathbf{x} = (1 - t)\mathbf{x}_0 + t\mathbf{x}_1 \quad (0 \le t \le 1)$$
represents the line segment in 2-space or 3-space that extends from $\mathbf{x}_0$ to $\mathbf{x}_1$. In Exercises 41 and 42, sketch the line segment represented by the vector equation.

41. (a) $\mathbf{x} = (1 - t)(1, 0) + t(0, 1) \quad (0 \le t \le 1)$
   (b) $\mathbf{x} = (1 - t)(1, 1, 0) + t(0, 0, 1) \quad (0 \le t \le 1)$

42. (a) $\mathbf{x} = (1 - t)(1, 1) + t(1, -1) \quad (0 \le t \le 1)$
   (b) $\mathbf{x} = (1 - t)(1, 1, 1) + t(1, 1, 0) \quad (0 \le t \le 1)$

In Exercises 43 and 44, write a vector equation for the line segment from P to Q.

43. $P(-2, 4, 1), \ Q(0, 4, 7)$ **44.** $P(0, -6, 5), \ Q(3, -1, 9)$

45. Let $P = (2, 3, -2)$ and $Q = (7, -4, 1)$.
   (a) Find the midpoint of the line segment connecting the points P and Q.
   (b) Find the point on the line segment connecting P and Q that is $\frac{3}{4}$ of the way from the point P to the point Q.

Discussion and Discovery

D1. Given that a, b, and c are not all zero, find parametric equations for a line in R^3 that passes through the point (x_0, y_0, z_0) and is perpendicular to the line

$$x = x_0 + at, \; y = y_0 + bt, \; z = z_0 + ct$$

D2. (a) How can you tell whether the line $\mathbf{x} = \mathbf{x}_0 + t\mathbf{v}$ in R^3 is parallel to the plane $\mathbf{x} = \mathbf{x}_0 + t_1\mathbf{v}_1 + t_2\mathbf{v}_2$?

(b) Invent a reasonable definition of what it means for a line to be parallel to a plane in R^n.

D3. (a) Let $\mathbf{v}$, $\mathbf{w}_1$, and $\mathbf{w}_2$ be vectors in R^n. Show that if $\mathbf{v}$ is orthogonal to both $\mathbf{w}_1$ and $\mathbf{w}_2$, then $\mathbf{v}$ is orthogonal to $\mathbf{x} = k_1\mathbf{w}_1 + k_2\mathbf{w}_2$ for all scalars k_1 and k_2.

(b) Give a geometric interpretation of this result in R^3.

D4. (a) The equation $Ax + By = 0$ represents a line through the origin in R^2 if A and B are not both zero. What does this equation represent in R^3 if you think of it as $Ax + By + 0z = 0$? Explain.

(b) Do you think that the equation $Ax_1 + Bx_2 + Cx_3 = 0$ represents a plane in R^4 if A, B, and C are not all zero? Explain.

D5. Indicate whether the statement is true (T) or false (F). Justify your answer.

(a) If a, b, and c are not all zero, then the line $x = at$, $y = bt$, $z = ct$ is perpendicular to the plane $ax + by + cz = 0$.

(b) Two nonparallel lines in R^3 must intersect in at least one point.

(c) If $\mathbf{u}$, $\mathbf{v}$, and $\mathbf{w}$ are vectors in R^3 such that $\mathbf{u} + \mathbf{v} + \mathbf{w} = \mathbf{0}$, then the three vectors lie in some plane.

(d) The equation $\mathbf{x} = t\mathbf{v}$ represents a line for every vector $\mathbf{v}$ in R^2.

Technology Exercises

T1. (*Parametric lines*) Many graphing utilities can graph parametric curves. If you have such a utility, then determine how to do this and generate the line $x = 5 + 5t$, $y = -7t$ (see Figure 1.3.3).

T2. Generate the line L through the point $(1, 2)$ that is parallel to $\mathbf{v} = (1, 1)$; in the same window, generate the line through the point $(1, 2)$ that is perpendicular to L. If your lines do not look perpendicular, explain why.

T3. Two intersecting planes in 3-space determine two angles of intersection, an acute angle $(0 \le \theta \le 90°)$ and its supplement $180° - \theta$ (see the accompanying figure). If $\mathbf{n}_1$ and $\mathbf{n}_2$ are normals to the planes, then the angle between $\mathbf{n}_1$ and $\mathbf{n}_2$ is θ or $180° - \theta$ depending on the directions of the normals. In each part, find the acute angle of intersection of the planes to the nearest degree.

(a) $x = 0$ and $2x - y + z - 4 = 0$

(b) $x + 2y - 2z = 5$ and $6x - 3y + 2z = 8$

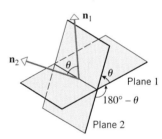

Figure Ex-T3

T4. Find the acute angle of intersection between the plane $x - y - 3z = 5$ and the line

$$x = 2 - t, \; y = 2t, \; z = 3t - 1$$

(See Exercise T3.)

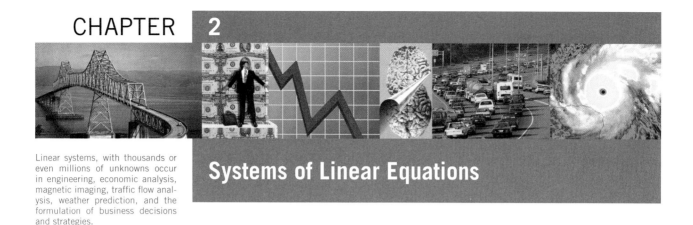

CHAPTER 2

Systems of Linear Equations

Linear systems, with thousands or even millions of unknowns occur in engineering, economic analysis, magnetic imaging, traffic flow analysis, weather prediction, and the formulation of business decisions and strategies.

Section 2.1 Introduction to Systems of Linear Equations

The study of systems of linear equations and their solutions is one of the major topics in linear algebra. In this introductory section we will discuss some ways in which systems of linear equations arise, what it means to solve them, and how their solutions can be interpreted geometrically. Our focus here will be on general ideas, and in the next section we will discuss computational methods for finding solutions.

LINEAR SYSTEMS

Recall that a line in R^2 can be represented by an equation of the form

$$a_1 x + a_2 y = b \quad (a_1, a_2 \text{ not both } 0) \tag{1}$$

and a plane in R^3 by an equation of the form

$$a_1 x + a_2 y + a_3 z = b \quad (a_1, a_2, a_3 \text{ not all } 0) \tag{2}$$

These are examples of *linear equations*. In general, we define a ***linear equation*** in the n variables $x_1, x_2, \ldots, x_n$ to be one that can be expressed in the form

$$a_1 x_1 + a_2 x_2 + \cdots + a_n x_n = b \tag{3}$$

where $a_1, a_2, \ldots, a_n$ and b are constants and the a's are not all zero. In the cases where $n = 2$ and $n = 3$ we will often use variables without subscripts, as in (1) and (2). In the special case where $b = 0$, Equation (3) has the form

$$a_1 x_1 + a_2 x_2 + \cdots + a_n x_n = 0 \tag{4}$$

which is called a ***homogeneous linear equation***.

EXAMPLE 1
Linear
Equations

Observe that a linear equation does not involve any products or roots of variables. All variables occur only to the first power and do not appear, for example, as arguments of trigonometric, logarithmic, or exponential functions. The following are linear equations:

$$x + 3y = 7 \qquad\qquad x_1 - 2x_2 - 3x_3 + x_4 = 0$$
$$\tfrac{1}{2}x - y + 3z = -1 \qquad\qquad x_1 + x_2 + \cdots + x_n = 1$$

The following are not linear equations:

$$x + 3y^2 = 4 \qquad\qquad 3x + 2y - xy = 5$$
$$\sin x + y = 0 \qquad\qquad \sqrt{x_1} + 2x_2 + x_3 = 1$$

■

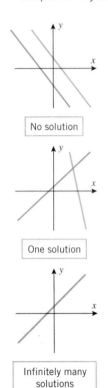

No solution

One solution

Infinitely many
solutions
(coincident lines)

Figure 2.1.1

A finite set of linear equations is called a *system of linear equations* or a *linear system*. The variables in a linear system are called the *unknowns*. For example,

$$4x_1 - x_2 + 3x_3 = -1$$
$$3x_1 + x_2 + 9x_3 = -4$$

(5)

is a linear system of two equations in the three unknowns x_1, x_2, and x_3. A general linear system of m equations in the n unknowns $x_1, x_2, \ldots, x_n$ can be written as

$$a_{11}x_1 + a_{12}x_2 + \cdots + a_{1n}x_n = b_1$$
$$a_{21}x_1 + a_{22}x_2 + \cdots + a_{2n}x_n = b_2$$
$$\vdots \qquad \vdots \qquad \qquad \vdots \qquad \vdots$$
$$a_{m1}x_1 + a_{m2}x_2 + \cdots + a_{mn}x_n = b_m$$

(6)

The double subscripting on the coefficients a_{ij} of the unknowns is used to specify their location—the first subscript indicates the equation in which the coefficient occurs, and the second indicates which unknown it multiplies. Thus, a_{12} is in the first equation and multiplies unknown x_2.

A *solution* of a linear system in the unknowns $x_1, x_2, \ldots, x_n$ is a sequence of n numbers $s_1, s_2, \ldots, s_n$ that when substituted for $x_1, x_2, \ldots, x_n$, respectively, makes every equation in the system a true statement. For example, $x_1 = 1, x_2 = 2, x_3 = -1$ is a solution of (5), but $x_1 = 1, x_2 = 8, x_3 = 1$ is not (verify). The set of all solutions of a linear system is called its *solution set*.

It will be useful to express solutions of linear systems as ordered n-tuples. Thus, for example, the solution $x_1 = 1, x_2 = 2, x_3 = -1$ of (5) might be expressed as the n-tuple $(1, 2, -1)$; and more generally, a solution $x_1 = s_1, x_2 = s_2, \ldots, x_n = s_n$ of (6) might be expressed as the n-tuple $(s_1, s_2, \ldots, s_n)$. By thinking of solutions as n-tuples, we interpret them as vectors or points in R^n, thereby opening up the possibility that the solution set may have some recognizable geometric form.

LINEAR SYSTEMS WITH TWO AND THREE UNKNOWNS

Linear systems in two unknowns arise in connection with intersections of lines in R^2. For example, consider the linear system

$$a_1x + b_1y = c_1$$
$$a_2x + b_2y = c_2$$

The graphs of these equations are lines in the xy-plane, so each solution (x, y) of this system corresponds to a point of intersection of these lines. Thus, there are three possibilities (Figure 2.1.1):

1. The lines may be parallel and distinct, in which case there is no intersection and consequently no solution.

2. The lines may intersect at only one point, in which case the system has exactly one solution.

3. The lines may coincide, in which case there are infinitely many points of intersection (the points on the common line) and consequently infinitely many solutions.

In general, we say that a linear system is *consistent* if it has at least one solution and *inconsistent* if it has no solutions. Thus, a *consistent* linear system of two equations in two unknowns has either one solution or infinitely many solutions—there are no other possibilities. The same is true for a linear system of three equations in three unknowns:

$$a_1x + b_1y + c_1z = d_1$$
$$a_2x + b_2y + c_2z = d_2$$
$$a_3x + b_3y + c_3z = d_3$$

In this case, the graph of each equation is a plane, so the solutions of the system, if any, correspond to points where all three planes intersect; and again we see that there are only three possibilities—no solutions, one solution, or infinitely many solutions (Figure 2.1.2).

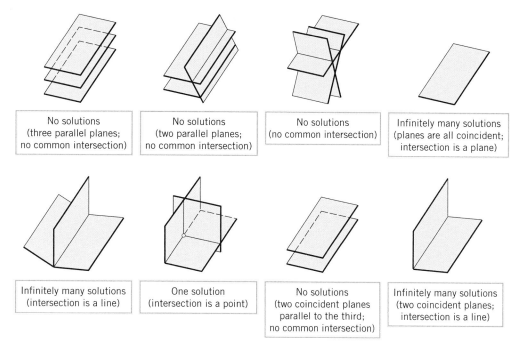

No solutions
(three parallel planes;
no common intersection)

No solutions
(two parallel planes;
no common intersection)

No solutions
(no common intersection)

Infinitely many solutions
(planes are all coincident;
intersection is a plane)

Infinitely many solutions
(intersection is a line)

One solution
(intersection is a point)

No solutions
(two coincident planes
parallel to the third;
no common intersection)

Infinitely many solutions
(two coincident planes;
intersection is a line)

Figure 2.1.2

We will prove later that our observations about the number of solutions of linear systems of two equations in two unknowns and three equations in three unknowns holds for *all* linear systems:

Theorem 2.1.1   *Every system of linear equations has zero, one, or infinitely many solutions; there are no other possibilities.*

EXAMPLE 2
A Linear
System with
One Solution

Solve the linear system

$$x - y = 1$$
$$2x + y = 6$$

Solution   We can eliminate x from the second equation by adding -2 times the first equation to the second. This yields the simplified system

$$x - y = 1$$
$$3y = 4$$

From the second equation we obtain $y = \frac{4}{3}$, and on substituting this value in the first equation we obtain $x = 1 + y = \frac{7}{3}$. Thus, the system has the unique solution $x = \frac{7}{3}$, $y = \frac{4}{3}$. Geometrically, this means that the lines represented by the equations in the system intersect at the single point $\left(\frac{7}{3}, \frac{4}{3}\right)$. We leave it for you to check this by graphing the lines.   ∎

EXAMPLE 3
A Linear
System with No
Solutions

Solve the linear system

$$x + y = 4$$
$$3x + 3y = 6$$

Solution   We can eliminate x from the second equation by adding -3 times the first equation to the second equation. This yields the simplified system

$$x + y = 4$$
$$0 = -6$$

The second equation is contradictory, so the given system has no solution. Geometrically, this means that the lines corresponding to the equations in the original system are parallel and distinct. We leave it for you to check this by graphing the lines or by showing that they have the same slope but different y-intercepts. ∎

EXAMPLE 4

A Linear System with Infinitely Many Solutions

Solve the linear system

$$4x - 2y = 1$$
$$16x - 8y = 4$$

Solution We can eliminate x from the second equation by adding -4 times the first equation to the second. This yields the simplified system

$$4x - 2y = 1$$
$$0 = 0$$

The second equation does not impose any restrictions on x and y and hence can be eliminated. Thus, the solutions of the system are those values of x and y that satisfy the single equation

$$4x - 2y = 1 \tag{7}$$

Geometrically, this means the lines corresponding to the two equations in the original system coincide. The most convenient way to describe the solution set in this case is to express (7) parametrically. We can do this by letting $y = t$ and solving for x in terms of t, or by letting $x = t$ and solving for y in terms of t. The first approach yields the following parametric equations:

$$x = \tfrac{1}{4} + \tfrac{1}{2}t, \ \ y = t$$

We can now obtain specific solutions by substituting numerical values for the parameter. For example, $t = 0$ yields the solution $\left(\tfrac{1}{4}, 0\right)$, $t = 1$ yields the solution $\left(\tfrac{3}{4}, 1\right)$, and $t = -1$ yields the solution $\left(-\tfrac{1}{4}, -1\right)$. You can confirm that these are solutions by substituting the coordinates into the given equation. ∎

CONCEPT PROBLEM Find parametric equations for the solution set in the last example by letting $x = t$ and solving for y. Then determine the value of the parameter t in those equations that produces the solution $\left(\tfrac{3}{4}, 1\right)$ obtained in the example.

EXAMPLE 5

A Linear System with Infinitely Many Solutions

Solve the linear system

$$x - \ \ y + 2z = \ \ 5$$
$$2x - 2y + 4z = 10$$
$$3x - 3y + 6z = 15$$

Solution This system can be solved by inspection, since the second and third equations are multiples of the first. Geometrically, this means that the three planes coincide and that those values of x, y, and z that satisfy the equation

$$x - y + 2z = 5 \tag{8}$$

automatically satisfy all three equations. Using the method of Example 7 in Section 1.3, we can express the solution set parametrically as

$$x = 5 + t_1 - 2t_2, \ \ y = t_1, \ \ z = t_2$$

Specific solutions can be obtained by choosing numerical values for the parameters. ∎

AUGMENTED MATRICES AND ELEMENTARY ROW OPERATIONS

As the number of equations and unknowns in a linear system increases, so does the complexity of the algebra involved in finding solutions. The required computations can be made more manageable by simplifying notation and standardizing procedures. For example, by mentally keeping track of the location of the $+$'s, the x's, and the $=$'s in the linear system

$$a_{11}x_1 + a_{12}x_2 + \cdots + a_{1n}x_n = b_1$$
$$a_{21}x_1 + a_{22}x_2 + \cdots + a_{2n}x_n = b_2$$
$$\vdots \qquad \vdots \qquad \qquad \vdots \qquad \vdots$$
$$a_{m1}x_1 + a_{m2}x_2 + \cdots + a_{mn}x_n = b_m$$

we can abbreviate the system by writing only the rectangular array of numbers

$$\begin{bmatrix} a_{11} & a_{12} & \cdots & a_{1n} & b_1 \\ a_{21} & a_{22} & \cdots & a_{2n} & b_2 \\ \vdots & \vdots & & \vdots & \vdots \\ a_{m1} & a_{m2} & \cdots & a_{mn} & b_m \end{bmatrix}$$

This is called the ***augmented matrix*** for the system. For example, the augmented matrix for the system of equations

$$\begin{aligned} x_1 + \ x_2 + 2x_3 &= 9 \\ 2x_1 + 4x_2 - 3x_3 &= 1 \\ 3x_1 + 6x_2 - 5x_3 &= 0 \end{aligned} \qquad \text{is} \qquad \begin{bmatrix} 1 & 1 & 2 & 9 \\ 2 & 4 & -3 & 1 \\ 3 & 6 & -5 & 0 \end{bmatrix}$$

Linear Algebra in History

The use of the term *augmented matrix* appears to have been introduced by the American mathematician Maxime Bôcher in his book *Introduction to Higher Algebra*, published in 1907. In addition to being an outstanding research mathematician and an expert in Latin, chemistry, philosophy, zoology, geography, meteorology, art, and music, Bôcher was an outstanding expositor of mathematics whose elementary textbooks were greatly appreciated by students and are still in demand today.

Maxime Bôcher
(1867–1918)

REMARK The term *matrix* is used in mathematics to denote any rectangular array of numbers. Later we will study matrices in more detail, but for now we will only be concerned with augmented matrices. When constructing the augmented matrix for a linear system, the unknowns in the system must be written in the same order in each equation, and the constant term in each equation must be on the right side by itself.

The basic method for solving a linear system is to perform appropriate algebraic operations on the equations to produce a succession of increasingly simpler systems that have the same solution set as the original system until a point is reached where it is apparent whether the system is consistent, and if so, what the solutions are. The succession of simpler systems can be obtained by eliminating unknowns systematically using three types of operations:

1. Multiply an equation through by a nonzero constant.

2. Interchange two equations.

3. Add a multiple of one equation to another.

Since the rows (horizontal lines) of an augmented matrix correspond to the equations in the associated system, these three operations correspond to the following operations on the rows of the augmented matrix:

1. Multiply a row through by a nonzero constant.

2. Interchange two rows.

3. Add a multiple of one row to another.

These are called ***elementary row operations*** on a matrix.

In the next example we will illustrate how to use elementary row operations and augmented matrices to solve a linear system in three unknowns. Since a systematic procedure for finding solutions will be developed in the next section, it is not necessary to worry about how the steps in this example were selected. Your objective here should be simply to understand the computations.

EXAMPLE 6
Using
Elementary
Row Operations
and Augmented
Matrices

In the left column we solve a linear system by operating on the equations, and in the right column we solve it by operating on the rows of the augmented matrix.

System	**Augmented Matrix**

$$x + y + 2z = 9$$
$$2x + 4y - 3z = 1$$
$$3x + 6y - 5z = 0$$

$$\begin{bmatrix} 1 & 1 & 2 & 9 \\ 2 & 4 & -3 & 1 \\ 3 & 6 & -5 & 0 \end{bmatrix}$$

Add -2 times the first equation to the second to obtain

Add -2 times the first row to the second to obtain

$$x + y + 2z = 9$$
$$2y - 7z = -17$$
$$3x + 6y - 5z = 0$$

$$\begin{bmatrix} 1 & 1 & 2 & 9 \\ 0 & 2 & -7 & -17 \\ 3 & 6 & -5 & 0 \end{bmatrix}$$

Add -3 times the first equation to the third to obtain

Add -3 times the first row to the third to obtain

$$x + y + 2z = 9$$
$$2y - 7z = -17$$
$$3y - 11z = -27$$

$$\begin{bmatrix} 1 & 1 & 2 & 9 \\ 0 & 2 & -7 & -17 \\ 0 & 3 & -11 & -27 \end{bmatrix}$$

Multiply the second equation by $\frac{1}{2}$ to obtain

Multiply the second row by $\frac{1}{2}$ to obtain

$$x + y + 2z = 9$$
$$y - \tfrac{7}{2}z = -\tfrac{17}{2}$$
$$3y - 11z = -27$$

$$\begin{bmatrix} 1 & 1 & 2 & 9 \\ 0 & 1 & -\frac{7}{2} & -\frac{17}{2} \\ 0 & 3 & -11 & -27 \end{bmatrix}$$

Add -3 times the second equation to the third to obtain

Add -3 times the second row to the third to obtain

$$x + y + 2z = 9$$
$$y - \tfrac{7}{2}z = -\tfrac{17}{2}$$
$$-\tfrac{1}{2}z = -\tfrac{3}{2}$$

$$\begin{bmatrix} 1 & 1 & 2 & 9 \\ 0 & 1 & -\frac{7}{2} & -\frac{17}{2} \\ 0 & 0 & -\frac{1}{2} & -\frac{3}{2} \end{bmatrix}$$

Multiply the third equation by -2 to obtain

Multiply the third row by -2 to obtain

$$x + y + 2z = 9$$
$$y - \tfrac{7}{2}z = -\tfrac{17}{2}$$
$$z = 3$$

$$\begin{bmatrix} 1 & 1 & 2 & 9 \\ 0 & 1 & -\frac{7}{2} & -\frac{17}{2} \\ 0 & 0 & 1 & 3 \end{bmatrix}$$

Add -1 times the second equation to the first to obtain

Add -1 times the second row to the first to obtain

$$x + \tfrac{11}{2}z = \tfrac{35}{2}$$
$$y - \tfrac{7}{2}z = -\tfrac{17}{2}$$
$$z = 3$$

$$\begin{bmatrix} 1 & 0 & \frac{11}{2} & \frac{35}{2} \\ 0 & 1 & -\frac{7}{2} & -\frac{17}{2} \\ 0 & 0 & 1 & 3 \end{bmatrix}$$

Add $-\frac{11}{2}$ times the third equation to the first and $\frac{7}{2}$ times the third equation to the second to obtain

Add $-\frac{11}{2}$ times the third row to the first and $\frac{7}{2}$ times the third row to the second to obtain

$$x = 1$$
$$y = 2$$
$$z = 3$$

$$\begin{bmatrix} 1 & 0 & 0 & 1 \\ 0 & 1 & 0 & 2 \\ 0 & 0 & 1 & 3 \end{bmatrix}$$

Linear Algebra in History

The first known example of using augmented matrices to describe linear systems appeared in a Chinese manuscript entitled *Nine Chapters of Mathematical Art* that was published between 200 B.C. and 100 B.C. during the Han Dynasty. The following problem was posed in the manuscript:

There are three types of corn, of which three bundles of the first, two of the second, and one of the third make 39 measures. Two of the first, three of the second, and one of the third make 34 measures. And one of the first, two of the second, and three of the third make 26 measures. How many measures of corn are contained in one bundle of each type?

The problem leads to a linear system of three equations in three unknowns, which the author sets up as

$$\begin{matrix} 1 & 2 & 3 \\ 2 & 3 & 2 \\ 3 & 1 & 1 \\ 26 & 34 & 39 \end{matrix}$$

Except for the arrangement of the coefficients by columns rather than rows and the omission of brackets, this is the augmented matrix for the system. Remarkably, the author then goes on to describe a succession of operations on the columns that leads to the solution of the system.

The solution

$$x = 1, \quad y = 2, \quad z = 3$$

is now evident. Geometrically, this means that the planes represented by the equations in the system intersect at the single point $(1, 2, 3)$ in R^3. ∎

EXAMPLE 7
Linear
Combinations

Determine whether the vector $\mathbf{w} = (9, 1, 0)$ can be expressed as a linear combination of the vectors

$$\mathbf{v}_1 = (1, 2, 3), \quad \mathbf{v}_2 = (1, 4, 6), \quad \mathbf{v}_3 = (2, -3, -5)$$

and, if so, find such a linear combination.

Solution For $\mathbf{w}$ to be a linear combination of $\mathbf{v}_1$, $\mathbf{v}_2$, and $\mathbf{v}_3$ there must exist scalars c_1, c_2, and c_3 such that

$$\mathbf{w} = c_1\mathbf{v}_1 + c_2\mathbf{v}_2 + c_3\mathbf{v}_3$$

or in component form,

$$(9, 1, 0) = c_1(1, 2, 3) + c_2(1, 4, 6) + c_3(2, -3, -5)$$

Thus, our problem reduces to determining whether there are values of c_1, c_2, and c_3 such that

$$(9, 1, 0) = (c_1 + c_2 + 2c_3, \ 2c_1 + 4c_2 - 3c_3, \ 3c_1 + 6c_2 - 5c_3)$$

By equating corresponding components, we see that this is equivalent to determining whether the linear system

$$\begin{aligned} c_1 + c_2 + 2c_3 &= 9 \\ 2c_1 + 4c_2 - 3c_3 &= 1 \\ 3c_1 + 6c_2 - 5c_3 &= 0 \end{aligned} \tag{9}$$

is consistent. However, this is the same as the system in Example 6 (except for a difference in the names of the unknowns), so we know from that example that the system is consistent and that it has the unique solution

$$c_1 = 1, \quad c_2 = 2, \quad c_3 = 3$$

Thus, $\mathbf{w}$ can be expressed as the linear combination

$$\mathbf{w} = \mathbf{v}_1 + 2\mathbf{v}_2 + 3\mathbf{v}_3$$

or in component form,

$$(9, 1, 0) = (1, 2, 3) + 2(1, 4, 6) + 3(2, -3, -5)$$

We leave it for you to confirm this vector equality. ∎

Exercise Set 2.1

In Exercises 1 and 2, determine whether the given equation is linear.

1. (a) $x_1 + 5x_2 - \sqrt{2}x_3 = 1$ (b) $x_1 + 3x_2 + x_1x_3 = 2$
   (c) $x_1 = -7x_2 + 3x_3$ (d) $x_1^{-2} + x_2 + 8x_3 = 5$

2. (a) $3x + 28y - \frac{1}{2}z = 0$ (b) $xyz = x + 2y$
   (c) $x_1^{3/5} - 2x_2 + x_3 = 4$ (d) $\pi x_1 - \sqrt{2}x_2 + \frac{1}{3}x_3 = 7^{1/3}$

In Exercises 3 and 4, determine whether the equation is linear, given that k and m are constants (the result may depend on the constant).

3. (a) $x_1 - x_2 + x_3 = \sin(k)$ (b) $kx_1 - \frac{1}{k}x_2 = 9$
   (c) $x_1^k + 7x_2 - x_3 = 0$

4. (a) $x - y + mz = m^2$ (b) $mx - my = 23$
   (c) $\sqrt[m]{x} + 7y + 2z = 6$

In Exercises 5 and 6, determine whether the vector is a solution of the linear system.

5. $2x_1 - 4x_2 - x_3 = 1$
$\quad\ \ x_1 - 3x_2 + x_3 = 1$
$\quad 3x_1 - 5x_2 - 3x_3 = 1$

(a) $(3, 1, 1)$ (b) $(3, -1, 1)$ (c) $(13, 5, 2)$
(d) $\left(\frac{13}{2}, \frac{5}{2}, 2\right)$ (e) $(17, 7, 5)$

6. $\quad x_1 + 2x_2 - 2x_3 = 3$
$\quad 3x_1 - x_2 + x_3 = 1$
$\quad -x_1 + 5x_2 - 5x_3 = 5$

(a) $\left(\frac{5}{7}, \frac{8}{7}, 1\right)$ (b) $\left(\frac{5}{7}, \frac{8}{7}, 0\right)$ (c) $(5, 8, 1)$
(d) $\left(\frac{5}{7}, \frac{10}{7}, \frac{2}{7}\right)$ (e) $\left(\frac{5}{7}, \frac{22}{7}, 2\right)$

In Exercises 7 and 8, graph the three equations in the linear system by hand or with a graphing utility, and use the graphs to make a statement about the number of solutions of the system. Confirm your conclusion algebraically.

7. $\ x + 2y = 1$
$\quad 2x + y = 2$
$\quad 3x - 3y = 3$

8. $\ x - y = 3$
$\quad x + y = 1$
$\quad 2x + 3y = 6$

In Exercises 9 and 10, find the solution set of the given linear equation.

9. (a) $7x - 5y = 3$
(b) $3x_1 - 5x_2 + 4x_3 = 7$
(c) $-8x_1 + 2x_2 - 5x_3 + 6x_4 = 1$
(d) $3v - 8w + 2x - y + 4z = 0$

10. (a) $x + 10y = 2$
(b) $x_1 + 3x_2 - 12x_3 = 3$
(c) $4x_1 + 2x_2 + 3x_3 + x_4 = 20$
(d) $v + w + x - 5y + 7z = 0$

11. (a) Find a linear equation in x and y whose solution set is given by the equations $x = 5 + 2t$, $y = t$.
(b) Show that the solution set is also given by the equations $x = t$, $y = \frac{1}{2}t - \frac{5}{2}$.

12. (a) Find a linear equation in the variables x_1 and x_2 whose solution set is given by the equations $x_1 = -3 + t$, $x_2 = 2t$.
(b) Show that the solution set is also given by the equations $x_1 = t$, $x_2 = 2t + 6$.

In Exercises 13 and 14, find vector and parametric equations for the line of intersection of the planes in R^3 represented by the given equations.

13. $x + y - z = 3$ and $2x + y + 3z = 4$

14. $x + 2y + 3z = 1$ and $3x - 2y + z = 2$

In Exercises 15 and 16, for which value(s) of the constant k does the system of linear equations have no solutions? Exactly one solution? Infinitely many solutions?

15. $\ x - y = 3$
$\quad 2x - 2y = k$

16. $\ x_1 + 4x_2 = 1$
$\quad 3x_1 + 12x_2 = k$

In Exercises 17–20, find the augmented matrix for the linear system.

17. $3x_1 - 2x_2 = -1$
$\quad 4x_1 + 5x_2 = \ \ \ 3$
$\quad 7x_1 + 3x_2 = \ \ \ 2$

18. $2x_1 \qquad\quad + 2x_3 = 1$
$\quad 3x_1 - x_2 + 4x_3 = 7$
$\quad 6x_1 + x_2 - \ x_3 = 0$

19. $x_1 + 2x_2 \qquad - x_4 + x_5 = 1$
$\quad 3x_2 + 2x_3 \qquad - x_5 = 2$
$\quad 3x_2 + \ x_3 + 7x_4 \qquad = 1$

20. $x_1 \qquad\qquad = 1$
$\quad\ \ x_2 \qquad = 2$
$\quad\qquad x_3 = 3$

In Exercises 21–24, find a system of linear equations corresponding to the given augmented matrix.

21. $\begin{bmatrix} 2 & 0 & 0 \\ 3 & -4 & 0 \\ 0 & 1 & 1 \end{bmatrix}$

22. $\begin{bmatrix} 3 & 0 & -2 & 5 \\ 7 & 1 & 4 & -3 \\ 0 & -2 & 1 & 7 \end{bmatrix}$

23. $\begin{bmatrix} 7 & 2 & 1 & -3 & 5 \\ 1 & 2 & 4 & 0 & 1 \end{bmatrix}$

24. $\begin{bmatrix} 1 & 0 & 0 & 0 & 7 \\ 0 & 1 & 0 & 0 & -2 \\ 0 & 0 & 1 & 0 & 3 \\ 0 & 0 & 0 & 1 & 4 \end{bmatrix}$

In Exercises 25 and 26, describe an elementary row operation that produces B from A, and then describe an elementary row operation that recovers A from B.

25. (a) $A = \begin{bmatrix} 1 & 3 & 5 \\ -2 & 2 & 1 \\ 3 & 4 & -2 \end{bmatrix}$, $B = \begin{bmatrix} 1 & 3 & 5 \\ 0 & 8 & 11 \\ 3 & 4 & -2 \end{bmatrix}$

(b) $A = \begin{bmatrix} 2 & 0 & -4 \\ -3 & -2 & 6 \\ 2 & 5 & 1 \end{bmatrix}$, $B = \begin{bmatrix} 1 & 0 & -2 \\ -3 & -2 & 6 \\ 2 & 5 & 1 \end{bmatrix}$

26. (a) $A = \begin{bmatrix} 2 & 0 & -4 \\ -3 & -2 & 6 \\ 2 & 5 & 1 \end{bmatrix}$, $B = \begin{bmatrix} 2 & 5 & 1 \\ -3 & -2 & 6 \\ 2 & 0 & -4 \end{bmatrix}$

(b) $A = \begin{bmatrix} 2 & 0 & -4 \\ -3 & -2 & 1 \\ 2 & 0 & 3 \end{bmatrix}$, $B = \begin{bmatrix} 2 & 0 & -4 \\ -3 & -2 & 1 \\ 10 & 0 & 15 \end{bmatrix}$

27. Suppose that a certain diet calls for 7 units of fat, 9 units of protein, and 16 units of carbohydrates for the main meal, and suppose that an individual has three possible foods to choose from to meet these requirements:

Food 1: Each ounce contains 2 units of fat, 2 units of protein, and 4 units of carbohydrates.

Food 2: Each ounce contains 3 units of fat, 1 unit of protein, and 2 units of carbohydrates.

Food 3: Each ounce contains 1 unit of fat, 3 units of protein, and 5 units of carbohydrates.

Let x, y, and z denote the number of ounces of the first, second, and third foods that the dieter will consume at the main meal. Find (but do not solve) a linear system in x, y, and z whose solution tells how many ounces of each food must be consumed to meet the diet requirements.

28. A manufacturer produces custom steel products from recycled steel in which each ton of the product must contain 4 pounds of chromium, 8 pounds of tungsten, and 7 pounds of carbon. The manufacturer has three sources of recycled steel:

Source 1: Each ton contains 2 pounds of chromium, 8 pounds of tungsten, and 6 pounds of carbon.

Source 2: Each ton contains 3 pounds of chromium, 9 pounds of tungsten, and 6 pounds of carbon.

Source 3: Each ton contains 12 pounds of chromium, 6 pounds of tungsten, and 12 pounds of carbon.

Let x, y, and z denote the percentages of the first, second, and third recycled steel sources that will be melted down for one ton of the product. Find (but do not solve) a linear system in x, y, and z whose solution tells the percentage of each source that must be used to meet the requirements for the finished product.

29. Suppose you are asked to find three real numbers such that the sum of the numbers is 12, the sum of two times the first plus the second plus two times the third is 5, and the third number is one more than the first. Find (but do not solve) a linear system whose equations describe the three conditions.

30. Suppose you are asked to find three real numbers such that the sum of the numbers is 3, the sum of the second and third numbers is 10, and the third number is six more than the second number. Find (but do not solve) a system of linear equations that represents these three conditions.

> In each part of Exercises 31 and 32, find (but do not solve) a system of linear equations whose consistency or inconsistency will determine whether the given vector $\mathbf{v}$ is a linear combination of $\mathbf{u}_1$, $\mathbf{u}_2$, $\mathbf{u}_3$, and $\mathbf{u}_4$.

31. $\mathbf{u}_1 = (3, 0, -1, 2)$, $\mathbf{u}_2 = (1, 1, 1, 1)$, $\mathbf{u}_3 = (2, 3, 0, 2)$, $\mathbf{u}_4 = (-1, 2, 5, 0)$
    (a) $\mathbf{v} = (5, 6, 5, 5)$ (b) $\mathbf{v} = (8, 3, -2, 6)$
    (c) $\mathbf{v} = (4, 4, 6, 2)$

32. $\mathbf{u}_1 = (1, 2, -4, 0, 5)$, $\mathbf{u}_2 = (1, 0, 2, 2, -1)$, $\mathbf{u}_3 = (2, -2, -1, 1, 3)$, $\mathbf{u}_4 = (0, 5, 4, -1, 1)$
    (a) $\mathbf{v} = (2, -2, -8, 0, 12)$ (b) $\mathbf{v} = (5, -3, -9, 4, 11)$
    (c) $\mathbf{v} = (4, -4, 2, 0, 24)$

Discussion and Discovery

D1. Consider the system of equations
$$ax + by = k$$
$$cx + dy = l$$
$$ex + fy = m$$
Discuss the relative position of each of the lines $ax + by = k$, $cx + dy = l$, and $ex + fy = m$ when
(a) the system has no solutions;
(b) the system has exactly one solution;
(c) the system has infinitely many solutions.

D2. Show that if the system of equations in the preceding exercise is consistent, then at least one equation can be discarded from the system without altering the solution set. What does this say about the three lines represented by those equations?

D3. If a matrix B results from a matrix A by applying an elementary row operation, is there always an elementary row operation that can be applied to B to recover A?

D4. If $k = l = m = 0$ in Exercise D1, explain why the system must be consistent. What can be said about the point of intersection of the three lines if the system has exactly one solution?

D5. Suppose that you want to find values for a, b, and c such that the parabola $y = ax^2 + bx + c$ passes through the points $(1, 1)$, $(2, 4)$, and $(-1, 1)$. Find (but do not solve) a system of linear equations whose solutions provide values for a, b, and c. How many solutions would you expect this system of equations to have, and why?

D6. Assume that the parabola $y = ax^2 + bx + c$ passes through the points (x_1, y_1), (x_2, y_2), and (x_3, y_3). Show that a, b, and c satisfy the linear system whose augmented matrix is
$$\begin{bmatrix} x_1^2 & x_1 & 1 & y_1 \\ x_2^2 & x_2 & 1 & y_2 \\ x_3^2 & x_3 & 1 & y_3 \end{bmatrix}$$

D7. Show that if the linear equations $x_1 + kx_2 = c$ and $x_1 + lx_2 = d$ have the same solution set, then the equations are identical.

D8. Indicate whether the statement is true (T) or false (F). Justify your answer.
(a) Every matrix with two or more columns is an augmented matrix for some system of linear equations.

(b) If two linear systems have exactly the same solutions, then they have the same augmented matrix.

(c) A row of an augmented matrix can be multiplied by zero without affecting the solution set of the corresponding linear system.

(d) A linear system of two equations in three unknowns cannot have exactly one solution.

D9. Indicate whether the statement is true (T) or false (F). Justify your answer.

(a) Four planes in 3-space can intersect in a line.

(b) The solution set of a linear system in three unknowns is not changed if the first two columns of its augmented matrix are interchanged.

(c) A linear system with 100 equations and two unknowns must be inconsistent.

(d) If the last column of an augmented matrix consists entirely of zeros, then the corresponding linear system must be consistent.

Technology Exercises

T1. (*Linear systems with unique solutions*) Solve the system in Example 6.

T2. Use your technology utility to determine whether the vector $\mathbf{u} = (8, -1, -7, 4, 0)$ is a linear combination of the vectors $(1, -1, 0, 1, 3)$, $(2, 3, -2, 2, 1)$, and $(-4, 2, 5, 0, 7)$, and if so, express $\mathbf{u}$ as such a linear combination.

Section 2.2 Solving Linear Systems by Row Reduction

In this section we will discuss a systematic procedure for solving systems of linear equations. The methods that we will discuss here are useful for solving small systems by hand and also form the foundation for most of the algorithms that are used to solve large linear systems by computer.

CONSIDERATIONS IN SOLVING LINEAR SYSTEMS

When considering methods for solving systems of linear equations, it is important to distinguish between large systems that must be solved by computer and small systems that can be solved by hand. For example, there are many applications that lead to linear systems in thousands or even millions of unknowns. Large systems require special techniques to deal with issues of memory size, roundoff errors, solution time, and so forth. Such techniques are studied in the field of *numerical analysis* and will only be touched on in this text. However, almost all of the methods that are used for large systems are based on the ideas that we will develop in this section.

ECHELON FORMS

In Example 6 of the last section, we solved a linear system in the unknowns x, y, and z by reducing the augmented matrix to the form

$$\begin{bmatrix} 1 & 0 & 0 & 1 \\ 0 & 1 & 0 & 2 \\ 0 & 0 & 1 & 3 \end{bmatrix}$$

from which the solution $x = 1$, $y = 2$, $z = 3$ became evident. This is an example of a matrix that is in *reduced row echelon form*. To be of this form a matrix must have the following properties:

1. If a row does not consist entirely of zeros, then the first nonzero number in the row is a 1. We call this a *leading* **1**.

2. If there are any rows that consist entirely of zeros, then they are grouped together at the bottom of the matrix.

3. In any two successive rows that do not consist entirely of zeros, the leading 1 in the lower row occurs farther to the right than the leading 1 in the higher row.

4. Each column that contains a leading 1 has zeros everywhere else.

A matrix that has the first three properties is said to be in ***row echelon form***. (Thus, a matrix in reduced row echelon form is of necessity in row echelon form, but not conversely.)

EXAMPLE 1
Row Echelon and Reduced Row Echelon Form

The following matrices are in reduced row echelon form:

$$\begin{bmatrix} 1 & 0 & 0 & 4 \\ 0 & 1 & 0 & 7 \\ 0 & 0 & 1 & -1 \end{bmatrix}, \quad \begin{bmatrix} 1 & 0 & 0 \\ 0 & 1 & 0 \\ 0 & 0 & 1 \end{bmatrix}, \quad \begin{bmatrix} 0 & 1 & -2 & 0 & 1 \\ 0 & 0 & 0 & 1 & 3 \\ 0 & 0 & 0 & 0 & 0 \\ 0 & 0 & 0 & 0 & 0 \end{bmatrix}, \quad \begin{bmatrix} 0 & 0 \\ 0 & 0 \end{bmatrix}$$

The following matrices are in row echelon form:

$$\begin{bmatrix} 1 & 4 & -3 & 7 \\ 0 & 1 & 6 & 2 \\ 0 & 0 & 1 & 5 \end{bmatrix}, \quad \begin{bmatrix} 1 & 1 & 0 \\ 0 & 1 & 0 \\ 0 & 0 & 0 \end{bmatrix}, \quad \begin{bmatrix} 0 & 1 & 2 & 6 & 0 \\ 0 & 0 & 1 & -1 & 0 \\ 0 & 0 & 0 & 0 & 1 \end{bmatrix}$$

We leave it for you to confirm that each of the matrices in this example satisfies all of the requirements for its stated form. ■

EXAMPLE 2
More on Row Echelon and Reduced Row Echelon Form

As the last example illustrates, a matrix in row echelon form has zeros below each leading 1, whereas a matrix in reduced row echelon form has zeros below *and above* each leading 1. Thus, with any real numbers substituted for the $*$'s, all matrices of the following types are in row echelon form:

$$\begin{bmatrix} 1 & * & * & * \\ 0 & 1 & * & * \\ 0 & 0 & 1 & * \\ 0 & 0 & 0 & 1 \end{bmatrix}, \quad \begin{bmatrix} 1 & * & * & * \\ 0 & 1 & * & * \\ 0 & 0 & 1 & * \\ 0 & 0 & 0 & 0 \end{bmatrix}, \quad \begin{bmatrix} 1 & * & * & * \\ 0 & 1 & * & * \\ 0 & 0 & 0 & 0 \\ 0 & 0 & 0 & 0 \end{bmatrix}, \quad \begin{bmatrix} 0 & 1 & * & * & * & * & * & * & * \\ 0 & 0 & 0 & 1 & * & * & * & * & * \\ 0 & 0 & 0 & 0 & 1 & * & * & * & * \\ 0 & 0 & 0 & 0 & 0 & 1 & * & * & * \\ 0 & 0 & 0 & 0 & 0 & 0 & 0 & 1 & * \end{bmatrix}$$

Moreover, all matrices of the following types are in reduced row echelon form:

$$\begin{bmatrix} 1 & 0 & 0 & 0 \\ 0 & 1 & 0 & 0 \\ 0 & 0 & 1 & 0 \\ 0 & 0 & 0 & 1 \end{bmatrix}, \quad \begin{bmatrix} 1 & 0 & 0 & * \\ 0 & 1 & 0 & * \\ 0 & 0 & 1 & * \\ 0 & 0 & 0 & 0 \end{bmatrix}, \quad \begin{bmatrix} 1 & 0 & * & * \\ 0 & 1 & * & * \\ 0 & 0 & 0 & 0 \\ 0 & 0 & 0 & 0 \end{bmatrix}, \quad \begin{bmatrix} 0 & 1 & * & 0 & 0 & * & * & 0 & * \\ 0 & 0 & 0 & 1 & 0 & 0 & * & * & 0 & * \\ 0 & 0 & 0 & 0 & 1 & 0 & * & * & 0 & * \\ 0 & 0 & 0 & 0 & 0 & 1 & * & * & 0 & * \\ 0 & 0 & 0 & 0 & 0 & 0 & 0 & 1 & * \end{bmatrix}$$

■

If by a sequence of elementary row operations, the augmented matrix for a system of linear equations is put in *reduced* row echelon form, then the solution set can be obtained either by inspection, or by converting certain linear equations to parametric form. Here are some examples.

EXAMPLE 3
Unique Solution

Suppose that the augmented matrix for a linear system in the unknowns x_1, x_2, x_3, and x_4 has been reduced by elementary row operations to

$$\begin{bmatrix} 1 & 0 & 0 & 0 & 3 \\ 0 & 1 & 0 & 0 & -1 \\ 0 & 0 & 1 & 0 & 0 \\ 0 & 0 & 0 & 1 & 5 \end{bmatrix}$$

This matrix is in reduced row echelon form and corresponds to the equations

$$
\begin{array}{cccc}
x_1 & & & = & 3 \\
& x_2 & & = & -1 \\
& & x_3 & = & 0 \\
& & & x_4 = & 5
\end{array}
$$

Thus, the system has a unique solution, namely, $x_1 = 3$, $x_2 = -1$, $x_3 = 0$, $x_4 = 5$. ∎

EXAMPLE 4
Linear Systems
in Three
Unknowns

In each part, suppose that the augmented matrix for a linear system in the unknowns x, y, and z has been reduced by elementary row operations to the given reduced row echelon form. Solve the system.

$$
\text{(a)} \begin{bmatrix} 1 & 0 & 0 & 0 \\ 0 & 1 & 2 & 0 \\ 0 & 0 & 0 & 1 \end{bmatrix}
\qquad
\text{(b)} \begin{bmatrix} 1 & 0 & 3 & -1 \\ 0 & 1 & -4 & 2 \\ 0 & 0 & 0 & 0 \end{bmatrix}
\qquad
\text{(c)} \begin{bmatrix} 1 & -5 & 1 & 4 \\ 0 & 0 & 0 & 0 \\ 0 & 0 & 0 & 0 \end{bmatrix}
$$

Solution (a) The equation that corresponds to the last row of the augmented matrix is

$$0x + 0y + 0z = 1$$

Since this equation is not satisfied by any values of x, y, and z, the system is inconsistent.

Solution (b) The equation that corresponds to the last row of the augmented matrix is

$$0x + 0y + 0z = 0$$

This equation can be omitted since it imposes no restrictions on x, y, and z; hence, the linear system corresponding to the augmented matrix is

$$
\begin{array}{rcr}
x & + 3z = & -1 \\
y - 4z = & 2
\end{array}
$$

Since x and y correspond to the leading 1's in the augmented matrix, we call these the **leading variables**. The remaining variables (in this case z) are called **free variables**. Solving for the leading variables in terms of the free variables gives

$$
\begin{array}{l}
x = -1 - 3z \\
y = 2 + 4z
\end{array}
$$

From these equations we see that the free variable z can be assigned an arbitrary value, say t, which then determines x and y. Thus, the solution set of the system can be represented by the parametric equations

$$x = -1 - 3t, \ y = 2 + 4t, \ z = t \tag{1}$$

By substituting various values for t in these equations we can obtain various solutions of the system. For example, setting $t = 0$ in (1) yields the solution

$$x = -1, \quad y = 2, \quad z = 0$$

and setting $t = 1$ yields the solution

$$x = -4, \quad y = 6, \quad z = 1$$

However, rather than thinking of (1) as a collection of individual solutions, we can think of the entire set of solutions as a geometric object by writing (1) in the vector form

$$(x, y, z) = (-1, 2, 0) + t(-3, 4, 1)$$

Now we see that the solutions form a line in R^3 that passes through the point $(-1, 2, 0)$ and is parallel to the vector $\mathbf{v} = (-3, 4, 1)$.

Solution (c) As explained in part (b), we can omit the equations corresponding to the zero rows, in which case the linear system associated with the augmented matrix consists of the single equation

$$x - 5y + z = 4 \tag{2}$$

from which we see that the solution set is a plane in 3-space. Although (2) is a valid form of the solution set, there are many applications in which it is preferable to express the solution set in parametric or vector form. We can convert (2) to parametric form by solving for the leading variable x in terms of the free variables y and z to obtain

$$x = 4 + 5y - z$$

From this equation we see that the free variables can be assigned arbitrary values, say $y = s$ and $z = t$, which then determine the value of x. Thus, the solution set can be expressed parametrically as

$$x = 4 + 5s - t, \ y = s, \ z = t \tag{3}$$

To obtain the solution set in vector form we can use the method of Example 6 in Section 1.3 (with s and t in place of t_1 and t_2) to rewrite (3) as

$$(x, y, z) = (4, 0, 0) + s(5, 1, 0) + t(-1, 0, 1)$$

This conveys that the solution set is the plane that passes through the point $(4, 0, 0)$ and is parallel to the vectors $(5, 1, 0)$ and $(-1, 0, 1)$. ∎

REMARK A set of parametric equations (or their vector equivalent) for the solution set of a linear system is commonly called a ***general solution*** of the system. As in the last example, we will usually denote the parameters in a general solution by $r, s, t, \ldots$, but any letters that do not conflict with the names of the unknowns can be used. For systems with more than three unknowns, subscripted letters such as $t_1, t_2, t_3, \ldots$ are convenient.

GENERAL SOLUTIONS AS LINEAR COMBINATIONS OF COLUMN VECTORS

For many purposes, it is desirable to express a general solution of a linear system as a linear combination of column vectors. As in Example 4, the basic idea for doing this is to separate the constant terms and the terms involving the parameters. For example, the parametric equations in (1) can be expressed as

$$\begin{bmatrix} x \\ y \\ z \end{bmatrix} = \begin{bmatrix} -1 - 3t \\ 2 + 4t \\ t \end{bmatrix} = \begin{bmatrix} -1 \\ 2 \\ 0 \end{bmatrix} + \begin{bmatrix} -3t \\ 4t \\ t \end{bmatrix} = \begin{bmatrix} -1 \\ 2 \\ 0 \end{bmatrix} + t \begin{bmatrix} -3 \\ 4 \\ 1 \end{bmatrix} \tag{4}$$

and similarly, the parametric equations in (3) can be expressed as

$$\begin{bmatrix} x \\ y \\ z \end{bmatrix} = \begin{bmatrix} 4 + 5s - t \\ s \\ t \end{bmatrix} = \begin{bmatrix} 4 \\ 0 \\ 0 \end{bmatrix} + \begin{bmatrix} 5s - t \\ s \\ t \end{bmatrix}$$

$$= \begin{bmatrix} 4 \\ 0 \\ 0 \end{bmatrix} + \begin{bmatrix} 5s \\ s \\ 0 \end{bmatrix} + \begin{bmatrix} -t \\ 0 \\ t \end{bmatrix} = \begin{bmatrix} 4 \\ 0 \\ 0 \end{bmatrix} + s \begin{bmatrix} 5 \\ 1 \\ 0 \end{bmatrix} + t \begin{bmatrix} -1 \\ 0 \\ 1 \end{bmatrix} \tag{5}$$

GAUSS–JORDAN AND GAUSSIAN ELIMINATION

We have seen how easy it is to solve a linear system once its augmented matrix is in reduced row echelon form. Now we will give a step-by-step procedure that can be used to reduce any matrix to reduced row echelon form by elementary row operations. As we state each step, we will illustrate the idea by reducing the following matrix to reduced row echelon form:

$$\begin{bmatrix} 0 & 0 & -2 & 0 & 7 & 12 \\ 2 & 4 & -10 & 6 & 12 & 28 \\ 2 & 4 & -5 & 6 & -5 & -1 \end{bmatrix} \tag{6}$$

It can be proved that elementary row operations, when applied to the augmented matrix of a linear system, do not change the solution set of the system. Thus, we are assured that the linear system corresponding to the reduced row echelon form of (6) will have the same solutions as the original system. Here are the steps for reducing (6) to reduced row echelon form:

Step 1. Locate the leftmost column that does not consist entirely of zeros.

$$\begin{bmatrix} 0 & 0 & -2 & 0 & 7 & 12 \\ 2 & 4 & -10 & 6 & 12 & 28 \\ 2 & 4 & -5 & 6 & -5 & -1 \end{bmatrix}$$

$\uparrow$ **Leftmost nonzero column**

Step 2. Interchange the top row with another row, if necessary, to bring a nonzero entry to the top of the column found in Step 1.

$$\begin{bmatrix} 2 & 4 & -10 & 6 & 12 & 28 \\ 0 & 0 & -2 & 0 & 7 & 12 \\ 2 & 4 & -5 & 6 & -5 & -1 \end{bmatrix}$$

> The first and second rows in the preceding matrix were interchanged.

Step 3. If the entry that is now at the top of the column found in Step 1 is a, multiply the first row by $1/a$ in order to introduce a leading 1.

$$\begin{bmatrix} 1 & 2 & -5 & 3 & 6 & 14 \\ 0 & 0 & -2 & 0 & 7 & 12 \\ 2 & 4 & -5 & 6 & -5 & -1 \end{bmatrix}$$

> The first row of the preceding matrix was multiplied by $\frac{1}{2}$.

Step 4. Add suitable multiples of the top row to the rows below so that all entries below the leading 1 become zeros.

$$\begin{bmatrix} 1 & 2 & -5 & 3 & 6 & 14 \\ 0 & 0 & -2 & 0 & 7 & 12 \\ 0 & 0 & 5 & 0 & -17 & -29 \end{bmatrix}$$

> -2 times the first row of the preceding matrix was added to the third row.

Step 5. Now cover the top row in the matrix and begin again with Step 1 applied to the submatrix that remains. Continue in this way until the *entire* matrix is in row echelon form.

$$\begin{bmatrix} 1 & 2 & -5 & 3 & 6 & 14 \\ 0 & 0 & -2 & 0 & 7 & 12 \\ 0 & 0 & 5 & 0 & -17 & -29 \end{bmatrix}$$

$\uparrow$ **Leftmost nonzero column**
in the submatrix

$$\begin{bmatrix} 1 & 2 & -5 & 3 & 6 & 14 \\ 0 & 0 & 1 & 0 & -\frac{7}{2} & -6 \\ 0 & 0 & 5 & 0 & -17 & -29 \end{bmatrix}$$

> The first row in the submatrix was multiplied by $-\frac{1}{2}$ to introduce a leading 1.

$$\begin{bmatrix} 1 & 2 & -5 & 3 & 6 & 14 \\ 0 & 0 & 1 & 0 & -\frac{7}{2} & -6 \\ 0 & 0 & 0 & 0 & \frac{1}{2} & 1 \end{bmatrix}$$

> -5 times the first row of the submatrix was added to the second row of the submatrix to introduce a zero below the leading 1.

$$\begin{bmatrix} 1 & 2 & -5 & 3 & 6 & 14 \\ 0 & 0 & 1 & 0 & -\frac{7}{2} & -6 \\ 0 & 0 & 0 & 0 & \frac{1}{2} & 1 \end{bmatrix}$$

> The top row in the submatrix was covered, and we returned again to Step 1.

$\uparrow$ **Leftmost nonzero column**
in the new submatrix

$$\begin{bmatrix} 1 & 2 & -5 & 3 & 6 & 14 \\ 0 & 0 & 1 & 0 & -\frac{7}{2} & -6 \\ 0 & 0 & 0 & 0 & 1 & 2 \end{bmatrix}$$

> The first (and only) row in the new submatrix was multiplied by 2 to introduce a leading 1.

The *entire* matrix is now in row echelon form. To find the reduced row echelon form we need the following additional step.

Step 6. Beginning with the last nonzero row and working upward, add suitable multiples of each row to the rows above to introduce zeros above the leading 1's.

$$\begin{bmatrix} 1 & 2 & -5 & 3 & 6 & 14 \\ 0 & 0 & 1 & 0 & 0 & 1 \\ 0 & 0 & 0 & 0 & 1 & 2 \end{bmatrix}$$

> $\frac{7}{2}$ times the third row of the preceding matrix was added to the second row.

$$\begin{bmatrix} 1 & 2 & -5 & 3 & 0 & 2 \\ 0 & 0 & 1 & 0 & 0 & 1 \\ 0 & 0 & 0 & 0 & 1 & 2 \end{bmatrix}$$

> -6 times the third row was added to the first row.

$$\begin{bmatrix} 1 & 2 & 0 & 3 & 0 & 7 \\ 0 & 0 & 1 & 0 & 0 & 1 \\ 0 & 0 & 0 & 0 & 1 & 2 \end{bmatrix}$$

> 5 times the second row was added to the first row.

The last matrix is in reduced row echelon form.

The procedure (or algorithm) we have just described for reducing a matrix to reduced row echelon form is called **_Gauss–Jordan elimination_**. This algorithm consists of two parts, a **_forward phase_** in which zeros are introduced below the leading 1's and then a **_backward phase_** in which zeros are introduced above the leading 1's. If only the forward phase is used, then the procedure produces a row echelon form and is called **_Gaussian elimination_**. For example, in the preceding computations a row echelon form was obtained at the end of Step 5.

SOME FACTS ABOUT ECHELON FORMS

There are two facts about row echelon forms and reduced row echelon forms that are important to know, but which we will state without proof.

1. Every matrix has a unique reduced row echelon form; that is, regardless of whether one uses Gauss–Jordan elimination or some other sequence of elementary row operations, the same reduced row echelon form will result in the end.[*]

2. Row echelon forms are not unique; that is, different sequences of elementary row operations may result in different row echelon forms for a given matrix. However, all of the row echelon forms have their leading 1's in the same positions and all have the same number of zero rows at the bottom. The positions that have the leading 1's are called the **_pivot positions_** in the augmented matrix, and the columns that contain the leading 1's are called the **_pivot columns_**.

EXAMPLE 5

Solving a Linear System by Gauss–Jordan Elimination

Solve the following linear system by Gauss–Jordan elimination:

$$\begin{aligned} x_1 + 3x_2 - 2x_3 \quad\quad + 2x_5 \quad\quad &= \quad 0 \\ 2x_1 + 6x_2 - 5x_3 - \; 2x_4 + 4x_5 - \; 3x_6 &= -1 \\ 5x_3 + 10x_4 \quad\quad + 15x_6 &= \quad 5 \\ 2x_1 + 6x_2 \quad\quad + \; 8x_4 + 4x_5 + 18x_6 &= \quad 6 \end{aligned}$$

---

[*]A proof of this result can be found in the article "The Reduced Row Echelon Form of a Matrix Is Unique: A Simple Proof," by Thomas Yuster, *Mathematics Magazine*, Vol. 57, No. 2, 1984, pp. 93–94.

Solution The augmented matrix for the system is

$$
\begin{bmatrix}
1 & 3 & -2 & 0 & 2 & 0 & 0 \\
2 & 6 & -5 & -2 & 4 & -3 & -1 \\
0 & 0 & 5 & 10 & 0 & 15 & 5 \\
2 & 6 & 0 & 8 & 4 & 18 & 6
\end{bmatrix}
$$

Adding -2 times the first row to the second and fourth rows gives

$$
\begin{bmatrix}
1 & 3 & -2 & 0 & 2 & 0 & 0 \\
0 & 0 & -1 & -2 & 0 & -3 & -1 \\
0 & 0 & 5 & 10 & 0 & 15 & 5 \\
0 & 0 & 4 & 8 & 0 & 18 & 6
\end{bmatrix}
$$

Multiplying the second row by -1 and then adding -5 times the new second row to the third row and -4 times the new second row to the fourth row gives

$$
\begin{bmatrix}
1 & 3 & -2 & 0 & 2 & 0 & 0 \\
0 & 0 & 1 & 2 & 0 & 3 & 1 \\
0 & 0 & 0 & 0 & 0 & 0 & 0 \\
0 & 0 & 0 & 0 & 0 & 6 & 2
\end{bmatrix}
$$

Interchanging the third and fourth rows and then multiplying the third row of the resulting matrix by $\frac{1}{6}$ gives the row echelon form

$$
\begin{bmatrix}
1 & 3 & -2 & 0 & 2 & 0 & 0 \\
0 & 0 & 1 & 2 & 0 & 3 & 1 \\
0 & 0 & 0 & 0 & 0 & 1 & \frac{1}{3} \\
0 & 0 & 0 & 0 & 0 & 0 & 0
\end{bmatrix}
$$

Adding -3 times the third row to the second row and then adding 2 times the second row of the resulting matrix to the first row yields the reduced row echelon form

$$
\begin{bmatrix}
1 & 3 & 0 & 4 & 2 & 0 & 0 \\
0 & 0 & 1 & 2 & 0 & 0 & 0 \\
0 & 0 & 0 & 0 & 0 & 1 & \frac{1}{3} \\
0 & 0 & 0 & 0 & 0 & 0 & 0
\end{bmatrix}
$$

The corresponding system of equations is

$$
\begin{aligned}
x_1 + 3x_2 \quad\;\; + 4x_4 + 2x_5 \quad\;\; &= 0 \\
x_3 + 2x_4 \qquad\qquad &= 0 \\
x_6 &= \tfrac{1}{3}
\end{aligned}
$$

Solving for the leading variables we obtain

$$
\begin{aligned}
x_1 &= -3x_2 - 4x_4 - 2x_5 \\
x_3 &= -2x_4 \\
x_6 &= \tfrac{1}{3}
\end{aligned}
$$

If we now assign the free variables x_2, x_4, and x_5 arbitrary values r, s, and t, respectively, then we can express the solution set parametrically as

$$
x_1 = -3r - 4s - 2t, \; x_2 = r, \; x_3 = -2s, \; x_4 = s, \; x_5 = t, \; x_6 = \tfrac{1}{3}
$$

Linear Algebra in History

A version of Gaussian elimination first appeared around 200 B.C. in the Chinese text *Nine Chapters of Mathematical Art.* However, the power of the method was not recognized until the great German mathematician Carl Friedrich Gauss used it to compute the orbit of the asteroid Ceres from limited data. What happened was this: On January 1, 1801 the Sicilian astronomer Giuseppe Piazzi (1746–1826) noticed a dim celestial object that he believed might be a "missing planet." He named the object Ceres and made a limited number of positional observations but then lost the object as it neared the Sun. Gauss undertook the problem of computing the orbit from the limited data using least squares and the procedure that we now call Gaussian elimination. The work of Gauss caused a sensation when Ceres reappeared a year later in the constellation Virgo at almost the precise position that Gauss predicted! The method was further popularized by the German engineer Wilhelm Jordan in his handbook on geodesy (the science of measuring Earth shapes) entitled *Handbuch der Vermessungskunde* and published in 1888.

Carl Friedrich Gauss (1777–1855)   *Wilhelm Jordan* (1842–1899)

Alternatively, we can express the solution set as a linear combination of column vectors by writing

$$\begin{bmatrix} x_1 \\ x_2 \\ x_3 \\ x_4 \\ x_5 \\ x_6 \end{bmatrix} = \begin{bmatrix} -3r - 4s - 2t \\ r \\ -2s \\ s \\ t \\ \frac{1}{3} \end{bmatrix} = \begin{bmatrix} 0 \\ 0 \\ 0 \\ 0 \\ 0 \\ \frac{1}{3} \end{bmatrix} + \begin{bmatrix} -3r - 4s - 2t \\ r \\ -2s \\ s \\ t \\ 0 \end{bmatrix}$$

$$= \begin{bmatrix} 0 \\ 0 \\ 0 \\ 0 \\ 0 \\ \frac{1}{3} \end{bmatrix} + r \begin{bmatrix} -3 \\ 1 \\ 0 \\ 0 \\ 0 \\ 0 \end{bmatrix} + s \begin{bmatrix} -4 \\ 0 \\ -2 \\ 1 \\ 0 \\ 0 \end{bmatrix} + t \begin{bmatrix} -2 \\ 0 \\ 0 \\ 0 \\ 1 \\ 0 \end{bmatrix} \tag{7}$$

∎

BACK SUBSTITUTION In the examples given thus far we solved various linear systems by first transforming the augmented matrix to reduced row echelon form (Gauss–Jordan elimination) and then solving the corresponding linear system. However, it is possible to use only the forward phase of the reduction algorithm (Gaussian elimination) and solve the system that corresponds to the resulting row echelon form. With this approach the backward phase of the Gauss–Jordan algorithm is replaced by an algebraic procedure, called *back substitution*, in which each equation corresponding to the row echelon form is systematically substituted into the equations above, starting at the bottom and working up.

EXAMPLE 6
Gaussian Elimination and Back Substitution

We will solve the linear system in Example 5 using the row echelon form of the augmented matrix produced by Gaussian elimination. In the forward phase of the computations in Example 5, we obtained the following row echelon form of the augmented matrix:

$$\begin{bmatrix} 1 & 3 & -2 & 0 & 2 & 0 & 0 \\ 0 & 0 & 1 & 2 & 0 & 3 & 1 \\ 0 & 0 & 0 & 0 & 0 & 1 & \frac{1}{3} \\ 0 & 0 & 0 & 0 & 0 & 0 & 0 \end{bmatrix}$$

To solve the corresponding system of equations

$$\begin{aligned} x_1 + 3x_2 - 2x_3 \quad\quad + 2x_5 \quad\quad &= 0 \\ x_3 + 2x_4 \quad\quad + 3x_6 &= 1 \\ x_6 &= \tfrac{1}{3} \end{aligned}$$

we proceed as follows:

Step 1. Solve the equations for the leading variables.

$$\begin{aligned} x_1 &= -3x_2 + 2x_3 - 2x_5 \\ x_3 &= 1 - 2x_4 - 3x_6 \\ x_6 &= \tfrac{1}{3} \end{aligned}$$

Step 2. Beginning with the bottom equation and working upward, successively substitute each equation into all the equations above it.

Substituting $x_6 = \tfrac{1}{3}$ into the second equation yields

$$\begin{aligned} x_1 &= -3x_2 + 2x_3 - 2x_5 \\ x_3 &= -2x_4 \\ x_6 &= \tfrac{1}{3} \end{aligned}$$

Substituting $x_3 = -2x_4$ into the first equation yields

$$x_1 = -3x_2 - 4x_4 - 2x_5$$
$$x_3 = -2x_4$$
$$x_6 = \tfrac{1}{3}$$

Step 3. Assign arbitrary values to the free variables, if any.

If we now assign x_2, x_4, and x_5 the arbitrary values r, s, and t, respectively, we obtain

$$x_1 = -3r - 4s - 2t, \ \ x_2 = r, \ \ x_3 = -2s, \ \ x_4 = s, \ \ x_5 = t, \ \ x_6 = \tfrac{1}{3}$$

which agrees with the solution obtained in Example 5 by Gauss–Jordan elimination.   ■

REMARK For hand calculation, it is sometimes possible to avoid annoying fractions by varying the order of the steps in the right way (Exercise 48). Thus, once you have mastered Gauss–Jordan elimination, you may wish to vary the steps in specific problems to avoid fractions.

HOMOGENEOUS LINEAR SYSTEMS

Recall from Section 2.1 that a linear equation is said to be *homogeneous* if its constant term is zero; that is, the equation is of the form

$$a_1x_1 + a_2x_2 + \cdots + a_nx_n = 0$$

More generally, we say that a linear system is **homogeneous** if each of its equations is homogeneous. Thus, a homogeneous linear system of m equations in n unknowns has the form

$$
\begin{aligned}
a_{11}x_1 \ + \ a_{12}x_2 \ + \cdots + \ a_{1n}x_n &= 0 \\
a_{21}x_1 \ + \ a_{22}x_2 \ + \cdots + \ a_{2n}x_n &= 0 \\
\vdots \qquad\quad \vdots \qquad\qquad\quad \vdots \qquad\quad \vdots \\
a_{m1}x_1 \ + \ a_{m2}x_2 \ + \cdots + \ a_{mn}x_n &= 0
\end{aligned}
\tag{8}
$$

Observe that every homogeneous linear system is consistent, since

$$x_1 = 0, \quad x_2 = 0, \ldots, \quad x_n = 0$$

is a solution of (8). This is called the **trivial solution**. All other solutions, if any, are called **nontrivial solutions**. If the homogeneous linear system (8) has some nontrivial solution

$$x_1 = s_1, \ x_2 = s_2, \ldots, x_n = s_n$$

then it must have infinitely many solutions, since

$$x_1 = ts_1, \ x_2 = ts_2, \ldots, x_n = ts_n$$

is also a solution for any scalar t (verify). Thus, we have the following result.

> **Theorem 2.2.1** *A homogeneous linear system has only the trivial solution or it has infinitely many solutions; there are no other possibilities.*

Since the graph of a homogeneous linear equation $ax + by = 0$ is a line through the origin, one can see geometrically from Figure 2.2.1 why Theorem 2.2.1 is true for a homogeneous linear system of two equations in x and y: If the lines have different slopes, then their only intersection is at the origin, which corresponds to the trivial solution $x = 0$, $y = 0$. If the lines have the same slope, then they must coincide, so there are infinitely many solutions.

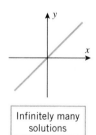

Only the trivial solution

Infinitely many solutions

Figure 2.2.1

EXAMPLE 7
Homogeneous System with Nontrivial Solutions

Use Gauss–Jordan elimination to solve the homogeneous linear system

$$
\begin{aligned}
x_1 + 3x_2 - 2x_3 \qquad\quad + 2x_5 \qquad\qquad &= 0 \\
2x_1 + 6x_2 - 5x_3 - \ 2x_4 + 4x_5 - \ 3x_6 &= 0 \\
5x_3 + 10x_4 \qquad + 15x_6 &= 0 \\
2x_1 + 6x_2 \qquad\quad + \ 8x_4 + 4x_5 + 18x_6 &= 0
\end{aligned}
$$

Solution Observe first that the coefficients of the unknowns in this system are the same as those in Example 5; that is, the two systems differ only in the constants on the right side. The augmented matrix for the given homogeneous system is

$$\begin{bmatrix} 1 & 3 & -2 & 0 & 2 & 0 & 0 \\ 2 & 6 & -5 & -2 & 4 & -3 & 0 \\ 0 & 0 & 5 & 10 & 0 & 15 & 0 \\ 2 & 6 & 0 & 8 & 4 & 18 & 0 \end{bmatrix} \tag{9}$$

which is the same as the augmented matrix for the system in Example 5, except for zeros in the last column. Thus, the reduced row echelon form of this matrix will be the same as the reduced row echelon form of the augmented matrix in Example 5, except for the last column. However, a moment's reflection will make it evident that a column of zeros is not changed by an elementary row operation, so the reduced row echelon form of (9) is

$$\begin{bmatrix} 1 & 3 & 0 & 4 & 2 & 0 & 0 \\ 0 & 0 & 1 & 2 & 0 & 0 & 0 \\ 0 & 0 & 0 & 0 & 0 & 1 & 0 \\ 0 & 0 & 0 & 0 & 0 & 0 & 0 \end{bmatrix} \tag{10}$$

The corresponding system of equations is

$$\begin{aligned} x_1 + 3x_2 \quad\quad + 4x_4 + 2x_5 \quad\quad &= 0 \\ x_3 + 2x_4 \quad\quad\quad &= 0 \\ x_6 &= 0 \end{aligned} \tag{11}$$

Solving for the leading variables we obtain

$$\begin{aligned} x_1 &= -3x_2 - 4x_4 - 2x_5 \\ x_3 &= -2x_4 \\ x_6 &= 0 \end{aligned} \tag{12}$$

If we now assign the free variables x_2, x_4, and x_5 arbitrary values r, s, and t, respectively, then we can express the solution set parametrically as

$$x_1 = -3r - 4s - 2t, \ x_2 = r, \ x_3 = -2s, \ x_4 = s, \ x_5 = t, \ x_6 = 0 \tag{13}$$

We leave it for you to show that the solution set can be expressed in vector form as

$$(x_1, x_2, x_3, x_4, x_5, x_6) = r(-3, 1, 0, 0, 0, 0) + s(-4, 0, -2, 1, 0, 0) + t(-2, 0, 0, 0, 1, 0) \tag{14}$$

or alternatively, as

$$\begin{bmatrix} x_1 \\ x_2 \\ x_3 \\ x_4 \\ x_5 \\ x_6 \end{bmatrix} = r \begin{bmatrix} -3 \\ 1 \\ 0 \\ 0 \\ 0 \\ 0 \end{bmatrix} + s \begin{bmatrix} -4 \\ 0 \\ -2 \\ 1 \\ 0 \\ 0 \end{bmatrix} + t \begin{bmatrix} -2 \\ 0 \\ 0 \\ 0 \\ 1 \\ 0 \end{bmatrix} \tag{15}$$

THE DIMENSION THEOREM FOR HOMOGENEOUS LINEAR SYSTEMS

Example 7 illustrates two important points about solving homogeneous linear systems:

1. Elementary row operations do not alter columns of zeros in a matrix, so the reduced row echelon form of the augmented matrix for a homogeneous linear system has a final column of zeros. This implies that the linear system corresponding to the reduced row echelon form is homogeneous, just like the original system.

2. When we constructed the homogeneous system of equations corresponding to (10) we ignored the row of zeros because the corresponding equation

$$0x_1 + 0x_2 + 0x_3 + 0x_4 + 0x_5 + 0x_6 = 0$$

does not impose any conditions on the unknowns. Thus, depending on whether or not the reduced row echelon form of the augmented matrix for a homogeneous linear system has zero rows, the linear system corresponding to the reduced row echelon form of the augmented matrix will either have the same number of equations as the original system or it will have fewer.

Now consider a general homogeneous linear system with n unknowns, and suppose that the reduced row echelon form of the augmented matrix has r nonzero rows. Since each nonzero row has a leading 1, and since each leading 1 corresponds to a leading variable, the homogeneous system corresponding to the reduced row echelon form of the augmented matrix must have r leading variables and $n - r$ free variables. Thus, this system is of the form

$$
\begin{aligned}
x_{k_1} \quad\quad\quad + \sum(\ ) &= 0 \\
x_{k_2} \quad\quad + \sum(\ ) &= 0 \\
\ddots \qquad \vdots \qquad\quad & \\
x_{k_r} + \sum(\ ) &= 0
\end{aligned}
\tag{16}
$$

where in each equation the expression $\sum(\ )$ denotes a sum that involves the free variables, if any [see (11)]. In summary, we have the following result.

Theorem 2.2.2 (*Dimension Theorem for Homogeneous Systems*) *If a homogeneous linear system has n unknowns, and if the reduced row echelon form of its augmented matrix has r nonzero rows, then the system has n − r free variables.*

This theorem has an important implication for homogeneous linear systems with more unknowns than equations. Specifically, if a homogeneous linear system has m equations and n unknowns, and if $m < n$, then it must also be true that $r < n$ (why?). Thus, Theorem 2.2.2 implies that there is at least one free variable, and this means that the system must have infinitely many solutions (see Example 7). Thus, we have the following result.

Theorem 2.2.3 *A homogeneous linear system with more unknowns than equations has infinitely many solutions.*

In retrospect, we could have anticipated that the homogeneous system in Example 7 would have infinitely many solutions, since it has four equations and six unknowns.

REMARK   It is important to keep in mind that this theorem is only applicable to homogeneous linear systems. Indeed, there exist nonhomogeneous linear systems with more unknowns than equations that have no solutions (Exercise 47).

STABILITY, ROUNDOFF ERROR, PARTIAL PIVOTING

There is often a gap between mathematical theory and its practical implementation, Gauss–Jordan elimination and Gaussian elimination being good examples. The source of the difficulty is that large-scale linear systems are solved on computers, which, by their nature, represent exact numbers by decimal (or more precisely, binary) approximations, thereby introducing ***roundoff error***; and unless precautions are taken, the accumulation of roundoff error in successive calculations may degrade the answer to the point of rendering it useless. Algorithms in which this tends to happen are said to be ***unstable***.

In practice, various techniques are used to minimize the instability that is inherent in Gauss–Jordan and Gaussian elimination. For example, it can be shown that division by numbers near zero tends to magnify roundoff error—the closer the denominator is to zero, the greater the magnification. Thus, in the practical implementation of Gauss–Jordan and Gaussian elimination, it is standard practice to perform a row interchange at each step to put the entry with the largest absolute value into the pivot position before dividing to introduce a leading 1. This procedure is called ***partial pivoting*** (Exercises 53 and 54).

Exercise Set 2.2

In Exercises 1 and 2, determine which of the matrices, if any, are in reduced row echelon form.

1. (a) $\begin{bmatrix} 1 & 0 & 0 \\ 0 & 1 & 0 \\ 0 & 0 & 1 \end{bmatrix}$ (b) $\begin{bmatrix} 1 & 1 & 0 \\ 0 & 1 & 0 \\ 0 & 0 & 0 \end{bmatrix}$ (c) $\begin{bmatrix} 0 & 1 & 0 \\ 0 & 0 & 1 \\ 0 & 0 & 0 \end{bmatrix}$

(d) $\begin{bmatrix} 1 & 0 & 0 \\ 0 & 0 & 1 \\ 0 & 0 & 0 \end{bmatrix}$ (e) $\begin{bmatrix} 0 & 0 & 0 \\ 1 & 0 & 0 \\ 0 & 0 & 0 \end{bmatrix}$

2. (a) $\begin{bmatrix} 0 & 1 & 0 \\ 1 & 0 & 0 \\ 0 & 0 & 0 \end{bmatrix}$ (b) $\begin{bmatrix} 1 & 1 & 0 \\ 0 & 1 & 0 \\ 0 & 0 & 0 \end{bmatrix}$ (c) $\begin{bmatrix} 1 & 0 & 2 \\ 0 & 1 & 3 \\ 0 & 0 & 0 \end{bmatrix}$

(d) $\begin{bmatrix} 0 & 0 & 1 \\ 0 & 0 & 0 \\ 0 & 0 & 0 \end{bmatrix}$ (e) $\begin{bmatrix} 0 & 0 & 0 \\ 0 & 0 & 0 \\ 0 & 0 & 0 \end{bmatrix}$

In Exercises 3 and 4, determine which of the matrices, if any, are in row echelon form.

3. (a) $\begin{bmatrix} 1 & 0 & 0 \\ 0 & 1 & 0 \\ 0 & 0 & 1 \end{bmatrix}$ (b) $\begin{bmatrix} 1 & 2 & 0 \\ 0 & 1 & 0 \\ 0 & 0 & 0 \end{bmatrix}$ (c) $\begin{bmatrix} 1 & 0 & 0 \\ 0 & 1 & 0 \\ 0 & 2 & 0 \end{bmatrix}$

4. (a) $\begin{bmatrix} 1 & 3 & 4 \\ 0 & 0 & 1 \\ 0 & 0 & 0 \end{bmatrix}$ (b) $\begin{bmatrix} 1 & 5 & -3 \\ 0 & 1 & 1 \\ 0 & 0 & 0 \end{bmatrix}$ (c) $\begin{bmatrix} 1 & 2 & 3 \\ 0 & 0 & 0 \\ 0 & 0 & 1 \end{bmatrix}$

In Exercises 5 and 6, determine whether the matrix is in row echelon form, reduced row echelon form, both, or neither.

5. (a) $\begin{bmatrix} 1 & 2 & 0 & 3 & 0 \\ 0 & 0 & 1 & 1 & 0 \\ 0 & 0 & 0 & 0 & 1 \\ 0 & 0 & 0 & 0 & 0 \end{bmatrix}$ (b) $\begin{bmatrix} 1 & 0 & 0 & 5 \\ 0 & 0 & 1 & 3 \\ 0 & 1 & 0 & 4 \end{bmatrix}$

(c) $\begin{bmatrix} 1 & 0 & 3 & 1 \\ 0 & 1 & 2 & 4 \end{bmatrix}$

6. (a) $\begin{bmatrix} 1 & -7 & 5 & 5 \\ 0 & 1 & 3 & 2 \end{bmatrix}$ (b) $\begin{bmatrix} 1 & 3 & 0 & 2 & 0 \\ 1 & 0 & 2 & 2 & 0 \\ 0 & 0 & 0 & 0 & 1 \\ 0 & 0 & 0 & 0 & 0 \end{bmatrix}$

(c) $\begin{bmatrix} 0 & 0 \\ 0 & 0 \\ 0 & 0 \end{bmatrix}$

7. Describe all possible 2×2 matrices that are in reduced row echelon form.

8. Describe all possible 3×3 matrices that are in reduced row echelon form.

In Exercises 9–14, suppose that the augmented matrix for a system of linear equations has been reduced by row operations to the given reduced row echelon form. Solve the system. Assume that the variables are named $x_1, x_2, \ldots$ from left to right.

9. $\begin{bmatrix} 1 & 0 & 0 & -3 \\ 0 & 1 & 0 & 0 \\ 0 & 0 & 1 & 7 \end{bmatrix}$

10. $\begin{bmatrix} 1 & 2 & 0 & 2 & -1 \\ 0 & 0 & 1 & 3 & 4 \end{bmatrix}$

11. $\begin{bmatrix} 1 & -6 & 0 & 0 & 3 & -2 \\ 0 & 0 & 1 & 0 & 4 & 7 \\ 0 & 0 & 0 & 1 & 5 & 8 \\ 0 & 0 & 0 & 0 & 0 & 0 \end{bmatrix}$

12. $\begin{bmatrix} 1 & -3 & 0 & 0 \\ 0 & 0 & 1 & 0 \\ 0 & 0 & 0 & 1 \end{bmatrix}$

13. $\begin{bmatrix} 1 & 0 & 0 & -7 & 8 \\ 0 & 1 & 0 & 3 & 2 \\ 0 & 0 & 1 & 1 & -5 \end{bmatrix}$

14. $[\begin{matrix} 1 & 2 & 0 & 2 & -1 & 3 \end{matrix}]$

In Exercises 15 and 16, suppose that the augmented matrix for a system of linear equations has been reduced by row operations to the given row echelon form. Solve the system by back substitution, and then solve it again by reducing the matrix to reduced row echelon form (Gauss–Jordan elimination). Assume that the variables are named $x_1, x_2, \ldots$ from left to right.

15. $\begin{bmatrix} 1 & -3 & 4 & 7 \\ 0 & 1 & 2 & 2 \\ 0 & 0 & 1 & 5 \end{bmatrix}$

16. $\begin{bmatrix} 1 & 0 & 8 & -5 & 6 \\ 0 & 1 & 4 & -9 & 3 \\ 0 & 0 & 1 & 1 & 2 \end{bmatrix}$

In Exercises 17–20, suppose that the augmented matrix for a system of linear equations has been reduced by row operations to the given row echelon form. Solve the system by back substitution. Assume that the variables are named $x_1, x_2, \ldots$ from left to right.

17. $\begin{bmatrix} 1 & 7 & -2 & 0 & -8 & -3 \\ 0 & 0 & 1 & 1 & 6 & 5 \\ 0 & 0 & 0 & 1 & 3 & 9 \\ 0 & 0 & 0 & 0 & 0 & 0 \end{bmatrix}$

18. $\begin{bmatrix} 1 & -3 & 7 & 1 \\ 0 & 1 & 4 & 0 \\ 0 & 0 & 0 & 1 \end{bmatrix}$ **19.** $\begin{bmatrix} 1 & 1 & -3 & 2 & 1 \\ 0 & 1 & 4 & 0 & 3 \\ 0 & 0 & 0 & 1 & 2 \end{bmatrix}$

20. $\begin{bmatrix} 1 & 0 & 5 & 3 & 2 \\ 0 & 1 & -2 & 4 & -7 \\ 0 & 0 & 1 & 1 & 3 \end{bmatrix}$

In Exercises 21 and 22, solve the linear system by solving for x_1 in the first equation, then using this result to find x_2 in the second equation, and finally using the values of x_1 and x_2 to find x_3 in the third equation. (This is called *forward substitution*, as compared to back substitution.)

21. $\begin{aligned} 2x_1 & & & = 4 \\ x_1 + 3x_2 & & & = 5 \\ 3x_1 + 2x_2 + 3x_3 & & & = 12 \end{aligned}$ **22.** $\begin{aligned} 3x_1 & & & = -3 \\ 2x_1 + 3x_2 & & & = 4 \\ x_1 + 4x_2 + x_3 & & & = 5 \end{aligned}$

In Exercises 23–28, solve the linear system by Gauss–Jordan elimination.

23. $\begin{aligned} x_1 + x_2 + 2x_3 &= 8 \\ -x_1 - 2x_2 + 3x_3 &= 1 \\ 3x_1 - 7x_2 + 4x_3 &= 10 \end{aligned}$ **24.** $\begin{aligned} 2x_1 + 2x_2 + 2x_3 &= 0 \\ -2x_1 + 5x_2 + 2x_3 &= 1 \\ 8x_1 + x_2 + 4x_3 &= -1 \end{aligned}$

25. $\begin{aligned} x - y + 2z - w &= -1 \\ 2x + y - 2z - 2w &= -2 \\ -x + 2y - 4z + w &= 1 \\ 3x \qquad\quad - 3w &= -3 \end{aligned}$

26. $\begin{aligned} -2b + 3c &= 2 \\ 3a + 6b - 3c &= -2 \\ 6a + 6b + 3c &= 5 \end{aligned}$ **27.** $\begin{aligned} 2x_1 - 3x_2 &= -2 \\ 2x_1 + x_2 &= 1 \\ 3x_1 + 2x_2 &= 1 \end{aligned}$

28. $\begin{aligned} 3x_1 + 2x_2 - x_3 &= -15 \\ 5x_1 + 3x_2 + 2x_3 &= 0 \\ 3x_1 + x_2 + 3x_3 &= 11 \\ 6x_1 - 4x_2 + 2x_3 &= 30 \end{aligned}$

In Exercises 29–32, solve the linear system in the referenced exercise by Gaussian elimination.

29. Exercise 23   **30.** Exercise 24

31. Exercise 25   **32.** Exercise 26

In Exercises 33 and 34, determine whether the homogeneous system has nontrivial solutions by inspection.

33. (a) $\begin{aligned} 2x_1 - 3x_2 + 4x_3 - x_4 &= 0 \\ 7x_1 + x_2 - 8x_3 + 9x_4 &= 0 \\ 2x_1 + 8x_2 + x_3 - x_4 &= 0 \end{aligned}$

   (b) $\begin{aligned} x_1 + 2x_2 - x_3 &= 0 \\ 3x_2 + 2x_3 &= 0 \\ 4x_3 &= 0 \end{aligned}$

34. (a) $\begin{aligned} a_{11}x_1 + a_{12}x_2 + a_{13}x_3 &= 0 \\ a_{21}x_1 + a_{22}x_2 + a_{23}x_3 &= 0 \end{aligned}$

   (b) $\begin{aligned} 3x_1 - 2x_2 &= 0 \\ 6x_1 - 4x_2 &= 0 \end{aligned}$

In Exercises 35–42, solve the given homogeneous system of linear equations by any method.

35. $\begin{aligned} 2x_1 + x_2 + 3x_3 &= 0 \\ x_1 + 2x_2 &= 0 \\ x_2 + 2x_3 &= 0 \end{aligned}$ **36.** $\begin{aligned} 3x_1 + x_2 + x_3 + x_4 &= 0 \\ 5x_1 - x_2 + x_3 - x_4 &= 0 \end{aligned}$

37. $\begin{aligned} 2x + 2y + 4z &= 0 \\ w - y - 3z &= 0 \\ -2w + x + 3y - 2z &= 0 \end{aligned}$

38. $\begin{aligned} 2x - y - 3z &= 0 \\ -x + 2y - 3z &= 0 \\ x + y + 4z &= 0 \end{aligned}$ **39.** $\begin{aligned} v + 3w - 2x &= 0 \\ 2u + v - 4w + 3x &= 0 \\ 2u + 3v + 2w - x &= 0 \\ -4u - 3v + 5w - 4x &= 0 \end{aligned}$

40. $\begin{aligned} x_1 + 3x_2 + x_4 &= 0 \\ x_1 + 4x_2 + 2x_3 &= 0 \\ -2x_2 - 2x_3 - x_4 &= 0 \\ 2x_1 - 4x_2 + x_3 + x_4 &= 0 \\ x_1 - 2x_2 - x_3 + x_4 &= 0 \end{aligned}$

41. $\begin{aligned} 2I_1 - I_2 + 3I_3 + 4I_4 &= 0 \\ I_1 - 2I_3 + 7I_4 &= 0 \\ 3I_1 - 3I_2 + I_3 + 5I_4 &= 0 \\ 2I_1 + I_2 + 4I_3 + 4I_4 &= 0 \end{aligned}$

42. $\begin{aligned} Z_3 + Z_4 + Z_5 &= 0 \\ -Z_1 - Z_2 + 2Z_3 - 3Z_4 + Z_5 &= 0 \\ Z_1 + Z_2 - 2Z_3 - Z_5 &= 0 \\ 2Z_1 + 2Z_2 - Z_3 + Z_5 &= 0 \end{aligned}$

In Exercises 43–46, determine the values of a for which the system has no solutions, exactly one solution, or infinitely many solutions.

43. $\begin{aligned} x + 2y + 3z &= 4 \\ 3x - y + 5z &= 2 \\ 4x + y - 14z &= a + 2 \end{aligned}$ **44.** $\begin{aligned} x + 2y + z &= 2 \\ 2x - 2y + 3z &= 1 \\ x + 2y + (a^2 - 3)z &= a \end{aligned}$

45. $\begin{aligned} x + 2y &= 1 \\ 2x + (a^2 - 5)y &= a - 1 \end{aligned}$

46. $\begin{aligned} x + y + 7z &= -7 \\ 2x + 3y + 17z &= -16 \\ x + 2y + (a^2 + 1)z &= 3a \end{aligned}$

47. Consider the linear systems

$\begin{aligned} x + y + z &= 1 \\ 2x + 2y + 2z &= 4 \end{aligned}$ and $\begin{aligned} x + y + z &= 0 \\ 2x + 2y + 2z &= 0 \end{aligned}$

   (a) Show that the first system has no solutions, and state what this tells you about the planes represented by these equations.

(b) Show that the second system has infinitely many solutions, and state what this tells you about the planes represented by these equations.

48. Reduce the matrix

$$\begin{bmatrix} 2 & 1 & 3 \\ 0 & -2 & 7 \\ 3 & 4 & 5 \end{bmatrix}$$

to reduced row echelon form without introducing fractions at any intermediate stage.

49. Solve the following system of nonlinear equations for the unknown angles α, β, and γ, where $0 \le \alpha \le 2\pi$, $0 \le \beta \le 2\pi$, and $0 \le \gamma \le \pi$. [*Hint:* Begin by making the substitutions $x = \sin \alpha$, $y = \cos \beta$, and $z = \tan \gamma$.]

$$2 \sin \alpha - \cos \beta + 3 \tan \gamma = 3$$
$$4 \sin \alpha + 2 \cos \beta - 2 \tan \gamma = 2$$
$$6 \sin \alpha - 3 \cos \beta + \tan \gamma = 9$$

50. Solve the following system of nonlinear equations for x, y, and z. [*Hint:* Begin by making the substitutions $X = x^2$, $Y = y^2$, $Z = z^2$.]

$$x^2 + y^2 + z^2 = 6$$
$$x^2 - y^2 + 2z^2 = 2$$
$$2x^2 + y^2 - z^2 = 3$$

51. Consider the linear system

$$2x_1 - x_2 \qquad = \lambda x_1$$
$$2x_1 + x_2 + x_3 = \lambda x_2$$
$$-2x_1 - 2x_2 + x_3 = \lambda x_3$$

in which λ is a constant. Solve the system taking $\lambda = 1$ and then taking $\lambda = 2$.

52. What relationship must exist between a, b, and c for the linear system

$$x + y + 2z = a$$
$$x \qquad + z = b$$
$$2x + y + 3z = c$$

to be consistent?

Computations that are performed using exact numbers are said to be performed in *exact arithmetic*, and those that are performed using numbers that have been rounded off to a certain number of significant digits are said to be performed in *finite-precision arithmetic*. Although computer algebra systems, such as *Mathematica*, *Maple*, and *Derive*, are capable of performing computations using either exact or finite-precision arithmetic, most large-scale problems are solved using finite-precision arithmetic. Thus, methods for minimizing roundoff error are extremely important in practical applications of linear algebra. Exercises 53 and 54 illustrate a method for reducing roundoff error when finite-precision arithmetic is used in Gauss–Jordan elimination or Gaussian elimination.

---

53. (a) Consider the linear system

$$\tfrac{1}{10,000}x + y = 1$$
$$x - y = 0$$

whose exact solution is $x = \frac{10,000}{10,001}$, $y = \frac{10,000}{10,001}$ (verify), and assume that this system is stored in a calculator or computer as

$$0.0001x + 1.000y = 1.000$$
$$1.000x - 1.000y = 0.000$$

Solve this system by Gauss–Jordan elimination using finite-precision arithmetic in which you perform each computation to the full accuracy of your calculating utility and then round to four significant digits at each step. This should yield $x \approx 0.000$ and $y \approx 1.000$, which is a poor approximation to the exact solution. [*Note:* One way to round a decimal number to four significant digits is to first write it in the exponential notation

$$M \times 10^k$$

where M is of the form $M = 0.d_1 d_2 d_3 d_4 \ldots$ and d_1 is nonzero; then round off M to four decimal places and express the result in decimal form. For example, the number 26.87643 rounds to four significant digits as 26.88, the number 0.0002687643 rounds to four significant digits as 0.0002688, and the number 10,001 rounds to four significant digits as 10,000.]

(b) The inaccuracy in part (a) is due to the fact that round-off error tends to be magnified by large multipliers (or, equivalently, small divisors), and a large multiplier of 10,000 was required to introduce a leading 1 in the first row of the augmented matrix. However this large multiplier can be avoided by first interchanging the equations to obtain

$$1.000x - 1.000y = 0.000$$
$$0.0001x + 1.000y = 1.000$$

If you now solve the system using the finite-precision procedure described in part (a), then you should obtain $x \approx 1.000$ and $y \approx 1.000$, which is a big improvement in accuracy. Perform these computations.

(c) In part (b) we were able to avoid the large multiplier needed to create the leading 1 in the first row of the augmented matrix by performing a preliminary row interchange. In general, when you are implementing Gauss–Jordan elimination or Gaussian elimination in finite-precision arithmetic, you can reduce the effect of roundoff error if, when you are ready to introduce a leading 1 in a certain row, you interchange that row with an appropriate *lower* row to make the number in the pivot position as large as possible in absolute value. This procedure is called *partial pivoting*. Solve the system

$$\tfrac{1}{50000}x + y = 1$$
$$x + y = 3$$

with and without partial pivoting using finite-precision arithmetic with four significant digits, and compare your answers to the exact solution.

54. Solve the system by Gauss–Jordan elimination with partial pivoting using finite-precision arithmetic with two significant digits.

(a) $0.21x + 0.33y = 0.54$
$0.70x + 0.24y = 0.94$

(b) $0.11x_1 - 0.13x_2 + 0.20x_3 = -0.02$
$0.10x_1 + 0.36x_2 + 0.45x_3 = 0.25$
$0.50x_1 - 0.01x_2 + 0.30x_3 = -0.70$

Discussion and Discovery

D1. If the linear system
$$a_1x + b_1y + c_1z = 0$$
$$a_2x + b_2y + c_2z = 0$$
$$a_3x + b_3y + c_3z = 0$$
has only the trivial solution, what can be said about the solutions of the following system?
$$a_1x + b_1y + c_1z = 3$$
$$a_2x + b_2y + c_2z = 7$$
$$a_3x + b_3y + c_3z = 11$$

D2. Consider the system of equations
$$ax + by = 0$$
$$cx + dy = 0$$
$$ex + fy = 0$$
Discuss the relative positions of the lines represented by the three equations when
(a) the system has only the trivial solution;
(b) the system has nontrivial solutions.

D3. Consider the system of equations
$$ax + by = 0$$
$$cx + dy = 0$$
(a) If $x = x_0$, $y = y_0$ is any solution of the system and k is any constant, do you think it is true that $x = kx_0$, $y = ky_0$ is also a solution? Justify your answer.
(b) If $x = x_0$, $y = y_0$ and $x = x_1$, $y = y_1$ are any two solutions of the system, do you think it is true that $x = x_0 + x_1$, $y = y_0 + y_1$ is also a solution? Justify your answer.
(c) Would your answers in parts (a) and (b) change if the system were not homogeneous?

D4. What geometric relationship exists between the solution sets of the following linear systems?
$$\begin{array}{l} ax + by = k \\ cx + dy = l \end{array} \quad \text{and} \quad \begin{array}{l} ax + by = 0 \\ cx + dy = 0 \end{array}$$

D5. (a) If A is a 3×5 matrix, then the number of leading 1's in its reduced row echelon form is at most _____.
(b) If B is a 3×6 matrix whose last column has all zeros, then the number of parameters in the general solution of the linear system with augmented matrix B is at most _____.

(c) If A is a 3×5 matrix, then the number of leading 1's in any of its row echelon forms is at most _____.

D6. (a) If A is a 5×3 matrix, then the number of leading 1's in its reduced row echelon form is at most _____.
(b) If B is a 5×4 matrix whose last column has all zeros, then the number of parameters in the general solution of the linear system with augmented matrix B is at most _____.
(c) If A is a 5×3 matrix, then the number of leading 1's in any of its row echelon forms is at most _____.

D7. Indicate whether the statement is true (T) or false (F). Justify your answer.
(a) A linear system that has more unknowns than equations is consistent.
(b) A linear system can have exactly two solutions.
(c) A linear system of two equations in three unknowns can have a unique solution.
(d) A homogeneous system of equations is consistent.

D8. Indicate whether the statement is true (T) or false (F). Justify your answer.
(a) A matrix may be reduced to more than one row echelon form.
(b) A matrix may be reduced to more than one reduced row echelon form.
(c) If the reduced row echelon form of the augmented matrix of a system of equations has a row consisting entirely of zeros, then the system of equations has infinitely many solutions.
(d) A nonhomogeneous system of equations with more equations than unknowns must be inconsistent.

D9. Show that the system
$$\sin\alpha + 2\cos\beta + 3\tan\gamma = 0$$
$$2\sin\alpha + 5\cos\beta + 3\tan\gamma = 0$$
$$-\sin\alpha - 5\cos\beta + 5\tan\gamma = 0$$
has 18 solutions if $0 \le \alpha \le 2\pi, 0 \le \beta \le 2\pi, 0 \le \gamma \le 2\pi$. Does this contradict Theorem 2.2.1?

Working with Proofs

P1. (a) Prove that if $ad - bc \neq 0$, then the reduced row echelon form of

$$\begin{bmatrix} a & b \\ c & d \end{bmatrix} \quad \text{is} \quad \begin{bmatrix} 1 & 0 \\ 0 & 1 \end{bmatrix}$$

(b) Use the result in part (a) to prove that if $ad - bc \neq 0$, then the linear system

$$ax + by = k$$
$$cx + dy = l$$

has exactly one solution.

Technology Exercises

T1. (**Reduced row echelon form**) Read your documentation on how to enter matrices and how to produce reduced row echelon forms. Check your understanding of these commands by finding the reduced row echelon form of the matrix

$$\begin{bmatrix} 2 & -3 & 1 & 0 & 4 \\ 1 & 1 & 2 & 2 & 0 \\ 3 & 0 & -1 & 4 & 5 \\ 1 & 6 & 5 & 6 & -4 \end{bmatrix}$$

T2. (**Inconsistent linear systems**) Technology utilities will often successfully identify inconsistent linear systems, but they can sometimes be fooled into reporting an inconsistent system as consistent or vice versa. This typically occurs when some of the numbers that occur in the computations are so small that roundoff error makes it difficult for the utility to determine whether or not they are equal to zero. Create some inconsistent linear systems and see how your utility handles them.

T3. (**Linear systems with infinitely many solutions**) Technology utilities used for solving linear systems all handle systems with infinitely many solutions differently. See what happens when you solve the system in Example 5.

T4. Solve the linear system

$$\tfrac{1}{5}x + \tfrac{1}{4}y + \tfrac{1}{2}z = \tfrac{37}{120}$$
$$\tfrac{1}{3}x + \tfrac{1}{7}y + \tfrac{1}{4}z = \tfrac{93}{336}$$
$$\tfrac{1}{4}x + \tfrac{1}{6}y + \tfrac{1}{3}z = \tfrac{43}{180}$$

T5. In each part find values of the constants that make the equation an identity.

(a) $\dfrac{3x^3 + 4x^2 - 6x}{(x^2 + 2x + 2)(x^2 - 1)}$

$$= \dfrac{Ax + B}{x^2 + 2x + 2} + \dfrac{C}{x - 1} + \dfrac{D}{x + 1}$$

(b) $\dfrac{3x^4 + 4x^3 + 16x^2 + 20x + 9}{(x + 2)(x^2 + 3)^2}$

$$= \dfrac{A}{x + 2} + \dfrac{Bx + C}{x^2 + 3} + \dfrac{Dx + E}{(x^2 + 3)^2}$$

[*Hint:* Obtain a common denominator on the right, and then equate corresponding coefficients of the various powers of x in the two resulting numerators. Students of calculus may recognize this as a problem in partial fractions.]

Section 2.3 Applications of Linear Systems

In this section we will discuss various applications of linear systems, including global positioning and the network analysis of traffic flow, electrical circuits, and chemical reactions. We will also show how linear systems can be used to find polynomial curves that pass through specified points.

GLOBAL POSITIONING GPS (**Global Positioning System**) is the system used by the military, ships, airplane pilots, surveyors, utility companies, automobiles, and hikers to locate current positions by communicating with a system of satellites. The system, which is operated by the U.S. Department of Defense, nominally uses 24 satellites that orbit the Earth every 12 hours at a height of about 11,000 miles. These satellites move in six orbital planes that have been chosen to make between five and eight satellites visible from any point on Earth.

To explain how the system works, assume that the Earth is a sphere, and suppose that there is an xyz-coordinate system with its origin at the Earth's center and its z-axis through the North Pole (Figure 2.3.1). Let us assume that relative to this coordinate system a ship is at an unknown

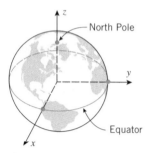

Figure 2.3.1

point (x, y, z) at some time t. For simplicity, assume that distances are measured in units equal to the Earth's radius, so that the coordinates of the ship always satisfy the equation

$$x^2 + y^2 + z^2 = 1$$

The GPS identifies the ship's coordinates (x, y, z) at a time t using a triangulation system and computed distances from four satellites. These distances are computed using the speed of light (approximately 0.469 Earth radii per hundredth of a second) and the time it takes for the signal to travel from the satellite to the ship. For example, if the ship receives the signal at time t and the satellite indicates that it transmitted the signal at time t_0, then the distance d traveled by the signal will be

$$d = 0.469(t - t_0) \tag{1}$$

In theory, knowing three ship-to-satellite distances would suffice to determine the three unknown coordinates of the ship. However, the problem is that the ships (or other GPS users) do not generally have clocks that can compute t with sufficient accuracy for global positioning. Thus, the variable t in (1) must be regarded as a fourth unknown, and hence the need for the distance to a fourth satellite. Suppose that in addition to transmitting the time t_0, each satellite also transmits its coordinates (x_0, y_0, z_0) at that time, thereby allowing d to be computed as

$$d = \sqrt{(x - x_0)^2 + (y - y_0)^2 + (z - z_0)^2} \tag{2}$$

If we now equate the squares of (1) and (2) and round off to three decimal places, then we obtain the second-degree equation

$$(x - x_0)^2 + (y - y_0)^2 + (z - z_0)^2 = 0.22(t - t_0)^2 \tag{3}$$

Since there are four different satellites, and we can get an equation like this for each one, we can produce four equations in the unknowns x, y, z, and t_0. Although these are second-degree equations, it is possible to use these equations and some algebra to produce a system of linear equations that can be solved for the unknowns. We will illustrate this with an example.

EXAMPLE 1
Global
Positioning

Suppose that a ship with unknown coordinates (x, y, z) at an unknown time t receives the following data from four satellites in which coordinates are measured in Earth radii and time in hundredths of a second after midnight:

Satellite	Satellite Position	Time
1	(1.12, 2.10, 1.40)	1.06
2	(0.00, 1.53, 2.30)	0.56
3	(1.40, 1.12, 2.10)	1.16
4	(2.30, 0.00, 1.53)	0.75

Substituting data from the first satellite into Formula (3) yields the equation

$$(x - 1.12)^2 + (y - 2.10)^2 + (z - 1.40)^2 = 0.22(t - 1.06)^2 \tag{4}$$

which, on squaring out, we can rewrite as

$$2.24x + 4.2y + 2.8z - 0.466t = x^2 + y^2 + z^2 - 0.22t^2 + 7.377$$

(All of our calculations are performed to the full accuracy of our calculating utility but rounded to three decimal places when displayed.) Similar calculations using the data from the other three satellites yields the following nonlinear system of equations:

$$\begin{aligned}
2.24x + 4.2y + 2.8z - 0.466t &= x^2 + y^2 + z^2 - 0.22t^2 + 7.377 \\
3.06y + 4.6z - 0.246t &= x^2 + y^2 + z^2 - 0.22t^2 + 7.562 \\
2.8x + 2.24y + 4.2z - 0.510t &= x^2 + y^2 + z^2 - 0.22t^2 + 7.328 \\
4.6x + 3.06z - 0.33t &= x^2 + y^2 + z^2 - 0.22t^2 + 7.507
\end{aligned}$$

The quadratic terms in all of these equations are the same, so if we subtract each of the last three equations from the first one, we obtain the linear system

$$
\begin{aligned}
2.24x + 1.14y - 1.8z - 0.22t &= -0.185 \\
-0.56x + 1.96y - 1.4z + 0.044t &= 0.049 \\
-2.36x + 4.2y - 0.26z - 0.136t &= -0.13
\end{aligned}
\tag{5}
$$

Since there are three equations in four unknowns, we can reasonably expect that the general solution of (5) will involve one parameter, say s. By solving for x, y, z, and t in terms of s and substituting into any one of the original quadratic equations, we can obtain a quadratic equation in s, a solution of which can then be used to calculate x, y, and z. To carry out this plan, we leave it for you to verify that the reduced row echelon form of the augmented matrix for (5) is

$$
\begin{bmatrix}
1 & 0 & 0 & -0.153 & -0.139 \\
0 & 1 & 0 & -0.128 & -0.118 \\
0 & 0 & 1 & -0.149 & -0.144
\end{bmatrix}
$$

from which we obtain

$$
\begin{aligned}
x &= -0.139 + 0.153s \\
y &= -0.118 + 0.128s \\
z &= -0.144 + 0.149s \\
t &= s
\end{aligned}
\tag{6}
$$

To find s we can substitute these expressions into any of the four quadratic equations from the satellites. To be specific, if we substitute into (4) and simplify, we obtain

$$
8.639 - 0.945s - 0.158s^2 = 0
$$

This yields the solutions

$$
s = -10.959 \quad \text{and} \quad s = 4.985
$$

(verify). Since $t \geq 0$, and $s = t$, we choose the positive value of s. Substituting this value into (6) yields

$$
(x, y, z) = (0.624, 0.519, 0.598)
$$

which are the coordinates of the ship. You may want to confirm that this is a point on the surface of the Earth by showing that it is one unit from the Earth's center, allowing for the roundoff error. For navigational purposes one would usually convert these coordinates to latitude and longitude, but we will not discuss this here. ∎

NETWORK ANALYSIS

The concept of a *network* appears in a variety of applications. Loosely stated, a **network** is a set of **branches** through which something "flows." For example, the branches might be electrical wires through which electricity flows, pipes through which water or oil flows, traffic lanes through which vehicular traffic flows, or economic linkages through which money flows, to name a few possibilities.

In most networks, the branches meet at points, called **nodes** or **junctions**, where the flow divides. For example, in an electrical network, nodes occur where three or more wires join, in a traffic network they occur at street intersections, and in a financial network they occur at banking centers where incoming money is distributed to individuals or other institutions.

In the study of networks, there is generally some numerical measure of the rate of flow through a branch. For example, the flow rate of electricity is often measured in amperes (A), the flow rate of water or oil in gallons per minute (gal/min), the flow rate of traffic in vehicles per hour, and so forth. Networks are of two basic types: **open**, in which the flow medium can enter and leave the network, and **closed**, in which the flow medium circulates continuously through the network, with none entering or leaving. Many of the most important kinds of networks have three basic properties:

1. **One-Directional Flow**—At any instant, the flow in a branch is in one direction only.
2. **Flow Conservation at a Node**—The rate of flow into a node is equal to the rate of flow out of the node.
3. **Flow Conservation in the Network**—The rate of flow into the network is equal to the rate of flow out of the network.

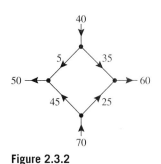

Figure 2.3.2

The second property ensures that the flow medium does not build up at the nodes, and the third ensures that it does not build up anywhere in the network. (In a closed network, the third property holds automatically, since the flow rate in and out of the network is zero.)

These properties are illustrated in Figure 2.3.2, which shows an open network with four nodes. For simplicity, the flow rates in the figure are indicated without units. Observe that the total rate of flow into the network is $70+40 = 110$ and the total rate of flow out is $50+60 = 110$, so there is conservation of flow in the network. Moreover, at each of the four nodes the rate of flow in is equal to the rate of flow out (verify), so there is conservation of flow at all nodes.

A common problem in network analysis is to use known flow rates in certain branches to find the flow rates in all of the branches. Here is an example.

EXAMPLE 2
Network Analysis Using Linear Systems

Figure 2.3.3a shows a network in which the flow rate and direction of flow in certain branches are known. Find the flow rates and directions of flow in the remaining branches.

Solution As illustrated in Figure 2.3.3b, we have assigned arbitrary directions to the unknown flow rates x_1, x_2, and x_3. We need not be concerned if some of the directions are incorrect, since an incorrect direction will be signaled by a negative value for the flow rate when we solve for the unknowns.

It follows from the conservation of flow at node A that

$$x_1 + x_2 = 30$$

Similarly, at the other nodes we have

$$x_2 + x_3 = 35 \qquad \text{(node } B\text{)}$$
$$x_3 + 15 = 60 \qquad \text{(node } C\text{)}$$
$$x_1 + 15 = 55 \qquad \text{(node } D\text{)}$$

These four conditions produce the linear system

$$
\begin{aligned}
x_1 + x_2 \quad\;\; &= 30 \\
x_2 + x_3 &= 35 \\
x_3 &= 45 \\
x_1 \qquad\;\; &= 40
\end{aligned}
$$

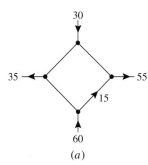

(a)

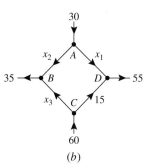

(b)

Figure 2.3.3

which we can now try to solve for the unknown flow rates. In this particular case the system is sufficiently simple that it can be solved by inspection (work from the bottom up). We leave it for you to confirm that the solution is

$$x_1 = 40, \quad x_2 = -10, \quad x_3 = 45$$

The fact that x_2 is negative tells us that the direction assigned to that flow in Figure 2.3.3b is incorrect; that is, the flow in that branch is *into* node A. ∎

EXAMPLE 3
Design of Traffic Patterns

The network in Figure 2.3.4 shows a proposed plan for the traffic flow around a new park that will house the Liberty Bell in Philadelphia, Pennsylvania. The plan calls for a computerized traffic light at the north exit on Third Street, and the diagram indicates the average number of vehicles per hour that are expected to flow in and out of the streets that border the complex. All streets are one-way.

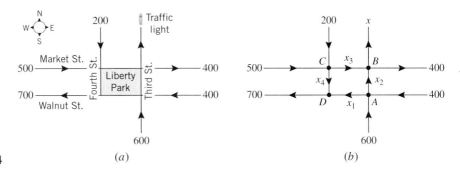

Figure 2.3.4
(a)   (b)

(a) How many vehicles per hour should the traffic light let through to ensure that the average number of vehicles per hour flowing into the complex is the same as the average number of vehicles flowing out?

(b) Assuming that the traffic light has been set to balance the total flow in and out of the complex, what can you say about the average number of vehicles per hour that will flow along the streets that border the complex?

Solution (a)   If we let x denote the number of vehicles per hour that the traffic light must let through, then the total number of vehicles per hour that flow in and out of the complex will be

Flowing in: $500 + 400 + 600 + 200 = 1700$

Flowing out: $x + 700 + 400$

Equating the flows in and out shows that the traffic light should let $x = 600$ vehicles per hour pass through.

Solution (b)   To avoid traffic congestion, the flow in must equal the flow out at each intersection. For this to happen, the following conditions must be satisfied:

Intersection	Flow In		Flow Out
A	$400 + 600$	$=$	$x_1 + x_2$
B	$x_2 + x_3$	$=$	$400 + x$
C	$500 + 200$	$=$	$x_3 + x_4$
D	$x_1 + x_4$	$=$	700

Thus, with $x = 600$, as computed in part (a), we obtain the following linear system:

$$\begin{aligned} x_1 + x_2 \qquad\qquad &= 1000 \\ x_2 + x_3 \qquad &= 1000 \\ x_3 + x_4 &= 700 \\ x_1 \qquad\qquad + x_4 &= 700 \end{aligned}$$

We leave it for you to show that the system has infinitely many solutions and that these are given by the parametric equations

$$x_1 = 700 - t, \ x_2 = 300 + t, \ x_3 = 700 - t, \ x_4 = t \tag{7}$$

However, the parameter t is not completely arbitrary here, since there are physical constraints to be considered. For example, the average flow rates must be nonnegative since we have assumed the streets to be one-way, and a negative flow rate would indicate a flow in the wrong direction. This being the case, we see from (7) that t can be any real number that satisfies $0 \leq t \leq 700$, which implies that the average flow rates along the streets will fall in the ranges

$$0 \leq x_1 \leq 700, \quad 300 \leq x_2 \leq 1000, \quad 0 \leq x_3 \leq 700, \quad 0 \leq x_4 \leq 700$$ ■

ELECTRICAL CIRCUITS

Figure 2.3.5

Next, we will show how network analysis can be used to analyze electrical circuits consisting of batteries and resistors. A ***battery*** is a source of electric energy, and a ***resistor***, such as a light-bulb, is an element that dissipates electric energy. Figure 2.3.5 shows a schematic diagram of a circuit with one battery (represented by the symbol ⊣⊢), one resistor (represented by the symbol –∿∿–), and a switch. The battery has a ***positive pole*** (+) and a ***negative pole*** (−). When the switch is closed, electrical current is considered to flow from the positive pole of the battery, through the resistor, and back to the negative pole (indicated by the arrowhead in the figure).

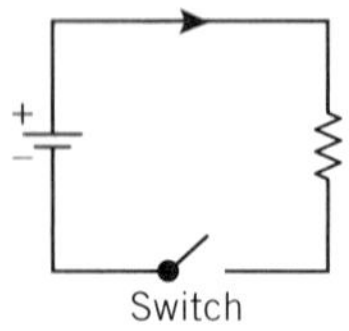

Electrical current, which is a flow of electrons through wires, behaves much like the flow of water through pipes. A battery acts like a pump that creates "electrical pressure" to increase the flow rate of electrons, and a resistor acts like a restriction in a pipe that reduces the flow rate of electrons. The technical term for electrical pressure is ***electrical potential***; it is commonly measured in ***volts*** (V). The degree to which a resistor reduces the electrical potential is called its ***resistance*** and is commonly measured in ***ohms*** (Ω). The rate of flow of electrons in a wire is called ***current*** and is commonly measured in ***amperes*** (also called ***amps***) (A). The precise effect of a resistor is given by the following law:

> **Ohm's Law**   If a current of I amperes passes through a resistor with a resistance of R ohms, then there is a resulting drop of E volts in electrical potential that is the product of the current and resistance; that is,
>
> $$E = IR$$

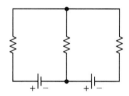

Figure 2.3.6

A typical electrical network will have multiple batteries and resistors joined by some configuration of wires. A point at which three or more wires in a network are joined is called a ***node*** (or ***junction point***). A ***branch*** is a wire connecting two nodes, and a ***closed loop*** is a succession of connected branches that begin and end at the same node. For example, the electrical network in Figure 2.3.6 has two nodes and three closed loops—two inner loops and one outer loop. As current flows through an electrical network, it undergoes increases and decreases in electrical potential, called ***voltage rises*** and ***voltage drops***, respectively. The behavior of the current at the nodes and around closed loops is governed by two fundamental laws:

> **Kirchhoff's Current Law**   The sum of the currents flowing into any node is equal to the sum of the currents flowing out.

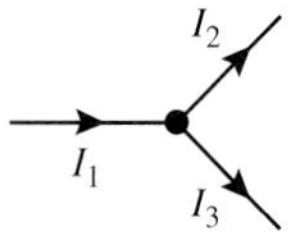

Figure 2.3.7

> **Kirchhoff's Voltage Law**   In one traversal of any closed loop, the sum of the voltage rises equals the sum of the voltage drops.

Kirchhoff's current law is a restatement of the principle of flow conservation at a node that was stated for general networks. Thus, for example, the currents at the top node in Figure 2.3.7 satisfy the equation $I_1 = I_2 + I_3$.

In circuits with multiple loops and batteries there is usually no way to tell in advance which way the currents are flowing, so the usual procedure in circuit analysis is to assign *arbitrary* directions to the current flows in the branches and let the mathematical computations determine whether the assignments are correct. In addition to assigning directions to the current flows, Kirchhoff's voltage law requires a direction of travel for each closed loop. The choice is arbitrary, but for consistency we will always take this direction to be *clockwise* (Figure 2.3.8). We also make the following conventions:

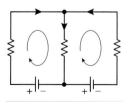

Clockwise closed-loop convention with arbitrary direction assignments to currents in the branches

Figure 2.3.8

- A voltage drop occurs at a resistor if the direction assigned to the current through the resistor is the same as the direction assigned to the loop, and a voltage rise occurs at a

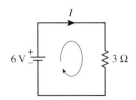

Figure 2.3.9

resistor if the direction assigned to the current through the resistor is the opposite to that assigned to the loop.

- A voltage rise occurs at a battery if the direction assigned to the loop is from $-$ to $+$ through the battery, and a voltage drop occurs at a battery if the direction assigned to the loop is from $+$ to $-$ through the battery.

If you follow these conventions when calculating currents, then those currents whose directions were assigned correctly will have positive values and those whose directions were assigned incorrectly will have negative values.

EXAMPLE 4
A Circuit with
One Closed
Loop

Determine the current I in the circuit shown in Figure 2.3.9.

Solution Since the direction assigned to the current through the resistor is the same as the direction of the loop, there is a voltage drop at the resistor. By Ohm's law this voltage drop is $E = IR = 3I$. Also, since the direction assigned to the loop is from $-$ to $+$ through the battery, there is a voltage rise of 6 volts at the battery. Thus, it follows from Kirchhoff's voltage law that

$$3I = 6$$

from which we conclude that the current is $I = 2$ A. Since I is positive, the direction assigned to the current flow is correct. ∎

EXAMPLE 5
A Circuit with
Three Closed
Loops

Determine the currents I_1, I_2, and I_3 in the circuit shown in Figure 2.3.10.

Solution Using the assigned directions for the currents, Kirchhoff's current law provides one equation for each node:

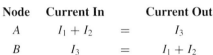

Figure 2.3.10

Node	Current In		Current Out
A	$I_1 + I_2$	$=$	I_3
B	I_3	$=$	$I_1 + I_2$

However, these equations are really the same, since both can be expressed as

$$I_1 + I_2 - I_3 = 0 \tag{8}$$

To find unique values for the currents we will need two more equations, which we will obtain from Kirchhoff's voltage law. We can see from the network diagram that there are three closed loops, a left inner loop containing the 50-V battery, a right inner loop containing the 30-V battery, and an outer loop that contains both batteries. Thus, Kirchhoff's voltage law will actually produce three equations. With a clockwise traversal of the loops, the voltage rises and drops in these loops are as follows:

	Voltage Rises	Voltage Drops
Left Inside Loop	50	$5I_1 + 20I_3$
Right Inside Loop	$30 + 10I_2 + 20I_3$	0
Outside Loop	$30 + 50 + 10I_2$	$5I_1$

These conditions can be rewritten as

$$
\begin{aligned}
5I_1 \quad\quad + 20I_3 &= 50 \\
10I_2 + 20I_3 &= -30 \\
5I_1 - 10I_2 \quad\quad &= 80
\end{aligned}
\tag{9}
$$

However, the last equation is superfluous, since it is the difference of the first two. Thus, if we combine (8) and the first two equations in (9), we obtain the following linear system of three equations in the three unknown currents:

$$
\begin{aligned}
I_1 + I_2 - I_3 &= 0 \\
5I_1 \qquad\;\; + 20I_3 &= 50 \\
10I_2 + 20I_3 &= -30
\end{aligned}
$$

We leave it for you to solve this system and show that $I_1 = 6$ A, $I_2 = -5$ A, and $I_3 = 1$ A. The fact that I_2 is negative tells us that the direction of this current is opposite to that indicated in Figure 2.3.10. ∎

BALANCING CHEMICAL EQUATIONS

Chemical compounds are represented by ***chemical formulas*** that describe the atomic makeup of their molecules. For example, water is composed of two hydrogen atoms and one oxygen atom, so its chemical formula is H_2O; and stable oxygen is composed of two oxygen atoms, so its chemical formula is O_2.

When chemical compounds are combined under the right conditions, the atoms in their molecules rearrange to form new compounds. For example, in methane burning, methane (CH_4) and stable oxygen (O_2) react to form carbon dioxide (CO_2) and water (H_2O). This is indicated by the ***chemical equation***

$$CH_4 + O_2 \longrightarrow CO_2 + H_2O \tag{10}$$

The molecules to the left of the arrow are called the ***reactants*** and those to the right the ***products***. In this equation the plus signs serve to separate the molecules and are not intended as algebraic operations. However, this equation does not tell the whole story, since it fails to account for the proportions of molecules required for a ***complete reaction*** (no reactants left over). For example, we can see from the right side of (10) that to produce one molecule of carbon dioxide and one molecule of water, one needs *three* oxygen atoms for each carbon atom. However, from the left side of (10) we see that one molecule of methane and one molecule of stable oxygen have only *two* oxygen atoms for each carbon atom. Thus, on the reactant side the ratio of methane to stable oxygen cannot be one-to-one in a complete reaction.

A chemical equation is said to be ***balanced*** if for each type of atom in the reaction, the same number of atoms appears on each side of the arrow. For example, the balanced version of Equation (10) is

$$CH_4 + 2O_2 \longrightarrow CO_2 + 2H_2O \tag{11}$$

by which we mean that one methane molecule combines with two stable oxygen molecules to produce one carbon dioxide molecule and two water molecules. In theory, one could multiply this equation through by any positive integer. For example, multiplying through by 2 yields the balanced chemical equation

$$2CH_4 + 4O_2 \longrightarrow 2CO_2 + 4H_2O$$

However, the standard convention is to use the smallest positive integers that will balance the equation.

Equation (10) is sufficiently simple that it could have been balanced by trial and error, but for more complicated chemical equations we will need a systematic method. There are various methods that can be used, but we will give one that uses systems of linear equations. To illustrate the method let us reexamine Equation (10). To balance this equation we must find positive integers, $x_1, x_2, x_3,$ and x_4 such that

$$x_1\,(CH_4) + x_2\,(O_2) \longrightarrow x_3\,(CO_2) + x_4\,(H_2O) \tag{12}$$

However, for each of the atoms in the equation, the number of atoms on the left must be equal to the number of atoms on the right. Expressing this in tabular form we have

Linear Algebra in History

The German physicist Gustav Kirchhoff was a student of Gauss. His work on Kirchhoff's laws, announced in 1854, was a major advance in the calculation of currents, voltages, and resistances of electrical circuits. Kirchhoff was severely disabled and spent most of his life on crutches or in a wheelchair.

Gustav Kirchhoff
(1824–1887)

	Left Side		**Right Side**
Carbon	x_1	$=$	x_3
Hydrogen	$4x_1$	$=$	$2x_4$
Oxygen	$2x_2$	$=$	$2x_3 + x_4$

from which we obtain the homogeneous linear system

$$
\begin{aligned}
x_1 \quad\; - \;\, x_3 \qquad\quad &= 0 \\
4x_1 \qquad\qquad\; - 2x_4 &= 0 \\
2x_2 - 2x_3 - \;\, x_4 &= 0
\end{aligned}
$$

The augmented matrix for this system is

$$
\begin{bmatrix}
1 & 0 & -1 & 0 & 0 \\
4 & 0 & 0 & -2 & 0 \\
0 & 2 & -2 & -1 & 0
\end{bmatrix}
$$

We leave it for you to show that the reduced row echelon form of this matrix is

$$
\begin{bmatrix}
1 & 0 & 0 & -\frac{1}{2} & 0 \\
0 & 1 & 0 & -1 & 0 \\
0 & 0 & 1 & -\frac{1}{2} & 0
\end{bmatrix}
$$

from which we conclude that the general solution of the system is

$$
x_1 = t/2, \quad x_2 = t, \quad x_3 = t/2, \quad x_4 = t
$$

where t is arbitrary. The smallest positive integer values for the unknowns occur when we let $t = 2$, so the equation can be balanced by letting $x_1 = 1, x_2 = 2, x_3 = 1, x_4 = 2$. This agrees with our earlier conclusions, since substituting these values into (12) yields (11).

EXAMPLE 6
Balancing
Chemical
Equations
Using Linear
Systems

Balance the chemical equation

$$
\text{HCl} \quad + \quad \text{Na}_3\text{PO}_4 \quad \longrightarrow \quad \text{H}_3\text{PO}_4 \quad + \quad \text{NaCl}
$$
$$
[\text{hydrochloric acid}] + [\text{sodium phosphate}] \longrightarrow [\text{phosphoric acid}] + [\text{sodium chloride}]
$$

Solution Let $x_1, x_2, x_3,$ and x_4 be positive integers that balance the equation

$$
x_1\,(\text{HCl}) + x_2\,(\text{Na}_3\text{PO}_4) \longrightarrow x_3\,(\text{H}_3\text{PO}_4) + x_4\,(\text{NaCl}) \tag{13}
$$

Equating the number of atoms of each type on the two sides yields

$$
\begin{aligned}
1x_1 &= 3x_3 && \text{Hydrogen (H)} \\
1x_1 &= 1x_4 && \text{Chlorine (Cl)} \\
3x_2 &= 1x_4 && \text{Sodium (Na)} \\
1x_2 &= 1x_3 && \text{Phosphorous (P)} \\
4x_2 &= 4x_3 && \text{Oxygen (O)}
\end{aligned}
$$

from which we obtain the homogeneous linear system

$$
\begin{aligned}
x_1 \qquad\; - 3x_3 \qquad\quad &= 0 \\
x_1 \qquad\qquad\quad - x_4 &= 0 \\
3x_2 \qquad\; - x_4 &= 0 \\
x_2 - \;\, x_3 \qquad\quad &= 0 \\
4x_2 - 4x_3 \qquad\quad &= 0
\end{aligned}
$$

We leave it for you to show that the reduced row echelon form of the augmented matrix for this system is

$$\begin{bmatrix} 1 & 0 & 0 & -1 & 0 \\ 0 & 1 & 0 & -\frac{1}{3} & 0 \\ 0 & 0 & 1 & -\frac{1}{3} & 0 \\ 0 & 0 & 0 & 0 & 0 \\ 0 & 0 & 0 & 0 & 0 \end{bmatrix}$$

from which we conclude that the general solution of the system is

$$x_1 = t, \quad x_2 = t/3, \quad x_3 = t/3, \quad x_4 = t$$

where t is arbitrary. To obtain the smallest positive integers that balance the equation, we let $t = 3$, in which case we obtain $x_1 = 3$, $x_2 = 1$, $x_3 = 1$, and $x_4 = 3$. Substituting these values in (13) produces the balanced equation

$$3HCl + Na_3PO_4 \longrightarrow H_3PO_4 + 3NaCl \qquad \blacksquare$$

POLYNOMIAL INTERPOLATION

An important problem in various applications is to find a polynomial whose graph passes through a specified set of points in the plane; this is called an *interpolating polynomial* for the points. The simplest example of such a problem is to find a linear polynomial

$$p(x) = ax + b \tag{14}$$

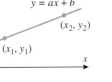

Figure 2.3.11

whose graph passes through two known distinct points, (x_1, y_1) and (x_2, y_2), in the xy-plane (Figure 2.3.11). You have probably encountered various methods in analytic geometry for finding the equation of a line through two points, but here we will give a method based on linear systems that can be adapted to general polynomial interpolation.

The graph of (14) is the line $y = ax + b$, and for this line to pass through the points (x_1, y_1) and (x_2, y_2), we must have

$$y_1 = ax_1 + b \quad \text{and} \quad y_2 = ax_2 + b$$

Therefore, the unknown coefficients a and b can be obtained by solving the linear system

$$ax_1 + b = y_1$$
$$ax_2 + b = y_2$$

We don't need any fancy methods to solve this system—the value of a can be obtained by subtracting the equations to eliminate b, and then the value of a can be substituted into either equation to find b. We leave it as an exercise for you to find a and b and then show that they can be expressed in the form

$$a = \frac{y_2 - y_1}{x_2 - x_1} \quad \text{and} \quad b = \frac{y_1 x_2 - y_2 x_1}{x_2 - x_1} \tag{15}$$

provided $x_1 \neq x_2$. Thus, for example, the line $y = ax + b$ that passes through the points

$$(2, 1) \quad \text{and} \quad (5, 4)$$

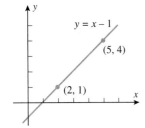

Figure 2.3.12

can be obtained by taking $(x_1, y_1) = (2, 1)$ and $(x_2, y_2) = (5, 4)$, in which case (15) yields

$$a = \frac{4 - 1}{5 - 2} = 1 \quad \text{and} \quad b = \frac{(1)(5) - (4)(2)}{5 - 2} = -1$$

Therefore, the equation of the line is

$$y = x - 1$$

(Figure 2.3.12).

Now let us consider the more general problem of finding a polynomial whose graph passes through n distinct points

$$(x_1, y_1), (x_2, y_2), (x_3, y_3), \ldots, (x_n, y_n) \tag{16}$$

Since there are n conditions to be satisfied, intuition suggests that we should begin by looking for a polynomial of the form

$$p(x) = a_0 + a_1 x + a_2 x^2 + \cdots + a_{n-1} x^{n-1} \tag{17}$$

since a polynomial of this form has n coefficients that are at our disposal to satisfy the n conditions. However, we want to allow for cases where the points may lie on a line or have some other configuration that would make it possible to use a polynomial whose degree is less than $n - 1$; thus, we allow for the possibility that a_{n-1} and other coefficients in (17) may be zero.

The following theorem, which we will prove in a later section, is the basic result on polynomial interpolation.

Theorem 2.3.1 (*Polynomial Interpolation*) *Given any n points in the xy-plane that have distinct x-coordinates, there is a unique polynomial of degree $n - 1$ or less whose graph passes through those points.*

Let us now consider how we might go about finding the interpolating polynomial (17) whose graph passes through the points in (16). Since the graph of this polynomial is the graph of the equation

$$y = a_0 + a_1 x + a_2 x^2 + \cdots + a_{n-1} x^{n-1}$$

it follows that the coordinates of the points must satisfy

$$
\begin{aligned}
a_0 + a_1 x_1 + a_2 x_1^2 + \cdots + a_{n-1} x_1^{n-1} &= y_1 \\
a_0 + a_1 x_2 + a_2 x_2^2 + \cdots + a_{n-1} x_2^{n-1} &= y_2 \\
\vdots \quad \vdots \quad \vdots \qquad\qquad \vdots \qquad\quad \vdots \\
a_0 + a_1 x_n + a_2 x_n^2 + \cdots + a_{n-1} x_n^{n-1} &= y_n
\end{aligned}
\tag{18}
$$

In these equations the values of x's and y's are assumed to be known, so we can view this as a linear system in the unknowns $a_0, a_1, \ldots, a_{n-1}$. From this point of view the augmented matrix for the system is

$$
\begin{bmatrix}
1 & x_1 & x_1^2 & \cdots & x_1^{n-1} & y_1 \\
1 & x_2 & x_2^2 & \cdots & x_2^{n-1} & y_2 \\
\vdots & \vdots & \vdots & & \vdots & \vdots \\
1 & x_n & x_n^2 & \cdots & x_n^{n-1} & y_n
\end{bmatrix}
\tag{19}
$$

and hence the interpolating polynomial can be found by reducing this matrix to reduced row echelon form (Gauss–Jordan elimination).

EXAMPLE 7
Polynomial
Interpolation by
Gauss–Jordan
Elimination

Find a cubic polynomial whose graph passes through the points

$$(1, 3), \quad (2, -2), \quad (3, -5), \quad (4, 0)$$

Solution Denote the interpolating polynomial by

$$p(x) = a_0 + a_1 x + a_2 x^2 + a_3 x^3$$

and denote the x- and y-coordinates of the given points by

$$x_1 = 1, \; x_2 = 2, \; x_3 = 3, \; x_4 = 4 \quad \text{and} \quad y_1 = 3, \; y_2 = -2, \; y_3 = -5, \; y_4 = 0$$

Thus, it follows from (19) that the augmented matrix for the linear system in the unknowns a_0, a_1, a_2, and a_3 is

$$\begin{bmatrix} 1 & x_1 & x_1^2 & x_1^3 & y_1 \\ 1 & x_2 & x_2^2 & x_2^3 & y_2 \\ 1 & x_3 & x_3^2 & x_3^3 & y_3 \\ 1 & x_4 & x_4^2 & x_4^3 & y_4 \end{bmatrix} = \begin{bmatrix} 1 & 1 & 1 & 1 & 3 \\ 1 & 2 & 4 & 8 & -2 \\ 1 & 3 & 9 & 27 & -5 \\ 1 & 4 & 16 & 64 & 0 \end{bmatrix}$$

We leave it for you to confirm that the reduced row echelon form of this matrix is

$$\begin{bmatrix} 1 & 0 & 0 & 0 & 4 \\ 0 & 1 & 0 & 0 & 3 \\ 0 & 0 & 1 & 0 & -5 \\ 0 & 0 & 0 & 1 & 1 \end{bmatrix}$$

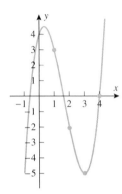

Figure 2.3.13

from which it follows that $a_0 = 4$, $a_1 = 3$, $a_2 = -5$, $a_3 = 1$. Thus, the interpolating polynomial is

$$p(x) = 4 + 3x - 5x^2 + x^3$$

The graph of this polynomial and the given points are shown in Figure 2.3.13. ∎

REMARK Later we will give a more efficient method for finding interpolating polynomials that is better suited for problems in which the number of data points is large.

EXAMPLE 8
Approximate
Integration
(*Calculus
Required*)

There is no way to evaluate the integral

$$\int_0^1 \sin\left(\frac{\pi x^2}{2}\right) dx$$

directly since there is no way to express an antiderivative of the integrand in terms of elementary functions. This integral could be approximated by Simpson's rule or some comparable method, but an alternative approach is to approximate the integrand by an interpolating polynomial and integrate the approximating polynomial. For example, let us consider the five points

$$x_0 = 0, \quad x_1 = 0.25, \quad x_2 = 0.5, \quad x_3 = 0.75, \quad x_4 = 1$$

that divide the interval [0, 1] into four equally spaced subintervals. The values of

$$f(x) = \sin\left(\frac{\pi x^2}{2}\right)$$

at these points are approximately

$$f(0) = 0, \quad f(0.25) = 0.098017, \quad f(0.5) = 0.382683, \quad f(0.75) = 0.77301, \quad f(1) = 1$$

In the Technology Exercises we will ask you to show that the interpolating polynomial is

$$p(x) = 0.098796x + 0.762356x^2 + 2.14429x^3 - 2.00544x^4 \tag{20}$$

and that

$$\int_0^1 p(x)\,dx \approx 0.438501 \tag{21}$$

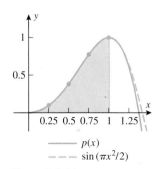

$$\text{———} \quad p(x)$$
$$\text{- - -} \quad \sin(\pi x^2/2)$$

Figure 2.3.14

As shown in Figure 2.3.14, the graphs of f and p match very closely over the interval [0, 1], so the approximation is quite good. ∎

Exercise Set 2.3

1. The accompanying figure shows a network in which the flow rate and direction of flow in certain branches are known. Find the flow rates and directions of flow in the remaining branches.

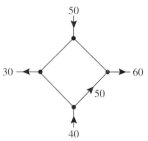

Figure Ex-1

2. The accompanying figure shows known flow rates of hydrocarbons into and out of a network of pipes at an oil refinery.
   (a) Set up a linear system whose solution provides the unknown flow rates.
   (b) Solve the system for the unknown flow rates.
   (c) Find the flow rates and directions of flow if $x_4 = 50$ and $x_6 = 0$.

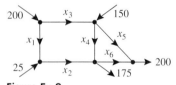

Figure Ex-2

3. The accompanying figure shows a network of one-way streets with traffic flowing in the directions indicated. The flow rates along the streets are measured as the average number of vehicles per hour.
   (a) Set up a linear system whose solution provides the unknown flow rates.
   (b) Solve the system for the unknown flow rates.
   (c) If the flow along the road from A to B must be reduced for construction, what is the minimum flow that is required to keep traffic flowing on all roads?

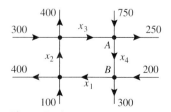

Figure Ex-3

4. The accompanying figure shows a network of one-way streets with traffic flowing in the directions indicated. The flow rates along the streets are measured as the average number of vehicles per hour.
   (a) Set up a linear system whose solution provides the unknown flow rates.
   (b) Solve the system for the unknown flow rates.
   (c) Is it possible to close the road from A to B for construction and keep traffic flowing on the other streets? Explain.

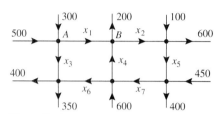

Figure Ex-4

In Exercises 5–8, analyze the given electrical circuits by finding the unknown currents.

5.

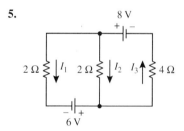

6.

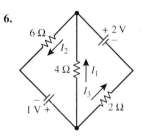

7.

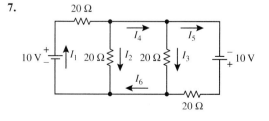

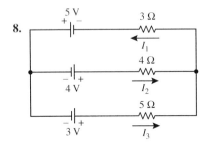

8.

In Exercises 9–12, write a balanced equation for the given chemical reaction.

9. $C_3H_8 + O_2 \rightarrow CO_2 + H_2O$ (propane combustion)

10. $C_6H_{12}O_6 \rightarrow CO_2 + C_2H_5OH$ (fermentation of sugar)

11. $CH_3COF + H_2O \rightarrow CH_3COOH + HF$

12. $CO_2 + H_2O \rightarrow C_6H_{12}O_6 + O_2$ (photosynthesis)

13. Find the quadratic polynomial whose graph passes through the points $(1, 1)$, $(2, 2)$, and $(3, 5)$.

14. Find the quadratic polynomial whose graph passes through the points $(0, 0)$, $(-1, 1)$, and $(1, 1)$.

15. Find the cubic polynomial whose graph passes through the points $(-1, -1)$, $(0, 1)$, $(1, 3)$, $(4, -1)$.

16. The accompanying figure shows the graph of a cubic polynomial. Find the polynomial.

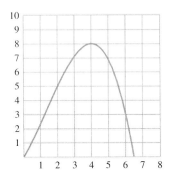

Figure Ex-16

Discussion and Discovery

D1. (a) Find an equation that represents the family of second-degree polynomials that pass through the points $(0, 1)$ and $(1, 2)$. [*Hint:* The equation will involve one arbitrary parameter that produces the members of the family when varied.]

 (b) By hand, or with the help of a graphing utility, sketch four curves in the family.

D2. In this section we have selected only a few applications of linear systems. Using the Internet as a search tool, try to find some more real-world applications of such systems. Select one that is of interest to you, and write a paragraph about it.

Technology Exercises

T1. Investigate your technology's commands for finding interpolating polynomials, and then confirm your understanding of these commands by checking the result obtained in Example 7.

T2. (a) Use your technology utility to find the polynomial of degree 5 that passes through the points $(1, 1)$, $(2, 3)$, $(3, 5)$, $(4, -2)$, $(5, 11)$, $(6, -12)$.

 (b) Follow the directions in part (a) for the points $(1, 1)$, $(2, 4)$, $(3, 9)$, $(4, 16)$, $(5, 25)$, $(6, 36)$. Give an explanation for what happens.

T3. Find integer values of the coefficients for which the equation
$$A(x^2 + y^2 + z^2) + Bx + Cy + Dz + E = 0$$
is satisfied by all of the following ordered triples (x, y, z):
$$\left(\tfrac{1}{2}, -\tfrac{1}{3}, 1\right), \quad \left(\tfrac{1}{3}, -\tfrac{1}{2}, \tfrac{2}{3}\right), \quad \left(2, -\tfrac{1}{3}, \tfrac{1}{2}\right), \quad (1, 0, 1)$$

T4. In an experiment for the design of an aircraft wing, the lifting force on the wing is measured at various forward velocities as follows:

Velocity (100 ft/s)	1	2	4	8	16	32
Lifting Force (100 lb)	0	3.12	15.86	33.7	81.5	123.0

Find an interpolating polynomial of degree 5 that models the data, and use your polynomial to estimate the lifting force at 2000 ft/s.

T5. (a) Devise a method for approximating $\sin x$ on the interval $0 \leq x \leq \pi/2$ by a cubic polynomial.

 (b) Compare the value of $\sin(0.5)$ to the approximation produced by your polynomial.

 (c) Generate the graphs of $\sin x$ and your polynomial over the interval $0 \leq x \leq \pi/2$ to see how they compare.

T6. Obtain Formula (20) in Example 8.

T7. Use the method of Example 8 to approximate the integral

$$\int_0^1 e^{x^2}\, dx$$

by subdividing the interval into five equal parts and using an interpolating polynomial to approximate the integrand. Compare your answer to that produced using the numerical integration capability of your calculating utility.

T8. Suppose that a ship with unknown coordinates (x, y, z) at an unknown time t, shortly after midnight, receives the following data from four satellites. We suppose that distances are measured in Earth radii and that time is listed in hundredths of a second after midnight.

Satellite	Satellite Position	Time
1	(0.94, 1.2, 2.3)	1.23
2	(0, 1.35, 2.41)	1.08
3	(2.3, 0.94, 1.2)	0.74
4	(2.41, 0, 1.35)	0.23

Using the technique of Example 1, find the coordinates of the ship, and verify that the ship is approximately 1 unit of distance from the Earth's center.

Matrices are used, for example, for solving linear systems, storing and manipulating tabular information, and as tools for transmitting digitized sound and visual images over the Internet.

Matrices and Matrix Algebra

Section 3.1 Operations on Matrices

In the last chapter we used matrices to simplify the notation for systems of linear equations. In this section we will consider matrices as mathematical objects in their own right and define some basic algebraic operations on them.

MATRIX NOTATION AND TERMINOLOGY

Recall that a *matrix* is a rectangular array of numbers, called *entries*, that a matrix with m rows and n columns is said to have *size* $m \times n$ (read, "m by n"), and that the number of rows is always written first. Here are some examples:

$$\begin{bmatrix} 1 & 2 \\ 3 & 0 \\ -1 & 4 \end{bmatrix}, \quad \begin{bmatrix} 1 & 2 & -1 \\ 3 & 0 & 4 \end{bmatrix}, \quad [2 \quad 1 \quad 0 \quad 3], \quad \begin{bmatrix} \pi & -\sqrt{2} & \frac{1}{2} \\ 0.5 & 1 & 7 \\ 0 & 0 & 0 \end{bmatrix}, \quad \begin{bmatrix} 1 \\ 3 \end{bmatrix}, \quad [4]$$

$$\boxed{3 \times 2} \qquad \boxed{2 \times 3} \qquad \boxed{1 \times 4} \qquad \boxed{3 \times 3} \qquad \boxed{2 \times 1} \quad \boxed{1 \times 1}$$

We will usually use capital letters to denote matrices and lowercase letters to denote entries. Thus, a general $m \times n$ matrix might be denoted as

$$A = \begin{bmatrix} a_{11} & a_{12} & \cdots & a_{1n} \\ a_{21} & a_{22} & \cdots & a_{2n} \\ \vdots & \vdots & & \vdots \\ a_{m1} & a_{m2} & \cdots & a_{mn} \end{bmatrix} \tag{1}$$

A matrix with n rows and n columns is called a ***square matrix of order n***, and the entries $a_{11}, a_{22}, \ldots, a_{nn}$ are said to form the main diagonal of the matrix (Figure 3.1.1).

When a more compact notation is needed, (1) can be written as

$$A = [a_{ij}]_{m \times n} \quad \text{or as} \quad A = [a_{ij}]$$

where the first notation would be used when the size of the matrix is important to the discussion, the second when it is not. Usually, we will match the letter denoting a matrix with the letter used for its entries. Thus, for example,

$$A = [a_{ij}], \quad B = [b_{ij}], \quad C = [c_{ij}]$$

Also, we will use the symbol $(A)_{ij}$ to stand for the entry in row i and column j of a matrix A. For example, if

$$A = \begin{bmatrix} 2 & -3 \\ 7 & 0 \end{bmatrix}$$

then $(A)_{11} = 2$, $(A)_{12} = -3$, $(A)_{21} = 7$, and $(A)_{22} = 0$.

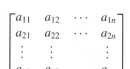

$$\begin{bmatrix} a_{11} & a_{12} & \cdots & a_{1n} \\ a_{21} & a_{22} & \cdots & a_{2n} \\ \vdots & \vdots & & \vdots \\ a_{n1} & a_{n2} & \cdots & a_{nn} \end{bmatrix}$$

Figure 3.1.1

REMARK It is common practice to omit brackets on 1×1 matrices, so we would usually write 4 rather than [4]. Although this makes it impossible to distinguish between the number "4" and the 1×1 matrix whose entry is "4," the appropriate interpretation will usually be clear from the context in which the symbol occurs.

OPERATIONS ON MATRICES

For many applications it is desirable to have an "arithmetic" of matrices in which matrices can be added, subtracted, and multiplied in a useful way. The remainder of this section will be devoted to developing this arithmetic.

Definition 3.1.1 Two matrices are defined to be *equal* if they have the same size and their corresponding entries are equal.

In matrix notation, if $A = [a_{ij}]$ and $B = [b_{ij}]$ have the same size, then $A = B$ if and only if $(A)_{ij} = (B)_{ij}$ (or equivalently, $a_{ij} = b_{ij}$) for all values of i and j.

EXAMPLE 1
Equality of Matrices

Consider the matrices

$$A = \begin{bmatrix} 2 & 1 \\ 3 & x+1 \end{bmatrix}, \quad B = \begin{bmatrix} 2 & 1 \\ 3 & 5 \end{bmatrix}, \quad C = \begin{bmatrix} 2 & 1 & 0 \\ 3 & 4 & 0 \end{bmatrix}$$

The matrices A and B are equal if and only if $x = 4$. There is no value of x for which $A = C$, since A and C have different sizes. ∎

Definition 3.1.2 If A and B are matrices with the same size, then we define the *sum $A + B$* to be the matrix obtained by adding the entries of B to the corresponding entries of A, and we define the *difference $A - B$* to be the matrix obtained by subtracting the entries of B from the corresponding entries of A.

If $A = [a_{ij}]$ and $B = [b_{ij}]$ have the same size, then this definition states that

$$(A + B)_{ij} = (A)_{ij} + (B)_{ij} = a_{ij} + b_{ij} \tag{2}$$

$$(A - B)_{ij} = (A)_{ij} - (B)_{ij} = a_{ij} - b_{ij} \tag{3}$$

EXAMPLE 2
Adding and Subtracting Matrices

Consider the matrices

$$A = \begin{bmatrix} 2 & 1 & 0 & 3 \\ -1 & 0 & 2 & 4 \\ 4 & -2 & 7 & 0 \end{bmatrix}, \quad B = \begin{bmatrix} -4 & 3 & 5 & 1 \\ 2 & 2 & 0 & -1 \\ 3 & 2 & -4 & 5 \end{bmatrix}, \quad C = \begin{bmatrix} 1 & 1 \\ 2 & 2 \end{bmatrix}$$

Then

$$A + B = \begin{bmatrix} -2 & 4 & 5 & 4 \\ 1 & 2 & 2 & 3 \\ 7 & 0 & 3 & 5 \end{bmatrix} \quad \text{and} \quad A - B = \begin{bmatrix} 6 & -2 & -5 & 2 \\ -3 & -2 & 2 & 5 \\ 1 & -4 & 11 & -5 \end{bmatrix}$$

The expressions $A + C$, $B + C$, $A - C$, and $B - C$ are undefined because the sizes of the matrices A, B, and C are not compatible for performing the operations. ∎

Definition 3.1.3 If A is any matrix and c is any scalar, then the *product cA* is defined to be the matrix obtained by multiplying each entry of A by c.

In matrix notation, if $A = [a_{ij}]$, then

$$(cA)_{ij} = c(A)_{ij} = ca_{ij} \tag{4}$$

EXAMPLE 3
Multiplying
Matrices by
Scalars

Consider the matrices

$$A = \begin{bmatrix} 2 & 3 & 4 \\ 1 & 3 & 1 \end{bmatrix}, \quad B = \begin{bmatrix} 0 & 2 & 7 \\ -1 & 3 & -5 \end{bmatrix}, \quad C = \begin{bmatrix} 9 & -6 & 3 \\ 3 & 0 & 12 \end{bmatrix}$$

Then

$$2A = \begin{bmatrix} 4 & 6 & 8 \\ 2 & 6 & 2 \end{bmatrix}, \quad (-1)B = \begin{bmatrix} 0 & -2 & -7 \\ 1 & -3 & 5 \end{bmatrix}, \quad \tfrac{1}{3}C = \begin{bmatrix} 3 & -2 & 1 \\ 1 & 0 & 4 \end{bmatrix}$$ ∎

REMARK As might be expected, we define the ***negative*** of a matrix A as $-A = (-1)A$.

ROW AND COLUMN VECTORS

Recall that a matrix with one row is called a *row vector*, a matrix with one column is called a *column vector*, and that row vectors and column vectors are denoted by lowercase boldface letters. Thus, a general $1 \times n$ row vector $\mathbf{r}$ and a general $m \times 1$ column vector $\mathbf{c}$ have the forms

$$\mathbf{r} = [r_1 \quad r_2 \quad \cdots \quad r_n] \quad \text{and} \quad \mathbf{c} = \begin{bmatrix} c_1 \\ c_2 \\ \vdots \\ c_m \end{bmatrix}$$

Sometimes it is desirable to think of a matrix as a list of row vectors or column vectors. For example, if

$$A = \begin{bmatrix} a_{11} & a_{12} & a_{13} & a_{14} \\ a_{21} & a_{22} & a_{23} & a_{24} \\ a_{31} & a_{32} & a_{33} & a_{34} \end{bmatrix} \tag{5}$$

then A can be subdivided into column vectors as

$$A = \begin{bmatrix} a_{11} & a_{12} & a_{13} & a_{14} \\ a_{21} & a_{22} & a_{23} & a_{24} \\ a_{31} & a_{32} & a_{33} & a_{34} \end{bmatrix} = [\mathbf{c}_1 \quad \mathbf{c}_2 \quad \mathbf{c}_3 \quad \mathbf{c}_4] \tag{6}$$

or into row vectors as

$$A = \begin{bmatrix} a_{11} & a_{12} & a_{13} & a_{14} \\ a_{21} & a_{22} & a_{23} & a_{24} \\ a_{31} & a_{32} & a_{33} & a_{34} \end{bmatrix} = \begin{bmatrix} \mathbf{r}_1 \\ \mathbf{r}_2 \\ \mathbf{r}_3 \end{bmatrix} \tag{7}$$

The dashed lines in (6) and (7) serve to emphasize that the rows and columns are being regarded as single entities. The lines are said to ***partition*** A into column vectors in (6) and into row vectors in (7). When convenient, we will use the symbol $\mathbf{r}_i(A)$ to denote the ith row vector of a matrix A and $\mathbf{c}_j(A)$ to denote the jth column vector. Thus, for the matrix A in (5) we have

$$\mathbf{r}_1(A) = [a_{11} \quad a_{12} \quad a_{13} \quad a_{14}]$$
$$\mathbf{r}_2(A) = [a_{21} \quad a_{22} \quad a_{23} \quad a_{24}]$$
$$\mathbf{r}_3(A) = [a_{31} \quad a_{32} \quad a_{33} \quad a_{34}]$$

and

$$\mathbf{c}_1(A) = \begin{bmatrix} a_{11} \\ a_{21} \\ a_{31} \end{bmatrix}, \quad \mathbf{c}_2(A) = \begin{bmatrix} a_{12} \\ a_{22} \\ a_{32} \end{bmatrix}, \quad \mathbf{c}_3(A) = \begin{bmatrix} a_{13} \\ a_{23} \\ a_{33} \end{bmatrix}, \quad \mathbf{c}_4(A) = \begin{bmatrix} a_{14} \\ a_{24} \\ a_{34} \end{bmatrix}$$

Linear Algebra in History

The term *matrix* was first used by the English mathematician (and lawyer) James Sylvester, who defined the term in 1850 to be an "oblong arrangement of terms." Sylvester communicated his work on matrices to a fellow English mathematician and lawyer named Arthur Cayley, who then introduced some of the basic operations on matrices in a book entitled *Memoir on the Theory of Matrices* that was published in 1858. As a matter of interest, Sylvester, who was Jewish, did not get his college degree because he refused to sign a required oath to the Church of England. He was appointed to a chair at the University of Virginia in the United States but resigned after swatting a student with a stick because he was reading a newspaper in class. Sylvester, thinking he had killed the student, fled back to England on the first available ship. Fortunately, the student was not dead, just in shock!

James Sylvester
(1814–1897)

Arthur Cayley
(1821–1895)

THE PRODUCT Ax

Having already discussed how to multiply a matrix by a scalar, we will now consider how to multiply two matrices. Since matrices with the same size are added by adding corresponding entries, it seems reasonable that two matrices should be multiplied in a similar way, that is, by requiring that they have the same size and multiplying corresponding entries. Although this is a perfectly acceptable definition, it turns out that it is not very useful. A definition that is better suited for applications can be motivated by thinking of matrix multiplication in the context of linear systems. Specifically, consider the linear system

$$
\begin{aligned}
a_{11}x_1 + a_{12}x_2 + \cdots + a_{1n}x_n &= b_1 \\
a_{21}x_1 + a_{22}x_2 + \cdots + a_{2n}x_n &= b_2 \\
\vdots \qquad\qquad \vdots \qquad\qquad \vdots \qquad &\ \ \vdots \\
a_{m1}x_1 + a_{m2}x_2 + \cdots + a_{mn}x_n &= b_m
\end{aligned}
\tag{8}
$$

and consider the following three matrices that are associated with this system:

$$
A = \begin{bmatrix} a_{11} & a_{12} & \cdots & a_{1n} \\ a_{21} & a_{22} & \cdots & a_{2n} \\ \vdots & \vdots & & \vdots \\ a_{m1} & a_{m2} & \cdots & a_{mn} \end{bmatrix}, \quad
\mathbf{x} = \begin{bmatrix} x_1 \\ x_2 \\ \vdots \\ x_n \end{bmatrix}, \quad
\mathbf{b} = \begin{bmatrix} b_1 \\ b_2 \\ \vdots \\ b_m \end{bmatrix}
$$

The matrix A is called the ***coefficient matrix*** of the system. Our objective is to define the product $A\mathbf{x}$ in such a way that the m linear equations in (8) can be rewritten as the single matrix equation

$$
A\mathbf{x} = \mathbf{b}
\tag{9}
$$

thereby creating a matrix analog of the algebraic linear equation $ax = b$.

As a first step, observe that the m individual equations in (8) can be replaced by the single matrix equation

$$
\begin{bmatrix} a_{11}x_1 + a_{12}x_2 + \cdots + a_{1n}x_n \\ a_{21}x_1 + a_{22}x_2 + \cdots + a_{2n}x_n \\ \vdots \qquad\qquad \vdots \qquad\qquad \vdots \\ a_{m1}x_1 + a_{m2}x_2 + \cdots + a_{mn}x_n \end{bmatrix} = \begin{bmatrix} b_1 \\ b_2 \\ \vdots \\ b_m \end{bmatrix}
$$

which is justified because two matrices are equal if and only if their corresponding entries are equal. This matrix equation can now be rewritten as

$$
x_1 \begin{bmatrix} a_{11} \\ a_{21} \\ \vdots \\ a_{m1} \end{bmatrix} + x_2 \begin{bmatrix} a_{12} \\ a_{22} \\ \vdots \\ a_{m2} \end{bmatrix} + \cdots + x_n \begin{bmatrix} a_{1n} \\ a_{2n} \\ \vdots \\ a_{mn} \end{bmatrix} = \mathbf{b}
$$

(verify). From (9), we want the left side of this equation to be the product $A\mathbf{x}$, so it is now clear that we must define $A\mathbf{x}$ as

$$
A\mathbf{x} = \begin{bmatrix} a_{11} & a_{12} & \cdots & a_{1n} \\ a_{21} & a_{22} & \cdots & a_{2n} \\ \vdots & \vdots & & \vdots \\ a_{m1} & a_{m2} & \cdots & a_{mn} \end{bmatrix} \begin{bmatrix} x_1 \\ x_2 \\ \vdots \\ x_n \end{bmatrix} = x_1 \begin{bmatrix} a_{11} \\ a_{21} \\ \vdots \\ a_{m1} \end{bmatrix} + x_2 \begin{bmatrix} a_{12} \\ a_{22} \\ \vdots \\ a_{m2} \end{bmatrix} + \cdots + x_n \begin{bmatrix} a_{1n} \\ a_{2n} \\ \vdots \\ a_{mn} \end{bmatrix}
$$

The expression on the right side of this equation is a linear combination of the column vectors of A with the entries of $\mathbf{x}$ as coefficients, so we are led to the following definition.

> **Definition 3.1.4** If A is an $m \times n$ matrix and $\mathbf{x}$ is an $n \times 1$ column vector, then the ***product*** $A\mathbf{x}$ is defined to be the $m \times 1$ column vector that results by forming the linear combination of the column vectors of A that has the entries of $\mathbf{x}$ as coefficients. More precisely, if the column vectors of A are $\mathbf{a}_1, \mathbf{a}_2, \ldots, \mathbf{a}_n$, then
>
> $$
> A\mathbf{x} = [\mathbf{a}_1 \ \ \mathbf{a}_2 \ \ \cdots \ \ \mathbf{a}_n] \begin{bmatrix} x_1 \\ x_2 \\ \vdots \\ x_n \end{bmatrix} = x_1\mathbf{a}_1 + x_2\mathbf{a}_2 + \cdots + x_n\mathbf{a}_n
> \tag{10}
> $$

For example,

$$\begin{bmatrix} 1 & -3 & 2 \\ 4 & 0 & -5 \end{bmatrix} \begin{bmatrix} 3 \\ 1 \\ 2 \end{bmatrix} = 3 \begin{bmatrix} 1 \\ 4 \end{bmatrix} + 1 \begin{bmatrix} -3 \\ 0 \end{bmatrix} + 2 \begin{bmatrix} 2 \\ -5 \end{bmatrix} = \begin{bmatrix} 4 \\ 2 \end{bmatrix}$$

$$\begin{bmatrix} 3 & 0 \\ -1 & 2 \\ 4 & 1 \end{bmatrix} \begin{bmatrix} 2 \\ -5 \end{bmatrix} = 2 \begin{bmatrix} 3 \\ -1 \\ 4 \end{bmatrix} + (-5) \begin{bmatrix} 0 \\ 2 \\ 1 \end{bmatrix} = \begin{bmatrix} 6 \\ -12 \\ 3 \end{bmatrix}$$

REMARK Observe that Formula (10) requires the number of entries in $\mathbf{x}$ to be the same as the number of columns of A. If this condition is not satisfied, then the product $A\mathbf{x}$ is not defined.

EXAMPLE 4
Writing a
Linear System
as $A\mathbf{x} = \mathbf{b}$

The linear system

$$\begin{aligned} x_1 + 2x_2 + 3x_3 &= 5 \\ 2x_1 + 5x_2 + 3x_3 &= 3 \\ x_1 \qquad\quad + 8x_3 &= 17 \end{aligned}$$

can be written in matrix form as $A\mathbf{x} = \mathbf{b}$, where

$$A = \begin{bmatrix} 1 & 2 & 3 \\ 2 & 5 & 3 \\ 1 & 0 & 8 \end{bmatrix}, \quad \mathbf{x} = \begin{bmatrix} x_1 \\ x_2 \\ x_3 \end{bmatrix}, \quad \mathbf{b} = \begin{bmatrix} 5 \\ 3 \\ 17 \end{bmatrix}$$

The following algebraic properties of matrix multiplication will play a fundamental role in our subsequent work.

> **Theorem 3.1.5** *If A is an $m \times n$ matrix, then the following relationships hold for all column vectors $\mathbf{u}$ and $\mathbf{v}$ in R^n and for every scalar c:*
>
> *(a)* $A(c\mathbf{u}) = c(A\mathbf{u})$
> *(b)* $A(\mathbf{u} + \mathbf{v}) = A\mathbf{u} + A\mathbf{v}$

Proof Suppose that A is partitioned into column vectors as

$$A = [\mathbf{a}_1 \quad \mathbf{a}_2 \quad \cdots \quad \mathbf{a}_n]$$

and that the vectors $\mathbf{u}$ and $\mathbf{v}$ are given in component form as

$$\mathbf{u} = \begin{bmatrix} u_1 \\ u_2 \\ \vdots \\ u_n \end{bmatrix} \quad \text{and} \quad \mathbf{v} = \begin{bmatrix} v_1 \\ v_2 \\ \vdots \\ v_n \end{bmatrix}$$

Then

$$c\mathbf{u} = \begin{bmatrix} cu_1 \\ cu_2 \\ \vdots \\ cu_n \end{bmatrix} \quad \text{and} \quad \mathbf{u} + \mathbf{v} = \begin{bmatrix} u_1 + v_1 \\ u_2 + v_2 \\ \vdots \\ u_n + v_n \end{bmatrix}$$

so it follows from Definition 3.1.4 that

$$\begin{aligned} A(c\mathbf{u}) = [\mathbf{a}_1 \quad \mathbf{a}_2 \quad \cdots \quad \mathbf{a}_n] \begin{bmatrix} cu_1 \\ cu_2 \\ \vdots \\ cu_n \end{bmatrix} &= (cu_1)\mathbf{a}_1 + (cu_2)\mathbf{a}_2 + \cdots + (cu_n)\mathbf{a}_n \\ &= c(u_1\mathbf{a}_1) + c(u_2\mathbf{a}_2) + \cdots + c(u_n\mathbf{a}_n) \\ &= c(A\mathbf{u}) \end{aligned}$$

which proves part (*a*). Also,

$$A(\mathbf{u} + \mathbf{v}) = [\mathbf{a}_1 \quad \mathbf{a}_2 \quad \cdots \quad \mathbf{a}_n] \begin{bmatrix} u_1 + v_1 \\ u_2 + v_2 \\ \vdots \\ u_n + v_n \end{bmatrix}$$

$$= (u_1 + v_1)\mathbf{a}_1 + (u_2 + v_2)\mathbf{a}_2 + \cdots + (u_n + v_n)\mathbf{a}_n$$
$$= (u_1\mathbf{a}_1 + u_2\mathbf{a}_2 + \cdots + u_n\mathbf{a}_n) + (v_1\mathbf{a}_1 + v_2\mathbf{a}_2 + \cdots + v_n\mathbf{a}_n)$$
$$= A\mathbf{u} + A\mathbf{v}$$

which proves part (*b*). ■

REMARK Part (*b*) of this theorem can be extended to sums with more than two terms, and the two parts of the theorem can then be used in combination to show that

$$A(c_1\mathbf{u}_1 + c_2\mathbf{u}_2 + \cdots + c_n\mathbf{u}_n) = c_1(A\mathbf{u}_1) + c_2(A\mathbf{u}_2) + \cdots + c_n(A\mathbf{u}_n) \tag{11}$$

THE PRODUCT *AB* Our next goal is to build on the definition of the product $A\mathbf{x}$ to define a more general matrix product AB in which B can have more than one column. To ensure that the product AB behaves in a reasonable way algebraically, we will require that it satisfy the *associativity condition*

$$A(B\mathbf{x}) = (AB)\mathbf{x} \tag{12}$$

for every column vector $\mathbf{x}$ in R^n (Figure 3.1.2).

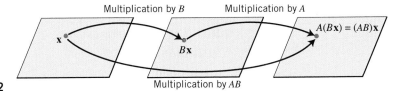

Multiplication by *B*   Multiplication by *A*

$A(B\mathbf{x}) = (AB)\mathbf{x}$

$\mathbf{x}$

$B\mathbf{x}$

Multiplication by *AB*

Figure 3.1.2

The condition in (12) imposes a size restriction on A and B. Specifically, since $\mathbf{x}$ is a column vector in R^n it has n rows, and hence the matrix B must have n columns for $B\mathbf{x}$ to be defined. This means that the size of B must be of the form $s \times n$, which implies that the size of $B\mathbf{x}$ must be $s \times 1$. This being the case, the matrix A must have s columns for $A(B\mathbf{x})$ to be defined, so we have shown that *the number of columns of A must be the same as the number of rows of B*. Assuming this to be so, and assuming that the column vectors of B are $\mathbf{b}_1, \mathbf{b}_2, \ldots, \mathbf{b}_n$, we obtain

$$B\mathbf{x} = [\mathbf{b}_1 \quad \mathbf{b}_2 \quad \cdots \quad \mathbf{b}_n] \begin{bmatrix} x_1 \\ x_2 \\ \vdots \\ x_n \end{bmatrix} = x_1\mathbf{b}_1 + x_2\mathbf{b}_2 + \cdots + x_n\mathbf{b}_n \tag{13}$$

and hence

$$A(B\mathbf{x}) = A(x_1\mathbf{b}_1 + x_2\mathbf{b}_2 + \cdots + x_n\mathbf{b}_n) = x_1(A\mathbf{b}_1) + x_2(A\mathbf{b}_2) + \cdots + x_n(A\mathbf{b}_n)$$

Consequently, for (12) to hold, it would have to be true that

$$(AB)\mathbf{x} = A(B\mathbf{x}) = x_1(A\mathbf{b}_1) + x_2(A\mathbf{b}_2) + \cdots + x_n(A\mathbf{b}_n)$$

and by Definition 3.1.4 we can achieve this by taking the column vectors of AB to be

$$A\mathbf{b}_1, A\mathbf{b}_2, \ldots, A\mathbf{b}_n$$

Thus we are led to the following definition.

Definition 3.1.6 If A is an $m \times s$ matrix and B is an $s \times n$ matrix, and if the column vectors of B are $\mathbf{b}_1, \mathbf{b}_2, \ldots, \mathbf{b}_n$, then the **product** AB is the $m \times n$ matrix defined as

$$AB = [A\mathbf{b}_1 \quad A\mathbf{b}_2 \quad \cdots \quad A\mathbf{b}_n] \tag{14}$$

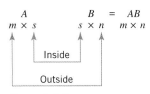

Figure 3.1.3

It is important to keep in mind that this definition requires the number of columns of the first factor A to be the same as the number of rows of the second factor B. When this condition is satisfied, the sizes of A and B are said to **conform** for the product AB. If the sizes of A and B do not conform for the product AB, then this product is **undefined**. A convenient way to determine whether A and B conform for the product AB and, if so, to find the size of the product is to write the sizes of the factors side by side as in Figure 3.1.3 (the size of the first factor on the left and the size of the second factor on the right). If the inside numbers are the same, then the product AB is defined, and the outside numbers then give the size of the product.

EXAMPLE 5
Computing a Matrix Product AB

Find the product AB for

$$A = \begin{bmatrix} 1 & 2 & 4 \\ 2 & 6 & 0 \end{bmatrix} \quad \text{and} \quad B = \begin{bmatrix} 4 & 1 & 4 & 3 \\ 0 & -1 & 3 & 1 \\ 2 & 7 & 5 & 2 \end{bmatrix}$$

Solution It follows from (14) that the product AB is formed in a column-by-column manner by multiplying the successive columns of B by A. The computations are

$$\begin{bmatrix} 1 & 2 & 4 \\ 2 & 6 & 0 \end{bmatrix} \begin{bmatrix} 4 \\ 0 \\ 2 \end{bmatrix} = (4)\begin{bmatrix} 1 \\ 2 \end{bmatrix} + (0)\begin{bmatrix} 2 \\ 6 \end{bmatrix} + (2)\begin{bmatrix} 4 \\ 0 \end{bmatrix} = \begin{bmatrix} 12 \\ 8 \end{bmatrix}$$

$$\begin{bmatrix} 1 & 2 & 4 \\ 2 & 6 & 0 \end{bmatrix} \begin{bmatrix} 1 \\ -1 \\ 7 \end{bmatrix} = (1)\begin{bmatrix} 1 \\ 2 \end{bmatrix} + (-1)\begin{bmatrix} 2 \\ 6 \end{bmatrix} + (7)\begin{bmatrix} 4 \\ 0 \end{bmatrix} = \begin{bmatrix} 27 \\ -4 \end{bmatrix}$$

$$\begin{bmatrix} 1 & 2 & 4 \\ 2 & 6 & 0 \end{bmatrix} \begin{bmatrix} 4 \\ 3 \\ 5 \end{bmatrix} = (4)\begin{bmatrix} 1 \\ 2 \end{bmatrix} + (3)\begin{bmatrix} 2 \\ 6 \end{bmatrix} + (5)\begin{bmatrix} 4 \\ 0 \end{bmatrix} = \begin{bmatrix} 30 \\ 26 \end{bmatrix}$$

$$\begin{bmatrix} 1 & 2 & 4 \\ 2 & 6 & 0 \end{bmatrix} \begin{bmatrix} 3 \\ 1 \\ 2 \end{bmatrix} = (3)\begin{bmatrix} 1 \\ 2 \end{bmatrix} + (1)\begin{bmatrix} 2 \\ 6 \end{bmatrix} + (2)\begin{bmatrix} 4 \\ 0 \end{bmatrix} = \begin{bmatrix} 13 \\ 12 \end{bmatrix}$$

Thus,

$$AB = \begin{bmatrix} 1 & 2 & 4 \\ 2 & 6 & 0 \end{bmatrix} \begin{bmatrix} 4 & 1 & 4 & 3 \\ 0 & -1 & 3 & 1 \\ 2 & 7 & 5 & 2 \end{bmatrix} = \begin{bmatrix} 12 & 27 & 30 & 13 \\ 8 & -4 & 26 & 12 \end{bmatrix} \qquad \blacksquare$$

EXAMPLE 6
An Undefined Product

Find the product BA for the matrices in Example 5.

Solution The first factor has size 3×4, and the second has size 2×3. The "inside" numbers are not the same, so the product is undefined. $\blacksquare$

FINDING SPECIFIC ENTRIES IN A MATRIX PRODUCT

Sometimes we will be interested in finding a specific entry in a matrix product without going through the work of computing the entire column that contains the entry. To see how this can be done, suppose that we want to find the entry $(AB)_{ij}$ in the ith row and jth column of a product AB, where $A = [a_{ij}]$ is an $m \times s$ matrix and $B = [b_{ij}]$ is an $s \times n$ matrix with column vectors

$\mathbf{b}_1, \mathbf{b}_2, \ldots, \mathbf{b}_n$. It follows from Definition 3.1.6 that the jth column vector of AB is

$$
A\mathbf{b}_j = \begin{bmatrix} a_{11} & a_{12} & \cdots & a_{1s} \\ a_{21} & a_{22} & \cdots & a_{2s} \\ \vdots & \vdots & & \vdots \\ a_{i1} & a_{i2} & \cdots & a_{is} \\ \vdots & \vdots & & \vdots \\ a_{m1} & a_{m2} & \cdots & a_{ms} \end{bmatrix} \begin{bmatrix} b_{1j} \\ b_{2j} \\ \vdots \\ b_{sj} \end{bmatrix} = b_{1j} \begin{bmatrix} a_{11} \\ a_{21} \\ \vdots \\ a_{i1} \\ \vdots \\ a_{m1} \end{bmatrix} + b_{2j} \begin{bmatrix} a_{12} \\ a_{22} \\ \vdots \\ a_{i2} \\ \vdots \\ a_{m2} \end{bmatrix} + \cdots + b_{sj} \begin{bmatrix} a_{1s} \\ a_{2s} \\ \vdots \\ a_{is} \\ \vdots \\ a_{ms} \end{bmatrix}
$$

(15)

Since the entry $(AB)_{ij}$ in the ith row and jth column of AB is the entry in the ith row of $A\mathbf{b}_j$, it follows from (15) and Figure 3.1.4 that

$$(AB)_{ij} = a_{i1}b_{1j} + a_{i2}b_{2j} + a_{i3}b_{3j} + \cdots + a_{is}b_{sj}$$

(16)

which is the dot product of the ith row vector of A with the jth column vector of B. We can also write (16) in the alternative notation

$$(AB)_{ij} = \mathbf{r}_i(A)\mathbf{c}_j(B) = \mathbf{r}_i(A) \cdot \mathbf{c}_j(B)$$

(17)

Thus, we have established the following result, called the **row-column rule** or the **dot product rule** for matrix multiplication.

Theorem 3.1.7 (*The Row-Column Rule or Dot Product Rule*)   *The entry in row i and column j of a matrix product AB is the ith row vector of A times the jth column vector of B, or equivalently, the dot product of the ith row vector of A and the jth column vector of B.*

$$
AB = \begin{bmatrix} a_{11} & a_{12} & \cdots & a_{1s} \\ a_{21} & a_{22} & \cdots & a_{2s} \\ \vdots & \vdots & & \vdots \\ a_{i1} & a_{i2} & \cdots & a_{is} \\ \vdots & \vdots & & \vdots \\ a_{m1} & a_{m2} & \cdots & a_{ms} \end{bmatrix} \begin{bmatrix} b_{11} & b_{12} & \cdots & b_{1j} & \cdots & b_{1n} \\ b_{21} & b_{22} & \cdots & b_{2j} & \cdots & b_{2n} \\ \vdots & \vdots & & \vdots & & \vdots \\ b_{s1} & b_{s2} & \cdots & b_{sj} & \cdots & b_{sn} \end{bmatrix}
$$

Figure 3.1.4

EXAMPLE 7
Example 5
Revisited

Use the dot product rule to compute the individual entries in the product of Example 5.

Solution   Since A has size 2×3 and B has size 3×4, the product AB is a 2×4 matrix of the form

$$
AB = \begin{bmatrix} \mathbf{r}_1(A) \cdot \mathbf{c}_1(B) & \mathbf{r}_1(A) \cdot \mathbf{c}_2(B) & \mathbf{r}_1(A) \cdot \mathbf{c}_3(B) & \mathbf{r}_1(A) \cdot \mathbf{c}_4(B) \\ \mathbf{r}_2(A) \cdot \mathbf{c}_1(B) & \mathbf{r}_2(A) \cdot \mathbf{c}_2(B) & \mathbf{r}_2(A) \cdot \mathbf{c}_3(B) & \mathbf{r}_2(A) \cdot \mathbf{c}_4(B) \end{bmatrix}
$$

where $\mathbf{r}_1(A)$ and $\mathbf{r}_2(A)$ are the row vectors of A and $\mathbf{c}_1(B)$, $\mathbf{c}_2(B)$, $\mathbf{c}_3(B)$, and $\mathbf{c}_4(B)$ are the column vectors of B. For example, the entry in row 2 and column 3 of AB can be computed as

$$
\begin{bmatrix} 1 & 2 & 4 \\ 2 & 6 & 0 \end{bmatrix} \begin{bmatrix} 4 & 1 & 4 & 3 \\ 0 & -1 & 3 & 1 \\ 2 & 7 & 5 & 2 \end{bmatrix} = \begin{bmatrix} \square & \square & \square & \square \\ \square & \square & 26 & \square \end{bmatrix}
$$

$$(2 \cdot 4) + (6 \cdot 3) + (0 \cdot 5) = 26$$

and the entry in row 1 and column 4 of AB can be computed as

$$\begin{bmatrix} 1 & 2 & 4 \\ 2 & 6 & 0 \end{bmatrix} \begin{bmatrix} 4 & 1 & 4 & 3 \\ 0 & -1 & 3 & 1 \\ 2 & 7 & 5 & 2 \end{bmatrix} = \begin{bmatrix} \square & \square & \square & \boxed{13} \\ \square & \square & \square & \square \end{bmatrix}$$

$$(1 \cdot 3) + (2 \cdot 1) + (4 \cdot 2) = 13$$

Here is the complete set of computations:

$$(AB)_{11} = (1 \cdot 4) + (2 \cdot 0) + (4 \cdot 2) = 12$$
$$(AB)_{12} = (1 \cdot 1) + (2 \cdot -1) + (4 \cdot 7) = 27$$
$$(AB)_{13} = (1 \cdot 4) + (2 \cdot 3) + (4 \cdot 5) = 30$$
$$(AB)_{14} = (1 \cdot 3) + (2 \cdot 1) + (4 \cdot 2) = 13$$
$$(AB)_{21} = (2 \cdot 4) + (6 \cdot 0) + (0 \cdot 2) = 8$$
$$(AB)_{22} = (2 \cdot 1) + (6 \cdot -1) + (0 \cdot 7) = -4$$
$$(AB)_{23} = (2 \cdot 4) + (6 \cdot 3) + (0 \cdot 5) = 26$$
$$(AB)_{24} = (2 \cdot 3) + (6 \cdot 1) + (0 \cdot 2) = 12$$ ∎

FINDING SPECIFIC ROWS AND COLUMNS OF A MATRIX PRODUCT

The row-column rule is useful for finding specific entries in a matrix product AB, but if you want to find a specific column of AB, then the formula

$$AB = A[\mathbf{b}_1 \quad \mathbf{b}_2 \quad \cdots \quad \mathbf{b}_n] = [A\mathbf{b}_1 \quad A\mathbf{b}_2 \quad \cdots \quad A\mathbf{b}_n]$$

(from Definition 3.1.6) is the one to focus on. It follows from this formula that the jth column of AB is $A\mathbf{b}_j$; that is,

$$\mathbf{c}_j(AB) = A\mathbf{c}_j(B) \tag{18}$$

Similarly, the ith row of AB is given by the formula

$$\mathbf{r}_i(AB) = \mathbf{r}_i(A)B \tag{19}$$

We call (18) the **column rule** for matrix multiplication and (19) the **row rule** for matrix multiplication. In words, *the column rule states that the jth column of a matrix product is the first factor times the jth column of the second factor, and the row rule states that the ith row of a product is the ith row of the first factor times the second factor.*

EXAMPLE 8
Finding a Specific Row and Column of AB

Let A and B be the matrices in Example 5. Use the column rule to find the second column of AB and the row rule to find the first row of AB.

Solution

$$\mathbf{c}_2(AB) = A\mathbf{c}_2(B) = \begin{bmatrix} 1 & 2 & 4 \\ 2 & 6 & 0 \end{bmatrix} \begin{bmatrix} 1 \\ -1 \\ 7 \end{bmatrix} = \begin{bmatrix} 27 \\ -4 \end{bmatrix}$$

$$\mathbf{r}_1(AB) = \mathbf{r}_1(A)B = [1 \quad 2 \quad 4] \begin{bmatrix} 4 & 1 & 4 & 3 \\ 0 & -1 & 3 & 1 \\ 2 & 7 & 5 & 2 \end{bmatrix} = [12 \quad 27 \quad 30 \quad 13]$$

both of which agree with the result obtained in Example 5. ∎

MATRIX PRODUCTS AS LINEAR COMBINATIONS

Sometimes it is useful to view the rows and columns of a product AB as linear combinations. For example, if A is an $m \times s$ matrix, and $\mathbf{x}$ is a column vector with entries $x_1, x_2, \ldots, x_s$, then Definition 3.1.4 implies that

$$A\mathbf{x} = x_1\mathbf{c}_1(A) + x_2\mathbf{c}_2(A) + \cdots + x_s\mathbf{c}_s(A) \tag{20}$$

which is a linear combination of the column vectors of A with coefficients from $\mathbf{x}$. Similarly, if B is an $s \times n$ matrix, and $\mathbf{y}$ is a row vector with entries $y_1, y_2, \ldots, y_s$, then

$$\mathbf{y}B = y_1\mathbf{r}_1(B) + y_2\mathbf{r}_2(B) + \cdots + y_s\mathbf{r}_s(B) \tag{21}$$

which is a linear combination of the row vectors of B with coefficients from $\mathbf{y}$ (Exercise P1). Since the column rule for AB states that $\mathbf{c}_j(AB) = A\mathbf{c}_j(B)$ and the row rule for AB states that $\mathbf{r}_i(AB) = \mathbf{r}_i(A)B$, Formulas (20) and (21) imply the following result.

> **Theorem 3.1.8**
>
> (a) *The jth column vector of a matrix product AB is a linear combination of the column vectors of A with the coefficients coming from the jth column of B.*
>
> (b) *The ith row vector of a matrix product AB is a linear combination of the row vectors of B with the coefficients coming from the ith row of A.*

EXAMPLE 9
Rows and Columns of AB as Linear Combinations

Here are the computations in Example 8 performed using linear combinations:

$$\begin{bmatrix} 1 & 2 & 4 \\ 2 & 6 & 0 \end{bmatrix} \begin{bmatrix} 1 \\ -1 \\ 7 \end{bmatrix} = (1)\begin{bmatrix} 1 \\ 2 \end{bmatrix} + (-1)\begin{bmatrix} 2 \\ 6 \end{bmatrix} + 7\begin{bmatrix} 4 \\ 0 \end{bmatrix} = \begin{bmatrix} 27 \\ -4 \end{bmatrix}$$

$$\begin{bmatrix} 1 & 2 & 4 \end{bmatrix} \begin{bmatrix} 4 & 1 & 4 & 3 \\ 0 & -1 & 3 & 1 \\ 2 & 7 & 5 & 2 \end{bmatrix} = (1)\begin{bmatrix} 4 & 1 & 4 & 3 \end{bmatrix} + (2)\begin{bmatrix} 0 & -1 & 3 & 1 \end{bmatrix} + (4)\begin{bmatrix} 2 & 7 & 5 & 2 \end{bmatrix}$$

$$= \begin{bmatrix} 12 & 27 & 30 & 13 \end{bmatrix} \qquad \blacksquare$$

TRANSPOSE OF A MATRIX

Next we will define a matrix operation that has no analog in the algebra of real numbers.

> **Definition 3.1.9** If A is an $m \times n$ matrix, then the ***transpose*** of A, denoted by A^T, is defined to be the $n \times m$ matrix that is obtained by making the rows of A into columns; that is, the first column of A^T is the first row of A, the second column of A^T is the second row of A, and so forth.

EXAMPLE 10
Transpose of a Matrix

The following are some examples of matrices and their transposes.

$$A = \begin{bmatrix} a_{11} & a_{12} & a_{13} & a_{14} \\ a_{21} & a_{22} & a_{23} & a_{24} \\ a_{31} & a_{32} & a_{33} & a_{34} \end{bmatrix}, \quad B = \begin{bmatrix} 2 & 3 \\ 1 & 4 \\ 5 & 6 \end{bmatrix}, \quad C = \begin{bmatrix} 1 & 3 & -5 \end{bmatrix}, \quad D = \begin{bmatrix} 4 \end{bmatrix}$$

$$A^T = \begin{bmatrix} a_{11} & a_{21} & a_{31} \\ a_{12} & a_{22} & a_{32} \\ a_{13} & a_{23} & a_{33} \\ a_{14} & a_{24} & a_{34} \end{bmatrix}, \quad B^T = \begin{bmatrix} 2 & 1 & 5 \\ 3 & 4 & 6 \end{bmatrix}, \quad C^T = \begin{bmatrix} 1 \\ 3 \\ -5 \end{bmatrix}, \quad D^T = \begin{bmatrix} 4 \end{bmatrix} \qquad \blacksquare$$

This example illustrates that the process of forming A^T from A by converting rows into columns automatically converts the columns of A into the rows of A^T. Thus, the entry in row i and column j of A becomes the entry in row j and column i of A^T; that is,

$$(A)_{ij} = (A^T)_{ji} \tag{22}$$

If A is a square matrix, then the transpose of A can be obtained by interchanging entries that are symmetrically positioned about the main diagonal, or stated another way, by "reflecting" A about its main diagonal (Figure 3.1.5).

$$A = \begin{bmatrix} -1 & 2 & 4 \\ 3 & 7 & 0 \\ 5 & 8 & -6 \end{bmatrix} \rightarrow A^T = \begin{bmatrix} -1 & 3 & 5 \\ 2 & 7 & 8 \\ 4 & 0 & -6 \end{bmatrix}$$

Interchange entries that are symmetrically positioned about the main diagonal.

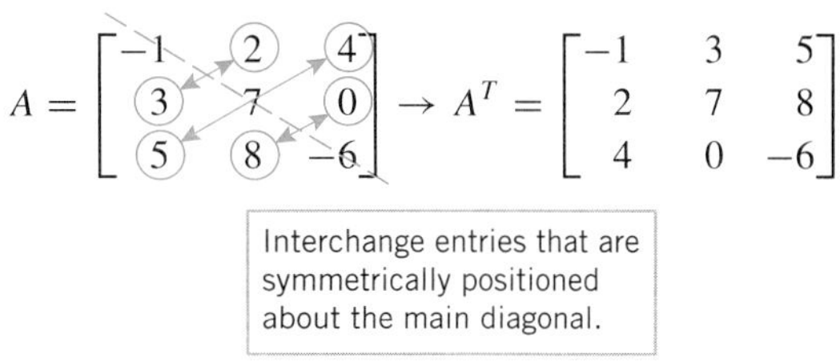

Figure 3.1.5

TRACE   The following definition, which we will need later in this text, applies only to square matrices.

Definition 3.1.10   If A is a square matrix, then the *trace* of A, denoted by tr(A), is defined to be the sum of the entries on the main diagonal of A.

For example, if

$$A = \begin{bmatrix} 3 & 2 \\ -1 & -8 \end{bmatrix} \quad \text{and} \quad B = \begin{bmatrix} b_{11} & b_{12} & b_{13} \\ b_{21} & b_{22} & b_{23} \\ b_{31} & b_{32} & b_{33} \end{bmatrix}$$

then

$$\text{tr}(A) = 3 + (-8) = -5 \quad \text{and} \quad \text{tr}(B) = b_{11} + b_{22} + b_{33}$$

Note that the trace of a matrix is the sum of the entries whose row and column numbers are the same.

INNER AND OUTER MATRIX PRODUCTS   Matrix products involving row and column vectors have some special terminology associated with them.

Definition 3.1.11   If $\mathbf{u}$ and $\mathbf{v}$ are column vectors with the same size, then the product $\mathbf{u}^T\mathbf{v}$ is called the *matrix inner product* of $\mathbf{u}$ with $\mathbf{v}$; and if $\mathbf{u}$ and $\mathbf{v}$ are column vectors of any size, then the product $\mathbf{u}\mathbf{v}^T$ is called the *matrix outer product* of $\mathbf{u}$ with $\mathbf{v}$.

REMARK   To help remember the distinction between inner and outer products, think of a matrix inner product as a "row vector times a column vector" and a matrix outer product as a "column vector times a row vector." If we follow our usual convention of not distinguishing between the number a and the 1×1 matrix $[a]$, then the matrix inner product of $\mathbf{u}$ with $\mathbf{v}$ is the same as $\mathbf{u} \cdot \mathbf{v}$. This is illustrated in the next example.

EXAMPLE 11
Matrix Inner and
Outer Products

If

$$\mathbf{u} = \begin{bmatrix} -1 \\ 3 \end{bmatrix} \quad \text{and} \quad \mathbf{v} = \begin{bmatrix} 2 \\ 5 \end{bmatrix}$$

then the matrix inner product of $\mathbf{u}$ with $\mathbf{v}$ (row times column) is

$$\mathbf{u}^T\mathbf{v} = \begin{bmatrix} -1 & 3 \end{bmatrix} \begin{bmatrix} 2 \\ 5 \end{bmatrix} = [13] = 13 = \mathbf{u} \cdot \mathbf{v}$$

The matrix outer product of $\mathbf{u}$ with $\mathbf{v}$ (column times row) is

$$\mathbf{u}\mathbf{v}^T = \begin{bmatrix} -1 \\ 3 \end{bmatrix} [2 \quad 5] = \begin{bmatrix} -2 & -5 \\ 6 & 15 \end{bmatrix}$$

In general, if

$$\mathbf{u} = \begin{bmatrix} u_1 \\ u_2 \\ \vdots \\ u_n \end{bmatrix} \quad \text{and} \quad \mathbf{v} = \begin{bmatrix} v_1 \\ v_2 \\ \vdots \\ v_n \end{bmatrix}$$

then the matrix inner product of $\mathbf{u}$ with $\mathbf{v}$ is

$$\mathbf{u}^T\mathbf{v} = [u_1 \quad u_2 \quad \cdots \quad u_n] \begin{bmatrix} v_1 \\ v_2 \\ \vdots \\ v_n \end{bmatrix} = [u_1v_1 + u_2v_2 + \cdots + u_nv_n] = \mathbf{u} \cdot \mathbf{v} \tag{23}$$

and the matrix outer product of $\mathbf{u}$ with $\mathbf{v}$ is

$$\mathbf{u}\mathbf{v}^T = \begin{bmatrix} u_1 \\ u_2 \\ \vdots \\ u_n \end{bmatrix} [v_1 \quad v_2 \quad \cdots \quad v_n] = \begin{bmatrix} u_1v_1 & u_1v_2 & \cdots & u_1v_n \\ u_2v_1 & u_2v_2 & \cdots & u_2v_n \\ \vdots & \vdots & & \vdots \\ u_nv_1 & u_nv_2 & \cdots & u_nv_n \end{bmatrix} \tag{24}$$

We will conclude this section with some useful relationships that tie together the notions of inner product, outer product, dot product, and trace. First, by comparing the sum in Formula (23) to the diagonal entries in (24), we see that the inner and outer products of two column vectors are related by the trace formula

$$\mathbf{u}^T\mathbf{v} = \text{tr}(\mathbf{u}\mathbf{v}^T) \tag{25}$$

Also, it follows from (23) and the symmetry property of the dot product that

$$\mathbf{u}^T\mathbf{v} = \mathbf{u} \cdot \mathbf{v} = \mathbf{v} \cdot \mathbf{u} = \mathbf{v}^T\mathbf{u} \tag{26}$$

Moreover, interchanging $\mathbf{u}$ and $\mathbf{v}$ in (25) and applying (26) yields

$$\text{tr}(\mathbf{u}\mathbf{v}^T) = \text{tr}(\mathbf{v}\mathbf{u}^T) = \mathbf{u} \cdot \mathbf{v} \tag{27}$$

Keep in mind, however, that these formulas apply only when $\mathbf{u}$ and $\mathbf{v}$ are expressed as column vectors.

Exercise Set 3.1

In Exercises 1 and 2, solve the matrix equation for a, b, c, and d.

1. $\begin{bmatrix} a - b & b + a \\ 3d + c & 2d - c \end{bmatrix} = \begin{bmatrix} 8 & 1 \\ 7 & 6 \end{bmatrix}$

2. $\begin{bmatrix} a & 3 \\ -1 & a + b \end{bmatrix} = \begin{bmatrix} 4 & d - 2c \\ d + 2c & -2 \end{bmatrix}$

In Exercises 3 and 4, let

$$A = [a_{ij}] = \begin{bmatrix} 3 & 1 & -7 & 8 \\ 2 & 4 & 11 & 0 \\ 3 & 3 & 9 & -4 \end{bmatrix}$$

and

$$B = [b_{ij}] = \begin{bmatrix} 11 & \frac{1}{2} & -2 & 3 \\ 4 & 1 & 1 & 0 \end{bmatrix}$$

3. Fill in the blanks.
   (a) A has size _____ by _____, and B^T has size _____ by _____.
   (b) $a_{32} = $ _____ and $a_{23} = $ _____.
   (c) $a_{ij} = 3$ for $(i, j) = $ _____.
   (d) The third column vector of A^T is _____.
   (e) The second row vector of $(2B)^T$ is _____.

4. Fill in the blanks.
   (a) B has size _____ by _____, and A^T has size _____ by _____.
   (b) $b_{12} = $ _____ and $b_{21} = $ _____.
   (c) $b_{ij} = 1$ for $(i, j) = $ _____.
   (d) The second row vector of A^T is _____.
   (e) The second column vector of $3B^T$ is _____.

Use the following matrices in Exercises 5–8:

$$A = \begin{bmatrix} 3 & 0 \\ -1 & 2 \\ 1 & 1 \end{bmatrix}, \quad B = \begin{bmatrix} 2 & 1 \\ -3 & 1 \\ 4 & 0 \end{bmatrix}, \quad C = \begin{bmatrix} 1 & 0 \\ 3 & -1 \end{bmatrix}$$

$$D = \begin{bmatrix} 1 & 1 \\ -3 & 3 \end{bmatrix}, \quad E = \begin{bmatrix} 1 & 4 & 2 \\ 3 & 1 & 5 \end{bmatrix}, \quad F = \begin{bmatrix} 1 & 5 & 2 \\ -1 & 0 & 1 \\ 3 & 2 & 4 \end{bmatrix}$$

$$G = \begin{bmatrix} 6 & 1 & 3 \\ -1 & 1 & 2 \\ 4 & 1 & 3 \end{bmatrix}$$

5. Compute the following matrices, where possible.
(a) $A + 2B$
(b) $A - B^T$
(c) $4D - 3C^T$
(d) $D - D^T$
(e) $G + (2F)^T$
(f) $(7A - B) + E$

6. Find the following matrices, where possible.
(a) $3C + D$
(b) $E - A^T$
(c) $4C - 5D^T$
(d) $F - F^T$
(e) $B + (4E)^T$
(f) $(7C - D) + B$

7. Compute the following matrices, where possible.
(a) CD
(b) AE
(c) FG
(d) $B^T F$
(e) BB^T
(f) GE

8. Compute the following matrices, where possible.
(a) GA
(b) FB
(c) GF
(d) $A^T G$
(e) EE^T
(f) DA

In Exercises 9 and 10, compute $A\mathbf{x}$ using Definition 3.1.4.

9. $A = \begin{bmatrix} 1 & 5 & 2 \\ -4 & 9 & 1 \\ 2 & 0 & 3 \end{bmatrix}; \quad \mathbf{x} = \begin{bmatrix} 2 \\ -1 \\ 3 \end{bmatrix}$

10. $A = \begin{bmatrix} 2 & 4 & 1 \\ -3 & 9 & 0 \\ 3 & -1 & 2 \end{bmatrix}; \quad \mathbf{x} = \begin{bmatrix} 4 \\ 5 \\ -2 \end{bmatrix}$

In Exercises 11 and 12, express the system in the form $A\mathbf{x} = \mathbf{b}$.

11. (a) $2x_1 - 3x_2 + 5x_3 = 7$
$9x_1 - x_2 + x_3 = -1$

(b) $x_1 + x_2 + x_3 = 4$
$2x_1 - 3x_2 + 4x_3 = 3$
$x_1 + 5x_2 - 2x_3 = -2$

12. (a) $x_1 - 2x_2 + 3x_3 = -3$
$2x_1 + x_2 = 0$
$-3x_2 + 4x_3 = 1$
$x_1 + x_3 = 5$

(b) $3x_1 + 3x_2 + 3x_3 = -3$
$-x_1 - 5x_2 - 2x_3 = 3$
$-4x_2 + x_3 = 0$

In Exercises 13 and 14, write out the equations of the linear system in x_1, x_2, x_3 whose matrix form is $A\mathbf{x} = \mathbf{b}$.

13. $A = \begin{bmatrix} 5 & 6 & -7 \\ -1 & -2 & 3 \\ 0 & 4 & -1 \end{bmatrix}; \quad \mathbf{b} = \begin{bmatrix} 2 \\ 0 \\ 3 \end{bmatrix}$

14. $A = \begin{bmatrix} 1 & 1 & 1 \\ 2 & 3 & 0 \\ 5 & -3 & -6 \end{bmatrix}; \quad \mathbf{b} = \begin{bmatrix} 2 \\ 2 \\ -9 \end{bmatrix}$

In Exercises 15–20, let

$$A = \begin{bmatrix} 3 & -2 & 7 \\ 6 & 5 & 4 \\ 0 & 4 & 9 \end{bmatrix} \quad \text{and} \quad B = \begin{bmatrix} 6 & -2 & 4 \\ 0 & 1 & 3 \\ 7 & 7 & 5 \end{bmatrix}$$

15. Find $(AB)_{23}$ with as few additions and multiplications as possible.

16. Find $(BA)_{21}$ with as few additions and multiplications as possible.

17. Use the method of Example 8 to find
(a) the first row vector of AB;
(b) the third row vector of AB;
(c) the second column vector of AB.

18. Use the method of Example 8 to find
(a) the first row vector of BA;
(b) the third row vector of BA;
(c) the second column vector of BA.

19. Find
(a) tr(A)
(b) tr(A^T)
(c) tr(AB) − tr(A)tr(B)

20. Find
(a) tr(B)
(b) tr(B^T)
(c) tr(BA) − tr(B)tr(A)

21. Let $\mathbf{u} = \begin{bmatrix} -2 \\ 3 \end{bmatrix}$ and $\mathbf{v} = \begin{bmatrix} 4 \\ 5 \end{bmatrix}$.

(a) Find the matrix inner product of $\mathbf{u}$ with $\mathbf{v}$.
(b) Find the matrix outer product of $\mathbf{u}$ with $\mathbf{v}$.
(c) Confirm Formula (25) for the vectors $\mathbf{u}$ and $\mathbf{v}$.
(d) Confirm Formula (26) for the vectors $\mathbf{u}$ and $\mathbf{v}$.
(e) Confirm Formula (27) for the vectors $\mathbf{u}$ and $\mathbf{v}$.

22. Let $\mathbf{u} = \begin{bmatrix} 3 \\ -4 \\ 5 \end{bmatrix}$ and $\mathbf{v} = \begin{bmatrix} 2 \\ 7 \\ 0 \end{bmatrix}$.

(a) Find the matrix inner product of $\mathbf{u}$ with $\mathbf{v}$.
(b) Find the matrix outer product of $\mathbf{u}$ with $\mathbf{v}$.
(c) Confirm Formula (25) for the vectors $\mathbf{u}$ and $\mathbf{v}$
(d) Confirm Formula (26) for the vectors $\mathbf{u}$ and $\mathbf{v}$.
(e) Confirm Formula (27) for the vectors $\mathbf{u}$ and $\mathbf{v}$.

In Exercises 23 and 24, find all values of k, if any, that satisfy the equation.

23. $\begin{bmatrix} k & 1 & 1 \end{bmatrix} \begin{bmatrix} 1 & 1 & 0 \\ 1 & 0 & 2 \\ 0 & 2 & -3 \end{bmatrix} \begin{bmatrix} k \\ 1 \\ 1 \end{bmatrix} = 0$

24. $\begin{bmatrix} 2 & 2 & k \end{bmatrix} \begin{bmatrix} 1 & 2 & 0 \\ 2 & 0 & 3 \\ 0 & 3 & 1 \end{bmatrix} \begin{bmatrix} 2 \\ 2 \\ k \end{bmatrix} = 0$

25. Let C, D, and E be the matrices used in Exercises 5–8. Using as few operations as possible, find the entry in row 2 and column 3 of $C(DE)$.

26. Let C, D, and E be the matrices used in Exercises 5–8. Using as few operations as possible, determine the entry in row 1 and column 2 of $(CD)E$.

27. Show that if AB and BA are both defined, then AB and BA are square matrices.

28. Show that if A is an $m \times n$ matrix and $A(BA)$ is defined, then B is an $n \times m$ matrix.

29. (a) Show that if A has a row of zeros and B is any matrix for which AB is defined, then AB also has a row of zeros.
(b) Find a similar result involving a column of zeros.

30. (a) Show that if B and C have two equal columns, and A is any matrix for which AB and AC are defined, then AB and AC also have two equal columns.
(b) Find a similar result involving matrices with two equal rows.

31. In each part, describe the form of a 6×6 matrix $A = [a_{ij}]$ that satisfies the stated condition. Make your answer as general as possible by using letters rather than specific numbers for the nonzero entries.
(a) $a_{ij} = 0$ if $i \neq j$ (b) $a_{ij} = 0$ if $i > j$
(c) $a_{ij} = 0$ if $i < j$ (d) $a_{ij} = 0$ if $|i - j| > 1$

32. In each part, find the 4×4 matrix $A = [a_{ij}]$ whose entries satisfy the stated condition.
(a) $a_{ij} = i + j$ (b) $a_{ij} = (-1)^{i+j}$
(c) $a_{ij} = \begin{cases} 1 & \text{if } |i - j| > 1 \\ -1 & \text{if } |i - j| \leq 1 \end{cases}$

33. Suppose that type I items cost $1 each, type II items cost $2 each, and type III items cost $3 each. Also, suppose that the accompanying table describes the number of items of each type purchased during the first four months of the year.

	Type I	Type II	Type III
Jan.	3	4	3
Feb.	5	6	0
Mar.	2	9	4
Apr.	1	1	7

Table Ex-33

What information is represented by the following matrix product?

$$\begin{bmatrix} 3 & 4 & 3 \\ 5 & 6 & 0 \\ 2 & 9 & 4 \\ 1 & 1 & 7 \end{bmatrix} \begin{bmatrix} 1 \\ 2 \\ 3 \end{bmatrix}$$

34. The accompanying table shows a record of May and June unit sales for a clothing store. Let M denote the 4×3 matrix of May sales and J the 4×3 matrix of June sales.
(a) What does the matrix $M + J$ represent?
(b) What does the matrix $M - J$ represent?
(c) Find a column vector $\mathbf{x}$ for which $M\mathbf{x}$ provides a list of the number of shirts, jeans, suits, and raincoats sold in May.
(d) Find a row vector $\mathbf{y}$ for which $\mathbf{y}M$ provides a list of the number of small, medium, and large items sold in May.
(e) Using the matrices $\mathbf{x}$ and $\mathbf{y}$ that you found in parts (c) and (d), what does $\mathbf{y}M\mathbf{x}$ represent?

	May Sales		
	Small	**Medium**	**Large**
Shirts	45	60	75
Jeans	30	30	40
Suits	12	65	45
Raincoats	15	40	35

	June Sales		
	Small	**Medium**	**Large**
Shirts	30	33	40
Jeans	21	23	25
Suits	9	12	11
Raincoats	8	10	9

Table Ex-34

Discussion and Discovery

D1. Given that AB is a matrix whose size is 6×8, what can you say about the sizes of A and B?

D2. Find a nonzero 2×2 matrix A such that AA has all zero entries.

D3. Describe three different methods for computing a matrix product, and illustrate the different methods by computing a product AB of your choice.

D4. How many 3×3 matrices A can you find such that A has constant entries and

$$A \begin{bmatrix} x \\ y \\ z \end{bmatrix} = \begin{bmatrix} x + y \\ x - y \\ 0 \end{bmatrix}$$

for all real values of x, y, and z?

D5. How many 3×3 matrices A can you find such that A has constant entries and

$$A \begin{bmatrix} x \\ y \\ z \end{bmatrix} = \begin{bmatrix} xy \\ 0 \\ 0 \end{bmatrix}$$

for all real values of x, y, and z?

D6. Let

$$A = \begin{bmatrix} 2 & 2 \\ 2 & 2 \end{bmatrix} \quad \text{and} \quad B = \begin{bmatrix} 5 & 0 \\ 0 & 9 \end{bmatrix}$$

(a) A matrix S is said to be a ***square root*** of a matrix M if $SS = M$. Find two square roots of A.

(b) How many different square roots of B can you find?

(c) Do you think that every matrix has a square root? Explain your reasoning.

D7. Is there a 3×3 matrix A such that AB has three equal rows for every 3×3 matrix B?

D8. Is there a 3×3 matrix A for which $AB = 2B$ for every 3×3 matrix B?

D9. Indicate whether the statement is true (T) or false (F). Justify your answer.

(a) If AB and BA are both defined, then A and B are square matrices.

(b) If $AB + BA$ is defined, then A and B are square matrices of the same size.

(c) If B has a column of zeros, then so does AB if this product is defined.

(d) If B has a column of zeros, then so does BA if this product is defined.

(e) The expressions $\text{tr}(A^T A)$ and $\text{tr}(AA^T)$ are defined for every matrix A.

(f) If $\mathbf{u}$ and $\mathbf{v}$ are row vectors, then $\mathbf{u}^T \mathbf{v} = \mathbf{u} \cdot \mathbf{v}$.

D10. If A and B are 3×3 matrices, and the second column of B is the sum of the first and third columns, what can be said about the second column of AB?

D11. (***Sigma notation***) Suppose that

$$A = [a_{ij}]_{m \times s} \quad \text{and} \quad B = [b_{ij}]_{s \times n}$$

(a) Write out the sum $\sum_{k=1}^{s} (a_{ik} b_{kj})$.

(b) What does this sum represent?

Working with Proofs

P1. Prove Formula (21) by multiplying matrices and comparing entries.

P2. Prove that a linear system $A\mathbf{x} = \mathbf{b}$ is consistent if and only if $\mathbf{b}$ can be expressed as a linear combination of the column vectors of A.

Technology Exercises

T1. (***Matrix addition and multiplication by scalars***) Perform the calculations in Examples 2 and 3.

T2. (***Matrix multiplication***) Compute the product AB from Example 5.

T3. (***Trace and transpose***) Find the trace and the transpose of the matrix

$$A = \begin{bmatrix} 3 & 1 & -7 \\ 2 & 4 & 11 \\ 3 & 3 & 9 \end{bmatrix}$$

T4. (***Extracting row vectors, column vectors, and entries***)

(a) Extract the row vectors and column vectors of the matrix in Exercise T3.

(b) Find the sum of the row vectors and the sum of the column vectors of the matrix in Exercise T3.

(c) Extract the diagonal entries of the matrix in Exercise T3 and compute their sum.

T5. See what happens when you try to multiply matrices whose sizes do not conform for the product.

T6. (*Linear combinations by matrix multiplication*) One way to obtain a linear combination $c_1\mathbf{v}_1 + c_2\mathbf{v}_2 + \cdots + c_k\mathbf{v}_k$ of vectors in R^n is to compute the matrix $A\mathbf{c}$ in which the successive column vectors of A are $\mathbf{v}_1, \mathbf{v}_2, \ldots, \mathbf{v}_k$ and $\mathbf{c}$ is the column vector whose successive entries are $c_1, c_2, \ldots, c_k$. Use this method to compute the linear combination

$$6(8, 2, 1, 4) + 17(3, 9, 11, 6) + 9(0, 1, 2, 4)$$

T7. Use the idea in Exercise T6 to compute the following linear combinations with a single matrix multiplication.

$$3(7, 1, 0, 3) - 4(-1, 5, 7, 0) + 2(6, 3, -2, 1)$$
$$5(7, 1, 0, 3) - (-1, 5, 7, 0) + (6, 3, -2, 1)$$
$$2(7, 1, 0, 3) + 4(-1, 5, 7, 0) + 7(6, 3, -2, 1)$$

Section 3.2   Inverses; Algebraic Properties of Matrices

In this section we will discuss some of the algebraic properties of matrix operations. We will see that many of the basic rules of arithmetic for real numbers also hold for matrices, but we will also see that some do not.

PROPERTIES OF MATRIX ADDITION AND SCALAR MULTIPLICATION

The following theorem lists some of the basic properties of matrix addition and scalar multiplication. All of these results are analogs of familiar rules for the arithmetic of real numbers.

Theorem 3.2.1   *If a and b are scalars, and if the sizes of the matrices A, B, and C are such that the indicated operations can be performed, then:*

(a) $A + B = B + A$ [Commutative law for addition]

(b) $A + (B + C) = (A + B) + C$ [Associative law for addition]

(c) $(ab)A = a(bA)$

(d) $(a + b)A = aA + bA$

(e) $(a - b)A = aA - bA$

(f) $a(A + B) = aA + aB$

(g) $a(A - B) = aA - aB$

In each part we must show that the left side has the same size as the right side and that corresponding entries on the two sides are equal. To prove that corresponding entries are equal, we can work with the individual entries, or we can show that corresponding column vectors (or row vectors) on the two sides are equal. We will prove part (b) by considering the individual entries. Some of the other proofs will be left as exercises.

Proof (b)   In order for $A + (B + C)$ to be defined, the matrices A, B, and C must have the same size, say $m \times n$. It follows from this that $A + (B + C)$ and $(A + B) + C$ also have size $m \times n$, so the expressions on both sides of the equation have the same size. Thus, it only remains to show that the corresponding entries on the two sides are equal; that is,

$$[A + (B + C)]_{ij} = [(A + B) + C]_{ij}$$

for all values of i and j. For this purpose, let

$$A = [a_{ij}], \quad B = [b_{ij}], \quad C = [c_{ij}]$$

Thus,

$$[A + (B + C)]_{ij} = a_{ij} + (b_{ij} + c_{ij}) \quad \text{[Definition of matrix addition]}$$
$$= (a_{ij} + b_{ij}) + c_{ij} \quad \text{[Associative law for addition of real numbers]}$$
$$= [(A + B) + C]_{ij} \quad \text{[Definition of matrix addition]} \quad \blacksquare$$

PROPERTIES OF MATRIX MULTIPLICATION

Do not let Theorem 3.2.1 lull you into believing that all of the laws of arithmetic for real numbers carry over to matrices. For example, you know that in the arithmetic of real numbers it is always true that $ab = ba$, which is called the ***commutative law for multiplication***. However, the commutative law does not hold for matrix multiplication; that is, AB and BA need not be equal matrices. Equality can fail to hold for three reasons:

1. AB may be defined and BA may not (for example, if A is 2×3 and B is 3×4).

2. AB and BA may both be defined, but they may have different sizes (for example, if A is 2×3 and B is 3×2).

3. AB and BA may both be defined and have the same size, but the two matrices may be different (as illustrated in the next example).

EXAMPLE 1
Order Matters in Matrix Multiplication

Consider the matrices

$$A = \begin{bmatrix} -1 & 0 \\ 2 & 3 \end{bmatrix} \quad \text{and} \quad B = \begin{bmatrix} 1 & 2 \\ 3 & 0 \end{bmatrix}$$

Multiplying gives

$$AB = \begin{bmatrix} -1 & -2 \\ 11 & 4 \end{bmatrix} \quad \text{and} \quad BA = \begin{bmatrix} 3 & 6 \\ -3 & 0 \end{bmatrix}$$

Thus, $AB \neq BA$. ■

REMARK Do not conclude from this example that AB and BA are never equal. In specific cases it may be true that $AB = BA$, in which case we say that the matrices A and B ***commute***. We will give some examples later in this section.

Although the commutative law is not valid for matrix multiplication, many familiar properties of multiplication do carry over to matrix arithmetic.

> **Theorem 3.2.2** *If a is a scalar, and if the sizes of the matrices A, B, and C are such that the indicated operations can be performed, then:*
>
> (a) $A(BC) = (AB)C$ **[Associative law for multiplication]**
>
> (b) $A(B + C) = AB + AC$ **[Left distributive law]**
>
> (c) $(B + C)A = BA + CA$ **[Right distributive law]**
>
> (d) $A(B - C) = AB - AC$
>
> (e) $(B - C)A = BA - CA$
>
> (f) $a(BC) = (aB)C = B(aC)$

In each part we must show that the left side has the same size as the right side and that corresponding entries on the two sides are equal. As observed in our discussion of Theorem 3.2.1, we can prove this equality by working with the individual entries, or by showing that the corresponding column vectors (or row vectors) on the two sides are equal. We will prove parts (a) and (b). Some of the remaining proofs are given as exercises.

Proof (a) We must show first that the matrices $A(BC)$ and $(AB)C$ have the same size. For BC to be defined, the number of columns of B must be the same as the number of rows of C. Thus, assume that B has size $k \times s$ and C has size $s \times n$. This implies that BC has size $k \times n$. For $A(BC)$ to be defined, the matrix A must have the same number of columns as BC has rows, so assume that A has size $m \times k$. It now follows that $A(BC)$ and $(AB)C$ have the same size, namely $m \times n$.

Next we want to show that the corresponding column vectors of $A(BC)$ and $(AB)C$ are equal. For this purpose, let $\mathbf{c}_j$ be the jth column vector of C. Thus, the jth column vector of $(AB)C$ is

$$(AB)\mathbf{c}_j \tag{1}$$

Also, the jth column vector of BC is $B\mathbf{c}_j$, which implies that the jth column vector of $A(BC)$ is

$$A(B\mathbf{c}_j) \tag{2}$$

However, the definition of matrix multiplication was created specifically to make it true that $A(B\mathbf{x}) = (AB)\mathbf{x}$ for every conforming column vector $\mathbf{x}$ [see Formula (12) of Section 3.1]. In particular, this holds for the column vector $\mathbf{c}_j$, so it follows that (1) and (2) are equal, which completes the proof.

Proof (*b*) We must show first that $A(B + C)$ and $AB + AC$ have the same size. To perform the operation $B + C$ the matrices B and C must have the same size, say $s \times n$. The matrix A must then have s columns to conform for the product $A(B + C)$, so its size must be of the form $m \times s$. This implies that $A(B + C)$, AB, and AC are $m \times n$ matrices, from which it follows that $A(B + C)$ and $AB + AC$ have the same size, namely $m \times n$.

Next we want to show that corresponding column vectors of $A(B + C)$ and $AB + AC$ are equal. For this purpose, let $\mathbf{b}_j$ and $\mathbf{c}_j$ be the jth column vectors of B and C, respectively. Then the jth column vector of $A(B + C)$ is

$$A(\mathbf{b}_j + \mathbf{c}_j) \tag{3}$$

and the jth column vector of $AB + AC$ is

$$A\mathbf{b}_j + A\mathbf{c}_j \tag{4}$$

But part (*b*) of Theorem 3.1.5 implies that (3) and (4) are equal, which completes the proof. ∎

REMARK Although the matrix addition and multiplication were only defined for pairs of matrices, the associative laws $A + (B + C) = (A + B) + C$ and $A(BC) = (AB)C$ allow us to use the expressions $A + B + C$ and ABC without ambiguity because the same result is obtained no matter how the matrices are grouped. In general, *given any sum or any product of matrices, pairs of parentheses can be inserted or deleted anywhere in the expression without affecting the end result.*

ZERO MATRICES   A matrix whose entries are all zero is called a *zero matrix*. Some examples are

$$\begin{bmatrix} 0 & 0 \\ 0 & 0 \end{bmatrix}, \quad \begin{bmatrix} 0 & 0 & 0 \\ 0 & 0 & 0 \\ 0 & 0 & 0 \end{bmatrix}, \quad \begin{bmatrix} 0 & 0 & 0 & 0 \\ 0 & 0 & 0 & 0 \end{bmatrix}, \quad \begin{bmatrix} 0 \\ 0 \\ 0 \\ 0 \end{bmatrix}, \quad [0]$$

We will denote a zero matrix by 0 unless it is important to give the size, in which case we will denote the $m \times n$ zero matrix by $0_{m \times n}$.

It should be evident that if A and 0 are matrices with the same size, then

$$A + 0 = 0 + A = A$$

Thus, 0 plays the same role in this matrix equation that the number 0 plays in the numerical equation $a + 0 = 0 + a = a$.

The following theorem lists the basic properties of zero matrices. Since the results should be self-evident, we will omit the formal proofs.

Theorem 3.2.3 *If c is a scalar, and if the sizes of the matrices are such that the operations can be performed, then:*

(*a*) $A + 0 = 0 + A = A$

(*b*) $A - 0 = A$

(*c*) $A - A = A + (-A) = 0$

(*d*) $0A = 0$

(*e*) *If $cA = 0$, then $c = 0$ or $A = 0$.*

You should not conclude from this theorem that all properties of the number zero in ordinary arithmetic carry over to zero matrices in matrix arithmetic. For example, consider the *cancellation law* for real numbers: If $ab = ac$, and if $a \neq 0$, then $b = c$. The following example shows that this result does not carry over to matrix arithmetic.

EXAMPLE 2

The Cancellation Law Is Not True for Matrices

Consider the matrices

$$A = \begin{bmatrix} 0 & 1 \\ 0 & 2 \end{bmatrix}, \quad B = \begin{bmatrix} 1 & 1 \\ 3 & 4 \end{bmatrix}, \quad C = \begin{bmatrix} 2 & 5 \\ 3 & 4 \end{bmatrix}$$

We leave it for you to confirm that

$$AB = AC = \begin{bmatrix} 3 & 4 \\ 6 & 8 \end{bmatrix}$$

Although $A \neq 0$, canceling A from both sides of the equation $AB = AC$ would lead to the incorrect conclusion that $B = C$. Thus, the cancellation law does not hold, in general, for matrix multiplication. ∎

EXAMPLE 3

Nonzero Matrices Can Have a Zero Product

Recall that if c and a are real numbers such that $ca = 0$, then $c = 0$ or $a = 0$; analogously, Theorem 3.2.3(e) states that if c is a scalar and A is a matrix such that $cA = 0$, then $c = 0$ or $A = 0$. However, this result does not extend to matrix products. For example, let

$$C = \begin{bmatrix} 0 & 1 \\ 0 & 2 \end{bmatrix} \quad \text{and} \quad A = \begin{bmatrix} 3 & 7 \\ 0 & 0 \end{bmatrix}$$

Here $CA = 0$, but $C \neq 0$ and $A \neq 0$. ∎

IDENTITY MATRICES

A square matrix with 1's on the main diagonal and zeros elsewhere is called an ***identity matrix***. Some examples are

$$[1], \quad \begin{bmatrix} 1 & 0 \\ 0 & 1 \end{bmatrix}, \quad \begin{bmatrix} 1 & 0 & 0 \\ 0 & 1 & 0 \\ 0 & 0 & 1 \end{bmatrix}, \quad \begin{bmatrix} 1 & 0 & 0 & 0 \\ 0 & 1 & 0 & 0 \\ 0 & 0 & 1 & 0 \\ 0 & 0 & 0 & 1 \end{bmatrix}$$

We will denote an identity matrix by the letter I unless it is important to give the size, in which case we will write I_n for the $n \times n$ identity matrix.

To explain the role of identity matrices in matrix arithmetic, let us consider the effect of multiplying a general 2×3 matrix A on each side by an identity matrix. Multiplying on the right by the 3×3 identity matrix yields

$$AI_3 = \begin{bmatrix} a_{11} & a_{12} & a_{13} \\ a_{21} & a_{22} & a_{23} \end{bmatrix} \begin{bmatrix} 1 & 0 & 0 \\ 0 & 1 & 0 \\ 0 & 0 & 1 \end{bmatrix} = \begin{bmatrix} a_{11} & a_{12} & a_{13} \\ a_{21} & a_{22} & a_{23} \end{bmatrix} = A$$

and multiplying on the left by the 2×2 identity matrix yields

$$I_2 A = \begin{bmatrix} 1 & 0 \\ 0 & 1 \end{bmatrix} \begin{bmatrix} a_{11} & a_{12} & a_{13} \\ a_{21} & a_{22} & a_{23} \end{bmatrix} = \begin{bmatrix} a_{11} & a_{12} & a_{13} \\ a_{21} & a_{22} & a_{23} \end{bmatrix} = A$$

The same result holds in general; that is, if A is any $m \times n$ matrix, then

$$AI_n = A \quad \text{and} \quad I_m A = A$$

Thus, the identity matrices play the same role in these matrix equations that the number 1 plays in the numerical equation $a \cdot 1 = 1 \cdot a = a$.

Identity matrices arise naturally in the process of reducing square matrices to reduced row echelon form by elementary row operations. For example, consider what might happen when a 3×3 matrix is put in reduced row echelon form. There are two possibilities—either the reduced row echelon form has a row of zeros or it does not. If it does not, then the three rows have leading

1's. However, in reduced row echelon form there are zeros above and below each leading 1, so this forces the reduced row echelon form to be the identity matrix if there are no zero rows. The same argument holds for any square matrix, so we have the following result.

Theorem 3.2.4 *If R is the reduced row echelon form of an n × n matrix A, then either R has a row of zeros or R is the identity matrix I_n.*

INVERSE OF A MATRIX

In ordinary arithmetic every nonzero number a has a reciprocal $a^{-1} (= 1/a)$ with the property

$$a \cdot a^{-1} = a^{-1} \cdot a = 1$$

The number a^{-1} is sometimes called the ***multiplicative inverse*** of a. Our next objective is to look for an analog of this result in matrix arithmetic. Toward this end we make the following definition.

Definition 3.2.5 If A is a square matrix, and if there is a matrix B with the same size as A such that $AB = BA = I$, then A is said to be ***invertible*** (or ***nonsingular***), and B is called an ***inverse*** of A. If there is no matrix B with this property, then A is said to be ***singular***.

REMARK Observe that the condition that $AB = BA = I$ is not altered by interchanging A and B. Thus, if A is invertible and B is an inverse of A, then it is also true that B is invertible and A is an inverse of B. Accordingly, when the condition $AB = BA = I$ holds, it is correct to say that A and B are *inverses of one another.*

EXAMPLE 4
An Invertible Matrix

Let

$$A = \begin{bmatrix} 2 & -5 \\ -1 & 3 \end{bmatrix} \quad \text{and} \quad B = \begin{bmatrix} 3 & 5 \\ 1 & 2 \end{bmatrix}$$

Then

$$AB = \begin{bmatrix} 2 & -5 \\ -1 & 3 \end{bmatrix} \begin{bmatrix} 3 & 5 \\ 1 & 2 \end{bmatrix} = \begin{bmatrix} 1 & 0 \\ 0 & 1 \end{bmatrix} = I$$

$$BA = \begin{bmatrix} 3 & 5 \\ 1 & 2 \end{bmatrix} \begin{bmatrix} 2 & -5 \\ -1 & 3 \end{bmatrix} = \begin{bmatrix} 1 & 0 \\ 0 & 1 \end{bmatrix} = I$$

Thus, A and B are invertible and each is an inverse of the other. ■

EXAMPLE 5
A Class of Singular Matrices

In general, a square matrix with a row or column of zeros is singular. To help understand why this is so, consider the matrix

$$A = \begin{bmatrix} 1 & 4 & 0 \\ 2 & 5 & 0 \\ 3 & 6 & 0 \end{bmatrix}$$

To prove that A is singular we must show that there is no 3×3 matrix B such that $AB = BA = I$. For this purpose let $\mathbf{c}_1, \mathbf{c}_2, \mathbf{0}$ be the column vectors of A. Thus, for any 3×3 matrix B we can express the product BA as

$$BA = B[\mathbf{c}_1 \quad \mathbf{c}_2 \quad \mathbf{0}] = [B\mathbf{c}_1 \quad B\mathbf{c}_2 \quad \mathbf{0}]$$

The column of zeros shows that $BA \neq I$ and hence that A is singular. ■

PROPERTIES OF INVERSES

We know that if a is a nonzero real number, then there is a *unique* real number b such that $ab = ba = 1$, namely $b = a^{-1}$. The next theorem shows that matrix inverses are also unique.

Theorem 3.2.6 *If A is an invertible matrix, and if B and C are both inverses of A, then $B = C$; that is, an invertible matrix has a unique inverse.*

Proof Since B is an inverse of A, we have $BA = I$. Multiplying both sides of this equation on the right by C and keeping in mind that $IC = C$ yields

$$(BA)C = C$$

Since C is also an inverse of A, we have $AC = I$. Thus, the left side of the above equation can be rewritten as

$$(BA)C = B(AC) = BI = B$$

which implies that $B = C$. ■

REMARK Because an invertible matrix A can have only one inverse, we are entitled to talk about "the" inverse of A. Motivated by the notation a^{-1} for the multiplicative inverse of a nonzero real number a, we will denote the inverse of an invertible matrix A by A^{-1}. Thus,

$$AA^{-1} = I \quad \text{and} \quad A^{-1}A = I$$

Later in this chapter we will discuss a general method for finding the inverse of an invertible matrix. However, in the simple case of an invertible 2×2 matrix, the inverse can be obtained using the formula in the next theorem.

Linear Algebra in History

The formula for A^{-1} given in Theorem 3.2.7 first appeared (in a more general form) in Arthur Cayley's 1858 *Memoir on the Theory of Matrices* (see p. 81). The more general result that Cayley discovered will be studied later.

Theorem 3.2.7 *The matrix*

$$A = \begin{bmatrix} a & b \\ c & d \end{bmatrix}$$

is invertible if and only if $ad - bc \neq 0$, in which case the inverse is given by the formula

$$A^{-1} = \frac{1}{ad - bc}\begin{bmatrix} d & -b \\ -c & a \end{bmatrix} = \begin{bmatrix} \dfrac{d}{ad - bc} & -\dfrac{b}{ad - bc} \\ -\dfrac{c}{ad - bc} & \dfrac{a}{ad - bc} \end{bmatrix} \tag{5}$$

Proof The heart of the proof is to show that $AA^{-1} = A^{-1}A = I$. We leave the computations to you. (Also, see Exercise P7.) ■

REMARK The quantity $ad - bc$ in this theorem is called the ***determinant*** of the 2×2 matrix A and is denoted by the symbol $\det(A)$ or, alternatively, by replacing the brackets around the matrix A with vertical bars as shown in Figure 3.2.1. That figure illustrates that the determinant of A is the product of the diagonal entries of A minus the product of the "off-diagonal" entries of A. In determinant terminology, Theorem 3.2.7 states that a 2×2 matrix A is invertible if and only if its determinant is nonzero, and in that case the inverse can be obtained by interchanging the diagonal entries of A, reversing the signs of the off-diagonal entries, and then dividing the entries by the determinant.

$$\det(A) = \begin{vmatrix} a & b \\ c & d \end{vmatrix} = ad - bc$$

Figure 3.2.1

EXAMPLE 6 Calculating the Inverse of a 2×2 Matrix

In each part, determine whether the matrix is invertible. If so, find its inverse.

(a) $A = \begin{bmatrix} 6 & 1 \\ 5 & 2 \end{bmatrix}$ (b) $A = \begin{bmatrix} -1 & 2 \\ 3 & -6 \end{bmatrix}$

Solution (a) The determinant of A is $\det(A) = (6)(2) - (1)(5) = 7$, which is nonzero. Thus, A is invertible, and its inverse is

$$A^{-1} = \frac{1}{7}\begin{bmatrix} 2 & -1 \\ -5 & 6 \end{bmatrix} = \begin{bmatrix} \frac{2}{7} & -\frac{1}{7} \\ -\frac{5}{7} & \frac{6}{7} \end{bmatrix}$$

We leave it for you to confirm that $AA^{-1} = A^{-1}A = I$.

Solution (b) Since $\det(A) = (-1)(-6) - (2)(3) = 0$, the matrix A is not invertible.   ∎

EXAMPLE 7
Solution of a
Linear System
by Matrix
Inversion

A problem that arises in many applications is to solve equations of the form

$$u = ax + by$$
$$v = cx + dy$$

for x and y in terms of u and v. One approach is to treat this as a linear system of two equations in the unknowns x and y and use Gauss–Jordan elimination to solve for x and y. However, because the coefficients of the unknowns are *literal* rather than *numerical*, this procedure is a little clumsy. As an alternative approach, let us replace the two equations by the single matrix equation

$$\begin{bmatrix} u \\ v \end{bmatrix} = \begin{bmatrix} ax + by \\ cx + dy \end{bmatrix}$$

which we can rewrite as

$$\begin{bmatrix} u \\ v \end{bmatrix} = \begin{bmatrix} a & b \\ c & d \end{bmatrix}\begin{bmatrix} x \\ y \end{bmatrix}$$

If we assume that the 2×2 matrix is invertible (i.e., $ad - bc \neq 0$), then we can multiply through on the left by the inverse and rewrite the equation as

$$\begin{bmatrix} a & b \\ c & d \end{bmatrix}^{-1}\begin{bmatrix} u \\ v \end{bmatrix} = \begin{bmatrix} a & b \\ c & d \end{bmatrix}^{-1}\begin{bmatrix} a & b \\ c & d \end{bmatrix}\begin{bmatrix} x \\ y \end{bmatrix}$$

which simplifies to

$$\begin{bmatrix} a & b \\ c & d \end{bmatrix}^{-1}\begin{bmatrix} u \\ v \end{bmatrix} = \begin{bmatrix} x \\ y \end{bmatrix}$$

Using Theorem 3.2.7, we can rewrite this equation as

$$\frac{1}{ad - bc}\begin{bmatrix} d & -b \\ -c & a \end{bmatrix}\begin{bmatrix} u \\ v \end{bmatrix} = \begin{bmatrix} x \\ y \end{bmatrix}$$

from which we obtain

$$x = \frac{du - bv}{ad - bc}, \quad y = \frac{av - cu}{ad - bc}$$ ∎

REMARK Readers who are familiar with determinants and Cramer's rule may want to check that the solution obtained in this example is consistent with that rule. Readers who are not familiar with Cramer's rule will learn about it in Chapter 4.

EXAMPLE 8
An Application
to Robotics

Figure 3.2.2 shows a diagram of a simplified industrial robot. The robot consists of two arms that can be rotated independently through angles α and β and that can be "telescoped" independently to lengths l_1 and l_2. For fixed angles α and β, what should the lengths of the arms be in order to position the tip of the working arm at the point (x, y) shown in the figure?

Solution We leave it for you to use basic trigonometry to show that

$$x = l_1 \cos\alpha + l_2 \cos\beta$$
$$y = l_1 \sin\alpha + l_2 \sin\beta$$

Thus, the problem is to solve these equations for l_1 and l_2 in terms of x and y. Proceeding as in the last example, we can rewrite the two equations as the single matrix equation

$$\begin{bmatrix} x \\ y \end{bmatrix} = \begin{bmatrix} \cos\alpha & \cos\beta \\ \sin\alpha & \sin\beta \end{bmatrix} \begin{bmatrix} l_1 \\ l_2 \end{bmatrix} \tag{6}$$

The determinant of the 2×2 matrix is $\cos\alpha\sin\beta - \sin\alpha\cos\beta = \sin(\beta - \alpha)$. Thus, if $\beta - \alpha$ is not a multiple of π (radians), then the determinant will be nonzero, and the 2×2 matrix will be invertible. In this case we can rewrite (6) as

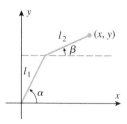

$$\begin{bmatrix} l_1 \\ l_2 \end{bmatrix} = \begin{bmatrix} \cos\alpha & \cos\beta \\ \sin\alpha & \sin\beta \end{bmatrix}^{-1} \begin{bmatrix} x \\ y \end{bmatrix} = \frac{1}{\sin(\beta - \alpha)} \begin{bmatrix} \sin\beta & -\cos\beta \\ -\sin\alpha & \cos\alpha \end{bmatrix} \begin{bmatrix} x \\ y \end{bmatrix}$$

from which it follows that

$$l_1 = \frac{\sin\beta}{\sin(\beta - \alpha)}x - \frac{\cos\beta}{\sin(\beta - \alpha)}y \quad \text{and} \quad l_2 = -\frac{\sin\alpha}{\sin(\beta - \alpha)}x + \frac{\cos\alpha}{\sin(\beta - \alpha)}y \qquad \blacksquare$$

Figure 3.2.2

The next theorem is concerned with inverses of matrix products.

Theorem 3.2.8   *If A and B are invertible matrices with the same size, then AB is invertible and*

$$(AB)^{-1} = B^{-1}A^{-1}$$

Proof We can establish the invertibility and obtain the formula at the same time by showing that

$$(AB)(B^{-1}A^{-1}) = (B^{-1}A^{-1})(AB) = I$$

But

$$(AB)(B^{-1}A^{-1}) = A(BB^{-1})A^{-1} = AIA^{-1} = AA^{-1} = I$$

and similarly, $(B^{-1}A^{-1})(AB) = I$.   ∎

Although we will not prove it, this result can be extended to three or more factors:

A product of any number of invertible matrices is invertible, and the inverse of the product is the product of the inverses in the reverse order.

REMARK It follows logically from this statement that if a product of matrices is singular, then at least one of the factors is singular.

Linear Algebra in History

Sometimes even brilliant mathematicians have feet of clay. For example, the great English mathematician Arthur Cayley (p. 81), whom some call the "father" of matrix theory, asserted that if the product of two nonzero square matrices, A and B, is zero, then *at least one* of the factors must be singular. Cayley was correct, but he surprisingly overlooked an important point, namely that if $AB = 0$, then A and B must *both* be singular. Why?

EXAMPLE 9
The Inverse of a Product

Consider the matrices

$$A = \begin{bmatrix} 1 & 2 \\ 1 & 3 \end{bmatrix}, \quad B = \begin{bmatrix} 3 & 2 \\ 2 & 2 \end{bmatrix}$$

We leave it for you to show that

$$AB = \begin{bmatrix} 7 & 6 \\ 9 & 8 \end{bmatrix}, \quad (AB)^{-1} = \begin{bmatrix} 4 & -3 \\ -\frac{9}{2} & \frac{7}{2} \end{bmatrix}$$

and also that

$$A^{-1} = \begin{bmatrix} 3 & -2 \\ -1 & 1 \end{bmatrix}, \quad B^{-1} = \begin{bmatrix} 1 & -1 \\ -1 & \frac{3}{2} \end{bmatrix}, \quad B^{-1}A^{-1} = \begin{bmatrix} 1 & -1 \\ -1 & \frac{3}{2} \end{bmatrix}\begin{bmatrix} 3 & -2 \\ -1 & 1 \end{bmatrix} = \begin{bmatrix} 4 & -3 \\ -\frac{9}{2} & \frac{7}{2} \end{bmatrix}$$

Thus, $(AB)^{-1} = B^{-1}A^{-1}$ as guaranteed by Theorem 3.2.8.   ∎

POWERS OF A MATRIX   If A is a *square* matrix, then we define the nonnegative integer powers of A to be

$$A^0 = I \quad \text{and} \quad A^n = AA \cdots A \qquad \text{[n factors]}$$

and if A is invertible, then we define the negative integer powers of A to be

$$A^{-n} = (A^{-1})^n = A^{-1}A^{-1} \cdots A^{-1} \qquad \text{[n factors]}$$

Because these definitions parallel those for real numbers, the usual laws of nonnegative exponents hold; for example,

$$A^r A^s = A^{r+s} \quad \text{and} \quad (A^r)^s = A^{rs}$$

In addition, we have the following properties of negative exponents.

> **Theorem 3.2.9** *If A is invertible and n is a nonnegative integer, then:*
>   (a) A^{-1} *is invertible and* $(A^{-1})^{-1} = A$.
>   (b) A^n *is invertible and* $(A^n)^{-1} = A^{-n} = (A^{-1})^n$.
>   (c) kA *is invertible for any nonzero scalar k, and* $(kA)^{-1} = k^{-1}A^{-1}$.

We will prove part (c) and leave the proofs of parts (a) and (b) as exercises.

Proof (c) Property (c) in Theorem 3.2.1 and property (f) in Theorem 3.2.2 imply that

$$(kA)(k^{-1}A^{-1}) = k^{-1}(kA)A^{-1} = (k^{-1}k)AA^{-1} = (1)I = I$$

and similarly, $(k^{-1}A^{-1})(kA) = I$. Thus, kA is invertible and $(kA)^{-1} = k^{-1}A^{-1}$. ∎

EXAMPLE 10
Properties of
Exponents

Let A and A^{-1} be the matrices in Example 9; that is,

$$A = \begin{bmatrix} 1 & 2 \\ 1 & 3 \end{bmatrix} \quad \text{and} \quad A^{-1} = \begin{bmatrix} 3 & -2 \\ -1 & 1 \end{bmatrix}$$

Then

$$A^{-3} = (A^{-1})^3 = \begin{bmatrix} 3 & -2 \\ -1 & 1 \end{bmatrix}\begin{bmatrix} 3 & -2 \\ -1 & 1 \end{bmatrix}\begin{bmatrix} 3 & -2 \\ -1 & 1 \end{bmatrix} = \begin{bmatrix} 41 & -30 \\ -15 & 11 \end{bmatrix}$$

Also,

$$A^3 = \begin{bmatrix} 1 & 2 \\ 1 & 3 \end{bmatrix}\begin{bmatrix} 1 & 2 \\ 1 & 3 \end{bmatrix}\begin{bmatrix} 1 & 2 \\ 1 & 3 \end{bmatrix} = \begin{bmatrix} 11 & 30 \\ 15 & 41 \end{bmatrix}$$

so, as expected from Theorem 3.2.9(b),

$$(A^3)^{-1} = \frac{1}{(11)(41) - (30)(15)}\begin{bmatrix} 41 & -30 \\ -15 & 11 \end{bmatrix} = \begin{bmatrix} 41 & -30 \\ -15 & 11 \end{bmatrix} = (A^{-1})^3 \qquad ■$$

EXAMPLE 11
The Square of a
Matrix Sum

In the arithmetic of real numbers, we can write

$$(a + b)^2 = a^2 + ab + ba + b^2 = a^2 + ab + ab + b^2 = a^2 + 2ab + b^2$$

However, in the arithmetic of matrices the commutative law for multiplication does not hold, so for general square matrices with the same size the best we can do is write

$$(A + B)^2 = A^2 + AB + BA + B^2$$

It is only in the special case where A and B commute (i.e., $AB = BA$) that we can go a step further and write

$$(A + B)^2 = A^2 + 2AB + B^2 \qquad ■$$

MATRIX POLYNOMIALS

If A is a square matrix, say $n \times n$, and if

$$p(x) = a_0 + a_1 x + a_2 x^2 + \cdots + a_m x^m$$

is any polynomial, then we define the $n \times n$ matrix $p(A)$ to be

$$p(A) = a_0 I + a_1 A + a_2 A^2 + \cdots + a_m A^m \tag{7}$$

where I is the $n \times n$ identity matrix; that is, $p(A)$ is obtained by substituting A for x and replacing the constant term a_0 by the matrix $a_0 I$. An expression of form (7) is called a ***matrix polynomial in A***.

EXAMPLE 12
A Matrix
Polynomial

Find $p(A)$ for

$$p(x) = x^2 - 2x - 3 \quad \text{and} \quad A = \begin{bmatrix} -1 & 2 \\ 0 & 3 \end{bmatrix}$$

Solution

$$
\begin{aligned}
p(A) &= A^2 - 2A - 3I \\
&= \begin{bmatrix} -1 & 2 \\ 0 & 3 \end{bmatrix}^2 - 2\begin{bmatrix} -1 & 2 \\ 0 & 3 \end{bmatrix} - 3\begin{bmatrix} 1 & 0 \\ 0 & 1 \end{bmatrix} \\
&= \begin{bmatrix} 1 & 4 \\ 0 & 9 \end{bmatrix} - \begin{bmatrix} -2 & 4 \\ 0 & 6 \end{bmatrix} - \begin{bmatrix} 3 & 0 \\ 0 & 3 \end{bmatrix} = \begin{bmatrix} 0 & 0 \\ 0 & 0 \end{bmatrix}
\end{aligned}
$$

or more briefly, $p(A) = 0$. ∎

REMARK It follows from the fact that $A^r A^s = A^{r+s} = A^{s+r} = A^s A^r$ that powers of a square matrix commute, and since a matrix polynomial in A is built up from powers of A, any two matrix polynomials in A also commute; that is, for any polynomials p_1 and p_2 we have

$$p_1(A) p_2(A) = p_2(A) p_1(A) \tag{8}$$

Moreover, it can be proved that if $p(x) = p_1(x) p_2(x)$, then

$$p(A) = p_1(A) p_2(A) \tag{9}$$

We omit the proof.

PROPERTIES OF THE TRANSPOSE

The following theorem lists the main properties of the transpose.

> **Theorem 3.2.10** *If the sizes of the matrices are such that the stated operations can be performed, then:*
> (a) $(A^T)^T = A$
> (b) $(A + B)^T = A^T + B^T$
> (c) $(A - B)^T = A^T - B^T$
> (d) $(kA)^T = kA^T$
> (e) $(AB)^T = B^T A^T$

If you keep in mind that transposing a matrix interchanges its rows and columns, then you should have little trouble visualizing the results in parts (a)–(d). For example, part (a) states the obvious fact that interchanging rows and columns twice leaves a matrix unchanged; and part (b) states that adding two matrices and then interchanging the rows and columns produces the same result as interchanging the rows and columns before adding. We will omit the formal proofs. The result in part (e) is not so obvious, so we will prove it.

Proof (e) For AB to be defined, the number of columns of A must be the same as the number of rows of B. Thus, assume that A has size $m \times s$ and B has size $s \times n$. This implies that A^T has size $s \times m$, and B^T has size $n \times s$, so the sizes that conform for AB also conform for $B^T A^T$, and the products $(AB)^T$ and $B^T A^T$ both have size $n \times m$.

The key to the proof is to establish the following relationship between row-column products of AB and row-column products of $B^T A^T$:

$$\mathbf{r}_i(A)\mathbf{c}_j(B) = \mathbf{r}_j(B^T)\mathbf{c}_i(A^T) \tag{10}$$

Once we have proved this, the equality $(AB)^T = B^T A^T$ can be established by the following argument, which shows that the corresponding entries on the two sides of this equation are the same:

$$
\begin{aligned}
\left((AB)^T\right)_{ji} &= (AB)_{ij} && \textbf{[Formula (22), Section 3.1]} \\
&= \mathbf{r}_i(A)\mathbf{c}_j(B) && \textbf{[Row-column rule]} \\
&= \mathbf{r}_j(B^T)\mathbf{c}_i(A^T) && \textbf{[Formula (10)]} \\
&= (B^T A^T)_{ji} && \textbf{[Row-column rule]}
\end{aligned}
$$

To prove (10) we will write out the products, keeping in mind that we have assumed that A has size $m \times s$ and B has size $s \times n$:

$$\mathbf{r}_i(A)\mathbf{c}_j(B) = [a_{i1} \quad a_{i2} \quad \cdots \quad a_{is}] \begin{bmatrix} b_{1j} \\ b_{2j} \\ \vdots \\ b_{sj} \end{bmatrix} = a_{i1}b_{1j} + a_{i2}b_{2j} + \cdots + a_{is}b_{sj}$$

$$\mathbf{r}_j(B^T)\mathbf{c}_i(A^T) = [b_{1j} \quad b_{2j} \quad \cdots \quad b_{sj}] \begin{bmatrix} a_{i1} \\ a_{i2} \\ \vdots \\ a_{is} \end{bmatrix} = a_{i1}b_{1j} + a_{i2}b_{2j} + \cdots + a_{is}b_{sj}$$

from which (10) follows.   ∎

REMARK Although we will not prove it, part (*e*) of this theorem can be extended to three or more factors:

> *The transpose of a product of any number of matrices is the product of the transposes in the reverse order.*

The following theorem establishes a relationship between the inverse of a matrix and the inverse of its transpose.

Theorem 3.2.11 *If A is an invertible matrix, then A^T is also invertible and*

$$(A^T)^{-1} = (A^{-1})^T$$

Proof We can establish the invertibility and obtain the formula at the same time by showing that

$$A^T(A^{-1})^T = (A^{-1})^T A^T = I$$

But from part (*e*) of Theorem 3.2.10 and the fact that $I^T = I$, we have

$$A^T(A^{-1})^T = (A^{-1}A)^T = I^T = I$$
$$(A^{-1})^T A^T = (AA^{-1})^T = I^T = I$$

which completes the proof.   ∎

EXAMPLE 13
Inverse of a
Transpose

Consider a general 2×2 invertible matrix and its transpose:

$$A = \begin{bmatrix} a & b \\ c & d \end{bmatrix} \quad \text{and} \quad A^T = \begin{bmatrix} a & c \\ b & d \end{bmatrix}$$

Since A is invertible, its determinant $ad - bc$ is nonzero. But the determinant of A^T is also $ad - bc$ (verify), so A^T is also invertible. It follows from Theorem 3.2.7 that

$$(A^T)^{-1} = \begin{bmatrix} \dfrac{d}{ad - bc} & -\dfrac{c}{ad - bc} \\ -\dfrac{b}{ad - bc} & \dfrac{a}{ad - bc} \end{bmatrix}$$

which is the same matrix that results if A^{-1} is transposed (verify). Thus, $(A^T)^{-1} = (A^{-1})^T$ as guaranteed by Theorem 3.2.10. ■

PROPERTIES OF THE TRACE

The following theorem lists the main properties of the trace.

> **Theorem 3.2.12** *If A and B are square matrices with the same size, then:*
> (a) $\text{tr}(A^T) = \text{tr}(A)$
> (b) $\text{tr}(cA) = c\,\text{tr}(A)$
> (c) $\text{tr}(A + B) = \text{tr}(A) + \text{tr}(B)$
> (d) $\text{tr}(A - B) = \text{tr}(A) - \text{tr}(B)$
> (e) $\text{tr}(AB) = \text{tr}(BA)$

The result in part (a) is evident because transposing a square matrix reflects the entries about the main diagonal, leaving the main diagonal fixed. Thus, A and A^T have the same main diagonal and hence the same trace. Parts (b) through (d) follow easily from properties of the matrix operations. Part (e) can be proved by writing out the sums on the two sides (Exercise P8), but we will also give a more insightful proof later in this chapter.

EXAMPLE 14
Trace of a Product

Part (e) of Theorem 3.2.12 is rather interesting because it states that for square matrices A and B of the same size, the products AB and BA have the same trace, even if $AB \neq BA$. For example, if A and B are the matrices in Example 1, then $AB \neq BA$, yet

$$\text{tr}(AB) = \text{tr}(BA) = 3$$

(verify). ■

The following result about products of row and column vectors will be useful in our later work.

> **Theorem 3.2.13** *If $\mathbf{r}$ is a $1 \times n$ row vector and $\mathbf{c}$ is an $n \times 1$ column vector, then*
>
> $$\mathbf{rc} = \text{tr}(\mathbf{cr}) \tag{11}$$

Proof Since $\mathbf{r}$ is a row vector, it follows that $\mathbf{r}^T$ is a column vector, so we can apply Formula (25) of Section 3.1 with $\mathbf{u} = \mathbf{r}^T$ and $\mathbf{v} = \mathbf{c}$. This yields

$$
\begin{aligned}
\mathbf{rc} &= \mathbf{u}^T\mathbf{v} && \textbf{[Theorem 3.2.10(a)]} \\
&= \text{tr}(\mathbf{uv}^T) && \textbf{[Formula (25), Section 3.1]} \\
&= \text{tr}(\mathbf{r}^T\mathbf{c}^T) && \textbf{[Definition of u and v]} \\
&= \text{tr}((\mathbf{cr})^T) && \textbf{[Theorem 3.2.10(e)]} \\
&= \text{tr}(\mathbf{cr}) && \textbf{[Theorem 3.2.12(a)]}
\end{aligned}
$$

■

EXAMPLE 15
Trace of a
Column Vector
Times a Row
Vector

Let $\mathbf{r} = [1 \quad 2]$ and $\mathbf{c} = \begin{bmatrix} 3 \\ 4 \end{bmatrix}$. Then

$$\mathbf{rc} = [1 \quad 2] \begin{bmatrix} 3 \\ 4 \end{bmatrix} = (1)(3) + (2)(4) = 11 \quad \text{and} \quad \mathbf{cr} = \begin{bmatrix} 3 \\ 4 \end{bmatrix}[1 \quad 2] = \begin{bmatrix} 3 & 6 \\ 4 & 8 \end{bmatrix}$$

Thus, $\text{tr}(\mathbf{cr}) = 3 + 8 = 11 = \mathbf{rc}$ as guaranteed by Theorem 3.2.13.   ■

TRANSPOSE AND DOT PRODUCT

We conclude this section with two formulas that provide an important link between multiplication by A and multiplication by A^T.

Recall from Formula (26) of Section 3.1 that if $\mathbf{u}$ and $\mathbf{v}$ are *column vectors*, then their dot product can be expressed as the matrix product $\mathbf{u} \cdot \mathbf{v} = \mathbf{v}^T\mathbf{u}$. Thus, if A is an $n \times n$ matrix, then

$$A\mathbf{u} \cdot \mathbf{v} = \mathbf{v}^T(A\mathbf{u}) = (\mathbf{v}^TA)\mathbf{u} = (A^T\mathbf{v})^T\mathbf{u} = \mathbf{u} \cdot A^T\mathbf{v}$$

$$\mathbf{u} \cdot A\mathbf{v} = (A\mathbf{v})^T\mathbf{u} = (\mathbf{v}^TA^T)\mathbf{u} = \mathbf{v}^T(A^T\mathbf{u}) = A^T\mathbf{u} \cdot \mathbf{v}$$

so we have established the formulas

$$A\mathbf{u} \cdot \mathbf{v} = \mathbf{u} \cdot A^T\mathbf{v} \tag{12}$$

$$\mathbf{u} \cdot A\mathbf{v} = A^T\mathbf{u} \cdot \mathbf{v} \tag{13}$$

In words, these formulas tell us that *in expressions of the form $A\mathbf{u} \cdot \mathbf{v}$ or $\mathbf{u} \cdot A\mathbf{v}$ the matrix A can be moved across the dot product sign by transposing A.* Some problems that use these formulas are given in the exercises.

Exercise Set 3.2

In Exercises 1–8, use the following matrices and scalars:

$$A = \begin{bmatrix} 2 & -1 & 3 \\ 0 & 4 & 5 \\ -2 & 1 & 4 \end{bmatrix}, \quad B = \begin{bmatrix} 8 & -3 & -5 \\ 0 & 1 & 2 \\ 4 & -7 & 6 \end{bmatrix}$$

$$C = \begin{bmatrix} 0 & -2 & 3 \\ 1 & 7 & 4 \\ 3 & 5 & 9 \end{bmatrix}, \quad a = 4, \quad b = -7$$

1. Confirm the following statements from Theorems 3.2.1 and 3.2.2.
   (a) $A + (B + C) = (A + B) + C$
   (b) $(AB)C = A(BC)$
   (c) $(a + b)C = aC + bC$
   (d) $a(B - C) = aB - aC$

2. Confirm the following statements from Theorems 3.2.1 and 3.2.2.
   (a) $a(BC) = (aB)C = B(aC)$
   (b) $A(B - C) = AB - AC$
   (c) $(B + C)A = BA + CA$
   (d) $a(bC) = (ab)C$

3. Confirm the following statements from Theorem 3.2.10.
   (a) $(A^T)^T = A$
   (b) $(A + B)^T = A^T + B^T$
   (c) $(3C)^T = 3C^T$
   (d) $(AB)^T = B^TA^T$

4. Confirm the following statements from Theorem 3.2.10.
   (a) $(B^T)^T = B$
   (b) $(B - C)^T = B^T - C^T$
   (c) $(4B)^T = 4B^T$
   (d) $(BC)^T = C^TB^T$

5. Confirm the following statements from Theorem 3.2.12.
   (a) $\text{tr}(A^T) = \text{tr}(A)$
   (b) $\text{tr}(3A) = 3\,\text{tr}(A)$
   (c) $\text{tr}(A + B) = \text{tr}(A) + \text{tr}(B)$
   (d) $\text{tr}(AB) = \text{tr}(BA)$

6. Confirm the following statements from Theorem 3.2.12.
   (a) $\text{tr}(C^T) = \text{tr}(C)$
   (b) $\text{tr}(3C) = 3\,\text{tr}(C)$
   (c) $\text{tr}(A - B) = \text{tr}(A) - \text{tr}(B)$
   (d) $\text{tr}(BC) = \text{tr}(CB)$

7. In each part, find a matrix X that satisfies the equation.
   (a) $\text{tr}(B)A + 3X = BC$ (b) $B + (A + X)^T = C$

8. In each part, find a matrix X that satisfies the equation.
   (a) $\text{tr}(2C)C + 2X = B$ (b) $B + (\text{tr}(A)X)^T = C$

In Exercises 9–18, use Theorem 3.2.7 and the following matrices:

$$A = \begin{bmatrix} 3 & 1 \\ 5 & 2 \end{bmatrix}, \quad B = \begin{bmatrix} 2 & -3 \\ 4 & 4 \end{bmatrix}$$

$$C = \begin{bmatrix} 6 & 4 \\ -2 & -1 \end{bmatrix}, \quad D = \begin{bmatrix} 2 & 0 \\ 0 & 3 \end{bmatrix}$$

9. (a) Find A^{-1}.
  (b) Confirm that $(A^{-1})^{-1} = A$.
  (c) Confirm that $(A^T)^{-1} = (A^{-1})^T$.
  (d) Confirm that $(2A)^{-1} = \frac{1}{2}A^{-1}$.

10. (a) Find B^{-1}.
  (b) Confirm that $(B^{-1})^{-1} = B$.
  (c) Confirm that $(B^T)^{-1} = (B^{-1})^T$.
  (d) Confirm that $(3B)^{-1} = \frac{1}{3}B^{-1}$.

11. (a) Confirm that $(AB)^{-1} = B^{-1}A^{-1}$.
  (b) Confirm that $(ABC)^{-1} = C^{-1}B^{-1}A^{-1}$.

12. (a) Confirm that $(BC)^{-1} = C^{-1}B^{-1}$.
  (b) Confirm that $(BCD)^{-1} = D^{-1}C^{-1}B^{-1}$.

13. Find a matrix X, if any, that satisfies the equation $AX + B = BC$.

14. Find a matrix X, if any, that satisfies the equation $BX + AB = CX$.

15. (a) Find A^{-2}.
  (b) Find $p(A)$ for $p(x) = x + 2$.
  (c) Find $p(A)$ for $p(x) = x^2 - 2x + 1$.

16. (a) Find A^{-3}.
  (b) Find $p(A)$ for $p(x) = 3x - 1$.
  (c) Find $p(A)$ for $p(x) = x^3 - 2x + 4$.

17. Find $(AB)^{-1}(AC^{-1})(D^{-1}C^{-1})^{-1}D^{-1}$ by first simplifying as much as possible.

18. Find $(AC^{-1})^{-1}(AC^{-1})(AC^{-1})^{-1}AD^{-1}$ by first simplifying as much as possible.

In Exercises 19 and 20, use the given information to find A.

19. (a) $A^{-1} = \begin{bmatrix} 2 & -1 \\ 3 & 5 \end{bmatrix}$ (b) $(7A)^{-1} = \begin{bmatrix} -3 & 7 \\ 1 & -2 \end{bmatrix}$

20. (a) $(5A^T)^{-1} = \begin{bmatrix} -3 & -1 \\ 5 & 2 \end{bmatrix}$ (b) $(I + 2A)^{-1} = \begin{bmatrix} -1 & 2 \\ 4 & 5 \end{bmatrix}$

In Exercises 21 and 22, find all values of c, if any, for which A is invertible.

21. $A = \begin{bmatrix} c & 1 \\ c & c \end{bmatrix}$ **22.** $A = \begin{bmatrix} -c & -1 \\ 1 & c \end{bmatrix}$

23. Find a nonzero 3×3 matrix A such that $A^T = A$.

24. Find a nonzero 3×3 matrix A such that $A^T = -A$.

In Exercises 25 and 26, determine whether the matrix is invertible by investigating the equation $AX = I$. If it is, then find the inverse.

25. $A = \begin{bmatrix} 1 & 0 & 1 \\ 1 & 1 & 0 \\ 0 & 1 & 1 \end{bmatrix}$ **26.** $A = \begin{bmatrix} 1 & 0 & 0 \\ 0 & 1 & 0 \\ 2 & 0 & 1 \end{bmatrix}$

In Exercises 27 and 28, use the following matrices:

$$A = \begin{bmatrix} 3 & 2 & -1 \\ 1 & 5 & 0 \\ -2 & 4 & 6 \end{bmatrix}, \quad \mathbf{u} = \begin{bmatrix} 3 \\ -2 \\ 4 \end{bmatrix}$$

$$\mathbf{v} = \begin{bmatrix} 1 \\ -1 \\ 3 \end{bmatrix}, \quad \mathbf{c} = \begin{bmatrix} 8 \\ 0 \\ 4 \end{bmatrix}, \quad \mathbf{r} = [0 \; -2 \; 3]$$

27. (a) Confirm the relationship $\mathbf{rc} = \text{tr}(\mathbf{cr})$ stated in Theorem 3.2.13.
  (b) Confirm the relationship $A\mathbf{u} \cdot \mathbf{v} = \mathbf{u} \cdot A^T\mathbf{v}$ given in Formula (12).

28. (a) Use Theorem 3.2.13 to find a matrix whose trace is $\mathbf{u}^T\mathbf{v}$.
  (b) Confirm the relationship $\mathbf{u} \cdot A\mathbf{v} = A^T\mathbf{u} \cdot \mathbf{v}$ given in Formula (13).

29. We showed in Example 2 that the cancellation law does not hold for matrix multiplication. However, show that if A is invertible and $AB = AC$, then $B = C$.

30. We showed in Example 3 that it is possible to have nonzero matrices A and C for which $AC = 0$. However, show that if A is invertible and $AC = 0$, then $C = 0$; similarly, if C is invertible and $AC = 0$, then $A = 0$.

31. (a) Find all values of θ for which

$$A = \begin{bmatrix} \cos\theta & \sin\theta \\ -\sin\theta & \cos\theta \end{bmatrix}$$

is invertible, and find its inverse for those values.
  (b) Use the inverse obtained in part (a) to solve the following system of equations for x' and y' in terms of x and y:

$$x = x'\cos\theta + y'\sin\theta$$
$$y = -x'\sin\theta + y'\cos\theta$$

32. A square matrix A is said to be ***idempotent*** if $A^2 = A$.
  (a) Show that if A is idempotent, then so is $I - A$.
  (b) Show that if A is idempotent, then $2A - I$ is invertible and is its own inverse.

33. (a) Show that if A, B, and $A + B$ are invertible matrices with the same size, then

$$A(A^{-1} + B^{-1})B(A + B)^{-1} = I$$

  (b) What does the result in part (a) tell you about the matrix $A^{-1} + B^{-1}$?

34. Let $\mathbf{u}$ and $\mathbf{v}$ be column vectors in R^n, and let $A = I + \mathbf{uv}^T$. Show that if $\mathbf{u}^T\mathbf{v} \neq -1$, then A is invertible and

$$A^{-1} = I - \frac{1}{1 + \mathbf{u}^T\mathbf{v}}\mathbf{uv}^T$$

35. Show that if $p(x) = x^2 - (a + d)x + (ad - bc)$ and

$$A = \begin{bmatrix} a & b \\ c & d \end{bmatrix}$$

then $p(A) = 0$.

36. Show that the relationship $AB - BA = I_n$ is impossible for two $n \times n$ matrices A and B. [*Hint:* Consider traces.]

37. Let A be the adjacency matrix of the directed graph shown in the accompanying figure. Compute A^2 and verify that the entry in the ij position of A^2 is the number of different ways of traveling from i to j, following the arrows, and having one intermediate step. Find a similar interpretation for the powers A^n, $n > 2$.

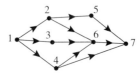

Figure Ex-37

Discussion and Discovery

D1. (a) Give an example of two matrices with the same size such that $(A + B)(A - B) \neq A^2 - B^2$.

    (b) State a valid formula for multiplying out

$$(A + B)(A - B)$$

    (c) What condition can you impose on A and B that will allow you to write $(A + B)(A - B) = A^2 - B^2$?

D2. The numerical equation $a^2 = 1$ has exactly two solutions. Find at least eight solutions of the matrix equation $A^2 = I_3$. [*Hint:* Look for solutions in which all entries off the main diagonal are zero.]

D3. (a) Show that if a square matrix A satisfies the equation $A^2 + 2A + I = 0$, then A must be invertible. What is the inverse?

    (b) Show that if $p(x)$ is a polynomial with a nonzero constant term, and if A is a square matrix for which $p(A) = 0$, then A is invertible.

D4. Is it possible for A^3 to be an identity matrix without A being invertible? Explain.

D5. Indicate whether the statement is true (T) or false (F). Justify your answer.

    (a) If A and B are square matrices of the same size, then $(AB)^2 = A^2 B^2$.

    (b) If A and B are square matrices with the same size, then $(A - B)^2 = (B - A)^2$.

    (c) If A is an invertible matrix and n is a positive integer, then $(A^{-n})^T = (A^T)^{-n}$.

    (d) If A and B are square matrices with the same size, then $\text{tr}(AB) = \text{tr}(A)\text{tr}(B)$.

    (e) If A and B are invertible matrices with the same size, then $A + B$ is invertible.

D6. (a) Let $\mathbf{e}_1$, $\mathbf{e}_2$, and $\mathbf{e}_3$ be the standard unit vectors in R^3 expressed in column form, and let A be an invertible matrix. Show that the linear systems $A\mathbf{x} = \mathbf{e}_1$, $A\mathbf{x} = \mathbf{e}_2$, and $A\mathbf{x} = \mathbf{e}_3$ are consistent and that their solutions are the successive column vectors of A^{-1}.

    (b) Use the result in part (a) to find the inverse of the matrix

$$A = \begin{bmatrix} 1 & 2 & 0 \\ 1 & 3 & 0 \\ 0 & 0 & 4 \end{bmatrix}$$

D7. There are sixteen 2×2 matrices that can be made using only the entries 0 and 1. How many of them have inverses?

D8. If A and B are invertible $n \times n$ matrices for which $AB = BA$, must it also be true that $A^{-1}B^{-1} = B^{-1}A^{-1}$? Justify your answer.

Working with Proofs

P1. Prove Theorem 3.2.1(*c*). [*Suggestion:* See the proof of part (*b*) given in the text.]

P2. Prove Theorem 3.2.1(*f*).

P3. Prove Theorem 3.2.2(*d*).

P4. Prove Theorem 3.2.2(*f*).

P5. Prove Theorem 3.2.3(*e*).

P6. Prove parts (*a*) and (*b*) of Theorem 3.2.9.

P7. (a) Confirm the validity of Formula (5) by computing AA^{-1} and $A^{-1}A$.

    (b) The computation in part (a) establishes that A is invertible if $ad - bc \neq 0$. Prove that if $ad - bc = 0$,

then A is not invertible. [*Suggestion:* First treat the case where $a, b, c,$ and d are all nonzero, and then turn to cases where one or more of these entries is zero.]

P8. (*Summation notation*) Prove that if A and B are any two $n \times n$ matrices, then $\text{tr}(AB) = \text{tr}(BA)$. [*Hint:* Show that $\text{tr}(AB)$ can be expressed in sigma notation as

$$\text{tr}(AB) = \sum_{i=1}^{n} \left(\sum_{s=1}^{n} a_{is} b_{si} \right)$$

and find a similar expression for $\text{tr}(BA)$. Show that the two expressions are equal.]

Technology Exercises

T1. (*Matrix powers*) Compute various positive powers of the matrix

$$A = \begin{bmatrix} 1 & 2 & -3 & 0 \\ 1 & 1 & -2 & 1 \\ 2 & 1 & 3 & 4 \\ -3 & 2 & 2 & -8 \end{bmatrix}$$

T2. Compute $A^5 - 3A^3 + 7A - 4I$ for the matrix A in Exercise T1.

T3. Confirm Formulas (12) and (13) for

$$A = \begin{bmatrix} 1 & 2 & -3 & 0 \\ 1 & 1 & -2 & 1 \\ 2 & 1 & 3 & 4 \\ -3 & 2 & 2 & -8 \end{bmatrix}, \quad \mathbf{u} = \begin{bmatrix} -1 \\ 2 \\ 3 \\ 5 \end{bmatrix}, \quad \mathbf{v} = \begin{bmatrix} 2 \\ 0 \\ -4 \\ 1 \end{bmatrix}$$

T4. Let

$$A = \begin{bmatrix} 0 & \frac{1}{2} & \frac{1}{3} \\ \frac{1}{4} & 0 & \frac{1}{5} \\ \frac{1}{6} & \frac{1}{7} & 0 \end{bmatrix}$$

Discuss the behavior of A^k as k increases indefinitely, that is, as $k \to \infty$.

T5. (a) Show that the following matrix is idempotent (see Exercise 32):

$$A = \begin{bmatrix} \frac{5}{9} & -\frac{4}{9} & \frac{2}{9} \\ -\frac{4}{9} & \frac{5}{9} & \frac{2}{9} \\ \frac{2}{9} & \frac{2}{9} & \frac{8}{9} \end{bmatrix}$$

(b) Confirm the statements in parts (a) and (b) of Exercise 32.

T6. A square matrix A is said to be *nilpotent* if $A^k = 0$ for some positive integer k. The smallest value of k for which this equation holds is called the *index of nilpotency*. In each part, confirm that the matrix is nilpotent, and find the index of nilpotency.

(a) $A = \begin{bmatrix} 0 & 0 & 0 & 0 \\ 1 & 0 & 0 & 0 \\ 2 & 1 & 0 & 0 \\ -3 & 2 & 2 & 0 \end{bmatrix}$ (b) $\begin{bmatrix} 0 & 0 & 0 & 0 \\ 1 & 0 & 0 & 0 \\ 2 & 1 & 0 & 0 \\ 0 & 2 & 0 & 0 \end{bmatrix}$

T7. (**CAS**) Make a conjecture about the form of A^n for positive integer powers of n.

(a) $A = \begin{bmatrix} a & 1 \\ 0 & a \end{bmatrix}$ (b) $A = \begin{bmatrix} \cos\theta & \sin\theta \\ -\sin\theta & \cos\theta \end{bmatrix}$

Section 3.3   Elementary Matrices; A Method for Finding A^{-1}

In the last section we showed how to find the inverse of a 2×2 matrix. In this section we will develop an algorithm that can be used to find the inverse of an invertible matrix of any order, and we will discuss some basic properties of invertible matrices.

ELEMENTARY MATRICES   Recall from Section 2.1 that there are three types of elementary row operations that can be performed on a matrix:

1. Interchange two rows

2. Multiply a row by a nonzero constant

3. Add a multiple of one row to another

We define an *elementary matrix* to be a matrix that results from applying a *single* elementary row operation to an identity matrix. Here are some examples:

$$\begin{bmatrix} 1 & 0 \\ 0 & -3 \end{bmatrix} \quad \begin{bmatrix} 1 & 0 & 0 & 0 \\ 0 & 0 & 0 & 1 \\ 0 & 0 & 1 & 0 \\ 0 & 1 & 0 & 0 \end{bmatrix} \quad \begin{bmatrix} 1 & 0 & 3 \\ 0 & 1 & 0 \\ 0 & 0 & 1 \end{bmatrix} \quad \begin{bmatrix} 1 & 0 & 0 \\ 0 & 1 & 0 \\ 0 & 0 & 1 \end{bmatrix}$$

| Multiply the second row of I_2 by -3. | Interchange the second and fourth rows of I_4. | Add 3 times the third row of I_3 to the first row. | Multiply the first row of I_3 by 1. |

Observe that elementary matrices are always square.

Elementary matrices are important because they can be used to execute elementary row operations by matrix multiplication. This is the content of the following theorem whose proof is left for the exercises.

Theorem 3.3.1   *If A is an $m \times n$ matrix, and if the elementary matrix E results by performing a certain row operation on the $m \times m$ identity matrix, then the product EA is the matrix that results when the same row operation is performed on A.*

In short, this theorem states that an elementary row operation can be performed on a matrix A using a *left* multiplication by an appropriate elementary matrix.

EXAMPLE 1
Performing
Row Operations
by Matrix
Multiplication

Consider the matrix

$$A = \begin{bmatrix} 1 & 0 & 2 & 3 \\ 2 & -1 & 3 & 6 \\ 1 & 4 & 4 & 0 \end{bmatrix}$$

Find an elementary matrix E such that EA is the matrix that results by adding 4 times the first row of A to the third row.

Solution The matrix E must be 3×3 to conform for the product EA. Thus, we obtain E by adding 4 times the first row of I_3 to the third row. This yields

$$E = \begin{bmatrix} 1 & 0 & 0 \\ 0 & 1 & 0 \\ 4 & 0 & 1 \end{bmatrix}$$

As a check, the product EA is

$$EA = \begin{bmatrix} 1 & 0 & 0 \\ 0 & 1 & 0 \\ 4 & 0 & 1 \end{bmatrix} \begin{bmatrix} 1 & 0 & 2 & 3 \\ 2 & -1 & 3 & 6 \\ 1 & 4 & 4 & 0 \end{bmatrix} = \begin{bmatrix} 1 & 0 & 2 & 3 \\ 2 & -1 & 3 & 6 \\ 5 & 4 & 12 & 12 \end{bmatrix}$$

so left multiplication by E does, in fact, add 4 times the first row of A to the third row. ∎

REMARK Theorem 3.3.1 is primarily a tool for studying matrices and linear systems and is not intended as a computational procedure for hand calculations. It is better to perform row operations directly, rather than multiplying by elementary matrices.

If an elementary row operation is applied to an identity matrix I to produce an elementary matrix E, then there is a second row operation that, when applied to E, produces I back again. For example, if E is obtained by multiplying the ith row of I by a nonzero scalar c, then I can be recovered by multiplying the ith row of E by $1/c$. The following table explains how to recover the identity matrix from an elementary matrix for each of the three elementary row operations. The operations on the right side of this table are called the ***inverse operations*** of the corresponding operations on the left side.

Row Operation on *I* That Produces *E*	Row Operation on *E* That Reproduces *I*
Multiply row i by $c \neq 0$	Multiply row i by $1/c$
Interchange rows i and j	Interchange rows i and j
Add c times row i to row j	Add $-c$ times row i to row j

EXAMPLE 2
Recovering
Identity
Matrices from
Elementary
Matrices

Here are three examples that use inverses of row operations to recover the identity matrix from an elementary matrix.

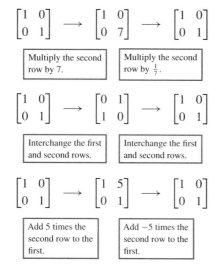

The next theorem is the basic result on the invertibility of elementary matrices.

Theorem 3.3.2 *An elementary matrix is invertible, and the inverse is also an elementary matrix.*

Proof If E is an elementary matrix, then E results from performing some row operation on I. Let E_0 be the elementary matrix that results when the inverse of this operation is performed on I. Applying Theorem 3.3.1 and using the fact that inverse row operations cancel the effect of one another, it follows that

$$E_0 E = I \quad \text{and} \quad E E_0 = I$$

which proves that the elementary matrix E_0 is the inverse of E. ∎

CHARACTERIZATIONS OF INVERTIBILITY

The next theorem establishes fundamental relationships between invertibility, reduced row echelon forms, and elementary matrices. This theorem will lead to a general method for inverting matrices.

Theorem 3.3.3 *If A is an n × n matrix, then the following statements are equivalent; that is, they are all true or all false.*

(*a*) *The reduced row echelon form of A is I_n.*

(*b*) *A is expressible as a product of elementary matrices.*

(*c*) *A is invertible.*

Proof We can prove the equivalence of all three statements by establishing the chain of implications $(a) \Rightarrow (b) \Rightarrow (c) \Rightarrow (a)$.

$(a) \Rightarrow (b)$ Since the reduced row echelon form of A is assumed to be I_n, there is a sequence of elementary row operations that reduces A to I_n. From Theorem 3.3.1, each of these elementary row operations can be performed with a left multiplication by an elementary matrix. Thus, there exists a sequence of elementary matrices $E_1, E_2, \ldots, E_k$ such that

$$E_k \cdots E_2 E_1 A = I_n \tag{1}$$

By Theorem 3.3.2, each of these elementary matrices is invertible. Thus, we can solve this equation for A by left multiplying both sides of (1) successively by $E_k^{-1}, \ldots, E_2^{-1}, E_1^{-1}$. This yields

$$A = E_1^{-1} E_2^{-1} \cdots E_k^{-1} I_n = E_1^{-1} E_2^{-1} \cdots E_k^{-1} \tag{2}$$

By Theorem 3.3.2, the factors on the right are elementary matrices, so this equation expresses A as a product of elementary matrices.

$(b) \Rightarrow (c)$ Suppose that A is expressible as a product of elementary matrices. Since a product of invertible matrices is invertible, and since elementary matrices are invertible, it follows that A is invertible.

$(c) \Rightarrow (a)$ Suppose that A is invertible and that its reduced row echelon form is R. Since R is obtained from A by a sequence of elementary row operations, it follows that there exists a sequence of elementary matrices $E_1, E_2, \ldots, E_k$ such that

$$E_k \cdots E_2 E_1 A = R$$

Since all of the factors on the left side are invertible, and since a product of invertible matrices is invertible, it follows that R must be invertible. Furthermore, it follows from Theorem 3.2.4 that there are only two possibilities for the form of R—either R has a row of zeros or $R = I_n$. However, a matrix with a row of zeros is not invertible, so R must be I_n.   ∎

ROW EQUIVALENCE   If a matrix B can be obtained from a matrix A by performing a finite sequence of elementary row operations, then there exists a sequence of elementary matrices $E_1, E_2, \ldots, E_k$ such that

$$E_k \cdots E_2 E_1 A = B \tag{3}$$

Since elementary matrices are invertible, this equation can be rewritten as

$$A = E_1^{-1} E_2^{-1} \cdots E_k^{-1} B \tag{4}$$

which tells us that A can be recovered from B by performing the inverses of the operations that produced B from A in the reverse order. In general, two matrices that can be obtained from one another by finite sequences of elementary row operations are said to be **row equivalent**. With this terminology, it follows from parts (a) and (c) of Theorem 3.3.3 that:

> *A square matrix A is invertible if and only if it is row equivalent to the identity matrix of the same size.*

The following useful theorem can be proved by multiplying out the k elementary matrices in (3) or (4).

> **Theorem 3.3.4** *If A and B are square matrices of the same size, then the following are equivalent:*
>   *(a) A and B are row equivalent.*
>   *(b) There is an invertible matrix E such that $B = EA$.*
>   *(c) There is an invertible matrix F such that $A = FB$.*

AN ALGORITHM FOR INVERTING MATRICES   We will now show how the ideas in the proof of Theorem 3.3.3 can be used to obtain a general method for finding the inverse of an invertible matrix A. For this purpose, suppose that A is reduced to I_n by a sequence of elementary row operations and that the corresponding sequence of elementary matrices is $E_1, E_2, \ldots, E_k$. Then from (2) we can express A as

$$A = E_1^{-1} E_2^{-1} \cdots E_k^{-1}$$

Taking the inverse of both sides yields

$$A^{-1} = E_k \cdots E_2 E_1$$

which we can also write as

$$A^{-1} = E_k \cdots E_2 E_1 I_n$$

This tells us that the same sequence of elementary row operations that reduces A to I_n will produce A^{-1} from I_n. Thus, we have the following result.

The Inversion Algorithm   *To find the inverse of an invertible matrix A, find a sequence of elementary row operations that reduces A to I, and then perform the same sequence of operations on I to obtain A^{-1}.*

The next example illustrates an efficient procedure for implementing this algorithm.

EXAMPLE 3
Applying the Inversion Algorithm

Find the inverse of

$$A = \begin{bmatrix} 1 & 2 & 3 \\ 2 & 5 & 3 \\ 1 & 0 & 8 \end{bmatrix}$$

Solution   We want to reduce A to the identity matrix by row operations and then apply the same sequence of operations to I to produce A^{-1}. A way of performing both tasks simultaneously is to adjoin the identity matrix to the right side of A, thereby creating a partitioned matrix of the form

$$[A \mid I]$$

and then apply row operations to this partitioned matrix until the left side is reduced to I. Those operations will convert the right side to A^{-1}, and the final matrix will have the form

$$[I \mid A^{-1}]$$

from which A^{-1} can then be read off. Here are the computations (with dashed rules inserted for clarity):

$$\left[\begin{array}{ccc|ccc} 1 & 2 & 3 & 1 & 0 & 0 \\ 2 & 5 & 3 & 0 & 1 & 0 \\ 1 & 0 & 8 & 0 & 0 & 1 \end{array}\right]$$

$$\left[\begin{array}{ccc|ccc} 1 & 2 & 3 & 1 & 0 & 0 \\ 0 & 1 & -3 & -2 & 1 & 0 \\ 0 & -2 & 5 & -1 & 0 & 1 \end{array}\right]$$

> We added -2 times the first row to the second and -1 times the first row to the third.

$$\left[\begin{array}{ccc|ccc} 1 & 2 & 3 & 1 & 0 & 0 \\ 0 & 1 & -3 & -2 & 1 & 0 \\ 0 & 0 & -1 & -5 & 2 & 1 \end{array}\right]$$

> We added 2 times the second row to the third.

$$\left[\begin{array}{ccc|ccc} 1 & 2 & 3 & 1 & 0 & 0 \\ 0 & 1 & -3 & -2 & 1 & 0 \\ 0 & 0 & 1 & 5 & -2 & -1 \end{array}\right]$$

> We multiplied the third row by -1.

$$\left[\begin{array}{ccc|ccc} 1 & 2 & 0 & -14 & 6 & 3 \\ 0 & 1 & 0 & 13 & -5 & -3 \\ 0 & 0 & 1 & 5 & -2 & -1 \end{array}\right]$$

> We added 3 times the third row to the second and -3 times the third row to the first.

$$\left[\begin{array}{ccc|ccc} 1 & 0 & 0 & -40 & 16 & 9 \\ 0 & 1 & 0 & 13 & -5 & -3 \\ 0 & 0 & 1 & 5 & -2 & -1 \end{array}\right]$$

> We added -2 times the second row to the first.

Thus,

$$A^{-1} = \begin{bmatrix} -40 & 16 & 9 \\ 13 & -5 & -3 \\ 5 & -2 & -1 \end{bmatrix}$$ ∎

REMARK Observe that in this example the invertibility of A was not known in advance—it was only when we succeeded in reducing A to I that the invertibility was established. If the inversion algorithm is attempted on a matrix that is not invertible, then somewhere during the computations a row of zeros will occur on the *left side* (why?). If this happens, you can conclude that the matrix is not invertible and stop the computations.

EXAMPLE 4
The Inversion Algorithm Will Reveal When a Matrix Is Singular

Consider the matrix

$$A = \begin{bmatrix} 1 & 6 & 4 \\ 2 & 4 & -1 \\ -1 & 2 & 5 \end{bmatrix}$$

If we apply the inversion algorithm, we obtain

$$\begin{bmatrix} 1 & 6 & 4 & \vdots & 1 & 0 & 0 \\ 2 & 4 & -1 & \vdots & 0 & 1 & 0 \\ -1 & 2 & 5 & \vdots & 0 & 0 & 1 \end{bmatrix}$$

$$\begin{bmatrix} 1 & 6 & 4 & \vdots & 1 & 0 & 0 \\ 0 & -8 & -9 & \vdots & -2 & 1 & 0 \\ 0 & 8 & 9 & \vdots & 1 & 0 & 1 \end{bmatrix}$$ We added -2 times the first row to the second and we added the first row to the third row.

$$\begin{bmatrix} 1 & 6 & 4 & \vdots & 1 & 0 & 0 \\ 0 & -8 & -9 & \vdots & -2 & 1 & 0 \\ 0 & 0 & 0 & \vdots & -1 & 1 & 1 \end{bmatrix}$$ We added the second row to the third row.

Since we have obtained a row of zeros on the left side, A is not invertible. ∎

SOLVING LINEAR SYSTEMS BY MATRIX INVERSION

In Sections 2.1 and 2.2 we showed how to solve a linear system

$$\begin{aligned} a_{11}x_1 + a_{12}x_2 + \cdots + a_{1n}x_n &= b_1 \\ a_{21}x_1 + a_{22}x_2 + \cdots + a_{2n}x_n &= b_2 \\ &\vdots \\ a_{m1}x_1 + a_{m2}x_2 + \cdots + a_{mn}x_n &= b_m \end{aligned}$$ (5)

by reducing the augmented matrix to reduced row echelon form (Gauss–Jordan elimination). However, there are other important methods for solving linear systems that are based on the idea of expressing the m equations in (5) as the single matrix equation

$$\begin{bmatrix} a_{11}x_1 + a_{12}x_2 + \cdots + a_{1n}x_n \\ a_{21}x_1 + a_{22}x_2 + \cdots + a_{2n}x_n \\ \vdots \\ a_{m1}x_1 + a_{m2}x_2 + \cdots + a_{mn}x_n \end{bmatrix} = \begin{bmatrix} b_1 \\ b_2 \\ \vdots \\ b_m \end{bmatrix}$$

This equation can be written as

$$\begin{bmatrix} a_{11} & a_{12} & \cdots & a_{1n} \\ a_{21} & a_{22} & \cdots & a_{2n} \\ \vdots & \vdots & & \vdots \\ a_{m1} & a_{m2} & \cdots & a_{mn} \end{bmatrix} \begin{bmatrix} x_1 \\ x_2 \\ \vdots \\ x_n \end{bmatrix} = \begin{bmatrix} b_1 \\ b_2 \\ \vdots \\ b_m \end{bmatrix}$$

or more briefly as

$$A\mathbf{x} = \mathbf{b}$$ (6)

Thus, we have replaced the problem of solving (5) for the unknowns $x_1, x_2, \ldots, x_n$ with the problem of solving (6) for the unknown vector **x**.

When working with Equation (6), it is important to keep in mind how the sizes of A, **x**, and **b** relate to the number of equations and unknowns in system (5). The matrix A, which is called the **coefficient matrix** for the system, has size $m \times n$, where m is the number of equations and n is the number of unknowns; the vector **x** has size $n \times 1$ and hence is a column vector in R^n; and the vector **b** has size $m \times 1$ and hence is a column vector in R^m. Finally, observe that the augmented matrix $[A \mid \mathbf{b}]$ is obtained by adjoining **b** to the coefficient matrix, so it has size $m \times (n + 1)$.

In the rest of this section we will be concerned primarily with the situation in which the number of equations is the same as the number of unknowns, in which case the coefficient matrix A in (5) is square. If, in addition, A is invertible, then we can solve (6) for **x** by left multiplying both sides of this equation by A^{-1} to obtain the unique solution

$$\mathbf{x} = A^{-1}\mathbf{b}$$

Thus, we are led to the following result.

Theorem 3.3.5   *If $A\mathbf{x} = \mathbf{b}$ is a linear system of n equations in n unknowns, and if the coefficient matrix A is invertible, then the system has a unique solution, namely $\mathbf{x} = A^{-1}\mathbf{b}$.*

EXAMPLE 5
Solution of a
Linear System
by Matrix
Inversion

Consider the linear system

$$\begin{array}{rcr} x_1 + 2x_2 + 3x_3 &=& 5 \\ 2x_1 + 5x_2 + 3x_3 &=& 3 \\ x_1 \quad\quad + 8x_3 &=& 17 \end{array}$$

This system can be written in matrix form as $A\mathbf{x} = \mathbf{b}$, where

$$A = \begin{bmatrix} 1 & 2 & 3 \\ 2 & 5 & 3 \\ 1 & 0 & 8 \end{bmatrix}, \quad \mathbf{x} = \begin{bmatrix} x_1 \\ x_2 \\ x_3 \end{bmatrix}, \quad \mathbf{b} = \begin{bmatrix} 5 \\ 3 \\ 17 \end{bmatrix}$$

We showed in Example 3 that A is invertible and

$$A^{-1} = \begin{bmatrix} -40 & 16 & 9 \\ 13 & -5 & -3 \\ 5 & -2 & -1 \end{bmatrix}$$

Thus, the solution of the linear system is

$$\mathbf{x} = A^{-1}\mathbf{b} = \begin{bmatrix} -40 & 16 & 9 \\ 13 & -5 & -3 \\ 5 & -2 & -1 \end{bmatrix} \begin{bmatrix} 5 \\ 3 \\ 17 \end{bmatrix} = \begin{bmatrix} 1 \\ -1 \\ 2 \end{bmatrix}$$

or, equivalently, $x_1 = 1$, $x_2 = -1$, $x_3 = 2$.   ■

REMARK   This method of solution is almost never used in professional computer programs, since there are more efficient methods available, some of which we will discuss later. However, the basic idea of the method provides an important way of thinking about solutions of linear systems that will prove invaluable in our subsequent work.

The following theorem establishes a fundamental relationship between the invertibility of a matrix A and the solutions of the homogeneous linear system $A\mathbf{x} = \mathbf{0}$ that has A as its coefficient matrix.

Theorem 3.3.6   *If $A\mathbf{x} = \mathbf{0}$ is a homogeneous linear system of n equations in n unknowns, then the system has only the trivial solution if and only if the coefficient matrix A is invertible.*

Proof If A is invertible, then it follows from Theorem 3.3.5 that the unique solution of the system is $\mathbf{x} = A^{-1}\mathbf{0} = \mathbf{0}$; that is, the system has only the trivial solution.

Conversely, suppose that the homogeneous linear system

$$
\begin{aligned}
a_{11}x_1 + a_{12}x_2 + \cdots + a_{1n}x_n &= 0 \\
a_{21}x_1 + a_{22}x_2 + \cdots + a_{2n}x_n &= 0 \\
\vdots \qquad \vdots \qquad\qquad \vdots \quad\ \ \vdots & \\
a_{n1}x_1 + a_{n2}x_2 + \cdots + a_{nn}x_n &= 0
\end{aligned}
\tag{7}
$$

has only the trivial solution. Then the system of equations corresponding to the reduced row echelon form of the augmented matrix for (7) is

$$
\begin{aligned}
x_1 \qquad\qquad\qquad &= 0 \\
x_2 \qquad\qquad &= 0 \\
\ddots \qquad \vdots & \\
x_n &= 0
\end{aligned}
$$

Thus, it follows that there is a sequence of elementary row operations that reduces the augmented matrix

$$
\begin{bmatrix}
a_{11} & a_{12} & \cdots & a_{1n} & 0 \\
a_{21} & a_{22} & \cdots & a_{2n} & 0 \\
\vdots & \vdots & & \vdots & \vdots \\
a_{n1} & a_{n2} & \cdots & a_{nn} & 0
\end{bmatrix}
$$

to the augmented matrix

$$
\begin{bmatrix}
1 & 0 & \cdots & 0 & 0 \\
0 & 1 & \cdots & 0 & 0 \\
\vdots & \vdots & & \vdots & \vdots \\
0 & 0 & \cdots & 1 & 0
\end{bmatrix}
$$

If we disregard the last column in each of these matrices, then we can conclude that the reduced row echelon form of A is I_n.   ∎

According to Theorem 3.3.6, a square matrix A is invertible if and only if $A\mathbf{x} = \mathbf{0}$ has only the trivial solution, and hence we can add a fourth statement to Theorem 3.3.3.

> **Theorem 3.3.7**   *If A is an $n \times n$ matrix, then the following statements are equivalent.*
>
> (a) *The reduced row echelon form of A is I_n.*
>
> (b) *A is expressible as a product of elementary matrices.*
>
> (c) *A is invertible.*
>
> (d) *$A\mathbf{x} = \mathbf{0}$ has only the trivial solution.*

EXAMPLE 6
Homogeneous
System with an
Invertible
Coefficient
Matrix

In Example 3 we showed that

$$
A = \begin{bmatrix}
1 & 2 & 3 \\
2 & 5 & 3 \\
1 & 0 & 8
\end{bmatrix}
$$

is an invertible matrix. Thus, we can conclude from Theorem 3.3.6 without any computation that the homogeneous linear system

$$
\begin{aligned}
x_1 + 2x_2 + 3x_3 &= 0 \\
2x_1 + 5x_2 + 3x_3 &= 0 \\
x_1 \qquad\ + 8x_3 &= 0
\end{aligned}
$$

has only the trivial solution.   ∎

According to the definition of invertibility, a square matrix A is invertible if and only if there is a matrix B such that $AB = I$ and $BA = I$. The first part of the next theorem shows that if we can find a matrix B satisfying *either* condition, then the other condition holds automatically. We also already know that a product of invertible matrices is invertible. The second part of the theorem provides a converse.

Theorem 3.3.8

(a) *If A and B are square matrices such that $AB = I$ or $BA = I$, then A and B are both invertible, and each is the inverse of the other.*

(b) *If A and B are square matrices whose product AB is invertible, then A and B are invertible.*

Proof (a) Suppose that $BA = I$. If we can show that A is invertible, then we can multiply both sides of this equation on the right by A^{-1} to obtain $B = A^{-1}$, from which it follows that $B^{-1} = (A^{-1})^{-1} = A$. This establishes that B is invertible and that A and B are inverses of one another. To prove the invertibility of A, it suffices to show that the homogeneous system $A\mathbf{x} = \mathbf{0}$ has only the trivial solution. However, if $\mathbf{x}$ is any solution of this system, then the assumption that $BA = I$ implies that

$$\mathbf{x} = I\mathbf{x} = BA\mathbf{x} = B(A\mathbf{x}) = B\mathbf{0} = \mathbf{0}$$

Thus, the system $A\mathbf{x} = \mathbf{0}$ has only the trivial solution, which establishes that A is invertible. The proof for the case where $AB = I$ can be obtained by interchanging A and B in the preceding argument.

Proof (b) If AB is invertible, then we can write

$$I = (AB)(AB)^{-1} = A(B(AB)^{-1}) \quad \text{and} \quad I = (AB)^{-1}(AB) = ((AB)^{-1}A)B$$

By part (a), the first set of equalities implies that A is invertible, and the second set implies that B is invertible. ∎

We are now in a position to add two more statements to Theorem 3.3.7.

Theorem 3.3.9 *If A is an $n \times n$ matrix, then the following statements are equivalent.*

(a) *The reduced row echelon form of A is I_n.*

(b) *A is expressible as a product of elementary matrices.*

(c) *A is invertible.*

(d) *$A\mathbf{x} = \mathbf{0}$ has only the trivial solution.*

(e) *$A\mathbf{x} = \mathbf{b}$ is consistent for every vector $\mathbf{b}$ in R^n.*

(f) *$A\mathbf{x} = \mathbf{b}$ has exactly one solution for every vector $\mathbf{b}$ in R^n.*

Proof We already know that (a), (b), (c), and (d) are equivalent, so we can complete the proof by showing that statements (c), (e), and (f) are equivalent, since this will automatically imply that (e) and (f) are equivalent to (a), (b), and (d). To show that (c), (e), and (f) are equivalent, we will prove that (f) $\Rightarrow$ (e) $\Rightarrow$ (c) $\Rightarrow$ (f).

(f) $\Rightarrow$ (e) If $A\mathbf{x} = \mathbf{b}$ has exactly one solution for every vector $\mathbf{b}$ in R^n, then it follows logically that $A\mathbf{x} = \mathbf{b}$ has at least one solution for every $n \times 1$ matrix $\mathbf{b}$.

(e) $\Rightarrow$ (c) If $A\mathbf{x} = \mathbf{b}$ is consistent for every vector $\mathbf{b}$ in R^n, then, in particular, this is true of the n linear systems

$$A\mathbf{x} = \mathbf{e}_1, \quad A\mathbf{x} = \mathbf{e}_2, \dots, \quad A\mathbf{x} = \mathbf{e}_n$$

where $\mathbf{e}_1, \mathbf{e}_2, \dots, \mathbf{e}_n$ are the standard unit vectors in R^n written in column form [see Formula (7) of Section 1.2]. Let $\mathbf{x}_1, \mathbf{x}_2, \dots, \mathbf{x}_n$ be solutions of the respective systems and form the partitioned matrix

$$C = [\mathbf{x}_1 \quad \mathbf{x}_2 \quad \cdots \quad \mathbf{x}_n]$$

Thus,

$$AC = [A\mathbf{x}_1 \quad A\mathbf{x}_2 \quad \cdots \quad A\mathbf{x}_n] = I_n$$

It now follows from Theorem 3.3.8 that A is invertible.

$(c) \Rightarrow (f)$ This is the statement of Theorem 3.3.5. ∎

SOLVING MULTIPLE LINEAR SYSTEMS WITH A COMMON COEFFICIENT MATRIX

In many applications one is concerned with solving a sequence of linear systems

$$A\mathbf{x} = \mathbf{b}_1, \quad A\mathbf{x} = \mathbf{b}_2, \ldots, \quad A\mathbf{x} = \mathbf{b}_k \tag{8}$$

each of which has the same coefficient matrix A. A *poor* method for solving the k systems is to apply Gauss–Jordan elimination or Gaussian elimination separately to each system, since the reduction operations on A are the same in all cases and there is no need to perform them over and over. We will consider some better procedures.

If the coefficient matrix A in (8) is invertible, then each system has a unique solution, and all k solutions can be obtained with one matrix inversion and k matrix multiplications:

$$\mathbf{x}_1 = A^{-1}\mathbf{b}_1, \quad \mathbf{x}_2 = A^{-1}\mathbf{b}_2, \ldots, \quad \mathbf{x}_k = A^{-1}\mathbf{b}_k$$

However, this procedure cannot be used unless A is invertible. An alternative approach that is more efficient and also applies when A is not square or not invertible is to create the augmented matrix

$$[A \mid \mathbf{b}_1 \mid \mathbf{b}_2 \mid \cdots \mid \mathbf{b}_k]$$

in which $\mathbf{b}_1, \mathbf{b}_2, \ldots, \mathbf{b}_k$ are adjoined to A, and reduce this matrix to reduced row echelon form, thereby solving all k systems at once by Gauss–Jordan elimination. Here is an example.

EXAMPLE 7
Solving Multiple Linear Systems by Gauss–Jordan Elimination

Solve the systems

(a) $x_1 + 2x_2 + 3x_3 = 4$
$2x_1 + 5x_2 + 3x_3 = 5$
$x_1 \qquad + 8x_3 = 9$

(b) $x_1 + 2x_2 + 3x_3 = 1$
$2x_1 + 5x_2 + 3x_3 = 6$
$x_1 \qquad + 8x_3 = -6$

Solution The two systems have the same coefficient matrix. If we augment this coefficient matrix with the columns of constants on the right sides of these systems, we obtain

$$\begin{bmatrix} 1 & 2 & 3 & 4 & 1 \\ 2 & 5 & 3 & 5 & 6 \\ 1 & 0 & 8 & 9 & -6 \end{bmatrix}$$

Reducing this matrix to reduced row echelon form yields (verify)

$$\begin{bmatrix} 1 & 0 & 0 & 1 & 2 \\ 0 & 1 & 0 & 0 & 1 \\ 0 & 0 & 1 & 1 & -1 \end{bmatrix}$$

It follows from the last two columns that the solution of system (a) is $x_1 = 1, x_2 = 0, x_3 = 1$ and of system (b) is $x_1 = 2, x_2 = 1, x_3 = -1$. ∎

CONSISTENCY OF LINEAR SYSTEMS

As we progress through this text, the following problem will occur in various contexts.

3.3.10 The Consistency Problem For a given matrix A, find all vectors $\mathbf{b}$ for which the linear system $A\mathbf{x} = \mathbf{b}$ is consistent.

If A is an invertible $n \times n$ matrix, then it follows from Theorem 3.3.9 that the system $A\mathbf{x} = \mathbf{b}$ is consistent for *every* vector $\mathbf{b}$ in R^n. If A is not square, or if A is square but not invertible, then the system will typically be consistent for some vectors but not others, and the problem is to determine which vectors produce a consistent system.

REMARK A linear system $A\mathbf{x} = \mathbf{b}$ is always consistent for at least one vector $\mathbf{b}$. Why?

The following example illustrates how Gaussian elimination can sometimes be used to solve the consistency problem for a given matrix.

EXAMPLE 8
Solving a
Consistency
Problem by
Gaussian
Elimination

What conditions must b_1, b_2, and b_3 satisfy for the following linear system to be consistent?

$$\begin{aligned} x_1 + x_2 + 2x_3 &= b_1 \\ x_1 \qquad + x_3 &= b_2 \\ 2x_1 + x_2 + 3x_3 &= b_3 \end{aligned}$$

Solution The augmented matrix is

$$\begin{bmatrix} 1 & 1 & 2 & b_1 \\ 1 & 0 & 1 & b_2 \\ 2 & 1 & 3 & b_3 \end{bmatrix}$$

which can be reduced to row echelon form as follows:

$$\begin{bmatrix} 1 & 1 & 2 & b_1 \\ 0 & -1 & -1 & b_2 - b_1 \\ 0 & -1 & -1 & b_3 - 2b_1 \end{bmatrix}$$

-1 times the first row was added to the second and -2 times the first row was added to the third.

$$\begin{bmatrix} 1 & 1 & 2 & b_1 \\ 0 & 1 & 1 & b_1 - b_2 \\ 0 & -1 & -1 & b_3 - 2b_1 \end{bmatrix}$$

The second row was multiplied by -1.

$$\begin{bmatrix} 1 & 1 & 2 & b_1 \\ 0 & 1 & 1 & b_1 - b_2 \\ 0 & 0 & 0 & b_3 - b_2 - b_1 \end{bmatrix}$$

The second row was added to the third.

It is now evident from the third row in the matrix that the system has a solution if and only if b_1, b_2, and b_3 satisfy the condition

$$b_3 - b_2 - b_1 = 0 \quad \text{or} \quad b_3 = b_1 + b_2$$

Thus, $A\mathbf{x} = \mathbf{b}$ is consistent if and only if $\mathbf{b}$ can be expressed in the form

$$\mathbf{b} = \begin{bmatrix} b_1 \\ b_2 \\ b_1 + b_2 \end{bmatrix} = \begin{bmatrix} b_1 \\ 0 \\ b_1 \end{bmatrix} + \begin{bmatrix} 0 \\ b_2 \\ b_2 \end{bmatrix} = b_1 \begin{bmatrix} 1 \\ 0 \\ 1 \end{bmatrix} + b_2 \begin{bmatrix} 0 \\ 1 \\ 1 \end{bmatrix}$$

where b_1 and b_2 are arbitrary; that is, the set of vectors in R^3 for which $A\mathbf{x} = \mathbf{b}$ is consistent is the subspace of R^3 that consists of all linear combinations of the vectors

$$\begin{bmatrix} 1 \\ 0 \\ 1 \end{bmatrix} \quad \text{and} \quad \begin{bmatrix} 0 \\ 1 \\ 1 \end{bmatrix}$$

This is the plane that passes through the origin and the points $(1, 0, 1)$ and $(0, 1, 1)$. ■

Exercise Set 3.3

In Exercises 1 and 2, determine whether the given matrix is elementary.

1. (a) $\begin{bmatrix} 1 & 0 \\ -5 & 1 \end{bmatrix}$ (b) $\begin{bmatrix} -5 & 1 \\ 1 & 0 \end{bmatrix}$ (c) $\begin{bmatrix} 0 & 0 & 1 \\ 0 & 1 & 0 \\ 1 & 0 & 0 \end{bmatrix}$ (d) $\begin{bmatrix} 0 & 1 & 0 \\ 1 & 0 & 0 \\ 0 & 0 & 0 \end{bmatrix}$

2. (a) $\begin{bmatrix} 1 & 1 \\ 1 & 0 \end{bmatrix}$ (b) $\begin{bmatrix} 2 & 0 \\ 0 & 3 \end{bmatrix}$ (c) $\begin{bmatrix} 1 & 0 & 0 \\ 0 & 1 & 9 \\ 0 & 0 & 1 \end{bmatrix}$ (d) $\begin{bmatrix} 0 & 2 & 0 \\ 1 & 0 & 0 \\ 0 & 0 & 1 \end{bmatrix}$

In Exercises 3 and 4, confirm that the matrix is elementary, and find a row operation that will restore it to an identity matrix.

3. (a) $\begin{bmatrix} 1 & 0 \\ -3 & 1 \end{bmatrix}$
(b) $\begin{bmatrix} 1 & 0 & 0 \\ 0 & 1 & 0 \\ 0 & 0 & 3 \end{bmatrix}$

(c) $\begin{bmatrix} 0 & 0 & 0 & 1 \\ 0 & 1 & 0 & 0 \\ 0 & 0 & 1 & 0 \\ 1 & 0 & 0 & 0 \end{bmatrix}$
(d) $\begin{bmatrix} 1 & 0 & -\frac{1}{7} & 0 \\ 0 & 1 & 0 & 0 \\ 0 & 0 & 1 & 0 \\ 0 & 0 & 0 & 1 \end{bmatrix}$

4. (a) $\begin{bmatrix} 1 & 2 \\ 0 & 1 \end{bmatrix}$
(b) $\begin{bmatrix} -1 & 0 & 0 \\ 0 & 1 & 0 \\ 0 & 0 & 1 \end{bmatrix}$

(c) $\begin{bmatrix} 0 & 0 & 1 & 0 \\ 0 & 1 & 0 & 0 \\ 1 & 0 & 0 & 0 \\ 0 & 0 & 0 & 1 \end{bmatrix}$
(d) $\begin{bmatrix} 1 & 0 & 0 & 0 \\ 0 & 1 & 0 & 0 \\ 0 & 0 & 1 & 0 \\ 0 & 12 & 0 & 1 \end{bmatrix}$

5. Use the row operations you obtained in Exercise 3 to find the inverses of the matrices.

6. Use the row operations you obtained in Exercise 4 to find the inverses of the matrices.

In Exercises 7 and 8, use the matrices

$$A = \begin{bmatrix} 3 & 4 & 1 \\ 2 & -7 & -1 \\ 8 & 1 & 5 \end{bmatrix}, \quad B = \begin{bmatrix} 8 & 1 & 5 \\ 2 & -7 & -1 \\ 3 & 4 & 1 \end{bmatrix}$$

$$C = \begin{bmatrix} 3 & 4 & 1 \\ 2 & -7 & -1 \\ 2 & -7 & 3 \end{bmatrix}, \quad D = \begin{bmatrix} 8 & 1 & 5 \\ -6 & 21 & 3 \\ 3 & 4 & 1 \end{bmatrix}$$

$$F = \begin{bmatrix} 8 & 1 & 5 \\ 8 & 1 & 1 \\ 3 & 4 & 1 \end{bmatrix}$$

7. Find an elementary matrix E that satisfies the equation.
(a) $EA = B$ (b) $EB = A$
(c) $EA = C$ (d) $EC = A$

8. Find an elementary matrix E that satisfies the equation.
(a) $EB = D$ (b) $ED = B$
(c) $EB = F$ (d) $EF = B$

In Exercises 9 and 10, use the method of Example 3 to find the inverse of A, and check your answer using Theorem 3.2.7.

9. $A = \begin{bmatrix} 1 & 5 \\ 2 & 20 \end{bmatrix}$ **10.** $A = \begin{bmatrix} 2 & -3 \\ 4 & 1 \end{bmatrix}$

In Exercises 11 and 12, use the method of Example 3 to find A^{-1} if A is invertible.

11. (a) $A = \begin{bmatrix} 3 & 4 & -1 \\ 1 & 0 & 3 \\ 2 & 5 & -4 \end{bmatrix}$
(b) $A = \begin{bmatrix} -1 & 3 & -4 \\ 2 & 4 & 1 \\ -4 & 2 & -9 \end{bmatrix}$

(c) $A = \begin{bmatrix} 1 & 0 & 1 \\ 0 & 1 & 1 \\ 1 & 1 & 0 \end{bmatrix}$

12. (a) $A = \begin{bmatrix} 1 & 2 & 0 \\ 2 & 1 & 2 \\ 0 & 2 & 1 \end{bmatrix}$
(b) $A = \begin{bmatrix} 2 & 0 & 0 \\ 0 & 4 & 3 \\ 0 & 1 & 1 \end{bmatrix}$

(c) $A = \begin{bmatrix} 1 & 2 & 3 \\ 0 & 2 & 3 \\ 0 & 0 & 3 \end{bmatrix}$

In Exercises 13 and 14, find the reduced row echelon form R of the matrix A, and then find a matrix B for which $BA = R$.

13. $A = \begin{bmatrix} 1 & 2 & 3 \\ 0 & 0 & 1 \\ 1 & 2 & 4 \end{bmatrix}$ **14.** $A = \begin{bmatrix} 1 & 0 & 0 \\ 2 & 4 & 0 \\ 4 & 8 & 0 \end{bmatrix}$

In Exercises 15 and 16, find all values of c, if any, for which the matrix is invertible.

15. $\begin{bmatrix} c & c & c \\ 1 & c & c \\ 1 & 1 & c \end{bmatrix}$ **16.** $\begin{bmatrix} c & 1 & 0 \\ 1 & c & 1 \\ 0 & 1 & c \end{bmatrix}$

17. Let A be a 3×3 matrix. Find a matrix B for which BA is the matrix that results from A by interchanging the first two rows and then multiplying the third row by six.

18. Let A be a 3×3 matrix. Find a matrix B for which BA is the matrix that results from A by adding four times the second row to the third and then interchanging the first and third rows of the result.

In Exercises 19 and 20, state conditions on the constants under which A will be invertible, and find A^{-1}.

19. $A = \begin{bmatrix} 0 & 0 & 0 & k_1 \\ 0 & 0 & k_2 & 0 \\ 0 & k_3 & 0 & 0 \\ k_4 & 0 & 0 & 0 \end{bmatrix}$ **20.** $A = \begin{bmatrix} k & 0 & 0 & 0 \\ 1 & k & 0 & 0 \\ 0 & 1 & k & 0 \\ 0 & 0 & 1 & k \end{bmatrix}$

21. Consider the matrix $A = \begin{bmatrix} 1 & 0 \\ -5 & 2 \end{bmatrix}$.

(a) Find elementary matrices E_1 and E_2 such that $E_2 E_1 A = I$.
(b) Write A^{-1} as a product of two elementary matrices.
(c) Write A as a product of two elementary matrices.

22. Consider the matrix $A = \begin{bmatrix} 1 & 0 & -2 \\ 0 & 1 & 0 \\ 0 & 0 & 2 \end{bmatrix}$.

   (a) Find elementary matrices E_1 and E_2 such that $E_2 E_1 A = I$.
   (b) Write A^{-1} as a product of two elementary matrices.
   (c) Write A as a product of two elementary matrices.

In Exercises 23 and 24, express A and A^{-1} as products of elementary matrices.

23. $A = \begin{bmatrix} 2 & 1 & 1 \\ 1 & 2 & 1 \\ 1 & 1 & 2 \end{bmatrix}$ **24.** $A = \begin{bmatrix} 1 & 1 & 0 \\ 1 & 1 & 1 \\ 0 & 1 & 1 \end{bmatrix}$

In Exercises 25 and 26, use the method of Example 7 to solve the two systems at once by row reduction.

25. $\begin{aligned} x_1 + 2x_2 + x_3 &= -1 \\ x_1 + 3x_2 + 2x_3 &= 3 \\ x_2 + 2x_3 &= 4 \end{aligned}$ and $\begin{aligned} x_1 + 2x_2 + x_3 &= 0 \\ x_1 + 3x_2 + 2x_3 &= 0 \\ x_2 + 2x_3 &= 4 \end{aligned}$

26. $\begin{aligned} x_1 + 2x_2 + 5x_3 &= -1 \\ x_2 - 3x_3 &= -1 \\ x_3 &= 3 \end{aligned}$ and $\begin{aligned} x_1 + 2x_2 + 5x_3 &= 2 \\ x_2 - 3x_3 &= 1 \\ x_3 &= -1 \end{aligned}$

27. (a) Write the systems in Exercise 25 as $A\mathbf{x} = \mathbf{b}_1$ and $A\mathbf{x} = \mathbf{b}_2$, and then solve each of them by the method of Example 5.
   (b) Obtain the two solutions at once by computing $A^{-1}[\mathbf{b}_1 \quad \mathbf{b}_2]$.

Discussion and Discovery

D1. Suppose that A is an unknown invertible matrix, but a sequence of elementary row operations is known that produces the identity matrix when applied in succession to A. Explain how you can use the known information to find A.

D2. Do you think that there is a 2×2 matrix A such that

$$A \begin{bmatrix} a & b \\ c & d \end{bmatrix} = \begin{bmatrix} b & d \\ a & c \end{bmatrix}$$

for all values of a, b, c, and d? Explain your reasoning.

D3. Determine by inspection (no pencil and paper) whether the given homogeneous system has a nontrivial solution, and then state whether the coefficient matrix is invertible.

$$\begin{aligned} 2x_1 + x_2 - 3x_3 + x_4 &= 0 \\ 5x_2 + 4x_3 + 3x_4 &= 0 \\ x_3 + 2x_4 &= 0 \\ 3x_4 &= 0 \end{aligned}$$

28. (a) Write the systems in Exercise 26 as $A\mathbf{x} = \mathbf{b}_1$ and $A\mathbf{x} = \mathbf{b}_2$, and then solve each of them by the method of Example 5.
   (b) Obtain the two solutions at once by computing $A^{-1}[\mathbf{b}_1 \quad \mathbf{b}_2]$.

In Exercises 29–32, find conditions on the b's that will ensure that the system is consistent.

29. $\begin{aligned} 6x_1 - 4x_2 &= b_1 \\ 3x_1 - 2x_2 &= b_2 \end{aligned}$ **30.** $\begin{aligned} x_1 - 2x_2 + 5x_3 &= b_1 \\ 4x_1 - 5x_2 + 8x_3 &= b_2 \\ -3x_1 + 3x_2 - 3x_3 &= b_3 \end{aligned}$

31. $\begin{aligned} x_1 - 2x_2 - x_3 &= b_1 \\ -2x_1 + 3x_2 + 2x_3 &= b_2 \\ -4x_1 + 7x_2 + 4x_3 &= b_3 \end{aligned}$

32. $\begin{aligned} x_1 - x_2 + 3x_3 + 2x_4 &= b_1 \\ -2x_1 + x_2 + 5x_3 + x_4 &= b_2 \\ -3x_1 + 2x_2 + 2x_3 - x_4 &= b_3 \\ 4x_1 - 3x_2 + x_3 + 3x_4 &= b_4 \end{aligned}$

33. Factor the matrix

$$A = \begin{bmatrix} 0 & 1 & 7 & 8 \\ 1 & 3 & 3 & 8 \\ -2 & -5 & 1 & -8 \end{bmatrix}$$

as $A = EFGR$, where E, F, and G are elementary matrices and R is in row echelon form.

34. Show that if

$$A = \begin{bmatrix} 1 & 0 & 0 \\ 0 & 1 & 0 \\ a & b & 1 \end{bmatrix}$$

is an elementary matrix, then $ab = 0$.

D4. (a) Find a matrix B for which

$$B \begin{bmatrix} 1 & 2 \\ 0 & -2 \\ 1 & 4 \end{bmatrix} = \begin{bmatrix} 1 & 0 \\ 0 & 1 \end{bmatrix}$$

   (b) Is there more than one such B?
   (c) Does B have an inverse?

D5. Indicate whether the statement is true (T) or false (F). Justify your answer.
   (a) Every square matrix can be expressed as the product of elementary matrices.
   (b) The product of two elementary matrices is an elementary matrix.
   (c) If A is invertible, and if a multiple of the first row is added to the second row, then the resulting matrix is invertible.
   (d) If A is invertible and $AB = 0$, then $B = 0$.

(e) If A is an $n \times n$ matrix, and if the homogeneous linear system $A\mathbf{x} = \mathbf{0}$ has infinitely many solutions, then A is singular.

D6. Indicate whether the statement is true (T) or false (F). Justify your answer.

(a) Every invertible matrix can be factored into a product of elementary matrices.

(b) If A is a singular $n \times n$ matrix, then $A\mathbf{x} = \mathbf{0}$ has infinitely many solutions.

(c) If A is a singular $n \times n$ matrix, then the reduced row echelon form of A has at least one row of zeros.

(d) If A is expressible as a product of elementary matrices, then the homogeneous linear system $A\mathbf{x} = \mathbf{0}$ has only the trivial solution.

(e) If A is a singular $n \times n$ matrix, and if B results by interchanging two rows of A, then B must be singular.

D7. Are there any values of the constants for which the following matrix is invertible?

$$A = \begin{bmatrix} 0 & a & 0 & 0 & 0 \\ b & 0 & c & 0 & 0 \\ 0 & d & 0 & e & 0 \\ 0 & 0 & f & 0 & g \\ 0 & 0 & 0 & h & 0 \end{bmatrix}$$

Working with Proofs

P1. Prove that if A and B are square matrices of the same size, and if AB is invertible, then A and B are invertible.

P2. Let A, B, and C be $n \times n$ matrices for which $A = BC$. Prove that if B is invertible, then any sequence of elementary row operations that reduces B to I_n will reduce A to C.

P3. Let $A\mathbf{x} = \mathbf{0}$ be a homogeneous system of n linear equations in n unknowns that has only the trivial solution. Prove that if k is any positive integer, then the system $A^k\mathbf{x} = \mathbf{0}$ also has only the trivial solution.

P4. Let $A\mathbf{x} = \mathbf{0}$ be a homogeneous linear system of n equations in n unknowns, and let B be an invertible $n \times n$ matrix. Prove that $A\mathbf{x} = \mathbf{0}$ has only the trivial solution if and only if $(BA)\mathbf{x} = \mathbf{0}$ has only the trivial solution.

P5. Prove Theorem 3.3.1. [*Hint:* Consider each of the three elementary operations separately.]

P6. Let A be a 2×2 matrix such that $AB = BA$ for every 2×2 matrix B. Prove that A must be a scalar multiple of the identity matrix. [*Hint:* Take B to be appropriate elementary matrices.]

P7. Prove that if A is an $m \times n$ matrix, then there is an invertible matrix C such that CA is in reduced row echelon form.

P8. Let A be an invertible $n \times n$ matrix, let B be any $n \times n$ matrix, and let $[A \mid B]$ denote the $n \times (2n)$ matrix that results by adjoining B to A. Prove that if elementary row operations are applied to $[A \mid B]$ until A is reduced to the $n \times n$ identity matrix I, then the resulting matrix is $[I \mid A^{-1}B]$.

Technology Exercises

T1. (*Inverses*) Compute the inverse of the matrix in Example 3, and then see what happens when you try to compute the inverse of the singular matrix in Example 4.

T2. (*Augmented matrices*) Many technology utilities provide methods for building up new matrices from a set of specified matrices. Determine whether your utility provides for this, and if so, form the augmented matrix for the system in Example 5 from the matrices A and $\mathbf{b}$.

T3. See what happens when you try to compute a negative power of the singular matrix in Example 4.

T4. Compute the inverse of the matrix A in Example 3 by adjoining the 3×3 identity matrix to A and reducing the 3×6 matrix to reduced row echelon form.

T5. Solve the following matrix equation for X:

$$\begin{bmatrix} 1 & -1 & 1 \\ 2 & 3 & 0 \\ 0 & 2 & -1 \end{bmatrix} X = \begin{bmatrix} 2 & -1 & 5 & 7 & 8 \\ 4 & 0 & -3 & 0 & 1 \\ 3 & 5 & -7 & 2 & 1 \end{bmatrix}$$

T6. (**CAS**) Obtain Formula (5) of Theorem 3.2.7.

T7. (a) Use matrix inversion to solve the linear system

$$\begin{bmatrix} 3 & 3 & -4 & -3 \\ 0 & 6 & 1 & 1 \\ 5 & 4 & 2 & 1 \\ 2 & 3 & 3 & 2 \end{bmatrix} \begin{bmatrix} x_1 \\ x_2 \\ x_3 \\ x_4 \end{bmatrix} = \begin{bmatrix} -2 \\ 3 \\ 5 \\ 1 \end{bmatrix}$$

(b) Solve the system in part (a) using the system-solving capability of your utility, and compare the result to that obtained in part (a).

T8. By experimenting with different values of n, find an expression for the inverse of an $n \times n$ matrix of the form

$$A = \begin{bmatrix} 1 & 2 & 3 & 4 & \cdots & n-1 & n \\ 0 & 1 & 2 & 3 & \cdots & n-2 & n-1 \\ 0 & 0 & 1 & 2 & \cdots & n-3 & n-2 \\ \vdots & \vdots & \vdots & \vdots & & \vdots & \vdots \\ 0 & 0 & 0 & 0 & \cdots & 1 & 2 \\ 0 & 0 & 0 & 0 & \cdots & 0 & 1 \end{bmatrix}$$

T9. (CAS) The $n \times n$ matrix $H_n = [h_{ij}]$ for which

$$h_{ij} = 1/(i + j - 1)$$

is called the nth-order ***Hilbert matrix***.

(a) Write out the Hilbert matrices H_2, H_3, H_4, and H_5.

(b) Hilbert matrices are invertible, and their inverses can be found exactly using computer algebra systems (if n is not too large). Find the exact inverses of the Hilbert matrices in part (a). [*Note:* Some programs have a command for automatically entering Hilbert matrices by specifying n. If your program has this feature, it will save some typing.]

(c) Hilbert matrices are notoriously difficult to invert numerically because of their sensitivity to slight roundoff errors. To illustrate this, create an approximation H to H_5 by converting the fractions

in H_5 to decimals with six decimal places. Invert H and compare the result to the exact inverse you obtained in part (b).

T10. Let

$$A = \begin{bmatrix} 1 & 3 & 2 \\ 4 & 5 & 1 \\ 3 & 7 & 2 \end{bmatrix}$$

Find a quadratic polynomial $f(x) = ax^2 + bx + c$ for which $f(A) = A^{-1}$.

T11. If A is a square matrix, and if $f(x) = p(x)/q(x)$, where $p(x)$ and $q(x)$ are polynomials for which $q(A)$ is invertible, then we define $f(A) = p(A)q(A)^{-1}$. Find $f(A)$ for $f(x) = (x^3 + 2)/(x^2 + 1)$ and the matrix in Exercise T10.

Section 3.4 Subspaces and Linear Independence

In the first chapter we extended the concepts of line and plane to R^n. In this section we will consider other kinds of geometric objects in R^n, and we will show how systems of linear equations can be used to study them.

SUBSPACES OF R^n

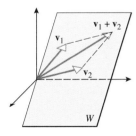

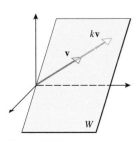

Figure 3.4.1

Recall from Section 1.3 that lines and planes through the origin in R^n are given by equations of the form $\mathbf{x} = t\mathbf{v}$ and $\mathbf{x} = t_1\mathbf{v}_1 + t_2\mathbf{v}_2$, respectively. Now we will turn our attention to geometric objects represented by equations of the general form

$$\mathbf{x} = t_1\mathbf{v}_1 + t_2\mathbf{v}_2 + \cdots + t_s\mathbf{v}_s$$

in which $\mathbf{v}_1, \mathbf{v}_2, \ldots, \mathbf{v}_s$ are vectors in R^n. We will begin with some observations about lines and planes through the origin of R^n.

It is evident geometrically that if $\mathbf{v}_1$ and $\mathbf{v}_2$ are vectors that lie in a plane W through the origin of R^2, then $\mathbf{v}_1 + \mathbf{v}_2$ is also in W, and that if $\mathbf{v}$ is a vector in W and k is any scalar, then $k\mathbf{v}$ is also in W (Figure 3.4.1). We describe these facts by saying that planes in R^2 are *closed under addition* and *closed under scalar multiplication*. More generally, these closure properties hold for planes in R^n. To see why this is so, let W be the plane through the origin of R^n whose equation is

$$\mathbf{x} = t_1\mathbf{v}_1 + t_2\mathbf{v}_2$$

If $\mathbf{x}$ is any vector in W and k is any scalar, then

$$k\mathbf{x} = k(t_1\mathbf{v}_1 + t_2\mathbf{v}_2) = (kt_1)\mathbf{v}_1 + (kt_2)\mathbf{v}_2 \tag{1}$$

This shows that $k\mathbf{x}$ is a linear combination of $\mathbf{v}_1$ and $\mathbf{v}_2$ and hence is a vector in W. Also, if

$$\mathbf{x} = t_1\mathbf{v}_1 + t_2\mathbf{v}_2 \quad \text{and} \quad \mathbf{x}' = t_1'\mathbf{v}_1 + t_2'\mathbf{v}_2$$

are vectors in W, then

$$\mathbf{x} + \mathbf{x}' = (t_1\mathbf{v}_1 + t_2\mathbf{v}_2) + (t_1'\mathbf{v}_1 + t_2'\mathbf{v}_2) = (t_1 + t_1')\mathbf{v}_1 + (t_2 + t_2')\mathbf{v}_2 \tag{2}$$

This shows that $\mathbf{x} + \mathbf{x}'$ is a linear combination of $\mathbf{v}_1$ and $\mathbf{v}_2$ and hence is a vector in W.

In general, if W is a nonempty set of vectors in R^n, then we say that W is ***closed under scalar multiplication*** if any scalar multiple of a vector in W is also in W, and we say that W is ***closed*** *under addition* if the sum of any two vectors in W is also in W. We also make the following definition to describe sets that have these two closure properties.

Definition 3.4.1   A nonempty set of vectors in R^n is called a *subspace* of R^n if it is closed under scalar multiplication and addition.

In addition to lines and planes through the origin of R^n, we can immediately identify two other subspaces of R^n, the *zero subspace*, which is the set $\{0\}$ consisting of the zero vector alone, and the set R^n itself. The set $\{0\}$ is a subspace because $\mathbf{0} + \mathbf{0} = \mathbf{0}$ (closure under addition) and $k\mathbf{0} = \mathbf{0}$ for all scalars k (closure under scalar multiplication). The set R^n is a subspace of R^n because adding two vectors in R^n or multiplying a vector in R^n by a scalar produces another vector in R^n. The zero subspace and R^n are called the *trivial subspaces* of R^n.

CONCEPT PROBLEM   Every subspace of R^n must contain the vector $\mathbf{0}$. Why?

EXAMPLE 1
A Subset of R^2
That Is Not a
Subspace

Let W be the set of all points (x, y) in R^2 such that $x > 0$ and $y > 0$ (points in the first quadrant). This set is closed under addition (why?), but it is not closed under scalar multiplication (Figure 3.4.2). Thus, W is not a subspace of R^2.   ■

CONCEPT PROBLEM   Is the RGB color cube in Figure 1.1.19 closed under addition? Under scalar multiplication? Explain.

REMARK   If W is a subspace of R^n, then the two closure properties can be used in combination to show that if $\mathbf{v}_1$ and $\mathbf{v}_2$ are vectors in W and c_1 and c_2 are scalars, then the linear combination

$$c_1 \mathbf{v}_1 + c_2 \mathbf{v}_2$$

is also a vector in W. More generally, any linear combination of vectors in W will be a vector in W, so we say that subspaces of R^n are *closed under linear combinations*.

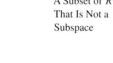

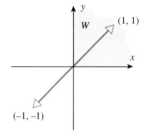

The vector $\mathbf{v} = (1, 1)$ lies in W, but the vector $(-1)\mathbf{v} = (-1, -1)$ does not.

Figure 3.4.2

To identify more subspaces of R^n, let $\mathbf{v}_1, \mathbf{v}_2, \ldots, \mathbf{v}_s$ be any vectors in R^n, and let W be the set of all vectors $\mathbf{x}$ that satisfy the equation

$$\mathbf{x} = t_1 \mathbf{v}_1 + t_2 \mathbf{v}_2 + \cdots + t_s \mathbf{v}_s$$

for appropriate choices of $t_1, t_2, \ldots, t_s$. The set W is nonempty because it contains the zero vector (take $t_1 = t_2 = \cdots = t_s = 0$) and it contains each of the vectors $\mathbf{v}_1, \mathbf{v}_2, \ldots, \mathbf{v}_s$ (why?). The set W is closed under scalar multiplication and addition, because multiplying a linear combination of $\mathbf{v}_1, \mathbf{v}_2, \ldots, \mathbf{v}_s$ by a scalar or adding two linear combinations of $\mathbf{v}_1, \mathbf{v}_2, \ldots, \mathbf{v}_s$ produces another linear combination of $\mathbf{v}_1, \mathbf{v}_2, \ldots, \mathbf{v}_s$ [see the computations in (1) and (2), for example]. Thus, we have the following result.

Theorem 3.4.2   *If $\mathbf{v}_1, \mathbf{v}_2, \ldots, \mathbf{v}_s$ are vectors in R^n, then the set of all linear combinations*

$$\mathbf{x} = t_1 \mathbf{v}_1 + t_2 \mathbf{v}_2 + \cdots + t_s \mathbf{v}_s \tag{3}$$

is a subspace of R^n.

The subspace W of R^n whose vectors satisfy (3) is called the *span* of $\mathbf{v}_1, \mathbf{v}_2, \ldots, \mathbf{v}_s$ and is denoted by

$$W = \text{span}\{\mathbf{v}_1, \mathbf{v}_2, \ldots, \mathbf{v}_s\} \tag{4}$$

We also say that the vectors $\mathbf{v}_1, \mathbf{v}_2, \ldots, \mathbf{v}_s$ *span* W. The scalars in (3) are called *parameters*, and you can think of $\text{span}\{\mathbf{v}_1, \mathbf{v}_2, \ldots, \mathbf{v}_s\}$ as the geometric object in R^n that results when the parameters in (3) are allowed to vary independently from $-\infty$ to $+\infty$. Alternatively, you can think of W as the set of all possible linear combinations of the vectors $\mathbf{v}_1, \mathbf{v}_2, \ldots, \mathbf{v}_s$; and more generally, if S is any nonempty set of vectors in R^n [not necessarily finite, as in (3)], then we define span(S) to be the set of all possible linear combinations that can be formed using vectors in S.

EXAMPLE 2
Spanning the
Trivial
Subspaces

Since every scalar multiple of the zero vector in R^n is zero, it follows that span$\{0\} = \{0\}$; that is, the zero subspace is spanned by the vector $\mathbf{0}$. Also, every vector $\mathbf{x} = (x_1, x_2, \ldots, x_n)$ in R^n can be expressed as a linear combination of the standard unit vectors $\mathbf{e}_1, \mathbf{e}_2, \ldots, \mathbf{e}_n$ by writing

$$\mathbf{x} = (x_1, x_2, \ldots, x_n) = x_1(1, 0, \ldots, 0) + x_2(0, 1, \ldots, 0) + x_n(0, 0, \ldots, 1)$$
$$= x_1\mathbf{e}_1 + x_2\mathbf{e}_2 + \cdots + x_n\mathbf{e}_n$$

Thus, span$\{\mathbf{e}_1, \mathbf{e}_2, \ldots, \mathbf{e}_n\} = R^n$; that is, R^n is spanned by the standard unit vectors. ∎

EXAMPLE 3
Spanning Lines
and Planes
Through the
Origin

Sometimes it is useful to express lines and planes through the origin of R^n in span notation. Thus, a line $\mathbf{x} = t\mathbf{v}$ can be written as span$\{\mathbf{v}\}$, and a plane $\mathbf{x} = t_1\mathbf{v}_1 + t_2\mathbf{v}_2$ as span$\{\mathbf{v}_1, \mathbf{v}_2\}$. For example, the line

$$(x_1, x_2, x_3, x_4) = t(1, 3, -2, 5)$$

can be expressed as span$\{\mathbf{v}\}$, where $\mathbf{v} = (1, 3, -2, 5)$. ∎

LOOKING AHEAD Two important tasks lie ahead in our study of subspaces: to identify all of the subspaces of R^n and to study their algebraic and geometric properties. With regard to the first task, we know at this point that span$\{\mathbf{v}_1, \mathbf{v}_2, \ldots, \mathbf{v}_s\}$ is a subspace of R^n for any choice of the vectors $\mathbf{v}_1, \mathbf{v}_2, \ldots, \mathbf{v}_s$, but we do not know whether these are *all* of the subspaces of R^n—it might be, for example, that there is some nonempty set in R^n that is closed under scalar multiplication and addition but which cannot be obtained by forming all possible linear combinations of some finite set of vectors. It will take some work to answer this question, and the details will have to wait until later; however, we will eventually show that *every* subspace of R^n is the span of some finite set of vectors, and, in fact, is the span of at most n vectors. Accepting this to be so, it follows that every subspace of R^2 is the span of at most two vectors, and every subspace of R^3 is the span of at most three vectors. The following example gives a complete list of all the subspaces of R^2 and R^3.

EXAMPLE 4
A Complete List
of Subspaces in
R^2 and in R^3

All subspaces of R^2 fall into one of three categories:

1. The zero subspace
2. Lines through the origin
3. All of R^2

All subspaces of R^3 fall into one of four categories:

1. The zero subspace
2. Lines through the origin
3. Planes through the origin
4. All of R^3 ∎

SOLUTION SPACE OF A LINEAR SYSTEM

Subspaces arise naturally in the course of solving homogeneous linear systems.

> **Theorem 3.4.3** *If $A\mathbf{x} = \mathbf{0}$ is a homogeneous linear system with n unknowns, then its solution set is a subspace of R^n.*

Proof Since $\mathbf{x} = \mathbf{0}$ is a solution of the system, we are assured that the solution set is nonempty. We must show that the solution set is closed under scalar multiplication and addition. To prove closure under scalar multiplication we must show that if $\mathbf{x}_0$ is any solution of the system and if

k is any scalar, then $k\mathbf{x}_0$ is also a solution of the system. But this is so since

$$A(k\mathbf{x}_0) = k(A\mathbf{x}_0) = k\mathbf{0} = \mathbf{0}$$

To prove closure under addition we must show that if $\mathbf{x}_1$ and $\mathbf{x}_2$ are solutions of the system, then $\mathbf{x}_1 + \mathbf{x}_2$ is also a solution of the system. But this is so since

$$A(\mathbf{x}_1 + \mathbf{x}_2) = A\mathbf{x}_1 + A\mathbf{x}_2 = \mathbf{0} + \mathbf{0} = \mathbf{0} \qquad ■$$

When we want to emphasize that the solution set of a homogeneous linear system is a subspace, we will refer to it as the *solution space* of the system. The solution space, being a subspace of R^n, must be expressible in the form

$$\mathbf{x} = t_1\mathbf{v}_1 + t_2\mathbf{v}_2 + \cdots + t_s\mathbf{v}_s \qquad (5)$$

which we call a *general solution* of the system. The usual procedure for obtaining a general solution of a homogeneous linear system is to solve the system, and then use the method of Example 7 in Section 2.2 to write $\mathbf{x}$ in form (5).

EXAMPLE 5
Finding a General Solution of a Homogeneous Linear System

We showed in Example 7 of Section 2.2 that the solutions of the homogeneous linear system

$$\begin{bmatrix} 1 & 3 & -2 & 0 & 2 & 0 \\ 2 & 6 & -5 & -2 & 4 & -3 \\ 0 & 0 & 5 & 10 & 0 & 15 \\ 2 & 6 & 0 & 8 & 4 & 18 \end{bmatrix} \begin{bmatrix} x_1 \\ x_2 \\ x_3 \\ x_4 \\ x_5 \\ x_6 \end{bmatrix} = \begin{bmatrix} 0 \\ 0 \\ 0 \\ 0 \end{bmatrix}$$

can be expressed in column form as

$$\begin{bmatrix} x_1 \\ x_2 \\ x_3 \\ x_4 \\ x_5 \\ x_6 \end{bmatrix} = r \begin{bmatrix} -3 \\ 1 \\ 0 \\ 0 \\ 0 \\ 0 \end{bmatrix} + s \begin{bmatrix} -4 \\ 0 \\ -2 \\ 1 \\ 0 \\ 0 \end{bmatrix} + t \begin{bmatrix} -2 \\ 0 \\ 0 \\ 0 \\ 1 \\ 0 \end{bmatrix} \qquad (6)$$

This is a general solution of the linear system. When convenient this general solution can be expressed in comma-delimited form as

$$(x_1, x_2, x_3, x_4, x_5, x_6) = r(-3, 1, 0, 0, 0, 0) + s(-4, 0, -2, 1, 0, 0) + t(-2, 0, 0, 0, 1, 0) \qquad (7)$$

or in parametric form as

$$x_1 = -3r - 4s - 2t, \ x_2 = r, \ x_3 = -2s, \ x_4 = s, \ x_5 = t, \ x_6 = 0$$

The solution space can also be denoted by span$\{\mathbf{v}_1, \mathbf{v}_2, \mathbf{v}_3\}$, where

$$\mathbf{v}_1 = (-3, 1, 0, 0, 0, 0), \quad \mathbf{v}_2 = (-4, 0, -2, 1, 0, 0), \quad \mathbf{v}_3 = (-2, 0, 0, 0, 1, 0) \qquad ■$$

EXAMPLE 6
Geometry of Homogeneous Systems in Two Unknowns

The solution space of a homogeneous linear system in two unknowns is a subspace of R^2 and hence must either be the origin $\mathbf{0}$, a line through the origin, or all of R^2 (see Figure 2.2.1, for example). The solution space will be all of R^2 if all of the coefficients in the equations are zero; for example, the system

$$0x + 0y = 0$$
$$0x + 0y = 0$$

is satisfied by all real values of x and y, so its solution space is all of R^2. $\qquad ■$

EXAMPLE 7

Geometry of
Homogeneous
Systems in
Three
Unknowns

The solution space of a homogeneous linear system in three unknowns is a subspace of R^3 and hence must either be the origin **0**, a line through the origin, a plane through the origin, or all of R^3. Keeping in mind that the graph of an equation of the form $ax + by + cz = 0$ is a plane through the origin if and only if a, b, and c are not all zero, the first three possibilities above are illustrated in Figure 3.4.3 for three equations in three unknowns. The solution space will be all of R^3 if all of the coefficients in the system are zero.   ■

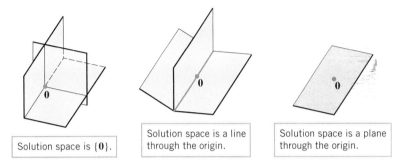

Figure 3.4.3

Solution space is {**0**}.

Solution space is a line through the origin.

Solution space is a plane through the origin.

In Examples 6 and 7 we were led to consider linear systems in which all of the coefficients are zero. Here is a theorem about such systems that will prove to be very useful in our later work.

Theorem 3.4.4

(a) *If A is a matrix with n columns, then the solution space of the homogeneous system $A\mathbf{x} = \mathbf{0}$ is all of R^n if and only if $A = 0$.*

(b) *If A and B are matrices with n columns, then $A = B$ if and only if $A\mathbf{x} = B\mathbf{x}$ for every $\mathbf{x}$ in R^n.*

Proof (a) If $A = 0$, then $A\mathbf{x} = 0\mathbf{x} = \mathbf{0}$ for every vector $\mathbf{x}$ in R^n, so the solution space of $A\mathbf{x} =$ is all of R^n. Conversely, assume that the solution space of $A\mathbf{x} = \mathbf{0}$ is all of R^n, and consider t equation

$$A = AI$$

where I is the $n \times n$ identity matrix. If we denote the successive column vectors of I by $\mathbf{e}_1, \mathbf{e}_2, \ldots, \mathbf{e}_n$, then we can rewrite this equation as

$$A = A[\mathbf{e}_1 \quad \mathbf{e}_2 \quad \cdots \quad \mathbf{e}_n] = [A\mathbf{e}_1 \quad A\mathbf{e}_2 \quad \cdots \quad A\mathbf{e}_n] = [\mathbf{0} \quad \mathbf{0} \quad \cdots \quad \mathbf{0}]$$

which implies that $A = 0$.

Proof (b) $A\mathbf{x} = B\mathbf{x}$ for all $\mathbf{x}$ in R^n if and only if $(A - B)\mathbf{x} = \mathbf{0}$ for all $\mathbf{x}$ in R^n. By part (a), this is true if and only if $A - B = 0$, or equivalently, if and only if $A = B$.   ■

REMARK The importance of part (b) of this theorem rests with the fact that it provides a way of showing that two matrices are equal without examining the individual entries of the matrices.

LINEAR INDEPENDENCE   Informally stated, we think of points as zero-dimensional, lines as one-dimensional, planes as two-dimensional, and space around us as three-dimensional. Thus, we see from Example 4 that R^2 has subspaces of dimension 0, 1, and 2, but none with dimension greater than 2, and R^3 has subspaces of dimension 0, 1, 2, and 3, but none with dimension greater than 3. Thus, it seems logical that with a reasonable definition of "dimension" the space R^n will have subspaces of dimension $0, 1, 2, \ldots, n$, but none with dimension greater than n. Some of these dimensions are already accounted for by the zero subspace (dimension 0), lines through the origin (dimension 1), and planes through the origin (dimension 2), but the subspaces of higher dimension remain to be identified. This will be done later, but we will begin developing some of the groundwork here.

We know that planes through the origin of R^n are represented by equations of the form

$$\mathbf{x} = t_1\mathbf{v}_1 + t_2\mathbf{v}_2 \tag{8}$$

in which $\mathbf{v}_1$ and $\mathbf{v}_2$ are nonzero vectors that are not scalar multiples of one another. These conditions on $\mathbf{v}_1$ and $\mathbf{v}_2$ are essential, for if either of the vectors is zero, or if either vector is a scalar multiple of the other, then (8) does not represent a plane. For example, if $\mathbf{v}_1 = \mathbf{v}_2 = \mathbf{0}$, then (8) reduces to $\mathbf{x} = \mathbf{0}$ and hence represents the zero subspace $\{\mathbf{0}\}$. As another example, if $\mathbf{v}_2 = c\mathbf{v}_1$, then (8) can be rewritten as

$$\mathbf{x} = t_1\mathbf{v}_1 + t_2\mathbf{v}_2 = t_1\mathbf{v}_1 + t_2(c\mathbf{v}_1) = (t_1 + ct_2)\mathbf{v}_1$$

This is an equation of the form $\mathbf{x} = t\mathbf{v}_1$ and hence represents the line through the origin parallel to $\mathbf{v}_1$. Thus, we see that the geometric properties of a subspace

$$\mathbf{x} = t_1\mathbf{v}_1 + t_2\mathbf{v}_2 + \cdots + t_s\mathbf{v}_s$$

are affected by interrelationships among the vectors $\mathbf{v}_1, \mathbf{v}_2, \ldots, \mathbf{v}_s$. The following definition will help us to explore this idea in more depth.

> **Definition 3.4.5**   A nonempty set of vectors $S = \{\mathbf{v}_1, \mathbf{v}_2, \ldots, \mathbf{v}_s\}$ in R^n is said to be **linearly independent** if the only scalars $c_1, c_2, \ldots, c_s$ that satisfy the equation
>
> $$c_1\mathbf{v}_1 + c_2\mathbf{v}_2 + \cdots + c_s\mathbf{v}_s = \mathbf{0} \tag{9}$$
>
> are $c_1 = 0, c_2 = 0, \ldots, c_s = 0$. If there are scalars, not all zero, that satisfy this equation, then the set is said to be **linearly dependent**.

REMARK Strictly speaking, the terms "linearly dependent" and "linearly independent" apply to nonempty finite *sets* of vectors; however, we will also find it convenient to apply them to the vectors themselves. Thus, we will say that the vectors $\mathbf{v}_1, \mathbf{v}_2, \ldots, \mathbf{v}_s$ are linearly independent or dependent in accordance with whether the set $S = \{\mathbf{v}_1, \mathbf{v}_2, \ldots, \mathbf{v}_s\}$ is linearly independent or dependent. Also, if S is a set with infinitely many vectors, then we will say that S is linearly independent if every finite subset is linearly independent and is linearly dependent if some finite subset is linearly dependent.

EXAMPLE 8
Linear Independence of One Vector

A single vector $\mathbf{v}$ is linearly dependent if it is zero and linearly independent if it is not—the vector $\mathbf{0}$ is linearly dependent because the equation $c\mathbf{0} = \mathbf{0}$ is satisfied by any nonzero scalar c, and a nonzero vector $\mathbf{v}$ is linearly independent because the only scalar c satisfying the equation $c\mathbf{v} = \mathbf{0}$ is $c = 0$. ∎

EXAMPLE 9
Sets Containing Zero Are Linearly Dependent

A nonempty set of vectors in R^n that *contains* the zero vector must be linearly dependent. For example, if $S = \{\mathbf{0}, \mathbf{v}_2, \ldots, \mathbf{v}_s\}$, then

$$1(\mathbf{0}) + 0\mathbf{v}_2 + \cdots + 0\mathbf{v}_s = \mathbf{0}$$

This implies that S is linearly dependent, since there are scalars, not all zero, that satisfy (9). ∎

The terminology *linearly dependent vectors* suggests that the vectors *depend* on each other in some way. The following theorem shows that this is, in fact, the case.

Theorem 3.4.6 *A set $S = \{\mathbf{v}_1, \mathbf{v}_2, \ldots, \mathbf{v}_s\}$ in R^n with two or more vectors is linearly dependent if and only if at least one of the vectors in S is expressible as a linear combination of the other vectors in S.*

Proof Assume that S is linearly dependent. This implies that there are scalars $c_1, c_2, \ldots, c_s$, not all zero, such that

$$c_1\mathbf{v}_1 + c_2\mathbf{v}_2 + \cdots + c_s\mathbf{v}_s = \mathbf{0} \tag{10}$$

To be specific, suppose that $c_1 \neq 0$. Then (10) can be rewritten as

$$\mathbf{v}_1 = \left(-\frac{c_2}{c_1}\right)\mathbf{v}_2 + \cdots + \left(-\frac{c_s}{c_1}\right)\mathbf{v}_s$$

which expresses $\mathbf{v}_1$ as a linear combination of the other vectors in the set. A similar argument holds if one of the other coefficients is nonzero.

Conversely, assume that at least one of the vectors in S is expressible as a linear combination of the other vectors in the set. To be specific, suppose that

$$\mathbf{v}_1 = c_2\mathbf{v}_2 + c_3\mathbf{v}_3 + \cdots + c_s\mathbf{v}_s$$

We can rewrite this as

$$\mathbf{v}_1 + (-c_2)\mathbf{v}_2 + (-c_3)\mathbf{v}_3 + \cdots + (-c_s)\mathbf{v}_s = \mathbf{0}$$

which is an equation of form (9) in which the scalars are not all zero. Thus, the vectors in S are linearly dependent. A similar argument holds for any vector that is a linear combination of the other vectors in the set. ∎

EXAMPLE 10
Linear Independence of Two Vectors

It follows from Definition 3.4.5 that two vectors $\mathbf{v}_1$ and $\mathbf{v}_2$ in R^n are linearly dependent if and only if there are scalars c_1 and c_2, not both zero, such that

$$c_1\mathbf{v}_1 + c_2\mathbf{v}_2 = \mathbf{0}$$

This equation can be rewritten as

$$\mathbf{v}_1 = -\left(\frac{c_2}{c_1}\right)\mathbf{v}_2 \quad \text{or} \quad \mathbf{v}_2 = -\left(\frac{c_1}{c_2}\right)\mathbf{v}_1$$

the first form being possible if $c_1 \neq 0$ and the second if $c_2 \neq 0$. Thus, we see that two vectors in R^n are linearly dependent if and only if at least one of the vectors is a scalar multiple of the other. Geometrically, this implies that two vectors in R^n are linearly dependent if they are collinear and linearly independent if they are not (Figure 3.4.4). ∎

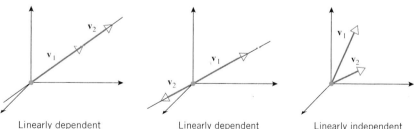

Figure 3.4.4   Linearly dependent   Linearly dependent   Linearly independent

EXAMPLE 11
Linear Independence of Three Vectors

Theorem 3.4.6 tells us that three vectors in R^n are linearly dependent if and only if at least one of them is a linear combination of the other two. But if one of them is a linear combination of the other two, then the three vectors must lie in a common plane through the origin (why?). Thus, three vectors in R^n are linearly dependent if they lie in a plane through the origin and are linearly independent if they do not (Figure 3.4.5). ∎

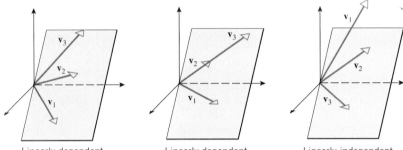

Figure 3.4.5   Linearly dependent   Linearly dependent   Linearly independent

LINEAR INDEPENDENCE AND HOMOGENEOUS LINEAR SYSTEMS

We will now show how to use a homogeneous system of linear equations to determine whether a set of vectors $S = \{\mathbf{v}_1, \mathbf{v}_2, \ldots, \mathbf{v}_s\}$ in R^n is linearly independent. For this purpose consider the $n \times s$ matrix

$$A = [\mathbf{v}_1 \quad \mathbf{v}_2 \quad \cdots \quad \mathbf{v}_s]$$

that has the vectors in S as its columns. Using this matrix and Formula (10) of Section 3.1, we can rewrite (9) as

$$[\mathbf{v}_1 \quad \mathbf{v}_2 \quad \cdots \quad \mathbf{v}_s] \begin{bmatrix} c_1 \\ c_2 \\ \vdots \\ c_s \end{bmatrix} = \begin{bmatrix} 0 \\ 0 \\ \vdots \\ 0 \end{bmatrix} \qquad (11)$$

which is a homogeneous linear system whose coefficient matrix is A and whose unknowns are the scalars in (9). Thus, the problem of determining whether $\mathbf{v}_1, \mathbf{v}_2, \ldots, \mathbf{v}_s$ are linearly independent reduces to determining whether (11) has nontrivial solutions—if the system has nontrivial solutions, then the vectors are linearly dependent, and if it has only the trivial solution, then they are linearly independent.

Theorem 3.4.7 *A homogeneous linear system $A\mathbf{x} = \mathbf{0}$ has only the trivial solution if and only if the column vectors of A are linearly independent.*

EXAMPLE 12

Linear Independence and Homogeneous Linear Systems

In each part, determine whether the vectors are linearly independent.

(a) $\mathbf{v}_1 = (1, 2, 1)$, $\mathbf{v}_2 = (2, 5, 0)$, $\mathbf{v}_3 = (3, 3, 8)$
(b) $\mathbf{v}_1 = (1, 2, -1)$, $\mathbf{v}_2 = (6, 4, 2)$, $\mathbf{v}_3 = (4, -1, 5)$
(c) $\mathbf{v}_1 = (2, -4, 6)$, $\mathbf{v}_2 = (0, 7, -5)$, $\mathbf{v}_3 = (6, 9, 8)$, $\mathbf{v}_4 = (5, 0, 1)$

Solution (a) We must determine whether the homogeneous linear system

$$\begin{bmatrix} 1 & 2 & 3 \\ 2 & 5 & 3 \\ 1 & 0 & 8 \end{bmatrix} \begin{bmatrix} c_1 \\ c_2 \\ c_3 \end{bmatrix} = \begin{bmatrix} 0 \\ 0 \\ 0 \end{bmatrix}$$

has nontrivial solutions. This system was considered in Example 6 of Section 3.3 (with different names for the unknowns), where we showed that it has only the trivial solution. Thus, the vectors are linearly independent.

Solution (b) We must determine whether the homogeneous linear system

$$\begin{bmatrix} 1 & 6 & 4 \\ 2 & 4 & -1 \\ -1 & 2 & 5 \end{bmatrix} \begin{bmatrix} c_1 \\ c_2 \\ c_3 \end{bmatrix} = \begin{bmatrix} 0 \\ 0 \\ 0 \end{bmatrix}$$

has nontrivial solutions. However, we showed in Example 4 of Section 3.3 that the coefficient matrix for this system is not invertible. This implies that the system has nontrivial solutions (Theorem 3.3.7), and hence that the vectors are linearly dependent.

Solution (*c*) We must determine whether the homogeneous linear system

$$\begin{bmatrix} 2 & 0 & 6 & 5 \\ -4 & 7 & 9 & 0 \\ 6 & -5 & 8 & 1 \end{bmatrix} \begin{bmatrix} c_1 \\ c_2 \\ c_3 \\ c_4 \end{bmatrix} = \begin{bmatrix} 0 \\ 0 \\ 0 \\ 0 \end{bmatrix}$$

has nontrivial solutions. However, this system has more unknowns than equations, so it must have nontrivial solutions by Theorem 2.2.3. This implies that the vectors are linearly dependent. ∎

Part (c) of this example illustrates an important fact: A finite set with more than *n* vectors in R^n must be linearly dependent because the homogeneous linear system whose coefficient matrix has those vectors as columns has more unknowns than equations and hence has nontrivial solutions. Thus, we have the following theorem.

> **Theorem 3.4.8**   *A set with more than n vectors in R^n is linearly dependent.*

TRANSLATED SUBSPACES

If $\mathbf{x} = t\mathbf{v}$ is a line though the origin of R^2 or R^3, then $\mathbf{x} = \mathbf{x}_0 + t\mathbf{v}$ is a parallel line that passes through the point $\mathbf{x}_0$; and, similarly, if $\mathbf{x} = t_1\mathbf{v}_1 + t_2\mathbf{v}_2$ is a plane through the origin of R^3, then $\mathbf{x} = \mathbf{x}_0 + t_1\mathbf{v}_1 + t_2\mathbf{v}_2$ is a parallel plane that passes through the point $\mathbf{x}_0$. Thus, we can view the line $\mathbf{x} = \mathbf{x}_0 + t\mathbf{v}$ as the translation of $\mathbf{x} = t\mathbf{v}$ by $\mathbf{x}_0$ and the plane $\mathbf{x} = \mathbf{x}_0 + t_1\mathbf{v}_1 + t_2\mathbf{v}_2$ as the translation of $\mathbf{x} = t_1\mathbf{v}_1 + t_2\mathbf{v}_2$ by $\mathbf{x}_0$. More generally, if $\mathbf{x}_0, \mathbf{v}_1, \mathbf{v}_2, \ldots, \mathbf{v}_s$ are vectors in R^n, then the set of vectors of the form

$$\mathbf{x} = \mathbf{x}_0 + t_1\mathbf{v}_1 + t_2\mathbf{v}_2 + \cdots + t_s\mathbf{v}_s \tag{12}$$

can be viewed as a translation by $\mathbf{x}_0$ of the subspace

$$\mathbf{x} = t_1\mathbf{v}_1 + t_2\mathbf{v}_2 + \cdots + t_s\mathbf{v}_s \tag{13}$$

Since (13) is an equation for $W = \text{span}\{\mathbf{v}_1, \mathbf{v}_2, \ldots, \mathbf{v}_s\}$, we call (12) the ***translation of W by*** $\mathbf{x}_0$ and denote it by

$$\mathbf{x}_0 + W \quad \text{or} \quad \mathbf{x}_0 + \text{span}\{\mathbf{v}_1, \mathbf{v}_2, \ldots, \mathbf{v}_s\}$$

Translations of subspaces have various names in the literature, the most common being ***linear manifolds***, ***flats***, and ***affine spaces***. We will call them linear manifolds. It follows from Example 4 that linear manifolds in R^2 fall into three categories (points, lines, and all of R^2) and that linear manifolds in R^3 fall into four categories (points, lines, planes, and all of R^3).

A UNIFYING THEOREM

Theorem 3.4.7 allows us to add two more statements to Theorem 3.3.9.

> **Theorem 3.4.9**   *If A is an n × n matrix, then the following statements are equivalent.*
> (*a*) *The reduced row echelon form of A is I_n.*
> (*b*) *A is expressible as a product of elementary matrices.*
> (*c*) *A is invertible.*
> (*d*) *A$\mathbf{x} = \mathbf{0}$ has only the trivial solution.*
> (*e*) *A$\mathbf{x} = \mathbf{b}$ is consistent for every vector $\mathbf{b}$ in R^n.*
> (*f*) *A$\mathbf{x} = \mathbf{b}$ has exactly one solution for every vector $\mathbf{b}$ in R^n.*
> (*g*) *The column vectors of A are linearly independent.*
> (*h*) *The row vectors of A are linearly independent.*

The equivalence of (*g*) and (*d*) follows from Theorem 3.4.7, and the equivalence of (*g*) and (*h*) follows from the fact that a matrix is invertible if and only if its transpose is invertible.

Exercise Set 3.4

In Exercises 1 and 2, find vector and parametric equations of span{**v**}.

1. (a) $\mathbf{v} = (1, -1)$ (b) $\mathbf{v} = (2, 1, -4)$

    (c) $\mathbf{v} = (1, 1, -2, 3)$

2. (a) $\mathbf{v} = (0, -4)$ (b) $\mathbf{v} = (0, -1, 5)$

    (c) $\mathbf{v} = (2, 0, 5, -4)$

In Exercises 3 and 4, find vector and parametric equations of span{$\mathbf{v}_1, \mathbf{v}_2$}.

3. (a) $\mathbf{v}_1 = (4, -4, 2), \mathbf{v}_2 = (-3, 5, 7)$

    (b) $\mathbf{v}_1 = (1, 2, 1, -3), \mathbf{v}_2 = (3, 4, 5, 0)$

4. (a) $\mathbf{v}_1 = (0, -1, 0), \mathbf{v}_2 = (-2, 1, 9)$

    (b) $\mathbf{v}_1 = (1, 5, -1, 4, 2), \mathbf{v}_2 = (2, 2, 0, 1, -4)$

In Exercises 5 and 6, determine by inspection whether **u** is in the subspace span{**v**}.

5. (a) $\mathbf{u} = (-2, 8, 4), \mathbf{v} = (1, -4, -2)$

    (b) $\mathbf{u} = (6, -9, 3, -3), \mathbf{v} = (2, -3, 1, 0)$

6. (a) $\mathbf{u} = (1, 0, -3), \mathbf{v} = (2, 0, 0)$

    (b) $\mathbf{u} = (1, -2, -3, 0), \mathbf{v} = (2, -4, -6, 0)$

7. In each part of Exercise 5, determine whether the vectors **u** and **v** are linearly independent.

8. In each part of Exercise 6, determine whether the vectors **u** and **v** are linearly independent.

In Exercises 9 and 10, show that the vectors are linearly dependent by expressing one of the vectors as a linear combination of the other two.

9. $\mathbf{u} = (1, 4, 4), \mathbf{v} = (-1, 3, 6), \mathbf{w} = (-3, 2, 8)$

10. $\mathbf{u} = (8, 10, -6), \mathbf{v} = (2, 3, 1), \mathbf{w} = (0, 1, 5)$

In Exercises 11 and 12, what kind of geometric object is represented by the equation?

11. (a) $(x_1, x_2, x_3, x_4) = (2, -3, 1, 4)t_1 + (4, -6, 2, 8)t_2$

    (b) $(x_1, x_2, x_3, x_4) = (3, -2, 2, 5)t_1 + (6, -4, 4, 0)t_2$

12. (a) $(x_1, x_2, x_3, x_4) = (6, -2, -4, 8)t_1 + (3, 0, 2, -4)t_2$

    (b) $(x_1, x_2, x_3, x_4) = (6, -2, -4, 8)t_1 + (-3, 1, 2, -4)t_2$

In Exercises 13–16, find a general solution of the linear system, and list a set of vectors that span the solution space.

13. $\begin{aligned} x_1 + 6x_2 + 2x_3 - 5x_4 &= 0 \\ -x_1 - 6x_2 - x_3 - 3x_4 &= 0 \\ 2x_1 + 12x_2 + 5x_3 - 18x_4 &= 0 \end{aligned}$

14. $\begin{aligned} x_1 - x_2 + x_3 + x_4 - 2x_5 &= 0 \\ -2x_1 + 2x_2 - x_3 \quad\quad + x_5 &= 0 \\ x_1 - x_2 + 2x_3 + 3x_4 - 5x_5 &= 0 \end{aligned}$

15. $\begin{aligned} x_1 + 2x_2 + x_3 + x_4 + x_5 &= 0 \\ 2x_1 + 4x_2 - x_3 \quad\quad + x_5 &= 0 \end{aligned}$

16. $\begin{aligned} x_1 + 2x_2 - 2x_3 - x_4 \quad\quad &= 0 \\ 2x_1 + 3x_2 - 5x_3 + x_4 - 7x_5 &= 0 \\ x_2 + x_3 - x_4 + x_5 &= 0 \\ -x_1 + x_2 + 5x_3 - x_4 \quad\quad &= 0 \end{aligned}$

In Exercises 17 and 18, explain why the given sets of vectors are linearly dependent.

17. (a) $\mathbf{v}_1 = (-1, 2, 4), \mathbf{v}_2 = (5, -10, -20)$

    (b) $\mathbf{v}_1 = (3, -1), \mathbf{v}_2 = (4, 5), \mathbf{v}_3 = (-4, 7)$

18. (a) $\mathbf{v}_1 = (-1, 2, 4), \mathbf{v}_2 = (0, 0, 0)$

    (b) $\mathbf{v}_1 = (3, -1, 2), \mathbf{v}_2 = (4, 5, 1), \mathbf{v}_3 = (-4, 7, 3),$
       $\mathbf{v}_4 = (-4, 7, 3)$

In Exercises 19 and 20, determine whether the vectors are linearly independent.

19. (a) $\mathbf{v}_1 = (4, -1, 2), \mathbf{v}_2 = (-4, 10, 2)$

    (b) $\mathbf{v}_1 = (3, 0, 6), \mathbf{v}_2 = (-1, 0, -2)$

    (c) $\mathbf{v}_1 = (-3, 0, 4), \mathbf{v}_2 = (5, -1, 2), \mathbf{v}_3 = (1, 1, 3)$

    (d) $\mathbf{v}_1 = (2, 5, 4), \mathbf{v}_2 = (0, -6, 2), \mathbf{v}_3 = (2, 1, 8),$
       $\mathbf{v}_4 = (4, -3, 7)$

20. (a) $\mathbf{v}_1 = (3, 8, 7, -3), \mathbf{v}_2 = (1, 5, 3, -1),$
       $\mathbf{v}_3 = (2, -1, 2, 6), \mathbf{v}_4 = (1, 4, 0, 3)$

    (b) $\mathbf{v}_1 = (0, 0, 2, 2), \mathbf{v}_2 = (3, 3, 0, 0), \mathbf{v}_3 = (1, 1, 0, -1)$

    (c) $\mathbf{v}_1 = (0, 0, 2, 2), \mathbf{v}_2 = (3, 3, 0, 0),$
       $\mathbf{v}_3 = (1, 1, 0, -1), \mathbf{v}_4 = (4, 4, 2, 1)$

    (d) $\mathbf{v}_1 = (4, 7, 6, -2), \mathbf{v}_2 = (9, 7, 2, -1),$
       $\mathbf{v}_3 = (4, -3, 1, 5), \mathbf{v}_4 = (6, 4, 9, 1), \mathbf{v}_5 = (4, -7, 0, 6)$

In Exercises 21 and 22, determine whether the vectors lie in a plane, and if so, whether they lie on a line.

21. (a) $\mathbf{v}_1 = (-2, -2, 0), \mathbf{v}_2 = (6, 1, 4), \mathbf{v}_3 = (2, 0, -4)$

    (b) $\mathbf{v}_1 = (-6, 7, 2), \mathbf{v}_2 = (3, 2, 4), \mathbf{v}_3 = (4, -1, 2)$

    (c) $\mathbf{v}_1 = (2, -4, 8), \mathbf{v}_2 = (1, -2, 4), \mathbf{v}_3 = (3, -6, 12)$

22. (a) $\mathbf{v}_1 = (1, 2, 1), \mathbf{v}_2 = (-7, 1, 3), \mathbf{v}_3 = (-14, -1, -4)$

    (b) $\mathbf{v}_1 = (2, 1, -1), \mathbf{v}_2 = (1, 3, 4), \mathbf{v}_3 = (1, 1, 0)$

    (c) $\mathbf{v}_1 = (6, -4, 8), \mathbf{v}_2 = (-3, 2, -4), \mathbf{v}_3 = (9, -6, -12)$

In Exercises 23 and 24, use the closure properties of subspaces to determine whether the set is a subspace of R^3. If it is not, indicate which closure properties fail.

23. (a) All vectors of the form $(a, 0, 0)$.

    (b) All vectors with integer components.

    (c) All vectors (a, b, c) for which $b = a + c$.

    (d) All vectors (a, b, c) for which $a + b + c = 1$.

24. (a) All vectors with exactly one nonzero component.

    (b) All **x** for which $\|\mathbf{x}\| \le 1$.

    (c) All vectors of the form $(a, -a, c)$.

    (d) All vectors (a, b, c) for which $a + b + c = 0$.

25. Show that the set W of all vectors that are of the form $\mathbf{x} = (a, 0, a, 0)$, in which a is any real number, is a subspace of R^4 by finding a set of spanning vectors for W. What kind of geometric object is W?

26. Show that the set W of all vectors that are of the form $\mathbf{x} = (a, b, 2a, 3b, -a)$ in which a and b are any real numbers is a subspace of R^5 by finding a set of spanning vectors for W. What kind of geometric object is W?

27. (a) Show that the vectors $\mathbf{v}_1 = (0, 3, 1, -1)$, $\mathbf{v}_2 = (6, 0, 5, 1)$, and $\mathbf{v}_3 = (4, -7, 1, 3)$ form a linearly dependent set in R^4.
(b) Express each vector as a linear combination of the other two.

28. For which real values of λ do the following vectors form a linearly dependent set in R^3?
$$\mathbf{v}_1 = \left(\lambda, -\tfrac{1}{2}, -\tfrac{1}{2}\right), \quad \mathbf{v}_2 = \left(-\tfrac{1}{2}, \lambda, -\tfrac{1}{2}\right), \quad \mathbf{v}_3 = \left(-\tfrac{1}{2}, -\tfrac{1}{2}, \lambda\right)$$

29. (a) Show that if $S = \{\mathbf{v}_1, \mathbf{v}_2, \mathbf{v}_3\}$ is a linearly independent set of vectors in R^n, then so is every nonempty subset of S.
(b) Show that if $S = \{\mathbf{v}_1, \mathbf{v}_2, \mathbf{v}_3\}$ is a linearly dependent set of vectors in R^n, then so is the set $\{\mathbf{v}_1, \mathbf{v}_2, \mathbf{v}_3, \mathbf{v}\}$ for every vector $\mathbf{v}$ in R^n.

30. (a) Show that if $S = \{\mathbf{v}_1, \mathbf{v}_2, \ldots, \mathbf{v}_r\}$ is a linearly independent set of vectors in R^n, then so is every nonempty subset of S.
(b) Show that if $S = \{\mathbf{v}_1, \mathbf{v}_2, \ldots, \mathbf{v}_r\}$ is a linearly dependent set of vectors in R^n, then so is the set $\{\mathbf{v}_1, \mathbf{v}_2, \ldots, \mathbf{v}_r, \mathbf{v}\}$ for every vector $\mathbf{v}$ in R^n.

31. Show that if $\mathbf{u}$, $\mathbf{v}$, and $\mathbf{w}$ are any vectors in R^n, then the vectors $\mathbf{u} - \mathbf{v}$, $\mathbf{v} - \mathbf{w}$, and $\mathbf{w} - \mathbf{u}$ form a linearly dependent set.

32. High fidelity sound can be recorded digitally by sampling a sound wave at the rate of 44,100 times a second. Thus, a 10-second segment of sound can be represented by a vector in R^{441000}. A sound technician at a jazz festival plans to record sound vectors with two microphones, one sound vector $\mathbf{s}$ from a microphone next to the saxophone player and a second concurrent sound vector $\mathbf{g}$ from a microphone next to the guitar player. A linear combination of the two sound vectors will then be created by a "mixer" in a sound studio to produce the desired result. Suppose that each microphone picks up all of the sound from its adjacent instrument and a small amount of sound from the other instrument, so the actual recorded vectors are $\mathbf{u} = \mathbf{s} + 0.06\mathbf{g}$ for the saxophone and $\mathbf{v} = \mathbf{g} + 0.12\mathbf{s}$ for the guitar.
(a) What linear combination of $\mathbf{u}$ and $\mathbf{v}$ will recover the saxophone vector $\mathbf{s}$?
(b) What linear combination of $\mathbf{u}$ and $\mathbf{v}$ will recover the guitar vector $\mathbf{g}$?
(c) What linear combination of $\mathbf{u}$ and $\mathbf{v}$ will produce an equal mix of $\mathbf{s}$ and $\mathbf{g}$, that is, $\tfrac{1}{2}(\mathbf{s} + \mathbf{g})$?

33. Color magazines and books are printed using what is called a **CMYK** *color model*. Colors in this model are created using four colored inks: cyan (C), magenta (M), yellow (Y), and black (K). The colors can be created either by mix-

ing inks of the four types and printing with the mixed inks (the *spot color method*) or by printing dot patterns (called *rosettes*) with the four colors and allowing the reader's eye and perception process to create the desired color combination (the *process color method*). There is a numbering system for commercial inks, called the *Pantone Matching System*, that assigns every commercial ink color a number in accordance with its percentages of cyan, magenta, yellow, and black. One way to represent a Pantone color is by associating the four base colors with the vectors

$$\mathbf{c} = (1, 0, 0, 0) \quad \text{(pure cyan)}$$
$$\mathbf{m} = (0, 1, 0, 0) \quad \text{(pure magenta)}$$
$$\mathbf{y} = (0, 0, 1, 0) \quad \text{(pure yellow)}$$
$$\mathbf{k} = (0, 0, 0, 1) \quad \text{(pure black)}$$

in R^4 and describing the ink color as a linear combination of these using coefficients between 0 and 1, inclusive. Thus, an ink color $\mathbf{p}$ is represented as a linear combination of the form

$$\mathbf{p} = c_1\mathbf{c} + c_2\mathbf{m} + c_3\mathbf{y} + c_4\mathbf{k} \quad (c_1, c_2, c_3, c_4)$$

where $0 \le c_i \le 1$. The set of all such linear combinations is called **CMYK** *space*.
(a) Is CMYK space a subspace of R^4? Explain.
(b) Pantone color 876CVC is a mixture of 38% cyan, 59% magenta, 73% yellow, and 7% black; Pantone color 216CVC is a mixture of 0% cyan, 83% magenta, 34% yellow, and 47% black; and Pantone color 328CVC is a mixture of 100% cyan, 0% magenta, 47% yellow, and 30% black. Denote these colors by $\mathbf{p}_{876} = (0.38, 0.59, 0.73, 0.07)$, $\mathbf{p}_{216} = (0, 0.83, 0.34, 0.47)$, and $\mathbf{p}_{328} = (1, 0, 0.47, 0.30)$, respectively, and express these vectors as linear combinations of $\mathbf{c}$, $\mathbf{m}$, $\mathbf{y}$, and $\mathbf{k}$.
(c) What CMYK vector do you think you would get if you mixed $\mathbf{p}_{876}$ and $\mathbf{p}_{216}$ in equal proportions? Why?

34. The following table shows the test scores of seven students on three tests.

	Test 1	Test 2	Test 3
Jones	90	75	60
Chan	54	92	70
Rocco	63	70	81
Johnson	70	71	72
Stein	46	90	63
Rio	87	72	69
Smith	50	77	83

View the columns in the body of the table as vectors $\mathbf{c}_1$, $\mathbf{c}_2$, and $\mathbf{c}_3$ in R^7, and view the rows in the body of the table as vectors $\mathbf{r}_1, \mathbf{r}_2, \ldots, \mathbf{r}_7$ in R^3.
(a) Find scalars k_1, k_2, and k_3 such that the components of the vector $\mathbf{x} = k_1\mathbf{c}_1 + k_2\mathbf{c}_2 + k_3\mathbf{c}_3$ are the average test scores for the students.

(b) Find scalars $k_1, k_2, \ldots, k_7$ such that the components of
$\mathbf{x} = k_1\mathbf{r}_1 + k_2\mathbf{r}_2 + \cdots + k_7\mathbf{r}_7$ are the average scores
for all of the students on each test.

(c) Give an interpretation of the vector
$\mathbf{x} = \frac{1}{4}\mathbf{c}_1 + \frac{1}{4}\mathbf{c}_2 + \frac{1}{2}\mathbf{c}_3$.

35. The following table shows the populations of five Pennsylvania counties in four different years.

	1950	1980	1992	1998
Philadelphia	408,762	847,170	1,552,572	1,436,287
Bucks	144,620	479,211	556,279	587,942
Delaware	414,234	555,007	549,506	542,593
Adams	44,197	68,292	81,232	86,537
Potter	16,810	17,726	16,863	17,184

Source: Population abstract of the United States, and Population Estimates
Program, Population Division, U.S. Bureau of the Census.

View the columns in the body of the table as vectors
$\mathbf{c}_1, \mathbf{c}_2, \mathbf{c}_3$, and $\mathbf{c}_4$ in R^5, and view the rows in the body of
the table as vectors $\mathbf{r}_1, \mathbf{r}_2, \mathbf{r}_3, \mathbf{r}_4$, and $\mathbf{r}_5$ in R^4.

(a) Find scalars k_1, k_2, k_3, and k_4 such that the components
of the vector $\mathbf{x} = k_1\mathbf{c}_1 + k_2\mathbf{c}_2 + k_3\mathbf{c}_3 + k_4\mathbf{c}_4$ are the
average populations of the counties over the four
sampled years.

(b) Find scalars k_1, k_2, k_3, k_4, and k_5 such that the
components of $\mathbf{x} = k_1\mathbf{r}_1 + k_2\mathbf{r}_2 + k_3\mathbf{r}_3 + k_4\mathbf{r}_4 + k_5\mathbf{r}_5$
are the average populations of the five counties in each
sampled year.

(c) Give an interpretation of the vector
$\mathbf{x} = \frac{1}{3}\mathbf{r}_1 + \frac{1}{3}\mathbf{r}_2 + \frac{1}{3}\mathbf{r}_3$.

36. (*Sigma notation*) Express the linear combination
$$\mathbf{v} = c_1\mathbf{v}_1 + c_2\mathbf{v}_2 + \cdots + c_n\mathbf{v}_n$$
in sigma notation.

Discussion and Discovery

D1. (a) What geometric property must a set of two vectors
have if they are to span R^2?

(b) What geometric property must a set of three vectors
in R^3 have if they are to span R^3?

D2. (a) Under what conditions will a set of two vectors in R^n
span a plane?

(b) Under what conditions will a set of two vectors in R^n
span a line?

(c) If $\mathbf{u}$ and $\mathbf{v}$ are vectors in R^n, under what conditions
will it be true that span$\{\mathbf{u}\}$ = span$\{\mathbf{v}\}$?

D3. (a) Do you think that every set of three nonzero mutually
orthogonal vectors in R^3 is linearly independent?
Justify your answer with a geometric argument.

(b) Justify your answer in part (a) with an algebraic
argument. [*Hint:* Use dot products.]

D4. Determine whether the vectors $\mathbf{v}_1, \mathbf{v}_2$, and $\mathbf{v}_3$ in each part
of the accompanying figure are linearly independent, and
explain your reasoning.

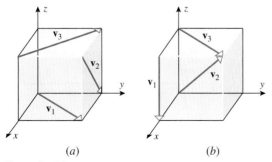

(a)   (b)

Figure Ex-D4

D5. The vectors corresponding to points in the shaded region of
the accompanying figure do not form a subspace of R^2 (see
Example 1), so one or both of the closure axioms must fail.
Which one(s)?

Figure Ex-D5

D6. Indicate whether the statement is true (T) or false (F).
Justify your answer.

(a) If three nonzero vectors form a linearly dependent set,
then each vector in the set can be expressed as a linear
combination of the other two.

(b) The set of all linear combinations of two vectors $\mathbf{v}$
and $\mathbf{w}$ in R^n is a plane.

(c) If $\mathbf{u}$ cannot be expressed as a linear combination of $\mathbf{v}$
and $\mathbf{w}$, then the three vectors are linearly independent.

(d) A set of vectors in R^n that contains $\mathbf{0}$ is linearly
dependent.

(e) If $\{\mathbf{v}_1, \mathbf{v}_2, \mathbf{v}_3\}$ is a linearly independent set, then so is
the set $\{k\mathbf{v}_1, k\mathbf{v}_2, k\mathbf{v}_3\}$ for every nonzero scalar k.

D7. Indicate whether the statement is true (T) or false (F). Justify your answer.

    (a) If $\mathbf{v}$ is a vector in R^n, then $\{\mathbf{v}, k\mathbf{v}\}$ is a linearly independent set if $k \neq \mathbf{0}$.

    (b) If $A\mathbf{x} = \mathbf{b}$ is any consistent linear system of m equations in n unknowns, then the solution set is a subspace of R^n.

    (c) If W is a subspace of R^n, then span$(W) = W$.

    (d) If span$(S_1) = $ span(S_2), then $S_1 = S_2$.

D8. State a relationship between span(S) and span(span(S)), and justify your answer.

Working with Proofs

P1. Let $\mathbf{u}$ and $\mathbf{v}$ be nonzero vectors in R^2 or R^3, and let $k = \|\mathbf{u}\|$ and $l = \|\mathbf{v}\|$. Prove that the linear combination $\mathbf{w} = l\mathbf{u}+k\mathbf{v}$ bisects the angle between $\mathbf{u}$ and $\mathbf{v}$ if these vectors have their initial points at the origin.

P2. Let W_1 and W_2 be subspaces of R^n. Prove that the intersection $W_1 \cap W_2$ is a subspace of R^n.

P3. If W_1 and W_2 are subspaces of R^n, then their ***sum***, written $W_1 + W_2$, is defined to be the set of all vectors of the form $\mathbf{x} + \mathbf{y}$, where $\mathbf{x}$ is in W_1 and $\mathbf{y}$ is in W_2. Prove that the sum $W_1 + W_2$ is a subspace of R^n.

Technology Exercises

T1. (***Sigma notation***) Use your technology utility to compute the linear combination

$$\mathbf{v} = \sum_{j=1}^{25} c_j\mathbf{v}_j$$

    for $c_j = 1/j$ and $\mathbf{v}_j = (\sin j, \cos j)$.

T2. Devise a procedure for using your technology utility to determine whether a set of vectors in R^n is linearly independent, and use it to determine whether the following vectors are linearly independent.

$$\mathbf{v}_1 = (4, -5, 2, 6), \quad \mathbf{v}_2 = (2, -2, 1, 3)$$
$$\mathbf{v}_3 = (6, -3, 3, 9), \quad \mathbf{v}_4 = (4, -1, 5, 6)$$

T3. Let $\mathbf{v}_1 = (4, 3, 2, 1)$, $\mathbf{v}_2 = (5, 1, 2, 4)$, $\mathbf{v}_3 = (7, 1, 5, 3)$, $\mathbf{x} = (16, 5, 9, 8)$, and $\mathbf{y} = (3, 1, 2, 7)$. Determine whether $\mathbf{x}$ and $\mathbf{y}$ lie in span$\{\mathbf{v}_1, \mathbf{v}_2, \mathbf{v}_3\}$.

Section 3.5   The Geometry of Linear Systems

In this section we will use some of the results we have obtained about matrices to explore geometric properties of solution sets of linear systems.

THE RELATIONSHIP BETWEEN $A\mathbf{x} = \mathbf{b}$ AND $A\mathbf{x} = \mathbf{0}$

Our first objective in this section is to establish a relationship between the solutions of a consistent nonhomogeneous system $A\mathbf{x} = \mathbf{b}$ and the solutions of the homogeneous system $A\mathbf{x} = \mathbf{0}$ that has the same coefficient matrix. For this purpose we will call $A\mathbf{x} = \mathbf{0}$ the homogeneous system ***associated*** with $A\mathbf{x} = \mathbf{b}$.

To motivate the result we are looking for, let us consider the nonhomogeneous system

$$\begin{bmatrix} 1 & 3 & -2 & 0 & 2 & 0 \\ 2 & 6 & -5 & -2 & 4 & -3 \\ 0 & 0 & 5 & 10 & 0 & 15 \\ 2 & 6 & 0 & 8 & 4 & 18 \end{bmatrix} \begin{bmatrix} x_1 \\ x_2 \\ x_3 \\ x_4 \\ x_5 \\ x_6 \end{bmatrix} = \begin{bmatrix} 0 \\ -1 \\ 5 \\ 6 \end{bmatrix} \qquad (1)$$

and its associated homogeneous system

$$
\begin{bmatrix}
1 & 3 & -2 & 0 & 2 & 0 \\
2 & 6 & -5 & -2 & 4 & -3 \\
0 & 0 & 5 & 10 & 0 & 15 \\
2 & 6 & 0 & 8 & 4 & 18
\end{bmatrix}
\begin{bmatrix}
x_1 \\ x_2 \\ x_3 \\ x_4 \\ x_5 \\ x_6
\end{bmatrix}
=
\begin{bmatrix}
0 \\ 0 \\ 0 \\ 0
\end{bmatrix}
\tag{2}
$$

We showed in Examples 5 and 7 of Section 2.2 that the solution sets of these systems can be expressed as

$$
\begin{bmatrix}
x_1 \\ x_2 \\ x_3 \\ x_4 \\ x_5 \\ x_6
\end{bmatrix}
=
\begin{bmatrix}
0 \\ 0 \\ 0 \\ 0 \\ 0 \\ \frac{1}{3}
\end{bmatrix}
+ r
\begin{bmatrix}
-3 \\ 1 \\ 0 \\ 0 \\ 0 \\ 0
\end{bmatrix}
+ s
\begin{bmatrix}
-4 \\ 0 \\ -2 \\ 1 \\ 0 \\ 0
\end{bmatrix}
+ t
\begin{bmatrix}
-2 \\ 0 \\ 0 \\ 0 \\ 1 \\ 0
\end{bmatrix}
$$

$$\boxed{\text{Solution set of (1)}}$$

and

$$
\begin{bmatrix}
x_1 \\ x_2 \\ x_3 \\ x_4 \\ x_5 \\ x_6
\end{bmatrix}
= r
\begin{bmatrix}
-3 \\ 1 \\ 0 \\ 0 \\ 0 \\ 0
\end{bmatrix}
+ s
\begin{bmatrix}
-4 \\ 0 \\ -2 \\ 1 \\ 0 \\ 0
\end{bmatrix}
+ t
\begin{bmatrix}
-2 \\ 0 \\ 0 \\ 0 \\ 1 \\ 0
\end{bmatrix}
$$

$$\boxed{\text{Solution set of (2)}}$$

Thus, we see that the solution set of the nonhomogeneous system (1) is the translation of the solution space of the associated homogeneous system (2) by the vector

$$
\mathbf{x}_0 =
\begin{bmatrix}
0 \\ 0 \\ 0 \\ 0 \\ 0 \\ \frac{1}{3}
\end{bmatrix}
$$

where $\mathbf{x}_0$ is a specific solution of (1) (set $r = s = t = 0$ in the above formula for its solution set). This illustrates the following general result.

Theorem 3.5.1 *If $A\mathbf{x} = \mathbf{b}$ is a consistent nonhomogeneous linear system, and if W is the solution space of the associated homogeneous system $A\mathbf{x} = \mathbf{0}$, then the solution set of $A\mathbf{x} = \mathbf{b}$ is the translated subspace $\mathbf{x}_0 + W$, where $\mathbf{x}_0$ is any solution of the nonhomogeneous system $A\mathbf{x} = \mathbf{b}$ (Figure 3.5.1).*

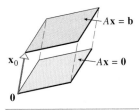

The solution set of $A\mathbf{x} = \mathbf{b}$ is a translation of the solution space of $A\mathbf{x} = \mathbf{0}$.

Figure 3.5.1

Proof We must show that if $\mathbf{x}$ is a vector in $\mathbf{x}_0 + W$, then $\mathbf{x}$ is a solution of $A\mathbf{x} = \mathbf{b}$, and, conversely, that every solution of $A\mathbf{x} = \mathbf{b}$ is in the set $\mathbf{x}_0 + W$.

Assume first that $\mathbf{x}$ is a vector in $\mathbf{x}_0 + W$. This implies that $\mathbf{x}$ is expressible in the form $\mathbf{x} = \mathbf{x}_0 + \mathbf{w}$, where $A\mathbf{x}_0 = \mathbf{b}$ and $A\mathbf{w} = \mathbf{0}$. Thus,

$$A\mathbf{x} = A(\mathbf{x}_0 + \mathbf{w}) = A\mathbf{x}_0 + A\mathbf{w} = \mathbf{b} + \mathbf{0} = \mathbf{b}$$

which shows that $\mathbf{x}$ is a solution of $A\mathbf{x} = \mathbf{b}$.

Conversely, let $\mathbf{x}$ be any solution of $A\mathbf{x} = \mathbf{b}$. To show that $\mathbf{x}$ is in the set $\mathbf{x}_0 + W$ we must show that $\mathbf{x}$ is expressible in the form

$$\mathbf{x} = \mathbf{x}_0 + \mathbf{w} \tag{3}$$

where $\mathbf{w}$ is in W (i.e., $A\mathbf{w} = \mathbf{0}$). We can do this by taking $\mathbf{w} = \mathbf{x} - \mathbf{x}_0$. This vector obviously satisfies (3), and it is in W, since

$$A\mathbf{w} = A(\mathbf{x} - \mathbf{x}_0) = A\mathbf{x} - A\mathbf{x}_0 = \mathbf{b} - \mathbf{b} = \mathbf{0} \qquad\blacksquare$$

It follows from Theorem 3.5.1 that the solution set of a consistent nonhomogeneous linear system is expressible in the form

$$\mathbf{x} = \mathbf{x}_0 + t_1\mathbf{v}_1 + t_2\mathbf{v}_2 + \cdots + t_s\mathbf{v}_s = \mathbf{x}_0 + \mathbf{x}_h \tag{4}$$

where

$$\mathbf{x}_h = t_1\mathbf{v}_1 + t_2\mathbf{v}_2 + \cdots + t_s\mathbf{v}_s$$

is a general solution of the associated homogeneous equation. We will call $\mathbf{x}_0$ a **_particular solution_** of $A\mathbf{x} = \mathbf{b}$, and we will call (4) a **_general solution_** of $A\mathbf{x} = \mathbf{b}$. The following theorem summarizes these ideas.

Theorem 3.5.2 *A general solution of a consistent linear system $A\mathbf{x} = \mathbf{b}$ can be obtained by adding a particular solution of $A\mathbf{x} = \mathbf{b}$ to a general solution of $A\mathbf{x} = \mathbf{0}$.*

REMARK It now follows from this theorem and Theorem 2.2.1 that every linear system has zero, one, or infinitely many solutions, as stated in Theorem 2.1.1.

EXAMPLE 1
The Geometry of Nonhomogeneous Linear Systems in Two or Three Unknowns

Since the solution set of a consistent nonhomogeneous linear system is the translation of the solution space of the associated homogeneous system, Example 4 of Section 3.4 implies that the solution set of a consistent nonhomogeneous linear system in two or three unknowns must be one of the following:

Solution Sets in R^2	Solution Sets in R^3
A point A line All of R^2	A point A line A plane All of R^3

$\blacksquare$

CONCEPT PROBLEM Explain why translating R^2 (or R^3) by a vector $\mathbf{x}_0$ produces R^2 (or R^3).

The following consequence of Theorem 3.5.2 relates the number of solutions of a nonhomogeneous system to the number of solutions of the associated homogeneous system in the general case where the number of equations need not be the same as the number of unknowns.

Theorem 3.5.3 *If A is an $m \times n$ matrix, then the following statements are equivalent.*

(a) $A\mathbf{x} = \mathbf{0}$ has only the trivial solution.

(b) $A\mathbf{x} = \mathbf{b}$ has at most one solution for every $\mathbf{b}$ in R^m (i.e., is inconsistent or has a unique solution).

It follows from this theorem that if a homogeneous linear system $A\mathbf{x} = \mathbf{0}$ has infinitely many solutions, then $A\mathbf{x} = \mathbf{b}$ is either inconsistent or has infinitely many solutions. In particular, if $A\mathbf{x} = \mathbf{0}$ has more unknowns than equations, then Theorem 2.2.3 implies the following result about nonhomogeneous linear systems.

Theorem 3.5.4 *A nonhomogeneous linear system with more unknowns than equations is either inconsistent or has infinitely many solutions.*

CONSISTENCY OF A LINEAR SYSTEM FROM THE VECTOR POINT OF VIEW

The consistency or inconsistency of a linear system $A\mathbf{x} = \mathbf{b}$ is determined by the relationship between the vector $\mathbf{b}$ and the column vectors of A. To see why this so, suppose that the successive column vectors of A are $\mathbf{a}_1, \mathbf{a}_2, \ldots, \mathbf{a}_n$, and use Formula (10) of Section 3.1 to rewrite the system as

$$x_1\mathbf{a}_1 + x_2\mathbf{a}_2 + \cdots + x_n\mathbf{a}_n = \mathbf{b} \tag{5}$$

We see from this expression that $A\mathbf{x} = \mathbf{b}$ is consistent if and only if $\mathbf{b}$ can be expressed as a linear combination of the column vectors of A, and if so, the solutions of the system are given by the coefficients in (5).

This idea can be expressed in a slightly different way: If A is an $m \times n$ matrix, then to say that $\mathbf{b}$ is a linear combination of the column vectors of A is the same as saying that $\mathbf{b}$ is in the subspace of R^m spanned by the column vectors of A. This subspace is called the ***column space*** of A and is denoted by $\mathrm{col}(A)$. The following theorem summarizes this discussion.

Theorem 3.5.5 *A linear system $A\mathbf{x} = \mathbf{b}$ is consistent if and only if $\mathbf{b}$ is in the column space of A.*

EXAMPLE 2
Linear Combinations Revisited

Determine whether the vector $\mathbf{w} = (9, 1, 0)$ can be expressed as a linear combination of the vectors

$$\mathbf{v}_1 = (1, 2, 3), \quad \mathbf{v}_2 = (1, 4, 6), \quad \mathbf{v}_3 = (2, -3, -5)$$

and, if so, find such a linear combination.

Solution We solved this problem in Example 7 of Section 2.1, but by using Theorem 3.5.5 with an appropriate adjustment in notation we can get to the heart of that solution more directly. First let us rewrite the vectors in column form, and consider the 3×3 matrix A whose successive column vectors are $\mathbf{v}_1$, $\mathbf{v}_2$, and $\mathbf{v}_3$. Our problem now becomes one of determining whether $\mathbf{w}$ is in the column space of A, and Theorem 3.5.5 tells us that this will be so if and only if the linear system

$$\begin{bmatrix} 1 & 1 & 2 \\ 2 & 4 & -3 \\ 3 & 6 & -5 \end{bmatrix} \begin{bmatrix} c_1 \\ c_2 \\ c_3 \end{bmatrix} = \begin{bmatrix} 9 \\ 1 \\ 0 \end{bmatrix} \tag{6}$$

is consistent. Thus, we have reached system (9) in Example 7 of Section 2.1 immediately, and we can now proceed as in that example. We showed in that example that the system is consistent and has the unique solution

$$c_1 = 1, \quad c_2 = 2, \quad c_3 = 3 \tag{7}$$

The consistency tells us that $\mathbf{w}$ is expressible as a linear combination of $\mathbf{v}_1$, $\mathbf{v}_2$, and $\mathbf{v}_3$, and (7) tells us that

$$\mathbf{w} = \mathbf{v}_1 + 2\mathbf{v}_2 + 3\mathbf{v}_3 \qquad\blacksquare$$

HYPERPLANES

We know that a linear equation $a_1x + a_2y = b$ in which a_1 and a_2 are not both zero represents a line in the xy-plane, and a linear equation $a_1x + a_2y + a_3z = b$ in which a_1, a_2, and a_3 are not all zero represents a plane in an xyz-coordinate system. More generally, the set of points

$(x_1, x_2, \ldots, x_n)$ in R^n that satisfy a linear equation of the form

$$a_1x_1 + a_2x_2 + \cdots + a_nx_n = b \qquad (a_1, a_2, \ldots, a_n \text{ not all zero}) \tag{8}$$

is called a ***hyperplane*** in R^n. Thus, for example, lines are hyperplanes in R^2 and planes are hyperplanes in R^3. If $b = 0$ in (8), then the equation simplifies to

$$a_1x_1 + a_2x_2 + \cdots + a_nx_n = 0 \qquad (a_1, a_2, \ldots, a_n \text{ not all zero}) \tag{9}$$

and we say that the hyperplane ***passes through the origin***.

When convenient, Equations (8) and (9) can be expressed in dot product notation as

$$\mathbf{a} \cdot \mathbf{x} = b \qquad (\mathbf{a} \neq \mathbf{0}) \tag{10}$$

and

$$\mathbf{a} \cdot \mathbf{x} = 0 \qquad (\mathbf{a} \neq \mathbf{0}) \tag{11}$$

respectively, where $\mathbf{a} = (a_1, a_2, \ldots, a_n)$ and $\mathbf{x} = (x_1, x_2, \ldots, x_n)$. The form of Equation (11) reveals that for a fixed nonzero vector $\mathbf{a}$, the hyperplane $\mathbf{a} \cdot \mathbf{x} = 0$ consists of all vectors $\mathbf{x}$ in R^n that are orthogonal to the vector $\mathbf{a}$. Accordingly, we call (11) the ***hyperplane through the origin with normal*** $\mathbf{a}$ or the ***orthogonal complement of*** $\mathbf{a}$. We denote this hyperplane by the symbol $\mathbf{a}^\perp$ (read, "$\mathbf{a}$ perp").

EXAMPLE 3
Finding an
Equation for a
Hyperplane

Let $\mathbf{a} = (1, -2, 4)$. Find an equation in the variables x, y, and z and also parametric equations for the hyperplane $\mathbf{a}^\perp$.

Solution The orthogonal complement of the vector $\mathbf{a}$ consists of all vectors $\mathbf{x} = (x, y, z)$ such that $\mathbf{a} \cdot \mathbf{x} = 0$. Writing this equation in component form yields

$$x - 2y + 4z = 0 \tag{12}$$

which is an equation in the variables x, y, and z for the hyperplane (a plane in this case). To find parametric equations for this plane, we could view (12) as a linear system of one equation in three unknowns and solve it by Gauss–Jordan elimination. However, because the system is so simple, we do not need the formality of matrices—we can simply solve (12) for the leading variable x and assign arbitrary values $y = t_1$ and $z = t_2$ to the free variables y and z. This yields the parametric equations

$$x = 2t_1 - 4t_2, \;\; y = t_1, \;\; z = t_2 \qquad\qquad \blacksquare$$

GEOMETRIC INTERPRETATIONS OF SOLUTION SPACES

Geometrically, the equations of a homogeneous linear system

$$\begin{aligned}
a_{11}x_1 + a_{12}x_2 + \cdots + a_{1n}x_n &= 0 \\
a_{21}x_1 + a_{22}x_2 + \cdots + a_{2n}x_n &= 0 \\
\vdots \qquad\quad \vdots \qquad\qquad \vdots \quad\;\; \vdots & \\
a_{m1}x_1 + a_{m2}x_2 + \cdots + a_{mn}x_n &= 0
\end{aligned} \tag{13}$$

represent hyperplanes through the origin in R^n and hence the solution space of the system can be viewed as the intersection of these hyperplanes. This is an extension to R^n of the familiar facts that the solution space of a homogeneous linear system in two unknowns is an intersection of lines through the origin of R^2, and the solution space of a homogeneous linear system in three unknowns is an intersection of planes through the origin of R^3.

Another geometric interpretation of the solution space of (13) can be obtained by writing these equations as dot products,

$$\begin{aligned}
\mathbf{a}_1 \cdot \mathbf{x} &= 0 \\
\mathbf{a}_2 \cdot \mathbf{x} &= 0 \\
\vdots \quad\;\; \vdots & \\
\mathbf{a}_m \cdot \mathbf{x} &= 0
\end{aligned} \tag{14}$$

and observing that $\mathbf{a}_1, \mathbf{a}_2, \ldots, \mathbf{a}_m$ are the row vectors of the coefficient matrix A when the system is expressed as $A\mathbf{x} = \mathbf{0}$. The form of the equations in (14) tells us that the solution space of the system consists of all vectors $\mathbf{x}$ in R^n that are orthogonal to every row vector of A. Thus, we have the following result.

Theorem 3.5.6   *If A is an $m \times n$ matrix, then the solution space of the homogeneous linear system $A\mathbf{x} = \mathbf{0}$ consists of all vectors in R^n that are orthogonal to every row vector of A.*

EXAMPLE 4
Orthogonality of Solutions and Row Vectors

As an illustration of Theorem 3.5.6, let us reexamine the system $A\mathbf{x} = \mathbf{0}$ that was solved in Example 7 of Section 2.2. The coefficient matrix of the system is

$$\begin{bmatrix} 1 & 3 & -2 & 0 & 2 & 0 \\ 2 & 6 & -5 & -2 & 4 & -3 \\ 0 & 0 & 5 & 10 & 0 & 15 \\ 2 & 6 & 0 & 8 & 4 & 18 \end{bmatrix}$$

and we showed in that example that a general solution of the system is

$$x_1 = -3r - 4s - 2t, \quad x_2 = r, \quad x_3 = -2s, \quad x_4 = s, \quad x_5 = t, \quad x_6 = 0$$

If we rewrite this general solution in the vector form

$$\mathbf{x} = (-3r - 4s - 2t, r, -2s, s, t, 0)$$

and if we rewrite the first row vector of A in comma-delimited form

$$\mathbf{a}_1 = (1, 3, -2, 0, 2, 0)$$

then we see that

$$\mathbf{a}_1 \cdot \mathbf{x} = 1(-3r - 4s - 2t) + 3(r) + (-2)(-2s) + 0(s) + 2(t) + 0(0) = 0$$

as guaranteed by Theorem 3.5.6. We leave it for you to confirm that the dot product of $\mathbf{x}$ with each of the other row vectors of A is also zero. ∎

LOOKING AHEAD You may have observed that we use the phrase "*a* general solution" and not "*the* general solution." This is because spanning vectors for the solution space of a homogeneous linear system are not unique. However, it can be shown that Gaussian elimination with back substitution and Gauss–Jordan elimination always produce the same general solution,

$$\mathbf{x} = t_1\mathbf{v}_1 + t_2\mathbf{v}_2 + \cdots + t_s\mathbf{v}_s$$

and we will show later that the vectors $\mathbf{v}_1, \mathbf{v}_2, \ldots, \mathbf{v}_s$ in this solution are linearly independent. The number of such vectors, or equivalently, the number of parameters, is called the ***dimension*** of the solution space. Thus, for example, in Example 7 of Section 2.2 Gauss–Jordan elimination produced a general solution with three parameters, so the solution space is three-dimensional.

The following table shows the relationship between the dimension of the solution space and its geometric form for a homogeneous linear system in three unknowns.

A Homogeneous Linear System in Three Unknowns

General Solution by Gauss–Jordan	Dimension of Solution Space	Description of Solution Space
$\mathbf{x} = \mathbf{0}$	0	A point (the origin)
$\mathbf{x} = t_1\mathbf{v}_1$	1	A line through the origin
$\mathbf{x} = t_1\mathbf{v}_1 + t_2\mathbf{v}_2$	2	A plane through the origin
$\mathbf{x} = t_1\mathbf{v}_1 + t_2\mathbf{v}_2 + t_3\mathbf{v}_3$	3	All of R^3

Exercise Set 3.5

1. Consider the linear systems

$$\begin{bmatrix} 3 & 2 & -1 \\ 6 & 4 & -2 \\ -3 & -2 & 1 \end{bmatrix} \begin{bmatrix} x_1 \\ x_2 \\ x_3 \end{bmatrix} = \begin{bmatrix} 0 \\ 0 \\ 0 \end{bmatrix}$$

and

$$\begin{bmatrix} 3 & 2 & -1 \\ 6 & 4 & -2 \\ -3 & -2 & 1 \end{bmatrix} \begin{bmatrix} x_1 \\ x_2 \\ x_3 \end{bmatrix} = \begin{bmatrix} 2 \\ 4 \\ -2 \end{bmatrix}$$

(a) Find a general solution of the homogeneous system.
(b) Confirm that $x_1 = 1, x_2 = 0, x_3 = 1$ is a solution of the nonhomogeneous system.
(c) Use the results in parts (a) and (b) to find a general solution of the nonhomogeneous system.
(d) Check your result in part (c) by solving the nonhomogeneous system directly.

2. Consider the linear systems

$$\begin{bmatrix} 1 & -2 & 3 \\ 2 & 1 & 4 \\ 1 & -7 & 5 \end{bmatrix} \begin{bmatrix} x_1 \\ x_2 \\ x_3 \end{bmatrix} = \begin{bmatrix} 0 \\ 0 \\ 0 \end{bmatrix}$$

and

$$\begin{bmatrix} 1 & -2 & 3 \\ 2 & 1 & 4 \\ 1 & -7 & 5 \end{bmatrix} \begin{bmatrix} x_1 \\ x_2 \\ x_3 \end{bmatrix} = \begin{bmatrix} 2 \\ 7 \\ -1 \end{bmatrix}$$

(a) Find a general solution of the homogeneous system.
(b) Confirm that $x_1 = 1, x_2 = 1, x_3 = 1$ is a solution of the nonhomogeneous system.
(c) Use the results in parts (a) and (b) to find a general solution of the nonhomogeneous system.
(d) Check your result in part (c) by solving the nonhomogeneous system directly.

In Exercises 3 and 4, (a) find a general solution of the given system, and (b) use the general solution obtained in part (a) to find a general solution of the associated homogeneous system and a particular solution of the given nonhomogeneous system.

3. $$\begin{bmatrix} 3 & 4 & 1 & 2 \\ 6 & 8 & 2 & 5 \\ 9 & 12 & 3 & 10 \end{bmatrix} \begin{bmatrix} x_1 \\ x_2 \\ x_3 \\ x_4 \end{bmatrix} = \begin{bmatrix} 3 \\ 7 \\ 13 \end{bmatrix}$$

4. $$\begin{bmatrix} 9 & -3 & 5 & 6 \\ 6 & -2 & 3 & 1 \\ 3 & -1 & 3 & 14 \end{bmatrix} \begin{bmatrix} x_1 \\ x_2 \\ x_3 \\ x_4 \end{bmatrix} = \begin{bmatrix} 4 \\ 5 \\ -8 \end{bmatrix}$$

In Exercises 5 and 6, use the method of Example 2 to confirm that $\mathbf{w}$ can be expressed as a linear combination of $\mathbf{v}_1, \mathbf{v}_2$, and $\mathbf{v}_3$, and find such a linear combination.

5. $\mathbf{w} = (-2, 0, 1); \mathbf{v}_1 = (2, 3, 1), \mathbf{v}_2 = (4, 9, 5),$
   $\mathbf{v}_3 = (-10, -21, -12)$

6. $\mathbf{w} = (-18, 14, -4); \mathbf{v}_1 = (3, 2, 2), \mathbf{v}_2 = (8, 1, 4),$
   $\mathbf{v}_3 = (1, 5, 2)$

In Exercises 7 and 8, use the method of Example 2 to determine whether $\mathbf{w}$ is in the span of the other vectors.

7. $\mathbf{w} = (1, 5, -2); \mathbf{v}_1 = (1, -1, 1), \mathbf{v}_2 = (1, 1, 0),$
   $\mathbf{v}_3 = (1, -3, 2), \mathbf{v}_4 = (1, 3, -1)$

8. $\mathbf{w} = (3, 5, 1); \mathbf{v}_1 = (1, 1, 1), \mathbf{v}_2 = (2, 3, 1),$
   $\mathbf{v}_3 = (0, -1, 1)$

In Exercises 9 and 10, find parametric equations for the hyperplane $\mathbf{a}^{\perp}$.

9. (a) $\mathbf{a} = (-2, 3)$ in R^2
   (b) $\mathbf{a} = (4, 0, -5)$ in R^3
   (c) $\mathbf{a} = (1, 2, -3, 7)$ in R^4

10. (a) $\mathbf{a} = (1, -4)$ in R^2
   (b) $\mathbf{a} = (3, 1, -6)$ in R^3
   (c) $\mathbf{a} = (2, -2, -4, 0)$ in R^4

In Exercises 11–14, find a general solution of the system, state the dimension of the solution space, and confirm that the row vectors of A are orthogonal to the solution vectors.

11. $x_1 + x_2 + x_3 = 0$
   $2x_1 + 2x_2 + 2x_3 = 0$
   $3x_1 + 3x_2 + 3x_3 = 0$

12. $x_1 + 3x_2 - 4x_3 = 0$
   $2x_1 + 6x_2 - 8x_3 = 0$

13. $x_1 + 5x_2 + x_3 + 2x_4 - x_5 = 0$
   $x_1 - 2x_2 - x_3 + 3x_4 + 2x_5 = 0$

14. $x_1 + 3x_2 - 4x_3 = 0$
   $x_1 + 2x_2 + 3x_3 = 0$

15. (a) The equation $x + y + z = 1$ can be viewed as a linear system of one equation in three unknowns. Express a general solution of this equation as a particular solution plus a general solution of the associated homogeneous system.
   (b) Give a geometric interpretation of the result in part (a).

16. (a) The equation $x + y = 1$ can be viewed as a linear system of one equation in two unknowns. Express a general solution of this equation as a particular solution plus a general solution of the associated homogeneous system.
   (b) Give a geometric interpretation of the result in part (a).

17. (a) Find a homogeneous linear system of two equations in three unknowns whose solution space consists of those vectors in R^3 that are orthogonal to $\mathbf{a} = (1, 1, 1)$ and $\mathbf{b} = (-2, 3, 0)$.

(b) What kind of geometric object is the solution space?
(c) Find a general solution of the system you obtained in part (a), and confirm that the solution space has the orthogonality and geometric properties of parts (a) and (b).

18. (a) Find a homogeneous linear system of two equations in three unknowns whose solution space consists of those vectors in R^3 that are orthogonal to
    $\mathbf{a} = (-3, 2, -1)$ and $\mathbf{b} = (0, -2, -2)$.
    (b) What kind of geometric object is the solution space?
    (c) Find a general solution of the system you obtained in part (a), and confirm that the solution space has the orthogonality and geometric properties of parts (a) and (b).

19. (a) Find a homogeneous linear system of two equations in four unknowns whose solution space consists of all vectors in R^4 that are orthogonal to the vectors
    $\mathbf{v}_1 = (1, 1, 2, 2)$ and $\mathbf{v}_2 = (5, 4, 3, 4)$.

(b) What kind of geometric object is the solution space?
(c) Find a general solution of the system you obtained in part (a), and confirm that the solution space has the orthogonality and geometric properties of parts (a) and (b).

20. (a) Find a homogeneous linear system of three equations in five unknowns whose solution space consists of all vectors in R^5 that are orthogonal to the vectors
    $\mathbf{v}_1 = (1, 3, 4, 4, -1)$, $\mathbf{v}_2 = (3, 2, 1, 2, -3)$, and
    $\mathbf{v}_3 = (1, 5, 0, -2, -4)$.
    (b) What kind of geometric object is the solution space?
    (c) Find a general solution of the system you obtained in part (a), and confirm that the solution space has the orthogonality and geometric properties of parts (a) and (b).

Discussion and Discovery

D1. If $A\mathbf{x} = \mathbf{b}$ is a consistent linear system, what is the relationship between its solution set and the solution space of $A\mathbf{x} = \mathbf{0}$?

D2. If A is an invertible $n \times n$ matrix, and if $\mathbf{v}$ is a vector in R^n that is orthogonal to every row vector of A, then $\mathbf{v} = \underline{\hspace{1cm}}$. Why?

D3. If $A\mathbf{x} = \mathbf{0}$ is a linear system of 4 equations in 7 unknowns, what can you say about the dimension of the solution space?

D4. Indicate whether the statement is true (T) or false (F). Justify your answer.

(a) If $A\mathbf{x} = \mathbf{b}$ has infinitely many solutions, then so does $A\mathbf{x} = \mathbf{0}$.
(b) If $A\mathbf{x} = \mathbf{b}$ is inconsistent, then $A\mathbf{x} = \mathbf{0}$ has only the trivial solution.
(c) The fewest number of hyperplanes in R^4 that can intersect in a single point is four.
(d) If W is any plane in R^3, there is a linear system in three unknowns whose solution set is that plane.
(e) If $A\mathbf{x} = \mathbf{b}$ is consistent, then every vector in the solution set is orthogonal to every row vector of A.

Working with Proofs

P1. Prove that $(a) \Rightarrow (b)$ in Theorem 3.5.3. [*Hint:* Proceed by contradiction, assuming that the nonhomogeneous system $A\mathbf{x} = \mathbf{b}$ has two distinct solutions, $\mathbf{x}_1$ and $\mathbf{x}_2$, and using those solutions to find a nontrivial solution of $A\mathbf{x} = \mathbf{0}$.]

P2. Prove that $(b) \Rightarrow (a)$ in Theorem 3.5.3.

P3. Prove that if $A\mathbf{x} = \mathbf{b}$ and $A\mathbf{x} = \mathbf{c}$ are consistent, then so is $A\mathbf{x} = \mathbf{b} + \mathbf{c}$. Create and prove a generalization of this result.

P4. Prove: If $\mathbf{a}$ is a nonzero vector in R^n and k is a nonzero scalar, then $\mathbf{a}^{\perp} = (k\mathbf{a})^{\perp}$. [*Note:* This shows that parallel vectors determine the same hyperplane through the origin.]

Technology Exercises

T1. (a) Show that the vector $\mathbf{v} = (-21, -60, -3, 108, 84)$ is in span$\{\mathbf{v}_1, \mathbf{v}_2, \mathbf{v}_3\}$, where $\mathbf{v}_1 = (1, -1, 3, 11, 20)$, $\mathbf{v}_2 = (10, 5, 15, 20, 11)$, and $\mathbf{v}_3 = (3, 3, 4, 4, 9)$.

(b) Express $\mathbf{v}$ as a linear combination of $\mathbf{v}_1$, $\mathbf{v}_2$, and $\mathbf{v}_3$.

Section 3.6   Matrices with Special Forms

In this section we will discuss matrices that have various special forms. The matrices that we will consider arise in a wide variety of applications and will also play an important role in our subsequent work.

DIAGONAL MATRICES   A square matrix in which all entries off the main diagonal are zero is called a ***diagonal matrix***. Thus, a general $n \times n$ diagonal matrix has the form

$$D = \begin{bmatrix} d_1 & 0 & \cdots & 0 \\ 0 & d_2 & \cdots & 0 \\ \vdots & \vdots & & \vdots \\ 0 & 0 & \cdots & d_n \end{bmatrix} \tag{1}$$

where $d_1, d_2, \ldots, d_n$ are any real numbers.

We leave it as an exercise for you to confirm that a diagonal matrix is invertible if and only if all of its diagonal entries are nonzero, in which case the inverse of (1) is

$$D^{-1} = \begin{bmatrix} 1/d_1 & 0 & \cdots & 0 \\ 0 & 1/d_2 & \cdots & 0 \\ \vdots & \vdots & & \vdots \\ 0 & 0 & \cdots & 1/d_n \end{bmatrix} \tag{2}$$

We also leave it as an exercise to confirm that if k is a positive integer, then the kth power of (1) is

$$D^k = \begin{bmatrix} d_1^k & 0 & \cdots & 0 \\ 0 & d_2^k & \cdots & 0 \\ \vdots & \vdots & & \vdots \\ 0 & 0 & \cdots & d_n^k \end{bmatrix} \tag{3}$$

The result is also true if $k < 0$ and D is invertible.

EXAMPLE 1
Inverses and Powers of Diagonal Matrices

Consider the diagonal matrix

$$A = \begin{bmatrix} 1 & 0 & 0 \\ 0 & -3 & 0 \\ 0 & 0 & 2 \end{bmatrix}$$

It follows from the preceding discussion that

$$A^{-1} = \begin{bmatrix} 1 & 0 & 0 \\ 0 & -\frac{1}{3} & 0 \\ 0 & 0 & \frac{1}{2} \end{bmatrix}, \quad A^5 = \begin{bmatrix} 1 & 0 & 0 \\ 0 & -243 & 0 \\ 0 & 0 & 32 \end{bmatrix}, \quad A^{-5} = \begin{bmatrix} 1 & 0 & 0 \\ 0 & \frac{1}{243} & 0 \\ 0 & 0 & \frac{1}{32} \end{bmatrix} \quad ∎$$

Matrix products involving diagonal matrices are easy to compute. For example,

$$\begin{bmatrix} d_1 & 0 & 0 \\ 0 & d_2 & 0 \\ 0 & 0 & d_3 \end{bmatrix} \begin{bmatrix} a_{11} & a_{12} & a_{13} & a_{14} \\ a_{21} & a_{22} & a_{23} & a_{24} \\ a_{31} & a_{32} & a_{33} & a_{34} \end{bmatrix} = \begin{bmatrix} d_1 a_{11} & d_1 a_{12} & d_1 a_{13} & d_1 a_{14} \\ d_2 a_{21} & d_2 a_{22} & d_2 a_{23} & d_2 a_{24} \\ d_3 a_{31} & d_3 a_{32} & d_3 a_{33} & d_3 a_{34} \end{bmatrix}$$

$$\begin{bmatrix} a_{11} & a_{12} & a_{13} \\ a_{21} & a_{22} & a_{23} \\ a_{31} & a_{32} & a_{33} \\ a_{41} & a_{42} & a_{43} \end{bmatrix} \begin{bmatrix} d_1 & 0 & 0 \\ 0 & d_2 & 0 \\ 0 & 0 & d_3 \end{bmatrix} = \begin{bmatrix} d_1 a_{11} & d_2 a_{12} & d_3 a_{13} \\ d_1 a_{21} & d_2 a_{22} & d_3 a_{23} \\ d_1 a_{31} & d_2 a_{32} & d_3 a_{33} \\ d_1 a_{41} & d_2 a_{42} & d_3 a_{43} \end{bmatrix}$$

That is, to multiply a matrix A on the left by a diagonal matrix D, multiply successive row vectors of A by the successive diagonal entries of D, and to multiply a matrix A on the right by a diagonal matrix D, multiply successive column vectors of A by the successive diagonal entries of D.

TRIANGULAR MATRICES   A square matrix in which all entries above the main diagonal are zero is called *lower triangular*, and a square matrix in which all the entries below the main diagonal are zero is called *upper triangular*. A matrix that is either upper triangular or lower triangular is called *triangular*.

EXAMPLE 2
Triangular
Matrices

General 4×4 triangular matrices have the forms

$$\begin{bmatrix} a_{11} & a_{12} & a_{13} & a_{14} \\ 0 & a_{22} & a_{23} & a_{24} \\ 0 & 0 & a_{33} & a_{34} \\ 0 & 0 & 0 & a_{44} \end{bmatrix} \qquad \begin{bmatrix} a_{11} & 0 & 0 & 0 \\ a_{21} & a_{22} & 0 & 0 \\ a_{31} & a_{32} & a_{33} & 0 \\ a_{41} & a_{42} & a_{43} & a_{44} \end{bmatrix}$$

Upper triangular   Lower triangular

This example illustrates the following facts about triangular matrices:

- A square matrix $A = [a_{ij}]$ is upper triangular if and only if in each row all entries to the left of the diagonal entry are zero; that is, the ith row starts with at least $i - 1$ zeros for every i.

- A square matrix $A = [a_{ij}]$ is lower triangular if and only if in each column all entries above the diagonal entry are zero; that is, the jth column starts with at least $j - 1$ zeros for every j.

- A square matrix $A = [a_{ij}]$ is upper triangular if and only if all entries to the left of the main diagonal are zero; that is, $a_{ij} = 0$ if $i > j$ (Figure 3.6.1).

- A square matrix $A = [a_{ij}]$ is lower triangular if and only if all entries to the right of the main diagonal are zero; that is, $a_{ij} = 0$ if $i < j$ (Figure 3.6.1).

Figure 3.6.1

We omit the formal proofs.

REMARK It is possible for a matrix to be both upper triangular and lower triangular. Can you find some examples?

EXAMPLE 3
Row Echelon
Forms of
Square Matrices
Are Upper
Triangular

Because a row echelon form of a square matrix has zeros below the main diagonal, it follows that row echelon forms of square matrices are upper triangular. For example, here are some typical row echelon forms of 4×4 matrices in which the $*$'s can be any real numbers:

$$\begin{bmatrix} 1 & * & * & * \\ 0 & 1 & * & * \\ 0 & 0 & 1 & * \\ 0 & 0 & 0 & 1 \end{bmatrix}, \begin{bmatrix} 1 & * & * & * \\ 0 & 1 & * & * \\ 0 & 0 & 1 & * \\ 0 & 0 & 0 & 0 \end{bmatrix}, \begin{bmatrix} 1 & * & * & * \\ 0 & 1 & * & * \\ 0 & 0 & 0 & 0 \\ 0 & 0 & 0 & 0 \end{bmatrix}, \begin{bmatrix} 1 & * & * & * \\ 0 & 0 & 1 & * \\ 0 & 0 & 0 & 1 \\ 0 & 0 & 0 & 0 \end{bmatrix}, \begin{bmatrix} 0 & 0 & 0 & 0 \\ 0 & 0 & 0 & 0 \\ 0 & 0 & 0 & 0 \\ 0 & 0 & 0 & 0 \end{bmatrix}$$

These are all upper triangular.

LINEAR SYSTEMS WITH TRIANGULAR COEFFICIENT MATRICES   Thus far, we have described two basic methods for solving linear systems, Gauss–Jordan elimination, in which the augmented matrix is taken to reduced row echelon form, and Gaussian elimination with back substitution, in which the augmented matrix is taken to row echelon form. In both methods the reduced matrix has leading 1's in each nonzero row. However, there are many computer algorithms for solving linear systems in which the augmented matrix is reduced without forcing the leading entries in the nonzero rows to be 1's. In such algorithms the divisions that we used to produce the leading 1's are simply incorporated into the process of solving for the leading variables.

**PROPERTIES OF
TRIANGULAR MATRICES**

Theorem 3.6.1

(a) *The transpose of a lower triangular matrix is upper triangular, and the transpose of an upper triangular matrix is lower triangular.*

(b) *A product of lower triangular matrices is lower triangular, and a product of upper triangular matrices is upper triangular.*

(c) *A triangular matrix is invertible if and only if its diagonal entries are all nonzero.*

(d) *The inverse of an invertible lower triangular matrix is lower triangular, and the inverse of an invertible upper triangular matrix is upper triangular.*

We will prove parts (a) and (b), but we will defer the proofs of parts (c) and (d) until Chapter 4, at which point we will have more tools to work with.

Proof (a) Transposing a square matrix reflects the entries about the main diagonal, so the transpose of a matrix with zeros below the main diagonal will have zeros above the main diagonal, and conversely. Thus, the transpose of an upper triangular matrix is lower triangular, and conversely.

Proof (b) We will prove that a product of lower triangular matrices, A and B, is lower triangular. The proof for upper triangular matrices is similar.

The fact that A is lower triangular means that its ith row vector is of the form

$$\mathbf{r}_i(A) = [a_{i1} \quad a_{i2} \quad \cdots \quad a_{ii} \quad 0 \quad \cdots \quad 0]$$

and the fact that B is lower triangular means that its jth column vector is of the form

$$\mathbf{c}_j(B) = \begin{bmatrix} 0 \\ \vdots \\ 0 \\ b_{jj} \\ b_{(j+1)j} \\ \vdots \\ b_{nj} \end{bmatrix}$$

A moment's reflection should make it evident that

$$(AB)_{ij} = \mathbf{r}_i(A)\mathbf{c}_j(B) = 0$$

if $i < j$, so all entries of AB above the main diagonal are zero (Figure 3.6.1). Thus, AB is lower triangular. ∎

EXAMPLE 4
Computations
Involving
Triangular
Matrices

Consider the upper triangular matrices

$$A = \begin{bmatrix} 1 & 3 & -1 \\ 0 & 2 & 4 \\ 0 & 0 & 5 \end{bmatrix} \quad \text{and} \quad B = \begin{bmatrix} 3 & -2 & 2 \\ 0 & 0 & -1 \\ 0 & 0 & 1 \end{bmatrix}$$

According to Theorem 3.6.1, the matrix A is invertible, but the matrix B is not. Moreover, the theorem also implies that A^{-1}, AB, and BA must be upper triangular. We leave it for you to confirm these three statements by showing that

$$A^{-1} = \begin{bmatrix} 1 & -\frac{3}{2} & \frac{7}{5} \\ 0 & \frac{1}{2} & -\frac{2}{5} \\ 0 & 0 & \frac{1}{5} \end{bmatrix}, \quad AB = \begin{bmatrix} 3 & -2 & -2 \\ 0 & 0 & 2 \\ 0 & 0 & 5 \end{bmatrix}, \quad BA = \begin{bmatrix} 3 & 5 & -1 \\ 0 & 0 & -5 \\ 0 & 0 & 5 \end{bmatrix}$$

Observe that AB and BA both have a zero on the main diagonal. You could have predicted this without performing any computations by using Theorem 3.6.1. How?   ■

SYMMETRIC AND SKEW-SYMMETRIC MATRICES

A square matrix A is called **symmetric** if $A^T = A$ and **skew-symmetric** if $A^T = -A$. Since the transpose of a matrix can be obtained by reflecting its entries about the main diagonal (Figure 3.1.5), entries that are symmetrically positioned across the main diagonal are equal in a symmetric matrix and negatives of one another in a skew-symmetric matrix. In a symmetric matrix the entries on the main diagonal are unrestricted, but in a skew-symmetric matrix they must be zero (why?). Here are some examples:

Symmetric: $\begin{bmatrix} 7 & -3 \\ -3 & 5 \end{bmatrix}$, $\begin{bmatrix} 1 & 4 & 5 \\ 4 & -3 & -6 \\ 5 & -6 & 7 \end{bmatrix}$, $\begin{bmatrix} 0 & 0 & 0 \\ 0 & 0 & 0 \\ 0 & 0 & 0 \end{bmatrix}$, $\begin{bmatrix} d_1 & 0 & 0 \\ 0 & d_2 & 0 \\ 0 & 0 & d_3 \end{bmatrix}$

Skew-Symmetric: $\begin{bmatrix} 0 & 3 \\ -3 & 0 \end{bmatrix}$, $\begin{bmatrix} 0 & -4 & 5 \\ 4 & 0 & -9 \\ -5 & 9 & 0 \end{bmatrix}$, $\begin{bmatrix} 0 & 0 & 0 \\ 0 & 0 & 0 \\ 0 & 0 & 0 \end{bmatrix}$

Stated algebraically, a matrix $A = [a_{ij}]$ is symmetric if and only if

$$(A)_{ij} = (A)_{ji} \quad \text{or equivalently,} \quad a_{ij} = a_{ji} \tag{4}$$

and is skew-symmetric if and only if

$$(A)_{ij} = -(A)_{ji} \quad \text{or equivalently,} \quad a_{ij} = -a_{ji} \tag{5}$$

The following theorem lists the main algebraic properties of symmetric matrices. The proofs are all direct consequences of Theorem 3.2.10 and will be left for the exercises, as will a discussion of the corresponding theorem for skew-symmetric matrices.

Theorem 3.6.2 *If A and B are symmetric matrices with the same size, and if k is any scalar, then:*

(a) A^T *is symmetric.*

(b) $A + B$ *and* $A - B$ *are symmetric.*

(c) kA *is symmetric.*

It is not true, in general, that a product of symmetric matrices is symmetric. To see why this is so, let A and B be symmetric matrices with the same size. Then

$$(AB)^T = B^T A^T = BA$$

Thus, it follows that $(AB)^T = AB$ if and only if $AB = BA$; that is, if and only if A and B commute. In summary, we have the following result.

Theorem 3.6.3 *The product of two symmetric matrices is symmetric if and only if the matrices commute.*

EXAMPLE 5
Products of Symmetric Matrices

Consider the product

$$\begin{bmatrix} 1 & 2 \\ 2 & 3 \end{bmatrix} \begin{bmatrix} -4 & 1 \\ 1 & 0 \end{bmatrix} = \begin{bmatrix} -2 & 1 \\ -5 & 2 \end{bmatrix} \tag{6}$$

The factors are symmetric, but the product is not, so we can conclude that the factors do not commute. We leave it for you to confirm this by showing that

$$\begin{bmatrix} -4 & 1 \\ 1 & 0 \end{bmatrix} \begin{bmatrix} 1 & 2 \\ 2 & 3 \end{bmatrix} = \begin{bmatrix} -2 & -5 \\ 1 & 2 \end{bmatrix}$$

You can do this by performing the multiplication or by transposing both sides of (6).   ■

CONCEPT PROBLEM   Do you think that the product of two skew-symmetric matrices is skew-symmetric if and only if they commute? Explain your reasoning.

INVERTIBILITY OF SYMMETRIC MATRICES

In general, a symmetric matrix need not be invertible. For example, a diagonal matrix with a zero on the main diagonal is symmetric but not invertible. However, the following theorem shows that if a symmetric matrix happens to be invertible, then its inverse must also be symmetric.

Theorem 3.6.4   *If A is an invertible symmetric matrix, then A^{-1} is symmetric.*

Proof   Assume that A is symmetric and invertible. To prove the A^{-1} is symmetric we must show that A^{-1} is equal to its transpose. However, it follows from Theorem 3.2.11 and the symmetry of A that

$$(A^{-1})^T = (A^T)^{-1} = A^{-1}$$

which completes the proof.   ∎

MATRICES OF THE FORM AA^T AND A^TA

Matrix products of the form AA^T and A^TA arise in many applications, so we will consider some of their properties. If A is an $m \times n$ matrix, then A^T is an $n \times m$ matrix, so the products AA^T and A^TA are both square, the matrix AA^T having size $m \times m$ and the matrix A^TA having size $n \times n$. The products AA^T and A^TA are always symmetric, since

$$(AA^T)^T = (A^T)^T A^T = AA^T \quad \text{and} \quad (A^TA)^T = A^T(A^T)^T = A^TA \tag{7}$$

The symmetry of AA^T and A^TA can be seen explicitly by computing the entries in these matrices. For example, if the column vectors of A are

$$\mathbf{a}_1, \mathbf{a}_2, \ldots, \mathbf{a}_n$$

then the row vectors of A^T are

$$\mathbf{a}_1^T, \mathbf{a}_2^T, \ldots, \mathbf{a}_n^T$$

so the row-column rule for matrix products (Theorem 3.1.7) implies that

$$A^TA = \begin{bmatrix} \mathbf{a}_1^T \\ \mathbf{a}_2^T \\ \vdots \\ \mathbf{a}_n^T \end{bmatrix} [\mathbf{a}_1 \ \ \mathbf{a}_2 \ \cdots \ \mathbf{a}_n] = \begin{bmatrix} \mathbf{a}_1^T\mathbf{a}_1 & \mathbf{a}_1^T\mathbf{a}_2 & \cdots & \mathbf{a}_1^T\mathbf{a}_n \\ \mathbf{a}_2^T\mathbf{a}_1 & \mathbf{a}_2^T\mathbf{a}_2 & \cdots & \mathbf{a}_2^T\mathbf{a}_n \\ \vdots & \vdots & & \vdots \\ \mathbf{a}_n^T\mathbf{a}_1 & \mathbf{a}_n^T\mathbf{a}_2 & \cdots & \mathbf{a}_n^T\mathbf{a}_n \end{bmatrix} \tag{8}$$

which, by Formula (23) of Section 3.1, can also be expressed as

$$A^TA = \begin{bmatrix} \mathbf{a}_1 \cdot \mathbf{a}_1 & \mathbf{a}_1 \cdot \mathbf{a}_2 & \cdots & \mathbf{a}_1 \cdot \mathbf{a}_n \\ \mathbf{a}_2 \cdot \mathbf{a}_1 & \mathbf{a}_2 \cdot \mathbf{a}_2 & \cdots & \mathbf{a}_2 \cdot \mathbf{a}_n \\ \vdots & \vdots & & \vdots \\ \mathbf{a}_n \cdot \mathbf{a}_1 & \mathbf{a}_n \cdot \mathbf{a}_2 & \cdots & \mathbf{a}_n \cdot \mathbf{a}_n \end{bmatrix} = \begin{bmatrix} \|\mathbf{a}_1\|^2 & \mathbf{a}_1 \cdot \mathbf{a}_2 & \cdots & \mathbf{a}_1 \cdot \mathbf{a}_n \\ \mathbf{a}_1 \cdot \mathbf{a}_2 & \|\mathbf{a}_2\|^2 & \cdots & \mathbf{a}_2 \cdot \mathbf{a}_n \\ \vdots & \vdots & & \vdots \\ \mathbf{a}_1 \cdot \mathbf{a}_n & \mathbf{a}_2 \cdot \mathbf{a}_n & \cdots & \|\mathbf{a}_n\|^2 \end{bmatrix} \tag{9}$$

As anticipated, A^TA is a symmetric matrix. In the exercises we will ask you to obtain an analogous formula for AA^T (Exercise 30).

Later in the text we will obtain general conditions under which AA^T and A^TA are invertible. However, in the case where A is square we have the following result.

Theorem 3.6.5   *If A is a square matrix, then the matrices A, AA^T, and A^TA are either all invertible or all singular.*

We will leave the complete proof as an exercise; however, observe that if A is invertible, then Theorem 3.2.11 implies that A^T is invertible, so the products AA^T and A^TA are invertible.

Linear Algebra in History

The following statement appeared in an article that was published in 1858 in the *Philosophical Transactions of the Royal Society:* "A matrix compounded with the transposed matrix gives rise to a symmetrical matrix."

32   MR. A. CAYLEY ON THE THEORY OF MATRICES.

which shows that a matrix compounded with the transposed matrix gives rise to a symmetrical matrix. It does not however follow, nor is it the fact, that the matrix and transposed matrix are convertible. And also

$$\begin{pmatrix} a, c \\ b, d \end{pmatrix}\begin{pmatrix} a, b \\ c, d \end{pmatrix}\begin{pmatrix} a, c \\ b, d \end{pmatrix} = \begin{pmatrix} a^2+bcd+a(b^2+c^2), & c^2+abd+c(a^2+d^2) \\ b^2+acd+b(a^2+d^2), & d^2+abc+d(b^2+c^2) \end{pmatrix}$$

which is a remarkably symmetrical form. It is needless to proceed further, since it is clear that

$$\begin{pmatrix} a, c \\ b, d \end{pmatrix}\begin{pmatrix} a, b \\ c, d \end{pmatrix}\begin{pmatrix} a, b \\ c, d \end{pmatrix} = \left(\begin{pmatrix} a, c \\ b, d \end{pmatrix}\begin{pmatrix} a, b \end{pmatrix}\right)^3$$

44. In all that precedes, the matrix of the order 2 has frequently been considered, but chiefly by way of illustration of the general theory; but it is worth while to develope more particularly the theory of such matrix. I call to mind the fundamental properties which have been obtained, viz. it was shown that the matrix

$$M = \begin{pmatrix} a, b \\ c, d \end{pmatrix}$$

satisfies the equation

$$M^2 - (a+d)M + ad - bc = 0,$$

and that the two matrices

$$\begin{pmatrix} a, b \\ c, d \end{pmatrix}, \quad \begin{pmatrix} a', b' \\ c', d' \end{pmatrix}$$

will be convertible if

$$a' - d' : b' : c' = a - d : b : c,$$

and that they will be skew convertible if

$$a+d=0, \quad a'+d'=0, \quad aa'+bc'+b'c+dd'=0,$$

the first two of these equations being the conditions in order that the two matrices may be respectively periodic of the second order to a factor *près*.

45. It may be noticed in passing, that if L, M are skew convertible matrices of the order 2, and if these matrices are also such that L² = −1, M² = −1, then putting N = LM = −ML, we obtain

$$L^2 = -1, \quad M^2 = -1, \quad N^2 = -1,$$

$$L = MN = -NM, \quad M = NL = -LN, \quad N = LM = -ML,$$

which is a system of relations precisely similar to that in the theory of quaternions.

46. The integer powers of the matrix

$$M = \begin{pmatrix} a, b \\ c, d \end{pmatrix},$$

are obtained with great facility from the quadratic equation; thus we have, attending first to the positive powers,

REMARK Observe that Equation (7) and Theorems 3.6.4 and 3.6.5 in combination imply that if A is invertible, then $(AA^T)^{-1}$ and $(A^TA)^{-1}$ are symmetric matrices.

FIXED POINTS OF A MATRIX

Square matrices of the form $I - A$ arise in a variety of applications, most notably in economics, where they occur as coefficient matrices of *large* linear systems of the form $(I - A)\mathbf{x} = \mathbf{b}$. Matrices of this type also occur in geometry and in various engineering problems in which the entries of $(I - A)^{-1}$ have important physical significance.

Matrices of the form $I - A$ occur in problems in which one is looking for vectors that remain unchanged when they are multiplied by a specified matrix. More precisely, suppose that A is an $n \times n$ matrix and we are looking for all vectors in R^n that satisfy the equation $A\mathbf{x} = \mathbf{x}$. The solutions of this equation, if any, are called the ***fixed points*** of A since they remain unchanged when multiplied by A. The vector $\mathbf{x} = \mathbf{0}$ is a fixed point of every matrix A since $A\mathbf{0} = \mathbf{0}$, so the problem of interest is to determine whether there are others. To do this we can rewrite the equation $A\mathbf{x} = \mathbf{x}$ as $\mathbf{x} - A\mathbf{x} = \mathbf{0}$, or equivalently, as

$$(I - A)\mathbf{x} = \mathbf{0} \tag{10}$$

which is a homogeneous linear system whose solutions are the fixed points of A.

EXAMPLE 6
Fixed Points of a Matrix

Find the fixed points of the matrix

$$A = \begin{bmatrix} 1 & 0 \\ 2 & 1 \end{bmatrix}$$

Solution As discussed above, the fixed points are the solutions of the homogeneous linear system

$$(I - A)\mathbf{x} = \mathbf{0}$$

which we can write as

$$\begin{bmatrix} 0 & 0 \\ -2 & 0 \end{bmatrix} \begin{bmatrix} x_1 \\ x_2 \end{bmatrix} = \begin{bmatrix} 0 \\ 0 \end{bmatrix}$$

(verify). We leave it for you to confirm that the general solution of this system is

$$x_1 = 0, \quad x_2 = t$$

Thus, the fixed points of A are all vectors of the form

$$\mathbf{x} = \begin{bmatrix} 0 \\ t \end{bmatrix}$$

As a check,

$$A\mathbf{x} = \begin{bmatrix} 1 & 0 \\ 2 & 1 \end{bmatrix} \begin{bmatrix} 0 \\ t \end{bmatrix} = \begin{bmatrix} 0 \\ t \end{bmatrix} = \mathbf{x} \quad \blacksquare$$

A TECHNIQUE FOR INVERTING $I - A$ WHEN A IS NILPOTENT

Since matrix algebra for polynomials is much like ordinary algebra, many polynomial identities for real numbers continue to hold for matrix polynomials. For example, if x is any real number and k is a positive integer, then the algebraic identity

$$(1 - x)(1 + x + x^2 + \cdots + x^{k-1}) = 1 - x^k$$

translates into the following identity for square matrices:

$$(I - A)(I + A + A^2 + \cdots + A^{k-1}) = I - A^k \tag{11}$$

This can be confirmed by multiplying out and simplifying the left side.

If it happens that $A^k = 0$ for some positive integer k, then (11) simplifies to

$$(I - A)(I + A + A^2 + \cdots + A^{k-1}) = I$$

in which case we can conclude that $I - A$ is invertible and that the second factor on the left side is its inverse. In summary, we have the following result.

Theorem 3.6.6 *If A is a square matrix, and if there is a positive integer k such that $A^k = 0$, then the matrix $I - A$ is invertible and*

$$(I - A)^{-1} = I + A + A^2 + \cdots + A^{k-1} \tag{12}$$

A square matrix A with the property that $A^k = 0$ for some positive integer k is said to be **nilpotent**, and the smallest positive power for which $A^k = 0$ is called the **index of nilpotency**. It is important to keep in mind that Theorem 3.6.6 only applies when A is nilpotent.

Formula (12) can sometimes be used to deduce properties of the matrix $(I - A)^{-1}$ from properties of powers of A, and in certain circumstances may involve less computation than the method of row reduction discussed in Section 3.3.

EXAMPLE 7
Inverting $I - A$
When A Is
Nilpotent

Consider the matrix

$$A = \begin{bmatrix} 0 & 2 & 1 \\ 0 & 0 & 3 \\ 0 & 0 & 0 \end{bmatrix}$$

We leave it for you to confirm that

$$A^2 = \begin{bmatrix} 0 & 0 & 6 \\ 0 & 0 & 0 \\ 0 & 0 & 0 \end{bmatrix} \quad \text{and} \quad A^3 = \begin{bmatrix} 0 & 0 & 0 \\ 0 & 0 & 0 \\ 0 & 0 & 0 \end{bmatrix}$$

Thus, A is nilpotent with an index of nilpotency of 3. Thus, it follows from Theorem 3.6.6 that the inverse of

$$I - A = \begin{bmatrix} 1 & -2 & -1 \\ 0 & 1 & -3 \\ 0 & 0 & 1 \end{bmatrix}$$

can be expressed as

$$(I - A)^{-1} = I + A + A^2 = \begin{bmatrix} 1 & 0 & 0 \\ 0 & 1 & 0 \\ 0 & 0 & 1 \end{bmatrix} + \begin{bmatrix} 0 & 2 & 1 \\ 0 & 0 & 3 \\ 0 & 0 & 0 \end{bmatrix} + \begin{bmatrix} 0 & 0 & 6 \\ 0 & 0 & 0 \\ 0 & 0 & 0 \end{bmatrix} = \begin{bmatrix} 1 & 2 & 7 \\ 0 & 1 & 3 \\ 0 & 0 & 1 \end{bmatrix} \quad \blacksquare$$

REMARK Observe that the matrix A in this example is upper triangular and has zeros on the main diagonal; such matrices are said to be **strictly upper triangular**. Similarly, a lower triangular matrix with zeros on the main diagonal is said to be **strictly lower triangular**. A matrix that is either strictly upper triangular or strictly lower triangular is said to be **strictly triangular**. It can be shown that every strictly triangular matrix is nilpotent. We omit the proof (but see Exercise T5).

INVERTING $I - A$ BY
POWER SERIES
(*Calculus Required*)

Our next objective is to investigate the invertibility of $I - A$ in cases where the nilpotency requirement of Theorem 3.6.6 is not met. To motivate the idea, let us consider what happens to the equation

$$(1 - x)(1 + x + x^2 + x^3 + \cdots + x^k) = 1 - x^{k+1} \tag{13}$$

if $0 < x < 1$ and k increases indefinitely; that is, $k \to +\infty$.

Since raising a positive fraction to higher and higher powers pushes its value toward zero, it follows that $x^{k+1} \to 0$ as $k \to +\infty$, so (13) suggests that the error in the approximation

$$(1 - x)(1 + x + x^2 + x^3 + \cdots + x^k) \approx 1 \tag{14}$$

approaches zero as $k \to +\infty$. We denote this by writing

$$\lim_{k \to +\infty} (1-x)(1 + x + x^2 + x^3 + \cdots + x^k) = 1$$

or more simply as

$$(1-x)(1 + x + x^2 + x^3 + \cdots) = 1$$

The matrix analog of this equation for a square matrix A is

$$(I - A)(I + A + A^2 + A^3 + \cdots) = I \tag{15}$$

which suggests that if $I - A$ is invertible, then

$$(I - A)^{-1} = I + A + A^2 + A^3 + \cdots \tag{16}$$

We interpret this to mean that the error in the approximation

$$(I - A)^{-1} \approx I + A + A^2 + A^3 + \cdots + A^k \tag{17}$$

approaches zero as $k \to +\infty$ in the sense that as k increases indefinitely the entries on the right side of (17) approach the corresponding entries on the left side. In this case we say that the infinite series on the right side of (16) converges to $(I - A)^{-1}$.

REMARK Students of calculus will recognize that (16) is a matrix analog of the formula

$$1 + x + x^2 + x^3 + \cdots = \frac{1}{1-x} = (1-x)^{-1}$$

for the sum of a geometric series.

The following theorem, which we state without proof, provides a condition on the entries of A that guarantees the invertibility of $I - A$ and the validity of (16).

Theorem 3.6.7 *If A is an $n \times n$ matrix for which the sum of the absolute values of the entries in each column (or each row) is less than 1, then $I - A$ is invertible and can be expressed as*

$$(I - A)^{-1} = I + A + A^2 + A^3 + \cdots \tag{18}$$

Formula (18) is called a ***power series representation*** of $(I - A)^{-1}$.

REMARK Observe that if A is nilpotent with nilpotency index k, then A^k and all subsequent powers of A are zero, in which case Formula (18) reduces to (12).

EXAMPLE 8
Approximating $(I - A)^{-1}$ by a Power Series

The matrix

$$A = \begin{bmatrix} 0 & \frac{1}{4} & \frac{1}{8} \\ \frac{1}{4} & \frac{1}{5} & \frac{1}{6} \\ \frac{1}{7} & \frac{1}{8} & \frac{1}{9} \end{bmatrix}$$

satisfies the condition in Theorem 3.6.7, so $I - A$ is invertible and can be approximated by (17). With the help of a computer program or a calculator that can calculate inverses, one can show that

$$(I - A)^{-1} = \begin{bmatrix} 1 & -\frac{1}{4} & -\frac{1}{8} \\ -\frac{1}{4} & \frac{4}{5} & -\frac{1}{6} \\ -\frac{1}{7} & -\frac{1}{8} & \frac{8}{9} \end{bmatrix}^{-1} \approx \begin{bmatrix} 1.1305 & 0.3895 & 0.2320 \\ 0.4029 & 1.4266 & 0.3241 \\ 0.2384 & 0.2632 & 1.2079 \end{bmatrix} \tag{19}$$

to four decimal place accuracy (verify). Here are some approximations to $(I - A)^{-1}$ using (17):

$k = 2$	$k = 5$	$k = 10$	$k = 12$
$\begin{bmatrix} 1.0804 & 0.3156 & 0.1806 \\ 0.3238 & 1.3233 & 0.2498 \\ 0.1900 & 0.1996 & 1.1621 \end{bmatrix}$	$\begin{bmatrix} 1.1248 & 0.3819 & 0.2265 \\ 0.3947 & 1.4154 & 0.3162 \\ 0.2333 & 0.2564 & 1.2030 \end{bmatrix}$	$\begin{bmatrix} 1.1304 & 0.3894 & 0.2319 \\ 0.4027 & 1.4263 & 0.3240 \\ 0.2382 & 0.2631 & 1.2078 \end{bmatrix}$	$\begin{bmatrix} 1.1305 & 0.3895 & 0.2320 \\ 0.4029 & 1.4265 & 0.3241 \\ 0.2383 & 0.2632 & 1.2078 \end{bmatrix}$

Note that with $k = 12$, the approximation agrees with (19) to three decimal places. ■

Exercise Set 3.6

In Exercises 1 and 2, determine whether the matrix is invertible; if so, find the inverse by inspection.

1. (a) $\begin{bmatrix} 4 & 0 & 0 \\ 0 & 0 & 0 \\ 0 & 0 & 5 \end{bmatrix}$ (b) $\begin{bmatrix} -1 & 0 & 0 \\ 0 & 2 & 0 \\ 0 & 0 & \frac{1}{3} \end{bmatrix}$

2. (a) $\begin{bmatrix} 1 & 0 & 0 \\ 0 & -2 & 0 \\ 0 & 0 & 3 \end{bmatrix}$ (b) $\begin{bmatrix} 4 & 0 & 0 \\ 0 & -7 & 0 \\ 0 & 0 & 0 \end{bmatrix}$

In Exercises 3 and 4, compute the products by inspection.

3. (a) $\begin{bmatrix} 3 & 0 & 0 \\ 0 & -1 & 0 \\ 0 & 0 & 2 \end{bmatrix} \begin{bmatrix} 2 & 1 \\ -4 & 1 \\ 2 & 5 \end{bmatrix}$

(b) $\begin{bmatrix} 2 & 1 \\ -4 & 1 \\ 2 & 5 \end{bmatrix} \begin{bmatrix} 5 & 0 \\ 0 & -2 \end{bmatrix}$

(c) $\begin{bmatrix} 3 & 0 & 0 \\ 0 & -1 & 0 \\ 0 & 0 & 2 \end{bmatrix} \begin{bmatrix} 2 & 1 \\ -4 & 1 \\ 2 & 5 \end{bmatrix} \begin{bmatrix} 5 & 0 \\ 0 & -2 \end{bmatrix}$

4. (a) $\begin{bmatrix} 2 & 0 & 0 \\ 0 & -1 & 0 \\ 0 & 0 & 4 \end{bmatrix} \begin{bmatrix} 4 & -1 & 3 \\ 1 & 2 & 0 \\ -5 & 1 & -2 \end{bmatrix}$

(b) $\begin{bmatrix} 4 & -1 & 3 \\ 1 & 2 & 0 \\ -5 & 1 & -2 \end{bmatrix} \begin{bmatrix} -3 & 0 & 0 \\ 0 & 5 & 0 \\ 0 & 0 & 2 \end{bmatrix}$

(c) $\begin{bmatrix} 2 & 0 & 0 \\ 0 & -1 & 0 \\ 0 & 0 & 4 \end{bmatrix} \begin{bmatrix} 4 & -1 & 3 \\ 1 & 2 & 0 \\ -5 & 1 & -2 \end{bmatrix} \begin{bmatrix} -3 & 0 & 0 \\ 0 & 5 & 0 \\ 0 & 0 & 2 \end{bmatrix}$

In Exercises 5 and 6, find A^2, A^{-2}, and A^{-k} by inspection.

5. $A = \begin{bmatrix} 1 & 0 & 0 \\ 0 & -2 & 0 \\ 0 & 0 & 4 \end{bmatrix}$ **6.** $A = \begin{bmatrix} \frac{1}{2} & 0 & 0 \\ 0 & \frac{1}{3} & 0 \\ 0 & 0 & \frac{1}{4} \end{bmatrix}$

In Exercises 7 and 8, find all values of x for which the matrix is invertible.

7. $A = \begin{bmatrix} x-1 & x^2 & x^4 \\ 0 & x+2 & x^3 \\ 0 & 0 & x-4 \end{bmatrix}$

8. $A = \begin{bmatrix} x-\frac{1}{2} & 0 & 0 \\ x & x-\frac{1}{3} & 0 \\ x^2 & x^3 & x+\frac{1}{4} \end{bmatrix}$

In Exercises 9 and 10, verify part (d) of Theorem 3.6.1 for the matrix A.

9. $A = \begin{bmatrix} 1 & 2 & 3 \\ 0 & 1 & -2 \\ 0 & 0 & 1 \end{bmatrix}$ **10.** $A = \begin{bmatrix} 1 & 0 & 0 \\ 2 & -1 & 0 \\ 3 & 4 & 1 \end{bmatrix}$

In Exercises 11 and 12, fill in the missing entries (marked with ×) to produce a skew-symmetric matrix.

11. $A = \begin{bmatrix} \times & \times & 4 \\ 0 & \times & \times \\ \times & -1 & \times \end{bmatrix}$ **12.** $A = \begin{bmatrix} \times & 0 & \times \\ \times & \times & -4 \\ 8 & \times & \times \end{bmatrix}$

13. Find all values of a, b, and c for which A is symmetric.

$$A = \begin{bmatrix} 2 & a-2b+2c & 2a+b+c \\ 3 & 5 & a+c \\ 0 & -2 & 7 \end{bmatrix}$$

14. Find all values of a, b, c, and d for which A is skew-symmetric.

$$A = \begin{bmatrix} 0 & 2a-3b+c & 3a-5b+5c \\ -2 & 0 & 5a-8b+6c \\ -3 & -5 & d \end{bmatrix}$$

In Exercises 15 and 16, verify Theorem 3.6.4 for the given matrix.

15. $A = \begin{bmatrix} 2 & -1 \\ -1 & 3 \end{bmatrix}$ **16.** $A = \begin{bmatrix} 1 & -2 & 3 \\ -2 & 1 & -7 \\ 3 & -7 & 4 \end{bmatrix}$

17. Verify Theorem 3.6.1(b) for the matrices

$$A = \begin{bmatrix} -1 & 2 & 5 \\ 0 & 1 & 3 \\ 0 & 0 & -4 \end{bmatrix} \quad \text{and} \quad B = \begin{bmatrix} 2 & -8 & 0 \\ 0 & 2 & 1 \\ 0 & 0 & 3 \end{bmatrix}$$

18. Use the given equation to determine by inspection whether the factors in the product commute. Explain your reasoning.

(a) $\begin{bmatrix} 1 & -3 \\ -3 & 2 \end{bmatrix} \begin{bmatrix} 4 & 1 \\ 1 & 2 \end{bmatrix} = \begin{bmatrix} 1 & -5 \\ -10 & 1 \end{bmatrix}$

(b) $\begin{bmatrix} 2 & -1 \\ -1 & 3 \end{bmatrix} \begin{bmatrix} 3 & 2 \\ 2 & 1 \end{bmatrix} = \begin{bmatrix} 4 & 3 \\ 3 & 1 \end{bmatrix}$

In Exercises 19 and 20, find a diagonal matrix A that satisfies the equation.

19. $A^{-5} = \begin{bmatrix} 1 & 0 & 0 \\ 0 & -1 & 0 \\ 0 & 0 & -1 \end{bmatrix}$ **20.** $A^{-2} = \begin{bmatrix} 9 & 0 & 0 \\ 0 & 4 & 0 \\ 0 & 0 & 1 \end{bmatrix}$

In Exercises 21 and 22, explain why A^TA and AA^T are invertible and have symmetric inverses (no computations are needed for this), and then confirm these facts by computing the inverses.

21. $A = \begin{bmatrix} 1 & 0 & 0 \\ 3 & 1 & 0 \\ 1 & 3 & 1 \end{bmatrix}$ **22.** $A = \begin{bmatrix} 1 & 0 & 0 \\ 3 & 1 & 0 \\ 1 & 2 & 1 \end{bmatrix}$

In Exercises 23 and 24, find the fixed points of the given matrix.

23. $\begin{bmatrix} 2 & 1 \\ 1 & 2 \end{bmatrix}$ **24.** $\begin{bmatrix} 3 & 2 & 0 \\ 3 & 4 & 0 \\ 0 & 0 & 2 \end{bmatrix}$

In Exercises 25 and 26, show that A is nilpotent, state the index of nilpotency, and then apply Theorem 3.6.6 to find $(I - A)^{-1}$.

25. (a) $A = \begin{bmatrix} 0 & 1 \\ 0 & 0 \end{bmatrix}$ (b) $A = \begin{bmatrix} 0 & 0 & 0 \\ 1 & 0 & 0 \\ 8 & 1 & 0 \end{bmatrix}$

26. (a) $A = \begin{bmatrix} 0 & 0 \\ 1 & 0 \end{bmatrix}$ (b) $A = \begin{bmatrix} 0 & 2 & 1 \\ 0 & 0 & 3 \\ 0 & 0 & 0 \end{bmatrix}$

27. (a) Show that if A is an invertible skew-symmetric matrix, then A^{-1} is skew-symmetric.

(b) Show that if A and B are skew-symmetric, then so are $A + B$, $A - B$, A^T, and kA for any scalar k.

28. (a) Show that if A is any $n \times n$ matrix, then $A + A^T$ is symmetric and $A - A^T$ is skew-symmetric.

(b) Show that every square matrix A can be expressed as the sum of a symmetric matrix and a skew-symmetric matrix.

(c) Express the matrix in Example 6 as the sum of a symmetric matrix and a skew-symmetric matrix.

29. Show that if $\mathbf{u}$ is a column vector in R^n, then $H = I_n - 2\mathbf{u}\mathbf{u}^T$ is symmetric.

30. (a) Let A be an $m \times n$ matrix, and find a formula for AA^T in terms of the row vectors $\mathbf{r}_1, \mathbf{r}_2, \ldots, \mathbf{r}_m$ of A that is analogous to Formula (8).

(b) Find a formula for AA^T that is analogous to Formula (9) for A^TA.

31. Find $\text{tr}(A^TA)$ given that A is an $m \times n$ matrix whose column vectors have length 1.

Discussion and Discovery

D1. (a) Factor the matrix

$$A = \begin{bmatrix} 3a_{11} & 5a_{12} & 7a_{13} \\ 3a_{21} & 5a_{22} & 7a_{23} \\ 3a_{31} & 5a_{32} & 7a_{33} \end{bmatrix}$$

into the form $A = BD$, where D is diagonal.

(b) Is your factorization the only one possible? Explain.

D2. Let $A = [a_{ij}]$ be an $n \times n$ matrix ($n > 1$). In each part, determine whether A is symmetric, and then devise a general test that can be applied to the formula for a_{ij} to determine whether $A = [a_{ij}]$ is symmetric.

(a) $a_{ij} = i^2 + j^2$ (b) $a_{ij} = i - j$

(c) $a_{ij} = 2i + 2j$ (d) $a_{ij} = 2i^2 + 2j^3$

D3. We showed in the text that the product of commuting symmetric matrices is symmetric. Do you think that the product of commuting skew-symmetric matrices is skew-symmetric? Explain.

D4. What can you say about a matrix for which $A^TA = 0$?

D5. (a) Find all 3×3 diagonal matrices A with the property that $A^2 = A$.

(b) How many $n \times n$ diagonal matrices A have the property that $A^2 = A$?

D6. Find all 2×2 diagonal matrices A with the property that $A^2 + 5A + 6I_2 = 0$.

D7. What can be said about a matrix A that is both symmetric and skew-symmetric?

D8. What is the maximum number of distinct entries that an $n \times n$ symmetric matrix can have? What about a skew-symmetric matrix? Explain your reasoning.

D9. Suppose that A is a square matrix and D is a diagonal matrix such that $AD = I$. What can you say about the matrix A? Explain your reasoning.

D10. Indicate whether the statement is true (T) or false (F). Justify your answer.

(a) If AA^T is invertible, then so is A.

(b) If $A + B$ is symmetric, then so are A and B.

(c) If A is symmetric and triangular, then every polynomial $p(A)$ in A is symmetric and triangular.

(d) An $n \times n$ matrix A can be written as $A = L + U + D$, where L is lower triangular with zeros on the main diagonal, U is upper triangular with zeros on the main diagonal, and D is diagonal.

(e) If A is an $n \times n$ matrix for which the system $A\mathbf{x} = 0$ has only the trivial solution, then $A^T\mathbf{x} = 0$ has only the trivial solution.

D11. Indicate whether the statement is true (T) or false (F). Justify your answer.

(a) If A^2 is symmetric, then so is A.

(b) A nilpotent matrix does not have an inverse.

(c) If $A^3 = A$, then A cannot be nilpotent.

(d) If A is invertible, then $(A^{-1})^T = (A^T)^{-1}$.

(e) If A is invertible, then so is $I - A$.

Working with Proofs

P1. Prove Theorem 3.6.2 using the properties given in Theorem 3.2.10, rather than working with individual entries.

P2. Prove that if

$$D = \begin{bmatrix} d_1 & \cdots & 0 \\ \vdots & \ddots & \vdots \\ 0 & \cdots & d_n \end{bmatrix}$$

is a diagonal matrix, then

$$D^k = \begin{bmatrix} d_1^k & \cdots & 0 \\ \vdots & \ddots & \vdots \\ 0 & \cdots & d_n^k \end{bmatrix}$$

P3. Prove that a diagonal matrix is invertible if and only if all diagonal entries are nonzero.

P4. (***For readers familiar with mathematical induction***) Use the method of mathematical induction to prove that if A is any symmetric matrix and n is a positive integer ($n \geq 2$), then A^n is symmetric.

P5. Prove Theorem 3.6.5.

Technology Exercises

T1. (***Special types of matrices***) Typing in the entries of a matrix can be tedious, so many technology utilities provide shortcuts for entering identity matrices, zero matrices, diagonal matrices, triangular matrices, symmetric matrices, and skew-symmetric matrices. Determine whether your utility has this feature, and if so, practice entering matrices of various special types.

T2. Confirm the results in Theorem 3.6.1 for some triangular matrices of your choice.

T3. Show that the matrix

$$A = \begin{bmatrix} 2 & 11 & 3 \\ -2 & -11 & -3 \\ 8 & 35 & 9 \end{bmatrix}$$

is nilpotent, and then use Formula (12) to compute $(I - A)^{-1}$. Check your answer by computing the inverse directly.

T4. (a) Use Theorem 3.6.7 to confirm that if

$$A = \begin{bmatrix} 0 & \frac{1}{4} & \frac{1}{8} \\ \frac{1}{4} & \frac{1}{8} & \frac{1}{10} \\ \frac{1}{8} & \frac{1}{10} & \frac{1}{10} \end{bmatrix}$$

then the inverse of $I - A$ can be expressed by the series in Formula (18).

(b) Compute the approximation $(I - A)^{-1} \approx I + A + A^2 + A^3 + \cdots + A^{10}$, and compare it to the inverse of $I - A$ produced directly by your utility. To how many decimal places do the results agree?

T5. (CAS) We stated in the text that every strictly triangular matrix is nilpotent. Show that this is true for matrices of size 2×2, 3×3, and 4×4, and make a conjecture about the index of nilpotency of an $n \times n$ strictly triangular matrix. Confirm your conjecture for matrices of size 5×5.

Section 3.7   Matrix Factorizations; *LU*-Decomposition

In this section we will discuss a method for factoring a square matrix into a product of upper and lower triangular matrices. Such factorizations provide the foundation for many of the most widely used algorithms for inverting matrices and solving linear systems. We will see that factorization methods for solving linear systems have certain advantages over Gauss–Jordan and Gaussian elimination.

SOLVING LINEAR SYSTEMS BY FACTORIZATION

Our primary goal in this section is to develop a method for factoring a square matrix A in the form

$$A = LU \tag{1}$$

where L is lower triangular and U is upper triangular. To motivate why one might be interested in doing this, let us assume that we want to solve a linear system $A\mathbf{x} = \mathbf{b}$ of n equations in n unknowns, and suppose that the coefficient matrix has somehow been factored in form (1), where L is an $n \times n$ lower triangular matrix and U is an $n \times n$ upper triangular matrix. Starting from this factorization we can solve the system $A\mathbf{x} = \mathbf{b}$ by the following procedure:

Step 1. Rewrite the system $A\mathbf{x} = \mathbf{b}$ as

$$LU\mathbf{x} = \mathbf{b} \tag{2}$$

Step 2. Define a new unknown $\mathbf{y}$ by letting

$$U\mathbf{x} = \mathbf{y} \tag{3}$$

and rewrite (2) as $L\mathbf{y} = \mathbf{b}$.

Step 3. Solve the system $L\mathbf{y} = \mathbf{b}$ for the unknown $\mathbf{y}$.

Step 4. Substitute the now-known vector $\mathbf{y}$ into (3) and solve for $\mathbf{x}$.

This procedure is called the method of ***LU-decomposition***. Although LU-decomposition converts the problem of solving the single system $A\mathbf{x} = \mathbf{b}$ into the problem of solving the two systems, $L\mathbf{y} = \mathbf{b}$ and $U\mathbf{x} = \mathbf{y}$, these systems are easy to solve because their coefficient matrices are triangular. Thus, it turns out to be no more work to solve the two systems than it is to solve the original system directly. Here is an example.

EXAMPLE 1
Solving $A\mathbf{x} = \mathbf{b}$ by LU-Decomposition

Later in this section we will derive the factorization

$$\begin{bmatrix} 2 & 6 & 2 \\ -3 & -8 & 0 \\ 4 & 9 & 2 \end{bmatrix} = \begin{bmatrix} 2 & 0 & 0 \\ -3 & 1 & 0 \\ 4 & -3 & 7 \end{bmatrix} \begin{bmatrix} 1 & 3 & 1 \\ 0 & 1 & 3 \\ 0 & 0 & 1 \end{bmatrix} \tag{4}$$

$$A \qquad = \qquad L \qquad \quad U$$

but for now do not worry about how it was derived—our only objective here is to illustrate how

Linear Algebra in History

Although the ideas were known earlier, credit for popularizing the matrix formulation of the *LU*-decomposition is often given to the British mathematician and logician Alan Turing for his work on the subject in 1948. Turing, one of the great geniuses of the twentieth century, is the founder of the field of artificial intelligence. Among his many accomplishments in that field, he developed the concept of an internally programmed computer before the practical technology had reached the point where the construction of such a machine was possible. During World War II Turing was secretly recruited by the British government's Code and Cypher School at Bletchley Park to help break the Nazi Enigma codes; it was Turing's statistical approach that provided the breakthrough. In addition to being a brilliant mathematician, Turing was a world-class runner who competed successfully with Olympic-level competition. Sadly, Turing, a homosexual, was tried and convicted of "gross indecency" in 1952, in violation of the then-existing British statutes. Although spared prison, he was subjected to hormone injections to "dampen his lust." Depressed, he committed suicide at age 41 by eating an apple laced with cyanide.

Alan Mathison Turing
(1912–1954)

this factorization can be used to solve the linear system

$$\underset{A}{\begin{bmatrix} 2 & 6 & 2 \\ -3 & -8 & 0 \\ 4 & 9 & 2 \end{bmatrix}} \underset{\mathbf{x}}{\begin{bmatrix} x_1 \\ x_2 \\ x_3 \end{bmatrix}} = \underset{\mathbf{b}}{\begin{bmatrix} 2 \\ 2 \\ 3 \end{bmatrix}}$$

From (4) we can rewrite this system as

$$\underset{L}{\begin{bmatrix} 2 & 0 & 0 \\ -3 & 1 & 0 \\ 4 & -3 & 7 \end{bmatrix}} \underset{U}{\begin{bmatrix} 1 & 3 & 1 \\ 0 & 1 & 3 \\ 0 & 0 & 1 \end{bmatrix}} \underset{\mathbf{x}}{\begin{bmatrix} x_1 \\ x_2 \\ x_3 \end{bmatrix}} = \underset{\mathbf{b}}{\begin{bmatrix} 2 \\ 2 \\ 3 \end{bmatrix}} \tag{5}$$

As specified in Step 2 above, let us define y_1, y_2, and y_3 by the equation

$$\underset{U}{\begin{bmatrix} 1 & 3 & 1 \\ 0 & 1 & 3 \\ 0 & 0 & 1 \end{bmatrix}} \underset{\mathbf{x}}{\begin{bmatrix} x_1 \\ x_2 \\ x_3 \end{bmatrix}} = \underset{\mathbf{y}}{\begin{bmatrix} y_1 \\ y_2 \\ y_3 \end{bmatrix}} \tag{6}$$

which allows us to rewrite (5) as

$$\underset{L}{\begin{bmatrix} 2 & 0 & 0 \\ -3 & 1 & 0 \\ 4 & -3 & 7 \end{bmatrix}} \underset{\mathbf{y}}{\begin{bmatrix} y_1 \\ y_2 \\ y_3 \end{bmatrix}} = \underset{\mathbf{b}}{\begin{bmatrix} 2 \\ 2 \\ 3 \end{bmatrix}} \tag{7}$$

or equivalently, as

$$\begin{aligned} 2y_1 & & & = 2 \\ -3y_1 + {}& y_2 & & = 2 \\ 4y_1 - {}& 3y_2 + 7y_3 & & = 3 \end{aligned}$$

This system can be solved by a procedure that is similar to back substitution, except that we solve the equations from the top down instead of from the bottom up. This procedure, called ***forward substitution***, yields

$$y_1 = 1, \quad y_2 = 5, \quad y_3 = 2$$

(verify). As indicated in Step 4 above, we substitute these values into (6), which yields the linear system

$$\begin{bmatrix} 1 & 3 & 1 \\ 0 & 1 & 3 \\ 0 & 0 & 1 \end{bmatrix} \begin{bmatrix} x_1 \\ x_2 \\ x_3 \end{bmatrix} = \begin{bmatrix} 1 \\ 5 \\ 2 \end{bmatrix}$$

or equivalently,

$$\begin{aligned} x_1 + 3x_2 + {}& x_3 = 1 \\ x_2 + {}& 3x_3 = 5 \\ & x_3 = 2 \end{aligned}$$

Solving this system by back substitution yields

$$x_1 = 2, \quad x_2 = -1, \quad x_3 = 2$$

(verify). ∎

FINDING
***LU*-DECOMPOSITIONS**

Example 1 makes it clear that after A is factored into lower and upper triangular matrices, the system $A\mathbf{x} = \mathbf{b}$ can be solved by one forward substitution and one back substitution. We will now show how to obtain such factorizations.

We begin with some terminology.

> **Definition 3.7.1** A factorization of a square matrix A as $A = LU$, where L is lower triangular and U is upper triangular, is called an ***LU*-decomposition** or ***LU*-factorization** of A.

Linear Algebra in History

In the late 1970s the National Science Foundation and the Department of Energy in the United States supported the development of computer routines for inverting matrices and analyzing and solving systems of linear equations. That research led to a set of Fortran programs, known as LINPACK, which has set the standard for many of today's computer algorithms, including those used by MATLAB. The LINPACK routines are organized around four matrix factorizations, of which the *LU*-decomposition is one. The primary developers of LINPACK, C. B. Moler, J. J. Dongarra, G. W. Stewart, and J. R. Bunch, based many of their ideas on the work of James Boyle and Kenneth Dritz at the Argonne National Laboratories.

In general, not every square matrix A has an LU-decomposition, nor is an LU-decomposition unique if it exists. However, we will now show that if A can be reduced to row echelon form by Gaussian elimination *without row interchanges*, then A must have an LU-decomposition. Moreover, as a by-product of this discussion, we will see how the factors can be obtained. Toward this end, suppose that A is an $n \times n$ matrix that has been reduced by elementary row operations without row interchanges to the row echelon form U. It follows from Theorem 3.3.1 that there is a sequence of elementary matrices $E_1, E_2, \ldots, E_k$ such that

$$E_k \cdots E_2 E_1 A = U \tag{8}$$

Since elementary matrices are invertible, we can solve (8) for A as

$$A = E_1^{-1} E_2^{-1} \cdots E_k^{-1} U$$

or more briefly as

$$A = LU \tag{9}$$

where

$$L = E_1^{-1} E_2^{-1} \cdots E_k^{-1} \tag{10}$$

If we can show that U is upper triangular and L is lower triangular, then (9) will be an LU-decomposition of A. However, U is upper triangular because it is a row echelon form of the square matrix A (see Example 3 of Section 3.6). To see that L is lower triangular, we will use the fact that no row interchanges are used to obtain U from A and that in Gaussian elimination zeros are introduced by adding multiples of rows to *lower* rows. This being the case, it follows that each elementary matrix in (8) arises either by multiplying a row of the $n \times n$ identity matrix by a scalar or by adding a multiple of a row to a lower row. In either case the resulting elementary matrix is lower triangular. Moreover, you can check directly that each of the matrices on the right side of (10) is lower triangular, so their product L is also lower triangular by part (*b*) of Theorem 3.6.1.

In summary, we have established the following result.

> **Theorem 3.7.2** *If a square matrix A can be reduced to row echelon form by Gaussian elimination with no row interchanges, then A has an LU-decomposition.*

The discussion that led us to this theorem actually provides a procedure for finding an LU-decomposition of the matrix A:

- Reduce A to a row echelon form U without using any row interchanges.
- Keep track of the sequence of row operations performed, and let $E_1, E_2, \ldots, E_k$ be the sequence of elementary matrices that corresponds to those operations.

- Let

$$L = E_1^{-1} E_2^{-1} \cdots E_k^{-1} = (E_k \cdots E_2 E_1)^{-1} \qquad (11)$$

- $A = LU$ is an *LU*-decomposition of A.

As a practical matter, there is no need to calculate L from Formula (11); a better approach is to observe that this formula can be rewritten as

$$(E_k \cdots E_2 E_1)L = I$$

which by comparison to (8) tells us that the same sequence of row operations that reduces A to U reduces L to I. This suggests that we may be able to construct L with some clever bookkeeping, the idea being to track the operations that reduce A to U and at each step try to figure what entry to put into L, so that L will be reduced to I by the sequence of row operations that reduces A to U. Here is a four-step procedure for doing this:

Step 1. Reduce A to row echelon form U without using row interchanges, keeping track of the multipliers used to introduce the leading 1's and the multipliers used to introduce zeros below the leading 1's.

Step 2. In each position along the main diagonal of L, place the reciprocal of the multiplier that introduced the leading 1 in that position in U.

Step 3. In each position below the main diagonal of L, place the negative of the multiplier used to introduce the zero in that position in U.

Step 4. Form the decomposition $A = LU$.

EXAMPLE 2
Constructing an *LU*-Decomposition

Find an *LU*-decomposition of

$$A = \begin{bmatrix} 6 & -2 & 0 \\ 9 & -1 & 1 \\ 3 & 7 & 5 \end{bmatrix}$$

Solution We will reduce A to a row echelon form U and at each step we will fill in an entry of L in accordance with the four-step procedure above.

Thus, we have constructed the LU-decomposition

$$A = LU = \begin{bmatrix} 6 & 0 & 0 \\ 9 & 2 & 0 \\ 3 & 8 & 1 \end{bmatrix} \begin{bmatrix} 1 & -\frac{1}{3} & 0 \\ 0 & 1 & \frac{1}{2} \\ 0 & 0 & 1 \end{bmatrix}$$

We leave it for you to confirm this end result by multiplying the factors. ∎

THE RELATIONSHIP BETWEEN GAUSSIAN ELIMINATION AND *LU*-DECOMPOSITION

There is a close relationship between Gaussian elimination and LU-decomposition that we will now explain. For this purpose, assume that $A\mathbf{x} = \mathbf{b}$ is a linear system of n equations in n unknowns, that A can be reduced to row echelon form without row interchanges, and that $A = LU$ is the corresponding LU-decomposition.

In the method of LU-decomposition, the system $A\mathbf{x} = \mathbf{b}$ is solved by first solving the system $L\mathbf{y} = \mathbf{b}$ for $\mathbf{y}$, and then solving the system $U\mathbf{x} = \mathbf{y}$ for $\mathbf{x}$. However, most of the work required to solve $L\mathbf{y} = \mathbf{b}$ is done in the course of constructing the LU-decomposition of A. To see why this is so, suppose that

$$E_1, E_2, \ldots, E_k$$

is the sequence of elementary matrices that corresponds to the sequence of row operations that reduces A to U; that is,

$$E_k \cdots E_2 E_1 A = U$$

Thus, if we multiply both sides of the equation $A\mathbf{x} = \mathbf{b}$ by $E_k \cdots E_2 E_1$, we obtain

$$U\mathbf{x} = E_k \cdots E_2 E_1 \mathbf{b}$$

which we can rewrite as

$$\mathbf{y} = E_k \cdots E_2 E_1 \mathbf{b}$$

This equation tells us that the sequence of row operations that reduces A to U produces the vector $\mathbf{y}$ when applied to $\mathbf{b}$. Accordingly, the process of Gaussian elimination reduces the augmented matrix $[A \mid \mathbf{b}]$ to $[U \mid \mathbf{y}]$ and hence produces the solution of the equation $L\mathbf{y} = \mathbf{b}$ in the last column. This shows that LU-decomposition and Gaussian elimination (the forward phase of Gauss–Jordan elimination) differ in organization and bookkeeping, but otherwise involve the same operations.

EXAMPLE 3
Gaussian Elimination Performed as an LU-Decomposition

In Example 1 we showed how to solve the linear system

$$\begin{bmatrix} 2 & 6 & 2 \\ -3 & -8 & 0 \\ 4 & 9 & 2 \end{bmatrix} \begin{bmatrix} x_1 \\ x_2 \\ x_3 \end{bmatrix} = \begin{bmatrix} 2 \\ 2 \\ 3 \end{bmatrix} \tag{12}$$

using an LU-decomposition of the coefficient matrix, but we did not discuss how the factorization was derived. In the course of solving the system we obtained the intermediate vector

$$\mathbf{y} = \begin{bmatrix} 1 \\ 5 \\ 2 \end{bmatrix}$$

by using forward substitution to solve system (7). We will now use the procedure discussed above to find both the LU-decomposition and the vector $\mathbf{y}$ by row operations on the augmented matrix for (12).

$$[A \mid \mathbf{b}] = \begin{bmatrix} 2 & 6 & 2 & \vdots & 2 \\ -3 & -8 & 0 & \vdots & 2 \\ 4 & 9 & 2 & \vdots & 3 \end{bmatrix} \quad \begin{bmatrix} \bullet & 0 & 0 \\ \bullet & \bullet & 0 \\ \bullet & \bullet & \bullet \end{bmatrix} = L \text{ (dots = unknown entries)}$$

$$\begin{bmatrix} 1 & 3 & 1 & \vdots & 1 \\ -3 & -8 & 0 & \vdots & 2 \\ 4 & 9 & 2 & \vdots & 3 \end{bmatrix} \quad \begin{bmatrix} 2 & 0 & 0 \\ \bullet & \bullet & 0 \\ \bullet & \bullet & \bullet \end{bmatrix}$$

$$
\begin{bmatrix} 1 & 3 & 1 & \vdots & 1 \\ 0 & 1 & 3 & \vdots & 5 \\ 0 & -3 & -2 & \vdots & -1 \end{bmatrix}
\begin{bmatrix} 2 & 0 & 0 \\ -3 & \bullet & 0 \\ 4 & \bullet & \bullet \end{bmatrix}
$$

$$
\begin{bmatrix} 1 & 3 & 1 & \vdots & 1 \\ 0 & 1 & 3 & \vdots & 5 \\ 0 & 0 & 7 & \vdots & 14 \end{bmatrix}
\begin{bmatrix} 2 & 0 & 0 \\ -3 & 1 & 0 \\ 4 & -3 & \bullet \end{bmatrix}
$$

$$
[U \mid \mathbf{y}] = \begin{bmatrix} 1 & 3 & 1 & \vdots & 1 \\ 0 & 1 & 3 & \vdots & 5 \\ 0 & 0 & 1 & \vdots & 2 \end{bmatrix}
\begin{bmatrix} 2 & 0 & 0 \\ -3 & 1 & 0 \\ 4 & -3 & 7 \end{bmatrix} = L
$$

These results agree with those in Example 1, so we have found an *LU*-decomposition of the coefficient matrix and simultaneously have completed the forward substitution required to find **y**. All that remains to solve the given system is to solve the system $U\mathbf{x} = \mathbf{y}$ by back substitution. The computations were performed in Example 1. ∎

MATRIX INVERSION BY *LU*-DECOMPOSITION

Many of the best algorithms for inverting matrices use *LU*-decomposition. To understand how this can be done, let A be an invertible $n \times n$ matrix, let $A^{-1} = [\mathbf{x}_1 \quad \mathbf{x}_2 \quad \cdots \quad \mathbf{x}_n]$ be its unknown inverse partitioned into column vectors, and let $I = [\mathbf{e}_1 \quad \mathbf{e}_2 \quad \cdots \quad \mathbf{e}_n]$ be the $n \times n$ identity matrix partitioned into column vectors. The matrix equation $AA^{-1} = I$ can be expressed as

$$
A[\mathbf{x}_1 \quad \mathbf{x}_2 \quad \cdots \quad \mathbf{x}_n] = [\mathbf{e}_1 \quad \mathbf{e}_2 \quad \cdots \quad \mathbf{e}_n]
$$

or alternatively as

$$
[A\mathbf{x}_1 \quad A\mathbf{x}_2 \quad \cdots \quad A\mathbf{x}_n] = [\mathbf{e}_1 \quad \mathbf{e}_2 \quad \cdots \quad \mathbf{e}_n]
$$

which tells us that the unknown column vectors of A^{-1} can be obtained by solving the n linear systems

$$
A\mathbf{x} = \mathbf{e}_1, \quad A\mathbf{x} = \mathbf{e}_2, \ldots, \quad A\mathbf{x} = \mathbf{e}_n \tag{13}
$$

As discussed above, this can be done by finding an *LU*-decomposition of A, and then using that decomposition to solve each of the n systems in (13).

LDU-DECOMPOSITIONS

The method we have described for computing *LU*-decompositions may result in an "asymmetry" in that the matrix U has 1's on the main diagonal but L need not. However, if it is preferred to have 1's on the main diagonal of the lower triangular factor, then we can "shift" the diagonal entries of L to a diagonal matrix D and write L as

$$
L = L'D
$$

where L' is a lower triangular matrix with 1's on the main diagonal. For example, a general 3×3 lower triangular matrix with nonzero entries on the main diagonal can be factored as

$$
\underset{L}{\begin{bmatrix} a_{11} & 0 & 0 \\ a_{21} & a_{22} & 0 \\ a_{31} & a_{32} & a_{33} \end{bmatrix}} = \underset{L'}{\begin{bmatrix} 1 & 0 & 0 \\ a_{21}/a_{11} & 1 & 0 \\ a_{31}/a_{11} & a_{32}/a_{22} & 1 \end{bmatrix}} \underset{D}{\begin{bmatrix} a_{11} & 0 & 0 \\ 0 & a_{22} & 0 \\ 0 & 0 & a_{33} \end{bmatrix}}
$$

Note that the columns of L' are obtained by dividing each entry in the column by the diagonal entry in the column. Thus, for example, we can rewrite (4) as

$$
\begin{bmatrix} 2 & 6 & 2 \\ -3 & -8 & 0 \\ 4 & 9 & 2 \end{bmatrix} = \begin{bmatrix} 2 & 0 & 0 \\ -3 & 1 & 0 \\ 4 & -3 & 7 \end{bmatrix} \begin{bmatrix} 1 & 3 & 1 \\ 0 & 1 & 3 \\ 0 & 0 & 1 \end{bmatrix}
$$

$$
= \begin{bmatrix} 1 & 0 & 0 \\ -3/2 & 1 & 0 \\ 2 & -3 & 1 \end{bmatrix} \begin{bmatrix} 2 & 0 & 0 \\ 0 & 1 & 0 \\ 0 & 0 & 7 \end{bmatrix} \begin{bmatrix} 1 & 3 & 1 \\ 0 & 1 & 3 \\ 0 & 0 & 1 \end{bmatrix}
$$

In general, one can prove that if A is a square matrix that can be reduced to row echelon form without row interchanges, then A can be factored *uniquely* as

$$A = LDU$$

where L is a lower triangular matrix with 1's on the main diagonal, D is a diagonal matrix, and U is an upper triangular matrix with 1's on the main diagonal. This is called the **LDU-decomposition** (or **LDU-factorization**) of A.

REMARK The procedure that we described for finding an LU-decomposition of a matrix A produces 1's on the main diagonal of U because the matrix U in our procedure is a row echelon form of A. Many programs for computing LU-decompositions do not introduce leading 1's; rather, they leave the leading entries, called **pivots**, in their original form and simply add suitable multiples of the pivots to the entries below to obtain the required zeros. This produces an upper triangular matrix U with the pivots, rather than 1's, on the main diagonal. There are certain advantages to preserving the pivots that we will discuss later in this text.

USING PERMUTATION MATRICES TO DEAL WITH ROW INTERCHANGES

Partial pivoting or some comparable procedure is used in most real-world applications to reduce roundoff error, so row interchanges almost always occur when numerically stable algorithms are used to solve a linear system $A\mathbf{x} = \mathbf{b}$. When row interchanges occur, LU-decomposition cannot be used directly; however, it is possible to circumvent this difficulty by "preprocessing" A so all of the row operations are performed *prior* to starting the LU-decomposition. More precisely, the idea is to form a matrix P by multiplying in sequence those elementary matrices that correspond to the row interchanges, and then execute all of these row interchanges on A by forming the product PA. Since all of the row interchanges are out of the way, the matrix PA can be reduced to row echelon form *without* row interchanges and hence has an LU-decomposition

$$PA = LU \tag{14}$$

Since the matrix P is invertible (being a product of elementary matrices), the systems $A\mathbf{x} = \mathbf{b}$ and $PA\mathbf{x} = P\mathbf{b}$ have the same solutions (why?), and the latter system can be solved by LU-decomposition.

REMARK If A has size $n \times n$, then the matrix P in the preceding discussion results by reordering the rows of I_n in some way. A matrix of this type is called a **permutation matrix**. Some writers use P^{-1} for the matrix we have denoted by P, in which case (14) can be rewritten as

$$A = PLU \tag{15}$$

This is called a **PLU-decomposition** of A.

FLOPS AND THE COST OF SOLVING A LINEAR SYSTEM

There is an old saying that "time is money." This is especially true in industry, where the cost of solving a problem is often determined by the time it takes for a computer to perform its computational tasks. In general, the time required for a computer to solve a problem depends on two factors—the speed of its processor and the number of operations it has to perform. The speed of computer processors is increasing all the time, so the natural advance of technology works toward reducing the cost of problem solving. However, the number of operations required to solve a problem is not a matter of technology, but rather of the algorithms that are used to solve the problem. Thus, choosing the right algorithm to solve a large linear system can have important financial implications. The rest of this section will be devoted to discussing factors that affect the choices of algorithms for solving linear systems.

In computer jargon, an arithmetic operation $(+, -, *, \div)$ on two real numbers is called a **flop**, which is an acronym for "floating-point operation."[*] The total number of flops required to solve a problem, which is called the **cost** of the solution, provides a convenient way of choosing

---

[*]Real numbers are stored in computers as numerical approximations called **floating-point numbers**. In base 10, a floating-point number has the form $\pm .d_1 d_2 \cdots d_n \times 10^m$, where m is an integer, called the **mantissa**, and n is the number of digits to the right of the decimal point. The value of n varies with the computer. In some literature the term *flop* is used as a measure of processing speed and stands for "floating-point operations *per second*." In our usage it is interpreted as a counting unit.

between various algorithms for solving the problem. When needed, the cost in flops can be converted to units of time or money if the speed of the computer processor and the financial aspects of its operation are known. For example, many of today's PCs are capable of performing nearly 10 gigaflops per second (1 gigaflop $= 10^9$ flops). Thus, an algorithm that costs 1,000,000 flops would be executed in 0.0001 s.

To illustrate how costs (in flops) can be computed, let us count the number of flops required to solve a linear system of n equations in n unknowns by Gauss–Jordan elimination. For this purpose we will need the following formulas for the sum of the first n positive integers and the sum of the squares of the first n positive integers:

$$1 + 2 + 3 + \cdots + n = \frac{n(n + 1)}{2} \tag{16}$$

$$1^2 + 2^2 + 3^2 + \cdots + n^2 = \frac{n(n + 1)(2n + 1)}{6} \tag{17}$$

Let $A\mathbf{x} = \mathbf{b}$ be a linear system of n equations in n unknowns to be solved by Gauss–Jordan elimination (or equivalently, by Gaussian elimination with back substitution). For simplicity, let us assume that A is invertible and that no row interchanges are required to reduce the augmented matrix $[A \mid \mathbf{b}]$ to row echelon form. The diagrams that accompany the analysis provide a convenient way of counting the operations required to introduce a leading 1 in the first row and then zeros below it. In our operation counts, we will lump divisions and multiplications together as "multiplications," and we will lump additions and subtractions together as "additions."

Step 1. It requires n flops (multiplications) to introduce the leading 1 in the first row.

$$\begin{bmatrix} 1 & \times & \times & \cdots & \times & \times & \vdots & \times \\ \bullet & \bullet & \bullet & \cdots & \bullet & \bullet & \vdots & \bullet \\ \bullet & \bullet & \bullet & \cdots & \bullet & \bullet & \vdots & \bullet \\ \vdots & \vdots & \vdots & & \vdots & \vdots & \vdots & \vdots \\ \bullet & \bullet & \bullet & \cdots & \bullet & \bullet & \vdots & \bullet \\ \bullet & \bullet & \bullet & \cdots & \bullet & \bullet & \vdots & \bullet \end{bmatrix}$$

$\begin{bmatrix} \times \text{ \textbf{denotes a quantity that is being computed.}} \\ \bullet \text{ \textbf{denotes a quantity that is not being computed.}} \\ \textbf{The matrix size is } n \times (n + 1). \end{bmatrix}$

Step 2. It requires n multiplications and n additions to introduce a zero below the leading 1, and there are $n - 1$ rows below the leading 1, so the number of flops required to introduce zeros below the leading 1 is $2n(n - 1)$.

$$\begin{bmatrix} 1 & \bullet & \bullet & \cdots & \bullet & \bullet & \vdots & \bullet \\ 0 & \times & \times & \cdots & \times & \times & \vdots & \times \\ 0 & \times & \times & \cdots & \times & \times & \vdots & \times \\ \vdots & \vdots & \vdots & & \vdots & \vdots & \vdots & \vdots \\ 0 & \times & \times & \cdots & \times & \times & \vdots & \times \\ 0 & \times & \times & \cdots & \times & \times & \vdots & \times \end{bmatrix}$$

Column 1. Combining Steps 1 and 2, the number of flops required for column 1 is

$$n + 2n(n - 1) = 2n^2 - n$$

Column 2. The procedure for column 2 is the same as for column 1, except that now we are dealing with one less row and one less column. Thus, the number of flops required to introduce the leading 1 in row 2 and the zeros below it can be obtained by replacing n by $n - 1$ in the flop count for the first column. Thus, the number of flops required for column 2 is

$$2(n - 1)^2 - (n - 1)$$

Column 3. By the argument for column 2, the number of flops required for column 3 is

$$2(n-2)^2 - (n-2)$$

Total for all columns. The pattern should now be clear. The total number of flops required to create the n leading 1's and the associated zeros is

$$(2n^2 - n) + [2(n-1)^2 - (n-1)] + [2(n-2)^2 - (n-2)] + \cdots + (2-1)$$

which we can rewrite as

$$2[n^2 + (n-1)^2 + \cdots + 1] - [n + (n-1) + \cdots + 1]$$

or on applying Formulas (16) and (17) as

$$2\frac{n(n+1)(2n+1)}{6} - \frac{n(n+1)}{2} = \frac{2}{3}n^3 + \frac{1}{2}n^2 - \frac{1}{6}n$$

Next, let us count the number of operations required to complete the backward phase (the back substitution).

Column n. It requires $n-1$ multiplications and $n-1$ additions to introduce zeros above the leading 1 in the nth column, so the total number of flops required for the column is $2(n-1)$.

$$\begin{bmatrix} 1 & \bullet & \bullet & \cdots & \bullet & 0 & \vdots & \times \\ 0 & 1 & \bullet & \cdots & \bullet & 0 & \vdots & \times \\ 0 & 0 & 1 & \cdots & \bullet & 0 & \vdots & \times \\ \vdots & \vdots & \vdots & & \vdots & \vdots & \vdots & \vdots \\ 0 & 0 & 0 & \cdots & 1 & 0 & \vdots & \times \\ 0 & 0 & 0 & \cdots & 0 & 1 & \vdots & \bullet \end{bmatrix}$$

Column $(n-1)$. The procedure is the same as for Step 1, except that now we are dealing with one less row. Thus, the number of flops required for the $(n-1)$st column is $2(n-2)$.

$$\begin{bmatrix} 1 & \bullet & \bullet & \cdots & 0 & 0 & \vdots & \times \\ 0 & 1 & \bullet & \cdots & 0 & 0 & \vdots & \times \\ 0 & 0 & 1 & \cdots & 0 & 0 & \vdots & \times \\ \vdots & \vdots & \vdots & & \vdots & \vdots & \vdots & \vdots \\ 0 & 0 & 0 & \cdots & 1 & 0 & \vdots & \bullet \\ 0 & 0 & 0 & \cdots & 0 & 1 & \vdots & \bullet \end{bmatrix}$$

Column $(n-2)$. By the argument for column $(n-1)$, the number of flops required for column $(n-2)$ is $2(n-3)$.

Total. The pattern should now be clear. The total number of flops to complete the backward phase is

$$2(n-1) + 2(n-2) + 2(n-3) + \cdots + 2(n-n) = 2[n^2 - (1 + 2 + \cdots + n)]$$

which we can rewrite using Formula (16) as

$$2\left(n^2 - \frac{n(n+1)}{2}\right) = n^2 - n$$

In summary, we have shown that for Gauss–Jordan elimination the number of flops required for the forward and backward phases is

$$\text{flops for forward phase} = \tfrac{2}{3}n^3 + \tfrac{1}{2}n^2 - \tfrac{1}{6}n \tag{18}$$

$$\text{flops for backward phase} = n^2 - n \tag{19}$$

Thus, the total cost of solving a linear system by Gauss–Jordan elimination is

$$\text{flops for both phases} = \tfrac{2}{3}n^3 + \tfrac{3}{2}n^2 - \tfrac{7}{6}n \tag{20}$$

COST ESTIMATES FOR SOLVING LARGE LINEAR SYSTEMS

It is a property of polynomials that for large values of the independent variable the term of highest power makes the major contribution to the value of the polynomial. Thus, for *large* linear systems we can use (18) and (19) to approximate the number of flops in the forward and backward phases as

$$\text{flops for forward phase} \approx \tfrac{2}{3}n^3 \tag{21}$$

$$\text{flops for backward phase} \approx n^2 \tag{22}$$

This shows that it is more costly to execute the forward phase than the backward phase for large linear systems. Indeed, the cost difference between the forward and backward phases can be enormous, as the next example shows.

EXAMPLE 4
Cost of Solving a Large Linear System

Approximate the time required to execute the forward and backward phases of Gauss–Jordan elimination for a system of 10,000 equations in 10,000 unknowns using a computer that can execute 10 gigaflops per second.

Solution We have $n = 10^4$ for the given system, so from (21) and (22) the number of gigaflops required for the forward and backward phases is

$$\text{gigaflops for forward phase} \approx \tfrac{2}{3}n^3 \times 10^{-9} = \tfrac{2}{3}(10^4)^3 \times 10^{-9} = \tfrac{2}{3} \times 10^3$$

$$\text{gigaflops for backward phase} \approx n^2 \times 10^{-9} = (10^4)^2 \times 10^{-9} = 10^{-1}$$

Thus, at 10 gigaflops/s the execution times for the forward and backward phases are

$$\text{time for forward phase} \approx \left(\tfrac{2}{3} \times 10^3\right) \times 10^{-1} \text{ s} \approx 66.67 \text{ s}$$

$$\text{time for backward phase} \approx (10^{-1}) \times 10^{-1} \text{ s} \approx 0.01 \text{ s} \qquad \blacksquare$$

We leave it as an exercise for you to confirm the results in Table 3.7.1.

Table 3.7.1

Approximate Cost for an $n \times n$ Matrix A with Large n	
Algorithm	**Cost in Flops**
Gauss–Jordan elimination (forward phase) [Gaussian elimination]	$\approx \tfrac{2}{3}n^3$
Gauss–Jordan elimination (backward phase) [back substitution]	$\approx n^2$
LU-decomposition of A	$\approx \tfrac{2}{3}n^3$
Forward substitution to solve $L\mathbf{y} = \mathbf{b}$	$\approx n^2$
Backward substitution to solve $U\mathbf{x} = \mathbf{y}$	$\approx n^2$
A^{-1} by reducing $[A \mid I]$ to $[I \mid A^{-1}]$	$\approx 2n^3$
Compute $A^{-1}\mathbf{b}$	$\approx 2n^3$

CONSIDERATIONS IN CHOOSING AN ALGORITHM FOR SOLVING A LINEAR SYSTEM

For a *single* linear system $A\mathbf{x} = \mathbf{b}$ of n equations in n unknowns, the methods of *LU*-decomposition and Gauss–Jordan elimination differ in bookkeeping but otherwise involve the same number of flops. Thus, neither method has a cost advantage over the other. However, *LU*-decomposition has other advantages that make it the method of choice:

- Gauss–Jordan elimination (or Gaussian elimination) uses the augmented matrix $[A \mid \mathbf{b}]$, so $\mathbf{b}$ must be known. In contrast, LU-decomposition uses only the matrix A, so once that decomposition is known it can be used with as many right-hand sides as are required, one at a time.

- The LU-decomposition that is computed to solve $A\mathbf{x} = \mathbf{b}$ can be used to compute A^{-1}, if needed—a real bonus in certain problems.

- For large linear systems in which computer memory is at a premium, one can dispense with the storage of the 1's and zeros that appear on or below the main diagonal of U, since those entries are known from the form of U. The space that this opens up can then be used to store the entries of L, thereby reducing the amount of memory required to solve the system.

- If A is a large matrix consisting mostly of zeros, and if the nonzero entries are concentrated in a "band" around the main diagonal, then there are techniques that can be used to reduce the cost of LU-decomposition, giving it an advantage over Gauss–Jordan elimination.

Exercise Set 3.7

In Exercises 1–4, use the given LU-decomposition to solve the system $A\mathbf{x} = \mathbf{b}$ by forward substitution followed by back substitution, as in Example 1.

1. $A = \begin{bmatrix} 3 & -6 \\ -2 & 5 \end{bmatrix} = \begin{bmatrix} 3 & 0 \\ -2 & 1 \end{bmatrix} \begin{bmatrix} 1 & -2 \\ 0 & 1 \end{bmatrix} = LU;$

$\mathbf{b} = \begin{bmatrix} 0 \\ 1 \end{bmatrix}$

2. $A = \begin{bmatrix} 2 & 5 \\ -3 & -4 \end{bmatrix} = \begin{bmatrix} 1 & 0 \\ -\frac{3}{2} & 1 \end{bmatrix} \begin{bmatrix} 2 & 5 \\ 0 & \frac{7}{2} \end{bmatrix} = LU;$

$\mathbf{b} = \begin{bmatrix} 3 \\ -4 \end{bmatrix}$

3. $\mathbf{b} = \begin{bmatrix} -3 \\ -22 \\ 3 \end{bmatrix}; \ A = \begin{bmatrix} 3 & -6 & -3 \\ 2 & 0 & 6 \\ -4 & 7 & 4 \end{bmatrix}$

$= \begin{bmatrix} 3 & 0 & 0 \\ 2 & 4 & 0 \\ -4 & -1 & 2 \end{bmatrix} \begin{bmatrix} 1 & -2 & -1 \\ 0 & 1 & 2 \\ 0 & 0 & 1 \end{bmatrix} = LU$

4. $\mathbf{b} = \begin{bmatrix} 1 \\ 5 \\ -7 \end{bmatrix}; \ A = \begin{bmatrix} 1 & -3 & 0 \\ 0 & 1 & 3 \\ 2 & -10 & 2 \end{bmatrix}$

$= \begin{bmatrix} 1 & 0 & 0 \\ 0 & 1 & 0 \\ 2 & -4 & 1 \end{bmatrix} \begin{bmatrix} 1 & -3 & 0 \\ 0 & 1 & 3 \\ 0 & 0 & 14 \end{bmatrix} = LU$

In Exercises 5–10, find an LU-decomposition of the coefficient matrix A, and use it to solve the system $A\mathbf{x} = \mathbf{b}$ by forward substitution followed by back substitution, as in Example 1.

5. $\begin{bmatrix} 2 & 8 \\ -1 & -1 \end{bmatrix} \begin{bmatrix} x_1 \\ x_2 \end{bmatrix} = \begin{bmatrix} -2 \\ -2 \end{bmatrix}$

6. $\begin{bmatrix} -5 & -10 \\ 6 & 5 \end{bmatrix} \begin{bmatrix} x_1 \\ x_2 \end{bmatrix} = \begin{bmatrix} -10 \\ 19 \end{bmatrix}$

7. $\begin{bmatrix} 2 & -2 & -2 \\ 0 & -2 & 2 \\ -1 & 5 & 2 \end{bmatrix} \begin{bmatrix} x_1 \\ x_2 \\ x_3 \end{bmatrix} = \begin{bmatrix} -4 \\ -2 \\ 6 \end{bmatrix}$

8. $\begin{bmatrix} 5 & 5 & 10 \\ -8 & -7 & -9 \\ 0 & 4 & 26 \end{bmatrix} \begin{bmatrix} x_1 \\ x_2 \\ x_3 \end{bmatrix} = \begin{bmatrix} 0 \\ 1 \\ 4 \end{bmatrix}$

9. $\begin{bmatrix} -1 & 0 & 1 & 0 \\ 2 & 3 & -2 & 6 \\ 0 & -1 & 2 & 0 \\ 0 & 0 & 1 & 5 \end{bmatrix} \begin{bmatrix} x_1 \\ x_2 \\ x_3 \\ x_4 \end{bmatrix} = \begin{bmatrix} 5 \\ -1 \\ 3 \\ 7 \end{bmatrix}$

10. $\begin{bmatrix} 2 & -4 & 0 & 0 \\ 1 & 2 & 12 & 0 \\ 0 & 0 & 2 & 0 \\ 0 & -1 & -4 & -5 \end{bmatrix} \begin{bmatrix} x_1 \\ x_2 \\ x_3 \\ x_4 \end{bmatrix} = \begin{bmatrix} 8 \\ 0 \\ 1 \\ 0 \end{bmatrix}$

11. Referring to the matrices in Exercise 3, use the given LU-decomposition to find A^{-1} by solving three appropriate linear systems.

12. Referring to the matrices in Exercise 4, use the given LU-decomposition to find A^{-1} by solving three appropriate linear systems.

In Exercises 13 and 14, find an LDU-decomposition of A.

13. $A = \begin{bmatrix} 2 & 1 & -1 \\ -2 & 0 & 2 \\ 2 & 2 & 1 \end{bmatrix}$ **14.** $A = \begin{bmatrix} -1 & -3 & -4 \\ 3 & 10 & -10 \\ -2 & -4 & 11 \end{bmatrix}$

In Exercises 15 and 16, determine whether A is a permutation matrix.

15. (a) $A = \begin{bmatrix} 0 & 1 \\ 1 & 0 \end{bmatrix}$ (b) $A = \begin{bmatrix} 0 & 1 & 0 \\ 1 & 0 & 1 \\ 0 & 0 & 1 \end{bmatrix}$

(c) $A = \begin{bmatrix} 0 & 0 & 1 & 0 \\ 0 & 1 & 0 & 0 \\ 0 & 0 & 0 & 1 \\ 1 & 0 & 0 & 0 \end{bmatrix}$

16. (a) $A = \begin{bmatrix} 0 & 1 \\ 0 & 1 \end{bmatrix}$ (b) $A = \begin{bmatrix} 0 & 0 & 1 \\ 0 & 1 & 0 \\ 1 & 0 & 0 \end{bmatrix}$

(c) $A = \begin{bmatrix} 0 & 0 & 1 & 0 \\ 0 & 1 & 1 & 0 \\ 0 & 0 & 0 & 1 \\ 1 & 0 & 0 & 0 \end{bmatrix}$

In Exercises 17 and 18, use the given PLU-decomposition of A to solve the linear system $A\mathbf{x} = \mathbf{b}$ by rewriting it as $P^{-1}A\mathbf{x} = P^{-1}\mathbf{b}$ and solving this system by LU-decomposition.

17. $\mathbf{b} = \begin{bmatrix} 2 \\ 1 \\ 5 \end{bmatrix}$; $A = \begin{bmatrix} 0 & 1 & 4 \\ 1 & 2 & 2 \\ 3 & 1 & 3 \end{bmatrix}$;

$A = \begin{bmatrix} 0 & 1 & 0 \\ 1 & 0 & 0 \\ 0 & 0 & 1 \end{bmatrix} \begin{bmatrix} 1 & 0 & 0 \\ 0 & 1 & 0 \\ 3 & -5 & 1 \end{bmatrix} \begin{bmatrix} 1 & 2 & 2 \\ 0 & 1 & 4 \\ 0 & 0 & 17 \end{bmatrix} = PLU$

18. $\mathbf{b} = \begin{bmatrix} 3 \\ 0 \\ 6 \end{bmatrix}$; $A = \begin{bmatrix} 4 & 1 & 2 \\ 0 & 2 & 1 \\ 8 & 1 & 8 \end{bmatrix}$;

$A = \begin{bmatrix} 1 & 0 & 0 \\ 0 & 0 & 1 \\ 0 & 1 & 0 \end{bmatrix} \begin{bmatrix} 1 & 0 & 0 \\ 2 & 1 & 0 \\ 0 & -2 & 1 \end{bmatrix} \begin{bmatrix} 4 & 1 & 2 \\ 0 & -1 & 4 \\ 0 & 0 & 9 \end{bmatrix} = PLU$

In Exercises 19 and 20, find a PLU-decomposition of A, and use it to solve the linear system $A\mathbf{x} = \mathbf{b}$ using the method of Exercises 17 and 18.

19. $A = \begin{bmatrix} 3 & -1 & 0 \\ 3 & -1 & 1 \\ 0 & 2 & 1 \end{bmatrix}$; $\mathbf{b} = \begin{bmatrix} -2 \\ 1 \\ 4 \end{bmatrix}$

20. $A = \begin{bmatrix} 0 & 3 & -2 \\ 1 & 1 & 4 \\ 2 & 2 & 5 \end{bmatrix}$; $\mathbf{b} = \begin{bmatrix} 7 \\ 5 \\ -2 \end{bmatrix}$

21. (a) Approximate the time required to execute the forward phase of Gauss–Jordan elimination for a system of 100,000 equations in 100,000 unknowns using a computer that can execute 1 gigaflop per second. Do the same for the backward phase. (See Table 3.7.1.)

(b) How many gigaflops per second must a computer be able to execute to find the LU-decomposition of a matrix of size $10{,}000 \times 10{,}000$ in less than 0.5 s? (See Table 3.7.1.)

22. Let $A = \begin{bmatrix} a & b \\ c & d \end{bmatrix}$.

(a) Show that if $a \neq 0$, then the matrix A has a unique LU-decomposition with 1's along the main diagonal of L.

(b) Find the LU-decomposition described in part (a).

Discussion and Discovery

D1. The first factor in the product

$$PA = \begin{bmatrix} 0 & 0 & 0 & 1 \\ 0 & 0 & 1 & 0 \\ 0 & 1 & 0 & 0 \\ 1 & 0 & 0 & 0 \end{bmatrix} \begin{bmatrix} 2 & 1 & 3 & 5 \\ 0 & 7 & 7 & 4 \\ -11 & 12 & 6 & 9 \\ 3 & -3 & 6 & -6 \end{bmatrix}$$

is a permutation matrix. Confirm this, and then use your observations to find PA by making appropriate row interchanges.

D2. If A is symmetric and invertible, what relationship exists between the factors L and U in the LDU-decomposition of A?

D3. Show that the following matrix A is invertible but has no LU-decomposition.

$$A = \begin{bmatrix} 0 & 1 \\ 1 & 0 \end{bmatrix}$$

Explain.

D4. Show that LU-decompositions are not unique by modifying the third diagonal entries of the matrices L and U appropriately in Example 1.

Technology Exercises

T1. (*LU-decomposition*) Technology utilities vary widely on how they handle LU-decompositions. For example, some

programs perform row interchanges to reduce roundoff error and hence produce a PLU-decomposition, even if an

LU-decomposition without row interchanges is possible. Determine how your utility handles *LU*-decompositions, and use it to find an *LU*- or *PLU*-decomposition of the matrix *A* in Example 1.

T2. (***Back and forward substitution***) *LU*-decomposition breaks up the process of solving linear systems into back and forward substitution. Some utilities have commands for solving linear systems with upper triangular coefficient matrices by back substitution, some have commands for solving linear systems with lower triangular coefficient matrices by forward substitution, and some have commands for using both back and forward substitution to solve linear systems whose coefficient matrices have previously been factored into *LU*- or *PLU*-form. Determine whether your utility has any or all of these capabilities, and experiment with them by solving the linear system in Example 1.

T3. Use *LU*- or *PLU*-decomposition (whichever is more convenient for your utility) to solve the linear systems $A\mathbf{x} = \mathbf{b}_1$, $A\mathbf{x} = \mathbf{b}_2$, and $A\mathbf{x} = \mathbf{b}_3$, where

$$A = \begin{bmatrix} 6 & 2 & -1 & 1 \\ 2 & 7 & 1 & -1 \\ 3 & -1 & 5 & 2 \\ 4 & 3 & 2 & -8 \end{bmatrix}; \quad \mathbf{b}_1 = \begin{bmatrix} 4 \\ 5 \\ 1 \\ 2 \end{bmatrix}$$

$$\mathbf{b}_2 = \begin{bmatrix} 0 \\ 4 \\ 2 \\ 3 \end{bmatrix}, \quad \mathbf{b}_3 = \begin{bmatrix} 6 \\ 7 \\ 8 \\ 4 \end{bmatrix}$$

T4. See what happens when you use your utility to find an *LU*-decomposition of a singular matrix.

Section 3.8   Partitioned Matrices and Parallel Processing

In earlier sections of this chapter we found it useful to subdivide a matrix into row vectors or column vectors. In this section we will consider other ways of subdividing matrices. This is typically done to isolate parts of a matrix that may be important in a particular problem or to break up a large matrix into smaller pieces that may be more manageable in large-scale computations.

GENERAL PARTITIONING   A matrix can be ***partitioned*** (subdivided) into ***submatrices*** (also called ***blocks***) in various ways by inserting lines between selected rows and columns. For example, a general 3×4 matrix *A* might be partitioned into four submatrices $A_{11}, A_{12}, A_{21},$ and A_{22} as

$$A = \left[\begin{array}{ccc|c} a_{11} & a_{12} & a_{13} & a_{14} \\ a_{21} & a_{22} & a_{23} & a_{24} \\ \hline a_{31} & a_{32} & a_{33} & a_{34} \end{array} \right] = \begin{bmatrix} A_{11} & A_{12} \\ A_{21} & A_{22} \end{bmatrix}$$

where

$$A_{11} = \begin{bmatrix} a_{11} & a_{12} & a_{13} \\ a_{21} & a_{22} & a_{23} \end{bmatrix}, \quad A_{12} = \begin{bmatrix} a_{14} \\ a_{24} \end{bmatrix}$$

$$A_{21} = [a_{31} \quad a_{32} \quad a_{33}], \quad A_{22} = [a_{34}]$$

With this partitioning *A* is being viewed as a 2×2 matrix whose entries are themselves matrices. Operations can be performed on partitioned matrices by treating the blocks as if they were numerical entries. For example, if

$$A = \begin{bmatrix} A_{11} & A_{12} \\ A_{21} & A_{22} \\ A_{31} & A_{32} \end{bmatrix} \quad \text{and} \quad B = \begin{bmatrix} B_{11} & B_{12} \\ B_{21} & B_{22} \end{bmatrix}$$

and if the sizes of the blocks conform for the required operations, then the block version of the row-column rule of Theorem 3.1.7 yields

$$AB = \begin{bmatrix} A_{11} & A_{12} \\ A_{21} & A_{22} \\ A_{31} & A_{32} \end{bmatrix} \begin{bmatrix} B_{11} & B_{12} \\ B_{21} & B_{22} \end{bmatrix} = \begin{bmatrix} A_{11}B_{11} + A_{12}B_{21} & A_{11}B_{12} + A_{12}B_{22} \\ A_{21}B_{11} + A_{22}B_{21} & A_{21}B_{12} + A_{22}B_{22} \\ A_{31}B_{11} + A_{32}B_{21} & A_{31}B_{12} + A_{32}B_{22} \end{bmatrix} \quad (1)$$

We call this procedure ***block multiplication***. Keep in mind, however, that for (1) to be a valid formula the partitioning of A and B must be such that the sizes of the blocks conform for the 12 products and 6 sums.

EXAMPLE 1
Block
Multiplication

The block version of the row-column rule for the product AB of the partitioned matrices

$$A = \begin{bmatrix} 3 & -4 & 1 & \vdots & 0 & 2 \\ -1 & 5 & -3 & \vdots & 1 & 4 \\ \cdots & \cdots & \cdots & \cdots & \cdots & \cdots \\ 2 & 0 & -2 & \vdots & 1 & 6 \end{bmatrix} = \begin{bmatrix} A_{11} & A_{12} \\ A_{21} & A_{22} \end{bmatrix}, \quad B = \begin{bmatrix} 2 & -1 \\ 3 & 0 \\ -5 & 1 \\ \cdots & \cdots \\ 4 & -3 \\ 0 & 2 \end{bmatrix} = \begin{bmatrix} B_{11} \\ B_{21} \end{bmatrix}$$

is

$$AB = \begin{bmatrix} A_{11} & A_{12} \\ A_{21} & A_{22} \end{bmatrix} \begin{bmatrix} B_{11} \\ B_{21} \end{bmatrix} = \begin{bmatrix} A_{11}B_{11} + A_{12}B_{21} \\ A_{21}B_{11} + A_{22}B_{21} \end{bmatrix}$$

This is a valid formula because the sizes of the blocks are such that all of the operations can be performed:

$$A_{11}B_{11} + A_{12}B_{21} = \begin{bmatrix} 3 & -4 & 1 \\ -1 & 5 & -3 \end{bmatrix} \begin{bmatrix} 2 & -1 \\ 3 & 0 \\ -5 & 1 \end{bmatrix} + \begin{bmatrix} 0 & 2 \\ 1 & 4 \end{bmatrix} \begin{bmatrix} 4 & -3 \\ 0 & 2 \end{bmatrix} = \begin{bmatrix} -11 & 2 \\ 32 & 3 \end{bmatrix}$$

$$A_{21}B_{11} + A_{22}B_{21} = \begin{bmatrix} 2 & 0 & -2 \end{bmatrix} \begin{bmatrix} 2 & -1 \\ 3 & 0 \\ -5 & 1 \end{bmatrix} + \begin{bmatrix} 1 & 6 \end{bmatrix} \begin{bmatrix} 4 & -3 \\ 0 & 2 \end{bmatrix} = \begin{bmatrix} 18 & 5 \end{bmatrix}$$

Thus,

$$AB = \begin{bmatrix} A_{11}B_{11} + A_{12}B_{21} \\ A_{21}B_{11} + A_{22}B_{21} \end{bmatrix} = \begin{bmatrix} -11 & 2 \\ 32 & 3 \\ \cdots & \cdots \\ 18 & 5 \end{bmatrix}$$

We leave it for you to confirm this result by performing the computation

$$AB = \begin{bmatrix} 3 & -4 & 1 & 0 & 2 \\ -1 & 5 & -3 & 1 & 4 \\ 2 & 0 & -2 & 1 & 6 \end{bmatrix} \begin{bmatrix} 2 & -1 \\ 3 & 0 \\ -5 & 1 \\ 4 & -3 \\ 0 & 2 \end{bmatrix} = \begin{bmatrix} -11 & 2 \\ 32 & 3 \\ 18 & 5 \end{bmatrix}$$

without partitioning.   ∎

CONCEPT PROBLEM   Devise a different way of partitioning the matrices A and B in this example that will allow you to compute AB by block multiplication. Perform the computations using your partitioning.

The following special case of block multiplication, which we will call the ***column-row rule*** for block multiplication, is particularly useful (compare to the row-column rule for matrix multiplication in Theorem 3.1.7).

> **Theorem 3.8.1** (***Column-Row Rule***) *If A has size $m \times s$ and B has size $s \times n$, and if these matrices are partitioned into column and row vectors as*
>
> $$A = \begin{bmatrix} \mathbf{c}_1 & \mathbf{c}_2 & \cdots & \mathbf{c}_s \end{bmatrix} \quad and \quad B = \begin{bmatrix} \mathbf{r}_1 \\ \mathbf{r}_2 \\ \vdots \\ \mathbf{r}_s \end{bmatrix}$$
>
> *then*
> $$AB = \mathbf{c}_1\mathbf{r}_1 + \mathbf{c}_2\mathbf{r}_2 + \cdots + \mathbf{c}_s\mathbf{r}_s \tag{2}$$

REMARK Formula (2) is sometimes called the ***outer product rule*** because it expresses AB as a sum of column vectors times row vectors (outer products). This formula is more important for theoretical analyses than for numerical computations, and we will use it many times as we progress through the text. Some computations involving this formula are given as exercises.

EXAMPLE 2
$\text{tr}(AB) = \text{tr}(BA)$

Here is the proof of Theorem 3.2.12(e) that we promised in Section 3.2. Assume that A and B are $m \times m$ matrices and that they are partitioned as in Theorem 3.8.1. Then it follows from the definition of the trace that

$$\text{tr}(BA) = \mathbf{r}_1\mathbf{c}_1 + \mathbf{r}_2\mathbf{c}_2 + \cdots + \mathbf{r}_s\mathbf{c}_s \qquad \textbf{[Row-column rule for matrix multiplication]}$$

$$= \text{tr}(\mathbf{c}_1\mathbf{r}_1) + \text{tr}(\mathbf{c}_2\mathbf{r}_2) + \cdots + \text{tr}(\mathbf{c}_s\mathbf{r}_s) \qquad \textbf{[Theorem 3.2.13]}$$

$$= \text{tr}(\mathbf{c}_1\mathbf{r}_1 + \mathbf{c}_2\mathbf{r}_2 + \cdots + \mathbf{c}_s\mathbf{r}_s) \qquad \textbf{[Theorem 3.2.12(c)]}$$

$$= \text{tr}(AB) \qquad \textbf{[Theorem 3.8.1]}$$ ∎

BLOCK DIAGONAL MATRICES

A partitioned matrix A is said to be ***block diagonal*** if the matrices on the main diagonal are square and all matrices off the main diagonal are zero; that is, the matrix is partitioned as

$$A = \begin{bmatrix} D_1 & 0 & \cdots & 0 \\ 0 & D_2 & \cdots & 0 \\ \vdots & \vdots & & \vdots \\ 0 & 0 & \cdots & D_k \end{bmatrix} \tag{3}$$

where the matrices $D_1, D_2, \ldots, D_k$ are square. It can be shown that the matrix A in (3) is invertible if and only if each matrix on the diagonal is invertible, in which case

$$A^{-1} = \begin{bmatrix} D_1^{-1} & 0 & \cdots & 0 \\ 0 & D_2^{-1} & \cdots & 0 \\ \vdots & \vdots & & \vdots \\ 0 & 0 & \cdots & D_k^{-1} \end{bmatrix} \tag{4}$$

EXAMPLE 3
Inverting a Block Diagonal Matrix

Consider the block diagonal matrix

$$A = \begin{bmatrix} 8 & -7 & 0 & 0 & 0 \\ 1 & -1 & 0 & 0 & 0 \\ 0 & 0 & 3 & 1 & 0 \\ 0 & 0 & 5 & 2 & 0 \\ 0 & 0 & 0 & 0 & 4 \end{bmatrix}$$

There are three matrices on the main diagonal—two 2×2 matrices and one 1×1 matrix. The 2×2 matrices are invertible by Theorem 3.2.7 because they have nonzero determinants (verify), and the 1×1 matrix is invertible because it is nonzero (its inverse is $\frac{1}{4}$). We leave it for you to apply Theorem 3.2.7 to invert the 2×2 matrices and show that

$$A^{-1} = \begin{bmatrix} 1 & -7 & 0 & 0 & 0 \\ 1 & -8 & 0 & 0 & 0 \\ 0 & 0 & 2 & -1 & 0 \\ 0 & 0 & -5 & 3 & 0 \\ 0 & 0 & 0 & 0 & \frac{1}{4} \end{bmatrix}$$ ∎

BLOCK UPPER TRIANGULAR MATRICES

A partitioned square matrix A is said to be ***block upper triangular*** if the matrices on the main diagonal are square and all matrices below the main diagonal are zero; that is, the matrix is partitioned as

$$
A = \begin{bmatrix} A_{11} & A_{12} & \cdots & A_{1k} \\ 0 & A_{22} & \cdots & A_{2k} \\ \vdots & \vdots & & \vdots \\ 0 & 0 & \cdots & A_{kk} \end{bmatrix} \tag{5}
$$

where the matrices $A_{11}, A_{22}, \ldots, A_{kk}$ are square. The definition of a ***block lower triangular matrix*** is similar.

Many computer algorithms for operating on large matrices exploit block structures to break the computations down into smaller pieces. For example, consider a block upper triangular matrix of the form

$$
A = \begin{bmatrix} A_{11} & A_{12} \\ 0 & A_{22} \end{bmatrix}
$$

in which A_{11} and A_{22} are square matrices. In the exercises we will ask you to show that if A_{11} and A_{22} are invertible, then the matrix A is invertible and

$$
A^{-1} = \begin{bmatrix} A_{11} & A_{12} \\ 0 & A_{22} \end{bmatrix}^{-1} = \begin{bmatrix} A_{11}^{-1} & -A_{11}^{-1} A_{12} A_{22}^{-1} \\ 0 & A_{22}^{-1} \end{bmatrix} \tag{6}
$$

This formula allows the work of inverting A to be accomplished by ***parallel processing***, that is, by using two individual processors working *simultaneously* to compute the inverses of the smaller matrices, A_{11} and A_{22}. Once the smaller matrices are inverted, the results can be combined using Formula (6) to construct the inverse of A. Parallel processing is not only fast, since many of the computations are performed simultaneously, but sometimes additional speed can also be gained by inverting the smaller matrices in high-speed memory that might not be large enough to accommodate the entire matrix A.

EXAMPLE 4   Confirm that

$$
A = \begin{bmatrix} 4 & 7 & -5 & 3 \\ 3 & 5 & 3 & -2 \\ 0 & 0 & 7 & 2 \\ 0 & 0 & 3 & 1 \end{bmatrix}
$$

is an invertible block upper triangular matrix, and then find its inverse by using Formula (6).

Solution   The matrix is block upper triangular because it can be partitioned into form (5) as

$$
A = \left[\begin{array}{cc:cc} 4 & 7 & 5 & 3 \\ 3 & 5 & 3 & -2 \\ \hdashline 0 & 0 & 7 & 2 \\ 0 & 0 & 3 & 1 \end{array} \right] = \begin{bmatrix} A_{11} & A_{12} \\ 0 & A_{22} \end{bmatrix}
$$

where

$$
A_{11} = \begin{bmatrix} 4 & 7 \\ 3 & 5 \end{bmatrix}, \quad A_{12} = \begin{bmatrix} -5 & 3 \\ 3 & -2 \end{bmatrix}, \quad A_{22} = \begin{bmatrix} 7 & 2 \\ 3 & 1 \end{bmatrix}
$$

It follows from Theorem 3.2.7 that A_{11} and A_{22} are invertible and their inverses are

$$
A_{11}^{-1} = \begin{bmatrix} -5 & 7 \\ 3 & -4 \end{bmatrix} \quad \text{and} \quad A_{22}^{-1} = \begin{bmatrix} 1 & -2 \\ -3 & 7 \end{bmatrix}
$$

Linear Algebra in History

In 1990 a cooperative international effort, called ***The Human Genome Project***, was undertaken to identify the roughly 100,000 genes in human DNA and to determine the sequences of the 3 billion chemical base pairs that make up human DNA. The project is due for completion by the end of 2003, two years ahead of the original schedule, because of advances in technology. The first part of the project focused on optimizing computational methods to increase the DNA mapping and sequencing efficiency by 10- to 20-fold. Parallel processing played a major role in that aspect of the project. By parallel processing it is possible to achieve performances in the petaflop per second range (1 petaflop = 10^{15} flops). A working draft of the human sequence was produced in the year 2000; it contains the functional blueprint and evolutionary history of the human species.

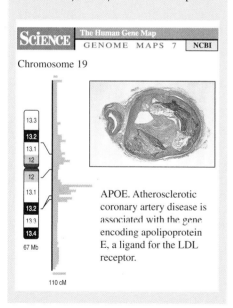

Chromosome 19

APOE. Atherosclerotic coronary artery disease is associated with the gene encoding apolipoprotein E, a ligand for the LDL receptor.

Moreover,

$$-A_{11}^{-1}A_{12}A_{22}^{-1} = -\begin{bmatrix} -5 & 7 \\ 3 & -4 \end{bmatrix}\begin{bmatrix} -5 & 3 \\ 3 & -2 \end{bmatrix}\begin{bmatrix} 1 & -2 \\ -3 & 7 \end{bmatrix} = \begin{bmatrix} -133 & 295 \\ 78 & -173 \end{bmatrix}$$

so it follows from (6) that the inverse of A is

$$A^{-1} = \begin{bmatrix} -5 & 7 & -133 & 295 \\ 3 & -4 & 78 & -173 \\ 0 & 0 & 1 & -2 \\ 0 & 0 & -3 & 7 \end{bmatrix}$$

■

Exercise Set 3.8

In Exercises 1 and 2, determine whether the product

$$\begin{bmatrix} 2 & 1 & -1 \\ 3 & 0 & 5 \\ 4 & 2 & 1 \end{bmatrix}\begin{bmatrix} 1 & -2 & -1 \\ 3 & 0 & -2 \\ -1 & 3 & 2 \end{bmatrix} = \begin{bmatrix} 6 & -7 & -6 \\ -2 & 9 & 7 \\ 9 & -5 & -6 \end{bmatrix}$$

can be computed using the indicated partitioning. If so, use it to compute the product by block multiplication.

1. (a) $\begin{bmatrix} 2 & 1 & -1 \\ 3 & 0 & 5 \\ \hline 4 & 2 & 1 \end{bmatrix}\begin{bmatrix} 1 & -2 & -1 \\ 3 & 0 & -2 \\ \hline -1 & 3 & 2 \end{bmatrix}$

(b) $\begin{bmatrix} 2 & 1 & -1 \\ \hline 3 & 0 & 5 \\ 4 & 2 & 1 \end{bmatrix}\begin{bmatrix} 1 & -2 & -1 \\ 3 & 0 & -2 \\ \hline -1 & 3 & 2 \end{bmatrix}$

(c) $\begin{bmatrix} 2 & 1 & -1 \\ \hline 3 & 0 & 5 \\ 4 & 2 & 1 \end{bmatrix}\begin{bmatrix} 1 & -2 & -1 \\ 3 & 0 & -2 \\ -1 & 3 & 2 \end{bmatrix}$

(d) $\begin{bmatrix} 2 & 1 & -1 \\ \hline 3 & 0 & 5 \\ 4 & 2 & 1 \end{bmatrix}\begin{bmatrix} 1 & -2 & -1 \\ 3 & 0 & -2 \\ -1 & 3 & 2 \end{bmatrix}$

2. (a) $\begin{bmatrix} 2 & 1 & -1 \\ 3 & 0 & 5 \\ 4 & 2 & 1 \end{bmatrix}\begin{bmatrix} 1 & -2 & -1 \\ 3 & 0 & -2 \\ -1 & 3 & 2 \end{bmatrix}$

(b) $\begin{bmatrix} 2 & 1 & -1 \\ \hline 3 & 0 & 5 \\ 4 & 2 & 1 \end{bmatrix}\begin{bmatrix} 1 & -2 & -1 \\ 3 & 0 & -2 \\ -1 & 3 & 2 \end{bmatrix}$

(c) $\begin{bmatrix} 2 & 1 & -1 \\ \hline 3 & 0 & 5 \\ 4 & 2 & 1 \end{bmatrix}\begin{bmatrix} 1 & -2 & -1 \\ 3 & 0 & -2 \\ -1 & 3 & 2 \end{bmatrix}$

(d) $\begin{bmatrix} 2 & 1 & -1 \\ 3 & 0 & 5 \\ 4 & 2 & 1 \end{bmatrix}\begin{bmatrix} 1 & -2 & -1 \\ 3 & 0 & -2 \\ -1 & 3 & 2 \end{bmatrix}$

In Exercises 3–6, determine whether block multiplication can be used to compute the product using the partitions shown. If so, compute the product by block multiplication.

3. (a) $\begin{bmatrix} -1 & 2 & 1 & 5 \\ 0 & -3 & 4 & 2 \\ \hline 1 & 5 & 6 & 1 \end{bmatrix}\begin{bmatrix} -2 & 1 & 4 \\ -3 & 5 & 2 \\ \hline 7 & -1 & 5 \\ 0 & 3 & -3 \end{bmatrix}$

(b) $\begin{bmatrix} -1 & 2 & 1 & 5 \\ \hline 0 & -3 & 4 & 2 \\ 1 & 5 & 6 & 1 \end{bmatrix}\begin{bmatrix} -2 & 1 & 4 \\ -3 & 5 & 2 \\ \hline 7 & -1 & 5 \\ 0 & 3 & -3 \end{bmatrix}$

4. (a) $\begin{bmatrix} 2 & 0 & 5 & 5 \\ \hline 0 & 2 & 3 & -2 \\ -1 & 1 & 4 & 4 \end{bmatrix}\begin{bmatrix} 6 & 3 & -2 \\ 2 & 4 & -2 \\ \hline 0 & 2 & 1 \\ 0 & 3 & -1 \end{bmatrix}$

(b) $\begin{bmatrix} 2 & 0 & 5 & 5 \\ 0 & 2 & 3 & -2 \\ -1 & 1 & 4 & 4 \end{bmatrix}\begin{bmatrix} 6 & 3 & -2 \\ 2 & 4 & -2 \\ \hline 0 & 2 & 1 \\ 0 & 3 & -1 \end{bmatrix}$

5. (a) $\begin{bmatrix} 3 & -1 & 0 & -3 \\ 2 & 1 & 4 & 5 \end{bmatrix}\begin{bmatrix} 2 & -4 & 1 \\ 3 & 0 & 2 \\ \hline 1 & -3 & 5 \\ 2 & 1 & 4 \end{bmatrix}$

(b) $\begin{bmatrix} 2 & -5 \\ 1 & 3 \\ 0 & 5 \\ \hline 1 & 4 \end{bmatrix}\begin{bmatrix} 2 & -1 & 3 & -4 \\ 0 & 1 & 5 & 7 \end{bmatrix}$

6. (a) $\begin{bmatrix} 1 & 0 & 0 & 0 & 0 \\ 0 & 1 & 0 & 0 & 0 \\ 0 & 0 & 1 & 0 & 0 \\ \hline 0 & 0 & 0 & 2 & 0 \\ 0 & 0 & 0 & -1 & 2 \end{bmatrix}\begin{bmatrix} 3 & 3 \\ -1 & 4 \\ 1 & 5 \\ 2 & -2 \\ 1 & 6 \end{bmatrix}$

(b) $\begin{bmatrix} 1 & \vdots & 2 & 0 & 0 \\ \hline -1 & \vdots & 3 & 0 & 0 \\ 0 & \vdots & 0 & 4 & 2 \\ 0 & \vdots & 0 & 2 & 1 \end{bmatrix} \begin{bmatrix} 0 & 0 & \vdots & 2 & 1 \\ 0 & 0 & \vdots & 5 & 3 \\ \hline -2 & -2 & \vdots & 0 & 0 \\ 1 & 3 & \vdots & 0 & 0 \end{bmatrix}$

In Exercises 7 and 8, compute the product using the column-row rule stated in Theorem 3.8.1, and check your answer by calculating the product directly.

7. $\begin{bmatrix} 1 & 2 \\ 3 & -5 \end{bmatrix} \begin{bmatrix} -1 & 2 \\ -4 & 5 \end{bmatrix}$ **8.** $\begin{bmatrix} 3 & 1 \\ 2 & -4 \end{bmatrix} \begin{bmatrix} 1 & 2 & 0 \\ -1 & 6 & 2 \end{bmatrix}$

In Exercises 9 and 10, find the inverse of the block diagonal matrix A using the method of Example 3.

9. (a) $A = \begin{bmatrix} 2 & 1 & 0 & 0 \\ 3 & 2 & 0 & 0 \\ 0 & 0 & 3 & 4 \\ 0 & 0 & 1 & -1 \end{bmatrix}$

(b) $A = \begin{bmatrix} 5 & 2 & 0 & 0 & 0 \\ 3 & 1 & 0 & 0 & 0 \\ 0 & 0 & 5 & 0 & 0 \\ 0 & 0 & 0 & 2 & 7 \\ 0 & 0 & 0 & 1 & 4 \end{bmatrix}$

10. (a) $A = \begin{bmatrix} 5 & 1 & 0 & 0 \\ 4 & 1 & 0 & 0 \\ 0 & 0 & 2 & -3 \\ 0 & 0 & -3 & 5 \end{bmatrix}$

(b) $A = \begin{bmatrix} 2 & 0 & 0 & 0 & 0 \\ 0 & 1 & 2 & 0 & 0 \\ 0 & 3 & 7 & 0 & 0 \\ 0 & 0 & 0 & 4 & 9 \\ 0 & 0 & 0 & 1 & 2 \end{bmatrix}$

In Exercises 11 and 12, use the method of Example 4 to find the inverse of the block upper triangular matrix A.

11. $A = \begin{bmatrix} 2 & 1 & 3 & -6 \\ 1 & 1 & 7 & 4 \\ 0 & 0 & 3 & 5 \\ 0 & 0 & 2 & 3 \end{bmatrix}$

12. $A = \begin{bmatrix} -1 & -1 & 2 & 5 \\ 2 & 1 & -3 & 8 \\ 0 & 0 & 4 & 1 \\ 0 & 0 & 7 & 2 \end{bmatrix}$

In Exercises 13 and 14, let M and N be partitioned matrices in which A and B are 2×2 matrices with A being invertible. Write a formula for the partitioned matrix MN in terms of A and B.

13. $M = \begin{bmatrix} A & B \\ 0 & A \end{bmatrix}$ and $N = \begin{bmatrix} A^{-1} & B \\ 0 & A \end{bmatrix}$

14. $M = \begin{bmatrix} A & A^{-1} \\ 0 & B \end{bmatrix}$ and $N = \begin{bmatrix} A & B \\ B & 0 \end{bmatrix}$

15. Find B_1, given that

$$\begin{bmatrix} A_1 & B_1 \\ 0 & C_1 \end{bmatrix} \begin{bmatrix} A_2 & B_2 \\ 0 & C_2 \end{bmatrix} = \begin{bmatrix} A_3 & B_3 \\ 0 & C_3 \end{bmatrix}$$

and

$$A_1 = \begin{bmatrix} 2 & 0 \\ 0 & 1 \end{bmatrix}, \quad B_2 = \begin{bmatrix} 1 & 1 \\ 1 & 2 \end{bmatrix}, \quad C_1 = \begin{bmatrix} 1 & 1 \\ 1 & -1 \end{bmatrix}$$

$$A_2 = \begin{bmatrix} 2 & 0 \\ 0 & 2 \end{bmatrix}, \quad C_2 = \begin{bmatrix} 1 & 0 \\ 0 & 2 \end{bmatrix}, \quad B_3 = \begin{bmatrix} 2 & 1 \\ 1 & 3 \end{bmatrix}$$

16. Given that I_k is the $k \times k$ identity matrix, that A, B, X, Y, and Z are $k \times k$ matrices, and that

$$\begin{bmatrix} A & B \\ 0 & I_k \end{bmatrix} \begin{bmatrix} X & Y & 3Z \\ 0 & 0 & I_k \end{bmatrix} = \begin{bmatrix} I_k & 0 & 0 \\ 0 & 0 & I_k \end{bmatrix}$$

find formulas for X, Y, and Z in terms of A and B.

17. Consider the partitioned linear system

$$\begin{bmatrix} 5 & 2 & \vdots & 2 & 3 \\ 2 & 1 & \vdots & -3 & 1 \\ \hline 1 & 0 & \vdots & 4 & 1 \\ 0 & 1 & \vdots & 0 & 2 \end{bmatrix} \begin{bmatrix} x_1 \\ x_2 \\ x_3 \\ x_4 \end{bmatrix} = \begin{bmatrix} 2 \\ 6 \\ 0 \\ 0 \end{bmatrix}$$

Solve this system by first expressing it as

$$\begin{bmatrix} A & B \\ I & D \end{bmatrix} \begin{bmatrix} \mathbf{u} \\ \mathbf{v} \end{bmatrix} = \begin{bmatrix} \mathbf{b} \\ 0 \end{bmatrix} \quad \text{or equivalently,} \quad \begin{matrix} A\mathbf{u} + B\mathbf{v} = \mathbf{b} \\ \mathbf{u} + D\mathbf{v} = 0 \end{matrix}$$

next solving the second equation for $\mathbf{u}$ in terms of $\mathbf{v}$, and then substituting in the first equation. Check your answer by solving the system directly.

18. Express the product

$$\begin{bmatrix} A_1 & A_2 \\ A_3 & A_4 \end{bmatrix} \begin{bmatrix} I & 0 \\ 0 & B \end{bmatrix}$$

of partitioned matrices as a single partitioned matrix in terms of A_1, A_2, A_3, A_4, and B. (Assume that A_1 and A_3 have as many columns as I has rows and that A_2 and A_4 have as many columns as B has rows.)

Discussion and Discovery

D1. Suppose that $A^2 = I_m$, $B^2 = I_k$, and

$$M = \begin{bmatrix} A & 0 \\ 0 & B \end{bmatrix}$$

What can you say about M^2?

D2. (a) Let A be a square matrix that can be partitioned as

$$A = \begin{bmatrix} A_{11} & A_{12} \\ 0 & A_{22} \end{bmatrix}$$

in which A_{11} and A_{22} are invertible. Confirm that A is invertible and that its inverse is given by Formula (6)

by computing the product

$$\begin{bmatrix} A_{11} & A_{12} \\ 0 & A_{22} \end{bmatrix} \begin{bmatrix} A_{11}^{-1} & -A_{11}^{-1} A_{12} A_{22}^{-1} \\ 0 & A_{22}^{-1} \end{bmatrix}$$

(b) Show that

$$A = \begin{bmatrix} A_{11} & 0 \\ A_{21} & A_{22} \end{bmatrix}$$

is also invertible by discovering a formula for its inverse that is analogous to Formula (6).

Working with Proofs

P1. Let M be a matrix that is partitioned as

$$M = \begin{bmatrix} A & B \end{bmatrix}$$

in which A is invertible. Suppose that there is a sequence of elementary row operations that reduces A to I. Prove that if that sequence of row operations is applied to M, then the

resulting matrix will be

$$M' = \begin{bmatrix} I & A^{-1}B \end{bmatrix}$$

[*Hint:* A can be expressed as a product of elementary matrices.]

Technology Exercises

T1. (***Extracting submatrices***) Many technology utilities provide methods for extracting rows, columns, and other submatrices of a given matrix. Determine whether your utility has this feature, and if so, extract the row vectors, column vectors, and the four submatrices of A that are indicated by the partitioning:

$$A = \begin{bmatrix} 1 & 2 & 3 & 4 \\ -3 & 9 & -6 & 12 \\ 7 & 6 & -5 & 2 \\ \hline 0 & 2 & -2 & 3 \end{bmatrix}$$

T2. (***Constructing matrices from submatrices***) Many technology utilities provide methods for building up new matrices from a set of specified matrices. Determine whether your utility provides for this, and if so, do the following:

(a) Have your utility construct the matrix A in Exercise T1 from the row vectors of A.

(b) Have your utility construct the matrix A in Exercise T1 from the column vectors of A.

(c) Have your utility construct the matrix A in Exercise T1 from the submatrices A_{11}, A_{12}, A_{21}, and A_{22} indicated by the partitioning in Exercise T1.

T3. (***Constructing block diagonal matrices***) Many technology utilities provide methods for constructing block diagonal matrices from a set of specified matrices. Determine

whether your utility has this feature, and if so, use it to construct the block diagonal matrix

$$\begin{bmatrix} A & 0 \\ 0 & A^2 \end{bmatrix}$$

from the matrix in Exercise T1.

T4. Compute the product

$$AB = \begin{bmatrix} 1 & 2 & 3 & 4 \\ 0 & 2 & -1 & 6 \\ 5 & 0 & 3 & 1 \\ -7 & 1 & 3 & 2 \end{bmatrix} \begin{bmatrix} 3 & 3 & -4 & -5 \\ 1 & 0 & 2 & 3 \\ 0 & 1 & 4 & 5 \\ 4 & -4 & -1 & 0 \end{bmatrix}$$

directly and by using the column-row rule of Theorem 3.8.1.

T5. Let A be the 9×9 block diagonal matrix whose successive diagonal blocks are

$$D_1 = \begin{bmatrix} -2 & 3 & 4 \\ 4 & -3 & -3 \\ 4 & -1 & 0 \end{bmatrix}, \quad D_2 = \begin{bmatrix} 0 & 1 & 2 \\ 1 & 0 & 3 \\ 4 & -3 & 8 \end{bmatrix}$$

$$D_3 = \begin{bmatrix} -1 & 0 & -5 \\ 1 & 1 & 0 \\ 3 & 2 & 6 \end{bmatrix}$$

Find the inverse of A using Formula (4), and check your result by constructing the matrix A and finding its inverse directly.

T6. Referring to the matrices in Exercise T5, use Formula (6) to find the inverse of the 6×6 matrix

$$A = \begin{bmatrix} D_1 & D_2 \\ 0 & D_3 \end{bmatrix}$$

and check your result by finding the inverse directly.

T7. If A is an $n \times n$ matrix, $\mathbf{u}$ and $\mathbf{v}$ are $n \times 1$ column vectors, and q is a scalar, then the $(n+1) \times (n+1)$ matrix

$$B = \begin{bmatrix} A & \mathbf{u} \\ \mathbf{v}^T & q \end{bmatrix}$$

is said to result by ***bordering*** A with $\mathbf{u}$, $\mathbf{v}$, and q. Border the matrix A in Exercise T1 with

$$\mathbf{u} = \begin{bmatrix} 1 \\ 0 \\ 5 \\ -2 \end{bmatrix}, \quad \mathbf{v} = \begin{bmatrix} 3 \\ 6 \\ -1 \\ 4 \end{bmatrix}, \quad q = 8$$

T8. In many applications it is important to know the effect on the inverse of an invertible matrix A of changing a single entry in A. To explain this, suppose that $A = [a_{ij}]$ is an invertible $n \times n$ matrix whose inverse $A^{-1} = [\gamma_{ij}]$ has column vectors $\mathbf{c}_1, \mathbf{c}_2, \ldots, \mathbf{c}_n$ and row vectors $\mathbf{r}_1, \mathbf{r}_2, \ldots, \mathbf{r}_n$. It can be shown that if a constant λ is added to the entry a_{ij} of A to obtain a matrix B, and if $\lambda \neq -1/\lambda_{ji}$, then the matrix B is invertible, and

$$B^{-1} = A^{-1} - \left(\frac{\lambda}{1 + \lambda \gamma_{ji}} \right) \mathbf{c}_i \mathbf{r}_j$$

Consider the matrices

$$A = \begin{bmatrix} 1 & -1 & 1 \\ 0 & 2 & -1 \\ 2 & 3 & 0 \end{bmatrix} \quad \text{and} \quad B = \begin{bmatrix} 1 & -1 & 1 \\ 0 & 2 & -1 \\ 2+\gamma & 3 & 0 \end{bmatrix}$$

(a) Find A^{-1} and B^{-1} for $\lambda = 2$ directly.

(b) Extract the appropriate row and column vectors from A^{-1}, and use them to find B^{-1} using the formula stated above. Confirm that your result is consistent with part (a).

(c) Suppose that an $n \times n$ electronic grid of indicators displays a rectangular array of numbers that forms a matrix $A(t)$ at time t, and suppose that the indicators change one entry at a time at times $t_1, t_2, t_3, \ldots$ in such a way that the matrices $A(t_1), A(t_2), A(t_3), \ldots$ are invertible. Compare the number of flops required to compute the inverse of $A(t_k)$ from the inverse of $A(t_{k-1})$ using the formula stated above as opposed to computing it by row reduction. [*Note:* Assume that n is large and see Table 3.7.1.]

Determinants are important in geometry and the theory of linear algebra. They also provide a way of distinguishing left-handedness from right-handedness in higher dimensions.

Determinants

Section 4.1 Determinants; Cofactor Expansion

In Section 3.2 we introduced the concept of a determinant as a convenience for writing a general formula for the inverse of a 2×2 invertible matrix. In this section we will extend the concept of a determinant in a way that will eventually produce formulas for inverses of invertible matrices of higher order as well as formulas for the solutions of certain linear systems.

DETERMINANTS OF 2×2 AND 3×3 MATRICES

Recall from Section 3.2 that the determinant of a 2×2 matrix

$$A = \begin{bmatrix} a & b \\ c & d \end{bmatrix}$$

is defined to be the product of the entries on the main diagonal minus the product of the entries off the main diagonal; that is, $\det(A) = ad - bc$. Alternatively, the determinant can be written as

$$\begin{vmatrix} a & b \\ c & d \end{vmatrix} = ad - bc \tag{1}$$

Historically, determinants first arose in the context of solving systems of linear equations for one set of variables in terms of another. For example, in Example 7 of Section 3.2 we showed that if the coefficient matrix of the system

$$u = ax + by$$
$$v = cx + dy$$

is invertible, then the equations can be solved for x and y in terms u and v as

$$x = \frac{du - bv}{ad - bc}, \quad y = \frac{av - cu}{ad - bc}$$

which we can write in determinant notation as

$$x = \frac{\begin{vmatrix} u & b \\ v & d \end{vmatrix}}{\begin{vmatrix} a & b \\ c & d \end{vmatrix}}, \quad y = \frac{\begin{vmatrix} a & u \\ c & v \end{vmatrix}}{\begin{vmatrix} a & b \\ c & d \end{vmatrix}} \tag{2}$$

(verify). In the late seventeenth and early eighteenth centuries these formulas were extended to higher-order systems by laboriously solving the systems directly and then searching for common patterns in the solutions. Once those patterns were discovered, they were used to define higher-order determinants in a way that would allow the solutions of the higher-order systems to be expressed as ratios of determinants, just as in (2). We will now discuss those definitions.

To extend the definition of $\det(A)$ to matrices of higher order, it will be helpful to use subscripted entries for A, in which case Formula (1) becomes

$$\det(A) = \begin{vmatrix} a_{11} & a_{12} \\ a_{21} & a_{22} \end{vmatrix} = a_{11}a_{22} - a_{12}a_{21} \tag{3}$$

This is called a **2 × 2 determinant**.

The determinant of a 3×3 matrix A, also called a **3 × 3 determinant**, is defined by the formula

$$\det(A) = \begin{vmatrix} a_{11} & a_{12} & a_{13} \\ a_{21} & a_{22} & a_{23} \\ a_{31} & a_{32} & a_{33} \end{vmatrix} = \begin{array}{l} a_{11}a_{22}a_{33} + a_{12}a_{23}a_{31} + a_{13}a_{21}a_{32} \\ - a_{13}a_{22}a_{31} - a_{12}a_{21}a_{33} - a_{11}a_{23}a_{32} \end{array} \tag{4}$$

Although this formula seems to be pulled out of thin air, it is devised so the solution of the system

$$u = a_{11}x + a_{12}y + a_{13}z$$
$$v = a_{21}x + a_{22}y + a_{23}z$$
$$w = a_{31}x + a_{32}y + a_{33}z$$

for x, y, and z in terms of u, v, and w can be expressed as a ratio of appropriate 3×3 determinants when the coefficient matrix is invertible. We will show that this is so later.

Linear Algebra in History

The term *determinant* was first introduced by the German mathematician Carl Friedrich Gauss in 1801 (see p. 54), who used them to "determine" properties of certain kinds of functions. Interestingly, the term *matrix* is derived from a Latin word for "womb" because it was viewed as a container of determinants.

There is no need to memorize Formulas (3) and (4), since they can both be obtained using the diagrams in Figure 4.1.1. As indicated in the left part of the figure, Formula (3) can be obtained by subtracting the product of the entries on the leftward arrow from the product of the entries on the rightward arrow; and as shown in the right part of the figure, Formula (4) can be obtained by duplicating the first two columns of the matrix, as illustrated, and then subtracting the sum of the products along the leftward arrows from the sum of the products along the rightward arrows (verify).

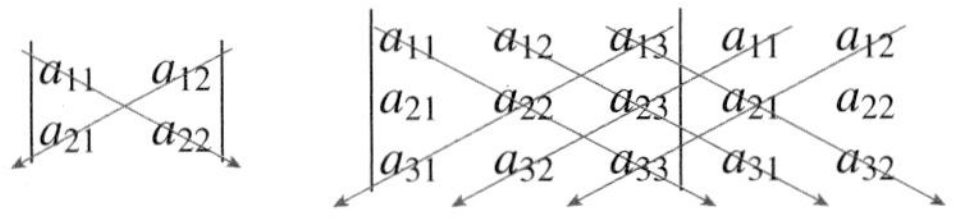

Figure 4.1.1

EXAMPLE 1
Evaluating
Determinants

$$\begin{vmatrix} 3 & 1 \\ 4 & -2 \end{vmatrix} = (3)(-2) - (1)(4) = -10$$

$$\begin{vmatrix} 1 & 2 & 3 \\ -4 & 5 & 6 \\ 7 & -8 & 9 \end{vmatrix} = \begin{array}{ccccc} 1 & 2 & 3 & 1 & 2 \\ -4 & 5 & 6 & -4 & 5 \\ 7 & -8 & 9 & 7 & -8 \end{array}$$

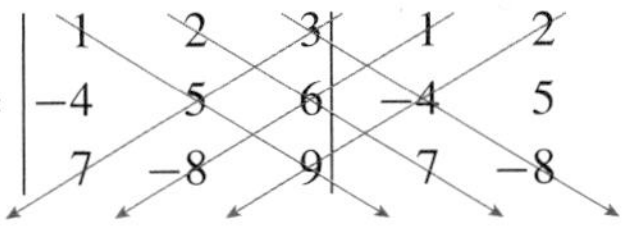

$$= [45 + 84 + 96] - [105 - 48 - 72] = 240 \qquad ■$$

ELEMENTARY PRODUCTS   To extend the definition of a determinant to general $n \times n$ matrices, it will be helpful to examine the structure of Formulas (3) and (4) in more detail. In both formulas the determinant is the sum of products, each containing exactly one entry from each row and one entry from each column of the matrix. Half of these products are preceded by a plus sign (not shown explicitly in the first term) and half by a minus sign. The products are called the **elementary products** of the matrix, and the elementary products with their associated + or − signs are called the **signed elementary products**. In Formulas (3) and (4), the row indices of the elementary products are in numerical order, but the column indices are mixed up. For example, in Formula (4) each elementary product is of the form

$$a_1\_a_2\_a_3\_$$

where the blanks contain some permutation of the column indices $\{1, 2, 3\}$. By filling in the blanks with all possible permutations of the column indices, you can obtain all six elementary products that appear in Formula (4) (verify). Similarly, the two elementary products that appear in Formula (3) can be obtained by filling in the blanks of

$$a_1\_a_2\_$$

with the two possible permutations of the column indices $\{1, 2\}$.

Although this may not be evident, the signs that precede the elementary products in Formulas (3) and (4) are related to the permutations of the column indices. More precisely, in each signed elementary product the sign can be determined by counting the minimum number of interchanges in the permutation of the column indices required to put those indices into their natural order: the sign is $+$ if the number is even and $-$ if it is odd. For example, in the formula

$$\det(A) = a_{11}a_{22} - a_{12}a_{21}$$

for a 2×2 determinant, the elementary product $a_{11}a_{22}$ takes a plus because the permutation $\{1, 2\}$ of its column indices is already in natural order (so the minimum number of interchanges required to put the indices in natural order is 0, which is an even integer). Similarly, the elementary product $a_{12}a_{21}$ takes a minus because the permutation $\{2, 1\}$ of the column indices requires 1 interchange to put them in natural order.

Finding the signs associated with the signed elementary products of a 3×3 determinant is also not difficult. A typical permutation of the column indices has the form

$$\{j_1, j_2, j_3\}$$

This can be put in natural order in at most two interchanges as follows:

1. Interchange the first index with the second or third index, if necessary, to bring the 1 to the first position.

2. Interchange the new second index with the third index, if necessary, to bring the 2 to the second position.

We leave it for you to use this procedure to obtain the results in the following table and then confirm that these results are consistent with Formula (4).

Permutation of Column Indices	Minimum Number of Interchanges to Put Permutation in Natural Order	Signed Elementary Product
$\{1, 2, 3\}$	0	$+a_{11}a_{22}a_{33}$
$\{1, 3, 2\}$	1	$-a_{11}a_{23}a_{32}$
$\{2, 1, 3\}$	1	$-a_{12}a_{21}a_{33}$
$\{2, 3, 1\}$	2	$+a_{12}a_{23}a_{31}$
$\{3, 1, 2\}$	2	$+a_{13}a_{21}a_{32}$
$\{3, 2, 1\}$	1	$-a_{13}a_{22}a_{31}$

CONCEPT PROBLEM Devise a general procedure for putting a permutation $\{j_1, j_2, j_3, j_4\}$ of the integers 1, 2, 3, and 4 in natural order with the minimum number of interchanges. Extend your procedure to permutations $\{j_1, j_2, \ldots, j_n\}$ of $1, 2, \ldots, n$.

GENERAL DETERMINANTS

To define the determinant of a general $n \times n$ matrix, we need some terminology. Motivated by our discussion of 2×2 and 3×3 determinants, we define an ***elementary product*** from an $n \times n$ matrix A to be a product of n entries from A, no two of which come from the same row or same column. Thus, if $A = [a_{ij}]$, then each elementary product is expressible in the form

$$a_{1j_1} a_{2j_2} \cdots a_{nj_n} \tag{5}$$

where the column indices form a permutation $\{j_1, j_2, \ldots, j_n\}$ of the integers from 1 to n and the row indices are in natural order. We will define this permutation to be ***even*** or ***odd*** in accordance with whether the minimum number of interchanges required to put the permutation in natural order is even or odd. The ***signed elementary product*** corresponding to (5) is defined to be

$$\pm a_{1j_1} a_{2j_2} \cdots a_{nj_n}$$

where the sign is $+$ if the permutation $\{j_1, j_2, \ldots, j_n\}$ is even and $-$ if it is odd.

> **Definition 4.1.1** The ***determinant*** of a square matrix A is denoted by $\det(A)$ and is defined to be the sum of all signed elementary products from A.

An $n \times n$ determinant can also be written in vertical bar notation as

$$\det(A) = |A| = \begin{vmatrix} a_{11} & a_{12} & \cdots & a_{1n} \\ a_{21} & a_{22} & \cdots & a_{2n} \\ \vdots & \vdots & & \vdots \\ a_{n1} & a_{n2} & \cdots & a_{nn} \end{vmatrix}$$

We will call this an ***$n \times n$ determinant*** or an ***nth-order determinant***. When convenient, Definition 4.1.1 can be expressed in summation notation as

$$\det(A) = \sum \pm a_{1j_1} a_{2j_2} \cdots a_{nj_n} \tag{6}$$

where the $\sum$ and the $\pm$ are intended to suggest that the signed elementary products are to be summed over all possible permutations $\{j_1, j_2, \ldots, j_n\}$ of the column indices.

REMARK For completeness, we note that the determinant of a 1×1 matrix $A = [a_{11}]$ is a_{11}.

EVALUATION DIFFICULTIES FOR HIGHER-ORDER DETERMINANTS

Evaluating a determinant from its definition has computational difficulties, the problem being that the amount of computation required gets out of control very quickly as n increases. This happens because the number of signed elementary products for an $n \times n$ determinant is

$$n! = n \cdot (n-1) \cdot (n-2) \cdots \cdot 1$$

(Exercise P2), which increases dramatically with n. For example, a 3×3 determinant has $3! = 6$ signed elementary products, a 4×4 has $4! = 24$, a 5×5 has $5! = 120$, and a 10×10 has $10! = 3,628,800$. A 30×30 determinant has so many signed elementary products that today's typical PC would require more than 10^{10} years to evaluate a determinant of this size—making it likely that the Sun would burn out first! Fortunately, there are other methods for evaluating determinants that require much less calculation.

WARNING The methods described in Figure 4.1.1 and used in Example 1 do not work for 4×4 determinants or higher.

DETERMINANTS OF MATRICES WITH ROWS OR COLUMNS THAT HAVE ALL ZEROS

Determinants of matrices with a row or column of zeros are easy to evaluate, regardless of size.

> **Theorem 4.1.2** *If A is a square matrix with a row or a column of zeros, then $\det(A) = 0$.*

Proof Assume that A has a row of zeros. Since every signed elementary product has an entry from each row, each such product has a factor of zero from the zero row. Thus, all signed elementary products are zero, and hence so is their sum. The same argument works for columns. ∎

DETERMINANTS OF TRIANGULAR MATRICES

As the following theorem shows, determinants of triangular matrices are also easy to evaluate, regardless of size.

Theorem 4.1.3 *If A is a triangular matrix, then* $\det(A)$ *is the product of the entries on the main diagonal.*

We will illustrate the idea of the proof for a 4×4 lower triangular matrix. The proofs in the upper triangular case and for general triangular matrices are similar in spirit.

Proof (**4 × 4** *lower triangular case*) Consider a general 4×4 lower triangular matrix

$$A = \begin{bmatrix} a_{11} & 0 & 0 & 0 \\ a_{21} & a_{22} & 0 & 0 \\ a_{31} & a_{32} & a_{33} & 0 \\ a_{41} & a_{42} & a_{43} & a_{44} \end{bmatrix}$$

Keeping in mind that an elementary product must have exactly one factor from each row and one factor from each column, the only elementary product that does not have one of the six zeros as a factor is $a_{11}a_{22}a_{33}a_{44}$ (verify). The column indices of this elementary product are in natural order, so the associated signed elementary product takes a $+$. Thus, $\det(A) = a_{11}a_{22}a_{33}a_{44}$. ∎

EXAMPLE 2
Determinant of a Triangular Matrix

By inspection,

$$\begin{vmatrix} -2 & 5 & 7 \\ 0 & 3 & 8 \\ 0 & 0 & 5 \end{vmatrix} = (-2)(3)(5) = -30 \quad \text{and} \quad \begin{vmatrix} 1 & 0 & 0 & 0 \\ 4 & 9 & 0 & 0 \\ -7 & 6 & -1 & 0 \\ 3 & 8 & -5 & -2 \end{vmatrix} = (1)(9)(-1)(-2) = 18 \quad ∎$$

MINORS AND COFACTORS

We will now develop a procedure for evaluating determinants that is based on the idea of expressing a determinant in terms of determinants of lower order. Although we will discuss better procedures for evaluating determinants in later sections, this procedure plays an important role in various applications of determinants. We begin with some terminology.

Definition 4.1.4 If A is a square matrix, then the ***minor*** of entry a_{ij} (also called the ijth minor of A) is denoted by M_{ij} and is defined to be the determinant of the submatrix that remains when the ith row and jth column of A are deleted. The number $C_{ij} = (-1)^{i+j} M_{ij}$ is called the ***cofactor*** of entry a_{ij} (or the ijth cofactor of A).

REMARK We have followed the tradition of denoting minors and cofactors by capital letters, even though they are numbers (scalars). For a 1×1 matrix $[a_{11}]$, the entry a_{11} itself is defined to be the minor and cofactor of a_{11}.

EXAMPLE 3
Minors and Cofactors

Let

$$A = \begin{bmatrix} 3 & 1 & -4 \\ 2 & 5 & 6 \\ 1 & 4 & 8 \end{bmatrix}$$

The minor of entry a_{11} is

$$M_{11} = \begin{vmatrix} 3 & 1 & -4 \\ 2 & 5 & 6 \\ 1 & 4 & 8 \end{vmatrix} = \begin{vmatrix} 5 & 6 \\ 4 & 8 \end{vmatrix} = 16$$

and the corresponding cofactor is

$$C_{11} = (-1)^{1+1} M_{11} = M_{11} = 16$$

The minor of entry a_{32} is

$$M_{32} = \begin{vmatrix} 3 & 1 & -4 \\ 2 & 5 & 6 \\ 1 & 4 & 8 \end{vmatrix} = \begin{vmatrix} 3 & -4 \\ 2 & 6 \end{vmatrix} = 26$$

and the corresponding cofactor is

$$C_{32} = (-1)^{3+2} M_{32} = -M_{32} = -26$$

REMARK Notice in Definition 4.1.4 that a minor and its associated cofactor are either the same or negatives of one another and that the relating sign $(-1)^{i+j}$ is either $+1$ or -1 in accordance with the pattern in the "checkerboard" array

$$\begin{bmatrix} + & - & + & - & + & \cdots \\ - & + & - & + & - & \cdots \\ + & - & + & - & + & \cdots \\ - & + & - & + & - & \cdots \\ \vdots & \vdots & \vdots & \vdots & \vdots & \end{bmatrix}$$

Thus, it is never really necessary to compute $(-1)^{i+j}$ to find C_{ij}—you can simply compute the minor M_{ij} and adjust the sign, if necessary, in accordance with the checkerboard. Try this in Example 3.

COFACTOR EXPANSIONS We will now show how a 3×3 determinant can be expressed in terms of 2×2 determinants. For this purpose, recall that the determinant of a 3×3 matrix A was defined in Formula (4) as

$$\det(A) = a_{11}a_{22}a_{33} + a_{12}a_{23}a_{31} + a_{13}a_{21}a_{32} - a_{13}a_{22}a_{31} - a_{12}a_{21}a_{33} - a_{11}a_{23}a_{32}$$

which we can rewrite as

$$\det(A) = a_{11}(a_{22}a_{33} - a_{23}a_{32}) + a_{21}(a_{13}a_{32} - a_{12}a_{33}) + a_{31}(a_{12}a_{23} - a_{13}a_{22})$$

However, the expressions in parentheses are the cofactors C_{11}, C_{21}, and C_{31} (verify), so we have shown that

$$\det(A) = a_{11}C_{11} + a_{21}C_{21} + a_{31}C_{31}$$

In words, this formula states that $\det(A)$ can be obtained by multiplying each entry in the first column of A by its cofactor and adding the resulting products. There is nothing special about the first column—by grouping terms in (4) appropriately, you should be able to show that there are actually six companion formulas:

$$\begin{aligned} \det(A) &= a_{11}C_{11} + a_{12}C_{12} + a_{13}C_{13} \\ &= a_{11}C_{11} + a_{21}C_{21} + a_{31}C_{31} \\ &= a_{21}C_{21} + a_{22}C_{22} + a_{23}C_{23} \\ &= a_{12}C_{12} + a_{22}C_{22} + a_{32}C_{32} \\ &= a_{31}C_{31} + a_{32}C_{32} + a_{33}C_{33} \\ &= a_{13}C_{13} + a_{23}C_{23} + a_{33}C_{33} \end{aligned} \tag{7}$$

These are called *cofactor expansions* of A. Note that in each cofactor expansion, the entries and cofactors all come from the same row or the same column. This shows that a 3×3 determinant can be evaluated by multiplying the entries in any row or column by their cofactors and adding the resulting products.

EXAMPLE 4
Cofactor
Expansions of a
3×3
Determinant

Here is the 3×3 determinant in Example 1 evaluated by a cofactor expansion along the first column:

$$\begin{vmatrix} 1 & 2 & 3 \\ -4 & 5 & 6 \\ 7 & -8 & 9 \end{vmatrix} = (1) \begin{vmatrix} 5 & 6 \\ -8 & 9 \end{vmatrix} + (-4)(-1) \begin{vmatrix} 2 & 3 \\ -8 & 9 \end{vmatrix} + (7) \begin{vmatrix} 2 & 3 \\ 5 & 6 \end{vmatrix}$$

$$= (1)(93) + (4)(42) + (7)(-3) = 240$$

And here is the same determinant evaluated by a cofactor expansion along the second column:

$$\begin{vmatrix} 1 & 2 & 3 \\ -4 & 5 & 6 \\ 7 & -8 & 9 \end{vmatrix} = (2)(-1) \begin{vmatrix} -4 & 6 \\ 7 & 9 \end{vmatrix} + (5) \begin{vmatrix} 1 & 3 \\ 7 & 9 \end{vmatrix} + (-8)(-1) \begin{vmatrix} 1 & 3 \\ -4 & 6 \end{vmatrix}$$

$$= (-2)(-78) + (5)(-12) + (8)(18) = 240 \qquad \blacksquare$$

CONCEPT PROBLEM To check your understanding of the method in Example 4, evaluate the determinant by cofactor expansions along the first and second rows.

The cofactor expansions for 3×3 determinants are special cases of the following general theorem, which we state without proof.

Theorem 4.1.5 *The determinant of an $n \times n$ matrix A can be computed by multiplying the entries in any row (or column) by their cofactors and adding the resulting products; that is, for each $1 \leq i \leq n$ and $1 \leq j \leq n$,*

$$\det(A) = a_{1j}C_{1j} + a_{2j}C_{2j} + \cdots + a_{nj}C_{nj}$$

(cofactor expansion along the jth column)

and

$$\det(A) = a_{i1}C_{i1} + a_{i2}C_{i2} + \cdots + a_{in}C_{in}$$

(cofactor expansion along the ith row)

Linear Algebra in History

Cofactor expansion is not the only method for expressing the determinant of a matrix in terms of determinants of lower order. For example, although it is not well known, the English mathematician Charles Dodgson, who was the author of *Alice's Adventures in Wonderland* and *Through the Looking Glass* under the pen name of Lewis Carroll, invented such a method, called *condensation*. That method has recently been resurrected from obscurity because of its suitability for parallel processing on computers.

Charles Lutwidge Dodgson
(Lewis Carroll)
(1832–1898)

EXAMPLE 5
Cofactor
Expansion of a
4×4
Determinant

Use a cofactor expansion to find the determinant of

$$A = \begin{bmatrix} 2 & 0 & 0 & 5 \\ -1 & 2 & 4 & 1 \\ 3 & 0 & 0 & 3 \\ 8 & 6 & 0 & 0 \end{bmatrix}$$

Solution We are free to expand along any row or column, but the third column is the best choice since it contains three zeros, each of which eliminates the need to calculate the corresponding cofactor. That is, by expanding along the third column we obtain

$$\det(A) = (0)C_{13} + (4)C_{23} + (0)C_{33} + (0)C_{43} = (4)C_{23}$$

which requires only one cofactor calculation. Thus,

$$\det(A) = \begin{vmatrix} 2 & 0 & 0 & 5 \\ -1 & 2 & 4 & 1 \\ 3 & 0 & 0 & 3 \\ 8 & 6 & 0 & 0 \end{vmatrix} = (-4)\begin{vmatrix} 2 & 0 & 5 \\ 3 & 0 & 3 \\ 8 & 6 & 0 \end{vmatrix} = (-4)(-6)\begin{vmatrix} 2 & 5 \\ 3 & 3 \end{vmatrix} = -216$$

where the 3×3 determinant was evaluated by a cofactor expansion along the second column. ∎

REMARK The computations in Example 5 were particularly simple because of the zeros. In the worst situation there will be no zeros, in which case a direct cofactor expansion will require the evaluation of four 3×3 determinants. However, in the next section we will discuss methods for introducing zeros to simplify cofactor expansions.

Exercise Set 4.1

In Exercises 1–10, evaluate the determinant by the method of Example 1.

1. $\begin{bmatrix} 3 & 5 \\ -2 & 4 \end{bmatrix}$

2. $\begin{bmatrix} 4 & 1 \\ 8 & 2 \end{bmatrix}$

3. $\begin{bmatrix} -5 & 7 \\ -7 & -2 \end{bmatrix}$

4. $\begin{bmatrix} \sqrt{2} & \sqrt{6} \\ 4 & \sqrt{3} \end{bmatrix}$

5. $\begin{bmatrix} a-3 & 5 \\ -3 & a-2 \end{bmatrix}$

6. $\begin{bmatrix} -2 & 7 & 6 \\ 5 & 1 & -2 \\ 3 & 8 & 4 \end{bmatrix}$

7. $\begin{bmatrix} -2 & 1 & 4 \\ 3 & 5 & -7 \\ 1 & 6 & 2 \end{bmatrix}$

8. $\begin{bmatrix} -1 & 1 & 2 \\ 3 & 0 & -5 \\ 1 & 7 & 2 \end{bmatrix}$

9. $\begin{bmatrix} 3 & 0 & 0 \\ 2 & -1 & 5 \\ 1 & 9 & -4 \end{bmatrix}$

10. $\begin{bmatrix} c & -4 & 3 \\ 2 & 1 & c^2 \\ 4 & c-1 & 2 \end{bmatrix}$

In Exercises 11 and 12, write down the permutation of column indices associated with the elementary product, and then find the sign of that elementary product by determining whether the permutation is even or odd.

11. (a) $a_{14}a_{21}a_{33}a_{45}a_{52}$
(b) $a_{15}a_{23}a_{34}a_{42}a_{51}$
(c) $a_{14}a_{22}a_{35}a_{43}a_{51}$
(d) $a_{15}a_{24}a_{33}a_{42}a_{51}$
(e) $a_{11}a_{22}a_{33}a_{44}a_{55}$
(f) $a_{11}a_{24}a_{32}a_{43}a_{55}$

12. (a) $a_{11}a_{22}a_{33}a_{44}$
(b) $a_{12}a_{21}a_{34}a_{43}$
(c) $a_{11}a_{23}a_{34}a_{42}$
(d) $a_{14}a_{23}a_{32}a_{41}$
(e) $a_{14}a_{21}a_{32}a_{43}$
(f) $a_{13}a_{21}a_{34}a_{42}$

In Exercises 13–16, find all values of λ for which $\det(A) = 0$.

13. $A = \begin{bmatrix} \lambda-2 & 1 \\ -5 & \lambda+4 \end{bmatrix}$

14. $A = \begin{bmatrix} \lambda-4 & 0 & 0 \\ 0 & \lambda & 2 \\ 0 & 3 & \lambda-1 \end{bmatrix}$

15. $A = \begin{bmatrix} \lambda-1 & 0 \\ 2 & \lambda+1 \end{bmatrix}$

16. $A = \begin{bmatrix} \lambda-4 & 4 & 0 \\ -1 & \lambda & 0 \\ 0 & 0 & \lambda-5 \end{bmatrix}$

17. Solve for x.

$$\begin{vmatrix} x & -1 \\ 3 & 1-x \end{vmatrix} = \begin{vmatrix} 1 & 0 & -3 \\ 2 & x & -6 \\ 1 & 3 & x-5 \end{vmatrix}$$

18. Solve for y.

$$\begin{vmatrix} 2y & y \\ -4 & 2+y \end{vmatrix} = \begin{vmatrix} 2 & -1 & 0 \\ 0 & -y & -6 \\ -1 & 3 & 1-y \end{vmatrix}$$

In Exercises 19 and 20, find the determinants of the matrices by inspection.

19. (a) $\begin{bmatrix} 1 & 0 & 0 \\ 0 & -1 & 0 \\ 0 & 0 & 1 \end{bmatrix}$
(b) $\begin{bmatrix} 0 & 0 & 0 & 0 \\ 1 & 2 & 0 & 0 \\ 0 & 4 & 3 & 0 \\ 1 & 2 & 3 & 8 \end{bmatrix}$

(c) $\begin{bmatrix} 1 & 2 & 7 & -3 \\ 0 & 1 & -4 & 1 \\ 0 & 0 & 2 & 7 \\ 0 & 0 & 0 & 3 \end{bmatrix}$

20. (a) $\begin{bmatrix} 2 & 0 & 0 \\ 0 & 2 & 0 \\ 0 & 0 & 2 \end{bmatrix}$
(b) $\begin{bmatrix} 1 & 1 & 1 & 1 \\ 0 & 2 & 2 & 2 \\ 0 & 0 & 3 & 3 \\ 0 & 0 & 0 & 4 \end{bmatrix}$

(c) $\begin{bmatrix} -3 & 0 & 0 & 0 \\ 1 & 2 & 0 & 0 \\ 40 & 10 & -1 & 0 \\ 100 & 200 & -23 & 3 \end{bmatrix}$

In Exercises 21 and 22, (a) find all the minors of A, and (b) find all the cofactors of A.

21. $A = \begin{bmatrix} 1 & -2 & 3 \\ 6 & 7 & -1 \\ -3 & 1 & 4 \end{bmatrix}$ **22.** $A = \begin{bmatrix} 1 & 1 & 2 \\ 3 & 3 & 6 \\ 0 & 1 & 4 \end{bmatrix}$

23. Let

$$A = \begin{bmatrix} 4 & -1 & 1 & 6 \\ 0 & 0 & -3 & 3 \\ 4 & 1 & 0 & 14 \\ 4 & 1 & 3 & 2 \end{bmatrix}$$

Find
- (a) M_{13} and C_{13}
- (b) M_{23} and C_{23}
- (c) M_{22} and C_{22}
- (d) M_{21} and C_{21}

24. Let

$$A = \begin{bmatrix} 2 & 3 & -1 & 1 \\ -3 & 2 & 0 & 3 \\ 3 & -2 & 1 & 0 \\ 3 & -2 & 1 & 4 \end{bmatrix}$$

Find
- (a) M_{32} and C_{32}
- (b) M_{44} and C_{44}
- (c) M_{41} and C_{41}
- (d) M_{24} and C_{24}

25. Evaluate the determinant of the matrix in Exercise 21 by a cofactor expansion along
- (a) the first row
- (b) the first column
- (c) the second row
- (d) the second column
- (e) the third row
- (f) the third column

26. Evaluate the determinant of the matrix in Exercise 22 by a cofactor expansion along
- (a) the first row
- (b) the first column
- (c) the second row
- (d) the second column
- (e) the third row
- (f) the third column

In Exercises 27–32, evaluate $\det(A)$ by a cofactor expansion along a row or column of your choice.

27. $A = \begin{bmatrix} -3 & 0 & 7 \\ 2 & 5 & 1 \\ -1 & 0 & 5 \end{bmatrix}$ **28.** $A = \begin{bmatrix} 3 & 3 & 1 \\ 1 & 0 & -4 \\ 1 & -3 & 5 \end{bmatrix}$

29. $A = \begin{bmatrix} 1 & k & k^2 \\ 1 & k & k^2 \\ 1 & k & k^2 \end{bmatrix}$ **30.** $A = \begin{bmatrix} k+1 & k-1 & 7 \\ 2 & k-3 & 4 \\ 5 & k+1 & k \end{bmatrix}$

31. $A = \begin{bmatrix} 3 & 3 & 0 & 5 \\ 2 & 2 & 0 & -2 \\ 4 & 1 & -3 & 0 \\ 2 & 10 & 3 & 2 \end{bmatrix}$

32. $A = \begin{bmatrix} 4 & 0 & 0 & 1 & 0 \\ 3 & 3 & 3 & -1 & 0 \\ 1 & 2 & 4 & 2 & 3 \\ 9 & 4 & 6 & 2 & 3 \\ 2 & 2 & 4 & 2 & 3 \end{bmatrix}$

33. Show that the value of the determinant

$$\begin{vmatrix} \sin(\theta) & \cos(\theta) & 0 \\ -\cos(\theta) & \sin(\theta) & 0 \\ \sin(\theta) - \cos(\theta) & \sin(\theta) + \cos(\theta) & 1 \end{vmatrix}$$

is independent of θ.

34. Show that the matrices

$$A = \begin{bmatrix} a & b \\ 0 & c \end{bmatrix} \quad \text{and} \quad B = \begin{bmatrix} d & e \\ 0 & f \end{bmatrix}$$

commute if and only if

$$\begin{vmatrix} b & a-c \\ e & d-f \end{vmatrix} = 0$$

35. By inspection, what is the relationship between the following determinants?

$$d_1 = \begin{vmatrix} a & b & c \\ d & 1 & f \\ g & 0 & 1 \end{vmatrix} \quad \text{and} \quad d_2 = \begin{vmatrix} a+\lambda & b & c \\ d & 1 & f \\ g & 0 & 1 \end{vmatrix}$$

36. Show that

$$\det(A) = \frac{1}{2} \begin{vmatrix} \operatorname{tr}(A) & 1 \\ \operatorname{tr}(A^2) & \operatorname{tr}(A) \end{vmatrix}$$

for any 2×2 matrix A.

Discussion and Discovery

D1. Explain why the determinant of a matrix with integer entries must be an integer.

D2. What can you say about an nth-order determinant all of whose entries are 1? Explain your reasoning.

D3. What is the maximum number of zeros that a 3×3 matrix can have without having a zero determinant? Explain your reasoning.

D4. What can you say about the graph of the equation

$$\begin{vmatrix} x & y & 1 \\ a_1 & b_1 & 1 \\ a_2 & b_2 & 1 \end{vmatrix} = 0$$

if a_1, a_2, b_1, and b_2 are constants?

D5. What can you say about the six elementary products of $A = \mathbf{u}\mathbf{v}^T$ if $\mathbf{u}$ and $\mathbf{v}$ are column vectors in R^3?

Working with Proofs

P1. Prove that (x_1, y_1), (x_2, y_2), and (x_3, y_3) are collinear points if and only if

$$\begin{vmatrix} x_1 & y_1 & 1 \\ x_2 & y_2 & 1 \\ x_3 & y_3 & 1 \end{vmatrix} = 0$$

P2. (*For readers familiar with proof by induction*) Prove by induction that there are $n!$ permutations of n distinct integers and hence $n!$ elementary products for an $n \times n$ matrix.

Technology Exercises

T1. Compute the determinants of the matrices in Exercises 23 and 24.

T2. Compute the cofactors of the matrix in Example 3, and calculate the determinant of the matrix by a cofactor expansion along the second row and also along the second column.

T3. Choose random 3×3 matrices A and B (possibly using your technology) and then compute $\det(AB) = \det(A) \det(B)$. Repeat this process as often as needed to be reasonably certain about the relationship between these quantities. What is the relationship?

T4. (**CAS**) Confirm Formula (4).

T5. (**CAS**) Use the determinant and simplification capabilities of a CAS to show that

$$\begin{vmatrix} \frac{1}{a+x} & \frac{1}{a+y} & 1 \\ \frac{1}{b+x} & \frac{1}{b+y} & 1 \\ \frac{1}{c+x} & \frac{1}{c+y} & 1 \end{vmatrix}$$

$$= \frac{(a-b)(a-c)(b-c)(x-y)}{(a+x)(b+x)(c+x)(a+y)(b+y)(c+y)}$$

T6. (**CAS**) Use the determinant and simplification capabilities of a CAS to show that

$$\begin{vmatrix} a & b & c & d \\ -b & a & d & -c \\ -c & -d & a & b \\ -d & c & -b & a \end{vmatrix} = (a^2 + b^2 + c^2 + d^2)^2$$

T7. (**CAS**) Find a simple formula for the determinant

$$\begin{vmatrix} (a+b)^2 & c^2 & c^2 \\ a^2 & (b+c)^2 & a^2 \\ b^2 & b^2 & (c+a)^2 \end{vmatrix}$$

T8. The nth-order **Fibonacci matrix** [named for the Italian mathematician (circa 1170–1250)] is the $n \times n$ matrix F_n that has 1's on the main diagonal, 1's along the diagonal immediately above the main diagonal, -1's along the diagonal immediately below the main diagonal, and zeros everywhere else. Construct the sequence

$$\det(F_1), \ \det(F_2), \ \det(F_3), \ldots, \ \det(F_7)$$

Make a conjecture about the relationship between a term in the sequence and its two immediate predecessors, and then use your conjecture to make a guess at $\det(F_8)$. Check your guess by calculating this number.

T9. Let A_n be the $n \times n$ matrix that has 2's along the main diagonal, 1's along the diagonals immediately above and below the main diagonal, and zeros everywhere else. Make a conjecture about the relationship between n and $\det(A_n)$.

Section 4.2   Properties of Determinants

In this section we will discuss some basic properties of determinants that will lead to an improved procedure for their evaluation and also shed some new light on solutions of linear systems and invertibility of matrices.

DETERMINANT OF A^T   Our first goal in this section is to establish a relationship between the determinant of a matrix and the determinant of its transpose. Because every elementary product from a square matrix A has one factor from each row and one from each column, there is an inherent symmetry between rows and columns in the definition of a determinant that naturally leads one to suspect that $\det(A) = \det(A^T)$. Indeed, this is true for 2×2 determinants, since

$$\det(A) = \begin{vmatrix} a & b \\ c & d \end{vmatrix} = ad - bc = ad - cb = \begin{vmatrix} a & c \\ b & d \end{vmatrix} = \det(A^T)$$

One way to extend this result to general $n \times n$ matrices is to use certain theorems on permutations to show that A and A^T have the same *signed* elementary products and hence the same determinant. An alternative proof, using induction and cofactor expansion, is given at the end of this section.

Theorem 4.2.1 *If A is a square matrix, then $\det(A) = \det(A^T)$.*

It follows from this that every theorem about rows of a determinant has a companion version for columns, and vice versa (see Theorem 4.1.2, for example).

CONCEPT PROBLEM Find the transpose of the matrix A in Example 4 of the last section, and confirm that $\det(A) = \det(A^T)$ using a cofactor expansion along any row or column of A^T.

EFFECT OF ELEMENTARY ROW OPERATIONS ON A DETERMINANT

The next theorem explains how elementary row operations and ***elementary column operations*** affect a determinant (elementary column operations being the same as elementary row operations, except performed on columns).

Theorem 4.2.2 *Let A be an $n \times n$ matrix.*

(a) *If B is the matrix that results when a single row or single column of A is multiplied by a scalar k, then $\det(B) = k \det(A)$.*

(b) *If B is the matrix that results when two rows or two columns of A are interchanged, then $\det(B) = -\det(A)$.*

(c) *If B is the matrix that results when a multiple of one row of A is added to another row or when a multiple of one column is added to another column, then $\det(B) = \det(A)$.*

We will prove parts (a) and (b) and leave part (c) for the exercises.

Proof (a) To be specific, suppose that the ith row of A is multiplied by the scalar k to produce the matrix B, and let

$$\det(A) = a_{i1}C_{i1} + a_{i2}C_{i2} + \cdots + a_{in}C_{in}$$

be the cofactor expansion of $\det(A)$ along the ith row. Since the ith row is deleted when the cofactors along that row are computed, the cofactors in this formula are unchanged when the ith row is multiplied by k. Thus, the cofactor expansion of $\det(B)$ along the ith row is

$$\det(B) = ka_{i1}C_{i1} + ka_{i2}C_{i2} + \cdots + ka_{in}C_{in} = k(a_{i1}C_{i1} + a_{i2}C_{i2} + \cdots + a_{in}C_{in}) = k \det(A)$$

The proof for columns can be obtained in a similar manner or by applying Theorem 4.2.1.

Proof (b) Here it will be easier to work with the determinant definition and consider a column interchange. If two columns of A are interchanged, then it can be shown that in each elementary product the minimum number of interchanges required to put the column indices in natural order is either increased or decreased by 1. This has the effect of reversing the sign of each signed elementary product, thereby reversing the sign of their sum, which is $\det(A)$. The proof for rows can now be obtained by applying Theorem 4.2.1. ■

EXAMPLE 1
Effect of Elementary Row Operations on a 3×3 Determinant

The following equations illustrate Theorem 4.2.2 for 3×3 determinants.

$$\overset{B}{\begin{vmatrix} ka_{11} & ka_{12} & ka_{13} \\ a_{21} & a_{22} & a_{23} \\ a_{31} & a_{32} & a_{33} \end{vmatrix}} = k \overset{A}{\begin{vmatrix} a_{11} & a_{12} & a_{13} \\ a_{21} & a_{22} & a_{23} \\ a_{31} & a_{32} & a_{33} \end{vmatrix}}$$

The first row of A was multiplied by k to obtain B. $\det(B) = k \det(A)$

$$\overset{B}{\begin{vmatrix} a_{21} & a_{22} & a_{23} \\ a_{11} & a_{12} & a_{13} \\ a_{31} & a_{32} & a_{33} \end{vmatrix}} = - \overset{A}{\begin{vmatrix} a_{11} & a_{12} & a_{13} \\ a_{21} & a_{22} & a_{23} \\ a_{31} & a_{32} & a_{33} \end{vmatrix}}$$

The first and second rows of A were interchanged to obtain B. $\det(B) = -\det(A)$

$$\overset{B}{\begin{vmatrix} a_{11} + ka_{21} & a_{12} + ka_{22} & a_{13} + ka_{23} \\ a_{21} & a_{22} & a_{23} \\ a_{31} & a_{32} & a_{33} \end{vmatrix}} = \overset{A}{\begin{vmatrix} a_{11} & a_{12} & a_{13} \\ a_{21} & a_{22} & a_{23} \\ a_{31} & a_{32} & a_{33} \end{vmatrix}}$$

k times the second row of A was added to the first to obtain B. $\det(B) = \det(A)$ ∎

REMARK Theorem 4.2.2(a) implies that a common factor from any row (or column) can be moved through the determinant symbol (see the first part of Example 1).

The next theorem is a direct consequence of Theorem 4.2.2.

> **Theorem 4.2.3** *Let A be an $n \times n$ matrix.*
> *(a) If A has two identical rows or columns, then $\det(A) = 0$.*
> *(b) If A has two proportional rows or columns, then $\det(A) = 0$.*
> *(c) $\det(kA) = k^n \det(A)$.*

Proof (a) If A has two identical rows or columns, then interchanging them does not change A, but it reverses the sign of its determinant. Thus, $\det(A) = -\det(A)$, which can only mean that $\det(A) = 0$.

Proof (b) If A has two proportional rows or columns, then one of the rows or columns is a scalar multiple of the other, say k times the other. If we move the scalar k through the determinant symbol, then the matrix B that remains has two identical rows or columns. Thus, applying part (a) to the matrix B yields $\det(A) = k \det(B) = 0$.

Proof (c) Multiplying A by k is the same as multiplying each row vector (or each column vector) by k. For each of the n rows (or columns) we can move the factor k through the determinant symbol, thereby multiplying $\det(A)$ by n factors of k; that is, $\det(kA) = k^n \det(A)$. ∎

EXAMPLE 2
Some Determinants That Can Be Evaluated by Inspection

Here are some determinants that can be evaluated by inspection if you think about them in the right way.

$$\begin{vmatrix} 3 & -1 & 4 & -5 \\ 6 & -2 & 5 & 2 \\ 5 & 8 & 1 & 4 \\ -9 & 3 & -12 & 15 \end{vmatrix} = 0, \qquad \begin{vmatrix} 0 & 0 & 0 & 1 \\ 0 & 0 & 1 & 0 \\ 0 & 1 & 0 & 0 \\ 1 & 0 & 0 & 0 \end{vmatrix} = 1, \qquad \begin{vmatrix} 1 & -2 & 7 \\ -4 & 8 & 5 \\ 2 & -4 & 3 \end{vmatrix} = 0$$

Two proportional rows

Two row interchanges produce I_4.

Two proportional columns ∎

EXAMPLE 3
Determinant of the Negative of a Matrix

What is the relationship between $\det(A)$ and $\det(-A)$?

Solution Since $-A = (-1)A$, it follows from Theorem 4.2.3(c) that if A has size $n \times n$, then

$$\det(-A) = (-1)^n \det(A)$$

Thus, $\det(-A) = \det(A)$ if n is even, and $\det(-A) = -\det(A)$ if n is odd. ∎

CONCEPT PROBLEM   See if you can obtain the result in Example 3 directly from the definition of a determinant as a sum of signed elementary products.

SIMPLIFYING COFACTOR EXPANSIONS

We observed in the last section that the work involved in a cofactor expansion can be minimized by expanding along a row or column with the maximum number of zeros (if any). Since zeros can be introduced into a matrix by adding appropriate multiples of one row (or column) to another, and since this operation does not change the determinant of the matrix, it is possible to combine operations of this type with cofactor expansions to produce a computational technique that is often better than cofactor expansion alone. This is illustrated in the next example.

EXAMPLE 4
Using Row Operations to Simplify a Cofactor Expansion

Use a cofactor expansion to find the determinant of the matrix

$$A = \begin{bmatrix} 3 & 5 & -2 & 6 \\ 1 & 2 & -1 & 1 \\ 2 & 4 & 1 & 5 \\ 3 & 7 & 5 & 3 \end{bmatrix}$$

Solution   This matrix has no zeros, so a direct cofactor expansion would require the evaluation of four 3×3 determinants. However, we can introduce zeros without altering the determinant by adding suitable multiples of one row (or column) to another. For example, one convenient possibility is to exploit the 1 in the second row of the first column and introduce three zeros into the first column by adding suitable multiples of the second row to the other three. This yields (verify)

$$\det(A) = \begin{vmatrix} 0 & -1 & 1 & 3 \\ 1 & 2 & -1 & 1 \\ 0 & 0 & 3 & 3 \\ 0 & 1 & 8 & 0 \end{vmatrix}$$

$$= - \begin{vmatrix} -1 & 1 & 3 \\ 0 & 3 & 3 \\ 1 & 8 & 0 \end{vmatrix} \qquad \boxed{\text{Cofactor expansion along the first column}}$$

$$= - \begin{vmatrix} -1 & 1 & 3 \\ 0 & 3 & 3 \\ 0 & 9 & 3 \end{vmatrix} \qquad \boxed{\text{We added the first row to the third row.}}$$

$$= -(-1) \begin{vmatrix} 3 & 3 \\ 9 & 3 \end{vmatrix} \qquad \boxed{\text{Cofactor expansion along the first column}}$$

$$= -18 \qquad\qquad\qquad\qquad\qquad\qquad \blacksquare$$

CONCEPT PROBLEM   Evaluate the determinant in Example 4 by first exploiting the 1 in the second row of the first column to introduce three zeros into the second row and then performing a cofactor expansion along the second row.

DETERMINANTS BY GAUSSIAN ELIMINATION

In practical applications, determinants are usually evaluated using some variation of Gaussian elimination (usually LU-decomposition). We will discuss this in more detail shortly, but here is an example that illustrates how $\det(A)$ can be evaluated by reducing A to row echelon form.

EXAMPLE 5
Evaluating a Determinant by Gaussian Elimination

In this example we will give a procedure for evaluating the determinant of

$$A = \begin{bmatrix} 0 & 1 & 5 \\ 3 & -6 & 9 \\ 2 & 6 & 1 \end{bmatrix}$$

by reducing A to row echelon form. The procedure we will give differs slightly from the usual method of row reduction in that the leading 1's are introduced by moving a common factor through the determinant symbol:

$$\det(A) = \begin{vmatrix} 0 & 1 & 5 \\ 3 & -6 & 9 \\ 2 & 6 & 1 \end{vmatrix} = - \begin{vmatrix} 3 & -6 & 9 \\ 0 & 1 & 5 \\ 2 & 6 & 1 \end{vmatrix}$$

The first and second rows of A were interchanged.

$$= -3 \begin{vmatrix} 1 & -2 & 3 \\ 0 & 1 & 5 \\ 2 & 6 & 1 \end{vmatrix}$$

A common factor of 3 from the first row was taken through the det symbol.

$$= -3 \begin{vmatrix} 1 & -2 & 3 \\ 0 & 1 & 5 \\ 0 & 10 & -5 \end{vmatrix}$$

-2 times the first row was added to the third row.

$$= -3 \begin{vmatrix} 1 & -2 & 3 \\ 0 & 1 & 5 \\ 0 & 0 & -55 \end{vmatrix}$$

-10 times the second row was added to the third row.

$$= (-3)(-55) \begin{vmatrix} 1 & -2 & 3 \\ 0 & 1 & 5 \\ 0 & 0 & 1 \end{vmatrix}$$

A common factor of -55 from the last row was taken through the det symbol.

$$= (-3)(-55)(1) = 165$$

$\blacksquare$

A DETERMINANT TEST FOR INVERTIBILITY

In many respects the next theorem is the most important in this section.

Theorem 4.2.4 *A square matrix A is invertible if and only if* $\det(A) \neq 0$.

Proof As a preliminary step we will show that the determinant of A and the determinant of its reduced row echelon form R are both zero or both nonzero. For this purpose, consider the effect of an elementary row operation on the determinant of a matrix: If the operation is a row interchange, then the effect is to multiply the determinant by -1; if a multiple of one row is added to another, then the value of the determinant is unchanged; and if a row of the matrix is multiplied by a *nonzero* constant k, then the effect is to multiply the determinant by k. In all three cases the determinants of the matrices before and after the operation are both zero or both nonzero. Since R is derived from A by a sequence of elementary row operations, it follows that $\det(A)$ and $\det(R)$ are both zero or both nonzero. Now to the main part of the proof.

Assume that A is invertible, in which case the reduced row echelon form of A is I (Theorem 3.3.3). Since $\det(I) \neq 0$, it follows that $\det(A) \neq 0$. Conversely, assume that $\det(A) \neq 0$. Thus, if R is the reduced row echelon form of A, then $\det(R) \neq 0$. This implies that R does not have any zero rows and hence that $R = I$ (Theorem 3.2.4). Thus, A is invertible (again by Theorem 3.3.3). $\blacksquare$

REMARK This theorem now makes it evident that a matrix with two proportional rows or columns cannot be invertible.

DETERMINANT OF A PRODUCT OF MATRICES

Since the definition of matrix multiplication and the definition of a determinant are both fairly complicated, it would seem unlikely that any simple relationship should exist between them. Thus, the following beautiful theorem, whose proof is given at the end of this section, should come as a remarkable surprise.

Theorem 4.2.5 *If A and B are square matrices of the same size, then*

$$\det(AB) = \det(A)\det(B) \tag{1}$$

The result in this theorem can be extended to three or more factors; and, in particular, if we apply the extended theorem to $A^n = AA \cdots A$ (n factors), then we obtain the relationship

$$\det(A^n) = [\det(A)]^n \tag{2}$$

EXAMPLE 6
An Illustration of Theorem 4.2.5

Verify the result in Theorem 4.2.5 for the matrices

$$A = \begin{bmatrix} 3 & 1 \\ 2 & 3 \end{bmatrix} \quad \text{and} \quad B = \begin{bmatrix} -1 & 3 \\ 5 & -4 \end{bmatrix}$$

Solution We leave it for you to confirm that

$$AB = \begin{bmatrix} 2 & 5 \\ 13 & -6 \end{bmatrix}$$

and that $\det(A) = 7$, $\det(B) = -11$, and $\det(AB) = -77$. Thus, $\det(AB) = \det(A)\det(B)$. ∎

DETERMINANT EVALUATION BY LU-DECOMPOSITION

In Example 5 we showed how to evaluate a determinant by Gaussian elimination. In practice, computer programs for evaluating determinants use LU-decomposition, which is essentially just an efficient way of performing Gaussian elimination. In the case where A has an LU-decomposition, say $A = LU$, the determinant of A can be expressed as $\det(A) = \det(L)\det(U)$, which is easy to compute since L and U are triangular matrices. Thus, nearly all of the computational work in evaluating $\det(A)$ is expended in obtaining the LU-decomposition. Now recall from Table 3.7.1 that the number of flops required to obtain the LU-decomposition of an $n \times n$ matrix is on the order of $\frac{2}{3}n^3$ for large values of n. This is an enormous improvement over the determinant definition, which involves the computation of $n!$ signed elementary products. For example, today's typical PC can evaluate a 30×30 determinant in less than one-thousandth of a second by LU-decomposition compared to the roughly 10^{10} years that would be required for it to evaluate $30!$ signed elementary products.

Recall from Section 3.7 that if partial pivoting is used, or if row interchanges are required to reduce A to row echelon form, then A has a factorization of the form $A = PLU$, where P^{-1} is the permutation matrix corresponding to the row interchanges performed on A. In this case $\det(A) = \det(P)\det(L)\det(U)$. In the exercises we will ask you to show that $\det(P) = \pm 1$, where the plus occurs if the number of row interchanges is even and the minus if it is odd. As a practical matter, the factor $\det(P)$ does not cause a complication in evaluating $\det(A)$, since computer programs for obtaining PLU-decompositions can easily be designed to track the number of row interchanges that are used.

Linear Algebra in History

In 1815 the French mathematician Augustin Cauchy (see p. 23) published a landmark paper in which he give the first systematic and modern treatment of determinants. It was in that paper that Theorem 4.2.5 was stated and proved for the first time in its full generality. Special cases of the theorem had been stated and proved earlier, but it was Cauchy who made the final jump.

CONCEPT PROBLEM Evaluate the determinant of the matrix A in Example 2 of Section 3.7 by inspection from its LU-decomposition, and then check your answer by calculating $\det(A)$ directly.

DETERMINANT OF THE INVERSE OF A MATRIX

The following theorem provides a fundamental relationship between the determinant of an invertible matrix and the determinant of its inverse.

Theorem 4.2.6 *If A is invertible, then*

$$\det(A^{-1}) = \frac{1}{\det(A)} \qquad (3)$$

Proof Since $AA^{-1} = I$, we have $\det(AA^{-1}) = \det(I) = 1$. This equation can be rewritten as $\det(A)\det(A^{-1}) = 1$, from which the result follows. ∎

EXAMPLE 7

Verifying Theorem 4.2.6 for 2×2 Matrices

Recall that if $A = \begin{bmatrix} a & b \\ c & d \end{bmatrix}$ is invertible, then

$$A^{-1} = \frac{1}{ad - bc} \begin{bmatrix} d & -b \\ -c & a \end{bmatrix}$$

Thus, from part (c) of Theorem 4.2.3 we have

$$\det(A^{-1}) = \frac{1}{(ad - bc)^2} \det \begin{bmatrix} d & -b \\ -c & a \end{bmatrix} = \frac{1}{(ad - bc)^2}(ad - bc) = \frac{1}{ad - bc} = \frac{1}{\det(A)} \blacksquare$$

DETERMINANT OF $A + B$

Given that $\det(AB) = \det(A)\det(B)$, you might be tempted to believe that $\det(A + B) = \det(A) + \det(B)$. However, the following example shows that this is *false*.

EXAMPLE 8

$\det(A + B) \neq \det(A) + \det(B)$

Consider the matrices

$$A = \begin{bmatrix} 1 & 2 \\ 2 & 5 \end{bmatrix}, \quad B = \begin{bmatrix} 3 & 1 \\ 1 & 3 \end{bmatrix}, \quad A + B = \begin{bmatrix} 4 & 3 \\ 3 & 8 \end{bmatrix}$$

We have $\det(A) = 1$, $\det(B) = 8$, and $\det(A + B) = 23$, so $\det(A + B) \neq \det(A) + \det(B)$. For a useful result about expressions of the form $\det(A) + \det(B)$ see Exercise P1. ∎

A UNIFYING THEOREM

In Theorem 3.4.9 we gave a theorem that tied together most of the major concepts developed at that point in the text. Theorem 4.2.4 now allows us to weave a property of determinants into that theorem.

Theorem 4.2.7 *If A is an $n \times n$ matrix, then the following statements are equivalent.*

(a) *The reduced row echelon form of A is I_n.*

(b) *A is expressible as a product of elementary matrices.*

(c) *A is invertible.*

(d) *$A\mathbf{x} = \mathbf{0}$ has only the trivial solution.*

(e) *$A\mathbf{x} = \mathbf{b}$ is consistent for every vector $\mathbf{b}$ in R^n.*

(f) *$A\mathbf{x} = \mathbf{b}$ has exactly one solution for every vector $\mathbf{b}$ in R^n.*

(g) *The column vectors of A are linearly independent.*

(h) *The row vectors of A are linearly independent.*

(i) *$\det(A) \neq 0$.*

OPTIONAL *Proof of Theorem 4.2.1*

The following proof of Theorem 4.2.1 is for readers who are familiar with the principle of induction.

Proof There are two parts to the induction proof: First we must prove that the result is true for matrices of size 1×1 (the case $n = 1$), and then we must prove that the result is true for matrices of size $n \times n$ under the hypothesis that it is true for matrices of size $(n-1) \times (n-1)$ (the induction hypothesis).

The proof is trivial for a 1×1 matrix $A = [a_{11}]$, since $A^T = A$ in this case. Next, assume that the theorem is true for matrices of size $(n-1) \times (n-1)$, and let A be a matrix of size $n \times n$. We can evaluate $\det(A)$ and $\det(A^T)$ by cofactor expansions along any row or column. In particular, if we evaluate $\det(A)$ by a cofactor expansion along the first row and $\det(A^T)$ by a cofactor expansion along the first column, we obtain

$$\det(A) = a_{11}C_{11} + a_{12}C_{12} + \cdots + a_{1n}C_{1n}$$

$$\det(A^T) = a_{11}C'_{11} + a_{12}C'_{21} + \cdots + a_{1n}C'_{n1}$$

where the C_{ij}'s and the C'_{ij}'s are the appropriate cofactors and the corresponding coefficients of the cofactors in the two equations are the same because the entries along the first column of A^T are the entries along the first row of A. We will now prove that

$$C_{11} = C'_{11}, \quad C_{12} = C'_{21}, \ldots, \quad C_{1n} = C'_{n1} \tag{4}$$

from which it will follow that $\det(A) = \det(A^T)$. To prove (4) observe that the matrices used to compute the cofactors in these equations all have size $(n-1) \times (n-1)$, so the induction hypothesis implies that transposing A does not affect the values of these cofactors; however, transposing does interchange the row and column indices, which establishes (4) and completes the proof. ∎

OPTIONAL *Proof of Theorem 4.2.5*

To prove Theorem 4.2.5 we will need two preliminary results, which we will refer to as lemmas. The first lemma follows directly from Theorem 4.2.2 and the fact that multiplying a matrix A on the left by an elementary matrix performs the row operation on A to which the elementary matrix corresponds.

Lemma 4.2.8 *Let E be an $n \times n$ elementary matrix and I_n the $n \times n$ identity matrix.*

    (a) If E results by multiplying a row of I_n by k, then $\det(E) = k$.

    (b) If E results by interchanging two rows of I_n, then $\det(E) = -1$.

    (c) If E results by adding a multiple of one row of I_n to another, then $\det(E) = 1$.

The next lemma is the special case of Theorem 4.2.5 in which A is an elementary matrix.

Lemma 4.2.9 *If B is an $n \times n$ matrix and E is an $n \times n$ elementary matrix, then*

$$\det(EB) = \det(E)\det(B)$$

Proof The proof is simply a matter of analyzing what happens for each of the three possible types of elementary row operations. For example, if E results from multiplying a row of I_n by k, then EB is the matrix that results when the same row of B is multiplied by k, and hence $\det(EB) = k\det(B)$ by Theorem 4.2.2(a). But Lemma 4.2.8(a) implies that $\det(E)\det(B) = k\det(B)$, so $\det(EB) = \det(E)\det(B)$. The argument is similar for the other two elementary row operations (verify). ∎

Lemma 4.2.9 can be applied repeatedly to prove the more general result that if $E_1, E_2, \ldots, E_r$ is a sequence of elementary matrices, then

$$\det(E_1 E_2 \cdots E_r B) = \det(E_1)\det(E_2)\cdots\det(E_r)\det(B) \tag{5}$$

In the special case where $B = I$ is the identity matrix, it follows from this equation that

$$\det(E_1 E_2 \cdots E_r) = \det(E_1)\det(E_2)\cdots\det(E_r) \tag{6}$$

Now to the general proof that $\det(AB) = \det(A)\det(B)$.

Proof of Theorem 4.2.5 First we will consider the case where the matrix A is singular and then the case where it is invertible. It follows from part (b) of Theorem 3.3.8 that if A is singular, then so is AB. Thus, in the case where A is singular we have $\det(A) = 0$ and $\det(AB) = 0$, from which it follows that $\det(A)\det(B) = \det(AB)$.

Now assume that A is invertible, in which case it can be factored into a product of elementary matrices, say

$$A = E_1 E_2 \cdots E_r \tag{7}$$

Thus, from (5)

$$\det(AB) = \det(E_1 E_2 \cdots E_r B) = \det(E_1)\det(E_2)\cdots\det(E_r)\det(B)$$

From (6) and (7) we can rewrite this as

$$\det(AB) = \det(E_1 E_2 \cdots E_r)\det(B) = \det(A)\det(B)$$

which completes the proof. ∎

Exercise Set 4.2

In Exercises 1 and 2, verify that $\det(A) = \det(A^T)$.

1. (a) $A = \begin{bmatrix} -2 & 3 \\ 1 & 4 \end{bmatrix}$ (b) $A = \begin{bmatrix} 2 & -1 & 3 \\ 1 & 2 & 4 \\ 5 & -3 & 6 \end{bmatrix}$

2. (a) $A = \begin{bmatrix} 1 & 2 \\ 2 & 6 \end{bmatrix}$ (b) $A = \begin{bmatrix} 1 & 3 & 2 \\ 3 & 1 & 5 \\ 2 & -1 & 1 \end{bmatrix}$

In Exercises 3 and 4, evaluate the determinants by inspection.

3. (a) $\begin{vmatrix} 3 & 1 & 3 & 33 \\ 0 & \frac{1}{3} & 9 & 22 \\ 0 & 0 & -2 & 12 \\ 0 & 0 & 0 & 2 \end{vmatrix}$ (b) $\begin{vmatrix} 3 & 1 & 9 \\ -1 & 2 & -3 \\ 1 & 5 & 3 \end{vmatrix}$

(c) $\begin{vmatrix} 3 & -17 & 4 \\ 0 & 5 & 1 \\ 0 & 0 & -2 \end{vmatrix}$

4. (a) $\begin{vmatrix} \sqrt{2} & 0 & 0 & 0 \\ -8 & \sqrt{2} & 0 & 0 \\ 7 & 0 & -1 & 0 \\ 9 & 5 & 6 & 1 \end{vmatrix}$ (b) $\begin{vmatrix} -2 & 1 & 3 \\ 1 & -7 & 4 \\ -2 & 1 & 3 \end{vmatrix}$

(c) $\begin{vmatrix} 1 & -2 & 3 \\ 2 & -4 & 6 \\ 5 & -8 & 1 \end{vmatrix}$

In Exercises 5 and 6, find the determinant given that

$$\begin{vmatrix} a & b & c \\ d & e & f \\ g & h & i \end{vmatrix} = -6$$

5. (a) $\begin{vmatrix} d & e & f \\ g & h & i \\ a & b & c \end{vmatrix}$ (b) $\begin{vmatrix} 3a & 3b & 3c \\ -d & -e & -f \\ 4g & 4h & 4i \end{vmatrix}$

(c) $\begin{vmatrix} a+g & b+h & c+i \\ d & e & f \\ g & h & i \end{vmatrix}$ (d) $\begin{vmatrix} -3a & -3b & -3c \\ d & e & f \\ g-4d & h-4e & i-4f \end{vmatrix}$

6. (a) $\begin{vmatrix} g & h & i \\ d & e & f \\ a & b & c \end{vmatrix}$ (b) $\begin{vmatrix} a & b & c \\ d & e & f \\ 2a & 2b & 2c \end{vmatrix}$

(c) $\begin{vmatrix} a+d & b+e & c+f \\ -d & -e & -f \\ g & h & i \end{vmatrix}$ (d) $\begin{vmatrix} a & b & c \\ 2d & 2e & 2f \\ g+3a & h+3b & i+3c \end{vmatrix}$

In Exercises 7 and 8, verify that $\det(kA) = k^n \det(A)$ for the given $n \times n$ matrix A.

7. (a) $A = \begin{bmatrix} -1 & 2 \\ 3 & 4 \end{bmatrix}$; $k = 2$

(b) $A = \begin{bmatrix} 2 & -1 & 3 \\ 3 & 2 & 1 \\ 1 & 4 & 5 \end{bmatrix}$; $k = -2$

8. (a) $A = \begin{bmatrix} 2 & 2 \\ 5 & -2 \end{bmatrix}; k = -4$

   (b) $A = \begin{bmatrix} 1 & 1 & 1 \\ 0 & 2 & 3 \\ 0 & 1 & -2 \end{bmatrix}; k = 3$

9. Without directly evaluating the determinant, explain why $x = 0$ and $x = 2$ satisfy the equation

$$\begin{vmatrix} x^2 & x & 2 \\ 2 & 1 & 1 \\ 0 & 0 & -5 \end{vmatrix} = 0$$

10. Without directly evaluating the determinant, explain why

$$\det \begin{bmatrix} b+c & c+a & b+a \\ a & b & c \\ 1 & 1 & 1 \end{bmatrix} = 0$$

In Exercises 11–18, evaluate the determinant of the matrix by reducing it to row echelon form.

11. $\begin{bmatrix} 3 & 6 & -9 \\ 0 & 0 & -2 \\ -2 & 1 & 5 \end{bmatrix}$ **12.** $\begin{bmatrix} 0 & 3 & 1 \\ 1 & 1 & 2 \\ 3 & 2 & 4 \end{bmatrix}$

13. $\begin{bmatrix} 1 & -3 & 0 \\ -2 & 4 & 1 \\ 5 & -2 & 2 \end{bmatrix}$ **14.** $\begin{bmatrix} 3 & -6 & 9 \\ -2 & 7 & -2 \\ 0 & 1 & 5 \end{bmatrix}$

15. $\begin{bmatrix} 1 & -2 & 3 & 1 \\ 5 & -9 & 6 & 3 \\ -1 & 2 & -6 & -2 \\ 2 & 8 & 6 & 1 \end{bmatrix}$ **16.** $\begin{bmatrix} 2 & 1 & 3 & 1 \\ 1 & 0 & 1 & 1 \\ 0 & 2 & 1 & 0 \\ 0 & 1 & 2 & 3 \end{bmatrix}$

17. $\begin{bmatrix} 0 & 1 & 1 & 1 \\ \frac{1}{2} & \frac{1}{2} & 1 & \frac{1}{2} \\ \frac{2}{3} & \frac{1}{3} & \frac{1}{3} & 0 \\ -\frac{1}{3} & \frac{2}{3} & 0 & 0 \end{bmatrix}$ **18.** $\begin{bmatrix} 1 & 3 & 1 & 5 & 3 \\ -2 & -7 & 0 & -4 & 2 \\ 0 & 0 & 1 & 0 & 1 \\ 0 & 0 & 2 & 1 & 1 \\ 0 & 0 & 0 & 1 & 1 \end{bmatrix}$

In Exercises 19 and 20, confirm the identities without evaluating the determinants directly.

19. (a) $\begin{vmatrix} a_1 & b_1 & a_1 + b_1 + c_1 \\ a_2 & b_2 & a_2 + b_2 + c_2 \\ a_3 & b_3 & a_3 + b_3 + c_3 \end{vmatrix} = \begin{vmatrix} a_1 & b_1 & c_1 \\ a_2 & b_2 & c_2 \\ a_3 & b_3 & c_3 \end{vmatrix}$

   (b) $\begin{vmatrix} a_1 + b_1 & a_1 - b_1 & c_1 \\ a_2 + b_2 & a_2 - b_2 & c_2 \\ a_3 + b_3 & a_3 - b_3 & c_3 \end{vmatrix} = -2 \begin{vmatrix} a_1 & b_1 & c_1 \\ a_2 & b_2 & c_2 \\ a_3 & b_3 & c_3 \end{vmatrix}$

20. (a) $\begin{vmatrix} a_1 + b_1 t & a_2 + b_2 t & a_3 + b_3 t \\ a_1 t + b_1 & a_2 t + b_2 & a_3 t + b_3 \\ c_1 & c_2 & c_3 \end{vmatrix} = (1 - t^2) \begin{vmatrix} a_1 & a_2 & a_3 \\ b_1 & b_2 & b_3 \\ c_1 & c_2 & c_3 \end{vmatrix}$

   (b) $\begin{vmatrix} a_1 & b_1 + ta_1 & c_1 + rb_1 + sa_1 \\ a_2 & b_2 + ta_2 & c_2 + rb_2 + sa_2 \\ a_3 & b_3 + ta_3 & c_3 + rb_3 + sa_3 \end{vmatrix} = \begin{vmatrix} a_1 & a_2 & a_3 \\ b_1 & b_2 & b_3 \\ c_1 & c_2 & c_3 \end{vmatrix}$

21. Show that

$$\begin{vmatrix} 1 & x & x^2 \\ 1 & y & y^2 \\ 1 & z & z^2 \end{vmatrix} = (y - x)(z - x)(z - y)$$

22. Find the determinant of the matrix

$$\begin{bmatrix} a & b & b & b \\ b & a & b & b \\ b & b & a & b \\ b & b & b & a \end{bmatrix}$$

In Exercises 23 and 24, show that $\det(A) = 0$ without directly evaluating the determinant.

23. $A = \begin{bmatrix} -2 & 8 & 1 & 4 \\ 3 & 2 & 5 & 1 \\ 1 & 10 & 6 & 5 \\ 4 & -6 & 4 & -3 \end{bmatrix}$

24. $A = \begin{bmatrix} -4 & 1 & 1 & 1 & 1 \\ 1 & -4 & 1 & 1 & 1 \\ 1 & 1 & -4 & 1 & 1 \\ 1 & 1 & 1 & -4 & 1 \\ 1 & 1 & 1 & 1 & -4 \end{bmatrix}$

In Exercises 25 and 26, find the values of k for which A is invertible.

25. (a) $A = \begin{bmatrix} k - 3 & -2 \\ -2 & k - 2 \end{bmatrix}$ (b) $A = \begin{bmatrix} 1 & 2 & 4 \\ 3 & 1 & 6 \\ k & 3 & 2 \end{bmatrix}$

26. (a) $A = \begin{bmatrix} k & 2 \\ 2 & k \end{bmatrix}$ (b) $A = \begin{bmatrix} 1 & 2 & 0 \\ k & 1 & k \\ 0 & 2 & 1 \end{bmatrix}$

27. In each part, find the determinant given that A is a 3×3 matrix for which $\det(A) = 7$.
   (a) $\det(3A)$ (b) $\det(A^{-1})$
   (c) $\det(2A^{-1})$ (d) $\det((2A)^{-1})$

28. In each part, find the determinant given that A is a 4×4 matrix for which $\det(A) = -2$.
   (a) $\det(-A)$ (b) $\det(A^{-1})$
   (c) $\det(2A^T)$ (d) $\det(A^3)$

In Exercises 29 and 30, verify that

$$\det(AB) = \det(A)\det(B)$$

29. $A = \begin{bmatrix} 1 & 0 & 0 \\ 3 & 3 & 0 \\ 5 & 2 & -2 \end{bmatrix}$, $B = \begin{bmatrix} 2 & 4 & -5 \\ 0 & 1 & 3 \\ 0 & 0 & 2 \end{bmatrix}$

30. $A = \begin{bmatrix} 2 & 1 & 0 \\ 3 & 4 & 0 \\ 0 & 0 & 2 \end{bmatrix}$, $B = \begin{bmatrix} 1 & -1 & 3 \\ 7 & 1 & 2 \\ 5 & 0 & 1 \end{bmatrix}$

In Exercises 31 and 32, evaluate the determinant of the matrix by forming its LU-decomposition and using Theorem 4.2.5.

31. The matrix in Exercise 15.

32. The matrix in Exercise 16.

33. Use a determinant to show that the matrix
$$\begin{bmatrix} \sin^2 \alpha & \sin^2 \beta & \sin^2 \gamma \\ \cos^2 \alpha & \cos^2 \beta & \cos^2 \gamma \\ 1 & 1 & 1 \end{bmatrix}$$
is not invertible for any values of α, β, and γ.

34. Find all real values of λ, if any, for which the system has a nontrivial solution.

(a) $\begin{aligned} x_1 + 2x_2 &= \lambda x_1 \\ 2x_1 + x_2 &= \lambda x_2 \end{aligned}$ (b) $\begin{aligned} 2x_1 + 3x_2 &= \lambda x_1 \\ 4x_1 + 3x_2 &= \lambda x_2 \end{aligned}$

(c) $\begin{aligned} 3x_1 + x_2 &= \lambda x_1 \\ -5x_1 - 3x_2 &= \lambda x_2 \end{aligned}$

35. (a) Show that if A is a square matrix, then $\det(A^T A) = \det(A A^T)$.

(b) Show that A is invertible if and only if $A^T A$ is invertible.

36. Let A and B be matrices. Show that if A is invertible, then $\det(A^{-1} B A) = \det(B)$.

37. Show that if the vectors $\mathbf{x} = (x_1, x_2, x_3)$ and $\mathbf{y} = (y_1, y_2, y_3)$ are any vectors in R^3, then
$$\|\mathbf{x}\|^2 \|\mathbf{y}\|^2 - (\mathbf{x} \cdot \mathbf{y})^2 = \begin{vmatrix} x_1 & x_2 \\ y_1 & y_2 \end{vmatrix}^2 + \begin{vmatrix} x_1 & x_3 \\ y_1 & y_3 \end{vmatrix}^2 + \begin{vmatrix} x_2 & x_3 \\ y_2 & y_3 \end{vmatrix}^2$$

It can be proved that if a square matrix M is partitioned into *block triangular form* as
$$M = \begin{bmatrix} A & 0 \\ C & B \end{bmatrix} \quad \text{or} \quad M = \begin{bmatrix} A & C \\ 0 & B \end{bmatrix}$$
in which A and B are square, then $\det(M) = \det(A) \det(B)$. Use this result to compute the determinants of the matrices in Exercises 38 and 39.

38. (a) $M = \begin{bmatrix} 1 & 2 & 0 & 0 \\ 2 & 5 & 0 & 0 \\ 5 & 12 & 3 & 3 \\ 11 & -8 & 2 & 1 \end{bmatrix}$

(b) $M = \begin{bmatrix} 1 & 2 & 0 & 8 & 6 & -9 \\ 2 & 5 & 0 & 4 & 7 & 5 \\ -1 & 3 & 2 & 6 & 9 & -2 \\ 0 & 0 & 0 & 3 & 0 & 0 \\ 0 & 0 & 0 & 2 & 1 & 0 \\ 0 & 0 & 0 & -3 & 8 & -4 \end{bmatrix}$

39. (a) $M = \begin{bmatrix} 2 & -1 & 2 & 5 & 6 \\ 4 & 3 & -1 & 3 & 4 \\ 0 & 0 & 1 & 3 & 5 \\ 0 & 0 & -2 & 6 & 2 \\ 0 & 0 & 3 & 5 & 2 \end{bmatrix}$

(b) $M = \begin{bmatrix} 1 & 2 & 0 & 0 & 0 \\ 0 & 1 & 2 & 0 & 0 \\ 0 & 0 & 1 & 0 & 0 \\ 0 & 0 & 0 & 1 & 2 \\ 2 & 0 & 0 & 0 & 1 \end{bmatrix}$

40. Show that if P is a permutation matrix, then $\det(P) = \pm 1$.

Discussion and Discovery

D1. What can you say about the values of s and t if the matrices
$$\begin{bmatrix} 1 & 2 & s \\ 2 & 3 & t \\ 4 & 5 & 7 \end{bmatrix} \quad \text{and} \quad \begin{bmatrix} 4 & 5 & 8 \\ s & 2 & 3 \\ t & 1 & 8 \end{bmatrix}$$
are both singular?

D2. Let A and B be $n \times n$ matrices. You know from earlier work that AB and BA need not be equal. Is this also true for $\det(AB)$ and $\det(BA)$? Explain your reasoning.

D3. Let A and B be $n \times n$ matrices. We know that AB is invertible if A and B are invertible. What can you say about the invertibility of AB if A or B is not invertible? Explain your reasoning.

D4. What can you say about the determinant of an $n \times n$ matrix with the following form?
$$\begin{bmatrix} 0 & 0 & \cdots & 0 & 1 \\ 0 & 0 & \cdots & 1 & 0 \\ \vdots & \vdots & & \vdots & \vdots \\ 0 & 1 & \cdots & 0 & 0 \\ 1 & 0 & \cdots & 0 & 0 \end{bmatrix}$$

D5. What can you say about the determinant of an $n \times n$ skew-symmetric matrix A if n is odd?

D6. Let A be an $n \times n$ matrix, and let B be the matrix that results when the rows of A are written in reverse order. State a theorem that describes how $\det(A)$ and $\det(B)$ are related.

D7. Indicate whether the statement is true (T) or false (F). Justify your answer.

    (a) $\det(I + A) = 1 + \det(A)$

    (b) $\det(A^4) = (\det(A))^4$

    (c) $\det(3A) = 3\det(A)$

    (d) If $\det(A) = 0$, then the homogeneous system $A\mathbf{x} = \mathbf{0}$ has infinitely many solutions.

D8. Indicate whether the statement is true (T) or false (F). Justify your answer.

    (a) If A is invertible and $\det(ABA) = 0$, then $\det(B) = 0$.

    (b) If $A = A^{-1}$, then $\det(A) = \pm 1$.

    (c) If the reduced row echelon form of A has a row of zeros, then the determinant of A is 0.

    (d) There is no square matrix A such that $\det(AA^T) = -1$.

    (e) If $\det(A) \neq 0$, then A is expressible as a product of elementary matrices.

D9. If $A = A^2$, what can you say about the determinant of A? What can you say if $A = A^3$?

D10. Let A be a matrix of the form

$$A = \begin{bmatrix} * & * & 0 & 0 & 0 \\ * & * & 0 & 0 & 0 \\ * & * & 0 & 0 & 0 \\ * & * & * & * & * \\ * & * & * & * & * \end{bmatrix}$$

How many different values can you obtain for $\det(A)$ by substituting numerical values (not necessarily all the same) for the $*$'s? Explain your reasoning.

D11. How will the value of an nth-order determinant change if the first column is made the last column, and all other columns are moved left by one position?

Working with Proofs

P1. Let A, B, and C be $n \times n$ matrices of the form

$$A = [\mathbf{c}_1 \cdots \mathbf{x} \cdots \mathbf{c}_n], \quad B = [\mathbf{c}_1 \cdots \mathbf{y} \cdots \mathbf{c}_n]$$
$$C = [\mathbf{c}_1 \cdots \mathbf{x} + \mathbf{y} \cdots \mathbf{c}_n]$$

where $\mathbf{x}$, $\mathbf{y}$, and $\mathbf{x} + \mathbf{y}$ are the jth column vectors. Use cofactor expansions to prove that $\det(C) = \det(A) + \det(B)$.

P2. Prove Theorem 4.2.2(c). [*Hint:* Theorem 4.2.3 uses only parts (a) and (b) of Theorem 4.2.2, so you can use those results in this proof. Assume that k times row i is added to row j, then expand along the new row j.]

Technology Exercises

T1. Find the LU-decomposition of the matrix A, and then use it to find $\det(A)$ by inspection.

$$A = \begin{bmatrix} -2 & 2 & -4 & -6 \\ -3 & 6 & 3 & -15 \\ 5 & -8 & -1 & 17 \\ 1 & 1 & 11 & 7 \end{bmatrix}$$

Check your result by computing $\det(A)$ directly.

T2. Confirm the formulas in Theorem 4.2.2 for a 5×5 matrix of your choice.

T3. Let

$$A = \begin{bmatrix} 1 & 3 & 5 & -8 & 9 \\ 11 & 21 & 7 & -3 & 6 \\ -12 & 0 & 3 & 7 & 8 \\ 0 & -3 & 7 & 21 & 3 \\ \varepsilon & -3 & 7 & 21 & 3 \end{bmatrix}$$

    (a) See if you can find a small nonzero value of ε for which your technology utility tells you that $\det(A) \neq 0$ and A is not invertible.

    (b) Do you think that this contradicts Theorem 4.2.4? Explain.

T4. We know from Exercise 35 that if A is a square matrix, then $\det(A^T A) = \det(AA^T)$. By experimentation, see if you think that the equality always holds if A is not square.

T5. (CAS) Use a determinant to show that if a, b, c, and d are not all zero, then the vectors

$$\mathbf{v}_1 = (a, b, c, d) \qquad \mathbf{v}_2 = (-b, a, d, -c)$$
$$\mathbf{v}_3 = (-c, -d, a, b) \qquad \mathbf{v}_4 = (-d, c, -b, a)$$

are linearly independent.

Section 4.3   Cramer's Rule; Formula for A^{-1}; Applications of Determinants

In this section we will use determinants to derive formulas for the inverse of a matrix and for the solution of a consistent linear system of n equations in n unknowns. We will also discuss some important applications of determinants.

ADJOINT OF A MATRIX   In a cofactor expansion we compute $\det(A)$ by multiplying the entries in any row or column of A by their cofactors and adding the resulting products. We now inquire as to what happens when the entries of any row (column) of A are multiplied by the corresponding cofactors from a *different* row (column) and the resulting products added. Remarkably, the result is always zero. To see why this so, consider the 3×3 matrix

$$A = \begin{bmatrix} a_{11} & a_{12} & a_{13} \\ a_{21} & a_{22} & a_{23} \\ a_{31} & a_{32} & a_{33} \end{bmatrix}$$

and suppose, for example, that we multiply the entries in the first row by the corresponding cofactors from the third row and form the sum

$$a_{11}C_{31} + a_{12}C_{32} + a_{13}C_{33} \tag{1}$$

To see that this sum is zero consider the matrix A' that results when the third row of A is replaced by a duplicate of the first row:

$$A' = \begin{bmatrix} a_{11} & a_{12} & a_{13} \\ a_{21} & a_{22} & a_{23} \\ a_{11} & a_{12} & a_{13} \end{bmatrix}$$

Without any computation we know that $\det(A') = 0$ because of the duplicate rows. However, let us evaluate $\det(A')$ by a cofactor expansion along the third row. The cofactors of the entries in this row are the same as the cofactors in the third row of A since the third row is crossed out in both matrices when these cofactors are computed. Thus,

$$\det(A') = a_{11}C_{31} + a_{12}C_{32} + a_{13}C_{33} = 0$$

which is what we wanted to show. This argument can be adapted to produce a proof of the following general theorem.

Theorem 4.3.1   *If the entries in any row (column) of a square matrix are multiplied by the cofactors of the corresponding entries in a different row (column), then the sum of the products is zero.*

CONCEPT PROBLEM   Confirm the result in this theorem for the matrix in Example 3 of Section 4.1 using one pair of rows and one pair of columns of your choice.

To derive a formula for the inverse of an invertible matrix we will need Theorem 4.3.1 and the following concept.

Definition 4.3.2   If A is an $n \times n$ matrix and C_{ij} is the cofactor of a_{ij}, then the matrix

$$C = \begin{bmatrix} C_{11} & C_{12} & \cdots & C_{1n} \\ C_{21} & C_{22} & \cdots & C_{2n} \\ \vdots & \vdots & & \vdots \\ C_{n1} & C_{n2} & \cdots & C_{nn} \end{bmatrix}$$

is called the ***matrix of cofactors*** from A. The transpose of this matrix is called the ***adjoint*** (or sometimes the ***adjugate***) of A and is denoted by adj(A).

EXAMPLE 1   The cofactors of the matrix

$$A = \begin{bmatrix} 3 & 2 & -1 \\ 1 & 6 & 3 \\ 2 & -4 & 0 \end{bmatrix}$$

are

$$\begin{array}{lll} C_{11} = 12 & C_{12} = 6 & C_{13} = -16 \\ C_{21} = 4 & C_{22} = 2 & C_{23} = 16 \\ C_{31} = 12 & C_{32} = -10 & C_{33} = 16 \end{array}$$

(verify) so the matrix of cofactors and the adjoint are

$$C = \begin{bmatrix} 12 & 6 & -16 \\ 4 & 2 & 16 \\ 12 & -10 & 16 \end{bmatrix} \quad \text{and} \quad \text{adj}(A) = C^T = \begin{bmatrix} 12 & 4 & 12 \\ 6 & 2 & -10 \\ -16 & 16 & 16 \end{bmatrix}$$

respectively.   ■

A FORMULA FOR THE INVERSE OF A MATRIX

We are now in position to derive a formula for the inverse of an invertible matrix.

Theorem 4.3.3   *If A is an invertible matrix, then*

$$A^{-1} = \frac{1}{\det(A)} \, \text{adj}(A) \tag{2}$$

Proof   We will show first that $A \, \text{adj}(A) = \det(A)I$. For this purpose, consider the product

$$A \, \text{adj}(A) = \begin{bmatrix} a_{11} & a_{12} & \cdots & a_{1n} \\ a_{21} & a_{22} & \cdots & a_{2n} \\ \vdots & \vdots & & \vdots \\ a_{i1} & a_{i2} & \cdots & a_{in} \\ \vdots & \vdots & & \vdots \\ a_{n1} & a_{n2} & \cdots & a_{nn} \end{bmatrix} \begin{bmatrix} C_{11} & C_{21} & \cdots & C_{j1} & \cdots & C_{n1} \\ C_{12} & C_{22} & \cdots & C_{j2} & \cdots & C_{n2} \\ \vdots & \vdots & & \vdots & & \vdots \\ C_{1n} & C_{2n} & \cdots & C_{jn} & \cdots & C_{nn} \end{bmatrix}$$

Referring to the shaded row and column in the figure, we see that the entry in the ith row and

jth column of this product is

$$a_{i1}C_{j1} + a_{i2}C_{j2} + \cdots + a_{in}C_{jn} \tag{3}$$

In the case where $i = j$ the entries and cofactors come from the same row of A, so (3) is the cofactor expansion of $\det(A)$ along that row; and in the case where $i \neq j$ the entries and cofactors come from different rows, so the sum is zero by Theorem 4.3.1. Thus,

$$A \operatorname{adj}(A) = \begin{bmatrix} \det(A) & 0 & \cdots & 0 \\ 0 & \det(A) & \cdots & 0 \\ \vdots & \vdots & & \vdots \\ 0 & 0 & \cdots & \det(A) \end{bmatrix} = \det(A)I$$

Since A is invertible, it follows that $\det(A) \neq 0$, so this equation can be rewritten as

$$A \left[\frac{1}{\det(A)} \operatorname{adj}(A) \right] = I$$

from which Formula (2) now follows.   ■

EXAMPLE 2
Using the Adjoint Formula to Calculate an Inverse

Use Formula (2) to find the inverse of the matrix A in Example 1.

Solution We leave it for you to confirm that $\det(A) = 64$. Using this determinant and the adjoint computed in Example 1, we obtain

$$A^{-1} = \frac{1}{\det(A)} \operatorname{adj}(A) = \begin{bmatrix} \frac{12}{64} & \frac{4}{64} & \frac{12}{64} \\ \frac{6}{64} & \frac{2}{64} & -\frac{10}{64} \\ -\frac{16}{64} & \frac{16}{64} & \frac{16}{64} \end{bmatrix}$$

■

CONCEPT PROBLEM Show that the formula for the inverse of a 2×2 matrix that was given in Theorem 3.2.7 is a special case of Formula (2).

HOW THE INVERSE FORMULA IS USED

Formula (2) provides a reasonable way of inverting 3×3 matrices by hand, but for 4×4 matrices or higher the row reduction algorithm discussed in Section 3.3 is usually better. Computer programs usually use LU-decomposition (as discussed in Section 3.7) and not Formula (2) to invert matrices. Thus, the value of Formula (2) is not for numerical computations, but rather as a tool in theoretical analysis.

Here is a simple example of how this formula can be used to deduce properties of A^{-1} that are not immediately evident from the algorithms for calculating the inverse.

EXAMPLE 3
Working with the Adjoint Formula for A^{-1}

If an invertible matrix A has integer entries, then its inverse may or may not have integer entries (verify for 2×2 matrices). However, if an $n \times n$ matrix A has integer entries and $\det(A) = 1$, then A^{-1} must have integer entries. To see why this so, we apply Formula (2) to write

$$A^{-1} = \frac{1}{\det(A)} \operatorname{adj}(A) = \operatorname{adj}(A)$$

But the cofactors that appear in $\operatorname{adj}(A)$ are derived from the minors by sign adjustment and hence involve only additions, subtractions, and multiplications (no divisions). Thus, the cofactors must be integers.   ■

CONCEPT PROBLEM If A is an invertible matrix with integer entries and $\det(A) \neq 1$, is it possible for A^{-1} to have integer entries? Explain.

In parts (*c*) and (*d*) of Theorem 3.6.1 we stated without proof that a triangular matrix *A* is invertible if and only if its diagonal entries are all nonzero and that in that case the inverse is also triangular (more precisely, upper triangular if *A* is upper triangular, and lower triangular if *A* is lower triangular). The proof of the first statement follows from the fact that det(*A*) is the product of the diagonal entries, so det(*A*) $\neq$ 0 if and only if the diagonal entries are all nonzero. A proof that A^{-1} is triangular can be based on Formula (2) (Exercises 55 and 56).

CRAMER'S RULE

When we began our study of determinants in Section 4.1, we stated that determinants are defined in a way that allows solutions of certain linear systems to be expressed as ratios of them. We will now explain how this can be done for linear systems of *n* equations in *n* unknowns. For motivation, let us start with a system $A\mathbf{x} = \mathbf{b}$ of two equations in two unknowns, say

$$\begin{aligned} a_{11}x + a_{12}y &= b_1 \\ a_{21}x + a_{22}y &= b_2 \end{aligned} \quad \text{or in matrix form,} \quad \begin{bmatrix} a_{11} & a_{12} \\ a_{21} & a_{22} \end{bmatrix} \begin{bmatrix} x \\ y \end{bmatrix} = \begin{bmatrix} b_1 \\ b_2 \end{bmatrix}$$

If the coefficient matrix *A* is invertible, then by making the appropriate notation adjustments to Formula (2) in Section 4.1, the solution of the system can be expressed as

$$x = \frac{\begin{vmatrix} b_1 & a_{12} \\ b_2 & a_{22} \end{vmatrix}}{\begin{vmatrix} a_{11} & a_{12} \\ a_{21} & a_{22} \end{vmatrix}}, \quad y = \frac{\begin{vmatrix} a_{11} & b_1 \\ a_{21} & b_2 \end{vmatrix}}{\begin{vmatrix} a_{11} & a_{12} \\ a_{21} & a_{22} \end{vmatrix}} \tag{4}$$

(verify). Note the pattern here: In both formulas the denominator is the determinant of the coefficient matrix, in the formula for *x* the numerator is the determinant that results when the coefficients of *x* in the denominator are replaced by the entries of **b**, and in the formula for *y* the numerator is the determinant that results when the coefficients of *y* in the denominator are replaced by the entries of **b**. This formula is called ***Cramer's rule*** for two equations in two unknowns. Here is an example.

EXAMPLE 4
Cramer's Rule for Two Equations in Two Unknowns

Use Cramer's rule to solve the system

$$\begin{aligned} 2x - 6y &= 1 \\ 3x - 4y &= 5 \end{aligned}$$

Solution From (4),

$$x = \frac{\begin{vmatrix} 1 & -6 \\ 5 & -4 \end{vmatrix}}{\begin{vmatrix} 2 & -6 \\ 3 & -4 \end{vmatrix}} = \frac{26}{10} = \frac{13}{5} \quad \text{and} \quad y = \frac{\begin{vmatrix} 2 & 1 \\ 3 & 5 \end{vmatrix}}{\begin{vmatrix} 2 & -6 \\ 3 & -4 \end{vmatrix}} = \frac{7}{10} \qquad \blacksquare$$

Cramer's rule is especially useful for linear systems that involve symbolic rather than numerical coefficients.

EXAMPLE 5
Solving Symbolic Equations by Cramer's Rule

Use Cramer's rule to solve the equations

$$\begin{aligned} x' &= x\cos\theta + y\sin\theta \\ y' &= -x\sin\theta + y\cos\theta \end{aligned}$$

for *x* and *y* in terms of x' and y'.

Solution The determinant of the coefficient matrix is

$$\begin{vmatrix} \cos\theta & \sin\theta \\ -\sin\theta & \cos\theta \end{vmatrix} = \cos^2\theta + \sin^2\theta = 1$$

Thus, Cramer's rule yields

$$x = \begin{vmatrix} x' & \sin\theta \\ y' & \cos\theta \end{vmatrix} = x'\cos\theta - y'\sin\theta \quad \text{and} \quad y = \begin{vmatrix} \cos\theta & x' \\ -\sin\theta & y' \end{vmatrix} = x'\sin\theta + y'\cos\theta \qquad \blacksquare$$

The following theorem is the extension of Cramer's rule to linear systems of n equations in n unknowns.

Theorem 4.3.4 (*Cramer's Rule*) *If* $A\mathbf{x} = \mathbf{b}$ *is a linear system of n equations in n unknowns, then the system has a unique solution if and only if* $\det(A) \neq 0$, *in which case the solution is*

$$x_1 = \frac{\det(A_1)}{\det(A)}, \quad x_2 = \frac{\det(A_2)}{\det(A)}, \ldots, \quad x_n = \frac{\det(A_n)}{\det(A)}$$

where A_j *is the matrix that results when the jth column of A is replaced by* **b**.

Proof We already know from Theorem 4.2.7 that if $\det(A) \neq 0$, then $A\mathbf{x} = \mathbf{b}$ has a unique solution. Conversely, if $A\mathbf{x} = \mathbf{b}$ has a unique solution, then $A\mathbf{x} = \mathbf{0}$ has only the trivial solution, for otherwise we could add a nontrivial solution of $A\mathbf{x} = \mathbf{0}$ to the solution of $A\mathbf{x} = \mathbf{b}$ and get another solution of this system—contradicting the uniqueness. Since $A\mathbf{x} = \mathbf{0}$ has only the trivial solution, Theorem 4.2.7 implies that $\det(A) \neq 0$. In the case where $\det(A) \neq 0$, we can use Formula (2) to rewrite the unique solution of $A\mathbf{x} = \mathbf{b}$ as

$$\mathbf{x} = A^{-1}\mathbf{b} = \frac{1}{\det(A)}\text{adj}(A)\mathbf{b} = \frac{1}{\det(A)} \begin{bmatrix} C_{11} & C_{21} & \cdots & C_{n1} \\ C_{12} & C_{22} & \cdots & C_{n2} \\ \vdots & \vdots & & \vdots \\ C_{1n} & C_{2n} & \cdots & C_{nn} \end{bmatrix} \begin{bmatrix} b_1 \\ b_2 \\ \vdots \\ b_n \end{bmatrix}$$

Therefore, the entry in the jth row of $\mathbf{x}$ is

$$x_j = \frac{b_1 C_{1j} + b_2 C_{2j} + \cdots + b_n C_{nj}}{\det(A)} \qquad (5)$$

where $b_1, b_2, \ldots, b_n$ are the entries of **b**. The cofactors in this expression come from the jth column of A and hence remain unchanged if we replace the jth column of A by **b** (the jth column is crossed out when the cofactors are computed). Since this substitution produces the matrix A_j, the numerator in (5) can be interpreted as the cofactor expansion along the jth column of A_j. Thus,

$$x_j = \frac{\det(A_j)}{\det(A)}$$

which completes the proof. $\blacksquare$

EXAMPLE 6   Solve the system

$$\begin{aligned} x_1 + + 2x_3 &= 6 \\ -3x_1 + 4x_2 + 6x_3 &= 30 \\ -x_1 - 2x_2 + 3x_3 &= 8 \end{aligned}$$

Solution

$$A = \begin{bmatrix} 1 & 0 & 2 \\ -3 & 4 & 6 \\ -1 & -2 & 3 \end{bmatrix}, \quad A_1 = \begin{bmatrix} 6 & 0 & 2 \\ 30 & 4 & 6 \\ 8 & -2 & 3 \end{bmatrix}, \quad A_2 = \begin{bmatrix} 1 & 6 & 2 \\ -3 & 30 & 6 \\ -1 & 8 & 3 \end{bmatrix}, \quad A_3 = \begin{bmatrix} 1 & 0 & 6 \\ -3 & 4 & 30 \\ -1 & -2 & 8 \end{bmatrix}$$

Linear Algebra in History

Variations of Cramer's rule were fairly well known before the Swiss mathematician Gabriel Cramer discussed it in his 1750 publication entitled *Introduction à l'analyse des lignes courbes algébriques*. It was Cramer's superior notation that popularized the method and led mathematicians to attach his name to it.

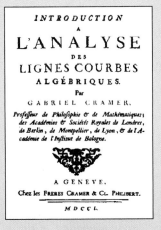

Gabriel Cramer
(1704–1752)

Therefore,

$$x_1 = \frac{\det(A_1)}{\det(A)} = \frac{-40}{44} = \frac{-10}{11}, \quad x_2 = \frac{\det(A_2)}{\det(A)} = \frac{72}{44} = \frac{18}{11}$$

$$x_3 = \frac{\det(A_3)}{\det(A)} = \frac{152}{44} = \frac{38}{11}$$

■

GEOMETRIC INTERPRETATION OF DETERMINANTS

The following theorem provides a geometric interpretation of 2×2 and 3×3 determinants (Figure 4.3.1).

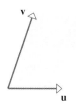

Theorem 4.3.5

(a) *If A is a 2×2 matrix, then $|\det(A)|$ represents the area of the parallelogram determined by the two column vectors of A when they are positioned so their initial points coincide.*

(b) *If A is a 3×3 matrix, then $|\det(A)|$ represents the volume of the parallelepiped determined by the three column vectors of A when they are positioned so their initial points coincide.*

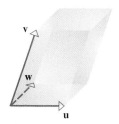

REMARK This theorem is intended to allow for the ***degenerate*** cases of parallelograms and parallelepipeds. (A degenerate parallelogram occurs if one of the vectors that determine the parallelogram has length zero or if the vectors lie on the same line; and a degenerate parallelepiped occurs if any of the vectors that determine the parallelepiped have length zero, or if two of those vectors lie on the same line, or if the three vectors lie in the same plane.) We define the area of a degenerate parallelogram and the volume of a degenerate parallelepiped to be zero, thereby making Theorem 4.3.5 valid in the degenerate cases.

Figure 4.3.1

We will prove part (*a*) of this theorem and omit the proof of part (*b*).

Proof (*a*) Suppose that the matrix A is partitioned into columns as

$$A = \begin{bmatrix} x_1 & x_2 \\ y_1 & y_2 \end{bmatrix} = \begin{bmatrix} \mathbf{u} & \mathbf{v} \end{bmatrix}$$

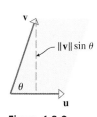

and let us assume that the parallelogram with adjacent sides $\mathbf{u}$ and $\mathbf{v}$ is not degenerate. We know from elementary geometry that the area of a parallelogram is its base times its height. Thus, we see from Figure 4.3.2 that this area can be expressed as

$$\text{area} = \text{base} \times \text{height} = \|\mathbf{u}\|\|\mathbf{v}\| \sin\theta$$

Figure 4.3.2

Thus, the square of the area can be expressed as

$$(\text{area})^2 = \|\mathbf{u}\|^2\|\mathbf{v}\|^2 \sin^2\theta = \|\mathbf{u}\|^2\|\mathbf{v}\|^2(1 - \cos^2\theta) = \|\mathbf{u}\|^2\|\mathbf{v}\|^2 - (\|\mathbf{u}\|\|\mathbf{v}\| \cos\theta)^2 \quad (6)$$

Thus, it follows from Formula (15) of Theorem 1.2.8 that

$$\begin{aligned}(\text{area})^2 &= \|\mathbf{u}\|^2\|\mathbf{v}\|^2 - (\mathbf{u} \cdot \mathbf{v})^2 \\ &= (x_1^2 + y_1^2)(x_2^2 + y_2^2) - (x_1x_2 + y_1y_2)^2 \\ &= x_1^2 y_2^2 + x_2^2 y_1^2 - 2(x_1 x_2)(y_1 y_2) \\ &= (x_1 y_2 - x_2 y_1)^2 = [\det(A)]^2\end{aligned}$$

Thus, it follows on taking square roots that the area of the parallelogram is area $= |\det(A)|$, which is what we wanted to show. In the degenerate cases where one or both columns of A are **0** or in which one of the columns is a scalar multiple of the other, we have $\det(A) = 0$, so the theorem is valid in these cases as well. ■

EXAMPLE 7
Area of a
Parallelogram
Using
Determinants

Find the area of the parallelogram with vertices $P_1(-1, 2)$, $P_2(1, 7)$, $P_3(7, 8)$, and $P_4(5, 3)$.

Solution From Figure 4.3.3 we see that the vectors $\overrightarrow{P_1P_2} = (2, 5)$ and $\overrightarrow{P_1P_4} = (6, 1)$ form adjacent sides of the parallelogram. Thus, it follows from Theorem 4.3.5(a) that area $= \pm \det(A)$, where the sign is chosen to produce a nonnegative value for the area. Thus,

$$\text{area of parallelogram} = \pm \begin{vmatrix} 2 & 6 \\ 5 & 1 \end{vmatrix} = \pm(-28) = 28 \qquad \blacksquare$$

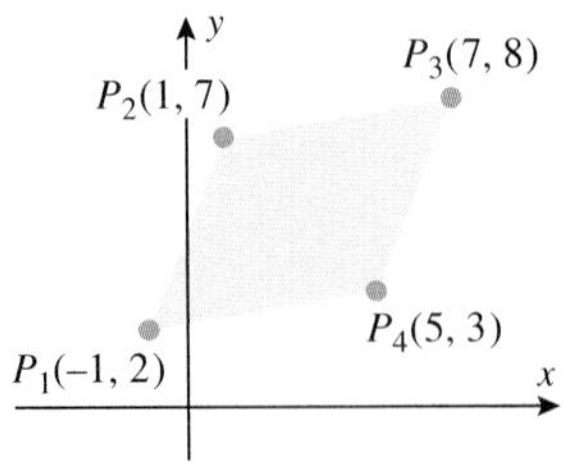

Figure 4.3.3

You know from geometry that the area of a triangle is one-half the base times the height. Thus, if you know that a triangle in the xy-plane has vertices $P_1(x_1, y_1)$, $P_2(x_2, y_2)$, and $P_3(x_3, y_3)$, you could find the area by using these vertices to calculate the base and height. However, the following theorem, whose proof is left for the exercises, provides a more efficient way of finding the area.

Theorem 4.3.6 *Suppose that a triangle in the xy-plane has vertices $P_1(x_1, y_1)$, $P_2(x_2, y_2)$, and $P_3(x_3, y_3)$ and that the labeling is such that the triangle is traversed counterclockwise from P_1 to P_2 to P_3. Then the area of the triangle is given by*

$$\text{area } \triangle P_1 P_2 P_3 = \frac{1}{2} \begin{vmatrix} x_1 & y_1 & 1 \\ x_2 & y_2 & 1 \\ x_3 & y_3 & 1 \end{vmatrix} \tag{7}$$

REMARK If the triangle in this theorem is traversed clockwise from P_1 to P_2 to P_3, then the right side in Formula (7) represents the *negative* of the area rather than the area itself. Thus, to apply this theorem you need not be concerned about how the vertices are labeled—you can label them arbitrarily and adjust the sign at the end to produce a positive area.

CONCEPT PROBLEM Show that if (x_3, y_3) is the origin, then the area of the triangle in Theorem 4.3.6 can be expressed as a 2×2 determinant:

$$\text{area } \triangle P_1 P_2 P_3 = \frac{1}{2} \begin{vmatrix} x_1 & y_1 \\ x_2 & y_2 \end{vmatrix} \tag{8}$$

EXAMPLE 8
Area of a
Triangle Using
Determinants

Find the area of the triangle with vertices $A(-5, 4)$, $B(3, 2)$, and $C(-2, -3)$.

Solution Rather than worry about the order of the vertices, let us just insert a $\pm$ in (7) for sign adjustment and write

$$\text{area } \triangle ABC = \pm\frac{1}{2} \begin{vmatrix} -5 & 4 & 1 \\ 3 & 2 & 1 \\ -2 & -3 & 1 \end{vmatrix} = \pm\frac{1}{2}(-50) = 25 \qquad \blacksquare$$

**POLYNOMIAL
INTERPOLATION AND
THE VANDERMONDE
DETERMINANT**

In Theorem 2.3.1 we stated without proof that if n points in the xy-plane have distinct x-coordinates, then there is a unique polynomial of degree $n - 1$ or less whose graph passes through those points. We are now in a position to give a proof of that theorem. If the n points are

$$(x_1, y_1), (x_2, y_2), (x_3, y_3), \ldots, (x_n, y_n)$$

and if the interpolating polynomial has the equation

$$y = a_0 + a_1 x + a_2 x^2 + \cdots + a_{n-1} x^{n-1}$$

then, as indicated in (18) of Section 2.3, the coefficients in this polynomial satisfy

$$a_0 + a_1 x_1 + a_2 x_1^2 + \cdots + a_{n-1} x_1^{n-1} = y_1$$
$$a_0 + a_1 x_2 + a_2 x_2^2 + \cdots + a_{n-1} x_2^{n-1} = y_2$$
$$\vdots \qquad \vdots \qquad \vdots \qquad \qquad \vdots \qquad \qquad \vdots$$
$$a_0 + a_1 x_n + a_2 x_n^2 + \cdots + a_{n-1} x_n^{n-1} = y_n$$

This is a linear system of n equations in n unknowns, so Theorem 4.2.7 implies that the system has a unique solution if and only if

$$\begin{vmatrix} 1 & x_1 & x_1^2 & \cdots & x_1^{n-1} \\ 1 & x_2 & x_2^2 & \cdots & x_2^{n-1} \\ \vdots & \vdots & \vdots & & \vdots \\ 1 & x_n & x_n^2 & \cdots & x_n^{n-1} \end{vmatrix} \neq 0$$

The $n \times n$ determinant on the left side of this equation is called the **_Vandermonde determinant_** after the French mathematician Alexandre Théophile Vandermonde (1735–1796). Thus, the proof of Theorem 2.3.1 reduces to showing that the Vandermonde determinant is nonzero if $x_1, x_2, \ldots, x_n$ are distinct. For simplicity, we will show this for the case where $n = 3$ and describe the procedure in the general case.

The Vandermonde determinant for $n = 3$ is

$$\begin{vmatrix} 1 & x_1 & x_1^2 \\ 1 & x_2 & x_2^2 \\ 1 & x_3 & x_3^2 \end{vmatrix}$$

We could evaluate this determinant by duplicating the first two columns and applying the "arrow method," but the result we are looking for will be more apparent if we use a combination of row operations and cofactor expansion. We write

$$\begin{vmatrix} 1 & x_1 & x_1^2 \\ 1 & x_2 & x_2^2 \\ 1 & x_3 & x_3^2 \end{vmatrix} = \begin{vmatrix} 1 & x_1 & x_1^2 \\ 0 & x_2 - x_1 & x_2^2 - x_1^2 \\ 0 & x_3 - x_1 & x_3^2 - x_1^2 \end{vmatrix} = (x_2 - x_1)(x_3 - x_1) \begin{vmatrix} 1 & x_1 & x_1^2 \\ 0 & 1 & x_2 + x_1 \\ 0 & 1 & x_3 + x_1 \end{vmatrix}$$

$$= (x_2 - x_1)(x_3 - x_1) \begin{vmatrix} 1 & x_2 + x_1 \\ 1 & x_3 + x_1 \end{vmatrix} = (x_2 - x_1)(x_3 - x_1)(x_3 - x_2)$$

Since we assumed that x_1, x_2, and x_3 are distinct, the last expression makes it evident that the determinant is nonzero, which is what we wanted to show.

The preceding computations show that the 3×3 Vandermonde determinant is expressible as the product of all possible factors of the form $(x_j - x_i)$, where $1 \leq i < j \leq 3$. In general, it can be proved that the $n \times n$ Vandermonde determinant is expressible as the product of all possible factors of the form $(x_j - x_i)$, where $1 \leq i < j \leq n$. You will often see this written as

$$\begin{vmatrix} 1 & x_1 & x_1^2 & \cdots & x_1^{n-1} \\ 1 & x_2 & x_2^2 & \cdots & x_2^{n-1} \\ \vdots & \vdots & \vdots & & \vdots \\ 1 & x_n & x_n^2 & \cdots & x_n^{n-1} \end{vmatrix} = \prod_{1 \leq i < j \leq n} (x_j - x_i) \tag{9}$$

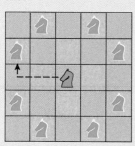

where the symbol Π (capital Greek pi) directs you to form the product of all factors $(x_j - x_i)$ whose subscripts satisfy the specified inequalities. As in the 3×3 case, this product is nonzero if $x_1, x_2, \ldots, x_n$ are distinct, which proves Theorem 2.3.1.

CROSS PRODUCTS

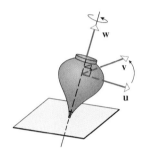

Figure 4.3.4

A basic problem in the study of rotational motion in 3-space is to find the axis of rotation of a spinning object and to identify whether the rotation is clockwise or counterclockwise from a specified point of view along the axis. To formulate this problem in vector terms, suppose that some rotation about an axis through the origin of R^3 causes a nonzero vector $\mathbf{u}$ to rotate into a nonzero vector $\mathbf{v}$. Since the axis of rotation must be perpendicular to the plane of $\mathbf{u}$ and $\mathbf{v}$, any nonzero vector $\mathbf{w}$ that is orthogonal to the plane of $\mathbf{u}$ and $\mathbf{v}$ will serve to identify the orientation of the rotational axis (Figure 4.3.4). Moreover, if the direction of $\mathbf{w}$ can be chosen so the rotation of $\mathbf{u}$ into $\mathbf{v}$ appears counterclockwise looking toward the origin from the terminal point of $\mathbf{w}$, then the vector $\mathbf{w}$ will carry all of the information needed to identify both the orientation of the axis and the direction of rotation. Accordingly, our goal in this subsection is to define a new kind of vector multiplication that will produce $\mathbf{w}$ when $\mathbf{u}$ and $\mathbf{v}$ are known.

Definition 4.3.7   If $\mathbf{u} = (u_1, u_2, u_3)$ and $\mathbf{v} = (v_1, v_2, v_3)$ are vectors in R^3, then the **cross product of u with v**, denoted by $\mathbf{u} \times \mathbf{v}$, is the vector in R^3 defined by

$$\mathbf{u} \times \mathbf{v} = (u_2 v_3 - u_3 v_2,\, u_3 v_1 - u_1 v_3,\, u_1 v_2 - u_2 v_1) \tag{10}$$

or equivalently,

$$\mathbf{u} \times \mathbf{v} = \left(\begin{vmatrix} u_2 & u_3 \\ v_2 & v_3 \end{vmatrix},\, -\begin{vmatrix} u_1 & u_3 \\ v_1 & v_3 \end{vmatrix},\, \begin{vmatrix} u_1 & u_2 \\ v_1 & v_2 \end{vmatrix} \right) \tag{11}$$

REMARK   Note that the cross product of vectors is a vector, whereas the dot product of vectors is a scalar.

A good way to remember Formula (10) is to express $\mathbf{u} \times \mathbf{v}$ in terms of the standard unit vectors $\mathbf{i} = (1, 0, 0)$, $\mathbf{j} = (0, 1, 0)$, $\mathbf{k} = (0, 0, 1)$ and to write (10) in the form of a 3×3 determinant as

$$\mathbf{u} \times \mathbf{v} = \begin{vmatrix} \mathbf{i} & \mathbf{j} & \mathbf{k} \\ u_1 & u_2 & u_3 \\ v_1 & v_2 & v_3 \end{vmatrix} = \begin{vmatrix} u_2 & u_3 \\ v_2 & v_3 \end{vmatrix} \mathbf{i} - \begin{vmatrix} u_1 & u_3 \\ v_1 & v_3 \end{vmatrix} \mathbf{j} + \begin{vmatrix} u_1 & u_2 \\ v_1 & v_2 \end{vmatrix} \mathbf{k} \tag{12}$$

You should confirm that the cross product formula on the right side results by expanding the 3×3 determinant[*] by cofactors along the first row.

EXAMPLE 9
Calculating a
Cross Product

Let $\mathbf{u} = (1, 2, -2)$ and $\mathbf{v} = (3, 0, 1)$. Find

   (a) $\mathbf{u} \times \mathbf{v}$ (b) $\mathbf{v} \times \mathbf{u}$ (c) $\mathbf{u} \times \mathbf{u}$

Solution (a)

$$\mathbf{u} \times \mathbf{v} = \begin{vmatrix} \mathbf{i} & \mathbf{j} & \mathbf{k} \\ 1 & 2 & -2 \\ 3 & 0 & 1 \end{vmatrix} = \begin{vmatrix} 2 & -2 \\ 0 & 1 \end{vmatrix} \mathbf{i} - \begin{vmatrix} 1 & -2 \\ 3 & 1 \end{vmatrix} \mathbf{j} + \begin{vmatrix} 1 & 2 \\ 3 & 0 \end{vmatrix} \mathbf{k} = 2\mathbf{i} - 7\mathbf{j} - 6\mathbf{k} = (2, -7, -6)$$

---

[*]This is not a determinant in the usual sense, since true determinants have scalar entries. Thus, you should think of this formula as a convenient mnemonic device.

Solution (*b*) We could proceed as in part (a), but a simpler approach is to observe that interchanging **u** and **v** in a cross product interchanges the rows of the 2×2 determinants on the right side of (12), and hence reverses the sign of each component. Thus, it follows from part (a) that

$$\mathbf{v} \times \mathbf{u} = -(\mathbf{u} \times \mathbf{v}) = -(2\mathbf{i} - 7\mathbf{j} - 6\mathbf{k}) = (-2, 7, 6)$$

Solution (*c*) If $\mathbf{u} = \mathbf{v}$, then each of the 2×2 determinants on the right side of (11) is zero because its rows are identical. Thus,

$$\mathbf{u} \times \mathbf{u} = 0\mathbf{i} - 0\mathbf{j} + 0\mathbf{k} = (0, 0, 0) = \mathbf{0} \qquad \blacksquare$$

The following theorem summarizes some basic properties of cross products that can be derived from properties of determinants. Some of the proofs are given as exercises.

> **Theorem 4.3.8** *If* **u**, **v**, *and* **w** *are vectors in* R^3 *and k is a scalar, then:*
>
> (*a*) $\mathbf{u} \times \mathbf{v} = -\mathbf{v} \times \mathbf{u}$
>
> (*b*) $\mathbf{u} \times (\mathbf{v} + \mathbf{w}) = (\mathbf{u} \times \mathbf{v}) + (\mathbf{u} \times \mathbf{w})$
>
> (*c*) $(\mathbf{u} + \mathbf{v}) \times \mathbf{w} = (\mathbf{u} \times \mathbf{w}) + (\mathbf{v} \times \mathbf{w})$
>
> (*d*) $k(\mathbf{u} \times \mathbf{v}) = (k\mathbf{u}) \times \mathbf{v} = \mathbf{u} \times (k\mathbf{v})$
>
> (*e*) $\mathbf{u} \times \mathbf{0} = \mathbf{0} \times \mathbf{u} = \mathbf{0}$
>
> (*f*) $\mathbf{u} \times \mathbf{u} = \mathbf{0}$

Recall that one of our goals in defining the cross product of **u** with **v** was to create a vector that is orthogonal to the plane of **u** and **v**. The following theorem shows that $\mathbf{u} \times \mathbf{v}$ has this property.

> **Theorem 4.3.9** *If* **u** *and* **v** *are vectors in* R^3, *then:*
>
> (*a*) $\mathbf{u} \cdot (\mathbf{u} \times \mathbf{v}) = 0$ [**u** x **v** is orthogonal to **u**]
>
> (*b*) $\mathbf{v} \cdot (\mathbf{u} \times \mathbf{v}) = 0$ [**u** x **v** is orthogonal to **v**]

We will prove part (*a*); the proof of part (*b*) is similar.

Proof (*a*) If $\mathbf{u} = (u_1, u_2, u_3)$ and $\mathbf{v} = (v_1, v_2, v_3)$, then it follows from Formula (10) that

$$\mathbf{u} \cdot (\mathbf{u} \times \mathbf{v}) = u_1(u_2 v_3 - u_3 v_2) + u_2(u_3 v_1 - u_1 v_3) + u_3(u_1 v_2 - u_2 v_1) = 0 \qquad \blacksquare$$

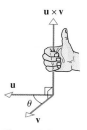

Figure 4.3.5

In general, if **u** and **v** are nonzero and nonparallel vectors, then the direction of $\mathbf{u} \times \mathbf{v}$ in relation to **u** and **v** can be determined by the following right-hand rule:[*] If the fingers of the right hand are cupped so they curl in the direction of rotation that takes **u** into **v** in at most $180°$, then the thumb will point (roughly) in the direction of $\mathbf{u} \times \mathbf{v}$ (Figure 4.3.5). It should be evident from this rule that the direction of rotation from **u** to **v** will appear to be counterclockwise to an observer looking toward the origin from the terminal point of $\mathbf{u} \times \mathbf{v}$. We will not prove this fact, but we will illustrate it with the six possible cross products of the standard unit vectors **i**, **j**, and **k**:

$$\begin{array}{ccc} \mathbf{i} \times \mathbf{j} = \mathbf{k} & \mathbf{j} \times \mathbf{k} = \mathbf{i} & \mathbf{k} \times \mathbf{i} = \mathbf{j} \\ \mathbf{j} \times \mathbf{i} = -\mathbf{k} & \mathbf{k} \times \mathbf{j} = -\mathbf{i} & \mathbf{i} \times \mathbf{k} = -\mathbf{j} \end{array} \qquad (13)$$

---

[*]Recall that we agreed to consider only right-handed coordinate systems in this text. Had we used left-handed coordinate systems instead, then a left-hand rule would apply here.

These products are easy to derive; for example,

$$\mathbf{i} \times \mathbf{j} = (1, 0, 0) \times (0, 1, 0) = \begin{vmatrix} \mathbf{i} & \mathbf{j} & \mathbf{k} \\ 1 & 0 & 0 \\ 0 & 1 & 0 \end{vmatrix} = \begin{vmatrix} 0 & 0 \\ 1 & 0 \end{vmatrix} \mathbf{i} - \begin{vmatrix} 1 & 0 \\ 0 & 0 \end{vmatrix} \mathbf{j} + \begin{vmatrix} 1 & 0 \\ 0 & 1 \end{vmatrix} \mathbf{k} = \mathbf{k}$$

As predicted by the right-hand rule, the 90° rotation of $\mathbf{i}$ into $\mathbf{j}$ appears to be counterclockwise looking toward the origin from the terminal point of $\mathbf{i} \times \mathbf{j} = \mathbf{k}$, and the 90° rotation of $\mathbf{j}$ into $\mathbf{k}$ appears to be counterclockwise looking toward the origin from the terminal point of $\mathbf{j} \times \mathbf{k} = \mathbf{i}$.

CONCEPT PROBLEM Confirm that the remaining four cross products in (13) satisfy the right-hand rule.

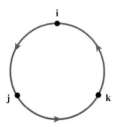

Figure 4.3.6

REMARK A useful way of remembering the six cross products in (13) is to use the diagram in Figure 4.3.6. In this diagram, the cross product of two consecutive vectors in the counterclockwise direction is the next vector around, and the cross product of two consecutive vectors in the clockwise direction is the negative of the next vector around.

WARNING We can write a product of three real numbers as abc and the product of three matrices as ABC because the associative laws $a(bc) = (ab)c$ and $A(BC) = (AB)C$ ensure that the same result is obtained no matter how parentheses are inserted. However, the associative law does *not* hold for cross products; for example,

$$\mathbf{i} \times (\mathbf{j} \times \mathbf{j}) = \mathbf{i} \times \mathbf{0} = \mathbf{0} \quad \text{whereas} \quad (\mathbf{i} \times \mathbf{j}) \times \mathbf{j} = \mathbf{k} \times \mathbf{j} = -\mathbf{i}$$

so $\mathbf{i} \times (\mathbf{j} \times \mathbf{j}) \neq (\mathbf{i} \times \mathbf{j}) \times \mathbf{j}$. Accordingly, expressions such as $\mathbf{u} \times \mathbf{v} \times \mathbf{w}$ should never be used because they are ambiguous.

The next theorem is concerned with the length of a cross product.

> **Theorem 4.3.10** *Let $\mathbf{u}$ and $\mathbf{v}$ be nonzero vectors in R^3, and let θ be the angle between these vectors.*
>
> (a) $\|\mathbf{u} \times \mathbf{v}\| = \|\mathbf{u}\| \|\mathbf{v}\| \sin \theta$
>
> (b) *The area A of the parallelogram that has $\mathbf{u}$ and $\mathbf{v}$ as adjacent sides is*
>
> $$A = \|\mathbf{u} \times \mathbf{v}\| \tag{14}$$

Proof (a) Since $0 \leq \theta \leq \pi$, it follows that $\sin \theta \geq 0$ and hence that

$$\sin \theta = \sqrt{1 - \cos^2 \theta}$$

Thus,

$$\|\mathbf{u}\| \|\mathbf{v}\| \sin \theta = \|\mathbf{u}\| \|\mathbf{v}\| \sqrt{1 - \cos^2 \theta}$$

$$= \|\mathbf{u}\| \|\mathbf{v}\| \sqrt{1 - \frac{(\mathbf{u} \cdot \mathbf{v})^2}{\|\mathbf{u}\|^2 \|\mathbf{v}\|^2}} \quad \text{[Theorem 1.2.8]}$$

$$= \sqrt{\|\mathbf{u}\|^2 \|\mathbf{v}\|^2 - (\mathbf{u} \cdot \mathbf{v})^2}$$

$$= \sqrt{\left(u_1^2 + u_2^2 + u_3^2\right)\left(v_1^2 + v_2^2 + v_3^2\right) - (u_1 v_1 + u_2 v_2 + u_3 v_3)^2}$$

$$= \sqrt{(u_2 v_3 - u_3 v_2)^2 + (u_3 v_1 - u_1 v_3)^2 + (u_1 v_2 - u_2 v_1)^2}$$

$$= \|\mathbf{u} \times \mathbf{v}\| \quad \text{[Formula (10)]}$$

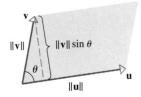

Figure 4.3.7

Proof (b) Referring to Figure 4.3.7, the parallelogram that has $\mathbf{u}$ and $\mathbf{v}$ as adjacent sides can be viewed as having a base of length $\|\mathbf{u}\|$ and an altitude of length $\|\mathbf{v}\| \sin \theta$. Thus, its area A is

$$A = (\text{base})(\text{altitude}) = \|\mathbf{u}\| \|\mathbf{v}\| \sin \theta = \|\mathbf{u} \times \mathbf{v}\| \qquad \blacksquare$$

EXAMPLE 10
Area of a Triangle
in 3-Space

Find the area of the triangle in R^3 that has vertices $P_1(2, 2, 0)$, $P_2(-1, 0, 2)$, and $P_3(0, 4, 3)$.

Solution The area A of the triangle is half the area of the parallelogram that has adjacent sides $\overrightarrow{P_1P_2}$ and $\overrightarrow{P_1P_3}$ (Figure 4.3.8). Thus,

$$A = \tfrac{1}{2} \left\| \overrightarrow{P_1P_2} \times \overrightarrow{P_1P_3} \right\|$$

We will leave it for you to show that $\overrightarrow{P_1P_2} = (-3, -2, 2)$ and $\overrightarrow{P_1P_3} = (-2, 2, 3)$ and also that

$$\overrightarrow{P_1P_2} \times \overrightarrow{P_1P_3} = \begin{vmatrix} \mathbf{i} & \mathbf{j} & \mathbf{k} \\ -3 & -2 & 2 \\ -2 & 2 & 3 \end{vmatrix} = -10\mathbf{i} + 5\mathbf{j} - 10\mathbf{k} = (-10, 5, -10)$$

Thus,

$$A = \tfrac{1}{2} \left\| \overrightarrow{P_1P_2} \times \overrightarrow{P_1P_3} \right\| = \tfrac{1}{2}\sqrt{225} = \tfrac{15}{2}$$ ∎

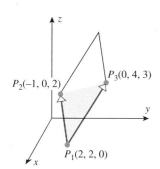

Figure 4.3.8

Exercise Set 4.3

In Exercises 1–4, find the adjoint of A, and then compute A^{-1} using Theorem 4.3.3.

1. $A = \begin{bmatrix} 2 & 5 & 5 \\ -1 & -1 & 0 \\ 2 & 4 & 3 \end{bmatrix}$ **2.** $A = \begin{bmatrix} 2 & 0 & 3 \\ 0 & 3 & 2 \\ -2 & 0 & -4 \end{bmatrix}$

3. $A = \begin{bmatrix} 2 & -3 & 5 \\ 0 & 1 & -3 \\ 0 & 0 & 2 \end{bmatrix}$ **4.** $A = \begin{bmatrix} 2 & 0 & 0 \\ 8 & 1 & 0 \\ -5 & 3 & 6 \end{bmatrix}$

In Exercises 5–10, solve the equations by Cramer's rule, where applicable.

5. $7x_1 - 3x_2 = 3$
$3x_1 + x_2 = 5$

6. $4x + 5y = -8$
$11x + y = 29$

7. $x - 4y + z = 6$
$4x - y + 2z = -1$
$2x + 2y - 3z = -20$

8. $x_1 - 3x_2 + x_3 = 4$
$2x_1 - x_2 = -2$
$4x_1 - 3x_3 = 0$

9. $-x_1 - 4x_2 + 2x_3 + x_4 = -32$
$2x_1 - x_2 + 7x_3 + 9x_4 = 14$
$-x_1 + x_2 + 3x_3 + x_4 = 11$
$x_1 - 2x_2 + x_3 - 4x_4 = -4$

10. $2x_1 + 2x_2 - x_3 + x_4 = 4$
$4x_1 + 3x_2 - x_3 + 2x_4 = 6$
$8x_1 + 5x_2 - 3x_3 + 4x_4 = 12$
$3x_1 + 3x_2 - 2x_3 + 2x_4 = 6$

11. Find x without solving for y and z:

$2x + 3y + 4z = 1$
$x - 2y - z = 2$
$3x + y + z = 4$

12. Find y without solving for x and z:

$x + 2y + 3z = -2$
$3x - y + z = 1$
$-x + 4y - 2z = -3$

13. Show that the matrix

$$A = \begin{bmatrix} \cos\theta & \sin\theta & 0 \\ -\sin\theta & \cos\theta & 0 \\ 0 & 0 & 1 \end{bmatrix}$$

is invertible for all values of θ, and use Theorem 4.3.3 to find A^{-1}.

14. Show that the matrix

$$A = \begin{bmatrix} \tan\alpha & -1 & 0 \\ 1 & \tan\alpha & 0 \\ 0 & 0 & \cos^2\alpha \end{bmatrix}$$

is invertible for $\alpha \neq \frac{\pi}{2} + n\pi$, where n is an integer. Then use Theorem 4.3.3 to find A^{-1}.

In Exercises 15 and 16, use determinants to find all values of k for which the system has a unique solution, and for those cases use Cramer's rule to find the solution.

15. $3x + 3y + z = 1$
$4x + ky + 2z = 2$
$2kx + 2ky + kz = 1$

16. $2x + 3ky - kz = 1$
$x - y + 2z = -1$
$3kx + 2y - z = 3$

In Exercises 17 and 18, state when the inverse of the matrix A exists in terms of the parameters that the matrix contains.

17. $A = \begin{bmatrix} x & y & 0 \\ 0 & x & y \\ y & x & 0 \end{bmatrix}$ **18.** $A = \begin{bmatrix} s & 0 & t & 0 \\ 0 & s & 0 & t \\ t & 0 & s & 0 \\ 0 & t & 0 & s \end{bmatrix}$

In Exercises 19 and 20, find the area of the parallelogram determined by the columns of A.

19. $A = \begin{bmatrix} 1 & -2 \\ 1 & 1 \end{bmatrix}$ **20.** $A = \begin{bmatrix} 0 & 3 \\ 2 & -4 \end{bmatrix}$

In Exercises 21 and 22, find the volume of the parallelepiped determined by the columns of A.

21. $A = \begin{bmatrix} 1 & 2 & 3 \\ 1 & 0 & 2 \\ -1 & 2 & 1 \end{bmatrix}$ **22.** $A = \begin{bmatrix} -1 & 1 & 0 \\ 2 & -1 & 1 \\ 1 & 3 & 1 \end{bmatrix}$

In Exercises 23 and 24, find the area of the parallelogram with the given vertices.

23. $P_1(1, 2)$, $P_2(4, 4)$, $P_3(7, 5)$, $P_4(4, 3)$

24. $P_1(3, 2)$, $P_2(5, 4)$, $P_3(9, 4)$, $P_4(7, 2)$

In Exercises 25 and 26, find the area of the triangle with the given vertices.

25. $A(2, 0)$, $B(3, 4)$, $C(-1, 2)$

26. $A(1, 1)$, $B(2, 2)$, $C(3, -3)$

In Exercises 27 and 28, find the volume of the parallelepiped with sides $\mathbf{u}$, $\mathbf{v}$, and $\mathbf{w}$.

27. $\mathbf{u} = (2, -6, 2)$, $\mathbf{v} = (0, 4, -2)$, $\mathbf{w} = (2, 2, -4)$

28. $\mathbf{u} = (3, 1, 2)$, $\mathbf{v} = (4, 5, 1)$, $\mathbf{w} = (1, 2, 4)$

In Exercises 29 and 30, determine whether $\mathbf{u}$, $\mathbf{v}$, and $\mathbf{w}$ lie in the same plane when positioned so that their initial points coincide.

29. $\mathbf{u} = (-1, -2, 1)$, $\mathbf{v} = (3, 0, -2)$, $\mathbf{w} = (5, -4, 0)$

30. $\mathbf{u} = (5, -2, 1)$, $\mathbf{v} = (4, -1, 1)$, $\mathbf{w} = (1, -1, 0)$

31. Find all unit vectors parallel to the yz-plane that are orthogonal to the vector $(3, -1, 2)$.

32. Find all unit vectors in the plane determined by $\mathbf{u} = (3, 0, 1)$ and $\mathbf{v} = (1, -1, 1)$ that are orthogonal to the vector $\mathbf{w} = (1, 2, 0)$.

33. Use the cross product to find the sine of the angle between the vectors $\mathbf{u} = (2, 3, -6)$ and $\mathbf{v} = (2, 3, 6)$.

34. (a) Find the area of the triangle having vertices $A(1, 0, 1)$, $B(0, 2, 3)$, and $C(2, 1, 0)$.
  (b) Use the result of part (a) to find the length of the altitude from vertex C to side AB.

In Exercises 35 and 36, let $\mathbf{u} = (3, 2, -1)$, $\mathbf{v} = (0, 2, -3)$, and $\mathbf{w} = (2, 6, 7)$. Compute the indicated vectors.

35. (a) $\mathbf{v} \times \mathbf{w}$ (b) $\mathbf{u} \times (\mathbf{v} \times \mathbf{w})$
  (c) $(\mathbf{u} \times \mathbf{v}) \times \mathbf{w}$

36. (a) $(\mathbf{u} \times \mathbf{v}) \times (\mathbf{v} \times \mathbf{w})$ (b) $\mathbf{u} \times (\mathbf{v} - 2\mathbf{w})$
  (c) $(\mathbf{u} \times \mathbf{v}) - 2\mathbf{w}$

In Exercises 37 and 38, find a vector that is orthogonal to both $\mathbf{u}$ and $\mathbf{v}$.

37. (a) $\mathbf{u} = (-6, 4, 2)$, $\mathbf{v} = (3, 1, 5)$
  (b) $\mathbf{u} = (-2, 1, 5)$, $\mathbf{v} = (3, 0, -3)$

38. (a) $\mathbf{u} = (1, 1, -2)$, $\mathbf{v} = (2, -1, 2)$
  (b) $\mathbf{u} = (3, 3, 1)$, $\mathbf{v} = (0, 4, 2)$

In Exercises 39–42, show that the given identities hold for any $\mathbf{u}$, $\mathbf{v}$, and $\mathbf{w}$ in R^3, and any scalar k.

39. $\mathbf{u} \times \mathbf{v} = -\mathbf{v} \times \mathbf{u}$

40. $\mathbf{u} \times (\mathbf{v} + \mathbf{w}) = (\mathbf{u} \times \mathbf{v}) + (\mathbf{u} \times \mathbf{w})$

41. $k(\mathbf{u} \times \mathbf{v}) = (k\mathbf{u}) \times \mathbf{v} = \mathbf{u} \times (k\mathbf{v})$

42. $\mathbf{u} \times \mathbf{0} = \mathbf{0}$ and $\mathbf{u} \times \mathbf{u} = \mathbf{0}$

In Exercises 43 and 44, find the area of the parallelogram determined by the given vectors $\mathbf{u}$ and $\mathbf{v}$.

43. (a) $\mathbf{u} = (1, -1, 2)$, $\mathbf{v} = (0, 3, 1)$
  (b) $\mathbf{u} = (2, 3, 0)$, $\mathbf{v} = (-1, 2, -2)$

44. (a) $\mathbf{u} = (3, -1, 4)$, $\mathbf{v} = (6, -2, 8)$
  (b) $\mathbf{u} = (1, 1, 1)$, $\mathbf{v} = (3, 2, -5)$

In Exercises 45 and 46, find the area of the triangle in 3-space that has the given vertices.

45. $P_1(2, 6, -1)$, $P_2(1, 1, 1)$, $P_3(4, 6, 2)$

46. $P(1, -1, 2)$, $Q(0, 3, 4)$, $R(6, 1, 8)$

47. Show that if $\mathbf{a}$, $\mathbf{b}$, $\mathbf{c}$, and $\mathbf{d}$ are any vectors in 3-space, then

$$(\mathbf{a} + \mathbf{d}) \cdot (\mathbf{b} \times \mathbf{c}) = \mathbf{a} \cdot (\mathbf{b} \times \mathbf{c}) + \mathbf{d} \cdot (\mathbf{b} \times \mathbf{c})$$

48. Simplify $(\mathbf{u} + \mathbf{v}) \times (\mathbf{u} - \mathbf{v})$.

49. Find a vector that is perpendicular to the plane determined by the points $A(0, -2, 1)$, $B(1, -1, -2)$, and $C(-1, 1, 0)$.

50. Consider the vectors $\mathbf{u} = (u_1, u_2, u_3)$, $\mathbf{v} = (v_1, v_2, v_3)$, and $\mathbf{w} = (w_1, w_2, w_3)$ in R^3. The expression $\mathbf{u} \cdot (\mathbf{v} \times \mathbf{w})$ is called the *scalar triple product* of $\mathbf{u}$, $\mathbf{v}$, and $\mathbf{w}$.
  (a) Show that

$$\mathbf{u} \cdot (\mathbf{v} \times \mathbf{w}) = \begin{vmatrix} u_1 & u_2 & u_3 \\ v_1 & v_2 & v_3 \\ w_1 & w_2 & w_3 \end{vmatrix}$$

  (b) Give a geometric interpretation of $|\mathbf{u} \cdot (\mathbf{v} \times \mathbf{w})|$ (vertical bars denote absolute value).

51. Let

$$A = \begin{bmatrix} 1 & 3 & 1 \\ 2 & 5 & 2 \\ 1 & 3 & 8 \end{bmatrix}$$

(a) Evaluate A^{-1} using Theorem 4.3.3.

(b) Evaluate A^{-1} using the method of Example 3 in Section 3.3.

(c) Which method involves less computation?

52. Suppose that A is nilpotent, that is, $A^k = 0$ for some k. Use properties of the determinant to show that A is not invertible.

53. Show that if **u**, **v**, and **w** are vectors in R^3, with no two of them collinear, then $\mathbf{u} \times (\mathbf{v} \times \mathbf{w})$ lies in the plane determined by **v** and **w**.

54. Show that if **u**, **v**, and **w** are vectors in R^3, with no two of them collinear, then $(\mathbf{u} \times \mathbf{v}) \times \mathbf{w}$ lies in the plane determined by **u** and **v**.

55. Use Formula (2) to show that the inverse of an invertible upper triangular matrix is upper triangular. [*Hint:* Examine which terms in the adjoint matrix of an upper triangular matrix must be zero.]

56. Use Formula (2) to show that the inverse of an invertible lower triangular matrix is lower triangular. [*Hint:* Examine which terms in the adjoint matrix of a lower triangular matrix must be zero.]

57. Use Cramer's rule to find a polynomial of degree 3 that passes through the points $(0, 1)$, $(1, -1)$, $(2, -1)$, and $(3, 7)$.

Discussion and Discovery

D1. Suppose that **u** and **v** are noncollinear vectors with their initial points at the origin in 3-space.

(a) Make a sketch that illustrates how $\mathbf{w} = \mathbf{v} \times (\mathbf{u} \times \mathbf{v})$ is oriented in relation to **u** and **v**.

(b) What can you say about the values of $\mathbf{u} \cdot \mathbf{w}$ and $\mathbf{v} \cdot \mathbf{w}$? Explain your reasoning.

D2. If $\mathbf{u} \neq \mathbf{0}$, is it valid to cancel **u** from both sides of the equation $\mathbf{u} \times \mathbf{v} = \mathbf{u} \times \mathbf{w}$ and conclude that $\mathbf{v} = \mathbf{w}$? Explain your reasoning.

D3. Something is wrong with one of the following expressions. Which one is it and what is wrong?

$$\mathbf{u} \cdot (\mathbf{v} \times \mathbf{w}), \quad \mathbf{u} \times (\mathbf{v} \times \mathbf{w}), \quad (\mathbf{u} \cdot \mathbf{v}) \times \mathbf{w}$$

D4. What can you say about the vectors **u** and **v** if $\mathbf{u} \times \mathbf{v} = \mathbf{0}$?

D5. Give some examples of other algebraic rules that hold for multiplication of real numbers but not for the cross product of vectors.

D6. Solve the following system by Cramer's rule:

$$cx_1 - (1-c)x_2 = \ \ 3$$
$$(1-c)x_1 + \ \ \ \ cx_2 = -4$$

For what values of c is this solution valid? Explain.

D7. Let $A\mathbf{x} = \mathbf{b}$ be the system in Exercise 12.

(a) Solve by Cramer's rule.

(b) Solve by Gauss–Jordan elimination.

(c) Which method involves less computation?

D8. Indicate whether the statement is true (T) or false (F). Justify your answer.

(a) If A is a square matrix, then $A \operatorname{adj}(A)$ is a diagonal matrix.

(b) Cramer's rule can be used to solve any system of linear equations if the number of equations is the same as the number of unknowns.

(c) If A is invertible, then $\operatorname{adj}(A)$ must also be invertible.

(d) If A has a row of zeros, then so does $\operatorname{adj}(A)$.

(e) If **u**, **v**, and **w** are vectors in R^3, then
$$\mathbf{u} \cdot (\mathbf{v} \times \mathbf{w}) = (\mathbf{u} \times \mathbf{v}) \cdot \mathbf{w}.$$

Working with Proofs

P1. Prove that if **u** and **v** are nonzero, nonorthogonal vectors in R^3, and θ is the angle between them, then

$$\tan \theta = \frac{\|\mathbf{u} \times \mathbf{v}\|}{(\mathbf{u} \cdot \mathbf{v})}$$

P2. Prove that if **u** and **v** are nonzero vectors in R^2 and α and β are the angles in the accompanying figure, then

$$\cos(\alpha - \beta) = \frac{\mathbf{u} \cdot \mathbf{v}}{\|\mathbf{u}\| \|\mathbf{v}\|}$$

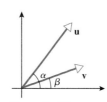

Figure Ex-P2

P3. Prove the identity for vectors in R^3.

(a) $(\mathbf{u} + k\mathbf{v}) \times \mathbf{v} = \mathbf{u} \times \mathbf{v}$

(b) $\mathbf{u} \cdot (\mathbf{v} \times \mathbf{w}) = -(\mathbf{u} \times \mathbf{w}) \cdot \mathbf{v}$ [*Hint:* See Exercise 50.]

P4. Prove that if $\mathbf{a}$, $\mathbf{b}$, $\mathbf{c}$, and $\mathbf{d}$ are vectors in R^3 that lie in the same plane, then $(\mathbf{a} \times \mathbf{b}) \times (\mathbf{c} \times \mathbf{d}) = \mathbf{0}$.

P5. Prove Theorem 4.3.6.

Technology Exercises

T1. Compute the cross products in Example 9.

T2. Compute the adjoint in Example 1, and confirm the computations in Example 2.

T3. Use Cramer's rule to solve for y without solving for x, z, and w, and then check your result by using any method to solve the system.

$$\begin{aligned} 4x + y + z + w &= 6 \\ 3x + 7y - z + w &= 1 \\ 7x + 3y - 5z + 8w &= -3 \\ x + y + z + 2w &= 3 \end{aligned}$$

T4. (CAS) Confirm some of the statements in Theorems 4.3.8 and 4.3.9.

T5. (CAS) Confirm Formula (9) for the fourth- and fifth-order Vandermonde determinants.

Section 4.4   A First Look at Eigenvalues and Eigenvectors

In this section we will discuss linear equations of the form $A\mathbf{x} = \mathbf{x}$ and, more generally, equations of the form $A\mathbf{x} = \lambda\mathbf{x}$, where λ is a scalar. Such equations arise in a wide variety of important applications and will be a recurring theme in the rest of this text.

FIXED POINTS   Recall that a *fixed point* of an $n \times n$ matrix A is a vector $\mathbf{x}$ in R^n such that $A\mathbf{x} = \mathbf{x}$ (see the discussion preceding Example 6 of Section 3.6). Every square matrix A has at least one fixed point, namely $\mathbf{x} = \mathbf{0}$. We call this the ***trivial fixed point*** of A.

The general procedure for finding the fixed points of a matrix A is to rewrite the equation $A\mathbf{x} = \mathbf{x}$ as $A\mathbf{x} = I\mathbf{x}$ or, alternatively, as

$$(I - A)\mathbf{x} = \mathbf{0} \tag{1}$$

Since this can be viewed as a homogeneous linear system of n equations in n unknowns with coefficient matrix $I - A$, we see that the set of fixed points of an $n \times n$ matrix is a subspace of R^n that can be obtained by solving (1).

The following theorem will be useful for ascertaining whether a matrix has nontrivial fixed points.

> **Theorem 4.4.1** *If A is an $n \times n$ matrix, then the following statements are equivalent.*
>
> (*a*) *A has nontrivial fixed points.*
>
> (*b*) *$I - A$ is singular.*
>
> (*c*) *$\det(I - A) = 0$.*

The proof follows by applying parts (*d*), (*c*), and (*i*) of Theorem 4.2.7 to the matrix $I - A$. We omit the details.

EXAMPLE 1
Fixed Points of
a 2 × 2 Matrix

In each part, determine whether the matrix has nontrivial fixed points; and, if so, graph the subspace of fixed points in an xy-coordinate system.

(a) $A = \begin{bmatrix} 3 & 6 \\ 1 & 2 \end{bmatrix}$ (b) $A = \begin{bmatrix} 0 & 2 \\ 0 & 1 \end{bmatrix}$

Solution (*a*) The matrix has only the trivial fixed point since

$$\det(I - A) = \begin{vmatrix} -2 & -6 \\ -1 & -1 \end{vmatrix} = -4 \neq 0$$

Solution (*b*) The matrix has nontrivial fixed points since

$$\det(I - A) = \begin{vmatrix} 1 & -2 \\ 0 & 0 \end{vmatrix} = 0$$

The fixed points $\mathbf{x} = (x, y)$ are the solutions of the linear system $(I - A)\mathbf{x} = \mathbf{0}$, which we can express in component form as

$$\begin{bmatrix} 1 & -2 \\ 0 & 0 \end{bmatrix} \begin{bmatrix} x \\ y \end{bmatrix} = \begin{bmatrix} 0 \\ 0 \end{bmatrix}$$

A general solution of this system is

$$x = 2t, \quad y = t \tag{2}$$

(verify), which are parametric equations of the line $y = \frac{1}{2}x$ (Figure 4.4.1). It follows from the corresponding vector form of this line that the fixed points are

$$\mathbf{x} = \begin{bmatrix} x \\ y \end{bmatrix} = \begin{bmatrix} 2t \\ t \end{bmatrix} = t \begin{bmatrix} 2 \\ 1 \end{bmatrix} \tag{3}$$

As a check,

$$A\mathbf{x} = \begin{bmatrix} 0 & 2 \\ 0 & 1 \end{bmatrix} \begin{bmatrix} 2t \\ t \end{bmatrix} = \begin{bmatrix} 2t \\ t \end{bmatrix} = \mathbf{x}$$

so every vector of form (3) is a fixed point of A. ∎

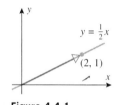

Figure 4.4.1

**EIGENVALUES AND
EIGENVECTORS**

In a fixed point problem one looks for nonzero vectors that satisfy the equation $A\mathbf{x} = \mathbf{x}$. One might also consider whether there are nonzero vectors that satisfy such equations as

$$A\mathbf{x} = 2\mathbf{x}, \quad A\mathbf{x} = -3\mathbf{x}, \quad A\mathbf{x} = \sqrt{2}\mathbf{x}$$

or, more generally, equations of the form $A\mathbf{x} = \lambda\mathbf{x}$ in which λ is a scalar. Thus, we pose the following problem.

Problem 4.4.2 If A is an $n \times n$ matrix, for what values of the scalar λ, if any, are there nonzero vectors in R^n such that $A\mathbf{x} = \lambda\mathbf{x}$?

Before we discuss how to solve such a problem, it will be helpful to introduce some additional terminology.

Definition 4.4.3 If A is an $n \times n$ matrix, then a scalar λ is called an ***eigenvalue*** of A if there is a nonzero vector $\mathbf{x}$ such that $A\mathbf{x} = \lambda\mathbf{x}$. If λ is an eigenvalue of A, then every nonzero vector $\mathbf{x}$ such that $A\mathbf{x} = \lambda\mathbf{x}$ is called an ***eigenvector*** of A corresponding to λ.

The most direct way of finding the eigenvalues of an $n \times n$ matrix A is to rewrite the equation $A\mathbf{x} = \lambda\mathbf{x}$ as $A\mathbf{x} = \lambda I\mathbf{x}$, or equivalently, as

$$(\lambda I - A)\mathbf{x} = \mathbf{0} \tag{4}$$

and then try to determine those values of λ, if any, for which this system has nontrivial solutions. Since (4) has nontrivial solutions if and only if the coefficient matrix $\lambda I - A$ is singular, we see that the eigenvalues of A are the solutions of the equation

$$\det(\lambda I - A) = 0 \tag{5}$$

This is called the **characteristic equation** of A. Also, if λ is an eigenvalue of A, then (4) has a nonzero solution space, which we call the **eigenspace** of A corresponding to λ. It is the *nonzero* vectors in the eigenspace of A corresponding to λ that are the eigenvectors of A corresponding to λ.

The preceding discussion is summarized by the following theorem.

Theorem 4.4.4 *If A is an $n \times n$ matrix and λ is a scalar, then the following statements are equivalent.*

  (a) λ *is an eigenvalue of A.*

  (b) λ *is a solution of the equation* $\det(\lambda I - A) = 0$.

  (c) *The linear system* $(\lambda I - A)\mathbf{x} = \mathbf{0}$ *has nontrivial solutions.*

EXAMPLE 2

  (a) Find the eigenvalues and corresponding eigenvectors of the matrix

$$A = \begin{bmatrix} 1 & 3 \\ 4 & 2 \end{bmatrix}$$

  (b) Graph the eigenspaces of A in an xy-coordinate system.

Solution (*a*) To find the eigenvalues we will solve the characteristic equation of A. Since

$$\lambda I - A = \lambda \begin{bmatrix} 1 & 0 \\ 0 & 1 \end{bmatrix} - \begin{bmatrix} 1 & 3 \\ 4 & 2 \end{bmatrix} = \begin{bmatrix} \lambda - 1 & -3 \\ -4 & \lambda - 2 \end{bmatrix}$$

the characteristic equation $\det(\lambda I - A) = 0$ is

$$\begin{vmatrix} \lambda - 1 & -3 \\ -4 & \lambda - 2 \end{vmatrix} = 0$$

Expanding and simplifying the determinant yields

$$\lambda^2 - 3\lambda - 10 = 0 \quad \text{or equivalently,} \quad (\lambda + 2)(\lambda - 5) = 0 \tag{6}$$

(verify), so the eigenvalues of A are $\lambda = -2$ and $\lambda = 5$.

To find the eigenspaces corresponding to these eigenvalues we must solve the system

$$\begin{bmatrix} \lambda - 1 & -3 \\ -4 & \lambda - 2 \end{bmatrix} \begin{bmatrix} x \\ y \end{bmatrix} = \begin{bmatrix} 0 \\ 0 \end{bmatrix} \tag{7}$$

with $\lambda = -2$ and then with $\lambda = 5$. Here are the computations in the two cases.

Case $\lambda = -2$ In this case (7) becomes

$$\begin{bmatrix} -3 & -3 \\ -4 & -4 \end{bmatrix} \begin{bmatrix} x \\ y \end{bmatrix} = \begin{bmatrix} 0 \\ 0 \end{bmatrix}$$

Solving this system yields (verify)

$$x = -t, \quad y = t \tag{8}$$

so the eigenvectors corresponding to $\lambda = -2$ are the nonzero vectors of the form

$$\mathbf{x} = \begin{bmatrix} x \\ y \end{bmatrix} = \begin{bmatrix} -t \\ t \end{bmatrix} = t \begin{bmatrix} -1 \\ 1 \end{bmatrix} \tag{9}$$

As a check,

$$A\mathbf{x} = \begin{bmatrix} 1 & 3 \\ 4 & 2 \end{bmatrix} \begin{bmatrix} -t \\ t \end{bmatrix} = \begin{bmatrix} 2t \\ -2t \end{bmatrix} = -2 \begin{bmatrix} -t \\ t \end{bmatrix} = -2\mathbf{x}$$

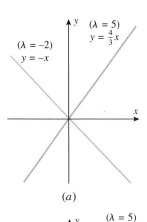

Case $\lambda = 5$ In this case (7) becomes

$$\begin{bmatrix} 4 & -3 \\ -4 & 3 \end{bmatrix} \begin{bmatrix} x \\ y \end{bmatrix} = \begin{bmatrix} 0 \\ 0 \end{bmatrix}$$

Solving this system yields (verify)

$$x = \tfrac{3}{4}t, \quad y = t \tag{10}$$

so the eigenvectors corresponding to $\lambda = 5$ are the nonzero vectors of the form

$$\mathbf{x} = \begin{bmatrix} x \\ y \end{bmatrix} = \begin{bmatrix} \tfrac{3}{4}t \\ t \end{bmatrix} = t \begin{bmatrix} \tfrac{3}{4} \\ 1 \end{bmatrix} \tag{11}$$

As a check,

$$A\mathbf{x} = \begin{bmatrix} 1 & 3 \\ 4 & 2 \end{bmatrix} \begin{bmatrix} \tfrac{3}{4}t \\ t \end{bmatrix} = \begin{bmatrix} \tfrac{15}{4}t \\ 5t \end{bmatrix} = 5 \begin{bmatrix} \tfrac{3}{4}t \\ t \end{bmatrix} = 5\mathbf{x}$$

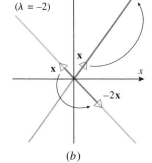

Figure 4.4.2

Solution (*b*) The eigenspaces corresponding to $\lambda = -2$ and $\lambda = 5$ can be graphed from the parametric equations in (8) and (10) or from the vector equations (9) and (11). These graphs are shown in Figure 4.4.2. When an eigenvector $\mathbf{x}$ in the eigenspace for $\lambda = 5$ is multiplied by A, the resulting vector has the same direction as $\mathbf{x}$ but the length is increased by a factor of 5; and when an eigenvector $\mathbf{x}$ in the eigenspace for $\lambda = -2$ is multiplied by A, the resulting vector is oppositely directed to $\mathbf{x}$ and the length is increased by a factor of 2. In both cases, multiplying an eigenvector by A produces a vector in the same eigenspace. ■

EXAMPLE 3
Eigenvalues of
a 3 × 3 Matrix

Find the eigenvalues of the matrix

$$A = \begin{bmatrix} 0 & -1 & 0 \\ 0 & 0 & 1 \\ -4 & -17 & 8 \end{bmatrix}$$

Solution We leave it for you to confirm that

$$\det(\lambda I - A) = \begin{vmatrix} \lambda & 1 & 0 \\ 0 & \lambda & -1 \\ 4 & 17 & \lambda - 8 \end{vmatrix} = \lambda^3 - 8\lambda^2 + 17\lambda - 4 \tag{12}$$

from which we obtain the characteristic equation

$$\lambda^3 - 8\lambda^2 + 17\lambda - 4 = 0 \tag{13}$$

To solve this equation we will begin by searching for integer solutions. This can be done by using the fact that if a polynomial equation has integer coefficients, then its integer solutions, if any, must be divisors of the constant term. Thus, the only possible integer solutions of (13) are the divisors of -4, namely ± 1, ± 2, and ± 4. Substituting these values successively into (13) shows that $\lambda = 4$ is an integer solution. This implies that $\lambda - 4$ is a factor of (12), so we divide $\lambda - 4$ into (12) and rewrite (13) as

$$(\lambda - 4)(\lambda^2 - 4\lambda + 1) = 0$$

Thus, the remaining solutions of the characteristic equation satisfy the quadratic equation

$$\lambda^2 - 4\lambda + 1 = 0$$

which we can solve by the quadratic formula to conclude that the eigenvalues of A are

$$\lambda = 4, \quad \lambda = 2 + \sqrt{3}, \quad \lambda = 2 - \sqrt{3} \qquad \blacksquare$$

EIGENVALUES OF TRIANGULAR MATRICES

If A is an $n \times n$ triangular matrix with diagonal entries $a_{11}, a_{22}, \ldots, a_{nn}$, then $\lambda I - A$ is a triangular matrix with diagonal entries $\lambda - a_{11}, \lambda - a_{22}, \ldots, \lambda - a_{nn}$ (verify). Thus, the characteristic polynomial of A is

$$\det(\lambda I - A) = (\lambda - a_{11})(\lambda - a_{22}) \cdots (\lambda - a_{nn})$$

which implies that the eigenvalues of A are

$$\lambda_1 = a_{11}, \quad \lambda_2 = a_{22}, \ldots, \quad \lambda_n = a_{nn}$$

Thus, we have the following theorem.

> **Theorem 4.4.5** *If A is a triangular matrix (upper triangular, lower triangular, or diagonal), then the eigenvalues of A are the entries on the main diagonal of A.*

EXAMPLE 4
Eigenvalues of Triangular Matrices

By inspection, the characteristic polynomial of the matrix

$$A = \begin{bmatrix} \frac{1}{2} & 0 & 0 & 0 \\ -1 & -\frac{2}{3} & 0 & 0 \\ 7 & \frac{5}{8} & 6 & 0 \\ \frac{4}{9} & -4 & 3 & 6 \end{bmatrix}$$

is

$$p(\lambda) = \left(\lambda - \tfrac{1}{2}\right)\left(\lambda + \tfrac{2}{3}\right)(\lambda - 6)^2$$

so the *distinct* eigenvalues of A are $\lambda = \frac{1}{2}, \lambda = -\frac{2}{3}$, and $\lambda = 6$. $\qquad \blacksquare$

EIGENVALUES OF POWERS OF A MATRIX

Once the eigenvalues and eigenvectors of a matrix A are found, it is a simple matter to find the eigenvalues and eigenvectors of any positive integer power of A. For example, if λ is an eigenvalue of A and $\mathbf{x}$ is a corresponding eigenvector, then

$$A^2\mathbf{x} = A(A\mathbf{x}) = A(\lambda\mathbf{x}) = \lambda(A\mathbf{x}) = \lambda(\lambda\mathbf{x}) = \lambda^2\mathbf{x}$$

which shows that λ^2 is an eigenvalue of A^2 and $\mathbf{x}$ is a corresponding eigenvector. In general, we have the following result.

> **Theorem 4.4.6** *If λ is an eigenvalue of a matrix A and $\mathbf{x}$ is a corresponding eigenvector, and if k is any positive integer, then λ^k is an eigenvalue of A^k and $\mathbf{x}$ is a corresponding eigenvector.*

Some problems that use this theorem are given in the exercises.

A UNIFYING THEOREM   Since λ is an eigenvalue of a square matrix A if and only if there is a nonzero vector $\mathbf{x}$ such that $A\mathbf{x} = \lambda\mathbf{x}$, it follows that $\lambda = 0$ is an eigenvalue of A if and only if there is a nonzero vector $\mathbf{x}$ such that $A\mathbf{x} = \mathbf{0}$. However, this is true if and only if $\det(A) = 0$, so we can add another statement to the list in Theorem 4.2.7.

> **Theorem 4.4.7** *If A is an $n \times n$ matrix, then the following statements are equivalent.*
>
> (a) *The reduced row echelon form of A is I_n.*
> (b) *A is expressible as a product of elementary matrices.*
> (c) *A is invertible.*
> (d) *$A\mathbf{x} = \mathbf{0}$ has only the trivial solution.*
> (e) *$A\mathbf{x} = \mathbf{b}$ is consistent for every vector $\mathbf{b}$ in R^n.*
> (f) *$A\mathbf{x} = \mathbf{b}$ has exactly one solution for every vector $\mathbf{b}$ in R^n.*
> (g) *The column vectors of A are linearly independent.*
> (h) *The row vectors of A are linearly independent.*
> (i) *$\det(A) \neq 0$.*
> (j) *$\lambda = 0$ is not an eigenvalue of A.*

REMARK In the exercises we will ask you to prove that if A is an invertible matrix with eigenvalues $\lambda_1, \lambda_2, \ldots, \lambda_k$, then $1/\lambda_1, 1/\lambda_2, \ldots, 1/\lambda_k$ are eigenvalues of A^{-1}. Parts (c) and (j) of Theorem 4.4.7 ensure that these reciprocals are defined.

COMPLEX EIGENVALUES   It is possible for the characteristic equation of a matrix A with real entries to have imaginary solutions. For example, the characteristic equation of the matrix

$$A = \begin{bmatrix} -2 & -1 \\ 5 & 2 \end{bmatrix} \tag{14}$$

is

$$\begin{vmatrix} \lambda + 2 & 1 \\ -5 & \lambda - 2 \end{vmatrix} = \lambda^2 + 1 = 0 \tag{15}$$

Thus, the roots of the characteristic equation are the imaginary numbers $\lambda = i$ and $\lambda = -i$.

This raises some important issues that need to be addressed. Specifically, up to now we have required scalars to be real numbers, so the imaginary numbers arising from (15) cannot be regarded as eigenvalues unless we are willing to extend the concept of a scalar to allow complex numbers. However, if we decide to do this, then the linear system $(\lambda I - A)\mathbf{x} = \mathbf{0}$ will have a coefficient matrix with complex entries and the solutions of this system may also involve complex numbers. Thus, by opening the door to complex eigenvalues we are forced to consider matrices with complex entries and vectors with complex components. It turns out that complex eigenvalues have important applications, so for a complete discussion of eigenvalues and eigenvectors it is necessary to allow complex scalars and vectors with complex entries. Such matters will be taken up in more detail later, so for now we will continue to assume that matrices have real entries and vectors have real components. We will, however, refer to complex solutions of the characteristic equation as ***complex eigenvalues***.

ALGEBRAIC MULTIPLICITY   If A is an $n \times n$ matrix, then the expanded form of the determinant $\det(\lambda I - A)$ is a polynomial of degree n in which the coefficient of λ^n is 1 (can you see why?); that is, $\det(\lambda I - A)$ is of the form

$$\det(\lambda I - A) = \lambda^n + c_1\lambda^{n-1} + \cdots + c_n \tag{16}$$

The polynomial

$$p(\lambda) = \lambda^n + c_1\lambda^{n-1} + \cdots + c_n \tag{17}$$

that appears on the right side of (16) is called the ***characteristic polynomial*** of A. For example, the characteristic polynomial of the 2×2 matrix A in Example 2 is the second-degree polynomial

$$p(\lambda) = \lambda^2 - 3\lambda - 10$$

[see (6)], and the characteristic polynomial of the 3×3 matrix A in Example 3 is the third-degree polynomial

$$p(\lambda) = \lambda^3 - 8\lambda^2 + 17\lambda - 4$$

[see (13)].

When you try to factor a characteristic polynomial

$$\det(\lambda I - A) = \lambda^n + c_1 \lambda^{n-1} + \cdots + c_n$$

one of three things can happen:

1. It may be possible to factor the polynomial completely into distinct linear factors using only real numbers; for example,

$$\lambda^3 + \lambda^2 - 2\lambda = \lambda(\lambda^2 + \lambda - 2) = \lambda(\lambda - 1)(\lambda + 2)$$

2. It may be possible to factor the polynomial completely into linear factors using only real numbers, but some of the factors may be repeated; for example,

$$\lambda^6 - 3\lambda^4 + 2\lambda^3 = \lambda^3(\lambda^3 - 3\lambda + 2) = \lambda^3(\lambda - 1)^2(\lambda + 2)$$

3. It may be possible to factor the polynomial completely into linear and quadratic factors using only real numbers, but it may not be possible to decompose the quadratic factors into linear factors without using imaginary numbers (such quadratic factors are said to be ***irreducible*** over the real numbers); for example,

$$\lambda^4 - 1 = (\lambda^2 - 1)(\lambda^2 + 1) = (\lambda - 1)(\lambda + 1)(\lambda - i)(\lambda + i)$$

Here the factor $\lambda^2 + 1$ is irreducible over the real numbers.

It can be proved that if imaginary eigenvalues are allowed, then the characteristic polynomial of an $n \times n$ matrix A can be factored as

$$\det(\lambda I - A) = \lambda^n + c_1 \lambda^{n-1} + \cdots + c_n = (\lambda - \lambda_1)(\lambda - \lambda_2) \cdots (\lambda - \lambda_n) \tag{18}$$

where $\lambda_1, \lambda_2, \ldots, \lambda_n$ are eigenvalues of A. This is called the ***complete linear factorization*** of the characteristic polynomial.[*] If some of the factors in (18) are repeated, then they can be combined; for example, if the first k factors are distinct and the rest are repetitions of the first k, then (18) can be rewritten in the form

$$\det(\lambda I - A) = \lambda^n + c_1 \lambda^{n-1} + \cdots + c_n = (\lambda - \lambda_1)^{m_1}(\lambda - \lambda_2)^{m_2} \cdots (\lambda - \lambda_k)^{m_k} \tag{19}$$

where $\lambda_1, \lambda_2, \ldots, \lambda_k$ are the *distinct* eigenvalues of A. The exponent m_i, called the ***algebraic multiplicity*** of the eigenvalue λ_i, tells how many times that eigenvalue is repeated in the complete factorization of the characteristic polynomial.

The sum of the algebraic multiplicities of the eigenvalues in (19) must be n, since the characteristic polynomial has degree n. For example, if A is a 6×6 matrix whose characteristic polynomial is

$$\lambda^6 - 3\lambda^4 + 2\lambda^3 = \lambda^3(\lambda^3 - 3\lambda + 2) = \lambda^3(\lambda - 1)^2(\lambda + 2)$$

then the distinct eigenvalues of A are $\lambda = 0$, $\lambda = 1$, and $\lambda = -2$. The algebraic multiplicities of these eigenvalues are 3, 2, and 1, respectively, which sum up to 6 as required.

---

[*]The *fundamental theorem of algebra* states that if $p(\lambda)$ is a nonconstant polynomial of degree n with real or complex coefficients, then the polynomial equation $p(\lambda) = 0$ has at least one real or imaginary root. If λ_1 is any such root, then the *factor theorem* in algebra states that $p(\lambda)$ can be factored as $p(\lambda) = (\lambda - \lambda_1)p_1(\lambda)$, where $p_1(\lambda)$ is a polynomial of degree $n - 1$. By applying this same factorization process to $p_1(\lambda)$ and then to the subsequent polynomials of lower degree, one can eventually obtain the factorization in (18).

The following theorem summarizes this discussion.

> **Theorem 4.4.8** *If A is an n × n matrix, then the characteristic polynomial of A can be expressed as*
>
> $$\det(\lambda I - A) = (\lambda - \lambda_1)^{m_1}(\lambda - \lambda_2)^{m_2} \cdots (\lambda - \lambda_k)^{m_k}$$
>
> *where $\lambda_1, \lambda_2, \ldots, \lambda_k$ are the distinct eigenvalues of A and $m_1 + m_2 + \cdots + m_k = n$.*

REMARK This theorem implies that an $n \times n$ matrix has n eigenvalues if we agree to count repetitions and allow complex eigenvalues, but the number of distinct eigenvalues may be less than n.

CONCEPT PROBLEM By setting $\lambda = 0$ in (18), deduce that the constant term in the characteristic polynomial of an $n \times n$ matrix A is

$$c_n = (-1)^n(\lambda_1 \lambda_2 \cdots \lambda_n) \tag{20}$$

EIGENVALUE ANALYSIS OF 2 × 2 MATRICES

Next we will derive formulas for the eigenvalues of 2×2 matrices, and we will discuss some geometric properties of their eigenspaces. The characteristic polynomial of a general 2×2 matrix

$$A = \begin{bmatrix} a & b \\ c & d \end{bmatrix}$$

is

$$\det(\lambda I - A) = \begin{vmatrix} \lambda - a & -b \\ -c & \lambda - d \end{vmatrix} = (\lambda - a)(\lambda - d) - bc = \lambda^2 - (a + d)\lambda + (ad - bc)$$

We can express this in terms of the trace and determinant of A as

$$\det(\lambda I - A) = \lambda^2 - \operatorname{tr}(A)\lambda + \det(A) \tag{21}$$

from which it follows that the characteristic equation of A is

$$\lambda^2 - \operatorname{tr}(A)\lambda + \det(A) = 0 \tag{22}$$

Now recall from algebra that if $ax^2 + bx + c = 0$ is a quadratic equation with real coefficients, then the **discriminant** $b^2 - 4ac$ determines the nature of the roots:

$b^2 - 4ac > 0$ **[Two distinct real roots]**
$b^2 - 4ac = 0$ **[One repeated real root]**
$b^2 - 4ac < 0$ **[Two conjugate imaginary roots]**

Applying this to (22) with $a = 1$, $b = -\operatorname{tr}(A)$, and $c = \det(A)$ yields the following theorem.

> **Theorem 4.4.9** *If A is a 2 × 2 matrix with real entries, then the characteristic equation of A is*
>
> $$\lambda^2 - \operatorname{tr}(A)\lambda + \det(A) = 0$$
>
> *and*
>
> (a) *A has two distinct real eigenvalues if* $\operatorname{tr}(A)^2 - 4\det(A) > 0$;
>
> (b) *A has one repeated real eigenvalue if* $\operatorname{tr}(A)^2 - 4\det(A) = 0$;
>
> (c) *A has two conjugate imaginary eigenvalues if* $\operatorname{tr}(A)^2 - 4\det(A) < 0$.

REMARK This theorem is not valid if A has complex entries, so we have emphasized in its statement that A must have real entries, even though these are the only kinds of matrices we are considering now. Later we will consider matrices with complex entries, in which case this theorem will not be applicable.

EXAMPLE 5
Eigenvalues of
a 2 × 2 Matrix

In each part, use Formula (22) for the characteristic equation to find the eigenvalues of

(a) $A = \begin{bmatrix} 2 & 2 \\ -1 & 5 \end{bmatrix}$ (b) $A = \begin{bmatrix} 0 & -1 \\ 1 & 2 \end{bmatrix}$ (c) $A = \begin{bmatrix} 2 & 3 \\ -3 & 2 \end{bmatrix}$

Solution (a) We have $\text{tr}(A) = 7$ and $\det(A) = 12$, so the characteristic equation of A is

$$\lambda^2 - 7\lambda + 12 = 0$$

Factoring yields $(\lambda - 4)(\lambda - 3) = 0$, so the eigenvalues of A are $\lambda = 4$ and $\lambda = 3$.

Solution (b) We have $\text{tr}(A) = 2$ and $\det(A) = 1$, so the characteristic equation of A is

$$\lambda^2 - 2\lambda + 1 = 0$$

Factoring yields $(\lambda - 1)^2 = 0$, so $\lambda = 1$ is the only eigenvalue of A; it has algebraic multiplicity 2.

Solution (c) We have $\text{tr}(A) = 4$ and $\det(A) = 13$, so the characteristic equation of A is

$$\lambda^2 - 4\lambda + 13 = 0$$

Solving this equation by the quadratic formula yields

$$\lambda = \frac{4 \pm \sqrt{(-4)^2 - 4(13)}}{2} = \frac{4 \pm \sqrt{-36}}{2} = 2 \pm 3i$$

Thus, the eigenvalues of A are $\lambda = 2 + 3i$ and $\lambda = 2 - 3i$. ∎

**EIGENVALUE ANALYSIS
OF 2 × 2 SYMMETRIC
MATRICES**

Later in the text we will show that *all* symmetric matrices with real entries have real eigenvalues. However, we already have the mathematical tools to prove this result in the 2 × 2 case.

> **Theorem 4.4.10** *A symmetric 2 × 2 matrix with real entries has real eigenvalues. Moreover, if A is of the form*
>
> $$A = \begin{bmatrix} a & 0 \\ 0 & a \end{bmatrix} \tag{23}$$
>
> *then A has one repeated eigenvalue, namely $\lambda = a$; otherwise it has two distinct eigenvalues.*

Proof If the 2 × 2 symmetric matrix is

$$A = \begin{bmatrix} a & b \\ b & d \end{bmatrix}$$

then

$$\text{tr}(A)^2 - 4\det(A) = (a + d)^2 - 4(ad - b^2) = (a - d)^2 + 4b^2 \geq 0$$

so Theorem 4.4.9 implies that A has real eigenvalues. It also follows from that theorem that A has one repeated eigenvalue if and only if

$$\text{tr}(A)^2 - 4\det(A) = (a - d)^2 + 4b^2 = 0$$

Since this holds if and only if $a = d$ and $b = 0$, it follows that the only 2 × 2 symmetric matrices with one repeated eigenvalue are those of form (23). ∎

In the exercises we will guide you through the steps in proving the following general result about the eigenspaces of 2×2 symmetric matrices.

Theorem 4.4.11

(a) *If a 2×2 symmetric matrix with real entries has one repeated eigenvalue, then the eigenspace corresponding to that eigenvalue is R^2.*

(b) *If a 2×2 symmetric matrix with real entries has two distinct eigenvalues, then the eigenspaces corresponding to those eigenvalues are perpendicular lines through the origin of R^2.*

EXAMPLE 6
Eigenspaces of a Symmetric 2×2 Matrix

Graph the eigenspaces of the symmetric matrix

$$A = \begin{bmatrix} 3 & 2 \\ 2 & 3 \end{bmatrix}$$

in an xy-coordinate system.

Solution Since $\operatorname{tr}(A) = 6$ and $\det(A) = 5$, it follows from (21) that the characteristic polynomial of A is

$$\lambda^2 - 6\lambda + 5 = (\lambda - 1)(\lambda - 5)$$

so the eigenvalues of A are $\lambda = 1$ and $\lambda = 5$. To find the corresponding eigenspaces we must solve the system

$$\begin{bmatrix} \lambda - 3 & -2 \\ -2 & \lambda - 3 \end{bmatrix} \begin{bmatrix} x \\ y \end{bmatrix} = \begin{bmatrix} 0 \\ 0 \end{bmatrix} \tag{24}$$

first with $\lambda = 1$ and then with $\lambda = 5$.

Case $\lambda = 1$ In this case (24) becomes

$$\begin{bmatrix} -2 & -2 \\ -2 & -2 \end{bmatrix} \begin{bmatrix} x \\ y \end{bmatrix} = \begin{bmatrix} 0 \\ 0 \end{bmatrix}$$

Solving this system yields (verify)

$$x = -t, \quad y = t \tag{25}$$

which are parametric equations of the line $y = -x$. This is the eigenspace corresponding to $\lambda = 1$ (Figure 4.4.3).

Case $\lambda = 5$ In this case (24) becomes

$$\begin{bmatrix} 2 & -2 \\ -2 & 2 \end{bmatrix} \begin{bmatrix} x \\ y \end{bmatrix} = \begin{bmatrix} 0 \\ 0 \end{bmatrix}$$

Solving this system yields (verify)

$$x = t, \quad y = t \tag{26}$$

which are parametric equations of the line $y = x$. This is the eigenspace corresponding to $\lambda = 5$ (Figure 4.4.3).

Note that the lines $y = -x$ and $y = x$ are perpendicular, as guaranteed by Theorem 4.4.11. From a vector point of view, we can write (25) and (26) in the vector forms

$$\mathbf{x} = \begin{bmatrix} x \\ y \end{bmatrix} = \begin{bmatrix} -t \\ t \end{bmatrix} = t \begin{bmatrix} -1 \\ 1 \end{bmatrix} \quad \text{and} \quad \mathbf{x} = \begin{bmatrix} x \\ y \end{bmatrix} = \begin{bmatrix} t \\ t \end{bmatrix} = t \begin{bmatrix} 1 \\ 1 \end{bmatrix}$$

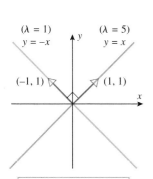

$(\lambda = 1)$
$y = -x$

$(\lambda = 5)$
$y = x$

$(-1, 1)$ $(1, 1)$

The eigenspaces of the symmetric matrix A are perpendicular.

Figure 4.4.3

from which we see that the spanning vectors

$$\mathbf{v}_1 = \begin{bmatrix} -1 \\ 1 \end{bmatrix} \quad \text{and} \quad \mathbf{v}_2 = \begin{bmatrix} 1 \\ 1 \end{bmatrix} \tag{27}$$

for the two eigenspaces are orthogonal.   ■

EXPRESSIONS FOR DETERMINANT AND TRACE IN TERMS OF EIGENVALUES

The following theorem provides expressions for the determinant and trace of a square matrix in terms of its eigenvalues.

Theorem 4.4.12   *If A is an $n \times n$ matrix with eigenvalues $\lambda_1, \lambda_2, \ldots, \lambda_n$ (repeated according to multiplicity), then:*

(a) $\det(A) = \lambda_1 \lambda_2 \cdots \lambda_n$

(b) $\operatorname{tr}(A) = \lambda_1 + \lambda_2 + \cdots + \lambda_n$

Proof (a) Write the characteristic polynomial in factored form:

$$p(\lambda) = \det(\lambda I - A) = (\lambda - \lambda_1)(\lambda - \lambda_2) \cdots (\lambda - \lambda_n) \tag{28}$$

Setting $\lambda = 0$ yields

$$\det(-A) = (-1)^n (\lambda_1 \lambda_2 \cdots \lambda_n)$$

But $\det(-A) = (-1)^n \det(A)$, so it follows that

$$\det(A) = \lambda_1 \lambda_2 \cdots \lambda_n \tag{29}$$

Proof (b) Assume that $A = [a_{ij}]$, so we can write $p(\lambda)$ as

$$p(\lambda) = \det(\lambda I - A) = \begin{vmatrix} \lambda - a_{11} & -a_{12} & \cdots & -a_{1n} \\ -a_{21} & \lambda - a_{22} & \cdots & -a_{2n} \\ \vdots & \vdots & & \vdots \\ -a_{n1} & -a_{n2} & \cdots & \lambda - a_{nn} \end{vmatrix} \tag{30}$$

If we compute $p(\lambda)$ from this determinant by forming the sum of the signed elementary products, then any elementary product that contains an entry that is *off* the main diagonal of (30) as a factor will contain at most $n - 2$ factors that involve λ (why?). Thus, the coefficient of λ^{n-1} in $p(\lambda)$ is the same as the coefficient of λ^{n-1} in the product

$$(\lambda - a_{11})(\lambda - a_{22}) \cdots (\lambda - a_{nn})$$

Expanding this product we see that it has the form

$$\lambda^n - (a_{11} + a_{22} + \cdots + a_{nn})\lambda^{n-1} + \cdots \tag{31}$$

and expanding the expression for $p(\lambda)$ in (28) we see that it has the form

$$\lambda^n - (\lambda_1 + \lambda_2 + \cdots + \lambda_n)\lambda^{n-1} + \cdots$$

Thus, we must have

$$\operatorname{tr}(A) = a_{11} + a_{22} + \cdots + a_{nn} = \lambda_1 + \lambda_2 + \cdots + \lambda_n \qquad ■$$

EXAMPLE 7
Determinant and Trace from Eigenvalues

Find the determinant and trace of a 3×3 matrix whose characteristic polynomial is

$$p(\lambda) = \lambda^3 - 3\lambda + 2 \tag{32}$$

Solution This polynomial can be factored as

$$p(\lambda) = (\lambda - 1)^2(\lambda + 2)$$

so the eigenvalues, repeated according to multiplicity, are $\lambda_1 = 1$, $\lambda_2 = 1$, and $\lambda_3 = -2$. Thus,

$$\det(A) = \lambda_1\lambda_2\lambda_3 = -2 \quad \text{and} \quad \text{tr}(A) = \lambda_1 + \lambda_2 + \lambda_3 = 0$$

Alternative Solution It follows from (31) that if $p(\lambda)$ is the characteristic polynomial of an $n \times n$ matrix A, then $\text{tr}(A)$ is the negative of the coefficient of λ^{n-1}, and it follows from (28) and (29) that $\det(A)$ is the constant term in $p(\lambda)$ if n is even and the negative of the constant term if n is odd (why?). Thus, we see directly from (32) that $\text{tr}(A) = 0$ and $\det(A) = -2$. ■

EIGENVALUES BY NUMERICAL METHODS Eigenvalues are rarely obtained by solving the characteristic equation in real-world applications primarily for two reasons:

1. In order to construct the characteristic equation of an $n \times n$ matrix A, it is necessary to expand the determinant $\det(\lambda I - A)$. Although computer algebra systems such as *Mathematica*, *Maple*, and *Derive* can do this for matrices of small size, the computations are prohibitive for matrices of the size that occur in typical applications.

2. There is no algebraic formula or finite algorithm that can be used to obtain the exact solutions of the characteristic equation of a general $n \times n$ matrix when $n \geq 5$.

Given these impediments, various algorithms have been developed for producing numerical approximations to the eigenvalues and eigenvectors. Some of these will be discussed later in the text.

Exercise Set 4.4

In Exercises 1 and 2, determine whether the matrix has non-trivial fixed points, and if so, find them.

1. (a) $\begin{bmatrix} 1 & 0 \\ 1 & 0 \end{bmatrix}$ (b) $\begin{bmatrix} 0 & 1 \\ 1 & 0 \end{bmatrix}$

2. (a) $\begin{bmatrix} 3 & 1 \\ 4 & 2 \end{bmatrix}$ (b) $\begin{bmatrix} 1 & 4 \\ 1 & 1 \end{bmatrix}$

In Exercises 3 and 4, confirm by multiplication that $\mathbf{x}$ is an eigenvector of A, and find the corresponding eigenvalue.

3. $A = \begin{bmatrix} 4 & 0 & 1 \\ 2 & 3 & 2 \\ 1 & 0 & 4 \end{bmatrix}$; $\mathbf{x} = \begin{bmatrix} 1 \\ 2 \\ 1 \end{bmatrix}$

4. $A = \begin{bmatrix} 2 & -1 & -1 \\ -1 & 2 & -1 \\ -1 & -1 & 2 \end{bmatrix}$; $\mathbf{x} = \begin{bmatrix} 1 \\ 1 \\ 1 \end{bmatrix}$

In Exercises 5–8, find the characteristic equation of the matrix, and then find the eigenvalues and their algebraic multiplicities.

5. (a) $\begin{bmatrix} 3 & 0 \\ 8 & -1 \end{bmatrix}$ (b) $\begin{bmatrix} 10 & -9 \\ 4 & -2 \end{bmatrix}$ (c) $\begin{bmatrix} 2 & 0 \\ 1 & 2 \end{bmatrix}$

6. (a) $\begin{bmatrix} -2 & 4 \\ 3 & 2 \end{bmatrix}$ (b) $\begin{bmatrix} 0 & 0 \\ 0 & 0 \end{bmatrix}$ (c) $\begin{bmatrix} 1 & 0 \\ 2 & 1 \end{bmatrix}$

7. (a) $\begin{bmatrix} 4 & 0 & 1 \\ -2 & 1 & 0 \\ -2 & 0 & 1 \end{bmatrix}$ (b) $\begin{bmatrix} 4 & -5 & -5 \\ \frac{2}{5} & 1 & -1 \\ \frac{6}{5} & -3 & -1 \end{bmatrix}$

(c) $\begin{bmatrix} 3 & 4 & -1 \\ -1 & -2 & 1 \\ 3 & 9 & 0 \end{bmatrix}$

8. (a) $\begin{bmatrix} 15 & 6 & -4 \\ -40 & -16 & 11 \\ 0 & 0 & -1 \end{bmatrix}$ (b) $\begin{bmatrix} 5 & 0 & 1 \\ 1 & 1 & 0 \\ -7 & 1 & 0 \end{bmatrix}$

(c) $\begin{bmatrix} 5 & 6 & 2 \\ 0 & -1 & -8 \\ 1 & 0 & -2 \end{bmatrix}$

In Exercises 9–12, find the eigenspaces of the matrix, and describe them geometrically.

9. The matrices in Exercise 5.

10. The matrices in Exercise 6.

11. The matrices in Exercise 7.

12. The matrices in Exercise 8.

13. Find the characteristic polynomial and the eigenvalues by inspection.

(a) $\begin{bmatrix} -1 & 6 \\ 0 & 5 \end{bmatrix}$

(b) $\begin{bmatrix} 3 & 0 & 0 \\ -2 & 7 & 0 \\ 4 & 8 & 1 \end{bmatrix}$

(c) $\begin{bmatrix} -\frac{1}{3} & 0 & 0 & 0 \\ 0 & -\frac{1}{3} & 0 & 0 \\ 0 & 0 & 1 & 0 \\ 0 & 0 & 0 & \frac{1}{2} \end{bmatrix}$

14. Find some matrices whose characteristic polynomial is $p(\lambda) = \lambda(\lambda - 2)^2(\lambda + 1)$.

In Exercises 15 and 16, use *block triangular partitioning* to find the characteristic polynomial and the eigenvalues. [*Note:* See the discussion preceding Exercises 38 and 39 of Section 4.2.]

15. $\begin{bmatrix} 2 & 3 & 0 & 0 \\ -1 & 6 & 0 & 0 \\ 0 & 0 & -2 & 5 \\ 0 & 0 & 1 & 2 \end{bmatrix}$

16. $\begin{bmatrix} 0 & 0 & 7 & -5 \\ 1 & 0 & 1 & 0 \\ 0 & 0 & -2 & 0 \\ 0 & 0 & 0 & 1 \end{bmatrix}$

In Exercises 17 and 18, find the eigenvalues and corresponding eigenvectors of A, and then use them to find the eigenvalues and corresponding eigenvectors of the stated power of A.

17. $A = \begin{bmatrix} -1 & -2 & -2 \\ 1 & 2 & 1 \\ -1 & -1 & 0 \end{bmatrix}$; A^{25}

18. $A = \begin{bmatrix} 1 & 3 & 7 & 11 \\ 0 & \frac{1}{2} & 3 & 8 \\ 0 & 0 & 0 & 4 \\ 0 & 0 & 0 & 2 \end{bmatrix}$; A^9

In Exercises 19 and 20, find the eigenvalues of the matrix, and confirm the statements in Theorem 4.4.12.

19. $A = \begin{bmatrix} 2 & 1 & 1 \\ 2 & 1 & -2 \\ -1 & 0 & -2 \end{bmatrix}$

20. $A = \begin{bmatrix} 1 & 1 & -3 \\ 2 & 0 & 6 \\ 1 & -1 & 5 \end{bmatrix}$

In Exercises 21 and 22, graph the eigenspaces of the given symmetric matrix in an xy-coordinate system, and use slopes to confirm that they are perpendicular lines.

21. $\begin{bmatrix} 1 & 2 \\ 2 & 4 \end{bmatrix}$

22. $\begin{bmatrix} 1 & \sqrt{2} \\ \sqrt{2} & 0 \end{bmatrix}$

23. Let A be a 2×2 matrix, and call a line through the origin of R^2 *invariant under A* if $A\mathbf{x}$ lies on the line when $\mathbf{x}$ does.

Find equations for all lines in R^2, if any, that are invariant under the given matrix.

(a) $\begin{bmatrix} 4 & -1 \\ 2 & 1 \end{bmatrix}$

(b) $\begin{bmatrix} 0 & 1 \\ -1 & 0 \end{bmatrix}$

(c) $\begin{bmatrix} 2 & 3 \\ 0 & 2 \end{bmatrix}$

In Exercises 24 and 25, find a and b so that A has the stated eigenvalues.

24. $A = \begin{bmatrix} 1 & 6 \\ a & b \end{bmatrix}$; $\lambda = 4, -3$

25. $A = \begin{bmatrix} 3 & 2 \\ a & b \end{bmatrix}$; $\lambda = 2, 5$

26. For what value(s) of x, if any, will the matrix

$$A = \begin{bmatrix} 3 & 0 & 0 \\ 0 & x & 2 \\ 0 & 2 & x \end{bmatrix}$$

have at least one repeated eigenvalue?

27. Show that if A is a 2×2 matrix such that $A^2 = I$ and if $\mathbf{x}$ is any vector in R^2, then $\mathbf{y} = \mathbf{x} + A\mathbf{x}$ and $\mathbf{z} = \mathbf{x} - A\mathbf{x}$ are eigenvectors of A. Find the corresponding eigenvalues.

28. Show that if A is a square matrix, then the constant term in the characteristic polynomial of A is $(-1)^n \det(A)$.

Exercises 29 and 30 are concerned with formulas for the eigenvectors and eigenvalues of a general 2×2 matrix

$$A = \begin{bmatrix} a & b \\ c & d \end{bmatrix}$$

29. (a) Use Formula (22) to show that the eigenvalues of A are

$$\lambda = \tfrac{1}{2}\left[(a + d) \pm \sqrt{(a - d)^2 + 4bc}\right]$$

(b) Show that A has two distinct real eigenvalues if $(a - d)^2 + 4bc > 0$.

(c) Show that A has one repeated real eigenvalue if $(a - d)^2 + 4bc = 0$.

(d) Show that A has no real eigenvalues if $(a - d)^2 + 4bc < 0$.

30. Show that if $(a - d)^2 + 4bc > 0$ and $b \neq 0$, then eigenvectors corresponding to the eigenvalues of A (obtained in the preceding exercise) are

$$\lambda_1 = \tfrac{1}{2}\left[(a + d) + \sqrt{(a - d)^2 + 4bc}\right]; \ \mathbf{x}_1 = \begin{bmatrix} -b \\ a - \lambda_1 \end{bmatrix}$$

$$\lambda_2 = \tfrac{1}{2}\left[(a + d) - \sqrt{(a - d)^2 + 4bc}\right]; \ \mathbf{x}_2 = \begin{bmatrix} -b \\ a - \lambda_2 \end{bmatrix}$$

31. Suppose that the characteristic polynomial of a matrix A is $p(\lambda) = \lambda^2 + 3\lambda - 4$. Use Theorem 4.4.6 and the statements in Exercises P2, P3, P4, and P5 below to find the eigenvalues of the matrix.

(a) A^{-1}

(b) A^{-3}

(c) $A - 4I$

(d) $5A$

(e) $4A^T + 2I$

32. Show that if λ is an eigenvalue of a matrix A and $\mathbf{x}$ is a corresponding eigenvector, then

$$\lambda = \frac{(A\mathbf{x}) \cdot \mathbf{x}}{\|\mathbf{x}\|^2}$$

33. (a) Show that the characteristic polynomial of the matrix

$$C = \begin{bmatrix} 0 & 0 & 0 & \cdots & 0 & -c_0 \\ 1 & 0 & 0 & \cdots & 0 & -c_1 \\ 0 & 1 & 0 & \cdots & 0 & -c_2 \\ \vdots & \vdots & \vdots & & \vdots & \vdots \\ 0 & 0 & 0 & \cdots & 1 & -c_{n-1} \end{bmatrix}$$

is $p(\lambda) = c_0 + c_1\lambda + \cdots + c_{n-1}\lambda^{n-1} + \lambda^n$. [*Hint:* Evaluate all required determinants by adding

a suitable multiple of the second row to the first to introduce a zero at the top of the first column, and then expand by cofactors along the first column. Then repeat the process.]

(b) The matrix in part (a) is called the **companion matrix** of the polynomial

$$p(\lambda) = c_0 + c_1\lambda + \cdots + c_{n-1}\lambda^{n-1} + \lambda^n$$

Thus we see that if $p(\lambda)$ is any polynomial whose highest power has a coefficient of 1, then there is some matrix whose characteristic polynomial is $p(\lambda)$, namely its companion matrix. Use this observation to find a matrix whose characteristic polynomial is $p(\lambda) = 2 - 3\lambda + \lambda^2 - 5\lambda^3 + \lambda^4$.

Discussion and Discovery

D1. Suppose that the characteristic polynomial of A is $p(\lambda) = (\lambda - 1)(\lambda - 3)^2(\lambda - 4)^3$.
  (a) What is the size of A?
  (b) Is A invertible? Explain your reasoning.

D2. A square matrix whose size is 2×2 or greater and whose entries are all the same must have _____ as one of its eigenvalues.

D3. If A is a 2×2 matrix with $\text{tr}(A) = \det(A) = 4$, then the eigenvalues of A are _____.

D4. If $p(\lambda) = (\lambda - 3)^2(\lambda + 2)^3$ is the characteristic polynomial of a matrix A, then $\det(A) = $ _____ and $\text{tr}(A) = $ _____.

D5. Find all 2×2 matrices for which $\text{tr}(A) = \det(A)$, if any.

D6. If the characteristic polynomial of A is

$$p(\lambda) = \lambda^4 + 5\lambda^3 + 6\lambda^2 - 4\lambda - 8$$

then the eigenvalues of A^2 are _____.

D7. Indicate whether the statement is true (T) or false (F). Justify your answer. In each part, assume that A is square.

(a) If $A\mathbf{x} = \lambda\mathbf{x}$ for some nonzero scalar λ, then $\mathbf{x}$ is an eigenvector of A.
(b) If λ is an eigenvalue of A, then the linear system $(\lambda^2 I - A^2)\mathbf{x} = 0$ has nontrivial solutions.
(c) If $\lambda = 0$ is an eigenvalue of A, then the row vectors and column vectors of A are linearly independent.
(d) A 2×2 matrix with real eigenvalues is symmetric.

D8. Indicate whether the statement is true (T) or false (F). Justify your answer. In each part, A is assumed to be square.

(a) The eigenvalues of A are the same as the eigenvalues of the reduced row echelon form of A.
(b) If eigenvectors corresponding to distinct eigenvalues are added, then the resulting vector is not an eigenvector of A.
(c) A 3×3 matrix has at least one real eigenvalue.
(d) If the characteristic polynomial of A is $p(\lambda) = \lambda^n + 1$, then A is invertible.

Working with Proofs

P1. Use Formula (22) to show that if $p(\lambda)$ is the characteristic polynomial of a 2×2 matrix, then $p(A) = 0$; that is, A satisfies its own characteristic equation. This result, called the **Cayley–Hamilton theorem**, will be revisited for $n \times n$ matrices later in the text. [*Hint:* Substitute the matrix $\begin{bmatrix} a & b \\ c & d \end{bmatrix}$ into the left side of (22).]

P2. (a) Prove that if A is a square matrix, then A and A^T have the same characteristic polynomial. [*Hint.* Consider the characteristic equation $\det(\lambda I - A) = 0$ and use properties of the determinant.]
  (b) Show that A and A^T need not have the same eigenspaces by considering the matrix

$$\begin{bmatrix} 2 & 0 \\ 2 & 3 \end{bmatrix}$$

P3. Prove that if λ is an eigenvalue of an invertible matrix A, and $\mathbf{x}$ is a corresponding eigenvector, then $1/\lambda$ is an eigenvalue of A^{-1}, and $\mathbf{x}$ is a corresponding eigenvector. [*Hint:* Begin with the equation $A\mathbf{x} = \lambda\mathbf{x}$.]

P4. Prove that if λ is an eigenvalue of A and $\mathbf{x}$ is a corresponding eigenvector, then $\lambda - s$ is an eigenvalue of $A - sI$ for any scalar s, and $\mathbf{x}$ is a corresponding eigenvector.

P5. Prove that if λ is an eigenvalue of A and $\mathbf{x}$ is a corresponding eigenvector, then $s\lambda$ is an eigenvalue of sA for every scalar s, and $\mathbf{x}$ is a corresponding eigenvector.

P6. Prove Theorem 4.4.11 by using the results of Exercise 30.

P7. Suppose that A and B are square matrices with the same size, and λ is an eigenvalue of A with a corresponding eigenvector $\mathbf{x}$ that is a fixed point of B. Prove that λ is an eigenvalue of both AB and BA and that $\mathbf{x}$ is a corresponding eigenvector.

Technology Exercises

T1. Find the eigenvalues and corresponding eigenvectors of the matrix in Example 2. If it happens that your eigenvectors are different from those obtained in the example, resolve the discrepancy.

T2. Find eigenvectors corresponding to the eigenvalues of the matrix in Example 4.

T3. Define an nth-order ***checkerboard matrix*** C_n to be a matrix that has a 1 in the upper left corner and alternates between 1 and 0 along rows and columns (see the accompanying figure for an example).

   (a) Find the eigenvalues of C_1, C_2, C_3, C_4, C_5 and make a conjecture about the eigenvalues of C_6. Check your conjecture by finding the eigenvalues of C_6.

   (b) In general, what can you say about the eigenvalues of C_n?

1	0	1	0
0	1	0	1
1	0	1	0
0	1	0	1

Figure Ex-T3

T4. Confirm the statement in Theorem 4.4.6 for the matrix in Example 2 with $n = 2, 3, 4,$ and 5.

T5. (CAS)

   (a) Use the command for finding characteristic polynomials to find the characteristic polynomial of the matrix in Example 3, and check the result by using a determinant operation.

   (b) Find the exact eigenvalues by solving the characteristic equation.

T6. (CAS) Obtain the formulas in Exercise 29.

T7. Graph the characteristic polynomial of the matrix

$$A = \begin{bmatrix} 1 & 1 & 2 & 1 & 1 \\ 1 & 2 & 3 & 2 & 1 \\ 2 & 3 & 1 & 2 & 1 \\ 1 & 2 & 2 & 3 & 1 \\ 1 & 1 & 1 & 1 & 7 \end{bmatrix}$$

and estimate the roots of the characteristic equation. Compare your estimates to the eigenvalues produced directly by your utility.

T8. Select some pairs of 3×3 matrices A and B, and compute the eigenvalues of AB and BA. Make an educated guess about the relationship between the eigenvalues of AB and BA in general.

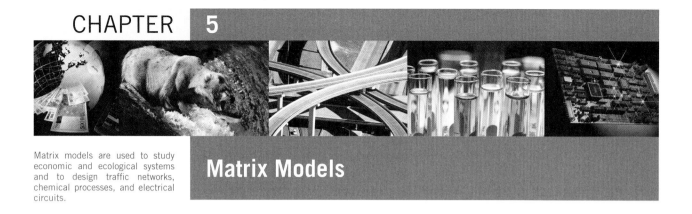

CHAPTER 5

Matrix models are used to study economic and ecological systems and to design traffic networks, chemical processes, and electrical circuits.

Matrix Models

Section 5.1 Dynamical Systems and Markov Chains

In this section we will show how matrix methods can be used to analyze the behavior of physical systems that evolve over time. The methods that we will study here have been applied to problems in business, ecology, demographics, sociology, and most of the physical sciences.

DYNAMICAL SYSTEMS

A ***dynamical system*** is a finite set of variables whose values change with time. The value of a variable at a point in time is called the ***state of the variable*** at that time, and the vector formed from these states is called the ***state of the dynamical system*** at that time. Our primary objective in this section is to analyze how the state of a dynamical system changes with time. Let us begin with an example.

EXAMPLE 1
Market Share as a Dynamical System

Suppose that two competing television news channels, channel 1 and channel 2, each have 50% of the viewer market at some initial point in time. Assume that over each one-year period channel 1 captures 10% of channel 2's share, and channel 2 captures 20% of channel 1's share (see Figure 5.1.1). What is each channel's market share after one year?

Solution Let us begin by introducing the time-dependent variables

$x_1(t)$ = fraction of the market held by channel 1 at time t
$x_2(t)$ = fraction of the market held by channel 2 at time t

and the column vector

$$\mathbf{x}(t) = \begin{bmatrix} x_1(t) \\ x_2(t) \end{bmatrix} \quad \begin{matrix} \leftarrow \textbf{Channel 1's fraction of the market at time } t \\ \leftarrow \textbf{Channel 2's fraction of the market at time } t \end{matrix}$$

The variables $x_1(t)$ and $x_2(t)$ form a dynamical system whose state at time t is the vector $\mathbf{x}(t)$. If we take $t = 0$ to be the starting point at which the two channels had 50% of the market, then the state of the system at that time is

$$\mathbf{x}(0) = \begin{bmatrix} x_1(0) \\ x_2(0) \end{bmatrix} = \begin{bmatrix} 0.5 \\ 0.5 \end{bmatrix} \quad \begin{matrix} \leftarrow \textbf{Channel 1's fraction of the market at time } t = 0 \\ \leftarrow \textbf{Channel 2's fraction of the market at time } t = 0 \end{matrix} \tag{1}$$

Now let us try to find the state of the system at time $t = 1$ (one year later). Over the one-year period, channel 1 retains 80% of its initial 50% (it loses 20% to channel 2), and it gains 10% of channel 2's initial 50%. Thus,

$$\mathbf{x}_1(1) = 0.8(0.5) + 0.1(0.5) = 0.45 \tag{2}$$

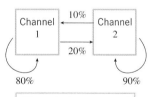

Channel 1 ← 10% ← Channel 2
Channel 1 → 20% → Channel 2
80%   90%

Channel 1 loses 20% and holds 80%.
Channel 2 loses 10% and holds 90%.

Figure 5.1.1

225

Similarly, channel 2 gains 20% of channel 1's initial 50%, and retains 90% of its initial 50% (it loses 10% to channel 1). Thus,

$$x_2(1) = 0.2(0.5) + 0.9(0.5) = 0.55 \tag{3}$$

Therefore, the state of the system at time $t = 1$ is

$$\mathbf{x}(1) = \begin{bmatrix} x_1(1) \\ x_2(1) \end{bmatrix} = \begin{bmatrix} 0.45 \\ 0.55 \end{bmatrix} \quad \begin{matrix} \leftarrow \textbf{Channel 1's fraction of the market at time } t = 1 \\ \leftarrow \textbf{Channel 2's fraction of the market at time } t = 1 \end{matrix} \tag{4}$$

■

EXAMPLE 2
Evolution of
Market Share
over Five Years

Track the market shares of channels 1 and 2 in Example 1 over a five-year period.

Solution To solve this problem suppose that we have already computed the market share of each channel at time $t = k$ and we are interested in using the known values of $x_1(k)$ and $x_2(k)$ to compute the market shares $x_1(k + 1)$ and $x_2(k + 1)$ one year later. The analysis is exactly the same as that used to obtain Equations (2) and (3). Over the one-year period, channel 1 retains 80% of its starting fraction $x_1(k)$ and gains 10% of channel 2's starting fraction $x_2(k)$. Thus,

$$x_1(k + 1) = (0.8)x_1(k) + (0.1)x_2(k) \tag{5}$$

Similarly, channel 2 gains 20% of channel 1's starting fraction $x_1(k)$ and retains 90% of its own starting fraction $x_2(k)$. Thus,

$$x_2(k + 1) = (0.2)x_1(k) + (0.9)x_2(k) \tag{6}$$

Equations (5) and (6) can be expressed in matrix form as

$$\begin{bmatrix} x_1(k + 1) \\ x_2(k + 1) \end{bmatrix} = \begin{bmatrix} 0.8 & 0.1 \\ 0.2 & 0.9 \end{bmatrix} \begin{bmatrix} x_1(k) \\ x_2(k) \end{bmatrix} \tag{7}$$

which provides a way of using matrix multiplication to compute the state of the system at time $t = k + 1$ from the state at time $t = k$. For example, using (1) and (7) we obtain

$$\mathbf{x}(1) = \begin{bmatrix} 0.8 & 0.1 \\ 0.2 & 0.9 \end{bmatrix} \mathbf{x}(0) = \begin{bmatrix} 0.8 & 0.1 \\ 0.2 & 0.9 \end{bmatrix} \begin{bmatrix} 0.5 \\ 0.5 \end{bmatrix} = \begin{bmatrix} 0.45 \\ 0.55 \end{bmatrix}$$

which agrees with (4). Similarly,

$$\mathbf{x}(2) = \begin{bmatrix} 0.8 & 0.1 \\ 0.2 & 0.9 \end{bmatrix} \mathbf{x}(1) = \begin{bmatrix} 0.8 & 0.1 \\ 0.2 & 0.9 \end{bmatrix} \begin{bmatrix} 0.45 \\ 0.55 \end{bmatrix} = \begin{bmatrix} 0.415 \\ 0.585 \end{bmatrix}$$

We can now continue this process, using Formula (7) to compute $\mathbf{x}(3)$ from $\mathbf{x}(2)$, then $\mathbf{x}(4)$ from $\mathbf{x}(3)$, and so on. This yields (verify)

$$\mathbf{x}(3) = \begin{bmatrix} 0.3905 \\ 0.6095 \end{bmatrix}, \quad \mathbf{x}(4) = \begin{bmatrix} 0.37335 \\ 0.62665 \end{bmatrix}, \quad \mathbf{x}(5) = \begin{bmatrix} 0.361345 \\ 0.638655 \end{bmatrix} \tag{8}$$

Thus, after five years, channel 1 will hold about 36% of the market and channel 2 will hold about 64% of the market. ■

If desired, we can continue the market analysis in the last example beyond the five-year period and explore what happens to the market share over the long term. We did so, using a computer, and obtained the following state vectors (rounded to six decimal places):

$$\mathbf{x}(10) \approx \begin{bmatrix} 0.338041 \\ 0.661959 \end{bmatrix}, \quad \mathbf{x}(20) \approx \begin{bmatrix} 0.333466 \\ 0.666534 \end{bmatrix}, \quad \mathbf{x}(40) \approx \begin{bmatrix} 0.333333 \\ 0.666667 \end{bmatrix} \tag{9}$$

All subsequent state vectors, when rounded to six decimal places, are the same as $\mathbf{x}(40)$, so we see that the market shares eventually stabilize with channel 1 holding about one-third of the market and channel 2 holding about two-thirds. Later in this section, we will explain why this stabilization occurs.

MARKOV CHAINS   In many dynamical systems the states of the variables are not known with certainty but can be expressed as probabilities; such dynamical systems are called ***stochastic processes*** (from the Greek word *stokastikos*, meaning "proceeding by guesswork"). A detailed study of stochastic processes requires a precise definition of the term *probability*, which is outside the scope of this course. However, the following interpretation of this term will suffice for our present purposes:

> Stated informally, the ***probability*** that an experiment or observation will have a certain outcome is approximately the fraction of the time that the outcome would occur if the experiment were to be repeated many times under constant conditions— the greater the number of repetitions, the more accurately the probability describes the fraction of occurrences.

For example, when we say that the probability of tossing heads with a fair coin is $\frac{1}{2}$, we mean that if the coin were tossed many times under constant conditions, then we would expect about half of the outcomes to be heads. Probabilities are often expressed as decimals or percentages. Thus, the probability of tossing heads with a fair coin can also be expressed as 0.5 or 50%.

If an experiment or observation has n possible outcomes, then the probabilities of those outcomes must be nonnegative fractions whose sum is 1. The probabilities are nonnegative because each describes the fraction of occurrences of an outcome over the long term, and the sum is 1 because they account for all possible outcomes. For example, if a box contains one red ball, three green balls, and six yellow balls, and if a ball is drawn at random from the box, then the probabilities of the possible outcomes are

$$p_1 = \text{prob(red)} = 1/10 = 0.1$$
$$p_2 = \text{prob(green)} = 3/10 = 0.3$$
$$p_3 = \text{prob(yellow)} = 6/10 = 0.6$$

Each probability is a nonnegative fraction and

$$p_1 + p_2 + p_3 = 0.1 + 0.3 + 0.6 = 1$$

In a stochastic process with n possible states, the state vector at each time t has the form

$$\mathbf{x}(t) = \begin{bmatrix} x_1(t) \\ x_2(t) \\ \vdots \\ x_n(t) \end{bmatrix} \quad \begin{array}{l} \textbf{Probability that the system is in state 1} \\ \textbf{Probability that the system is in state 2} \\ \vdots \\ \textbf{Probability that the system is in state } n \end{array}$$

The entries in this vector must add up to 1 since they account for all n possibilities. In general, a vector with nonnegative entries that add up to 1 is called a ***probability vector***.

EXAMPLE 3

Example 1 Revisited from the Probability Viewpoint

Observe that the state vectors in Examples 1 and 2 are all probability vectors. This is to be expected since the entries in each state vector are the fractional market shares of the channels, and together they account for the entire market. Moreover, it is actually preferable to interpret the entries in the state vectors as probabilities rather than exact market fractions, since market information is usually obtained by statistical sampling procedures with intrinsic uncertainties. Thus, for example, the state vector

$$\mathbf{x}(1) = \begin{bmatrix} x_1(1) \\ x_2(1) \end{bmatrix} = \begin{bmatrix} 0.45 \\ 0.55 \end{bmatrix}$$

which we interpreted in Example 1 to mean that channel 1 has 45% of the market and channel 2 has 55%, can also be interpreted to mean that an individual picked at random from the market will be a channel 1 viewer with probability 0.45 and a channel 2 viewer with probability 0.55. ∎

A square matrix, each of whose columns is a probability vector, is called a ***stochastic matrix***. Such matrices commonly occur in formulas that relate successive states of a stochastic process. For example, the state vectors $\mathbf{x}(k + 1)$ and $\mathbf{x}(k)$ in (7) are related by an equation of the form

Linear Algebra in History

Markov chains are named in honor of the Russian mathematician A. A. Markov, a lover of poetry, who used them to analyze the alternation of vowels and consonants in the poem *Eugene Onegin* by Pushkin. Markov believed that the only applications of his chains were to the analysis of literary works, so he would be astonished to learn that his discovery is used today in the social sciences, quantum theory, and genetics!

Andrei Andreyevich Markov
(1856–1922)

$\mathbf{x}(k + 1) = P\mathbf{x}(k)$ in which

$$P = \begin{bmatrix} 0.8 & 0.1 \\ 0.2 & 0.9 \end{bmatrix} \tag{10}$$

is a stochastic matrix. It should not be surprising that the column vectors of P are probability vectors, since the entries in each column provide a breakdown of what happens to each channel's market share over the year—the entries in column 1 convey that each year channel 1 retains 80% of its market share and loses 20%; and the entries in column 2 convey that each year channel 2 retains 90% of its market share and loses 10%. The entries in (10) can also be viewed as probabilities:

$p_{11} = 0.8 =$ probability that a channel 1 viewer remains a channel 1 viewer
$p_{21} = 0.2 =$ probability that a channel 1 viewer becomes a channel 2 viewer
$p_{12} = 0.1 =$ probability that a channel 2 viewer becomes a channel 1 viewer
$p_{22} = 0.9 =$ probability that a channel 2 viewer remains a channel 2 viewer

Example 1 is a special case of a large class of stochastic processes, called *Markov chains*.

Definition 5.1.1 A *Markov chain* is a dynamical system whose state vectors at a succession of time intervals are probability vectors and for which the state vectors at successive time intervals are related by an equation of the form

$$\mathbf{x}(k + 1) = P\mathbf{x}(k)$$

in which $P = [p_{ij}]$ is a stochastic matrix and p_{ij} is the probability that the system will be in state i at time $t = k + 1$ if it is in state j at time $t = k$. The matrix P is called the *transition matrix* for the system.

REMARK Note that in this definition the row index i corresponds to the later state and the column index j to the earlier state (Figure 5.1.2).

EXAMPLE 4
Wildlife Migration as a Markov Chain

Suppose that a tagged lion can migrate over three adjacent game reserves in search of food, reserve 1, reserve 2, and reserve 3. Based on data about the food resources, researchers conclude that the monthly migration pattern of the lion can be modeled by a Markov chain with transition matrix

Reserve at Time $t = k$

$$P = \begin{matrix} & \begin{matrix} 1 & \ \ 2 & \ \ 3 \end{matrix} \\ \begin{bmatrix} 0.5 & 0.4 & 0.6 \\ 0.2 & 0.2 & 0.3 \\ 0.3 & 0.4 & 0.1 \end{bmatrix} & \begin{matrix} 1 \\ 2 \\ 3 \end{matrix} \end{matrix} \quad \textbf{Reserve at Time } t = k + 1$$

(see Figure 5.1.3). That is,

$p_{11} = 0.5 =$ probability that the lion will stay in reserve 1 when it is in reserve 1
$p_{12} = 0.4 =$ probability that the lion will move from reserve 2 to reserve 1
$p_{13} = 0.6 =$ probability that the lion will move from reserve 3 to reserve 1
$p_{21} = 0.2 =$ probability that the lion will move from reserve 1 to reserve 2
$p_{22} = 0.2 =$ probability that the lion will stay in reserve 2 when it is in reserve 2
$p_{23} = 0.3 =$ probability that the lion will move from reserve 3 to reserve 2
$p_{31} = 0.3 =$ probability that the lion will move from reserve 1 to reserve 3
$p_{32} = 0.4 =$ probability that the lion will move from reserve 2 to reserve 3
$p_{33} = 0.1 =$ probability that the lion will stay in reserve 3 when it is in reserve 3

State at time $t = k$

$$\begin{bmatrix} & \downarrow & \\ & \vdots & \\ \text{—} & p_{ij} & \text{—} \\ & \vdots & \end{bmatrix} \xleftarrow[\ t = k + 1\ ]{\text{State at time}}$$

The entry p_{ij} is the probability that the system is in state i at time $t = k + 1$ if it is in state j at time $t = k$.

Figure 5.1.2

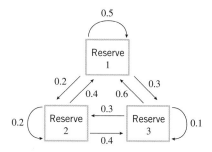

Figure 5.1.3

Assuming that t is in months and the lion is released in reserve 2 at time $t = 0$, track its probable locations over a six-month period.

Solution Let $x_1(k)$, $x_2(k)$, and $x_3(k)$ be the probabilities that the lion is in reserve 1, 2, or 3, respectively, at time $t = k$, and let

$$\mathbf{x}(k) = \begin{bmatrix} x_1(k) \\ x_2(k) \\ x_3(k) \end{bmatrix}$$

be the state vector at that time. Since we know with certainty that the lion is in reserve 2 at time $t = 0$, the initial state vector is

$$\mathbf{x}(0) = \begin{bmatrix} 0 \\ 1 \\ 0 \end{bmatrix}$$

We leave it for you to show that the state vectors over a six-month period are

$$\mathbf{x}(1) = P\mathbf{x}(0) = \begin{bmatrix} 0.400 \\ 0.200 \\ 0.400 \end{bmatrix}, \quad \mathbf{x}(2) = P\mathbf{x}(1) = \begin{bmatrix} 0.520 \\ 0.240 \\ 0.240 \end{bmatrix}, \quad \mathbf{x}(3) = P\mathbf{x}(2) = \begin{bmatrix} 0.500 \\ 0.224 \\ 0.276 \end{bmatrix}$$

$$\mathbf{x}(4) = P\mathbf{x}(3) \approx \begin{bmatrix} 0.505 \\ 0.228 \\ 0.267 \end{bmatrix}, \quad \mathbf{x}(5) = P\mathbf{x}(4) \approx \begin{bmatrix} 0.504 \\ 0.227 \\ 0.269 \end{bmatrix}, \quad \mathbf{x}(6) = P\mathbf{x}(5) \approx \begin{bmatrix} 0.504 \\ 0.227 \\ 0.269 \end{bmatrix}$$

As in Example 2, the state vectors here seem to stabilize over time with a probability of approximately 0.504 that the lion is in reserve 1, a probability of approximately 0.227 that it is in reserve 2, and a probability of approximately 0.269 that it is in reserve 3. ∎

MARKOV CHAINS IN TERMS OF POWERS OF THE TRANSITION MATRIX

In a Markov chain with an initial state of $\mathbf{x}(0)$, the successive state vectors are

$$\mathbf{x}(1) = P\mathbf{x}(0), \quad \mathbf{x}(2) = P\mathbf{x}(1), \quad \mathbf{x}(3) = P\mathbf{x}(2), \quad \mathbf{x}(4) = P\mathbf{x}(3), \ldots$$

For brevity, it is common to denote $\mathbf{x}(k)$ by $\mathbf{x}_k$, which allows us to write the successive state vectors more briefly as

$$\mathbf{x}_1 = P\mathbf{x}_0, \quad \mathbf{x}_2 = P\mathbf{x}_1, \quad \mathbf{x}_3 = P\mathbf{x}_2, \quad \mathbf{x}_4 = P\mathbf{x}_3, \ldots \tag{11}$$

Alternatively, these state vectors can be expressed in terms of the initial state vector $\mathbf{x}_0$ as

$$\mathbf{x}_1 = P\mathbf{x}_0, \quad \mathbf{x}_2 = P(P\mathbf{x}_0) = P^2\mathbf{x}_0, \quad \mathbf{x}_3 = P(P^2\mathbf{x}_0) = P^3\mathbf{x}_0, \quad \mathbf{x}_4 = P(P^3\mathbf{x}_0) = P^4\mathbf{x}_0, \ldots$$

from which it follows that

$$\mathbf{x}_k = P^k\mathbf{x}_0 \tag{12}$$

without computing all of the intermediate states. Later in the text we will discuss efficient methods for calculating powers of a matrix that will make this formula even more useful.

EXAMPLE 5
Finding a State
Vector Directly
from $\mathbf{x_0}$

Use Formula (12) to find the state vector $\mathbf{x}(3)$ in Example 2.

Solution From (1) and (7), the initial state vector and transition matrix are

$$\mathbf{x}_0 = \mathbf{x}(0) = \begin{bmatrix} 0.5 \\ 0.5 \end{bmatrix} \quad \text{and} \quad P = \begin{bmatrix} 0.8 & 0.1 \\ 0.2 & 0.9 \end{bmatrix}$$

We leave it for you to calculate P^3 and show that

$$\mathbf{x}(3) = \mathbf{x}_3 = P^3\mathbf{x}_0 = \begin{bmatrix} 0.562 & 0.219 \\ 0.438 & 0.781 \end{bmatrix} \begin{bmatrix} 0.5 \\ 0.5 \end{bmatrix} = \begin{bmatrix} 0.3905 \\ 0.6095 \end{bmatrix}$$

which agrees with the result in (8). ∎

LONG-TERM BEHAVIOR OF A MARKOV CHAIN

We have seen two examples of Markov chains in which the state vectors seem to stabilize after a period of time. Thus, it is reasonable to ask whether all Markov chains have this property. The following example shows that this is not the case.

EXAMPLE 6
A Markov
Chain That
Does Not
Stabilize

The matrix

$$P = \begin{bmatrix} 0 & 1 \\ 1 & 0 \end{bmatrix}$$

is stochastic and hence can be regarded as the transition matrix for a Markov chain. A simple calculation shows that $P^2 = I$, from which it follows that

$$I = P^2 = P^4 = P^6 = \cdots \quad \text{and} \quad P = P^3 = P^5 = P^7 = \cdots$$

Thus, the successive states in the Markov chain with initial vector $\mathbf{x}_0$ are

$$\mathbf{x}_0, \ P\mathbf{x}_0, \ \mathbf{x}_0, \ P\mathbf{x}_0, \ \mathbf{x}_0, \ldots$$

which oscillate between $\mathbf{x}_0$ and $P\mathbf{x}_0$. Thus, the Markov chain does not stabilize unless both components of $\mathbf{x}_0$ are $\frac{1}{2}$ (verify). ∎

A precise definition of what it means for a sequence of numbers or vectors to stabilize by approaching a limiting value is given in calculus; however, that level of precision will not be needed here. Stated informally, we will say that a sequence of vectors

$$\mathbf{x}_1, \mathbf{x}_2, \ldots, \mathbf{x}_k, \ldots$$

approaches a *limit* $\mathbf{q}$ or that it *converges* to $\mathbf{q}$ if all entries in $\mathbf{x}_k$ can be made as close as we like to the corresponding entries in the vector $\mathbf{q}$ by taking k sufficiently large. We denote this by writing $\mathbf{x}_k \to \mathbf{q}$ as $k \to \infty$.

We saw in Example 6 that the state vectors of a Markov chain need not approach a limit in all cases. However, by imposing a mild condition on the transition matrix of a Markov chain, we can guarantee that the state vectors will approach a limit.

Definition 5.1.2 A stochastic matrix P is said to be *regular* if P or some positive power of P has all positive entries, and a Markov chain whose transition matrix is regular is said to be a *regular Markov chain*.

EXAMPLE 7
Regular
Stochastic
Matrices

The transition matrices in Examples 2 and 4 are regular because their entries are positive. The matrix

$$P = \begin{bmatrix} 0.5 & 1 \\ 0.5 & 0 \end{bmatrix}$$

is regular because

$$P^2 = \begin{bmatrix} 0.75 & 0.5 \\ 0.25 & 0.5 \end{bmatrix}$$

has positive entries. The matrix P in Example 6 is not regular because P and every positive power of P have some zero entries (verify). ∎

The following theorem, which we state without proof, is the fundamental result about the long-term behavior of Markov chains.

> **Theorem 5.1.3** *If P is the transition matrix for a regular Markov chain, then:*
>
> (a) *There is a unique probability vector $\mathbf{q}$ such that $P\mathbf{q} = \mathbf{q}$.*
>
> (b) *For any initial probability vector $\mathbf{x}_0$, the sequence of state vectors*
>
> $$\mathbf{x}_0, P\mathbf{x}_0, \ldots, P^k\mathbf{x}_0, \ldots$$
>
> *approaches $\mathbf{q}$ as a limit; that is, $P^k\mathbf{x}_0 \to \mathbf{q}$ as $k \to \infty$.*

The vector $\mathbf{q}$, which is called the **steady-state vector** of the Markov chain, is a fixed point of the transition matrix P and hence can be found by solving the homogeneous linear system

$$(I - P)\mathbf{q} = \mathbf{0}$$

subject to the requirement that the solution be a probability vector.

EXAMPLE 8
Examples 1 and 2 Revisited

The transition matrix for the Markov chain in Example 2 is

$$P = \begin{bmatrix} 0.8 & 0.1 \\ 0.2 & 0.9 \end{bmatrix}$$

Since the entries of P are positive, the Markov chain is regular and hence has a unique steady-state vector $\mathbf{q}$. To find $\mathbf{q}$ we will solve the system $(I - P)\mathbf{q} = \mathbf{0}$, which we can write as

$$\begin{bmatrix} 0.2 & -0.1 \\ -0.2 & 0.1 \end{bmatrix} \begin{bmatrix} q_1 \\ q_2 \end{bmatrix} = \begin{bmatrix} 0 \\ 0 \end{bmatrix}$$

The general solution of this system is

$$q_1 = 0.5s, \quad q_2 = s$$

(verify), which we can write in vector form as

$$\mathbf{q} = \begin{bmatrix} q_1 \\ q_2 \end{bmatrix} = \begin{bmatrix} 0.5s \\ s \end{bmatrix} = \begin{bmatrix} \frac{1}{2}s \\ s \end{bmatrix} \tag{13}$$

For $\mathbf{q}$ to be a probability vector, we must have

$$1 = q_1 + q_2 = \tfrac{3}{2}s$$

which implies that $s = \frac{2}{3}$. Substituting this value in (13) yields the steady-state vector

$$\mathbf{q} = \begin{bmatrix} \frac{1}{3} \\ \frac{2}{3} \end{bmatrix}$$

which is consistent with the numerical results obtained in (9). ∎

EXAMPLE 9
Example 4 Revisited

The transition matrix for the Markov chain in Example 4 is

$$P = \begin{bmatrix} 0.5 & 0.4 & 0.6 \\ 0.2 & 0.2 & 0.3 \\ 0.3 & 0.4 & 0.1 \end{bmatrix}$$

Since the entries of P are positive, the Markov chain is regular and hence has a unique steady-state vector $\mathbf{q}$. To find $\mathbf{q}$ we will solve the system $(I - P)\mathbf{q} = \mathbf{0}$, which we can write (using fractions) as

$$
\begin{bmatrix} \frac{1}{2} & -\frac{2}{5} & -\frac{3}{5} \\ -\frac{1}{5} & \frac{4}{5} & -\frac{3}{10} \\ -\frac{3}{10} & -\frac{2}{5} & \frac{9}{10} \end{bmatrix} \begin{bmatrix} q_1 \\ q_2 \\ q_3 \end{bmatrix} = \begin{bmatrix} 0 \\ 0 \\ 0 \end{bmatrix} \tag{14}
$$

(We have converted to fractions to avoid roundoff error in this illustrative example.) We leave it for you to confirm that the reduced row echelon form of the coefficient matrix is

$$
\begin{bmatrix} 1 & 0 & -\frac{15}{8} \\ 0 & 1 & -\frac{27}{32} \\ 0 & 0 & 0 \end{bmatrix}
$$

and that the general solution of (14) is

$$
q_1 = \tfrac{15}{8}s, \quad q_2 = \tfrac{27}{32}s, \quad q_3 = s \tag{15}
$$

For $\mathbf{q}$ to be a probability vector we must have $q_1 + q_2 + q_3 = 1$, from which it follows that $s = \frac{32}{119}$ (verify). Substituting this value in (15) yields the steady-state vector

$$
\mathbf{q} = \begin{bmatrix} \frac{60}{119} \\ \frac{27}{119} \\ \frac{32}{119} \end{bmatrix} \approx \begin{bmatrix} 0.5042 \\ 0.2269 \\ 0.2689 \end{bmatrix}
$$

(verify), which is consistent with the results obtained in Example 4. ∎

REMARK Readers who are interested in exploring more on the theory of Markov chains are referred to specialized books on the subject such as the classic *Finite Markov Chains* by J. Kemeny and J. Snell, Springer-Verlag, New York, 1976.

Exercise Set 5.1

In Exercises 1 and 2, determine whether A is a stochastic matrix. If not, explain why not.

1. (a) $A = \begin{bmatrix} 0.4 & 0.3 \\ 0.6 & 0.7 \end{bmatrix}$ (b) $A = \begin{bmatrix} 0.4 & 0.6 \\ 0.3 & 0.7 \end{bmatrix}$

(c) $A = \begin{bmatrix} 1 & \frac{1}{2} & \frac{1}{3} \\ 0 & 0 & \frac{1}{3} \\ 0 & \frac{1}{2} & \frac{1}{3} \end{bmatrix}$ (d) $A = \begin{bmatrix} \frac{1}{3} & \frac{1}{3} & \frac{1}{2} \\ \frac{1}{6} & \frac{1}{3} & -\frac{1}{2} \\ \frac{1}{2} & \frac{1}{3} & 1 \end{bmatrix}$

2. (a) $A = \begin{bmatrix} 0.2 & 0.9 \\ 0.8 & 0.1 \end{bmatrix}$ (b) $A = \begin{bmatrix} 0.2 & 0.8 \\ 0.9 & 0.1 \end{bmatrix}$

(c) $A = \begin{bmatrix} \frac{1}{12} & \frac{1}{9} & \frac{1}{6} \\ \frac{1}{2} & 0 & \frac{5}{6} \\ \frac{5}{12} & \frac{8}{9} & 0 \end{bmatrix}$ (d) $A = \begin{bmatrix} -1 & \frac{1}{3} & \frac{1}{2} \\ 0 & \frac{1}{3} & \frac{1}{2} \\ 2 & \frac{1}{3} & 0 \end{bmatrix}$

In Exercises 3 and 4, use Formulas (11) and (12) to compute the state vector $\mathbf{x}_4$ in two different ways.

3. $P = \begin{bmatrix} 0.5 & 0.6 \\ 0.5 & 0.4 \end{bmatrix}$; $\mathbf{x}_0 = \begin{bmatrix} 0.5 \\ 0.5 \end{bmatrix}$

4. $P = \begin{bmatrix} 0.8 & 0.5 \\ 0.2 & 0.5 \end{bmatrix}$; $\mathbf{x}_0 = \begin{bmatrix} 1 \\ 0 \end{bmatrix}$

In Exercises 5 and 6, determine whether P is a regular stochastic matrix.

5. (a) $P = \begin{bmatrix} \frac{1}{5} & \frac{1}{7} \\ \frac{4}{5} & \frac{6}{7} \end{bmatrix}$ (b) $P = \begin{bmatrix} \frac{1}{5} & 0 \\ \frac{4}{5} & 1 \end{bmatrix}$ (c) $P = \begin{bmatrix} \frac{1}{5} & 1 \\ \frac{4}{5} & 0 \end{bmatrix}$

6. (a) $P = \begin{bmatrix} \frac{1}{2} & 1 \\ \frac{1}{2} & 0 \end{bmatrix}$ (b) $P = \begin{bmatrix} 1 & \frac{2}{3} \\ 0 & \frac{1}{3} \end{bmatrix}$ (c) $P = \begin{bmatrix} \frac{3}{4} & \frac{1}{3} \\ \frac{1}{4} & \frac{2}{3} \end{bmatrix}$

In Exercises 7–10, confirm that P is a regular transition matrix and find its steady-state vector.

7. $P = \begin{bmatrix} \frac{1}{4} & \frac{2}{3} \\ \frac{3}{4} & \frac{1}{3} \end{bmatrix}$

8. $P = \begin{bmatrix} 0.2 & 0.6 \\ 0.8 & 0.4 \end{bmatrix}$

9. $P = \begin{bmatrix} \frac{1}{2} & \frac{1}{2} & 0 \\ \frac{1}{4} & \frac{1}{2} & \frac{1}{3} \\ \frac{1}{4} & 0 & \frac{2}{3} \end{bmatrix}$

10. $P = \begin{bmatrix} \frac{1}{3} & \frac{1}{4} & \frac{2}{5} \\ 0 & \frac{3}{4} & \frac{2}{5} \\ \frac{2}{3} & 0 & \frac{1}{5} \end{bmatrix}$

11. Consider a Markov process with transition matrix

$$\begin{array}{cc} & \begin{array}{cc} \textbf{State 1} & \textbf{State 2} \end{array} \\ \begin{array}{c} \textbf{State 1} \\ \textbf{State 2} \end{array} & \begin{bmatrix} 0.2 & 0.1 \\ 0.8 & 0.9 \end{bmatrix} \end{array}$$

(a) What does the entry 0.2 represent?

(b) What does the entry 0.1 represent?

(c) If the system is in state 1 initially, what is the probability that it will be in state 2 at the next observation?

(d) If the system has a 50% chance of being in state 1 initially, what is the probability that it will be in state 2 at the next observation?

12. Consider a Markov process with transition matrix

$$\begin{array}{cc} & \begin{array}{cc} \textbf{State 1} & \textbf{State 2} \end{array} \\ \begin{array}{c} \textbf{State 1} \\ \textbf{State 2} \end{array} & \begin{bmatrix} 0 & \frac{1}{7} \\ 1 & \frac{6}{7} \end{bmatrix} \end{array}$$

(a) What does the entry $\frac{6}{7}$ represent?

(b) What does the entry 0 represent?

(c) If the system is in state 1 initially, what is the probability that it will be in state 1 at the next observation?

(d) If the system has a 50% chance of being in state 1 initially, what is the probability that it will be in state 2 at the next observation?

13. On a given day the air quality in a certain city is either good or bad. Records show that when the air quality is good on one day, then there is a 95% chance that it will be good the next day, and when the air quality is bad on one day, then there is a 45% chance that it will be bad the next day.

(a) Find a transition matrix for this phenomenon.

(b) If the air quality is good today, what is the probability that it will be good two days from now?

(c) If the air quality is bad today, what is the probability that it will be bad three days from now?

(d) If there is 20% chance that the air quality will be good today, what is the probability that it will be good tomorrow?

14. In a laboratory experiment, a mouse can choose one of two food types each day, type I or type II. Records show that if the mouse chooses type I on a given day, then there is a 75% chance that it will choose type I the next day, and if it

chooses type II on one day, then there is a 50% chance that it will choose type II the next day.

(a) Find a transition matrix for this phenomenon.

(b) If the mouse chooses type I today, what is the probability that it will choose type I two days from now?

(c) If the mouse chooses type II today, what is the probability that it will choose type II three days from now?

(d) If there is 10% chance that the mouse will choose type I today, what is the probability that it will choose type I tomorrow?

15. Suppose that at some initial point in time 100,000 people live in a certain city and 25,000 people live in its suburbs. The Regional Planning Commission determines that each year 5% of the city population moves to the suburbs and 3% of the suburban population moves to the city.

(a) Assuming that the total population remains constant, make a table that shows the populations of the city and its suburbs over a five-year period (round to the nearest integer).

(b) Over the long term, how will the population be distributed between the city and its suburbs?

16. Suppose that two competing television stations, station 1 and station 2, each have 50% of the viewer market at some initial point in time. Assume that over each one-year period station 1 captures 5% of station 2's market share and station 2 captures 10% of station 1's market share.

(a) Make a table that shows the market share of each station over a five-year period.

(b) Over the long term, how will the market share be distributed between the two stations?

17. Suppose that a car rental agency has three locations, numbered 1, 2, and 3. A customer may rent a car from any of the three locations and return it to any of the three locations. Records show that cars are rented and returned in accordance with the following probabilities:

		Rented from Location		
		1	2	3
Returned to Location	1	$\frac{1}{10}$	$\frac{1}{5}$	$\frac{3}{5}$
	2	$\frac{4}{5}$	$\frac{3}{10}$	$\frac{1}{5}$
	3	$\frac{1}{10}$	$\frac{1}{2}$	$\frac{1}{5}$

(a) Assuming that a car is rented from location 1, what is the probability that it will be at location 1 after two rentals?

(b) Assuming that this dynamical system can be modeled as a Markov chain, find the steady-state vector.

(c) If the rental agency owns 120 cars, how many parking spaces should it allocate at each location to be reasonably certain that it will have enough spaces for the cars over the long term? Explain your reasoning.

18. Physical traits are determined by the genes that an offspring receives from its parents. In the simplest case a trait in the offspring is determined by one pair of genes, one member of the pair inherited from the male parent and the other from the female parent. Typically, each gene in a pair can assume one of two forms, called **alleles**, denoted by A and a. This leads to three possible pairings:

   AA, Aa, aa

called **genotypes** (the pairs Aa and aA determine the same trait and hence are not distinguished from one another). It is shown in the study of heredity that if a parent of known genotype is crossed with a random parent of unknown genotype, then the offspring will have the genotype probabilities given in the following table, which can be viewed as a transition matrix for a Markov process:

		Genotype of Parent	
	AA	Aa	aa
AA	$\frac{1}{2}$	$\frac{1}{4}$	0
Genotype of Offspring — Aa	$\frac{1}{2}$	$\frac{1}{2}$	$\frac{1}{2}$
aa	0	$\frac{1}{4}$	$\frac{1}{2}$

Thus, for example, the offspring of a parent of genotype AA that is crossed at random with a parent of unknown genotype will have a 50% chance of being AA, a 50% chance of being Aa, and no chance of being aa.

(a) Show that the transition matrix is regular.
(b) Find the steady-state vector, and discuss its physical interpretation.

Discussion and Discovery

D1. Fill in the missing entries of the stochastic matrix

$$P = \begin{bmatrix} \frac{7}{10} & * & \frac{1}{5} \\ * & \frac{3}{10} & * \\ \frac{1}{10} & \frac{3}{5} & \frac{3}{10} \end{bmatrix}$$

and find its steady-state vector.

D2. If P is an $n \times n$ stochastic matrix, and if M is a $1 \times n$ matrix whose entries are all 1's, then $MP =$ _____.

D3. If P is a regular stochastic matrix with steady-state vector $\mathbf{q}$, what can you say about the sequence of products

$$P\mathbf{q}, P^2\mathbf{q}, P^3\mathbf{q}, \ldots, P^k\mathbf{q}, \ldots$$

as $k \to \infty$?

D4. (a) If P is a regular $n \times n$ stochastic matrix with steady-state vector $\mathbf{q}$, and if $\mathbf{e}_1, \mathbf{e}_2, \ldots, \mathbf{e}_n$ are the standard unit vectors in column form, what can you say about the behavior of the sequence

$$P\mathbf{e}_i, P^2\mathbf{e}_i, P^3\mathbf{e}_i, \ldots, P^k\mathbf{e}_i, \ldots$$

as $k \to \infty$ for each $i = 1, 2, \ldots, n$?

(b) What does this tell you about the behavior of the column vectors of P^k as $k \to \infty$?

(c) What can you say about the steady-state vector of a stochastic matrix that is regular and symmetric? Prove your assertion.

Working with Proofs

P1. Prove that the product of two stochastic matrices is a stochastic matrix. [*Hint:* Write each column of the product as a linear combination of the columns of the first factor.]

P2. (a) Let P be a regular $k \times k$ stochastic matrix. Prove that the row sums of P are all equal to 1 if and only if all entries in the steady-state vector are equal to $1/k$. [*Hint:* Use part (*a*) of Theorem 5.1.3.]

(b) Use the result in part (*a*) to find the steady-state vector of

$$P = \begin{bmatrix} 0 & \frac{1}{3} & \frac{2}{3} \\ \frac{1}{3} & \frac{2}{3} & 0 \\ \frac{2}{3} & 0 & \frac{1}{3} \end{bmatrix}$$

P3. Prove that if P is a stochastic matrix whose entries are all greater than or equal to ρ, then the entries of P^2 are greater than or equal to ρ.

Technology Exercises

T1. By calculating sufficiently high powers of P, confirm the result in part (b) of Exercise D4 for the matrix

$$P = \begin{bmatrix} 0.2 & 0.4 & 0.5 \\ 0.1 & 0.3 & 0.1 \\ 0.7 & 0.3 & 0.4 \end{bmatrix}$$

T2. Each night a guard patrols an art gallery with seven rooms connected by corridors, as shown in the accompanying figure. The guard spends 10 minutes in a room and then moves to a neighboring room that is chosen at random, each possible choice being equally likely.

(a) Find the 7×7 transition matrix for the surveillance pattern.

(b) Assuming that the guard (or a replacement) follows the surveillance pattern indefinitely, what proportion of time does the guard spend in each room?

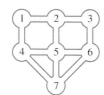

Figure Ex-T2

T3. Acme trucking rents trucks in New York, Boston, and Chicago, and the trucks are returned to those cities in accordance with the accompanying table. Determine the distribution of the trucks over the long run.

		Trucks Rented At		
		New York	**Boston**	**Chicago**
Trucks Returned To	**New York**	0.721	0.05	0.211
	Boston	0.122	0.92	0.095
	Chicago	0.157	0.03	0.694

Table Ex-T3

Section 5.2 Leontief Input-Output Models

In 1973 the economist Wassily Leontief was awarded the Nobel prize for his work on economic modeling in which he used matrix methods to study the relationships between different sectors in an economy. In this section we will discuss some of the ideas developed by Leontief.

INPUTS AND OUTPUTS IN AN ECONOMY

One way to analyze an economy is to divide it into **sectors** and study how the sectors interact with one another. For example, a simple economy might be divided into three sectors—manufacturing, agriculture, and utilities. Typically, a sector will produce certain **outputs** but will require **inputs** from the other sectors and itself. For example, the agricultural sector may produce wheat as an output but will require inputs of farm machinery from the manufacturing sector, electrical power from the utilities sector, and food from its own sector to feed its workers. Thus, we can imagine an economy to be a network in which inputs and outputs flow in and out of the sectors; the study of such flows is called **input-output analysis**. Inputs and outputs are commonly measured in monetary units (dollars or millions of dollars, for example) but other units of measurement are also possible.

The flows between sectors of a real economy are not always obvious. For example, in World War II the United States had a demand for 50,000 new airplanes that required the construction of many new aluminum manufacturing plants. This produced an unexpectedly large demand for certain copper electrical components, which in turn produced a copper shortage. The problem was eventually resolved by using silver borrowed from Fort Knox as a copper substitute. In all likelihood modern input-output analysis would have anticipated the copper shortage.

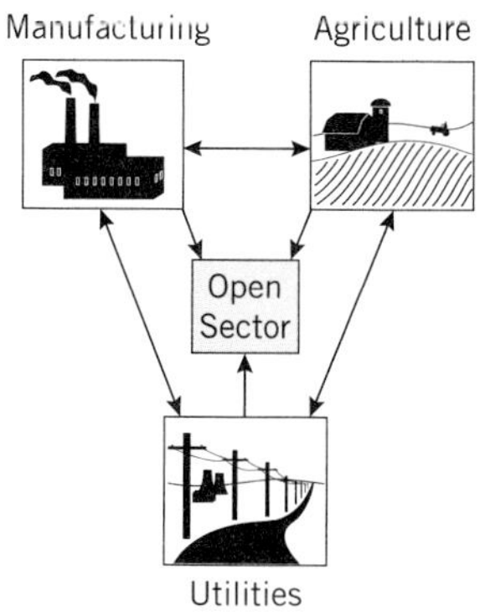

Most sectors of an economy will produce outputs, but there may exist sectors that consume outputs without producing anything themselves (the consumer market, for example). Those sectors that do not produce outputs are called **open sectors**. Economies with no open sectors are called **closed economies**, and economies with one or more open sectors are called **open economies** (Figure 5.2.1). In this section we will be concerned with economies with one open

Figure 5.2.1

sector, and our primary goal will be to determine the output levels that are required for the productive sectors to sustain themselves and satisfy the demand of the open sector.

LEONTIEF MODEL OF AN OPEN ECONOMY

Let us consider a simple open economy with one open sector and three product-producing sectors: manufacturing, agriculture, and utilities. Assume that inputs and outputs are measured in dollars and that the inputs required by the productive sectors to produce one dollar's worth of output are in accordance with the following table.

Table 5.2.1

		Output		
		Manufacturing	**Agriculture**	**Utilities**
Input	**Manufacturing**	$ 0.50	$ 0.10	$ 0.10
	Agriculture	$ 0.20	$ 0.50	$ 0.30
	Utilities	$ 0.10	$ 0.30	$ 0.40

Usually, one would suppress the labeling and express this matrix as

$$C = \begin{bmatrix} 0.5 & 0.1 & 0.1 \\ 0.2 & 0.5 & 0.3 \\ 0.1 & 0.3 & 0.4 \end{bmatrix} \tag{1}$$

This is called the **consumption matrix** (or sometimes the **technology matrix**) for the economy. The column vectors

$$\mathbf{c}_1 = \begin{bmatrix} 0.5 \\ 0.2 \\ 0.1 \end{bmatrix}, \quad \mathbf{c}_2 = \begin{bmatrix} 0.1 \\ 0.5 \\ 0.3 \end{bmatrix}, \quad \mathbf{c}_3 = \begin{bmatrix} 0.1 \\ 0.3 \\ 0.4 \end{bmatrix}$$

in C list the inputs required by the manufacturing, agricultural, and utilities sectors, respectively, to produce $1.00 worth of output. These are called the **consumption vectors** of the sectors. For example, $\mathbf{c}_1$ tells us that to produce $1.00 worth of output the manufacturing sector needs $0.50 worth of manufacturing output, $0.20 worth of agricultural output, and $0.10 worth of utilities output.

Continuing with the above example, suppose that the open sector wants the economy to supply it manufactured goods, agricultural products, and utilities with dollar values:

d_1 dollars of manufactured goods

d_2 dollars of agricultural products

d_3 dollars of utilities

The column vector **d** that has these numbers as successive components is called the **outside demand vector**. Since the product-producing sectors consume some of their own output, the dollar value of their output must cover their own needs plus the outside demand. Suppose that the dollar values required to do this are

x_1 dollars of manufactured goods

x_2 dollars of agricultural products

x_3 dollars of utilities

The column vector **x** that has these numbers as successive components is called the **production vector** for the economy. For the economy with consumption matrix (1), that portion of the production vector **x** that will be consumed by the

three productive sectors is

$$x_1 \begin{bmatrix} 0.5 \\ 0.2 \\ 0.1 \end{bmatrix} + x_2 \begin{bmatrix} 0.1 \\ 0.5 \\ 0.3 \end{bmatrix} + x_3 \begin{bmatrix} 0.1 \\ 0.3 \\ 0.4 \end{bmatrix} = \begin{bmatrix} 0.5 & 0.1 & 0.1 \\ 0.2 & 0.5 & 0.3 \\ 0.1 & 0.3 & 0.4 \end{bmatrix} \begin{bmatrix} x_1 \\ x_2 \\ x_3 \end{bmatrix} = C\mathbf{x}$$

| Fractions consumed by manufacturing | Fractions consumed by agriculture | Fractions consumed by utilities |

The vector $C\mathbf{x}$ is called the ***intermediate demand vector*** for the economy. Once the intermediate demand is met, the portion of the production that is left to satisfy the outside demand is $\mathbf{x} - C\mathbf{x}$. Thus, if the outside demand vector is $\mathbf{d}$, then $\mathbf{x}$ must satisfy the equation

$$\mathbf{x} \quad - \quad C\mathbf{x} \quad = \quad \mathbf{d}$$

| Amount produced | Intermediate demand | Outside demand |

which we will find convenient to rewrite as

$$(I - C)\mathbf{x} = \mathbf{d} \tag{2}$$

The matrix $I - C$ is called the ***Leontief matrix*** and (2) is called the ***Leontief equation***.

EXAMPLE 1
Satisfying
Outside
Demand

Consider the economy described in Table 5.2.1. Suppose that the open sector has a demand for $7900 worth of manufacturing products, $3950 worth of agricultural products, and $1975 worth of utilities.

(a) Can the economy meet this demand?

(b) If so, find a production vector $\mathbf{x}$ that will meet it exactly.

Solution The consumption matrix, production vector, and outside demand vector are

$$C = \begin{bmatrix} 0.5 & 0.1 & 0.1 \\ 0.2 & 0.5 & 0.3 \\ 0.1 & 0.3 & 0.4 \end{bmatrix}, \quad \mathbf{x} = \begin{bmatrix} x_1 \\ x_2 \\ x_3 \end{bmatrix}, \quad \mathbf{d} = \begin{bmatrix} 7900 \\ 3950 \\ 1975 \end{bmatrix} \tag{3}$$

To meet the outside demand, the vector $\mathbf{x}$ must satisfy the Leontief equation (2), so the problem reduces to solving the linear system

$$\underbrace{\begin{bmatrix} 0.5 & -0.1 & -0.1 \\ -0.2 & 0.5 & -0.3 \\ -0.1 & -0.3 & 0.6 \end{bmatrix}}_{I - C} \underbrace{\begin{bmatrix} x_1 \\ x_2 \\ x_3 \end{bmatrix}}_{\mathbf{x}} = \underbrace{\begin{bmatrix} 7900 \\ 3950 \\ 1975 \end{bmatrix}}_{\mathbf{d}} \tag{4}$$

(if consistent). We leave it for you to show that the reduced row echelon form of the augmented matrix for this system is

$$\begin{bmatrix} 1 & 0 & 0 & | & 27{,}500 \\ 0 & 1 & 0 & | & 33{,}750 \\ 0 & 0 & 1 & | & 24{,}750 \end{bmatrix}$$

This tells us that (4) is consistent, and the economy can satisfy the demand of the open sector exactly by producing $27,500 worth of manufacturing output, $33,750 worth of agricultural output, and $24,750 worth of utilities output.   ■

**PRODUCTIVE OPEN
ECONOMIES**

In the preceding discussion we considered an open economy with three product-producing sectors; the same ideas apply to an open economy with n product-producing sectors. In this case, the consumption matrix, production vector, and outside demand vector have the form

$$C = \begin{bmatrix} c_{11} & c_{12} & \cdots & c_{1n} \\ c_{21} & c_{22} & \cdots & c_{2n} \\ \vdots & \vdots & & \vdots \\ c_{n1} & c_{n2} & \cdots & c_{nn} \end{bmatrix}, \quad \mathbf{x} = \begin{bmatrix} x_1 \\ x_2 \\ \vdots \\ x_n \end{bmatrix}, \quad \mathbf{d} = \begin{bmatrix} d_1 \\ d_2 \\ \vdots \\ d_n \end{bmatrix}$$

where all entries are nonnegative and

$c_{ij} = $ the monetary value of the output of the ith sector that is needed by the jth sector to produce one unit of output

$x_i = $ the monetary value of the output of the ith sector

$d_i = $ the monetary value of the output of the ith sector that is required to meet the demand of the open sector

REMARK Note that the jth column vector of C contains the monetary values that the jth sector requires of the other sectors to produce one monetary unit of output, and the ith row vector of C contains the monetary values required of the ith sector by the other sectors for each of them to produce one monetary unit of output.

As discussed in our example above, a production vector $\mathbf{x}$ that meets the demand $\mathbf{d}$ of the outside sector must satisfy the Leontief equation

$$(I - C)\mathbf{x} = \mathbf{d}$$

If the matrix $I - C$ is invertible, then this equation has the unique solution

$$\mathbf{x} = (I - C)^{-1}\mathbf{d} \tag{5}$$

for every demand vector $\mathbf{d}$. However, for $\mathbf{x}$ to be a valid production vector it must have nonnegative entries, so the problem of importance in economics is to determine conditions under which the Leontief equation has a solution with nonnegative entries.

In the case where $I - C$ is invertible, it is evident from the form of (5) that if $(I - C)^{-1}$ has nonnegative entries, then for every demand vector $\mathbf{d}$ the corresponding $\mathbf{x}$ will have nonnegative entries and hence will be a valid production vector for the economy. An economy for which $(I - C)^{-1}$ has nonnegative entries is said to be ***productive***. Such economies are particularly nice because every demand can be met by some appropriate level of production.

Since C has nonnegative entries, it follows from Theorem 3.6.7 that if the column sums of C are all less than 1, then $I - C$ will be invertible and its inverse will have nonnegative entries. Thus, we have the following result about open economies.

Theorem 5.2.1 *If all of the column sums of the consumption matrix C of an open economy are less than 1, then the economy is productive.*

REMARK The jth column sum of C represents the total dollar value of input that the jth sector requires to produce \$1 of output, so if the jth column sum is less than 1, then the jth sector requires less than \$1 of input to produce \$1 of output; in this case we say that the jth sector is ***profitable***. Thus, Theorem 5.2.1 states that if all product-producing sectors of an open economy are profitable, then the economy is productive. In the exercises we will ask you to show that an open economy is productive if all of the row sums of C are less than 1. Thus, an open economy is productive if *either* all of the column sums or all of the row sums of C are less than 1.

CONCEPT PROBLEM What is the economic significance of the row sums of the consumption matrix?

EXAMPLE 2
An Open
Economy
Whose Sectors
Are All
Profitable

The column sums of the consumption matrix C in (1) are less than 1, so $(I - C)^{-1}$ exists and has nonnegative entries. Use a calculating utility to confirm this, and use this inverse to solve Equation (4) in Example 1.

Solution We leave it for you to show that

$$(I - C)^{-1} \approx \begin{bmatrix} 2.65823 & 1.13924 & 1.01266 \\ 1.89873 & 3.67089 & 2.15190 \\ 1.39241 & 2.02532 & 2.91139 \end{bmatrix}$$

This matrix has nonnegative entries, and

$$\mathbf{x} = (I - C)^{-1}\mathbf{d} \approx \begin{bmatrix} 2.65823 & 1.13924 & 1.01266 \\ 1.89873 & 3.67089 & 2.15190 \\ 1.39241 & 2.02532 & 2.91139 \end{bmatrix} \begin{bmatrix} 7900 \\ 3950 \\ 1975 \end{bmatrix} \approx \begin{bmatrix} 27,500 \\ 33,750 \\ 24,750 \end{bmatrix}$$

which is consistent with the solution in Example 1.   ∎

Exercise Set 5.2

1. An automobile mechanic (M) and a body shop (B) use each other's services. For each $1.00 of business that M does, it uses $0.50 of its own services and $0.25 of B's services, and for each $1.00 of business that B does it uses $0.10 of its own services and $0.25 of M's services.

(a) Construct a consumption matrix for this economy.

(b) How much must M and B each produce to provide customers with $7000 worth of mechanical work and $14,000 worth of body work?

2. A simple economy produces food (F) and housing (H). The production of $1.00 worth of food requires $0.30 worth of food and $0.10 worth of housing, and the production of $1.00 worth of housing requires $0.20 worth of food and $0.60 worth of housing.

(a) Construct a consumption matrix for this economy.

(b) What dollar value of food and housing must be produced for the economy to provide consumers $130,000 worth of food and $130,000 worth of housing?

3. Consider the open economy described by the accompanying table, where the input is in dollars needed for $1.00 of output.

Output

	Housing	Food	Utilities
Housing	$ 0.10	$ 0.60	$ 0.40
Food	$ 0.30	$ 0.20	$ 0.30
Utilities	$ 0.40	$ 0.10	$ 0.20

Input

Table Ex-3

(a) Find the consumption matrix for the economy.

(b) Suppose that the open sector has a demand for $1930 worth of housing, $3860 worth of food, and $5790 worth of utilities. Use row reduction to find a production vector that will meet this demand exactly.

4. A company produces Web design, software, and networking services. View the company as an open economy described by the accompanying table, where input is in dollars needed for $1.00 of output.

Output

	Web Design	Software	Networking
Web Design	$ 0.40	$ 0.20	$ 0.45
Software	$ 0.30	$ 0.35	$ 0.30
Networking	$ 0.15	$ 0.10	$ 0.20

Input

Table Ex-4

(a) Find the consumption matrix for the company.

(b) Suppose that the customers (the open sector) have a demand for $5400 worth of Web design, $2700 worth of software, and $900 worth of networking. Use row reduction to find a production vector that will meet this demand exactly.

In Exercises 5 and 6, use matrix inversion to find the production vector $\mathbf{x}$ that meets the demand $\mathbf{d}$ for the consumption matrix C.

5. $C = \begin{bmatrix} 0.1 & 0.3 \\ 0.5 & 0.4 \end{bmatrix}$; $\mathbf{d} = \begin{bmatrix} 50 \\ 60 \end{bmatrix}$

6. $C = \begin{bmatrix} 0.3 & 0.1 \\ 0.3 & 0.7 \end{bmatrix}$; $\mathbf{d} = \begin{bmatrix} 22 \\ 14 \end{bmatrix}$

> We know from Theorem 5.2.1 that if C is an $n \times n$ matrix with nonnegative entries and column sums that are all less than 1, then $I - C$ is invertible and has nonnegative entries. In Exercise P2 below we will ask you to prove that the same conclusion is true if C has nonnegative entries and its *row sums* are less than 1. Use this result and Theorem 5.2.1 in Exercises 7 and 8.

7. In each part, show that the open economy with consumption matrix C is productive.

(a) $C = \begin{bmatrix} 0.4 & 0.3 & 0.5 \\ 0.2 & 0.5 & 0.2 \\ 0.2 & 0.1 & 0.1 \end{bmatrix}$ (b) $C = \begin{bmatrix} 0.1 & 0.3 & 0.4 \\ 0.2 & 0.2 & 0.2 \\ 0.8 & 0.1 & 0 \end{bmatrix}$

8. In each part, show that the open economy with consumption matrix C is productive.

(a) $C = \begin{bmatrix} 0.4 & 0.1 & 0.7 \\ 0.3 & 0.5 & 0.1 \\ 0.1 & 0.2 & 0.1 \end{bmatrix}$ (b) $C = \begin{bmatrix} 0.5 & 0.2 & 0.2 \\ 0.3 & 0.3 & 0.3 \\ 0.4 & 0.4 & 0.1 \end{bmatrix}$

9. Consider an open economy with consumption matrix

$$C = \begin{bmatrix} 0.50 & 0 & 0.25 \\ 0.20 & 0.80 & 0.10 \\ 1 & 0.40 & 0 \end{bmatrix}$$

Show that the economy is productive even though some of the column sums and row sums are greater than 1. Does this violate Theorem 5.2.1?

Discussion and Discovery

D1. Consider an open economy with consumption matrix

$$C = \begin{bmatrix} \frac{1}{2} & 0 \\ 0 & 1 \end{bmatrix}$$

(a) Show that the economy can meet a demand of $d_1 = 2$ units from the first sector and $d_2 = 0$ units from the second sector, but it cannot meet a demand of $d_1 = 2$ units from the first sector and $d_2 = 1$ unit from the second sector.

(b) Give both a mathematical and an economic explanation of the result in part (a).

D2. Consider an open economy with consumption matrix

$$C = \begin{bmatrix} \frac{1}{2} & \frac{1}{4} & \frac{1}{4} \\ \frac{1}{2} & \frac{1}{8} & \frac{1}{4} \\ \frac{1}{2} & \frac{1}{4} & \frac{1}{8} \end{bmatrix}$$

If the open sector demands the same dollar value from each product-producing sector, which such sector must produce the greatest dollar value to meet the demand?

D3. Consider an open economy with consumption matrix

$$C = \begin{bmatrix} c_{11} & c_{12} \\ c_{21} & 0 \end{bmatrix}$$

Show that the Leontief equation $\mathbf{x} - C\mathbf{x} = \mathbf{d}$ has a unique solution for every demand vector $\mathbf{d}$ if $c_{21}c_{12} < 1 - c_{11}$.

Working with Proofs

P1. (a) Consider an open economy with a consumption matrix C whose column sums are less than 1, and let $\mathbf{x}$ be the production vector that satisfies an outside demand $\mathbf{d}$; that is, $(I - C)^{-1}\mathbf{d} = \mathbf{x}$. Let $\mathbf{d}_j$ be the demand vector that is obtained by increasing the jth entry of $\mathbf{d}$ by 1 and leaving the other entries fixed. Prove that the production vector $\mathbf{x}_j$ that meets this demand is

$$\mathbf{x}_j = \mathbf{x} + j\text{th column vector of } (I - C)^{-1}$$

(b) In words, what is the economic significance of the jth column vector of $(I - C)^{-1}$? [*Suggestion:* Look at $\mathbf{x}_j - \mathbf{x}$.]

P2. Prove that if C is an $n \times n$ matrix whose entries are nonnegative and whose row sums are less than 1, then $I - C$ is invertible and has nonnegative entries. [*Hint:* $(A^T)^{-1} = (A^{-1})^T$ for any invertible matrix A.]

P3. Prove that an open economy with a nilpotent consumption matrix is productive. [*Hint:* See Theorem 3.6.6.]

Technology Exercises

T1. Suppose that the consumption matrix for an open economy is

$$C = \begin{bmatrix} 0.29 & 0.05 & 0.04 & 0.01 \\ 0.02 & 0.31 & 0.01 & 0.03 \\ 0.04 & 0.02 & 0.44 & 0.01 \\ 0.01 & 0.03 & 0.04 & 0.32 \end{bmatrix}$$

   (a) Confirm that the economy is productive, and then show by direct computation that $(I - C)^{-1}$ has positive entries.

   (b) Use matrix inversion or row reduction to find the production vector $\mathbf{x}$ that satisfies the demand

$$\mathbf{d} = \begin{bmatrix} 200 \\ 100 \\ 350 \\ 275 \end{bmatrix}$$

T2. The Leontief equation $\mathbf{x} - C\mathbf{x} = \mathbf{d}$ can be rewritten as
$$\mathbf{x} = C\mathbf{x} + \mathbf{d}$$
and then solved approximately by substituting an arbitrary initial approximation $\mathbf{x}_0$ into the right side of this equation and using the resulting vector $\mathbf{x}_1 = C\mathbf{x}_0 + \mathbf{d}$ as a new (and often better) approximation to the solution. By repeating this process, you can generate a succession of approximations $\mathbf{x}_1, \mathbf{x}_2, \mathbf{x}_3, \ldots, \mathbf{x}_k, \ldots$ recursively from the relationship $\mathbf{x}_k = C\mathbf{x}_{k-1} + \mathbf{d}$. We will see in the next section that this sequence converges to the exact solution under fairly general conditions. Take $\mathbf{x}_0 = \mathbf{0}$, and use this method to generate a succession of ten approximations, $\mathbf{x}_1, \mathbf{x}_2, \ldots, \mathbf{x}_{10}$, to the solution of the problem in Exercise T1. Compare $\mathbf{x}_{10}$ to the result obtained in that exercise.

T3. Consider an open economy described by the accompanying table.

   (a) Show that the sectors are all profitable.

   (b) Find the production levels that will satisfy the following demand by the open sector (units in millions of dollars): agriculture, $1.2; manufacturing, $3.4; trade, $2.7; services, $4.3; and energy, $2.9.

   (c) If the demand for services doubles from the level in part (b), which sector will be affected the most? Explain your reasoning.

Output

		Agriculture	Manufacturing	Trade	Services	Energy
	Agriculture	$ 0.27	$ 0.39	$ 0.03	$ 0.02	$ 0.23
	Manufacturing	$ 0.15	$ 0.15	$ 0.10	$ 0.01	$ 0.22
Input	**Trade**	$ 0.06	$ 0.07	$ 0.36	$ 0.15	$ 0.35
	Services	$ 0.27	$ 0.08	$ 0.07	$ 0.41	$ 0.09
	Energy	$ 0.23	$ 0.19	$ 0.36	$ 0.24	$ 0.10

Table Ex-T3

Section 5.3 Gauss–Seidel and Jacobi Iteration; Sparse Linear Systems

*Many mathematical models lead to large linear systems in which the coefficient matrix has a high proportion of zeros. In such cases the system and its coefficient matrix are said to be **sparse**. Although Gauss–Jordan elimination and LU-decomposition can be applied to sparse linear systems, those methods tend to destroy the zeros, thereby failing to take computational advantage of their presence. In this section we will discuss two methods that are appropriate for linear systems with sparse coefficient matrices.*

ITERATIVE METHODS   Let $A\mathbf{x} = \mathbf{b}$ be a linear system of n equations in n unknowns with an invertible coefficient matrix A (so the system has a unique solution). An ***iterative method*** for solving such a system is an algorithm that generates a sequence of vectors

$$\mathbf{x}_1, \mathbf{x}_2, \ldots, \mathbf{x}_k, \ldots$$

called ***iterates***, that converge to the exact solution $\mathbf{x}$ in the sense that the entries of $\mathbf{x}_k$ can be made as close as we like to the entries of $\mathbf{x}$ by making k sufficiently large. Whereas Gauss–Jordan elimination and LU-decomposition can produce the exact solution if there is no roundoff error, iterative methods are specifically designed to produce an approximation to the exact solution.

The basic procedure for designing iterative methods is to devise matrices B and $\mathbf{c}$ that allow the system $A\mathbf{x} = \mathbf{b}$ to be rewritten in the form

$$\mathbf{x} = B\mathbf{x} + \mathbf{c} \tag{1}$$

This modified equation is then solved by forming the **recurrence relation**

$$\mathbf{x}_{k+1} = B\mathbf{x}_k + \mathbf{c} \tag{2}$$

and proceeding as follows:

Step 1. Choose an arbitrary **initial approximation** $\mathbf{x}_0$.

Step 2. Substitute $\mathbf{x}_0$ into the right side of (2) and compute the **first iterate**
$$\mathbf{x}_1 = B\mathbf{x}_0 + \mathbf{c}$$

Step 3. Substitute $\mathbf{x}_1$ into the right side of (2) and compute the **second iterate**
$$\mathbf{x}_2 = B\mathbf{x}_1 + \mathbf{c}$$

Step 4. Keep repeating the procedure of Steps 2 and 3, substituting each new iterate into the right side of (2), thereby producing a **third iterate**, a **fourth iterate**, and so on. Generate as many iterates as may be required to achieve the desired accuracy.

Whether the sequence of iterates produced by (2) will actually converge to the exact solution will depend on how B and $\mathbf{c}$ are chosen and properties of the matrix A. In this section we will consider two of the most basic iterative methods.

JACOBI ITERATION Let $A\mathbf{x} = \mathbf{b}$ be a linear system of n equations in n unknowns in which A is invertible and has nonzero diagonal entries. Let D be the $n \times n$ diagonal matrix formed from the diagonal entries of A. The matrix D is invertible since its diagonal entries are nonzero [Formula (2) of Section 3.6], and hence we can rewrite the system $A\mathbf{x} = \mathbf{b}$ as

$$(A - D)\mathbf{x} + D\mathbf{x} = \mathbf{b}$$
$$D\mathbf{x} = (D - A)\mathbf{x} + \mathbf{b}$$
$$\mathbf{x} = D^{-1}(D - A)\mathbf{x} + D^{-1}\mathbf{b} \tag{3}$$

Equation (3) is of form (1) with $B = D^{-1}(D - A)$ and $\mathbf{c} = D^{-1}\mathbf{b}$, and the corresponding recursion formula is

$$\mathbf{x}_{k+1} = D^{-1}(D - A)\mathbf{x}_k + D^{-1}\mathbf{b} \tag{4}$$

The iteration algorithm that uses this formula is called **Jacobi iteration** or **the method of simultaneous displacements**.

If $A\mathbf{x} = \mathbf{b}$ is the linear system

$$
\begin{array}{l}
a_{11}x_1 + a_{12}x_2 + \cdots + a_{1n}x_n = b_1 \\
a_{21}x_1 + a_{22}x_2 + \cdots + a_{2n}x_n = b_2 \\
\quad\vdots \qquad\quad \vdots \qquad\qquad \vdots \qquad \vdots \\
a_{n1}x_1 + a_{n2}x_2 + \cdots + a_{nn}x_n = b_n
\end{array} \tag{5}
$$

then the individual equations in (3) are

$$x_1 = \frac{1}{a_{11}}(b_1 - a_{12}x_2 - a_{13}x_3 - \cdots - a_{1n}x_n)$$

$$x_2 = \frac{1}{a_{22}}(b_2 - a_{21}x_1 - a_{23}x_3 - \cdots - a_{2n}x_n) \tag{6}$$

$$\vdots \qquad\qquad\qquad \vdots$$

$$x_n = \frac{1}{a_{nn}}(b_n - a_{n1}x_1 - a_{n2}x_2 - \cdots - a_{n(n-1)}x_{n-1})$$

These can be obtained by solving the first equation in (5) for x_1 in terms of the remaining unknowns, the second equation for x_2 in terms of the remaining unknowns, and so forth.

Jacobi iteration can be programmed for computer implementation using Formula (4), but for hand computation (with computer or calculator assistance) you can proceed as follows:

Jacobi Iteration

Step 1. Rewrite the system $A\mathbf{x} = \mathbf{b}$ in form (6), and choose arbitrary initial values for the unknowns. The column vector $\mathbf{x}_0$ formed from these values is the initial approximation. If no better choice is available, you can take $\mathbf{x}_0 = \mathbf{0}$; that is,

$$x_1 = 0, \quad x_2 = 0, \dots, \quad x_n = 0$$

Step 2. Substitute the entries of the initial approximation into the right side of (6) to produce new values of $x_1, x_2, \dots, x_n$ on the left side. The column vector $\mathbf{x}_1$ with these new entries is the first iterate.

Step 3. Substitute the entries in the first iterate into the right side of (6) to produce new values of $x_1, x_2, \dots, x_n$ on the left side. The column vector $\mathbf{x}_2$ with these new entries is the second iterate.

Step 4. Substitute the entries in the second iterate into the right side of (6) to produce the third iterate $\mathbf{x}_3$, and continue the process to produce as many iterates as may be required to achieve the desired accuracy.

EXAMPLE 1   Use Jacobi iteration to approximate the solution of the system

$$
\begin{aligned}
20x_1 + x_2 - x_3 &= 17 \\
x_1 - 10x_2 + x_3 &= 13 \\
-x_1 + x_2 + 10x_3 &= 18
\end{aligned}
\tag{7}
$$

Stop the process when the entries in two successive iterates are the same when rounded to four decimal places.

Solution   As required for Jacobi iteration, we begin by solving the first equation for x_1, the second for x_2, and the third for x_3. This yields

$$
\begin{array}{ll}
x_1 = \frac{17}{20} - \frac{1}{20}x_2 + \frac{1}{20}x_3 & \qquad x_1 = 0.85 - 0.05x_2 + 0.05x_3 \\
x_2 = -\frac{13}{10} + \frac{1}{10}x_1 + \frac{1}{10}x_3 \quad \text{or} & \qquad x_2 = -1.3 + 0.1x_1 + 0.1x_3 \\
x_3 = \frac{18}{10} + \frac{1}{10}x_1 - \frac{1}{10}x_2 & \qquad x_3 = 1.8 + 0.1x_1 - 0.1x_2
\end{array}
\tag{8}
$$

which we can write in matrix form as

$$
\begin{bmatrix} x_1 \\ x_2 \\ x_3 \end{bmatrix} =
\begin{bmatrix} 0 & -0.05 & 0.05 \\ 0.1 & 0 & 0.1 \\ 0.1 & -0.1 & 0 \end{bmatrix}
\begin{bmatrix} x_1 \\ x_2 \\ x_3 \end{bmatrix} +
\begin{bmatrix} 0.85 \\ -1.3 \\ 1.8 \end{bmatrix}
\tag{9}
$$

Since we have no special information about the solution, we will take the initial approximation to be $x_1 = x_2 = x_3 = 0$. To obtain the first iterate, we substitute these values into the right side of (9). This yields

$$
\mathbf{x}_1 = \begin{bmatrix} 0.85 \\ -1.3 \\ 1.8 \end{bmatrix}
$$

To obtain the second iterate, we substitute the entries of $\mathbf{x}_1$ into the right side of (9). This yields

$$
\mathbf{x}_2 = \begin{bmatrix} x_1 \\ x_2 \\ x_3 \end{bmatrix} =
\begin{bmatrix} 0 & -0.05 & 0.05 \\ 0.1 & 0 & 0.1 \\ 0.1 & -0.1 & 0 \end{bmatrix}
\begin{bmatrix} 0.85 \\ -1.3 \\ 1.8 \end{bmatrix} +
\begin{bmatrix} 0.85 \\ -1.3 \\ 1.8 \end{bmatrix} =
\begin{bmatrix} 1.005 \\ -1.035 \\ 2.015 \end{bmatrix}
$$

Repeating this process until two successive iterates match to four decimal places yields the results in Table 5.3.1. This agrees with the exact solution $x_1 = 1$, $x_2 = -1$, $x_3 = 2$ to four decimal places. ■

Table 5.3.1

	x_0	x_1	x_2	x_3	x_4	x_5	x_6	x_7
x_1	0	0.8500	1.0050	1.0025	1.0001	1.0000	1.0000	1.0000
x_2	0	−1.3000	−1.0350	−0.9980	−0.9994	−1.0000	−1.0000	−1.0000
x_3	0	1.8000	2.0150	2.0040	2.0000	1.9999	2.0000	2.0000

REMARK As a rule of thumb, if you want to round an iterate to m decimal places, then you should use at least $m + 1$ decimal places in all computations (the more the better). Thus, in the above example, all computations should be carried out using at least five decimal places, even though the iterates are rounded to four decimal places for display in the table. Your calculating utility may produce slightly different values from those in the table, depending on what rounding conventions it uses.

GAUSS–SEIDEL ITERATION

Jacobi iteration is reasonable for small linear systems but the convergence tends to be too slow for large systems. We will now consider a procedure that can be used to speed up the convergence.

In each step of the Jacobi method the new x-values are obtained by substituting the previous x-values into the right side of (6). These new x-values are not all computed simultaneously— first x_1 is obtained from the top equation, then x_2 from the second equation, then x_3 from the third equation, and so on. Since the new x-values are expected to be more accurate than their predecessors, it seems reasonable that better accuracy might be obtained by using the new x-values as soon as they become available. If this is done, then the resulting algorithm is called *Gauss–Seidel iteration* or the *method of successive displacements*. Here is an example.

EXAMPLE 2
Gauss–Seidel Iteration

Use Gauss–Seidel iteration to approximate the solution of the linear system in Example 1 to four decimal places.

Solution As before, we will take $x_1 = x_2 = x_3 = 0$ as the initial approximation. First we will substitute $x_2 = 0$ and $x_3 = 0$ into the right side of the first equation in (8) to obtain the new x_1, then we will substitute $x_3 = 0$ and the new x_1 into the right side of the second equation to obtain the new x_2, and finally we will substitute the new x_1 and new x_2 into the right side of the third equation to obtain the new x_3. The computations are as follows:

$$x_1 = 0.85 - (0.05)(0) + (0.05)(0) = 0.85$$
$$x_2 = -1.3 + (0.1)(0.85) + (0.1)(0) = -1.215$$
$$x_3 = 1.8 + (0.1)(0.85) - (0.1)(-1.215) = 2.0065$$

Thus, the first Gauss–Seidel iterate is

$$\mathbf{x}_1 = \begin{bmatrix} 0.8500 \\ -1.2150 \\ 2.0065 \end{bmatrix}$$

Similarly, the computations for the second iterate are

$$x_1 = 0.85 - (0.05)(-1.215) + (0.05)(2.0065) = 1.011075$$
$$x_2 = -1.3 + (0.1)(1.011075) + (0.1)(2.0065) = -0.9982425$$
$$x_3 = 1.8 + (0.1)(1.011075) - (0.1)(-0.9982425) = 2.00093175$$

Thus, the second Gauss–Seidel iterate to four decimal places is

$$\mathbf{x}_2 \approx \begin{bmatrix} 1.0111 \\ -0.9982 \\ 2.0009 \end{bmatrix}$$

Linear Algebra in History

Ludwig Philipp von Seidel (1821–1896) was a German physicist and mathematician who studied under a student of Gauss (see p. 54). Seidel published the method in 1874, but it is unclear how Gauss's name became associated with it, since Gauss, who was aware of the method much earlier, declared the method to be worthless! Gauss's criticism notwithstanding, adaptations of the method are commonly used for solving certain kinds of sparse linear systems.

Table 5.3.2 shows the first four Gauss–Seidel iterates to four decimal places. Comparing Tables 5.3.1 and 5.3.2, we see that the Gauss–Seidel method produced the solution to four decimal places in four iterations, whereas the Jacobi method required six. ■

Table 5.3.2

	$\mathbf{x}_0$	$\mathbf{x}_1$	$\mathbf{x}_2$	$\mathbf{x}_3$	$\mathbf{x}_4$
x_1	0	0.8500	1.0111	1.0000	1.0000
x_2	0	−1.2150	−0.9982	−0.9999	−1.0000
x_3	0	2.0065	2.0009	2.0000	2.0000

REMARK To program Gauss–Seidel iteration for computer implementation, it is desirable to have a recursion formula comparable to (4). Such a formula can be obtained by first writing A as

$$A = D - L - U$$

where D, $-L$, and $-U$ are the diagonal, lower triangular, and upper triangular matrices suggested in Figure 5.3.1. (The matrices L and U are not those from the LU-decomposition of A.) It can be shown that if $D - L$ is invertible, then a recursion formula for Gauss–Seidel iteration can be expressed in terms of these matrices as

$$\mathbf{x}_{k+1} = (D - L)^{-1}U\mathbf{x}_k + (D - L)^{-1}\mathbf{b} \tag{10}$$

We omit the proof.

$$\begin{bmatrix} \bullet & \bullet & \bullet & \bullet & \bullet \\ \bullet & \bullet & \bullet & \bullet & \bullet \\ \bullet & \bullet & \bullet & \bullet & \bullet \\ \bullet & \bullet & \bullet & \bullet & \bullet \\ \bullet & \bullet & \bullet & \bullet & \bullet \end{bmatrix} = \begin{bmatrix} \bullet & 0 & 0 & 0 & 0 \\ 0 & \bullet & 0 & 0 & 0 \\ 0 & 0 & \bullet & 0 & 0 \\ 0 & 0 & 0 & \bullet & 0 \\ 0 & 0 & 0 & 0 & \bullet \end{bmatrix} - \begin{bmatrix} 0 & 0 & 0 & 0 & 0 \\ -\bullet & 0 & 0 & 0 & 0 \\ -\bullet & \bullet & 0 & 0 & 0 \\ -\bullet & -\bullet & -\bullet & 0 & 0 \\ -\bullet & -\bullet & -\bullet & -\bullet & 0 \end{bmatrix} - \begin{bmatrix} 0 & -\bullet & -\bullet & -\bullet & -\bullet \\ 0 & 0 & -\bullet & -\bullet & -\bullet \\ 0 & 0 & 0 & -\bullet & -\bullet \\ 0 & 0 & 0 & 0 & -\bullet \\ \bullet & 0 & 0 & 0 & 0 \end{bmatrix}$$

$$A \quad = \quad D \quad - \quad L \quad - \quad U$$

$-\bullet$ denotes the negative of entry $\bullet$ in A

Figure 5.3.1

CONVERGENCE

There are situations in which the iterates produced by the Jacobi and Gauss–Seidel methods fail to converge to the solution of the system. Our next objective is to discuss conditions under which convergence is ensured. For this purpose we define a square matrix

$$A = \begin{bmatrix} a_{11} & a_{12} & \cdots & a_{1n} \\ a_{21} & a_{22} & \cdots & a_{2n} \\ \vdots & \vdots & & \vdots \\ a_{n1} & a_{n2} & \cdots & a_{nn} \end{bmatrix}$$

to be ***strictly diagonally dominant*** if the absolute value of each diagonal entry is greater than the sum of the absolute values of the remaining entries in the same row; that is,

$$\begin{aligned} |a_{11}| &> |a_{12}| + |a_{13}| + \cdots + |a_{1n}| \\ |a_{22}| &> |a_{21}| + |a_{23}| + \cdots + |a_{2n}| \\ &\vdots \\ |a_{nn}| &> |a_{n1}| + |a_{n2}| + \cdots + |a_{n(n-1)}| \end{aligned} \tag{11}$$

EXAMPLE 3   The matrix

A Strictly
Diagonally
Dominant
Matrix

$$\begin{bmatrix} 7 & -2 & 3 \\ 4 & 1 & -6 \\ 5 & 12 & -4 \end{bmatrix}$$

is not strictly diagonally dominant because the required condition fails to hold in both the second and third rows. In the second row $|1|$ is not greater than $|4| + |-6|$, and in the third row $|-4|$ is not greater than $|5| + |12|$. However, if the second and third rows are interchanged, then the resulting matrix

$$\begin{bmatrix} 7 & -2 & 3 \\ 5 & 12 & -4 \\ 4 & 1 & -6 \end{bmatrix}$$

is strictly diagonally dominant since

$$|7| > |-2| + |3|$$
$$|12| > |5| + |-4|$$
$$|-6| > |4| + |1|$$

Strictly diagonally dominant matrices are important because of the following theorem whose proof can be found in books on numerical methods of linear algebra. (Also, see Exercise P2.)

> **Theorem 5.3.1** *If A is strictly diagonally dominant, then* $A\mathbf{x} = \mathbf{b}$ *has a unique solution, and for any choice of the initial approximation the iterates in the Gauss–Seidel and Jacobi methods converge to that solution.*

EXAMPLE 4   The calculations in Examples 1 and 2 strongly suggest that both the Jacobi and Gauss–Seidel

Convergence of
Iterates

iterates converge to the exact solution of system (7). Theorem 5.3.1 guarantees that this is so, since the coefficient matrix of the system is strictly diagonally dominant (verify). ■

SPEEDING UP CONVERGENCE   In applications one is concerned not only with the convergence of iterative methods but also with *how fast* they converge. One can show that in the case where A is strictly diagonally dominant, the rate of convergence is determined by *how much* the left sides of (11) dominate the right sides—the more dominant the left sides are, the more rapid the convergence. Thus, even if the inequalities in (11) hold, the two sides of the inequalities may be so close that the convergence of the Jacobi and Gauss–Seidel iterates is too slow to make the algorithms practical. To deal with this problem numerical analysts have devised various methods for improving the rate of convergence. One of the most important of these methods is known as ***extrapolated Gauss–Seidel iteration*** or the ***method of successive overrelaxation*** (abbreviated SOR in the literature). Readers interested in this topic are referred to the following standard reference on the subject: G. H. Golub and C. F. Van Loan, *Matrix Computations*, Johns Hopkins University Press, Baltimore, 1996.

Exercise Set 5.3

In Exercises 1 and 2, approximate the solution of the system by Jacobi iteration and by Gauss–Seidel iteration, starting with $x_1 = 0$, $x_2 = 0$, and using three iterations. Compare your results to the exact solution.

1. $2x_1 + x_2 = 7$
$\quad\ x_1 - 2x_2 = 1$

2. $3x_1 - x_2 = 5$
$\quad\ 2x_1 + 3x_2 = -4$

In Exercises 3 and 4, approximate the solution of the system by Jacobi iteration and by Gauss–Seidel iteration, starting with $x_1 = 0$, $x_2 = 0$, $x_3 = 0$, and using three iterations. Compare your results to the exact solution.

3. $10x_1 + x_2 + 2x_3 = 3$
$\quad\ \ x_1 + 10x_2 - x_3 = \frac{3}{2}$
$\quad\ 2x_1 + x_2 + 10x_3 = -9$

4. $20x_1 - x_2 + x_3 = 20$
$\quad\ 2x_1 + 10x_2 - x_3 = 11$
$\quad\ \ x_1 + x_2 - 20x_3 = -18$

In Exercises 5 and 6, determine whether the matrix is strictly diagonally dominant.

5. (a) $\begin{bmatrix} 4 & -3 \\ 2 & -2 \end{bmatrix}$
  (b) $\begin{bmatrix} 4 & -1 & -2 \\ 0 & -7 & 2 \\ 3 & -1 & 5 \end{bmatrix}$

  (c) $\begin{bmatrix} 6 & 2 & 3 & 0 \\ 3 & 5 & 0 & 1 \\ 1 & 2 & -8 & 3 \\ -1 & 0 & 4 & 6 \end{bmatrix}$

6. (a) $\begin{bmatrix} -5 & -3 \\ 4 & 6 \end{bmatrix}$
  (b) $\begin{bmatrix} 4 & 1 & 1 \\ -2 & 5 & 2 \\ 2 & -4 & 6 \end{bmatrix}$

  (c) $\begin{bmatrix} 4 & 3 & 0 & -2 \\ 1 & 5 & 1 & -3 \\ -2 & 2 & 6 & 1 \\ 1 & 2 & 3 & -7 \end{bmatrix}$

In Exercises 7–10, show that the matrix is not strictly diagonally dominant, and determine whether it can be made strictly diagonally dominant by interchanging appropriate rows.

7. $\begin{bmatrix} -1 & 4 & -6 \\ 5 & 2 & 2 \\ 3 & -7 & -3 \end{bmatrix}$
8. $\begin{bmatrix} 1 & -2 & 3 & 7 \\ -6 & 1 & 0 & -4 \\ 1 & -2 & 5 & 1 \\ 3 & -8 & -1 & 1 \end{bmatrix}$

9. $\begin{bmatrix} 1 & -2 & 3 \\ 4 & 1 & 1 \\ 1 & -4 & 2 \end{bmatrix}$
10. $\begin{bmatrix} 1 & 3 & -1 & 2 \\ 5 & 1 & 4 & 0 \\ 2 & -1 & 5 & 1 \\ 0 & 3 & 6 & -2 \end{bmatrix}$

Working with Proofs

Define the ***infinity norm*** of an $m \times n$ matrix A to be the maximum of the absolute row sums, that is,

$$\|A\|_\infty = \max_{1 \le i \le m} \left[\sum_{j=1}^{n} |a_{ij}| \right]$$

P1. (a) Prove that $\|aA\|_\infty = |a|\,\|A\|_\infty$.
  (b) Prove that $\|A + B\|_\infty \le \|A\|_\infty + \|B\|_\infty$.
  (c) Prove that if $\mathbf{x}$ is a $1 \times n$ row vector, then
  $\|\mathbf{x}B\|_\infty \le \|B\|_\infty\|\mathbf{x}\|_\infty$.
  (d) Prove that $\|AB\|_\infty \le \|A\|_\infty\|B\|_\infty$.

P2. (a) Let M be an $n \times n$ matrix, and consider the sequence defined recursively by $\mathbf{x}_k = M\mathbf{x}_{k-1} + \mathbf{y}$, where $\mathbf{x}_0$ is arbitrary. Show that
$$\mathbf{x}_k = M^k\mathbf{x}_0 + (I + M + M^2 + \cdots + M^{k-1})\mathbf{y}$$

  (b) Use part (a) together with part (d) of Exercise P1 for the infinity norm and Theorem 3.6.7 to prove that if $\|M\|_\infty < 1$, then $\mathbf{x}_k \to (I - M)^{-1}\mathbf{y}$ as $k \to \infty$.

P3. Formula (4) states that if A is an $n \times n$ invertible matrix with nonzero diagonal entries, and if $A\mathbf{x} = \mathbf{b}$ is a system of equations, then the Jacobi iteration sequence approximating $\mathbf{x}$ can be written in the form $\mathbf{x}_{k+1} = M\mathbf{x}_k + \mathbf{y}$, where $M = D^{-1}(D - A)$ and $\mathbf{y} = D^{-1}\mathbf{b}$. ($D$ is the diagonal matrix whose diagonal agrees with that of A.)
  (a) Show that if A is strictly diagonally dominant, then $\|M\|_\infty = \|D^{-1}(D - A)\|_\infty < 1$. [*Hint:* Write out the entries in the kth row of $D^{-1}(D - A)$ and then add them.]
  (b) Use part (a) and Exercise P2 to complete the proof of Theorem 5.3.1 in the Jacobi case.

P4. Suppose that A is an invertible $n \times n$ matrix with the property that the absolute value of each diagonal entry is greater than the sum of the absolute values of the remaining entries in the same column (instead of row). Show that the Jacobi method of approximating the solution to $A\mathbf{x} = \mathbf{b}$ works in this case. [*Hint:* Define the **1-*norm*** of an $m \times n$ matrix A to be the maximum of the absolute column sums; that is,

$$\|A\|_1 = \max_{1 \le j \le n} \sum_{i=1}^{m} |a_{ij}|$$

It can be proved that the results stated for the infinity norm in Exercises P1 and P2 also hold for the 1-norm. Accepting this to be so, proceed as in Exercise P3 with the 1-norm in place of the infinity norm.]

Technology Exercises

In Exercises T1 and T2, approximate the solutions using Jacobi iteration, starting with $\mathbf{x} = \mathbf{0}$ and continuing until two successive iterations agree to three decimal places. Repeat the process using Gauss–Seidel iteration and compare the number of iterations required by each method.

T1. The system in Exercise 3.

T2. The system in Exercise 4.

T3. Consider the linear system

$$\begin{aligned} 2x_1 - 4x_2 + 7x_3 &= 8 \\ -2x_1 + 5x_2 + 2x_3 &= 0 \\ 4x_1 + x_2 + x_3 &= 5 \end{aligned}$$

(a) The coefficient matrix of the system is not strictly diagonally dominant, so Theorem 5.3.1 does not guarantee convergence of either Jacobi iteration or Gauss–Seidel iteration. Compute the first five Gauss–Seidel iterates to illustrate a lack of convergence.

(b) Reorder the equations to produce a linear system with a strictly diagonally dominant coefficient matrix, and compute the first five Gauss–Seidel iterates to illustrate convergence.

T4. *Heat* is energy that flows from a point with a higher temperature to a point with a lower temperature. This energy flow causes a temperature decrease at the point with higher temperature and a temperature increase at the point with lower temperature until the temperatures at the two points are the same and the energy flow stops; the temperatures are then said to have reached a ***steady state***. We will consider the problem of approximating the steady-state temperature at the interior points of a rectangular metal plate whose edges are kept at fixed temperatures. Methods for finding steady-state temperatures at all points of a plate generally require calculus, but if we limit the problem to finding the steady-state temperature at a finite number of points, then we can use linear algebra. The idea is to overlay the plate with a rectangular grid and approximate the steady-state temperatures at the interior grid points. For example, the last part of the accompanying figure shows a rectangular plate with fixed temperatures of $0°C, 0°C, 4°C,$ and $16°C$ on the edges and unknown steady-state temperatures $t_1, t_2, \ldots, t_9$ at nine interior grid points. Our approach will use the ***discrete averaging*** model from thermodynamics, which states that steady-state temperature at an interior grid point is approximately the average of the temperatures at the four adjacent grid points. Thus, for example, the steady-state temperatures t_5 and t_1 are

$$t_5 = \tfrac{1}{4}(t_4 + t_2 + t_6 + t_8) \quad \text{and} \quad t_1 = \tfrac{1}{4}(0 + 16 + t_2 + t_4)$$

(a) Write out all nine discrete averaging equations for the steady-state temperatures $t_1, t_2, \ldots, t_9$.

(b) Rewrite the equations in part (a) in the matrix form $\mathbf{t} = B\mathbf{t} + \mathbf{c}$, where $\mathbf{t}$ is the column vector of unknown steady-state vectors, B is a 9×9 matrix, and $\mathbf{c}$ is a 9×1 column vector of constants.

(c) The form of the equation $\mathbf{t} = B\mathbf{t} + \mathbf{c}$ is the same as Equation (1), so Equation (2) suggests that it can be solved iteratively using the recurrence relation $\mathbf{t}_{k+1} = B\mathbf{t}_k + \mathbf{c}$, as with Jacobi iteration. Use this method to approximate the steady-state temperatures, starting with $t_1 = 0, t_2 = 0, \ldots, t_9 = 0$ and using 10 iterations. [*Comment:* This is a good example of how sparse matrices arise—each unknown depends on only a few of the other unknowns, so the resulting matrix has many zeros.]

(d) Approximate the steady-state temperatures using as many iterations as are required until the difference between two successive iterations is less than 0.001 in each entry.

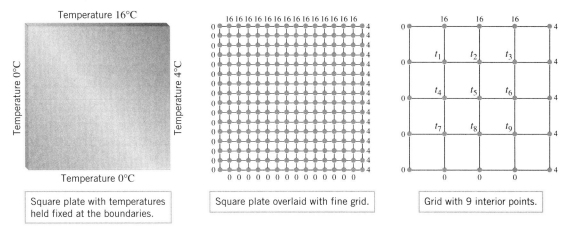

| Square plate with temperatures held fixed at the boundaries. | Square plate overlaid with fine grid. | Grid with 9 interior points. |

Figure Ex-T4

Section 5.4   The Power Method; Application to Internet Search Engines

The eigenvalues of a square matrix can, in theory, be found by solving the characteristic equation. However, this procedure has so many computational difficulties that it is almost never used in applications. In this section we will discuss an algorithm that can be used to approximate the eigenvalue with greatest absolute value and a corresponding eigenvector. This particular eigenvalue and its corresponding eigenvectors are important because they arise naturally in many iterative processes. The methods that we will study in this section have recently been applied to produce extremely fast Internet search engines, and we will explain how that is done.

THE POWER METHOD

There are many applications in which some vector $\mathbf{x}_0$ in R^n is multiplied repeatedly by an $n \times n$ matrix A to produce a sequence

$$\mathbf{x}_0, \ A\mathbf{x}_0, \ A^2\mathbf{x}_0, \dots, \ A^k\mathbf{x}_0, \dots$$

We call a sequence of this form a ***power sequence generated by*** A. In this section we will be concerned with the convergence of power sequences and their application to the study of eigenvalues and eigenvectors. For this purpose, we make the following definition.

Definition 5.4.1   If the *distinct* eigenvalues of a matrix A are $\lambda_1, \lambda_2, \dots, \lambda_k$, and if $|\lambda_1|$ is larger than $|\lambda_2|, \dots, |\lambda_k|$, then λ_1 is called a ***dominant eigenvalue*** of A. Any eigenvector corresponding to a dominant eigenvalue is called a ***dominant eigenvector*** of A.

EXAMPLE 1
Dominant
Eigenvalues

Some matrices have dominant eigenvalues and some do not. For example, if the distinct eigenvalues of a matrix are

$$\lambda_1 = -4, \quad \lambda_2 = -2, \quad \lambda_3 = 1, \quad \lambda_4 = 3$$

then $\lambda_1 = -4$ is dominant since $|\lambda_1| = 4$ is greater than the absolute values of all the other eigenvalues, but if the distinct eigenvalues of a matrix are

$$\lambda_1 = 7, \quad \lambda_2 = -7, \quad \lambda_3 = -2, \quad \lambda_4 = 5$$

then $|\lambda_1| = |\lambda_2| = 7$, so there is no eigenvalue whose absolute value is greater than the absolute value of all the other eigenvalues.   ■

The most important theorems about convergence of power sequences apply to $n \times n$ matrices that have n linearly independent eigenvectors. We will show later in the text that symmetric matrices have this property, and since many of the most important applications involve symmetric matrices, we will limit our discussion to that case in this section.

Theorem 5.4.2 *Let A be a symmetric $n \times n$ matrix with a positive* [*] *dominant eigenvalue λ. If $\mathbf{x}_0$ is a unit vector in R^n that is not orthogonal to the eigenspace corresponding to λ, then the normalized power sequence*

$$\mathbf{x}_0, \quad \mathbf{x}_1 = \frac{A\mathbf{x}_0}{\|A\mathbf{x}_0\|}, \quad \mathbf{x}_2 = \frac{A\mathbf{x}_1}{\|A\mathbf{x}_1\|}, \dots, \quad \mathbf{x}_k = \frac{A\mathbf{x}_{k-1}}{\|A\mathbf{x}_{k-1}\|}, \dots \tag{1}$$

converges to a unit dominant eigenvector, and the sequence

$$A\mathbf{x}_1 \cdot \mathbf{x}_1, \quad A\mathbf{x}_2 \cdot \mathbf{x}_2, \quad A\mathbf{x}_3 \cdot \mathbf{x}_3, \dots, \quad A\mathbf{x}_k \cdot \mathbf{x}_k, \dots \tag{2}$$

converges to the dominant eigenvalue λ.

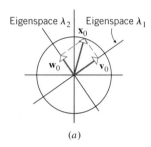

Eigenspace λ_2 Eigenspace λ_1

(*a*)

REMARK In the exercises we will ask you to show that (1) can also be expressed as

$$\mathbf{x}_0, \quad \mathbf{x}_1 = \frac{A\mathbf{x}_0}{\|A\mathbf{x}_0\|}, \quad \mathbf{x}_2 = \frac{A^2\mathbf{x}_0}{\|A^2\mathbf{x}_0\|}, \dots, \quad \mathbf{x}_k = \frac{A^k\mathbf{x}_0}{\|A^k\mathbf{x}_0\|}, \dots \tag{3}$$

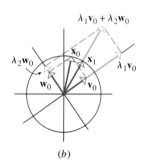

(*b*)

This form of the power sequence expresses each iterate in terms of the starting vector $\mathbf{x}_0$, rather than in terms of its predecessor.

We will not prove Theorem 5.4.2, but we can make it plausible geometrically in the 2×2 case where A is a symmetric matrix with distinct positive eigenvalues, λ_1 and λ_2, one of which is dominant. To be specific, assume that λ_1 is dominant and

$$\lambda_1 > \lambda_2 > 0$$

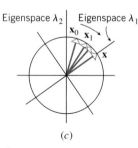

Eigenspace λ_2 Eigenspace λ_1

(*c*)

Figure 5.4.1

Since we are assuming that A is symmetric and has distinct eigenvalues, Theorem 4.4.11 tells us that the eigenspaces corresponding to λ_1 and λ_2 are perpendicular lines through the origin. Thus, the assumption that $\mathbf{x}_0$ is a unit vector that is not orthogonal to the eigenspace corresponding to λ_1 implies that $\mathbf{x}_0$ does not lie in the eigenspace corresponding to λ_2. To help understand the geometric effect of multiplying $\mathbf{x}_0$ by A, it will be useful to split $\mathbf{x}_0$ into the sum

$$\mathbf{x}_0 = \mathbf{v}_0 + \mathbf{w}_0 \tag{4}$$

where $\mathbf{v}_0$ and $\mathbf{w}_0$ are the orthogonal projections of $\mathbf{x}_0$ on the eigenspaces of λ_1 and λ_2, respectively (Figure 5.4.1*a*). This enables us to express $A\mathbf{x}_0$ as

$$A\mathbf{x}_0 = A\mathbf{v}_0 + A\mathbf{w}_0 = \lambda_1\mathbf{v}_0 + \lambda_2\mathbf{w}_0 \tag{5}$$

---

[*] If the dominant eigenvalue is not positive, sequence (2) will still converge to the dominant eigenvalue, but sequence (1) may not converge to a *specific* dominant eigenvector because of *alternation* (see Exercise 9). Nevertheless, each term of (1) will closely approximate *some* dominant eigenvector for sufficiently large values of k.

which tells us that multiplying $\mathbf{x}_0$ by A "scales" the terms $\mathbf{v}_0$ and $\mathbf{w}_0$ in (4) by λ_1 and λ_2, respectively. However, λ_1 is larger than λ_2, so the scaling is greater in the direction of $\mathbf{v}_0$ than in the direction of $\mathbf{w}_0$. Thus, multiplying $\mathbf{x}_0$ by A "pulls" $\mathbf{x}_0$ toward the eigenspace of λ_1, and normalizing produces a vector $\mathbf{x}_1 = A\mathbf{x}_0/\|A\mathbf{x}_0\|$, which is on the unit circle and is closer to the eigenspace of λ_1 than $\mathbf{x}_0$ (Figure 5.4.1b). Similarly, multiplying $\mathbf{x}_1$ by A and normalizing produces a unit vector $\mathbf{x}_2$ that is closer to the eigenspace of λ_1 than $\mathbf{x}_1$. Thus, it seems reasonable that by repeatedly multiplying by A and normalizing we will produce a sequence of vectors $\mathbf{x}_k$ that lie on the unit circle and converge to a unit vector $\mathbf{x}$ in the eigenspace of λ_1 (Figure 5.4.1c). Moreover, if $\mathbf{x}_k$ converges to $\mathbf{x}$, then it also seems reasonable that $A\mathbf{x}_k \cdot \mathbf{x}_k$ will converge to

$$A\mathbf{x} \cdot \mathbf{x} = \lambda_1 \mathbf{x} \cdot \mathbf{x} = \lambda_1 \|\mathbf{x}\|^2 = \lambda_1$$

which is the dominant eigenvalue of A.

THE POWER METHOD WITH EUCLIDEAN SCALING

Theorem 5.4.2 provides us with an algorithm for approximating the dominant eigenvalue and a corresponding unit eigenvector of a symmetric matrix A, provided the dominant eigenvalue is positive. This algorithm, called the ***power method with Euclidean scaling***, is as follows:

The Power Method with Euclidean Scaling

Step 1. Choose an arbitrary nonzero vector and normalize it, if need be, to obtain a unit vector $\mathbf{x}_0$.

Step 2. Compute $A\mathbf{x}_0$ and normalize it to obtain the first approximation $\mathbf{x}_1$ to a dominant unit eigenvector. Compute $A\mathbf{x}_1 \cdot \mathbf{x}_1$ to obtain the first approximation to the dominant eigenvalue.

Step 3. Compute $A\mathbf{x}_1$ and normalize it to obtain the second approximation $\mathbf{x}_2$ to a dominant unit eigenvector. Compute $A\mathbf{x}_2 \cdot \mathbf{x}_2$ to obtain the second approximation to the dominant eigenvalue.

Step 4. Compute $A\mathbf{x}_2$ and normalize it to obtain the third approximation $\mathbf{x}_3$ to a dominant unit eigenvector. Compute $A\mathbf{x}_3 \cdot \mathbf{x}_3$ to obtain the third approximation to the dominant eigenvalue.

Continuing in this way will usually generate a sequence of better and better approximations to the dominant eigenvalue and a corresponding unit eigenvector.[*]

EXAMPLE 2
The Power Method with Euclidean Scaling

Apply the power method with Euclidean scaling to

$$A = \begin{bmatrix} 3 & 2 \\ 2 & 3 \end{bmatrix} \quad \text{with} \quad \mathbf{x}_0 = \begin{bmatrix} 1 \\ 0 \end{bmatrix}$$

Stop at $\mathbf{x}_5$ and compare the resulting approximations to the exact values of the dominant eigenvalue and eigenvector.

Solution In Example 6 of Section 4.4 we found the eigenvalues of A to be $\lambda = 1$ and $\lambda = 5$, so the dominant eigenvalue of A is $\lambda = 5$. We also showed that the eigenspace corresponding to $\lambda = 5$ is the line represented by the parametric equations $x_1 = t$, $x_2 = t$, which can be written in vector form as

$$\mathbf{x} = \begin{bmatrix} t \\ t \end{bmatrix} = t \begin{bmatrix} 1 \\ 1 \end{bmatrix} \tag{6}$$

---

[*] If the vector $\mathbf{x}_0$ happens to be orthogonal to the eigenspace of the dominant eigenvalue, then the hypotheses of Theorem 5.4.2 will be violated and the method may fail. However, the reality is that computer roundoff errors usually perturb $\mathbf{x}_0$ enough to destroy any orthogonality and make the algorithm work. This is one instance in which errors help to obtain correct results!

Thus, taking $t = 1/\sqrt{2}$ and $t = -1/\sqrt{2}$ yields two dominant unit eigenvectors:

$$\mathbf{v}_1 = \begin{bmatrix} \frac{1}{\sqrt{2}} \\ \frac{1}{\sqrt{2}} \end{bmatrix} \quad \text{and} \quad \mathbf{v}_2 = \begin{bmatrix} -\frac{1}{\sqrt{2}} \\ -\frac{1}{\sqrt{2}} \end{bmatrix} \tag{7}$$

Now let us see what happens when we use the power method, starting with the unit vector $\mathbf{x}_0$:

$$A\mathbf{x}_0 = \begin{bmatrix} 3 & 2 \\ 2 & 3 \end{bmatrix} \begin{bmatrix} 1 \\ 0 \end{bmatrix} = \begin{bmatrix} 3 \\ 2 \end{bmatrix} \qquad \mathbf{x}_1 = \frac{A\mathbf{x}_0}{\|A\mathbf{x}_0\|} = \frac{1}{\sqrt{13}} \begin{bmatrix} 3 \\ 2 \end{bmatrix} \approx \frac{1}{3.60555} \begin{bmatrix} 3 \\ 2 \end{bmatrix} \approx \begin{bmatrix} 0.83205 \\ 0.55470 \end{bmatrix}$$

$$A\mathbf{x}_1 \approx \begin{bmatrix} 3 & 2 \\ 2 & 3 \end{bmatrix} \begin{bmatrix} 0.83205 \\ 0.55470 \end{bmatrix} \approx \begin{bmatrix} 3.60555 \\ 3.32820 \end{bmatrix} \qquad \mathbf{x}_2 = \frac{A\mathbf{x}_1}{\|A\mathbf{x}_1\|} \approx \frac{1}{4.90682} \begin{bmatrix} 3.60555 \\ 3.32820 \end{bmatrix} \approx \begin{bmatrix} 0.73480 \\ 0.67828 \end{bmatrix}$$

$$A\mathbf{x}_2 \approx \begin{bmatrix} 3 & 2 \\ 2 & 3 \end{bmatrix} \begin{bmatrix} 0.73480 \\ 0.67828 \end{bmatrix} \approx \begin{bmatrix} 3.56097 \\ 3.50445 \end{bmatrix} \qquad \mathbf{x}_3 = \frac{A\mathbf{x}_2}{\|A\mathbf{x}_2\|} \approx \frac{1}{4.99616} \begin{bmatrix} 3.56097 \\ 3.50445 \end{bmatrix} \approx \begin{bmatrix} 0.71274 \\ 0.70143 \end{bmatrix}$$

$$A\mathbf{x}_3 \approx \begin{bmatrix} 3 & 2 \\ 2 & 3 \end{bmatrix} \begin{bmatrix} 0.71274 \\ 0.70143 \end{bmatrix} \approx \begin{bmatrix} 3.54108 \\ 3.52976 \end{bmatrix} \qquad \mathbf{x}_4 = \frac{A\mathbf{x}_3}{\|A\mathbf{x}_3\|} \approx \frac{1}{4.99985} \begin{bmatrix} 3.54108 \\ 3.52976 \end{bmatrix} \approx \begin{bmatrix} 0.70824 \\ 0.70597 \end{bmatrix}$$

$$A\mathbf{x}_4 \approx \begin{bmatrix} 3 & 2 \\ 2 & 3 \end{bmatrix} \begin{bmatrix} 0.70824 \\ 0.70597 \end{bmatrix} \approx \begin{bmatrix} 3.53666 \\ 3.53440 \end{bmatrix} \qquad \mathbf{x}_5 = \frac{A\mathbf{x}_4}{\|A\mathbf{x}_4\|} \approx \frac{1}{4.99999} \begin{bmatrix} 3.53666 \\ 3.53440 \end{bmatrix} \approx \begin{bmatrix} 0.70733 \\ 0.70688 \end{bmatrix}$$

$$\lambda^{(1)} = (A\mathbf{x}_1) \cdot \mathbf{x}_1 = (A\mathbf{x}_1)^T \mathbf{x}_1 \approx \begin{bmatrix} 3.60555 & 3.32820 \end{bmatrix} \begin{bmatrix} 0.83205 \\ 0.55470 \end{bmatrix} \approx 4.84615$$

$$\lambda^{(2)} = (A\mathbf{x}_2) \cdot \mathbf{x}_2 = (A\mathbf{x}_2)^T \mathbf{x}_2 \approx \begin{bmatrix} 3.56097 & 3.50445 \end{bmatrix} \begin{bmatrix} 0.73480 \\ 0.67828 \end{bmatrix} \approx 4.99361$$

$$\lambda^{(3)} = (A\mathbf{x}_3) \cdot \mathbf{x}_3 = (A\mathbf{x}_3)^T \mathbf{x}_3 \approx \begin{bmatrix} 3.54108 & 3.52976 \end{bmatrix} \begin{bmatrix} 0.71274 \\ 0.70143 \end{bmatrix} \approx 4.99974$$

$$\lambda^{(4)} = (A\mathbf{x}_4) \cdot \mathbf{x}_4 = (A\mathbf{x}_4)^T \mathbf{x}_4 \approx \begin{bmatrix} 3.53666 & 3.53440 \end{bmatrix} \begin{bmatrix} 0.70824 \\ 0.70597 \end{bmatrix} \approx 4.99999$$

$$\lambda^{(5)} = (A\mathbf{x}_5) \cdot \mathbf{x}_5 = (A\mathbf{x}_5)^T \mathbf{x}_5 \approx \begin{bmatrix} 3.53576 & 3.53531 \end{bmatrix} \begin{bmatrix} 0.70733 \\ 0.70688 \end{bmatrix} \approx 5.00000$$

Thus, $\lambda^{(5)}$ approximates the dominant eigenvalue to five decimal place accuracy (the two "fives" are accidental) and $\mathbf{x}_5$ approximates the dominant eigenvector

$$\mathbf{v}_1 = \begin{bmatrix} \frac{1}{\sqrt{2}} \\ \frac{1}{\sqrt{2}} \end{bmatrix} \approx \begin{bmatrix} 0.707106781187\ldots \\ 0.707106781187\ldots \end{bmatrix}$$

correctly to three decimal place accuracy. ∎

THE POWER METHOD WITH MAXIMUM ENTRY SCALING

Next we will consider a variation of the power method in which each iterate is scaled to make its largest entry a 1, rather than being normalized. To describe the method we will use the notation $\max(\mathbf{x})$ to denote the maximum *absolute value* of the entries in $\mathbf{x}$. For example, if

$$\mathbf{x} = \begin{bmatrix} 5 \\ 3 \\ -7 \\ 2 \end{bmatrix}$$

then $\max(\mathbf{x}) = 7$. The following theorem is a useful variation of Theorem 5.4.2.

Theorem 5.4.3 *Let A be a symmetric $n \times n$ matrix with a positive dominant* * *eigenvalue λ. If $\mathbf{x}_0$ is a nonzero vector in R^n that is not orthogonal to the eigenspace corresponding to λ, then the sequence*

$$\mathbf{x}_0, \quad \mathbf{x}_1 = \frac{A\mathbf{x}_0}{\max(A\mathbf{x}_0)}, \quad \mathbf{x}_2 = \frac{A\mathbf{x}_1}{\max(A\mathbf{x}_1)}, \ldots, \quad \mathbf{x}_k = \frac{A\mathbf{x}_{k-1}}{\max(A\mathbf{x}_{k-1})}, \ldots \qquad (8)$$

converges to an eigenvector corresponding to λ, and the sequence

$$\frac{A\mathbf{x}_1 \cdot \mathbf{x}_1}{\mathbf{x}_1 \cdot \mathbf{x}_1}, \quad \frac{A\mathbf{x}_2 \cdot \mathbf{x}_2}{\mathbf{x}_2 \cdot \mathbf{x}_2}, \quad \frac{A\mathbf{x}_3 \cdot \mathbf{x}_3}{\mathbf{x}_3 \cdot \mathbf{x}_3}, \ldots, \quad \frac{A\mathbf{x}_k \cdot \mathbf{x}_k}{\mathbf{x}_k \cdot \mathbf{x}_k}, \ldots \qquad (9)$$

converges to λ.

Linear Algebra in History

The British physicist John Rayleigh won the Nobel prize in physics in 1904 for his discovery of the inert gas argon. Rayleigh also made fundamental discoveries in acoustics and optics, and his work in wave phenomena enabled him to give the first accurate explanation of why the sky is blue.

John William Strutt Rayleigh (1842–1919)

REMARK In the exercises we will ask you to show that (8) can be expressed in the alternative form

$$\mathbf{x}_0, \quad \mathbf{x}_1 = \frac{A\mathbf{x}_0}{\max(A\mathbf{x}_0)}, \quad \mathbf{x}_2 = \frac{A^2\mathbf{x}_0}{\max(A^2\mathbf{x}_0)}, \ldots, \quad \mathbf{x}_k = \frac{A^k\mathbf{x}_0}{\max(A^k\mathbf{x}_0)}, \ldots \ (10)$$

which expresses the iterates in terms of the initial vector $\mathbf{x}_0$.

We will omit the proof of this theorem, but if we accept that (8) converges to an eigenvector of A, then it is not hard to see why (9) converges to the dominant eigenvalue. For this purpose we note that each term in (9) is of the form

$$\frac{A\mathbf{x} \cdot \mathbf{x}}{\mathbf{x} \cdot \mathbf{x}} \qquad (11)$$

which is called a ***Rayleigh quotient*** of A. In the case where λ is an eigenvalue of A and $\mathbf{x}$ is a corresponding eigenvector, the Rayleigh quotient is

$$\frac{A\mathbf{x} \cdot \mathbf{x}}{\mathbf{x} \cdot \mathbf{x}} = \frac{\lambda\mathbf{x} \cdot \mathbf{x}}{\mathbf{x} \cdot \mathbf{x}} = \frac{\lambda(\mathbf{x} \cdot \mathbf{x})}{\mathbf{x} \cdot \mathbf{x}} = \lambda$$

Thus, if $\mathbf{x}_k$ converges to a dominant eigenvector $\mathbf{x}$, then it seems reasonable that

$$\frac{A\mathbf{x}_k \cdot \mathbf{x}_k}{\mathbf{x}_k \cdot \mathbf{x}_k} \quad \text{converges to} \quad \frac{A\mathbf{x} \cdot \mathbf{x}}{\mathbf{x} \cdot \mathbf{x}} = \lambda$$

which is the dominant eigenvalue.

Theorem 5.4.3 produces the following algorithm, called the ***power method with maximum entry scaling***.

The Power Method with Maximum Entry Scaling

Step 1. Choose an arbitrary nonzero vector $\mathbf{x}_0$.

Step 2. Compute $A\mathbf{x}_0$ and multiply it by the factor $1/\max(A\mathbf{x}_0)$ to obtain the first approximation $\mathbf{x}_1$ to a dominant eigenvector. Compute the Rayleigh quotient of $\mathbf{x}_1$ to obtain the first approximation to the dominant eigenvalue.

Step 3. Compute $A\mathbf{x}_1$ and scale it by the factor $1/\max(A\mathbf{x}_1)$ to obtain the second approximation $\mathbf{x}_2$ to a dominant eigenvector. Compute the Rayleigh quotient of $\mathbf{x}_2$ to obtain the second approximation to the dominant eigenvalue.

---

*As in Theorem 5.4.2, if the dominant eigenvalue is not positive, sequence (9) will still converge to the dominant eigenvalue, but sequence (8) may not converge to a *specific* dominant eigenvector. Nevertheless, each term of (8) will closely approximate *some* dominant eigenvector for sufficiently large values of k.

Step 4. Compute $A\mathbf{x}_2$ and scale it by the factor $1/\max(A\mathbf{x}_2)$ to obtain the third approximation $\mathbf{x}_3$ to a dominant eigenvector. Compute the Rayleigh quotient of $\mathbf{x}_3$ to obtain the third approximation to the dominant eigenvalue.

Continuing in this way will generate a sequence of better and better approximations to the dominant eigenvalue and a corresponding eigenvector.

REMARK One difference between the power method with Euclidean scaling and the power method with maximum entry scaling is that Euclidean scaling produces a sequence that approaches a *unit* dominant eigenvector, whereas maximum entry scaling produces a sequence that approaches a dominant eigenvector whose largest component is 1.

EXAMPLE 3
Example 2
Revisited Using
Maximum
Entry Scaling

Apply the power method with maximum entry scaling to

$$A = \begin{bmatrix} 3 & 2 \\ 2 & 3 \end{bmatrix} \quad \text{with} \quad \mathbf{x}_0 = \begin{bmatrix} 1 \\ 0 \end{bmatrix}$$

Stop at $\mathbf{x}_5$ and compare the resulting approximations to the exact values and to the approximations obtained in Example 2.

Solution We leave it for you to confirm that

$$A\mathbf{x}_0 = \begin{bmatrix} 3 & 2 \\ 2 & 3 \end{bmatrix}\begin{bmatrix} 1 \\ 0 \end{bmatrix} = \begin{bmatrix} 3 \\ 2 \end{bmatrix} \qquad \mathbf{x}_1 = \frac{A\mathbf{x}_0}{\max(A\mathbf{x}_0)} = \frac{1}{3}\begin{bmatrix} 3 \\ 2 \end{bmatrix} \approx \begin{bmatrix} 1.00000 \\ 0.66667 \end{bmatrix}$$

$$A\mathbf{x}_1 \approx \begin{bmatrix} 3 & 2 \\ 2 & 3 \end{bmatrix}\begin{bmatrix} 1.00000 \\ 0.66667 \end{bmatrix} \approx \begin{bmatrix} 4.33333 \\ 4.00000 \end{bmatrix} \qquad \mathbf{x}_2 = \frac{A\mathbf{x}_1}{\max(A\mathbf{x}_1)} \approx \frac{1}{4.33333}\begin{bmatrix} 4.33333 \\ 4.00000 \end{bmatrix} \approx \begin{bmatrix} 1.00000 \\ 0.92308 \end{bmatrix}$$

$$A\mathbf{x}_2 \approx \begin{bmatrix} 3 & 2 \\ 2 & 3 \end{bmatrix}\begin{bmatrix} 1.00000 \\ 0.92308 \end{bmatrix} \approx \begin{bmatrix} 4.84615 \\ 4.76923 \end{bmatrix} \qquad \mathbf{x}_3 = \frac{A\mathbf{x}_2}{\max(A\mathbf{x}_2)} \approx \frac{1}{4.84615}\begin{bmatrix} 4.84615 \\ 4.76923 \end{bmatrix} \approx \begin{bmatrix} 1.00000 \\ 0.98413 \end{bmatrix}$$

$$A\mathbf{x}_3 \approx \begin{bmatrix} 3 & 2 \\ 2 & 3 \end{bmatrix}\begin{bmatrix} 1.00000 \\ 0.98413 \end{bmatrix} \approx \begin{bmatrix} 4.96825 \\ 4.95238 \end{bmatrix} \qquad \mathbf{x}_4 = \frac{A\mathbf{x}_3}{\max(A\mathbf{x}_3)} \approx \frac{1}{4.96825}\begin{bmatrix} 4.96825 \\ 4.95238 \end{bmatrix} \approx \begin{bmatrix} 1.00000 \\ 0.99681 \end{bmatrix}$$

$$A\mathbf{x}_4 \approx \begin{bmatrix} 3 & 2 \\ 2 & 3 \end{bmatrix}\begin{bmatrix} 1.00000 \\ 0.99681 \end{bmatrix} \approx \begin{bmatrix} 4.99361 \\ 4.99042 \end{bmatrix} \qquad \mathbf{x}_5 = \frac{A\mathbf{x}_4}{\max(A\mathbf{x}_4)} \approx \frac{1}{4.99361}\begin{bmatrix} 4.99361 \\ 4.99042 \end{bmatrix} \approx \begin{bmatrix} 1.00000 \\ 0.99936 \end{bmatrix}$$

$$\lambda^{(1)} = \frac{A\mathbf{x}_1 \cdot \mathbf{x}_1}{\mathbf{x}_1 \cdot \mathbf{x}_1} = \frac{(A\mathbf{x}_1)^T\mathbf{x}_1}{\mathbf{x}_1^T\mathbf{x}_1} \approx \frac{7.00000}{1.44444} \approx 4.84615$$

$$\lambda^{(2)} = \frac{A\mathbf{x}_2 \cdot \mathbf{x}_2}{\mathbf{x}_2 \cdot \mathbf{x}_2} = \frac{(A\mathbf{x}_2)^T\mathbf{x}_2}{\mathbf{x}_2^T\mathbf{x}_2} \approx \frac{9.24852}{1.85207} \approx 4.99361$$

$$\lambda^{(3)} = \frac{A\mathbf{x}_3 \cdot \mathbf{x}_3}{\mathbf{x}_3 \cdot \mathbf{x}_3} = \frac{(A\mathbf{x}_3)^T\mathbf{x}_3}{\mathbf{x}_3^T\mathbf{x}_3} \approx \frac{9.84203}{1.96851} \approx 4.99974$$

$$\lambda^{(4)} = \frac{A\mathbf{x}_4 \cdot \mathbf{x}_4}{\mathbf{x}_4 \cdot \mathbf{x}_4} = \frac{(A\mathbf{x}_4)^T\mathbf{x}_4}{\mathbf{x}_4^T\mathbf{x}_4} \approx \frac{9.96808}{1.99362} \approx 4.99999$$

$$\lambda^{(5)} = \frac{A\mathbf{x}_5 \cdot \mathbf{x}_5}{\mathbf{x}_5 \cdot \mathbf{x}_5} = \frac{(A\mathbf{x}_5)^T\mathbf{x}_5}{\mathbf{x}_5^T\mathbf{x}_5} \approx \frac{9.99360}{1.99872} \approx 5.00000$$

Thus, $\lambda^{(5)}$ approximates the dominant eigenvalue correctly to five decimal places and $\mathbf{x}_5$ closely approximates the dominant eigenvector

$$\mathbf{x} = \begin{bmatrix} 1 \\ 1 \end{bmatrix}$$

that results by taking $t = 1$ in (6). ∎

RATE OF CONVERGENCE   If A is a symmetric matrix whose distinct eigenvalues can be arranged so that

$$|\lambda_1| > |\lambda_2| \geq |\lambda_3| \geq \cdots \geq |\lambda_k|$$

then the "rate" at which the Rayleigh quotients converge to the dominant eigenvalue λ_1 depends on the ratio $|\lambda_1|/|\lambda_2|$; that is, the convergence is slow when this ratio is near 1 and rapid when it is large—the greater the ratio, the more rapid the convergence. For example, if A is a 2×2 symmetric matrix, then the greater the ratio $|\lambda_1|/|\lambda_2|$, the greater the disparity between the scaling effects of λ_1 and λ_2 in Figure 5.4.1, and hence the greater the effect that multiplication by A has on pulling the iterates toward the eigenspace of λ_1. Indeed, the rapid convergence in Example 3 is due to the fact that $|\lambda_1|/|\lambda_2| = 5/1 = 5$, which is quite a large ratio. In cases where the ratio is close to 1, the convergence of the power method may be so slow that other methods have to be used.

STOPPING PROCEDURES   If λ is the exact value of the dominant eigenvalue, and if a power method produces the approximation $\lambda^{(k)}$ at the kth iteration, then we call

$$\left| \frac{\lambda - \lambda^{(k)}}{\lambda} \right| \tag{12}$$

the ***relative error*** in $\lambda^{(k)}$. If this is expressed as a percentage, then it is called the ***percentage error*** in $\lambda^{(k)}$. For example, if $\lambda = 5$ and the approximation after three iterations is $\lambda^{(3)} = 5.1$, then

$$\text{relative error in } \lambda^{(3)} = \left| \frac{\lambda - \lambda^{(3)}}{\lambda} \right| = \left| \frac{5 - 5.1}{5} \right| = |-0.02| = 0.02$$

$$\text{percentage error in } \lambda^{(3)} = 0.02 \times 100\% = 2\%$$

In applications one usually knows the relative error E that can be tolerated in the dominant eigenvalue, and the objective is to stop computing iterates once the relative error in the approximation to that eigenvalue is less than E. However, there is a problem in computing the relative error from (12) since the eigenvalue λ is unknown. To circumvent this problem, it is usual to estimate λ by $\lambda^{(k)}$ and stop the computations when

$$\left| \frac{\lambda^{(k)} - \lambda^{(k-1)}}{\lambda^{(k)}} \right| < E \tag{13}$$

(where $k \geq 2$). The quantity on the left side of (13) is called the ***estimated relative error*** in $\lambda^{(k)}$ and its percentage form is called the ***estimated percentage error*** in $\lambda^{(k)}$.

EXAMPLE 4
Estimated
Relative Error

For the computations in Example 3, find the smallest value of k for which the estimated percentage error in $\lambda^{(k)}$ is less than 0.1%.

Solution   The percentage errors of the approximations in Example 3 are as follows:

	Approximation	**Relative Error**	**Percentage Error**
$\lambda^{(2)}$:	$\left\| \dfrac{\lambda^{(2)} - \lambda^{(1)}}{\lambda^{(2)}} \right\| \approx \left\| \dfrac{4.99361 - 4.84615}{4.99361} \right\|$	≈ 0.02953	$= 2.953\%$
$\lambda^{(3)}$:	$\left\| \dfrac{\lambda^{(3)} - \lambda^{(2)}}{\lambda^{(3)}} \right\| \approx \left\| \dfrac{4.99974 - 4.99361}{4.99974} \right\|$	≈ 0.00123	$= 0.123\%$
$\lambda^{(4)}$:	$\left\| \dfrac{\lambda^{(4)} - \lambda^{(3)}}{\lambda^{(4)}} \right\| \approx \left\| \dfrac{4.99999 - 4.99974}{4.99999} \right\|$	≈ 0.00005	$= 0.005\%$
$\lambda^{(5)}$:	$\left\| \dfrac{\lambda^{(5)} - \lambda^{(4)}}{\lambda^{(5)}} \right\| \approx \left\| \dfrac{5.00000 - 4.99999}{5.00000} \right\|$	≈ 0.00000	$= 0\%$

Thus, $\lambda^{(4)} = 4.99999$ is the first approximation whose estimated percentage error is less than 0.1%. ∎

REMARK A rule for deciding when to stop an iterative process is called a *stopping procedure*. In the exercises, we will discuss stopping procedures for the power method that are based on the dominant eigenvector rather than the dominant eigenvalue.

AN APPLICATION OF THE POWER METHOD TO INTERNET SEARCHES

The power method has recently been used to develop a new kind of algorithm for searching the Internet. This algorithm, called the *PageRank algorithm*, is implemented in the *Google* [*] *search engine*, developed by Stanford graduates Larry Page and Sergey Brin. Whereas most search engines check page titles and content to determine a list of relevant sites, PageRank determines which pages are referenced by other pages, creates a *link matrix* to describe the referencing structure, and then uses the dominant eigenvector of a related matrix to create an ordered list of the sites that appear to be most relevant to the search.

Here is the basic idea behind the Google search engine:

- When a user asks Google to search for a word or phrase, Google's first step is to use a standard text-based search engine to find an initial collection S_0 of relevant sites, usually a few hundred or so.

- Since words can have multiple meanings, the set S_0 will typically contain some irrelevant sites, and since words can have synonyms, the set S_0 will likely omit important sites that use different terminology for the search words. Recognizing this, Google looks for sites that reference (or link to) those in S_0 and then expands S_0 to a larger set S that includes these sites. The underlying assumption is that the set S will contain the most important sites that are related to the search words; we call this the *search set*.

- Since the search set may contain thousands of sites, the main task of the search engine is to order those sites according to their likely relevance to the search words. It is in this part of the Google search that the power method and the PageRank algorithm come into play.

To explain the PageRank algorithm, suppose that the search set S contains n sites, and define the *adjacency matrix* for S to be the $n \times n$ matrix $A = [a_{ij}]$ in which $a_{ij} = 1$ if site i references site j and $a_{ij} = 0$ if it does not. We will make the convention that no site references itself, so the diagonal entries of A are zeros.

EXAMPLE 5 Here is a typical adjacency matrix for a search set with four Internet sites:

$$
A = \begin{array}{c} \quad \\ \text{Referenced Site} \\ \begin{array}{cccc} 1 & 2 & 3 & 4 \end{array} \\ \begin{bmatrix} 0 & 0 & 1 & 1 \\ 1 & 0 & 0 & 0 \\ 1 & 0 & 0 & 1 \\ 1 & 1 & 1 & 0 \end{bmatrix} \begin{array}{c} 1 \\ 2 \\ 3 \\ 4 \end{array} \end{array} \quad \textbf{Referencing Site} \tag{14}
$$

Thus, site 1 references sites 3 and 4, site 2 references site 1, and so forth. ∎

---

[*] The term *google* is a variation of the word *googol*, which stands for the number 10^{100} (1 followed by 100 zeros). This term was invented by the American mathematician Edward Kasner (1878–1955) in 1938, and the story goes that it came about when Kasner asked his eight-year-old nephew to give a name to a really big number—he responded with "googol." Kasner then went on to define a *googolplex* to be 10^{googol} (1 followed by googol zeros).

There are two basic roles that a site can play in the search process—the site may be a ***hub***, meaning that it *references* many other sites, or it may be an ***authority***, meaning that it is *referenced by* many other sites. A given site will typically have both hub and authority properties in that it will both reference and be referenced.

In general, if A is an adjacency matrix for n Internet sites, then the column sums of A measure the authority aspect of the sites and the row sums of A measure their hub aspect. For example, the column sums of (14) are 3, 1, 2, and 2, which means that site 1 is referenced by three other sites, site 2 is referenced by one other site, and so forth. Similarly, the row sums of (14) are 2, 1, 2, and 3, so site 1 references two other sites, site 2 references one other site, and so forth.

Accordingly, if A is an adjacency matrix, then we call the vector $\mathbf{h}_0$ of row sums of A the ***initial hub vector*** of A, and we call the vector $\mathbf{a}_0$ of column sums of A the ***initial authority vector*** of A. Alternatively, we can think of $\mathbf{a}_0$ as the vector of *row* sums of A^T, which turns out to be more convenient for computations. The entries in the hub vector are called ***hub weights*** and those in the authority vector ***authority weights***.

EXAMPLE 6
Initial Hub and Authority Vectors of an Adjacency Matrix

Find the initial hub and authority vectors for the adjacency matrix A in Example 5.

Solution   The row sums of A yield the initial hub vector

$$
\mathbf{h}_0 = \begin{bmatrix} 2 \\ 1 \\ 2 \\ 3 \end{bmatrix} \begin{matrix} \textbf{Site 1} \\ \textbf{Site 2} \\ \textbf{Site 3} \\ \textbf{Site 4} \end{matrix} \tag{15}
$$

and the row sums of A^T (the column sums of A) yield the initial authority vector

$$
\mathbf{a}_0 = \begin{bmatrix} 3 \\ 1 \\ 2 \\ 2 \end{bmatrix} \begin{matrix} \textbf{Site 1} \\ \textbf{Site 2} \\ \textbf{Site 3} \\ \textbf{Site 4} \end{matrix} \tag{16}
$$

■

The link counting in Example 6 suggests that site 4 is the major hub and site 1 is the greatest authority. However, counting links does not tell the whole story; for example, it seems reasonable that if site 1 is to be considered the greatest authority, then more weight should be given to hubs that link to that site, and if site 4 is to be considered the major hub, then more weight should be given to sites that it links to. Thus, there is an interaction between hubs and authorities that needs to be accounted for in the search process. Accordingly, once Google has calculated the initial authority vector $\mathbf{a}_0$, it then uses the information in that vector to create new hub and authority vectors $\mathbf{h}_1$ and $\mathbf{a}_1$ using the formulas

$$
\mathbf{h}_1 = \frac{A\mathbf{a}_0}{\|A\mathbf{a}_0\|} \quad \text{and} \quad \mathbf{a}_1 = \frac{A^T\mathbf{h}_1}{\|A^T\mathbf{h}_1\|} \tag{17}
$$

The numerators in these formulas do the weighting, and the normalization serves to control the size of the entries. To understand how the numerators accomplish the weighting, view the product $A\mathbf{a}_0$ as a linear combination of the column vectors of A with coefficients from $\mathbf{a}_0$. For example, with the adjacency matrix in Example 5 and the authority vector calculated in Example 6 we have

$$
\begin{matrix} & \textbf{Referenced Site} \\ & \textbf{1\ \ 2\ \ 3\ \ 4} \end{matrix}
$$

$$
A\mathbf{a}_0 = \begin{bmatrix} 0 & 0 & 1 & 1 \\ 1 & 0 & 0 & 0 \\ 1 & 0 & 0 & 1 \\ 1 & 1 & 1 & 0 \end{bmatrix} \begin{bmatrix} 3 \\ 1 \\ 2 \\ 2 \end{bmatrix} = 3\begin{bmatrix} 0 \\ 1 \\ 1 \\ 1 \end{bmatrix} + 1\begin{bmatrix} 0 \\ 0 \\ 0 \\ 1 \end{bmatrix} + 2\begin{bmatrix} 1 \\ 0 \\ 0 \\ 1 \end{bmatrix} + 2\begin{bmatrix} 1 \\ 0 \\ 1 \\ 0 \end{bmatrix} = \begin{bmatrix} 4 \\ 3 \\ 5 \\ 6 \end{bmatrix} \begin{matrix} \textbf{Site 1} \\ \textbf{Site 2} \\ \textbf{Site 3} \\ \textbf{Site 4} \end{matrix}
$$

Thus, we see that the links to each referenced site are weighted by the authority values in $\mathbf{a}_0$. To control the size of the entries, Google normalizes $A\mathbf{a}_0$ to produce the updated hub vector

$$\mathbf{h}_1 = \frac{A\mathbf{a}_0}{\|A\mathbf{a}_0\|} = \frac{1}{\sqrt{86}} \begin{bmatrix} 4 \\ 3 \\ 5 \\ 6 \end{bmatrix} \approx \begin{bmatrix} 0.43133 \\ 0.32350 \\ 0.53916 \\ 0.64700 \end{bmatrix} \begin{matrix} \text{Site 1} \\ \text{Site 2} \\ \text{Site 3} \\ \text{Site 4} \end{matrix} \quad \textbf{New Hub Weights}$$

The new hub vector $\mathbf{h}_1$ can now be used to update the authority vector using Formula (17). The product $A^T\mathbf{h}_1$ performs the weighting, and the normalization controls the size:

$$A^T\mathbf{h}_1 \approx \overset{\begin{matrix}\text{Referencing Site}\\ 1\ \ 2\ \ 3\ \ 4\end{matrix}}{\begin{bmatrix} 0 & 1 & 1 & 1 \\ 0 & 0 & 0 & 1 \\ 1 & 0 & 0 & 1 \\ 1 & 0 & 1 & 0 \end{bmatrix}} \begin{bmatrix} 0.43133 \\ 0.32350 \\ 0.53916 \\ 0.64700 \end{bmatrix} \approx 0.43133 \begin{bmatrix} 0 \\ 0 \\ 1 \\ 1 \end{bmatrix} + 0.32350 \begin{bmatrix} 1 \\ 0 \\ 0 \\ 0 \end{bmatrix} + 0.53916 \begin{bmatrix} 1 \\ 0 \\ 0 \\ 1 \end{bmatrix} + 0.64700 \begin{bmatrix} 1 \\ 1 \\ 1 \\ 0 \end{bmatrix} \approx \begin{bmatrix} 1.50966 \\ 0.64700 \\ 1.07833 \\ 0.97049 \end{bmatrix} \begin{matrix} \text{Site 1} \\ \text{Site 2} \\ \text{Site 3} \\ \text{Site 4} \end{matrix}$$

$$\mathbf{a}_1 = \frac{A^T\mathbf{h}_1}{\|A^T\mathbf{h}_1\|} \approx \frac{1}{2.19142} \begin{bmatrix} 1.50966 \\ 0.64700 \\ 1.07833 \\ 0.97049 \end{bmatrix} \approx \begin{bmatrix} 0.68889 \\ 0.29524 \\ 0.49207 \\ 0.44286 \end{bmatrix} \begin{matrix} \text{Site 1} \\ \text{Site 2} \\ \text{Site 3} \\ \text{Site 4} \end{matrix} \quad \textbf{New Authority Weights}$$

Once the updated hub and authority vectors, $\mathbf{h}_1$ and $\mathbf{a}_1$, are obtained, the Google engine repeats the process and computes a succession of hub and authority vectors, thereby generating the interrelated sequences

$$\mathbf{h}_1 = \frac{A\mathbf{a}_0}{\|A\mathbf{a}_0\|}, \quad \mathbf{h}_2 = \frac{A\mathbf{a}_1}{\|A\mathbf{a}_1\|}, \quad \mathbf{h}_3 = \frac{A\mathbf{a}_2}{\|A\mathbf{a}_2\|}, \ldots, \quad \mathbf{h}_k = \frac{A\mathbf{a}_{k-1}}{\|A\mathbf{a}_{k-1}\|}, \ldots \quad (18)$$

$$\mathbf{a}_0, \quad \mathbf{a}_1 = \frac{A^T\mathbf{h}_1}{\|A^T\mathbf{h}_1\|}, \quad \mathbf{a}_2 = \frac{A^T\mathbf{h}_2}{\|A^T\mathbf{h}_2\|}, \quad \mathbf{a}_3 = \frac{A^T\mathbf{h}_3}{\|A^T\mathbf{h}_3\|}, \ldots, \quad \mathbf{a}_k = \frac{A^T\mathbf{h}_k}{\|A^T\mathbf{h}_k\|}, \ldots \quad (19)$$

However, each of these sequences is a power sequence in disguise. For example, if we substitute the expression for $\mathbf{h}_k$ into the expression for $\mathbf{a}_k$, then we obtain

$$\mathbf{a}_k = \frac{A^T\mathbf{h}_k}{\|A^T\mathbf{h}_k\|} = \frac{A^T\left(\dfrac{A\mathbf{a}_{k-1}}{\|A\mathbf{a}_{k-1}\|}\right)}{\left\|A^T\left(\dfrac{A\mathbf{a}_{k-1}}{\|A\mathbf{a}_{k-1}\|}\right)\right\|} = \frac{(A^TA)\mathbf{a}_{k-1}}{\|(A^TA)\mathbf{a}_{k-1}\|}$$

which means that we can rewrite (19) as

$$\mathbf{a}_0, \quad \mathbf{a}_1 = \frac{(A^TA)\mathbf{a}_0}{\|(A^TA)\mathbf{a}_0\|}, \quad \mathbf{a}_2 = \frac{(A^TA)\mathbf{a}_1}{\|(A^TA)\mathbf{a}_1\|}, \ldots, \quad \mathbf{a}_k = \frac{(A^TA)\mathbf{a}_{k-1}}{\|(A^TA)\mathbf{a}_{k-1}\|}, \ldots \quad (20)$$

Similarly, sequence (18) can be expressed as

$$\mathbf{h}_1 = \frac{A\mathbf{a}_0}{\|A\mathbf{a}_0\|}, \quad \mathbf{h}_2 = \frac{(AA^T)\mathbf{h}_1}{\|(AA^T)\mathbf{h}_1\|}, \ldots, \quad \mathbf{h}_k = \frac{(AA^T)\mathbf{h}_{k-1}}{\|(AA^T)\mathbf{h}_1\|}, \ldots \quad (21)$$

REMARK   The matrices AA^T and A^TA are symmetric, and in the exercises we will ask you to show that they have positive dominant eigenvalues (Exercise P1). Thus, Theorem 5.4.2 ensures that (20) and (21) will converge to dominant eigenvectors of A^TA and AA^T, respectively. The entries in those eigenvectors are the authority and hub weights that Google uses to rank the search sites in order of importance as hubs and authorities.

EXAMPLE 7
A Google
Search Using
the PageRank
Algorithm

Suppose that the Google search engine produces 10 Internet sites in its search set and that the adjacency matrix for those sites is

Referenced Site

$$A = \begin{bmatrix} 0 & 1 & 0 & 0 & 1 & 0 & 0 & 1 & 0 & 0 \\ 0 & 0 & 0 & 0 & 1 & 0 & 0 & 0 & 0 & 0 \\ 0 & 0 & 0 & 0 & 1 & 0 & 0 & 0 & 0 & 0 \\ 0 & 0 & 0 & 0 & 0 & 1 & 1 & 0 & 0 & 0 \\ 0 & 0 & 0 & 0 & 0 & 0 & 0 & 1 & 0 & 0 \\ 0 & 1 & 1 & 1 & 1 & 0 & 0 & 1 & 0 & 1 \\ 0 & 0 & 0 & 0 & 0 & 0 & 0 & 0 & 0 & 1 \\ 0 & 0 & 0 & 0 & 1 & 0 & 0 & 0 & 0 & 0 \\ 0 & 0 & 0 & 0 & 0 & 1 & 0 & 0 & 0 & 0 \\ 0 & 0 & 0 & 0 & 0 & 1 & 0 & 0 & 0 & 0 \end{bmatrix} \begin{matrix} 1 \\ 2 \\ 3 \\ 4 \\ 5 \\ 6 \\ 7 \\ 8 \\ 9 \\ 10 \end{matrix}$$

with columns labeled $1\ 2\ 3\ 4\ 5\ 6\ 7\ 8\ 9\ 10$ and **Referencing Site** at the right.

Use the PageRank algorithm to rank the sites in decreasing order of authority for the Google search engine.

Solution   We will take $\mathbf{a}_0$ to be the normalized vector of column sums of A, and then we will compute the iterates in (20) until the authority vectors seem to stabilize. We leave it for you to show that

$$\mathbf{a}_0 = \frac{1}{\sqrt{54}} \begin{bmatrix} 0 \\ 2 \\ 1 \\ 1 \\ 5 \\ 3 \\ 1 \\ 3 \\ 0 \\ 2 \end{bmatrix} \approx \begin{bmatrix} 0 \\ 0.27217 \\ 0.13608 \\ 0.13608 \\ 0.68041 \\ 0.40825 \\ 0.13608 \\ 0.40825 \\ 0 \\ 0.27217 \end{bmatrix}$$

and that

$$(A^T A)\mathbf{a}_0 \approx \begin{bmatrix} 0 & 0 & 0 & 0 & 0 & 0 & 0 & 0 & 0 & 0 \\ 0 & 2 & 1 & 1 & 2 & 0 & 0 & 2 & 0 & 1 \\ 0 & 1 & 1 & 1 & 1 & 0 & 0 & 1 & 0 & 1 \\ 0 & 1 & 1 & 1 & 1 & 0 & 0 & 1 & 0 & 1 \\ 0 & 2 & 1 & 1 & 5 & 0 & 0 & 2 & 0 & 1 \\ 0 & 0 & 0 & 0 & 0 & 3 & 1 & 0 & 0 & 0 \\ 0 & 0 & 0 & 0 & 0 & 1 & 1 & 0 & 0 & 0 \\ 0 & 2 & 1 & 1 & 2 & 0 & 0 & 3 & 0 & 1 \\ 0 & 0 & 0 & 0 & 0 & 0 & 0 & 0 & 0 & 0 \\ 0 & 1 & 1 & 1 & 1 & 0 & 0 & 1 & 0 & 2 \end{bmatrix} \begin{bmatrix} 0 \\ 0.27217 \\ 0.13608 \\ 0.13608 \\ 0.68041 \\ 0.40825 \\ 0.13608 \\ 0.40825 \\ 0 \\ 0.27217 \end{bmatrix} \approx \begin{bmatrix} 0 \\ 3.26599 \\ 1.90516 \\ 1.90516 \\ 5.30723 \\ 1.36083 \\ 0.54433 \\ 3.67423 \\ 0 \\ 2.17732 \end{bmatrix}$$

Thus,

$$
\mathbf{a}_1 = \frac{(A^TA)\mathbf{a}_0}{\left\|(A^TA)\mathbf{a}_0\right\|} \approx \frac{1}{8.15362}
\begin{bmatrix}
0 \\
3.26599 \\
1.90516 \\
1.90516 \\
5.30723 \\
1.36083 \\
0.54433 \\
3.67423 \\
0 \\
2.17732
\end{bmatrix}
\approx
\begin{bmatrix}
0 \\
0.40056 \\
0.23366 \\
0.23366 \\
0.65090 \\
0.16690 \\
0.06676 \\
0.45063 \\
0 \\
0.26704
\end{bmatrix}
$$

Continuing in this way yields the following authority iterates:

$$
\mathbf{a}_0 \quad \mathbf{a}_1 = \frac{(A^TA)\mathbf{a}_0}{\left\|(A^TA)\mathbf{a}_0\right\|} \quad \mathbf{a}_2 = \frac{(A^TA)\mathbf{a}_1}{\left\|(A^TA)\mathbf{a}_1\right\|} \quad \mathbf{a}_3 = \frac{(A^TA)\mathbf{a}_2}{\left\|(A^TA)\mathbf{a}_2\right\|} \quad \mathbf{a}_4 = \frac{(A^TA)\mathbf{a}_3}{\left\|(A^TA)\mathbf{a}_3\right\|} \cdots \mathbf{a}_9 = \frac{(A^TA)\mathbf{a}_8}{\left\|(A^TA)\mathbf{a}_8\right\|} \quad \mathbf{a}_{10} = \frac{(A^TA)\mathbf{a}_9}{\left\|(A^TA)\mathbf{a}_9\right\|}
$$

$\mathbf{a}_0$	$\mathbf{a}_1$	$\mathbf{a}_2$	$\mathbf{a}_3$	$\mathbf{a}_4$	$\mathbf{a}_9$	$\mathbf{a}_{10}$	
0	0	0	0	0	0	0	**Site 1**
0.27217	0.40056	0.41652	0.41918	0.41973	0.41990	0.41990	**Site 2**
0.13608	0.23366	0.24917	0.25233	0.25309	0.25337	0.25337	**Site 3**
0.13608	0.23366	0.24917	0.25233	0.25309	0.25337	0.25337	**Site 4**
0.68041	0.65090	0.63407	0.62836	0.62665	0.62597	0.62597	**Site 5**
0.40825	0.16690	0.06322	0.02372	0.00889	0.00007	0.00002	**Site 6**
0.13608	0.06676	0.02603	0.00981	0.00368	0.00003	0.00001	**Site 7**
0.40825	0.45063	0.46672	0.47050	0.47137	0.47165	0.47165	**Site 8**
0	0	0	0	0	0	0	**Site 9**
0.27217	0.26704	0.27892	0.28300	0.28416	0.28460	0.28460	**Site 10**

The small changes between $\mathbf{a}_9$ and $\mathbf{a}_{10}$ suggest that the iterates have stabilized near a dominant eigenvector of A^TA. From the entries in $\mathbf{a}_{10}$ we conclude that sites 1, 6, 7, and 9 are probably irrelevant to the search and that the remaining sites should be searched in the order

site 5, site 8, site 2, site 10, sites 3 and 4 (a tie) ∎

VARIATIONS OF THE POWER METHOD

Although Theorems 5.4.2 and 5.4.3 are stated for symmetric matrices, they hold for certain nonsymmetric matrices as well. For example, we will show later in the text that these theorems are true for any $n \times n$ matrix A that has n linearly independent eigenvectors and a dominant eigenvalue and for stochastic matrices.

Indeed, Theorem 5.1.3 is essentially a statement about convergence of power sequences. More precisely, it can be proved that every stochastic matrix has a dominant eigenvalue of $\lambda = 1$ and that every regular stochastic matrix has a unique dominant eigenvector that is also a probability vector. Thus, part (*b*) of Theorem 5.1.3 for Markov chains is just a statement of the fact that the power sequence

$$\mathbf{x}_0, \ P\mathbf{x}_0, \ P^2\mathbf{x}_0, \ldots, \ P^k\mathbf{x}_0, \ldots$$

converges to the (probability) eigenvector $\mathbf{q}$ associated with the eigenvalue $\lambda = 1$.

In the exercises we will outline a technique, called the ***inverse power method***, that can be used to approximate the eigenvalue of smallest absolute value, and a technique, called ***shifting***, that can be used to approximate intermediate eigenvalues. In applications, the power method, inverse power method, and shifting are sometimes used for sparse matrices and in simple problems, but the most widely used methods are based on a result called the ***QR-algorithm***. A description of that algorithm can be found in many numerical analysis texts.

Exercise Set 5.4

In Exercises 1 and 2, the distinct eigenvalues of a matrix A are given. Determine whether A has a dominant eigenvalue, and if so, find it.

1. (a) $\lambda_1 = 7, \lambda_2 = 3, \lambda_3 = -8, \lambda_4 = 1$
   (b) $\lambda_1 = -5, \lambda_2 = 3, \lambda_3 = 2, \lambda_4 = 5$

2. (a) $\lambda_1 = 1, \lambda_2 = 0, \lambda_3 = -3, \lambda_4 = 2$
   (b) $\lambda_1 = -3, \lambda_2 = -2, \lambda_3 = -1, \lambda_4 = 3$

In Exercises 3 and 4, a matrix A and several terms in a normalized power sequence for A are given. Show that A has a dominant eigenvalue, and compare its exact value to the approximations in sequence (2) of Theorem 5.4.2. Find the exact unit eigenvector that the terms in sequence (1) of Theorem 5.4.2 are approaching.

3. $A = \begin{bmatrix} -7 & -12 \\ 8 & 13 \end{bmatrix}$; $\mathbf{x}_0 = \begin{bmatrix} 1 \\ 0 \end{bmatrix}$, $\mathbf{x}_1 \approx \begin{bmatrix} -0.6585 \\ 0.7526 \end{bmatrix}$,

$\mathbf{x}_2 \approx \begin{bmatrix} -0.6996 \\ 0.7145 \end{bmatrix}$, $\mathbf{x}_3 \approx \begin{bmatrix} -0.7057 \\ 0.7085 \end{bmatrix}$, $\mathbf{x}_4 \approx \begin{bmatrix} -0.7068 \\ 0.7074 \end{bmatrix}$

4. $A = \begin{bmatrix} 1 & -3 \\ -3 & 5 \end{bmatrix}$; $\mathbf{x}_0 = \begin{bmatrix} \frac{1}{\sqrt{2}} \\ \frac{1}{\sqrt{2}} \end{bmatrix}$, $\mathbf{x}_1 \approx \begin{bmatrix} -0.7071 \\ 0.7071 \end{bmatrix}$,

$\mathbf{x}_2 \approx \begin{bmatrix} -0.4472 \\ 0.8944 \end{bmatrix}$, $\mathbf{x}_3 \approx \begin{bmatrix} -0.4741 \\ 0.8805 \end{bmatrix}$, $\mathbf{x}_4 \approx \begin{bmatrix} -0.4717 \\ 0.8818 \end{bmatrix}$

In Exercises 5 and 6, a matrix A and several terms in a power sequence for A with maximum entry scaling are given. Show that A has a dominant eigenvalue, and compare its exact value to the approximations produced by the Rayleigh quotients in sequence (9) of Theorem 5.4.3. Find the exact eigenvector that the terms in sequence (8) of Theorem 5.4.3 are approaching.

5. $A = \begin{bmatrix} 5 & -1 \\ -1 & -1 \end{bmatrix}$; $\mathbf{x}_0 = \begin{bmatrix} 0 \\ 1 \end{bmatrix}$, $\mathbf{x}_1 = \begin{bmatrix} 1 \\ 1 \end{bmatrix}$, $\mathbf{x}_2 = \begin{bmatrix} 1 \\ -0.5 \end{bmatrix}$,

$\mathbf{x}_3 = \begin{bmatrix} 1 \\ -0.091 \end{bmatrix}$, $\mathbf{x}_4 = \begin{bmatrix} 1 \\ -0.179 \end{bmatrix}$

6. $A = \begin{bmatrix} 3.300 & -2.200 \\ -2.200 & 3.300 \end{bmatrix}$; $\mathbf{x}_0 = \begin{bmatrix} 1 \\ 0 \end{bmatrix}$, $\mathbf{x}_1 \approx \begin{bmatrix} 1 \\ -0.667 \end{bmatrix}$,

$\mathbf{x}_2 \approx \begin{bmatrix} 1 \\ -0.923 \end{bmatrix}$, $\mathbf{x}_3 \approx \begin{bmatrix} 1 \\ -0.984 \end{bmatrix}$, $\mathbf{x}_4 \approx \begin{bmatrix} 1 \\ -0.997 \end{bmatrix}$

In Exercises 7 and 8, a matrix A (with a dominant eigenvalue) and a sequence $\mathbf{x}_0, A\mathbf{x}_0, \ldots, A^5\mathbf{x}_0$ are given. Use (9) and (10) to approximate the dominant eigenvalue and a corresponding eigenvector.

7. $A = \begin{bmatrix} 1 & 2 \\ 2 & 1 \end{bmatrix}$; $\mathbf{x}_0 = \begin{bmatrix} 1 \\ 0 \end{bmatrix}$, $A\mathbf{x}_0 = \begin{bmatrix} 1 \\ 2 \end{bmatrix}$, $A^2\mathbf{x}_0 = \begin{bmatrix} 5 \\ 4 \end{bmatrix}$,

$A^3\mathbf{x}_0 = \begin{bmatrix} 13 \\ 14 \end{bmatrix}$, $A^4\mathbf{x}_0 = \begin{bmatrix} 41 \\ 40 \end{bmatrix}$, $A^5\mathbf{x}_0 = \begin{bmatrix} 121 \\ 122 \end{bmatrix}$

8. $A = \begin{bmatrix} 1 & 2 \\ 2 & 1 \end{bmatrix}$; $\mathbf{x}_0 = \begin{bmatrix} 0 \\ 1 \end{bmatrix}$, $A\mathbf{x}_0 = \begin{bmatrix} 2 \\ 1 \end{bmatrix}$, $A^2\mathbf{x}_0 = \begin{bmatrix} 4 \\ 5 \end{bmatrix}$,

$A^3\mathbf{x}_0 = \begin{bmatrix} 14 \\ 13 \end{bmatrix}$, $A^4\mathbf{x}_0 = \begin{bmatrix} 40 \\ 41 \end{bmatrix}$, $A^5\mathbf{x}_0 = \begin{bmatrix} 122 \\ 121 \end{bmatrix}$

9. Consider matrices

$$A = \begin{bmatrix} -1 & 0 \\ 0 & 0 \end{bmatrix} \quad \text{and} \quad \mathbf{x}_0 = \begin{bmatrix} a \\ b \end{bmatrix}$$

where $\mathbf{x}_0$ is a unit vector and $a \neq 0$. Show that even though the matrix A is symmetric and has a dominant eigenvalue, the power sequence (1) in Theorem 5.4.2 does not converge. This shows that the requirement in that theorem that the dominant eigenvalue be positive is essential.

Discussion and Discovery

D1. Consider the symmetric matrix

$$A = \begin{bmatrix} 0 & 1 \\ 1 & 0 \end{bmatrix}$$

Discuss the behavior of the power sequence $\mathbf{x}_0, \mathbf{x}_1, \ldots, \mathbf{x}_k, \ldots$ with Euclidean scaling for a general

nonzero vector $\mathbf{x}_0$. What is it about the matrix that causes the observed behavior?

D2. Suppose that a symmetric matrix A has distinct eigenvalues $\lambda_1 = 8, \lambda_2 = 1.4, \lambda_3 = 2.3$, and $\lambda_4 = -8.1$. What can you say about the convergence of the Rayleigh quotients?

Working with Proofs

P1. Prove that if A is a nonzero $n \times n$ matrix, then A^TA and AA^T have positive dominant eigenvalues.

P2. (***For readers familiar with proof by induction***) Let A be an $n \times n$ matrix, let $\mathbf{x}_0$ be a unit vector in R^n, and define the sequence $\mathbf{x}_1, \mathbf{x}_2, \ldots, \mathbf{x}_k, \ldots$ by

$$\mathbf{x}_1 = \frac{A\mathbf{x}_0}{\|A\mathbf{x}_0\|}, \quad \mathbf{x}_2 = \frac{A\mathbf{x}_1}{\|A\mathbf{x}_1\|}, \ldots, \quad \mathbf{x}_k = \frac{A\mathbf{x}_{k-1}}{\|A\mathbf{x}_{k-1}\|}, \ldots$$

Prove by induction that $\mathbf{x}_k = A^k\mathbf{x}_0/\|A^k\mathbf{x}_0\|$.

P3. (***For readers familiar with proof by induction***) Let A be an $n \times n$ matrix, let $\mathbf{x}_0$ be a nonzero vector in R^n, and define the sequence $\mathbf{x}_1, \mathbf{x}_2, \ldots, \mathbf{x}_k, \ldots$ by

$$\mathbf{x}_1 = \frac{A\mathbf{x}_0}{\max(A\mathbf{x}_0)}, \quad \mathbf{x}_2 = \frac{A\mathbf{x}_1}{\max(A\mathbf{x}_1)}, \ldots,$$

$$\mathbf{x}_k = \frac{A\mathbf{x}_{k-1}}{\max(A\mathbf{x}_{k-1})}, \ldots$$

Prove by induction that $\mathbf{x}_k = A^k\mathbf{x}_0/\max(A^k\mathbf{x}_0)$.

Technology Exercises

T1. Use the power method with Euclidean scaling to approximate the dominant eigenvalue and a corresponding eigenvector of A. Choose your own starting vector, and stop when the estimated percentage error in the eigenvalue approximation is less than 0.1%.

(a) $\begin{bmatrix} 1 & 3 & 3 \\ 3 & 4 & -1 \\ 3 & -1 & 10 \end{bmatrix}$
(b) $\begin{bmatrix} 1 & 0 & 1 & 1 \\ 0 & 2 & -1 & 1 \\ 1 & -1 & 4 & 1 \\ 1 & 1 & 1 & 8 \end{bmatrix}$

T2. Repeat Exercise T1, but this time stop when all corresponding entries in two successive eigenvector approximations differ by less than 0.01 in absolute value.

T3. Repeat Exercise T1 using maximum entry scaling.

T4. Suppose that the Google search engine produces 10 Internet sites in the search set and that the adjacency matrix for those sites is

Referenced Site

$$A = \begin{bmatrix} 0 & 1 & 1 & 0 & 1 & 1 & 0 & 0 & 0 & 1 \\ 0 & 0 & 1 & 0 & 0 & 0 & 0 & 0 & 0 & 0 \\ 0 & 0 & 0 & 0 & 0 & 0 & 0 & 0 & 0 & 1 \\ 0 & 1 & 1 & 0 & 0 & 1 & 1 & 0 & 0 & 1 \\ 0 & 0 & 0 & 1 & 0 & 0 & 0 & 0 & 0 & 0 \\ 0 & 1 & 0 & 0 & 0 & 0 & 0 & 0 & 0 & 0 \\ 0 & 0 & 0 & 0 & 0 & 0 & 0 & 0 & 1 & 0 \\ 0 & 0 & 0 & 0 & 0 & 1 & 0 & 0 & 0 & 0 \\ 0 & 1 & 1 & 0 & 0 & 1 & 0 & 1 & 0 & 1 \\ 0 & 0 & 0 & 0 & 0 & 1 & 0 & 0 & 0 & 0 \end{bmatrix} \begin{matrix} 1 \\ 2 \\ 3 \\ 4 \\ 5 \\ 6 \\ 7 \\ 8 \\ 9 \\ 10 \end{matrix}$$

(columns labeled 1 2 3 4 5 6 7 8 9 10; rows 5 and 6 labeled **Referencing Site**)

Use the PageRank algorithm to rank the sites in decreasing order of authority for the Google search engine.

There is a way of using the power method to approximate the eigenvalue of A with smallest absolute value when A is an invertible $n \times n$ matrix, and the eigenvalues of A can be ordered according to the size of their absolute values as

$$|\lambda_1| \geq |\lambda_2| \geq \cdots \geq |\lambda_{n-1}| > |\lambda_n|$$

The method uses the fact (proved in Exercise P3 of Section 4.4) that if $\lambda_1, \lambda_2, \ldots, \lambda_n$ are the eigenvalues of A, then $1/\lambda_1, 1/\lambda_2, \ldots, 1/\lambda_n$ are the eigenvalues of A^{-1}, and the eigenvectors of A^{-1} corresponding to $1/\lambda_k$ are the same as the eigenvectors of A corresponding to λ_k. The above inequalities imply that A^{-1} has a dominant eigenvalue of $1/\lambda_n$ (why?), which together with a corresponding eigenvector $\mathbf{x}$ can be approximated by applying the power method to A^{-1}. Once obtained, the reciprocal of this approximation will provide an approximation to the eigenvalue of A that has smallest absolute value, and $\mathbf{x}$ will be a corresponding eigenvector. This technique is called the ***inverse power method***. In practice, the inverse power method is rarely implemented by finding A^{-1} and computing successive iterates as

$$\mathbf{x}_k = \frac{A^{-1}\mathbf{x}_{k-1}}{\max(A^{-1}\mathbf{x}_{k-1})} \quad \text{or} \quad \mathbf{x}_k = \frac{A^{-1}\mathbf{x}_{k-1}}{\|A^{-1}\mathbf{x}_{k-1}\|}$$

Rather, it is usual to let $\mathbf{y}_k = A^{-1}\mathbf{x}_{k-1}$, solve the equation $\mathbf{x}_{k-1} = A\mathbf{y}_k$ for $\mathbf{y}_k$, say by LU-decomposition, and then scale to obtain $\mathbf{x}_k$. The LU-decomposition only needs to be computed once, after which it can be reused to find each new iterate. Use the inverse power method in Exercises T5 and T6.

T5. In Example 6 of Section 4.4, we found the eigenvalues of

$$A = \begin{bmatrix} 3 & 2 \\ 2 & 3 \end{bmatrix}$$

to be $\lambda = 1$ and $\lambda = 5$, and in Example 2 of this section we approximated the eigenvalue $\lambda = 5$ and a corresponding eigenvector using the power method with Euclidean scaling. Use the inverse power method with Euclidean scaling to approximate the eigenvalue $\lambda = 1$ and a corresponding eigenvector. Start with the vector $\mathbf{x}_0$ used in Example 2, and stop when the estimated relative error in the eigenvalue is less than 0.001.

T6. Use the inverse power method with Euclidean scaling to approximate the eigenvalue of

$$A = \begin{bmatrix} 0 & 0 & 2 \\ 0 & 2 & 0 \\ 2 & 0 & 3 \end{bmatrix}$$

that has the smallest absolute value and approximate a corresponding eigenvector. Choose your own starting vector, and stop when the estimated relative error in the eigenvalue is less than 0.001.

There is a way of using the inverse power method to approximate any eigenvalue λ of a symmetric $n \times n$ matrix A provided one can find a scalar s that is closer to λ than to any other eigenvalue of A. The method is based on the result in Exercise P4 of Section 4.4, which states that if the eigenvalues of A are $\lambda_1, \lambda_2, \ldots, \lambda_n$, then the eigenvalues of $A - sI$ are $\lambda_1 - s, \lambda_2 - s, \ldots, \lambda_n - s$, and the eigenvectors of $A - sI$ corresponding to $\lambda - s$ are the same as the eigenvectors of A corresponding to λ. Since we are assuming that s is closer to λ than to any other eigenvalue of A, it follows that $1/(\lambda - s)$ is a dominant eigenvalue of the matrix $(A - sI)^{-1}$, so $\lambda - s$ and a corresponding eigenvector $\mathbf{x}$ can be approximated by applying the inverse power method to $A - sI$. Adding s to this approximation will yield an approximation to the eigenvalue λ of A, and $\mathbf{x}$ will be a corresponding eigenvector. This technique is called the **shifted inverse power** method. Use this method in Exercise T7.

T7. Given that the matrix

$$A = \begin{bmatrix} 2.0800 & 0.8818 & 0.6235 \\ 0.8818 & 4.0533 & -2.7907 \\ 0.6235 & -2.7907 & 6.0267 \end{bmatrix}$$

has an eigenvalue λ near 3, use the shifted inverse power method with $s = 3$ to approximate λ and a corresponding eigenvector. Choose your own starting vector and stop when the estimated relative error is less than 0.001.

CHAPTER 6

Linear Transformations

Linear transformations are used in the study of chaotic processes and the design of engineering control systems. They are also important in such applications as filtration of noise from electrical and acoustical signals and computer graphics.

Section 6.1 Matrices as Transformations

Up to now our work with matrices has been in the context of linear systems. In this section we will consider matrices from an "operational" point of view, meaning that we will be concerned with how matrix multiplication affects algebraic and geometric relationships between vectors.

A REVIEW OF FUNCTIONS

We will begin by briefly reviewing some ideas about functions. Recall first that if a variable y depends on a variable x in such a way that each allowable value of x determines exactly one value of y, then we say that *y is a function of x*. For example, if x is any real number, then the equation $y = x^2$ defines y as a function of x, since every real number x has a unique square y.

In the mid eighteenth century mathematicians began denoting functions by letters, thereby making it possible to talk about functions without stating specific formulas. To understand this idea, think of a function as a computer program that takes an *input* x, operates on it in some way, and produces exactly one *output* y. The computer program is an object in its own right, so we can give it a name, say f. Thus, f (the computer program) can be viewed as a procedure for associating a unique output y with each input x (Figure 6.1.1).

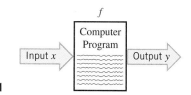

Figure 6.1.1

In general, the inputs and outputs of a function can be any kinds of objects (though they will usually be numbers, vectors, or matrices in this text)—all that is important is that a set of allowable inputs be specified and that the function assign exactly one output to each allowable input. Here is a more precise definition of a function.

Definition 6.1.1 Given a set D of allowable inputs, a *function* f is a rule that associates a unique output with each input from D; the set D is called the *domain* of f. If the input is denoted by x, then the corresponding output is denoted by $f(x)$ (read, "f of x"). The output is also called the *value* of f at x or the *image* of x under f, and we say that f *maps* x into $f(x)$. It is common to denote the output by the single letter y and write $y = f(x)$. The set of all outputs y that results as x varies over the domain is called the *range* of f.

A function whose inputs and outputs are vectors is called a *transformation*, and it is standard to denote transformations by capital letters such as F, T, or L. If T is a transformation that maps

265

the vector $\mathbf{x}$ into the vector $\mathbf{w}$, then the relationship $\mathbf{w} = T(\mathbf{x})$ is sometimes written as

$$\mathbf{x} \xrightarrow{T} \mathbf{w}$$

which is read, "T maps $\mathbf{x}$ into $\mathbf{w}$."

EXAMPLE 1
A Scaling
Transformation

Let T be the transformation that maps a vector $\mathbf{x} = (x_1, x_2)$ in R^2 into the vector $2\mathbf{x} = (2x_1, 2x_2)$ in R^2. This relationship can be expressed in various ways:

$$T(\mathbf{x}) = 2\mathbf{x}, \qquad T(x_1, x_2) = (2x_1, 2x_2)$$
$$\mathbf{x} \xrightarrow{T} 2\mathbf{x}, \qquad (x_1, x_2) \xrightarrow{T} (2x_1, 2x_2)$$

In particular, if $\mathbf{x} = (-1, 3)$, then $T(\mathbf{x}) = 2\mathbf{x} = (-2, 6)$, which we can express as

$$T(-1, 3) = (-2, 6) \quad \text{or equivalently,} \quad (-1, 3) \xrightarrow{T} (-2, 6)$$ ∎

EXAMPLE 2
A Component-
Squaring
Transformation

Let T be the transformation that maps a vector $\mathbf{x}$ in R^3 into the vector in R^3 whose components are the squares of the components of $\mathbf{x}$. Thus, if $\mathbf{x} = (x_1, x_2, x_3)$, then

$$T(x_1, x_2, x_3) = \left(x_1^2, x_2^2, x_3^2\right) \quad \text{or equivalently,} \quad (x_1, x_2, x_3) \xrightarrow{T} \left(x_1^2, x_2^2, x_3^2\right)$$ ∎

EXAMPLE 3
A Matrix
Multiplication
Transformation

Consider the 3×2 matrix

$$A = \begin{bmatrix} 1 & -1 \\ 2 & 5 \\ 3 & 4 \end{bmatrix} \tag{1}$$

and let T_A be the transformation that maps a 2×1 column vector $\mathbf{x}$ in R^2 into the 3×1 column vector $A\mathbf{x}$ in R^3. This relationship can be expressed as

$$T_A(\mathbf{x}) = A\mathbf{x} \quad \text{or as} \quad \mathbf{x} \xrightarrow{T_A} A\mathbf{x}$$

If we write $A\mathbf{x}$ in component form as

$$A\mathbf{x} = \begin{bmatrix} 1 & -1 \\ 2 & 5 \\ 3 & 4 \end{bmatrix} \begin{bmatrix} x_1 \\ x_2 \end{bmatrix} = \begin{bmatrix} x_1 - x_2 \\ 2x_1 + 5x_2 \\ 3x_1 + 4x_2 \end{bmatrix}$$

then we can express the transformation T_A in component form as

$$T_A\left(\begin{bmatrix} x_1 \\ x_2 \end{bmatrix}\right) = \begin{bmatrix} x_1 - x_2 \\ 2x_1 + 5x_2 \\ 3x_1 + 4x_2 \end{bmatrix} \tag{2}$$

This formula can also be expressed more compactly in comma-delimited form as

$$T_A(x_1, x_2) = (x_1 - x_2, 2x_1 + 5x_2, 3x_1 + 4x_2) \tag{3}$$

which emphasizes T_A as a mapping from points into points. For example, Formulas (2) and (3) yield

$$T_A\left(\begin{bmatrix} -1 \\ 3 \end{bmatrix}\right) = \begin{bmatrix} -4 \\ 13 \\ 9 \end{bmatrix} \quad \text{or} \quad T_A(-1, 3) = (-4, 13, 9)$$ ∎

If T is a transformation whose domain *is* R^n and whose range is *in* R^m, then we will write

$$T : R^n \to R^m \tag{4}$$

(read, "T maps R^n into R^m") when we want to emphasize the spaces involved. Depending on your geometric point of view, you can think of a transformation $T : R^n \to R^m$ as mapping points into points or vectors into vectors (Figure 6.1.2). For example, the scaling transformation in

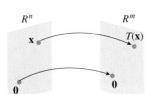

T maps points to points.

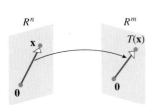

T maps vectors to vectors.

Figure 6.1.2

Example 1 maps the vectors (or points) in R^2 into vectors (or points) in R^2, so

$$T : R^2 \to R^2 \tag{5}$$

and the transformation T_A in Example 3 maps the vectors (or points) in R^2 into vectors (or points) in R^3, so

$$T_A : R^2 \to R^3 \tag{6}$$

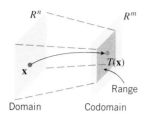

Figure 6.1.3

Keep in mind that the set R^n in (4) is the domain of T but that R^m may not be the range of T. The set R^m, which is called the **codomain** of T, is intended only to describe the space in which the image vectors lie and may actually be larger than the range of T (Figure 6.1.3).

Note that the transformation T in (5) maps the space R^2 back into itself, whereas the transformation T_A in (6) maps R^2 into a different space. In general, if $T : R^n \to R^n$, then we will refer to the transformation T as an **operator** on R^n to emphasize that it maps R^n back into R^n.

MATRIX TRANSFORMATIONS

Example 3 is a special case of a general class of transformations, called **matrix transformations**. Specifically, if A is an $m \times n$ matrix, and if $\mathbf{x}$ is a column vector in R^n, then the product $A\mathbf{x}$ is a vector in R^m, so multiplying $\mathbf{x}$ by A creates a transformation that maps vectors in R^n into vectors in R^m. We call this transformation **multiplication by** A or the **transformation** A and denote it by T_A to emphasize the matrix A. Thus,

$$T_A : R^n \to R^m$$

and

$$T_A(\mathbf{x}) = A\mathbf{x} \quad \text{or equivalently,} \quad \mathbf{x} \xrightarrow{T_A} A\mathbf{x}$$

In the special case where A is square, say $n \times n$, we have $T_A : R^n \to R^n$, and we call T_A a **matrix operator** on R^n.

EXAMPLE 4
Zero
Transformations

If 0 is the $m \times n$ zero matrix, then

$$T_0(\mathbf{x}) = 0\mathbf{x} = \mathbf{0}$$

so multiplication by 0 maps every vector in R^n into the zero vector in R^m. Accordingly, we call T_0 the **zero transformation** from R^n to R^m. ∎

EXAMPLE 5
Identity
Operators

If I is the $n \times n$ identity matrix, then for every vector $\mathbf{x}$ in R^n we have

$$T_I(\mathbf{x}) = I\mathbf{x} = \mathbf{x}$$

so multiplication by I maps every vector in R^n back into itself. Accordingly, we call T_I the **identity operator** on R^n. ∎

Thus far, much of our work with matrices has been in the context of solving a linear system

$$A\mathbf{x} = \mathbf{b} \tag{7}$$

Now we will focus on the transformation aspect of matrix multiplication. For example, if A has size $m \times n$, then multiplication by A defines a matrix transformation from R^n to R^m, so the problem of solving (7) can be viewed geometrically as the problem of finding a vector $\mathbf{x}$ in R^n whose image under the transformation T_A is the vector $\mathbf{b}$ in R^m.

EXAMPLE 6
A Matrix
Transformation

Let $T_A : R^2 \to R^3$ be the matrix transformation in Example 3.

(a) Find a vector $\mathbf{x}$ in R^2, if any, whose image under T_A is

$$\mathbf{b} = \begin{bmatrix} 7 \\ 0 \\ 7 \end{bmatrix}$$

(b) Find a vector $\mathbf{x}$ in R^2, if any, whose image under T_A is

$$\mathbf{b} = \begin{bmatrix} 9 \\ -3 \\ -1 \end{bmatrix}$$

Solution (a) The stated problem is equivalent to finding a solution $\mathbf{x}$ of the linear system

$$\begin{bmatrix} 1 & -1 \\ 2 & 5 \\ 3 & 4 \end{bmatrix} \begin{bmatrix} x_1 \\ x_2 \end{bmatrix} = \begin{bmatrix} 7 \\ 0 \\ 7 \end{bmatrix} \tag{8}$$

We leave it for you to show that the reduced row echelon form of the augmented matrix for this system is

$$\begin{bmatrix} 1 & 0 & 5 \\ 0 & 1 & -2 \\ 0 & 0 & 0 \end{bmatrix}$$

It follows from this that (8) has the unique solution

$$\mathbf{x} = \begin{bmatrix} 5 \\ -2 \end{bmatrix}$$

and hence that $T_A(\mathbf{x}) = \mathbf{b}$. This shows that the vector $\mathbf{b}$ is in the range of T_A.

Solution (b) The stated problem is equivalent to finding a solution $\mathbf{x}$ of the linear system

$$\begin{bmatrix} 1 & -1 \\ 2 & 5 \\ 3 & 4 \end{bmatrix} \begin{bmatrix} x_1 \\ x_2 \end{bmatrix} = \begin{bmatrix} 9 \\ -3 \\ -1 \end{bmatrix}$$

We leave it for you to show by Gaussian elimination that a row echelon form of the augmented matrix is

$$\begin{bmatrix} 1 & -1 & 9 \\ 0 & 1 & -3 \\ 0 & 0 & 1 \end{bmatrix}$$

The last row reveals that the system is inconsistent, and hence that there is no vector $\mathbf{x}$ in R^2 for which $T_A(\mathbf{x}) = \mathbf{b}$. This shows that the vector $\mathbf{b}$ is not in the range of T_A. ■

LINEAR TRANSFORMATIONS

According to the dictionary, the term *linear* is derived from the Latin word *linearis*, meaning "of or pertaining to a line or lines." The dictionary also gives a secondary meaning of the term: "having an effect or giving a response directly proportional to a stimulus, force, or input." Finally, the dictionary describes a *linear equation* as "an algebraic equation whose variable quantity or quantities are in the first power only." Thus, we have three categories of use for the term *linear*:

1. Algebraic (describing the form of an equation)
2. Geometric (describing the form of objects)
3. Operational (describing the way a system, function, or transformation responds to inputs)

In Chapter 2 we discussed the algebraic and geometric interpretations of linearity. For example, we defined a linear equation in the n variables $x_1, x_2, \ldots, x_n$ to be one that can be expressed in the form

$$a_1 x_1 + a_2 x_2 + \cdots + a_n x_n = b$$

where $a_1, a_2, \ldots, a_n$ and b are constants and the a's are not all zero. We also showed that linear equations correspond to lines in R^2, to planes in R^3, and to hyperplanes in R^n. We will now turn our attention to the operational interpretation of linearity.

Recall that a variable y is said to be **directly proportional** to a variable x if there is a constant k, called the **constant of proportionality**, such that

$$y = kx$$

Hooke's law in physics is a good example of a physical law that can be modeled by direct proportion. It follows from this law that a weight of x units suspended from a spring stretches the spring from its natural length by an amount y that is directly proportional to x (Figure 6.1.4); that is, the variables x and y are related by an equation of the form $y = kx$, where k is a positive constant. The constant k depends on the stiffness of the spring—the stiffer the spring, the smaller the value of k (why?).

For convenience, we will let $f(x) = kx$ and write the direct proportion equation $y = kx$ in the functional form $y = f(x)$. This equation has two important properties:

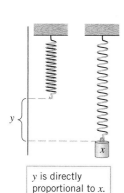

y is directly proportional to x.

Figure 6.1.4

1. **Homogeneity**—Changing the input by a multiplicative factor changes the output by the same factor; that is,

$$f(cx) = k(cx) = c(kx) = cf(x) \tag{9}$$

2. **Additivity**—Adding two inputs adds the corresponding outputs; that is,

$$f(x_1 + x_2) = k(x_1 + x_2) = kx_1 + kx_2 = f(x_1) + f(x_2) \tag{10}$$

These ideas can be illustrated physically with the weight-spring system in Figure 6.1.4. Since the amount y that the spring is stretched from its natural length is directly proportional to the weight x, it follows from the homogeneity that increasing the weight by a factor of c increases the amount the spring is stretched by the same factor. Thus, doubling the weight increases this amount by a factor of 2, tripling the weight increases it by a factor of 3, and so forth (Figure 6.1.5a). Moreover, it follows from the additivity that the amount the spring is stretched by a combined weight of $x_1 + x_2$ is equal to the amount of increase from x_1 alone plus the amount of increase from x_2 alone (Figure 6.1.5b).

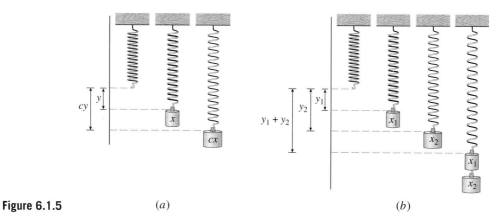

Figure 6.1.5   (a)   (b)

Motivated by Formulas (9) and (10), we make the following definition.

Definition 6.1.2   A function $T : R^n \to R^m$ is called a **linear transformation** from R^n to R^m if the following two properties hold for all vectors **u** and **v** in R^n and for all scalars c:

(i) $T(c\mathbf{u}) = cT(\mathbf{u})$ **[Homogeneity property]**
(ii) $T(\mathbf{u} + \mathbf{v}) = T(\mathbf{u}) + T(\mathbf{v})$ **[Additivity property]**

In the special case where $m = n$, the linear transformation T is called a **linear operator** on R^n.

The two properties in this definition can be used in combination to show that if $\mathbf{v}_1$ and $\mathbf{v}_2$ are vectors in R^n and c_1 and c_2 are any scalars, then

$$T(c_1\mathbf{v}_1 + c_2\mathbf{v}_2) = c_1 T(\mathbf{v}_1) + c_2 T(\mathbf{v}_2)$$

(verify). More generally, if $\mathbf{v}_1, \mathbf{v}_2, \ldots, \mathbf{v}_k$ are vectors in R^n and $c_1, c_2, \ldots, c_k$ are any scalars, then

$$T(c_1\mathbf{v}_1 + c_2\mathbf{v}_2 + \cdots + c_k\mathbf{v}_k) = c_1 T(\mathbf{v}_1) + c_2 T(\mathbf{v}_2) + \cdots + c_k T(\mathbf{v}_k) \tag{11}$$

Engineers and physicists sometimes call this the **superposition principle**.

EXAMPLE 7 Recall from Theorem 3.1.5 that if A is an $m \times n$ matrix, $\mathbf{u}$ and $\mathbf{v}$ are column vectors in R^n, and c is a scalar, then $A(c\mathbf{u}) = c(A\mathbf{u})$ and $A(\mathbf{u} + \mathbf{v}) = A\mathbf{u} + A\mathbf{v}$. Thus, the matrix transformation $T_A : R^n \to R^m$ is linear since

$$T_A(c\mathbf{u}) = A(c\mathbf{u}) = c(A\mathbf{u}) = cT_A(\mathbf{u})$$
$$T_A(\mathbf{u} + \mathbf{v}) = A(\mathbf{u} + \mathbf{v}) = A\mathbf{u} + A\mathbf{v} = T_A(\mathbf{u}) + T_A(\mathbf{v})$$

EXAMPLE 8 In Example 2 we considered the transformation

$$T(x_1, x_2, x_3) = \left(x_1^2, x_2^2, x_3^2\right)$$

This transformation is not linear since it violates both conditions in Definition 6.1.2 (although the violation of either condition would have been sufficient for nonlinearity). The homogeneity condition is violated since

$$T(c\mathbf{u}) = T(cu_1, cu_2, cu_3) = \left(c^2u_1^2, c^2u_2^2, c^2u_3^2\right) = c^2\left(u_1^2, u_2^2, u_3^2\right) = c^2 T(\mathbf{u})$$

which means that $T(c\mathbf{u}) \neq cT(\mathbf{u})$ for some scalars and vectors. The additivity condition is violated since

$$T(\mathbf{u} + \mathbf{v}) = T(u_1 + v_1, u_2 + v_2, u_3 + v_3) = ((u_1 + v_1)^2, (u_2 + v_2)^2, (u_3 + v_3)^2)$$

whereas

$$T(\mathbf{u}) + T(\mathbf{v}) = \left(u_1^2, u_2^2, u_3^2\right) + \left(v_1^2, v_2^2, v_3^2\right) = \left(u_1^2 + v_1^2, u_2^2 + v_2^2, u_3^2 + v_3^2\right)$$

Thus, $T(\mathbf{u} + \mathbf{v}) \neq T(\mathbf{u}) + T(\mathbf{v})$ for some vectors in R^3.

SOME PROPERTIES OF LINEAR TRANSFORMATIONS

The next theorem gives some basic properties of linear transformations.

Theorem 6.1.3 *If $T : R^n \to R^m$ is a linear transformation, then:*
(a) $T(\mathbf{0}) = \mathbf{0}$
(b) $T(-\mathbf{u}) = -T(\mathbf{u})$
(c) $T(\mathbf{u} - \mathbf{v}) = T(\mathbf{u}) - T(\mathbf{v})$

Proof To prove (a) set $c = 0$ in the formula $T(c\mathbf{u}) = cT(\mathbf{u})$, and to prove (b) set $c = -1$ in this formula. To prove part (c) replace $\mathbf{v}$ by $-\mathbf{v}$ in the formula $T(\mathbf{u} + \mathbf{v}) = T(\mathbf{u}) + T(\mathbf{v})$ and apply part (b) on the right side of the resulting equation.

CONCEPT PROBLEM We showed in Example 8 that the component-squaring operator on R^2 is not linear. Show that it fails to have at least one of the properties in Theorem 6.1.3. Which ones?

EXAMPLE 9
Translations Are Not Linear

Recall that adding $\mathbf{x}_0$ to $\mathbf{x}$ has the effect of translating the terminal point of $\mathbf{x}$ by $\mathbf{x}_0$ (Figure 6.1.6). Thus, the operator $T(\mathbf{x}) = \mathbf{x}_0 + \mathbf{x}$ on R^n has the effect of translating every point in R^n by $\mathbf{x}_0$. Show that T is not linear if $\mathbf{x}_0 \neq \mathbf{0}$.

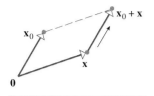

Adding $\mathbf{x}_0$ to $\mathbf{x}$ translates the terminal point of $\mathbf{x}$ by $\mathbf{x}_0$.

Figure 6.1.6

Solution We could proceed by showing that either the homogeneity or linearity property (or both) fails to hold. However, we can see that T is not linear, since $T(\mathbf{0}) = \mathbf{x}_0 + \mathbf{0} = \mathbf{x}_0 \neq \mathbf{0}$, in violation of part (*a*) of Theorem 6.1.3. As a check, we leave it for you to confirm that both the homogeneity and linearity properties fail to hold for this operator. ■

REMARK Sometimes we will need to consider transformations in which the domain V is a proper *subspace* of R^n, rather than all of R^n. Just as in Definition 6.1.2, a transformation $T : V \rightarrow R^m$ of this type is said to be ***linear*** if it has the additivity and homogeneity properties. Theorem 6.1.3 is valid for such linear transformations, since the proof carries over without change.

ALL LINEAR TRANSFORMATIONS FROM R^n TO R^m ARE MATRIX TRANSFORMATIONS

We saw in Example 7 that every matrix transformation from R^n to R^m is linear. We will now show that matrix transformations are the *only* linear transformations from R^n to R^m in the sense that if $T : R^n \rightarrow R^m$ is a linear transformation, then there is a unique $m \times n$ matrix A such that

$$T(\mathbf{x}) = A\mathbf{x}$$

for every vector $\mathbf{x}$ in R^n (assuming, of course, that $\mathbf{x}$ is expressed in column form). This is an extremely important result because it means *all* linear transformations from R^n to R^m can be performed by matrix multiplications, even if they don't arise in that way.

To prove this result, let us assume that $\mathbf{x}$ is written in column form and express it as a linear combination of the standard unit vectors by writing

$$\mathbf{x} = \begin{bmatrix} x_1 \\ x_2 \\ \vdots \\ x_n \end{bmatrix} = x_1 \begin{bmatrix} 1 \\ 0 \\ \vdots \\ 0 \end{bmatrix} + x_2 \begin{bmatrix} 0 \\ 1 \\ \vdots \\ 0 \end{bmatrix} + \cdots + x_n \begin{bmatrix} 0 \\ 0 \\ \vdots \\ 1 \end{bmatrix} = x_1\mathbf{e}_1 + x_2\mathbf{e}_2 + \cdots + x_n\mathbf{e}_n$$

Thus, it follows from (11) that

$$T(\mathbf{x}) = x_1 T(\mathbf{e}_1) + x_2 T(\mathbf{e}_2) + \cdots + x_n T(\mathbf{e}_n) \tag{12}$$

If we now create the matrix A that has $T(\mathbf{e}_1)$, $T(\mathbf{e}_2)$, ..., $T(\mathbf{e}_n)$ as successive column vectors, then it follows from Formula (10) of Section 3.1 that (12) can be expressed as

$$T(\mathbf{x}) = [T(\mathbf{e}_1) \quad T(\mathbf{e}_2) \quad \cdots \quad T(\mathbf{e}_n)] \begin{bmatrix} x_1 \\ x_2 \\ \vdots \\ x_n \end{bmatrix} = A\mathbf{x}$$

Thus, we have established the following result.

Theorem 6.1.4 *Let $T : R^n \rightarrow R^m$ be a linear transformation, and suppose that vectors are expressed in column form. If $\mathbf{e}_1, \mathbf{e}_2, \ldots, \mathbf{e}_n$ are the standard unit vectors in R^n, and if $\mathbf{x}$ is any vector in R^n, then $T(\mathbf{x})$ can be expressed as*

$$T(\mathbf{x}) = A\mathbf{x} \tag{13}$$

where

$$A = [T(\mathbf{e}_1) \quad T(\mathbf{e}_2) \quad \cdots \quad T(\mathbf{e}_n)]$$

The matrix A in this theorem is called the ***standard matrix for T***, and we say that T is the ***transformation corresponding to A***, or that T is the ***transformation represented by A***, or sometimes simply that ***T is the transformation A***.

When it is desirable to emphasize the relationship between T and its standard matrix, we will denote A by $[T]$; that is, we will write

$$[T] = [T(\mathbf{e}_1) \quad T(\mathbf{e}_2) \quad \cdots \quad T(\mathbf{e}_n)]$$

With this notation, the relationship in (13) becomes

$$T(\mathbf{x}) = [T]\mathbf{x} \tag{14}$$

REMARK Theorem 6.1.4 shows that *a linear transformation* $T : R^n \rightarrow R^m$ *is completely determined by its values at the standard unit vectors* in the sense that once the images of the standard unit vectors are known, the standard matrix $[T]$ can be constructed and then used to compute images of all other vectors using (14).

EXAMPLE 10
Standard Matrix
for a Scaling
Operator

In Example 1 we considered the scaling operator $T : R^2 \rightarrow R^2$ defined by $T(\mathbf{x}) = 2\mathbf{x}$. Show that this operator is linear, and find its standard matrix.

Solution The transformation T is homogeneous since

$$T(c\mathbf{u}) = 2(c\mathbf{u}) = c(2\mathbf{u}) = cT(\mathbf{u})$$

and it is additive since

$$T(\mathbf{u} + \mathbf{v}) = 2(\mathbf{u} + \mathbf{v}) = 2\mathbf{u} + 2\mathbf{v} = T(\mathbf{u}) + T(\mathbf{v})$$

From Theorem 6.1.4, the standard matrix for T is

$$[T] = \begin{bmatrix} T(\mathbf{e}_1) & T(\mathbf{e}_2) \end{bmatrix} = \begin{bmatrix} 2\mathbf{e}_1 & 2\mathbf{e}_2 \end{bmatrix} = \begin{bmatrix} 2 & 0 \\ 0 & 2 \end{bmatrix}$$

As a check,

$$[T]\mathbf{x} = \begin{bmatrix} 2 & 0 \\ 0 & 2 \end{bmatrix} \begin{bmatrix} x_1 \\ x_2 \end{bmatrix} = \begin{bmatrix} 2x_1 \\ 2x_2 \end{bmatrix} = 2 \begin{bmatrix} x_1 \\ x_2 \end{bmatrix} = 2\mathbf{x} = T(\mathbf{x}) \qquad \blacksquare$$

Transformations from R^n to R^m are often specified by formulas that relate the components of a vector $\mathbf{x} = (x_1, x_2, \ldots, x_n)$ in R^n with those of its image $\mathbf{w} = T(\mathbf{x}) = (w_1, w_2, \ldots, w_m)$ in R^m. It follows from Theorem 6.1.4 that such a transformation is linear if and only if the relationship between $\mathbf{w}$ and $\mathbf{x}$ is expressible as $\mathbf{w} = A\mathbf{x}$, where $A = [a_{ij}]$ is the standard matrix for T. If we write out the individual equations in this matrix equation, we obtain

$$w_1 = a_{11}x_1 + a_{12}x_2 + \cdots + a_{1n}x_n$$
$$w_2 = a_{21}x_1 + a_{22}x_2 + \cdots + a_{2n}x_n$$
$$\vdots \qquad \vdots \qquad \vdots \qquad \qquad \vdots$$
$$w_m = a_{m1}x_1 + a_{m2}x_2 + \cdots + a_{mn}x_n$$

Thus, it follows that $T(\mathbf{x}) = (w_1, w_2, \ldots, w_m)$ *is a linear transformation if and only if the equations relating the components of* $\mathbf{x}$ *and* $\mathbf{w}$ *are linear equations.*

EXAMPLE 11
Standard Matrix
for a Linear
Transformation

Show that the transformation $T : R^3 \rightarrow R^2$ defined by the formula

$$T(x_1, x_2, x_3) = (x_1 + x_2, x_2 - x_3) \tag{15}$$

is linear and find its standard matrix.

Solution Let $\mathbf{x} = (x_1, x_2, x_3)$ be a vector in R^3, and let $\mathbf{w} = (w_1, w_2)$ be its image under the transformation T. It follows from (15) that

$$w_1 = x_1 + x_2 \quad \text{and} \quad w_2 = x_2 - x_3$$

Since these are linear equations, the transformation T is linear. To find the standard matrix for T we compute the images of the standard unit vectors under the transformation. These are

$$T(\mathbf{e}_1) = T(1, 0, 0) = (1, 0)$$
$$T(\mathbf{e}_2) = T(0, 1, 0) = (1, 1)$$
$$T(\mathbf{e}_3) = T(0, 0, 1) = (0, -1)$$

Writing these vectors in column form yields the standard matrix

$$[T] = [T(\mathbf{e}_1) \quad T(\mathbf{e}_2) \quad T(\mathbf{e}_3)] = \begin{bmatrix} 1 & 1 & 0 \\ 0 & 1 & -1 \end{bmatrix}$$

As a check,

$$[T]\mathbf{x} = \begin{bmatrix} 1 & 1 & 0 \\ 0 & 1 & -1 \end{bmatrix} \begin{bmatrix} x_1 \\ x_2 \\ x_3 \end{bmatrix} = \begin{bmatrix} x_1 + x_2 \\ x_2 - x_3 \end{bmatrix}$$

which agrees with (15), except for the use of matrix notation.   ■

CONCEPT PROBLEM What can you say about the standard matrix for the identity operator on R^n? A zero transformation from R^n to R^m?

ROTATIONS ABOUT THE ORIGIN

Some of the most important linear operators on R^2 and R^3 are rotations, reflections, and projections. In this section we will show how to find standard matrices for such operators on R^2, and in the next section we will use these matrices to study the operators in more detail.

Let θ be a fixed angle, and consider the operator T that rotates each vector $\mathbf{x}$ in R^2 about the origin through the angle θ^* (Figure 6.1.7a). It is not hard to visualize that T is linear by drawing some appropriate pictures. For example, Figure 6.1.7b makes it evident that the rotation T is homogeneous because the same image of a vector $\mathbf{u}$ results whether one first multiplies $\mathbf{u}$ by c and then rotates or first rotates $\mathbf{u}$ and then multiplies by c; also, Figure 6.1.7c makes it evident that T is additive because the same image results whether one first adds $\mathbf{u}$ and $\mathbf{v}$ and then rotates the sum or first rotates the vectors $\mathbf{u}$ and $\mathbf{v}$ and then forms the sum of the rotated vectors.

Let us now try to find the standard matrix for the rotation operator. In keeping with standard usage, we will denote the standard matrix for the rotation about the origin through an angle θ by R_θ. From Figure 6.1.8 we see that this matrix is

$$R_\theta = [T(\mathbf{e}_1) \quad T(\mathbf{e}_2)] = \begin{bmatrix} \cos\theta & -\sin\theta \\ \sin\theta & \cos\theta \end{bmatrix} \tag{16}$$

so the image of a vector $\mathbf{x} = (x, y)$ under this rotation is

$$R_\theta \mathbf{x} = \begin{bmatrix} \cos\theta & -\sin\theta \\ \sin\theta & \cos\theta \end{bmatrix} \begin{bmatrix} x \\ y \end{bmatrix} \tag{17}$$

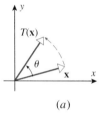

(a)

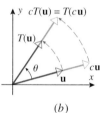

(b)

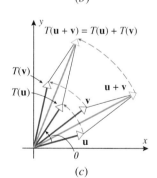

(c)

Figure 6.1.7

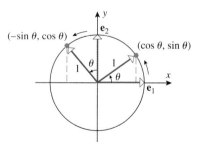

Figure 6.1.8

EXAMPLE 12
A Rotation
Operator

Find the image of $\mathbf{x} = (1, 1)$ under a rotation of $\pi/6$ radians $(= 30°)$ about the origin.

Solution It follows from (17) with $\theta = \pi/6$ that

$$R_{\pi/6}\mathbf{x} = \begin{bmatrix} \frac{\sqrt{3}}{2} & -\frac{1}{2} \\ \frac{1}{2} & \frac{\sqrt{3}}{2} \end{bmatrix} \begin{bmatrix} 1 \\ 1 \end{bmatrix} = \begin{bmatrix} \frac{\sqrt{3}-1}{2} \\ \frac{1+\sqrt{3}}{2} \end{bmatrix} \approx \begin{bmatrix} 0.37 \\ 1.37 \end{bmatrix}$$

or in comma-delimited notation this image is approximately $(0.37, 1.37)$.   ■

---

* In the plane, counterclockwise angles are positive and clockwise angles are negative.

REFLECTIONS ABOUT LINES THROUGH THE ORIGIN

Now let us consider the operator $T : R^2 \to R^2$ that reflects each vector $\mathbf{x}$ about a line through the origin that makes an angle θ with the positive x-axis (Figure 6.1.9). The same kind of geometric argument that we used to establish the linearity of rotation operators can be used to establish linearity of reflection operators. In keeping with a common convention of associating the letter H with reflections, we will denote the standard matrix for the reflection in Figure 6.1.9 by H_θ. From Figure 6.1.10 we see that this matrix is

$$H_\theta = [T(\mathbf{e}_1) \quad T(\mathbf{e}_2)] = \begin{bmatrix} \cos 2\theta & \cos\left(\frac{\pi}{2} - 2\theta\right) \\ \sin 2\theta & -\sin\left(\frac{\pi}{2} - 2\theta\right) \end{bmatrix} = \begin{bmatrix} \cos 2\theta & \sin 2\theta \\ \sin 2\theta & -\cos 2\theta \end{bmatrix} \tag{18}$$

so the image of a vector $\mathbf{x} = (x, y)$ under this reflection is

$$H_\theta \mathbf{x} = \begin{bmatrix} \cos 2\theta & \sin 2\theta \\ \sin 2\theta & -\cos 2\theta \end{bmatrix} \begin{bmatrix} x \\ y \end{bmatrix} \tag{19}$$

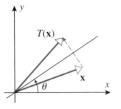

Figure 6.1.9

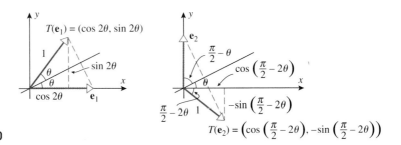

Figure 6.1.10

The most basic reflections in an xy-coordinate system are about the x-axis ($\theta = 0$), the y-axis ($\theta = \pi/2$), and the line $y = x$ ($\theta = \pi/4$). Some information about these reflections is given in Table 6.1.1.

Table 6.1.1

Operator	Illustration	Images of $\mathbf{e}_1$ and $\mathbf{e}_2$	Standard Matrix
Reflection about the y-axis $T(x, y) = (-x, y)$		$T(\mathbf{e}_1) = T(1, 0) = (-1, 0)$ $T(\mathbf{e}_2) = T(0, 1) = (0, 1)$	$\begin{bmatrix} -1 & 0 \\ 0 & 1 \end{bmatrix}$
Reflection about the x-axis $T(x, y) = (x, -y)$		$T(\mathbf{e}_1) = T(1, 0) = (1, 0)$ $T(\mathbf{e}_2) = T(0, 1) = (0, -1)$	$\begin{bmatrix} 1 & 0 \\ 0 & -1 \end{bmatrix}$
Reflection about the line $y = x$ $T(x, y) = (y, x)$		$T(\mathbf{e}_1) = T(1, 0) = (0, 1)$ $T(\mathbf{e}_2) = T(0, 1) = (1, 0)$	$\begin{bmatrix} 0 & 1 \\ 1 & 0 \end{bmatrix}$

EXAMPLE 13
A Reflection
Operator

Find the image of the vector $\mathbf{x} = (1, 1)$ under a reflection about the line through the origin that makes an angle of $\pi/6 (= 30°)$ with the positive x-axis.

Solution Substituting $\theta = \pi/6$ in (19) yields

$$H_{\pi/6}\mathbf{x} = \begin{bmatrix} \frac{1}{2} & \frac{\sqrt{3}}{2} \\ \frac{\sqrt{3}}{2} & -\frac{1}{2} \end{bmatrix} \begin{bmatrix} 1 \\ 1 \end{bmatrix} = \begin{bmatrix} \frac{1+\sqrt{3}}{2} \\ \frac{\sqrt{3}-1}{2} \end{bmatrix} \approx \begin{bmatrix} 1.37 \\ 0.37 \end{bmatrix}$$

or in comma-delimited notation this image is approximately $(1.37, 0.37)$. ∎

CONCEPT PROBLEM Obtain the standard matrices in Table 6.1.1 from (19).

ORTHOGONAL PROJECTIONS ONTO LINES THROUGH THE ORIGIN

Consider the operator $T : R^2 \rightarrow R^2$ that projects each vector $\mathbf{x}$ in R^2 onto a line through the origin by dropping a perpendicular to that line as shown in Figure 6.1.11; we call this operator the ***orthogonal projection*** of R^2 onto the line. One can show that orthogonal projections onto lines are linear and hence are matrix operators.

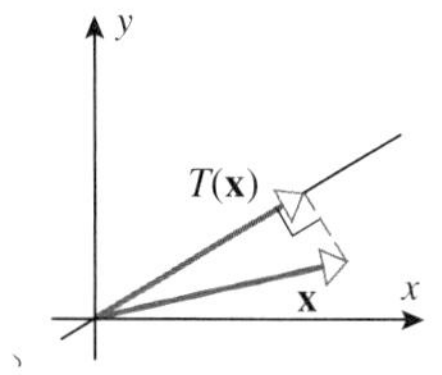

The standard matrix for an orthogonal projection onto a general line through the origin can be obtained using Theorem 6.1.4; however, it will be instructive to consider an alternative approach in which we will express the orthogonal projection in terms of a reflection and then use the known standard matrix for the reflection to obtain the matrix for the projection.

Consider a line through the origin that makes an angle θ with the positive x-axis, and denote the standard matrix for the orthogonal projection by P_θ. It is evident from Figure 6.1.12 that for each $\mathbf{x}$ in R^2 the vector $P_\theta\mathbf{x}$ is related to the vector $H_\theta\mathbf{x}$ by the equation

$$P_\theta\mathbf{x} - \mathbf{x} = \tfrac{1}{2}(H_\theta\mathbf{x} - \mathbf{x})$$

Solving for $P_\theta\mathbf{x}$ yields

$$P_\theta\mathbf{x} = \tfrac{1}{2}H_\theta\mathbf{x} + \tfrac{1}{2}\mathbf{x} = \tfrac{1}{2}H_\theta\mathbf{x} + \tfrac{1}{2}I\mathbf{x} = \tfrac{1}{2}(H_\theta + I)\mathbf{x}$$

so part (*b*) of Theorem 3.4.4 implies that

$$P_\theta = \tfrac{1}{2}(H_\theta + I) \tag{20}$$

We now leave it for you to use this equation and Formula (18) to show that

$$P_\theta = \begin{bmatrix} \frac{1}{2}(1 + \cos 2\theta) & \frac{1}{2}\sin 2\theta \\ \frac{1}{2}\sin 2\theta & \frac{1}{2}(1 - \cos 2\theta) \end{bmatrix} = \begin{bmatrix} \cos^2 \theta & \sin \theta \cos \theta \\ \sin \theta \cos \theta & \sin^2 \theta \end{bmatrix} \tag{21}$$

so the image of a vector $\mathbf{x} = (x, y)$ under this projection is

$$P_\theta\mathbf{x} = \begin{bmatrix} \cos^2 \theta & \sin \theta \cos \theta \\ \sin \theta \cos \theta & \sin^2 \theta \end{bmatrix} \begin{bmatrix} x \\ y \end{bmatrix} \tag{22}$$

The most basic orthogonal projections in R^2 are onto the coordinate axes. Information about these operators is given in Table 6.1.2.

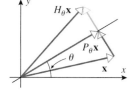

Orthogonal projection onto a line through the origin

Figure 6.1.11

Figure 6.1.12

EXAMPLE 14
An Orthogonal
Projection
Operator

Find the orthogonal projection of the vector $\mathbf{x} = (1, 1)$ on the line through the origin that makes an angle of $\pi/12 (= 15°)$ with the x-axis.

Solution Here it is easier to work with the first form of the standard matrix given in (21), since the angle $2\theta = \pi/6$ is nicer to work with than $\theta = \pi/12$. This yields the standard matrix

$$P_{\pi/12} = \begin{bmatrix} \frac{1}{2}\left(1 + \cos \frac{\pi}{6}\right) & \frac{1}{2}\sin \frac{\pi}{6} \\ \frac{1}{2}\sin \frac{\pi}{6} & \frac{1}{2}\left(1 - \cos \frac{\pi}{6}\right) \end{bmatrix} = \begin{bmatrix} \frac{1}{2}\left(1 + \frac{\sqrt{3}}{2}\right) & \frac{1}{4} \\ \frac{1}{4} & \frac{1}{2}\left(1 - \frac{\sqrt{3}}{2}\right) \end{bmatrix}$$

Table 6.1.2

Operator	Illustration	Images of e_1 and e_2	Standard Matrix
Orthogonal projection on the x-axis $T(x, y) = (x, 0)$		$T(e_1) = T(1, 0) = (1, 0)$ $T(e_2) = T(0, 1) = (0, 0)$	$\begin{bmatrix} 1 & 0 \\ 0 & 0 \end{bmatrix}$
Orthogonal projection on the y-axis $T(x, y) = (0, y)$		$T(e_1) = T(1, 0) = (0, 0)$ $T(e_2) = T(0, 1) = (0, 1)$	$\begin{bmatrix} 0 & 0 \\ 0 & 1 \end{bmatrix}$

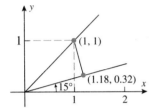

Figure 6.1.13

for the projection. Thus, the image of $x = (1, 1)$ under this projection is

$$P_{\pi/12}x = \begin{bmatrix} \frac{1}{2}\left(1 + \frac{\sqrt{3}}{2}\right) & \frac{1}{4} \\ \frac{1}{4} & \frac{1}{2}\left(1 - \frac{\sqrt{3}}{2}\right) \end{bmatrix} \begin{bmatrix} 1 \\ 1 \end{bmatrix} = \begin{bmatrix} \frac{1}{4}\left(3 + \sqrt{3}\right) \\ \frac{1}{4}\left(3 - \sqrt{3}\right) \end{bmatrix} \approx \begin{bmatrix} 1.18 \\ 0.32 \end{bmatrix}$$

or in comma-delimited notation this projection is approximately $(1.18, 0.32)$. The projection is shown in Figure 6.1.13. ∎

CONCEPT PROBLEM Use Formula (22) to derive the results in Table 6.1.2.

TRANSFORMATIONS OF THE UNIT SQUARE

The *unit square* in R^2 is the square that has e_1 and e_2 as adjacent sides; its vertices are $(0, 0)$, $(1, 0)$, $(1, 1)$, and $(0, 1)$. It is often possible to gain some insight into the geometric behavior of a linear operator on R^2 by graphing the images of these vertices (Figure 6.1.14).

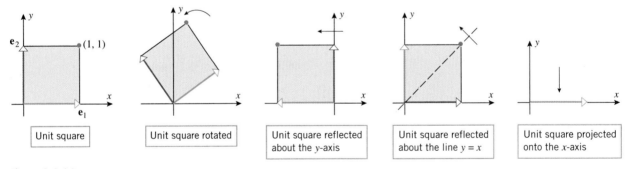

Figure 6.1.14

POWER SEQUENCES

There are many problems in which one is more interested in how successive applications of a linear transformation affect a specific vector than in how the transformation affects geometric objects. That is, if A is the standard matrix for a linear operator on R^n and x_0 is some fixed vector in R^n, then one might be interested in the behavior of the power sequence

$$x_0, Ax_0, A^2x_0, \ldots, A^kx_0, \ldots$$

For example, if

$$A = \begin{bmatrix} \frac{1}{2} & \frac{3}{4} \\ -\frac{3}{5} & \frac{11}{10} \end{bmatrix} \quad \text{and} \quad x_0 = \begin{bmatrix} 1 \\ 1 \end{bmatrix}$$

then with the help of a computer or calculator one can show that the first five terms in the power

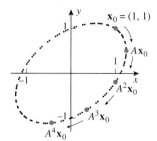

Figure 6.1.15

sequence are

$$\mathbf{x}_0 = \begin{bmatrix} 1 \\ 1 \end{bmatrix}, \quad A\mathbf{x}_0 = \begin{bmatrix} 1.25 \\ 0.5 \end{bmatrix}, \quad A^2\mathbf{x}_0 = \begin{bmatrix} 1.0 \\ -0.2 \end{bmatrix}, \quad A^3\mathbf{x}_0 = \begin{bmatrix} 0.35 \\ -0.82 \end{bmatrix}, \quad A^4\mathbf{x}_0 = \begin{bmatrix} -0.44 \\ -1.112 \end{bmatrix}$$

With the help of MATLAB or a computer algebra system one can show that if the first 100 terms are plotted as ordered pairs (x, y), then the points move along the elliptical orbit shown in Figure 6.1.15. An explanation of why this occurs requires a deeper analysis of the structure of matrices, a task that we will take up in subsequent chapters.

Exercise Set 6.1

In Exercises 1 and 2, find the domain and codomain of the transformation $T_A(\mathbf{x}) = A\mathbf{x}$.

1. (a) A has size 3×2.   (b) A has size 2×3.

   (c) A has size 3×3.

2. (a) A has size 4×5.   (b) A has size 5×4.

   (c) A has size 4×4.

3. If $T(x_1, x_2) = (x_1 + x_2, -x_2, 3x_1)$, then the domain of T is _____, the codomain of T is _____, and the image of $\mathbf{x} = (1, -2)$ under T is _____.

4. If $T(x_1, x_2, x_3) = (x_1 + 2x_2, x_1 - 2x_2)$, then the domain of T is _____, the codomain of T is _____, and the image of $\mathbf{x} = (0, -1, 4)$ under T is _____.

In Exercises 5 and 6, the standard matrix $[T]$ of a linear transformation T is given. Use it to find $T(\mathbf{x})$.

5. (a) $[T] = \begin{bmatrix} 1 & 2 \\ 3 & 4 \end{bmatrix}$; $\mathbf{x} = \begin{bmatrix} 3 \\ -2 \end{bmatrix}$

   (b) $[T] = \begin{bmatrix} -1 & 2 & 0 \\ 3 & 1 & 5 \end{bmatrix}$; $\mathbf{x} = \begin{bmatrix} -1 \\ 1 \\ 3 \end{bmatrix}$

6. (a) $[T] = \begin{bmatrix} -2 & 1 & 4 \\ 3 & 5 & 7 \\ 6 & 0 & -1 \end{bmatrix}$; $\mathbf{x} = \begin{bmatrix} x_1 \\ x_2 \\ x_3 \end{bmatrix}$

   (b) $[T] = \begin{bmatrix} -1 & 1 \\ 2 & 4 \\ 7 & 8 \end{bmatrix}$; $\mathbf{x} = \begin{bmatrix} x_1 \\ x_2 \end{bmatrix}$

In Exercises 7 and 8, let T_A be the transformation whose standard matrix is A. Find a vector $\mathbf{x}$, if any, whose image under T_A is $\mathbf{b}$.

7. (a) $A = \begin{bmatrix} 1 & 2 & 0 \\ 0 & -1 & 3 \\ 2 & 5 & -3 \end{bmatrix}$; $\mathbf{b} = \begin{bmatrix} 1 \\ -1 \\ 3 \end{bmatrix}$

   (b) $A = \begin{bmatrix} 1 & 2 & 0 \\ 0 & -1 & 3 \\ 2 & 5 & -3 \end{bmatrix}$; $\mathbf{b} = \begin{bmatrix} 2 \\ 1 \\ 1 \end{bmatrix}$

8. (a) $A = \begin{bmatrix} 2 & 1 & 5 & 0 \\ 3 & -2 & 4 & -7 \\ -1 & 2 & 0 & 5 \end{bmatrix}$; $\mathbf{b} = \begin{bmatrix} 7 \\ 0 \\ 4 \end{bmatrix}$

   (b) $A = \begin{bmatrix} 2 & 1 & 5 & 0 \\ 3 & -2 & 4 & -7 \\ -1 & 2 & 0 & 5 \end{bmatrix}$; $\mathbf{b} = \begin{bmatrix} -1 \\ 2 \\ 3 \end{bmatrix}$

In Exercises 9–12, determine whether T is linear. If not, state whether it is the additivity or the homogeneity (or both) that fails.

9. (a) $T(x, y) = (2x, y)$ (b) $T(x, y) = (x^2, y)$

   (c) $T(x, y) = (-y, x)$ (d) $T(x, y) = (x, 0)$

10. (a) $T(x, y) = (2x + y, x - y)$

   (b) $T(x, y) = (x + 1, y)$

   (c) $T(x, y) = (y, y)$

   (d) $T(x, y) = (\sqrt[3]{x}, \sqrt[3]{y})$

11. (a) $T(x, y, z) = (x, x + y + z)$

   (b) $T(x, y, z) = (1, 1)$

   (c) $T(x, y, z) = (0, x)$

12. (a) $T(x, y, z) = (0, 0)$

   (b) $T(x, y, z) = (y^2, z)$

   (c) $T(x, y, z) = (3x - 4y, 2x - 5z)$

In Exercises 13–16, find the domain and codomain of the transformation $\mathbf{x} \xrightarrow{T} \mathbf{w}$ defined by the equations, and determine whether T is linear.

13. $w_1 = 3x_1 - 2x_2 + 4x_3$
$w_2 = 5x_1 - 8x_2 + x_3$

14. $w_1 = 2x_1x_2 - x_2$
$w_2 = x_1 + 3x_1x_2x_3$
$w_3 = x_1 + x_2$

15. $w_1 = 5x_1 - x_2 + x_3$
$w_2 = -x_1 + x_2 + 7x_3$
$w_3 = 2x_1 - 4x_2 - x_3$

16. $w_1 = x_1^2 - 3x_2 + x_3 - 2x_4$
$w_2 = 3x_1 - 4x_2 - x_3^2 + x_4$

In Exercises 17 and 18, assume that T is linear, and use the given information to find $[T]$.

---

17. $T(1, 0) = (3, 2, 4)$ and $T(0, 1) = (-1, 3, 0)$

18. $T(1, 0, 0) = (1, 1, 2)$, $T(0, 1, 0) = (1, 1, 2)$, and $T(0, 0, 1) = (0, 3, 1)$

In Exercises 19 and 20, find the standard matrix for T, and use it to compute $T(\mathbf{x})$. Check your answer by calculating $T(\mathbf{x})$ directly.

---

19. (a) $T(x_1, x_2) = (-x_1 + x_2, x_2)$; $\mathbf{x} = (-1, 4)$
(b) $T(x_1, x_2, x_3) = (2x_1 - x_2 + x_3, x_2 + x_3, 0)$;
$\mathbf{x} = (2, 1, -3)$

20. (a) $T(x_1, x_2, x_3) = (4x_1, -x_2 + x_3)$; $\mathbf{x} = (2, 0, -5)$
(b) $T(x_1, x_2) = (3x_1, -5x_2)$; $\mathbf{x} = (-4, 0)$

21. (a) Find the standard matrix for the linear transformation $\mathbf{x} \xrightarrow{T} \mathbf{w}$ defined by the equations

$$w_1 = 3x_1 + 5x_2 - x_3$$
$$w_2 = 4x_1 - x_2 + x_3$$
$$w_3 = 3x_1 + 2x_2 - x_3$$

(b) Find the image of the vector $\mathbf{x} = (-1, 2, 4)$ under T by substituting in the equations and then by using the standard matrix.

22. (a) Find the standard matrix for the linear transformation $\mathbf{x} \xrightarrow{T} \mathbf{w}$ defined by the equations

$$w_1 = 2x_1 - 3x_2 + x_3$$
$$w_2 = 3x_1 + 5x_2 - x_3$$

(b) Find the image of the vector $\mathbf{x} = (-1, 2, 4)$ under T by substituting in the equations and then by using the standard matrix.

23. Use matrix multiplication to find the image of the vector $\mathbf{x} = (-2, 1)$ under the transformation.
(a) Reflection about the x-axis.
(b) Reflection about the line $y = -x$.
(c) Orthogonal projection on the x-axis.
(d) Orthogonal projection on the y-axis.

24. Use matrix multiplication to find the image of the vector $\mathbf{x} = (2, -1)$ under the transformation.
(a) Reflection about the y-axis.
(b) Reflection about the line $y = x$.

(c) Orthogonal projection on the x-axis.
(d) Orthogonal projection on the y-axis.

25. Use matrix multiplication to find the image of the vector $\mathbf{x} = (3, 4)$ under the rotation.
(a) Rotation of $45°$ about the origin.
(b) Rotation of $90°$ about the origin.
(c) Rotation of π radians about the origin.
(d) Rotation of $-30°$ about the origin.

26. Use matrix multiplication to find the image of the vector $\mathbf{x} = (4, 3)$ under the rotation.
(a) Rotation of $30°$ about the origin.
(b) Rotation of $-90°$ about the origin.
(c) Rotation of $3\pi/2$ radians about the origin.
(d) Rotation of $-60°$ about the origin.

27. Use matrix multiplication to find the image of the point $\mathbf{x} = (3, 4)$ under the transformation.
(a) Reflection about the line through the origin that makes an angle of $\theta = \pi/3$ with the positive x-axis.
(b) Orthogonal projection on the line in part (a).

28. Use matrix multiplication to find the image of the point $\mathbf{x} = (4, 3)$ under the transformation.
(a) Reflection about the line through the origin that makes an angle of $\theta = 120°$ with the positive x-axis.
(b) Orthogonal projection on the line in part (a).

29. Show that

$$A = \begin{bmatrix} -\frac{1}{\sqrt{2}} & -\frac{1}{\sqrt{2}} \\ \frac{1}{\sqrt{2}} & -\frac{1}{\sqrt{2}} \end{bmatrix}$$

is the standard matrix for some rotation about the origin, and find the smallest positive angle of rotation that it represents.

30. Show that

$$A = \begin{bmatrix} \frac{1}{\sqrt{2}} & \frac{1}{\sqrt{2}} \\ \frac{1}{\sqrt{2}} & -\frac{1}{\sqrt{2}} \end{bmatrix}$$

is the standard matrix for a reflection about a line through the origin. What line?

31. Let L be the line $y = mx$ that passes through the origin and has slope m, let H_L be the standard matrix for the reflection of R^2 about L, and let P_L be the standard matrix for the orthogonal projection of R^2 onto L. Use Formulas (18) and (21), the appropriate trigonometric identities, and the accompanying figure to show that

(a) $H_L = \dfrac{1}{1 + m^2} \begin{bmatrix} 1 - m^2 & 2m \\ 2m & m^2 - 1 \end{bmatrix}$

(b) $P_L = \dfrac{1}{1 + m^2} \begin{bmatrix} 1 & m \\ m & m^2 \end{bmatrix}$

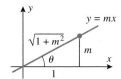

Figure Ex-31

32. Use the result in Exercise 31 to find
  (a) the reflection of the vector $\mathbf{x} = (3, 4)$ about the line $y = 2x$;
  (b) the orthogonal projection of the vector $\mathbf{x} = (3, 4)$ onto the line $y = 2x$.

33. Use the result in Exercise 31 to find
  (a) the reflection of the vector $\mathbf{x} = (4, 3)$ about the line $y = 3x$;
  (b) the orthogonal projection of the vector $\mathbf{x} = (4, 3)$ onto the line $y = 3x$.

34. Show that the formula $T(x, y) = (0, 0)$ defines a linear operator on R^2, but the formula $T(x, y) = (1, 1)$ does not.

35. Let $T_A : R^3 \rightarrow R^3$ be multiplication by

$$A = \begin{bmatrix} -1 & 3 & 0 \\ 2 & 1 & 2 \\ 4 & 5 & -3 \end{bmatrix}$$

and let $\mathbf{e}_1$, $\mathbf{e}_2$, and $\mathbf{e}_3$ be the standard basis vectors for R^3. Find the following vectors by inspection.
  (a) $T_A(\mathbf{e}_1), T_A(\mathbf{e}_2), T_A(\mathbf{e}_3)$ (b) $T_A(\mathbf{e}_1 + \mathbf{e}_2 + \mathbf{e}_3)$
  (c) $T_A(7\mathbf{e}_3)$

36. Suppose that $(x, y) \xrightarrow{T} (s, t)$ is the linear operator on R^2 defined by the equations

$$\begin{aligned} 2x + y &= s \\ 6x + 2y &= t \end{aligned}$$

Find the image of the line $x + y = 1$ under this operator.

37. Find the standard matrix for the linear operator $T : R^3 \rightarrow R^3$ described in the figure.

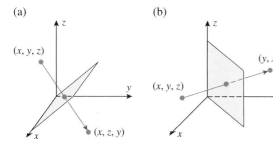

(a)   (b)

38. Sketch the image under the stated transformation of the rectangle in the xy-plane whose vertices are $(0, 0)$, $(1, 0)$, $(1, 2)$, $(0, 2)$.
  (a) Reflection about the x-axis.
  (b) Reflection about the y-axis.
  (c) Rotation about the origin through $\theta = \pi/4$.
  (d) Reflection about $y = x$.

39. If $T : R^2 \rightarrow R^2$ projects a vector orthogonally onto the x-axis and then reflects that vector about the y-axis, then the standard matrix for T is _____.

40. If $T : R^2 \rightarrow R^2$ reflects a vector about the line $y = x$ and then reflects that vector about the x-axis, then the standard matrix for T is _____.

Discussion and Discovery

D1. Indicate whether the statement is true (T) or false (F). Justify your answer.
  (a) If T maps R^n into R^m and $T(\mathbf{0}) = \mathbf{0}$, then T is linear.
  (b) If T maps R^n into R^m, and if

$$T(c_1\mathbf{u} + c_2\mathbf{v}) = c_1 T(\mathbf{u}) + c_2 T(\mathbf{v})$$

  for all scalars c_1 and c_2 and for all vectors $\mathbf{u}$ and $\mathbf{v}$ in R^n, then T is linear.
  (c) There is only one linear transformation $T : R^n \rightarrow R^n$ such that $T(-\mathbf{v}) = -T(\mathbf{v})$ for all $\mathbf{v}$ in R^n.
  (d) There is only one linear transformation $T : R^n \rightarrow R^n$ for which $T(\mathbf{u} + \mathbf{v}) = T(\mathbf{u} - \mathbf{v})$ for all vectors $\mathbf{u}$ and $\mathbf{v}$ in R^n.
  (e) If $\mathbf{v}_0$ is a nonzero vector in R^n, then the formula $T(\mathbf{v}) = \mathbf{v}_0 + \mathbf{v}$ defines a linear operator on V.

D2. Let L be the line through the origin of R^2 that makes an angle of $\pi/4$ with the positive x-axis, and let A be the standard matrix for the reflection of R^2 about that line. Make a conjecture about the eigenvalues and eigenvectors of A and confirm your conjecture by computing them in the usual way.

D3. In words, describe the geometric effect of multiplying a vector $\mathbf{x}$ by the matrix

$$A = \begin{bmatrix} \cos^2\theta - \sin^2\theta & -2\sin\theta\cos\theta \\ 2\sin\theta\cos\theta & \cos^2\theta - \sin^2\theta \end{bmatrix}$$

D4. If multiplication by A rotates a vector $\mathbf{x}$ in the xy-plane through an angle θ, what is the effect of multiplying $\mathbf{x}$ by A^T? Explain your reasoning.

D5. Let $\mathbf{x}_0$ be a nonzero column vector in R^2, and suppose that $T : R^2 \to R^2$ is the transformation defined by $T(\mathbf{x}) = \mathbf{x}_0 + R_\theta\mathbf{x}$, where R_θ is the standard matrix of the rotation of R^2 about the origin through the angle θ. Give a geometric description of this transformation. Is it a linear transformation? Explain.

D6. A function of the form $f(x) = mx + b$ is commonly called a "linear function" because the graph of $y = mx + b$ is a line. Is f additive and homogeneous?

D7. Let $\mathbf{x} = \mathbf{x}_0 + t\mathbf{v}$ be a line in R^n, and let $T : R^n \to R^n$ be a linear operator on R^n. What kind of geometric object is the image of this line under the operator T? Explain your reasoning.

Technology Exercises

T1. (a) Find the reflection of the point $(1, 3)$ about the line through the origin of the xy-plane that makes an angle of $12°$ with the positive x-axis.

(b) Given that the point $(2, -1)$ reflects into the point $(1, 2)$ about an unknown line L through the origin of the xy-plane, find the reflection of the point $(5, -2)$ about L.

T2. (a) Find the orthogonal projection of the point $(1, 3)$ onto the line through the origin of the xy-plane that makes an angle of $12°$ with the positive x-axis.

(b) Given that the point $(5, 1)$ projects orthogonally onto the point $(2, 3)$ on a line L through the origin of the xy-plane, find the orthogonal projection of the point $(5, -2)$ on L.

T3. Generate Figure 6.1.15 and explore the behavior of the power sequence for other choices of $\mathbf{x}_0$.

Section 6.2   Geometry of Linear Operators

In this section we will study some of the geometric properties of linear operators. The ideas that we will develop here have applications in computer graphics and are used in many important numerical algorithms.

NORM-PRESERVING LINEAR OPERATORS

In the last section we studied three kinds of operators on R^2: rotations about the origin, reflections about lines through the origin, and orthogonal projections onto lines through the origin; and we showed that the standard matrices for these operators are

$$R_\theta = \begin{bmatrix} \cos\theta & -\sin\theta \\ \sin\theta & \cos\theta \end{bmatrix}, \quad H_\theta = \begin{bmatrix} \cos 2\theta & \sin 2\theta \\ \sin 2\theta & -\cos 2\theta \end{bmatrix}, \quad P_\theta = \begin{bmatrix} \cos^2\theta & \sin\theta\cos\theta \\ \sin\theta\cos\theta & \sin^2\theta \end{bmatrix} \tag{1–3}$$

Rotation about the origin through an angle θ	Reflection about the line through the origin making an angle θ with the positive x-axis	Orthogonal projection onto the line through the origin making an angle θ with the positive x-axis

As suggested in Figure 6.2.1, rotations about the origin and reflections about lines through the origin do not change the lengths of vectors or the angles between vectors; thus, we say that these operators are *length preserving* and *angle preserving*. In contrast, an orthogonal projection onto a line through the origin can change the length of a vector and the angle between vectors (verify by drawing a picture).

In general, a linear operator $T : R^n \to R^n$ with the length-preserving property $\|T(\mathbf{x})\| = \|\mathbf{x}\|$ is called an ***orthogonal operator*** or a ***linear isometry*** (from the Greek *isometros*, meaning "equal measure"). Thus, for example, rotations about the origin and reflections about lines through the origin of R^2 are orthogonal operators. As we will see, the fact that these two operators preserve

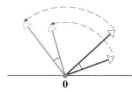

A rotation about the origin does not change lengths of vectors or angles between vectors.

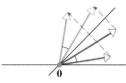

A reflection about a line through the origin does not change lengths of vectors or angles between vectors.

Figure 6.2.1

angles as well as lengths is a consequence of the following theorem, which shows that length-preserving linear operators are ***dot product preserving***, and conversely.

Theorem 6.2.1 *If $T: R^n \to R^n$ is a linear operator on R^n, then the following statements are equivalent.*

(a) $\|T(\mathbf{x})\| = \|\mathbf{x}\|$ *for all $\mathbf{x}$ in R^n.*   **[T orthogonal (i.e., length preserving)]**

(b) $T(\mathbf{x}) \cdot T(\mathbf{y}) = \mathbf{x} \cdot \mathbf{y}$ *for all $\mathbf{x}$ and $\mathbf{y}$ in R^n.*   **[T is dot product preserving.]**

Proof (a) $\Rightarrow$ (b) Suppose that T is length preserving, and let $\mathbf{x}$ and $\mathbf{y}$ be any two vectors in R^n. We leave it for you to derive the relationship

$$\mathbf{x} \cdot \mathbf{y} = \tfrac{1}{4}\left(\|\mathbf{x} + \mathbf{y}\|^2 - \|\mathbf{x} - \mathbf{y}\|^2\right) \tag{4}$$

by writing $\|\mathbf{x} + \mathbf{y}\|^2$ and $\|\mathbf{x} - \mathbf{y}\|^2$ as

$$\|\mathbf{x} + \mathbf{y}\|^2 = (\mathbf{x} + \mathbf{y}) \cdot (\mathbf{x} + \mathbf{y}) \quad \text{and} \quad \|\mathbf{x} - \mathbf{y}\|^2 = (\mathbf{x} - \mathbf{y}) \cdot (\mathbf{x} - \mathbf{y})$$

and then expanding the dot products. It now follows from (4) that

$$T(\mathbf{x}) \cdot T(\mathbf{y}) = \tfrac{1}{4}\left(\|T(\mathbf{x}) + T(\mathbf{y})\|^2 - \|T(\mathbf{x}) - T(\mathbf{y})\|^2\right)$$

$$= \tfrac{1}{4}\left(\|T(\mathbf{x} + \mathbf{y})\|^2 - \|T(\mathbf{x} - \mathbf{y})\|^2\right) \quad \textbf{[Additivity and Theorem 6.1.3]}$$

$$= \tfrac{1}{4}\left(\|\mathbf{x} + \mathbf{y}\|^2 - \|\mathbf{x} - \mathbf{y}\|^2\right) \quad \textbf{[T is length preserving.]}$$

$$= \mathbf{x} \cdot \mathbf{y} \quad \textbf{[Formula (4)]}$$

Proof (b) $\Rightarrow$ (a) Conversely, suppose that T is dot product preserving, and let $\mathbf{x}$ be any vector in R^n. Since we can express $\|\mathbf{x}\|$ as

$$\|\mathbf{x}\| = \sqrt{\mathbf{x} \cdot \mathbf{x}} \tag{5}$$

it follows that

$$\|T(\mathbf{x})\| = \sqrt{T(\mathbf{x}) \cdot T(\mathbf{x})} = \sqrt{\mathbf{x} \cdot \mathbf{x}} = \|\mathbf{x}\| \qquad \blacksquare$$

REMARK Formulas (4) and (5) are "flip sides of a coin" in that (5) provides a way of expressing norms in terms of dot products, whereas (4), which is sometimes called the ***polarization identity***, provides a way of expressing dot products in terms of norms.

ORTHOGONAL OPERATORS PRESERVE ANGLES AND ORTHOGONALITY

Recall from the remark following Theorem 1.2.12 that the angle between two nonzero vectors $\mathbf{x}$ and $\mathbf{y}$ in R^n is given by the formula

$$\theta = \cos^{-1}\left(\frac{\mathbf{x} \cdot \mathbf{y}}{\|\mathbf{x}\|\|\mathbf{y}\|}\right) \tag{6}$$

Thus, if $T: R^n \to R^n$ is an orthogonal operator, the fact that T is length preserving and dot product preserving implies that

$$\cos^{-1}\left(\frac{T(\mathbf{x}) \cdot T(\mathbf{y})}{\|T(\mathbf{x})\|\|T(\mathbf{y})\|}\right) = \cos^{-1}\left(\frac{\mathbf{x} \cdot \mathbf{y}}{\|\mathbf{x}\|\|\mathbf{y}\|}\right) \tag{7}$$

which implies that an orthogonal operator preserves angles. In particular, an orthogonal operator preserves orthogonality in the sense that the images of two vectors are orthogonal if and only if the original vectors are orthogonal.

ORTHOGONAL MATRICES

Our next goal is to explore the relationship between the orthogonality of an operator and properties of its standard matrix. As a first step, suppose that A is the standard matrix for an orthogonal linear operator $T: R^n \to R^n$. Since $T(\mathbf{x}) = A\mathbf{x}$ for all $\mathbf{x}$ in R^n, and since $\|T(\mathbf{x})\| = \|\mathbf{x}\|$, it follows that

$$\|A\mathbf{x}\| = \|\mathbf{x}\| \tag{8}$$

for all $\mathbf{x}$ in R^n. Thus, the problem of determining whether a linear operator is orthogonal reduces to determining whether its standard matrix satisfies (8) for all $\mathbf{x}$ in R^n. The following definition will be useful in our investigation of this problem.

Definition 6.2.2 A square matrix A is said to be **orthogonal** if $A^{-1} = A^T$.

REMARK As a practical matter, you can show that a square matrix A is orthogonal by showing that $A^T A = I$ or $AA^T = I$, since either condition implies the other by Theorem 3.3.8.

EXAMPLE 1
An Orthogonal
Matrix

The matrix

$$A = \begin{bmatrix} \frac{3}{7} & \frac{2}{7} & \frac{6}{7} \\ -\frac{6}{7} & \frac{3}{7} & \frac{2}{7} \\ \frac{2}{7} & \frac{6}{7} & -\frac{3}{7} \end{bmatrix}$$

is orthogonal since

$$A^T A = \begin{bmatrix} \frac{3}{7} & -\frac{6}{7} & \frac{2}{7} \\ \frac{2}{7} & \frac{3}{7} & \frac{6}{7} \\ \frac{6}{7} & \frac{2}{7} & -\frac{3}{7} \end{bmatrix} \begin{bmatrix} \frac{3}{7} & \frac{2}{7} & \frac{6}{7} \\ -\frac{6}{7} & \frac{3}{7} & \frac{2}{7} \\ \frac{2}{7} & \frac{6}{7} & -\frac{3}{7} \end{bmatrix} = \begin{bmatrix} 1 & 0 & 0 \\ 0 & 1 & 0 \\ 0 & 0 & 1 \end{bmatrix} = I$$

and hence

$$A^{-1} = A^T = \begin{bmatrix} \frac{3}{7} & -\frac{6}{7} & \frac{2}{7} \\ \frac{2}{7} & \frac{3}{7} & \frac{6}{7} \\ \frac{6}{7} & \frac{2}{7} & -\frac{3}{7} \end{bmatrix}$$

∎

The following theorem states some of the basic properties of orthogonal matrices.

Theorem 6.2.3

(a) *The transpose of an orthogonal matrix is orthogonal.*

(b) *The inverse of an orthogonal matrix is orthogonal.*

(c) *A product of orthogonal matrices is orthogonal.*

(d) *If A is orthogonal, then $\det(A) = 1$ or $\det(A) = -1$.*

Proof (a) If A is orthogonal, then $A^T A = I$. We can rewrite this as $A^T (A^T)^T = I$, which implies that $(A^T)^{-1} = (A^T)^T$. Thus, A^T is orthogonal.

Proof (b) If A is orthogonal, then $A^{-1} = A^T$. Transposing both sides of this equation yields

$$(A^{-1})^T = (A^T)^T = A = (A^{-1})^{-1}$$

which implies that A^{-1} is orthogonal.

Proof (c) We will give the proof for a product of two orthogonal matrices. If A and B are orthogonal matrices, then

$$(AB)^{-1} = B^{-1}A^{-1} = B^T A^T = (AB)^T$$

Thus, AB is orthogonal.

Proof (*d*) If A is orthogonal, then $A^TA = I$. Taking the determinant of both sides, and using properties of determinants yields

$$\det(A)\det(A^T) = \det(A)\det(A) = 1$$

which implies that $\det(A) = 1$ or $\det(A) = -1$. ∎

CONCEPT PROBLEM As a check, you may want to confirm that $\det(A) = -1$ for the orthogonal matrix A in Example 1.

The following theorem, which is applicable to general matrices, is important in its own right and will lead to an effective way of telling whether a square matrix is orthogonal.

Theorem 6.2.4 *If A is an $m \times n$ matrix, then the following statements are equivalent.*

(*a*) $A^TA = I$.

(*b*) $\|A\mathbf{x}\| = \|\mathbf{x}\|$ *for all $\mathbf{x}$ in R^n.*

(*c*) $A\mathbf{x} \cdot A\mathbf{y} = \mathbf{x} \cdot \mathbf{y}$ *for all $\mathbf{x}$ and $\mathbf{y}$ in R^n.*

(*d*) *The column vectors of A are orthonormal.*

We will prove the chain of implications $(a) \Rightarrow (b) \Rightarrow (c) \Rightarrow (d) \Rightarrow (a)$.

Proof $(a) \Rightarrow (b)$ It follows from Formula (12) of Section 3.2 that

$$\|A\mathbf{x}\|^2 = A\mathbf{x} \cdot A\mathbf{x} = \mathbf{x} \cdot A^TA\mathbf{x} = \mathbf{x} \cdot I\mathbf{x} = \mathbf{x} \cdot \mathbf{x} = \|\mathbf{x}\|^2$$

from which part (*b*) follows.

Proof $(b) \Rightarrow (c)$ This follows from Theorem 6.2.1 with $T(\mathbf{x}) = A\mathbf{x}$.

Proof $(c) \Rightarrow (d)$ Define $T : R^n \to R^n$ to be the matrix operator $T(\mathbf{x}) = A\mathbf{x}$. By hypothesis, $T(\mathbf{x}) \cdot T(\mathbf{y}) = \mathbf{x} \cdot \mathbf{y}$ for all $\mathbf{x}$ and $\mathbf{y}$ in R^n, so Theorem 6.2.1 implies that $\|T(\mathbf{x})\| = \|\mathbf{x}\|$ for all $\mathbf{x}$ in R^n. This tells us that T preserves lengths and orthogonality, so T must map every set of orthonormal vectors into another set of orthonormal vectors. This is true, in particular, for the set of standard unit vectors, so

$$T(\mathbf{e}_1) = A\mathbf{e}_1, \quad T(\mathbf{e}_2) = A\mathbf{e}_2, \ldots, \quad T(\mathbf{e}_n) = A\mathbf{e}_n$$

must be an orthonormal set. However, these are the column vectors of A (why?), which proves part (*d*).

Proof $(d) \Rightarrow (a)$ Assume that the column vectors of A are orthonormal, and denote these vectors by $\mathbf{a}_1, \mathbf{a}_2, \ldots, \mathbf{a}_n$. It follows from Formula (9) of Section 3.6 that $A^TA = I$ (verify). ∎

If A is square, then the condition $A^TA = I$ in part (*a*) of Theorem 6.2.4 is equivalent to saying that $A^{-1} = A^T$ (i.e., A is orthogonal). Thus, in the case of a square matrix, Theorems 6.2.4 and 6.2.3 together yield the following theorem about orthogonal matrices.

Theorem 6.2.5 *If A is an $n \times n$ matrix, then the following statements are equivalent.*

(*a*) *A is orthogonal.*

(*b*) $\|A\mathbf{x}\| = \|\mathbf{x}\|$ *for all $\mathbf{x}$ in R^n.*

(*c*) $A\mathbf{x} \cdot A\mathbf{y} = \mathbf{x} \cdot \mathbf{y}$ *for all $\mathbf{x}$ and $\mathbf{y}$ in R^n.*

(*d*) *The column vectors of A are orthonormal.*

(*e*) *The row vectors of A are orthonormal.*

Recall that a linear operator $T : R^n \to R^n$ is defined to be orthogonal if $\|T(\mathbf{x})\| = \|\mathbf{x}\|$ for all $\mathbf{x}$ in R^n. Thus, T is orthogonal if and only if its standard matrix has the property $\|A\mathbf{x}\| = \|\mathbf{x}\|$

for all **x** in R^n. This fact and parts (a) and (b) of Theorem 6.2.5 yield the following result about standard matrices of orthogonal operators.

Theorem 6.2.6   *A linear operator* $T : R^n \rightarrow R^n$ *is orthogonal if and only if its standard matrix is orthogonal.*

EXAMPLE 2
Standard
Matrices of
Rotations and
Reflections Are
Orthogonal

Since rotations about the origin and reflections about lines through the origin of R^2 are orthogonal operators, the standard matrices of these operators must be orthogonal. This is indeed the case, since Formula (1) implies that

$$R_\theta^T R_\theta = \begin{bmatrix} \cos\theta & \sin\theta \\ -\sin\theta & \cos\theta \end{bmatrix} \begin{bmatrix} \cos\theta & -\sin\theta \\ \sin\theta & \cos\theta \end{bmatrix}$$

$$= \begin{bmatrix} \cos^2\theta + \sin^2\theta & 0 \\ 0 & \sin^2\theta + \cos^2\theta \end{bmatrix} = \begin{bmatrix} 1 & 0 \\ 0 & 1 \end{bmatrix} = I$$

and, similarly, Formula (2) implies that $H_\theta^T H_\theta = I$ (verify).   ∎

EXAMPLE 3
Identifying
Orthogonal
Matrices

We showed in Example 1 that the matrix

$$A = \begin{bmatrix} \frac{3}{7} & \frac{2}{7} & \frac{6}{7} \\ -\frac{6}{7} & \frac{3}{7} & \frac{2}{7} \\ \frac{2}{7} & \frac{6}{7} & -\frac{3}{7} \end{bmatrix} \tag{9}$$

is orthogonal by confirming that $A^T A = I$. In light of Theorem 6.2.5, we can also establish the orthogonality of A by showing that the row vectors or the column vectors are orthonormal. We leave it for you to check both.   ∎

**ALL ORTHOGONAL
LINEAR OPERATORS ON
R^2 ARE ROTATIONS OR
REFLECTIONS**

We have seen that rotations about the origin and reflections about lines through the origin of R^2 are orthogonal (i.e., length preserving) operators. We will now show that these are the *only* orthogonal operators on R^2.

Theorem 6.2.7   *If* $T : R^2 \rightarrow R^2$ *is an orthogonal linear operator, then the standard matrix for* T *is expressible in the form*

$$R_\theta = \begin{bmatrix} \cos\theta & -\sin\theta \\ \sin\theta & \cos\theta \end{bmatrix} \quad or \quad H_{\theta/2} = \begin{bmatrix} \cos\theta & \sin\theta \\ \sin\theta & -\cos\theta \end{bmatrix} \tag{10}$$

That is, T *is either a rotation about the origin or a reflection about a line through the origin.*

Proof   Assume that T is an orthogonal linear operator on R^2 and that its standard matrix is

$$A = \begin{bmatrix} a & b \\ c & d \end{bmatrix}$$

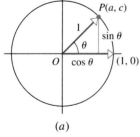

(a)

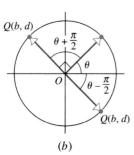

(b)

Figure 6.2.2

This matrix is orthogonal, so its column vectors are orthonormal. Thus, $a^2 + c^2 = 1$, which means that the point $P(a, c)$ lies on the unit circle (Figure 6.2.2a). If we now let θ be the angle from $\mathbf{e}_1 = (1, 0)$ to the vector $\overrightarrow{OP} = (a, c)$, then we can express a and c as

$$a = \cos\theta \quad \text{and} \quad c = \sin\theta$$

and we can rewrite A as

$$A = \begin{bmatrix} \cos\theta & b \\ \sin\theta & d \end{bmatrix} \tag{11}$$

The orthogonality of A also implies that the second column vector of A is orthogonal to the first, and hence the counterclockwise angle from $\mathbf{e}_1$ to the vector $\overrightarrow{OQ} = (b, d)$ must be $\theta + \pi/2$ or

$\theta - \pi/2$ (Figure 6.2.2*b*). In the first case we have

$$b = \cos(\theta + \pi/2) = -\sin\theta \quad \text{and} \quad d = \sin(\theta + \pi/2) = \cos\theta$$

and in the second case we have

$$b = \cos(\theta - \pi/2) = \sin\theta \quad \text{and} \quad d = \sin(\theta - \pi/2) = -\cos\theta$$

Substituting these expressions in (11) yields the two possible forms in (10). ∎

If A is an orthogonal 2×2 matrix, then we know from Theorem 6.2.7 that the corresponding linear operator is either a rotation about the origin or a reflection about a line through the origin. The determinant of A can be used to distinguish between the two cases, since it follows from (1) and (2) that

$$\det(R_\theta) = \begin{vmatrix} \cos\theta & -\sin\theta \\ \sin\theta & \cos\theta \end{vmatrix} = \cos^2\theta + \sin^2\theta = 1$$

$$\det(H_{\theta/2}) = \begin{vmatrix} \cos\theta & \sin\theta \\ \sin\theta & -\cos\theta \end{vmatrix} = -(\cos^2\theta + \sin^2\theta) = -1$$

Thus, a 2×2 orthogonal matrix represents a rotation if $\det(A) = 1$ and a reflection if $\det(A) = -1$.

EXAMPLE 4
Geometric
Properties of
Orthogonal
Matrices

In each part, describe the linear operator on R^2 corresponding to the matrix A.

(a) $A = \begin{bmatrix} 1/\sqrt{2} & -1/\sqrt{2} \\ 1/\sqrt{2} & 1/\sqrt{2} \end{bmatrix}$ (b) $A = \begin{bmatrix} 1/\sqrt{2} & 1/\sqrt{2} \\ 1/\sqrt{2} & -1/\sqrt{2} \end{bmatrix}$

Solution (*a*) The column vectors of A are orthonormal (verify), so the matrix A is orthogonal. This implies that the operator is either a rotation about the origin or a reflection about a line through the origin. Since $\det(A) = 1$, we know definitively that the operator is a rotation. We can determine the angle of rotation by comparing A to the general rotation matrix R_θ in (1). This yields

$$\cos\theta = 1/\sqrt{2} \quad \text{and} \quad \sin\theta = 1/\sqrt{2}$$

from which we conclude that the angle of rotation is $\theta = \pi/4\,(= 45°)$.

Solution (*b*) The matrix A is orthogonal and $\det(A) = -1$, so the corresponding operator is a reflection about a line through the origin. We can determine the angle that the line makes with the positive x-axis by comparing A to the general reflection matrix H_θ in (2). This yields

$$\cos 2\theta = 1/\sqrt{2} \quad \text{and} \quad \sin 2\theta = 1/\sqrt{2}$$

from which we conclude that $\theta = \pi/8\,(= 22.5°)$. ∎

**CONTRACTIONS AND
DILATIONS OF R^2**

Up to now we have focused primarily on length-preserving linear operators; now we will consider some important linear operators that are not length preserving.

If k is a nonnegative scalar, then the linear operator $T(x, y) = (kx, ky)$ is called the *scaling operator with factor k*. In particular, this operator is called a *contraction* if $0 \le k < 1$ and a *dilation* if $k > 1$. Contractions preserve the directions of vectors but reduce their lengths by the factor k, and dilations preserve the directions of vectors but increase their lengths by the factor k. Table 6.2.1 provides the basic information about scaling operators on R^2.

**VERTICAL AND
HORIZONTAL
COMPRESSIONS AND
EXPANSIONS OF R^2**

An operator $T(x, y) = (kx, y)$ that multiplies the x-coordinate of each point in the xy-plane by a nonnegative constant k has the effect of expanding or compressing every figure in the plane in the x-direction—it compresses if $0 \le k < 1$ and expands if $k > 1$. Accordingly, we call T the *expansion* (or *compression*) *in the x-direction with factor k*. Similarly, $T(x, y) = (x, ky)$ is the *expansion* (or *compression*) *in the y-direction with factor k*. Table 6.2.2 provides the basic information about expansion and compression operators on R^2.

Table 6.2.1

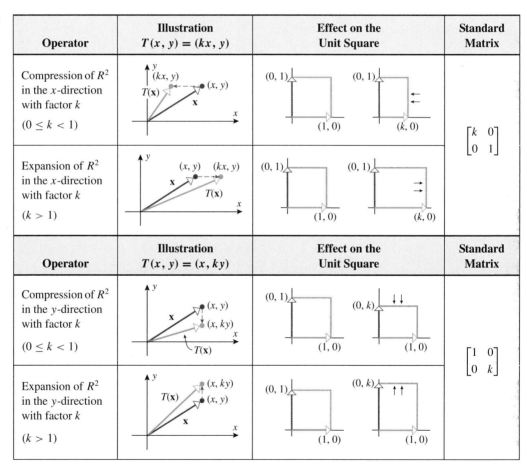

Operator	Illustration $T(x, y) = (kx, ky)$	Effect on the Unit Square	Standard Matrix
Contraction with factor k on R^2 ($0 \leq k < 1$)			$\begin{bmatrix} k & 0 \\ 0 & k \end{bmatrix}$
Dilation with factor k on R^2 ($k > 1$)			

Table 6.2.2

Operator	Illustration $T(x, y) = (kx, y)$	Effect on the Unit Square	Standard Matrix
Compression of R^2 in the x-direction with factor k ($0 \leq k < 1$)			$\begin{bmatrix} k & 0 \\ 0 & 1 \end{bmatrix}$
Expansion of R^2 in the x-direction with factor k ($k > 1$)			

Operator	Illustration $T(x, y) = (x, ky)$	Effect on the Unit Square	Standard Matrix
Compression of R^2 in the y-direction with factor k ($0 \leq k < 1$)			$\begin{bmatrix} 1 & 0 \\ 0 & k \end{bmatrix}$
Expansion of R^2 in the y-direction with factor k ($k > 1$)			

SHEARS   A linear operator of the form $T(x, y) = (x + ky, y)$ translates a point (x, y) in the xy-plane parallel to the x-axis by an amount ky that is proportional to the y-coordinate of the point. This operator leaves the points on the x-axis fixed (since $y = 0$), but as we progress away from the x-axis, the translation distance increases. We call this operator the ***shear in the x-direction with factor k***. Similarly, a linear operator of the form $T(x, y) = (x, y + kx)$ is called the ***shear in the y-direction with factor k***. Table 6.2.3 provides the basic information about shears in R^2.

Table 6.2.3

Operator	Effect on the Unit Square	Standard Matrix
Shear of R^2 in the x-direction with factor k $T(x, y) = (x + ky, y)$	$(0, 1)$ $(1, 0)$ $(k, 1)$ $(1, 0)$ $(k > 0)$ $(k, 1)$ $(1, 0)$ $(k < 0)$	$\begin{bmatrix} 1 & k \\ 0 & 1 \end{bmatrix}$
Shear of R^2 in the y-direction with factor k $T(x, y) = (x, y + kx)$	$(0, 1)$ $(1, 0)$ $(0, 1)$ $(1, k)$ $(k > 0)$ $(0, 1)$ $(1, k)$ $(k < 0)$	$\begin{bmatrix} 1 & 0 \\ k & 1 \end{bmatrix}$

EXAMPLE 5
Some Basic
Linear
Operators
on R^2

In each part describe the linear operator corresponding to A, and show its effect on the unit square.

(a) $A_1 = \begin{bmatrix} 1 & 2 \\ 0 & 1 \end{bmatrix}$ (b) $A_2 = \begin{bmatrix} 2 & 0 \\ 0 & 2 \end{bmatrix}$ (c) $A_3 = \begin{bmatrix} 2 & 0 \\ 0 & 1 \end{bmatrix}$

Solution By comparing the forms of these matrices to those in Tables 6.2.1, 6.2.2, and 6.2.3, we see that the matrix A_1 corresponds to a shear in the x-direction with factor 2, the matrix A_2 corresponds to a dilation with factor 2, and A_3 corresponds to an expansion in the x-direction with factor 2. The effects of these operators on the unit square are shown in Figure 6.2.3.   ■

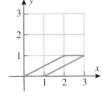 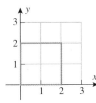 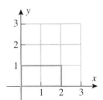

Figure 6.2.3

EXAMPLE 6
Application to
Computer
Graphics

Figure 6.2.4 shows a famous picture of Albert Einstein and three computer-generated linear transformations of that picture. The original picture was scanned and then digitized to decompose it into a rectangular array of pixels. The transformed picture was then obtained as follows:

- The program MATLAB was used to assign coordinates and a gray level to each pixel.
- The coordinates of the pixels were transformed by matrix multiplication.
- The images were then assigned their original gray levels to produce the transformed picture.   ■

LINEAR OPERATORS ON R^3

We now turn our attention to linear operators on R^3. As in R^2, we will want to distinguish between operators that preserve lengths (orthogonal operators) and those that do not. The most important linear operators that are not length preserving are orthogonal projections onto subspaces, and the simplest of these are the orthogonal projections onto the coordinate planes of an xyz-coordinate system. Table 6.2.4 provides the basic information about such operators. More general orthogonal projections will be studied in the next chapter, so we will devote the rest of this section to orthogonal operators.

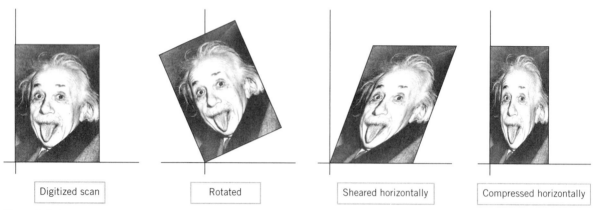

| Digitized scan | Rotated | Sheared horizontally | Compressed horizontally |

Figure 6.2.4

Table 6.2.4

Operator	Illustration	Standard Matrix
Orthogonal projection on the xy-plane $T(x, y, z) = (x, y, 0)$		$\begin{bmatrix} 1 & 0 & 0 \\ 0 & 1 & 0 \\ 0 & 0 & 0 \end{bmatrix}$
Orthogonal projection on the xz-plane $T(x, y, z) = (x, 0, z)$		$\begin{bmatrix} 1 & 0 & 0 \\ 0 & 0 & 0 \\ 0 & 0 & 1 \end{bmatrix}$
Orthogonal projection on the yz-plane $T(x, y, z) = (0, y, z)$		$\begin{bmatrix} 0 & 0 & 0 \\ 0 & 1 & 0 \\ 0 & 0 & 1 \end{bmatrix}$

We have seen that 2×2 orthogonal matrices correspond to rotations about the origin or reflections about lines through the origin in R^2. One can prove that all 3×3 orthogonal matrices correspond to linear operators on R^3 of the following types:

Type 1: Rotations about lines through the origin.

Type 2: Reflections about planes through the origin.

Type 3: A rotation about a line through the origin followed by a reflection about the plane through the origin that is perpendicular to the line.

(A proof of this result can be found in *Linear Algebra and Geometry*, by David Bloom, Cambridge University Press, New York, 1979.)

Recall that one call tell whether a 2×2 orthogonal matrix A represents a rotation or a reflection by its determinant—a rotation if $\det(A) = 1$ and a reflection if $\det(A) = -1$. Similarly, if A is a 3×3 orthogonal matrix, then A represents a rotation (i.e., is of type 1) if $\det(A) = 1$ and represents a type 2 or type 3 operator if $\det(A) = -1$. Accordingly, we will frequently refer to 2×2 or 3×3 orthogonal matrices with determinant 1 as ***rotation matrices***. To tell whether a 3×3 orthogonal matrix with determinant -1 represents a type 2 or a type 3 operator requires an analysis of eigenvectors and eigenvalues.

REFLECTIONS ABOUT COORDINATE PLANES

The most basic reflections in a rectangular xyz-coordinate system are those about the coordinate planes. Table 6.2.5 provides the basic information about such operators on R^3.

Table 6.2.5

Operator	Illustration	Standard Matrix
Reflection about the xy-plane $T(x, y, z) = (x, y, -z)$		$\begin{bmatrix} 1 & 0 & 0 \\ 0 & 1 & 0 \\ 0 & 0 & -1 \end{bmatrix}$
Reflection about the xz-plane $T(x, y, z) = (x, -y, z)$		$\begin{bmatrix} 1 & 0 & 0 \\ 0 & -1 & 0 \\ 0 & 0 & 1 \end{bmatrix}$
Reflection about the yz-plane $T(x, y, z) = (-x, y, z)$		$\begin{bmatrix} -1 & 0 & 0 \\ 0 & 1 & 0 \\ 0 & 0 & 1 \end{bmatrix}$

ROTATIONS IN R^3

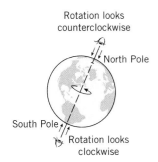

Rotation looks counterclockwise

North Pole

South Pole

Rotation looks clockwise

Figure 6.2.5

We will now turn our attention to rotations in R^3. To help understand some of the issues involved, we will begin with a familiar example—the rotation of the Earth about its axis through the North and South Poles. For simplicity, we will assume that the Earth is a sphere. Since the Sun rises in the east and sets in the west, we know that the Earth rotates from west to east. However, to an observer above the North Pole the rotation will appear counterclockwise, and to an observer below the South Pole it will appear clockwise (Figure 6.2.5). Thus, when a rotation in R^3 is described as clockwise or counterclockwise, a direction of view along the axis of rotation must also be stated.

There are some other facts about the Earth's rotation that are useful for understanding general rotations in R^3. For example, as the Earth rotates about its axis, the North and South Poles remain fixed, as do all other points that lie on the axis of rotation. Thus, the axis of rotation can be thought of as the line of fixed points in the Earth's rotation. Moreover, all points on the Earth that are not on the axis of rotation move in circular paths that are centered on the axis and lie in planes that are perpendicular to the axis. For example, the points in the Equatorial Plane move within the Equatorial Plane in circles about the Earth's center.

A rotation of R^3 is an orthogonal operator with a line of fixed points, called the ***axis of rotation***. In this section we will only be concerned with rotations about lines through the origin, and we will assume for simplicity that an angle of rotation is at most 180° (π radians). If $T : R^3 \to R^3$ is a rotation through an angle θ about a line through the origin, and if W is the

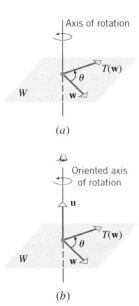

Axis of rotation

(a)

Oriented axis of rotation

(b)

Figure 6.2.6

plane through the origin that is perpendicular to the axis of rotation, then T rotates each nonzero vector $\mathbf{w}$ in W about the origin through the angle θ into a vector $T(\mathbf{w})$ in W (Figure 6.2.6a). Thus, within the plane W, the operator T behaves like a rotation of R^2 about the origin. To establish a direction of rotation in W for the angle θ, we need to establish a direction of view along the axis of rotation. We can do this by choosing a nonzero vector $\mathbf{u}$ on the axis of rotation with its initial point at the origin and agree to view W by looking from the terminal point of $\mathbf{u}$ toward the origin; we will call $\mathbf{u}$ an ***orientation*** of the axis of rotation (Figure 6.2.6b).

Now let us see how we might choose the orientation $\mathbf{u}$ so that rotations in the plane W appear counterclockwise when viewed from the terminal point of $\mathbf{u}$. If $\theta \neq 0$ and $\theta \neq \pi$,[*] then we can accomplish this by taking

$$\mathbf{u} = \mathbf{w} \times T(\mathbf{w}) \tag{12}$$

where $\mathbf{w}$ is any nonzero vector in W. With this choice of $\mathbf{u}$, the right-hand rule holds, and the rotation of $\mathbf{w}$ into $T(\mathbf{w})$ is counterclockwise looking from the terminal point of $\mathbf{u}$ toward the origin (Figure 6.2.7). If we now agree to follow the standard convention of making counterclockwise angles nonnegative, then the angle θ will satisfy the inequalities $0 \leq \theta \leq \pi$.

The most basic rotations in a rectangular xyz-coordinate system are those about the coordinate axes. Table 6.2.6 provides the basic information about these rotations. For each of these rotations, one of the standard unit vectors remains fixed and the images of the other two can be computed by adapting Figure 6.1.8 appropriately. For example, in a rotation about the positive y-axis through an angle θ, the vector $\mathbf{e}_2 = (0, 1, 0)$ along the positive y-axis remains fixed, and the vectors $\mathbf{e}_1 = (1, 0, 0)$ and $\mathbf{e}_3 = (0, 0, 1)$ undergo rotations through the angle θ in the zx-plane. Thus, if we denote the standard matrix for this rotation by $R_{y,\theta}$, then

$$\mathbf{e}_1 = (1, 0, 0) \xrightarrow{R_{y,\theta}} (\cos\theta, 0, -\sin\theta)$$

$$\mathbf{e}_2 = (0, 1, 0) \xrightarrow{R_{y,\theta}} (0, 1, 0)$$

$$\mathbf{e}_3 = (0, 0, 1) \xrightarrow{R_{y,\theta}} (\sin\theta, 0, \cos\theta)$$

(see Figure 6.2.8).

GENERAL ROTATIONS

A complete analysis of general rotations in R^3 involves too much detail to present here, so we will focus on the highlights and fill in the gaps with references to other sources of information. We will be concerned with two basic problems:

1. Find the standard matrix for a rotation whose axis of rotation and angle of rotation are known.

2. Given the standard matrix for a rotation, find the axis and angle of rotation.

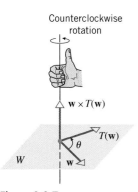

Counterclockwise rotation

$\mathbf{w} \times T(\mathbf{w})$

$T(\mathbf{w})$

W

Figure 6.2.7

The solution of the first problem is given by the following theorem, whose proof can be found, for example, in the book *Principles of Interactive Computer Graphics*, by W. M. Newman and R. F. Sproull, McGraw-Hill, New York, 1979, or in the paper "The Matrix of a Rotation," by Roger C. Alperin, *College Mathematics Journal*, Vol. 20, No. 3, May 1989.

Theorem 6.2.8 *If* $\mathbf{u} = (a, b, c)$ *is a unit vector, then the standard matrix* $R_{\mathbf{u},\theta}$ *for the rotation through the angle* θ *about an axis through the origin with orientation* $\mathbf{u}$ *is*

$$R_{\mathbf{u},\theta} = \begin{bmatrix} a^2(1-\cos\theta) + \cos\theta & ab(1-\cos\theta) - c\sin\theta & ac(1-\cos\theta) + b\sin\theta \\ ab(1-\cos\theta) + c\sin\theta & b^2(1-\cos\theta) + \cos\theta & bc(1-\cos\theta) - a\sin\theta \\ ac(1-\cos\theta) - b\sin\theta & bc(1-\cos\theta) + a\sin\theta & c^2(1-\cos\theta) + \cos\theta \end{bmatrix} \tag{13}$$

---

[*]In these cases the direction of view is not significant. For example, the same result is obtained regardless of whether a vector $\mathbf{w}$ is rotated clockwise 180° or counterclockwise 180°.

Table 6.2.6

Operator	Illustration	Standard Matrix
Rotation about the positive x-axis through an angle θ		$\begin{bmatrix} 1 & 0 & 0 \\ 0 & \cos\theta & -\sin\theta \\ 0 & \sin\theta & \cos\theta \end{bmatrix}$
Rotation about the positive y-axis through an angle θ		$\begin{bmatrix} \cos\theta & 0 & \sin\theta \\ 0 & 1 & 0 \\ -\sin\theta & 0 & \cos\theta \end{bmatrix}$
Rotation about the positive z-axis through an angle θ		$\begin{bmatrix} \cos\theta & -\sin\theta & 0 \\ \sin\theta & \cos\theta & 0 \\ 0 & 0 & 1 \end{bmatrix}$

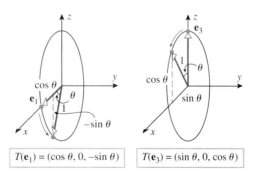

Figure 6.2.8 $T(\mathbf{e}_1) = (\cos\theta, 0, -\sin\theta)$ $T(\mathbf{e}_3) = (\sin\theta, 0, \cos\theta)$

You may also find it instructive to deduce the results in Table 6.2.6 from this more general result.

We now turn to the second problem posed above—given a rotation matrix A, find the axis and angle of rotation. Since the axis of rotation consists of the fixed points of A, we can determine this axis by solving the linear system

$$(I - A)\mathbf{x} = \mathbf{0}$$

(see the discussion in the beginning of Section 4.4). Once we know the axis of rotation, we can find a nonzero vector $\mathbf{w}$ in the plane W through the origin that is perpendicular to this axis and orient the axis using the vector

$$\mathbf{u} = \mathbf{w} \times A\mathbf{w}$$

Looking toward the origin from the terminal point of $\mathbf{u}$, the angle θ of rotation will be counterclockwise in W and hence can be computed from the formula

$$\cos\theta = \frac{\mathbf{w} \cdot A\mathbf{w}}{\|\mathbf{w}\| \|A\mathbf{w}\|} \tag{14}$$

Here is an example.

EXAMPLE 7
Axis and Angle
of Rotation

(a) Show that the matrix

$$A = \begin{bmatrix} 0 & 0 & 1 \\ 1 & 0 & 0 \\ 0 & 1 & 0 \end{bmatrix}$$

represents a rotation about a line through the origin of R^3.

(b) Find the axis and angle of rotation.

Solution (a) The matrix A is a rotation matrix since it is orthogonal and $\det(A) = 1$ (verify).

Solution (b) To find the axis of rotation we must solve the linear system $(I - A)\mathbf{x} = \mathbf{0}$. We leave it for you to show that this linear system is

$$\begin{bmatrix} 1 & 0 & -1 \\ -1 & 1 & 0 \\ 0 & -1 & 1 \end{bmatrix} \begin{bmatrix} x \\ y \\ z \end{bmatrix} = \begin{bmatrix} 0 \\ 0 \\ 0 \end{bmatrix}$$

and that a general solution is

$$\mathbf{x} = \begin{bmatrix} x \\ y \\ z \end{bmatrix} = t \begin{bmatrix} 1 \\ 1 \\ 1 \end{bmatrix}$$

Thus, the axis of rotation is the line through the origin that passes through the point $(1, 1, 1)$. The plane through the origin that is perpendicular to this line is given by the equation

$$x + y + z = 0$$

To find a nonzero vector in this plane, we can assign two of the variables arbitrary values (not both zero) and calculate the value of the third variable. Thus, for example, setting $x = 1$ and $y = -1$ produces the vector $\mathbf{w} = (1, -1, 0)$ in the plane W. Writing this vector in column form yields

$$\mathbf{w} = \begin{bmatrix} 1 \\ -1 \\ 0 \end{bmatrix}, \quad A\mathbf{w} = \begin{bmatrix} 0 & 0 & 1 \\ 1 & 0 & 0 \\ 0 & 1 & 0 \end{bmatrix} \begin{bmatrix} 1 \\ -1 \\ 0 \end{bmatrix} = \begin{bmatrix} 0 \\ 1 \\ -1 \end{bmatrix}, \quad \mathbf{w} \times A\mathbf{w} = \begin{bmatrix} 1 \\ 1 \\ 1 \end{bmatrix}$$

(verify). Thus, the rotation angle θ relative to the orientation $\mathbf{u} = \mathbf{w} \times A\mathbf{w} = (1, 1, 1)$ satisfies

$$\cos\theta = \frac{\mathbf{w} \cdot A\mathbf{w}}{\|\mathbf{w}\| \|A\mathbf{w}\|} = -\frac{1}{2} \tag{15}$$

Hence the angle of rotation is $2\pi/3$ ($= 120°$) (Figure 6.2.9). ∎

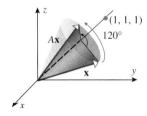

Figure 6.2.9

For applications that involve many rotations it is desirable to have formulas for computing axes and rotation angles. A formula for the cosine of the rotation angle in terms of the entries of A can be obtained from (13) by observing that

$$\mathrm{tr}(A) = (a^2 + b^2 + c^2)(1 - \cos\theta) + 3\cos\theta = 1 - \cos\theta + 3\cos\theta = 1 + 2\cos\theta$$

from which it follows that

$$\cos\theta = \frac{\mathrm{tr}(A) - 1}{2} \tag{16}$$

There also exist various formulas for the axis of rotation. One such formula appears in an article entitled "The Axis of Rotation: Analysis, Algebra, Geometry," by Dan Kalman, *Mathematics Magazine*, Vol. 62, No. 4, October 1989. It is shown in that article that if A is a rotation matrix, then for any nonzero vector $\mathbf{x}$ in R^3 that is not perpendicular to the axis of rotation, the vector

$$\mathbf{v} = A\mathbf{x} + A^T\mathbf{x} + [1 - \text{tr}(A)]\mathbf{x} \tag{17}$$

is nonzero and is along the axis of rotation when $\mathbf{x}$ has its initial point at the origin. Here is an application of Formulas (16) and (17) to Example 7.

EXAMPLE 8
Example 7
Revisited

Use Formulas (16) and (17) to solve the problem in part (b) of Example 7.

Solution In this case we have $\text{tr}(A) = 0$ (verify), so Formula (17) simplifies to

$$\mathbf{v} = A\mathbf{x} + A^T\mathbf{x} + \mathbf{x} = (A + A^T + I)\mathbf{x}$$

Let us take $\mathbf{x}$ to be the standard unit vector $\mathbf{e}_1 = (1, 0, 0)$, expressed in column form. Thus, an axis of rotation is determined by the vector

$$\mathbf{v} = (A + A^T + I)\mathbf{e}_1 = \begin{bmatrix} 1 & 1 & 1 \\ 1 & 1 & 1 \\ 1 & 1 & 1 \end{bmatrix} \begin{bmatrix} 1 \\ 0 \\ 0 \end{bmatrix} = \begin{bmatrix} 1 \\ 1 \\ 1 \end{bmatrix}$$

This implies that the axis of rotation passes through the point $(1, 1, 1)$, which is the same result we obtained in Example 7. Moreover, since $\text{tr}(A) = 0$, it follows from (16) that the angle of rotation satisfies

$$\cos\theta = \frac{\text{tr}(A) - 1}{2} = -\frac{1}{2}$$

which agrees with (15). ∎

Exercise Set 6.2

In Exercises 1–4, confirm that the matrix is orthogonal and find its inverse.

1. $\begin{bmatrix} \frac{3}{5} & -\frac{4}{5} \\ \frac{4}{5} & \frac{3}{5} \end{bmatrix}$

2. $\begin{bmatrix} \frac{20}{29} & \frac{21}{29} \\ -\frac{21}{29} & \frac{20}{29} \end{bmatrix}$

3. $\begin{bmatrix} \frac{4}{5} & 0 & -\frac{3}{5} \\ -\frac{9}{25} & \frac{4}{5} & -\frac{12}{25} \\ \frac{12}{25} & \frac{3}{5} & \frac{16}{25} \end{bmatrix}$

4. $\begin{bmatrix} \frac{1}{\sqrt{3}} & \frac{1}{\sqrt{2}} & \frac{1}{\sqrt{6}} \\ \frac{1}{\sqrt{3}} & -\frac{1}{\sqrt{2}} & \frac{1}{\sqrt{6}} \\ \frac{1}{\sqrt{3}} & 0 & -\frac{2}{\sqrt{6}} \end{bmatrix}$

In Exercises 5 and 6, confirm that the matrix A is orthogonal, and determine whether multiplication by A is a rotation about the origin or a reflection about a line through the origin. As appropriate, find the rotation angle or the angle that the line makes with the positive x-axis.

5. (a) $A = \begin{bmatrix} -\frac{1}{\sqrt{2}} & \frac{1}{\sqrt{2}} \\ -\frac{1}{\sqrt{2}} & -\frac{1}{\sqrt{2}} \end{bmatrix}$ (b) $A = \begin{bmatrix} -\frac{1}{2} & \frac{\sqrt{3}}{2} \\ \frac{\sqrt{3}}{2} & \frac{1}{2} \end{bmatrix}$

6. (a) $\begin{bmatrix} \frac{1}{2} & -\frac{\sqrt{3}}{2} \\ \frac{\sqrt{3}}{2} & \frac{1}{2} \end{bmatrix}$ (b) $\begin{bmatrix} \frac{1}{2} & \frac{\sqrt{3}}{2} \\ \frac{\sqrt{3}}{2} & -\frac{1}{2} \end{bmatrix}$

In Exercises 7 and 8, find the standard matrix for the stated linear operator on R^2.

7. (a) Contraction with factor $\frac{1}{5}$.
   (b) Compression in the x-direction with factor $\frac{1}{3}$.
   (c) Expansion in the y-direction with factor 6.
   (d) Shear in the x-direction with factor 3.

8. (a) Dilation with factor 5.
   (b) Expansion in the x-direction with factor 3.
   (c) Compression in the y-direction with factor $\frac{1}{6}$.
   (d) Shear in the y-direction with factor 2.

In Exercises 9 and 10, describe the geometric effect of multiplication by the given matrix.

9. (a) $\begin{bmatrix} 3 & 0 \\ 0 & 1 \end{bmatrix}$ (b) $\begin{bmatrix} \frac{1}{4} & 0 \\ 0 & \frac{1}{4} \end{bmatrix}$

   (c) $\begin{bmatrix} 1 & 4 \\ 0 & 1 \end{bmatrix}$ (d) $\begin{bmatrix} 1 & 0 \\ -4 & 1 \end{bmatrix}$

10. (a) $\begin{bmatrix} 1 & 0 \\ 0 & \frac{2}{3} \end{bmatrix}$ (b) $\begin{bmatrix} 8 & 0 \\ 0 & 8 \end{bmatrix}$

   (c) $\begin{bmatrix} 1 & -3 \\ 0 & 1 \end{bmatrix}$ (d) $\begin{bmatrix} 1 & 0 \\ 3 & 1 \end{bmatrix}$

In Exercises 11–14, use Theorem 6.1.4 to find the standard matrix for the linear transformation.

11. $T: R^2 \to R^2$ dilates a vector by a factor of 3, then reflects that vector about the line $y = x$, and then projects that vector orthogonally onto the y-axis.

12. $T: R^2 \to R^2$ reflects a vector about the line $y = -x$, then projects that vector onto the y-axis, and then compresses that vector by a factor of $\frac{1}{2}$ in the y-direction.

13. $T: R^3 \to R^3$ projects a vector orthogonally onto the xz-plane and then projects that vector orthogonally onto the xy-plane.

14. $T: R^3 \to R^3$ reflects a vector about the xy-plane, then reflects that vector about the xz-plane, and then reflects the vector about the yz-plane.

In Exercises 15 and 16, sketch the image of the rectangle with vertices $(0, 0)$, $(1, 0)$, $(1, 2)$, and $(0, 2)$ under the given transformation.

15. (a) Reflection about the x-axis.
   (b) Compression in the y-direction with factor $\frac{1}{4}$.
   (c) Shear in the x-direction with factor 3.

16. (a) Reflection about the y-axis.
   (b) Expansion in the x-direction with factor 3.
   (c) Shear in the y-direction with factor 2.

In Exercises 17 and 18, sketch the image of the square with vertices $(0, 0)$, $(1, 0)$, $(1, 1)$, and $(0, 1)$ under multiplication by A.

17. (a) $A = \begin{bmatrix} 2 & 0 \\ 0 & 1 \end{bmatrix}$ (b) $A = \begin{bmatrix} 1 & 0 \\ 0 & 2 \end{bmatrix}$

   (c) $A = \begin{bmatrix} 2 & 0 \\ 0 & 2 \end{bmatrix}$

18. (a) $A = \begin{bmatrix} \frac{2}{3} & 0 \\ 0 & 1 \end{bmatrix}$ (b) $A = \begin{bmatrix} 1 & 0 \\ 0 & \frac{2}{3} \end{bmatrix}$

   (c) $A = \begin{bmatrix} \frac{2}{3} & 0 \\ 0 & \frac{2}{3} \end{bmatrix}$

19. Use matrix multiplication to find the reflection of $(2, 5, 3)$ about the
   (a) xy-plane   (b) xz-plane   (c) yz-plane

20. Use matrix multiplication to find the orthogonal projection of $(-2, 1, 3)$ on the
   (a) xy-plane   (b) xz-plane   (c) yz-plane

21. Find the standard matrix for the linear operator that performs the stated rotation in R^3.
   (a) $90°$ about the positive x-axis.
   (b) $90°$ about the positive y-axis.
   (c) $-90°$ about the positive z-axis.

22. Find the standard matrix for the linear operator that performs the stated rotation in R^3.
   (a) $-90°$ about the positive x-axis.
   (b) $-90°$ about the positive y-axis.
   (c) $90°$ about the positive z-axis.

23. Use matrix multiplication to find the image of the vector $(-2, 1, 2)$ under the stated rotation.
   (a) Through an angle of $30°$ about the positive x-axis.
   (b) Through an angle of $45°$ about the positive y-axis.
   (c) Through an angle of $-60°$ about the positive z-axis.

24. Use matrix multiplication to find the image of the vector $(-2, 1, 2)$ under the stated rotation.
   (a) Through an angle of $60°$ about the positive x-axis.
   (b) Through an angle of $30°$ about the positive y-axis.
   (c) Through an angle of $-45°$ about the positive z-axis.

In Exercises 25 and 26, show that the matrix represents a rotation about the origin of R^3, and find the axis and angle of rotation using the method of Example 7.

25. $A = \begin{bmatrix} 1 & 0 & 0 \\ 0 & 0 & -1 \\ 0 & 1 & 0 \end{bmatrix}$ **26.** $A = \begin{bmatrix} 0 & 1 & 0 \\ 0 & 0 & 1 \\ 1 & 0 & 0 \end{bmatrix}$

27. Solve the problem in Exercise 25 using the method of Example 8.

28. Solve the problem in Exercise 26 using the method of Example 8.

29. The *orthogonal projections* on the x-axis, y-axis, and z-axis of a rectangular coordinate system in R^3 are defined by

$$T_1(x, y, z) = (x, 0, 0), \quad T_2(x, y, z) = (0, y, 0)$$
$$T_3(x, y, z) = (0, 0, z)$$

respectively.
   (a) Show that the orthogonal projections on the coordinate axes are linear operators, and find their standard matrices M_1, M_2, and M_3.
   (b) Show algebraically that if $T: R^3 \to R^3$ is an orthogonal projection on one of the coordinate axes, then $T(\mathbf{x})$ and $\mathbf{x} - T(\mathbf{x})$ are orthogonal for every vector $\mathbf{x}$ in R^3. Illustrate this with a sketch that shows $\mathbf{x}$ and $\mathbf{x} - T(\mathbf{x})$ in the case where T is the orthogonal projection on the x-axis.

30. As illustrated in the accompanying figure, the *shear in the xy-direction with factor k* in R^3 is the linear transformation that moves each point (x, y, z) parallel to the xy-plane to the new position $(x + kz, y + kz, z)$.
   (a) Find the standard matrix for the shear in the xy-direction with factor k.
   (b) How would you define the shear in the xz-direction with factor k and the shear in the yz-direction with factor k? Find the standard matrices for these shears.

Figure Ex-30

31. Deduce the standard matrices in Table 6.2.6 from Formula (13).

Discussion and Discovery

D1. In words, describe the geometric effect that multiplication by A has on the unit square.

(a) $A = \begin{bmatrix} 2 & 0 \\ 0 & 0 \end{bmatrix}$ 　　(b) $A = \begin{bmatrix} 0 & 0 \\ 0 & 3 \end{bmatrix}$

(c) $A = \begin{bmatrix} 2 & 0 \\ 0 & 3 \end{bmatrix}$ 　　(d) $A = \begin{bmatrix} \frac{\sqrt{3}}{2} & \frac{1}{2} \\ -\frac{1}{2} & \frac{\sqrt{3}}{2} \end{bmatrix}$

D2. Find a, b, and c for which the matrix

$$\begin{bmatrix} a & \frac{1}{\sqrt{2}} & -\frac{1}{\sqrt{2}} \\ b & \frac{1}{\sqrt{6}} & \frac{1}{\sqrt{6}} \\ c & \frac{1}{\sqrt{3}} & \frac{1}{\sqrt{3}} \end{bmatrix}$$

is orthogonal. Are the values of a, b, and c unique? Explain.

D3. What conditions must a and b satisfy for the matrix

$$\begin{bmatrix} a+b & b-a \\ a-b & b+a \end{bmatrix}$$

to be orthogonal?

D4. Given that orthogonal matrices are norm preserving, what can be said about an eigenvalue λ of an orthogonal matrix A?

D5. In each part, make a conjecture about the eigenvectors and eigenvalues of the matrix A corresponding to the given transformation by considering the geometric properties of multiplication by A. Confirm each of your conjectures with computations.

(a) Reflection about the line $y = x$.

(b) Contraction by a factor of $\frac{1}{2}$.

D6. Find the matrix for a shear in the x-direction that transforms the triangle with vertices $(0, 0)$, $(2, 1)$, and $(3, 0)$ into a right triangle with a right angle at the origin.

D7. Given that $\mathbf{x}$ and $\mathbf{y}$ are vectors in R^n such that $\|\mathbf{x} + \mathbf{y}\| = 4$ and $\|\mathbf{x} - \mathbf{y}\| = 2$, what can you say about the value of $\mathbf{x} \cdot \mathbf{y}$?

D8. It follows from the polarization identity (4) that if $\mathbf{x}$ and $\mathbf{y}$ are vectors in R^n such that $\|\mathbf{x} + \mathbf{y}\| = \|\mathbf{x} - \mathbf{y}\|$, then $\mathbf{x} \cdot \mathbf{y} = 0$. Illustrate this geometrically by drawing a picture in R^2.

Technology Exercises

T1. Use Formula (13) to find the image of $\mathbf{x} = (1, -2, 5)$ under a rotation of $\theta = \pi/4$ about an axis through the origin oriented in the direction of $\mathbf{u} = \left(\frac{2}{7}, \frac{3}{7}, \frac{6}{7}\right)$.

T2. Let

$$A = \begin{bmatrix} -\frac{3}{7} & -\frac{2}{7} & -\frac{6}{7} \\ \frac{6}{7} & -\frac{3}{7} & -\frac{2}{7} \\ -\frac{2}{7} & -\frac{6}{7} & \frac{3}{7} \end{bmatrix}$$

Show that A represents a rotation, and use Formulas (16) and (17) to find the axis and angle of rotation.

T3. Let

$$\mathbf{u} = \begin{bmatrix} \frac{1}{3} \\ \frac{2}{3} \\ \frac{2}{3} \end{bmatrix}$$

Use Theorem 6.2.8 to construct the standard matrix for the rotation through an angle of $\pi/6$ about an axis oriented in the direction of $\mathbf{u}$.

Section 6.3   Kernel and Range

In this section we will discuss the range of a linear transformation in more detail, and we will show that the set of vectors that a linear transformation maps into zero plays an important role in understanding the geometric effect that the transformation has on subspaces of its domain.

KERNEL OF A LINEAR TRANSFORMATION

If $\mathbf{x} = t\mathbf{v}$ is a line through the origin of R^n, and if T is a linear operator on R^n, then the image of the line under the transformation T is the set of vectors of the form

$$T(\mathbf{x}) = T(t\mathbf{v}) = tT(\mathbf{v})$$

Geometrically, there are two possibilities for this image:

1. If $T(\mathbf{v}) = \mathbf{0}$, then $T(\mathbf{x}) = \mathbf{0}$ for all $\mathbf{x}$, so the image is the single point $\mathbf{0}$.
2. If $T(\mathbf{v}) \neq \mathbf{0}$, then the image is the line through the origin determined by $T(\mathbf{v})$.

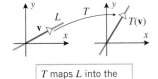

T maps L into the point $\mathbf{0}$ if $T(\mathbf{v}) = \mathbf{0}$.

(See Figure 6.3.1.) Similarly, if $\mathbf{x} = t_1\mathbf{v}_1 + t_2\mathbf{v}_2$ is a plane through the origin of R^n, then the image of this plane under the transformation T is the set of vectors of the form

$$T(\mathbf{x}) = T(t_1\mathbf{v}_1 + t_2\mathbf{v}_2) = t_1 T(\mathbf{v}_1) + t_2 T(\mathbf{v}_2)$$

There are three possibilities for this image:

1. If $T(\mathbf{v}_1) = \mathbf{0}$ and $T(\mathbf{v}_2) = \mathbf{0}$, then $T(\mathbf{x}) = \mathbf{0}$ for all $\mathbf{x}$, so the image is the single point $\mathbf{0}$.
2. If $T(\mathbf{v}_1) \neq \mathbf{0}$ and $T(\mathbf{v}_2) \neq \mathbf{0}$, and if $T(\mathbf{v}_1)$ and $T(\mathbf{v}_2)$ are not scalar multiples of one another, then the image is a plane through the origin.
3. The image is a line through the origin in the remaining cases.

T maps L into the line spanned by $T(\mathbf{v})$ if $T(\mathbf{v}) \neq \mathbf{0}$.

Figure 6.3.1

In light of the preceding discussion, we see that to understand the geometric effect of a linear transformation, one must know something about the set of vectors that the transformation maps into $\mathbf{0}$. This set of vectors is sufficiently important that there is some special terminology and notation associated with it.

Definition 6.3.1   If $T : R^n \rightarrow R^m$ is a linear transformation, then the set of vectors in R^n that T maps into $\mathbf{0}$ is called the **kernel** of T and is denoted by $\ker(T)$.

EXAMPLE 1
Kernels of Some Basic Operators

In each part, find the kernel of the stated linear operator on R^3.

(a) The zero operator $T_0(\mathbf{x}) = 0\mathbf{x} = \mathbf{0}$.
(b) The identity operator $T_I(\mathbf{x}) = I\mathbf{x} = \mathbf{x}$.
(c) The orthogonal projection T on the xy-plane.
(d) A rotation T about a line through the origin through an angle θ.

Solution (a)   The transformation maps every vector $\mathbf{x}$ into $\mathbf{0}$, so the kernel is all of R^3; that is, $\ker(T_0) = R^3$.

Solution (b)   Since $T_I(\mathbf{x}) = \mathbf{x}$, it follows that $T_I(\mathbf{x}) = \mathbf{0}$ if and only if $\mathbf{x} = \mathbf{0}$. This implies that $\ker(T_I) = \{\mathbf{0}\}$.

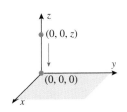

Figure 6.3.2

Solution (c)   The orthogonal projection on the xy-plane maps a general point $\mathbf{x} = (x, y, z)$ into $(x, y, 0)$, so the points that get mapped into $\mathbf{0} = (0, 0, 0)$ are those for which $x = 0$ and $y = 0$. Thus, the kernel of the projection T is the z-axis (Figure 6.3.2).

Solution (d)   The only vector whose image under the rotation is $\mathbf{0}$ is the vector $\mathbf{0}$ itself; that is, the kernel of the rotation T is $\{\mathbf{0}\}$.   ■

It is important to note that the kernel of a linear transformation always contains the vector $\mathbf{0}$ by Theorem 6.1.3; the following theorem shows that the kernel of a linear transformation is always a subspace.

Theorem 6.3.2   *If $T : R^n \to R^m$ is a linear transformation, then the kernel of T is a subspace of R^n.*

Proof   The kernel of T is a nonempty set since it contains the zero vector in R^n. To show that it is a subspace of R^n we must show that it is closed under scalar multiplication and addition. For this purpose, let $\mathbf{u}$ and $\mathbf{v}$ be any vectors in $\ker(T)$, and let c be any scalar. Then

$$T(c\mathbf{u}) = cT(\mathbf{u}) = c\mathbf{0} = \mathbf{0}$$

so $c\mathbf{u}$ is in $\ker(T)$, which shows that $\ker(T)$ is closed under scalar multiplication. Also,

$$T(\mathbf{u} + \mathbf{v}) = T(\mathbf{u}) + T(\mathbf{v}) = \mathbf{0} + \mathbf{0} = \mathbf{0}$$

so $\mathbf{u} + \mathbf{v}$ is in $\ker(T)$, which shows that $\ker(T)$ is closed under addition.   ∎

KERNEL OF A MATRIX TRANSFORMATION

If A is an $m \times n$ matrix and $T_A : R^n \to R^m$ is the corresponding linear transformation, then $T_A(\mathbf{x}) = A\mathbf{x}$, so that $\mathbf{x}$ is in the kernel of T_A if and only if $A\mathbf{x} = \mathbf{0}$. Thus, we have the following result.

Theorem 6.3.3   *If A is an $m \times n$ matrix, then the kernel of the corresponding linear transformation is the solution space of $A\mathbf{x} = \mathbf{0}$.*

EXAMPLE 2
Kernel of a
Matrix Operator

In part (c) of Example 1 we showed that the kernel of the orthogonal projection of R^3 onto the xy-plane is the z-axis. This can also be deduced from Theorem 6.3.3 by considering the standard matrix for this projection, namely

$$A = \begin{bmatrix} 1 & 0 & 0 \\ 0 & 1 & 0 \\ 0 & 0 & 0 \end{bmatrix}$$

It is evident from this matrix that a general solution of the system $A\mathbf{x} = \mathbf{0}$ is

$$x = 0, \quad y = 0, \quad z = t$$

which are parametric equations for the z-axis.   ∎

There are many instances in mathematics in which an object is given different names to emphasize different points of view. In the current context, for example, the *solution space* of $A\mathbf{x} = \mathbf{0}$ and the *kernel* of T_A are really the same thing, the choice of terminology depending on whether one wants to emphasize linear systems or linear transformations. A third possibility is to regard this subspace as an object associated with the *matrix A* rather than with the *system* $A\mathbf{x} = \mathbf{0}$ or with the *transformation* T_A; the following terminology emphasizes this point of view.

Definition 6.3.4   If A is an $m \times n$ matrix, then the solution space of the linear system $A\mathbf{x} = \mathbf{0}$, or, equivalently, the kernel of the transformation T_A, is called the ***null space*** of the matrix A and is denoted by $\text{null}(A)$.

EXAMPLE 3
Finding the
Null Space of a
Matrix

Find the null space of the matrix

$$A = \begin{bmatrix} 1 & 3 & -2 & 0 & 2 & 0 \\ 2 & 6 & -5 & -2 & 4 & -3 \\ 0 & 0 & 5 & 10 & 0 & 15 \\ 2 & 6 & 0 & 8 & 4 & 18 \end{bmatrix}$$

Solution We will solve the problem by producing a set of vectors that spans the subspace. The null space of A is the solution space of $A\mathbf{x} = \mathbf{0}$, so the stated problem boils down to solving this linear system. The computations were performed in Example 7 of Section 2.2, where we showed that the solution space consists of all linear combinations of the vectors

$$\mathbf{v}_1 = \begin{bmatrix} -3 \\ 1 \\ 0 \\ 0 \\ 0 \\ 0 \end{bmatrix}, \quad \mathbf{v}_2 = \begin{bmatrix} -4 \\ 0 \\ -2 \\ 1 \\ 0 \\ 0 \end{bmatrix}, \quad \mathbf{v}_3 = \begin{bmatrix} -2 \\ 0 \\ 0 \\ 0 \\ 1 \\ 0 \end{bmatrix}$$

Thus, $\operatorname{null}(A) = \operatorname{span}\{\mathbf{v}_1, \mathbf{v}_2, \mathbf{v}_3\}$. ∎

At the beginning of this section we showed that a linear transformation $T : R^n \to R^m$ maps a line through the origin into either another line through the origin or the single point $\mathbf{0}$, and we showed that it maps a plane through the origin into either another plane through the origin, a line through the origin, or the single point $\mathbf{0}$. In all cases the image is a subspace, which is in keeping with the next theorem.

Theorem 6.3.5 *If $T : R^n \to R^m$ is a linear transformation, then T maps subspaces of R^n into subspaces of R^m.*

Proof Let S be any subspace of R^n, and let $W = T(S)$ be its image under T. We want to show that W is closed under scalar multiplication and addition; so we must show that if $\mathbf{u}$ and $\mathbf{v}$ are any vectors in W, and if c is any scalar, then $c\mathbf{u}$ and $\mathbf{u} + \mathbf{v}$ are images under T of vectors in S. To find vectors with these images, suppose that $\mathbf{u}$ and $\mathbf{v}$ are the images of the vectors $\mathbf{u}_0$ and $\mathbf{v}_0$ in S, respectively; that is,

$$\mathbf{u} = T(\mathbf{u}_0) \quad \text{and} \quad \mathbf{v} = T(\mathbf{v}_0)$$

Since S is a subspace of R^n, it is closed under scalar multiplication and addition, so $c\mathbf{u}_0$ and $\mathbf{u}_0 + \mathbf{v}_0$ are also vectors in S. These are the vectors we are looking for, since

$$T(c\mathbf{u}_0) = cT(\mathbf{u}_0) = c\mathbf{u} \quad \text{and} \quad T(\mathbf{u}_0 + \mathbf{v}_0) = T(\mathbf{u}_0) + T(\mathbf{v}_0) = \mathbf{u} + \mathbf{v}$$

which shows that $c\mathbf{u}$ and $\mathbf{u} + \mathbf{v}$ are images of vectors in S. ∎

RANGE OF A LINEAR TRANSFORMATION

We will now shift our focus from the kernel to the range of a linear transformation. The following definition is a reformulation of Definition 6.1.1 in the context of transformations.

Definition 6.3.6 If $T : R^n \to R^m$ is a linear transformation, then the ***range*** of T, denoted by $\operatorname{ran}(T)$, is the set of all vectors in R^m that are images of at least one vector in R^n. Stated another way, $\operatorname{ran}(T)$ is the image of the domain R^n under the transformation T.

EXAMPLE 4
Ranges of Some Basic Operators on R^3

Describe the ranges of the following linear operators on R^3.

(a) The zero operator $T_0(\mathbf{x}) = 0\mathbf{x} = \mathbf{0}$.

(b) The identity operator $T_I(\mathbf{x}) = I\mathbf{x} = \mathbf{x}$.

(c) The orthogonal projection T on the xy-plane.

(d) A rotation T about a line through the origin through an angle θ.

Solution (a) This transformation maps every vector in R^3 into $\mathbf{0}$, so $\operatorname{ran}(T_0) = \{\mathbf{0}\}$.

Solution (b) This transformation maps every vector into itself, so every vector in R^3 is the image of some vector. Thus, $\operatorname{ran}(T_I) = R^3$.

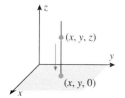

Figure 6.3.3

Solution (c) This transformation maps a general point $\mathbf{x} = (x, y, z)$ into $(x, y, 0)$, so the range consists of all points with a z-component of zero. Geometrically, $\text{ran}(T)$ is the xy-plane (Figure 6.3.3).

Solution (d) Every vector in R^3 is the image of some vector under the rotation T. For example, to find a vector whose image is $\mathbf{x}$, rotate $\mathbf{x}$ about the line through the angle $-\theta$ to obtain a vector $\mathbf{w}$; the image of $\mathbf{w}$, when rotated through the angle θ, will be $\mathbf{x}$. Thus, $\text{ran}(T) = R^3$. ∎

The range of a linear transformation $T : R^n \to R^m$ can be viewed as the image of R^n under T, so it follows as a special case of Theorem 6.3.5 that the range of T is a subspace of R^m. This is consistent with the results in Example 4.

> **Theorem 6.3.7** *If $T : R^n \to R^m$ is a linear transformation, then $\text{ran}(T)$ is a subspace of R^m.*

RANGE OF A MATRIX TRANSFORMATION

If A is an $m \times n$ matrix and $T_A : R^n \to R^m$ is the corresponding linear transformation, then $T_A(\mathbf{x}) = A\mathbf{x}$, so that a vector $\mathbf{b}$ in R^m is in the range of T_A if and only if there is a vector $\mathbf{x}$ such that $A\mathbf{x} = \mathbf{b}$. Stated another way, $\mathbf{b}$ is in the range of T_A if and only if the linear system $A\mathbf{x} = \mathbf{b}$ is consistent. Thus, Theorem 3.5.5 implies the following result.

> **Theorem 6.3.8** *If A is an $m \times n$ matrix, then the range of the corresponding linear transformation is the column space of A.*

If $T_A : R^n \to R^m$ is the linear transformation corresponding to the matrix A, then the range of T_A and the column space of A are the same object from different points of view—the first emphasizes the transformation and the second the matrix.

EXAMPLE 5
Range of a
Matrix Operator

In part (c) of Example 4 we showed that the range of the orthogonal projection of R^3 onto the xy-plane is the xy-plane. This can also be deduced from Theorem 6.3.8 by considering the standard matrix for this projection, namely

$$A = \begin{bmatrix} 1 & 0 & 0 \\ 0 & 1 & 0 \\ 0 & 0 & 0 \end{bmatrix}$$

The range of the projection is the column space of A, which consists of all vectors of the form

$$A\mathbf{x} = \begin{bmatrix} 1 & 0 & 0 \\ 0 & 1 & 0 \\ 0 & 0 & 0 \end{bmatrix} \begin{bmatrix} x \\ y \\ z \end{bmatrix} = x \begin{bmatrix} 1 \\ 0 \\ 0 \end{bmatrix} + y \begin{bmatrix} 0 \\ 1 \\ 0 \end{bmatrix} + z \begin{bmatrix} 0 \\ 0 \\ 0 \end{bmatrix} = \begin{bmatrix} x \\ y \\ 0 \end{bmatrix}$$

Thus, the range of the projection, in comma-delimited notation, is the set of points of the form $(x, y, 0)$, which is the xy-plane. ∎

It is important in many kinds of problems to be able to determine whether a given vector $\mathbf{b}$ in R^m is in the range of a linear transformation $T : R^n \to R^m$. If A is the standard matrix for T, then this problem reduces to determining whether $\mathbf{b}$ is in the column space of A. Here is an example.

EXAMPLE 6
Column Space

Suppose that

$$A = \begin{bmatrix} 1 & -8 & -7 & -4 \\ 2 & -3 & -1 & 5 \\ 3 & 2 & 5 & 14 \end{bmatrix} \quad \text{and} \quad \mathbf{b} = \begin{bmatrix} 8 \\ -10 \\ -28 \end{bmatrix}$$

Determine whether $\mathbf{b}$ is in the column space of A, and, if so, express it as a linear combination of the column vectors of A.

Solution The problem can be solved by determining whether the linear system $A\mathbf{x} = \mathbf{b}$ is consistent. If the answer is "yes," then $\mathbf{b}$ is in the column space, and the components of any solution $\mathbf{x}$ can be used as coefficients for the desired linear combination; if the answer is "no," then $\mathbf{b}$ is not in the column space of A. We leave it for you to confirm that the reduced row echelon form of the augmented matrix for the system is

$$\begin{bmatrix} 1 & 0 & 1 & 4 & -8 \\ 0 & 1 & 1 & 1 & -2 \\ 0 & 0 & 0 & 0 & 0 \end{bmatrix}$$

We can see from this matrix that the system is consistent, and we leave it for you to show that a general solution is

$$x_1 = -8 - s - 4t, \quad x_2 = -2 - s - t, \quad x_3 = s, \quad x_4 = t$$

Since the parameters s and t are arbitrary, there are infinitely many ways to express $\mathbf{b}$ as a linear combination of the column vectors of A. A particularly simple way is to take $s = 0$ and $t = 0$, in which case we obtain $x_1 = -8, x_2 = -2, x_3 = 0, x_4 = 0$. This yields the linear combination

$$\mathbf{b} = \begin{bmatrix} 8 \\ -10 \\ -28 \end{bmatrix} = -8 \begin{bmatrix} 1 \\ 2 \\ 3 \end{bmatrix} - 2 \begin{bmatrix} -8 \\ -3 \\ 2 \end{bmatrix} + 0 \begin{bmatrix} -7 \\ -1 \\ 5 \end{bmatrix} + 0 \begin{bmatrix} -4 \\ 5 \\ 14 \end{bmatrix}$$

You may find it instructive to express $\mathbf{b}$ as a linear combination of the column vectors of A in some other ways by choosing different values for the parameters s and t.   ■

EXISTENCE AND UNIQUENESS ISSUES

There are many problems in which one is concerned with the following questions about a linear transformation $T: R^n \to R^m$:

- **The Existence Question**—Is every vector in R^m the image of at least one vector in R^n; that is, is the range of T all of R^m? (See the schematic diagram in Figure 6.3.4.)
- **The Uniqueness Question**—Can two different vectors in R^n have the same image in R^m? (See the schematic diagram in Figure 6.3.5.)

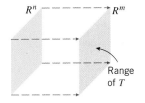

The range is R^m, so every vector in R^m is the image of at least one vector in R^n.

Figure 6.3.4

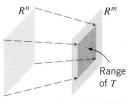

The range is not all of R^m, so there are vectors in R^m that are not images of any vectors in R^n.

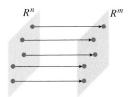

Distinct vectors in R^n have distinct images in R^m.

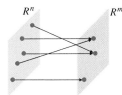

There are distinct vectors in R^n that have the same image in R^m.

Figure 6.3.5

The following terminology relates to these questions.

> **Definition 6.3.9** A transformation $T: R^n \to R^m$ is said to be ***onto*** if its range is the entire codomain R^m; that is, every vector in R^m is the image of at least one vector in R^n.

> **Definition 6.3.10** A transformation $T: R^n \to R^m$ is said to be ***one-to-one*** (sometimes written 1–1) if T maps distinct vectors in R^n into distinct vectors in R^m.

In general, a transformation can have both, neither, or just one of the properties in these definitions. Here are some examples.

EXAMPLE 7
One-to-One and
Onto

Let $T : R^2 \rightarrow R^2$ be the operator that rotates each vector in the xy-plane about the origin through an angle θ. This operator is one-to-one because rotating distinct vectors through the same angle produces distinct vectors; it is onto because any vector $\mathbf{x}$ in R^2 is the image of some vector $\mathbf{w}$ under the rotation (rotate $\mathbf{x}$ through the angle $-\theta$ to obtain $\mathbf{w}$). ∎

EXAMPLE 8
Neither
One-to-One nor
Onto

Let $T : R^3 \rightarrow R^3$ be the orthogonal projection on the xy-plane. This operator is not one-to-one because distinct points on any vertical line map into the same point in the xy-plane; it is not onto because its range (the xy-plane) is not all of R^3. ∎

EXAMPLE 9
One-to-One but
Not Onto

Let $T : R^2 \rightarrow R^3$ be the linear transformation defined by the formula $T(x, y) = (x, y, 0)$. To show that this linear transformation is one-to-one, consider the images of two points $\mathbf{x}_1 = (x_1, y_1)$ and $\mathbf{x}_2 = (x_2, y_2)$. If $T(\mathbf{x}_1) = T(\mathbf{x}_2)$, then $(x_1, y_1, 0) = (x_2, y_2, 0)$, which implies that $x_1 = y_1$ and $x_2 = y_2$. Thus if $\mathbf{x}_1 \neq \mathbf{x}_2$, then $T(\mathbf{x}_1) \neq T(\mathbf{x}_2)$, which means T maps distinct vectors into distinct vectors. The transformation is not onto because its range is not all of R^3. For example, there is no vector in R^2 that maps into $(0, 0, 1)$. ∎

EXAMPLE 10
Onto but Not
One-to-One

Let $T : R^3 \rightarrow R^2$ be the linear transformation defined by the formula $T(x, y, z) = (x, y)$. This transformation is onto because each vector $\mathbf{w} = (x, y)$ in R^2 is the image of at least one vector in R^3; in fact, it is the image of any vector $\mathbf{x} = (x, y, z)$ whose first two components are the same as those of $\mathbf{w}$. The transformation is not one-to-one because two distinct vectors of the form $\mathbf{x}_1 = (x, y, z_1)$ and $\mathbf{x}_2 = (x, y, z_2)$ map into the same point (x, y). ∎

The following theorem establishes an important relationship between the kernel of a linear transformation and the property of being one-to-one.

> **Theorem 6.3.11** *If $T : R^n \rightarrow R^m$ is a linear transformation, then the following statements are equivalent.*
>
> *(a) T is one-to-one.*
>
> *(b) $\ker(T) = \{\mathbf{0}\}$.*

Proof (a) $\Rightarrow$ (b) Assume that T is one-to-one. Since T is linear, we know that $T(\mathbf{0}) = \mathbf{0}$ by Theorem 6.1.3. The fact that T is one-to-one implies that $\mathbf{x} = \mathbf{0}$ is the only vector for which $T(\mathbf{x}) = \mathbf{0}$, so $\ker(T) = \{\mathbf{0}\}$.

Proof (b) $\Rightarrow$ (a) Assume that $\ker(T) = \{\mathbf{0}\}$. To prove that T is one-to-one we will show that if $\mathbf{x}_1 \neq \mathbf{x}_2$, then $T(\mathbf{x}_1) \neq T(\mathbf{x}_2)$. But if $\mathbf{x}_1 \neq \mathbf{x}_2$, then $\mathbf{x}_1 - \mathbf{x}_2 \neq \mathbf{0}$, which means that $\mathbf{x}_1 - \mathbf{x}_2$ is not in $\ker(T)$. This being the case,

$$T(\mathbf{x}_1 - \mathbf{x}_2) = T(\mathbf{x}_1) - T(\mathbf{x}_2) \neq \mathbf{0}$$

Thus, $T(\mathbf{x}_1) \neq T(\mathbf{x}_2)$. ∎

**ONE-TO-ONE AND ONTO
FROM THE VIEWPOINT
OF LINEAR SYSTEMS**

If A is an $m \times n$ matrix and $T_A : R^n \rightarrow R^m$ is the corresponding linear transformation, then $T_A(\mathbf{x}) = A\mathbf{x}$. Thus, to say that $\ker(T_A) = \{\mathbf{0}\}$ (i.e., that T_A is one-to-one) is the same as saying that the linear system $A\mathbf{x} = \mathbf{0}$ has only the trivial solution. Also, to say that T_A is onto is the same as saying that for each vector $\mathbf{b}$ in R^m there is at least one vector $\mathbf{x}$ in R^n such that $A\mathbf{x} = \mathbf{b}$. This establishes the following theorems.

> **Theorem 6.3.12** *If A is an $m \times n$ matrix, then the corresponding linear transformation $T_A : R^n \rightarrow R^m$ is one-to-one if and only if the linear system $A\mathbf{x} = \mathbf{0}$ has only the trivial solution.*

> **Theorem 6.3.13** *If A is an $m \times n$ matrix, then the corresponding linear transformation $T_A : R^n \rightarrow R^m$ is onto if and only if the linear system $A\mathbf{x} = \mathbf{b}$ is consistent for every $\mathbf{b}$ in R^n.*

EXAMPLE 11

Mapping "Bigger" Spaces into "Smaller" Spaces

Let $T : R^n \to R^m$ be a linear transformation, and suppose that $n > m$. If A is the standard matrix for T, then the linear system $A\mathbf{x} = \mathbf{0}$ has more unknowns than equations and hence has nontrivial solutions. Accordingly, it follows from Theorem 6.3.12 that T is not one-to-one, and hence we have shown that if a matrix transformation maps a space R^n of higher dimension into a space R^m of smaller dimension, then there must be distinct points in R^n that map into the same point in R^m. For example, the linear transformation

$$T(x_1, x_2, x_3) = (x_1 + x_2, x_1 - x_3)$$

maps the higher-dimensional space R^3 into the lower-dimensional space R^2, so you can tell without any computation that T is not one-to-one. ∎

We observed earlier in this section that a linear transformation $T : R^n \to R^m$ can be one-to-one and not onto or can be onto and not one-to-one (Examples 9 and 10). The next theorem shows that in the special case where T is a *linear operator*, the two properties go hand in hand—both hold or neither holds.

Theorem 6.3.14 *If* $T : R^n \to R^n$ *is a linear operator on* R^n, *then* T *is one-to-one if and only if it is onto.*

Proof Let A be the standard matrix for T. By parts (*d*) and (*e*) of Theorem 4.4.7, the system $A\mathbf{x} = \mathbf{0}$ has only the trivial solution if and only if the system $A\mathbf{x} = \mathbf{b}$ is consistent for every vector $\mathbf{b}$ in R^n. Combining this with Theorems 6.3.12 and 6.3.13 completes the proof. ∎

EXAMPLE 12

Examples 7 and 8 Revisited

We saw in Examples 7 and 8 that a rotation about the origin of R^2 is both one-to-one and onto and that the orthogonal projection on the xy-plane in R^3 is neither one-to-one nor onto. The "both" and "neither" are consistent with Theorem 6.3.14, since the rotation and the projection are both linear *operators*. ∎

A UNIFYING THEOREM

In Theorem 4.4.7 we tied together most of the major concepts developed at that point in the text. Theorems 6.3.12, 6.3.13, and 6.3.14 now enable us to add two more results to that theorem.

Theorem 6.3.15 *If* A *is an* $n \times n$ *matrix, and if* T_A *is the linear operator on* R^n *with standard matrix* A, *then the following statements are equivalent.*

(*a*) *The reduced row echelon form of* A *is* I_n.

(*b*) A *is expressible as a product of elementary matrices.*

(*c*) A *is invertible.*

(*d*) $A\mathbf{x} = \mathbf{0}$ *has only the trivial solution.* ~~ker(T_A) = 0~~

(*e*) $A\mathbf{x} = \mathbf{b}$ *is consistent for every vector* $\mathbf{b}$ *in* R^n.

(*f*) $A\mathbf{x} = \mathbf{b}$ *has exactly one solution for every vector* $\mathbf{b}$ *in* R^n.

(*g*) *The column vectors of* A *are linearly independent.*

(*h*) *The row vectors of* A *are linearly independent.*

(*i*) $\det(A) \neq 0$.

(*j*) $\lambda = 0$ *is not an eigenvalue of* A.

(*k*) T_A *is one-to-one.*

(*l*) T_A *is onto.*

EXAMPLE 13

Examples 7 and 8 Revisited Using Determinants

The fact that a rotation about the origin R^2 is one-to-one and onto can be established algebraically by showing that the determinant of its standard matrix is not zero. This can be confirmed using Formula (16) of Section 6.1 to obtain

$$\det(R_\theta) = \begin{vmatrix} \cos\theta & -\sin\theta \\ \sin\theta & \cos\theta \end{vmatrix} = \cos^2\theta + \sin^2\theta = 1 \neq 0$$

The fact that the orthogonal projection of R^3 on the xy-plane is neither one-to-one nor onto can be established by showing that the determinant of its standard matrix A is zero. This is, in fact, the case, since

$$\det(A) = \begin{vmatrix} 1 & 0 & 0 \\ 0 & 1 & 0 \\ 0 & 0 & 0 \end{vmatrix} = 0$$

■

Exercise Set 6.3

In Exercises 1 and 2, find the kernel and range of T without performing any computations. Which transformations, if any, are one-to-one? Onto?

1. (a) T is the orthogonal projection of R^2 on the x-axis.
　 (b) T is the orthogonal projection of R^3 on the yz-plane.
　 (c) $T : R^2 \to R^2$ is the dilation $T(\mathbf{x}) = 2\mathbf{x}$.
　 (d) $T : R^3 \to R^3$ is the reflection about the xy-plane.

2. (a) T is the orthogonal projection of R^2 on the y-axis.
　 (b) T is the orthogonal projection of R^3 on the xz-plane.
　 (c) $T : R^2 \to R^2$ is the contraction $T(\mathbf{x}) = \frac{1}{2}\mathbf{x}$.
　 (d) $T : R^3 \to R^3$ is the rotation about the z-axis through an angle of $\pi/4$.

In Exercises 3–6, find the kernel of the linear transformation whose standard matrix is A. Express your answer as the span of a set of vectors.

3. $A = \begin{bmatrix} 1 & 1 \\ 2 & 2 \end{bmatrix}$

4. $A = \begin{bmatrix} 1 & 0 & 2 \\ 2 & 1 & -1 \\ 1 & -1 & 7 \end{bmatrix}$

5. $A = \begin{bmatrix} 1 & 2 & -3 \\ -1 & -2 & 3 \\ 4 & 4 & 4 \end{bmatrix}$

6. $A = \begin{bmatrix} 2 & 1 & -1 \\ 1 & -2 & 1 \\ 1 & -7 & 4 \\ 3 & 4 & -3 \end{bmatrix}$

In Exercises 7 and 8, set up and solve a linear system whose solution space is the kernel of T, and then express the kernel as the span of some set of vectors.

7. $T : R^3 \to R^4$; $T(x, y, z) = (x - z, y - z, x - y, x + y + z)$

8. $T : R^3 \to R^2$; $T(x, y, z) = (x + 2y + z, x - y + z)$

In Exercises 9 and 10, determine whether $\mathbf{b}$ is in the column space of A, and if so, express it as a linear combination of the column vectors of A.

9. (a) $A = \begin{bmatrix} 1 & 2 & 0 \\ -2 & 2 & 1 \\ 1 & 8 & 1 \end{bmatrix}$; $\mathbf{b} = \begin{bmatrix} 1 \\ 1 \\ 1 \end{bmatrix}$

　 (b) $A = \begin{bmatrix} 1 & 2 & 0 \\ -2 & 2 & 1 \\ 1 & 8 & 1 \end{bmatrix}$; $\mathbf{b} = \begin{bmatrix} 3 \\ -1 \\ 8 \end{bmatrix}$

10. (a) $A = \begin{bmatrix} 3 & -2 & 1 & 5 \\ 1 & 4 & 5 & -3 \\ 0 & 1 & 1 & -1 \end{bmatrix}$; $\mathbf{b} = \begin{bmatrix} 2 \\ 1 \\ 0 \end{bmatrix}$

　 (b) $A = \begin{bmatrix} 3 & -2 & 1 & 5 \\ 1 & 4 & 5 & -3 \\ 0 & 1 & 1 & -1 \end{bmatrix}$; $\mathbf{b} = \begin{bmatrix} 4 \\ 6 \\ 1 \end{bmatrix}$

In Exercises 11 and 12, determine whether $\mathbf{w}$ is in the range of the linear operator T.

11. $T : R^3 \to R^3$; $T(x, y, z) = (2x - y, x + z, y - z)$;
　 $\mathbf{w} = (3, 3, 0)$

12. $T : R^3 \to R^3$; $T(x, y, z) = (x - y, x + y + z, x + 2z)$;
　 $\mathbf{w} = (1, 2, -1)$

In Exercises 13–16, find the standard matrix for the linear operator defined by the equations, and determine whether the operator is one-to-one and/or onto.

13. $w_1 = 2x_1 - 3x_2$
　　 $w_2 = 5x_1 + x_2$

14. $w_1 = 8x_1 + 4x_2$
　　 $w_2 = 2x_1 + x_2$

15. $w_1 = -x_1 + 3x_2 + 2x_3$
　　 $w_2 = 2x_1 + 4x_3$
　　 $w_3 = x_1 + 3x_2 + 6x_3$

16. $w_1 = x_1 + 2x_2 + 3x_3$
　　 $w_2 = 2x_1 + 5x_2 + 3x_3$
　　 $w_3 = x_1 + 8x_3$

In Exercises 17 and 18, show that the linear operator defined by the equations is not onto, and find a vector that is not in the range.

17. $w_1 = 4x_1 - 2x_2$
   $w_2 = 2x_1 - x_2$

18. $w_1 = x_1 - 2x_2 + x_3$
   $w_2 = 5x_1 - x_2 + 3x_3$
   $w_3 = 4x_1 + x_2 + 2x_3$

19. Determine whether multiplication by A is a one-to-one linear transformation.

(a) $A = \begin{bmatrix} 1 & -1 \\ 2 & 0 \\ 3 & -4 \end{bmatrix}$

(b) $A = \begin{bmatrix} 1 & 2 & 3 \\ -1 & 0 & -4 \end{bmatrix}$

20. Determine whether the linear transformations in Exercise 19 are onto.

21. Consider the linear system $A\mathbf{x} = \mathbf{b}$ given by

$$\begin{bmatrix} 1 & -2 & -1 & 3 \\ 2 & 4 & 6 & -2 \\ 3 & 0 & 3 & 3 \end{bmatrix} \begin{bmatrix} x_1 \\ x_2 \\ x_3 \\ x_4 \end{bmatrix} = \begin{bmatrix} b_1 \\ b_2 \\ b_3 \end{bmatrix}$$

(a) What conditions must b_1, b_2, and b_3 satisfy for this system to be consistent?

(b) Use the result in part (a) to express the range of the linear transformation T_A as a linear combination of a set of linearly independent vectors.

(c) Express the kernel of T_A as a linear combination of a set of linearly independent vectors.

Discussion and Discovery

D1. Indicate whether the statement is true (T) or false (F). Justify your answer.

(a) If the linear transformation $T : R^n \to R^n$ is one-to-one and $T(\mathbf{u} - \mathbf{v}) = \mathbf{0}$, then $\mathbf{u} = \mathbf{v}$.

(b) If the linear transformation $T : R^n \to R^n$ is onto and $T(\mathbf{u} - \mathbf{v}) = \mathbf{0}$, then $\mathbf{u} = \mathbf{v}$.

(c) If $\det(A) = 0$, then T_A is neither onto nor one-to-one.

(d) If $T : R^n \to R^m$ is not one-to-one, then $\ker(T)$ contains infinitely many vectors.

(e) Shears in R^2 are one-to-one linear operators.

D2. Let $\mathbf{a}$ be a fixed vector in R^3. Do you think that the formula $T(\mathbf{v}) = \mathbf{a} \times \mathbf{v}$ defines a one-to-one linear operator on R^3? Explain your reasoning.

D3. If A is an $m \times n$ matrix, and if the linear system $A\mathbf{x} = \mathbf{b}$ is consistent for every vector $\mathbf{b}$ in R^m, what can you say about the range of $T_A : R^n \to R^m$?

D4. If $\mathbf{x} = t\mathbf{v}$ is a line through the origin of R^n, and if $\mathbf{v}_0$ is a nonzero vector in R^n, is there some linear operator on R^n that maps $\mathbf{x} = t\mathbf{v}$ onto the line $\mathbf{x} = \mathbf{v}_0 + t\mathbf{v}$? Explain your reasoning.

Working with Proofs

P1. Prove that if A and B are $n \times n$ matrices and $\mathbf{x}$ is in the null space of B, then $\mathbf{x}$ is in the null space of AB.

Technology Exercises

T1. Consider the matrix

$$A = \begin{bmatrix} 2 & 5 & -3 & 7 & 1 & 3 \\ 5 & -2 & 9 & 8 & 4 & -2 \\ -4 & 3 & 8 & 11 & -5 & 2 \\ 11 & 0 & -2 & 4 & 10 & -1 \end{bmatrix}$$

(a) Find the null space of A. Express your answer as the span of a set of vectors.

(b) Determine whether the vector $\mathbf{w} = (5, -2, -3, 6)$ is in the range of the linear transformation T_A. If so, find a vector whose image under T_A is $\mathbf{w}$.

T2. Consider the matrix

$$A = \begin{bmatrix} 3 & -5 & -2 & 2 \\ -4 & 7 & 4 & 4 \\ 4 & -9 & -3 & 7 \\ 2 & -6 & -3 & 2 \end{bmatrix}$$

Show that $T_A : R^4 \to R^4$ is onto in three different ways.

Section 6.4   Composition and Invertibility of Linear Transformations

In this section we will investigate problems that involve two or more linear transformations performed in succession, and we will explore the relationship between the invertibility of a matrix A and the geometric properties of the corresponding linear operator.

COMPOSITIONS OF LINEAR TRANSFORMATIONS

There are many applications in which sequences of linear transformations are applied in succession, with each transformation acting on the output of its predecessor—for example, a rotation, followed by a reflection, followed by a projection. Our first goal in this section is to show how to combine a succession of linear transformations into a single linear transformation.

If $T_1: R^n \to R^k$ and $T_2: R^k \to R^m$ are linear transformations in which the codomain of T_1 is the same as the domain of T_2, then for each $\mathbf{x}$ in R^n we can first compute $T_1(\mathbf{x})$ to produce a vector in R^k, and then we can compute $T_2(T_1(\mathbf{x}))$ to produce a vector in R^m. Thus, first applying T_1 and then applying T_2 to the output of T_1 produces a transformation from R^n to R^m. This transformation, called the **composition of T_2 with T_1**, is denoted by $T_2 \circ T_1$ (read, "T_2 circle T_1"); that is,

$$(T_2 \circ T_1)(\mathbf{x}) = T_2(T_1(\mathbf{x})) \tag{1}$$

(Figure 6.4.1).

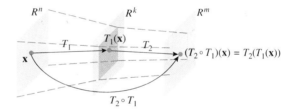

Figure 6.4.1

The following theorem shows that the composition of two linear transformations is itself a linear transformation.

Theorem 6.4.1   *If $T_1: R^n \to R^k$ and $T_2: R^k \to R^m$ are both linear transformations, then $(T_2 \circ T_1): R^n \to R^m$ is also a linear transformation.*

Proof To prove that the composition $T_2 \circ T_1$ is linear we must show that it is additive and homogeneous. Accordingly, let $\mathbf{u}$ and $\mathbf{v}$ be any vectors in R^n, and let c be a scalar. Then it follows from (1) and the linearity of T_1 and T_2 that

$$(T_2 \circ T_1)(\mathbf{u} + \mathbf{v}) = T_2(T_1(\mathbf{u} + \mathbf{v})) = T_2(T_1(\mathbf{u}) + T_1(\mathbf{v}))$$
$$= T_2(T_1(\mathbf{u})) + T_2(T_1(\mathbf{v}))$$
$$= (T_2 \circ T_1)(\mathbf{u}) + (T_2 \circ T_1)(\mathbf{v})$$

which proves the additivity. Also,

$$(T_2 \circ T_1)(c\mathbf{u}) = T_2(T_1(c\mathbf{u})) = T_2(cT_1(\mathbf{u})) = cT_2(T_1(\mathbf{u})) = c(T_2 \circ T_1)(\mathbf{u})$$

which proves the homogeneity. ∎

Now let us consider how the standard matrix for a composition of two linear transformations is related to the standard matrices for the individual transformations. For this purpose, suppose that $T_1: R^n \to R^k$ has standard matrix $[T_1]$ and that $T_2: R^k \to R^m$ has standard matrix $[T_2]$. Thus, for each standard unit vector $\mathbf{e}_i$ in R^n we have

$$(T_2 \circ T_1)(\mathbf{e}_i) = T_2(T_1(\mathbf{e}_i)) = T_2([T_1]\mathbf{e}_i) = [T_2]([T_1]\mathbf{e}_i) = ([T_2][T_1])\mathbf{e}_i$$

which implies that $[T_2][T_1]$ is the standard matrix for $T_2 \circ T_1$ (why?). Thus, we have shown that

$$[T_2 \circ T_1] = [T_2][T_1] \tag{2}$$

That is, *the standard matrix for the composition of two linear transformations is the product of their standard matrices in the appropriate order.*

Formula (2) can be expressed in an alternative form that is useful when specific letters are used to denote the standard matrices: If A is the standard matrix for a linear transformation $T_A : R^n \to R^k$ and B is the standard matrix for a linear transformation $T_B : R^k \to R^m$, then Formula (2) states that BA is the standard matrix for $T_B \circ T_A$; that is,

$$T_B \circ T_A = T_{BA} \tag{3}$$

In summary, we have the following transformation interpretation of matrix multiplication.

Theorem 6.4.2 *If A is a $k \times n$ matrix and B is an $m \times k$ matrix, then the $m \times n$ matrix BA is the standard matrix for the composition of the linear transformation corresponding to B with the linear transformation corresponding to A.*

EXAMPLE 1
Composing
Rotations in R^2

Let $T_1 : R^2 \to R^2$ be the rotation about the origin of R^2 through the angle θ_1, and let $T_2 : R^2 \to R^2$ be the rotation about the origin through the angle θ_2. The standard matrices for these rotations are

$$R_{\theta_1} = \begin{bmatrix} \cos\theta_1 & -\sin\theta_1 \\ \sin\theta_1 & \cos\theta_1 \end{bmatrix} \quad \text{and} \quad R_{\theta_2} = \begin{bmatrix} \cos\theta_2 & -\sin\theta_2 \\ \sin\theta_2 & \cos\theta_2 \end{bmatrix} \tag{4}$$

The composition

$$(T_2 \circ T_1)(\mathbf{x}) = T_2(T_1(\mathbf{x}))$$

first rotates $\mathbf{x}$ through the angle θ_1 and then rotates $T_1(\mathbf{x})$ through the angle θ_2, so the standard matrix for $T_2 \circ T_1$ should be

$$R_{\theta_1 + \theta_2} = \begin{bmatrix} \cos(\theta_1 + \theta_2) & -\sin(\theta_1 + \theta_2) \\ \sin(\theta_1 + \theta_2) & \cos(\theta_1 + \theta_2) \end{bmatrix} \tag{5}$$

To confirm that this is so let us apply Formula (2). According to that formula the standard matrix for $T_2 \circ T_1$ is

$$\begin{aligned}
R_{\theta_2} R_{\theta_1} &= \begin{bmatrix} \cos\theta_2 & -\sin\theta_2 \\ \sin\theta_2 & \cos\theta_2 \end{bmatrix} \begin{bmatrix} \cos\theta_1 & -\sin\theta_1 \\ \sin\theta_1 & \cos\theta_1 \end{bmatrix} \\
&= \begin{bmatrix} \cos\theta_2\cos\theta_1 - \sin\theta_2\sin\theta_1 & -\cos\theta_2\sin\theta_1 - \sin\theta_2\cos\theta_1 \\ \sin\theta_2\cos\theta_1 + \cos\theta_2\sin\theta_1 & -\sin\theta_2\sin\theta_1 + \cos\theta_2\cos\theta_1 \end{bmatrix} \\
&= \begin{bmatrix} \cos(\theta_1 + \theta_2) & -\sin(\theta_1 + \theta_2) \\ \sin(\theta_1 + \theta_2) & \cos(\theta_1 + \theta_2) \end{bmatrix}
\end{aligned}$$

which agrees with (5). ∎

In light of Example 1 you might think that a composition of two reflections in R^2 would be another reflection. The following example shows that this is never the case—the composition of two reflections about lines through the origin of R^2 is a rotation about the origin.

EXAMPLE 2
Composing
Reflections

By Formula (18) of Section 6.1, the matrices

$$H_{\theta_1} = \begin{bmatrix} \cos 2\theta_1 & \sin 2\theta_1 \\ \sin 2\theta_1 & -\cos 2\theta_1 \end{bmatrix} \quad \text{and} \quad H_{\theta_2} = \begin{bmatrix} \cos 2\theta_2 & \sin 2\theta_2 \\ \sin 2\theta_2 & -\cos 2\theta_2 \end{bmatrix}$$

represent reflections about lines through the origin of R^2 making angles of θ_1 and θ_2 with the x-axis, respectively. Accordingly, if we first reflect about the line making the angle θ_1 and then about the line making the angle θ_2, then we obtain a linear operator whose standard matrix is

$$
\begin{aligned}
H_{\theta_2} H_{\theta_1} &= \begin{bmatrix} \cos 2\theta_2 & \sin 2\theta_2 \\ \sin 2\theta_2 & -\cos 2\theta_2 \end{bmatrix} \begin{bmatrix} \cos 2\theta_1 & \sin 2\theta_1 \\ \sin 2\theta_1 & -\cos 2\theta_1 \end{bmatrix} \\
&= \begin{bmatrix} \cos 2\theta_2 \cos 2\theta_1 + \sin 2\theta_2 \sin 2\theta_1 & \cos 2\theta_2 \sin 2\theta_1 - \sin 2\theta_2 \cos 2\theta_1 \\ \sin 2\theta_2 \cos 2\theta_1 - \cos 2\theta_2 \sin 2\theta_1 & \sin 2\theta_2 \sin 2\theta_1 + \cos 2\theta_2 \cos 2\theta_1 \end{bmatrix} \\
&= \begin{bmatrix} \cos(2\theta_2 - 2\theta_1) & -\sin(2\theta_2 - 2\theta_1) \\ \sin(2\theta_2 - 2\theta_1) & \cos(2\theta_2 - 2\theta_1) \end{bmatrix}
\end{aligned}
$$

Comparing this matrix to the matrix R_θ in Formula (16) of Section 6.1, we see that this matrix represents a rotation about the origin through an angle of $2\theta_2 - 2\theta_1$. Thus, we have shown that

$$H_{\theta_2} H_{\theta_1} = R_{2(\theta_2 - \theta_1)}$$

This result is illustrated in Figure 6.4.2. ∎

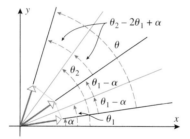

Figure 6.4.2

$\theta = 2(\theta_2 - 2\theta_1 + \alpha) + 2(\theta_1 - \alpha) = 2\theta_2 - 2\theta_1$

REMARK In the last two examples we saw that in R^2 the composition of two rotations about the origin or of two reflections about lines through the origin produces a rotation about the origin. We could have anticipated this from Theorem 6.2.7, since a rotation is represented by an orthogonal matrix with determinant $+1$ and a reflection by an orthogonal matrix with determinant -1. Thus, the product of two rotation matrices or two reflection matrices is an orthogonal matrix with determinant $+1$ and hence represents a rotation.

EXAMPLE 3
Composition Is
Not a
Commutative
Operation

(a) Find the standard matrix for the linear operator on R^2 that first shears by a factor of 2 in the x-direction and then reflects about the line $y = x$.

(b) Find the standard matrix for the linear operator on R^2 that first reflects about the line $y = x$ and then shears by a factor of 2 in the x-direction.

Solution Let A_1 and A_2 be the standard matrices for the shear and reflection, respectively. Then from Table 6.2.3 and Table 6.1.1 we have

$$A_1 = \begin{bmatrix} 1 & 2 \\ 0 & 1 \end{bmatrix} \quad \text{and} \quad A_2 = \begin{bmatrix} 0 & 1 \\ 1 & 0 \end{bmatrix}$$

Thus, the standard matrix for the shear followed by the reflection is

$$A_2 A_1 = \begin{bmatrix} 0 & 1 \\ 1 & 0 \end{bmatrix} \begin{bmatrix} 1 & 2 \\ 0 & 1 \end{bmatrix} = \begin{bmatrix} 0 & 1 \\ 1 & 2 \end{bmatrix}$$

and the standard matrix for the reflection followed by the shear is

$$A_1 A_2 = \begin{bmatrix} 1 & 2 \\ 0 & 1 \end{bmatrix} \begin{bmatrix} 0 & 1 \\ 1 & 0 \end{bmatrix} = \begin{bmatrix} 2 & 1 \\ 1 & 0 \end{bmatrix}$$

Since the matrices $A_2 A_1$ and $A_1 A_2$ are not the same, shearing and then reflecting is different from reflecting and then shearing (Figure 6.4.3). ■

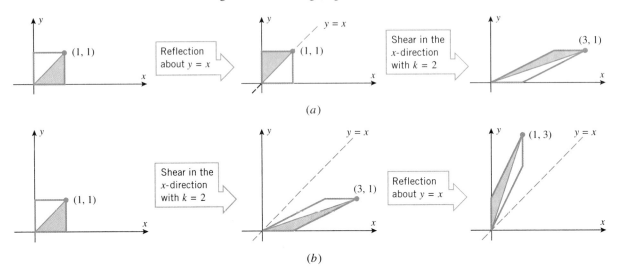

Figure 6.4.3

REMARK If T_A and T_B are linear operators whose standard matrices are A and B, then it follows from (3) that $T_A \circ T_B = T_B \circ T_A$ if and only if $AB = BA$. Thus, the composition of linear operators is the same in either order if and only if their standard matrices commute.

COMPOSITION OF THREE OR MORE LINEAR TRANSFORMATIONS

Compositions can be defined for three or more matrix transformations when the domains and codomains match up appropriately. Specifically, if

$$T_1 : R^n \to R^k, \quad T_2 : R^k \to R^l, \quad T_3 : R^l \to R^m$$

then we define the composition $(T_3 \circ T_2 \circ T_1) : R^n \to R^m$ by

$$(T_3 \circ T_2 \circ T_1)(\mathbf{x}) = T_3(T_2(T_1(\mathbf{x}))) \tag{6}$$

In this case the analog of Formula (2) is

$$[T_3 \circ T_2 \circ T_1] = [T_3][T_2][T_1] \tag{7}$$

Also, if we let A, B, and C denote the standard matrices for the linear transformations T_A, T_B, and T_C, respectively, then the analog of Formula (3) is

$$T_C \circ T_B \circ T_A = T_{CBA} \tag{8}$$

The extensions of (6), (7), and (8) to four or more linear transformations should be clear.

EXAMPLE 4
A Composition
of Three Matrix
Transformations

Find the standard matrix for the linear operator $T : R^3 \to R^3$ that first rotates a vector about the z-axis through an angle θ, then reflects the resulting vector about the yz-plane, and then projects that vector orthogonally onto the xy-plane.

Solution The operator T can be expressed as the composition

$$T = T_C \circ T_B \circ T_A = T_{CBA}$$

where A, B, and C are the standard matrices for the rotation, reflection, and projection, respectively. These matrices are

$$A = \begin{bmatrix} \cos\theta & -\sin\theta & 0 \\ \sin\theta & \cos\theta & 0 \\ 0 & 0 & 1 \end{bmatrix}, \quad B = \begin{bmatrix} -1 & 0 & 0 \\ 0 & 1 & 0 \\ 0 & 0 & 1 \end{bmatrix}, \quad C = \begin{bmatrix} 1 & 0 & 0 \\ 0 & 1 & 0 \\ 0 & 0 & 0 \end{bmatrix}$$

(verify) and hence the standard matrix for T is

$$CBA = \begin{bmatrix} 1 & 0 & 0 \\ 0 & 1 & 0 \\ 0 & 0 & 0 \end{bmatrix} \begin{bmatrix} -1 & 0 & 0 \\ 0 & 1 & 0 \\ 0 & 0 & 1 \end{bmatrix} \begin{bmatrix} \cos\theta & -\sin\theta & 0 \\ \sin\theta & \cos\theta & 0 \\ 0 & 0 & 1 \end{bmatrix}$$

$$= \begin{bmatrix} -\cos\theta & \sin\theta & 0 \\ \sin\theta & \cos\theta & 0 \\ 0 & 0 & 0 \end{bmatrix}$$ ∎

In many applications an object undergoes a succession of rotations about different axes through the origin. The following important theorem shows that the net effect of these rotations is the same as that of a single rotation about some appropriate axis through the origin.

> **Theorem 6.4.3** *If $T_1, T_2, \ldots, T_k$ is a succession of rotations about axes through the origin of R^3, then the k rotations can be accomplished by a single rotation about some appropriate axis through the origin of R^3.*

Proof Let $A_1, A_2, \ldots, A_k$ be the standard matrices for the rotations. Each matrix is orthogonal and has determinant 1, so the same is true for the product

$$A = A_k \cdots A_2 A_1$$

Thus, A represents a rotation about some axis through the origin of R^3. Since A is the standard matrix for the composition $T_k \circ \cdots \circ T_2 \circ T_1$, the result is proved. ∎

In aeronautics and astronautics, the orientation of an aircraft or space shuttle relative to an xyz-coordinate system is often described in terms of angles called *yaw*, *pitch*, and *roll*. If, for example, an aircraft is flying along the y-axis and the xy-plane defines the horizontal, then the aircraft's angle of rotation about the z-axis is called the *yaw*, its angle of rotation about the x-axis is called the *pitch*, and its angle of rotation about the y-axis is called the *roll* (Figure 6.4.4). As a result of Theorem 6.4.3, a combination of yaw, pitch, and roll can be achieved by a single rotation about some axis through the origin. This is, in fact, how a space shuttle makes attitude adjustments—it doesn't perform each rotation separately; it calculates one axis, and rotates about that axis to get the correct orientation. Such rotation maneuvers are used to align an antenna, point the nose toward a celestial object, or position a payload bay for docking.

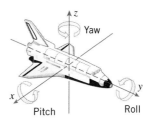

Figure 6.4.4

EXAMPLE 5
A Rotation
Problem

Suppose that a vector in R^3 is first rotated $45°$ about the positive x-axis, then the resulting vector is rotated $45°$ about the positive y-axis, and then that vector is rotated $45°$ about the positive z-axis. Find an appropriate axis and angle of rotation that achieves the same result in one rotation.

Solution Let R_x, R_y, and R_z denote the standard matrices for the rotations about the positive x-, y-, and z-axes, respectively. Referring to Table 6.2.6, these matrices are

$$R_x = \begin{bmatrix} 1 & 0 & 0 \\ 0 & \frac{1}{\sqrt{2}} & -\frac{1}{\sqrt{2}} \\ 0 & \frac{1}{\sqrt{2}} & \frac{1}{\sqrt{2}} \end{bmatrix}, \quad R_y = \begin{bmatrix} \frac{1}{\sqrt{2}} & 0 & \frac{1}{\sqrt{2}} \\ 0 & 1 & 0 \\ -\frac{1}{\sqrt{2}} & 0 & \frac{1}{\sqrt{2}} \end{bmatrix}, \quad R_z = \begin{bmatrix} \frac{1}{\sqrt{2}} & -\frac{1}{\sqrt{2}} & 0 \\ \frac{1}{\sqrt{2}} & \frac{1}{\sqrt{2}} & 0 \\ 0 & 0 & 1 \end{bmatrix}$$

Thus, the standard matrix for the composition of the rotations in the order stated in the problem is

$$A = R_z R_y R_x = \begin{bmatrix} \frac{1}{2} & \frac{\sqrt{2}}{4} - \frac{1}{2} & \frac{\sqrt{2}}{4} + \frac{1}{2} \\ \frac{1}{2} & \frac{\sqrt{2}}{4} + \frac{1}{2} & \frac{\sqrt{2}}{4} - \frac{1}{2} \\ -\frac{1}{\sqrt{2}} & \frac{1}{2} & \frac{1}{2} \end{bmatrix} \approx \begin{bmatrix} 0.5 & -0.1464 & 0.8536 \\ 0.5 & 0.8536 & -0.1464 \\ -0.7071 & 0.5 & 0.5 \end{bmatrix}$$

(verify). To find the axis of rotation $\mathbf{v}$ we will apply Formula (17) of Section 6.2, taking the arbitrary vector $\mathbf{x}$ to be $\mathbf{e}_1$. We leave it for you to confirm that

$$\mathbf{v} = \begin{bmatrix} \frac{1}{2} - \frac{\sqrt{2}}{4} \\ \frac{\sqrt{2}}{4} \\ \frac{1}{2} - \frac{\sqrt{2}}{4} \end{bmatrix} \approx \begin{bmatrix} 0.1464 \\ 0.3536 \\ 0.1464 \end{bmatrix}$$

Also, it follows from Formula (16) of Section 6.2 that the angle of rotation satisfies

$$\cos\theta = \frac{\text{tr}(A) - 1}{2} = \frac{2 + \sqrt{2}}{8} \approx 0.4268$$

from which it follows that $\theta \approx 64.74°$.   ■

FACTORING LINEAR OPERATORS INTO COMPOSITIONS

We have seen that a succession of linear transformations can be composed into a single linear transformation. Sometimes the geometric effect of a matrix transformation can best be understood by reversing this process and expressing the transformation as a composition of simpler transformations whose geometric effects are known. We will begin with some examples in R^2.

EXAMPLE 6
Transforming
with a Diagonal
Matrix

A diagonal matrix

$$D = \begin{bmatrix} \lambda_1 & 0 \\ 0 & \lambda_2 \end{bmatrix}$$

can be factored as

$$D = \begin{bmatrix} \lambda_1 & 0 \\ 0 & \lambda_2 \end{bmatrix} = \begin{bmatrix} 1 & 0 \\ 0 & \lambda_2 \end{bmatrix} \begin{bmatrix} \lambda_1 & 0 \\ 0 & 1 \end{bmatrix} = D_2 D_1$$

Multiplication by D_1 produces a compression in the x-direction if $0 \leq \lambda_1 < 1$, an expansion in the x-direction if $\lambda_1 > 1$, and has no effect if $\lambda_1 = 1$; multiplication by D_2 produces analogous results in the y-direction. Thus, for example, multiplication by

$$D = \begin{bmatrix} 3 & 0 \\ 0 & \frac{1}{2} \end{bmatrix} = \begin{bmatrix} 3 & 0 \\ 0 & 1 \end{bmatrix} \begin{bmatrix} 1 & 0 \\ 0 & \frac{1}{2} \end{bmatrix}$$

causes an expansion by a factor of 3 in the x-direction and a compression by a factor of $\frac{1}{2}$ in the y-direction. ∎

The result in the last example is a special case of a more general result about diagonal matrices. Specifically, if

$$D = \begin{bmatrix} \lambda_1 & 0 & \cdots & 0 \\ 0 & \lambda_2 & \cdots & 0 \\ \vdots & \vdots & \ddots & \vdots \\ 0 & 0 & \cdots & \lambda_n \end{bmatrix}$$

has nonnegative entries, then multiplication by D maps the standard unit vector $\mathbf{e}_i$ into the vector $\lambda_i \mathbf{e}_i$, so you can think of this operator as causing compressions or expansions in the directions of the standard unit vectors—it causes a compression in the direction of $\mathbf{e}_i$ if $0 \leq \lambda_i < 1$ and an expansion if $\lambda_i > 1$. Multiplication by D has no effect in the direction of $\mathbf{e}_i$ if $\lambda_i = 1$. Because of these geometric properties, diagonal matrices with nonnegative entries are called *scaling matrices*.

EXAMPLE 7
Transforming with 2×2 Elementary Matrices

The 2×2 elementary matrices have five possible forms (verify):

$$\begin{bmatrix} 1 & k \\ 0 & 1 \end{bmatrix} \qquad \begin{bmatrix} 1 & 0 \\ k & 1 \end{bmatrix} \qquad \begin{bmatrix} 0 & 1 \\ 1 & 0 \end{bmatrix} \qquad \begin{bmatrix} k & 0 \\ 0 & 1 \end{bmatrix} \qquad \begin{bmatrix} 1 & 0 \\ 0 & k \end{bmatrix}$$

Type 1   **Type 2**   **Type 3**   **Type 4**   **Type 5**

Type 1 represents a shear in the x-direction, type 2 a shear in the y-direction, and type 3 a reflection about the line $y = x$. If $k \geq 0$, then types 4 and 5 represent compressions or expansions in the x- and y-directions, respectively. If $k < 0$, then we can express k in the form $k = -k_1$, where $k_1 > 0$, and we can factor the type 4 and 5 matrices as

$$\begin{bmatrix} k & 0 \\ 0 & 1 \end{bmatrix} = \begin{bmatrix} -k_1 & 0 \\ 0 & 1 \end{bmatrix} = \begin{bmatrix} -1 & 0 \\ 0 & 1 \end{bmatrix} \begin{bmatrix} k_1 & 0 \\ 0 & 1 \end{bmatrix} \qquad (9)$$

$$\begin{bmatrix} 1 & 0 \\ 0 & k \end{bmatrix} = \begin{bmatrix} 1 & 0 \\ 0 & -k_1 \end{bmatrix} = \begin{bmatrix} 1 & 0 \\ 0 & -1 \end{bmatrix} \begin{bmatrix} 1 & 0 \\ 0 & k_1 \end{bmatrix} \qquad (10)$$

Thus, a type 4 matrix with negative k represents a compression or expansion in the x-direction, followed by a reflection about the y-axis; and a type 5 matrix with negative k represents an expansion or compression in the y-direction, followed by a reflection about the x-axis. ∎

Recall from Theorem 3.3.3 that an invertible matrix A can be expressed as a product of elementary matrices. Thus, Example 7 leads to the following result about the geometric effect of linear operators on R^2 whose standard matrices are invertible.

Theorem 6.4.4   *If A is an invertible 2×2 matrix, then the corresponding linear operator on R^2 is a composition of shears, compressions, and expansions in the directions of the coordinate axes, and reflections about the coordinate axes and about the line $y = x$.*

EXAMPLE 8
Transforming with an Invertible 2×2 Matrix

Describe the geometric effect of multiplication by

$$A = \begin{bmatrix} 1 & 2 \\ 3 & 4 \end{bmatrix}$$

in terms of shears, compressions, expansions, and reflections.

Solution Since $\det(A) \neq 0$, the matrix A is invertible and hence can be reduced to I by a sequence of elementary row operations; for example,

$$\begin{bmatrix} 1 & 2 \\ 3 & 4 \end{bmatrix} \rightarrow \begin{bmatrix} 1 & 2 \\ 0 & -2 \end{bmatrix} \rightarrow \begin{bmatrix} 1 & 2 \\ 0 & 1 \end{bmatrix} \rightarrow \begin{bmatrix} 1 & 0 \\ 0 & 1 \end{bmatrix}$$

| Add -3 times the first row to the second. | Multiply the second row by $-\frac{1}{2}$. | Add -2 times the second row to the first. |

The three successive row operations can be performed using multiplications by the elementary matrices

$$E_1 = \begin{bmatrix} 1 & 0 \\ -3 & 1 \end{bmatrix}, \quad E_2 = \begin{bmatrix} 1 & 0 \\ 0 & -\frac{1}{2} \end{bmatrix}, \quad E_3 = \begin{bmatrix} 1 & -2 \\ 0 & 1 \end{bmatrix}$$

Inverting these matrices and applying Formula (2) of Section 3.3 yields the factorization

$$A = E_1^{-1} E_2^{-1} E_3^{-1} = \begin{bmatrix} 1 & 0 \\ 3 & 1 \end{bmatrix} \begin{bmatrix} 1 & 0 \\ 0 & -2 \end{bmatrix} \begin{bmatrix} 1 & 2 \\ 0 & 1 \end{bmatrix} = \begin{bmatrix} 1 & 0 \\ 3 & 1 \end{bmatrix} \begin{bmatrix} 1 & 0 \\ 0 & -1 \end{bmatrix} \begin{bmatrix} 1 & 0 \\ 0 & 2 \end{bmatrix} \begin{bmatrix} 1 & 2 \\ 0 & 1 \end{bmatrix}$$

Thus, reading right to left, the geometric effect of A is to successively shear by a factor of 2 in the x-direction, expand by a factor of 2 in the y-direction, reflect about the x-axis, and shear by a factor of 3 in the y-direction. ∎

INVERSE OF A LINEAR TRANSFORMATION

Our next objective is to find a relationship between the linear operators represented by A and A^{-1} when A is invertible. We will begin with some terminology.

If $T : R^n \rightarrow R^m$ is a one-to-one linear transformation, then each vector $\mathbf{w}$ in the range of T is the image of a unique vector $\mathbf{x}$ in the domain of T (Figure 6.4.5a); we call $\mathbf{x}$ the *preimage* of $\mathbf{w}$. The uniqueness of the preimage allows us to create a new function that maps $\mathbf{w}$ into $\mathbf{x}$; we call this function the *inverse* of T and denote it by T^{-1}. Thus,

$$T^{-1}(\mathbf{w}) = \mathbf{x} \quad \text{if and only if} \quad T(\mathbf{x}) = \mathbf{w}$$

(Figure 6.4.5b). The domain of the function T^{-1} is the range of T and the range of T^{-1} is the domain of T. When we want to emphasize that the domain of T^{-1} is $\text{ran}(T)$ we will write

$$T^{-1} : \text{ran}(T) \rightarrow R^n \tag{11}$$

Stated informally, T and T^{-1} "cancel out" the effect of one another in the sense that if $\mathbf{w} = T(\mathbf{x})$, then

$$T(T^{-1}(\mathbf{w})) = T(\mathbf{x}) = \mathbf{w} \tag{12}$$

$$T^{-1}(T(\mathbf{x})) = T^{-1}(\mathbf{w}) = \mathbf{x} \tag{13}$$

We will leave it as an exercise for you to prove the following result.

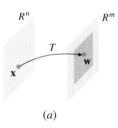

(a)

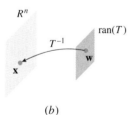

(b)

Figure 6.4.5

Theorem 6.4.5   *If T is a one-to-one linear transformation, then so is T^{-1}.*

INVERTIBLE LINEAR OPERATORS

In the special case where T is a one-to-one linear operator on R^n, it follows from Theorem 6.3.15 that T is onto and hence that the domain of T^{-1} is all of R^n. This, together with Theorem 6.4.5, implies that if T is a one-to-one linear operator on R^n, then so is T^{-1}. This being the case, we are naturally led to inquire as to what relationship might exist between the standard matrix for T and the standard matrix for T^{-1}.

Theorem 6.4.6   *If T is a one-to-one linear operator on R^n, then the standard matrix for T is invertible and its inverse is the standard matrix for T^{-1}.*

Proof Let A and B be the standard matrices for T and T^{-1}, respectively, and let $\mathbf{x}$ be any vector in R^n. We know from (13) that

$$T^{-1}(T(\mathbf{x})) = \mathbf{x}$$

which we can write in matrix form as

$$B(A\mathbf{x}) = \mathbf{x} \quad \text{or} \quad (BA)\mathbf{x} = \mathbf{x} = I\mathbf{x}$$

Since this holds for all $\mathbf{x}$ in R^n, it follows from Theorem 3.4.4 that $BA = I$. Thus, A is invertible and its inverse is B, which is what we wanted to prove. ∎

If T is a one-to-one linear operator on R^n, then the statement of Theorem 6.4.6 is captured by the formula

$$[T^{-1}] = [T]^{-1} \tag{14}$$

Alternatively, if we use the notation T_A to denote a one-to-one linear operator with standard matrix A, then (14) implies that

$$T_A^{-1} = T_{A^{-1}} \tag{15}$$

REMARK A one-to-one linear operator is also called an ***invertible linear operator*** when it is desired to emphasize the existence of the inverse operator.

EXAMPLE 9
Inverse of a
Rotation
Operator

Recall that the linear operator on R^2 corresponding to

$$A = \begin{bmatrix} \cos\theta & -\sin\theta \\ \sin\theta & \cos\theta \end{bmatrix} \tag{16}$$

is the rotation about the origin through the angle θ. It is evident that the inverse of this operator is the rotation through the angle $-\theta$, since rotating $\mathbf{x}$ through the angle θ and then rotating the image through the angle $-\theta$ produces the vector $\mathbf{x}$ back again. This is consistent with Theorem 6.4.6 since

$$A^{-1} = A^T = \begin{bmatrix} \cos\theta & \sin\theta \\ -\sin\theta & \cos\theta \end{bmatrix} = \begin{bmatrix} \cos(-\theta) & -\sin(-\theta) \\ \sin(-\theta) & \cos(-\theta) \end{bmatrix} \tag{17}$$

represents the rotation about the origin through the angle $-\theta$. ∎

EXAMPLE 10
Inverse of a
Compression
Operator

The linear operator on R^2 corresponding to

$$A = \begin{bmatrix} 1 & 0 \\ 0 & \frac{1}{2} \end{bmatrix}$$

is the compression in the y-direction by a factor of $\frac{1}{2}$. It is evident that the inverse of this operator is the expansion in the y-direction by a factor of 2. This is consistent with Theorem 6.4.6 since

$$A^{-1} = \begin{bmatrix} 1 & 0 \\ 0 & 2 \end{bmatrix}$$

is the expansion in the y-direction with factor 2. ∎

EXAMPLE 11
Inverse of a
Reflection
Operator

Recall from Formula (18) of Section 6.1 that the linear operator on R^2 corresponding to

$$A = \begin{bmatrix} \cos\theta & \sin\theta \\ \sin\theta & -\cos\theta \end{bmatrix}$$

is the reflection about the line through the origin that makes an angle of $\theta/2$ with the positive x-axis. It is evident geometrically that A must be its own inverse, since reflecting $\mathbf{x}$ about this line, and then reflecting the image of $\mathbf{x}$ about the line produces $\mathbf{x}$ back again. This is consistent with Theorem 6.4.6, since

$$A^{-1} = A^T = \begin{bmatrix} \cos\theta & \sin\theta \\ \sin\theta & -\cos\theta \end{bmatrix}$$ ■

EXAMPLE 12
Inverse of a
Linear Operator
Defined by a
Linear System

Show that the linear operator $T(x_1, x_2, x_3) = (w_1, w_2, w_3)$ that is defined by the linear equations

$$\begin{aligned} w_1 &= x_1 + 2x_2 + 3x_3 \\ w_2 &= 2x_1 + 5x_2 + 3x_3 \\ w_3 &= x_1 \qquad\quad + 8x_3 \end{aligned}$$ (18)

is one-to-one, and find a set of linear equations that define T^{-1}.

Solution The standard matrix for the operator is

$$A = \begin{bmatrix} 1 & 2 & 3 \\ 2 & 5 & 3 \\ 1 & 0 & 8 \end{bmatrix}$$

(verify). It was shown in Example 3 of Section 3.3 that this matrix is invertible and that

$$A^{-1} = \begin{bmatrix} -40 & 16 & 9 \\ 13 & -5 & -3 \\ 5 & -2 & -1 \end{bmatrix}$$ (19)

The invertibility of A implies that T is one-to-one and that $\mathbf{x} = T^{-1}(\mathbf{w}) = A^{-1}\mathbf{w}$. Thus, it follows from (19) that $T^{-1}(w_1, w_2, w_3) = (x_1, x_2, x_3)$, where

$$\begin{aligned} x_1 &= -40w_1 + 16w_2 + 9w_3 \\ x_2 &= 13w_1 - 5w_2 - 3w_3 \\ x_3 &= 5w_1 - 2w_2 - w_3 \end{aligned}$$

Note that these equations are simply the equations that result from solving (18) for x_1, x_2, and x_3 in terms of w_1, w_2, and w_3.   ■

GEOMETRIC PROPERTIES OF INVERTIBLE LINEAR OPERATORS ON R^2

The next theorem is concerned with the geometric effect that an invertible linear operator on R^2 has on lines. This result will help us to determine how such operators map regions that are bounded by polygons—triangles or rectangles, for example.

> **Theorem 6.4.7** *If $T: R^2 \to R^2$ is an invertible linear operator, then:*
>
> *(a) The image of a line is a line.*
>
> *(b) The image of a line passes through the origin if and only if the original line passes through the origin.*
>
> *(c) The images of two lines are parallel if and only if the original lines are parallel.*
>
> *(d) The images of three points lie on a line if and only if the original points lie on a line.*
>
> *(e) The image of the line segment joining two points is the line segment joining the images of those points.*

We will prove parts (a), (b), and (c).

Proof (a) Recall from Formula (4) of Section 1.3 that the line L through $\mathbf{x}_0$ that is parallel to a nonzero vector $\mathbf{v}$ is given by the vector equation $\mathbf{x} = \mathbf{x}_0 + t\mathbf{v}$. Thus, the linearity of T implies that the image of this line under T consists of all vectors of the form

$$T(\mathbf{x}) = T(\mathbf{x}_0) + tT(\mathbf{v})$$ (20)

Since T is invertible and $\mathbf{v} \neq \mathbf{0}$, it follows that $T(\mathbf{v}) \neq \mathbf{0}$ (why?). Thus, (20) is a vector equation of the line through $T(\mathbf{x}_0)$ that is parallel to $T(\mathbf{v})$.

***Proof** (b)* Since T is invertible, it follows that $T(\mathbf{x}) = \mathbf{0}$ if and only if $\mathbf{x} = \mathbf{0}$. Thus, (20) passes through the origin if and only if $\mathbf{x} = \mathbf{x}_0 + t\mathbf{v}$ passes through the origin.

***Proof** (c)* If L_1 and L_2 are parallel lines, then they are both parallel to some nonzero vector $\mathbf{v}$ and hence can be expressed in vector form as

$$\mathbf{x} = \mathbf{x}_1 + t\mathbf{v} \quad \text{and} \quad \mathbf{x} = \mathbf{x}_2 + t\mathbf{v}$$

The images of these lines are

$$T(\mathbf{x}) = T(\mathbf{x}_1) + tT(\mathbf{v}) \quad \text{and} \quad T(\mathbf{x}) = T(\mathbf{x}_2) + tT(\mathbf{v})$$

both of which are lines parallel to the nonzero vector $T(\mathbf{v})$. Thus, the images must be parallel lines. The same argument applied to T^{-1} can be used to prove the converse. ∎

IMAGE OF THE UNIT SQUARE UNDER AN INVERTIBLE LINEAR OPERATOR

Let us see what we can say about the image of the unit square under an invertible linear operator T on R^2. Since a linear operator maps $\mathbf{0}$ into $\mathbf{0}$, the vertex at the origin remains fixed under the transformation. The images of the other three vertices must be distinct, for otherwise they would lie on a line, and this is impossible by part (d) of Theorem 6.4.7. Finally, since the images of the parallel sides remain parallel, we can conclude that the image of the unit square is a nondegenerate parallelogram that has a vertex at the origin and whose adjacent sides are $T(\mathbf{e}_1)$ and $T(\mathbf{e}_2)$ (Figure 6.4.6). If

$$A = [T(\mathbf{e}_1) \quad T(\mathbf{e}_2)] = \begin{bmatrix} x_1 & x_2 \\ y_1 & y_2 \end{bmatrix}$$

denotes the standard matrix for T, then it follows from Theorem 4.3.5 that $|\det(A)|$ is the area of the parallelogram with adjacent sides $T(\mathbf{e}_1)$ and $T(\mathbf{e}_2)$. Since this parallelogram is the image of the unit square under T, we have established the following result.

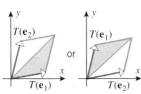

The image of the unit square under an invertible linear operator is a nondegenerate parallelogram.

Figure 6.4.6

Theorem 6.4.8 *If $T : R^2 \to R^2$ is an invertible linear operator, then T maps the unit square into a nondegenerate parallelogram that has a vertex at the origin and has adjacent sides $T(\mathbf{e}_1)$ and $T(\mathbf{e}_2)$. The area of this parallelogram is $|\det(A)|$, where $A = [T(\mathbf{e}_1) \quad T(\mathbf{e}_2)]$ is the standard matrix for T.*

CONCEPT PROBLEM If $T : R^2 \to R^2$ is a linear operator on R^2 that is not invertible, then its standard matrix A is singular and $\det(A) = 0$. What does this tell you about the image of the unit square in this case? Explain.

EXAMPLE 13
Determinants of Rotation and Reflection Operators

If R_θ is the standard matrix for the rotation about the origin of R^2 through the angle θ, and if H_θ is the standard matrix for the reflection about the line making an angle θ with the x-axis of R^2, then we must have $|\det(R_\theta)| = 1$ and $|\det(H_\theta)| = 1$, since the rotation and reflection do not change the area of the unit square. This is consistent with our observation in Section 6.2 that $\det(R_\theta) = 1$ and $\det(H_\theta) = -1$. ∎

Exercise Set 6.4

In Exercises 1 and 2, let T_A and T_B be the linear operators whose standard matrices are given. Find the standard matrices for $T_B \circ T_A$ and $T_A \circ T_B$.

1. $A = \begin{bmatrix} 1 & -2 & 0 \\ 4 & 1 & -3 \\ 5 & 2 & 4 \end{bmatrix}$, $B = \begin{bmatrix} 2 & -3 & 3 \\ 5 & 0 & 1 \\ 6 & 1 & 7 \end{bmatrix}$

2. $A = \begin{bmatrix} 6 & 3 & -1 \\ 2 & 0 & 1 \\ 4 & -3 & 6 \end{bmatrix}$, $B = \begin{bmatrix} 4 & 0 & 4 \\ -1 & 5 & 2 \\ 2 & -3 & 8 \end{bmatrix}$

3. Let $T_1(x_1, x_2) = (x_1 + x_2, x_1 - x_2)$ and
$T_2(x_1, x_2) = (3x_1, 2x_1 + 4x_2)$.
  (a) Find the standard matrices for T_1 and T_2.
  (b) Find the standard matrices for $T_2 \circ T_1$ and $T_1 \circ T_2$.
  (c) Use the matrices obtained in part (b) to find formulas for $T_1(T_2(x_1, x_2))$ and $T_2(T_1(x_1, x_2))$.

4. Let $T_1(x_1, x_2, x_3) = (4x_1, -2x_1 + x_2, -x_1 - 3x_2)$ and
$T_2(x_1, x_2, x_3) = (x_1 + 2x_2, -x_3, 4x_1 - x_3)$.
  (a) Find the standard matrices for T_1 and T_2.
  (b) Find the standard matrices for $T_2 \circ T_1$ and $T_1 \circ T_2$.
  (c) Use the matrices obtained in part (b) to find formulas for $T_1(T_2(x_1, x_2, x_3))$ and $T_2(T_1(x_1, x_2, x_3))$.

In Exercises 5 and 6, use matrix multiplication to find the standard matrix for the stated composition of linear operators on R^2.

5. (a) A counterclockwise rotation of $90°$, followed by a reflection about the line $y = x$.
  (b) An orthogonal projection on the y-axis, followed by a contraction with factor $k = \frac{1}{2}$.
  (c) A reflection about the x-axis, followed by a dilation with factor $k = 3$.

6. (a) A counterclockwise rotation of $60°$, followed by an orthogonal projection on the x-axis, followed by a reflection about the line $y = x$.
  (b) A dilation with factor $k = 2$, followed by a counterclockwise rotation of $45°$, followed by a reflection about the y-axis.
  (c) A counterclockwise rotation of $15°$, followed by a counterclockwise rotation of $105°$, followed by a counterclockwise rotation of $60°$.

In Exercises 7–10, use matrix multiplication to find the standard matrix for the stated composition of linear operators on R^3.

7. (a) A reflection about the yz-plane, followed by an orthogonal projection on the xz-plane.
  (b) A counterclockwise rotation of $45°$ about the y-axis, followed by a dilation with factor $k = \sqrt{2}$.
  (c) An orthogonal projection on the xy-plane, followed by a reflection about the yz-plane.

8. (a) A reflection about the xz-plane, followed by an orthogonal projection on the xy-plane.
  (b) A counterclockwise rotation of $30°$ about the x-axis, followed by a contraction with factor $k = \frac{1}{3}$.
  (c) An orthogonal projection on the xz-plane, followed by a reflection about the xy-plane.

9. (a) A counterclockwise rotation of $30°$ about the x-axis, followed by a counterclockwise rotation of $30°$ about the z-axis, followed by a contraction with factor $k = \frac{1}{4}$.
  (b) A reflection about the xy-plane, followed by a reflection about the xz-plane, followed by an orthogonal projection on the yz-plane.

10. (a) A counterclockwise rotation of $270°$ about the x-axis, followed by a counterclockwise rotation of $90°$ about the y-axis, followed by a rotation of $180°$ about the z-axis.
  (b) A counterclockwise rotation of $30°$ about the z-axis, followed by a reflection in the xy-plane, followed by an orthogonal projection on the yz-plane.

In Exercises 11 and 12, express the matrix as a product of elementary matrices, and then describe the effect of multiplication by A as a succession of compressions, expansions, reflections, and shears.

11. (a) $A = \begin{bmatrix} 2 & 0 \\ 0 & 3 \end{bmatrix}$ (b) $A = \begin{bmatrix} 2 & 1 \\ 1 & 0 \end{bmatrix}$

  (c) $A = \begin{bmatrix} 0 & -2 \\ 4 & 0 \end{bmatrix}$ (d) $A = \begin{bmatrix} 1 & -3 \\ 4 & 6 \end{bmatrix}$

12. (a) $A = \begin{bmatrix} \frac{1}{2} & 0 \\ 0 & \frac{1}{3} \end{bmatrix}$ (b) $A = \begin{bmatrix} 0 & 1 \\ 1 & 2 \end{bmatrix}$

  (c) $A = \begin{bmatrix} 0 & 2 \\ -5 & 0 \end{bmatrix}$ (d) $A = \begin{bmatrix} 1 & 4 \\ 2 & 9 \end{bmatrix}$

In Exercises 13 and 14, describe the inverse of the linear operator.

13. (a) Reflection of R^2 about the x-axis.
  (b) Rotation of R^2 about the origin through an angle of $\pi/4$.
  (c) Dilation of R^2 by a factor of 3.
  (d) Compression of R^2 in the y-direction with factor $\frac{1}{2}$.

14. (a) Reflection of R^2 about the y-axis.
  (b) Rotation of R^2 about the origin through an angle of $-\pi/6$.
  (c) Contraction of R^2 with factor $\frac{1}{5}$.
  (d) Expansion of R^2 in the x-direction with factor 7.

In Exercises 15–18, determine whether the linear operator $T : R^2 \to R^2$ defined by the equations is one-to-one; if so, find the standard matrix for the inverse operator, and find $T^{-1}(w_1, w_2)$.

15. $w_1 = x_1 + 2x_2$
  $w_2 = x_1 + x_2$

16. $w_1 = 4x_1 - 6x_2$
  $w_2 = 2x_1 + 3x_2$

17. $w_1 = - x_2$
  $w_2 = x_1$

18. $w_1 = 3x_1$
  $w_2 = 5x_1$

In Exercises 19–22, determine whether the linear operator $T : R^3 \to R^3$ defined by the equations is one-to-one; if so, find the standard matrix for the inverse operator, and find $T^{-1}(w_1, w_2, w_3)$.

19. $\begin{aligned} w_1 &= x_1 - 2x_2 + 2x_3 \\ w_2 &= 2x_1 + x_2 + x_3 \\ w_3 &= x_1 + x_2 \end{aligned}$

20. $\begin{aligned} w_1 &= x_1 - 3x_2 + 4x_3 \\ w_2 &= -x_1 + x_2 + x_3 \\ w_3 &= - 2x_2 + 5x_3 \end{aligned}$

21. $\begin{aligned} w_1 &= x_1 + 4x_2 - x_3 \\ w_2 &= 2x_1 + 7x_2 + x_3 \\ w_3 &= x_1 + 3x_2 \end{aligned}$

22. $\begin{aligned} w_1 &= x_1 + 2x_2 + x_3 \\ w_2 &= 2x_1 + x_2 + 4x_3 \\ w_3 &= 7x_1 + 4x_2 + 5x_3 \end{aligned}$

In Exercises 23 and 24, determine whether $T_1 \circ T_2 = T_2 \circ T_1$.

23. (a) $T_1 : R^2 \to R^2$ is the orthogonal projection on the x-axis, and $T_2 : R^2 \to R^2$ is the orthogonal projection on the y-axis.

(b) $T_1 : R^2 \to R^2$ is the rotation about the origin through an angle θ_1, and $T_2 : R^2 \to R^2$ is the rotation about the origin through an angle θ_2.

(c) $T_1 : R^3 \to R^3$ is the rotation about the x-axis through an angle θ_1, and $T_2 : R^3 \to R^3$ is the rotation about the z-axis through an angle θ_2.

24. (a) $T_1 : R^2 \to R^2$ is the reflection about the x-axis, and $T_2 : R^2 \to R^2$ is the reflection about the y-axis.

(b) $T_1 : R^2 \to R^2$ is the orthogonal projection on the x-axis, and $T_2 : R^2 \to R^2$ is the counterclockwise rotation through an angle θ.

(c) $T_1 : R^3 \to R^3$ is a dilation by a factor k and $T_2 : R^3 \to R^3$ is the counterclockwise rotation about the z-axis through an angle θ.

25. Let $H_{\pi/3}$ and $H_{\pi/6}$ be the standard matrices for the reflections of R^2 about lines through the origin making angles of $\pi/3$ and $\pi/6$, respectively, with the positive x-axis. Find the standard matrix for a rotation that has the same effect as the reflection $H_{\pi/3}$ followed by the reflection $H_{\pi/6}$.

26. Let $H_{\pi/4}$ and $H_{\pi/8}$ be the standard matrices for the reflections of R^2 about lines through the origin making angles of $\pi/4$ and $\pi/8$, respectively, with the positive x-axis. Find the standard matrix for a rotation that has the same effect as the reflection $H_{\pi/4}$ followed by the reflection $H_{\pi/8}$.

In Exercises 27 and 28, sketch the image of the unit square under multiplication by A, and use a determinant to find its area.

27. $A = \begin{bmatrix} 1 & -1 \\ 1 & 2 \end{bmatrix}$

28. $A = \begin{bmatrix} 2 & 3 \\ 3 & 4 \end{bmatrix}$

Discussion and Discovery

D1. Indicate whether the statement is true (T) or false (F). Justify your answer.

(a) If $T_1 : R^n \to R^m$ and $T_2 : R^m \to R^k$ are linear transformations, and if T_1 is not one-to-one, then neither is $T_2 \circ T_1$.

(b) If $T_1 : R^n \to R^m$ and $T_2 : R^m \to R^k$ are linear transformations, and if T_1 is not onto, then neither is $T_2 \circ T_1$.

(c) If $T_1 : R^n \to R^m$ and $T_2 : R^m \to R^k$ are linear transformations, and if T_2 is not one-to-one, then neither is $T_2 \circ T_1$.

(d) If $T_1 : R^n \to R^m$ and $T_2 : R^m \to R^k$ are linear transformations, and if T_2 is not onto, then neither is $T_2 \circ T_1$.

D2. Let L be the line through the origin that makes an angle β with the positive x-axis in R^2, let R_β be the standard matrix for the rotation about the origin through an angle β, and let H_0 be the standard matrix for the reflection about the x-axis. Describe the geometric effect of multiplication by $R_\beta H_0 R_\beta^{-1}$ in terms of L.

D3. Show that every rotation about the origin of R^2 can be expressed as a composition of two reflections about lines through the origin, and draw a picture to illustrate how one might execute a rotation of $120°$ by two such reflections.

Technology Exercises

T1. Consider successive rotations of R^3 by $30°$ about the z-axis, then by $60°$ about the x-axis, and then by $45°$ about the y-axis. If it is desired to execute the three rotations by a single rotation about an appropriate axis, what axis and angle should be used?

T2. (CAS) Consider successive rotations of R^3 through an angle θ_1 about the x-axis, then through an angle θ_2 about the y-axis, and then through an angle θ_3 about the z-axis. Find a single matrix that executes the three rotations.

Section 6.5   Computer Graphics

The field of computer graphics is concerned with displaying, transforming, and animating representations of two- and three-dimensional objects on a computer screen. A complete study of the subject would delve into such topics as color, lighting, and modeling for three-dimensional effects, but in this section we will focus only on using matrix transformations to move and transform objects composed of line segments.

WIREFRAMES

In this section we will be concerned with screen displays of graphical images that are composed of finitely many points joined by straight line segments. The points are called *vertices*, the line segments are called *wires*, and an object formed by joining vertices with wires is called a *wireframe*. For example, Figure 6.5.1*a* shows a wireframe of a "fallen house." Objects with curved boundaries can be approximated as wireframes by choosing closely spaced points along the curves and connecting those points by straight line segments.

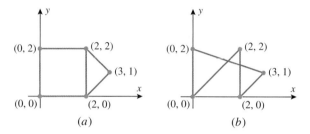

Figure 6.5.1   (*a*)   (*b*)

For a computer to draw a wireframe it must be given coordinates of the vertices in some coordinate system together with information that specifies which pairs of vertices are joined by wires. For example, the wireframe in Figure 6.5.1*b* has the same vertices as the house, but it looks different because the vertices are connected differently.

MATRIX REPRESENTATIONS OF WIREFRAMES

A convenient way to store the positions of the vertices in a wireframe is to form a *vertex matrix* V that has the coordinates of the vertices as column vectors. For example, the vertices of the wireframes in Figure 6.5.1 can be stored as

$$V = \begin{bmatrix} 0 & 2 & 3 & 2 & 0 \\ 0 & 0 & 1 & 2 & 2 \end{bmatrix} \tag{1}$$

The order in which the vertices are listed in a vertex matrix does not matter; however, once the order is chosen, the information on how to connect the vertices of an n-vertex wireframe can be stored in an $n \times n$ *connectivity matrix* C in which the entry in row i and column j is 1 if the vertex in column i of V is connected to the vertex in column j and is zero if it is not. (By agreement, every vertex is considered to be connected to itself, so the diagonal entries are all 1's.) For example, the connectivity information for the house in Figure 6.5.1*a* can be stored as

$$C = \begin{bmatrix} 1 & 1 & 0 & 0 & 1 \\ 1 & 1 & 1 & 1 & 0 \\ 0 & 1 & 1 & 1 & 0 \\ 0 & 1 & 1 & 1 & 1 \\ 1 & 0 & 0 & 1 & 1 \end{bmatrix} \tag{2}$$

CONCEPT PROBLEM Write down the connectivity matrix for the wireframe in Figure 6.5.1*b*.

REMARK Connectivity matrices are always symmetric (why?). Thus, for efficient use of computer storage space one need only store the entries above or below the main diagonal.

EXAMPLE 1
Constructing a
Wireframe

Draw the wireframe whose vertex and connectivity matrices are

$$V = \begin{bmatrix} 1 & 2 & 1 & 0 \\ 0 & 1 & 2 & 1 \end{bmatrix} \quad \text{and} \quad C = \begin{bmatrix} 1 & 1 & 0 & 1 \\ 1 & 1 & 1 & 1 \\ 0 & 1 & 1 & 1 \\ 1 & 1 & 1 & 1 \end{bmatrix}$$

The wireframe is shown in Figure 6.5.2. The connections were obtained from C by observing that all vertices are connected to each other except those in the first and third columns of V. ■

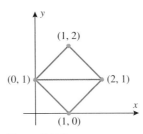

Figure 6.5.2

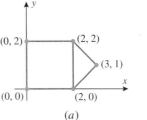
(a)

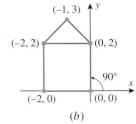
(b)

Figure 6.5.3

TRANSFORMING WIREFRAMES

Next we will consider how to transform a wireframe by applying an invertible linear transformation to its vertex matrix. We will assume that the underlying connectivity matrix of the wireframe is known and the connectivity of the transformed vertices is described by the same matrix.

EXAMPLE 2   The fallen house in Figure 6.5.3a can be brought to the upright position in Figure 6.5.3b by rotating the vectors in its vertex matrix counterclockwise by 90°; the standard matrix for this rotation is

$$R = \begin{bmatrix} \cos(90°) & -\sin(90°) \\ \sin(90°) & \cos(90°) \end{bmatrix} = \begin{bmatrix} 0 & -1 \\ 1 & 0 \end{bmatrix} \tag{3}$$

We can perform all of the rotations in one swoop by multiplying the vertex matrix V in (1) on the left by R. Thus, the vertex matrix V_1 for the rotated house is

$$V_1 = RV = \begin{bmatrix} 0 & -1 \\ 1 & 0 \end{bmatrix} \begin{bmatrix} 0 & 2 & 3 & 2 & 0 \\ 0 & 0 & 1 & 2 & 2 \end{bmatrix} = \begin{bmatrix} 0 & 0 & -1 & -2 & -2 \\ 0 & 2 & 3 & 2 & 0 \end{bmatrix} \tag{4}$$

which is consistent with Figure 6.5.3b. Similarly, one could reflect, project, compress, expand, or shear the house by multiplying its vertex matrix on the left by the appropriate transformation matrix. ■

EXAMPLE 3   Fonts used in computer displays usually have an upright (or **roman**) version and a slanted (or **italic**) version. The italic version is usually created by shearing the roman version; for example, Figure 6.5.4a shows the wireframe for a roman T and also the wireframe for the italic version that results by shearing the roman version in the positive x-direction to an angle that is 15° off the vertical. Find the vertices of the italic T to two decimal places.

Solution We leave it for you to show that a vertex matrix for the roman T is

$$V = \begin{bmatrix} 0.5 & 0.5 & 3 & 3 & -3 & -3 & -0.5 & -0.5 \\ 0 & 6.5 & 6.5 & 7.5 & 7.5 & 6.5 & 6.5 & 0 \end{bmatrix}$$

The column vectors of the shear matrix S are $S\mathbf{e}_1$ and $S\mathbf{e}_2$, where $\mathbf{e}_1$ and $\mathbf{e}_2$ are the standard unit vectors in R^2. Since the shear leaves points on the x-axis fixed,

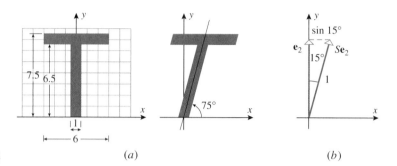

Figure 6.5.4   (*a*)   (*b*)

we have $S\mathbf{e}_1 = \mathbf{e}_1$. The effect of the shear on $\mathbf{e}_2$ is to move the terminal point of $\mathbf{e}_2$ horizontally to the right so its image $S\mathbf{e}_2$ leans at an angle of $15°$ off the vertical (Figure 6.5.4*b*). Thus,

$$S\mathbf{e}_1 = \mathbf{e}_1 = \begin{bmatrix} 1 \\ 0 \end{bmatrix} \quad \text{and} \quad S\mathbf{e}_2 = \begin{bmatrix} \sin 15° \\ 1 \end{bmatrix} \approx \begin{bmatrix} 0.26 \\ 1 \end{bmatrix}$$

and hence a vertex matrix for the sheared T is

$$SV \approx \begin{bmatrix} 1 & 0.26 \\ 0 & 1 \end{bmatrix} \begin{bmatrix} 0.5 & 0.5 & 3 & 3 & -3 & -3 & -0.5 & -0.5 \\ 0 & 6.5 & 6.5 & 7.5 & 7.5 & 6.5 & 6.5 & 0 \end{bmatrix}$$

$$= \begin{bmatrix} 0.5 & 2.19 & 4.69 & 4.95 & -1.05 & -1.31 & 1.19 & -0.5 \\ 0 & 6.5 & 6.5 & 7.5 & 7.5 & 6.5 & 6.5 & 0 \end{bmatrix}$$

We leave it for you to confirm that this result is consistent with the picture of the sheared T in Figure 6.5.4*a* by plotting the vertices.   ■

TRANSLATION USING HOMOGENEOUS COORDINATES

Although translation is an important operation in computer graphics, it presents a problem because it is not a linear operator and hence not a matrix operator (see Example 9 of Section 6.1). Thus, for example, there is no 2×2 matrix that will translate vectors in R^2 by matrix multiplication and, similarly, no 3×3 matrix that will translate vectors in R^3. Fortunately, there is a way to circumvent this problem by using the following theorem about partitioned matrices.

> **Theorem 6.5.1** *If $\mathbf{x}$ and $\mathbf{x}_0$ are vectors in R^n, and if I_n is the $n \times n$ identity matrix, then*
>
> $$\left[\begin{array}{c|c} I_n & \mathbf{x}_0 \\ \hline 0 & 1 \end{array}\right] \begin{bmatrix} \mathbf{x} \\ 1 \end{bmatrix} = \begin{bmatrix} \mathbf{x}_0 + \mathbf{x} \\ 1 \end{bmatrix} \tag{5}$$

Proof Following the usual convention of writing 1×1 matrices as scalars, we obtain

$$\left[\begin{array}{c|c} I_n & \mathbf{x}_0 \\ \hline 0 & 1 \end{array}\right] \begin{bmatrix} \mathbf{x} \\ 1 \end{bmatrix} = \begin{bmatrix} I_n\mathbf{x} + \mathbf{x}_0 \\ 0 + 1 \end{bmatrix} = \begin{bmatrix} \mathbf{x}_0 + \mathbf{x} \\ 1 \end{bmatrix} \qquad ■$$

    This theorem tells us that if we modify $\mathbf{x}$ and $\mathbf{x}_0 + \mathbf{x}$ by adjoining an additional component of 1, then there is an $(n + 1) \times (n + 1)$ matrix that will transform the modified $\mathbf{x}$ into the modified $\mathbf{x}_0 + \mathbf{x}$ by matrix multiplication. Once the modified $\mathbf{x}_0 + \mathbf{x}$ is computed, the final component of 1 can be dropped to produce the translated vector $\mathbf{x}_0 + \mathbf{x}$ in R^n. As a matter of terminology, if $\mathbf{x} = (x_1, x_2, \ldots, x_n)$ is a vector in R^n, then the modified vector $(x_1, x_2, \ldots, x_n, 1)$ in R^{n+1} is said to represent $\mathbf{x}$ in ***homogeneous coordinates***. Thus, for example, the point $\mathbf{x} = (x, y)$ in R^2 is represented in homogeneous coordinates by $(x, y, 1)$, and the point $\mathbf{x} = (x, y, z)$ in R^3 is represented in homogeneous coordinates by $(x, y, z, 1)$.

EXAMPLE 4
Translation by Matrix Multiplication

Translating $\mathbf{x} = (x, y)$ by $\mathbf{x}_0 = (h, k)$ produces the vector $\mathbf{x} + \mathbf{x}_0 = (x + h, y + k)$. Using Theorem 6.5.1, this computation can be performed in homogeneous coordinates as

$$\begin{bmatrix} 1 & 0 & | & h \\ 0 & 1 & | & k \\ \hline 0 & 0 & | & 1 \end{bmatrix} \begin{bmatrix} x \\ y \\ \hline 1 \end{bmatrix} = \begin{bmatrix} x + h \\ y + k \\ \hline 1 \end{bmatrix}$$

The translated vector can now be recovered by dropping the final 1. ∎

EXAMPLE 5
Translating a Wireframe by Matrix Multiplication

Use matrix multiplication to translate the upright house in Figure 6.5.5a to the position shown in Figure 6.5.5b.

Solution A vertex matrix for the upright house is

$$V_1 = \begin{bmatrix} 0 & 0 & -1 & -2 & -2 \\ 0 & 2 & 3 & 2 & 0 \end{bmatrix}$$

To translate the house to the desired position we must translate each column vector in V_1 by

$$\mathbf{x}_0 = \begin{bmatrix} 3 \\ 2 \end{bmatrix}$$

From Theorem 6.5.1, these translations can be obtained in homogeneous coordinates via matrix multiplication by

$$T = \begin{bmatrix} I_2 & | & \mathbf{x}_0 \\ \hline 0 & | & 1 \end{bmatrix} = \begin{bmatrix} 1 & 0 & | & 3 \\ 0 & 1 & | & 2 \\ \hline 0 & 0 & | & 1 \end{bmatrix} \tag{6}$$

If we first convert the column vectors of V_1 to homogeneous coordinates, then we can perform all of the translations by the single multiplication

$$\begin{bmatrix} 1 & 0 & 3 \\ 0 & 1 & 2 \\ 0 & 0 & 1 \end{bmatrix} \begin{bmatrix} 0 & 0 & -1 & -2 & -2 \\ 0 & 2 & 3 & 2 & 0 \\ 1 & 1 & 1 & 1 & 1 \end{bmatrix} = \begin{bmatrix} 3 & 3 & 2 & 1 & 1 \\ 2 & 4 & 5 & 4 & 2 \\ 1 & 1 & 1 & 1 & 1 \end{bmatrix}$$

If we now drop the 1's in the third row of the product, then we obtain the vertex matrix V_2 for the translated house; namely

$$V_2 = \begin{bmatrix} 3 & 3 & 2 & 1 & 1 \\ 2 & 4 & 5 & 4 & 2 \end{bmatrix}$$

You should check that this is consistent with Figure 6.5.5. ∎

We now know how to perform all of the basic transformations in computer graphics by matrix multiplication. However, translation still sticks out like a sore thumb because it requires a matrix of a different size; for example, a rotation in R^2 is executed by a 2×2 matrix, whereas a translation requires a 3×3 matrix. This is a problem because it makes it impossible to compose translations with other transformations by multiplying matrices. One way to eliminate this size discrepancy is to perform *all* of the basic transformations in homogeneous coordinates. The following theorem, which we leave for you to prove, will enable us to do this.

Theorem 6.5.2 *If A is an $n \times n$ matrix, and $\mathbf{x}$ is a vector in R^n that is expressed in column form, then*

$$\begin{bmatrix} A & | & 0 \\ \hline 0 & | & 1 \end{bmatrix} \begin{bmatrix} \mathbf{x} \\ \hline 1 \end{bmatrix} = \begin{bmatrix} A\mathbf{x} \\ \hline 1 \end{bmatrix}$$

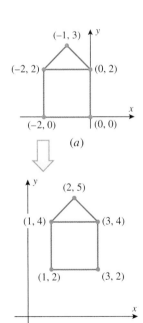

(a)

(b)

Figure 6.5.5

EXAMPLE 6
A Rotation in Homogeneous Coordinates

If the vector $\mathbf{x} = (x, y)$ is rotated about the origin through an angle θ, then the resulting vector in column form is

$$\begin{bmatrix} \cos\theta & -\sin\theta \\ \sin\theta & \cos\theta \end{bmatrix} \begin{bmatrix} x \\ y \end{bmatrix} = \begin{bmatrix} x\cos\theta - y\sin\theta \\ x\sin\theta + y\cos\theta \end{bmatrix} \tag{7}$$

Using Theorem 6.5.2, this computation can be performed in homogeneous coordinates as

$$\begin{bmatrix} \cos\theta & -\sin\theta & 0 \\ \sin\theta & \cos\theta & 0 \\ \hline 0 & 0 & 1 \end{bmatrix} \begin{bmatrix} x \\ y \\ 1 \end{bmatrix} = \begin{bmatrix} x\cos\theta - y\sin\theta \\ x\sin\theta + y\cos\theta \\ \hline 1 \end{bmatrix}$$

This is consistent with (7) after the final 1 is dropped. ∎

EXAMPLE 7
Composition in Homogeneous Coordinates

Figure 6.5.6, which is a combination of Figures 6.5.3 and 6.5.5, shows a wireframe for a fallen house that is first rotated 90° to an upright position and then translated to a new position. The rotation was performed in Example 2 using the matrix R in (3), and the translation was performed in Example 5 using the 3×3 matrix T in (6). To compose these transformations and execute the composition as a single matrix multiplication in homogeneous coordinates, we must first express the rotation matrix R in Example 2 as

$$R' = \begin{bmatrix} R & 0 \\ \hline 0 & 1 \end{bmatrix} = \begin{bmatrix} 0 & -1 & 0 \\ 1 & 0 & 0 \\ \hline 0 & 0 & 1 \end{bmatrix}$$

Since the matrix for the translation in homogeneous coordinates is

$$T = \begin{bmatrix} I_2 & \mathbf{x}_0 \\ \hline 0 & 1 \end{bmatrix} = \begin{bmatrix} 1 & 0 & 3 \\ 0 & 1 & 2 \\ \hline 0 & 0 & 1 \end{bmatrix}$$

the composition of the translation with the rotation can be performed in homogeneous coordinates by multiplying the vertex matrix in homogeneous coordinates by

$$TR' = \begin{bmatrix} 1 & 0 & 3 \\ 0 & 1 & 2 \\ 0 & 0 & 1 \end{bmatrix} \begin{bmatrix} 0 & -1 & 0 \\ 1 & 0 & 0 \\ 0 & 0 & 1 \end{bmatrix} = \begin{bmatrix} 0 & -1 & 3 \\ 1 & 0 & 2 \\ 0 & 0 & 1 \end{bmatrix}$$

This yields

$$TR'V = \begin{bmatrix} 0 & -1 & 3 \\ 1 & 0 & 2 \\ 0 & 0 & 1 \end{bmatrix} \begin{bmatrix} 0 & 2 & 3 & 2 & 0 \\ 0 & 0 & 1 & 2 & 2 \\ 1 & 1 & 1 & 1 & 1 \end{bmatrix} = \begin{bmatrix} 3 & 3 & 2 & 1 & 1 \\ 2 & 4 & 5 & 4 & 2 \\ 1 & 1 & 1 & 1 & 1 \end{bmatrix}$$

which is consistent with Figure 6.5.6 after the final 1's are dropped (verify). ∎

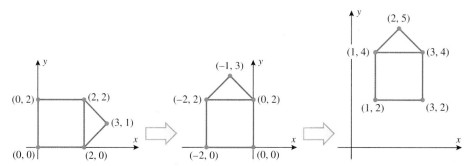

Figure 6.5.6

THREE-DIMENSIONAL GRAPHICS

A three-dimensional wireframe can be represented on a flat computer screen by projecting the vertices and wires onto the screen to obtain a two-dimensional representation of the object. More precisely, suppose that a three-dimensional wireframe is embedded in an xyz-coordinate system whose xy-plane coincides with the computer screen and whose z-axis is perpendicular to the screen. If, as in Figure 6.5.7, we imagine a viewer's eye to be positioned at a point $Q(0, 0, d)$ on the z-axis, then a vertex $P(x, y, z)$ of the wireframe can be represented on the computer screen by the point $(x^*, y^*, 0)$ at which the ray from Q through P intersects the screen. These are called the **screen coordinates** of P, and this procedure for obtaining screen coordinates is called the **perspective projection with viewpoint Q**.

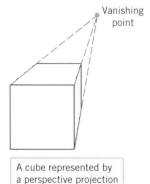

A cube represented by a perspective projection

(a)

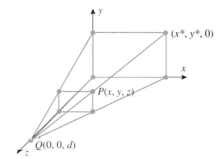

Figure 6.5.7

Perspective projections, combined with lighting and techniques for removing background lines, are useful for creating realistic images of three-dimensional solids. The illusion of realism occurs because perspective projections create a **vanishing point** in the image plane at which lines that are parallel to the z-axis in the actual object meet in the projected image (Figure 6.5.8a). However, there are many applications of computer graphics (engineering drawings, for example) in which perspective is not desirable. In such cases the screen image of the wireframe is typically created using the orthogonal projection on the xy-plane (Figure 6.5.8b).

REMARK Observe that if we allow d to increase indefinitely in Figure 6.5.7, then the screen point $(x^*, y^*, 0)$ produced by the perspective projection will approach the screen point $(x, y, 0)$ produced by the orthogonal projection. Thus, the orthogonal projection can be viewed as a perspective projection in which the viewer's eye is "infinitely far" from the screen.

A cube represented by an orthogonal projection

(b)

Figure 6.5.8

EXAMPLE 8
Transformations of a Three-Dimensional Wireframe

Figure 6.5.9a shows the orthogonal projection on the xy-plane of a three-dimensional wireframe with vertex matrix

$$V = \begin{bmatrix} 3 & 5 & 6 & 7 & 9 & 9 & 11 & 12 & 13 & 15 \\ 5 & 17 & 0 & 11 & 20 & 8 & 11 & 0 & 17 & 5 \\ -9 & -1 & 0 & 3 & -6 & -16 & 3 & 0 & -1 & -9 \end{bmatrix}$$

(a) Working in homogeneous coordinates, use an appropriate 4×4 matrix to rotate the wireframe counterclockwise $60°$ about the positive x-axis in R^3; then sketch the orthogonal projection of the rotated wireframe on the xy-plane.

(b) Working in homogeneous coordinates, use an appropriate 4×4 matrix to translate the rotated wireframe obtained in part (a) by

$$\mathbf{x}_0 = \begin{bmatrix} 1 \\ 2 \\ 5 \end{bmatrix} \tag{8}$$

and then sketch the orthogonal projection of the translated wireframe on the xy-plane.

(c) Find a single 4×4 matrix that will perform the rotation followed by the translation.

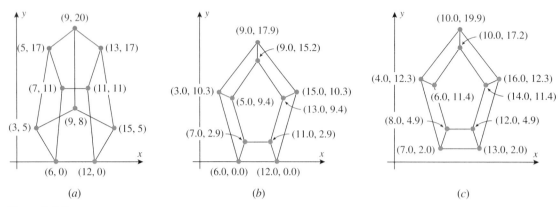

Figure 6.5.9

Solution (a) To find the 4×4 matrix that produces the desired rotation in homogeneous coordinates, we first use Table 6.2.6 to find the standard 3×3 matrix for the rotation and then apply Theorem 6.5.2. This yields the matrix

$$
R = \begin{bmatrix}
1 & 0 & 0 & 0 \\
0 & \frac{1}{2} & -\frac{1}{2}\sqrt{3} & 0 \\
0 & \frac{1}{2}\sqrt{3} & \frac{1}{2} & 0 \\
0 & 0 & 0 & 1
\end{bmatrix} \tag{9}
$$

(verify). To perform the rotation we write the column vectors of V in homogeneous coordinates and multiply by R. This yields

$$
V_1 = \begin{bmatrix}
1 & 0 & 0 & 0 \\
0 & \frac{1}{2} & -\frac{1}{2}\sqrt{3} & 0 \\
0 & \frac{1}{2}\sqrt{3} & \frac{1}{2} & 0 \\
0 & 0 & 0 & 1
\end{bmatrix}
\begin{bmatrix}
3 & 5 & 6 & 7 & 9 & 9 & 11 & 12 & 13 & 15 \\
5 & 17 & 0 & 11 & 20 & 8 & 11 & 0 & 17 & 5 \\
-9 & -1 & 0 & 3 & -6 & -16 & 3 & 0 & -1 & -9 \\
1 & 1 & 1 & 1 & 1 & 1 & 1 & 1 & 1 & 1
\end{bmatrix}
$$

$$
\approx \begin{bmatrix}
3.000 & 5.000 & 6.000 & 7.000 & 9.000 & 9.000 & 11.000 & 12.000 & 13.000 & 15.000 \\
10.294 & 9.366 & 0.000 & 2.902 & 15.196 & 17.856 & 2.902 & 0.000 & 9.366 & 10.294 \\
-0.170 & 14.222 & 0.000 & 11.026 & 14.321 & -1.072 & 11.026 & 0.000 & 14.222 & -0.170 \\
1 & 1 & 1 & 1 & 1 & 1 & 1 & 1 & 1 & 1
\end{bmatrix}
$$

The vertex matrix for the rotated wireframe in R^3 can now be obtained by dropping the 1's, and the vertex matrix for the orthogonal projection of the rotated wireframe on the xy-plane can be obtained by setting the z-coordinates equal to zero. The resulting wireframe is shown in Figure 6.5.9b with the coordinate labels rounded to one decimal place and the zero z-coordinates suppressed.

Solution (b) From Theorem 6.5.1 and (8), the matrix T that performs the translation in homogeneous coordinates is

$$
T = \begin{bmatrix}
1 & 0 & 0 & 1 \\
0 & 1 & 0 & 2 \\
0 & 0 & 1 & 5 \\
0 & 0 & 0 & 1
\end{bmatrix} \tag{10}
$$

Thus, the vertex matrix for the translated wireframe in homogeneous coordinates is

$$V_2 = TV_1 \approx \begin{bmatrix} 1 & 0 & 0 & 1 \\ 0 & 1 & 0 & 2 \\ 0 & 0 & 1 & 5 \\ 0 & 0 & 0 & 1 \end{bmatrix} \begin{bmatrix} 3.000 & 5.000 & 6.000 & 7.000 & 9.000 & 9.000 & 11.000 & 12.000 & 13.000 & 15.000 \\ 10.294 & 9.366 & 0.000 & 2.902 & 15.196 & 17.856 & 2.902 & 0.000 & 9.366 & 10.294 \\ -0.170 & 14.222 & 0.000 & 11.026 & 14.321 & -1.072 & 11.026 & 0.000 & 14.222 & -0.170 \\ 1 & 1 & 1 & 1 & 1 & 1 & 1 & 1 & 1 & 1 \end{bmatrix}$$

$$\approx \begin{bmatrix} 4.000 & 6.000 & 7.000 & 8.000 & 10.000 & 10.000 & 12.000 & 13.000 & 14.000 & 16.000 \\ 12.294 & 11.366 & 2.000 & 4.902 & 17.196 & 19.856 & 4.902 & 2.000 & 11.366 & 12.294 \\ 4.830 & 19.222 & 5.000 & 16.026 & 19.321 & 3.928 & 16.026 & 5.000 & 19.222 & 4.830 \\ 1 & 1 & 1 & 1 & 1 & 1 & 1 & 1 & 1 & 1 \end{bmatrix}$$

The vertex matrix for the translated wireframe in R^3 can now be obtained by dropping the 1's, and the vertex matrix for the orthogonal projection of the rotated wireframe on the xy-plane can be obtained by setting the z-coordinates equal to zero. The resulting wireframe is shown in Figure 6.5.9c with the coordinate labels rounded to one decimal place and the zero z-coordinates suppressed.

Solution (c) It follows from (9) and (10) that the rotation followed by the translation can be implemented by the matrix

$$TR = \begin{bmatrix} 1 & 0 & 0 & 1 \\ 0 & 1 & 0 & 2 \\ 0 & 0 & 1 & 5 \\ 0 & 0 & 0 & 1 \end{bmatrix} \begin{bmatrix} 1 & 0 & 0 & 0 \\ 0 & \frac{1}{2} & -\frac{1}{2}\sqrt{3} & 0 \\ 0 & \frac{1}{2}\sqrt{3} & \frac{1}{2} & 0 \\ 0 & 0 & 0 & 1 \end{bmatrix} = \begin{bmatrix} 1 & 0 & 0 & 1 \\ 0 & \frac{1}{2} & -\frac{1}{2}\sqrt{3} & 2 \\ 0 & \frac{1}{2}\sqrt{3} & \frac{1}{2} & 5 \\ 0 & 0 & 0 & 1 \end{bmatrix} \qquad \blacksquare$$

Exercise Set 6.5

In Exercises 1–4, draw the wireframe whose vertex and connectivity matrices are given.

1. $V = \begin{bmatrix} 1 & 1 & 2 & 2 \\ 0 & 1 & 1 & 0 \end{bmatrix}$; $C = \begin{bmatrix} 1 & 1 & 1 & 1 \\ 1 & 1 & 1 & 0 \\ 1 & 1 & 1 & 1 \\ 1 & 0 & 1 & 1 \end{bmatrix}$

2. $V = \begin{bmatrix} 0 & 1 & 2 & 3 & 2 & 1 \\ 1 & 2 & 2 & 1 & 0 & 0 \end{bmatrix}$; $C = \begin{bmatrix} 1 & 1 & 0 & 0 & 0 & 1 \\ 1 & 1 & 1 & 0 & 0 & 0 \\ 0 & 1 & 1 & 1 & 0 & 0 \\ 0 & 0 & 1 & 1 & 1 & 0 \\ 0 & 0 & 0 & 1 & 1 & 1 \\ 1 & 0 & 0 & 0 & 1 & 1 \end{bmatrix}$

3. $V = \begin{bmatrix} 0 & 1 & 2 & 3 \\ 0 & 1 & 0 & 1 \end{bmatrix}$; $C = \begin{bmatrix} 1 & 1 & 0 & 1 \\ 1 & 1 & 1 & 1 \\ 0 & 1 & 1 & 1 \\ 1 & 1 & 1 & 1 \end{bmatrix}$

4. $V = \begin{bmatrix} 0 & 1 & 2 & 3 & 4 \\ 1 & 0 & 1 & 0 & 1 \end{bmatrix}$; $C = \begin{bmatrix} 1 & 1 & 0 & 0 & 0 \\ 1 & 1 & 1 & 0 & 0 \\ 0 & 1 & 1 & 1 & 0 \\ 0 & 0 & 1 & 1 & 1 \\ 0 & 0 & 0 & 1 & 1 \end{bmatrix}$

In Exercises 5–8, find vertex and connectivity matrices for the wireframe.

5.

6.

7.

8.

In Exercises 9–12, use matrix multiplication to find the image of the wireframe under the transformation A, and sketch the image.

9. The wireframe in Exercise 7; $A = \begin{bmatrix} 2 & 0 \\ 0 & 3 \end{bmatrix}$.

10. The wireframe in Exercise 8; $A = \begin{bmatrix} \frac{1}{2} & 0 \\ 0 & 2 \end{bmatrix}$.

11. The wireframe in Exercise 7; $A = \begin{bmatrix} 1 & 2 \\ 0 & 1 \end{bmatrix}$.

12. The wireframe in Exercise 8; $A = \begin{bmatrix} 1 & 0 \\ 2 & 1 \end{bmatrix}$.

> In Exercises 13–16, use matrix multiplication in homogeneous coordinates to translate the wireframe by $\mathbf{x}_0$, and draw the image.

13. The wireframe in Exercise 7; $\mathbf{x}_0 = \begin{bmatrix} 1 \\ 1 \end{bmatrix}$.

14. The wireframe in Exercise 8; $\mathbf{x}_0 = \begin{bmatrix} -1 \\ 0 \end{bmatrix}$.

15. The wireframe in Exercise 7; $\mathbf{x}_0 = \begin{bmatrix} 2 \\ -1 \end{bmatrix}$.

16. The wireframe in Exercise 8; $\mathbf{x}_0 = \begin{bmatrix} 0 \\ 2 \end{bmatrix}$.

> In Exercises 17 and 18, find the 3×3 matrix in homogeneous coordinates that performs the given operation on R^2.

17. (a) Rotation of $30°$ about the origin.
   (b) Reflection about the x-axis.

18. (a) Compression with factor $\frac{1}{2}$ in the x-direction.
   (b) Dilation with factor 6.

> In Exercises 19 and 20, find the 4×4 matrix in homogeneous coordinates that performs the given operation on R^3.

19. (a) Rotation of $60°$ about the z-axis.
   (b) Reflection about the yz-plane.

20. (a) Rotation of $45°$ about the x-axis.
   (b) Reflection about the xy-plane.

21. (a) Find a 3×3 matrix in homogeneous coordinates that translates $\mathbf{x} = (x, y)$ by $\mathbf{x}_0 = (1, 2)$.
   (b) Use the matrix obtained in part (a) to find the image of the point $\mathbf{x} = (3, 4)$ under the translation.

22. (a) Find a 3×3 matrix in homogeneous coordinates that translates $\mathbf{x} = (x, y)$ by $\mathbf{x}_0 = (-2, 4)$.
   (b) Use the matrix obtained in part (a) to find the image of the point $\mathbf{x} = (1, 3)$ under the translation.

23. (a) Find a 4×4 matrix in homogeneous coordinates that translates $\mathbf{x} = (x, y, z)$ by $\mathbf{x}_0 = (5, 3, -1)$.
   (b) Use the matrix obtained in part (a) to find the image of the point $\mathbf{x} = (5, -4, 1)$ under the translation.

24. (a) Find a 4×4 matrix in homogeneous coordinates that translates $\mathbf{x} = (x, y, z)$ by $\mathbf{x}_0 = (4, 2, 0)$.
   (b) Use the matrix obtained in part (a) to find the image of the point $\mathbf{x} = (6, -3, 2)$ under the translation.

> In Exercises 25–28, find a matrix that performs the stated composition in homogeneous coordinates.

25. The rotation of R^2 about the origin through an angle of $60°$, followed by the translation by $(3, -1)$.

26. The translation of R^2 by $(1, 1)$, followed by the scaling transformation $(x, y) \rightarrow (2x, 7y)$.

27. The transformation $(x, y, z) \rightarrow (-x, 2y + z, x + y)$, followed by the translation by $(2, -3, 5)$.

28. The rotation of R^3 about the positive y-axis through an angle of $30°$, followed by the translation by $(1, -2, 3)$.

Discussion and Discovery

In Section 6.1 we considered rotations of R^2 about the origin and reflections of R^2 about lines through the origin [see Formulas (16) and (18) of that section]. However, there are many applications in which one is interested in rotations about points other than the origin or reflections about lines that do not pass through the origin. Such transformations are not linear, but they can be performed using matrix multiplication in homogeneous coordinates. This idea is explored in Exercises D1–D3.

D1. One way to rotate R^2 about the point $\mathbf{x}_0 = (x_0, y_0)$ through the angle θ is first to translate R^2 by $-\mathbf{x}_0$ to bring the point $\mathbf{x}_0$ to the origin, then rotate R^2 about the origin through the angle θ, and then translate R^2 by $\mathbf{x}_0$ to bring the origin back to the point $\mathbf{x}_0 = (x_0, y_0)$. Thus, a matrix R in homogeneous coordinates that performs the rotation of R^2 through the angle θ about the point $\mathbf{x}_0 = (x_0, y_0)$ is

$$R = \begin{bmatrix} 1 & 0 & x_0 \\ 0 & 1 & y_0 \\ 0 & 0 & 1 \end{bmatrix} \begin{bmatrix} \cos\theta & -\sin\theta & 0 \\ \sin\theta & \cos\theta & 0 \\ 0 & 0 & 1 \end{bmatrix} \begin{bmatrix} 1 & 0 & -x_0 \\ 0 & 1 & -y_0 \\ 0 & 0 & 1 \end{bmatrix}$$

Find the image of a general point (x, y) under a rotation of R^2 by $30°$ about the point $(2, -1)$.

D2. (a) Use the idea discussed in Exercise D1 to find a matrix H in homogeneous coordinates that reflects R^2 about the line through the point $\mathbf{x}_0 = (x_0, y_0)$ that makes an angle θ with the positive x-axis. (Express H as a product, as in Exercise D1.)
   (b) Use the matrix obtained in part (a) to find the image of a general point (x, y) under a reflection about the line through the point $(2, -1)$ that makes an angle of $30°$ with the positive x-axis.

D3. Find the image of the point $(3, 4)$ under a reflection about the line with slope $m = \frac{1}{2}$ that passes through the point

(1, 2). [*Suggestion:* See Exercise 31(a) of Section 6.1 and Exercise D2 above.]

D4. Recall that if λ_1 and λ_2 are positive scalars, then the scaling operator $(x, y) \rightarrow (\lambda_1 x, \lambda_2 y)$ on R^2 stretches or compresses x-coordinates by the factor λ_1 and stretches or compresses y-coordinates by the factor λ_2. Since such scaling operators leave the origin fixed, the origin is sometimes called the **center of scaling** for linear scaling operators. However, there also exist nonlinear scaling operators on R^2 that stretch or compress in the x- and y-directions but leave a point $\mathbf{x}_0 = (x_0, y_0)$ different from the origin fixed. Operators of this type can be created by first translating R^2 by $-\mathbf{x}_0$ to bring the point $\mathbf{x}_0$ to the origin, then applying the linear scaling operator $(x, y) \rightarrow (\lambda_1 x, \lambda_2 y)$, and then translating by $\mathbf{x}_0$ to bring the origin back to the point $\mathbf{x}_0$. Find a 3×3 matrix S in homogeneous coordinates that performs the three operations, and use it to show that the transformation can be expressed as

$$(x, y) \rightarrow (x_0 + \lambda_1(x - x_0), y_0 + \lambda_2(y - y_0))$$

[Note that (x_0, y_0) remains fixed under this transformation and that the transformation is nonlinear if $(x_0, y_0) \neq (0, 0)$, unless $\lambda_1 = \lambda_2 = 1$.]

D5. (a) Find a 3×3 matrix in homogeneous coordinates that first translates $\mathbf{x} = (x, y)$ by $\mathbf{x}_0 = (x_0, y_0)$ and then rotates the resulting vector about the origin through the angle θ.

   (b) Find a 3×3 matrix in homogeneous coordinates that performs the operations in part (a) in the opposite order.

   (c) What do the results in parts (a) and (b) tell you about the commutativity of rotation and translation? Draw a picture that illustrates your conclusion.

Technology Exercises

T1. (a) The accompanying figure shows the wireframe of a cube in a rectangular xyz-coordinate system. Find a vertex matrix for the wireframe, and sketch the orthogonal projection of the wireframe on the xy-plane.

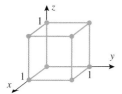

Figure Ex-T1

   (b) Find a 4×4 matrix in homogeneous coordinates that rotates the wireframe $30°$ about the positive z-axis and then translates the wireframe by one unit in the positive x-direction.

   (c) Compute the vertex matrix for the rotated and translated wireframe.

   (d) Use the matrix obtained in part (c) to sketch the orthogonal projection on the xy-plane of the rotated and translated wireframe. Does the result agree with your intuitive geometric sense of what the projection should look like?

T2. Repeat parts (b), (c), and (d) of Exercise T1 assuming that the wireframe is rotated $30°$ about the positive y-axis and then translated so the vertex at the origin moves to the point $(1, 1, 1)$.

T3. The accompanying figure shows a letter L in an xy-coordinate system and an italicized version of that letter created by shearing and translating. Use the method of Example 3 to find the vertices of the shifted italic L to two decimal places.

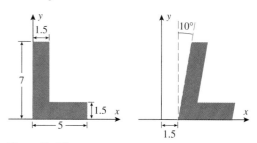

Figure Ex-T3

CHAPTER 7

Notions of dimension and structure in *n*-space make it possible to visualize and interpret data using familiar geometric ideas. Virtually all applications of linear algebra use these ideas in some way.

Dimension and Structure

Section 7.1   Basis and Dimension

In Section 3.5 we discussed the concept of dimension informally. The goal of this section is to make that idea mathematically precise.

BASES FOR SUBSPACES

If $V = \text{span}\{\mathbf{v}_1, \mathbf{v}_2, \ldots, \mathbf{v}_s\}$ is a subspace of R^n, and if the vectors in the set $S = \{\mathbf{v}_1, \mathbf{v}_2, \ldots, \mathbf{v}_s\}$ are linearly dependent, then at least one of the vectors in S can be deleted, and the remaining vectors will still span V. For example, suppose that $V = \text{span}\{\mathbf{v}_1, \mathbf{v}_2, \mathbf{v}_3\}$, where $S = \{\mathbf{v}_1, \mathbf{v}_2, \mathbf{v}_3\}$ is a linearly dependent set. The linear dependence of S implies that at least one vector in that set is a linear combination of the others, say

$$\mathbf{v}_3 = k_1\mathbf{v}_1 + k_2\mathbf{v}_2 \tag{1}$$

Thus, every vector $\mathbf{w}$ in V can be expressed as a linear combination of $\mathbf{v}_1$ and $\mathbf{v}_2$ alone by first writing it as a linear combination of $\mathbf{v}_1$, $\mathbf{v}_2$, and $\mathbf{v}_3$, say

$$\mathbf{w} = c_1\mathbf{v}_1 + c_2\mathbf{v}_2 + c_3\mathbf{v}_3$$

and then substituting (1) to obtain

$$\mathbf{w} = c_1\mathbf{v}_1 + c_2\mathbf{v}_2 + c_3(k_1\mathbf{v}_1 + k_2\mathbf{v}_2) = (c_1 + c_3k_1)\mathbf{v}_1 + (c_2 + c_3k_2)\mathbf{v}_2$$

This discussion suggests that spanning sets of linearly independent vectors are special in that they do not contain superfluous vectors. We make the following definition.

> **Definition 7.1.1** A set of vectors in a subspace V of R^n is said to be a ***basis*** for V if it is linearly independent and spans V.

Here are some examples.

EXAMPLE 1
Some Simple Bases

- If V is a line through the origin of R^n, then any nonzero vector on the line forms a basis for V.
- If V is a plane through the origin of R^n, then any two nonzero vectors in the plane that are not scalar multiples of one another form a basis for V.
- If $V = \{\mathbf{0}\}$ is the zero subspace of R^n, then V has no basis since it does not contain any linearly independent vectors.[*] ∎

---

[*] Some writers define the empty set to be a basis for $\{\mathbf{0}\}$, but we prefer to think of $\{\mathbf{0}\}$ as a subspace with no basis.

EXAMPLE 2
The Standard
Basis for R^n

The standard unit vectors $e_1, e_2, \ldots, e_n$ are linearly independent, for if we write

$$c_1 e_1 + c_2 e_2 + \cdots + c_n e_n = 0 \tag{2}$$

in component form, then we obtain $(c_1, c_2, \ldots, c_n) = (0, 0, \ldots, 0)$, which implies that all of the coefficients in (2) are 0. Furthermore, these vectors span R^n because an arbitrary vector $\mathbf{x} = (x_1, x_2, \ldots, x_n)$ in R^n can be expressed as

$$\mathbf{x} = x_1(1, 0, \ldots, 0) + x_2(0, 1, \ldots, 0) + \cdots + x_n(0, 0, \ldots, 1) = x_1 e_1 + x_2 e_2 + \cdots + x_n e_n$$

We call $\{e_1, e_2, \ldots, e_n\}$ the **standard basis** for R^n. ∎

We know that a set of two or more vectors $S = \{\mathbf{v}_1, \mathbf{v}_2, \ldots, \mathbf{v}_k\}$ in R^n is linearly dependent if and only if some vector in the set is a linear combination of other vectors in the set. The next theorem, which will be useful in our study of bases, takes this a step further by showing that regardless of the order in which the vectors in S are listed, at least one vector in the list will be a linear combination of those that come before it.

Theorem 7.1.2 *If $S = \{\mathbf{v}_1, \mathbf{v}_2, \ldots, \mathbf{v}_k\}$ is a set of two or more nonzero vectors in R^n, then S is linearly dependent if and only if some vector in S is a linear combination of its predecessors.*

Proof If some vector in S is a linear combination of predecessors in S, then the linear dependence of S follows from Theorem 3.4.6.

Conversely, assume that S is a linearly dependent set. This implies that there exist scalars, not all zero, such that

$$t_1 \mathbf{v}_1 + t_2 \mathbf{v}_2 + \cdots + t_k \mathbf{v}_k = 0 \tag{3}$$

so let t_j be that nonzero scalar that has the largest index. Since this implies that all terms in (3) beyond the jth (if any) are zero, we can rewrite this equation as

$$t_1 \mathbf{v}_1 + t_2 \mathbf{v}_2 + \cdots + t_j \mathbf{v}_j = 0$$

Since $t_j \neq 0$, we can multiply this equation through by $1/t_j$ and solve for $\mathbf{v}_j$ as a linear combination of its predecessors $\mathbf{v}_1, \mathbf{v}_2, \ldots, \mathbf{v}_{j-1}$, which completes the proof. ∎

EXAMPLE 3
Linear
Independence
Using Theorem
7.1.2

Show that the vectors

$$\mathbf{v}_1 = (0, 2, 0), \quad \mathbf{v}_2 = (3, 0, 3), \quad \mathbf{v}_3 = (-4, 0, 4)$$

are linearly independent by showing that no vector in the set $\{\mathbf{v}_1, \mathbf{v}_2, \mathbf{v}_3\}$ is a linear combination of predecessors.

Solution The vector $\mathbf{v}_2$ is not a scalar multiple of $\mathbf{v}_1$ and hence is not a linear combination of $\mathbf{v}_1$. The vector $\mathbf{v}_3$ is not a linear combination of $\mathbf{v}_1$ and $\mathbf{v}_2$, for if there were a relationship of the form $\mathbf{v}_3 = t_1 \mathbf{v}_1 + t_2 \mathbf{v}_2$, then we would have to have $t_1 = 0$ in order to produce the zero second coefficient of $\mathbf{v}_3$. This would then imply that $\mathbf{v}_3$ is a scalar multiple of $\mathbf{v}_2$, which it is not. Thus, no vector in the set $\{\mathbf{v}_1, \mathbf{v}_2, \mathbf{v}_3\}$ is a linear combination of predecessors. ∎

EXAMPLE 4
Independence
of Nonzero
Row Vectors in
a Row Echelon
Form

The nonzero row vectors of a matrix in row echelon form are linearly independent. To visualize why this is true, consider the following typical matrices in row echelon form, where the $*$'s denote arbitrary real numbers:

$$\begin{bmatrix} 1 & * & * & * \\ 0 & 1 & * & * \\ 0 & 0 & 1 & * \\ 0 & 0 & 0 & 1 \end{bmatrix}, \quad \begin{bmatrix} 1 & * & * & * \\ 0 & 1 & * & * \\ 0 & 0 & 1 & * \\ 0 & 0 & 0 & 0 \end{bmatrix}, \quad \begin{bmatrix} 1 & * & * & * \\ 0 & 1 & * & * \\ 0 & 0 & 0 & 0 \\ 0 & 0 & 0 & 0 \end{bmatrix}, \quad \begin{bmatrix} 0 & 1 & * & * & * & * & * & * & * \\ 0 & 0 & 0 & 1 & * & * & * & * & * \\ 0 & 0 & 0 & 0 & 1 & * & * & * & * \\ 0 & 0 & 0 & 0 & 0 & 1 & * & * & * \\ 0 & 0 & 0 & 0 & 0 & 0 & 0 & 1 & * \end{bmatrix}$$

If we list the nonzero row vectors of such matrices in the order $\mathbf{v}_1, \mathbf{v}_2, \ldots$, *starting at the bottom and working up*, then no row vector in the list can be expressed as a linear combination of predecessors in the list because there is no way to produce its leading 1 by such a linear combination. The linear independence of the nonzero row vectors now follows from Theorem 7.1.2. ∎

We now have all of the mathematical tools to prove one of the main theorems in this section.

Theorem 7.1.3 (***Existence of a Basis***) *If V is a nonzero subspace of R^n, then there exists a basis for V that has at most n vectors.*

Proof Let V be a nonzero subspace of R^n. We will give a procedure for constructing a set in V with at most n vectors that spans V and in which no vector is a linear combination of predecessors (and hence is linearly independent). Here is the construction:

- Let $\mathbf{v}_1$ be any nonzero vector in V. If $V = \text{span}\{\mathbf{v}_1\}$, then we have our linearly independent spanning set.

- If $V \neq \text{span}\{\mathbf{v}_1\}$, then choose any vector $\mathbf{v}_2$ in V that is not a linear combination of $\mathbf{v}_1$ (i.e., is not a scalar multiple of $\mathbf{v}_1$). If $V = \text{span}\{\mathbf{v}_1, \mathbf{v}_2\}$, then we have our linearly independent spanning set.

- If $V \neq \text{span}\{\mathbf{v}_1, \mathbf{v}_2\}$, then choose any vector $\mathbf{v}_3$ in V that is not a linear combination of $\mathbf{v}_1$ and $\mathbf{v}_2$. If $V = \text{span}\{\mathbf{v}_1, \mathbf{v}_2, \mathbf{v}_3\}$, then we have our linearly independent spanning set. If $V \neq \text{span}\{\mathbf{v}_1, \mathbf{v}_2, \mathbf{v}_3\}$, then choose any $\mathbf{v}_4$ in V that is not a linear combination of $\mathbf{v}_1, \mathbf{v}_2$, and $\mathbf{v}_3$, and repeat the process in the preceding steps.

- If we continue this construction process, then there are two logical possibilities: At some stage we will produce a linearly independent set that spans V or, if not, we will encounter a linearly independent set of $n + 1$ vectors. But the latter is impossible since a linearly independent set in R^n can contain at most n vectors (Theorem 3.4.8). ∎

In general, nonzero subspaces have many bases. For example, if V is a line through the origin of R^n, then any nonzero vector on that line forms a basis for it, and if V is a plane through the origin of R^n, then any two noncollinear vectors in that plane form a basis for it. It is not accidental that the bases for the line all have one vector and the bases for the plane all have two vectors, for the following theorem shows that all bases for a nonzero subspace of R^n must have the same number of vectors.

Theorem 7.1.4 *All bases for a nonzero subspace of R^n have the same number of vectors.*

Proof Let V be a nonzero subspace of R^n, and suppose that the sets $B_1 = \{\mathbf{v}_1, \mathbf{v}_2, \ldots, \mathbf{v}_k\}$ and $B_2 = \{\mathbf{w}_1, \mathbf{w}_2, \ldots, \mathbf{w}_m\}$ are bases for V. Our goal is to prove that $k = m$, which we will do by assuming that $k \neq m$ and obtaining a contradiction. Since the cases $k < m$ and $m < k$ differ only in notation, it will be sufficient to give the proof in the case where $k < m$.

Since B_1 spans V, and since the vectors in B_2 are in V, each $\mathbf{w}_i$ in B_2 can be expressed as a linear combination of the vectors in B_1, say

$$\mathbf{w}_1 = a_{11}\mathbf{v}_1 + a_{21}\mathbf{v}_2 + \cdots + a_{k1}\mathbf{v}_k$$
$$\mathbf{w}_2 = a_{12}\mathbf{v}_1 + a_{22}\mathbf{v}_2 + \cdots + a_{k2}\mathbf{v}_k$$
$$\vdots \qquad \vdots \qquad \vdots \qquad \qquad \vdots \tag{4}$$
$$\mathbf{w}_m = a_{1m}\mathbf{v}_1 + a_{2m}\mathbf{v}_2 + \cdots + a_{km}\mathbf{v}_k$$

Now consider the homogeneous linear system

$$\begin{bmatrix} a_{11} & a_{12} & \cdots & a_{1m} \\ a_{21} & a_{22} & \cdots & a_{2m} \\ \vdots & \vdots & & \vdots \\ a_{k1} & a_{k2} & \cdots & a_{km} \end{bmatrix} \begin{bmatrix} c_1 \\ c_2 \\ \vdots \\ c_m \end{bmatrix} = \begin{bmatrix} 0 \\ 0 \\ \vdots \\ 0 \end{bmatrix}$$

of k equations in the m unknowns $c_1, c_2, \ldots, c_m$. Since $k < m$, this system has more unknowns than equations and hence has a nontrivial solution (Theorem 2.2.3). This implies that there exist numbers $c_1, c_2, \ldots, c_m$, not all zero, such that

$$c_1 \begin{bmatrix} a_{11} \\ a_{21} \\ \vdots \\ a_{k1} \end{bmatrix} + c_2 \begin{bmatrix} a_{12} \\ a_{22} \\ \vdots \\ a_{k2} \end{bmatrix} + \cdots + c_m \begin{bmatrix} a_{1m} \\ a_{2m} \\ \vdots \\ a_{km} \end{bmatrix} = \begin{bmatrix} 0 \\ 0 \\ \vdots \\ 0 \end{bmatrix}$$

or equivalently,

$$\begin{aligned} c_1 a_{11} + c_2 a_{12} + \cdots + c_m a_{1m} &= 0 \\ c_1 a_{21} + c_2 a_{22} + \cdots + c_m a_{2m} &= 0 \\ \vdots \qquad \vdots \qquad\qquad \vdots \qquad \vdots \\ c_1 a_{k1} + c_2 a_{k2} + \cdots + c_m a_{km} &= 0 \end{aligned} \tag{5}$$

To complete the proof, we will show that

$$c_1 \mathbf{w}_1 + c_2 \mathbf{w}_2 + \cdots + c_m \mathbf{w}_m = \mathbf{0} \tag{6}$$

which will contradict the linear independence of $\mathbf{w}_1, \mathbf{w}_2, \ldots, \mathbf{w}_m$. For this purpose, we first use (4) to rewrite the left side of (6) as

$$\begin{aligned} c_1 \mathbf{w}_1 + c_2 \mathbf{w}_2 + \cdots + c_m \mathbf{w}_m = \\ c_1(a_{11}\mathbf{v}_1 + a_{21}\mathbf{v}_2 + \cdots + a_{k1}\mathbf{v}_k) \\ + c_2(a_{12}\mathbf{v}_1 + a_{22}\mathbf{v}_2 + \cdots + a_{k2}\mathbf{v}_k) \\ \ddots \\ + c_m(a_{1m}\mathbf{v}_1 + a_{2m}\mathbf{v}_2 + \cdots + a_{km}\mathbf{v}_k) \end{aligned} \tag{7}$$

Next we multiply out on the right side of (7) and regroup the terms to form a linear combination of $\mathbf{v}_1, \mathbf{v}_2, \ldots, \mathbf{v}_k$. The resulting coefficients in this linear combination match up with the expressions on the left side of (5) (verify), so it follows from (5) that

$$c_1 \mathbf{w}_1 + c_2 \mathbf{w}_2 + \cdots + c_m \mathbf{w}_m = 0\mathbf{v}_1 + 0\mathbf{v}_2 + \cdots + 0\mathbf{v}_k = \mathbf{0}$$

which is the contradiction we were looking for. ∎

EXAMPLE 5
Number of Vectors in a Basis

- Every basis for a line through the origin of R^n has one vector.
- Every basis for a plane through the origin of R^n has two vectors.
- Every basis for R^n has n vectors (since the standard basis has n vectors). ∎

We see in each part of Example 5 that the number of basis vectors for the subspace matches our intuitive concept of its dimension. Thus, we are naturally led to the following definition.

Definition 7.1.5 If V is a nonzero subspace of R^n, then the **dimension** of V, written $\dim(V)$, is defined to be the number of vectors in a basis for V. In addition, we define the zero subspace to have dimension 0.

EXAMPLE 6
Dimensions of
Subspaces of R^n

It follows from Definition 7.1.5 and Example 5 that:

- A line through the origin of R^n has dimension 1.
- A plane through the origin of R^n has dimension 2.
- R^n has dimension n.

DIMENSION OF A SOLUTION SPACE

At the end of Section 3.5 we stated that the general solution of a homogeneous linear system $A\mathbf{x} = \mathbf{0}$ that results from Gauss–Jordan elimination is of the form

$$\mathbf{x} = t_1\mathbf{v}_1 + t_2\mathbf{v}_2 + \cdots + t_s\mathbf{v}_s \tag{8}$$

in which the vectors $\mathbf{v}_1, \mathbf{v}_2, \ldots, \mathbf{v}_s$ are linearly independent. We will call these vectors the ***canonical solutions*** of $A\mathbf{x} = \mathbf{0}$. Note that these are the solutions that result from (8) by setting one of the parameters to 1 and the others to zero. Since the canonical solution vectors span the solution space and are linearly independent, they form a basis for the solution space; we call that basis the ***canonical basis*** for the solution space.

EXAMPLE 7
Basis and
Dimension of
the Solution
Space of
$A\mathbf{x} = \mathbf{0}$

Find the canonical basis for the solution space of the homogeneous system

$$
\begin{aligned}
x_1 + 3x_2 - 2x_3 &\phantom{{}+{}} + 2x_5 &= 0 \\
2x_1 + 6x_2 - 5x_3 - 2x_4 + 4x_5 - 3x_6 &= 0 \\
5x_3 + 10x_4 &\phantom{{}+{}} + 15x_6 &= 0 \\
2x_1 + 6x_2 \phantom{{}+{}} + 8x_4 + 4x_5 + 18x_6 &= 0
\end{aligned}
$$

and state the dimension of that solution space.

Solution We showed in Example 7 of Section 2.2 that the general solution produced by Gauss–Jordan elimination is

$$(x_1, x_2, x_3, x_4, x_5, x_6) = r(-3, 1, 0, 0, 0, 0) + s(-4, 0, -2, 1, 0, 0) + t(-2, 0, 0, 0, 1, 0)$$

Thus, the canonical basis vectors are

$$\mathbf{v}_1 = (-3, 1, 0, 0, 0, 0), \quad \mathbf{v}_2 = (-4, 0, -2, 1, 0, 0), \quad \mathbf{v}_3 = (-2, 0, 0, 0, 1, 0)$$

and the solution space is a three-dimensional subspace of R^6.

DIMENSION OF A HYPERPLANE

Recall from Section 3.5 that if $\mathbf{a} = (a_1, a_2, \ldots, a_n)$ is a nonzero vector in R^n, then the hyperplane $\mathbf{a}^\perp$ through the origin of R^n is given by the equation

$$a_1x_1 + a_2x_2 + \cdots + a_nx_n = 0$$

Let us view this as a linear system of one equation in n unknowns. Since this system has one leading variable and $n - 1$ free variables, its solution space has dimension $n - 1$, and this implies that $\dim(\mathbf{a}^\perp) = n - 1$. For example, hyperplanes through the origin of R^2 (lines) have dimension 1, and hyperplanes through the origin of R^3 (planes) have dimension 2.

> **Theorem 7.1.6** *If $\mathbf{a}$ is a nonzero vector in R^n, then $\dim(\mathbf{a}^\perp) = n - 1$.*

If we exclude R^n itself, then the hyperplanes in R^n are the subspaces of maximal dimension.

Exercise Set 7.1

In Exercises 1 and 2, show that the vectors are linearly dependent by finding a vector in the list that is a linear combination of predecessors. Specify the first vector in the list with this property. (You should be able to solve these problems by inspection.)

1. (a) $\mathbf{v}_1 = (1, -3, 4)$, $\mathbf{v}_2 = (2, -6, 8)$, $\mathbf{v}_3 = (1, 1, 0)$
   (b) $\mathbf{v}_1 = (1, 1, 1)$, $\mathbf{v}_2 = (1, 1, 0)$, $\mathbf{v}_3 = (0, 0, 1)$
   (c) $\mathbf{v}_1 = (1, 0, 0, 0)$, $\mathbf{v}_2 = (0, 1, 0, 0)$, $\mathbf{v}_3 = (0, 0, 0, 1)$, $\mathbf{v}_4 = (-1, 2, 0, 6)$

2. (a) $\mathbf{v}_1 = (8, -6, 2)$, $\mathbf{v}_2 = (4, -3, 1)$, $\mathbf{v}_3 = (0, 0, 7)$
   (b) $\mathbf{v}_1 = (1, 1, 1)$, $\mathbf{v}_2 = (1, 1, 0)$, $\mathbf{v}_3 = (0, 0, 5)$
   (c) $\mathbf{v}_1 = (0, 0, 1, 0)$, $\mathbf{v}_2 = (0, 1, 0, 0)$, $\mathbf{v}_3 = (0, 0, 0, 1)$, $\mathbf{v}_4 = (0, 4, 5, 8)$

In Exercises 3 and 4, show that the vectors in the list are linearly dependent by expressing one of the vectors as a linear combination of its predecessors.

3. $\mathbf{v}_1 = (0, 3, 1, -1)$, $\mathbf{v}_2 = (6, 0, 5, 1)$, $\mathbf{v}_3 = (4, -7, 1, 3)$
4. $\mathbf{v}_1 = (2, 4, 2, -8)$, $\mathbf{v}_2 = (6, 3, 9, -6)$, $\mathbf{v}_3 = (-1, 1, -2, -2)$
5. (a) Find three different bases for the line $y = 2x$ in R^2.
   (b) Find three different bases for the plane $x + y + 2z = 0$ in R^3.

6. (a) Find three different bases for the line $x = t$, $y = 3t$ in R^2.
   (b) Find three different bases for the plane $x = t_1 + t_2$, $y = t_1 - t_2$, $z = 3t_1 + 2t_2$ in R^3.

In Exercises 7–10, find the canonical basis for the solution space of the homogeneous system, and state the dimension of the space.

7. $3x_1 + x_2 + x_3 + x_4 = 0$
   $5x_1 - x_2 + x_3 - x_4 = 0$

8. $x_1 - x_2 + x_3 = 0$
   $2x_1 - x_2 + 4x_3 = 0$
   $3x_1 + x_2 + 11x_3 = 0$

9. $2x_1 + 2x_2 - x_3 \qquad + x_5 = 0$
   $-x_1 - x_2 + 2x_3 - 3x_4 + x_5 = 0$
   $x_1 + x_2 - 2x_3 \qquad - x_5 = 0$
   $x_3 + x_4 + x_5 = 0$

10. $x_1 + 2x_2 - 2x_3 + x_4 + 3x_5 = 0$
    $x_2 + 3x_3 - x_4 - x_5 = 0$
    $2x_1 + 3x_2 - 7x_3 + 3x_4 + 7x_5 = 0$
    $x_2 \qquad - x_4 + x_5 = 0$

In Exercises 11 and 12, find a basis for the hyperplane $\mathbf{a}^\perp$.

11. (a) $\mathbf{a} = (1, 2, -3)$ (b) $\mathbf{a} = (2, -1, 4, 1)$
12. (a) $\mathbf{a} = (-2, 1, 4)$ (b) $\mathbf{a} = (0, -3, 5, 7)$

Discussion and Discovery

D1. (a) If $A\mathbf{x} = \mathbf{0}$ is a linear system of m equations in n unknowns, then its solution space has dimension at most _____.
   (b) A hyperplane in R^6 has dimension _____.
   (c) The subspaces of R^5 have dimensions _____.
   (d) The dimension of the subspace of R^4 spanned by the vectors $\mathbf{v}_1 = (1, 0, 1, 0)$, $\mathbf{v}_2 = (1, 1, 0, 0)$, $\mathbf{v}_3 = (1, 1, 1, 0)$ is _____.

D2. Let $\mathbf{v}_1$, $\mathbf{v}_2$, $\mathbf{v}_3$, and $\mathbf{v}_4$ be vectors in R^6 of the form
$$\mathbf{v}_1 = (1, *, *, *, *, *), \quad \mathbf{v}_2 = (0, 0, 1, *, *, *)$$
$$\mathbf{v}_3 = (0, 0, 0, 1, *, *), \quad \mathbf{v}_4 = (0, 0, 0, 0, 1, *)$$
in which each entry denoted by $*$ can be an arbitrary real number. Is this set linearly independent? Justify your answer.

D3. True or false: If $S = \{\mathbf{v}_1, \mathbf{v}_2, \ldots, \mathbf{v}_k\}$ is a set of two or more nonzero vectors in R^n, then S is linearly independent if and only if some vector in S is a linear combination of its successors. Justify your answer.

D4. Let
$$A = \begin{bmatrix} 2 & 4 & t \\ t & -1 & 1 \\ 3 & 1 & -1 \end{bmatrix}$$

For what values of t, if any, does the solution space of $A\mathbf{x} = \mathbf{0}$ have positive dimension? Find the dimension for each such t.

Working with Proofs

P1. Use Theorem 7.1.2 to prove the following results:
   (a) If $S = \{\mathbf{v}_1, \mathbf{v}_2, \ldots, \mathbf{v}_k\}$ is a linearly dependent set in R^n, then so is the set
   $$S' = \{\mathbf{v}_1, \mathbf{v}_2, \ldots, \mathbf{v}_k, \mathbf{w}_1, \ldots, \mathbf{w}_r\}$$
   for any vectors $\mathbf{w}_1, \mathbf{w}_2, \ldots, \mathbf{w}_r$ in R^n.

   (b) If $S = \{\mathbf{v}_1, \mathbf{v}_2, \ldots, \mathbf{v}_k\}$ is a linearly independent set in R^n, then so is every nonempty subset of S.

P2. Prove that if k is a nonzero scalar and $\mathbf{a}$ is a nonzero vector in R^n, then $(k\mathbf{a})^\perp = \mathbf{a}^\perp$.

Technology Exercises

T1. Are any of the vectors in the set

$$S = \{(2, 6, 3, 4, 2), (3, 1, 5, 8, 3), (5, 1, 2, 6, 7),$$
$$(8, 4, 3, 2, 6), (5, 5, 6, 3, 4)\}$$

linear combinations of predecessors? Justify your answer.

T2. (CAS) Find the exact canonical basis (no decimal approximations) for the solution space of $A\mathbf{x} = \mathbf{0}$, given that

$$A = \begin{bmatrix} 1 & 5 & 2 & 4 & 4 & 7 \\ 3 & 2 & 4 & 9 & 1 & 3 \\ 5 & 2 & 4 & 8 & 5 & 7 \\ 9 & 9 & 10 & 21 & 10 & 17 \end{bmatrix}$$

Section 7.2   Properties of Bases

In this section we will continue our study of basis and dimension and will develop some important properties of subspaces.

PROPERTIES OF BASES

In absence of restrictive conditions, there will generally be many ways to express a vector in span$\{\mathbf{v}_1, \mathbf{v}_2, \ldots, \mathbf{v}_k\}$ as a linear combination of the spanning vectors. For example, let us consider how we might express the vector $\mathbf{v} = (3, 4, 5)$ as a linear combination of the vectors

$$\mathbf{v}_1 = (1, 0, 0), \quad \mathbf{v}_2 = (0, 1, 0), \quad \mathbf{v}_3 = (0, 0, 1), \quad \mathbf{v}_4 = (1, 1, 1)$$

One obvious possibility is to discount the presence of $\mathbf{v}_4$ and write

$$(3, 4, 5) = 3\mathbf{v}_1 + 4\mathbf{v}_2 + 5\mathbf{v}_3 + 0\mathbf{v}_4$$

Other ways can be discovered by expressing the vectors in column form and writing the vector equation

$$\mathbf{v} = c_1\mathbf{v}_1 + c_2\mathbf{v}_2 + c_3\mathbf{v}_3 + c_4\mathbf{v}_4 \tag{1}$$

as the linear system

$$\begin{bmatrix} 1 & 0 & 0 & 1 \\ 0 & 1 & 0 & 1 \\ 0 & 0 & 1 & 1 \end{bmatrix} \begin{bmatrix} c_1 \\ c_2 \\ c_3 \\ c_4 \end{bmatrix} = \begin{bmatrix} 3 \\ 4 \\ 5 \end{bmatrix}$$

Solving this system yields (verify)

$$c_1 = 3 - t, \quad c_2 = 4 - t, \quad c_3 = 5 - t, \quad c_4 = t$$

so substituting in (1) and writing the vectors in component form yields

$$(3, 4, 5) = (3 - t)(1, 0, 0) + (4 - t)(0, 1, 0) + (5 - t)(0, 0, 1) + t(1, 1, 1)$$

Thus, for example, taking $t = 1$ yields

$$(3, 4, 5) = 2(1, 0, 0) + 3(0, 1, 0) + 4(0, 0, 1) + (1, 1, 1)$$

and taking $t = -1$ yields

$$(3, 4, 5) = 4(1, 0, 0) + 5(0, 1, 0) + 6(0, 0, 1) - (1, 1, 1)$$

The following theorem shows that it was the linear dependence of $\mathbf{v}_1$, $\mathbf{v}_2$, $\mathbf{v}_3$, and $\mathbf{v}_4$ that made it possible to express $\mathbf{v}$ as a linear combination of these vectors in more than one way.

Theorem 7.2.1 *If $S = \{\mathbf{v}_1, \mathbf{v}_2, \ldots, \mathbf{v}_k\}$ is a basis for subspace V of R^n, then every vector $\mathbf{v}$ in V can be expressed in exactly one way as a linear combination of the vectors in S.*

Proof Let $\mathbf{v}$ be any vector in V. Since S spans V, there is at least one way to express $\mathbf{v}$ as a linear combination of the vectors in S. To see that there is exactly one way to do this, suppose that

$$\mathbf{v} = t_1\mathbf{v}_1 + t_2\mathbf{v}_2 + \cdots + t_k\mathbf{v}_k \quad \text{and} \quad \mathbf{v} = t_1'\mathbf{v}_1 + t_2'\mathbf{v}_2 + \cdots + t_k'\mathbf{v}_k \tag{2}$$

Subtracting the second equation from the first yields

$$\mathbf{0} = (t_1 - t_1')\mathbf{v}_1 + (t_2 - t_2')\mathbf{v}_2 + \cdots + (t_k - t_k')\mathbf{v}_k$$

Since the right side of this equation is a linear combination of the vectors in S, and since these vectors are linearly independent, each of the coefficients in the linear combination must be zero. Thus, the two linear combinations in (2) are the same. ■

The following theorem reveals two important facts about bases:

1. Every spanning set for a subspace is either a basis for that subspace or has a basis as a subset.

2. Every linearly independent set in a subspace is either a basis for the subspace or can be extended to a basis for the subspace.

Theorem 7.2.2 *Let S be a finite set of vectors in a nonzero subspace V of R^n.*

(a) *If S spans V, but is not a basis for V, then a basis for V can be obtained by removing appropriate vectors from S.*

(b) *If S is a linearly independent set, but is not a basis for V, then a basis for V can be obtained by adding appropriate vectors from V to S.*

Proof (a) If S spans V but is not a basis for V, then S must be a linearly dependent set. This means that some vector $\mathbf{v}$ in S is a linear combination of predecessors. Remove this vector from S to obtain a set S'. The set S' must still span V, since any linear combination of the vectors in S can be rewritten as a linear combination of the vectors in S' by expressing $\mathbf{v}$ in terms of its predecessors. If S' is linearly independent, then it is a basis for V, and we are done. If S' is not linearly independent, then some vector in S' can be expressed as a linear combination of predecessors. Remove this vector from S' to obtain a set S''. As before, this new set will still span V. If S'' is linearly independent, then it is a basis for V and we are done; otherwise, we continue the process of removing vectors until we reach a basis.

Proof (b) If $S = \{\mathbf{w}_1, \mathbf{w}_2, \ldots, \mathbf{w}_s\}$ is a linearly independent set of vectors in V but is not a basis for V, then S does not span V. Thus, there is some vector $\mathbf{v}_1$ in V that is not a linear combination of the vectors in S. Add this vector to S to obtain the set $S' = \{\mathbf{w}_1, \mathbf{w}_2, \ldots, \mathbf{w}_s, \mathbf{v}_1\}$. This set must still be linearly independent since none of the vectors in S' can be linear combinations of predecessors. If S' spans V, then it is a basis for V, and we are done. If S' does not span V, then there is some vector $\mathbf{v}_2$ in V that is not a linear combination of the vectors in S'. Add this vector to S' to obtain the set $S'' = \{\mathbf{w}_1, \mathbf{w}_2, \ldots, \mathbf{w}_s, \mathbf{v}_1, \mathbf{v}_2\}$. As before, this set will still be linearly independent. If S'' spans V, then it is a basis and we are done; otherwise we continue the process until we reach a basis or a linearly independent set with n vectors. But in the latter case the set also has to be a basis for V; otherwise, it would not span V, and the procedure we have been following would allow us to add another vector to the set and create a linearly independent set with $n + 1$ vectors—an impossibility by Theorem 3.4.8. Thus, the procedure eventually produces a basis in all cases. ■

By definition, the dimension of a nonzero subspace V of R^n is the number of vectors in a basis for V; however, the dimension can also be viewed as the maximum number of linearly independent vectors in V, for if $\dim(V) = k$, and if we could produce more than k linearly independent vectors, then by part (b) of Theorem 7.2.2, that set of vectors would either have to be a basis for V or part of a basis for V, contradicting the fact that all bases for V have k vectors.

Theorem 7.2.3   *If V is a nonzero subspace of R^n, then* $\dim(V)$ *is the maximum number of linearly independent vectors in V.*

REMARK   Engineers use the term ***degrees of freedom*** as a synonym for dimension, the idea being that a space with k degrees of freedom allows freedom of motion or variation in at most k independent directions.

SUBSPACES OF SUBSPACES   Up to now we have focused on subspaces of R^n. However, if V and W are subspaces of R^n, and if V is a subset of W, then we also say that V is a ***subspace*** of W. For example, in Figure 7.2.1 the space $\{0\}$ is a subspace of the line, which in turn is a subspace of the plane, which in turn is a subspace of R^3.

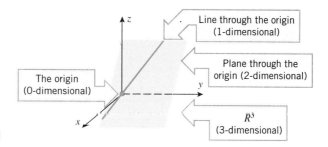

Figure 7.2.1

Since the dimension of a subspace of R^n is the maximum number of linearly independent vectors that the subspace can have, it follows that if V is a subspace of W, then the dimension of V cannot exceed the dimension of W. In particular, the dimension of a subspace of R^n can be at most n, just as you would suspect. Further, if V is a subspace of W, and if the two spaces have the same dimension, then they must be the same space (Exercise P8).

Theorem 7.2.4   *If V and W are subspaces of R^n, and if V is a subspace of W, then:*
(a) $0 \leq \dim(V) \leq \dim(W) \leq n$
(b) $V = W$ *if and only if* $\dim(V) = \dim(W)$

There will be occasions on which we are given some nonempty set S, and we will be interested in knowing how the subspace span(S) is affected by adding additional vectors to S. The following theorem deals with this question.

Theorem 7.2.5   *Let S be a nonempty set of vectors in R^n, and let S′ be a set that results by adding additional vectors in R^n to S.*
(a) *If the additional vectors are in* span(S), *then* span$(S') =$ span(S).
(b) *If* span$(S') =$ span(S), *then the additional vectors are in* span(S).
(c) *If* span(S') *and* span(S) *have the same dimension, then the additional vectors are in* span(S) *and* span$(S') =$ span(S).

We will not formally prove this theorem, but its statements should almost be self-evident. For example, part (a) tells you that if you add vectors to S that are already linear combinations of vectors in S, then you are not going to add anything new to the set of all possible linear combinations of vectors in S. Part (b) tells you that if the additional vectors do not add anything new to the set of all linear combinations of vectors in S, then the additional vectors must already be linear combinations of vectors in S. Finally, in part (c) the fact that S is a subset of S' means that span(S) is a subspace of span(S') (why?). Thus, if the two spaces have the same dimension, then they must be the same space by Theorem 7.2.4(b); and this means that the additional vectors must be in span(S).

SOMETIMES SPANNING IMPLIES LINEAR INDEPENDENCE, AND CONVERSELY

In general, when you want to show that a set of vectors is a basis for a subspace V of R^n you must show that the set is linearly independent *and* spans V. However, if you know a priori that the number of vectors in the set is the same as the dimension of V, then to show that the set is a basis it suffices to show either that it is linearly independent *or* that it spans V—the other condition will follow automatically.

Theorem 7.2.6

(a) *A set of k linearly independent vectors in a nonzero k-dimensional subspace of R^n is a basis for that subspace.*

(b) *A set of k vectors that span a nonzero k-dimensional subspace of R^n is a basis for that subspace.*

(c) *A set of fewer than k vectors in a nonzero k-dimensional subspace of R^n cannot span that subspace.*

(d) *A set with more than k vectors in a nonzero k-dimensional subspace of R^n is linearly dependent.*

Proof (a) Let S be a linearly independent set of k vectors in a nonzero k-dimensional subspace V of R^n. If S is not a basis for V, then S can be extended to a basis by adding appropriate vectors from V. However, this would produce a basis for V with more than k vectors, which is impossible. Thus, S must be a basis.

Proof (b) Let S be a set of k vectors in a nonzero k-dimensional subspace V of R^n that span V. If S is not a basis for V, then it can be pared down to a basis for V by removing appropriate vectors. However, this would produce a basis for V with fewer than k vectors, which is impossible. Thus, S must be a basis.

Proof (c) Let S be a set with fewer than k vectors that spans a nonzero k-dimensional subspace V of R^n. Then either S is a basis for V or can be made into a basis for V by removing appropriate vectors. In either case we have a basis for V with fewer than k vectors, which is impossible.

Proof (d) Let S be a linearly independent set with more than k vectors from a nonzero k-dimensional subspace of R^n. Then either S is a basis for V or can be made into a basis for V by adding appropriate vectors. In either case we have a basis for V with more than k vectors, which is impossible. ∎

EXAMPLE 1
Bases by Inspection

(a) Show that the vectors $\mathbf{v}_1 = (-3, 7)$ and $\mathbf{v}_2 = (5, 5)$ form a basis for R^2 by inspection.

(b) Show that $\mathbf{v}_1 = (2, 0, -1)$, $\mathbf{v}_2 = (4, 0, 7)$, and $\mathbf{v}_3 = (6, 1, -5)$ form a basis for R^3 by inspection.

Solution (a) We have two vectors in a two-dimensional space, so it suffices to show that the vectors are linearly independent. However, this is obvious, since neither vector is a scalar multiple of the other.

Solution (b) We have three vectors in a three-dimensional space, so again it suffices to show that the vectors are linearly independent. We can do this by showing that none of the vectors $\mathbf{v}_1$, $\mathbf{v}_2$, and $\mathbf{v}_3$ is a linear combination of predecessors. But the vector $\mathbf{v}_2$ is not a linear combination of $\mathbf{v}_1$, since it is not a scalar multiple of $\mathbf{v}_1$, and the vector $\mathbf{v}_3$ is not a linear combination of $\mathbf{v}_1$ and $\mathbf{v}_2$, since any such linear combination has a second component of zero, and $\mathbf{v}_3$ does not. ∎

EXAMPLE 2
A Determinant Test for Linear Independence

(a) Show that the vectors $\mathbf{v}_1 = (1, 2, 1)$, $\mathbf{v}_2 = (1, -1, 3)$, and $\mathbf{v}_3 = (1, 1, 4)$ form a basis for R^3.

(b) Express $\mathbf{w} = (4, 9, 8)$ as a linear combination of $\mathbf{v}_1$, $\mathbf{v}_2$, and $\mathbf{v}_3$.

Solution (*a*) We have three vectors in a three-dimensional space, so it suffices to show that the vectors are linearly independent. One way to do this is to form the matrix

$$A = \begin{bmatrix} 1 & 1 & 1 \\ 2 & -1 & 1 \\ 1 & 3 & 4 \end{bmatrix}$$

that has $\mathbf{v}_1$, $\mathbf{v}_2$, and $\mathbf{v}_3$ as its column vectors and apply Theorem 6.3.15. The determinant of the matrix A is nonzero [verify that $\det(A) = -7$], so parts (*i*) and (*g*) of that theorem imply that the column vectors are linearly independent.

Solution (*b*) The result in part (a) guarantees that $\mathbf{w}$ can be expressed as a unique linear combination of $\mathbf{v}_1$, $\mathbf{v}_2$, and $\mathbf{v}_3$, but the method used in that part does not tell us what that linear combination is. To find it, we rewrite the vector equation

$$(4, 9, 8) = c_1(1, 2, 1) + c_2(1, -1, 3) + c_3(1, 1, 4) \tag{3}$$

as the linear system

$$\begin{bmatrix} 1 & 1 & 1 \\ 2 & -1 & 1 \\ 1 & 3 & 4 \end{bmatrix} \begin{bmatrix} c_1 \\ c_2 \\ c_3 \end{bmatrix} = \begin{bmatrix} 4 \\ 9 \\ 8 \end{bmatrix}$$

This system has the unique solution $c_1 = 3$, $c_2 = -1$, $c_3 = 2$ (verify), and substituting these values in (3) yields

$$(4, 9, 8) = 3(1, 2, 1) - (1, -1, 3) + 2(1, 1, 4)$$

which expresses $\mathbf{w}$ as the linear combination $\mathbf{w} = 3\mathbf{v}_1 - \mathbf{v}_2 + 2\mathbf{v}_3$. ∎

A UNIFYING THEOREM By combining Theorem 7.2.6 with parts (*g*) and (*h*) of Theorem 6.3.15, we can add four more statements to the latter. (Note that we have reordered the parts that appeared in Theorem 6.3.15 to bring the row and column statements together.)

> **Theorem 7.2.7** *If A is an $n \times n$ matrix, and if T_A is the linear operator on R^n with standard matrix A, then the following statements are equivalent.*
>
> (*a*) *The reduced row echelon form of A is I_n.*
> (*b*) *A is expressible as a product of elementary matrices.*
> (*c*) *A is invertible.*
> (*d*) *$A\mathbf{x} = \mathbf{0}$ has only the trivial solution.*
> (*e*) *$A\mathbf{x} = \mathbf{b}$ is consistent for every vector $\mathbf{b}$ in R^n.*
> (*f*) *$A\mathbf{x} = \mathbf{b}$ has exactly one solution for every vector $\mathbf{b}$ in R^n.*
> (*g*) *$\det(A) \neq 0$.*
> (*h*) *$\lambda = 0$ is not an eigenvalue of A.*
> (*i*) *T_A is one-to-one.*
> (*j*) *T_A is onto.*
> (*k*) *The column vectors of A are linearly independent.*
> (*l*) *The row vectors of A are linearly independent.*
> (*m*) *The column vectors of A span R^n.*
> (*n*) *The row vectors of A span R^n.*
> (*o*) *The column vectors of A form a basis for R^n.*
> (*p*) *The row vectors of A form a basis for R^n.*

Exercise Set 7.2

In Exercises 1 and 2, explain why the vectors do not form a basis for the indicated vector space. (No computations needed.)

1. (a) $\mathbf{v}_1 = (1, 2)$, $\mathbf{v}_2 = (0, 3)$, $\mathbf{v}_3 = (2, 7)$ for R^2.
   (b) $\mathbf{v}_1 = (-1, 3, 2)$, $\mathbf{v}_2 = (6, 1, 1)$ for R^3.
   (c) $\mathbf{v}_1 = (4, 3)$, $\mathbf{v}_2 = (8, 6)$ for R^2.

2. (a) $\mathbf{v}_1 = (4, 3)$ for R^2.
   (b) $\mathbf{v}_1 = (1, 1, 1)$, $\mathbf{v}_2 = (-1, 2, 4)$, $\mathbf{v}_3 = (0, 7, 9)$,
      $\mathbf{v}_4 = (-9, 8, 6)$ for R^3.
   (c) $\mathbf{v}_1 = (1, 1, 1)$, $\mathbf{v}_2 = (-1, 2, 4)$, $\mathbf{v}_3 = (0, 0, 0)$ for R^3.

In Exercises 3 and 4, use the ideas in Example 1 to show that the vectors form a basis for the indicated vector space.

3. (a) $\mathbf{v}_1 = (2, 1)$, $\mathbf{v}_2 = (3, 0)$ for R^2.
   (b) $\mathbf{v}_1 = (4, 1, 0)$, $\mathbf{v}_2 = (-7, 8, 0)$, $\mathbf{v}_3 = (1, 1, 1)$ for R^3.

4. (a) $\mathbf{v}_1 = (7, 5)$, $\mathbf{v}_2 = (4, 8)$ for R^2.
   (b) $\mathbf{v}_1 = (0, 1, 3)$, $\mathbf{v}_2 = (0, 8, 0)$, $\mathbf{v}_3 = (1, 6, 0)$ for R^3.

In Exercises 5 and 6, use a determinant test to determine whether the vectors form a basis for R^3.

5. (a) $\mathbf{v}_1 = (3, 2, -4)$, $\mathbf{v}_2 = (4, 1, -2)$, $\mathbf{v}_3 = (5, 2, -3)$
   (b) $\mathbf{v}_1 = (3, 4, -5)$, $\mathbf{v}_2 = (8, 7, -2)$, $\mathbf{v}_3 = (2, -1, 8)$

6. (a) $\mathbf{v}_1 = (5, 1, -8)$, $\mathbf{v}_2 = (3, 0, 5)$, $\mathbf{v}_3 = (8, 1, -3)$
   (b) $\mathbf{v}_1 = (-1, 1, 1)$, $\mathbf{v}_2 = (1, -1, 1)$, $\mathbf{v}_3 = (1, 1, -1)$

7. (a) Show that the vectors

$$\mathbf{v}_1 = (1, 0, 0), \quad \mathbf{v}_2 = (0, 2, 0)$$
$$\mathbf{v}_3 = (0, 0, 3), \quad \mathbf{v}_4 = (1, 1, 1)$$

span R^3 but do not form a basis for R^3.
   (b) Find three different ways to express the vector $\mathbf{v} = (1, 2, 3)$ as a linear combination of the vectors in part (a).

8. (a) Show that the vectors

$$\mathbf{v}_1 = (1, 1, 1), \quad \mathbf{v}_2 = (1, 1, 0)$$
$$\mathbf{v}_3 = (1, 0, 0), \quad \mathbf{v}_4 = (3, 2, 0)$$

span R^3 but do not form a basis for R^3.
   (b) Find three different ways to express the vector $\mathbf{v} = (1, 2, 3)$ as a linear combination of the vectors in part (a).

In Exercises 9 and 10, show that $S = \{\mathbf{v}_1, \mathbf{v}_2\}$ is a linearly independent set in R^3, and extend S to a basis for R^3 by adding an appropriate standard unit vector to it.

9. $\mathbf{v}_1 = (-1, 2, 3)$, $\mathbf{v}_2 = (1, -2, -2)$

10. $\mathbf{v}_1 = (1, -1, 0)$, $\mathbf{v}_2 = (3, 1, -2)$

In Exercises 11 and 12, show that the set $S = \{\mathbf{v}_1, \mathbf{v}_2, \mathbf{v}_3, \mathbf{v}_4\}$ spans R^3, and create a basis for R^3 by removing vectors from S that are linear combinations of predecessors. Confirm that the resulting set is a basis.

11. $\mathbf{v}_1 = (1, 2, -2)$, $\mathbf{v}_2 = (3, -1, 1)$, $\mathbf{v}_3 = (4, 1, -1)$,
   $\mathbf{v}_4 = (1, 3, 6)$

12. $\mathbf{v}_1 = (3, 2, 2)$, $\mathbf{v}_2 = (6, 4, 4)$, $\mathbf{v}_3 = (1, 2, -3)$,
   $\mathbf{v}_4 = (0, 1, 4)$

In Exercises 13 and 14, show that $B = \{\mathbf{v}_1, \mathbf{v}_2, \mathbf{v}_3\}$ is a basis for R^3, and express $\mathbf{v}$ as a linear combination of the basis vectors.

13. $\mathbf{v}_1 = (1, 0, 0)$, $\mathbf{v}_2 = (1, 1, 0)$, $\mathbf{v}_3 = (1, 1, 1)$; $\mathbf{v} = (2, 5, 1)$

14. $\mathbf{v}_1 = (1, 1, 0)$, $\mathbf{v}_2 = (1, 0, 1)$, $\mathbf{v}_3 = (0, 1, 1)$;
   $\mathbf{v} = (-4, 2, 3)$

In Exercises 15 and 16, subspaces V and W of R^3 are given. Determine whether V is a subspace of W.

15. (a) V is the line through the origin spanned by
      $\mathbf{u} = (1, 2, -1)$, and W is the plane through the origin
      with normal $\mathbf{n} = (2, 0, 1)$.
   (b) V is the line with parametric equations $x = 2t$,
      $y = -t$, $z = t$, and W is the plane whose equation is
      $x + 3y + z = 0$.

16. (a) V is the line through the origin spanned by
      $\mathbf{u} = (1, 1, 3)$, and W is the plane through the origin
      with normal $\mathbf{n} = (2, 1, -1)$.
   (b) V is the line with parametric equations $x = t$, $y = 2t$,
      $z = -5t$, and W is the plane whose equation is
      $3x + 2y + z = 0$.

Suppose that $S = \{\mathbf{v}_1, \mathbf{v}_2, \ldots, \mathbf{v}_n\}$ is a basis for R^n and that $T : R^n \rightarrow R^m$ is a linear transformation. If the images $T(\mathbf{v}_1), T(\mathbf{v}_2), \ldots, T(\mathbf{v}_n)$ of the basis vectors are known, then these image vectors can be used to compute the image of any vector $\mathbf{x}$ in R^n by first writing $\mathbf{x}$ as a linear combination of the vectors in S, say $\mathbf{x} = c_1\mathbf{v}_1 + c_2\mathbf{v}_2 + \cdots + c_n\mathbf{v}_n$, and then using the linearity of the transformation T to express $T(\mathbf{x})$ as $T(\mathbf{x}) = c_1 T(\mathbf{v}_1) + c_2 T(\mathbf{v}_2) + \cdots + c_n T(\mathbf{v}_n)$. This fact is sometimes described by saying that a linear transformation is completely determined by its "values" at a basis. Use this idea in Exercises 17 and 18.

17. Suppose that $T : R^3 \rightarrow R^3$ is a linear transformation and we know that

$$T(1, 1, 0) = (2, 1, -1), \quad T(0, 1, 2) = (1, 0, 2)$$
$$T(2, 1, 3) = (4, 0, 1)$$

   (a) Find $T(3, 2, -1)$.   (b) Find $T(a, b, c)$.
   (c) Find the standard matrix for T.

18. Suppose that $T : R^3 \to R^4$ is a linear transformation and we know that
$$T(1, 1, 1) = (3, 2, 0, 1), \quad T(1, 1, 0) = (2, 1, 3, -1)$$
$$T(1, 0, 0) = (5, -2, 1, 0)$$
(a) Find $T(4, 3, 0)$.   (b) Find $T(a, b, c)$.
(c) Find the standard matrix for T.

19. Show that the vectors $\mathbf{v}_1 = (1, -7, -5)$, $\mathbf{v}_2 = (3, -2, 8)$, $\mathbf{v}_3 = (4, -3, 5)$ form a basis for R^3 and express a general vector $\mathbf{x} = (x, y, z)$ as a linear combination of these basis vectors.

Discussion and Discovery

D1. Indicate whether the statement is true (T) or false (F). Justify your answer.
(a) If $S = \{\mathbf{v}_1, \mathbf{v}_2, \ldots, \mathbf{v}_k\}$ is a linearly independent set in R^n, then $k \le n$.
(b) If $S = \{\mathbf{v}_1, \mathbf{v}_2, \ldots, \mathbf{v}_k\}$ spans R^n, then $k \ge n$.
(c) If $S = \{\mathbf{v}_1, \mathbf{v}_2, \ldots, \mathbf{v}_k\}$ is a set of vectors in R^n, and if every vector in R^n can be expressed in exactly one way as a linear combination of the vectors in S, then $k = n$.
(d) If $A\mathbf{x} = \mathbf{0}$ is a linear system of n equations in n unknowns with infinitely many solutions, then the row vectors of A do not form a basis for R^n.
(e) If V and W are distinct subspaces of R^n with the same dimension, then neither V nor W is a subspace of the other.

D2. If $S = \{\mathbf{v}_1, \mathbf{v}_2, \ldots, \mathbf{v}_n\}$ is a linearly dependent set in R^n, is it possible to create a basis for R^n by forming appropriate linear combinations of the vectors in S? Explain your reasoning.

D3. If $B = \{\mathbf{v}_1, \mathbf{v}_2, \ldots, \mathbf{v}_n\}$ is a basis for R^n, how many different one-to-one linear operators can be created that map each vector in B into a vector in B? (See Exercise P4.)

D4. For what values of α and β, if any, do the vectors
$$\mathbf{v}_1 = (\sin\alpha + \sin\beta, \cos\beta + \cos\alpha)$$
and
$$\mathbf{v}_2 = (\cos\beta - \cos\alpha, \sin\alpha - \sin\beta)$$
form a basis for R^2?

D5. Explain why the dimension of a subspace W of R^n is the minimum number of vectors in a spanning set for W.

Working with Proofs

P1. Prove: Every nonzero subspace of R^n has a basis.

P2. Prove: If k is any integer between 0 and n, then R^n contains a subspace of dimension k.

P3. Prove that if every vector in R^n can be written uniquely as a linear combination of the vectors in $S = \{\mathbf{v}_1, \mathbf{v}_2, \ldots, \mathbf{v}_n\}$, then S is a basis for R^n. [*Hint:* Use the uniqueness assumption to prove linear independence.]

P4. Prove: If $T : R^n \to R^n$ is a one-to-one linear operator, and if $\{\mathbf{v}_1, \mathbf{v}_2, \ldots, \mathbf{v}_n\}$ is a basis for R^n, then $\{T(\mathbf{v}_1), T(\mathbf{v}_2), \ldots, T(\mathbf{v}_n)\}$ is also a basis for R^n; that is, one-to-one linear operators map bases into bases. [*Hint:* Use the one-to-one assumption to prove linear independence.]

P5. Prove: If $B = \{\mathbf{v}_1, \mathbf{v}_2, \ldots, \mathbf{v}_n\}$ and $B' = \{\mathbf{w}_1, \mathbf{w}_2, \ldots, \mathbf{w}_n\}$ are bases for R^n, then there exists a unique linear operator on R^n such that $T(\mathbf{v}_1) = \mathbf{w}_1, T(\mathbf{v}_2) = \mathbf{w}_2, \ldots, T(\mathbf{v}_n) = \mathbf{w}_n$.

[*Hint:* Since B is a basis for R^n, every vector $\mathbf{x}$ in R^n can be expressed uniquely as $\mathbf{x} = c_1\mathbf{v}_1 + c_2\mathbf{v}_2 + \cdots + c_n\mathbf{v}_n$. Show that the formula $T(\mathbf{x}) = c_1\mathbf{w}_1 + c_2\mathbf{w}_2 + \cdots + c_n\mathbf{w}_n$ defines a linear operator with the required property.]

P6. (a) Prove that if $\{\mathbf{v}_1, \mathbf{v}_2, \mathbf{v}_3\}$ is a basis for R^3, then so is $\{\mathbf{u}_1, \mathbf{u}_2, \mathbf{u}_3\}$, where
$$\mathbf{u}_1 = \mathbf{v}_1, \quad \mathbf{u}_2 = \mathbf{v}_1 + \mathbf{v}_2, \quad \mathbf{u}_3 = \mathbf{v}_1 + \mathbf{v}_2 + \mathbf{v}_3$$
(b) State a generalization of the result in part (a).

P7. Prove that $\mathbf{x}$ is an eigenvector of an $n \times n$ matrix A if and only if the subspace of R^n spanned by $\mathbf{x}$ and $A\mathbf{x}$ has dimension 1.

P8. Prove: If V and W are subspaces of R^n such that $V \subset W$, and $\dim(V) = \dim(W)$, then $V = W$. [*Hint:* Use Theorem 7.2.2(b) to show that if $S = \{\mathbf{v}_1, \mathbf{v}_2, \ldots, \mathbf{v}_k\}$ is a basis for V, then S is also a basis for W. Then use this result to show that $W \subset V$.]

Technology Exercises

T1. Devise three different procedures for using your technology utility to find the dimension of the subspace spanned by a set of vectors in R^n, and use all of your procedures to find

the dimension of the subspace of R^5 spanned by the vectors
$$\mathbf{v}_1 = (2, 2, -1, 0, 1), \quad \mathbf{v}_2 = (-1, -1, 2, -3, 1)$$
$$\mathbf{v}_3 = (1, 1, -2, 0, -1), \quad \mathbf{v}_4 = (0, 0, 1, 1, 1)$$

T2. Let $S = \{\mathbf{v}_1, \mathbf{v}_2, \mathbf{v}_3, \mathbf{v}_4, \mathbf{v}_5\}$, where

$$\mathbf{v}_1 = (1, 2, 1), \quad \mathbf{v}_2 = (4, 4, 4), \quad \mathbf{v}_3 = (1, 0, 1)$$
$$\mathbf{v}_4 = (2, 4, 2), \quad \mathbf{v}_5 = (0, 1, 1)$$

Find all possible subsets of S that are bases for R^3.

T3. (CAS) Use a determinant test to find conditions on a, b, c, and d under which the vectors

$$\mathbf{v}_1 = (a, b, c, d), \quad \mathbf{v}_2 = (-b, a, d, -c)$$
$$\mathbf{v}_3 = (-c, -d, a, b), \quad \mathbf{v}_4 = (-d, c, -b, a)$$

form a basis for R^4.

Section 7.3   The Fundamental Spaces of a Matrix

The development of analytic geometry was one of the great milestones in mathematics in that it provided a way of studying properties of geometric curves using algebraic equations and conversely a way of using geometric curves to study properties of algebraic equations. In this section and those that follow we will see how algebraic properties of matrices can be used to study geometric properties of subspaces and conversely how geometric properties of subspaces can be used to study algebraic properties of matrices. In this section we will also consider various methods for finding bases for subspaces.

THE FUNDAMENTAL SPACES OF A MATRIX

If A is an $m \times n$ matrix, then there are three important spaces associated with A:

1. The **row space** of A, denoted by row(A), is the subspace of R^n that is spanned by the row vectors of A.

2. The **column space** of A, denoted by col(A), is the subspace of R^m that is spanned by the column vectors of A.

3. The **null space** of A, denoted by null(A), is the solution space of $A\mathbf{x} = \mathbf{0}$. This is a subspace of R^n.

If we consider A and A^T together, then there appear to be six such subspaces:

$$\text{row}(A), \quad \text{row}(A^T), \quad \text{col}(A), \quad \text{col}(A^T), \quad \text{null}(A), \quad \text{null}(A^T)$$

But transposing a matrix converts rows to columns, and columns to rows, so row(A^T) = col(A) and col(A^T) = row(A). Thus, of the six subspaces only the following four are distinct:

row(A)	null(A)
col(A)	null(A^T)

These are called the **fundamental spaces** of A. The dimensions of row(A) and null(A) are sufficiently important that there is some terminology associated with them.

> **Definition 7.3.1** The dimension of the row space of a matrix A is called the **rank** of A and is denoted by rank(A); and the dimension of the null space of A is called the **nullity** of A and is denoted by nullity(A).

REMARK Later in this chapter we will show that the row space and column space of a matrix always have the same dimension, so you can also think of the rank of A as the dimension of the column space. However, this will not be relevant to our current work.

Linear Algebra in History

The concept of rank appeared for the first time in an 1879 research paper by the German mathematician Ferdinand Frobenius, who used the German word *rang* to describe the idea.

Ferdinand Georg Frobenius
(1849–1917)

ORTHOGONAL COMPLEMENTS

One of the goals in this section is to develop some of the basic properties of the fundamental spaces. As a first step, we will need to establish some more results about orthogonality.

Recall from Section 3.5 that if **a** is a nonzero vector in R^n, then $\mathbf{a}^\perp$ is the set of all vectors in R^n that are orthogonal to **a**. We call this set the *orthogonal complement* of **a** (or the *hyperplane through the origin with normal* **a**). The following definition extends the idea of an orthogonal complement to sets with more than one vector.

Definition 7.3.2 If S is a nonempty set in R^n, then the **orthogonal complement** of S, denoted by $S^\perp$, is defined to be the set of all vectors in R^n that are orthogonal to every vector in S.

EXAMPLE 1
Orthogonal
Complements of
Subspaces of R^3

If L is a line through the origin of R^3, then $L^\perp$ is the plane through the origin that is perpendicular to L, and if W is a plane through the origin of R^3, then $W^\perp$ is the line through the origin that is perpendicular to W (Figure 7.3.1). ∎

EXAMPLE 2
Orthogonal
Complement of
Row Vectors

If S is the set of row vectors of an $m \times n$ matrix A, then it follows from Theorem 3.5.6 that $S^\perp$ is the solution space of $A\mathbf{x} = \mathbf{0}$. ∎

The set S in Definition 7.3.2 is not required to be a subspace of R^n. However, the following theorem shows that the orthogonal complement of S is always a subspace of R^n, regardless of whether S is or not.

Theorem 7.3.3 *If S is a nonempty set in R^n, then $S^\perp$ is a subspace of R^n.*

Proof The set $S^\perp$ contains the vector **0**, so we can be assured that it is nonempty. We will show that it is closed under scalar multiplication and addition. For this purpose, let **u** and **v** be vectors in $S^\perp$ and let c be a scalar. To show that $c\mathbf{u}$ and $\mathbf{u} + \mathbf{v}$ are vectors in $S^\perp$, we must show that $c\mathbf{u} \cdot \mathbf{x} = 0$ and $(\mathbf{u} + \mathbf{v}) \cdot \mathbf{x} = 0$ for every vector **x** in S. But **u** and **v** are vectors in $S^\perp$, so $\mathbf{u} \cdot \mathbf{x} = 0$ and $\mathbf{v} \cdot \mathbf{x} = 0$. Thus, using properties of the dot product we obtain

$$c\mathbf{u} \cdot \mathbf{x} = c(\mathbf{u} \cdot \mathbf{x}) = c(0) = 0 \quad \text{and} \quad (\mathbf{u} + \mathbf{v}) \cdot \mathbf{x} = (\mathbf{u} \cdot \mathbf{x}) + (\mathbf{v} \cdot \mathbf{x}) = 0 + 0 = 0$$

which completes the proof. ∎

Figure 7.3.1

EXAMPLE 3
Orthogonal
Complement of
Two Vectors

Find the orthogonal complement in an xyz-coordinate system of the set $S = \{\mathbf{v}_1, \mathbf{v}_2\}$, where

$$\mathbf{v}_1 = (1, -2, 1) \quad \text{and} \quad \mathbf{v}_2 = (3, -7, 5)$$

Solution It should be evident geometrically that $S^\perp$ is the line through the origin that is perpendicular to the plane determined by $\mathbf{v}_1$ and $\mathbf{v}_2$. One way to find this line is to use the result in Example 2 by letting A be the matrix with row vectors $\mathbf{v}_1$ and $\mathbf{v}_2$ and solving the system $A\mathbf{x} = \mathbf{0}$. This system is

$$\begin{bmatrix} 1 & -2 & 1 \\ 3 & -7 & 5 \end{bmatrix} \begin{bmatrix} x \\ y \\ z \end{bmatrix} = \begin{bmatrix} 0 \\ 0 \\ 0 \end{bmatrix}$$

and a general solution is (verify)

$$x = 3t, \quad y = 2t, \quad z = t \tag{1}$$

Thus, $S^\perp$ is the line through the origin that is parallel to the vector $\mathbf{w} = (3, 2, 1)$.

Alternative Solution A vector that is orthogonal to both $\mathbf{v}_1$ and $\mathbf{v}_2$ is

$$\mathbf{v}_1 \times \mathbf{v}_2 = \begin{vmatrix} \mathbf{i} & \mathbf{j} & \mathbf{k} \\ 1 & -2 & 1 \\ 3 & -7 & 5 \end{vmatrix} = -3\mathbf{i} - 2\mathbf{j} - \mathbf{k} = (-3, -2, -1)$$

The vector $\mathbf{w} = -(\mathbf{v}_1 \times \mathbf{v}_2) = (3, 2, 1)$ is also orthogonal to $\mathbf{v}_1$ and $\mathbf{v}_2$ and is actually a simpler choice because it has fewer minus signs. This is the vector **w** that we obtained in our first solution, so again we find that the orthogonal complement is the line through the origin given parametrically by (1). ∎

The following theorem lists three basic facts about orthogonal complements.

Theorem 7.3.4

    (a) *If W is a subspace of R^n, then $W^\perp \cap W = \{\mathbf{0}\}$.*

    (b) *If S is a nonempty subset of R^n, then $S^\perp = \text{span}(S)^\perp$.*

    (c) *If W is a subspace of R^n, then $(W^\perp)^\perp = W$.*

We will prove parts (a) and (b); the proof of part (c) will be deferred until a later section in which we will have more mathematical tools to use.

Proof (a) The set $W^\perp \cap W$ contains at least the zero vector, since W and $W^\perp$ are subspaces of R^n. But this is the only vector in $W^\perp \cap W$, for if **v** is any vector in this intersection, then $\mathbf{v} \cdot \mathbf{v} = 0$, which implies that $\mathbf{v} = 0$ by part (d) of Theorem 1.2.6.

Proof (b) We must show that every vector in $\text{span}(S)^\perp$ is in $S^\perp$ and conversely. Accordingly, let **v** be any vector in $\text{span}(S)^\perp$. This vector is orthogonal to every vector in $\text{span}(S)$, so it has to be orthogonal to every vector in S, since S is contained in $\text{span}(S)$. Thus, **v** is in $S^\perp$. Conversely, let **v** be any vector in $S^\perp$. To show that **v** is in $\text{span}(S)^\perp$, we must show that **v** is orthogonal to every linear combination of the vectors in S. Accordingly, let

$$\mathbf{w} = t_1\mathbf{v}_1 + t_2\mathbf{v}_2 + \cdots + t_k\mathbf{v}_k$$

be any such linear combination. Then using properties of the dot product we obtain

$$\mathbf{v} \cdot \mathbf{w} = \mathbf{v} \cdot (t_1\mathbf{v}_1 + t_2\mathbf{v}_2 + \cdots + t_k\mathbf{v}_k) = t_1(\mathbf{v} \cdot \mathbf{v}_1) + t_2(\mathbf{v} \cdot \mathbf{v}_2) + \cdots + t_k(\mathbf{v} \cdot \mathbf{v}_k)$$

But each dot product on the right side is zero, since **v** is orthogonal to every vector in S. Thus, $\mathbf{v} \cdot \mathbf{w} = 0$, which shows that **v** is orthogonal to every vector in $\text{span}(S)$. ∎

REMARK If W is a subspace of R^n and $W^\perp$ is its orthogonal complement, then the equation $(W^\perp)^\perp = W$ in part (c) of the last theorem states that W is the orthogonal complement of $W^\perp$. This establishes a symmetry that allows us to say that W and $W^\perp$ are orthogonal complements *of one another*. Note, however, that it is required that W be a subspace of R^n (not just a subset) for this to be true.

In words, part (b) of Theorem 7.3.4 states that *the orthogonal complement of a nonempty set and the orthogonal complement of the subspace spanned by that set are the same.* Thus, for example, the orthogonal complement of the set of row vectors of a matrix is the same as the orthogonal complement of the row space of that matrix. Thus, we have established the following stronger version of Theorem 3.5.6.

Theorem 7.3.5 *If A is an $m \times n$ matrix, then the row space of A and the null space of A are orthogonal complements.*

Moreover, if we apply this theorem to A^T, and use the fact that the row space of A^T is the column space of A, we obtain the following companion theorem.

Theorem 7.3.6 *If A is an $m \times n$ matrix, then the column space of A and the null space of A^T are orthogonal complements.*

The results in these two theorems are captured by the formulas

$$\text{row}(A)^{\perp} = \text{null}(A), \qquad \text{null}(A)^{\perp} = \text{row}(A)$$
$$\text{col}(A)^{\perp} = \text{null}(A^T), \qquad \text{null}(A^T)^{\perp} = \text{col}(A) \tag{2}$$

The following theorem provides an important computational tool for studying relationships between the fundamental spaces of a matrix. The first two statements in the theorem go to the heart of Gauss–Jordan elimination and Gaussian elimination and were simply accepted to be true when we developed those methods.

Theorem 7.3.7

(a) *Elementary row operations do not change the row space of a matrix.*

(b) *Elementary row operations do not change the null space of a matrix.*

(c) *The nonzero row vectors in any row echelon form of a matrix form a basis for the row space of the matrix.*

We will prove parts (a) and (b) and give an informal argument for part (c).

Proof (a) Observe first that when you multiply a row of a matrix A by a nonzero scalar or when you add a scalar multiple of one row to another, you are computing a linear combination of row vectors of A. Thus, if B is obtained from A by a succession of elementary row operations, then every vector in $\text{row}(B)$ must be in $\text{row}(A)$. However, if B is obtained from A by elementary row operations, then A can be obtained from B by performing the inverse operations in reverse order. Thus, every vector in $\text{row}(A)$ must also be in $\text{row}(B)$, from which we conclude that $\text{row}(A) = \text{row}(B)$.

Proof (b) By part (a), performing an elementary row operation on a matrix does not change the row space of the matrix and hence does not change the orthogonal complement of the row space. But the orthogonal complement of the row space of A is the null space of A (Theorem 7.3.5), so performing an elementary row operation on a matrix A does not change the null space of A.

Proof (c) The nonzero row vectors in a row echelon form of a matrix A form a basis for the row space of A because they span the row space by part (a) of this theorem, and they are linearly independent by Example 4 of Section 7.1. ■

CONCEPT PROBLEM Do you think that elementary row operations change the column space of a matrix? Justify your answer.

The following useful theorem deals with the relationships between the fundamental spaces of two matrices.

Theorem 7.3.8 *If A and B are matrices with the same number of columns, then the following statements are equivalent.*

(a) *A and B have the same row space.*

(b) *A and B have the same null space.*

(c) *The row vectors of A are linear combinations of the row vectors of B, and conversely.*

We will prove the equivalence of parts (a) and (b) and leave the proof of equivalence (a) ⇔ (c) as an exercise. The equivalence (b) ⇔ (c) will then follow as a logical consequence.

Proof (a) ⇔ (b) The row space and null space of a matrix are orthogonal complements of one another. Thus, if A and B have the same row space, then they must have the same null space; and conversely. ■

FINDING BASES BY ROW REDUCTION

We now turn to the problem of finding a basis for the subspace W of R^n that is spanned by a given set of vectors

$$S = \{\mathbf{v}_1, \mathbf{v}_2, \ldots, \mathbf{v}_s\}$$

There are two variations of this problem, each requiring different methods:

1. If *any* basis for W will suffice for the problem at hand, then we can start by forming a matrix A that has $\mathbf{v}_1, \mathbf{v}_2, \ldots, \mathbf{v}_s$ as row vectors. This makes W the row space of A, so a basis can be found by reducing A to row echelon form and extracting the nonzero rows.

2. If the basis must consist of vectors from the *original set S*, then the preceding method is not appropriate because elementary row operations usually alter row vectors. A method for solving this kind of basis problem will be discussed later.

EXAMPLE 4
Finding a Basis by Row Reduction

(a) Find a basis for the subspace W of R^5 that is spanned by the vectors

$$\mathbf{v}_1 = (1, 0, 0, 0, 2), \qquad \mathbf{v}_2 = (-2, 1, -3, -2, -4)$$
$$\mathbf{v}_3 = (0, 5, -14, -9, 0), \quad \mathbf{v}_4 = (2, 10, -28, -18, 4)$$

(b) Find a basis for $W^{\perp}$.

Solution (a) The subspace spanned by the given vectors is the row space of the matrix

$$A = \begin{bmatrix} 1 & 0 & 0 & 0 & 2 \\ -2 & 1 & -3 & -2 & -4 \\ 0 & 5 & -14 & -9 & 0 \\ 2 & 10 & -28 & -18 & 4 \end{bmatrix} \tag{3}$$

Reducing this matrix to row echelon form yields

$$U = \begin{bmatrix} 1 & 0 & 0 & 0 & 2 \\ 0 & 1 & -3 & -2 & 0 \\ 0 & 0 & 1 & 1 & 0 \\ 0 & 0 & 0 & 0 & 0 \end{bmatrix} \tag{4}$$

Extracting the nonzero rows yields the basis vectors

$$\mathbf{w}_1 = (1, 0, 0, 0, 2), \quad \mathbf{w}_2 = (0, 1, -3, -2, 0), \quad \mathbf{w}_3 = (0, 0, 1, 1, 0)$$

which we have expressed in comma-delimited notation for consistency with the form of the original vectors. Since there are three basis vectors, we have shown that $\dim(W) = 3$. Alternatively, we can take the matrix A all the way to the reduced row echelon form

$$R = \begin{bmatrix} 1 & 0 & 0 & 0 & 2 \\ 0 & 1 & 0 & 1 & 0 \\ 0 & 0 & 1 & 1 & 0 \\ 0 & 0 & 0 & 0 & 0 \end{bmatrix} \tag{5}$$

which yields the basis vectors

$$\mathbf{w}_1' = (1, 0, 0, 0, 2), \quad \mathbf{w}_2' = (0, 1, 0, 1, 0), \quad \mathbf{w}_3' = (0, 0, 1, 1, 0) \tag{6}$$

Although it is extra work to obtain the reduced row echelon form, the resulting basis vectors are usually simpler in that they have more zeros. Whether it justifies the extra work depends on the purpose for which the basis will be used.

Solution (b) It follows from Theorem 7.3.5 that $W^{\perp}$ is the null space of A, so our problem reduces to finding a basis for the solution space of the homogeneous system $A\mathbf{x} = \mathbf{0}$. We will use the canonical basis produced by Gauss–Jordan elimination. Most of the work has already been done, since R in (5) is the reduced row echelon form of A. We leave it for you to use R to obtain the general solution

$$\begin{bmatrix} x_1 \\ x_2 \\ x_3 \\ x_4 \\ x_5 \end{bmatrix} = s \begin{bmatrix} -2 \\ 0 \\ 0 \\ 0 \\ 1 \end{bmatrix} + t \begin{bmatrix} 0 \\ -1 \\ -1 \\ 1 \\ 0 \end{bmatrix} \tag{7}$$

Thus, the vectors

$$\mathbf{u}_1 = (-2, 0, 0, 0, 1) \quad \text{and} \quad \mathbf{u}_2 = (0, -1, -1, 1, 0)$$

form a basis for $W^{\perp}$. As a check, we leave it for you to confirm that $\mathbf{u}_1$ and $\mathbf{u}_2$ are orthogonal to all of the basis vectors for W that were obtained in part (a). ∎

EXAMPLE 5
Finding a
Linear System
with a Specified
Solution Space

Find a homogeneous linear system $B\mathbf{x} = \mathbf{0}$ whose solution space is the space W spanned by the vectors $\mathbf{v}_1$, $\mathbf{v}_2$, $\mathbf{v}_3$, and $\mathbf{v}_4$ in Example 4.

Solution In part (b) of Example 4 we found basis vectors $\mathbf{u}_1$ and $\mathbf{u}_2$ for $W^{\perp}$. Use these as row vectors to form the matrix

$$B = \begin{bmatrix} -2 & 0 & 0 & 0 & 1 \\ 0 & -1 & -1 & 1 & 0 \end{bmatrix}$$

The row space of B is $W^{\perp}$, so the null space of B is $(W^{\perp})^{\perp} = W$. Thus, the linear system $B\mathbf{x} = \mathbf{0}$, or equivalently,

$$\begin{aligned} -2x_1 \qquad\qquad\quad + x_5 &= 0 \\ -x_2 - x_3 + x_4 \quad\;\; &= 0 \end{aligned}$$

has W as its solution space. ∎

**DETERMINING WHETHER
A VECTOR IS IN A GIVEN
SUBSPACE**

Consider the following three problems:

Problem 1. Given a set of vectors $S = \{\mathbf{v}_1, \mathbf{v}_2, \ldots, \mathbf{v}_n\}$ in R^m, find conditions on the numbers $b_1, b_2, \ldots, b_m$ under which $\mathbf{b} = (b_1, b_2, \ldots, b_m)$ will lie in span(S).

Problem 2. Given an $m \times n$ matrix A, find conditions on the numbers $b_1, b_2, \ldots, b_m$ under which $\mathbf{b} = (b_1, b_2, \ldots, b_m)$ will lie in col(A).

Problem 3. Given a linear transformation $T : R^n \to R^m$, find conditions on the numbers $b_1, b_2, \ldots, b_m$ under which $\mathbf{b} = (b_1, b_2, \ldots, b_m)$ will lie in ran(T).

Although these problems look different at the surface, they are just different formulations of the same problem (why?). The following example illustrates three ways of attacking the first formulation of the problem.

EXAMPLE 6
Conditions for a
Vector to Lie in
a Given
Subspace

What conditions must a vector $\mathbf{b} = (b_1, b_2, b_3, b_4, b_5)$ satisfy to lie in the subspace of R^5 spanned by the vectors $\mathbf{v}_1$, $\mathbf{v}_2$, $\mathbf{v}_3$, and $\mathbf{v}_4$ in Example 4?

Solution 1 The most direct way to solve this problem is to look for conditions under which the vector equation

$$x_1\mathbf{v}_1 + x_2\mathbf{v}_2 + x_3\mathbf{v}_3 + x_4\mathbf{v}_4 = \mathbf{b} \tag{8}$$

has a solution for x_1, x_2, x_3, and x_4. This is the vector form of the linear system $C\mathbf{x} = \mathbf{b}$ in which $\mathbf{v}_1$, $\mathbf{v}_2$, $\mathbf{v}_3$, and $\mathbf{v}_4$ are the successive column vectors of C. The augmented matrix for this system is

$$\begin{bmatrix} 1 & -2 & 0 & 2 & \vdots & b_1 \\ 0 & 1 & 5 & 10 & \vdots & b_2 \\ 0 & -3 & -14 & -28 & \vdots & b_3 \\ 0 & -2 & -9 & -18 & \vdots & b_4 \\ 2 & -4 & 0 & 4 & \vdots & b_5 \end{bmatrix} \tag{9}$$

and our goal is to determine conditions on the b's under which this system is consistent. This is what we referred to as a "consistency problem" (see 3.3.10). As in Example 8 of that section, we reduce the left side of (9) to row echelon form, which yields (verify)

$$\begin{bmatrix} 1 & -2 & 0 & 2 & \vdots & b_1 \\ 0 & 1 & 5 & 10 & \vdots & b_2 \\ 0 & 0 & 1 & 2 & \vdots & b_3 + 3b_2 \\ 0 & 0 & 0 & 0 & \vdots & b_4 - b_3 - b_2 \\ 0 & 0 & 0 & 0 & \vdots & b_5 - 2b_1 \end{bmatrix}$$

Thus, for the system to be consistent the components of $\mathbf{b}$ must satisfy the two conditions

$$\begin{array}{ll} b_4 - b_3 - b_2 = 0 & \quad b_4 = b_2 + b_3 \\ b_5 - 2b_1 \quad = 0 & \text{or} \quad b_5 = \quad 2b_1 \end{array}$$

For example, the vector $(7, -2, 5, 3, 14)$ lies in W, but $(7, -2, 5, 3, 6)$ and $(0, -1, 3, -2, 0)$ do not (verify).

Solution 2 Here is a way to attack the same problem by focusing on rows rather than columns. It follows from Theorem 7.2.5 that $\mathbf{b}$ will lie in span$\{\mathbf{v}_1, \mathbf{v}_2, \mathbf{v}_3, \mathbf{v}_4\}$ if and only if this space has the same dimension as span$\{\mathbf{v}_1, \mathbf{v}_2, \mathbf{v}_3, \mathbf{v}_4, \mathbf{b}\}$, that is, if and only if the matrix A with row vectors $\mathbf{v}_1, \mathbf{v}_2, \mathbf{v}_3$, and $\mathbf{v}_4$ has the same rank as the matrix that results when $\mathbf{b}$ is adjoined to A as an additional row vector. The matrix A is given in (3), so that adjoining $\mathbf{b}$ as an additional row vector yields

$$\begin{bmatrix} 1 & 0 & 0 & 0 & 2 \\ -2 & 1 & -3 & -2 & -4 \\ 0 & 5 & -14 & -9 & 0 \\ 2 & 10 & -28 & -18 & 4 \\ \hdashline b_1 & b_2 & b_3 & b_4 & b_5 \end{bmatrix} \qquad (10)$$

To determine conditions under which (3) and (10) have the same rank, we will begin by reducing the A portion of (10) to reduced row echelon form to reveal a basis for the row space of A (a row echelon form will also work). This yields

$$\begin{bmatrix} 1 & 0 & 0 & 0 & 2 \\ 0 & 1 & 0 & 1 & 0 \\ 0 & 0 & 1 & 1 & 0 \\ 0 & 0 & 0 & 0 & 0 \\ \hdashline b_1 & b_2 & b_3 & b_4 & b_5 \end{bmatrix} \qquad (11)$$

For this matrix to have the same rank as A (rank 3) it would have to be possible to make the last row zero using elementary row operations. Accordingly, we will now start "zeroing out" those entries in the last row that lie in the pivot columns of A by adding suitable multiples of the pivot rows of A to the bottom row. We leave it for you to verify that this yields

$$\begin{bmatrix} 1 & 0 & 0 & 0 & 2 \\ 0 & 1 & 0 & 1 & 0 \\ 0 & 0 & 1 & 1 & 0 \\ 0 & 0 & 0 & 0 & 0 \\ \hdashline 0 & 0 & 0 & b_4 - b_3 - b_2 & b_5 - 2b_1 \end{bmatrix} \qquad (12)$$

Thus, for (11) to have rank 3 we must have $b_4 - b_3 - b_2 = 0$ and $b_5 - 2b_1 = 0$. These are the same conditions that we obtained by the first method.

Solution 3 Here is a third way to attack the same problem. To say that $\mathbf{b} = (b_1, b_2, b_3, b_4, b_5)$ lies in the subspace W spanned by the vectors $\mathbf{v}_1, \mathbf{v}_2, \mathbf{v}_3$, and $\mathbf{v}_4$ is the same as saying that $\mathbf{b}$ is orthogonal to every vector in $W^\perp$. But we showed in part (b) of Example 4 that the vectors

$$\mathbf{u}_1 = (-2, 0, 0, 0, 1) \quad \text{and} \quad \mathbf{u}_2 = (0, -1, -1, 1, 0)$$

form a basis for $W^\perp$. Thus, $\mathbf{b}$ will be orthogonal to every vector in $W^\perp$ if and only if it is orthogonal to $\mathbf{u}_1$ and $\mathbf{u}_2$. If we write the orthogonality conditions $\mathbf{u}_1 \cdot \mathbf{b} = 0$ and $\mathbf{u}_2 \cdot \mathbf{b} = 0$ in component form, we obtain

$$-2b_1 + b_5 = 0 \quad \text{and} \quad -b_2 - b_3 + b_4 = 0$$

which are the same conditions we obtained by the first two methods.   ■

EXAMPLE 7
A Useful
Algorithm

Determine which of the three vectors $\mathbf{b}_1 = (7, -2, 5, 3, 14)$, $\mathbf{b}_2 = (7, -2, 5, 3, 6)$, and $\mathbf{b}_3 = (0, -1, 3, -2, 0)$, if any, lie in the subspace of R^5 spanned by the vectors $\mathbf{v}_1$, $\mathbf{v}_2$, $\mathbf{v}_3$, and $\mathbf{v}_4$ in Example 4.

Solution One way to solve this problem is to consider the matrix C that has $\mathbf{v}_1$, $\mathbf{v}_2$, $\mathbf{v}_3$, and $\mathbf{v}_4$ as its successive column vectors, and determine which of the $\mathbf{b}$'s, if any, lie in the column space of C by investigating whether the systems $C\mathbf{x} = \mathbf{b}_1$, $C\mathbf{x} = \mathbf{b}_2$, and $C\mathbf{x} = \mathbf{b}_3$ are consistent. An efficient way of doing this was presented in Example 7 of Section 3.3. As in that example, we adjoin the column vectors $\mathbf{b}_1$, $\mathbf{b}_2$, or $\mathbf{b}_3$ to C and consider the partitioned matrix

$$\begin{bmatrix} 1 & -2 & 0 & 2 & | & 7 & | & 7 & | & 0 \\ 0 & 1 & 5 & 10 & | & -2 & | & -2 & | & -1 \\ 0 & -3 & -14 & -28 & | & 5 & | & 5 & | & 3 \\ 0 & -2 & -9 & -18 & | & 3 & | & 3 & | & -2 \\ 2 & -4 & 0 & 4 & | & 14 & | & 6 & | & 0 \end{bmatrix}$$

If we now apply row operations to this until the submatrix C is in row echelon form, we obtain (verify)

$$\begin{bmatrix} 1 & -2 & 0 & 2 & | & 7 & | & 7 & | & 0 \\ 0 & 1 & 5 & 10 & | & -2 & | & -2 & | & -1 \\ 0 & 0 & 1 & 2 & | & -1 & | & -1 & | & 0 \\ 0 & 0 & 0 & 0 & | & 0 & | & 0 & | & -4 \\ 0 & 0 & 0 & 0 & | & 0 & | & -8 & | & 0 \end{bmatrix}$$

At this point we can see that the system $C\mathbf{x} = \mathbf{b}_1$ is consistent but the systems $C\mathbf{x} = \mathbf{b}_2$ and $C\mathbf{x} = \mathbf{b}_3$ are not. Thus, vector $\mathbf{b}_1 = (7, -2, 5, 3, 14)$ lies in span$\{\mathbf{v}_1, \mathbf{v}_2, \mathbf{v}_3, \mathbf{v}_4\}$ but vectors $\mathbf{b}_2 = (7, -2, 5, 3, 6)$ and $\mathbf{b}_3 = (0, -1, 3, -2, 0)$ do not. This is consistent with the observation made at the end of the first solution in Example 6.   ■

Exercise Set 7.3

In Exercises 1 and 2, use the two methods of Example 3 to find the orthogonal complement of the set $S = \{\mathbf{v}_1, \mathbf{v}_2\}$ in an xyz-coordinate system. Confirm that the answers are consistent with one another.

1. $\mathbf{v}_1 = (1, 1, 3)$, $\mathbf{v}_2 = (0, 2, -1)$

2. $\mathbf{v}_1 = (2, 0, -1)$, $\mathbf{v}_2 = (1, 1, 5)$

In Exercises 3 and 4, show that $\mathbf{u}$ is in the orthogonal complement of $W = \text{span}(\mathbf{v}_1, \mathbf{v}_2, \mathbf{v}_3)$.

3. $\mathbf{u} = (-1, 1, 0, 2)$; $\mathbf{v}_1 = (6, 2, 7, 2)$, $\mathbf{v}_2 = (1, 1, 3, 0)$, $\mathbf{v}_3 = (4, 0, 9, 2)$

4. $\mathbf{u} = (0, 2, 1, -2)$; $\mathbf{v}_1 = (4, 1, 2, 2)$, $\mathbf{v}_2 = (3, 5, 0, 5)$, $\mathbf{v}_3 = (1, 2, 2, 3)$

5. Let W be the line in R^2 with the equation $y = 2x$. Find an equation for $W^\perp$.

6. Let W be the plane in R^3 with the equation $x - 2y - 3z = 0$. Find parametric equations for $W^\perp$.

7. Let W be the line in R^3 with parametric equations $x = 2t$, $y = -5t$, $z = 4t$. Find parametric equations for $W^\perp$.

8. Let W be the intersection of the planes $x + y + z = 0$ and $x - y + z = 0$ in R^3. Find parametric equations for $W^\perp$.

In Exercises 9–14, let W be the space spanned by the given vectors. Find bases for W and $W^\perp$.

9. $\mathbf{v}_1 = (1, -1, 3)$, $\mathbf{v}_2 = (5, -4, -4)$, $\mathbf{v}_3 = (7, -6, 2)$

10. $\mathbf{v}_1 = (2, 0, -1)$, $\mathbf{v}_2 = (4, 1, -2)$, $\mathbf{v}_3 = (8, 1, -4)$

11. $\mathbf{v}_1 = (1, 1, 0, 0)$, $\mathbf{v}_2 = (0, 0, 1, 1)$,
$\mathbf{v}_3 = (-2, 0, 2, 2)$, $\mathbf{v}_4 = (4, 2, -1, -1)$

12. $\mathbf{v}_1 = (2, 4, -5, 6)$, $\mathbf{v}_2 = (1, 2, -2, 3)$,
$\mathbf{v}_3 = (3, 6, -3, 9)$, $\mathbf{v}_4 = (5, 4, -1, 6)$

13. $\mathbf{v}_1 = (1, 0, 0, 2, 5)$, $\mathbf{v}_2 = (0, 1, 0, 3, 4)$,
$\mathbf{v}_3 = (0, 0, 1, 4, 7)$, $\mathbf{v}_4 = (2, -3, 4, 11, 12)$

14. $\mathbf{v}_1 = (1, 4, -2, 3, 5)$, $\mathbf{v}_2 = (0, 1, 6, -7, 1)$,
$\mathbf{v}_3 = (0, 0, 1, 2, -3)$, $\mathbf{v}_4 = (0, 0, 0, 1, 2)$,
$\mathbf{v}_5 = (1, 5, 5, -1, 5)$

In Exercises 15–18, find a linear system whose solution space is the span of the given vectors.

15. The vectors in Exercise 9.

16. The vectors in Exercise 10.

17. The vectors in Exercise 11.

18. The vectors in Exercise 12.

In Exercises 19–22, use the three methods in Example 6 to determine conditions that a vector $\mathbf{b}$ must satisfy to lie in the space spanned by the given vectors.

19. The vectors in Exercise 9.

20. The vectors in Exercise 10.

21. The vectors in Exercise 11.

22. The vectors in Exercise 12.

In Exercises 23 and 24, determine which of the $\mathbf{b}$'s, if any, lie in the space spanned by the $\mathbf{v}$'s.

23. $\mathbf{v}_1 = (1, 1, 0, -1, 2)$, $\mathbf{v}_2 = (-2, 0, 1, 1, 3)$,
$\mathbf{v}_3 = (-1, 1, 2, 1, -1)$, $\mathbf{v}_4 = (0, 2, -1, 1, 1)$;
$\mathbf{b}_1 = (-2, 4, 2, 2, 5)$, $\mathbf{b}_2 = (0, -2, -3, -1, 5)$,
$\mathbf{b}_3 = (-2, 2, -1, 1, 0)$

24. $\mathbf{v}_1 = (0, 1, 0, 2, 0)$, $\mathbf{v}_2 = (1, 1, 3, 1, -1)$,
$\mathbf{v}_3 = (-1, 0, 2, 1, 1)$, $\mathbf{v}_4 = (3, -2, 1, 0, 1)$;
$\mathbf{b}_1 = (3, -1, 7, 2, 1)$, $\mathbf{b}_2 = (-2, 0, -1, 2, 2)$,
$\mathbf{b}_3 = (3, 2, 6, 4, 1)$

In Exercises 25 and 26, confirm that null(A) and row(A) are orthogonal complements, as guaranteed by Theorem 7.3.5.

25. $A = \begin{bmatrix} 1 & 3 & -2 & 0 & 2 & 0 \\ 2 & 6 & -5 & -2 & 4 & -3 \\ 0 & 0 & 5 & 10 & 10 & 15 \\ 2 & 6 & 0 & 8 & 4 & 18 \end{bmatrix}$

26. $A = \begin{bmatrix} 2 & 7 & 4 & 5 & 8 \\ 4 & 4 & 8 & 5 & 4 \\ 1 & -9 & -3 & -5 & -14 \\ 3 & 5 & 7 & 5 & 6 \end{bmatrix}$

In Exercises 27 and 28, find a basis for the row space of the matrix.

27. The matrix in Exercise 25.

28. The matrix in Exercise 26.

In Exercises 29 and 30, show that the matrices A and B have the same row space.

29. $A = \begin{bmatrix} 1 & 3 & 3 & 2 \\ 2 & 0 & 6 & 1 \\ -2 & 4 & 2 & 4 \end{bmatrix}$, $B = \begin{bmatrix} 2 & -6 & -6 & -4 \\ 0 & 6 & 12 & 5 \\ 0 & -1 & -2 & 0 \\ 2 & -7 & -8 & -4 \end{bmatrix}$

30. $A = \begin{bmatrix} 1 & -2 & 3 & 8 \\ 2 & -3 & 2 & 7 \\ -1 & 1 & 2 & 4 \\ -1 & 1 & 1 & 1 \end{bmatrix}$, $B = \begin{bmatrix} 3 & -6 & 0 & -3 \\ 0 & 1 & 0 & 3 \\ 0 & 0 & 1 & 3 \end{bmatrix}$

31. Construct a matrix whose null space consists of all linear combinations of the vectors

$$\mathbf{v}_1 = \begin{bmatrix} 1 \\ -1 \\ 3 \\ 2 \end{bmatrix} \quad \text{and} \quad \mathbf{v}_2 = \begin{bmatrix} 2 \\ 0 \\ -2 \\ 4 \end{bmatrix}$$

32. (a) Show that in an xyz-coordinate system the null space of the matrix

$$A = \begin{bmatrix} 0 & 1 & 0 \\ 1 & 0 & 0 \\ 0 & 0 & 0 \end{bmatrix}$$

consists of all points on the z-axis and the column space consists of all points in the xy-plane.

(b) Find a 3×3 matrix whose null space is the x-axis and whose column space is the yz-plane.

Discussion and Discovery

D1. Indicate whether the statement is true (T) or false (F). Justify your answer.

(a) The nonzero row vectors of a matrix A form a basis for the row space of A.

(b) If E is an elementary matrix, then A and EA have the same null space.

(c) If A is an $n \times n$ matrix with rank n, then A is invertible.

(d) If A is a nonzero $m \times n$ matrix, then null(A) $\cap$ row(A) contains at least one nonzero vector.

(e) If S is a nonempty subset of R^n, then every linear combination of vectors in $S^{\perp}$ lies in $S^{\perp}$.

D2. Indicate whether the statement is true (T) or false (F). Justify your answer.

    (a) The row space and column space of an invertible matrix are the same.

    (b) If V is a subspace of R^n and W is a subspace of V, then $W^\perp$ is a subspace of $V^\perp$.

    (c) If each row of a matrix A is a linear combination of the rows of a matrix B, then A and B have the same null space.

    (d) If A and B are $n \times n$ matrices with the same row space, then A and B have the same column space.

    (e) If E is an elementary matrix, then A and EA have the same row space.

D3. Find all 2×2 matrices whose null space is the line $3x - 5y = 0$.

D4. If $T_A : R^n \to R^n$ is multiplication by A, then the null space of A is the same as the _____ of T_A, and the column space of A is the same as the _____ of T_A.

D5. If W is a hyperplane in R^n, what can you say about $W^\perp$?

D6. Let $A\mathbf{x} = \mathbf{0}$ be a homogeneous system of three equations in the unknowns x, y, and z.

    (a) If the solution space is a line through the origin in R^3, what kind of geometric object is the row space of A? Explain your reasoning.

    (b) If the column space of A is a line through the origin, what kind of geometric object is the solution space of the homogeneous system $A^T\mathbf{x} = \mathbf{0}$? Explain your reasoning.

D7. Sketch the null spaces of the following matrices:

$$A = \begin{bmatrix} 1 & 4 \\ 0 & 5 \end{bmatrix}, \quad A = \begin{bmatrix} 1 & 0 \\ 0 & 5 \end{bmatrix}$$

$$A = \begin{bmatrix} 6 & 2 \\ 3 & 1 \end{bmatrix}, \quad A = \begin{bmatrix} 0 & 0 \\ 0 & 0 \end{bmatrix}$$

D8. (a) Let S be the set of all vectors on the line $y = 3x$ in the xy-plane. Find equations for $S^\perp$ and $(S^\perp)^\perp$.

    (b) Let $\mathbf{v}$ be the vector $(1, 2)$ in the xy-plane, and let $S = \{\mathbf{v}\}$. Find equations for $S^\perp$ and $(S^\perp)^\perp$.

D9. Do you think that it is possible to find an invertible $n \times n$ matrix and a singular $n \times n$ matrix that have the same row space? Explain your reasoning.

Working with Proofs

P1. Prove the equivalence of parts (a) and (c) of Theorem 7.3.8. [*Hint:* Show that each row space contains the other.]

P2. Prove that the row vectors of an invertible $n \times n$ matrix form a basis for R^n.

P3. Prove: If P is an invertible $n \times n$ matrix, and A is any $n \times k$ matrix, then

$$\text{rank}(PA) = \text{rank}(A) \quad \text{and} \quad \text{nullity}(PA) = \text{nullity}(A)$$

[*Hint:* Use parts (a) and (b) of Theorem 7.3.8.]

P4. Prove: If S is a nonempty subset of R^n, then

$$(S^\perp)^\perp = \text{span}(S)$$

[*Hint:* Use Theorem 7.3.4.]

P5. Prove: If A is an invertible $n \times n$ matrix and k is an integer satisfying $1 \le k < n$, then the first k rows of A and the last $n - k$ columns of A^{-1} are orthogonal. [*Hint:* Partition A and A^{-1} appropriately.]

Technology Exercises

T1. Many technology utilities provide a command for finding a basis for the null space of a matrix.

    (a) Determine whether your utility has this capability; if so, use that command to find a basis for the null space of the matrix

$$A = \begin{bmatrix} 3 & 2 & 1 & 3 & 5 \\ 6 & 4 & 3 & 5 & 7 \\ 9 & 6 & 5 & 7 & 9 \\ 3 & 2 & 0 & 4 & 8 \end{bmatrix}$$

    (b) Confirm that the basis obtained in part (a) is consistent with the basis that results when your utility is used to find the general solution of the linear system $A\mathbf{x} = \mathbf{0}$.

T2. Some technology utilities provide a command for finding a basis for the row space of a matrix.

    (a) Determine whether your utility has this capability; if so, use that command to find a basis for the row space of the matrix

$$A = \begin{bmatrix} 2 & -1 & 3 & 5 \\ 4 & -3 & 1 & 3 \\ 3 & -2 & 3 & 4 \\ 4 & -1 & 15 & 17 \\ 7 & -6 & -7 & 0 \end{bmatrix}$$

    (b) Confirm that the basis obtained in part (a) is consistent with the basis that results when your utility is used to find the reduced row echelon form of A.

T3. Some technology utilities provide a command for finding a basis for the column space of a matrix.

    (a) Determine whether your utility has this capability; if so, use that command to find a basis for the column space of the matrix in Exercise T2.

(b) Confirm that the basis obtained is consistent with the basis that results when your utility is used to find a basis for the row space of A^T.

T4. Determine which of the vectors $\mathbf{b}_1 = (2, 6, -17, -11, 4)$, $\mathbf{b}_2 = (1, 16, -45, -29, 2)$, and $\mathbf{b}_3 = (7, 14, 2, 1, 14)$, if any, lie in the subspace of R^5 spanned by the vectors in Example 4.

Section 7.4   The Dimension Theorem and Its Implications

In this section we will derive a relationship between the rank and nullity of a matrix, and we will use that relationship to explore the geometric structure of R^n.

THE DIMENSION THEOREM FOR MATRICES

Recall from Theorem 2.2.2 that if $A\mathbf{x} = \mathbf{0}$ is a homogeneous linear system with n unknowns, and if the reduced row echelon form of the augmented matrix has r nonzero rows, then the system has $n - r$ free variables; we called this the *dimension theorem for homogeneous linear systems*. However, for a homogeneous system, the augmented matrix and the coefficient matrix have the same number of zero rows in their reduced row echelon forms (the number being the rank of A), so we can restate the dimension theorem for homogeneous linear systems as

$$\text{number of free variables} = n - \text{rank}(A)$$

or, alternatively, as

$$\text{rank}(A) + \text{number of free variables} = \text{number of columns of } A \tag{1}$$

But each free variable produces a parameter in a general solution of the system $A\mathbf{x} = \mathbf{0}$, so the number of free variables is the same as the dimension of the solution space of the system (which is the nullity of A). Thus, we can rewrite (1) as

$$\text{rank}(A) + \text{nullity}(A) = \text{number of columns of } A$$

and hence we have established the following matrix version of the dimension theorem.

> **Theorem 7.4.1** (*The Dimension Theorem for Matrices*)   *If A is an $m \times n$ matrix, then*
>
> $$\text{rank}(A) + \text{nullity}(A) = n \tag{2}$$

EXAMPLE 1
The Dimension Theorem for Matrices

In Example 4 of the last section we saw that the reduced row echelon form of the matrix

$$A = \begin{bmatrix} 1 & 0 & 0 & 0 & 2 \\ -2 & 1 & -3 & -2 & -4 \\ 0 & 5 & -14 & -9 & 0 \\ 2 & 10 & -28 & -18 & 4 \end{bmatrix}$$

has three nonzero rows [Formula (5) of Section 7.3], and we saw that the null space of A has dimension 2 [Formula (7) of Section 7.3]. Thus,

$$\text{rank}(A) + \text{nullity}(A) = 3 + 2 = 5$$

which is consistent with Formula (2) and the fact that A has five columns.   ■

EXTENDING A LINEARLY INDEPENDENT SET TO A BASIS

It follows from part (*b*) of Theorem 7.2.2 that every linearly independent set $\{\mathbf{v}_1, \mathbf{v}_2, \ldots, \mathbf{v}_k\}$ in R^n can be enlarged to a basis for R^n by adding appropriate linearly independent vectors to it. One way to find such vectors is to form the matrix A that has $\mathbf{v}_1, \mathbf{v}_2, \ldots, \mathbf{v}_k$ as row vectors, thereby making the subspace spanned by these vectors into the row space of A. By solving the homogeneous linear system $A\mathbf{x} = \mathbf{0}$, we can find a basis for the null space of A. This basis has $n - k$ vectors, say $\mathbf{w}_{k+1}, \ldots, \mathbf{w}_n$, by the dimension theorem for matrices, and each of the **w**'s is orthogonal to all of the **v**'s, since null(A) and row(A) are orthogonal. This orthogonality implies that the set $\{\mathbf{v}_1, \mathbf{v}_2, \ldots, \mathbf{v}_k, \mathbf{w}_{k+1}, \ldots, \mathbf{w}_n\}$ is linearly independent (Exercise P4) and hence forms a basis for R^n.

EXAMPLE 2
Extending a Linearly Independent Set to a Basis

The vectors

$$\mathbf{v}_1 = (1, 3, -1, 1) \quad \text{and} \quad \mathbf{v}_2 = (0, 1, 1, 6)$$

are linearly independent, since neither vector is a scalar multiple of the other. Enlarge the set $\{\mathbf{v}_1, \mathbf{v}_2\}$ to a basis for R^4.

Solution We will find a basis for the null space of the matrix

$$A = \begin{bmatrix} 1 & 3 & -1 & 1 \\ 0 & 1 & 1 & 6 \end{bmatrix}$$

by solving the linear system $A\mathbf{x} = \mathbf{0}$. The reduced row echelon form of the augmented matrix for the system is

$$\begin{bmatrix} 1 & 0 & -4 & -17 & \vdots & 0 \\ 0 & 1 & 1 & 6 & \vdots & 0 \end{bmatrix}$$

so a general solution is

$$\begin{bmatrix} x_1 \\ x_2 \\ x_3 \\ x_4 \end{bmatrix} = \begin{bmatrix} 4s + 17t \\ -s - 6t \\ s \\ t \end{bmatrix} = s \begin{bmatrix} 4 \\ -1 \\ 1 \\ 0 \end{bmatrix} + t \begin{bmatrix} 17 \\ -6 \\ 0 \\ 1 \end{bmatrix}$$

Thus, the vectors

$$\mathbf{v}_1 = (1, 3, -1, 1), \quad \mathbf{v}_2 = (0, 1, 1, 6), \quad \mathbf{w}_3 = (4, -1, 1, 0), \quad \mathbf{w}_4 = (17, -6, 0, 1)$$

form a basis for R^4. ∎

SOME CONSEQUENCES OF THE DIMENSION THEOREM FOR MATRICES

The following theorem lists some properties of an $m \times n$ matrix that you should be able to deduce from the dimension theorem for matrices and other results we have discussed.

> **Theorem 7.4.2** *If an $m \times n$ matrix A has rank k, then:*
> (a) *A has nullity $n - k$.*
> (b) *Every row echelon form of A has k nonzero rows.*
> (c) *Every row echelon form of A has $m - k$ zero rows.*
> (d) *The homogeneous system $A\mathbf{x} = \mathbf{0}$ has k pivot variables (leading variables) and $n - k$ free variables.*

EXAMPLE 3
Consequences of the Dimension Theorem for Matrices

State some facts about a 5×7 matrix A with nullity 3.

Solution Here are some possibilities:

- rank(A) $= 7 - 3 = 4$.
- Every row echelon form of A has $5 - 4 = 1$ zero row.
- The homogeneous system $A\mathbf{x} = \mathbf{0}$ has 4 pivot variables and $7 - 4 = 3$ free variables. ∎

EXAMPLE 4
Restrictions Imposed by the Dimension Theorem for Matrices

Can a 5×7 matrix A have a one-dimensional null space?

Solution No. Otherwise, the rank of A would be

$$\text{rank}(A) = 7 - \text{nullity}(A) = 7 - 1 = 6$$

which is impossible, since the five row vectors of A cannot span a six-dimensional space. ∎

THE DIMENSION THEOREM FOR SUBSPACES

The dimension theorem for matrices (Theorem 7.4.1) can be recast as the following theorem about subspaces of R^n.

Theorem 7.4.3 (*The Dimension Theorem for Subspaces*) *If W is a subspace of R^n, then*

$$\dim(W) + \dim(W^\perp) = n \tag{3}$$

Proof If $W = \{\mathbf{0}\}$, then $W^\perp = R^n$, in which case $\dim(W) + \dim(W^\perp) = 0 + n = n$. If $W \neq \{\mathbf{0}\}$, then choose a basis for W and form a matrix A that has these basis vectors as row vectors. The matrix A has n columns since its row vectors come from R^n. Moreover, the row space of A is W and the null space of A is $W^\perp$, so it follows from Theorem 7.4.1 that

$$\dim(W) + \dim(W^\perp) = \operatorname{rank}(A) + \operatorname{nullity}(A) = n \qquad \blacksquare$$

EXAMPLE 5
The Dimension Theorem for Subspaces

In Example 4 of the last section we considered a subspace W spanned by four vectors, $\mathbf{v}_1$, $\mathbf{v}_2$, $\mathbf{v}_3$, and $\mathbf{v}_4$ in R^5. In part (a) of that example we found a basis for W with three vectors, and in part (b) we found a basis for $W^\perp$ with two vectors. Thus,

$$\dim(W) + \dim(W^\perp) = 3 + 2 = 5$$

which is consistent with Formula (3) and the fact that R^5 is five-dimensional. $\qquad \blacksquare$

A UNIFYING THEOREM

Unifying Theorem 7.2.7 tied together all of the major concepts that had been discussed at that point. The dimension theorem for matrices enables us to add two additional results to Theorem 7.2.7.

Theorem 7.4.4 *If A is an $n \times n$ matrix, and if T_A is the linear operator on R^n with standard matrix A, then the following statements are equivalent.*

(a) *The reduced row echelon form of A is I_n.*

(b) *A is expressible as a product of elementary matrices.*

(c) *A is invertible.*

(d) *$A\mathbf{x} = \mathbf{0}$ has only the trivial solution.*

(e) *$A\mathbf{x} = \mathbf{b}$ is consistent for every vector $\mathbf{b}$ in R^n.*

(f) *$A\mathbf{x} = \mathbf{b}$ has exactly one solution for every vector $\mathbf{b}$ in R^n.*

(g) *$\det(A) \neq 0$.*

(h) *$\lambda = 0$ is not an eigenvalue of A.*

(i) *T_A is one-to-one.*

(j) *T_A is onto.*

(k) *The column vectors of A are linearly independent.*

(l) *The row vectors of A are linearly independent.*

(m) *The column vectors of A span R^n.*

(n) *The row vectors of A span R^n.*

(o) *The column vectors of A form a basis for R^n.*

(p) *The row vectors of A form a basis for R^n.*

(q) *$\operatorname{rank}(A) = n$.*

(r) *$\operatorname{nullity}(A) = 0$.*

Statements (q) and (r) are equivalent by the dimension theorem for matrices, and statements (r) and (d) are equivalent since $\operatorname{nullity}(A)$ is the dimension of the solution space of $A\mathbf{x} = \mathbf{0}$. Thus, statements (q) and (r) are equivalent to all others in the theorem by logical implication.

MORE ON HYPERPLANES

Recall from Theorem 7.1.6 that if $\mathbf{a}$ is a nonzero vector, then the hyperplane $\mathbf{a}^\perp$ is a subspace of dimension $n - 1$. The following theorem shows that the converse is also true.

> **Theorem 7.4.5** *If W is a subspace of R^n with dimension $n - 1$, then there is a nonzero vector $\mathbf{a}$ for which $W = \mathbf{a}^\perp$; that is, W is a hyperplane through the origin of R^n.*

Proof Let W be a subspace of R^n with dimension $n - 1$. It follows from the dimension theorem for subspaces that $\dim(W^\perp) = 1$; and this implies that $W^\perp$ is the span of some nonzero vector in R^n, say $W^\perp = \text{span}\{\mathbf{a}\}$. Thus, it follows from parts (*b*) and (*c*) of Theorem 7.3.4 that

$$W = (W^\perp)^\perp = (\text{span}\{\mathbf{a}\})^\perp = \mathbf{a}^\perp$$

which shows that W is the hyperplane through the origin of R^n with normal $\mathbf{a}$. ■

Since hyperplanes through the origin of R^n are the subspaces of dimension $n - 1$, their orthogonal complements are the subspaces of dimension 1, which are the lines through the origin of R^n. Thus, we have the following geometric result.

> **Theorem 7.4.6** *The orthogonal complement of a hyperplane through the origin of R^n is a line through the origin of R^n, and the orthogonal complement of a line through the origin of R^n is a hyperplane through the origin of R^n. Specifically, if $\mathbf{a}$ is a nonzero vector in R^n, then the line $\text{span}\{\mathbf{a}\}$ and the hyperplane $\mathbf{a}^\perp$ are orthogonal complements of one another.*

RANK 1 MATRICES

Matrices of rank 1 will play an important role in our later work, so we will conclude this section by discussing some basic results about them. Here are some facts about an $m \times n$ matrix A that follow immediately from our previous work:

- If $\text{rank}(A) = 1$, then $\text{nullity}(A) = n - 1$, so the row space of A is a line through the origin of R^n and the null space is a hyperplane through the origin of R^n. Conversely, if the row space of A is a line through the origin of R^n, or, equivalently, if the null space of A is a hyperplane through the origin of R^n; then A has rank 1.

- If $\text{rank}(A) = 1$, then the row space of A is spanned by some nonzero vector $\mathbf{a}$, so all row vectors of A are scalar multiples of $\mathbf{a}$ and the null space of A is $\mathbf{a}^\perp$. Conversely, if the row vectors of A are all scalar multiples of some nonzero vector $\mathbf{a}$, then A has rank 1 and the null space of A is the hyperplane $\mathbf{a}^\perp$.

EXAMPLE 6
Some Rank 1 Matrices

The following matrices have rank 1 because in each case the row vectors are expressible as scalar multiples of any nonzero row vector:

$$\begin{bmatrix} 2 & -4 & -6 & 0 \\ -3 & 6 & 9 & 0 \end{bmatrix}, \quad \begin{bmatrix} 3 & -1 \\ -6 & 2 \\ -3 & 1 \end{bmatrix}, \quad \begin{bmatrix} 1 & 1 & 1 \\ -2 & -2 & -2 \\ 0 & 0 & 0 \end{bmatrix}$$

Observe that in each case the column vectors are also scalar multiples of a nonzero vector. This is because the row space and column space of a matrix always have the same dimension—a result that will be proved in the next section. ■

Rank 1 matrices arise when outer products of nonzero column vectors are computed. To see why this is so, suppose that

$$\mathbf{u} = \begin{bmatrix} u_1 \\ u_2 \\ \vdots \\ u_m \end{bmatrix} \quad \text{and} \quad \mathbf{v} = \begin{bmatrix} v_1 \\ v_2 \\ \vdots \\ v_n \end{bmatrix}$$

are nonzero, and recall from Definition 3.1.11 that the outer product of $\mathbf{u}$ with $\mathbf{v}$ is

$$\mathbf{uv}^T = \begin{bmatrix} u_1 \\ u_2 \\ \vdots \\ u_m \end{bmatrix} [v_1 \quad v_2 \quad \cdots \quad v_n] = \begin{bmatrix} u_1 v_1 & u_1 v_2 & \cdots & u_1 v_n \\ u_2 v_1 & u_2 v_2 & \cdots & u_2 v_n \\ \vdots & \vdots & & \vdots \\ u_m v_1 & u_m v_2 & \cdots & u_m v_n \end{bmatrix} \tag{4}$$

This matrix has rank 1 since all row vectors are scalar multiples of the nonzero vector $\mathbf{v}^T$ and at least one of the components of $\mathbf{u}$ is nonzero.

EXAMPLE 7
A Rank 1
Matrix Arising
from a Product
$\mathbf{uv}^T$

Let

$$\mathbf{u} = \begin{bmatrix} -2 \\ 1 \\ 4 \end{bmatrix} \quad \text{and} \quad \mathbf{v} = \begin{bmatrix} 1 \\ 3 \\ -2 \\ -1 \end{bmatrix}$$

Then

$$\mathbf{uv}^T = \begin{bmatrix} -2 \\ 1 \\ 4 \end{bmatrix} [1 \quad 3 \quad -2 \quad -1] = \begin{bmatrix} -2 & -6 & 4 & 2 \\ 1 & 3 & -2 & -1 \\ 4 & 12 & -8 & -4 \end{bmatrix}$$

which is a matrix of rank 1. ∎

We saw in (4) that the outer product of nonzero column vectors has rank 1. The following theorem shows that all rank 1 matrices arise from outer products.

Theorem 7.4.7 *If $\mathbf{u}$ is a nonzero $m \times 1$ matrix and $\mathbf{v}$ is a nonzero $n \times 1$ matrix, then the outer product*

$$A = \mathbf{uv}^T$$

has rank 1. Conversely, if A is an $m \times n$ matrix with rank 1, then A can be factored into a product of the above form.

Proof Only the converse remains to be proved. Accordingly, let A be any $m \times n$ matrix of rank 1. The row vectors of A are all scalar multiples of some nonzero row vector $\mathbf{v}^T$, so we can express A in the form

$$A = \begin{bmatrix} u_1 \mathbf{v}^T \\ u_2 \mathbf{v}^T \\ \vdots \\ u_m \mathbf{v}^T \end{bmatrix} = \begin{bmatrix} u_1 \\ u_2 \\ \vdots \\ u_m \end{bmatrix} \mathbf{v}^T = \mathbf{uv}^T$$

where $\mathbf{u}$ is the column vector with components $u_1, u_2, \ldots, u_m$. These components cannot all be zero, otherwise A would have rank 0. ∎

The proof of this theorem suggests a method for factoring a rank 1 matrix into a product $\mathbf{uv}^T$ of a column vector times a row vector—take $\mathbf{v}^T$ to be any nonzero row of A and take the entries of the column vector $\mathbf{u}$ to be the scalars that produce the successive rows of A from $\mathbf{v}^T$. Here is an example.

EXAMPLE 8
Factoring a
Rank 1 Matrix
into the Form
$\mathbf{uv}^T$

We will factor the first matrix in Example 6, taking $\mathbf{v}^T$ to be the first row. This yields

$$\begin{bmatrix} 2 & -4 & -6 & 0 \\ -3 & 6 & 9 & 0 \end{bmatrix} = \begin{bmatrix} 1 \\ -\frac{3}{2} \end{bmatrix} [2 \quad -4 \quad -6 \quad 0] = \mathbf{uv}^T$$
∎

SYMMETRIC RANK 1 MATRICES

If $\mathbf{u}$ is a nonzero column vector, then

$$\mathbf{uu}^T = \begin{bmatrix} u_1 \\ u_2 \\ \vdots \\ u_n \end{bmatrix} [u_1 \quad u_2 \quad \cdots \quad u_n] = \begin{bmatrix} u_1^2 & u_1u_2 & \cdots & u_1u_n \\ u_2u_1 & u_2^2 & \cdots & u_2u_n \\ \vdots & \vdots & & \vdots \\ u_nu_1 & u_nu_2 & \cdots & u_n^2 \end{bmatrix} \quad (5)$$

which, in addition to having rank 1, is symmetric. This is part of the following theorem whose proof is outlined in the exercises.

Theorem 7.4.8 *If $\mathbf{u}$ is a nonzero $n \times 1$ column vector, then the outer product $\mathbf{uu}^T$ is a symmetric matrix of rank 1. Conversely, if A is a symmetric $n \times n$ matrix of rank 1, then it can be factored as $A = \mathbf{uu}^T$ or else as $A = -\mathbf{uu}^T$ for some nonzero $n \times 1$ column vector $\mathbf{u}$.*

EXAMPLE 9
A Symmetric
Matrix of Rank
One Arising
from $\mathbf{uu}^T$

If

$$\mathbf{u} = \begin{bmatrix} -2 \\ 1 \\ 3 \end{bmatrix}$$

then

$$\mathbf{uu}^T = \begin{bmatrix} -2 \\ 1 \\ 3 \end{bmatrix} [-2 \quad 1 \quad 3] = \begin{bmatrix} 4 & -2 & -6 \\ -2 & 1 & 3 \\ -6 & 3 & 9 \end{bmatrix}$$

which we see directly is symmetric and has rank 1.
∎

Exercise Set 7.4

In Exercises 1–6, confirm that the rank and nullity of the matrix satisfy Formula (2) in the dimension theorem.

1. $A = \begin{bmatrix} 1 & -1 & 3 \\ 5 & -4 & -4 \\ 7 & -6 & 2 \end{bmatrix}$

2. $A = \begin{bmatrix} 2 & 0 & -1 \\ 4 & 0 & -2 \\ 0 & 0 & 0 \end{bmatrix}$

3. $A = \begin{bmatrix} 1 & 4 & 5 & 2 \\ 2 & 1 & 3 & 0 \\ -1 & 3 & 2 & 2 \end{bmatrix}$

4. $A = \begin{bmatrix} 1 & 4 & 5 & 6 & 9 \\ 3 & -2 & 1 & 4 & -1 \\ -1 & 0 & -1 & -2 & -1 \\ 2 & 3 & 5 & 7 & 8 \end{bmatrix}$

5. $A = \begin{bmatrix} 1 & -3 & 2 & 2 & 1 \\ 0 & 3 & 6 & 0 & -3 \\ 2 & -3 & -2 & 4 & 4 \\ 3 & -6 & 0 & 6 & 5 \\ -2 & 9 & 2 & -4 & -5 \end{bmatrix}$

6. $A = \begin{bmatrix} 1 & -1 & 0 & 3 & 2 & 1 & -3 \\ 2 & 1 & 1 & -1 & 0 & 3 & -2 \\ 1 & -4 & -1 & 10 & 6 & 0 & -7 \\ 3 & 3 & 2 & -5 & -2 & 5 & -1 \end{bmatrix}$

In Exercises 7 and 8, use the given information to find the number of pivot variables and the number of parameters in a general solution of the system $A\mathbf{x} = \mathbf{0}$.

7. (a) A is a 5×8 matrix with rank 3.
  (b) A is a 7×4 matrix with nullity 2.
  (c) A is a 6×6 matrix whose row echelon forms have two nonzero rows.

8. (a) A is a 7×9 matrix with rank 5.
  (b) A is an 8×6 matrix with nullity 3.
  (c) A is a 7×7 matrix whose row echelon forms have three nonzero rows.

In Exercises 9 and 10, find the largest possible value for rank(A) and the smallest possible value for nullity(A).

9. (a) A is 5×3. (b) A is 3×5. (c) A is 4×4.

10. (a) A is 6×4. (b) A is 2×6. (c) A is 5×5.

11. Confirm that $\mathbf{v}_1 = (1, 1, 0, 0)$ and $\mathbf{v}_2 = (0, 3, 4, 5)$ are linearly independent vectors, and use the method of Example 2 to extend the set $\{\mathbf{v}_1, \mathbf{v}_2\}$ to a basis for R^4.

12. Confirm that $\mathbf{v}_1 = (1, 0, -2, 3, -5), \mathbf{v}_2 = (-2, 3, 3, 1, -1)$, and $\mathbf{v}_3 = (4, 1, -3, 0, 5)$ are linearly independent vectors, and use the method of Example 2 to extend the set $\{\mathbf{v}_1, \mathbf{v}_2, \mathbf{v}_3\}$ to a basis for R^5.

In Exercises 13 and 14, find the matrices, if any, that have rank 1, and express those matrices in the form $A = \mathbf{u}\mathbf{v}^T$, as guaranteed by Theorem 7.4.7.

13. (a) $A = \begin{bmatrix} 1 & -7 \\ -2 & 14 \end{bmatrix}$ (b) $A = \begin{bmatrix} 1 & 0 & 0 \\ 0 & 2 & 2 \\ 0 & 1 & 1 \end{bmatrix}$

  (c) $A = \begin{bmatrix} 1 & 1 & 3 & 3 & -9 \\ -2 & -2 & -6 & -6 & 18 \\ 3 & 3 & 9 & 9 & -27 \end{bmatrix}$

14. (a) $A = \begin{bmatrix} 1 & 0 \\ 0 & 1 \end{bmatrix}$ (b) $A = \begin{bmatrix} 1 & 6 & -3 \\ 0 & 0 & 0 \\ -4 & -24 & 12 \end{bmatrix}$

  (c) $A = \begin{bmatrix} 2 & -6 & 10 & 12 & -6 & 8 \\ 2 & -6 & 10 & 12 & -6 & 8 \\ 3 & -9 & 15 & 18 & -9 & 12 \\ -1 & 3 & -5 & -6 & 3 & -4 \end{bmatrix}$

In Exercises 15 and 16, express $\mathbf{u}$ in column form, and confirm that $\mathbf{u}\mathbf{u}^T$ is a symmetric matrix of rank 1, as guaranteed by Theorem 7.4.8.

15. $\mathbf{u} = (2, 3, 1, 1)$ **16.** $\mathbf{u} = (0, -4, 5, -7)$

In Exercises 17 and 18, express the rank of A in terms of t.

17. $A = \begin{bmatrix} 1 & 1 & t \\ 1 & t & 1 \\ t & 1 & 1 \end{bmatrix}$ **18.** $A = \begin{bmatrix} t & 3 & -1 \\ 3 & 6 & -2 \\ -1 & -3 & t \end{bmatrix}$

19. Show that if $\begin{bmatrix} x & y & z \\ 1 & x & y \end{bmatrix}$ has rank 1, then $x = t$, $y = t^2$, $z = t^3$ for some t.

20. Let W be the subspace of R^3 spanned by the vectors $\mathbf{v}_1 = (1, 1, 1)$, $\mathbf{v}_2 = (1, 2, -3)$, and $\mathbf{v}_3 = (4, 5, 0)$. Find bases for W and $W^\perp$, and verify Theorem 7.4.3.

21. Let W be the line in R^3 given parametrically by $x = 2t$, $y = -t, z = -3t$. Find a basis for $W^\perp$, and verify Theorem 7.4.3. Is W or $W^\perp$ a hyperplane in R^3? Explain.

22. Let $\mathbf{u}$ and $\mathbf{v}$ be nonzero column vectors in R^n, and let $T: R^n \to R^n$ be the linear operator whose standard matrix is $A = \mathbf{u}\mathbf{v}^T$. Show that $\ker(T) = \mathbf{v}^\perp$ and $\text{ran}(T) = \text{span}\{\mathbf{u}\}$.

23. (a) Show that if a matrix B results by changing one entry of a matrix A, then $A - B$ has rank 1.
  (b) Show that if a matrix B results by changing one column or one row of a matrix A, then $A - B$ has rank 1.
  (c) Show that if one entry, one column, or one row of an $m \times n$ matrix A is changed, then the resulting matrix B can be expressed in the form $B = A + \mathbf{u}\mathbf{v}^T$, where $\mathbf{u}$ is an $m \times 1$ column vector and $\mathbf{v}$ is an $n \times 1$ column vector.

24. Let $\mathbf{u}$ and $\mathbf{v}$ be nonzero column vectors in R^n, and let $A = \mathbf{u}\mathbf{v}^T$.
  (a) Show that $A^2 = (\mathbf{u} \cdot \mathbf{v})A$.
  (b) Use the result in part (a) to show that if $\mathbf{u} \cdot \mathbf{v} \neq 0$, then $\mathbf{u} \cdot \mathbf{v}$ is the only nonzero eigenvalue of A.
  (c) Use the results in parts (a) and (b) to show that if a matrix A has rank 1, then $I - A$ is invertible if and only if $A^2 \neq A$.

Discussion and Discovery

D1. Indicate whether the statement is true (T) or false (F). Justify your answer.

  (a) If A is not square, then the row vectors of A or the column vectors of A are linearly dependent.
  (b) Adding one additional nonzero row to a matrix A increases its rank by 1.
  (c) If A is a nonzero $m \times n$ matrix, then the nullity of A is at most m.
  (d) The nullity of a square matrix with linearly dependent rows is at least 1.
  (e) If A is square, and $A\mathbf{x} = \mathbf{b}$ is inconsistent for some vector $\mathbf{b}$, then nullity$(A) = 0$.
  (f) There is no 3×3 matrix whose row space and null space are both lines through the origin.

D2. If A is an $m \times n$ matrix, then rank$(A^T) + $ nullity$(A^T) = $ _____. Why?

D3. If A is a 3×5 matrix, then the number of leading 1's in the reduced row echelon form of A is at most _____, and the number of parameters in a general solution of $A\mathbf{x} = \mathbf{0}$ is at most _____.

D4. If A is a 5×3 matrix, then the number of leading 1's in the reduced row echelon form of A is at most _____, and the number of parameters in a general solution of $A\mathbf{x} = \mathbf{0}$ is at most _____.

D5. What are the possible values for the rank and nullity of a 3×5 matrix? A 5×3 matrix? A 5×5 matrix?

D6. If $\mathbf{u}$ and $\mathbf{v}$ are nonzero column vectors in R^n, then the nullity of the matrix $A = \mathbf{u}\mathbf{v}^T$ is _____.

D7. If $T: R^n \to R^n$ is a linear operator, and if the kernel of T is a line through the origin, what kind of geometric object is the range of T? Explain your reasoning.

D8. What can you say about the rank of the following matrix?

$$A = \begin{bmatrix} 1 & 0 & 0 \\ 0 & r-2 & 2 \\ 0 & s-1 & r+2 \\ 0 & 0 & 3 \end{bmatrix}$$

D9. Find the value(s) of λ for which the matrix

$$A = \begin{bmatrix} 3 & 1 & 1 & 4 \\ \lambda & 4 & 10 & 1 \\ 1 & 7 & 17 & 3 \\ 2 & 2 & 4 & 3 \end{bmatrix}$$

has lowest rank.

D10. Show by example that it is possible for two matrices A and B to have the same rank and A^2 and B^2 to have different ranks.

D11. Use Sylvester's inequalities in Exercise T4 below to show that if A and B are $n \times n$ matrices for which $AB = 0$, then $\text{rank}(A) + \text{rank}(B) \leq n$. What does this tell you about the relationship between the rank of A and the nullity of B? Between the rank of B and the nullity of A?

Working with Proofs

P1. Prove that the matrix

$$\begin{bmatrix} a_{11} & a_{12} & a_{13} \\ a_{21} & a_{22} & a_{23} \end{bmatrix}$$

has rank 2 if and only if one or more of the determinants

$$\begin{vmatrix} a_{11} & a_{12} \\ a_{21} & a_{22} \end{vmatrix}, \quad \begin{vmatrix} a_{11} & a_{13} \\ a_{21} & a_{23} \end{vmatrix}, \quad \begin{vmatrix} a_{12} & a_{13} \\ a_{22} & a_{23} \end{vmatrix}$$

is nonzero. [*Note:* This follows from the more general result stated in Exercise T6 below, but prove this result independently.]

P2. Prove that if A is an $n \times n$ symmetric matrix of rank 1, then A can be expressed as $A = \mathbf{u}\mathbf{u}^T$ or $-\mathbf{u}\mathbf{u}^T$, where $\mathbf{u}$ is a column vector in R^n. [*Hint:* Theorem 7.4.7 and the symmetry of A imply that $A = \mathbf{x}\mathbf{y}^T = \mathbf{y}\mathbf{x}^T$. Show that $\mathbf{x}^T\mathbf{y} \neq 0$, and then consider the cases $\mathbf{x}^T\mathbf{y} > 0$ and $\mathbf{x}^T\mathbf{y} < 0$ separately. Show that if $\mathbf{x}^T\mathbf{y} > 0$, then

$$\mathbf{u} = \sqrt{\frac{\mathbf{x}^T\mathbf{x}}{\mathbf{x}^T\mathbf{y}}}\, \mathbf{y}$$

has the required property, and find a similar formula in the case where $\mathbf{x}^T\mathbf{y} < 0$.]

P3. Let A and B be nonzero matrices, and partition A into column vectors and B into row vectors. Prove that multiplying A and B as partitioned matrices produces a decomposition of AB as a sum of rank 1 matrices.

P4. Prove: If $V = \{\mathbf{v}_1, \mathbf{v}_2, \ldots, \mathbf{v}_k\}$ is a linearly independent set of vectors in R^n, and if $W = \{\mathbf{w}_{k+1}, \ldots, \mathbf{w}_n\}$ is a basis for the null space of the matrix A that has the vectors $\mathbf{v}_1, \mathbf{v}_2, \ldots, \mathbf{v}_k$ as its successive rows, then $V \cup W = \{\mathbf{v}_1, \mathbf{v}_2, \ldots, \mathbf{v}_k, \mathbf{w}_{k+1}, \ldots, \mathbf{w}_n\}$ is a basis for R^n. [*Hint:* Since $V \cup W$ contains n vectors, it suffices to show that $V \cup W$ is linearly independent. As a first step, rewrite the equation

$$c_1\mathbf{v}_1 + c_2\mathbf{v}_2 + \cdots + c_k\mathbf{v}_k + d_1\mathbf{w}_1$$
$$+ d_2\mathbf{w}_2 + \cdots + d_{n-k}\mathbf{w}_{n-k} = \mathbf{0}$$

as

$$c_1\mathbf{v}_1 + c_2\mathbf{v}_2 + \cdots + c_k\mathbf{v}_k$$
$$= -d_1\mathbf{w}_1 - d_2\mathbf{w}_2 - \cdots - d_{n-k}\mathbf{w}_{n-k}$$

and use this to show that the expressions on each side are equal to $\mathbf{0}$.]

P5. Use Sylvester's rank inequalities in Exercise T4 below to prove the following results (known as *Sylvester's laws of nullity*): If A and B are square matrices for which the product AB is defined, then

$$\text{nullity}(A) \leq \text{nullity}(AB) \leq \text{nullity}(A) + \text{nullity}(B)$$
$$\text{nullity}(B) \leq \text{nullity}(AB) \leq \text{nullity}(A) + \text{nullity}(B)$$

P6. Suppose that A is an invertible $n \times n$ matrix whose inverse is known and that B is a matrix that results by changing one entry, one row, or one column of A. We know from Exercise 23(c) that B can be expressed as $B = A + \mathbf{u}\mathbf{v}^T$, where $\mathbf{u}$ and $\mathbf{v}$ are column vectors in R^n. Thus, one is led to inquire whether the invertibility of A implies the invertibility of $B = A + \mathbf{u}\mathbf{v}^T$, and if so, what relationship might exist between A^{-1} and B^{-1}. Prove that B is invertible if $1 + \mathbf{v}^T A^{-1}\mathbf{u} \neq 0$ and that in that case

$$B^{-1} = (A + \mathbf{u}\mathbf{v}^T)^{-1} = A^{-1} - \frac{A^{-1}\mathbf{u}\mathbf{v}^T A^{-1}}{1 + \mathbf{v}^T A^{-1}\mathbf{u}}$$

Technology Exercises

T1. (a) Some technology utilities provide a command for finding the rank of a matrix. Determine whether your utility has this capability; if so, use that command to find the rank of the matrix in Example 1.

(b) Confirm that the rank obtained in part (a) is consistent with the rank obtained by using your utility to find the number of nonzero rows in the reduced row echelon form of the matrix.

T2. Most technology utilities do not provide a direct command for finding the nullity of a matrix since the nullity can be computed using the rank command and Formula (2). Use that method to find the nullity of the matrix in Exercise T1 of Section 7.3, and confirm that the result obtained is consistent with the number of basis vectors obtained in that exercise.

T3. Confirm Formula (2) for some 5×7 matrices of your choice.

T4. *Sylvester's rank inequalities* (whose proofs are somewhat detailed) state that if A is a matrix with n columns and B is a matrix with n rows, then

$$\text{rank}(A) + \text{rank}(B) - n \le \text{rank}(AB) \le \text{rank}(A)$$
$$\text{rank}(A) + \text{rank}(B) - n \le \text{rank}(AB) \le \text{rank}(B)$$

Confirm these inequalities for some matrices of your choice.

T5. (a) Consider the matrices

$$A = \begin{bmatrix} 7 & 4 & -2 & 4 \\ 2 & -3 & 7 & -6 \\ 5 & 6 & 2 & -5 \\ 3 & 3 & -5 & 8 \end{bmatrix}, \quad A = \begin{bmatrix} 7.1 & 4 & -2 & 4 \\ 2 & -3 & 7 & -6 \\ 5 & 6 & 2 & -5 \\ 3 & 3 & -5 & 8 \end{bmatrix}$$

which differ only in one entry. Compute A^{-1} and use the result in Exercise P6 to compute B^{-1}.

(b) Check your result by computing B^{-1} directly.

T6. It can be proved that the rank of a matrix A is the order of the largest square submatrix of A (formed by deleting rows and columns of A) whose determinant is nonzero. Use this result to find the rank of the matrix

$$A = \begin{bmatrix} 3 & -1 & 3 & 2 & 5 \\ 5 & -3 & 2 & 3 & 4 \\ 1 & -3 & -5 & 0 & -7 \\ 7 & -5 & 1 & 4 & 1 \end{bmatrix}$$

and check your answer by using a different method to find the rank.

Section 7.5 The Rank Theorem and Its Implications

In this section we will prove that the row space and column space have the same dimension, and we will discuss some of the implications of this result.

THE RANK THEOREM

The following theorem, which is proved at the end of this section, is one of the most important in linear algebra.

> **Theorem 7.5.1** (*The Rank Theorem*) *The row space and column space of a matrix have the same dimension.*

EXAMPLE 1
Row Space and
Column Space
Have the Same
Dimension

In Example 4 of Section 7.3 we showed that the row space of the matrix

$$A = \begin{bmatrix} 1 & 0 & 0 & 0 & 2 \\ -2 & 1 & -3 & -2 & -4 \\ 0 & 5 & -14 & -9 & 0 \\ 2 & 10 & -28 & -18 & 4 \end{bmatrix} \tag{1}$$

is three-dimensional, so the rank theorem implies that the column space is also three-dimensional. Let us confirm this by finding a basis for the column space. One way to do this is to transpose A (which converts columns to rows) and then find a basis for the row space of A^T by reducing it to row echelon form and extracting the nonzero row vectors. Proceeding in this way, we first transpose A to obtain

$$A^T = \begin{bmatrix} 1 & -2 & 0 & 2 \\ 0 & 1 & 5 & 10 \\ 0 & -3 & -14 & -28 \\ 0 & -2 & -9 & -18 \\ 2 & -4 & 0 & 4 \end{bmatrix}$$

and then reduce this matrix to row echelon form to obtain

$$\begin{bmatrix} 1 & -2 & 0 & 2 \\ 0 & 1 & 5 & 10 \\ 0 & 0 & 1 & 2 \\ 0 & 0 & 0 & 0 \\ 0 & 0 & 0 & 0 \end{bmatrix} \tag{2}$$

(verify). The nonzero row vectors in this matrix form a basis for the row space of A^T, so the column space of A is three-dimensional as anticipated. If desired, a basis of column vectors for the column space of A can be obtained by transposing the row vectors in (2) to obtain

$$\mathbf{c}_1 = \begin{bmatrix} 1 \\ -2 \\ 0 \\ 2 \end{bmatrix}, \quad \mathbf{c}_2 = \begin{bmatrix} 0 \\ 1 \\ 5 \\ 10 \end{bmatrix}, \quad \mathbf{c}_3 = \begin{bmatrix} 0 \\ 0 \\ 1 \\ 2 \end{bmatrix} \qquad \blacksquare$$

Recall from Definition 7.3.1 that the rank of a matrix A is the dimension of its row space. As a result of Theorem 7.5.1, we can now also think of the rank of a matrix as the dimension of its column space. Moreover, since transposing a matrix converts columns to rows and rows to columns, it is evident that a matrix and its transpose must have the same rank.

> **Theorem 7.5.2**   *If A is an $m \times n$ matrix, then*
>
> $$\text{rank}(A) = \text{rank}(A^T) \tag{3}$$

This result has some important implications. For example, if A is an $m \times n$ matrix, then applying Theorem 7.4.1 to A^T yields

$$\text{rank}(A^T) + \text{nullity}(A^T) = m$$

which we can rewrite using (3) as

$$\text{rank}(A) + \text{nullity}(A^T) = m \tag{4}$$

This relationship now makes it possible to express the dimensions of all four fundamental spaces of a matrix in terms of the size and rank of the matrix. Specifically, if A is an $m \times n$ matrix with rank k, then

$$\begin{array}{ll} \dim(\text{row}(A)) = k, & \dim(\text{null}(A)) = n - k \\ \dim(\text{col}(A)) = k, & \dim(\text{null}(A^T)) = m - k \end{array} \tag{5}$$

EXAMPLE 2
Dimensions of the Fundamental Spaces from the Rank

Find the rank of

$$A = \begin{bmatrix} 1 & 2 & -3 & 1 & 1 \\ -3 & 1 & 7 & -1 & 1 \\ -2 & 3 & 4 & 0 & 2 \end{bmatrix}$$

and then use that result to compute the dimensions of the fundamental spaces of A.

Solution The rank of A is the number of nonzero rows in any row echelon form of A, so we will begin by reducing A to row echelon form. Introducing the required zeros in the first column yields

$$\begin{bmatrix} 1 & 2 & -3 & 1 & 1 \\ 0 & 7 & -2 & 2 & 4 \\ 0 & 7 & -2 & 2 & 4 \end{bmatrix}$$

At this point there is no need to go any further, since it is now evident that the row space is two-dimensional. Thus, A has rank 2 and

$$\dim(\text{row}(A)) = \text{rank} = 2, \quad \dim(\text{null}(A)) = \text{number of columns} - \text{rank} = 5 - 2 = 3$$
$$\dim(\text{col}(A)) = \text{rank} = 2, \quad \dim(\text{null}(A^T)) = \text{number of rows} - \text{rank} = 3 - 2 = 1 \quad \blacksquare$$

CONCEPT PROBLEM If A is an $m \times n$ matrix, what is the largest possible value for the rank of A? Explain.

RELATIONSHIP BETWEEN CONSISTENCY AND RANK

In the course of progressing through this text, we have developed a succession of unifying theorems involving linear systems of n equations in n unknowns, the last being Theorem 7.4.4. We will now turn our attention to linear systems in which the number of equations and unknowns need not be the same. The following theorem, which is an extension of Theorem 3.5.5, provides a relationship between the consistency of a linear system and the ranks of its coefficient and augmented matrices.

> **Theorem 7.5.3** (*The Consistency Theorem*) *If $A\mathbf{x} = \mathbf{b}$ is a linear system of m equations in n unknowns, then the following statements are equivalent.*
>
> (*a*) $A\mathbf{x} = \mathbf{b}$ *is consistent.*
>
> (*b*) $\mathbf{b}$ *is in the column space of A.*
>
> (*c*) *The coefficient matrix A and the augmented matrix $[A \mid \mathbf{b}]$ have the same rank.*

The equivalence of parts (*a*) and (*b*) was given in Theorem 3.5.5, so we need only prove that (*b*) $\Leftrightarrow$ (*c*). The equivalence (*a*) $\Leftrightarrow$ (*c*) will then follow as a logical consequence.

Proof (*b*) $\Leftrightarrow$ (*c*) If $\mathbf{b}$ is in the column space of A, then Theorem 7.2.5 implies that the column spaces of A and $[A \mid \mathbf{b}]$ have the same dimension; that is, the two matrices have the same rank. Conversely, if A and $[A \mid \mathbf{b}]$ have the same rank, then their column spaces have the same dimension, so Theorem 7.2.5 implies that $\mathbf{b}$ is a linear combination of the column vectors of A. $\blacksquare$

EXAMPLE 3
Visualizing the Consistency Theorem

To obtain a better understanding of the relationship between the ranks of the coefficient and augmented matrices of a linear system, consider the system

$$\begin{aligned} x_1 - 2x_2 - 3x_3 &= -4 \\ -3x_1 + 7x_2 - x_3 &= -3 \\ 2x_1 - 5x_2 + 4x_3 &= 7 \\ -3x_1 + 6x_2 + 9x_3 &= -1 \end{aligned}$$

The augmented matrix for the system is

$$\begin{bmatrix} 1 & -2 & -3 & -4 \\ -3 & 7 & -1 & -3 \\ 2 & -5 & 4 & 7 \\ -3 & 6 & 9 & -1 \end{bmatrix}$$

and the reduced row echelon form of this matrix is (verify)

$$\begin{bmatrix} 1 & 0 & -23 & 0 \\ 0 & 1 & -10 & 0 \\ 0 & 0 & 0 & 1 \\ 0 & 0 & 0 & 0 \end{bmatrix} \tag{6}$$

The "bad" third row in this matrix makes it evident that the system is inconsistent. However, this row also causes the corresponding row echelon form of the coefficient matrix to have smaller rank than the row echelon form of the augmented matrix [cover the last column of (6) to see this]. This example should make it evident that the augmented matrix and the coefficient matrix of a linear system have the same rank if and only if there are no bad rows in any row echelon form of the augmented matrix, or equivalently, if and only if the system is consistent. ∎

The following concept is an important tool in the study of linear systems in which the number of equations and number of unknowns need not be the same.

Definition 7.5.4   An $m \times n$ matrix A is said to have *full column rank* if its column vectors are linearly independent, and it is said to have *full row rank* if its row vectors are linearly independent.

Since the column vectors of a matrix span the column space and the row vectors span the row space, the column vectors of a matrix with full column rank must be a basis for the column space, and the row vectors of a matrix with full row rank must be a basis for the row space. Thus, we have the following alternative way of viewing the concepts of full column rank and full row rank.

Theorem 7.5.5   *Let A be an m × n matrix.*

(a) *A has full column rank if and only if the column vectors of A form a basis for the column space, that is, if and only if* rank$(A) = n$.

(b) *A has full row rank if and only if the row vectors of A form a basis for the row space, that is, if and only if* rank$(A) = m$.

EXAMPLE 4
Full Column
Rank and Full
Row Rank

The matrix

$$A = \begin{bmatrix} 1 & 0 \\ 2 & 1 \\ -3 & 1 \end{bmatrix}$$

has full column rank because the column vectors are not scalar multiples of one another; it does not have full row rank because three vectors in R^2 are linearly dependent. In contrast,

$$A^T = \begin{bmatrix} 1 & 2 & -3 \\ 0 & 1 & 1 \end{bmatrix}$$

has full row rank but not full column rank. ∎

CONCEPT PROBLEM   If A is an $m \times n$ matrix with full column rank, what can you say about the relative sizes of m and n? What if A has full row rank? Explain.

The following theorem is closely related to Theorem 3.5.3.

Theorem 7.5.6   *If A is an m × n matrix, then the following statements are equivalent.*

(a) $A\mathbf{x} = \mathbf{0}$ *has only the trivial solution.*

(b) $A\mathbf{x} = \mathbf{b}$ *has at most one solution for every* $\mathbf{b}$ *in* R^m.

(c) *A has full column rank.*

Since the equivalence of parts (a) and (b) is the content of Theorem 3.5.3, it suffices to show that parts (a) and (c) are equivalent to complete the proof.

Proof (a) ⇔ ***(c)***   Let $\mathbf{a}_1, \mathbf{a}_2, \ldots, \mathbf{a}_n$ be the column vectors of A, and write the system $A\mathbf{x} = \mathbf{0}$ in the vector form

$$x_1\mathbf{a}_1 + x_2\mathbf{a}_2 + \cdots + x_n\mathbf{a}_n = \mathbf{0} \tag{7}$$

Thus, to say that $A\mathbf{x} = \mathbf{0}$ has only the trivial solution is equivalent to saying that the n column vectors in (7) are linearly independent; that is, $A\mathbf{x} = \mathbf{0}$ has only the trivial solution if and only if A has full column rank.   ∎

EXAMPLE 5
Implications of
Full Column
Rank

We showed in Example 4 that

$$A = \begin{bmatrix} 1 & 0 \\ 2 & 1 \\ -3 & 1 \end{bmatrix}$$

has full column rank. Thus, Theorem 7.5.6 implies that the system $A\mathbf{x} = \mathbf{0}$ has only the trivial solution and that the system $A\mathbf{x} = \mathbf{b}$ has at most one solution for every $\mathbf{b}$ in R^3. We will leave it for you to confirm the first statement by solving the system $A\mathbf{x} = \mathbf{0}$; and we will show that $A\mathbf{x} = \mathbf{b}$ has at most one solution for every vector $\mathbf{b} = (b_1, b_2, b_3)$ in R^3.

Reducing the augmented matrix $[A \mid \mathbf{b}]$ until the left side is in reduced row echelon form yields

$$\begin{bmatrix} 1 & 0 & | & b_1 \\ 0 & 1 & | & b_2 - 2b_1 \\ 0 & 0 & | & b_3 - b_2 + 5b_1 \end{bmatrix}$$

(verify), so there are two possibilities: $b_3 - b_2 + 5b_1 \neq 0$ or $b_3 - b_2 + 5b_1 = 0$. In the first case the system is inconsistent, and in the second case the system has the unique solution $x_1 = b_1$, $x_2 = b_2 - 2b_1$. In either case it is correct to say that there is at most one solution.   ∎

OVERDETERMINED AND UNDERDETERMINED LINEAR SYSTEMS

In engineering applications, the equations in a linear system $A\mathbf{x} = \mathbf{b}$ are often mathematical formulations of physical constraints on a set of variables, and engineers generally try to match the number of variables and constraints. However, this is not always possible, so an engineer may be faced with a linear system that has more equations than unknowns (called an ***overdetermined system***) or a linear system that has fewer equations than unknowns (called an ***underdetermined system***). The occurrence of an overdetermined or underdetermined linear system in applications often signals that some undesirable physical phenomenon may occur. The following theorem explains why.

Theorem 7.5.7   *Let A be an m × n matrix.*

(a) (***Overdetermined Case***) *If $m > n$, then the system $A\mathbf{x} = \mathbf{b}$ is inconsistent for some vector $\mathbf{b}$ in R^m.*

(b) (***Underdetermined Case***) *If $m < n$, then for every vector $\mathbf{b}$ in R^m the system $A\mathbf{x} = \mathbf{b}$ is either inconsistent or has infinitely many solutions.*

Proof (a)   If $m > n$, then the column vectors of A cannot span R^m. Thus, there is at least one vector $\mathbf{b}$ in R^m that is not a linear combination of the column vectors of A, and for such a $\mathbf{b}$ the system $A\mathbf{x} = \mathbf{b}$ has no solution.

Proof (b)   If $m < n$, then the column vectors of A must be linearly dependent (n vectors in R^m). This implies that $A\mathbf{x} = \mathbf{0}$ has infinitely many solutions, so the result follows from Theorem 3.5.2.   ∎

EXAMPLE 6
A Misbehaving
Robot

To express Theorem 7.5.7 in transformation terms, think of $A\mathbf{x}$ as a matrix transformation from R^n to R^m, and think of the vector $\mathbf{b}$ in the equation $A\mathbf{x} = \mathbf{b}$ as some output that we would like the transformation to produce in response to some input $\mathbf{x}$. Part (a) of Theorem 7.5.7 states that if $m > n$, then there is some output that cannot be produced by any input, and part (b) states that if $m < n$, then for each possible output $\mathbf{b}$ there is either no input that produces that output or there are infinitely many inputs that produce that output. Thus, for example, if the input $\mathbf{x}$ is a vector of voltages to the driving motors of a robot, and if the output $\mathbf{b}$ is a vector of speeds and position coordinates that describe the action of the robot in response to the input, then an overdetermined system governs a robot that cannot achieve certain desired actions, and an underdetermined system governs a robot for which certain actions can be achieved in infinitely many ways, which may not be desirable. ∎

MATRICES OF THE FORM A^TA AND AA^T

Matrices of the form A^TA and AA^T play an important role in many applications, so we will now focus our attention on matrices of this form.

To start, recall from Formula (9) of Section 3.6 that if A is an $m \times n$ matrix with column vectors $\mathbf{a}_1, \mathbf{a}_2, \ldots, \mathbf{a}_n$, then

$$A^TA = \begin{bmatrix} \mathbf{a}_1 \cdot \mathbf{a}_1 & \mathbf{a}_1 \cdot \mathbf{a}_2 & \cdots & \mathbf{a}_1 \cdot \mathbf{a}_n \\ \mathbf{a}_2 \cdot \mathbf{a}_1 & \mathbf{a}_2 \cdot \mathbf{a}_2 & \cdots & \mathbf{a}_2 \cdot \mathbf{a}_n \\ \vdots & \vdots & & \vdots \\ \mathbf{a}_n \cdot \mathbf{a}_1 & \mathbf{a}_n \cdot \mathbf{a}_2 & \cdots & \mathbf{a}_n \cdot \mathbf{a}_n \end{bmatrix} \tag{8}$$

Since transposing a matrix converts columns to rows and rows to columns, it follows from (8) that if $\mathbf{r}_1, \mathbf{r}_2, \ldots, \mathbf{r}_m$ are the row vectors of A, then

$$AA^T = \begin{bmatrix} \mathbf{r}_1 \cdot \mathbf{r}_1 & \mathbf{r}_1 \cdot \mathbf{r}_2 & \cdots & \mathbf{r}_1 \cdot \mathbf{r}_m \\ \mathbf{r}_2 \cdot \mathbf{r}_1 & \mathbf{r}_2 \cdot \mathbf{r}_2 & \cdots & \mathbf{r}_2 \cdot \mathbf{r}_m \\ \vdots & \vdots & & \vdots \\ \mathbf{r}_m \cdot \mathbf{r}_1 & \mathbf{r}_m \cdot \mathbf{r}_2 & \cdots & \mathbf{r}_m \cdot \mathbf{r}_m \end{bmatrix} \tag{9}$$

The next theorem provides some important links between properties of a general matrix A, its transpose A^T, and the square symmetric matrix A^TA.

> **Theorem 7.5.8** *If A is an $m \times n$ matrix, then:*
> *(a) A and A^TA have the same null space.*
> *(b) A and A^TA have the same row space.*
> *(c) A^T and A^TA have the same column space.*
> *(d) A and A^TA have the same rank.*

We will prove part (a) and leave the remaining proofs for the exercises.

Proof (a) We must show that every solution of $A\mathbf{x} = \mathbf{0}$ is a solution of $A^TA\mathbf{x} = \mathbf{0}$, and conversely. If $\mathbf{x}_0$ is any solution of $A\mathbf{x} = \mathbf{0}$, then $\mathbf{x}_0$ is also a solution of $A^TA\mathbf{x} = \mathbf{0}$ since

$$A^TA\mathbf{x}_0 = A^T(A\mathbf{x}_0) = A^T\mathbf{0} = \mathbf{0}$$

Conversely, if $\mathbf{x}_0$ is any solution of $A^TA\mathbf{x} = \mathbf{0}$, then $\mathbf{x}_0$ is in the null space of A^TA and hence is orthogonal to every vector in the row space of A^TA by Theorem 3.5.6. However, A^TA is symmetric, so $\mathbf{x}_0$ is also orthogonal to every vector in the column space of A^TA. In particular, $\mathbf{x}_0$ must be orthogonal to the vector $A^TA\mathbf{x}_0$; that is, $\mathbf{x}_0 \cdot (A^TA\mathbf{x}_0) = 0$. From Formula (23) of Section 3.1, we can write this as

$$\mathbf{x}_0^T(A^TA\mathbf{x}_0) = 0 \quad \text{or, equivalently, as} \quad (A\mathbf{x}_0)^T(A\mathbf{x}_0) = 0$$

This implies that $A\mathbf{x}_0 \cdot A\mathbf{x}_0 = 0$, so $A\mathbf{x}_0 = \mathbf{0}$ by part (d) of Theorem 1.2.6. This proves that $\mathbf{x}_0$ is a solution of $A\mathbf{x} = \mathbf{0}$. ∎

The following companion to Theorem 7.5.8 follows on replacing A by A^T in that theorem and using the fact that A and A^T have the same rank for part (d).

> **Theorem 7.5.9** *If A is an $m \times n$ matrix, then:*
> *(a) A^T and AA^T have the same null space.*
> *(b) A^T and AA^T have the same row space.*
> *(c) A and AA^T have the same column space.*
> *(d) A and AA^T have the same rank.*

CONCEPT PROBLEM   What is the relationship between $\text{rank}(A^T A)$ and $\text{rank}(A A^T)$. Why?

SOME UNIFYING THEOREMS

The following unifying theorem adds another condition to those in Theorem 7.5.6.

Theorem 7.5.10   *If A is an m × n matrix, then the following statements are equivalent.*

(a) $A\mathbf{x} = \mathbf{0}$ *has only the trivial solution.*

(b) $A\mathbf{x} = \mathbf{b}$ *has at most one solution for every* $\mathbf{b}$ *in* R^m.

(c) *A has full column rank.*

(d) $A^T A$ *is invertible.*

It suffices to prove that statements (c) and (d) are equivalent, since the remaining equivalences follow immediately from Theorem 7.5.6.

Proof (c) ⇔ (d)   Since $A^T A$ is an $n \times n$ matrix, it follows from statements (c) and (q) of Theorem 7.4.4 that $A^T A$ is invertible if and only if $A^T A$ has rank n. However, $A^T A$ has the same rank as A by part (d) of Theorem 7.5.8, so $A^T A$ is invertible if and only if $\text{rank}(A) = n$, that is, if and only if A has full column rank.   ■

REMARK   We know from Theorem 3.6.5 that if A is square, then $A^T A$ is invertible if and only if A is invertible. That result is a special case of the equivalence of (c) and (d) in Theorem 7.5.10.

The following companion to Theorem 7.5.10 follows on replacing A by A^T.

Theorem 7.5.11   *If A is an m × n matrix, then the following statements are equivalent.*

(a) $A^T\mathbf{x} = \mathbf{0}$ *has only the trivial solution.*

(b) $A^T\mathbf{x} = \mathbf{b}$ *has at most one solution for every vector* $\mathbf{b}$ *in* R^n.

(c) *A has full row rank.*

(d) $A A^T$ *is invertible.*

Theorems 7.5.10 and 7.5.11 make it possible to use results about square matrices to deduce results about matrices that are not square. For example, we know that $A^T A$ is invertible if and only if $\det(A^T A) \neq 0$, and $A A^T$ is invertible if and only if $\det(A A^T) \neq 0$. Thus, it follows from Theorems 7.5.10 and 7.5.11 that A has full column rank if and only if $\det(A^T A) \neq 0$, and A has full row rank if and only if $\det(A A^T) \neq 0$.

EXAMPLE 7
A Determinant Test for Full Column Rank and Full Row Rank

We showed in Example 4 that the matrix

$$A = \begin{bmatrix} 1 & 0 \\ 2 & 1 \\ -3 & 1 \end{bmatrix}$$

has full column rank, but not full row rank. Confirm these results by evaluating appropriate determinants.

Solution   To test for full column rank we consider the matrix

$$A^T A = \begin{bmatrix} 1 & 2 & -3 \\ 0 & 1 & 1 \end{bmatrix} \begin{bmatrix} 1 & 0 \\ 2 & 1 \\ -3 & 1 \end{bmatrix} = \begin{bmatrix} 14 & -1 \\ -1 & 2 \end{bmatrix}$$

and to test for full row rank we consider the matrix

$$A A^T = \begin{bmatrix} 1 & 0 \\ 2 & 1 \\ -3 & 1 \end{bmatrix} \begin{bmatrix} 1 & 2 & -3 \\ 0 & 1 & 1 \end{bmatrix} = \begin{bmatrix} 1 & 2 & -3 \\ 2 & 5 & -5 \\ -3 & -5 & 10 \end{bmatrix}$$

Since $\det(A^T A) = 27 \neq 0$ (verify), the matrix A has full column rank, and since $\det(A A^T) = 0$ (verify), the matrix A does not have full row rank. ∎

APPLICATIONS OF RANK   The advent of the Internet has stimulated research on finding efficient methods for transmitting large amounts of digital data over communications lines with limited bandwidth. Digital data are commonly stored in matrix form, and many techniques for improving transmission speed use the rank of a matrix in some way. Rank plays a role because it measures the "redundancy" in a matrix in the sense that if A is an $m \times n$ matrix of rank k, then $n - k$ of the column vectors and $m - k$ of the row vectors can be expressed in terms of k linearly independent column or row vectors. The essential idea in many data compression schemes is to approximate the original data set by a data set with smaller rank that conveys nearly the same information, then eliminate redundant vectors in the approximating set to speed up the transmission time.

Linear Algebra in History

In 1924 the U.S. Federal Bureau of Investigation (FBI) began collecting fingerprints and handprints and now has more than 30 million such prints in its files. To reduce the storage cost, the FBI began working with the Los Alamos National Laboratory, the National Bureau of Standards, and other groups in 1993 to devise compression methods for storing prints in digital form. The following figure shows an original fingerprint and a reconstruction from digital data that was compressed at a ratio of 26:1.

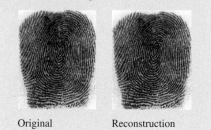

Original   Reconstruction

OPTIONAL *Proof of Theorem 7.5.1*

We want to prove that the row space and column space of an $m \times n$ matrix A have the same dimension. For this purpose, assume that A has rank k, which implies that the reduced row echelon form R has exactly k nonzero row vectors, say $\mathbf{r}_1, \mathbf{r}_2, \ldots, \mathbf{r}_k$. Since A and R have the same row space by Theorem 7.3.7, it follows that the row vectors $\mathbf{a}_1, \mathbf{a}_2, \ldots, \mathbf{a}_m$ of A can be expressed as linear combinations of the row vectors of R, say

$$
\begin{aligned}
\mathbf{a}_1 &= c_{11}\mathbf{r}_1 + c_{12}\mathbf{r}_2 + c_{13}\mathbf{r}_3 + \cdots + c_{1k}\mathbf{r}_k \\
\mathbf{a}_2 &= c_{21}\mathbf{r}_1 + c_{22}\mathbf{r}_2 + c_{23}\mathbf{r}_3 + \cdots + c_{2k}\mathbf{r}_k \\
&\;\;\vdots \qquad \vdots \qquad \vdots \qquad \vdots \qquad \vdots \\
\mathbf{a}_m &= c_{m1}\mathbf{r}_1 + c_{m2}\mathbf{r}_2 + c_{m3}\mathbf{r}_3 + \cdots + c_{mk}\mathbf{r}_k
\end{aligned}
\tag{10}
$$

Next we equate corresponding components on the two sides of each equation. For this purpose let a_{ij} be the jth component of $\mathbf{a}_i$, and let r_{ij} be the jth component of $\mathbf{r}_i$. Thus, the relationships between the jth components on the two sides of (10) are

$$
\begin{aligned}
a_{1j} &= c_{11}r_{1j} + c_{12}r_{2j} + c_{13}r_{3j} + \cdots + c_{1k}r_{kj} \\
a_{2j} &= c_{21}r_{1j} + c_{22}r_{2j} + c_{23}r_{3j} + \cdots + c_{2k}r_{kj} \\
&\;\;\vdots \qquad \vdots \qquad \vdots \qquad \vdots \qquad \vdots \\
a_{mj} &= c_{m1}r_{1j} + c_{m2}r_{2j} + c_{m3}r_{3j} + \cdots + c_{mk}r_{kj}
\end{aligned}
$$

which we can rewrite in matrix form as

$$
\begin{bmatrix} a_{1j} \\ a_{2j} \\ \vdots \\ a_{mj} \end{bmatrix}
= r_{1j} \begin{bmatrix} c_{11} \\ c_{21} \\ \vdots \\ c_{m1} \end{bmatrix}
+ r_{2j} \begin{bmatrix} c_{12} \\ c_{22} \\ \vdots \\ c_{m2} \end{bmatrix}
+ r_{3j} \begin{bmatrix} c_{13} \\ c_{23} \\ \vdots \\ c_{m3} \end{bmatrix}
+ \cdots + r_{kj} \begin{bmatrix} c_{1k} \\ c_{2k} \\ \vdots \\ c_{mk} \end{bmatrix}
$$

Since the left side of this equation is the jth column vector of A, we have shown that the k column vectors on the right side of the equation span the column space of A. Thus, the dimension of the column space of A is at most k; that is,

$$\dim(\mathrm{col}(A)) \leq \dim(\mathrm{row}(A)) \tag{11}$$

It follows from this that

$$\dim(\mathrm{col}(A^T)) \leq \dim(\mathrm{row}(A^T))$$

or

$$\dim(\mathrm{row}(A)) \leq \dim(\mathrm{col}(A)) \tag{12}$$

We can conclude from (11) and (12) that $\dim(\mathrm{row}(A)) = \dim(\mathrm{col}(A))$. ∎

Exercise Set 7.5

In Exercises 1 and 2, verify that the row space of A and the column space of A have the same dimension, and use that number to find the dimensions of the other two fundamental spaces of A and the number of parameters in a general solution of $A\mathbf{x} = \mathbf{0}$.

1. $A = \begin{bmatrix} 1 & -1 & 2 & -1 & 1 \\ 4 & -3 & -1 & 1 & -2 \\ 3 & -2 & -3 & 2 & -3 \\ 6 & -5 & 3 & -1 & 0 \end{bmatrix}$

2. $A = \begin{bmatrix} 1 & 1 & -2 & 3 & 2 \\ 2 & -1 & 3 & 4 & 1 \\ -1 & -2 & 3 & 1 & 0 \\ 3 & 0 & 1 & 7 & 3 \end{bmatrix}$

In Exercises 3 and 4, verify that rank$(A) = $ rank(A^T), in accordance with Theorem 7.5.2.

3. $A = \begin{bmatrix} 2 & 0 & -1 \\ 1 & 2 & 3 \\ 3 & -2 & -5 \\ 4 & 2 & 5 \end{bmatrix}$ **4.** $A = \begin{bmatrix} 1 & 2 & 4 & 0 \\ -3 & 1 & 5 & 2 \\ -2 & 3 & 9 & 2 \end{bmatrix}$

In each part of Exercises 5 and 6, use the information in the table to find the dimensions of the four fundamental spaces of A.

5.

	(a)	(b)	(c)	(d)	(e)
Size of A	3×3	3×3	3×3	5×9	9×5
Rank(A)	3	2	1	2	2

6.

	(a)	(b)	(c)	(d)	(e)
Size of A	3×4	4×3	6×3	5×7	7×4
Rank(A)	3	2	1	2	2

In each part of Exercises 7 and 8, use the information in the table to determine whether the linear system $A\mathbf{x} = \mathbf{b}$ is consistent. If so, state the number of parameters in a general solution.

7.

	(a)	(b)	(c)	(d)
Size of A	3×3	3×3	3×3	5×9
Rank(A)	3	2	1	2
Rank[$A \mid \mathbf{b}$]	3	3	1	2

8.

	(a)	(b)	(c)	(d)
Size of A	4×5	4×4	5×3	8×7
Rank(A)	3	2	3	4
Rank[$A \mid \mathbf{b}$]	4	2	3	4

In Exercises 9 and 10, determine whether A has full column rank, full row rank, both, or neither.

9. (a) $A = \begin{bmatrix} 1 & 3 \\ 2 & 0 \\ -1 & 4 \end{bmatrix}$ (b) $A = \begin{bmatrix} 1 & 4 & -2 \\ 2 & 8 & -4 \end{bmatrix}$

(c) $A = \begin{bmatrix} 3 & 1 & 5 \\ -1 & 0 & 1 \end{bmatrix}$ (d) $A = \begin{bmatrix} 2 & 0 \\ 1 & 4 \end{bmatrix}$

10. (a) $A = \begin{bmatrix} 1 & 3 \\ 2 & 6 \\ -1 & 0 \end{bmatrix}$ (b) $A = \begin{bmatrix} 2 & 5 & -1 \\ 0 & 3 & 6 \end{bmatrix}$

(c) $A = \begin{bmatrix} 4 & -1 & 3 \\ 8 & -2 & 6 \end{bmatrix}$ (d) $A = \begin{bmatrix} 5 & 7 \\ 6 & 1 \end{bmatrix}$

11. Check your answers in Exercise 9 using determinants to test for the invertibility of $A^T A$ and $A A^T$.

12. Check your answers in Exercise 10 using determinants to test for the invertibility of $A^T A$ and $A A^T$.

In Exercises 13 and 14, a matrix A with full column rank is given. In accordance with Theorem 7.5.6, verify that $A\mathbf{x} = \mathbf{0}$ has only the trivial solution and that the system $A\mathbf{x} = \mathbf{b}$ has at most one solution for every vector $\mathbf{b}$ in R^3.

13. $A = \begin{bmatrix} 1 & 2 \\ -1 & -2 \\ 3 & 0 \end{bmatrix}$ **14.** $A = \begin{bmatrix} 1 & 4 \\ 2 & 5 \\ 3 & 6 \end{bmatrix}$

In Exercises 15 and 16, verify that A and $A^T A$ have the same null space and row space, in accordance with Theorem 7.5.8.

15. $A = \begin{bmatrix} 1 & 2 \\ 2 & 4 \\ -1 & -2 \end{bmatrix}$ **16.** $A = \begin{bmatrix} 1 & 1 & 1 \\ 2 & 3 & -4 \end{bmatrix}$

17. According to Theorem 7.5.7, an overdetermined linear system $A\mathbf{x} = \mathbf{b}$ must be inconsistent for some vector $\mathbf{b}$. Find all values of b_1, b_2, b_3, b_4, b_5 for which the following overdetermined linear system is inconsistent:

$$\begin{aligned} x_1 - 3x_2 &= b_1 \\ x_1 - 2x_2 &= b_2 \\ x_1 + x_2 &= b_3 \\ x_1 - 4x_2 &= b_4 \\ x_1 + 5x_2 &= b_5 \end{aligned}$$

18. According to Theorem 7.5.7, an underdetermined linear system $A\mathbf{x} = \mathbf{b}$ is either inconsistent or has infinitely many solutions for each given vector $\mathbf{b}$. Find all values of b_1, b_2, and b_3 for which the following underdetermined linear system

has infinitely many solutions:

$$
\begin{aligned}
x_1 + 2x_2 + 3x_3 - x_4 &= b_1 \\
3x_1 - x_2 + x_3 + 2x_4 &= b_2 \\
4x_1 + x_2 + 4x_3 + x_4 &= b_3
\end{aligned}
$$

Discussion and Discovery

D1. If A is a 7×5 matrix with rank 3, then it follows that $\dim(\text{row}(A^T)) = $ _____, $\dim(\text{col}(A^T)) = $ _____, and $\dim(\text{null}(A^T)) = $ _____.

D2. If A is an $m \times n$ matrix with rank k, then it follows that $\dim(\text{row}(A^TA)) = $ _____ and $\dim(\text{row}(AA^T)) = $ _____.

D3. If the homogeneous system $A^T\mathbf{x} = \mathbf{0}$ has a unique solution, what can you say about the row space and column space of A? Explain your reasoning.

D4. Indicate whether the statement is true (T) or false (F). Justify your answer.

(a) If A has more rows than columns, then the dimension of the row space is greater than the dimension of the column space.

(b) If $\text{rank}(A) = \text{rank}(A^T)$, then A is square.

(c) If A is an invertible $n \times n$ matrix and $\mathbf{b}$ is any column vector in R^n, then the matrix A and the augmented matrix $[A \mid \mathbf{b}]$ have the same rank.

(d) If A has full row rank and full column rank, then A is square.

(e) If A^TA and AA^T are both invertible, then A is square.

(f) There is no 3×3 matrix whose rank and nullity are the same.

D5. The equation $x_1 + x_2 + x_3 = b$ can be viewed as an underdetermined linear system of one equation in three unknowns.

(a) Show that this system has infinitely many solutions for all values of b. Does this violate part (b) of Theorem 7.5.7?

(b) In accordance with Theorem 3.5.2, express a general solution of this system as the sum of a particular solution plus a general solution of the corresponding homogeneous system.

D6. (a) Show that if A is a 3×5 matrix, then the column vectors of A are linearly dependent.

(b) Show that if A is a 5×3 matrix, then the row vectors of A are linearly dependent.

(c) Generalize the results in parts (a) and (b) to an $m \times n$ matrix for which $m \neq n$.

Working with Proofs

P1. Prove that if A is an $m \times n$ matrix, then A^TA and AA^T have the same rank.

P2. Prove part (d) of Theorem 7.5.8 by using part (a) of the theorem and the fact that A and A^TA have n columns.

P3. (a) Prove part (b) of Theorem 7.5.8 by first showing that $\text{row}(A^TA)$ is a subspace of $\text{row}(A)$.

(b) Prove part (c) of Theorem 7.5.8 by using part (b).

P4. Prove: If A is a matrix that is not square, then either the row vectors of A or the column vectors of A are linearly dependent.

P5. Prove: If A is a square matrix for which A and A^2 have the same rank, then $\text{null}(A) \cap \text{col}(A) = \{\mathbf{0}\}$. [*Hint:* First show that $\text{null}(A) = \text{null}(A^2)$.]

P6. Prove: If A is a nonzero matrix with rank k, then A has at least one invertible $k \times k$ submatrix, and all square submatrices with larger size are singular. Conversely, if the size of the largest invertible submatrix of a nonzero matrix A is $k \times k$, then A has rank k. (Interpret a submatrix of A to be A itself or a matrix obtained from A by deleting rows and

columns.) For the first part of the theorem, assume that A has rank k and proceed as follows:

Step 1. First show that there is a submatrix with k linearly independent columns, and then show that there is a $k \times k$ invertible submatrix of that matrix.

Step 2. Show that if C is an $r \times r$ submatrix of A for which $r > k$, then those columns of A that contain the columns of C are linearly dependent.

For the converse, assume that A has rank r, that A has an invertible $k \times k$ submatrix, and that all square submatrices of A with larger size are singular. Use the first part of the theorem to show that $r = k$.

P7. Exercise P3 of Section 7.3 stated that if P is an invertible $n \times n$ matrix and A is any $n \times k$ matrix, then

$$\text{rank}(PA) = \text{rank}(A) \quad \text{and} \quad \text{nullity}(PA) = \text{nullity}(A)$$

Use these facts and Theorem 7.5.2 to prove that if P is an invertible $n \times n$ matrix and C is any $k \times n$ matrix, then

$$\text{rank}(CP) = \text{rank}(C) \quad \text{and} \quad \text{nullity}(CP) = \text{nullity}(C)$$

Technology Exercises

T1. (*Finding rank using determinants*) Since a square matrix is invertible if and only if its determinant is nonzero, it follows from Exercise P6 that the rank of a nonzero matrix A is the order of the largest square submatrix of A whose determinant is nonzero. Use this fact to find the rank of A, and check your answer by using a different method to find the rank.

$$A = \begin{bmatrix} 3 & -1 & 3 & 2 & 5 \\ 5 & -3 & 2 & 3 & 4 \\ 1 & -3 & -5 & 0 & -7 \\ 7 & -5 & 1 & 4 & 1 \end{bmatrix}$$

Section 7.6 The Pivot Theorem and Its Implications

In this section we will develop an important theorem about column spaces of matrices that will lead to a method for extracting bases from spanning sets.

BASIS PROBLEMS REVISITED

Let us reconsider the problem of finding a basis for a subspace W spanned by a set of vectors $S = \{\mathbf{v}_1, \mathbf{v}_2, \ldots, \mathbf{v}_s\}$. There are two variations of this problem:

1. Find any basis for W.

2. Find a basis for W consisting of vectors from S.

We have already seen that the first basis problem can be solved by making the vectors in S into row vectors of a matrix, reducing the matrix to row echelon form, and then extracting the nonzero row vectors (Example 4 of Section 7.3). One way to solve the second basis problem is to create a matrix A that has the vectors of S as *column vectors*. This makes W into the column space of A and converts the problem into one of finding a basis for the column space of A consisting of column vectors of A. To develop a method for doing this, we will need some preliminary ideas.

We know from Theorem 7.3.7 that elementary row operations do not change the row space or the null space of a matrix. However, elementary row operations *do* change the column space of a matrix. For example, the matrix

$$A = \begin{bmatrix} 1 & 2 \\ 1 & 2 \end{bmatrix}$$

can be reduced by one elementary row operation to

$$B = \begin{bmatrix} 1 & 2 \\ 0 & 0 \end{bmatrix}$$

But these matrices do not have the same column space—the column space of A is the span of the vector $\mathbf{v} = (1, 1)$ and the column space of B is the span of the vector $\mathbf{w} = (1, 0)$; and these are different lines through the origin of R^2. Note, however, that the column vectors of A, which we will denote by $\mathbf{c}_1$ and $\mathbf{c}_2$, satisfy the equation

$$\mathbf{c}_1 - \tfrac{1}{2}\mathbf{c}_2 = \mathbf{0}$$

and the column vectors of B, which we will denote by $\mathbf{c}'_1$ and $\mathbf{c}'_2$, satisfy the equation

$$\mathbf{c}'_1 - \tfrac{1}{2}\mathbf{c}'_2 = \mathbf{0}$$

The corresponding coefficients in these equations are the same; thus, even though the row operation that produced B from A did not preserve the column space, it *did* preserve the dependency relationship between the column vectors. More generally, suppose that A and B are row equivalent matrices that have been partitioned into column vectors as

$$A = [\mathbf{c}_1 \quad \mathbf{c}_2 \quad \cdots \quad \mathbf{c}_n] \quad \text{and} \quad B = [\mathbf{c}'_1 \quad \mathbf{c}'_2 \quad \cdots \quad \mathbf{c}'_n]$$

It follows from part (b) of Theorem 7.3.7 that the homogeneous linear systems $A\mathbf{x} = \mathbf{0}$ and $B\mathbf{x} = \mathbf{0}$ have the same solution set and hence the same is true of the vector equations

$$x_1\mathbf{c}_1 + x_2\mathbf{c}_2 + \cdots + x_n\mathbf{c}_n = \mathbf{0}$$

and

$$x_1\mathbf{c}_1' + x_2\mathbf{c}_2' + \cdots + x_n\mathbf{c}_n' = \mathbf{0}$$

since these are the vector forms of the two homogeneous systems. It follows from these equations that the column vectors of A are linearly independent if and only if the column vectors of B are linearly independent; and further, if the column vectors of A and B are linearly dependent, then those column vectors have the same dependency relationships. That is, *elementary row operations do not change linear independence or dependence of column vectors, and in the case of linear dependence they do not change dependency relationships among column vectors.* It can be proved that these conclusions also apply to any subset of the column vectors, which leads us to the following theorem.

> **Theorem 7.6.1** *Let A and B be row equivalent matrices.*
>
> (a) *If some subset of column vectors from A is linearly independent, then the corresponding column vectors from B are linearly independent, and conversely.*
>
> (b) *If some subset of column vectors from B is linearly dependent, then the corresponding column vectors from A are linearly dependent, and conversely. Moreover, the column vectors in the two matrices have the same dependency relationships.*

This theorem is the key to finding a set of column vectors of a matrix that forms a basis for its column space. Here is an example.

EXAMPLE 1
A Basis for
col(A)
Consisting of
Column Vectors
of A

Find a subset of the column vectors of

$$A = \begin{bmatrix} 1 & -3 & 4 & -2 & 5 & 4 \\ 2 & -6 & 9 & -1 & 8 & 2 \\ 2 & -6 & 9 & -1 & 9 & 7 \\ -1 & 3 & -4 & 2 & -5 & -4 \end{bmatrix}$$

that forms a basis for the column space of A.

Solution We leave it for you to confirm that reducing A to row echelon form by Gaussian elimination yields

$$U = \begin{bmatrix} 1 & -3 & 4 & -2 & 5 & 4 \\ 0 & 0 & 1 & 3 & -2 & -6 \\ 0 & 0 & 0 & 0 & 1 & 5 \\ 0 & 0 & 0 & 0 & 0 & 0 \end{bmatrix}$$

Since elementary row operations do not alter rank, and since U has three nonzero rows, it follows that A has rank 3 and hence that the column space of A is three-dimensional. Thus, if we can find three linearly independent column vectors in A, then those vectors will form a basis for the column space of A by Theorem 7.2.6. For this purpose, focus on the column vectors of U that have the leading 1's (columns 1, 3, and 5):

$$U = \begin{bmatrix} 1 & * & 4 & * & 5 & * \\ 0 & * & 1 & * & -2 & * \\ 0 & * & 0 & * & 1 & * \\ 0 & * & 0 & * & 0 & * \end{bmatrix}$$

If we progress from left to right through these three column vectors, we see that none of them is a linear combination of predecessors because there is no way to obtain its leading 1 by such a linear combination. This implies that these column vectors are linearly independent, and hence so are the corresponding column vectors of A by Theorem 7.6.1. Thus, the column vectors

$$\mathbf{c}_1 = \begin{bmatrix} 1 \\ 2 \\ 2 \\ -1 \end{bmatrix}, \quad \mathbf{c}_3 = \begin{bmatrix} 4 \\ 9 \\ 9 \\ -4 \end{bmatrix}, \quad \mathbf{c}_5 = \begin{bmatrix} 5 \\ 8 \\ 9 \\ -5 \end{bmatrix}$$

form a basis for the column space of A.   ∎

In this example we found that the column vectors of A occupying the column positions of the leading 1's in the row echelon form U form a basis for the column space of A. In general, the column vectors in these positions have a name associated with them.

Definition 7.6.2   The column vectors of a matrix A that lie in the column positions where the leading 1's occur in the row echelon forms of A are called the ***pivot columns*** of A.

It is a straightforward matter to convert the method of Example 1 into a proof of the following general result.

Theorem 7.6.3 (*The Pivot Theorem*)   *The pivot columns of a nonzero matrix A form a basis for the column space of A.*

REMARK At the end of Section 7.5 we gave a slightly tedious proof of the rank theorem (Theorem 7.5.1). Theorem 7.6.3 now provides us with a simpler way of seeing that result. We need only observe that the number of pivot columns in a nonzero matrix A is the same as the number of leading 1's in a row echelon form, which is the same as the number of nonzero rows in that row echelon form. Thus, Theorems 7.6.3 and part (c) of Theorem 7.3.7 imply that the column space and row space have the same dimension.

We now have all of the mathematical machinery required to solve the second basis problem posed at the beginning of this section:

Algorithm 1   If W is the subspace of R^n spanned by $S = \{\mathbf{v}_1, \mathbf{v}_2, \ldots, \mathbf{v}_s\}$, then the following procedure extracts a basis for W from S and expresses the vectors of S that are not in the basis as linear combinations of the basis vectors.

Step 1. Form the matrix A that has $\mathbf{v}_1, \mathbf{v}_2, \ldots, \mathbf{v}_s$ as successive column vectors.

Step 2. Reduce A to a row echelon form U, and identify the columns with the leading 1's to determine the pivot columns of A.

Step 3. Extract the pivot columns of A to obtain a basis for W. If appropriate, rewrite these basis vectors in comma-delimited form.

Step 4. If it is desired to express the vectors of S that are not in the basis as linear combinations of the basis vectors, then continue reducing U to obtain the reduced row echelon form R of A.

Step 5. By inspection, express each column vector of R that does not contain a leading 1 as a linear combination of preceding column vectors that contain leading 1's. Replace the column vectors in these linear combinations by the corresponding column vectors of A to obtain equations that express the column vectors of A that are not in the basis as linear combinations of basis vectors.

EXAMPLE 2
Extracting a
Basis from a
Set of Spanning
Vectors

Let W be the subspace of R^4 that is spanned by the vectors

$$\mathbf{v}_1 = (1, -2, 0, 3), \quad \mathbf{v}_2 = (2, -5, -3, 6), \quad \mathbf{v}_3 = (0, 1, 3, 0)$$
$$\mathbf{v}_4 = (2, -1, 4, -7), \quad \mathbf{v}_5 = (5, -8, 1, 2)$$

(a) Find a subset of these vectors that forms a basis for W.

(b) Express those vectors of $S = \{\mathbf{v}_1, \mathbf{v}_2, \mathbf{v}_3, \mathbf{v}_4, \mathbf{v}_5\}$ that are not in the basis as linear combinations of those vectors that are.

Solution (a) We start by creating a matrix whose column space is W. Such a matrix is

$$A = \begin{bmatrix} 1 & 2 & 0 & 2 & 5 \\ -2 & -5 & 1 & -1 & -8 \\ 0 & -3 & 3 & 4 & 1 \\ 3 & 6 & 0 & -7 & 2 \end{bmatrix}$$
$$\quad\; \uparrow \quad \uparrow \quad \uparrow \quad \uparrow \quad \uparrow$$
$$\quad\; \mathbf{v}_1 \quad \mathbf{v}_2 \quad \mathbf{v}_3 \quad \mathbf{v}_4 \quad \mathbf{v}_5$$

To find the pivot columns, we reduce A to row echelon form U by Gaussian elimination. We leave it for you to confirm that this yields

$$U = \begin{bmatrix} 1 & 2 & 0 & 2 & 5 \\ 0 & 1 & -1 & -3 & -2 \\ 0 & 0 & 0 & 1 & 1 \\ 0 & 0 & 0 & 0 & 0 \end{bmatrix}$$

The leading 1's occur in columns 1, 2, and 4, so the basis vectors for W (expressed in comma-delimited form) are

$$\mathbf{v}_1 = (1, -2, 0, 3), \quad \mathbf{v}_2 = (2, -5, -3, 6), \quad \mathbf{v}_4 = (2, -1, 4, -7)$$

Solution (b) For this problem it will be helpful to take A all the way to reduced row echelon form. We leave it for you to continue the reduction of U and confirm that the reduced row echelon form of A is

$$R = \begin{bmatrix} 1 & 0 & 2 & 0 & 1 \\ 0 & 1 & -1 & 0 & 1 \\ 0 & 0 & 0 & 1 & 1 \\ 0 & 0 & 0 & 0 & 0 \end{bmatrix}$$
$$\quad\; \uparrow \quad \uparrow \quad \uparrow \quad \uparrow \quad \uparrow$$
$$\quad\; \mathbf{v}'_1 \quad \mathbf{v}'_2 \quad \mathbf{v}'_3 \quad \mathbf{v}'_4 \quad \mathbf{v}'_5$$

where we have named the column vectors of R as $\mathbf{v}'_1$, $\mathbf{v}'_2$, $\mathbf{v}'_3$, $\mathbf{v}'_4$, and $\mathbf{v}'_5$. Our goal is to express $\mathbf{v}_3$ and $\mathbf{v}_5$ as linear combinations of $\mathbf{v}_1$, $\mathbf{v}_2$, and $\mathbf{v}_4$. However, we know that elementary row operations do not alter dependency relationships among column vectors. Thus, if we can express $\mathbf{v}'_3$ and $\mathbf{v}'_5$ as linear combinations of $\mathbf{v}'_1$, $\mathbf{v}'_2$, and $\mathbf{v}'_4$, then those same linear combinations will apply to the corresponding column vectors of A. By inspection from R,

$$\mathbf{v}'_3 = 2\mathbf{v}'_1 - \mathbf{v}'_2 \quad \text{and} \quad \mathbf{v}'_5 = \mathbf{v}'_1 + \mathbf{v}'_2 + \mathbf{v}'_4$$

Thus, it follows that

$$\mathbf{v}_3 = 2\mathbf{v}_1 - \mathbf{v}_2 \quad \text{and} \quad \mathbf{v}_5 = \mathbf{v}_1 + \mathbf{v}_2 + \mathbf{v}_4$$

As a check, you may want to confirm directly from the components of the vectors that these relationships are correct.   ∎

REMARK   To find a basis for the row space of a matrix A that consists of row vectors of A, you can apply Algorithm 1 in the following way: Transpose A, identify the pivot columns of A^T by row reduction, and then transpose those pivot columns to obtain row vectors of A that form a basis for the row space of A.

BASES FOR THE FUNDAMENTAL SPACES OF A MATRIX

We have already seen how to find bases for three of the four fundamental spaces of a matrix A by reducing the matrix to a row echelon form U or its reduced row echelon form R:

1. The nonzero rows of U form a basis for row(A).
2. The columns of U with leading 1's identify the pivot columns of A, and these form a basis for col(A).
3. The canonical solutions of $A\mathbf{x} = \mathbf{0}$ form a basis for null(A), and these are readily obtained from the system $R\mathbf{x} = \mathbf{0}$.

A basis for null(A^T) can be obtained by using row reduction of A^T to solve $A^T\mathbf{x} = \mathbf{0}$. However, it would be desirable to have an algorithm for finding a basis for null(A^T) by row reduction of A, since we would then have a common procedure for producing bases for all four fundamental spaces of A. We will now show how to do this.

Suppose that A is an $m \times n$ matrix with rank k, and we are interested in finding a basis for null(A^T) using elementary row operations on A. Recall that the dimension of null(A^T) is $m - k$ (number of rows − rank), so if $k = m$, then null(A^T) is the zero subspace of R^m, which has no basis. This being the case, we will assume that $k < m$. With this assumption, we are guaranteed that every row echelon form of A has at least one zero row (why?). Here is the procedure (which is justified by a proof at the end of this section):

Algorithm 2   If A is an $m \times n$ matrix with rank k, and if $k < m$, then the following procedure produces a basis for null(A^T) by elementary row operations on A.

Step 1. Adjoin the $m \times m$ identity matrix I_m to the right side of A to create the partitioned matrix $[A \mid I_m]$.

Step 2. Apply elementary row operations to $[A \mid I_m]$ until A is reduced to a row echelon form U, and let the resulting partitioned matrix be $[U \mid E]$.

Step 3. Repartition $[U \mid E]$ by adding a horizontal rule to split off the zero rows of U. This yields a matrix of the form

$$\begin{bmatrix} V & \vdots & E_1 \\ \hdashline 0 & \vdots & E_2 \end{bmatrix} \begin{matrix} k \\ m-k \end{matrix}$$
$$\quad n \quad m$$

where the margin entries indicate sizes.

Step 4. The row vectors of E_2 form a basis for null(A^T).

Here is an example.

EXAMPLE 3
A Basis for null(A^T) by Row Reduction of A

In Example 1 we found a basis for the column space of the matrix

$$A = \begin{bmatrix} 1 & -3 & 4 & -2 & 5 & 4 \\ 2 & -6 & 9 & -1 & 8 & 2 \\ 2 & -6 & 9 & -1 & 9 & 7 \\ -1 & 3 & -4 & 2 & -5 & -4 \end{bmatrix}$$

Apply Algorithm 2 to find a basis for null(A^T) by row reduction.

Solution In Example 1 we found that A has rank 3, so we know without performing any computations that null(A^T) has dimension 1 (number of rows − rank). Following the steps in the algorithm we obtain

$$[A \mid I_4] = \begin{bmatrix} 1 & -3 & 4 & -2 & 5 & 4 & | & 1 & 0 & 0 & 0 \\ 2 & -6 & 9 & -1 & 8 & 2 & | & 0 & 1 & 0 & 0 \\ 2 & -6 & 9 & -1 & 9 & 7 & | & 0 & 0 & 1 & 0 \\ -1 & 3 & -4 & 2 & -5 & -4 & | & 0 & 0 & 0 & 1 \end{bmatrix} \text{Step 1}$$

$$[U \mid E] = \begin{bmatrix} 1 & -3 & 4 & -2 & 5 & 4 & | & 1 & 0 & 0 & 0 \\ 0 & 0 & 1 & 3 & -2 & -6 & | & -2 & 1 & 0 & 0 \\ 0 & 0 & 0 & 0 & 1 & 5 & | & 0 & -1 & 1 & 0 \\ 0 & 0 & 0 & 0 & 0 & 0 & | & 1 & 0 & 0 & 1 \end{bmatrix} \text{Step 2}$$

$$\begin{bmatrix} V & | & E_1 \\ \hline 0 & | & E_2 \end{bmatrix} = \begin{bmatrix} 1 & -3 & 4 & -2 & 5 & 4 & | & 1 & 0 & 0 & 0 \\ 0 & 0 & 1 & 3 & -2 & -6 & | & -2 & 1 & 0 & 0 \\ 0 & 0 & 0 & 0 & 1 & 5 & | & 0 & -1 & 1 & 0 \\ \hline 0 & 0 & 0 & 0 & 0 & 0 & | & 1 & 0 & 0 & 1 \end{bmatrix} \text{Step 3}$$

As anticipated, the matrix E_2 has only one row vector, namely

$$\mathbf{w} = [1 \quad 0 \quad 0 \quad 1]$$

This vector is a basis for $\text{null}(A^T)$. Since we know that $\text{null}(A^T)$ and $\text{col}(A)$ are orthogonal complements, the vector $\mathbf{w}$ should be orthogonal to the basis vectors for $\text{col}(A)$ obtained in Example 1. We leave it for you confirm that this is so by showing that $\mathbf{w} \cdot \mathbf{c}_1 = 0$, $\mathbf{w} \cdot \mathbf{c}_3 = 0$, and $\mathbf{w} \cdot \mathbf{c}_5 = 0$. ∎

CONCEPT PROBLEM   Examples 1 and 3 provide bases for $\text{col}(A)$ and $\text{null}(A^T)$. Use the matrix in Step 3 above to find bases for $\text{row}(A)$ and $\text{null}(A)$.

A COLUMN-ROW FACTORIZATION

We have seen that the pivot columns of a matrix A form a basis for the column space of A (Theorem 7.6.3) and that the nonzero rows of any row echelon form of A form a basis for the row space of A (Theorem 7.3.7). The following beautiful theorem shows that every nonzero matrix A can be factored into a product of two matrices, the first factor consisting of the pivot columns of A and the second consisting of the nonzero rows in the reduced row echelon form of A.

> **Theorem 7.6.4 (Column-Row Factorization)**   *If A is a nonzero $m \times n$ matrix of rank k, then A can be factored as*
>
> $$A = CR \tag{1}$$
>
> *where C is the $m \times k$ matrix whose column vectors are the pivot columns of A and R is the $k \times n$ matrix whose row vectors are the nonzero rows in the reduced row echelon form of A.*

Proof   As in Algorithm 2, adjoin the $m \times m$ identity matrix I_m to the right side of A, and apply elementary row operations to $[A \mid I_m]$ until A is in its reduced row echelon form R_0. If the resulting partitioned matrix is $[R_0 \mid E]$, then E is the product of the elementary matrices that perform the row operations, so

$$EA = R_0 \tag{2}$$

Partition R_0 and E^{-1} as

$$R_0 = \begin{bmatrix} R \\ \hline 0 \end{bmatrix} \quad \text{and} \quad E^{-1} = [C \mid D]$$

where the matrix R consists of the nonzero row vectors of A, the matrix C consists of the first k column vectors of E, and the matrix D consists of the last $m - k$ columns of E. Thus, we can rewrite (2) as

$$A = E^{-1}R_0 = [C \mid D]\begin{bmatrix} R \\ \hline 0 \end{bmatrix} = CR + D0 = CR \tag{3}$$

It now remains to show that the successive columns of C are the successive pivot columns of A. For this purpose suppose that the pivot columns of A (and hence R_0) have column numbers

$$c_1, c_2, \ldots, c_k$$

A moment's reflection should make it evident that the column vectors of R in those positions are the standard unit vectors

$$\mathbf{e}_1, \mathbf{e}_2, \ldots, \mathbf{e}_k$$

in R^k. Thus, (3) implies that the jth pivot column of A is $C\mathbf{e}_j$, which is the jth column of C. ∎

EXAMPLE 4
Column-Row
Factorization

The reduced row echelon form of the matrix

$$A = \begin{bmatrix} 1 & 2 & 8 \\ -1 & -1 & -5 \\ 2 & 5 & 19 \end{bmatrix}$$

is

$$R_0 = \begin{bmatrix} 1 & 0 & 2 \\ 0 & 1 & 3 \\ 0 & 0 & 0 \end{bmatrix}$$

(verify), so A has the column-row factorization

$$\begin{bmatrix} 1 & 2 & 8 \\ -1 & -1 & -5 \\ 2 & 5 & 19 \end{bmatrix} = \begin{bmatrix} 1 & 2 \\ -1 & -1 \\ 2 & 5 \end{bmatrix}\begin{bmatrix} 1 & 0 & 2 \\ 0 & 1 & 3 \end{bmatrix}$$

$$\qquad\quad A \qquad\qquad\quad C \qquad\quad R$$

∎

COLUMN-ROW EXPANSION

We know from the column-row rule for matrix multiplication (Theorem 3.8.1) that a matrix product can be expressed as the sum of the outer products of the columns from the first factor times the corresponding rows of the second factor. Applying this to (1) yields the following result.

> **Theorem 7.6.5** (*Column-Row Expansion*) *If A is a nonzero matrix of rank k, then A can be expressed as*
>
> $$A = \mathbf{c}_1\mathbf{r}_1 + \mathbf{c}_2\mathbf{r}_2 + \cdots + \mathbf{c}_k\mathbf{r}_k \tag{4}$$
>
> *where $\mathbf{c}_1, \mathbf{c}_2, \ldots, \mathbf{c}_k$ are the successive pivot columns of A and $\mathbf{r}_1, \mathbf{r}_2, \ldots, \mathbf{r}_k$ are the successive nonzero row vectors in the reduced row echelon form of A.*

EXAMPLE 5
Column-Row
Expansion

From the column-row factorization obtained for the matrix A in Example 4, the corresponding column-row expansion of A is

$$\begin{bmatrix} 1 & 2 & 8 \\ -1 & -1 & -5 \\ 2 & 5 & 19 \end{bmatrix} = \begin{bmatrix} 1 \\ -1 \\ 2 \end{bmatrix}[1 \ \ 0 \ \ 2] + \begin{bmatrix} 2 \\ -1 \\ 5 \end{bmatrix}[0 \ \ 1 \ \ 3] = \begin{bmatrix} 1 & 0 & 2 \\ -1 & 0 & -2 \\ 2 & 0 & 4 \end{bmatrix} + \begin{bmatrix} 0 & 2 & 6 \\ 0 & -1 & -3 \\ 0 & 5 & 15 \end{bmatrix}$$

∎

OPTIONAL *Proof of Algorithm 2*

Assume that A is an $m \times n$ matrix with rank k, where $k < m$, and apply elementary row operations to $[A \mid I_m]$ until A is reduced to a row echelon form U. The row operations that reduce A to U can be performed by multiplying A on the left by an appropriate product E of elementary matrices. Thus,

$$EA = U \tag{5}$$

where E is invertible, since it is a product of elementary matrices, each of which is invertible. Multiplying $[A \mid I_m]$ on the left by E yields

$$E[A \mid I_m] = [EA \mid EI_m] = [U \mid E]$$

Now partition the matrix $[U \mid E]$ as

$$[U \mid E] = \begin{bmatrix} V & \vdots & E_1 \\ \cdots & + & \cdots \\ 0 & \vdots & E_2 \end{bmatrix} \tag{6}$$

where V is the $k \times n$ matrix of nonzero rows in U. It now follows from (5) and (6) that

$$\begin{bmatrix} V \\ 0 \end{bmatrix} = U = EA = \begin{bmatrix} E_1 \\ E_2 \end{bmatrix} A = \begin{bmatrix} E_1 A \\ E_2 A \end{bmatrix}$$

from which we see that $E_2 A = 0$. If we view the entries in $E_2 A$ as dot products of row vectors from E_2 with column vectors from A, then the equation $E_2 A = 0$ implies that each row vector of E_2 is orthogonal to each column vector of A. This places the row vectors of E_2 in the orthogonal complement of $\text{col}(A)$, which is $\text{null}(A^T)$. Thus, it only remains to show that the row vectors of E_2 form a basis for $\text{null}(A^T)$. Let us show first that the row vectors of E_2 are linearly independent. Since E is invertible, its row vectors are linearly independent by Theorem 7.4.4, and this implies that the row vectors of E_2 are linearly independent, since they are a subset of the row vectors of E. Moreover, there are $m - k$ row vectors in E_2, and the dimension of $\text{null}(A^T)$ is also $m - k$, so the row vectors of E_2 must be a basis for $\text{null}(A^T)$ by Theorem 7.2.6. ■

Exercise Set 7.6

In Exercises 1–6, find a basis for the column space of A that consists of column vectors of A, and find a basis for the row space of A that consists of row vectors of A.

1. $A = \begin{bmatrix} 1 & -1 & 3 \\ 5 & -4 & -4 \\ 7 & -6 & 2 \end{bmatrix}$

2. $A = \begin{bmatrix} 2 & 0 & -1 \\ 4 & 0 & -2 \\ 0 & 0 & 0 \end{bmatrix}$

3. $A = \begin{bmatrix} 1 & 4 & 5 & 2 \\ 2 & 1 & 3 & 0 \\ -1 & 3 & 2 & 2 \end{bmatrix}$

4. $A = \begin{bmatrix} 1 & 4 & 5 & 6 & 9 \\ 3 & -2 & 1 & 4 & -1 \\ -1 & 0 & -1 & -2 & -1 \\ 2 & 3 & 5 & 7 & 8 \end{bmatrix}$

5. $A = \begin{bmatrix} 1 & -3 & 2 & 2 & 1 \\ 0 & 3 & 6 & 0 & -3 \\ 2 & -3 & -2 & 4 & 4 \\ 3 & -6 & 0 & 6 & 5 \\ -2 & 9 & 2 & -4 & -5 \end{bmatrix}$

6. $A = \begin{bmatrix} 1 & 3 & 2 & 1 \\ -2 & -6 & 0 & -6 \\ 3 & 9 & 1 & 8 \\ -1 & -3 & -3 & -6 \\ 1 & 3 & 2 & 1 \\ 4 & 12 & 1 & 11 \end{bmatrix}$

In Exercises 7–10, find a subset of the vectors that forms a basis for the space spanned by the vectors; then express each of the remaining vectors in the set as a linear combination of the basis vectors.

7. $\mathbf{v}_1 = (1, 2, -1, 1)$, $\mathbf{v}_2 = (4, 0, -6, 2)$,
$\mathbf{v}_3 = (1, 10, -11, 3)$

8. $\mathbf{v}_1 = (1, 0, 1, 1)$, $\mathbf{v}_2 = (-3, 3, 7, 1)$, $\mathbf{v}_3 = (-1, 3, 9, 3)$,
$\mathbf{v}_4 = (-5, 3, 5, -1)$

9. $\mathbf{v}_1 = (1, -2, 0, 3)$, $\mathbf{v}_2 = (2, -4, 0, 6)$, $\mathbf{v}_3 = (-1, 1, 2, 0)$,
$\mathbf{v}_4 = (0, -1, 2, 3)$

10. $\mathbf{v}_1 = (1, -1, 5, 2)$, $\mathbf{v}_2 = (-2, 3, 1, 0)$, $\mathbf{v}_3 = (4, -5, 9, 4)$,
$\mathbf{v}_4 = (0, 4, 2, -3)$, $\mathbf{v}_5 = (-7, 18, 2, -8)$

In Exercises 11 and 12, use Algorithm 2 to find a basis for $\text{null}(A^T)$.

11. The matrix in Exercise 2.   **12.** The matrix in Exercise 3.

In Exercises 13 and 14, find bases for the four fundamental spaces of A.

13. $A = \begin{bmatrix} 1 & -1 & 2 & 3 \\ -2 & 1 & -1 & 1 \\ 4 & -3 & 5 & 5 \\ 0 & -1 & 3 & 7 \end{bmatrix}$

14. $A = \begin{bmatrix} 3 & 2 & 1 & 1 \\ -1 & 2 & 3 & -2 \\ 1 & 6 & 7 & -5 \\ 7 & 2 & -1 & 0 \end{bmatrix}$

15. The following is a matrix A, its reduced row echelon form R, and the reduced row echelon form C of its transpose:

$$A = \begin{bmatrix} 4 & 16 & 8 & 8 \\ 0 & 0 & 5 & 15 \\ 0 & 0 & 0 & 4 \\ 0 & 0 & 3 & 9 \\ 1 & 4 & 2 & 2 \end{bmatrix}, \quad R = \begin{bmatrix} 1 & 4 & 0 & 0 \\ 0 & 0 & 1 & 0 \\ 0 & 0 & 0 & 1 \\ 0 & 0 & 0 & 0 \\ 0 & 0 & 0 & 0 \end{bmatrix}$$

$$C = \begin{bmatrix} 1 & 0 & 0 & 0 & \frac{1}{4} \\ 0 & 1 & 0 & \frac{3}{5} & 0 \\ 0 & 0 & 1 & 0 & 0 \\ 0 & 0 & 0 & 0 & 0 \end{bmatrix}$$

(a) Find a basis for $\text{col}(A)$ consisting of column vectors of A.

(b) Find a basis for $\text{row}(A)$ consisting of row vectors of A.

(c) Find a basis for $\text{null}(A)$.

(d) Find a basis for $\text{null}(A^T)$.

In Exercises 16 and 17, find the column-row factorization and column-row expansion of the matrix.

16. $\begin{bmatrix} 1 & 1 & 3 \\ 2 & -3 & -1 \\ 1 & 6 & 10 \end{bmatrix}$ **17.** $\begin{bmatrix} 2 & -4 & 2 & 6 \\ 1 & 1 & -4 & -2 \\ 3 & -9 & 8 & 14 \\ 0 & 6 & -10 & -10 \end{bmatrix}$

Discussion and Discovery

D1. (a) If A is a 3×5 matrix, then the number of leading 1's in a row echelon form of A is at most _____, the number of parameters in a general solution of $A\mathbf{x} = \mathbf{0}$ is at most _____, the rank of A is at most _____, the rank of A^T is at most _____, and the nullity of A^T is at most _____.

(b) If A is a 5×3 matrix, then the number of leading 1's in a row echelon form of A is at most _____, the number of parameters in a general solution of $A\mathbf{x} = \mathbf{0}$ is at most _____, the rank of A is at most _____, the rank of A^T is at most _____, and the nullity of A^T is at most _____.

(c) If A is a 4×4 matrix, then the number of leading 1's in a row echelon form of A is at most _____, the number of parameters in a general solution of $A\mathbf{x} = \mathbf{0}$ is at most _____, the rank of A is at most _____, the rank of A^T is at most _____, and the nullity of A^T is at most _____.

(d) If A is an $m \times n$ matrix, then the number of leading 1's in a row echelon form of A is at most _____,

the number of parameters in a general solution of $A\mathbf{x} = \mathbf{0}$ is at most _____, the rank of A is at most _____, the rank of A^T is at most _____, and the nullity of A^T is at most _____.

D2. In words, what are the pivot columns of a matrix? Find the pivot columns of the matrix

$$A = \begin{bmatrix} 0 & 0 & -2 & 0 & 7 & 12 \\ 2 & 4 & -10 & 6 & 12 & 28 \\ 2 & 4 & -5 & 6 & -5 & -1 \end{bmatrix}$$

D3. By inspection, find a basis for $\text{null}(A^T)$ given that the following matrix results when elementary row operations are applied to $[A \mid I_5]$ until A is in row echelon form.

$$\begin{bmatrix} 1 & -3 & 4 & -2 & 5 & 4 & 3 & 1 & 0 & 0 & 1 \\ 0 & 0 & 1 & 3 & -2 & -6 & -5 & 2 & 0 & 0 & 0 \\ 0 & 0 & 0 & 0 & 1 & 5 & 0 & 3 & -4 & 2 & 2 \\ 0 & 0 & 0 & 0 & 0 & 0 & 4 & 0 & 1 & -4 & 5 \\ 0 & 0 & 0 & 0 & 0 & 0 & 0 & 1 & 0 & 0 & 1 \end{bmatrix}$$

D4. Suppose that $A = [\mathbf{a}_1 \quad \mathbf{a}_2 \quad \mathbf{a}_3 \quad \mathbf{a}_4 \quad \mathbf{a}_5]$ is a 5×5 matrix whose reduced row echelon form is

$$\begin{bmatrix} 1 & 0 & 4 & 0 & 6 \\ 0 & 1 & -3 & 0 & 7 \\ 0 & 0 & 0 & 1 & 2 \\ 0 & 0 & 0 & 0 & 0 \\ 0 & 0 & 0 & 0 & 0 \end{bmatrix}$$

(a) Find a subset of $S = \{\mathbf{a}_1, \mathbf{a}_2, \mathbf{a}_3, \mathbf{a}_4, \mathbf{a}_5\}$ that forms a basis for $\mathrm{col}(A)$.

(b) Express each vector of S that is not in the basis as a linear combination of the basis vectors.

Technology Exercises

T1. Consider the vectors

$$\mathbf{v}_1 = (1, 2, 4, -6, 11, 23, -14, 0, 2, 2)$$
$$\mathbf{v}_2 = (3, 1, -1, 7, 9, 13, -12, 8, 6, -30)$$
$$\mathbf{v}_3 = (5, 5, 7, -5, 31, 59, -40, 8, 10, -26)$$
$$\mathbf{v}_4 = (5, 0, -6, 20, 7, 3, -10, 16, 10, -62)$$

Use Algorithm 1 to find a subset of these vectors that forms a basis for $\mathrm{span}\{\mathbf{v}_1, \mathbf{v}_2, \mathbf{v}_3, \mathbf{v}_4\}$, and express those vectors not in the basis as linear combinations of basis vectors.

T2. Consider the matrix

$$A = \begin{bmatrix} 1 & 3 & 2 & 1 \\ -2 & -6 & 0 & -6 \\ 3 & 9 & 1 & 8 \\ -1 & -3 & -3 & -6 \\ 1 & 3 & 2 & 1 \\ 4 & 12 & 1 & 11 \end{bmatrix}$$

(a) Use Algorithm 1 to find a subset of the column vectors of A that forms a basis for the column space of A, and express each column vector of A that is not in that basis as a linear combination of the basis vectors.

(b) By applying Algorithm 1 to A^T, find a subset of the row vectors of A that forms a basis for the row space of A, and express each row vector that is not in the basis as a linear combination of the basis vectors.

(c) Use Algorithm 2 to find a basis for the null space of the matrix A^T.

Section 7.7 The Projection Theorem and Its Implications

In this chapter we have studied three theorems that are fundamental in the study of R^n: the dimension theorem, the rank theorem, and the pivot theorem. In this section we will add a fourth result to that list of fundamental theorems.

ORTHOGONAL PROJECTIONS ONTO LINES IN R^2

In Sections 6.1 and 6.2 we discussed orthogonal projections onto lines through the origin of R^2 and onto the coordinate planes of a rectangular coordinate system in R^3. In this section we will be concerned with the more general problem of defining and calculating orthogonal projections onto subspaces of R^n. To motivate the appropriate definition, we will revisit orthogonal projections onto lines through the origin of R^2 from another point of view.

Recall from Formula (21) of Section 6.1 that the standard matrix P_θ for the orthogonal projection of R^2 onto the line through the origin making an angle θ with the positive x-axis of a rectangular xy-coordinate system can be expressed as

$$P_\theta = \begin{bmatrix} \cos^2 \theta & \sin \theta \cos \theta \\ \sin \theta \cos \theta & \sin^2 \theta \end{bmatrix} \tag{1}$$

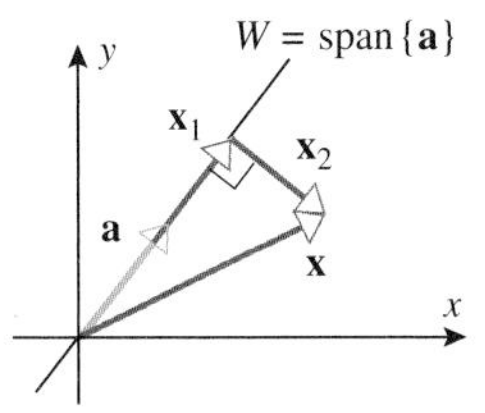

Now suppose that we are given a nonzero vector $\mathbf{a}$ in R^2, and let us consider how we might compute the orthogonal projection of a vector $\mathbf{x}$ onto the line $W = \mathrm{span}\{\mathbf{a}\}$ without explicitly computing θ.

Figure 7.7.1 suggests that the vector $\mathbf{x}$ can be expressed as

$$\mathbf{x} = \mathbf{x}_1 + \mathbf{x}_2 \tag{2}$$

Figure 7.7.1

where $\mathbf{x}_1$ is the orthogonal projection of $\mathbf{x}$ onto W, and

$$\mathbf{x}_2 = \mathbf{x} - \mathbf{x}_1$$

is the orthogonal projection onto the line through the origin that is perpendicular to W. The vector $\mathbf{x}_1$ is some scalar multiple of $\mathbf{a}$, say

$$\mathbf{x}_1 = k\mathbf{a} \tag{3}$$

and the vector $\mathbf{x}_2 = \mathbf{x} - \mathbf{x}_1 = \mathbf{x} - k\mathbf{a}$ is orthogonal to $\mathbf{a}$, so we must have

$$(\mathbf{x} - k\mathbf{a}) \cdot \mathbf{a} = 0$$

which we can rewrite as

$$\mathbf{x} \cdot \mathbf{a} - k(\mathbf{a} \cdot \mathbf{a}) = 0$$

Solving for k and substituting in (3) yields

$$\mathbf{x}_1 = \frac{\mathbf{x} \cdot \mathbf{a}}{\mathbf{a} \cdot \mathbf{a}}\mathbf{a} = \frac{\mathbf{x} \cdot \mathbf{a}}{\|\mathbf{a}\|^2}\mathbf{a} \tag{4}$$

which is a formula for the orthogonal projection of $\mathbf{x}$ onto the line span$\{\mathbf{a}\}$ in terms of $\mathbf{a}$ and $\mathbf{x}$. It is common to denote $\mathbf{x}_1$ by $\mathrm{proj}_\mathbf{a}\mathbf{x}$ and to express Formula (4) as

$$\mathrm{proj}_\mathbf{a}\mathbf{x} = \frac{\mathbf{x} \cdot \mathbf{a}}{\|\mathbf{a}\|^2}\mathbf{a} \tag{5}$$

The following example shows that this formula is consistent with Formula (1).

EXAMPLE 1
Orthogonal
Projection onto
a Line Through
the Origin of R^2

Use Formula (5) to obtain the standard matrix P_θ for the orthogonal projection of R^2 onto the line W through the origin that makes an angle θ with the positive x-axis of a rectangular xy-coordinate system.

Solution The vector $\mathbf{u} = (\cos\theta, \sin\theta)$ is a unit vector along W (Figure 7.7.2), so if we use this vector as the $\mathbf{a}$ in Formula (5) and use the fact that $\|\mathbf{u}\| = 1$, then we obtain

$$\mathrm{proj}_\mathbf{u}\mathbf{x} = (\mathbf{x} \cdot \mathbf{u})\mathbf{u}$$

In particular, the orthogonal projections of the standard unit vectors $\mathbf{e}_1 = (1, 0)$ and $\mathbf{e}_2 = (0, 1)$ onto the line are

$$\mathrm{proj}_\mathbf{u}\mathbf{e}_1 = (\mathbf{e}_1 \cdot \mathbf{u})\mathbf{u} = (\cos\theta)\mathbf{u} = (\cos^2\theta, \cos\theta\sin\theta) = (\cos^2\theta, \sin\theta\cos\theta)$$
$$\mathrm{proj}_\mathbf{u}\mathbf{e}_2 = (\mathbf{e}_2 \cdot \mathbf{u})\mathbf{u} = (\sin\theta)\mathbf{u} = (\sin\theta\cos\theta, \sin^2\theta)$$

Expressing these vectors in column form yields the standard matrix

$$P_\theta = \begin{bmatrix} \mathrm{proj}_\mathbf{u}\mathbf{e}_1 & \mathrm{proj}_\mathbf{u}\mathbf{e}_2 \end{bmatrix} = \begin{bmatrix} \cos^2\theta & \sin\theta\cos\theta \\ \sin\theta\cos\theta & \sin^2\theta \end{bmatrix}$$

which is consistent with Formula (1).   ∎

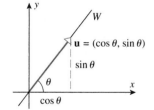

Figure 7.7.2

**ORTHOGONAL
PROJECTIONS ONTO
LINES THROUGH THE
ORIGIN OF R^n**

The following theorem, which extends Formula (2) to R^n, is the foundation for defining orthogonal projections onto lines through the origin of R^n (Figure 7.7.3).

Theorem 7.7.1 *If $\mathbf{a}$ is a nonzero vector in R^n, then every vector $\mathbf{x}$ in R^n can be expressed in exactly one way as*

$$\mathbf{x} = \mathbf{x}_1 + \mathbf{x}_2 \tag{6}$$

where $\mathbf{x}_1$ is a scalar multiple of $\mathbf{a}$ and $\mathbf{x}_2$ is orthogonal to $\mathbf{a}$ (and hence to $\mathbf{x}_1$). The vectors $\mathbf{x}_1$ and $\mathbf{x}_2$ are given by the formulas

$$\mathbf{x}_1 = \frac{\mathbf{x} \cdot \mathbf{a}}{\|\mathbf{a}\|^2}\mathbf{a} \quad and \quad \mathbf{x}_2 = \mathbf{x} - \mathbf{x}_1 = \mathbf{x} - \frac{\mathbf{x} \cdot \mathbf{a}}{\|\mathbf{a}\|^2}\mathbf{a} \tag{7}$$

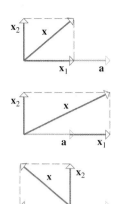

x_1 = vector component of x along a

x_2 = vector component of x orthogonal to a

Figure 7.7.3

Proof There are two parts to the proof, an *existence* part and a *uniqueness* part. The existence part is to show that there actually exist vectors x_1 and x_2 that satisfy (6) such that x_1 is a scalar multiple of a and x_2 is orthogonal to a; and the uniqueness part is to show that if x_1' and x_2' is a second pair of vectors that satisfy these conditions, then $x_1 = x_1'$ and $x_2 = x_2'$.

For the existence part, we will show that the vectors x_1 and x_2 in (7) satisfy the required conditions. It is obvious that x_1 is a scalar multiple of a and that (6) holds, so let us focus on proving that x_2 is orthogonal to a, that is, $x_2 \cdot a = 0$. The computations are as follows:

$$x_2 \cdot a = (x - x_1) \cdot a = \left(x - \frac{x \cdot a}{a \cdot a}a\right) \cdot a = (x \cdot a) - \frac{x \cdot a}{a \cdot a}(a \cdot a) = 0$$

For the uniqueness part, suppose that x can also be written as $x = x_1' + x_2'$, where x_1' is a scalar multiple of a and x_2' is orthogonal to a. Then

$$x_1 + x_2 = x_1' + x_2'$$

from which it follows that

$$x_1 - x_1' = x_2' - x_2 \tag{8}$$

Since x_1 and x_1' are both scalar multiples of a, so is their difference, and hence we can rewrite (8) in the form

$$ka = x_2' - x_2 \tag{9}$$

Moreover, since x_2' and x_2 are orthogonal to a, so is their difference, and hence it follows from (9) that

$$ka \cdot a = k(a \cdot a) = 0 \tag{10}$$

But $a \cdot a = \|a\|^2 \neq 0$, since we assumed that $a \neq 0$, so (10) implies that $k = 0$. Thus (9) implies that $x_2 = x_2'$ and (8), in turn, implies that $x_1 = x_1'$, which proves the uniqueness. ∎

In the special case where a and x are vectors in R^2, the formula for x_1 in (7) coincides with Formula (5) for the orthogonal projection of x onto span{a}, and this suggests that we use Formula (7) to *define* orthogonal projections onto lines through the origin in R^n.

Definition 7.7.2   If a is a nonzero vector in R^n, and if x is any vector in R^n, then the ***orthogonal projection of*** x ***onto*** span{a} is denoted by $\mathrm{proj}_a x$ and is defined as

$$\mathrm{proj}_a x = \frac{x \cdot a}{\|a\|^2}a \tag{11}$$

The vector $\mathrm{proj}_a x$ is also called the ***vector component of*** x ***along*** a, and $x - \mathrm{proj}_a x$ is called the ***vector component of*** x ***orthogonal to*** a.

EXAMPLE 2
Calculating Vector Components

Let $x = (2, -1, 3)$ and $a = (4, -1, 2)$. Find the vector components of x along a and orthogonal to a.

Solution Since

$$x \cdot a = (2)(4) + (-1)(-1) + (3)(2) = 15 \tag{12}$$

and

$$\|a\|^2 = a \cdot a = 4^2 + (-1)^2 + 2^2 = 21 \tag{13}$$

it follows from (7) and (11) that the vector component of x along a is

$$x_1 = \mathrm{proj}_a x = \frac{x \cdot a}{\|a\|^2}a = \tfrac{15}{21}(4, -1, 2) = \left(\tfrac{20}{7}, -\tfrac{5}{7}, \tfrac{10}{7}\right)$$

and that the vector component of x orthogonal to a is

$$x_2 = x - \mathrm{proj}_a x = (2, -1, 3) - \left(\tfrac{20}{7}, -\tfrac{5}{7}, \tfrac{10}{7}\right) = \left(-\tfrac{6}{7}, -\tfrac{2}{7}, \tfrac{11}{7}\right)$$

As a check, you may want to confirm that x_1 and x_2 are orthogonal and that $x = x_1 + x_2$. ∎

Sometimes we will be interested in finding the *length* of $\text{proj}_\mathbf{a}\mathbf{x}$ but will not need the projection itself. A formula for this length can be derived from (11) by writing

$$\|\text{proj}_\mathbf{a}\mathbf{x}\| = \left\| \frac{\mathbf{x} \cdot \mathbf{a}}{\|\mathbf{a}\|^2}\mathbf{a} \right\| = \frac{|\mathbf{x} \cdot \mathbf{a}|}{\|\mathbf{a}\|^2}\|\mathbf{a}\|$$

which simplifies to

$$\|\text{proj}_\mathbf{a}\mathbf{x}\| = \frac{|\mathbf{x} \cdot \mathbf{a}|}{\|\mathbf{a}\|} \tag{14}$$

EXAMPLE 3
Computing the
Length of an
Orthogonal
Projection

Use Formula (14) to compute the length of $\text{proj}_\mathbf{a}\mathbf{x}$ for the vectors $\mathbf{a}$ and $\mathbf{x}$ in Example 2.

Solution   Using the results in (12) and (13), we obtain

$$\|\text{proj}_\mathbf{a}\mathbf{x}\| = \frac{|\mathbf{x} \cdot \mathbf{a}|}{\|\mathbf{a}\|} = \frac{|15|}{\sqrt{21}} = \frac{15}{\sqrt{21}}$$

We leave it for you to check this result by calculating directly from the vector $\text{proj}_\mathbf{a}\mathbf{x}$ obtained in Example 2.   ∎

PROJECTION OPERATORS
ON R^n

Since the vector $\mathbf{x}$ in Definition 7.7.2 is arbitrary, we can use Formula (11) to define an operator $T : R^n \to R^n$ by

$$T(\mathbf{x}) = \text{proj}_\mathbf{a}\mathbf{x} = \frac{\mathbf{x} \cdot \mathbf{a}}{\|\mathbf{a}\|^2}\mathbf{a} \tag{15}$$

This is called the ***orthogonal projection of R^n onto*** **span**$\{\mathbf{a}\}$. We leave it as an exercise for you to show that this is a linear operator. The following theorem provides a formula for the standard matrix for T.

> **Theorem 7.7.3**   *If $\mathbf{a}$ is a nonzero vector in R^n, and if $\mathbf{a}$ is expressed in column form, then the standard matrix for the linear operator $T(\mathbf{x}) = \text{proj}_\mathbf{a}\mathbf{x}$ is*
>
> $$P = \frac{1}{\mathbf{a}^T\mathbf{a}}\mathbf{a}\mathbf{a}^T \tag{16}$$
>
> *This matrix is symmetric and has rank 1.*

Proof   The column vectors of the standard matrix for a linear transformation are the images of the standard basis vectors under the transformation. Thus, if we denote the jth entry of $\mathbf{a}$ by a_j, then the jth column of P is given by

$$T(\mathbf{e}_j) = \text{proj}_\mathbf{a}\mathbf{e}_j = \frac{\mathbf{e}_j \cdot \mathbf{a}}{\|\mathbf{a}\|^2}\mathbf{a} = \frac{a_j}{\|\mathbf{a}\|^2}\mathbf{a}$$

Accordingly, partitioning P into column vectors yields

$$P = \left[\frac{a_1}{\|\mathbf{a}\|^2}\mathbf{a} \,\middle|\, \frac{a_2}{\|\mathbf{a}\|^2}\mathbf{a} \,\middle|\, \cdots \,\middle|\, \frac{a_n}{\|\mathbf{a}\|^2}\mathbf{a} \right] = \frac{1}{\|\mathbf{a}\|^2}[a_1\mathbf{a} \mid a_2\mathbf{a} \mid \cdots \mid a_n\mathbf{a}]$$

$$= \frac{1}{\|\mathbf{a}\|^2}\mathbf{a}[a_1 \mid a_2 \mid \cdots \mid a_n] = \frac{1}{\mathbf{a}^T\mathbf{a}}\mathbf{a}\mathbf{a}^T$$

which proves (16). Finally, the matrix $\mathbf{a}\mathbf{a}^T$ is symmetric and has rank 1 (Theorem 7.4.7), so P, being a nonzero scalar multiple of $\mathbf{a}\mathbf{a}^T$, must also be symmetric and have rank 1.   ∎

CONCEPT PROBLEM   Explain geometrically why you would expect the standard matrix for the projection $T(\mathbf{x}) = \text{proj}_\mathbf{a}\mathbf{x}$ to have rank 1.

We leave it as an exercise for you to show that the matrix P in Formula (16) does not change if $\mathbf{a}$ is replaced by any nonzero scalar multiple of $\mathbf{a}$. This means that P is determined by the *line* onto which it projects and not by the particular basis vector $\mathbf{a}$ that is used to span the line. In particular, we can use a unit vector $\mathbf{u}$ along the line, in which case $\mathbf{u}^T\mathbf{u} = \|\mathbf{u}\|^2 = 1$, and the formula for P simplifies to

$$P = \mathbf{u}\mathbf{u}^T \tag{17}$$

Thus, we have shown that *the standard matrix for the orthogonal projection of R^n onto a line through the origin can be obtained by finding a unit vector along the line and forming the outer product of that vector with itself.*

EXAMPLE 4
Example 1
Revisited

Use Formula (17) to obtain the standard matrix P_θ for the orthogonal projection of R^2 onto the line W through the origin that makes an angle θ with the positive x-axis of a rectangular xy-coordinate system.

Solution Since $\mathbf{u} = (\cos\theta, \sin\theta)$ is a unit vector along the line, we write this vector in column form and apply Formula (17) to obtain

$$P_\theta = \mathbf{u}\mathbf{u}^T = \begin{bmatrix} \cos\theta \\ \sin\theta \end{bmatrix} [\cos\theta \quad \sin\theta] = \begin{bmatrix} \cos^2\theta & \sin\theta\cos\theta \\ \sin\theta\cos\theta & \sin^2\theta \end{bmatrix} \tag{18}$$

This agrees with the result in Example 1.   ∎

EXAMPLE 5
The Standard
Matrix for an
Orthogonal
Projection

(a) Find the standard matrix P for the orthogonal projection of R^3 onto the line spanned by the vector $\mathbf{a} = (1, -4, 2)$.

(b) Use the matrix to find the orthogonal projection of the vector $\mathbf{x} = (2, -1, 3)$ onto the line spanned by $\mathbf{a}$.

(c) Show that P has rank 1, and interpret this result geometrically.

Solution (a) Expressing $\mathbf{a}$ in column form we obtain

$$\mathbf{a}^T\mathbf{a} = [1 \quad -4 \quad 2] \begin{bmatrix} 1 \\ -4 \\ 2 \end{bmatrix} = 21 \quad \text{and} \quad \mathbf{a}\mathbf{a}^T = \begin{bmatrix} 1 \\ -4 \\ 2 \end{bmatrix} [1 \quad -4 \quad 2] = \begin{bmatrix} 1 & -4 & 2 \\ -4 & 16 & -8 \\ 2 & -8 & 4 \end{bmatrix}$$

Thus, it follows from (16) that the standard matrix P for the orthogonal projection is

$$P = \frac{1}{21} \begin{bmatrix} 1 & -4 & 2 \\ -4 & 16 & -8 \\ 2 & -8 & 4 \end{bmatrix} = \begin{bmatrix} \frac{1}{21} & -\frac{4}{21} & \frac{2}{21} \\ -\frac{4}{21} & \frac{16}{21} & -\frac{8}{21} \\ \frac{2}{21} & -\frac{8}{21} & \frac{4}{21} \end{bmatrix} \tag{19}$$

[We could also have obtained this result by normalizing $\mathbf{a}$ and applying Formula (17) using the normalized vector $\mathbf{u}$.] Note that P is symmetric, as expected.

Solution (b) The orthogonal projection of $\mathbf{x}$ onto the line spanned by $\mathbf{a}$ is the product $P\mathbf{x}$ with $\mathbf{x}$ expressed in column form. Thus,

$$\text{proj}_\mathbf{a}\mathbf{x} = P\mathbf{x} = \begin{bmatrix} \frac{1}{21} & -\frac{4}{21} & \frac{2}{21} \\ -\frac{4}{21} & \frac{16}{21} & -\frac{8}{21} \\ \frac{2}{21} & -\frac{8}{21} & \frac{4}{21} \end{bmatrix} \begin{bmatrix} 2 \\ -1 \\ 3 \end{bmatrix} = \begin{bmatrix} \frac{12}{21} \\ -\frac{48}{21} \\ \frac{24}{21} \end{bmatrix} = \begin{bmatrix} \frac{4}{7} \\ -\frac{16}{7} \\ \frac{8}{7} \end{bmatrix}$$

Solution (c) The matrix P in (19) has rank 1 since the second and third columns are scalar multiples of the first. This tells us that the column space of P is one-dimensional, which makes sense because the column space is the range of the linear operator represented by P, and we know that this is a line through the origin. ∎

ORTHOGONAL PROJECTIONS ONTO GENERAL SUBSPACES

We will now turn to the problem of defining and calculating orthogonal projections onto general subspaces of R^n. The following generalization of Theorem 7.7.1 will be our first step in that direction.

Theorem 7.7.4 (*Projection Theorem for Subspaces*) *If W is a subspace of R^n, then every vector $\mathbf{x}$ in R^n can be expressed in exactly one way as*

$$\mathbf{x} = \mathbf{x}_1 + \mathbf{x}_2 \tag{20}$$

where $\mathbf{x}_1$ is in W and $\mathbf{x}_2$ is in $W^{\perp}$.

Proof We will leave the case where $W = \{\mathbf{0}\}$ as an exercise, so we may assume that $W \neq \{\mathbf{0}\}$ and hence has a basis. Let $\{\mathbf{w}_1, \mathbf{w}_2, \ldots, \mathbf{w}_k\}$ be a basis for W, and form the matrix M that has these basis vectors as successive columns. This makes W the column space of M and $W^{\perp}$ the null space of M^T. Thus, the proof will be complete if we can show that every vector $\mathbf{x}$ in R^n can be expressed in exactly one way as

$$\mathbf{x} = \mathbf{x}_1 + \mathbf{x}_2$$

where $\mathbf{x}_1$ is in the column space of M and $M^T\mathbf{x}_2 = \mathbf{0}$. However, to say that $\mathbf{x}_1$ is in the column space of M is equivalent to saying that $\mathbf{x}_1 = M\mathbf{v}$ for some vector $\mathbf{v}$ in R^k, and to say that $M^T\mathbf{x}_2 = \mathbf{0}$ is equivalent to saying that $M^T(\mathbf{x} - \mathbf{x}_1) = \mathbf{0}$. Thus, if we can show that the equation

$$M^T(\mathbf{x} - M\mathbf{v}) = \mathbf{0} \tag{21}$$

has a unique solution for $\mathbf{v}$, then $\mathbf{x}_1 = M\mathbf{v}$ and $\mathbf{x}_2 = \mathbf{x} - \mathbf{x}_1$ will be uniquely determined vectors with the required properties. To do this, let us rewrite (21) as

$$M^TM\mathbf{v} = M^T\mathbf{x} \tag{22}$$

The matrix M in this equation has full column rank, since its column vectors are linearly independent. Thus, it follows from Theorem 7.5.10 that M^TM is invertible, so (22) has the unique solution

$$\mathbf{v} = (M^TM)^{-1}M^T\mathbf{x} \tag{23}$$

∎

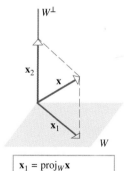

$\mathbf{x}_1 = \text{proj}_W\mathbf{x}$
$\mathbf{x}_2 = \mathbf{x} - \mathbf{x}_1 = \text{proj}_{W^{\perp}}\mathbf{x}$

Figure 7.7.4

In the special case where W is a line through the origin of R^n, the vectors $\mathbf{x}_1$ and $\mathbf{x}_2$ in this theorem are those given in Theorem 7.7.1; this suggests that we *define* the vector $\mathbf{x}_1$ in (20) to be the ***orthogonal projection of* $\mathbf{x}$ *on* W**. We will see later that the vector $\mathbf{x}_2$ is the ***orthogonal projection of* $\mathbf{x}$ *on* $W^{\perp}$**. We will denote these vectors by $\mathbf{x}_1 = \text{proj}_W\mathbf{x}$ and $\mathbf{x}_2 = \text{proj}_{W^{\perp}}\mathbf{x}$, respectively (Figure 7.7.4). Thus, Formula (20) can be expressed as

$$\mathbf{x} = \text{proj}_W\mathbf{x} + \text{proj}_{W^{\perp}}\mathbf{x} \tag{24}$$

The proof of the following result follows from Formula (23) in the proof of Theorem 7.7.4 and the relationship $\mathbf{x}_1 = \text{proj}_W\mathbf{x} = M\mathbf{v}$ that was established in that proof.

Theorem 7.7.5 *If W is a nonzero subspace of R^n, and if M is any matrix whose column vectors form a basis for W, then*

$$\text{proj}_W\mathbf{x} = M(M^TM)^{-1}M^T\mathbf{x} \tag{25}$$

for every column vector $\mathbf{x}$ in R^n.

Formula (25) can be used to define the linear operator

$$T(\mathbf{x}) = \text{proj}_W(\mathbf{x}) = M(M^TM)^{-1}M^T\mathbf{x} \tag{26}$$

on R^n whose standard matrix P is

$$P = M(M^TM)^{-1}M^T \tag{27}$$

We call this operator the ***orthogonal projection of R^n onto W***.

REMARK When working with Formulas (26) and (27), it is important to keep in mind that the matrix M is not unique, since its column vectors can be any basis vectors for W; that is, no matter what basis vectors you use to construct M, you will obtain the same operator T and the same matrix P.

EXAMPLE 6
Orthogonal
Projection of
R^3 onto a Plane
Through the
Origin

(a) Find the standard matrix P for the orthogonal projection of R^3 onto the plane $x - 4y + 2z = 0$.

(b) Use the matrix P to find the orthogonal projection of the vector $\mathbf{x} = (1, 0, 4)$ onto the plane.

Solution (a) Our strategy will be to find a basis for the plane, create a matrix M with the basis vectors as columns, and then use (27) to obtain P. To find a basis for the plane, we will view the equation $x - 4y + 2z = 0$ as a linear system of one equation in three unknowns and find a basis for the solution space. Solving the equation for its leading variable x and assigning arbitrary values to the free variables yields the general solution

$$\begin{bmatrix} x \\ y \\ z \end{bmatrix} = \begin{bmatrix} 4t_1 - 2t_2 \\ t_1 \\ t_2 \end{bmatrix} = t_1 \begin{bmatrix} 4 \\ 1 \\ 0 \end{bmatrix} + t_2 \begin{bmatrix} -2 \\ 0 \\ 1 \end{bmatrix}$$

The two column vectors on the right side form a basis for the solution space, so we take the matrix M to be

$$M = \begin{bmatrix} 4 & -2 \\ 1 & 0 \\ 0 & 1 \end{bmatrix}$$

Thus,

$$M^TM = \begin{bmatrix} 4 & 1 & 0 \\ -2 & 0 & 1 \end{bmatrix}\begin{bmatrix} 4 & -2 \\ 1 & 0 \\ 0 & 1 \end{bmatrix} = \begin{bmatrix} 17 & -8 \\ -8 & 5 \end{bmatrix} \quad \text{and} \quad (M^TM)^{-1} = \begin{bmatrix} \frac{5}{21} & \frac{8}{21} \\ \frac{8}{21} & \frac{17}{21} \end{bmatrix}$$

and therefore

$$P = M(M^TM)^{-1}M^T = \begin{bmatrix} 4 & -2 \\ 1 & 0 \\ 0 & 1 \end{bmatrix}\begin{bmatrix} \frac{5}{21} & \frac{8}{21} \\ \frac{8}{21} & \frac{17}{21} \end{bmatrix}\begin{bmatrix} 4 & 1 & 0 \\ -2 & 0 & 1 \end{bmatrix} = \begin{bmatrix} \frac{20}{21} & \frac{4}{21} & -\frac{2}{21} \\ \frac{4}{21} & \frac{5}{21} & \frac{8}{21} \\ -\frac{2}{21} & \frac{8}{21} & \frac{17}{21} \end{bmatrix} \tag{28}$$

Solution (b) The orthogonal projection of $\mathbf{x}$ onto the plane $x - 4y + 2z = 0$ is $P\mathbf{x}$ with $\mathbf{x}$ expressed in column form. Thus,

$$P\mathbf{x} = \begin{bmatrix} \frac{20}{21} & \frac{4}{21} & -\frac{2}{21} \\ \frac{4}{21} & \frac{5}{21} & \frac{8}{21} \\ -\frac{2}{21} & \frac{8}{21} & \frac{17}{21} \end{bmatrix}\begin{bmatrix} 1 \\ 0 \\ 4 \end{bmatrix} = \begin{bmatrix} \frac{12}{21} \\ \frac{36}{21} \\ \frac{66}{21} \end{bmatrix} = \begin{bmatrix} \frac{4}{7} \\ \frac{12}{7} \\ \frac{22}{7} \end{bmatrix}$$

If preferred, this can be written in the comma-delimited form $P\mathbf{x} = \left(\frac{4}{7}, \frac{12}{7}, \frac{22}{7}\right)$ for consistency with the comma-delimited notation that was originally used for $\mathbf{x}$. As a check, you may want to confirm that $P\mathbf{x}$ is in the plane $x - 4y + 2z = 0$ and $\mathbf{x} - P\mathbf{x}$ is orthogonal to $P\mathbf{x}$.   ■

WHEN DOES A MATRIX REPRESENT AN ORTHOGONAL PROJECTION?

We now turn to the problem of determining what properties an $n \times n$ matrix P must have in order to represent an orthogonal projection onto a k-dimensional subspace W of R^n. Some of the properties are clear. For example, since W is k-dimensional, the column space of P must be k-dimensional, and P must have rank k. We also know from (27) that if M is any $n \times k$ matrix whose column vectors form a basis for W, then

$$P^T = \left(M(M^TM)^{-1}M^T\right)^T = M(M^TM)^{-1}M^T = P$$

so P must be symmetric. Moreover,

$$\begin{aligned} P^2 &= \left(M(M^TM)^{-1}M^T\right)\left(M(M^TM)^{-1}M^T\right) \\ &= M(M^TM)^{-1}(M^TM)(M^TM)^{-1}M^T = M(M^TM)^{-1}M^T = P \end{aligned}$$

so P must be the same as its square. This makes sense intuitively, since the orthogonal projection of R^n onto W leaves vectors in W unchanged. In particular, it leaves $P\mathbf{x}$ unchanged for each $\mathbf{x}$ in R^n, so

$$P^2\mathbf{x} = P(P\mathbf{x}) = P\mathbf{x}$$

and this implies that $P^2 = P$.

A matrix that is the same as its square is said to be *idempotent*. Thus, we have shown that the standard matrix for an orthogonal projection of R^n onto a k-dimensional subspace has rank k, is symmetric, and is idempotent. In the exercises we will ask you to prove that the converse is also true, thereby establishing the following theorem.

Theorem 7.7.6 *An $n \times n$ matrix P is the standard matrix for an orthogonal projection of R^n onto a k-dimensional subspace of R^n if and only if P is symmetric, idempotent, and has rank k. The subspace W is the column space of P.*

EXAMPLE 7
Properties of Orthogonal Projections

In Example 5 we showed that the standard matrix P in (19) for the orthogonal projection of R^3 onto the line spanned by the vector $\mathbf{a} = (1, -4, 2)$ has rank 1, which is consistent with the fact that the line is one-dimensional. In accordance with Theorem 7.7.6, the matrix P is symmetric (verify) and is idempotent, since

$$P^2 = \begin{bmatrix} \frac{1}{21} & -\frac{4}{21} & \frac{2}{21} \\ -\frac{4}{21} & \frac{16}{21} & -\frac{8}{21} \\ \frac{2}{21} & -\frac{8}{21} & \frac{4}{21} \end{bmatrix} \begin{bmatrix} \frac{1}{21} & -\frac{4}{21} & \frac{2}{21} \\ -\frac{4}{21} & \frac{16}{21} & -\frac{8}{21} \\ \frac{2}{21} & -\frac{8}{21} & \frac{4}{21} \end{bmatrix} = \begin{bmatrix} \frac{1}{21} & -\frac{4}{21} & \frac{2}{21} \\ -\frac{4}{21} & \frac{16}{21} & -\frac{8}{21} \\ \frac{2}{21} & -\frac{8}{21} & \frac{4}{21} \end{bmatrix} = P$$

■

EXAMPLE 8
Identifying Orthogonal Projections

Show that

$$A = \begin{bmatrix} \frac{1}{9} & \frac{2}{9} & \frac{2}{9} \\ \frac{2}{9} & \frac{4}{9} & \frac{4}{9} \\ \frac{2}{9} & \frac{4}{9} & \frac{4}{9} \end{bmatrix}$$

is the standard matrix for an orthogonal projection of R^3 onto a line through the origin, and find the line.

Solution We leave it for you to confirm that A is symmetric, idempotent, and has rank 1. Thus, it follows from Theorem 7.7.6 that A represents an orthogonal projection of R^3 onto a line through the origin. That line is the column space of A, and since the second and third column vectors are scalar multiples of the first, we can take the first column vector of A as a basis for the line.

Moreover, since any scalar multiple of this column vector is also a basis for the line, we might as well multiply by 9 to simplify the components. Thus, the line can be expressed in comma-delimited form as the span of the vector $\mathbf{a} = (1, 2, 2)$, or it can be expressed parametrically in xyz-coordinates as

$$x = t, \quad y = 2t, \quad z = 2t \qquad \blacksquare$$

STRANG DIAGRAMS

Formula (24) is useful for studying systems of linear equations. To see why this is so, suppose that A is an $m \times n$ matrix, so that $A\mathbf{x} = \mathbf{b}$ is a linear system of m equations in n unknowns. Since $\mathbf{x}$ is a vector in R^n, we can apply Formula (24) with $W = \text{row}(A)$ and $W^{\perp} = \text{null}(A)$ to express $\mathbf{x}$ as a sum of two orthogonal terms

$$\mathbf{x} = \mathbf{x}_{\text{row}(A)} + \mathbf{x}_{\text{null}(A)} \qquad (29)$$

where $\mathbf{x}_{\text{row}(A)}$ is the orthogonal projection of $\mathbf{x}$ onto the row space of A, and $\mathbf{x}_{\text{null}(A)}$ is the orthogonal projection of $\mathbf{x}$ onto the null space of A; similarly, since $\mathbf{b}$ is a vector in R^m, we can apply Formula (24) to $\mathbf{b}$ with $W = \text{col}(A)$ and $W^{\perp} = \text{null}(A^T)$ to express $\mathbf{b}$ as a sum of two orthogonal terms

$$\mathbf{b} = \mathbf{b}_{\text{col}(A)} + \mathbf{b}_{\text{null}(A^T)} \qquad (30)$$

where $\mathbf{b}_{\text{col}(A)}$ is the orthogonal projection of $\mathbf{b}$ onto the column space of A, and $\mathbf{b}_{\text{null}(A^T)}$ is the orthogonal projection of $\mathbf{b}$ onto the null space of A^T. The decompositions in (29) and (30) can be pictured as in Figure 7.7.5 in which we have represented the fundamental spaces of A as perpendicular lines. We will call this a **Strang diagram**.[*]

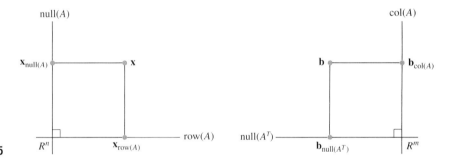

Figure 7.7.5

The fact that the fundamental spaces are represented by lines in a Strang diagram is pictorial only and is not intended to suggest that those spaces are one-dimensional. What we do know for sure is that

$$\dim(\text{row}(A)) + \dim(\text{null}(A)) = n \qquad (31)$$

$$\dim(\text{col}(A)) + \dim(\text{null}(A^T)) = m \qquad (32)$$

[see (5) in Section 7.5]. Also, we know from Theorem 3.5.5 that the system $A\mathbf{x} = \mathbf{b}$ is consistent if and only if $\mathbf{b}$ is in the column space of A, that is, if and only if $\mathbf{b}_{\text{null}(A^T)} = \mathbf{0}$ in (30). This is illustrated by the Strang diagrams in Figure 7.7.6.

FULL COLUMN RANK AND CONSISTENCY OF A LINEAR SYSTEM

The following theorem provides a deeper insight into the role that full column rank plays in the study of linear systems.

---

[*]We have taken the liberty of naming this kind of diagram for Gilbert Strang, currently a professor at MIT, who popularized them in a series of expository papers. In his papers, Strang used rectangles, rather than lines, to represent the fundamental spaces, but the idea is essentially the same.

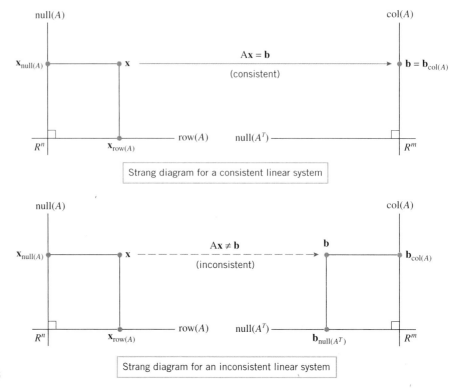

Figure 7.7.6

Theorem 7.7.7 *Suppose that A is an m × n matrix and* **b** *is in the column space of A.*

    *(a) If A has full column rank, then the system A***x** = **b** *has a unique solution, and that solution is in the row space of A.*

    *(b) If A does not have full column rank, then the system A***x** = **b** *has infinitely many solutions, but there is a unique solution in the row space of A. Moreover, among all solutions of the system, the solution in the row space of A has the smallest norm.*

Proof If A has full column rank, it follows from Theorem 7.5.6 that the system $A\mathbf{x} = \mathbf{b}$ is either inconsistent or has a unique solution. But $\mathbf{b}$ is in the column space of A, so the system must be consistent (Theorem 3.5.5) and hence has a unique solution. If A does not have full column rank, then Theorem 7.5.6 implies that $A\mathbf{x} = \mathbf{0}$ has infinitely many solutions and hence so does $A\mathbf{x} = \mathbf{b}$ by Theorem 3.5.2 and the consistency of $A\mathbf{x} = \mathbf{b}$. In either case, if $\mathbf{x}$ is a solution, then Theorem 7.7.4 implies that $\mathbf{x}$ can be split uniquely into a sum of two orthogonal terms

$$\mathbf{x} = \mathbf{x}_{\text{row}(A)} + \mathbf{x}_{\text{null}(A)} \tag{33}$$

where $\mathbf{x}_{\text{row}(A)}$ is in row(A) and $\mathbf{x}_{\text{null}(A)}$ is in null(A). Thus,

$$\mathbf{b} = A\mathbf{x} = A(\mathbf{x}_{\text{row}(A)} + \mathbf{x}_{\text{null}(A)}) = A\mathbf{x}_{\text{row}(A)} + A\mathbf{x}_{\text{null}(A)} = A\mathbf{x}_{\text{row}(A)} + \mathbf{0} = A\mathbf{x}_{\text{row}(A)} \tag{34}$$

which shows that $\mathbf{x}_{\text{row}(A)}$ is a solution of the system $A\mathbf{x} = \mathbf{b}$. In the case where A has full column rank, this is the only solution of the system [which proves part (a)], and in the case where A does not have full column rank, it shows that there exists at least one solution in the row space of A. To see in the latter case that there is only one solution in the row space of A, suppose that $\mathbf{x}_r$ and $\mathbf{x}'_r$ are two such solutions. Then

$$A(\mathbf{x}_r - \mathbf{x}'_r) = A\mathbf{x}_r - A\mathbf{x}'_r = \mathbf{b} - \mathbf{b} = \mathbf{0}$$

which implies that $\mathbf{x}_r - \mathbf{x}'_r$ is in null(A). However, $\mathbf{x}_r - \mathbf{x}'_r$ also lies in row$(A) = \text{null}(A)^\perp$, so part (a) of Theorem 7.3.4 implies that $\mathbf{x}_r - \mathbf{x}'_r = \mathbf{0}$ and hence that $\mathbf{x}_r = \mathbf{x}'_r$. Thus, there is a unique solution of $A\mathbf{x} = \mathbf{b}$ in the row space of A. Finally, if (33) is any solution of the system, then the theorem of Pythagoras (Theorem 1.2.11) implies that

$$\|\mathbf{x}\| = \sqrt{\|\mathbf{x}_{\text{row}(A)}\|^2 + \|\mathbf{x}_{\text{null}(A)}\|^2} \geq \|\mathbf{x}_{\text{row}(A)}\|$$

which shows that the solution $\mathbf{x}_{\text{row}(A)}$ in the row space of A has minimum norm. ∎

This theorem is illustrated by the Strang diagram in Figure 7.7.7. Part (a) of the figure illustrates the case where A is an $m \times n$ matrix with full column rank and $A\mathbf{x} = \mathbf{b}$ is consistent. In this case null$(A) = \{\mathbf{0}\}$, so the vertical line representing the null space of A collapses to a point, and $\mathbf{x} = \mathbf{x}_{\text{row}}$ is the unique solution of the system. Part (b) of the figure illustrates the case where the system is consistent but A does not have full column rank. In this case, if $\mathbf{x}_{\text{row}(A)}$ is the solution of the system that lies in the row space of A, and if $\mathbf{w}$ is any vector in the null space of A, then

$$\mathbf{x}' = \mathbf{x}_{\text{row}(A)} + \mathbf{w}$$

is also a solution of the system, since

$$A\mathbf{x}' = A\mathbf{x}_{\text{row}(A)} + A\mathbf{w} = \mathbf{b} + \mathbf{0} = \mathbf{b}$$

Thus, the solutions of $A\mathbf{x} = \mathbf{b}$ are represented by points on a vertical line through $\mathbf{x}_{\text{row}(A)}$ parallel to the "null(A) axis."

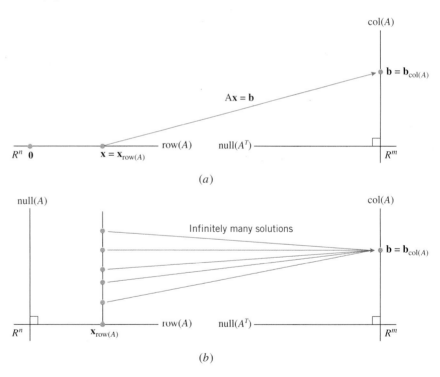

Figure 7.7.7

(a)

(b)

THE DOUBLE PERP THEOREM

In Theorem 7.3.4 we stated without proof that if W is a subspace of R^n, then $(W^\perp)^\perp = W$. Although this result may seem somewhat technical, it is important because it establishes that orthogonal complements occur in "companion pairs" in the sense that each of the spaces W and $W^\perp$ is the orthogonal complement of the other. Thus, for example, knowing that the null space of a matrix is the orthogonal complement of the row space automatically implies that the row space is the orthogonal complement of the null space.

We now have all of the mathematical machinery to prove this result.

Theorem 7.7.8 (*Double Perp Theorem*)   *If W is a subspace of R^n, then $(W^\perp)^\perp = W$.*

Proof Our strategy will be to show that every vector in W is in $(W^\perp)^\perp$, and conversely that every vector in $(W^\perp)^\perp$ is in W. Stated using set notation, we will be showing that $W \subset (W^\perp)^\perp$ and $(W^\perp)^\perp \subset W$, thereby proving that $W = (W^\perp)^\perp$.

Let $\mathbf{w}$ be any vector in W. By definition, $W^\perp$ consists of all vectors that are orthogonal to every vector in W. Thus, every vector in $W^\perp$ is orthogonal to $\mathbf{w}$, and this implies that $\mathbf{w}$ is in $(W^\perp)^\perp$. Conversely, let $\mathbf{w}$ be any vector in $(W^\perp)^\perp$. It follows from the projection theorem (Theorem 7.7.4) that $\mathbf{w}$ can be expressed uniquely as

$$\mathbf{w} = \mathbf{w}_1 + \mathbf{w}_2$$

where $\mathbf{w}_1$ is a vector in W and $\mathbf{w}_2$ is a vector in $W^\perp$. To show that $\mathbf{w}$ is a vector in W, we will show that $\mathbf{w}_2 = \mathbf{0}$ (which then makes $\mathbf{w} = \mathbf{w}_1$, which we know to be in W). Toward this end, observe that $\mathbf{w}_2$ is orthogonal to $\mathbf{w}$, so

$$\mathbf{w}_2 \cdot \mathbf{w} = 0$$

which implies that

$$(\mathbf{w}_2 \cdot \mathbf{w}_1) + (\mathbf{w}_2 \cdot \mathbf{w}_2) = 0$$

However, $\mathbf{w}_2$ is orthogonal to $\mathbf{w}_1$ (why?), so this equation simplifies to $\mathbf{w}_2 \cdot \mathbf{w}_2 = 0$, which implies that $\mathbf{w}_2 = \mathbf{0}$. ∎

ORTHOGONAL PROJECTIONS ONTO $W^\perp$

Given a subspace W of R^n, the standard matrix for the orthogonal projection $\text{proj}_{W^\perp}\mathbf{x}$ can be computed in one of two ways—you can either apply Formula (26) with the column vectors of M taken to be basis vectors for $W^\perp$, or you can use the fact that $\mathbf{x} = \text{proj}_W\mathbf{x} + \text{proj}_{W^\perp}\mathbf{x}$ to write $\text{proj}_{W^\perp}\mathbf{x}$ in terms of $\text{proj}_W\mathbf{x}$ as

$$\text{proj}_{W^\perp}\mathbf{x} = \mathbf{x} - \text{proj}_W\mathbf{x} = I\mathbf{x} - P\mathbf{x} = (I - P)\mathbf{x} \tag{35}$$

It now follows from (27) and (35) that the standard matrix for $\text{proj}_{W^\perp}\mathbf{x}$ can be expressed in terms of the standard matrix P for $\text{proj}_W\mathbf{x}$ as

$$I - P = I - M(M^TM)^{-1}M^T \tag{36}$$

where the column vectors of M form a basis for W.

EXAMPLE 9
Orthogonal
Projection onto
an Orthogonal
Complement

In Example 6 we showed that the standard matrix for the orthogonal projection of R^3 onto the plane $x - 4y + 2z = 0$ is

$$P = \begin{bmatrix} \frac{20}{21} & \frac{4}{21} & -\frac{2}{21} \\ \frac{4}{21} & \frac{5}{21} & \frac{8}{21} \\ -\frac{2}{21} & \frac{8}{21} & \frac{17}{21} \end{bmatrix}$$

In this case the orthogonal complement of the plane is the line through the origin that is perpendicular to the given plane, that is, the line spanned by the vector $\mathbf{a} = (1, -4, 2)$. It follows from (36) that the orthogonal projection of R^3 onto this line is

$$I - P = \begin{bmatrix} 1 & 0 & 0 \\ 0 & 1 & 0 \\ 0 & 0 & 1 \end{bmatrix} - \begin{bmatrix} \frac{20}{21} & \frac{4}{21} & -\frac{2}{21} \\ \frac{4}{21} & \frac{5}{21} & \frac{8}{21} \\ -\frac{2}{21} & \frac{8}{21} & \frac{17}{21} \end{bmatrix} = \begin{bmatrix} \frac{1}{21} & -\frac{4}{21} & \frac{2}{21} \\ -\frac{4}{21} & \frac{16}{21} & -\frac{8}{21} \\ \frac{2}{21} & -\frac{8}{21} & \frac{4}{21} \end{bmatrix}$$

Note that this is consistent with the result that we obtained in Example 5 using Theorem 7.7.3. ∎

Exercise Set 7.7

In Exercises 1 and 2, find the orthogonal projection of $\mathbf{x}$ on span$\{\mathbf{a}\}$, first using Formula (5) and then by finding and using the standard matrix for the projection.

1. $\mathbf{x} = (-1, 6)$, $\mathbf{a} = (1, 2)$ **2.** $\mathbf{x} = (2, 3)$, $\mathbf{a} = (-2, 5)$

In Exercises 3 and 4, find the orthogonal projection of $\mathbf{x}$ on the line l, first by using Formula (5) and then by finding and using the standard matrix for the projection.

3. $\mathbf{x} = (1, 1)$; l: $2x - y = 0$
4. $\mathbf{x} = (-3, 2)$; l: $x + 3y = 0$

In Exercises 5–8, find the vector components of $\mathbf{x}$ along $\mathbf{a}$ and orthogonal to $\mathbf{a}$.

5. $\mathbf{x} = (1, 1, 1)$, $\mathbf{a} = (0, 2, -1)$
6. $\mathbf{x} = (2, 0, 1)$, $\mathbf{a} = (1, 2, 3)$
7. $\mathbf{x} = (2, 1, 1, 2)$, $\mathbf{a} = (4, -4, 2, -2)$
8. $\mathbf{x} = (5, 0, -3, 7)$, $\mathbf{a} = (2, 1, -1, -1)$

In Exercises 9–12, use Formula (14) to find the length of the orthogonal projection without finding the orthogonal projection itself.

9. $\mathbf{x} = (1, -2, 4)$, $\mathbf{a} = (2, 3, 6)$
10. $\mathbf{x} = (4, -5, 1)$, $\mathbf{a} = (2, 2, 4)$
11. $\mathbf{x} = (2, 3, 1, 5)$, $\mathbf{a} = (-4, 2, -2, 3)$
12. $\mathbf{x} = (5, -3, 7, 1)$, $\mathbf{a} = (7, 1, 0, -1)$

In Exercises 13 and 14, find the standard matrix for the orthogonal projection of R^3 onto span$\{\mathbf{a}\}$, and confirm that the matrix is symmetric, idempotent, and has rank 1, in accordance with Theorem 7.7.6.

13. $\mathbf{a} = (-1, 5, 2)$ **14.** $\mathbf{a} = (7, -2, 2)$

In Exercises 15 and 16, find the standard matrix for the orthogonal projection of R^3 onto span$\{\mathbf{a}_1, \mathbf{a}_2\}$, and confirm that the matrix is symmetric, idempotent, and has rank 2, in accordance with Theorem 7.7.6.

15. $\mathbf{a}_1 = (3, -4, 1)$, $\mathbf{a}_2 = (2, 0, 3)$
16. $\mathbf{a}_1 = (1, -2, 5)$, $\mathbf{a}_2 = (4, -2, 3)$

In Exercises 17 and 18, you should be able to write down the standard matrix for the orthogonal projection of R^3 onto the stated plane without performing any computations. Write down this matrix and check your answer by finding a basis for the plane and applying Formula (27).

17. The xz-plane   **18.** The yz-plane

In Exercises 19 and 20, find the standard matrix for the orthogonal projection of R^3 onto the plane, and use that matrix to find the orthogonal projection of $\mathbf{v} = (2, 4, -1)$ on that plane.

19. The plane with equation $x + y + z = 0$.
20. The plane with equation $2x - y + 3z = 0$.

In Exercises 21 and 22, a linearly dependent set of vectors in R^4 is given. Find the standard matrix for the orthogonal projection of R^4 onto the subspace spanned by those vectors.

21. $\mathbf{a}_1 = (4, -6, 2, -4)$, $\mathbf{a}_2 = (2, -3, 1, -2)$,
     $\mathbf{a}_3 = (1, 0, -2, 5)$, $\mathbf{a}_4 = (5, -6, 0, 1)$
22. $\mathbf{a}_1 = (5, -3, 1, -1)$, $\mathbf{a}_2 = (3, -6, 0, 9)$,
     $\mathbf{a}_3 = (8, -9, 1, 8)$, $\mathbf{a}_4 = (1, -2, 0, 3)$

In Exercises 23 and 24, find the orthogonal projection of $\mathbf{v}$ on the solution space of the linear system $A\mathbf{x} = \mathbf{0}$.

23. $\mathbf{v} = (5, 6, 7, 2)$; $A = \begin{bmatrix} 1 & 1 & 1 & 0 \\ 0 & 2 & 1 & 1 \end{bmatrix}$

24. $\mathbf{v} = (1, 1, 2, 3)$; $A = \begin{bmatrix} 1 & 2 & 1 & 1 \\ 1 & 1 & -1 & 2 \\ 1 & -5 & 3 & 0 \end{bmatrix}$

In Exercises 25 and 26, use the reduced row echelon form of A to help find the standard matrices for the orthogonal projections onto the row and column spaces of A.

25. $A = \begin{bmatrix} 1 & 1 & -1 & -1 \\ 1 & 0 & 1 & 3 \\ 3 & 2 & -1 & 1 \end{bmatrix}$

26. $A = \begin{bmatrix} 1 & 4 & 5 & 6 & 9 \\ 3 & -2 & 1 & 4 & -1 \\ -1 & 0 & -1 & -2 & -1 \\ 2 & 3 & 5 & 7 & 8 \end{bmatrix}$

In Exercises 27 and 28, show that $A\mathbf{x} = \mathbf{b}$ is consistent and find the solution $\mathbf{x}_{\text{row}}$ that lies in the row space of A. [*Hint:* If $\mathbf{x}$ is any solution, then $\mathbf{x}_{\text{row}} = \text{proj}_{\text{row}(A)}\mathbf{x}$ (Figure 7.7.7).]

27. $A = \begin{bmatrix} 1 & 2 & -1 & 2 \\ 3 & 1 & 0 & 5 \\ -1 & 3 & -2 & -1 \end{bmatrix}$; $\mathbf{b} = \begin{bmatrix} 0 \\ -1 \\ 1 \end{bmatrix}$

28. $A = \begin{bmatrix} 3 & 0 & 2 & -1 \\ 1 & 4 & 1 & 2 \\ 4 & -8 & 2 & -6 \end{bmatrix}$; $\mathbf{b} = \begin{bmatrix} 4 \\ 0 \\ 8 \end{bmatrix}$

In Exercises 29 and 30, find the standard matrix for the orthogonal projection of R^3 onto the orthogonal complement of the indicated plane.

29. The plane in Exercise 19.   **30.** The plane in Exercise 20.

31. Find the orthogonal projection of the vector $\mathbf{v} = (1, 1, 1, 1)$ on the orthogonal complement of the subspace of R^4 spanned by $\mathbf{v}_1 = (1, -2, 3, 0)$ and $\mathbf{v}_2 = (3, 4, -1, 2)$.

Discussion and Discovery

D1. (a) The rank of the standard matrix for the orthogonal projection of R^n onto a line through the origin is _____ and onto its orthogonal complement is _____.

   (b) If $n \geq 2$, then the rank of the standard matrix for the orthogonal projection of R^n onto a plane through the origin is _____ and onto its orthogonal complement is _____.

D2. A 5×5 matrix P is the orthogonal projection of R^5 onto some three-dimensional subspace of R^5 if it has what properties?

D3. If $\mathbf{a}$ is a nonzero vector in R^n and $\mathbf{x}$ is any vector in R^n, what does the number

$$q = \sqrt{\|\mathbf{x}\|^2 - \frac{|\mathbf{x} \cdot \mathbf{a}|^2}{\|\mathbf{a}\|^2}}$$

represent geometrically?

D4. If P is the standard matrix for the orthogonal projection of R^n on a subspace W, what can you say about the matrix P^n?

D5. Indicate whether the statement is true (T) or false (F). Justify your answer.

   (a) If W is a subspace of R^n, then $\text{proj}_W \mathbf{u}$ is orthogonal to $\text{proj}_{W^\perp} \mathbf{u}$ for every vector $\mathbf{u}$ in R^n.

   (b) An $n \times n$ matrix P that satisfies the equation $P^2 = P$ is the standard matrix for an orthogonal projection onto some subspace of R^n.

   (c) If $\mathbf{x}$ is a solution of a linear system $A\mathbf{x} = \mathbf{b}$, then $\text{proj}_{\text{row}(A)}\mathbf{x}$ is also a solution.

   (d) If P is the standard matrix for the orthogonal projection of R^n onto a subspace W, then $I - P$ is idempotent.

   (e) If $A\mathbf{x} = \mathbf{b}$ is an inconsistent linear system, then so is $A\mathbf{x} = \text{proj}_{\text{col}(A)}\mathbf{b}$.

D6. If W is a subspace of R^n, what can you say about $((W^\perp)^\perp)^\perp$?

D7. Find a 3×3 symmetric matrix of rank 1 that is not the standard matrix for some orthogonal projection.

D8. Let A be an $n \times n$ matrix with linearly independent row vectors. Find the standard matrix for the orthogonal projection of R^n onto the row space of A.

D9. What are the possible eigenvalues of an $n \times n$ idempotent matrix?

D10. (*Calculus required*) For the given matrices, confirm that the linear system $A\mathbf{x} = \mathbf{b}$ is consistent, and find the solution $\mathbf{x}_{\text{row}}$ that lies in the row space A by using calculus to minimize the length of a general solution vector $\mathbf{x}$. Check your answer using an appropriate orthogonal projection. [*Suggestion:* You can simplify the calculations by minimizing $\|\mathbf{x}\|^2$ rather than $\|\mathbf{x}\|$.]

$$A = \begin{bmatrix} 1 & -2 & 1 \\ 2 & -5 & 1 \\ 3 & -7 & 2 \end{bmatrix}; \quad \mathbf{b} = \begin{bmatrix} 1 \\ -1 \\ 0 \end{bmatrix}$$

D11. Let R be the matrix formed from the nonzero rows of the reduced row echelon form of a nonzero matrix A, and let G be the transpose of R. What does the matrix $G(G^T G)^{-1}G^T$ represent? Explain your reasoning.

Working with Proofs

P1. Prove that Formula (15) defines a linear operator on R^n.

P2. Prove that multiplying $\mathbf{a}$ by a nonzero scalar does not change the matrix P in (16).

P3. Let P be a symmetric $n \times n$ matrix that is idempotent and has rank k. Prove that P is the standard matrix for an or-

thogonal projection onto some k-dimensional subspace W of R^n. [*Hint:* The subspace W is the column space of P. Use the fact that any vector $\mathbf{x}$ in R^n can be written as $\mathbf{x} = P\mathbf{x} + (I - P)\mathbf{x}$ to prove that $P(I - P)\mathbf{x} = \mathbf{0}$. To finish the proof show that $(I - P)\mathbf{x}$ is in $W^\perp$, and use Theorem 7.7.4 and Formula (24).]

Technology Exercises

T1. (*Standard matrix for an orthogonal projection*) Most technology utilities do not have a special command for computing orthogonal projections, so several commands must be used in combination. One way to find the standard matrix

for the orthogonal projection onto a subspace W spanned by a set of vectors $\{\mathbf{v}_1, \mathbf{v}_2, \ldots, \mathbf{v}_k\}$ is first to find a basis for W, then create a matrix A that has the basis vectors as columns, and then use Formula (27).

(a) Find the standard matrix for the orthogonal projection of R^4 onto the subspace W spanned by
$\mathbf{v}_1 = (1, 2, 3, -4)$, $\mathbf{v}_2 = (2, 3, -4, 1)$,
$\mathbf{v}_3 = (2, -5, 8, -3)$, $\mathbf{v}_4 = (5, 26, -9, -12)$,
$\mathbf{v}_5 = (3, -4, 1, 2)$.

(b) Use the matrix obtained in part (a) to find $\text{proj}_W \mathbf{x}$, where $\mathbf{x} = (1, 0, -3, 7)$.

(c) Find $\text{proj}_{W^\perp} \mathbf{x}$ for the vector in part (b).

T2. Confirm that the following linear system is consistent, and find the solution that lies in the row space of the coefficient matrix:

$$12x_1 + 14x_2 - 15x_3 + 23x_4 + 27x_5 = 5$$
$$16x_1 + 18x_2 - 22x_3 + 29x_4 + 37x_5 = 8$$
$$18x_1 + 20x_2 - 21x_3 + 32x_4 + 41x_5 = 9$$
$$10x_1 + 12x_2 - 16x_3 + 20x_4 + 23x_5 = 4$$

T3. (CAS) Use Formula (16) to compute the standard matrix for the orthogonal projection of R^3 onto the line spanned by the nonzero vector $\mathbf{a} = (a_1, a_2, a_3)$, and then confirm that the resulting matrix has the properties stated in Theorem 7.7.6.

Section 7.8 Best Approximation and Least Squares

The problem of finding the orthogonal projection of a vector onto a subspace of R^n is closely related to the problem of finding the distance between a point and a subspace. In this section we will explore this relationship and discuss its application to linear systems.

Minimum Distance Problems

We will be concerned here with the following problem.

The Minimum Distance Problem in R^n Given a subspace W and a vector $\mathbf{b}$ in R^n, find a vector $\hat{\mathbf{w}}$ in W that is closest to $\mathbf{b}$ in the sense that $\|\mathbf{b} - \hat{\mathbf{w}}\| < \|\mathbf{b} - \mathbf{w}\|$ for every vector $\mathbf{w}$ in W that is distinct from $\hat{\mathbf{w}}$. Such a vector $\hat{\mathbf{w}}$, if it exists, is called a **best approximation to b *from* W** (Figure 7.8.1).

To motivate a method for solving the minimum distance problem, let us focus on R^3. We know from geometry that if $\mathbf{b}$ is a point in R^3 and W is a plane through the origin, then the point $\hat{\mathbf{w}}$ in W that is closest to $\mathbf{b}$ is obtained by dropping a perpendicular from $\mathbf{b}$ to W; that is, $\hat{\mathbf{w}} = \text{proj}_W \mathbf{b}$. It follows from this that the distance from $\mathbf{b}$ to W is $d = \|\mathbf{b} - \text{proj}_W \mathbf{b}\|$, or equivalently, $d = \|\text{proj}_{W^\perp} \mathbf{b}\|$, where $W^\perp$ is the line through the origin that is perpendicular to W (Figure 7.8.2).

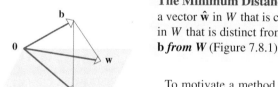

$\hat{\mathbf{w}}$ is closer to $\mathbf{b}$ than any other vector $\mathbf{w}$ in W.

Figure 7.8.1

EXAMPLE 1
Distance from a Point to a Plane in R^3

Use an appropriate orthogonal projection to find a formula for the distance d from the point (x_0, y_0, z_0) to the plane $ax + by + cz = 0$.

Solution Let $\mathbf{b} = (x_0, y_0, z_0)$, let W be the given plane, and let l be the line through the origin that is perpendicular to W (i.e., l is $W^\perp$). The line l is spanned by the normal $\mathbf{n} = (a, b, c)$ and hence it follows from Formula (14) of Section 7.7 that

$$d = \|\text{proj}_{W^\perp} \mathbf{b}\| = \|\text{proj}_{\mathbf{n}} \mathbf{b}\| = \frac{|\mathbf{n} \cdot \mathbf{b}|}{\|\mathbf{n}\|}$$

Substituting the components for $\mathbf{n}$ and $\mathbf{b}$ into this formula yields

$$d = \frac{|(a, b, c) \cdot (x_0, y_0, z_0)|}{\|(a, b, c)\|} = \frac{|ax_0 + by_0 + cz_0|}{\sqrt{a^2 + b^2 + c^2}} \tag{1}$$

Thus, for example, the distance from the point $(-1, 5, 4)$ to the plane $x - 2y + 3z = 0$ is

$$d = \frac{|ax_0 + by_0 + cz_0|}{\sqrt{a^2 + b^2 + c^2}} = \frac{|(1)(-1) + (-2)(5) + (3)(4)|}{\sqrt{1^2 + (-2)^2 + 3^2}} = \frac{1}{\sqrt{14}} \quad\blacksquare$$

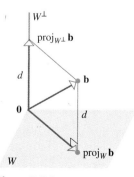

Figure 7.8.2

In light of the preceding discussion, the following theorem should not be surprising.

Theorem 7.8.1 (*Best Approximation Theorem*)   *If W is a subspace of R^n, and $\mathbf{b}$ is a point in R^n, then there is a unique best approximation to $\mathbf{b}$ from W, namely $\hat{\mathbf{w}} = \text{proj}_W \mathbf{b}$.*

Proof   For every vector $\mathbf{w}$ in W we can write

$$\mathbf{b} - \mathbf{w} = (\mathbf{b} - \text{proj}_W \mathbf{b}) + (\text{proj}_W \mathbf{b} - \mathbf{w})$$

The two terms on the right side of this equation are orthogonal (since the first term is in $W^\perp$, and the second term, being a difference of vectors in W, is in W). Thus, we can apply the theorem of Pythagoras to write

$$\|\mathbf{b} - \mathbf{w}\|^2 = \|\mathbf{b} - \text{proj}_W \mathbf{b}\|^2 + \|\text{proj}_W \mathbf{b} - \mathbf{w}\|^2$$

If $\mathbf{w} \neq \text{proj}_W \mathbf{b}$, then the second term on the right side of this equation is positive and hence

$$\|\mathbf{b} - \text{proj}_W \mathbf{b}\|^2 < \|\mathbf{b} - \mathbf{w}\|^2$$

This implies that $\|\mathbf{b} - \text{proj}_W \mathbf{b}\| < \|\mathbf{b} - \mathbf{w}\|$ if $\mathbf{w} \neq \text{proj}_W \mathbf{b}$, which tells us that $\hat{\mathbf{w}} = \text{proj}_W \mathbf{b}$ is a best approximation to $\mathbf{b}$ from W; we leave the proof of the uniqueness as an exercise.   ∎

Motivated by Figure 7.8.2, we define the *distance* from a point $\mathbf{b}$ to a subspace W in R^n to be

$$d = \|\mathbf{b} - \text{proj}_W \mathbf{b}\| \qquad \text{[Distance from b to } W\text{]} \tag{2}$$

or equivalently,

$$d = \|\text{proj}_{W^\perp} \mathbf{b}\| \qquad \text{[Distance from b to } W\text{]} \tag{3}$$

The following example extends the result in Example 1 to R^n.

EXAMPLE 2
Distance from a
Point to a
Hyperplane

Find a formula for the distance d from a point $\mathbf{b} = (b_1, b_2, \ldots, b_n)$ in R^n to the hyperplane $a_1 x_1 + a_2 x_2 + \cdots + a_n x_n = 0$.

Solution   Denote the hyperplane by W. This hyperplane is the orthogonal complement of $\mathbf{a} = (a_1, a_2, \ldots, a_n)$, so Theorem 7.4.6 implies that $W^\perp = \text{span}\{\mathbf{a}\}$. Thus, Formula (3) above and Formula (14) of Section 7.7 imply that

$$d = \|\text{proj}_{W^\perp} \mathbf{b}\| = \|\text{proj}_{\mathbf{a}} \mathbf{b}\| = \frac{|\mathbf{a} \cdot \mathbf{b}|}{\|\mathbf{a}\|} = \frac{|a_1 b_1 + a_2 b_2 + \cdots + a_n b_n|}{\sqrt{a_1^2 + a_2^2 + \cdots + a_n^2}} \tag{4}$$

With the appropriate change in notation, this reduces to Formula (1) in R^3.   ∎

LEAST SQUARES SOLUTIONS OF LINEAR SYSTEMS

There are many applications in which a linear system $A\mathbf{x} = \mathbf{b}$ should be consistent on theoretical grounds but fails to be so because of measurement errors in the entries of A or $\mathbf{b}$. In such cases, a common scientific procedure is to look for vectors that come as close as possible to being solutions in the sense that they minimize $\|\mathbf{b} - A\mathbf{x}\|$. Accordingly, we make the following definition.

Definition 7.8.2   If A is an $m \times n$ matrix and $\mathbf{b}$ is a vector in R^m, then a vector $\hat{\mathbf{x}}$ in R^n is called a *best approximate solution* or a *least squares solution* of $A\mathbf{x} = \mathbf{b}$ if

$$\|\mathbf{b} - A\hat{\mathbf{x}}\| \leq \|\mathbf{b} - A\mathbf{x}\| \tag{5}$$

for all $\mathbf{x}$ in R^n. The vector $\mathbf{b} - A\hat{\mathbf{x}}$ is called the *least squares error vector*, and the scalar $\|\mathbf{b} - A\hat{\mathbf{x}}\|$ is called the *least squares error*.

REMARK To understand the terminology in this definition, let $\mathbf{b} - A\mathbf{x} = (e_1, e_2, \ldots, e_n)$. The components of this vector can be interpreted as the errors that result in the individual components when $\mathbf{x}$ is used as an approximate solution. Since a best approximate solution minimizes

$$\|\mathbf{b} - A\mathbf{x}\| = \left(e_1^2 + e_2^2 + \cdots + e_n^2\right)^{1/2} \tag{6}$$

this solution also minimizes $e_1^2 + e_2^2 + \cdots + e_n^2$, which is the sum of the squares of the errors in the components, and hence the term *least squares solution*.

FINDING LEAST SQUARES SOLUTIONS OF LINEAR SYSTEMS

Our next objective is to develop a method for finding least squares solutions of a linear system $A\mathbf{x} = \mathbf{b}$ of m equations in n unknowns. To start, observe that $A\mathbf{x}$ is in the column space of A for all $\mathbf{x}$ in R^n, so $\|\mathbf{b} - A\mathbf{x}\|$ is minimized when

$$A\mathbf{x} = \mathrm{proj}_{\mathrm{col}(A)}\mathbf{b} \tag{7}$$

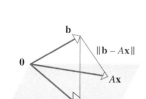

$\|\mathbf{b} - A\mathbf{x}\|$ is minimized when $A\mathbf{x} = \mathrm{proj}_{\mathrm{col}(A)}\mathbf{b}$.

Figure 7.8.3

(Figure 7.8.3). Since $\mathrm{proj}_{\mathrm{col}(A)}\mathbf{b}$ is a vector in the column space of A, system (7) is consistent and its solutions are the least squares solutions of $A\mathbf{x} = \mathbf{b}$. Thus, we are guaranteed that *every linear system $A\mathbf{x} = \mathbf{b}$ has at least one least squares solution.*

As a practical matter, least squares solutions are rarely obtained by solving (7), since this equation can be rewritten in an alternative form that eliminates the need to calculate the orthogonal projection. To see how this can be accomplished, rewrite (7) as

$$\mathbf{b} - A\mathbf{x} = \mathbf{b} - \mathrm{proj}_{\mathrm{col}(A)}\mathbf{b} \tag{8}$$

and multiply both sides of this equation by A^T to obtain

$$A^T(\mathbf{b} - A\mathbf{x}) = A^T(\mathbf{b} - \mathrm{proj}_{\mathrm{col}(A)}\mathbf{b}) \tag{9}$$

Since the orthogonal complement of $\mathrm{col}(A)$ is $\mathrm{null}(A^T)$, it follows from Formula (24) of Section 7.7 that

$$\mathbf{b} - \mathrm{proj}_{\mathrm{col}(A)}\mathbf{b} = \mathrm{proj}_{\mathrm{null}(A^T)}\mathbf{b}$$

This implies that $\mathbf{b} - \mathrm{proj}_{\mathrm{col}(A)}\mathbf{b}$ is in the null space of A^T, and hence that

$$A^T(\mathbf{b} - \mathrm{proj}_{\mathrm{col}(A)}\mathbf{b}) = \mathbf{0}$$

Thus, (9) can be rewritten as $A^T(\mathbf{b} - A\mathbf{x}) = \mathbf{0}$ or, alternatively, as

$$A^TA\mathbf{x} = A^T\mathbf{b} \tag{10}$$

This is called the ***normal equation*** or ***normal system*** associated with $A\mathbf{x} = \mathbf{b}$. The individual equations in (10) are called the ***normal equations*** associated with $A\mathbf{x} = \mathbf{b}$. Using this terminology, the problem of finding least squares solutions of $A\mathbf{x} = \mathbf{b}$ has been reduced to solving the associated normal system exactly. The following theorem provides the basic facts about solutions of the normal equation.

Theorem 7.8.3

   (a) *The least squares solutions of a linear system $A\mathbf{x} = \mathbf{b}$ are the exact solutions of the normal equation*

$$A^TA\mathbf{x} = A^T\mathbf{b} \tag{11}$$

   (b) *If A has full column rank, the normal equation has a unique solution, namely*

$$\hat{\mathbf{x}} = (A^TA)^{-1}A^T\mathbf{b} \tag{12}$$

   (c) *If A does not have full column rank, then the normal equation has infinitely many solutions, but there is a unique solution in the row space of A. Moreover, among all solutions of the normal equation, the solution in the row space of A has the smallest norm.*

Proof (a) We have already established that every least squares solution of $A\mathbf{x} = \mathbf{b}$ satisfies (11). Conversely, if $\mathbf{x}$ satisfies (11), then this vector also satisfies the equation

$$A^T(\mathbf{b} - A\mathbf{x}) = \mathbf{0}$$

This implies that $\mathbf{b} - A\mathbf{x}$ is orthogonal to the row vectors of A^T, and hence to the column vectors of A, and hence to the column space of A. It follows from this that the equation

$$\mathbf{b} = A\mathbf{x} + (\mathbf{b} - A\mathbf{x})$$

expresses $\mathbf{b}$ as the sum of a vector in $\text{col}(A)$ and a vector orthogonal to $\text{col}(A)$, which implies that $A\mathbf{x} = \text{proj}_{\text{col}(A)}\mathbf{b}$ by Theorem 7.7.4. Thus, $\mathbf{x}$ is a least squares solution of $A\mathbf{x} = \mathbf{b}$.

Proof (b) If A has full column rank, then Theorem 7.5.10 implies that A^TA is invertible, so (12) is the unique solution of (11).

Proof (c) If A does not have full column rank, then A^TA is not invertible (Theorem 7.5.10), so (11) is a consistent linear system whose coefficient matrix does not have full column rank. This being the case, it follows from part (*b*) of Theorem 7.7.7 that (11) has infinitely many solutions but has a unique solution in the row space of A^TA. Moreover, that theorem also tells us that the solution in the row space of A^TA is the solution with smallest norm. However, the row space of A^TA is the same as the row space of A (Theorem 7.5.8), so we have proved the final assertion of the theorem.   ■

CONCEPT PROBLEM   Show that if A is invertible, then (12) simplifies to $\mathbf{x} = A^{-1}\mathbf{b}$, and explain why this is to be expected.

EXAMPLE 3
Least Squares
Solutions

Find the least squares solutions of the linear system

$$\begin{aligned}
x_1 - x_2 &= 4 \\
3x_1 + 2x_2 &= 1 \\
-2x_1 + 4x_2 &= 3
\end{aligned}$$

Solution The matrix form of the system is $A\mathbf{x} = \mathbf{b}$, where

$$A = \begin{bmatrix} 1 & -1 \\ 3 & 2 \\ -2 & 4 \end{bmatrix} \quad \text{and} \quad \mathbf{b} = \begin{bmatrix} 4 \\ 1 \\ 3 \end{bmatrix}$$

Since the columns of A are not scalar multiples of one another, the matrix has full column rank. Thus, it follows from Theorem 7.8.3 that there is a unique least squares solution given by Formula (12). We leave it for you to confirm that

$$A^TA = \begin{bmatrix} 1 & 3 & -2 \\ -1 & 2 & 4 \end{bmatrix} \begin{bmatrix} 1 & -1 \\ 3 & 2 \\ -2 & 4 \end{bmatrix} = \begin{bmatrix} 14 & -3 \\ -3 & 21 \end{bmatrix}$$

$$(A^TA)^{-1} = \begin{bmatrix} 14 & -3 \\ -3 & 21 \end{bmatrix}^{-1} = \begin{bmatrix} \frac{21}{285} & \frac{3}{285} \\ \frac{3}{285} & \frac{14}{285} \end{bmatrix}$$

$$\hat{\mathbf{x}} = (A^TA)^{-1}A^T\mathbf{b} = \begin{bmatrix} \frac{21}{285} & \frac{3}{285} \\ \frac{3}{285} & \frac{14}{285} \end{bmatrix} \begin{bmatrix} 1 & 3 & -2 \\ -1 & 2 & 4 \end{bmatrix} \begin{bmatrix} 4 \\ 1 \\ 3 \end{bmatrix} = \begin{bmatrix} \frac{21}{285} & \frac{3}{285} \\ \frac{3}{285} & \frac{14}{285} \end{bmatrix} \begin{bmatrix} 1 \\ 10 \end{bmatrix} = \begin{bmatrix} \frac{17}{95} \\ \frac{143}{285} \end{bmatrix}$$

Thus, the least squares solution is

$$x_1 = \tfrac{17}{95}, \quad x_2 = \tfrac{143}{285}$$ ■

ORTHOGONALITY PROPERTY OF LEAST SQUARES ERROR VECTORS

Before considering another example, it will be helpful to develop some of the properties of least squares error vectors. For this purpose, consider a linear system $A\mathbf{x} = \mathbf{b}$, and recall from Formula (30) of Section 7.7 that $\mathbf{b}$ can be written as

$$\mathbf{b} = \text{proj}_{\text{col}(A)}\mathbf{b} + \text{proj}_{\text{null}(A^T)}\mathbf{b}$$

from which it follows that

$$\mathbf{b} - A\mathbf{x} = (\text{proj}_{\text{col}(A)}\mathbf{b} - A\mathbf{x}) + \text{proj}_{\text{null}(A^T)}\mathbf{b} \tag{13}$$

However, we know from (7) that $\mathbf{x}$ is a least squares solution of $A\mathbf{x} = \mathbf{b}$ if and only if

$$\text{proj}_{\text{col}(A)}\mathbf{b} - A\mathbf{x} = \mathbf{0}$$

which, together with (13), implies that $\hat{\mathbf{x}}$ is a least squares solution of $A\mathbf{x} = \mathbf{b}$ if and only if

$$\mathbf{b} - A\hat{\mathbf{x}} = \text{proj}_{\text{null}(A^T)}\mathbf{b} \tag{14}$$

This shows that every least squares solution $\hat{\mathbf{x}}$ of $A\mathbf{x} = \mathbf{b}$ has the same error vector, namely

$$\text{least squares error vector} = \mathbf{b} - A\hat{\mathbf{x}} = \text{proj}_{\text{null}(A^T)}\mathbf{b} \tag{15}$$

Thus, the least squares error can be written as

$$\text{least squares error} = \|\mathbf{b} - A\hat{\mathbf{x}}\| = \|\text{proj}_{\text{null}(A^T)}\mathbf{b}\| \tag{16}$$

Moreover, since the least squares error vector lies in $\text{null}(A^T)$, and since this space is orthogonal to $\text{col}(A)$, we have also established the following result.

> **Theorem 7.8.4** *A vector $\hat{\mathbf{x}}$ is a least squares solution of $A\mathbf{x} = \mathbf{b}$ if and only if the error vector $\mathbf{b} - A\hat{\mathbf{x}}$ is orthogonal to the column space of A.*

EXAMPLE 4
Least Squares
Solutions and
Their Error
Vector

Find the least squares solutions and least squares error for the linear system

$$\begin{array}{rrrcr} 3x_1 & + & 2x_2 & - & x_3 & = & 2 \\ x_1 & - & 4x_2 & + & 3x_3 & = & -2 \\ x_1 & + & 10x_2 & - & 7x_3 & = & 1 \end{array}$$

Solution   The matrix form of the system is $A\mathbf{x} = \mathbf{b}$, where

$$A = \begin{bmatrix} 3 & 2 & -1 \\ 1 & -4 & 3 \\ 1 & 10 & -7 \end{bmatrix} \quad \text{and} \quad \mathbf{b} = \begin{bmatrix} 2 \\ -2 \\ 1 \end{bmatrix}$$

Since it is not evident by inspection whether A has full column rank (in fact it does not), we will simply proceed by solving the associated normal system $A^TA\mathbf{x} = A^T\mathbf{b}$. We leave it for you to confirm that

$$A^TA = \begin{bmatrix} 11 & 12 & -7 \\ 12 & 120 & -84 \\ -7 & -84 & 59 \end{bmatrix} \quad \text{and} \quad A^T\mathbf{b} = \begin{bmatrix} 5 \\ 22 \\ -15 \end{bmatrix}$$

Thus, the augmented matrix for the normal system is

$$\begin{bmatrix} 11 & 12 & -7 & \vdots & 5 \\ 12 & 120 & -84 & \vdots & 22 \\ -7 & -84 & 59 & \vdots & -15 \end{bmatrix}$$

We leave it for you to confirm that the reduced row echelon form of this matrix is

$$\begin{bmatrix} 1 & 0 & \frac{1}{7} & \vdots & \frac{2}{7} \\ 0 & 1 & -\frac{5}{7} & \vdots & \frac{13}{84} \\ 0 & 0 & 0 & \vdots & 0 \end{bmatrix}$$

Thus, there are infinitely many least squares solutions, and they are given by

$$x_1 = \frac{2}{7} - \frac{1}{7}t$$
$$x_2 = \frac{13}{84} + \frac{5}{7}t$$
$$x_3 = t$$

As a check, let us verify that all least squares solutions produce the same least squares error vector and the same least squares error. To see that this is so, we first compute

$$\mathbf{b} - A\mathbf{x} = \begin{bmatrix} 2 \\ -2 \\ 1 \end{bmatrix} - \begin{bmatrix} 3 & 2 & -1 \\ 1 & -4 & 3 \\ 1 & 10 & -7 \end{bmatrix} \begin{bmatrix} \frac{2}{7} - \frac{1}{7}t \\ \frac{13}{84} + \frac{5}{7}t \\ t \end{bmatrix} = \begin{bmatrix} 2 \\ -2 \\ 1 \end{bmatrix} - \begin{bmatrix} \frac{7}{6} \\ -\frac{1}{3} \\ \frac{11}{6} \end{bmatrix} = \begin{bmatrix} \frac{5}{6} \\ -\frac{5}{3} \\ -\frac{5}{6} \end{bmatrix}$$

Since $\mathbf{b} - A\mathbf{x}$ does not depend on t, all least squares solutions produce the same error vector. The resulting least squares error is

$$\|\mathbf{b} - A\mathbf{x}\| = \sqrt{\left(\frac{5}{6}\right)^2 + \left(-\frac{5}{3}\right)^2 + \left(-\frac{5}{6}\right)^2} = \frac{5}{6}\sqrt{6}$$

We leave it for you to confirm that the error vector is orthogonal to the column vectors of the matrix

$$A = \begin{bmatrix} 3 & 2 & -1 \\ 1 & -4 & 3 \\ 1 & 10 & -7 \end{bmatrix}$$

in agreement with Theorem 7.8.4.   ■

STRANG DIAGRAMS FOR LEAST SQUARES PROBLEMS

The Strang diagrams in Figure 7.8.4 illustrate some of the ideas that we have been discussing about least squares solutions of a linear system $A\mathbf{x} = \mathbf{b}$. In that figure we have split $\mathbf{b}$ into the sum of orthogonal terms as

$$\mathbf{b} = \text{proj}_{\text{col}(A)}\mathbf{b} + \text{proj}_{\text{null}(A^T)}\mathbf{b}$$

The terms in this sum are significant in least squares problems because we know from (7) that each least squares solution $\hat{\mathbf{x}}$ satisfies

$$A\hat{\mathbf{x}} = \text{proj}_{\text{col}(A)}\mathbf{b}$$

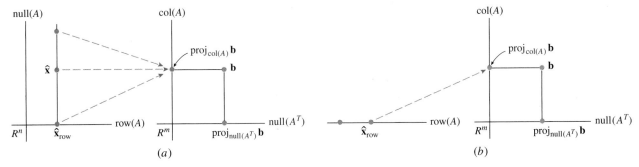

Figure 7.8.4

and from (15) that the error vector is

$$\mathbf{b} - A\hat{\mathbf{x}} = \text{proj}_{\text{null}(A^T)}\mathbf{b}$$

The Strang diagram in Figure 7.8.4a illustrates the case where A does not have full column rank. In this case there are infinitely many least squares solutions, but there is a unique least squares solution $\hat{\mathbf{x}}_{\text{row}}$ in the row space of A, this being the least squares solution with minimum norm. The Strang diagram in Figure 7.8.4b illustrates the case where A has full column rank. In this case null$(A) = \{\mathbf{0}\}$, so the vertical line representing null(A) collapses to a point, and $\hat{\mathbf{x}}_{\text{row}}$ is the unique least squares solution.

FITTING A CURVE TO EXPERIMENTAL DATA

A common problem in experimental work is to obtain a mathematical relationship between two variables x and y by "fitting" a curve $y = f(x)$ of a specified form to a set of points in the plane that correspond to experimentally determined values of x and y, say

$$(x_1, y_1), (x_2, y_2), \ldots, (x_n, y_n)$$

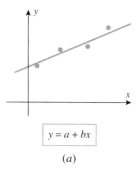

$y = a + bx$

(a)

The curve $y = f(x)$ is called a ***mathematical model*** for the data. The form of the function f is sometimes determined by a physical theory and sometimes by the pattern of the data. For example, Figure 7.8.5 shows some data patterns that suggest polynomial models. Once the form of the function has been decided on, the idea is to determine values of coefficients that make the graph of the function fit the data *as closely as possible*. In this section we will be concerned exclusively with polynomial models, but we will discuss some other kinds of mathematical models in the exercises.

We will begin with linear models (polynomials of degree 1). For this purpose let x and y be given variables, and assume that there is evidence to suggest that the variables are related by a linear equation

$$y = a + bx \tag{17}$$

where a and b are to be determined from two or more data points

$$(x_1, y_1), (x_2, y_2), \ldots, (x_n, y_n)$$

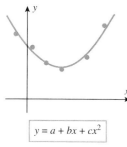

$y = a + bx + cx^2$

(b)

If the data happen to fall exactly on a line, then the coordinates of each point will satisfy (17), and the unknown coefficients will be a solution of the linear system

$$\begin{aligned} y_1 &= a + bx_1 \\ y_2 &= a + bx_2 \\ &\vdots \\ y_n &= a + bx_n \end{aligned} \tag{18}$$

We can write this system in matrix form as

$$\begin{bmatrix} 1 & x_1 \\ 1 & x_2 \\ \vdots & \vdots \\ 1 & x_n \end{bmatrix} \begin{bmatrix} a \\ b \end{bmatrix} = \begin{bmatrix} y_1 \\ y_2 \\ \vdots \\ y_n \end{bmatrix} \tag{19}$$

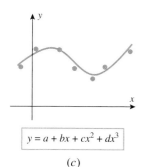

$y = a + bx + cx^2 + dx^3$

(c)

Figure 7.8.5

or more compactly as

$$M\mathbf{v} = \mathbf{y} \tag{20}$$

where

$$M = \begin{bmatrix} 1 & x_1 \\ 1 & x_2 \\ \vdots & \vdots \\ 1 & x_n \end{bmatrix}, \quad \mathbf{v} = \begin{bmatrix} a \\ b \end{bmatrix}, \quad \mathbf{y} = \begin{bmatrix} y_1 \\ y_1 \\ \vdots \\ y_n \end{bmatrix} \tag{21}$$

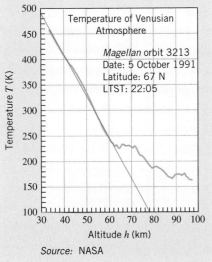
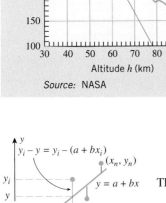

Figure 7.8.6

If there are measurement errors in the data, then the data points will typically not lie on a line, and (20) will be inconsistent. In this case we look for a least squares approximation to the values of a and b by solving the normal system

$$M^TM\mathbf{v} = M^T\mathbf{y} \tag{22}$$

If the x-coordinates of the data points are not all the same, then M will have rank 2 (full column rank), and the normal system will have a unique least squares solution

$$\mathbf{v} = (M^TM)^{-1}M^T\mathbf{y} \tag{23}$$

The line $y = a + bx$ that results from this solution is called the ***least squares line of best fit*** to the data or, alternatively, the ***regression line***. Referring to the equations in (18), we see that this line minimizes

$$S = [y_1 - (a + bx_1)]^2 + [y_2 - (a + bx_2)]^2 + \cdots + [y_n - (a + bx_n)]^2 \tag{24}$$

The differences in Equation (24) are called ***residuals***, so, in words, *the least squares line of best fit minimizes the sum of the squares of the residuals* (Figure 7.8.6).

EXAMPLE 5   Find the least squares line of best fit to the four points $(0, 1)$, $(1, 3)$, $(2, 4)$, and $(3, 4)$.

Solution   The first step is to use the data to build the matrices M and $\mathbf{y}$ in (21). This yields

$$M = \begin{bmatrix} 1 & 0 \\ 1 & 1 \\ 1 & 2 \\ 1 & 3 \end{bmatrix} \quad \text{and} \quad \mathbf{y} = \begin{bmatrix} 1 \\ 3 \\ 4 \\ 4 \end{bmatrix}$$

Since the x-coordinates of the data points are not all the same, the normal system has a unique solution, and the coefficients for the least squares line of best fit can be obtained from Formula (23). We leave it for you to confirm that

$$M^TM = \begin{bmatrix} 1 & 1 & 1 & 1 \\ 0 & 1 & 2 & 3 \end{bmatrix} \begin{bmatrix} 1 & 0 \\ 1 & 1 \\ 1 & 2 \\ 1 & 3 \end{bmatrix} = \begin{bmatrix} 4 & 6 \\ 6 & 14 \end{bmatrix} \quad \text{and} \quad (M^TM)^{-1} = \begin{bmatrix} \frac{7}{10} & -\frac{3}{10} \\ -\frac{3}{10} & \frac{2}{10} \end{bmatrix}$$

Thus, applying Formula (23) yields

$$\mathbf{v} = \begin{bmatrix} a \\ b \end{bmatrix} = (M^TM)^{-1}M^T\mathbf{y} = \begin{bmatrix} \frac{7}{10} & -\frac{3}{10} \\ -\frac{3}{10} & \frac{2}{10} \end{bmatrix} \begin{bmatrix} 1 & 1 & 1 & 1 \\ 0 & 1 & 2 & 3 \end{bmatrix} \begin{bmatrix} 1 \\ 3 \\ 4 \\ 4 \end{bmatrix} = \begin{bmatrix} \frac{3}{2} \\ 1 \end{bmatrix} = \begin{bmatrix} 1.5 \\ 1 \end{bmatrix}$$

Thus, the approximate values of a and b are $a = 1.5$ and $b = 1$, and the least squares straight line fit to the data is $y = 1.5 + x$. This line and the data points are shown in Figure 7.8.7.

Alternative Solution   In the exercises we will ask you to use (21) and (22) to show that the normal system can be expressed in terms of the coordinates of the data points as

$$\begin{bmatrix} n & \Sigma x_i \\ \Sigma x_i & \Sigma x_i^2 \end{bmatrix} \begin{bmatrix} a \\ b \end{bmatrix} = \begin{bmatrix} \Sigma y_i \\ \Sigma x_i y_i \end{bmatrix} \tag{25}$$

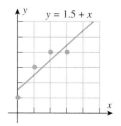

Figure 7.8.7

where the Σ indicates that the adjacent expression is to be summed as i varies from 1 to n. For the given data we have

$$\Sigma x_i = x_1 + x_2 + x_3 + x_4 = 0 + 1 + 2 + 3 = 6$$
$$\Sigma x_i^2 = x_1^2 + x_2^2 + x_3^2 + x_4^2 = 0 + 1 + 4 + 9 = 14$$
$$\Sigma y_i = y_1 + y_2 + y_3 + y_4 = 1 + 3 + 4 + 4 = 12$$
$$\Sigma x_i y_i = x_1 y_1 + x_2 y_2 + x_3 y_3 + x_4 y_4 = (0)(1) + (1)(3) + (2)(4) + (3)(4) = 23$$

so from (25) the normal system is

$$\begin{bmatrix} 4 & 6 \\ 6 & 14 \end{bmatrix} \begin{bmatrix} a \\ b \end{bmatrix} = \begin{bmatrix} 12 \\ 23 \end{bmatrix} \tag{26}$$

We leave it for you to show that the solution of this system is $a = 1.5$, $b = 1$, as before. ∎

EXAMPLE 6

An Application of Least Squares to Hooke's Law

It follows from Hooke's law in physics that if a mass with a weight of x units is suspended from a spring (as in Figure 7.8.8), then under appropriate conditions the spring will be stretched to a length y that is related to x by a linear equation $y = a + bx$. The constant a is called the ***natural length*** of the spring, since it is the length of the spring with no weight attached. The constant $k = 1/b$ is called the ***stiffness*** of the spring—the greater the stiffness, the smaller the value of b and hence the less the spring will stretch under a given weight.

The following table shows five measurements of weights in newtons and corresponding lengths in centimeters for a certain spring. Use a least squares line of best fit to the data to estimate the natural length of the spring and the stiffness.

Weight x (N)	1.0	2.0	4.0	6.0	8.0
Length y (cm)	6.9	7.6	8.7	10.4	11.6

Figure 7.8.8

Solution We will use (25) to find the normal system. Here is a convenient way to arrange the computations:

x_i	y_i	x_i^2	$x_i y_i$
1.0	6.9	1.0	6.9
2.0	7.6	4.0	15.2
4.0	8.7	16.0	34.8
6.0	10.4	36.0	62.4
8.0	11.6	64.0	92.8
$\Sigma x_i = 21.0$	$\Sigma y_i = 45.2$	$\Sigma x_i^2 = 121.0$	$\Sigma x_i y_i = 212.1$

Substituting the sums in (25) yields the normal system

$$\begin{bmatrix} 5 & 21.0 \\ 21.0 & 121.0 \end{bmatrix} \begin{bmatrix} a \\ b \end{bmatrix} = \begin{bmatrix} 45.2 \\ 212.1 \end{bmatrix}$$

We leave it for you to solve the system and show that to one decimal place the least squares estimate of the natural length is $a = 6.2$ cm, and the least squares estimate of the stiffness is $k = 1/b = 1.5$ N/cm. ∎

LEAST SQUARES FITS BY HIGHER-DEGREE POLYNOMIALS

The technique described for fitting a straight line to data generalizes easily to fitting a polynomial of any specified degree to a set of data points. To illustrate the idea, suppose that we want to find a polynomial of the form

$$y = a_0 + a_1 x + \cdots + a_m x^m \tag{27}$$

whose graph comes *as close as possible* to passing through n known data points

$$(x_1, y_1), (x_2, y_2), \ldots, (x_n, y_n)$$

We assume that m is specified and that the coefficients $a_0, a_1, \ldots, a_m$ are to be determined. For the polynomial to pass through the data points exactly, the coefficients would satisfy the n conditions

$$y_1 = a_0 + a_1 x_1 + a_2 x_1^2 + \cdots + a_m x_1^m$$
$$y_2 = a_0 + a_1 x_2 + a_2 x_2^2 + \cdots + a_m x_2^m$$
$$\vdots \qquad \vdots \qquad \vdots \qquad \vdots \qquad \qquad \vdots$$
$$y_n = a_0 + a_1 x_n + a_2 x_n^2 + \cdots + a_m x_n^m$$

which is a linear system of n equations in the $m + 1$ unknowns $a_0, a_1, \ldots, a_m$. We can write this system in matrix form as

$$\begin{bmatrix} 1 & x_1 & x_1^2 & \cdots & x_1^m \\ 1 & x_2 & x_2^2 & \cdots & x_2^m \\ \vdots & \vdots & \vdots & & \vdots \\ 1 & x_n & x_n^2 & \cdots & x_n^m \end{bmatrix} \begin{bmatrix} a_0 \\ a_1 \\ \vdots \\ a_m \end{bmatrix} = \begin{bmatrix} y_1 \\ y_2 \\ \vdots \\ y_n \end{bmatrix}$$

or more compactly as

$$M\mathbf{v} = \mathbf{y} \tag{28}$$

where

$$M = \begin{bmatrix} 1 & x_1 & x_1^2 & \cdots & x_1^m \\ 1 & x_2 & x_2^2 & \cdots & x_2^m \\ \vdots & \vdots & \vdots & & \vdots \\ 1 & x_n & x_n^2 & \cdots & x_n^m \end{bmatrix}, \quad \mathbf{v} = \begin{bmatrix} a_0 \\ a_1 \\ \vdots \\ a_m \end{bmatrix}, \quad \mathbf{y} = \begin{bmatrix} y_1 \\ y_2 \\ \vdots \\ y_n \end{bmatrix} \tag{29}$$

In the special case where $m = n - 1$, and the x-coordinates are distinct, Theorem 2.3.1 implies that there is a unique polynomial of degree m whose graph passes through the points exactly (the interpolating polynomial). In this case, (28) is consistent and has a unique solution. However, if $m < n - 1$, it will usually not be possible to find a curve of form (27) whose graph passes through all of the points. When that happens, system (28) will be inconsistent and we will have to settle for a least squares solution. As in the linear case, we can find the least squares solutions by solving the normal system

$$M^T M \mathbf{v} = M^T \mathbf{y} \tag{30}$$

In the exercises we will ask you to show that if $m < n$ and at least $m + 1$ of the x-coordinates of the data points are distinct, then M has full column rank and (30) has the unique solution

$$\mathbf{v} = (M^T M)^{-1} M^T \mathbf{y} \tag{31}$$

Linear Algebra in History

The technique of least squares was developed by the German mathematician Carl Friedrich Gauss in 1795 (see p. 54). Gauss's public application of the method was dramatic, to say the least. In 1801 the Italian astronomer Giuseppi Piazzi discovered the asteroid Ceres and observed it for 1/40 of its orbit before losing it in the Sun. Using three of Piazzi's observations and the method of least squares, Gauss computed the orbit, but his results differed dramatically from those of the leading astronomers. To the astonishment of all of the experts, Ceres reappeared a year later in virtually the exact position that Gauss had predicted. This achievement brought Gauss instant recognition as the premier mathematician in Europe.

EXAMPLE 7
Application of Least Squares to Newton's Second Law of Motion

According to Newton's second law of motion, a body near the surface of the Earth falls vertically downward according to the equation

$$y = y_0 + v_0 t + \tfrac{1}{2} g t^2 \tag{32}$$

where

y = the coordinate of the body relative to the origin of a vertical y-axis pointing down (Figure 7.8.9)

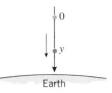

Figure 7.8.9

y_0 = the coordinate of the body at the initial time $t = 0$

v_0 = the velocity of the body at the initial time $t = 0$

g = a constant, called the ***acceleration due to gravity***[*]

Suppose that the initial displacement and velocity of a falling body and the local value of g are to be determined experimentally by measuring displacements of the body at various times. Find the least squares estimates of y_0, v_0, and g from the data in the following table:

Time t (seconds)	0.10	0.20	0.30	0.40	0.50
Displacement y (feet)	−0.18	0.31	1.03	2.48	3.73

Solution The first step is to use the data to build the matrices M, $\mathbf{v}$, and $\mathbf{y}$ in (29). Making the appropriate adjustments in the notation for the entries, we obtain

$$
M = \begin{bmatrix} 1 & t_1 & t_1^2 \\ 1 & t_2 & t_2^2 \\ 1 & t_3 & t_3^2 \\ 1 & t_4 & t_4^2 \\ 1 & t_5 & t_5^2 \end{bmatrix} = \begin{bmatrix} 1 & 0.10 & 0.01 \\ 1 & 0.20 & 0.04 \\ 1 & 0.30 & 0.09 \\ 1 & 0.40 & 0.16 \\ 1 & 0.50 & 0.25 \end{bmatrix}, \quad \mathbf{v} = \begin{bmatrix} y_0 \\ v_0 \\ \frac{1}{2}g \end{bmatrix}, \quad \mathbf{y} = \begin{bmatrix} y_1 \\ y_2 \\ y_3 \\ y_4 \\ y_5 \end{bmatrix} = \begin{bmatrix} -0.18 \\ 0.31 \\ 1.03 \\ 2.48 \\ 3.73 \end{bmatrix}
$$

We leave it for you to show that the normal system $M^T M \mathbf{v} = M^T \mathbf{y}$ is

$$
\begin{bmatrix} 5 & 1.5 & 0.55 \\ 1.5 & 0.55 & 0.225 \\ 0.55 & 0.225 & 0.0979 \end{bmatrix} \begin{bmatrix} y_0 \\ v_0 \\ \frac{1}{2}g \end{bmatrix} = \begin{bmatrix} 7.37 \\ 3.21 \\ 1.4326 \end{bmatrix}
$$

and we also leave it for you to solve this system and show that to two decimal places the least squares approximations of y_0, v_0, and g are

$$
y_0 = -0.40, \quad v_0 = 0.35, \quad g = 32.14
$$

The five data points and the graph of the equation $y = y_0 + v_0 t + \frac{1}{2}g t^2$ are shown in Figure 7.8.10. ∎

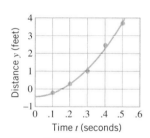

Figure 7.8.10

THEORY VERSUS PRACTICE

Procedures that are workable in theory often fail in practice because of computer roundoff error. For example, it can be shown that if there are slight errors in the entries of a matrix A, then those errors tend to become magnified into large errors in the entries of $A^T A$. This has an effect on least squares problems, since the normal equation $A^T A \mathbf{x} = A^T \mathbf{b}$ tends to be too sensitive to roundoff error to make it useful for the large-scale systems that typically occur in real-world applications. Accordingly, various algorithms for computing least squares solutions have been created for the specific purpose of circumventing the calculation of $A^T A$.

---

[*]Although the acceleration due to gravity is commonly taken to be 9.8 m/s^2 or 32.2 ft/s^2 in physics and engineering books, the elliptical shape of the Earth and other factors cause variations in this constant that make it latitude and altitude dependent. Appropriate local values of g have been determined experimentally and compiled by various international bodies for scientific reference.

Exercise Set 7.8

In Exercises 1 and 2, find the least squares solution of $A\mathbf{x} = \mathbf{b}$ by solving the associated normal system, and show that the least squares error vector is orthogonal to the column space of A (as guaranteed by Theorem 7.8.4).

1. $A = \begin{bmatrix} 1 & -1 \\ 2 & 3 \\ 4 & 5 \end{bmatrix}$; $\mathbf{b} = \begin{bmatrix} 2 \\ -1 \\ 5 \end{bmatrix}$

2. $A = \begin{bmatrix} 2 & -2 \\ 1 & 1 \\ 3 & 1 \end{bmatrix}$; $\mathbf{b} = \begin{bmatrix} 2 \\ -1 \\ 1 \end{bmatrix}$

3. For the matrices in Exercise 1, find $\text{proj}_{\text{col}(A)}\mathbf{b}$, and confirm that the least squares solution satisfies Equation (7).

4. For the matrices in Exercise 2, find $\text{proj}_{\text{col}(A)}\mathbf{b}$, and confirm that the least squares solution satisfies Equation (7).

In Exercises 5–8, find all least squares solutions of $A\mathbf{x} = \mathbf{b}$, and confirm that all of the solutions have the same error vector (and hence the same least squares error). Compute the least squares error.

5. $A = \begin{bmatrix} 2 & 1 \\ 4 & 2 \\ -2 & 1 \end{bmatrix}$; $\mathbf{b} = \begin{bmatrix} 3 \\ 2 \\ 1 \end{bmatrix}$

6. $A = \begin{bmatrix} 1 & 3 \\ -2 & -6 \\ 3 & 9 \end{bmatrix}$; $\mathbf{b} = \begin{bmatrix} 1 \\ 0 \\ 1 \end{bmatrix}$

7. $A = \begin{bmatrix} -1 & 3 & 2 \\ 2 & 1 & 3 \\ 0 & 1 & 1 \end{bmatrix}$; $\mathbf{b} = \begin{bmatrix} 7 \\ 0 \\ -7 \end{bmatrix}$

8. $A = \begin{bmatrix} 1 & 2 & 3 \\ -2 & -3 & -5 \\ 1 & 3 & 4 \end{bmatrix}$; $\mathbf{b} = \begin{bmatrix} 1 \\ 1 \\ 2 \end{bmatrix}$

In Exercises 9 and 10, find the least squares straight line fit $y = a + bx$ to the given points. Show that the result is reasonable by graphing the line and plotting the data in the same coordinate system.

9. $(2, 1)$, $(3, 2)$, $(5, 3)$, $(6, 4)$

10. $(0, 1)$, $(2, 0)$, $(3, 1)$, $(3, 2)$

In Exercises 11 and 12, find the least squares quadratic fit $y = a_0 + a_1x + a_2x^2$ to the given points. Show that the result is reasonable by graphing the curve and plotting the data in the same coordinate system.

11. $(0, 1)$, $(2, 0)$, $(3, 1)$, $(3, 2)$

12. $(1, -2)$, $(0, -1)$, $(1, 0)$, $(2, 4)$

In Exercises 13 and 14, set up but do not solve the normal system for finding the stated least squares fit. (The solution of the system is best found using a technology utility.)

13. The least squares cubic fit $y = a_0 + a_1x + a_2x^2 + a_3x^3$ to the data points $(1, 4.9)$, $(2, 10.8)$, $(3, 27.9)$, $(4, 60.2)$, $(5, 113)$.

14. Show that if M, $\mathbf{v}$, and $\mathbf{y}$ are the vectors in (21), then the normal system in (22) can be written as

$$\begin{bmatrix} n & \Sigma x_i \\ \Sigma x_i & \Sigma x_i^2 \end{bmatrix} \begin{bmatrix} a \\ b \end{bmatrix} = \begin{bmatrix} \Sigma y_i \\ \Sigma x_i y_i \end{bmatrix}$$

Discussion and Discovery

D1. (a) The distance in R^3 from the point $P_0 = (1, -2, 1)$ to the plane $x + y - z = 0$ is _____, and the point in the plane that is closest to P_0 is _____.

(b) The distance in R^4 from the point $\mathbf{b} = (1, 2, 0, -1)$ to the hyperplane $x_1 - x_2 + 2x_3 - 2x_4 = 0$ is _____, and the point in the hyperplane that is closest to $\mathbf{b}$ is

_____.

D2. Let A be an $m \times n$ matrix with linearly independent column vectors, and let $\mathbf{b}$ be a column vector in R^m. State a formula in terms of A and A^T for

(a) the vector in the column space of A that is closest to $\mathbf{b}$;

(b) the least squares solution of $A\mathbf{x} = \mathbf{b}$;

(c) the least squares error vector;

(d) the least squares error;

(e) the standard matrix for the orthogonal projection of R^m onto the column space of A.

D3. Is there any value of s for which $x_1 = 1$ and $x_2 = 2$ is the least squares solution of the following linear system?

$$\begin{aligned} x_1 - x_2 &= 1 \\ 2x_1 + 3x_2 &= 1 \\ 4x_1 + 5x_2 &= s \end{aligned}$$

Explain your reasoning.

D4. A corporation obtains the following data relating the number of sales representatives on its staff to annual sales:

Number of Sales Representatives	5	10	15	20	25	30
Annual Sales (millions)	3.4	4.3	5.2	6.1	7.2	8.3

Explain how you might use least squares methods to estimate the annual sales with 45 representatives, and discuss the assumptions that you are making. [You need not perform the actual computations.]

D5. Find a curve of the form $y = a + (b/x)$ that best fits the data points $(1, 7)$, $(3, 3)$, $(6, 1)$ by making the substitution $X = 1/x$. Draw the curve and plot the data points in the same coordinate system.

D6. Let A be an $m \times n$ matrix and $\mathbf{b}$ a column vector in R^m. What is the significance of vectors $\mathbf{x}$ and $\mathbf{r}$ for which

$$\begin{bmatrix} A & I_m \\ 0 & A^T \end{bmatrix} \begin{bmatrix} \mathbf{x} \\ \mathbf{r} \end{bmatrix} = \begin{bmatrix} \mathbf{b} \\ \mathbf{0} \end{bmatrix}?$$

Working with Proofs

P1. Prove: If A has linearly independent column vectors, and if $A\mathbf{x} = \mathbf{b}$ is consistent, then the least squares solution of $A\mathbf{x} = \mathbf{b}$ and the exact solution of $A\mathbf{x} = \mathbf{b}$ are the same.

P2. Prove: If A has linearly independent column vectors, and if $\mathbf{b}$ is orthogonal to the column space of A, then the least squares solution of $A\mathbf{x} = \mathbf{b}$ is $\mathbf{x} = \mathbf{0}$.

P3. Prove: The set of all least squares solutions of a linear system $A\mathbf{x} = \mathbf{b}$ is the translated subspace $\hat{\mathbf{x}} + W$, where $\hat{\mathbf{x}}$ is any least squares solution and W is the solution space of $A\mathbf{x} = \mathbf{0}$. [*Hint:* Model your proof after that of Theorem 3.5.1, and use the fact that A and A^TA have the same null space.]

P4. Prove that the best approximation to $\mathbf{b}$ in Theorem 7.8.1 is unique. [*Hint:* Use the fact that $\|\mathbf{b} - \text{proj}_W\mathbf{b}\| < \|\mathbf{b} - \mathbf{w}\|$ for any $\mathbf{w}$ in W other than $\text{proj}_W\mathbf{b}$.]

P5. Prove: If $m < n$ and at least $m + 1$ of the numbers $x_1, x_2, \ldots, x_n$ are distinct, then the column vectors of the $n \times (m + 1)$ matrix M in Equation (29) are linearly independent. [*Hint:* A nonzero polynomial of degree m has at most m distinct roots.]

P6. Let M be the matrix in Equation (29). Use the result in the preceding exercise to prove that if $m < n$ and at least $m + 1$ of the numbers $x_1, x_2, \ldots, x_n$ are distinct, then M^TM is invertible.

Technology Exercises

T1. (*Least squares fit to data*) Most technology utilities provide a direct command for fitting various kinds of functions to data by minimizing the sum of the squares of the deviations between the function values and the data. Use this command to check the result in Example 5.

T2. (*Least squares solutions of linear systems*) Some technology utilities provide a direct command for finding least squares solutions of linear systems, whereas others require that you set up and solve the associated normal system. Determine how your utility handles this problem and check the result in Example 3.

T3. Find the least squares fit in Exercise 13 by solving the normal system, and then compare the result to that obtained by using the direct command for a least squares fit.

T4. *Pathfinder* is an experimental, lightweight, remotely piloted, solar-powered aircraft that was used in a series of experiments by NASA to determine the feasibility of applying solar power for long-duration, high-altitude flight. In August 1997 *Pathfinder* recorded the data in the accompanying table relating altitude H and temperature T. Show that a linear model is reasonable by plotting the data, and then find the least squares line $H = H_0 + kT$ of best fit.

Three important models in applications are

    exponential models $(y = ae^{bx})$
    power function models $(y = ax^b)$
    logarithmic models $(y = a + b\ln x)$

where a and b are to be determined to fit experimental data as closely as possible. Exercises T5–T7 are concerned with a procedure, called *linearization*, by which the data are transformed to a form in which a straight-line fit by least squares can be used to approximate the constants.

Altitude H (thousands of feet)	15	20	25	30	35	40	45
Temperature T (°C)	4.5	−5.9	−16.1	−27.6	−39.8	−50.2	−62.9

Table Ex-T4

T5. (a) Show that making the substitution $Y = \ln y$ in the equation $y = ae^{bx}$ produces the equation $Y = bx + \ln a$ whose graph in the xY-plane is a line of slope b and Y-intercept $\ln a$.

(b) Part (a) suggests that a curve of the form $y = ae^{bx}$ can be fitted to n data points (x_i, y_i) by letting $Y_i = \ln y_i$, then fitting a straight line to the transformed data points (x_i, Y_i) by least squares to find b and $\ln a$, and then computing a from $\ln a$. Use this method to fit an exponential model to the following data, and graph the curve and data in the same coordinate system.

x	0	1	2	3	4	5	6	7
y	3.9	5.3	7.2	9.6	12	17	23	31

T6. (a) Show that making the substitutions $X = \ln x$ and $Y = \ln y$ in the equation $y = ax^b$ produces the equation $Y = bX + \ln a$ whose graph in the XY-plane is a line of slope b and Y-intercept $\ln a$.

(b) Part (a) suggests that a curve of the form $y = ax^b$ can be fitted to n data points (x_i, y_i) by letting $X_i = \ln x_i$ and $Y_i = \ln y_i$, then fitting a straight line to the transformed data points (X_i, Y_i) by least squares to find b and $\ln a$, and then computing a from $\ln a$. Use this method to fit a power function model to the following data, and graph the curve and data in the same coordinate system.

x	2	3	4	5	6	7	8	9
y	1.75	1.91	2.03	2.13	2.22	2.30	2.37	2.43

T7. (a) Show that making the substitution $X = \ln x$ in the equation $y = a + b \ln x$ produces the equation $y = a + bX$ whose graph in the Xy-plane is a line of slope b and y-intercept a.

(b) Part (a) suggests that a curve of the form $y = a + b \ln x$ can be fitted to n data points (x_i, y_i) by letting $X_i = \ln x_i$ and then fitting a straight line to the transformed data points (X_i, y_i) by least squares to find b and a. Use this method to fit a logarithmic model to the following data, and graph the curve and data in the same coordinate system.

x	2	3	4	5	6	7	8	9
y	4.07	5.30	6.21	6.79	7.32	7.91	8.23	8.51

T8. (*Center of a circle by least squares*) Least squares methods can be used to estimate the center (h, k) of a circle $(x - h)^2 + (y - k)^2 = r^2$ using measured data points on its circumference. To see how, suppose that the data points are

$$(x_1, y_1), (x_2, y_2), \ldots, (x_n, y_n)$$

and rewrite the equation of the circle in the form

$$2xh + 2yk + s = x^2 + y^2 \qquad (*)$$

where

$$s = r^2 - h^2 - k^2 \qquad (**)$$

Substituting the data points in $(*)$ yields a linear system in the unknowns h, k, and s, which can be solved by least squares to estimate their values. Equation $(**)$ can then be used to estimate r. Use this method to approximate the center and radius of a circle from the measured data points on the circumference given in the accompanying table. [*Note: The data in this problem are based on archaeological excavations of a circular starting line for a race track in the Greek city of Corinth dating from about 500 B.C. For a more detailed discussion of the problem and its history, see the article by C. Rorres and D. G. Romano entitled "Finding the Center of a Circular Starting Line in an Ancient Greek Stadium," SIAM Review, Vol. 39, No. 4, 1997.*]

x	19.880	20.919	21.735	23.375	24.361	25.375	25.979
y	68.874	67.676	66.692	64.385	62.908	61.292	60.277

Table Ex-T8

Section 7.9 Orthonormal Bases and the Gram–Schmidt Process

In this section we will show that every nonzero subspace of R^n has a basis of orthonormal vectors, and we will discuss a procedure for finding such bases. Bases of orthonormal vectors are important because they simplify many kinds of calculations.

ORTHOGONAL AND ORTHONORMAL BASES

Recall from Definitions 1.2.9 and 1.2.10 that a set of vectors in R^n is said to be *orthogonal* if each pair of distinct vectors in the set is orthogonal, and it is said to be *orthonormal* if it is orthogonal and each vector has length 1. In this section we will be concerned with **orthogonal bases** and **orthonormal bases** for subspaces of R^n. Here are some examples.

EXAMPLE 1
Converting an
Orthogonal
Basis to an
Orthonormal
Basis

Show that the vectors

$$\mathbf{v}_1 = (0, 2, 0), \quad \mathbf{v}_2 = (3, 0, 3), \quad \mathbf{v}_3 = (-4, 0, 4)$$

form an orthogonal basis for R^3, and convert it into an orthonormal basis by normalizing each vector.

Solution We showed in Example 3 of Section 7.1 that these vectors are linearly independent, so they must form a basis for R^3 by Theorem 7.2.6. We leave it for you to confirm that this is an orthogonal basis by showing that

$$\mathbf{v}_1 \cdot \mathbf{v}_2 = 0, \quad \mathbf{v}_1 \cdot \mathbf{v}_3 = 0, \quad \mathbf{v}_2 \cdot \mathbf{v}_3 = 0$$

To convert the orthogonal basis $\{\mathbf{v}_1, \mathbf{v}_2, \mathbf{v}_3\}$ to an orthonormal basis $\{\mathbf{q}_1, \mathbf{q}_2, \mathbf{q}_3\}$, we first compute $\|\mathbf{v}_1\| = 2$, $\|\mathbf{v}_2\| = 3\sqrt{2}$, and $\|\mathbf{v}_3\| = 4\sqrt{2}$, and then normalize to obtain

$$\mathbf{q}_1 = \frac{\mathbf{v}_1}{\|\mathbf{v}_1\|} = (0, 1, 0), \quad \mathbf{q}_2 = \frac{\mathbf{v}_2}{\|\mathbf{v}_2\|} = \left(\tfrac{1}{\sqrt{2}}, 0, \tfrac{1}{\sqrt{2}}\right), \quad \mathbf{q}_3 = \frac{\mathbf{v}_3}{\|\mathbf{v}_3\|} = \left(-\tfrac{1}{\sqrt{2}}, 0, \tfrac{1}{\sqrt{2}}\right) \quad ∎$$

EXAMPLE 2
The Standard
Basis for R^n is
an Orthonormal
Basis

Recall from Example 2 of Section 7.1 that the vectors

$$\mathbf{e}_1 = (1, 0, \ldots, 0), \quad \mathbf{e}_2 = (0, 1, \ldots, 0), \ldots, \quad \mathbf{e}_n = (0, 0, \ldots, 1)$$

form the standard basis for R^n. This is an orthonormal basis, since these are unit vectors and $\mathbf{e}_i \cdot \mathbf{e}_j = 0$ if $i \neq j$. ∎

It is clear geometrically that three nonzero mutually perpendicular vectors in R^3 must be linearly independent, for if one of them is a linear combination of the other two, then the three vectors would lie in a common plane, which they do not. This is a special case of the following more general result.

Theorem 7.9.1 *An orthogonal set of nonzero vectors in R^n is linearly independent.*

Proof Let $S = \{\mathbf{v}_1, \mathbf{v}_2, \ldots, \mathbf{v}_k\}$ be an orthogonal set of nonzero vectors in R^n. We must show that the only scalars that satisfy the vector equation

$$t_1\mathbf{v}_1 + t_2\mathbf{v}_2 + \cdots + t_k\mathbf{v}_k = \mathbf{0} \tag{1}$$

are $t_1 = 0$, $t_2 = 0$, $\ldots$, $t_k = 0$. To do this, let $\mathbf{v}_j$ be any vector in S; then (1) implies that

$$(t_1\mathbf{v}_1 + t_2\mathbf{v}_2 + \cdots + t_k\mathbf{v}_k) \cdot \mathbf{v}_j = \mathbf{0} \cdot \mathbf{v}_j = 0$$

which we can rewrite as

$$t_1(\mathbf{v}_1 \cdot \mathbf{v}_j) + t_2(\mathbf{v}_2 \cdot \mathbf{v}_j) + \cdots + t_k(\mathbf{v}_k \cdot \mathbf{v}_j) = 0 \tag{2}$$

But each pair of distinct vectors in S is orthogonal, so all of the dot products in this equation are zero, with the possible exception of $\mathbf{v}_j \cdot \mathbf{v}_j$. Thus, (2) can be simplified to

$$t_j(\mathbf{v}_j \cdot \mathbf{v}_j) = 0 \tag{3}$$

Since we have assumed that each vector in S is nonzero, this is true of $\mathbf{v}_j$, so it follows that $\mathbf{v}_j \cdot \mathbf{v}_j = \|\mathbf{v}_j\|^2 \neq 0$. Thus, (3) implies that $t_j = 0$, and since the choice of j is arbitrary, the proof is complete. ∎

EXAMPLE 3
An
Orthonormal
Basis for R^3

Show that the vectors

$$\mathbf{v}_1 = \left(\tfrac{3}{7}, -\tfrac{6}{7}, \tfrac{2}{7}\right), \quad \mathbf{v}_2 = \left(\tfrac{2}{7}, \tfrac{3}{7}, \tfrac{6}{7}\right), \quad \mathbf{v}_3 = \left(\tfrac{6}{7}, \tfrac{2}{7}, -\tfrac{3}{7}\right)$$

form an orthonormal basis for R^3.

Solution The vectors are orthonormal, since

$$\|\mathbf{v}_1\| = \|\mathbf{v}_2\| = \|\mathbf{v}_3\| = 1 \quad \text{and} \quad \mathbf{v}_1 \cdot \mathbf{v}_2 = \mathbf{v}_1 \cdot \mathbf{v}_3 = \mathbf{v}_2 \cdot \mathbf{v}_3 = 0$$

and hence they are linearly independent by Theorem 7.9.1. Since we have three linearly independent vectors in R^3, they must form a basis for R^3. ∎

ORTHOGONAL PROJECTIONS USING ORTHONORMAL BASES

Orthonormal bases are important because they simplify many formulas and numerical calculations. For example, we know from Theorem 7.7.5 that if W is a nonzero subspace of R^n, and if $\mathbf{x}$ is a vector in R^n that is expressed in column form, then

$$\text{proj}_W \mathbf{x} = M(M^T M)^{-1} M^T \mathbf{x} \tag{4}$$

for any matrix M whose column vectors form a basis for W. In particular, if the column vectors of M are orthonormal, then $M^T M = I$, so (4) simplifies to

$$\text{proj}_W \mathbf{x} = M M^T \mathbf{x} \tag{5}$$

and Formula (27) of Section 7.7 for the standard matrix of this orthogonal projection simplifies to

$$P = M M^T \tag{6}$$

Thus, using an orthonormal basis for W eliminates the matrix inversions in the projection formulas, and reduces the calculation of an orthogonal projection to matrix multiplication.

EXAMPLE 4
Standard Matrix for a Projection Using an Orthonormal Basis

Find the standard matrix P for the orthogonal projection of R^3 onto the plane through the origin that is spanned by the orthonormal vectors $\mathbf{v}_1 = (0, 1, 0)$ and $\mathbf{v}_2 = \left(-\frac{4}{5}, 0, \frac{3}{5}\right)$.

Solution Writing the vectors in column form and applying Formula (6) shows that the standard matrix for the projection is

$$P = M M^T = \begin{bmatrix} 0 & -\frac{4}{5} \\ 1 & 0 \\ 0 & \frac{3}{5} \end{bmatrix} \begin{bmatrix} 0 & 1 & 0 \\ -\frac{4}{5} & 0 & \frac{3}{5} \end{bmatrix} = \begin{bmatrix} \frac{16}{25} & 0 & -\frac{12}{25} \\ 0 & 1 & 0 \\ -\frac{12}{25} & 0 & \frac{9}{25} \end{bmatrix}$$ ∎

The following theorem expresses Formula (5) in terms of the basis vectors for the subspace W in the cases where the basis is orthonormal or orthogonal.

Theorem 7.9.2

(a) *If $\{\mathbf{v}_1, \mathbf{v}_2, \ldots, \mathbf{v}_k\}$ is an orthonormal basis for a subspace W of R^n, then the orthogonal projection of a vector $\mathbf{x}$ in R^n onto W can be expressed as*

$$\text{proj}_W \mathbf{x} = (\mathbf{x} \cdot \mathbf{v}_1)\mathbf{v}_1 + (\mathbf{x} \cdot \mathbf{v}_2)\mathbf{v}_2 + \cdots + (\mathbf{x} \cdot \mathbf{v}_k)\mathbf{v}_k \tag{7}$$

(b) *If $\{\mathbf{v}_1, \mathbf{v}_2, \ldots, \mathbf{v}_k\}$ is an orthogonal basis for a subspace W of R^n, then the orthogonal projection of a vector $\mathbf{x}$ in R^n onto W can be expressed as*

$$\text{proj}_W \mathbf{x} = \frac{\mathbf{x} \cdot \mathbf{v}_1}{\|\mathbf{v}_1\|^2}\mathbf{v}_1 + \frac{\mathbf{x} \cdot \mathbf{v}_2}{\|\mathbf{v}_2\|^2}\mathbf{v}_2 + \cdots + \frac{\mathbf{x} \cdot \mathbf{v}_k}{\|\mathbf{v}_k\|^2}\mathbf{v}_k \tag{8}$$

Proof (a) If we let

$$M = [\mathbf{v}_1 \quad \mathbf{v}_2 \quad \cdots \quad \mathbf{v}_k]$$

then it follows from Formula (5) that

$$\text{proj}_W \mathbf{x} = MM^T \mathbf{x}$$

Since the row vectors of M^T are the transposes of the column vectors of M, it follows from the row-column rule for matrix multiplication (Theorem 3.1.7) that

$$M^T \mathbf{x} = \begin{bmatrix} \mathbf{v}_1^T \mathbf{x} \\ \mathbf{v}_2^T \mathbf{x} \\ \vdots \\ \mathbf{v}_k^T \mathbf{x} \end{bmatrix} = \begin{bmatrix} \mathbf{x} \cdot \mathbf{v}_1 \\ \mathbf{x} \cdot \mathbf{v}_2 \\ \vdots \\ \mathbf{x} \cdot \mathbf{v}_k \end{bmatrix} \tag{9}$$

Thus, it follows from (5) and (9) that

$$\text{proj}_W \mathbf{x} = M(M^T \mathbf{x}) = \begin{bmatrix} \mathbf{v}_1 & \mathbf{v}_2 & \cdots & \mathbf{v}_k \end{bmatrix} \begin{bmatrix} \mathbf{x} \cdot \mathbf{v}_1 \\ \mathbf{x} \cdot \mathbf{v}_2 \\ \vdots \\ \mathbf{x} \cdot \mathbf{v}_k \end{bmatrix} = (\mathbf{x} \cdot \mathbf{v}_1)\mathbf{v}_1 + (\mathbf{x} \cdot \mathbf{v}_2)\mathbf{v}_2 + \cdots + (\mathbf{x} \cdot \mathbf{v}_k)\mathbf{v}_k$$

which proves (7).

Proof (b) Formula (8) can be derived by normalizing the orthogonal basis to obtain an orthonormal basis and applying (7). We leave the details for the exercises. ∎

EXAMPLE 5
An Orthogonal
Projection
Using an
Orthonormal
Basis

Find the orthogonal projection of $\mathbf{x} = (1, 1, 1)$ onto the plane W in R^3 that is spanned by the orthonormal vectors $\mathbf{v}_1 = (0, 1, 0)$ and $\mathbf{v}_2 = \left(-\frac{4}{5}, 0, \frac{3}{5}\right)$.

Solution One way to compute the orthogonal projection is to write $\mathbf{x}$ in column form and use the standard matrix P for the projection that was computed in Example 4. This yields

$$\text{proj}_W \mathbf{x} = P\mathbf{x} = \begin{bmatrix} \frac{16}{25} & 0 & -\frac{12}{25} \\ 0 & 1 & 0 \\ -\frac{12}{25} & 0 & \frac{9}{25} \end{bmatrix} \begin{bmatrix} 1 \\ 1 \\ 1 \end{bmatrix} = \begin{bmatrix} \frac{4}{25} \\ 1 \\ -\frac{3}{25} \end{bmatrix}$$

which we can write in comma-delimited form as $\text{proj}_W \mathbf{x} = \left(\frac{4}{25}, 1, -\frac{3}{25}\right)$.

Alternative Solution A second method for computing the orthogonal projection is to use Formula (7). This yields

$$\text{proj}_W \mathbf{x} = (\mathbf{x} \cdot \mathbf{v}_1)\mathbf{v}_1 + (\mathbf{x} \cdot \mathbf{v}_2)\mathbf{v}_2 = (1)(0, 1, 0) + \left(-\tfrac{1}{5}\right)\left(-\tfrac{4}{5}, 0, \tfrac{3}{5}\right) = \left(\tfrac{4}{25}, 1, -\tfrac{3}{25}\right)$$

which agrees with the result obtained using the standard matrix. ∎

EXAMPLE 6
An Orthogonal
Projection
Using an
Orthogonal
Basis

Find the orthogonal projection of $\mathbf{x} = (-5, 3, 1)$ onto the plane W in R^3 that is spanned by the orthogonal vectors

$$\mathbf{v}_1 = (0, 1, -1) \quad \text{and} \quad \mathbf{v}_2 = (1, 2, 2)$$

Solution We could normalize the basis vectors and apply Formula (7) to the resulting orthonormal basis for W, but let us apply (8) directly. We have

$$\|\mathbf{v}_1\|^2 = 0^2 + 1^2 + (-1)^2 = 2 \quad \text{and} \quad \|\mathbf{v}_2\|^2 = 1^2 + 2^2 + 2^2 = 9$$

so it follows from (8) that

$$\text{proj}_W \mathbf{x} = \frac{\mathbf{x} \cdot \mathbf{v}_1}{\|\mathbf{v}_1\|^2}\mathbf{v}_1 + \frac{\mathbf{x} \cdot \mathbf{v}_2}{\|\mathbf{v}_2\|^2}\mathbf{v}_2 = \tfrac{2}{2}(0, 1, -1) + \tfrac{3}{9}(1, 2, 2) = \left(\tfrac{1}{3}, \tfrac{5}{3}, -\tfrac{1}{3}\right)$$

∎

TRACE AND ORTHOGONAL PROJECTIONS

The following theorem provides a simple way of finding the dimension of the range of an orthogonal projection.

Theorem 7.9.3 *If P is the standard matrix for an orthogonal projection of R^n onto a subspace of R^n, then $\mathrm{tr}(P) = \mathrm{rank}(P)$.*

Proof Suppose that P is the standard matrix for an orthogonal projection of R^n onto a k-dimensional subspace W. If we let $\{\mathbf{v}_1, \mathbf{v}_2, \ldots, \mathbf{v}_k\}$ be an orthonormal basis for W, then it follows from Formula (6) and Theorem 3.8.1 that

$$P = MM^T = [\mathbf{v}_1 \quad \mathbf{v}_2 \quad \cdots \quad \mathbf{v}_k] \begin{bmatrix} \mathbf{v}_1^T \\ \mathbf{v}_2^T \\ \vdots \\ \mathbf{v}_k^T \end{bmatrix} = \mathbf{v}_1\mathbf{v}_1^T + \mathbf{v}_2\mathbf{v}_2^T + \cdots + \mathbf{v}_k\mathbf{v}_k^T \tag{10}$$

Using this result, the additive property of the trace, and Formula (27) of Section 3.1, we obtain

$$\mathrm{tr}(P) = \mathrm{tr}\left(\mathbf{v}_1\mathbf{v}_1^T\right) + \mathrm{tr}\left(\mathbf{v}_2\mathbf{v}_2^T\right) + \cdots + \mathrm{tr}\left(\mathbf{v}_k\mathbf{v}_k^T\right)$$
$$= (\mathbf{v}_1 \cdot \mathbf{v}_1) + (\mathbf{v}_2 \cdot \mathbf{v}_2) + \cdots + (\mathbf{v}_k \cdot \mathbf{v}_k)$$
$$= \|\mathbf{v}_1\|^2 + \|\mathbf{v}_2\|^2 + \cdots + \|\mathbf{v}_k\|^2$$
$$= 1 + 1 + \cdots + 1 = k = \dim(W)$$

But the range of a matrix transformation is the column space of the matrix, so it follows from this computation that $\mathrm{tr}(P) = \dim(\mathrm{col}(P)) = \mathrm{rank}(P)$. ∎

EXAMPLE 7
Using the Trace to Find the Rank of an Orthogonal Projection

We showed in Example 4 that the standard matrix P for the orthogonal projection onto the plane spanned by the vectors $\mathbf{v}_1 = (0, 1, 0)$ and $\mathbf{v}_2 = \left(-\frac{4}{5}, 0, \frac{3}{5}\right)$ is

$$P = \begin{bmatrix} \frac{16}{25} & 0 & -\frac{12}{25} \\ 0 & 1 & 0 \\ -\frac{12}{25} & 0 & \frac{9}{25} \end{bmatrix}$$

Since the plane is two-dimensional, the matrix P must have rank 2, which is confirmed by the computation

$$\mathrm{rank}(P) = \mathrm{tr}(P) = \frac{16}{25} + 1 + \frac{9}{25} = 2$$ ∎

LINEAR COMBINATIONS OF ORTHONORMAL BASIS VECTORS

If W is a subspace of R^n, and if $\mathbf{w}$ is a vector in W, then $\mathrm{proj}_W\mathbf{w} = \mathbf{w}$. Thus, we have the following special case of Theorem 7.9.2.

Theorem 7.9.4

(a) *If $\{\mathbf{v}_1, \mathbf{v}_2, \ldots, \mathbf{v}_k\}$ is an orthonormal basis for a subspace W of R^n, and if $\mathbf{w}$ is a vector in W, then*

$$\mathbf{w} = (\mathbf{w} \cdot \mathbf{v}_1)\mathbf{v}_1 + (\mathbf{w} \cdot \mathbf{v}_2)\mathbf{v}_2 + \cdots + (\mathbf{w} \cdot \mathbf{v}_k)\mathbf{v}_k \tag{11}$$

(b) *If $\{\mathbf{v}_1, \mathbf{v}_2, \ldots, \mathbf{v}_k\}$ is an orthogonal basis for a subspace W of R^n, and if $\mathbf{w}$ is a vector in W, then*

$$\mathbf{w} = \frac{\mathbf{w} \cdot \mathbf{v}_1}{\|\mathbf{v}_1\|^2}\mathbf{v}_1 + \frac{\mathbf{w} \cdot \mathbf{v}_2}{\|\mathbf{v}_2\|^2}\mathbf{v}_2 + \cdots + \frac{\mathbf{w} \cdot \mathbf{v}_k}{\|\mathbf{v}_k\|^2}\mathbf{v}_k \tag{12}$$

Recalling Formula (5) of Section 7.7, observe that the terms on the right side of (12) are the orthogonal projections onto the lines spanned by $\mathbf{v}_1, \mathbf{v}_2, \ldots, \mathbf{v}_k$, so that Formula (12) decomposes each vector $\mathbf{w}$ in the k-dimensional subspace W into a sum of k projections onto one-dimensional subspaces. Figure 7.9.1 illustrates this idea in the case where W is R^2.

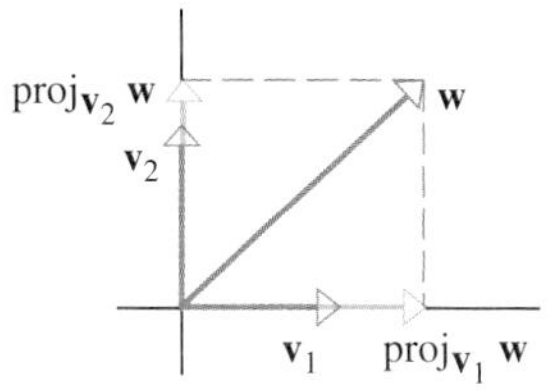

EXAMPLE 8

Linear Combinations of Orthonormal Basis Vectors

Express the vector $\mathbf{w} = (1, 1, 1)$ in R^3 as a linear combination of the orthonormal basis vectors

$$\mathbf{v}_1 = \left(\tfrac{3}{7}, -\tfrac{6}{7}, \tfrac{2}{7}\right), \quad \mathbf{v}_2 = \left(\tfrac{2}{7}, \tfrac{3}{7}, \tfrac{6}{7}\right), \quad \mathbf{v}_3 = \left(\tfrac{6}{7}, \tfrac{2}{7}, -\tfrac{3}{7}\right)$$

Solution We showed in Example 3 that the given vectors form an orthonormal basis for R^3. Thus, $\mathbf{w}$ can be expressed as a linear combination of $\mathbf{v}_1, \mathbf{v}_2$, and $\mathbf{v}_3$ using Formula (11). We leave it for you to confirm that

$$\mathbf{w} \cdot \mathbf{v}_1 = -\tfrac{1}{7}, \quad \mathbf{w} \cdot \mathbf{v}_2 = \tfrac{11}{7}, \quad \mathbf{w} \cdot \mathbf{v}_3 = \tfrac{5}{7}$$

Thus, it follows from Formula (11) that

$$\mathbf{w} = -\tfrac{1}{7}\mathbf{v}_1 + \tfrac{11}{7}\mathbf{v}_2 + \tfrac{5}{7}\mathbf{v}_3$$

or expressed in component form,

$$\mathbf{w} = -\tfrac{1}{7}\left(\tfrac{3}{7}, -\tfrac{6}{7}, \tfrac{2}{7}\right) + \tfrac{11}{7}\left(\tfrac{2}{7}, \tfrac{3}{7}, \tfrac{6}{7}\right) + \tfrac{5}{7}\left(\tfrac{6}{7}, \tfrac{2}{7}, -\tfrac{3}{7}\right)$$

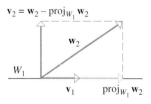

Figure 7.9.1

REMARK The general procedure for expressing a vector $\mathbf{w}$ as a linear combination

$$\mathbf{w} = t_1\mathbf{v}_1 + t_2\mathbf{v}_2 + \cdots + t_k\mathbf{v}_k$$

is to equate corresponding components on the two sides and solve the resulting linear system for the unknown coefficients $t_1, t_2, \ldots, t_k$ (see Example 7 of Section 2.1). The last example illustrates that if $\mathbf{v}_1, \mathbf{v}_2, \ldots, \mathbf{v}_k$ are orthonormal, then the coefficients can be obtained by computing appropriate dot products, thereby eliminating the need to solve a linear system.

FINDING ORTHOGONAL AND ORTHONORMAL BASES

The following theorem, which is the main result in this section, shows that every nonzero subspace of R^n has an orthonormal basis. The proof of this theorem is especially important because it provides a method for converting any basis for a subspace of R^n into an orthogonal basis.

Theorem 7.9.5 *Every nonzero subspace of R^n has an orthonormal basis.*

Proof Let W be a nonzero subspace of R^n, and let $\{\mathbf{w}_1, \mathbf{w}_2, \ldots, \mathbf{w}_k\}$ be *any* basis for W. To prove that W has an orthonormal basis, it suffices to show that W has an *orthogonal* basis, since such a basis can then be converted into an orthonormal basis by normalizing the vectors.

The following sequence of steps will produce an orthogonal basis $\{\mathbf{v}_1, \mathbf{v}_2, \ldots, \mathbf{v}_k\}$ for W:

Step 1. Let $\mathbf{v}_1 = \mathbf{w}_1$.

Step 2. As illustrated in Figure 7.9.2, we can obtain a vector $\mathbf{v}_2$ that is orthogonal to $\mathbf{v}_1$ by computing the component of $\mathbf{w}_2$ that is orthogonal to the subspace W_1 spanned by $\mathbf{v}_1$. By applying Formula (8) in Theorem 7.9.2 and Formula (24) of Section 7.7, we can express this component as

$$\mathbf{v}_2 = \mathbf{w}_2 - \text{proj}_{W_1}\mathbf{w}_2 = \mathbf{w}_2 - \frac{\mathbf{w}_2 \cdot \mathbf{v}_1}{\|\mathbf{v}_1\|^2}\mathbf{v}_1 \tag{13}$$

Of course, if $\mathbf{v}_2 = \mathbf{0}$, then $\mathbf{v}_2$ is not a basis vector. But this cannot happen, since it would then follow from (13) and Step 1 that

$$\mathbf{w}_2 = \frac{\mathbf{w}_2 \cdot \mathbf{v}_1}{\|\mathbf{v}_1\|^2}\mathbf{v}_1 = \frac{\mathbf{w}_2 \cdot \mathbf{v}_1}{\|\mathbf{v}_1\|^2}\mathbf{w}_1$$

which states that $\mathbf{w}_2$ is a scalar multiple of $\mathbf{w}_1$, contradicting the linear independence of the basis vectors $\{\mathbf{w}_1, \mathbf{w}_2, \ldots, \mathbf{w}_k\}$.

Figure 7.9.2

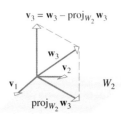

$$v_3 = w_3 - \text{proj}_{W_2} w_3$$

$$w_3$$

$$v_2$$

$$v_1 \qquad W_2$$

$$\text{proj}_{W_2} w_3$$

Figure 7.9.3

Step 3. To obtain a vector $\mathbf{v}_3$ that is orthogonal to $\mathbf{v}_1$ and $\mathbf{v}_2$, we will compute the component of $\mathbf{w}_3$ that is orthogonal to the subspace W_2 that is spanned by $\mathbf{v}_1$ and $\mathbf{v}_2$ (Figure 7.9.3). By applying Formula (8), we can express this component as

$$\mathbf{v}_3 = \mathbf{w}_3 - \text{proj}_{W_2} \mathbf{w}_3 = \mathbf{w}_3 - \frac{\mathbf{w}_3 \cdot \mathbf{v}_1}{\|\mathbf{v}_1\|^2}\mathbf{v}_1 - \frac{\mathbf{w}_3 \cdot \mathbf{v}_2}{\|\mathbf{v}_2\|^2}\mathbf{v}_2$$

As in Step 2, the linear independence of $\{\mathbf{w}_1, \mathbf{w}_2, \dots, \mathbf{w}_k\}$ ensures that $\mathbf{v}_3 \neq \mathbf{0}$. We leave the details as an exercise.

Step 4. To obtain a vector $\mathbf{v}_4$ that is orthogonal to $\mathbf{v}_1$, $\mathbf{v}_2$, and $\mathbf{v}_3$, we will compute the component of $\mathbf{w}_4$ that is orthogonal to the subspace W_3 spanned by $\mathbf{v}_1$, $\mathbf{v}_2$, and $\mathbf{v}_3$. By applying Formula (8) again, we can express this component as

$$\mathbf{v}_4 = \mathbf{w}_4 - \text{proj}_{W_3} \mathbf{w}_4 = \mathbf{w}_4 - \frac{\mathbf{w}_4 \cdot \mathbf{v}_1}{\|\mathbf{v}_1\|^2}\mathbf{v}_1 - \frac{\mathbf{w}_4 \cdot \mathbf{v}_2}{\|\mathbf{v}_2\|^2}\mathbf{v}_2 - \frac{\mathbf{w}_4 \cdot \mathbf{v}_3}{\|\mathbf{v}_3\|^2}\mathbf{v}_3$$

Steps 5 to k. Continuing in this way produces an orthogonal set $\{\mathbf{v}_1, \mathbf{v}_2, \dots, \mathbf{v}_k\}$ after k steps. Since W is k-dimensional, this set is an orthogonal basis for W, which completes the proof. ■

The proof of this theorem provides an algorithm, called the **Gram–Schmidt orthogonalization process**, for converting an arbitrary basis for a subspace of R^n into an orthogonal basis for the subspace. If the resulting orthogonal vectors are normalized to produce an orthonormal basis for the subspace, then the algorithm is called the **Gram–Schmidt process**.

EXAMPLE 9 The vectors $\mathbf{w}_1 = (1, 1, 1)$, $\mathbf{w}_2 = (0, 1, 1)$, and $\mathbf{w}_3 = (0, 0, 1)$ form a basis for R^3 (verify). Use the Gram–Schmidt orthogonalization process to transform this basis into an orthogonal basis, and then normalize the orthogonal basis vectors to obtain an orthonormal basis for R^3.

Solution Let $\{\mathbf{v}_1, \mathbf{v}_2, \mathbf{v}_3\}$ denote the orthogonal basis produced by the Gram–Schmidt orthogonalization process, and let $\{\mathbf{q}_1, \mathbf{q}_2, \mathbf{q}_3\}$ denote the orthonormal basis that results from normalizing $\mathbf{v}_1$, $\mathbf{v}_2$, and $\mathbf{v}_3$. To find the orthogonal basis we follow the steps in the proof of Theorem 7.9.5:

Step 1. Let $\mathbf{v}_1 = \mathbf{w}_1 = (1, 1, 1)$.

Step 2. Let $\mathbf{v}_2 = \mathbf{w}_2 - \text{proj}_{W_1} \mathbf{w}_2 = \mathbf{w}_2 - \frac{\mathbf{w}_2 \cdot \mathbf{v}_1}{\|\mathbf{v}_1\|^2}\mathbf{v}_1$

$$= (0, 1, 1) - \tfrac{2}{3}(1, 1, 1) = \left(-\tfrac{2}{3}, \tfrac{1}{3}, \tfrac{1}{3}\right)$$

Step 3. Let $\mathbf{v}_3 = \mathbf{w}_3 - \text{proj}_{W_2} \mathbf{w}_3 = \mathbf{w}_3 - \frac{\mathbf{w}_3 \cdot \mathbf{v}_1}{\|\mathbf{v}_1\|^2}\mathbf{v}_1 - \frac{\mathbf{w}_3 \cdot \mathbf{v}_2}{\|\mathbf{v}_2\|^2}\mathbf{v}_2$

$$= (0, 0, 1) - \tfrac{1}{3}(1, 1, 1) - \frac{1/3}{2/3}\left(-\tfrac{2}{3}, \tfrac{1}{3}, \tfrac{1}{3}\right) = \left(0, -\tfrac{1}{2}, \tfrac{1}{2}\right)$$

Thus, the vectors

$$\mathbf{v}_1 = (1, 1, 1), \quad \mathbf{v}_2 = \left(-\tfrac{2}{3}, \tfrac{1}{3}, \tfrac{1}{3}\right), \quad \mathbf{v}_3 = \left(0, -\tfrac{1}{2}, \tfrac{1}{2}\right)$$

form an orthogonal basis for R^3. The norms of these vectors are

$$\|\mathbf{v}_1\| = \sqrt{3}, \quad \|\mathbf{v}_2\| = \frac{\sqrt{6}}{3}, \quad \|\mathbf{v}_3\| = \frac{1}{\sqrt{2}}$$

so an orthonormal basis for R^3 is given by

$$\mathbf{q}_1 = \frac{\mathbf{v}_1}{\|\mathbf{v}_1\|} = \left(\tfrac{1}{\sqrt{3}}, \tfrac{1}{\sqrt{3}}, \tfrac{1}{\sqrt{3}}\right), \quad \mathbf{q}_2 = \frac{\mathbf{v}_2}{\|\mathbf{v}_2\|} = \left(-\tfrac{2}{\sqrt{6}}, \tfrac{1}{\sqrt{6}}, \tfrac{1}{\sqrt{6}}\right)$$

$$\mathbf{q}_3 = \frac{\mathbf{v}_3}{\|\mathbf{v}_3\|} = \left(0, -\tfrac{1}{\sqrt{2}}, \tfrac{1}{\sqrt{2}}\right)$$ ■

Linear Algebra in History

The names of Jörgen Pederson Gram, a Danish actuary, and Erhard Schmidt, a German mathematician, are as closely linked in the world of mathematics as the names of Gilbert and Sullivan in the world of musical theater. However, Gram and Schmidt probably never met, and the process that bears their names was not discovered by either one of them. Gram's name is linked to the process as the result of his Ph.D. thesis, in which he used it to solve least squares problems; and Schmidt's name is linked to it as the result of his studies of certain kinds of vector spaces.

Gram loved the interplay between theoretical mathematics and applications and wrote four mathematical treatises on forest management, whereas Schmidt was primarily a theoretician whose work significantly influenced the direction of mathematics in the twentieth century. During World War II Schmidt held positions of authority at the University of Berlin and had to carry out various Nazi resolutions against the Jews—a job that he apparently did not do well, since he was criticized at one point for not understanding the "Jewish question." At the celebration of Schmidt's 75th birthday in 1951 a prominent Jewish mathematician, who had survived the Nazi years, spoke of the difficulties that Schmidt faced during that period without criticism.

Jörgen Pederson Gram
(1850–1916)

Erhard Schmidt
(1876–1959)

REMARK In this example we first found an orthogonal basis and then normalized at the end to produce the orthonormal basis. Alternatively, we could have normalized each orthogonal basis vector as soon as it was calculated, thereby generating the orthonormal basis step by step. For hand calculation, it is usually better to do the normalization at the end, since this tends to postpone the introduction of bothersome square roots.

EXAMPLE 10
Orthonormal
Basis for a Plane
in R^3

Use the Gram–Schmidt process to construct an orthonormal basis for the plane $x + y + z = 0$ in R^3.

Solution First we will find a basis for the plane, and then we will apply the Gram–Schmidt process to that basis. Any two nonzero vectors in the plane that are not scalar multiples of one another will serve as a basis. One way to find such vectors is to use the method of Example 7 in Section 1.3 to write the plane in the parametric form

$$x = -t_1 - t_2, \ y = t_1, \ z = t_2$$

The parameter values $t_1 = 1$, $t_2 = 0$ and $t_1 = 0$, $t_2 = 1$ produce the vectors

$$\mathbf{w}_1 = (-1, 1, 0) \quad \text{and} \quad \mathbf{w}_2 = (-1, 0, 1)$$

in the given plane. Now we are ready to apply the Gram–Schmidt process. First we construct the orthogonal basis vectors

$$\mathbf{v}_1 = \mathbf{w}_1 = (-1, 1, 0)$$

$$\mathbf{v}_2 = \mathbf{w}_2 - \frac{\mathbf{w}_2 \cdot \mathbf{v}_1}{\|\mathbf{v}_1\|^2} \mathbf{v}_1 = (-1, 0, 1) - \tfrac{1}{2}(-1, 1, 0) = \left(-\tfrac{1}{2}, -\tfrac{1}{2}, 1\right)$$

and then normalize these to obtain the orthonormal basis vectors

$$\mathbf{q}_1 = \frac{\mathbf{v}_1}{\|\mathbf{v}_1\|} = \left(-\tfrac{1}{\sqrt{2}}, \tfrac{1}{\sqrt{2}}, 0\right)$$

$$\mathbf{q}_2 = \frac{\mathbf{v}_2}{\|\mathbf{v}_2\|} = \tfrac{2}{\sqrt{6}}\left(-\tfrac{1}{2}, -\tfrac{1}{2}, 1\right) = \left(-\tfrac{1}{\sqrt{6}}, -\tfrac{1}{\sqrt{6}}, \tfrac{2}{\sqrt{6}}\right)$$ ∎

A PROPERTY OF THE GRAM–SCHMIDT PROCESS

In the exercises we will ask you to show that the vector $\mathbf{v}_j$ that is produced at the jth step of the Gram–Schmidt process is expressible as a linear combination of $\mathbf{w}_1, \mathbf{w}_2, \ldots, \mathbf{w}_j$. Thus, not only does the Gram–Schmidt process produce an orthogonal basis $\{\mathbf{v}_1, \mathbf{v}_2, \ldots, \mathbf{v}_k\}$ for the subspace W spanned by $\{\mathbf{w}_1, \mathbf{w}_2, \ldots, \mathbf{w}_k\}$, but it also creates the basis in such a way that at each intermediate stage the vectors $\{\mathbf{v}_1, \mathbf{v}_2, \ldots, \mathbf{v}_j\}$ form an orthogonal basis for the subspace of R^n spanned by $\{\mathbf{w}_1, \mathbf{w}_2, \ldots, \mathbf{w}_j\}$. Moreover, since $\mathbf{v}_j$ is constructed to be orthogonal to span$\{\mathbf{v}_1, \mathbf{v}_2, \ldots, \mathbf{v}_{j-1}\}$, the vector $\mathbf{v}_j$ must also be orthogonal to span$\{\mathbf{w}_1, \mathbf{w}_2, \ldots, \mathbf{w}_{j-1}\}$, since the two subspaces are the same.

In summary, we have the following theorem.

Theorem 7.9.6 *If $S = \{\mathbf{w}_1, \mathbf{w}_2, \ldots, \mathbf{w}_k\}$ is a basis for a nonzero subspace of R^n, and if $S' = \{\mathbf{v}_1, \mathbf{v}_2, \ldots, \mathbf{v}_k\}$ is the corresponding orthogonal basis produced by the Gram–Schmidt process, then:*

(a) *$\{\mathbf{v}_1, \mathbf{v}_2, \ldots, \mathbf{v}_j\}$ is an orthogonal basis for span$\{\mathbf{w}_1, \mathbf{w}_2, \ldots, \mathbf{w}_j\}$ at the jth step.*

(b) *$\mathbf{v}_j$ is orthogonal to span$\{\mathbf{w}_1, \mathbf{w}_2, \ldots, \mathbf{w}_{j-1}\}$ at the jth step ($j \geq 2$).*

REMARK This theorem remains true if the orthogonal basis vectors are normalized at each step, rather than at the end of the process; that is, $\{\mathbf{q}_1, \mathbf{q}_2, \ldots, \mathbf{q}_j\}$ is an *orthonormal* basis for span$\{\mathbf{w}_1, \mathbf{w}_2, \ldots, \mathbf{w}_j\}$ and $\mathbf{q}_j$ is orthogonal to span$\{\mathbf{w}_1, \mathbf{w}_2, \ldots, \mathbf{w}_{j-1}\}$.

EXTENDING ORTHONORMAL SETS TO ORTHONORMAL BASES

Recall from part (*b*) of Theorem 7.2.2 that a linearly independent set in a nonzero subspace W of R^n can be enlarged to a basis for W. The following theorem is an analog of that result for orthogonal bases and orthonormal bases.

Theorem 7.9.7 *If W is a nonzero subspace of R^n, then:*

(a) *Every orthogonal set of nonzero vectors in W can be enlarged to an orthogonal basis for W.*

(b) *Every orthonormal set in W can be enlarged to an orthonormal basis for W.*

We will prove part (b); the proof of part (a) is similar.

Proof (b) Suppose that $S = \{\mathbf{v}_1, \mathbf{v}_2, \ldots, \mathbf{v}_s\}$ is an orthonormal set of vectors in W. Part (b) of Theorem 7.2.2 tells us that we can enlarge S to some basis $S' = \{\mathbf{v}_1, \mathbf{v}_2, \ldots, \mathbf{v}_s, \mathbf{w}_{s+1}, \ldots, \mathbf{w}_k\}$ for W. If we now apply the Gram–Schmidt process to the set S', then the vectors $\mathbf{v}_1, \mathbf{v}_2, \ldots, \mathbf{v}_s$ will not be altered since they are already orthonormal, and the resulting set

$$S'' = \{\mathbf{v}_1, \mathbf{v}_2, \ldots, \mathbf{v}_s, \mathbf{v}_{s+1}, \ldots, \mathbf{v}_k\}$$

will be an orthonormal basis for W. ∎

In the exercises we will ask you to use the method of Example 2 in Section 7.4 and the Gram–Schmidt process to develop a computational technique for extending orthonormal sets to orthonormal bases.

Exercise Set 7.9

In Exercises 1 and 2, determine whether the vectors form an orthogonal set. If so, convert it to an orthonormal set by normalizing.

1. (a) $\mathbf{v}_1 = (2, 3), \ \mathbf{v}_2 = (3, 2)$
   (b) $\mathbf{v}_1 = (-1, 1), \ \mathbf{v}_2 = (1, 1)$
   (c) $\mathbf{v}_1 = (-2, 1, 1), \ \mathbf{v}_2 = (1, 0, 2), \ \mathbf{v}_3 = (-2, -5, 1)$
   (d) $\mathbf{v}_1 = (-3, 4, -1), \ \mathbf{v}_2 = (1, 2, 5), \ \mathbf{v}_3 = (4, -3, 0)$

2. (a) $\mathbf{v}_1 = (2, 3), \ \mathbf{v}_2 = (-3, 2)$
   (b) $\mathbf{v}_1 = (1, -2), \ \mathbf{v}_2 = (-2, 1)$
   (c) $\mathbf{v}_1 = (1, 0, 1), \ \mathbf{v}_2 = (1, 1, 1), \ \mathbf{v}_3 = (-1, 0, 1)$
   (d) $\mathbf{v}_1 = (2, -2, 1), \ \mathbf{v}_2 = (2, 1, -2), \ \mathbf{v}_3 = (1, 2, 2)$

In Exercises 3 and 4, determine whether the vectors form an orthonormal set in R^3. If not, state which of the required properties fail to hold.

3. (a) $\mathbf{v}_1 = \left(\frac{2}{3}, -\frac{2}{3}, \frac{1}{3}\right), \ \mathbf{v}_2 = \left(\frac{2}{3}, \frac{1}{3}, -\frac{2}{3}\right), \ \mathbf{v}_3 = \left(\frac{1}{3}, \frac{2}{3}, \frac{2}{3}\right)$
   (b) $\mathbf{v}_1 = \left(\frac{1}{\sqrt{2}}, 0, \frac{1}{\sqrt{2}}\right), \ \mathbf{v}_2 = \left(\frac{1}{\sqrt{3}}, \frac{1}{\sqrt{3}}, -\frac{1}{\sqrt{3}}\right),$
   $\mathbf{v}_3 = \left(-\frac{1}{\sqrt{2}}, 0, \frac{1}{\sqrt{2}}\right)$
   (c) $\mathbf{v}_1 = \left(\frac{1}{\sqrt{2}}, 0, -\frac{1}{\sqrt{2}}\right), \ \mathbf{v}_2 = \left(\frac{1}{\sqrt{18}}, \frac{4}{\sqrt{18}}, \frac{1}{\sqrt{18}}\right),$
   $\mathbf{v}_3 = \left(-\frac{2}{\sqrt{3}}, \frac{1}{\sqrt{3}}, -\frac{2}{\sqrt{3}}\right)$

4. (a) $\mathbf{v}_1 = \left(\frac{1}{\sqrt{2}}, \frac{1}{\sqrt{3}}, \frac{1}{\sqrt{6}}\right), \ \mathbf{v}_2 = \left(0, \frac{1}{\sqrt{3}}, -\frac{2}{\sqrt{6}}\right),$
   $\mathbf{v}_3 = \left(-\frac{1}{\sqrt{2}}, \frac{1}{\sqrt{3}}, \frac{1}{\sqrt{6}}\right)$
   (b) $\mathbf{v}_1 = \left(\frac{1}{3}, \frac{1}{3}, \frac{1}{3}\right), \ \mathbf{v}_2 = (0, 0, 0), \ \mathbf{v}_3 = \left(-\frac{1}{\sqrt{2}}, 0, \frac{1}{\sqrt{2}}\right)$
   (c) $\mathbf{v}_1 = \left(\frac{1}{\sqrt{10}}, \frac{3}{\sqrt{20}}, \frac{3}{\sqrt{20}}\right), \ \mathbf{v}_2 = \left(\frac{3}{\sqrt{10}}, -\frac{1}{\sqrt{10}}, -\frac{1}{\sqrt{10}}\right),$
   $\mathbf{v}_3 = \left(0, -\frac{1}{\sqrt{3}}, \frac{1}{\sqrt{3}}\right)$

In Exercises 5 and 6, find the orthogonal projection of $\mathbf{x} = (1, 2, 0, -1)$ on the subspace of R^4 spanned by the given orthonormal vectors.

5. (a) $\mathbf{v}_1 = \left(0, \frac{1}{\sqrt{18}}, -\frac{4}{\sqrt{18}}, -\frac{1}{\sqrt{18}}\right), \ \mathbf{v}_2 = \left(\frac{1}{2}, \frac{5}{6}, \frac{1}{6}, \frac{1}{6}\right)$
   (b) $\mathbf{v}_1 = \left(0, \frac{1}{\sqrt{18}}, -\frac{4}{\sqrt{18}}, -\frac{1}{\sqrt{18}}\right), \ \mathbf{v}_2 = \left(\frac{1}{2}, \frac{5}{6}, \frac{1}{6}, \frac{1}{6}\right),$
   $\mathbf{v}_3 = \left(\frac{1}{\sqrt{18}}, 0, \frac{1}{\sqrt{18}}, -\frac{4}{\sqrt{18}}\right)$

6. (a) $\mathbf{v}_1 = \left(\frac{1}{2}, \frac{1}{2}, \frac{1}{2}, \frac{1}{2}\right), \ \mathbf{v}_2 = \left(\frac{1}{2}, \frac{1}{2}, -\frac{1}{2}, -\frac{1}{2}\right)$
   (b) $\mathbf{v}_1 = \left(\frac{1}{2}, \frac{1}{2}, \frac{1}{2}, \frac{1}{2}\right), \ \mathbf{v}_2 = \left(\frac{1}{2}, \frac{1}{2}, -\frac{1}{2}, -\frac{1}{2}\right),$
   $\mathbf{v}_3 = \left(\frac{1}{2}, -\frac{1}{2}, \frac{1}{2}, -\frac{1}{2}\right)$

In Exercises 7 and 8, find the orthogonal projection of $\mathbf{x} = (1, 2, 0, -2)$ on the subspace of R^4 spanned by the given orthogonal vectors.

7. (a) $\mathbf{v}_1 = (1, 1, 1, 1), \ \mathbf{v}_2 = (1, 1, -1, -1)$
   (b) $\mathbf{v}_1 = (1, 1, 1, 1), \ \mathbf{v}_2 = (1, 1, -1, -1),$
   $\mathbf{v}_3 = (1, -1, 1, -1)$

8. (a) $\mathbf{v}_1 = (0, 1, -4, -1), \ \mathbf{v}_2 = (3, 5, 1, 1)$
   (b) $\mathbf{v}_1 = (0, 1, -4, -1), \ \mathbf{v}_2 = (3, 5, 1, 1),$
   $\mathbf{v}_3 = (1, 0, 1, -4)$

In Exercises 9 and 10, confirm that $\{\mathbf{v}_1, \mathbf{v}_2, \mathbf{v}_3\}$ is an orthonormal basis for R^3, and express $\mathbf{w}$ as a linear combination of those vectors.

9. $\mathbf{v}_1 = \left(\frac{3}{\sqrt{11}}, \frac{1}{\sqrt{11}}, \frac{1}{\sqrt{11}}\right), \ \mathbf{v}_2 = \left(-\frac{1}{\sqrt{6}}, \frac{2}{\sqrt{6}}, \frac{1}{\sqrt{6}}\right),$
   $\mathbf{v}_3 = \left(-\frac{1}{\sqrt{66}}, -\frac{4}{\sqrt{66}}, \frac{7}{\sqrt{66}}\right); \ \mathbf{w} = (-1, 1, 2)$

10. $\mathbf{v}_1 = \left(\frac{2}{3}, \frac{1}{3}, \frac{2}{3}\right), \ \mathbf{v}_2 = \left(\frac{1}{3}, \frac{2}{3}, -\frac{2}{3}\right), \ \mathbf{v}_3 = \left(\frac{2}{3}, -\frac{2}{3}, -\frac{1}{3}\right);$
    $\mathbf{w} = (2, 0, 5)$

In Exercises 11 and 12, confirm that $\{\mathbf{v}_1, \mathbf{v}_2, \mathbf{v}_3, \mathbf{v}_4\}$ is an orthogonal basis for R^4, and express $\mathbf{w}$ as a linear combination of those vectors.

11. $\mathbf{v}_1 = (1, -2, 2, -3), \ \mathbf{v}_2 = (2, -3, 2, 4), \ \mathbf{v}_3 = (2, 2, 1, 0),$
    $\mathbf{v}_4 = (5, -2, -6, -1); \ \mathbf{w} = (1, 1, 1, 1)$

12. $\mathbf{v}_1 = (1, 1, 1, 2)$, $\mathbf{v}_2 = (1, 2, 3, -3)$, $\mathbf{v}_3 = (1, -2, 1, 0)$, $\mathbf{v}_4 = (25, 4, -17, -6)$; $\mathbf{w} = (1, 1, 1, 1)$

In Exercises 13 and 14, find the standard matrix for the orthogonal projection onto the subspace of R^3 spanned by the given orthonormal vectors.

13. $\mathbf{v}_1 = \left(\frac{2}{3}, \frac{1}{3}, \frac{2}{3}\right)$, $\mathbf{v}_2 = \left(\frac{1}{3}, \frac{2}{3}, -\frac{2}{3}\right)$

14. $\mathbf{v}_1 = \left(\frac{1}{\sqrt{2}}, \frac{1}{\sqrt{3}}, \frac{1}{\sqrt{6}}\right)$, $\mathbf{v}_2 = \left(0, \frac{1}{\sqrt{3}}, -\frac{2}{\sqrt{6}}\right)$

15. Use the matrix obtained in Exercise 13 to find the orthogonal projection of $\mathbf{w} = (0, 2, -3)$ onto span$\{\mathbf{v}_1, \mathbf{v}_2\}$, and check the result using Formula (7).

16. Use the matrix obtained in Exercise 13 to find the orthogonal projection of $\mathbf{w} = (4, -5, 1)$ onto span$\{\mathbf{v}_1, \mathbf{v}_2\}$, and check the result using Formula (7).

In Exercises 17 and 18, find the standard matrix for the orthogonal projection onto the subspace of R^3 spanned by the given orthogonal vectors.

17. $\mathbf{v}_1 = (1, 1, 1)$, $\mathbf{v}_2 = (1, 1, -2)$

18. $\mathbf{v}_1 = (2, 1, 2)$, $\mathbf{v}_2 = (1, 2, -2)$

19. Use the matrix obtained in Exercise 17 to find the orthogonal projection of $\mathbf{w} = (4, -5, 0)$ onto span$\{\mathbf{v}_1, \mathbf{v}_2\}$, and check the result using Formula (8).

20. Use the matrix obtained in Exercise 18 to find the orthogonal projection of $\mathbf{w} = (2, -3, 1)$ onto span$\{\mathbf{v}_1, \mathbf{v}_2\}$, and check the result using Formula (8).

21. The range of the projection in Exercise 17 is the plane through the origin spanned by $\mathbf{v}_1$ and $\mathbf{v}_2$, so the standard matrix for that projection should have rank 2. Confirm this by computing the trace of the matrix.

22. Follow the directions of Exercise 21 for the matrix in Exercise 18.

In Exercises 23–26, use Theorem 7.7.6 to confirm that P is the standard matrix for an orthogonal projection, and then use Theorem 7.9.3 to find the dimension of the range of that projection.

23. $P = \begin{bmatrix} \frac{20}{21} & \frac{4}{21} & -\frac{2}{21} \\ \frac{4}{21} & \frac{5}{21} & \frac{8}{21} \\ -\frac{2}{21} & \frac{8}{21} & \frac{17}{21} \end{bmatrix}$

24. $P = \begin{bmatrix} \frac{4}{49} & \frac{6}{49} & \frac{12}{49} \\ \frac{6}{49} & \frac{9}{49} & \frac{18}{49} \\ \frac{12}{49} & \frac{18}{49} & \frac{36}{49} \end{bmatrix}$

25. $P = \begin{bmatrix} \frac{1}{9} & \frac{2}{9} & \frac{2}{9} \\ \frac{2}{9} & \frac{4}{9} & \frac{4}{9} \\ \frac{2}{9} & \frac{4}{9} & \frac{4}{9} \end{bmatrix}$

26. $P = \begin{bmatrix} \frac{8}{9} & -\frac{2}{9} & \frac{2}{9} \\ -\frac{2}{9} & \frac{5}{9} & \frac{4}{9} \\ \frac{2}{9} & \frac{4}{9} & \frac{5}{9} \end{bmatrix}$

In Exercises 27 and 28, use the Gram–Schmidt process to transform the basis $\{\mathbf{w}_1, \mathbf{w}_2\}$ into an orthonormal basis. Draw both sets of basis vectors in the xy-plane.

27. $\mathbf{w}_1 = (1, -3)$, $\mathbf{w}_2 = (2, 2)$

28. $\mathbf{w}_1 = (1, 0)$, $\mathbf{w}_2 = (3, -5)$

In Exercises 29–32, use the Gram–Schmidt process to transform the given basis into an orthonormal basis.

29. $\mathbf{w}_1 = (1, 1, 1)$, $\mathbf{w}_2 = (-1, 1, 0)$, $\mathbf{w}_3 = (1, 2, 1)$

30. $\mathbf{w}_1 = (1, 0, 0)$, $\mathbf{w}_2 = (3, 7, -2)$, $\mathbf{w}_3 = (0, 4, 1)$

31. $\mathbf{w}_1 = (0, 2, 1, 0)$, $\mathbf{w}_2 = (1, -1, 0, 0)$, $\mathbf{w}_3 = (1, 2, 0, -1)$, $\mathbf{w}_4 = (1, 0, 0, 1)$

32. $\mathbf{w}_1 = (1, 2, 1, 0)$, $\mathbf{w}_2 = (1, 1, 2, 0)$, $\mathbf{w}_3 = (0, 1, 1, -2)$, $\mathbf{w}_4 = (1, 0, 3, 1)$

In Exercises 33 and 34, extend the given orthonormal set to an orthonormal basis for R^3 or R^4 (as appropriate) by using the method of Example 2 of Section 7.4 to extend the set to a basis and then applying the Gram–Schmidt process.

33. $\mathbf{w}_1 = \left(\frac{1}{\sqrt{2}}, \frac{1}{\sqrt{2}}, 0\right)$, $\mathbf{w}_2 = \left(\frac{1}{\sqrt{2}}, -\frac{1}{\sqrt{2}}, 0\right)$

34. $\mathbf{w}_1 = \left(\frac{2}{3}, \frac{2}{3}, 0, \frac{1}{3}\right)$, $\mathbf{w}_2 = \left(-\frac{1}{\sqrt{3}}, \frac{1}{\sqrt{3}}, \frac{1}{\sqrt{3}}, 0\right)$

35. Find an orthonormal basis for the subspace of R^3 spanned by the vectors $\mathbf{w}_1 = (0, 1, 2)$, $\mathbf{w}_2 = (-1, 0, 1)$, and $\mathbf{w}_3 = (-1, 1, 3)$.

36. Find an orthonormal basis for the subspace of R^4 spanned by the vectors $\mathbf{w}_1 = (-1, 2, 4, 7)$, $\mathbf{w}_2 = (-3, 0, 4, -2)$, $\mathbf{w}_3 = (2, 2, 7, -3)$, and $\mathbf{w}_4 = (4, 4, 7, 6)$.

37. Express $\mathbf{w} = (1, 2, 3)$ in the form $\mathbf{w} = \mathbf{w}_1 + \mathbf{w}_2$, where $\mathbf{w}_1$ lies in the plane W spanned by the vectors $\mathbf{u}_1 = \left(\frac{4}{5}, 0, -\frac{3}{5}\right)$ and $\mathbf{u}_2 = (0, 1, 0)$, and $\mathbf{w}_2$ is orthogonal to W.

38. Express $\mathbf{w} = (-1, 2, 6, 0)$ in the form $\mathbf{w} = \mathbf{w}_1 + \mathbf{w}_2$, where $\mathbf{w}_1$ is in the subspace W of R^4 spanned by the vectors $\mathbf{u}_1 = (-1, 0, 1, 2)$ and $\mathbf{u}_2 = (0, 1, 0, 1)$, and $\mathbf{w}_2$ is orthogonal to W.

39. Show that if $\mathbf{w} = (a, b, c)$ is a nonzero vector, then the standard matrix for the orthogonal projection of R^3 onto the line span$\{\mathbf{w}\}$ is

$$P = \frac{1}{a^2 + b^2 + c^2} \begin{bmatrix} a^2 & ab & ac \\ ab & b^2 & bc \\ ac & bc & c^2 \end{bmatrix}$$

Discussion and Discovery

D1. If a and b are nonzero, then an orthonormal basis for the plane $z = ax + by$ is _____ .

D2. (a) If $\mathbf{v}_1$, $\mathbf{v}_2$, and $\mathbf{v}_3$ are the orthogonal vectors that result by applying the Gram–Schmidt orthogonalization process to linearly independent vectors $\mathbf{w}_1$, $\mathbf{w}_2$, and $\mathbf{w}_3$, what relationships exist between span$\{\mathbf{v}_1\}$, span$\{\mathbf{v}_1, \mathbf{v}_2\}$, span$\{\mathbf{v}_1, \mathbf{v}_2, \mathbf{v}_3\}$ and span$\{\mathbf{w}_1\}$, span$\{\mathbf{w}_1, \mathbf{w}_2\}$, span$\{\mathbf{w}_1, \mathbf{w}_2, \mathbf{w}_3\}$?

(b) What relationship exists between $\mathbf{v}_3$ and span$\{\mathbf{w}_1, \mathbf{w}_2\}$?

D3. What would happen if you tried to apply the Gram–Schmidt process to a linearly dependent set of vectors?

D4. If A is a matrix whose column vectors are orthonormal, what relationship does AA^T have to the column space of A?

D5. (a) We know from Formula (6) that the standard matrix

for the orthogonal projection of R^n onto a subspace W can be expressed in the form $P = MM^T$. What is the relationship between the column spaces of M and P?

(b) If you know P, how might you find a matrix M for which $P = MM^T$?

(c) Is the matrix M unique?

D6. Indicate whether the statement is true (T) or false (F). Justify your answer.

(a) There are no linearly dependent orthonormal sets in R^n.

(b) There are no linearly dependent orthogonal sets in R^n.

(c) Every subspace of R^n has an orthonormal basis.

(d) If $\mathbf{q}_1$, $\mathbf{q}_2$, $\mathbf{q}_3$ are the orthonormal vectors that result by applying the Gram–Schmidt process to $\mathbf{w}_1$, $\mathbf{w}_2$, $\mathbf{w}_3$, then $\mathbf{q}_3 \cdot \mathbf{w}_1 = 0$ and $\mathbf{q}_3 \cdot \mathbf{w}_2 = 0$.

Working with Proofs

P1. Prove part (b) of Theorem 7.9.2 by normalizing the orthogonal vectors and using the result in part (a).

P2. Prove: If A is symmetric and idempotent, then A can be factored as $A = UU^T$, where U has orthonormal columns.

P3. Prove that the vector $\mathbf{v}_j$ that is produced at the jth step of the Gram–Schmidt orthogonalization process is expressible

as a linear combination of the starting linearly independent vectors $\mathbf{w}_1, \mathbf{w}_2, \ldots, \mathbf{w}_j$, thereby proving part (a) of Theorem 7.9.6. [*Hint:* The result is true for $j = 1$, since $\mathbf{v}_1 = \mathbf{w}_1$. Assume it is true for $k = j - 1$ and prove it is true for $k = j$, thereby proving the statement by mathematical induction.]

Technology Exercises

T1. (*Gram–Schmidt process*) Most technology utilities provide a command for performing some variation of the Gram–Schmidt process to produce either an orthogonal or orthonormal set. Some utilities require the starting vectors to be linearly independent, and others allow the set to be linearly dependent. In the latter case the utility eliminates linearly dependent vectors and produces an orthogonal or orthonormal basis for the space spanned by the original set. Determine how your utility performs the Gram–Schmidt process and use it to check the results that you obtained in Example 9.

T2. (*Normalization*) Some technology utilities have a command for normalizing a set of nonzero vectors. Determine whether your utility has such a command; if so, use it to convert the following set of orthogonal vectors to an orthonormal set:

$$\mathbf{v}_1 = (2, 1, 3, -1), \quad \mathbf{v}_2 = (3, 2, -3, -1)$$
$$\mathbf{v}_3 = (1, 5, 1, 10)$$

T3. Find an orthonormal basis for the subspace of R^7 spanned by the vectors

$$\mathbf{w}_1 = (1, 2, 3, 4, 5, 6, 7), \quad \mathbf{w}_2 = (1, 0, 3, 1, 1, 2, -2)$$
$$\mathbf{w}_3 = (1, 4, 3, 7, 9, 10, 1)$$

T4. Find orthonormal bases for the four fundamental spaces of the matrix

$$A = \begin{bmatrix} 2 & -1 & 3 & 5 \\ 4 & -3 & 1 & 3 \\ 3 & -2 & 3 & 4 \\ 4 & -1 & 15 & 17 \\ 7 & -6 & -7 & 0 \end{bmatrix}$$

T5. (**CAS**) Find the standard matrix for orthogonal projection onto the subspace of R^4 spanned by the nonzero vector $\mathbf{w} = (a, b, c, d)$. Confirm that the matrix is symmetric and idempotent, as guaranteed by Theorem 7.7.6, and use Theorem 7.9.3 to confirm that it has rank 1.

Section 7.10   *QR*-Decomposition; Householder Transformations

In this section we will show that the Gram–Schmidt process can be viewed as a method for factoring matrices, and we will show how these factorizations can be used to resolve various numerical difficulties that occur in the practical solution of least squares problems.

_QR_-DECOMPOSITION   We begin by posing the following question:

Suppose that A is an $m \times k$ matrix with full column rank whose successive column vectors are $\mathbf{w}_1, \mathbf{w}_2, \ldots, \mathbf{w}_k$. If the Gram–Schmidt process is applied to these vectors to produce an orthonormal basis $\{\mathbf{q}_1, \mathbf{q}_2, \ldots, \mathbf{q}_k\}$ for the column space of A, and if we form the matrix Q that has $\mathbf{q}_1, \mathbf{q}_2, \ldots, \mathbf{q}_k$ as successive columns, what relationship exists between the matrices A and Q?

To answer this question, let us write the matrices A and Q in the partitioned form

$$A = [\mathbf{w}_1 \quad \mathbf{w}_2 \quad \cdots \quad \mathbf{w}_k] \quad \text{and} \quad Q = [\mathbf{q}_1 \quad \mathbf{q}_2 \quad \cdots \quad \mathbf{q}_k]$$

It follows from Theorem 7.9.4 that the column vectors of A are expressible in terms of the column vectors of Q as

$$\mathbf{w}_1 = (\mathbf{w}_1 \cdot \mathbf{q}_1)\,\mathbf{q}_1 + (\mathbf{w}_1 \cdot \mathbf{q}_2)\,\mathbf{q}_2 + \cdots + (\mathbf{w}_1 \cdot \mathbf{q}_k)\,\mathbf{q}_k$$
$$\mathbf{w}_2 = (\mathbf{w}_2 \cdot \mathbf{q}_1)\,\mathbf{q}_1 + (\mathbf{w}_2 \cdot \mathbf{q}_2)\,\mathbf{q}_2 + \cdots + (\mathbf{w}_2 \cdot \mathbf{q}_k)\,\mathbf{q}_k$$
$$\vdots \qquad\qquad \vdots \qquad\qquad \vdots \qquad\qquad \vdots$$
$$\mathbf{w}_k = (\mathbf{w}_k \cdot \mathbf{q}_1)\,\mathbf{q}_1 + (\mathbf{w}_k \cdot \mathbf{q}_2)\,\mathbf{q}_2 + \cdots + (\mathbf{w}_k \cdot \mathbf{q}_k)\,\mathbf{q}_k$$

Moreover, it follows from part (*b*) of Theorem 7.9.6 that $\mathbf{q}_j$ is orthogonal to $\mathbf{w}_i$ whenever the index i is less than j, so this system can be rewritten more simply as

$$\mathbf{w}_1 = (\mathbf{w}_1 \cdot \mathbf{q}_1)\,\mathbf{q}_1$$
$$\mathbf{w}_2 = (\mathbf{w}_2 \cdot \mathbf{q}_1)\,\mathbf{q}_1 + (\mathbf{w}_2 \cdot \mathbf{q}_2)\,\mathbf{q}_2$$
$$\vdots \qquad\qquad \vdots \qquad\qquad \vdots \tag{1}$$
$$\mathbf{w}_k = (\mathbf{w}_k \cdot \mathbf{q}_1)\,\mathbf{q}_1 + (\mathbf{w}_k \cdot \mathbf{q}_2)\,\mathbf{q}_2 + \cdots + (\mathbf{w}_k \cdot \mathbf{q}_k)\,\mathbf{q}_k$$

Let us now form the upper triangular matrix

$$R = \begin{bmatrix} (\mathbf{w}_1 \cdot \mathbf{q}_1) & (\mathbf{w}_2 \cdot \mathbf{q}_1) & \cdots & (\mathbf{w}_k \cdot \mathbf{q}_1) \\ 0 & (\mathbf{w}_2 \cdot \mathbf{q}_2) & \cdots & (\mathbf{w}_k \cdot \mathbf{q}_2) \\ \vdots & \vdots & & \vdots \\ 0 & 0 & \cdots & (\mathbf{w}_k \cdot \mathbf{q}_k) \end{bmatrix} \tag{2}$$

and consider the product QR. It follows from Theorem 3.1.8 that the jth column vector of this product is a linear combination of the column vectors of Q with the coefficients coming from the jth column of R. But this is exactly the expression for $\mathbf{w}_j$ in (1), so the jth column of QR is the same as the jth column of A, and thus A and Q are related by the equation

$$\underbrace{[\mathbf{w}_1 \quad \mathbf{w}_2 \quad \cdots \quad \mathbf{w}_k]}_{A} = \underbrace{[\mathbf{q}_1 \quad \mathbf{q}_2 \quad \cdots \quad \mathbf{q}_k]}_{Q} \underbrace{\begin{bmatrix} (\mathbf{w}_1 \cdot \mathbf{q}_1) & (\mathbf{w}_2 \cdot \mathbf{q}_1) & \cdots & (\mathbf{w}_k \cdot \mathbf{q}_1) \\ 0 & (\mathbf{w}_2 \cdot \mathbf{q}_2) & \cdots & (\mathbf{w}_k \cdot \mathbf{q}_2) \\ \vdots & \vdots & \ddots & \vdots \\ 0 & 0 & \cdots & (\mathbf{w}_k \cdot \mathbf{q}_k) \end{bmatrix}}_{R} \tag{3}$$

In the exercises we will ask you to show that the matrix R is invertible by showing that its diagonal entries are nonzero. Thus, we have the following theorem.

> **Theorem 7.10.1** (*QR-Decomposition*)   *If A is an $m \times k$ matrix with full column rank, then A can be factored as*
>
> $$A = QR \tag{4}$$
>
> *where Q is an $m \times k$ matrix whose column vectors form an orthonormal basis for the column space of A and R is a $k \times k$ invertible upper triangular matrix.*

In general, a factorization of a matrix A as $A = QR$ in which the column vectors of Q are orthonormal and R is both invertible and upper triangular is called a **QR-decomposition** or a **QR-factorization** of A. Using this terminology, Theorem 7.10.1 guarantees that every matrix A with full column rank has a QR-factorization, and this is true, in particular, if A is invertible. The fact that Q has orthonormal columns implies that $Q^T Q = I$ (see Theorem 6.2.4), so multiplying both sides of (4) by Q^T on the left yields

$$R = Q^T A \tag{5}$$

Thus, one method for finding a QR-decomposition of a matrix A with full column rank is to apply the Gram–Schmidt process to the column vectors of A, then form the matrix Q from the resulting orthonormal basis vectors, and then find R from (5). Here is an example.

EXAMPLE 1

Finding a *QR*-Decomposition

Find a QR-decomposition of

$$A = \begin{bmatrix} 1 & 0 & 0 \\ 1 & 1 & 0 \\ 1 & 1 & 1 \end{bmatrix}$$

Solution The matrix A has full column rank (verify), so it is guaranteed to have a QR-decomposition. Applying the Gram–Schmidt process to

$$\mathbf{w}_1 = \begin{bmatrix} 1 \\ 1 \\ 1 \end{bmatrix}, \quad \mathbf{w}_2 = \begin{bmatrix} 0 \\ 1 \\ 1 \end{bmatrix}, \quad \mathbf{w}_3 = \begin{bmatrix} 0 \\ 0 \\ 1 \end{bmatrix}$$

and forming the matrix Q that has the resulting orthonormal basis vectors as columns yields

$$Q = \begin{bmatrix} \frac{1}{\sqrt{3}} & -\frac{2}{\sqrt{6}} & 0 \\ \frac{1}{\sqrt{3}} & \frac{1}{\sqrt{6}} & -\frac{1}{\sqrt{2}} \\ \frac{1}{\sqrt{3}} & \frac{1}{\sqrt{6}} & \frac{1}{\sqrt{2}} \end{bmatrix}$$

(see Example 9 of Section 7.9). It now follows from (5) that

$$R = Q^T A = \begin{bmatrix} \frac{1}{\sqrt{3}} & \frac{1}{\sqrt{3}} & \frac{1}{\sqrt{3}} \\ -\frac{2}{\sqrt{6}} & \frac{1}{\sqrt{6}} & \frac{1}{\sqrt{6}} \\ 0 & -\frac{1}{\sqrt{2}} & \frac{1}{\sqrt{2}} \end{bmatrix} \begin{bmatrix} 1 & 0 & 0 \\ 1 & 1 & 0 \\ 1 & 1 & 1 \end{bmatrix} = \begin{bmatrix} \sqrt{3} & \frac{2}{\sqrt{3}} & \frac{1}{\sqrt{3}} \\ 0 & \frac{2}{\sqrt{6}} & \frac{1}{\sqrt{6}} \\ 0 & 0 & \frac{1}{\sqrt{2}} \end{bmatrix}$$

from which we obtain the QR-decomposition

$$\underset{A}{\begin{bmatrix} 1 & 0 & 0 \\ 1 & 1 & 0 \\ 1 & 1 & 1 \end{bmatrix}} = \underset{Q}{\begin{bmatrix} \frac{1}{\sqrt{3}} & -\frac{2}{\sqrt{6}} & 0 \\ \frac{1}{\sqrt{3}} & \frac{1}{\sqrt{6}} & -\frac{1}{\sqrt{2}} \\ \frac{1}{\sqrt{3}} & \frac{1}{\sqrt{6}} & \frac{1}{\sqrt{2}} \end{bmatrix}} \underset{R}{\begin{bmatrix} \sqrt{3} & \frac{2}{\sqrt{3}} & \frac{1}{\sqrt{3}} \\ 0 & \frac{2}{\sqrt{6}} & \frac{1}{\sqrt{6}} \\ 0 & 0 & \frac{1}{\sqrt{2}} \end{bmatrix}}$$

∎

Recall from Theorem 7.8.3 that the least squares solutions of a linear system $A\mathbf{x} = \mathbf{b}$ are the exact solutions of the normal equation $A^T A\mathbf{x} = A^T \mathbf{b}$, and that if A has full column rank, then there is a unique least squares solution

$$\hat{\mathbf{x}} = (A^T A)^{-1} A^T \mathbf{b} \tag{6}$$

This suggests two possible procedures for computing the least squares solution when A has full column rank:

1. Solve the normal equation $A^T A\mathbf{x} = A^T \mathbf{b}$ directly, say by an LU-decomposition of $A^T A$.

2. Invert $A^T A$ and apply Formula (6).

Although fine in theory, neither of these methods works well in practice because slight roundoff errors in the entries of A are often magnified in computing the entries of $A^T A$. Thus, most computer algorithms for finding least squares solutions use methods that avoid computing the matrix $A^T A$. One way to do this when A has full column rank is to use a QR-decomposition $A = QR$ to rewrite the normal equation $A^T A\mathbf{x} = A^T \mathbf{b}$ as

$$(R^T Q^T)(QR)\mathbf{x} = R^T Q^T \mathbf{b} \tag{7}$$

and use the fact that $Q^T Q = I$ to rewrite this as

$$R^T R\mathbf{x} = R^T Q^T \mathbf{b} \tag{8}$$

It follows from the definition of QR-decomposition that R, and hence R^T, is invertible, so we can multiply both sides of (8) on the left by $(R^T)^{-1}$ to obtain the following result.

Theorem 7.10.2 *If A is an $m \times k$ matrix with full column rank, and if $A = QR$ is a QR-decomposition of A, then the normal system for $A\mathbf{x} = \mathbf{b}$ can be expressed as*

$$R\mathbf{x} = Q^T \mathbf{b} \tag{9}$$

and the least squares solution can be expressed as

$$\hat{\mathbf{x}} = R^{-1} Q^T \mathbf{b} \tag{10}$$

Since the matrix R in (9) is upper triangular, this form of the normal system can be solved readily by back substitution to obtain the least squares solution $\hat{\mathbf{x}}$. This procedure is important in practice because it avoids the troublesome $A^T A$ that appears in the normal equation $A^T A\mathbf{x} = A^T \mathbf{b}$.

EXAMPLE 2
Least Squares
Solutions Using
QR-Decomposition

Use a QR-decomposition to find the least squares solution of the linear system $A\mathbf{x} = \mathbf{b}$ given by the equations

$$\begin{aligned}
x_1 + 3x_2 + 5x_3 &= -2 \\
x_1 + x_2 \qquad\ &= 3 \\
x_1 + x_2 + 2x_3 &= -1 \\
x_1 + 3x_2 + 3x_3 &= 2
\end{aligned}$$

Solution We leave it for you to confirm that the coefficient matrix

$$A = \begin{bmatrix} 1 & 3 & 5 \\ 1 & 1 & 0 \\ 1 & 1 & 2 \\ 1 & 3 & 3 \end{bmatrix}$$

has full column rank, and to use the method of Example 1 to show that a QR-decomposition of this matrix is

$$A = QR = \begin{bmatrix} \frac{1}{2} & \frac{1}{2} & \frac{1}{2} \\ \frac{1}{2} & -\frac{1}{2} & -\frac{1}{2} \\ \frac{1}{2} & -\frac{1}{2} & \frac{1}{2} \\ \frac{1}{2} & \frac{1}{2} & -\frac{1}{2} \end{bmatrix} \begin{bmatrix} 2 & 4 & 5 \\ 0 & 2 & 3 \\ 0 & 0 & 2 \end{bmatrix}$$

This implies that

$$Q^T\mathbf{b} = \begin{bmatrix} \frac{1}{2} & \frac{1}{2} & \frac{1}{2} & \frac{1}{2} \\ \frac{1}{2} & -\frac{1}{2} & -\frac{1}{2} & \frac{1}{2} \\ \frac{1}{2} & -\frac{1}{2} & \frac{1}{2} & -\frac{1}{2} \end{bmatrix} \begin{bmatrix} -2 \\ 3 \\ -1 \\ 2 \end{bmatrix} = \begin{bmatrix} 1 \\ -1 \\ -4 \end{bmatrix}$$

so the normal system $R\mathbf{x} = Q^T\mathbf{b}$ is

$$\begin{bmatrix} 2 & 4 & 5 \\ 0 & 2 & 3 \\ 0 & 0 & 2 \end{bmatrix} \begin{bmatrix} x_1 \\ x_2 \\ x_3 \end{bmatrix} = \begin{bmatrix} 1 \\ -1 \\ -4 \end{bmatrix}$$

We leave it for you to verify that back substitution yields the least squares solution

$$x_1 = \tfrac{1}{2}, \quad x_2 = \tfrac{5}{2}, \quad x_3 = -2$$ ∎

OTHER NUMERICAL ISSUES

Although QR-decomposition avoids the troublesome A^TA in least squares problems, there are also numerical difficulties that arise when the Gram–Schmidt process is used to construct a QR-decomposition, the problem being that slight roundoff errors in the entries of A can produce a severe loss of orthogonality in the computed vectors of Q. There is a way of rearranging the order of the calculations in the Gram–Schmidt process (called the ***modified Gram–Schmidt process***) that reduces the effect of roundoff error, but the more common approach is to compute the QR-decomposition without using the Gram–Schmidt process at all. There are two basic methods for doing this, one based on reflections and one based on rotations. We will just touch on some of the basic ideas here and leave a detailed study of the subject to books on numerical methods of linear algebra.

HOUSEHOLDER REFLECTIONS

If $\mathbf{a}$ is a nonzero vector in R^2 or R^3, then there is a simple relationship between the orthogonal projection onto the line span$\{\mathbf{a}\}$ and the reflection about the hyperplane $\mathbf{a}^\perp$ that is illustrated by Figure 7.10.1 in R^3: If we denote the orthogonal projection of a vector $\mathbf{x}$ onto the line by $\text{proj}_{\mathbf{a}}\mathbf{x}$ and the reflection of $\mathbf{x}$ about the hyperplane by $\text{refl}_{\mathbf{a}^\perp}\mathbf{x}$, then the figure suggests that

$$\mathbf{x} - \text{refl}_{\mathbf{a}^\perp}\mathbf{x} = 2\text{proj}_{\mathbf{a}}\mathbf{x} \quad \text{or, equivalently, that} \quad \text{refl}_{\mathbf{a}^\perp}\mathbf{x} = \mathbf{x} - 2\text{proj}_{\mathbf{a}}\mathbf{x}$$

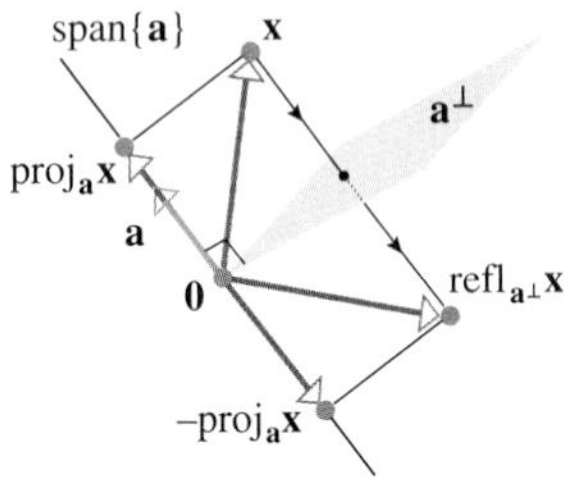

Motivated by this result, we make the following definition.

$$\mathbf{x} - \text{refl}_{\mathbf{a}\perp}\mathbf{x} = 2\,\text{proj}_{\mathbf{a}}\mathbf{x}$$

Figure 7.10.1

> **Definition 7.10.3** If $\mathbf{a}$ is a nonzero vector in R^n, and if $\mathbf{x}$ is any vector in R^n, then the ***reflection of*** $\mathbf{x}$ ***about the hyperplane*** $\mathbf{a}^\perp$ is denoted by $\text{refl}_{\mathbf{a}^\perp}\mathbf{x}$ and defined as
>
> $$\text{refl}_{\mathbf{a}^\perp}\mathbf{x} = \mathbf{x} - 2\text{proj}_{\mathbf{a}}\mathbf{x} \tag{11}$$
>
> The operator $T : R^n \to R^n$ defined by $T(\mathbf{x}) = \text{refl}_{\mathbf{a}^\perp}\mathbf{x}$ is called the ***reflection of R^n about the hyperplane*** $\mathbf{a}^\perp$.

CONCEPT PROBLEM Verify that $\text{refl}_{\mathbf{a}^\perp}$ is a linear operator on R^n.

It follows from Formula (11) of Section 7.7 that

$$\text{refl}_{\mathbf{a}^\perp}\mathbf{x} = \mathbf{x} - 2\frac{\mathbf{x} \cdot \mathbf{a}}{\|\mathbf{a}\|^2}\mathbf{a} \tag{12}$$

and from Theorem 7.7.3 that the standard matrix $H_{\mathbf{a}^{\perp}}$ for $\text{refl}_{\mathbf{a}^{\perp}}$ is

$$H_{\mathbf{a}^{\perp}} = I - \frac{2}{\mathbf{a}^T \mathbf{a}} \mathbf{a}\mathbf{a}^T \tag{13}$$

In the special case where the hyperplane is specified by a unit vector $\mathbf{u}$ we have $\mathbf{u}^T\mathbf{u} = \|\mathbf{u}\|^2 = 1$, so Formulas (12) and (13) simplify to

$$\text{refl}_{\mathbf{u}^{\perp}}\mathbf{x} = \mathbf{x} - 2(\mathbf{x} \cdot \mathbf{u})\mathbf{u} \quad \text{and} \quad H_{\mathbf{u}^{\perp}} = I - 2\mathbf{u}\mathbf{u}^T \tag{14–15}$$

EXAMPLE 3
Reflection
About a
Coordinate
Plane in R^3

Recall from Table 6.2.5 that the standard matrix for the reflection of R^3 about the yz-plane of an xyz-coordinate system is

$$\begin{bmatrix} -1 & 0 & 0 \\ 0 & 1 & 0 \\ 0 & 0 & 1 \end{bmatrix}$$

This is consistent with Formula (15) because the yz-plane is the orthogonal complement of the unit vector $\mathbf{u} = (1, 0, 0)$ along the positive x-axis. Thus, if we write $\mathbf{u}$ in column form and apply (13), we obtain

$$H_{\mathbf{u}^{\perp}} = I - 2\mathbf{u}\mathbf{u}^T = \begin{bmatrix} 1 & 0 & 0 \\ 0 & 1 & 0 \\ 0 & 0 & 1 \end{bmatrix} - 2\begin{bmatrix} 1 \\ 0 \\ 0 \end{bmatrix}\begin{bmatrix} 1 & 0 & 0 \end{bmatrix}$$

$$= \begin{bmatrix} 1 & 0 & 0 \\ 0 & 1 & 0 \\ 0 & 0 & 1 \end{bmatrix} - \begin{bmatrix} 2 & 0 & 0 \\ 0 & 0 & 0 \\ 0 & 0 & 0 \end{bmatrix} = \begin{bmatrix} -1 & 0 & 0 \\ 0 & 1 & 0 \\ 0 & 0 & 1 \end{bmatrix} \qquad ■$$

EXAMPLE 4
Reflection
About a Plane
Through the
Origin of R^3

(a) Find the standard matrix H for the reflection of R^3 about the plane $x - 4y + 2z = 0$.
(b) Use the matrix H to find the reflection of the vector $\mathbf{b} = (1, 0, 4)$ about the plane.

Solution (*a*) The vector $\mathbf{a} = (1, -4, 2)$ is perpendicular to the plane, so we can regard the plane as the hyperplane $\mathbf{a}^{\perp}$ and apply Formula (14) to find H. Alternatively, we can normalize $\mathbf{a}$ and apply Formula (15); this is the approach we will take. Normalizing $\mathbf{a}$ yields

$$\mathbf{u} = \left(\tfrac{1}{\sqrt{21}}, -\tfrac{4}{\sqrt{21}}, \tfrac{2}{\sqrt{21}} \right)$$

Writing this in column form and computing $\mathbf{u}\mathbf{u}^T$ yields

$$\mathbf{u}\mathbf{u}^T = \begin{bmatrix} \tfrac{1}{\sqrt{21}} \\ -\tfrac{4}{\sqrt{21}} \\ \tfrac{2}{\sqrt{21}} \end{bmatrix} \begin{bmatrix} \tfrac{1}{\sqrt{21}} & -\tfrac{4}{\sqrt{21}} & \tfrac{2}{\sqrt{21}} \end{bmatrix} = \begin{bmatrix} \tfrac{1}{21} & -\tfrac{4}{21} & \tfrac{2}{21} \\ -\tfrac{4}{21} & \tfrac{16}{21} & -\tfrac{8}{21} \\ \tfrac{2}{21} & -\tfrac{8}{21} & \tfrac{4}{21} \end{bmatrix}$$

from which we obtain

$$H = I - 2\mathbf{u}\mathbf{u}^T = \begin{bmatrix} 1 & 0 & 0 \\ 0 & 1 & 0 \\ 0 & 0 & 1 \end{bmatrix} - 2\begin{bmatrix} \tfrac{1}{21} & -\tfrac{4}{21} & \tfrac{2}{21} \\ -\tfrac{4}{21} & \tfrac{16}{21} & -\tfrac{8}{21} \\ \tfrac{2}{21} & -\tfrac{8}{21} & \tfrac{4}{21} \end{bmatrix} = \begin{bmatrix} \tfrac{19}{21} & \tfrac{8}{21} & -\tfrac{4}{21} \\ \tfrac{8}{21} & -\tfrac{11}{21} & \tfrac{16}{21} \\ -\tfrac{4}{21} & \tfrac{16}{21} & \tfrac{13}{21} \end{bmatrix}$$

Solution (b) The reflection of **b** about the plane $x - 4y + 2z = 0$ is the product $H\mathbf{b}$ with **b** expressed in column form. Thus,

$$H\mathbf{b} = \begin{bmatrix} \frac{19}{21} & \frac{8}{21} & -\frac{4}{21} \\ \frac{8}{21} & -\frac{11}{21} & \frac{16}{21} \\ -\frac{4}{21} & \frac{16}{21} & \frac{13}{21} \end{bmatrix} \begin{bmatrix} 1 \\ 0 \\ 4 \end{bmatrix} = \begin{bmatrix} \frac{3}{21} \\ \frac{72}{21} \\ \frac{48}{21} \end{bmatrix} = \begin{bmatrix} \frac{1}{7} \\ \frac{24}{7} \\ \frac{16}{7} \end{bmatrix}$$

or in comma-delimited notation, $H\mathbf{b} = \left(\frac{1}{7}, \frac{24}{7}, \frac{16}{7}\right)$. ∎

Linear Algebra in History

In the early 1950s, numerical linear algebra was a conglomeration of unrelated algorithms that had been developed ad hoc for solving various kinds of problems. It was the American mathematician Alston Scott Householder who, during his tenure as Director of the Oak Ridge National Laboratory, is generally credited with bringing order and precision to the field. Householder's interest in numerical mathematics blossomed late—he studied philosophy as an undergraduate at Northwestern University and obtained a Master's Degree in that field from Cornell University in 1927—it was not until 1947 that he received his Ph.D. in mathematics from the University of Chicago. Following that he worked in mathematical biology before finally settling into the field of numerical linear algebra, in which he gained his fame.

Alston Scott
Householder
(1904–1993)

Because of the pioneering work of the American mathematician A. S. Householder in applying reflections about hyperplanes to important numerical algorithms, reflections about hyperplanes are often called ***Householder reflections*** or ***Householder transformations*** in his honor. The terminology in the following definition is also common.

Definition 7.10.4 An $n \times n$ matrix of the form

$$H = I - \frac{2}{\mathbf{a}^T\mathbf{a}}\mathbf{a}\mathbf{a}^T \tag{16}$$

in which **a** is a nonzero vector in R^n is called a ***Householder matrix***. Geometrically, H is the standard matrix for the Householder reflection about the hyperplane $\mathbf{a}^\perp$.

The following theorem, whose proof is left as an exercise, shows that Householder reflections in R^n have the familiar properties of reflections in R^2 and R^3.

Theorem 7.10.5 *Householder matrices are symmetric and orthogonal.*

CONCEPT PROBLEM Show that if H is a Householder matrix, then $H^{-1} = H$, and explain why this makes sense geometrically.

The fact that Householder matrices are orthogonal means that Householder reflections preserve lengths. The following theorem is concerned with the reverse situation—it shows that two vectors with the same length are related by some Householder reflection.

Theorem 7.10.6 *If* **v** *and* **w** *are distinct vectors in* R^n *with the same length, then the Householder reflection about the hyperplane* $(\mathbf{v} - \mathbf{w})^\perp$ *maps* **v** *into* **w**, *and conversely.*

Proof It follows from (16) with $\mathbf{a} = \mathbf{v} - \mathbf{w}$ that the image of the vector **v** under the Householder reflection about the hyperplane $(\mathbf{v} - \mathbf{w})^\perp$ is

$$H\mathbf{v} = \mathbf{v} - \frac{2}{(\mathbf{v} - \mathbf{w})^T(\mathbf{v} - \mathbf{w})}(\mathbf{v} - \mathbf{w})(\mathbf{v} - \mathbf{w})^T\mathbf{v}$$

Since **v** and **w** have the same length, it follows that $\mathbf{v}^T\mathbf{v} = \mathbf{w}^T\mathbf{w}$. Using this and the symmetry property $\mathbf{v}^T\mathbf{w} = \mathbf{w}^T\mathbf{v}$ of the dot product, we can rewrite the denominator in the above formula for $H\mathbf{v}$ as

$$(\mathbf{v} - \mathbf{w})^T(\mathbf{v} - \mathbf{w}) = \mathbf{v}^T\mathbf{v} - \mathbf{w}^T\mathbf{v} - \mathbf{v}^T\mathbf{w} + \mathbf{w}^T\mathbf{w} = 2\mathbf{v}^T\mathbf{v} - 2\mathbf{w}^T\mathbf{v}$$

from which it follows that

$$Hv = v - \frac{1}{v^Tv - w^Tv}(v - w)(v - w)^Tv$$

$$= v - \frac{1}{v^Tv - w^Tv}(v - w)^Tv(v - w) \qquad \boxed{\begin{array}{l}\text{Since } (v - w)^T v \text{ is a} \\ 1 \times 1 \text{ matrix}\end{array}}$$

$$= v - (v - w) = w$$

This shows that H maps v into w. Conversely, H maps w into v, since $Hw = H^{-1}w = v$.   ∎

Theorem 7.10.6 is important because it provides a way of using Householder reflections to transform a given vector into a vector in which specified components are zero. For example, the vectors

$$v = \begin{bmatrix} v_1 \\ v_2 \\ \vdots \\ v_n \end{bmatrix} \quad \text{and} \quad w = \begin{bmatrix} \|v\| \\ 0 \\ \vdots \\ 0 \end{bmatrix}$$

have the same length, so Theorem 7.10.6 guarantees that there is a Householder reflection that maps v into w. Moreover, the scalar $\|v\|$ could have been placed anywhere in w, so there are Householder reflections that map v into a vector with zeros in any $n - 1$ selected positions. Here are some examples.

EXAMPLE 5
Creating Zero Entries Using Householder Reflections

Find a Householder reflection that maps the vector $v = (1, 2, 2)$ into a vector w that has zeros as its second and third components.

Solution Since $\|v\| = 3$, the vector $w = (3, 0, 0)$ has the same norm as v; thus, it follows from Theorem 7.10.6 that the Householder reflection about the hyperplane $(v - w)^\perp$ maps v into w. The Householder matrix H for this reflection can be obtained from (16) with the vector $a = v - w = (-2, 2, 2)$ written in column form. Since $a^Ta = \|a\|^2 = 12$, this yields

$$H = I - \frac{2}{a^Ta}aa^T = \begin{bmatrix} 1 & 0 & 0 \\ 0 & 1 & 0 \\ 0 & 0 & 1 \end{bmatrix} - \frac{1}{6}\begin{bmatrix} -2 \\ 2 \\ 2 \end{bmatrix}[-2 \quad 2 \quad 2]$$

$$= \begin{bmatrix} 1 & 0 & 0 \\ 0 & 1 & 0 \\ 0 & 0 & 1 \end{bmatrix} - \frac{1}{6}\begin{bmatrix} 4 & -4 & -4 \\ -4 & 4 & 4 \\ -4 & 4 & 4 \end{bmatrix} = \begin{bmatrix} \frac{1}{3} & \frac{2}{3} & \frac{2}{3} \\ \frac{2}{3} & \frac{1}{3} & -\frac{2}{3} \\ \frac{2}{3} & -\frac{2}{3} & \frac{1}{3} \end{bmatrix}$$

We leave it for you to write v and w in column form and confirm that $Hv = w$.   ∎

EXAMPLE 6
More on Householder Reflections

Show that if $v = (v_1, v_2, \ldots, v_n)$, then there exists a Householder reflection that maps v into a vector of the form $w = (v_1, v_2, \ldots, v_{k-1}, s, 0, \ldots, 0)$; that is, the reflection preserves the first $k - 1$ components of v, possibly modifies v_k, and converts the remaining components, if any, to zero.

Solution If we can find a value of s for which $\|v\| = \|w\|$, then Theorem 7.10.6 guarantees that there is a Householder reflection that maps v into w. We leave it for you to show that either of the values

$$s = \pm\sqrt{v_k^2 + v_{k+1}^2 + \cdots + v_n^2} \qquad (17)$$

will work.   ∎

REMARK In numerical algorithms, a judicious choice of the sign in (17) can help to reduce roundoff error.

**QR-DECOMPOSITION
USING HOUSEHOLDER
REFLECTIONS**

Our next goal is to illustrate a way in which Householder reflections can be used to construct *QR*-decompositions. Suppose, for example, that we are looking for a *QR*-decomposition of some 4×4 invertible matrix

$$A = \begin{bmatrix} \times & \times & \times & \times \\ \times & \times & \times & \times \\ \times & \times & \times & \times \\ \times & \times & \times & \times \end{bmatrix} \tag{18}$$

If we can find orthogonal matrices Q_1, Q_2, and Q_3 such that

$$Q_3 Q_2 Q_1 A = R$$

is upper triangular, then we can rewrite this equation as

$$A = Q_1^{-1} Q_2^{-1} Q_3^{-1} R = Q_1^T Q_2^T Q_3^T R = QR \tag{19}$$

which will be a QR-decomposition of A with $Q = Q_1^T Q_2^T Q_3^T$.

As a first step in implementing this idea, let Q_1 be the Householder matrix for a Householder reflection that "zeros out" the second, third, and fourth entries of the first column of A. Thus, the product $Q_1 A$ will be of the form

$$Q_1 A = \begin{bmatrix} \times & \times & \times & \times \\ 0 & \times & \times & \times \\ 0 & \times & \times & \times \\ 0 & \times & \times & \times \end{bmatrix} \tag{20}$$

where the entries represented by $\times$'s here need not be the same as those in (18).

Now we want to introduce zeros below the main diagonal in the second column without destroying the zeros already created in (20). To do this, focus on the 3×3 submatrix in the lower right corner of (20), and let H_2 be the 3×3 Householder matrix for a Householder reflection that zeros out the second and third entries in the first column of the submatrix. If we form the matrix

$$Q_2 = \begin{bmatrix} 1 & 0 & 0 & 0 \\ 0 & & & \\ 0 & & H_2 & \\ 0 & & & \end{bmatrix}$$

then Q_2 will be orthogonal (verify), and $Q_2 Q_1 A$ will be of the form

$$Q_2 Q_1 A = \begin{bmatrix} \times & \times & \times & \times \\ 0 & \times & \times & \times \\ 0 & 0 & \times & \times \\ 0 & 0 & \times & \times \end{bmatrix} \tag{21}$$

Next, focus on the 2×2 submatrix in the lower right corner of (21), and let H_3 be the 2×2 Householder matrix for a Householder reflection that zeros out the second entry in the first column of the submatrix. If we form the matrix

$$Q_3 = \begin{bmatrix} 1 & 0 & 0 & 0 \\ 0 & 1 & 0 & 0 \\ 0 & 0 & & \\ 0 & 0 & & H_3 \end{bmatrix}$$

then Q_3 will be orthogonal (verify), and $Q_3 Q_2 Q_1 A$ will be of the form

$$Q_3 Q_2 Q_1 A = \begin{bmatrix} \times & \times & \times & \times \\ 0 & \times & \times & \times \\ 0 & 0 & \times & \times \\ 0 & 0 & 0 & \times \end{bmatrix}$$

The matrix R on the right side of this equation is now upper triangular, so (19) provides a QR-decomposition of A. Here is an example.

EXAMPLE 7
QR Using
Householder
Reflections

Use Householder reflections to construct a QR-decomposition of

$$A = \begin{bmatrix} 1 & 3 & 1 \\ 2 & -5 & -2 \\ 2 & -4 & -3 \end{bmatrix}$$

Solution We showed in Example 5 that the second and third entries of the first column of A can be zeroed out by the Householder matrix

$$Q_1 = \begin{bmatrix} \frac{1}{3} & \frac{2}{3} & \frac{2}{3} \\ \frac{2}{3} & \frac{1}{3} & -\frac{2}{3} \\ \frac{2}{3} & -\frac{2}{3} & \frac{1}{3} \end{bmatrix}$$

Computing $Q_1 A$ yields

$$Q_1 A = \begin{bmatrix} \frac{1}{3} & \frac{2}{3} & \frac{2}{3} \\ \frac{2}{3} & \frac{1}{3} & -\frac{2}{3} \\ \frac{2}{3} & -\frac{2}{3} & \frac{1}{3} \end{bmatrix} \begin{bmatrix} 1 & 3 & 1 \\ 2 & -5 & -2 \\ 2 & -4 & -3 \end{bmatrix} = \begin{bmatrix} 3 & -5 & -3 \\ 0 & 3 & 2 \\ 0 & 4 & 1 \end{bmatrix} \qquad (22)$$

Now consider the 2×2 submatrix B in the lower right corner of (22), namely

$$B = \begin{bmatrix} 3 & 2 \\ 4 & 1 \end{bmatrix}$$

We leave it as an exercise to show that the second entry in the first column of B is zeroed out by the Householder matrix

$$H_2 = \begin{bmatrix} \frac{3}{5} & \frac{4}{5} \\ \frac{4}{5} & -\frac{3}{5} \end{bmatrix}$$

Now form the matrix

$$Q_2 = \begin{bmatrix} 1 & 0 & 0 \\ 0 & \frac{3}{5} & \frac{4}{5} \\ 0 & \frac{4}{5} & -\frac{3}{5} \end{bmatrix}$$

and multiply (22) on the left by Q_2 to obtain

$$Q_2 Q_1 A = \begin{bmatrix} 1 & 0 & 0 \\ 0 & \frac{3}{5} & \frac{4}{5} \\ 0 & \frac{4}{5} & -\frac{3}{5} \end{bmatrix} \begin{bmatrix} \frac{1}{3} & \frac{2}{3} & \frac{2}{3} \\ \frac{2}{3} & \frac{1}{3} & -\frac{2}{3} \\ \frac{2}{3} & -\frac{2}{3} & \frac{1}{3} \end{bmatrix} \begin{bmatrix} 1 & 3 & 1 \\ 2 & -5 & -2 \\ 2 & -4 & -3 \end{bmatrix} = \begin{bmatrix} 3 & -5 & -3 \\ 0 & 5 & 2 \\ 0 & 0 & 1 \end{bmatrix}$$

Thus,

$$A = \begin{bmatrix} 1 & 3 & 1 \\ 2 & -5 & -2 \\ 2 & -4 & -3 \end{bmatrix} = \begin{bmatrix} \frac{1}{3} & \frac{2}{3} & \frac{2}{3} \\ \frac{2}{3} & \frac{1}{3} & -\frac{2}{3} \\ \frac{2}{3} & -\frac{2}{3} & \frac{1}{3} \end{bmatrix} \begin{bmatrix} 1 & 0 & 0 \\ 0 & \frac{3}{5} & \frac{4}{5} \\ 0 & \frac{4}{5} & -\frac{3}{5} \end{bmatrix} \begin{bmatrix} 3 & -5 & -3 \\ 0 & 5 & 2 \\ 0 & 0 & 1 \end{bmatrix}$$

$$= \begin{bmatrix} \frac{1}{3} & \frac{14}{15} & \frac{2}{15} \\ \frac{2}{3} & -\frac{1}{3} & \frac{2}{3} \\ \frac{2}{3} & -\frac{2}{15} & -\frac{11}{15} \end{bmatrix} \begin{bmatrix} 3 & -5 & -3 \\ 0 & 5 & 2 \\ 0 & 0 & 1 \end{bmatrix}$$

which is a QR-decomposition of A. ∎

HOUSEHOLDER REFLECTIONS IN APPLICATIONS   In applications where one needs to compute the Householder reflection of a vector about a hyperplane, the Householder matrix need not be computed explicitly. Rather, if H is the standard matrix for a Householder reflection about the hyperplane $\mathbf{a}^\perp$, then it is usual to compute $H\mathbf{x}$ by using (16) to rewrite this product as

$$H\mathbf{x} = \left(I - \frac{2}{\mathbf{a}^T\mathbf{a}}\mathbf{a}\mathbf{a}^T \right)\mathbf{x} = \mathbf{x} - \beta\mathbf{a} \tag{23}$$

where $\beta = 2\mathbf{a}^T\mathbf{x}/\mathbf{a}^T\mathbf{a}$. Since this formula expresses $H\mathbf{x}$ directly in terms of $\mathbf{a}$ and $\mathbf{x}$, there is no need to find H explicitly for the purpose of computing $H\mathbf{x}$. In the special case where $\mathbf{a}$ is a unit vector, Formula (23) simplifies to

$$H\mathbf{x} = \mathbf{x} - \beta\mathbf{a} = \mathbf{x} - 2(\mathbf{a}^T\mathbf{x})\mathbf{a} = \mathbf{x} - 2(\mathbf{a} \cdot \mathbf{x})\mathbf{a} \tag{24}$$

Note that the term $2(\mathbf{a} \cdot \mathbf{x})\mathbf{a}$ in this formula is twice the orthogonal projection of $\mathbf{x}$ onto $\mathbf{a}$.

Exercise Set 7.10

In Exercises 1–6, a matrix A with full column rank is given. Find a QR-decomposition of A.

1. $A = \begin{bmatrix} 1 & -1 \\ 2 & 3 \end{bmatrix}$ **2.** $A = \begin{bmatrix} 1 & 2 \\ 0 & 1 \\ 1 & 4 \end{bmatrix}$

3. $A = \begin{bmatrix} 1 & 1 \\ -2 & 1 \\ 2 & 1 \end{bmatrix}$ **4.** $A = \begin{bmatrix} 1 & 0 & 2 \\ 0 & 1 & 1 \\ 1 & 2 & 0 \end{bmatrix}$

5. $A = \begin{bmatrix} 1 & 2 & 1 \\ 1 & 1 & 1 \\ 0 & 3 & 1 \end{bmatrix}$ **6.** $A = \begin{bmatrix} 1 & 0 & 1 \\ -1 & 1 & 1 \\ 1 & 0 & 1 \\ -1 & 1 & 0 \end{bmatrix}$

In Exercises 7–10, use the QR-decomposition of A and the method of Example 2 to find the least squares solution of the system $A\mathbf{x} = \mathbf{b}$.

7. The matrix A in Exercise 3; $\mathbf{b} = \begin{bmatrix} 1 \\ 1 \\ 0 \end{bmatrix}$.

8. The matrix A in Exercise 4; $\mathbf{b} = \begin{bmatrix} 1 \\ 2 \\ 3 \end{bmatrix}$.

9. The matrix A in Exercise 5; $\mathbf{b} = \begin{bmatrix} -2 \\ 3 \\ 1 \end{bmatrix}$.

10. The matrix A in Exercise 6; $\mathbf{b} = \begin{bmatrix} 3 \\ -1 \\ 2 \\ 1 \end{bmatrix}$.

In Exercises 11 and 12, find the standard matrix H for the reflection of R^3 about the given plane, and use H to find the reflection of $\mathbf{b}$ about that plane.

11. $2x - y + 3z = 0$; $\mathbf{b} = (1, 2, 2)$

12. $x + y - 4z = 0$; $\mathbf{b} = (1, 0, 1)$

In Exercises 13–16, find the standard matrix H for the reflection of R^3 (or R^4) about the hyperplane $\mathbf{a}^\perp$.

13. $\mathbf{a} = (1, -1, 1)$ **14.** $\mathbf{a} = (1, -1, 2)$

15. $\mathbf{a} = (0, 1, -1, 3)$ **16.** $\mathbf{a} = (-1, 2, 3, 1)$

In Exercises 17 and 18, vectors $\mathbf{v}$ and $\mathbf{w}$ with the same length are given. Find a Householder matrix H such that $H\mathbf{v} = \mathbf{w}$ when $\mathbf{v}$ and $\mathbf{w}$ are written in column form.

17. (a) $\mathbf{v} = (3, 4)$, $\mathbf{w} = (5, 0)$
   (b) $\mathbf{v} = (3, 4)$, $\mathbf{w} = (0, 5)$
   (c) $\mathbf{v} = (3, 4)$, $\mathbf{w} = \left(\frac{7\sqrt{2}}{2}, -\frac{\sqrt{2}}{2} \right)$

18. (a) $\mathbf{v} = (1, 1)$, $\mathbf{w} = (\sqrt{2}, 0)$
   (b) $\mathbf{v} = (1, 1)$, $\mathbf{w} = (0, \sqrt{2})$
   (c) $\mathbf{v} = (1, 1)$, $\mathbf{w} = \left(\frac{\sqrt{3}-1}{2}, \frac{1+\sqrt{3}}{2} \right)$

In Exercises 19 and 20, find the standard matrix for a Householder reflection that maps the vector $\mathbf{v}$ into a vector whose last two components are zero.

19. $\mathbf{v} = (-2, 1, 2)$ **20.** $\mathbf{v} = (1, -2, 2)$

In Exercises 21–24, use Householder reflections and the method of Example 7 to find a QR-decomposition of the given matrix.

21. $\begin{bmatrix} 1 & 2 \\ -1 & 3 \end{bmatrix}$ **22.** $\begin{bmatrix} 1 & 0 \\ 0 & 2 \\ 0 & 1 \end{bmatrix}$

23. $\begin{bmatrix} 1 & 2 & 1 \\ -1 & -2 & 3 \\ 0 & 4 & 5 \end{bmatrix}$ **24.** $\begin{bmatrix} 1 & 2 & 1 \\ 1 & 3 & -2 \\ 0 & -1 & 0 \\ 0 & 0 & 1 \end{bmatrix}$

25. Use the given QR-decomposition of A to solve the linear system $A\mathbf{x} = \mathbf{b}$ without computing A^{-1}:

$$A = \begin{bmatrix} \frac{1}{\sqrt{3}} & 0 & \frac{\sqrt{6}}{3} \\ \frac{1}{\sqrt{3}} & \frac{1}{\sqrt{2}} & -\frac{1}{\sqrt{6}} \\ \frac{1}{\sqrt{3}} & -\frac{1}{\sqrt{2}} & -\frac{1}{\sqrt{6}} \end{bmatrix} \begin{bmatrix} \sqrt{3} & -\sqrt{3} & \frac{5}{\sqrt{3}} \\ 0 & \sqrt{2} & 0 \\ 0 & 0 & \frac{2}{3}\sqrt{6} \end{bmatrix}; \quad \mathbf{b} = \begin{bmatrix} 3 \\ 2 \\ 0 \end{bmatrix}$$

26. (a) Confirm the validity of Formula (23). [*Hint:* $\mathbf{a}^T\mathbf{x} = \mathbf{a} \cdot \mathbf{x}$ is a scalar.]
(b) Let H be the standard matrix for the Householder reflection of R^3 about the hyperplane $\mathbf{a}^{\perp}$, where $\mathbf{a} = (1, 1, 1)$. Use Formula (23) to find the image of the vector $\mathbf{x} = (3, 4, 1)$ under this reflection without computing H. Check your result by finding H and computing $H\mathbf{x}$ with $\mathbf{x}$ in column form.

Discussion and Discovery

D1. If $\mathbf{e}_1$, $\mathbf{e}_2$, and $\mathbf{e}_3$ are the standard basis vectors for R^3, then the standard matrices for the reflections of R^3 about the hyperplanes $\mathbf{e}_1^{\perp}$, $\mathbf{e}_2^{\perp}$, and $\mathbf{e}_3^{\perp}$ are _____. Try to generalize your result to R^n.

D2. The standard matrix for the reflection of R^2 about the line $y = mx$ is _____.

D3. For what value(s) of s, if any, does there exist a Householder reflection that maps $\mathbf{v} = (3, 4, -7, 2)$ into $\mathbf{w} = (3, 4, s, 0)$?

D4. Find a line $y = mx$ such that the reflection of R^2 about the line maps $\mathbf{v} = (5, 12)$ into $\mathbf{w} = (13, 0)$.

D5. Find a plane $z = ax + by$ such that the reflection of R^3 about that plane maps $\mathbf{v} = (1, 2, 2)$ into $\mathbf{w} = (0, 0, 3)$.

Working with Proofs

P1. Use Definition 7.10.4 to prove that Householder matrices are symmetric.

P2. Use Definition 7.10.4 to prove that Householder matrices are orthogonal.

P3. Prove that the matrix R in (2) is invertible by showing that it has nonzero diagonal entries. [*Hint:* Theorem 7.9.6 im-

plies that $\mathbf{w}_j$ can be expressed as a linear combination of $\mathbf{q}_1, \mathbf{q}_2, \ldots, \mathbf{q}_j$.]

P4. Prove that if $A = QR$ is a QR-decomposition of A, then the column vectors of Q form an orthonormal basis for the column space of A.

Technology Exercises

T1. Most linear algebra technology utilities have a command for finding QR-decompositions. Use that command to find the QR-decompositions of the matrices given in Examples 1 and 2.

T2. Construct a QR-decomposition of the matrix

$$A = \begin{bmatrix} 1 & 1 & 1 \\ 1 & 0 & 2 \\ 0 & 1 & 2 \end{bmatrix}$$

by applying the Gram–Schmidt process to the column vectors to find Q and using Formula (5) to compute R. Compare your result to that produced by the command for computing QR-decompositions.

T3. Consider the linear system

$$\begin{aligned} x_1 + 5x_2 + 3x_3 &= 0.8 \\ x_1 + 3x_2 + 4x_3 &= 0.8 \\ x_1 + x_2 + 5x_3 &= 0.6 \\ x_1 + 2x_2 + x_3 &= 0.4 \end{aligned}$$

Find the least squares solution of the system using the command provided for that purpose. Compare your result to that obtained by finding a QR-decomposition of the coefficient matrix and applying the method of Example 2.

T4. Use Householder reflections and the method of Example 7 to find a QR-decomposition of the matrix

$$A = \begin{bmatrix} 1 & 2 & 1 \\ -1 & -2 & 3 \\ 0 & 4 & 5 \end{bmatrix}$$

Compare your answer to that produced by the command provided by your utility for finding QR-decompositions.

T5. **(CAS)** Show that if $\mathbf{a} = (a, b, c)$ is a nonzero vector in R^3, then the standard matrix H for the reflection of R^3 about the hyperplane $\mathbf{a}^\perp$ is

$$H = \begin{bmatrix} 1 - \dfrac{2a^2}{a^2 + b^2 + c^2} & -\dfrac{2ab}{a^2 + b^2 + c^2} & -\dfrac{2ac}{a^2 + b^2 + c^2} \\[3mm] -\dfrac{2ab}{a^2 + b^2 + c^2} & 1 - \dfrac{2b^2}{a^2 + b^2 + c^2} & -\dfrac{2bc}{a^2 + b^2 + c^2} \\[3mm] -\dfrac{2ac}{a^2 + b^2 + c^2} & -\dfrac{2bc}{a^2 + b^2 + c^2} & 1 - \dfrac{2c^2}{a^2 + b^2 + c^2} \end{bmatrix}$$

Section 7.11   Coordinates with Respect to a Basis

A basis that is convenient for one purpose may not be convenient for another, so it is not uncommon in various applications to be working with multiple bases in the same problem. In this section we will consider how results with respect to one basis can be converted to results with respect to another basis.

NONRECTANGULAR COORDINATE SYSTEMS IN R^2 AND R^3

Our first goal in this section is to extend the notion of a coordinate system from R^2 and R^3 to R^n by adopting a vector point of view. For this purpose, recall that in a rectangular xy-coordinate system in R^2 we associate an ordered pair of coordinates (a, b) with a point P by projecting the point onto the coordinate axes and finding the signed distances of the projections from the origin. This establishes a **one-to-one correspondence** between points in the plane and ordered pairs of real numbers. The same one-to-one correspondence can be obtained by considering the unit vectors $\mathbf{i}$ and $\mathbf{j}$ in the positive x- and y-directions, and expressing the vector $\overrightarrow{OP}$ as the linear combination

$$\overrightarrow{OP} = a\mathbf{i} + b\mathbf{j}$$

(Figure 7.11.1a). Since the coefficients in this linear combination are the same as the coordinates obtained using the coordinate axes, we can view the coordinates of a point P in a rectangular xy-coordinate system as the ordered pair of coefficients that result when $\overrightarrow{OP}$ is expressed in terms of the ordered basis* $B = \{\mathbf{i}, \mathbf{j}\}$. Similarly, the coordinates (a, b, c) of a point P in a rectangular xyz-coordinate system can be viewed as the coefficients in the linear combination

$$\overrightarrow{OP} = a\mathbf{i} + b\mathbf{j} + c\mathbf{k}$$

of the ordered basis $B = \{\mathbf{i}, \mathbf{j}, \mathbf{k}\}$ (Figure 7.11.1b).

For the purpose of establishing a one-to-one correspondence between points in the plane and ordered pairs of real numbers, it is not essential that the basis vectors be orthogonal or have length 1. For example, if $B = \{\mathbf{v}_1, \mathbf{v}_2\}$ is *any* ordered basis for R^2, then for each point P in the plane, there is exactly one way to express the vector $\overrightarrow{OP}$ as a linear combination

$$\overrightarrow{OP} = a\mathbf{v}_1 + b\mathbf{v}_2$$

Thus, the ordered basis B associates a unique ordered pair of numbers (a, b) with the point P, and conversely. Accordingly, we can think of the basis $B = \{\mathbf{v}_1, \mathbf{v}_2\}$ as defining a "generalized

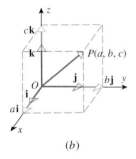

Figure 7.11.1

---

*The order in which the basis vectors are listed here is important, since changing the order of the vectors would change the order of the coordinates. In general, when the order in which the basis vectors are listed must be adhered to, the basis is called an **ordered basis.** In this section we will consider all bases to be ordered, even if not stated explicitly.

coordinate system" in which the coordinates (a, b) of a point P tell us how to reach the point P from the origin by a displacement $a\mathbf{v}_1$ followed by a displacement $b\mathbf{v}_2$ (Figure 7.11.2). More generally, we can think of an ordered basis $B = \{\mathbf{v}_1, \mathbf{v}_2, \ldots, \mathbf{v}_n\}$ for R^n as defining a generalized coordinate system in which each point

$$\mathbf{x} = a_1\mathbf{v}_1 + a_2\mathbf{v}_2 + \cdots + a_n\mathbf{v}_n$$

in R^n is represented by the n-tuple of "coordinates" $(a_1, a_2, \ldots, a_n)$.

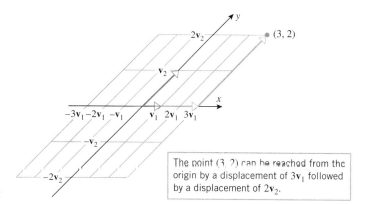

The point (3, 2) can be reached from the origin by a displacement of $3\mathbf{v}_1$ followed by a displacement of $2\mathbf{v}_2$.

Figure 7.11.2

Motivated by this discussion, we make the following definition.

> **Definition 7.11.1**   If $B = \{\mathbf{v}_1, \mathbf{v}_2, \ldots, \mathbf{v}_k\}$ is an ordered basis for a subspace W of R^n, and if
>
> $$\mathbf{w} = a_1\mathbf{v}_1 + a_2\mathbf{v}_2 + \cdots + a_k\mathbf{v}_k$$
>
> is the expression for a vector $\mathbf{w}$ in W as a linear combination of the vectors in B, then we call
>
> $$a_1, a_2, \ldots, a_n$$
>
> the ***coordinates of w with respect to*** B; and more specifically, we call a_j the $\mathbf{v}_j$***-coordinate of*** **w**. We denote the ordered k-tuple of coordinates by
>
> $$(\mathbf{w})_B = (a_1, a_2, \ldots, a_k)$$
>
> and call it the ***coordinate vector*** for $\mathbf{w}$ with respect to B; and we denote the column vector of coordinates by
>
> $$[\mathbf{w}]_B = \begin{bmatrix} a_1 \\ a_2 \\ \vdots \\ a_k \end{bmatrix}$$
>
> and call it the ***coordinate matrix*** for $\mathbf{w}$ with respect to B.

EXAMPLE 1
Finding
Coordinates

In Example 2(a) of Section 7.2 we showed that the vectors

$$\mathbf{v}_1 = (1, 2, 1), \quad \mathbf{v}_2 = (1, -1, 3), \quad \mathbf{v}_3 = (1, 1, 4)$$

form a basis for R^3.

(a) Find the coordinate vector and coordinate matrix for the vector $\mathbf{w} = (4, 9, 8)$ with respect to the ordered basis $B = \{\mathbf{v}_1, \mathbf{v}_2, \mathbf{v}_3\}$.

(b) Find the vector $\mathbf{w}$ in R^3 whose coordinate vector relative to B is $(\mathbf{w})_B = (1, 2, -3)$.

Solution (a) In Example 2(b) of Section 7.2 we showed that $\mathbf{w}$ can be expressed as

$$\mathbf{w} = 3\mathbf{v}_1 - \mathbf{v}_2 + 2\mathbf{v}_3$$

Thus,

$$(\mathbf{w})_B = (3, -1, 2) \quad \text{and} \quad [\mathbf{w}]_B = \begin{bmatrix} 3 \\ -1 \\ 2 \end{bmatrix}$$

Solution (b) The entries in the coordinate vector tell us how to express $\mathbf{w}$ as a linear combination of the basis vectors:

$$\mathbf{w} = \mathbf{v}_1 + 2\mathbf{v}_2 - 3\mathbf{v}_3 = (1, 2, 1) + 2(1, -1, 3) - 3(1, 1, 4) = (0, -3, -5) \qquad \blacksquare$$

EXAMPLE 2
Coordinates with Respect to the Standard Basis

If $S = \{\mathbf{e}_1, \mathbf{e}_2, \ldots, \mathbf{e}_n\}$ is the standard basis for R^n, and $\mathbf{w} = (w_1, w_2, \ldots, w_n)$, then $\mathbf{w}$ can be expressed as a linear combination of the standard basis vectors as

$$\mathbf{w} = w_1(1, 0, \ldots, 0) + w_2(0, 1, \ldots, 0) + \cdots + w_n(0, 0, \ldots, 1) = w_1\mathbf{e}_1 + w_2\mathbf{e}_2 + \cdots + w_n\mathbf{e}_n$$

Thus,

$$(\mathbf{w})_S = (w_1, w_2, \ldots, w_n) = \mathbf{w} \tag{1}$$

That is, the components of $\mathbf{w}$ are the same as its coordinates with respect to the standard basis. If $\mathbf{w}$ is written in column form, then

$$[\mathbf{w}]_S = \begin{bmatrix} w_1 \\ w_2 \\ \vdots \\ w_n \end{bmatrix} = \mathbf{w} \tag{2}$$

When we want to think of the components of a vector in R^n as coordinates with respect to the standard basis, we will call them the ***standard coordinates*** of the vector. $\qquad \blacksquare$

COORDINATES WITH RESPECT TO AN ORTHONORMAL BASIS

Recall from Theorem 7.9.4 that if $B = \{\mathbf{v}_1, \mathbf{v}_2, \ldots, \mathbf{v}_k\}$ is an orthonormal basis for a subspace W of R^n, and if $\mathbf{w}$ is a vector in W, then the expression for $\mathbf{w}$ as a linear combination of the vectors in B is

$$\mathbf{w} = (\mathbf{w} \cdot \mathbf{v}_1)\mathbf{v}_1 + (\mathbf{w} \cdot \mathbf{v}_2)\mathbf{v}_2 + \cdots + (\mathbf{w} \cdot \mathbf{v}_k)\mathbf{v}_k$$

Thus, the coordinate vector for $\mathbf{w}$ with respect to B is

$$(\mathbf{w})_B = \big((\mathbf{w} \cdot \mathbf{v}_1), (\mathbf{w} \cdot \mathbf{v}_2), \ldots, (\mathbf{w} \cdot \mathbf{v}_k)\big) \tag{3}$$

This result is noteworthy because it tells us that the components of a vector with respect to an orthonormal basis can be obtained by computing appropriate inner products, whereas for a general basis it is usually necessary to solve a linear system (as in Example 1). This is yet another computational advantage of orthonormal bases.

EXAMPLE 3
Finding Coordinates with Respect to an Orthonormal Basis

We showed in Example 3 of Section 7.9 that the vectors

$$\mathbf{v}_1 = \left(\tfrac{3}{7}, -\tfrac{6}{7}, \tfrac{2}{7}\right), \quad \mathbf{v}_2 = \left(\tfrac{2}{7}, \tfrac{3}{7}, \tfrac{6}{7}\right), \quad \mathbf{v}_3 = \left(\tfrac{6}{7}, \tfrac{2}{7}, -\tfrac{3}{7}\right)$$

form an orthonormal basis for R^3. Find the coordinate vector for $\mathbf{w} = (1, -1, 1)$ with respect to the basis $B = \{\mathbf{v}_1, \mathbf{v}_2, \mathbf{v}_3\}$.

Solution We leave it for you to show that

$$\mathbf{w} \cdot \mathbf{v}_1 = \tfrac{11}{7}, \quad \mathbf{w} \cdot \mathbf{v}_2 = \tfrac{5}{7}, \quad \mathbf{w} \cdot \mathbf{v}_3 = \tfrac{1}{7}$$

Thus, $(\mathbf{w})_B = \left(\frac{11}{7}, \frac{5}{7}, \frac{1}{7}\right)$, or in column form,

$$[\mathbf{w}]_B = \begin{bmatrix} \frac{11}{7} \\ \frac{5}{7} \\ \frac{1}{7} \end{bmatrix}$$

■

COMPUTING WITH COORDINATES WITH RESPECT TO AN ORTHONORMAL BASIS

The following theorem shows that if B is an orthonormal basis, then the norm of a vector is the same as the norm of its coordinate vector, and the dot product of two vectors is the same as the dot product of their coordinate vectors.

Theorem 7.11.2   *If B is an orthonormal basis for a k-dimensional subspace W of R^n, and if $\mathbf{u}$, $\mathbf{v}$, and $\mathbf{w}$ are vectors in W with coordinate vectors*

$$(\mathbf{u})_B = (u_1, u_2, \ldots, u_k), \quad (\mathbf{v})_B = (v_1, v_2, \ldots, v_k), \quad (\mathbf{w})_B = (w_1, w_2, \ldots, w_k)$$

then:

(*a*) $\|\mathbf{w}\| = \sqrt{w_1^2 + w_2^2 + \cdots + w_k^2} = \|(\mathbf{w})_B\|$

(*b*) $\mathbf{u} \cdot \mathbf{v} = u_1 v_1 + u_2 v_2 + \cdots + u_k v_k = (\mathbf{u})_B \cdot (\mathbf{v})_B$

CONCEPT PROBLEM   Do you think that this theorem is true if the basis is not orthonormal? Explain.

EXAMPLE 4
Computing with Coordinates

Let $B = \{\mathbf{v}_1, \mathbf{v}_2, \mathbf{v}_3\}$ be the orthonormal basis for R^3 that was given in Example 3, and let $\mathbf{w} = (1, -1, 1)$. We showed in Example 3 that $(\mathbf{w})_B = \left(\frac{11}{7}, \frac{5}{7}, \frac{1}{7}\right)$. Thus,

$$\|(\mathbf{w})_B\| = \sqrt{\left(\frac{11}{7}\right)^2 + \left(\frac{5}{7}\right)^2 + \left(\frac{1}{7}\right)^2} = \sqrt{3} = \|\mathbf{w}\|$$

as guaranteed by part (*a*) of Theorem 7.11.2.

■

CHANGE OF BASIS FOR R^n

We now come to the main problem in this section.

The Change of Basis Problem   If $\mathbf{w}$ is a vector in R^n, and if we change the basis for R^n from a basis B to a basis B', how are the coordinate matrices $[\mathbf{w}]_B$ and $[\mathbf{w}]_{B'}$ related?

To solve this problem, it will be convenient to refer to B as the "old basis" and B' as the "new basis." Thus, our objective is to find a relationship between the old and new coordinates of a fixed vector $\mathbf{w}$. For notational simplicity, we will solve the problem in R^2. For this purpose, let

$$B = [\mathbf{v}_1, \mathbf{v}_2] \quad \text{and} \quad B' = [\mathbf{v}_1', \mathbf{v}_2']$$

be the old and new bases, respectively, and suppose that the coordinate matrices for the old basis vectors with respect to the new basis are

$$[\mathbf{v}_1]_{B'} = \begin{bmatrix} a \\ b \end{bmatrix} \quad \text{and} \quad [\mathbf{v}_2]_{B'} = \begin{bmatrix} c \\ d \end{bmatrix} \tag{4}$$

That is,

$$\begin{aligned} \mathbf{v}_1 &= a\mathbf{v}_1' + b\mathbf{v}_2' \\ \mathbf{v}_2 &= c\mathbf{v}_1' + d\mathbf{v}_2' \end{aligned} \tag{5}$$

Now let $\mathbf{w}$ be any vector in W, and let

$$[\mathbf{w}]_B = \begin{bmatrix} k_1 \\ k_2 \end{bmatrix} \tag{6}$$

be the old coordinate matrix; that is,

$$\mathbf{w} = k_1 \mathbf{v}_1 + k_2 \mathbf{v}_2 \tag{7}$$

To find the new coordinate matrix for $\mathbf{w}$ we must express $\mathbf{w}$ in terms of the new basis B'. For this purpose we substitute (5) into (7) to obtain

$$\mathbf{w} = k_1(a\mathbf{v}'_1 + b\mathbf{v}'_2) + k_2(c\mathbf{v}'_1 + d\mathbf{v}'_2)$$

which we can rewrite as

$$\mathbf{w} = (k_1 a + k_2 c)\mathbf{v}'_1 + (k_1 b + k_2 d)\mathbf{v}'_2$$

Thus,

$$[\mathbf{w}]_{B'} = \begin{bmatrix} k_1 a + k_2 c \\ k_1 b + k_2 d \end{bmatrix}$$

Now using (6), we can express this as

$$[\mathbf{w}]_{B'} = \begin{bmatrix} k_1 a + k_2 c \\ k_1 b + k_2 d \end{bmatrix} = \begin{bmatrix} a & c \\ b & d \end{bmatrix} \begin{bmatrix} k_1 \\ k_2 \end{bmatrix} = \begin{bmatrix} a & c \\ b & d \end{bmatrix} [\mathbf{w}]_B \tag{8}$$

which is the relationship we were looking for, since it tells us that the new coordinate matrix can be obtained by multiplying the old coordinate matrix by

$$\begin{bmatrix} a & c \\ b & d \end{bmatrix} = \begin{bmatrix} [\mathbf{v}_1]_{B'} & | & [\mathbf{v}_2]_{B'} \end{bmatrix} \tag{9}$$

We will denote this matrix by $P_{B \to B'}$ to suggest that it transforms B-coordinates to B'-coordinates. Using this notation, we can express (8) as

$$[\mathbf{w}]_{B'} = P_{B \to B'}[\mathbf{w}]_B$$

Although this relationship was derived for R^2 for notational simplicity, the same relationship holds for R^n. Here is the general result.

Theorem 7.11.3 (*Solution of the Change of Basis Problem*) *If $\mathbf{w}$ is a vector in R^n, and if $B = \{\mathbf{v}_1, \mathbf{v}_2, \ldots, \mathbf{v}_n\}$ and $B' = \{\mathbf{v}'_1, \mathbf{v}'_2, \ldots, \mathbf{v}'_n\}$ are bases for R^n, then the coordinate matrices of $\mathbf{w}$ with respect to the two bases are related by the equation*

$$[\mathbf{w}]_{B'} = P_{B \to B'}[\mathbf{w}]_B \tag{10}$$

where

$$P_{B \to B'} = \begin{bmatrix} [\mathbf{v}_1]_{B'} & | & [\mathbf{v}_2]_{B'} & | & \cdots & | & [\mathbf{v}_n]_{B'} \end{bmatrix} \tag{11}$$

*This matrix is called the **transition matrix** (or the **change of coordinates matrix**) from B to B'.*

REMARK Formula (10) can be confusing when different letters are used for the bases or when the roles of B and B' are reversed, but you won't go wrong if you keep in mind that the columns in the transition matrix are coordinate matrices of the basis you are *transforming from* with respect to the basis you are *transforming to*.

EXAMPLE 5
Transition
Matrices

Consider the bases $B_1 = \{\mathbf{e}_1, \mathbf{e}_2\}$ and $B_2 = \{\mathbf{v}_1, \mathbf{v}_2\}$ for R^2, where

$$\mathbf{e}_1 = (1, 0), \quad \mathbf{e}_2 = (0, 1), \quad \mathbf{v}_1 = (1, 1), \quad \mathbf{v}_2 = (2, 1)$$

(a) Find the transition matrix from B_1 to B_2.

(b) Use the transition matrix from B_1 to B_2 to find $[\mathbf{w}]_{B_2}$ given that

$$[\mathbf{w}]_{B_1} = \begin{bmatrix} 7 \\ 2 \end{bmatrix} \tag{12}$$

(c) Find the transition matrix from B_2 to B_1.

(d) Use the transition matrix from B_2 to B_1 to recover the vector $[\mathbf{w}]_{B_1}$ from the vector $[\mathbf{w}]_{B_2}$.

Solution (*a*) Since we are transforming *to* B_2-coordinates, the form of the required transition matrix is

$$P_{B_1 \rightarrow B_2} = \left[[\mathbf{e}_1]_{B_2} \mid [\mathbf{e}_2]_{B_2} \right] \tag{13}$$

We leave it for you to show that

$$\mathbf{e}_1 = -\mathbf{v}_1 + \mathbf{v}_2$$
$$\mathbf{e}_2 = 2\mathbf{v}_1 - \mathbf{v}_2$$

from which we obtain

$$[\mathbf{e}_1]_{B_2} = \begin{bmatrix} -1 \\ 1 \end{bmatrix} \quad \text{and} \quad [\mathbf{e}_2]_{B_2} = \begin{bmatrix} 2 \\ -1 \end{bmatrix}$$

Thus, the transition matrix from B_1 to B_2 is

$$P_{B_1 \rightarrow B_2} = \left[[\mathbf{e}_1]_{B_2} \mid [\mathbf{e}_2]_{B_2} \right] = \begin{bmatrix} -1 & 2 \\ 1 & -1 \end{bmatrix} \tag{14}$$

Solution (*b*) Using (12) and (14) we obtain

$$[\mathbf{w}]_{B_2} = P_{B_1 \rightarrow B_2}[\mathbf{w}]_{B_1} = \begin{bmatrix} -1 & 2 \\ 1 & -1 \end{bmatrix} \begin{bmatrix} 7 \\ 2 \end{bmatrix} = \begin{bmatrix} -3 \\ 5 \end{bmatrix} \tag{15}$$

As a check, (12) and (15) should correspond to the same vector $\mathbf{w}$. This is in fact the case, since (12) yields

$$\mathbf{w} = 7\mathbf{e}_1 + 2\mathbf{e}_2 = 7\begin{bmatrix} 1 \\ 0 \end{bmatrix} + 2\begin{bmatrix} 0 \\ 1 \end{bmatrix} = \begin{bmatrix} 7 \\ 2 \end{bmatrix}$$

and (15) yields

$$\mathbf{w} = -3\mathbf{v}_1 + 5\mathbf{v}_2 = -3\begin{bmatrix} 1 \\ 1 \end{bmatrix} + 5\begin{bmatrix} 2 \\ 1 \end{bmatrix} = \begin{bmatrix} 7 \\ 2 \end{bmatrix}$$

Solution (*c*) Since we are transforming *to* B_1-coordinates, the form of the required transition matrix is

$$P_{B_2 \rightarrow B_1} = \left[[\mathbf{v}_1]_{B_1} \mid [\mathbf{v}_2]_{B_1} \right]$$

But B_1 is the standard basis, so if $\mathbf{v}_1$ and $\mathbf{v}_2$ are written in column form, then $[\mathbf{v}_1]_{B_1} = \mathbf{v}_1$ and $[\mathbf{v}_2]_{B_1} = \mathbf{v}_2$. Thus,

$$P_{B_2 \rightarrow B_1} = \left[[\mathbf{v}_1]_{B_1} \mid [\mathbf{v}_2]_{B_1} \right] = [\mathbf{v}_1 \mid \mathbf{v}_2] = \begin{bmatrix} 1 & 2 \\ 1 & 1 \end{bmatrix} \tag{16}$$

Solution (*d*) Using (15) and (16) we obtain

$$[\mathbf{w}]_{B_1} = P_{B_2 \rightarrow B_1}[\mathbf{w}]_{B_2} = \begin{bmatrix} 1 & 2 \\ 1 & 1 \end{bmatrix} \begin{bmatrix} -3 \\ 5 \end{bmatrix} = \begin{bmatrix} 7 \\ 2 \end{bmatrix}$$

which is consistent with (12).   ∎

INVERTIBILITY OF TRANSITION MATRICES

If B_1, B_2, and B_3 are bases for R^n, then it is reasonable, though we will not formally prove it, that

$$P_{B_2 \to B_3} P_{B_1 \to B_2} = P_{B_1 \to B_3} \tag{17}$$

This is because multiplication by $P_{B_1 \to B_2}$ maps B_1-coordinates into B_2-coordinates and multiplication by $P_{B_2 \to B_3}$ maps B_2-coordinates into B_3-coordinates, so the effect of first multiplying by $P_{B_1 \to B_2}$ and then by $P_{B_2 \to B_3}$ is to map B_1-coordinates into B_3-coordinates. In particular, if B and B' are two bases for W, then

$$P_{B' \to B} P_{B \to B'} = P_{B \to B} \tag{18}$$

But $P_{B \to B} = I$ (why?), so (18) implies that $P_{B' \to B}$ and $P_{B \to B'}$ are invertible and are inverses of one another.

Theorem 7.11.4 *If B and B' are bases for R^n, then the transition matrices $P_{B' \to B}$ and $P_{B \to B'}$ are invertible and are inverses of one another; that is,*

$$(P_{B' \to B})^{-1} = P_{B \to B'} \quad and \quad (P_{B \to B'})^{-1} = P_{B' \to B}$$

EXAMPLE 6
Inverse of a Transition Matrix

For the bases in Example 5 we found that

$$P_{B_2 \to B_1} = \begin{bmatrix} 1 & 2 \\ 1 & 1 \end{bmatrix} \quad \text{and} \quad P_{B_1 \to B_2} = \begin{bmatrix} -1 & 2 \\ 1 & -1 \end{bmatrix}$$

A simple multiplication will show that these matrices are inverses, as guaranteed by Theorem 7.11.4. ∎

A GOOD TECHNIQUE FOR FINDING TRANSITION MATRICES

Our next goal is to develop an efficient technique for finding transition matrices. For this purpose, let $B = \{\mathbf{v}_1, \mathbf{v}_2, \ldots, \mathbf{v}_n\}$ and $B' = \{\mathbf{v}'_1, \mathbf{v}'_2, \ldots, \mathbf{v}'_n\}$ be bases for R^n, and consider how the columns of the transition matrix

$$P_{B \to B'} = \left[[\mathbf{v}_1]_{B'} \mid [\mathbf{v}_2]_{B'} \mid \cdots \mid [\mathbf{v}_n]_{B'} \right] \tag{19}$$

are computed. The entries of $[\mathbf{v}_j]_{B'}$ are the coefficients that are required to express $\mathbf{v}_j$ as a linear combination of $\mathbf{v}'_1, \mathbf{v}'_2, \ldots, \mathbf{v}'_n$, and hence can be obtained by solving the linear system

$$\begin{bmatrix} \mathbf{v}'_1 & \mathbf{v}'_2 & \cdots & \mathbf{v}'_n \end{bmatrix} \mathbf{x} = \mathbf{v}_j \tag{20}$$

whose augmented matrix is

$$\begin{bmatrix} \mathbf{v}'_1 & \mathbf{v}'_2 & \cdots & \mathbf{v}'_n \mid \mathbf{v}_j \end{bmatrix} \tag{21}$$

Since $\mathbf{v}'_1, \mathbf{v}'_2, \ldots, \mathbf{v}'_n$ are linearly independent, the coefficient matrix in (20) has rank n, so its reduced row echelon form is the identity matrix; hence, the reduced row echelon form of (21) is

$$\begin{bmatrix} I \mid [\mathbf{v}_j]_{B'} \end{bmatrix}$$

Thus, we have shown that $[\mathbf{v}_j]_{B'}$ is the matrix that results on the right side when row operations are applied to (21) to reduce the left side to the identity matrix. However, rather than compute one column at a time, we can obtain all of the columns of (19) at once by applying row operations to the matrix

$$\begin{bmatrix} \mathbf{v}'_1 & \mathbf{v}'_2 & \cdots & \mathbf{v}'_n \mid \mathbf{v}_1 & \mathbf{v}_2 & \cdots & \mathbf{v}_n \end{bmatrix} \tag{22}$$

to reduce it to

$$\begin{bmatrix} I \mid [\mathbf{v}_1]_{B'} & [\mathbf{v}_2]_{B'} & \cdots & [\mathbf{v}_n]_{B'} \end{bmatrix} = \begin{bmatrix} I \mid P_{B \to B'} \end{bmatrix} \tag{23}$$

In summary, if we call $B = \{\mathbf{v}_1, \mathbf{v}_2, \ldots, \mathbf{v}_n\}$ the old basis and $B' = \{\mathbf{v}'_1, \mathbf{v}'_2, \ldots, \mathbf{v}'_n\}$ the new basis, then the process of obtaining (23) from (22) by row operations is captured in the diagram

$$[\text{new basis} \mid \text{old basis}] \xrightarrow{\text{row operations}} [I \mid \text{transition from old to new}] \qquad (24)$$

In summary, we have the following procedure for finding transition matrices by row reduction.

A Procedure for Computing $P_{B \to B'}$

Step 1. Form the matrix $[B' \mid B]$.

Step 2. Use elementary row operations to reduce the matrix in Step 1 to reduced row echelon form.

Step 3. The resulting matrix will be $[I \mid P_{B \to B'}]$.

Step 4. Extract the matrix $P_{B \to B'}$ from the right side of the matrix in Step 3.

EXAMPLE 7
Transition
Matrices by
Row Reduction

In Example 5 we found the transition matrices between the bases $B_1 = \{\mathbf{e}_1, \mathbf{e}_2\}$ and $B_2 = \{\mathbf{v}_1, \mathbf{v}_2\}$ for R^2, where

$$\mathbf{e}_1 = (1, 0), \quad \mathbf{e}_2 = (0, 1), \quad \mathbf{v}_1 = (1, 1), \quad \mathbf{v}_2 = (2, 1)$$

Find these transition matrices using (24).

Solution To find the transition matrix $P_{B_1 \to B_2}$ we must reduce the matrix

$$[B_2 \mid B_1] = \begin{bmatrix} 1 & 2 & 1 & 0 \\ 1 & 1 & 0 & 1 \end{bmatrix}$$

to make the left side the identity matrix. This yields (verify)

$$[I \mid P_{B_1 \to B_2}] = \begin{bmatrix} 1 & 0 & -1 & 2 \\ 0 & 1 & 1 & -1 \end{bmatrix}$$

which agrees with (14). To find the transition matrix $P_{B_2 \to B_1}$ we must reduce the matrix

$$[B_1 \mid B_2] = \begin{bmatrix} 1 & 0 & 1 & 2 \\ 0 & 1 & 1 & 1 \end{bmatrix}$$

to make the left side the identity matrix. However, it is already in that form, so there is nothing to do; we see immediately that

$$P_{B_2 \to B_1} = \begin{bmatrix} 1 & 2 \\ 1 & 1 \end{bmatrix}$$

which agrees with (16).   ■

REMARK The second part of this example illustrates the general fact that if $B = \{\mathbf{v}_1, \mathbf{v}_2, \dots, \mathbf{v}_n\}$ is a basis for R^n, then the transition matrix from B to the standard basis S for R^n is

$$P_{B \to S} = [\mathbf{v}_1 \mid \mathbf{v}_2 \mid \cdots \mid \mathbf{v}_n] \qquad (25)$$

COORDINATE MAPS

If B is a basis for R^n, then the transformation

$$\mathbf{x} \to (\mathbf{x})_B, \quad \text{or in column notation,} \quad \mathbf{x} \to [\mathbf{x}]_B$$

is called the ***coordinate map*** for B. In the exercises we will ask you to show that the following relationships hold for any scalar c and for any vectors $\mathbf{v}$ and $\mathbf{w}$ in R^n:

$$(c\mathbf{v})_B = c(\mathbf{v})_B \quad \text{and} \quad [c\mathbf{v}]_B = c[\mathbf{v}]_B \tag{26}$$

$$(\mathbf{v} + \mathbf{w})_B = (\mathbf{v})_B + (\mathbf{w})_B \quad \text{and} \quad [\mathbf{v} + \mathbf{w}]_B = [\mathbf{v}]_B + [\mathbf{w}]_B \tag{27}$$

It follows from these relationships that the coordinate map for B is a linear operator on R^n. Moreover, since distinct vectors in R^n have distinct coordinate vectors with respect to B (why?), it follows that the coordinate map is one-to-one (and hence also onto by Theorem 6.3.14). In the case where B is an orthonormal basis, it follows from Theorem 7.11.2 that the coordinate map is length preserving and inner product preserving; that is, it is an orthogonal operator on R^n.

Theorem 7.11.5   *If B is a basis for R^n, then the coordinate map $\mathbf{x} \to (\mathbf{x})_B$ (or $\mathbf{x} \to [\mathbf{x}]_B$) is a one-to-one linear operator on R^n. Moreover, if B is an orthonormal basis for R^n, then it is an orthogonal operator.*

We leave it as an exercise for you to use Theorem 3.4.4 and the fact that coordinate maps are onto to prove the following result.

Theorem 7.11.6   *If A and C are $m \times n$ matrices, and if B is any basis for R^n, then $A = C$ if and only if $A[\mathbf{x}]_B = C[\mathbf{x}]_B$ for every $\mathbf{x}$ in R^n.*

This theorem is useful because it provides a way of using coordinate matrices to determine whether two matrices are equal in cases where the entries of those matrices are not known explicitly.

TRANSITION BETWEEN ORTHONORMAL BASES

We will now consider a fundamental property of transition matrices between orthonormal bases.

Theorem 7.11.7   *If B and B' are orthonormal bases for R^n, then the transition matrices $P_{B \to B'}$ and $P_{B' \to B}$ are orthogonal.*

Proof Since $P_{B \to B'}$ and $P_{B' \to B}$ are inverses, we need only prove that $P_{B \to B'}$ is orthogonal, since the orthogonality of $P_{B' \to B}$ will then follow from Theorem 6.2.3. Accordingly, suppose that B and B' are orthonormal bases for W and that $B = \{\mathbf{v}_1, \mathbf{v}_2, \dots, \mathbf{v}_n\}$. To prove that

$$P_{B \to B'} = \left[[\mathbf{v}_1]_{B'} \mid [\mathbf{v}_2]_{B'} \mid \cdots \mid [\mathbf{v}_n]_{B'} \right] \tag{28}$$

is an orthogonal matrix, we will show that the column vectors are orthonormal (see Theorem 6.2.5). But this follows from Theorem 7.11.2, since

$$\| [\mathbf{v}_j]_{B'} \| = \| \mathbf{v}_j \| = 1 \quad \text{and} \quad [\mathbf{v}_i]_{B'} \cdot [\mathbf{v}_j]_{B'} = \mathbf{v}_i \cdot \mathbf{v}_j = 0 \qquad (i \neq j) \qquad \blacksquare$$

EXAMPLE 8

A Rotation of the Standard Basis for R^2

Let $S = \{\mathbf{e}_1, \mathbf{e}_2\}$ be the standard basis for R^2, and let $B = \{\mathbf{v}_1, \mathbf{v}_2\}$ be the basis that results when the vectors in S are rotated about the origin through the angle θ. From (25) and Figure 7.11.3,

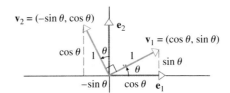

Figure 7.11.3

we see that the transition matrix from B to S is

$$P = P_{B \to S} = [\mathbf{v}_1 \mid \mathbf{v}_2] = \begin{bmatrix} \cos\theta & -\sin\theta \\ \sin\theta & \cos\theta \end{bmatrix}$$

This matrix is orthogonal, as guaranteed by Theorem 7.11.7, and hence the transition matrix from S to B is

$$P^T = P_{S \to B} = \begin{bmatrix} \cos\theta & \sin\theta \\ -\sin\theta & \cos\theta \end{bmatrix}$$ ∎

APPLICATION TO ROTATION OF COORDINATE AXES

In Chapter 6 we discussed rotations about the origin in R^2 and R^3. In that discussion the coordinate axes remained fixed and the vectors were rotated. However, there are many kinds of problems in which it is preferable to think of the vectors as being fixed and the axes rotated. Here is an example.

EXAMPLE 9
Rotation of Coordinate Axes in R^2

Suppose that a rectangular xy-coordinate system is given and an $x'y'$-coordinate system is obtained by rotating the xy-coordinate system about the origin through an angle θ. Since there are now two coordinate systems, each point Q in the plane has two pairs of coordinates, a pair of coordinates (x, y) with respect to the xy-system and a pair of coordinates (x', y') with respect to the $x'y'$-system (Figure 7.11.4a). To find a relationship between the two pairs of coordinates, we will treat the axis rotation as a change of basis from the standard basis $S = \{\mathbf{e}_1, \mathbf{e}_2\}$ to the basis $B = \{\mathbf{v}_1, \mathbf{v}_2\}$, where $\mathbf{e}_1$ and $\mathbf{e}_2$ run along the positive x- and y-axes, respectively, and $\mathbf{v}_1$ and $\mathbf{v}_2$ are the unit vectors along the positive x'- and y'-axes, respectively (Figure 7.11.4b). Since the vectors in B result from rotating the vectors in S about the origin through the angle θ, it follows from Example 8 that the transition matrices between the two bases are

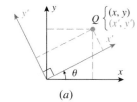

(a)

$$P_{B \to S} = \begin{bmatrix} \cos\theta & -\sin\theta \\ \sin\theta & \cos\theta \end{bmatrix} \quad \text{and} \quad P_{S \to B} = \begin{bmatrix} \cos\theta & \sin\theta \\ -\sin\theta & \cos\theta \end{bmatrix} \tag{29–30}$$

Thus, the relationship between two pairs of coordinates can be expressed as

$$\begin{bmatrix} x \\ y \end{bmatrix} = \begin{bmatrix} \cos\theta & -\sin\theta \\ \sin\theta & \cos\theta \end{bmatrix} \begin{bmatrix} x' \\ y' \end{bmatrix} \quad \text{or} \quad \begin{bmatrix} x' \\ y' \end{bmatrix} = \begin{bmatrix} \cos\theta & \sin\theta \\ -\sin\theta & \cos\theta \end{bmatrix} \begin{bmatrix} x \\ y \end{bmatrix} \tag{31–32}$$

If preferred, these matrix relationships can be expressed in equation form as

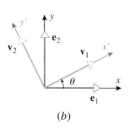

(b)

Figure 7.11.4

$$\begin{aligned} x &= x'\cos\theta - y'\sin\theta \\ y &= x'\sin\theta + y'\cos\theta \end{aligned} \quad \text{or} \quad \begin{aligned} x' &= x\cos\theta + y\sin\theta \\ y' &= -x\sin\theta + y\cos\theta \end{aligned} \tag{33–34}$$

These are sometimes called the ***rotation equations*** for the plane.   ∎

REMARK If you compare the transition matrix in (32) to the rotation matrix R_θ in Formula (16) of Section 6.1, you will see that they are inverses of one another. This is to be expected since rotating the coordinate axes through an angle θ with the vectors held fixed has the same effect as rotating the vectors through the angle $-\theta$ with the axes held fixed.

NEW WAYS TO THINK ABOUT MATRICES

Coordinates provide a new way of thinking about matrices. For example, although it is natural to think of

$$\mathbf{x} = \begin{bmatrix} 2 \\ 3 \end{bmatrix}$$

as a vector in R^2 with components $x_1 = 2$ and $x_2 = 3$, we can also think of $\mathbf{x}$ as a coordinate vector

$$\mathbf{x} = [\mathbf{w}]_B$$

in which $B = \{\mathbf{v}_1, \mathbf{v}_2\}$ is any basis for R^2 and $\mathbf{w}$ is the vector

$$\mathbf{w} = 2\mathbf{v}_1 + 4\mathbf{v}_2$$

Coordinates and transition matrices also provide a new way of thinking about invertible matrices. For example, although the invertible matrix

$$A = \begin{bmatrix} 1 & 3 \\ 2 & 5 \end{bmatrix} \tag{35}$$

might be viewed as the coefficient matrix for a linear system, it can also be viewed as the transition matrix from the basis

$$B = \left\{ \begin{bmatrix} 1 \\ 2 \end{bmatrix}, \begin{bmatrix} 3 \\ 5 \end{bmatrix} \right\}$$

to the standard basis $S = \{\mathbf{e}_1, \mathbf{e}_2\}$. More generally, we have the following result.

Theorem 7.11.8 *If P is an invertible $n \times n$ matrix with column vectors $\mathbf{p}_1, \mathbf{p}_2, \ldots, \mathbf{p}_n$, then P is the transition matrix from the basis $B = \{\mathbf{p}_1, \mathbf{p}_2, \ldots, \mathbf{p}_n\}$ for R^n to the standard basis $S = \{\mathbf{e}_1, \mathbf{e}_2, \ldots, \mathbf{e}_n\}$ for R^n.*

CONCEPT PROBLEM In what other ways might you interpret (35) as a transition matrix?

In the special case where P is a 2×2 or 3×3 orthogonal matrix and $\det(P) = 1$, the matrix P represents a rotation, which we can view either as a rotation of vectors or as a change in coordinates resulting from a rotation of coordinate axes. For example, if $P = [\mathbf{p}_1 \ \mathbf{p}_2]$ is 2×2, and if we view P as the standard matrix for a linear operator, then multiplication by P represents the rotation of R^2 that rotates $\mathbf{e}_1$ and $\mathbf{e}_2$ into $\mathbf{p}_1$ and $\mathbf{p}_2$, respectively. Alternatively, if we view the same matrix P as a transition matrix, then it follows from Theorem 7.11.3 that multiplication by P changes coordinates relative to the rectangular coordinate system whose positive axes are in the directions of $\mathbf{p}_1$ and $\mathbf{p}_2$ into those relative to the rotated coordinate system whose positive axes are in the directions of $\mathbf{e}_1$ and $\mathbf{e}_2$; hence, multiplication by $P^{-1} = P^T$ changes coordinates relative to the system whose positive axes are in the directions of $\mathbf{e}_1$ and $\mathbf{e}_2$ into those relative to the system whose positive axes are in the directions of $\mathbf{p}_1$ and $\mathbf{p}_2$. These two interpretations of

$$P = \begin{bmatrix} \frac{1}{\sqrt{2}} & -\frac{1}{\sqrt{2}} \\ \frac{1}{\sqrt{2}} & \frac{1}{\sqrt{2}} \end{bmatrix} \quad \text{and} \quad P^T = \begin{bmatrix} \frac{1}{\sqrt{2}} & \frac{1}{\sqrt{2}} \\ -\frac{1}{\sqrt{2}} & \frac{1}{\sqrt{2}} \end{bmatrix}$$

are illustrated in Figure 7.11.5.

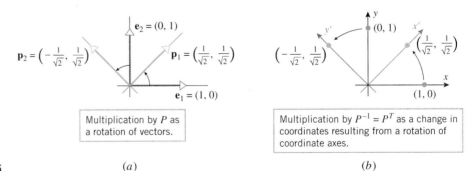

Multiplication by P as a rotation of vectors.

Multiplication by $P^{-1} = P^T$ as a change in coordinates resulting from a rotation of coordinate axes.

Figure 7.11.5

(a)

(b)

Exercise Set 7.11

1. Find $(\mathbf{w})_B$ and $[\mathbf{w}]_B$ with respect to the basis $B = \{\mathbf{v}_1, \mathbf{v}_2\}$ for R^2.

  (a) $\mathbf{w} = (3, -7)$; $\mathbf{v}_1 = (1, 0)$, $\mathbf{v}_2 = (0, 1)$

  (b) $\mathbf{w} = (1, 1)$; $\mathbf{v}_1 = (2, -4)$, $\mathbf{v}_2 = (3, 8)$

2. Find $(\mathbf{w})_B$ and $[\mathbf{w}]_B$ with respect to the basis $B = \{\mathbf{v}_1, \mathbf{v}_2\}$ for R^2.

  (a) $\mathbf{w} = (-2, 5)$; $\mathbf{v}_1 = (1, 0)$, $\mathbf{v}_2 = (0, 1)$

  (b) $\mathbf{w} = (7, 2)$; $\mathbf{v}_1 = (3, -5)$, $\mathbf{v}_2 = (4, 7)$

3. Find $(\mathbf{w})_B$ and $[\mathbf{w}]_B$ with respect to the basis $B = \{\mathbf{v}_1, \mathbf{v}_2, \mathbf{v}_3\}$ for R^3 given that $\mathbf{w} = (2, -1, 3)$; $\mathbf{v}_1 = (1, 0, 0)$, $\mathbf{v}_2 = (2, 2, 0)$, $\mathbf{v}_3 = (3, 3, 3)$.

4. Find $(\mathbf{w})_B$ and $[\mathbf{w}]_B$ with respect to the basis $B = \{\mathbf{v}_1, \mathbf{v}_2, \mathbf{v}_3\}$ for R^3 given that $\mathbf{w} = (5, -12, 3)$; $\mathbf{v}_1 = (1, 2, 3)$, $\mathbf{v}_2 = (-4, 5, 6)$, $\mathbf{v}_3 = (7, -8, 9)$.

5. Let B be the basis for R^3 in Exercise 3. Find the vector $\mathbf{u}$ for which $(\mathbf{u})_B = (7, -2, 1)$.

6. Let B be the basis for R^3 in Exercise 4. Find the vector $\mathbf{u}$ for which $(\mathbf{u})_B = (8, -5, 4)$.

In Exercises 7–10, find the coordinate vector of $\mathbf{w}$ with respect to the orthonormal basis formed by the $\mathbf{v}$'s.

7. $\mathbf{w} = (3, 7)$; $\mathbf{v}_1 = \left(\frac{1}{\sqrt{2}}, -\frac{1}{\sqrt{2}}\right)$, $\mathbf{v}_2 = \left(\frac{1}{\sqrt{2}}, \frac{1}{\sqrt{2}}\right)$

8. $\mathbf{w} = (-1, 0, 2)$; $\mathbf{v}_1 = \left(\frac{2}{3}, -\frac{2}{3}, \frac{1}{3}\right)$, $\mathbf{v}_2 = \left(\frac{2}{3}, \frac{1}{3}, -\frac{2}{3}\right)$, $\mathbf{v}_3 = \left(\frac{1}{3}, \frac{2}{3}, \frac{2}{3}\right)$

9. $\mathbf{w} = (2, -3, -1)$; $\mathbf{v}_1 = \left(\frac{1}{\sqrt{2}}, -\frac{1}{\sqrt{2}}, 0\right)$, $\mathbf{v}_2 = \left(\frac{1}{\sqrt{3}}, \frac{1}{\sqrt{3}}, \frac{1}{\sqrt{3}}\right)$, $\mathbf{v}_3 = \left(\frac{1}{\sqrt{6}}, \frac{1}{\sqrt{6}}, -\frac{2}{\sqrt{6}}\right)$

10. $\mathbf{w} = (4, 3, 0, -2)$; $\mathbf{v}_1 = \left(\frac{2}{3}, \frac{2}{3}, -\frac{1}{3}, 0\right)$, $\mathbf{v}_2 = \left(\frac{1}{\sqrt{3}}, -\frac{1}{\sqrt{3}}, 0, \frac{1}{\sqrt{3}}\right)$, $\mathbf{v}_3 = \left(\frac{1}{\sqrt{6}}, -\frac{1}{\sqrt{6}}, 0, -\frac{2}{\sqrt{6}}\right)$, $\mathbf{v}_4 = \left(\frac{1}{3\sqrt{2}}, \frac{1}{3\sqrt{2}}, \frac{4}{3\sqrt{2}}, 0\right)$

11. Let $B = \{\mathbf{v}_1, \mathbf{v}_2\}$ be the orthonormal basis for R^2 for which $\mathbf{v}_1 = \left(\frac{3}{5}, -\frac{4}{5}\right)$, $\mathbf{v}_2 = \left(\frac{4}{5}, \frac{3}{5}\right)$.
  (a) Find the vectors $\mathbf{u}$ and $\mathbf{v}$ that have coordinate vectors $(\mathbf{u})_B = (1, 1)$ and $(\mathbf{v})_B = (-1, 4)$.
  (b) Compute $\|\mathbf{u}\|$, $\|\mathbf{v}\|$, and $\mathbf{u} \cdot \mathbf{v}$ by applying Theorem 7.11.2 to the coordinate vectors $(\mathbf{u})_B$ and $(\mathbf{v})_B$, and then check the results by performing the computations directly with $\mathbf{u}$ and $\mathbf{v}$.

12. Let $B = \{\mathbf{v}_1, \mathbf{v}_2, \mathbf{v}_3\}$ be the orthonormal basis for R^3 for which
$$\mathbf{v}_1 = \left(\tfrac{1}{3}, \tfrac{2}{3}, \tfrac{2}{3}\right), \quad \mathbf{v}_2 = \left(\tfrac{2}{3}, \tfrac{1}{3}, -\tfrac{2}{3}\right), \quad \mathbf{v}_3 = \left(\tfrac{2}{3}, -\tfrac{2}{3}, \tfrac{1}{3}\right)$$
  (a) Find the vectors $\mathbf{u}$ and $\mathbf{v}$ that have coordinate vectors $(\mathbf{u})_B = (-2, 1, 2)$ and $(\mathbf{v})_B = (3, 0 - 2)$.
  (b) Compute $\|\mathbf{u}\|$, $\|\mathbf{v}\|$, and $\mathbf{u} \cdot \mathbf{v}$ by applying Theorem 7.11.2 to the coordinate vectors $(\mathbf{u})_B$ and $(\mathbf{v})_B$, and then check the results by performing the computations directly with $\mathbf{u}$ and $\mathbf{v}$.

In Exercises 13 and 14, find $\|\mathbf{u}\|$, $\|\mathbf{v}\|$, $\|\mathbf{w}\|$, $\|\mathbf{v}+\mathbf{w}\|$, $\|\mathbf{v}-\mathbf{w}\|$, and $\mathbf{v} \cdot \mathbf{w}$ assuming that B is an orthonormal basis for R^4.

13. $(\mathbf{u})_B = (-1, 2, 1, 3)$, $(\mathbf{v})_B = (0, -3, 1, 5)$, $(\mathbf{w})_B = (-2, -4, 3, 1)$

14. $(\mathbf{u})_B = (0, 0, -1, -1)$, $(\mathbf{v})_B = (5, 5, -2, -2)$, $(\mathbf{w})_B = (3, 0, -3, 0)$

15. The accompanying figure shows a rectangular xy-coordinate system and an $x'y'$-coordinate system with skewed axes. Assuming 1-unit scales are used on all axes, find the $x'y'$-

coordinates of the points whose xy-coordinates are given.
  (a) $(1, 1)$ (b) $(1, 0)$
  (c) $(0, 1)$ (d) (a, b)

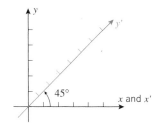

Figure Ex-15

16. The accompanying figure shows a rectangular xy-coordinate system determined by the unit basis vectors $\mathbf{i}$ and $\mathbf{j}$, and an $x'y'$-coordinate system determined by unit basis vectors $\mathbf{u}_1$ and $\mathbf{u}_2$. Find the $x'y'$-coordinates of the points whose xy-coordinates are given.
  (a) $(\sqrt{3}, 1)$ (b) $(1, 0)$
  (c) $(0, 1)$ (d) (a, b)

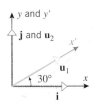

Figure Ex-16

17. Let S be the standard basis for R^2, and let $B = \{\mathbf{v}_1, \mathbf{v}_2\}$ be the basis in which $\mathbf{v}_1 = (2, 1)$ and $\mathbf{v}_2 = (-3, 4)$.
  (a) Find the transition matrix $P_{B \to S}$ by inspection.
  (b) Find the transition matrix $P_{S \to B}$ using row reduction (see Example 7).
  (c) Confirm that $P_{B \to S}$ and $P_{S \to B}$ are inverses of one another.
  (d) Find the coordinate matrix for $\mathbf{w} = (5, -3)$ with respect to B, and use the matrix $P_{B \to S}$ to compute $[\mathbf{w}]_S$ from $[\mathbf{w}]_B$.
  (e) Find the coordinate matrix for $\mathbf{w} = (3, -5)$ with respect to S, and use the matrix $P_{S \to B}$ to compute $[\mathbf{w}]_B$ from $[\mathbf{w}]_S$.

18. Let S be the standard basis for R^3, and let $B = \{\mathbf{v}_1, \mathbf{v}_2, \mathbf{v}_3\}$ be the basis in which $\mathbf{v}_1 = (1, 2, 1)$, $\mathbf{v}_2 = (2, 5, 0)$, and $\mathbf{v}_3 = (3, 3, 8)$.
  (a) Find the transition matrix $P_{B \to S}$ by inspection.
  (b) Find the transition matrix $P_{S \to B}$ using row reduction (see Example 7).
  (c) Confirm that $P_{B \to S}$ and $P_{S \to B}$ are inverses of one another.
  (d) Find the coordinate matrix for $\mathbf{w} = (5, -3, 1)$ with respect to B, and use the matrix $P_{B \to S}$ to compute $[\mathbf{w}]_S$ from $[\mathbf{w}]_B$.

(e) Find the coordinate matrix for $\mathbf{w} = (3, -5, 0)$ with respect to S, and use the matrix $P_{S \to B}$ to compute $[\mathbf{w}]_B$ from $[\mathbf{w}]_S$.

19. Let $B_1 = \{\mathbf{u}_1, \mathbf{u}_2\}$ and $B_2 = \{\mathbf{v}_1, \mathbf{v}_2\}$ be the bases for R^2 in which $\mathbf{u}_1 = (2, 2)$, $\mathbf{u}_2 = (4, -1)$, $\mathbf{v}_1 = (1, 3)$, and $\mathbf{v}_2 = (-1, -1)$.

(a) Find the transition matrix $P_{B_2 \to B_1}$ by row reduction.

(b) Find the transition matrix $P_{B_1 \to B_2}$ by row reduction.

(c) Confirm that $P_{B_2 \to B_1}$ and $P_{B_1 \to B_2}$ are inverses of one another.

(d) Find the coordinate matrix for $\mathbf{w} = (5, -3)$ with respect to B_1, and use the matrix $P_{B_1 \to B_2}$ to compute $[\mathbf{w}]_{B_2}$ from $[\mathbf{w}]_{B_1}$.

(e) Find the coordinate matrix for $\mathbf{w} = (3, -5)$ with respect to B_2, and use the matrix $P_{B_2 \to B_1}$ to compute $[\mathbf{w}]_{B_1}$ from $[\mathbf{w}]_{B_2}$.

20. Let $B_1 = \{\mathbf{u}_1, \mathbf{u}_2\}$ and $B_2 = \{\mathbf{v}_1, \mathbf{v}_2\}$ be the bases for R^2 in which $\mathbf{u}_1 = (1, 2)$, $\mathbf{u}_2 = (2, 3)$, $\mathbf{v}_1 = (1, 3)$, and $\mathbf{v}_2 = (1, 4)$.

(a) Find the transition matrix $P_{B_2 \to B_1}$ by row reduction.

(b) Find the transition matrix $P_{B_1 \to B_2}$ by row reduction.

(c) Confirm that $P_{B_2 \to B_1}$ and $P_{B_1 \to B_2}$ are inverses of one another.

(d) Find the coordinate matrix for $\mathbf{w} = (0, 1)$ with respect to B_1, and use the matrix $P_{B_1 \to B_2}$ to compute $[\mathbf{w}]_{B_2}$ from $[\mathbf{w}]_{B_1}$.

(e) Find the coordinate matrix for $\mathbf{w} = (2, 5)$ with respect to B_2, and use the matrix $P_{B_2 \to B_1}$ to compute $[\mathbf{w}]_{B_1}$ from $[\mathbf{w}]_{B_2}$.

21. Let $B_1 = \{\mathbf{u}_1, \mathbf{u}_2, \mathbf{u}_3\}$ and $B_2 = \{\mathbf{v}_1, \mathbf{v}_2, \mathbf{v}_3\}$ be the bases for R^3 in which $\mathbf{u}_1 = (-3, 0, -3)$, $\mathbf{u}_2 = (-3, 2, -1)$, $\mathbf{u}_3 = (1, 6, -1)$, $\mathbf{v}_1 = (-6, -6, 0)$, $\mathbf{v}_2 = (-2, -6, 4)$, and $\mathbf{v}_3 = (-2, -3, 7)$.

(a) Find the transition matrix $P_{B_1 \to B_2}$.

(b) Find the coordinate matrix with respect to B_1 of $\mathbf{w} = (-5, 8, -5)$, and then use the transition matrix obtained in part (a) to compute $[\mathbf{w}]_{B_2}$ by matrix multiplication.

(c) Check the result in part (b) by computing $[\mathbf{w}]_{B_2}$ directly.

22. Follow the directions of Exercise 21 with the same vector $\mathbf{w}$ but with $\mathbf{u}_1 = (2, 1, 1)$, $\mathbf{u}_2 = (2, -1, 1)$, $\mathbf{u}_3 = (1, 2, 1)$, $\mathbf{v}_1 = (3, 1, -5)$, $\mathbf{v}_2 = (1, 1, -3)$, and $\mathbf{v}_3 = (-1, 0, 2)$.

23. Let $B_1 = \{\mathbf{u}_1, \mathbf{u}_2, \mathbf{u}_3\}$ and $B_2 = \{\mathbf{v}_1, \mathbf{v}_2, \mathbf{v}_3\}$ be the orthonormal bases for R^3 in which $\mathbf{u}_1 = \left(\frac{1}{\sqrt{3}}, \frac{1}{\sqrt{3}}, \frac{1}{\sqrt{3}}\right)$, $\mathbf{u}_2 = \left(\frac{1}{\sqrt{2}}, -\frac{1}{\sqrt{2}}, 0\right)$, $\mathbf{u}_3 = \left(\frac{1}{\sqrt{6}}, \frac{1}{\sqrt{6}}, -\frac{2}{\sqrt{6}}\right)$, $\mathbf{v}_1 = \left(\frac{1}{\sqrt{2}}, 0, \frac{1}{\sqrt{2}}\right)$, $\mathbf{v}_2 = \left(\frac{1}{\sqrt{3}}, \frac{1}{\sqrt{3}}, -\frac{1}{\sqrt{3}}\right)$, and $\mathbf{v}_3 = \left(\frac{1}{\sqrt{6}}, -\frac{2}{\sqrt{6}}, -\frac{1}{\sqrt{6}}\right)$. Confirm that $P_{B_1 \to B_2}$ and $P_{B_2 \to B_1}$ are orthogonal matrices, as guaranteed by Theorem 7.11.7.

24. Let B_1 be the basis in Exercise 23. Confirm that the coordinate map $(\mathbf{x}) \to (\mathbf{x})_{B_1}$ is an orthogonal operator (as guaran-

teed by Theorem 7.11.5) by showing that its standard matrix is orthogonal.

25. Let $S = \{\mathbf{e}_1, \mathbf{e}_2\}$ be the standard basis for R^2, and let $B = \{\mathbf{v}_1, \mathbf{v}_2\}$ be the basis that results when the vectors in S are reflected about the line $y = x$.

(a) Find the transition matrix $P_{B \to S}$.

(b) Let $P = P_{B \to S}$ and show that $P^T = P_{S \to B}$. Give a geometric explanation of this.

26. Let $S = \{\mathbf{e}_1, \mathbf{e}_2\}$ be the standard basis for R^2, and let $B = \{\mathbf{v}_1, \mathbf{v}_2\}$ be the basis that results when the vectors in S are reflected about the line that makes an angle θ with the positive x-axis.

(a) Find the transition matrix $P_{B \to S}$.

(b) Let $P = P_{B \to S}$ and show that $P^T = P_{S \to B}$. Give a geometric explanation of this.

27. Suppose that a rectangular $x'y'$-coordinate system is obtained by rotating a rectangular xy-coordinate system about the origin through the angle $\theta = 3\pi/4$.

(a) Find the rotation equations that express $x'y'$-coordinates in terms of xy-coordinates, and use those equations to find the $x'y'$-coordinates of the point whose xy-coordinates are $(-2, 6)$.

(b) Find the rotation equations that express xy-coordinates in terms of $x'y'$-coordinates, and use those equations to find the xy-coordinates of the point whose $x'y'$-coordinates are $(5, 2)$.

28. Follow the directions of Exercise 27 with $\theta = \pi/3$.

29. Suppose that a rectangular $x'y'z'$-coordinate system is obtained by rotating a rectangular xyz-coordinate system counterclockwise about the positive z-axis through the angle $\theta = \pi/4$ (looking toward the origin along the positive z-axis).

(a) Find the $x'y'z'$-coordinates of the point whose xyz-coordinates are $(-1, 2, 5)$.

(b) Find the xyz-coordinates of the point whose $x'y'z'$-coordinates are $(1, 6, -3)$.

30. Follow the directions of Exercise 29 for a counterclockwise rotation about the positive y-axis through the angle $\theta = \pi/3$ (looking toward the origin along the positive y-axis).

31. Suppose that a rectangular $x''y''z''$-coordinate system is obtained by first rotating a rectangular xyz-coordinate system counterclockwise $60°$ about the positive z-axis (looking toward the origin along the positive z-axis) to obtain an $x'y'z'$-coordinate system, and then rotating the $x'y'z'$-coordinate system $45°$ counterclockwise about the positive y'-axis (looking toward the origin along the positive y'-axis). Find a matrix A that relates the xyz-coordinates and the $x''y''z''$-coordinates of a fixed point by

$$\begin{bmatrix} x'' \\ y'' \\ z'' \end{bmatrix} = A \begin{bmatrix} x \\ y \\ z \end{bmatrix}$$

Discussion and Discovery

D1. If B_1, B_2, and B_3 are bases for R^2, and if

$$P_{B_1 \to B_2} = \begin{bmatrix} 3 & 1 \\ 5 & 2 \end{bmatrix} \quad \text{and} \quad P_{B_2 \to B_3} = \begin{bmatrix} 7 & 2 \\ 4 & -1 \end{bmatrix}$$

then $P_{B_3 \to B_1} = \underline{\hspace{2cm}}$.

D2. Consider the matrix

$$P = \begin{bmatrix} 1 & 1 & 0 \\ 1 & 0 & 2 \\ 0 & 2 & 1 \end{bmatrix}$$

(a) P is the transition matrix from what basis B to the standard basis $S = \{\mathbf{e}_1, \mathbf{e}_2, \mathbf{e}_3\}$ for R^3?

(b) P is the transition matrix from the standard basis $S = \{\mathbf{e}_1, \mathbf{e}_2, \mathbf{e}_3\}$ to what basis B for R^3?

D3. The matrix

$$P = \begin{bmatrix} 1 & 0 & 0 \\ 0 & 3 & 2 \\ 0 & 1 & 1 \end{bmatrix}$$

is the transition matrix from what basis B to the basis $\{(1, 1, 1), (1, 1, 0), (1, 0, 0)\}$ for R^3?

D4. If $[\mathbf{w}]_B = \mathbf{w}$ holds for all vectors $\mathbf{w}$ in R^n, what can you say about the basis B?

D5. If $[\mathbf{x} - \mathbf{y}]_B = \mathbf{0}$, what can you say about $\mathbf{x}$ and $\mathbf{y}$?

Working with Proofs

P1. Let B be a basis for R^n. Prove that the vectors $\mathbf{v}_1, \mathbf{v}_2, \ldots, \mathbf{v}_k$ form a linearly independent set in R^n if and only if the vectors $(\mathbf{v}_1)_B, (\mathbf{v}_2)_B, \ldots, (\mathbf{v}_k)_B$ form a linearly independent set in R^n.

P2. Let B be a basis for R^n. Prove that the vectors $\mathbf{v}_1, \mathbf{v}_2, \ldots, \mathbf{v}_k$ span R^n if and only if the vectors $(\mathbf{v}_1)_B, (\mathbf{v}_2)_B, \ldots, (\mathbf{v}_k)_B$ span R^n.

P3. Use Theorem 3.4.4 and the fact that coordinate maps are onto to prove Theorem 7.11.6.

P4. Show that Formulas (25) and (27) hold for any scalar c and for any vectors $\mathbf{v}$ and $\mathbf{w}$ in R^n.

Technology Exercises

T1. (a) Confirm that $B_1 = \{\mathbf{u}_1, \mathbf{u}_2, \mathbf{u}_3, \mathbf{u}_4, \mathbf{u}_5\}$ and $B_2 = \{\mathbf{v}_1, \mathbf{v}_2, \mathbf{v}_3, \mathbf{v}_4, \mathbf{v}_5\}$ are bases for R^5, and find the transition matrices $P_{B_1 \to B_2}$ and $P_{B_2 \to B_1}$.

$$\mathbf{u}_1 = (3, 1, 3, 2, 6) \qquad \mathbf{v}_1 = (2, 6, 3, 4, 2)$$
$$\mathbf{u}_2 = (4, 5, 7, 2, 4) \qquad \mathbf{v}_2 = (3, 1, 5, 8, 3)$$
$$\mathbf{u}_3 = (3, 2, 1, 5, 4) \qquad \mathbf{v}_3 = (5, 1, 2, 6, 7)$$
$$\mathbf{u}_4 = (2, 9, 1, 4, 4) \qquad \mathbf{v}_4 = (8, 4, 3, 2, 6)$$
$$\mathbf{u}_5 = (3, 3, 6, 6, 7) \qquad \mathbf{v}_5 = (5, 5, 6, 3, 4)$$

(b) Find the coordinate matrices with respect to B_1 and B_2 of $\mathbf{w} = (1, 1, 1, 1, 1)$.

T2. An important problem in many applications is to perform a succession of rotations to align the positive axes of a right-handed xyz-coordinate system with the corresponding axes of a right-handed XYZ-coordinate system that has the same origin (see the accompanying figure). This can be accomplished by three successive rotations, which, if needed, can be composed into a single rotation about an appropriate axis. The three rotations involve angles θ, ϕ, and ψ, called **Euler angles**, and a vector $\mathbf{n}$, called the **axis of nodes**. As indicated in the figure, the axis of nodes is orthogonal to both the z-axis and Z-axis and hence is along the line of intersection of the xy- and XY-planes, θ is the angle from the positive z-axis to the positive Z-axis, ϕ is the angle from the positive x-axis to the axis of nodes, and ψ is the angle from the axis of nodes to the positive X-axis. The positive xyz-axes can be aligned with the positive XYZ-axes by first rotating the xyz-axes counterclockwise about the positive z-axis through the angle ϕ to carry the positive x-axis into the axis of nodes, then rotating the resulting axes counterclockwise about the axis of nodes through the angle θ to carry the positive z-axis into the positive Z-axis, and then rotating the resulting axes counterclockwise through the angle ψ about the positive Z-axis to carry the axis of nodes into the positive X-axis. Suppose that a rectangular xyz-coordinate system and a rectangular XYZ-coordinate system have the same origin and that $\theta = \pi/6$, $\phi = \pi/3$, and $\psi = \pi/4$ are Euler angles. Find a matrix A that relates the xyz-coordinates and XYZ-coordinates of a fixed point by

$$\begin{bmatrix} X \\ Y \\ Z \end{bmatrix} = A \begin{bmatrix} x \\ y \\ z \end{bmatrix}$$

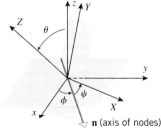

Figure Ex-T2

Transforming matrices to diagonal form is important mathematically as well as in such applications as vibration analysis, face and fingerprint recognition, statistics, and data compression.

Diagonalization

Section 8.1   Matrix Representations of Linear Transformations

Standard matrices for linear transformations provide a convenient way of using matrix operations to perform calculations with transformations. However, this is not the only role that matrices play in the study of linear transformations. In this section we will show how other kinds of matrices can be used to uncover geometric properties of a linear transformation that may not be evident from its standard matrix.

MATRIX OF A LINEAR OPERATOR WITH RESPECT TO A BASIS

We know that every linear transformation $T : R^n \to R^m$ has an associated standard matrix

$$[T] = [T(\mathbf{e}_1) \mid T(\mathbf{e}_2) \mid \cdots \mid T(\mathbf{e}_n)]$$

with the property that

$$T(\mathbf{x}) = [T]\mathbf{x}$$

for every vector $\mathbf{x}$ in R^n. For the moment we will focus on the case where T is a linear operator on R^n, so the standard matrix $[T]$ is a square matrix of size $n \times n$.

Sometimes the form of the standard matrix fully reveals the geometric properties of a linear operator and sometimes it does not. For example, we can tell by inspection of the matrix

$$[T_1] = \begin{bmatrix} \cos \frac{\pi}{4} & -\sin \frac{\pi}{4} & 0 \\ \sin \frac{\pi}{4} & \cos \frac{\pi}{4} & 0 \\ 0 & 0 & 1 \end{bmatrix} = \begin{bmatrix} \frac{1}{\sqrt{2}} & -\frac{1}{\sqrt{2}} & 0 \\ \frac{1}{\sqrt{2}} & \frac{1}{\sqrt{2}} & 0 \\ 0 & 0 & 1 \end{bmatrix} \tag{1}$$

that T_1 is a rotation through an angle of $\pi/4$ about the z-axis of an xyz-coordinate system. In contrast, a casual inspection of the matrix

$$[T_2] = \begin{bmatrix} 0 & 0 & 1 \\ 1 & 0 & 0 \\ 0 & 1 & 0 \end{bmatrix} \tag{2}$$

provides only partial geometric information about the operator T_2; we can tell that T_2 is a rotation since the matrix $[T_2]$ is orthogonal and has determinant 1, but, unlike (1), this matrix does not explicitly reveal the axis and angle of rotation.

The difference between (1) and (2) has to do with the orientation of the standard basis. In the case of the operator T_1, the standard basis vector $\mathbf{e}_3$ aligns with the axis of rotation, and the basis vectors $\mathbf{e}_1$ and $\mathbf{e}_2$ rotate in a plane perpendicular to $\mathbf{e}_3$, thereby making the axis and angle of rotation recognizable from the standard matrix. However, in the case of the operator T_2, none of the standard basis vectors aligns with the axis of rotation (see Example 7 and Figure 6.2.9 of Section 6.2), so the operator T_2 does not transform $\mathbf{e}_1$, $\mathbf{e}_2$, and $\mathbf{e}_3$ in a way that provides useful geometric information. Thus, although the standard basis is simple algebraically, it is not always the best basis from a geometric point of view.

Our primary goal in this section is to develop a way of using bases other than the standard basis to create matrices that describe the geometric behavior of a linear transformation better than the standard matrix. The key to doing this is to work with *coordinates* of vectors rather than with the vectors themselves, as we will now explain.

Suppose that

$$\mathbf{x} \xrightarrow{\;T\;} T(\mathbf{x})$$

is a linear operator on R^n and B is a basis for R^n. In the course of mapping $\mathbf{x}$ into $T(\mathbf{x})$ this operator creates a companion operator

$$[\mathbf{x}]_B \longrightarrow [T(\mathbf{x})]_B \tag{3}$$

that maps the coordinate matrix $[\mathbf{x}]_B$ into the coordinate matrix $[T(\mathbf{x})]_B$. In the exercises we will ask you to show that (3) is linear and hence must be a matrix transformation; that is, there must be a matrix A such that

$$A[\mathbf{x}]_B = [T(\mathbf{x})]_B$$

The following theorem shows how to find the matrix A.

Theorem 8.1.1   *Let $T : R^n \to R^n$ be a linear operator, let $B = \{\mathbf{v}_1, \mathbf{v}_2, \ldots, \mathbf{v}_n\}$ be a basis for R^n, and let*

$$A = \left[[T(\mathbf{v}_1)]_B \mid [T(\mathbf{v}_2)]_B \mid \cdots \mid [T(\mathbf{v}_n)]_B \right] \tag{4}$$

Then

$$[T(\mathbf{x})]_B = A[\mathbf{x}]_B \tag{5}$$

for every vector $\mathbf{x}$ in R^n. Moreover, the matrix A given by Formula (4) is the only matrix with property (5).

Proof Let $\mathbf{x}$ be any vector in R^n, and suppose that its coordinate matrix with respect to B is

$$[\mathbf{x}]_B = \begin{bmatrix} c_1 \\ c_2 \\ \vdots \\ c_n \end{bmatrix}$$

That is,

$$\mathbf{x} = c_1 \mathbf{v}_1 + c_2 \mathbf{v}_2 + \cdots + c_n \mathbf{v}_n$$

It now follows from the linearity of T that

$$T(\mathbf{x}) = c_1 T(\mathbf{v}_1) + c_2 T(\mathbf{v}_2) + \cdots + c_n T(\mathbf{v}_n)$$

and from the linearity of coordinate maps that

$$[T(\mathbf{x})]_B = c_1 [T(\mathbf{v}_1)]_B + c_2 [T(\mathbf{v}_2)]_B + \cdots + c_n [T(\mathbf{v}_n)]_B$$

which we can write in matrix form as

$$[T(\mathbf{x})]_B = \left[[T(\mathbf{v}_1)]_B \mid [T(\mathbf{v}_2)]_B \mid \cdots \mid [T(\mathbf{v}_n)]_B \right] \begin{bmatrix} c_1 \\ c_2 \\ \vdots \\ c_n \end{bmatrix} = A[\mathbf{x}]_B$$

This proves that the matrix A in (4) has property (5). Moreover, A is the only matrix with this property, for if there exists a matrix C such that

$$[T(\mathbf{x})]_B = A[\mathbf{x}]_B = C[\mathbf{x}]_B$$

for all $\mathbf{x}$ in R^n, then Theorem 7.11.6 implies that $A = C$. ■

The matrix A in (4) is called the ***matrix for T with respect to the basis B*** and is denoted by

$$[T]_B = \left[[T(\mathbf{v}_1)]_B \mid [T(\mathbf{v}_2)]_B \mid \cdots \mid [T(\mathbf{v}_n)]_B\right] \tag{6}$$

Using this notation we can write (5) as

$$[T(\mathbf{x})]_B = [T]_B[\mathbf{x}]_B \tag{7}$$

Recalling that the components of a vector in R^n are the same as its coordinates with respect to the standard basis S, it follows from (6) that

$$[T]_S = \left[[T(\mathbf{v}_1)]_S \mid [T(\mathbf{v}_2)]_S \mid \cdots \mid [T(\mathbf{v}_n)]_S\right] = [T(\mathbf{v}_1) \mid T(\mathbf{v}_2) \mid \cdots \mid T(\mathbf{v}_n)] = [T]$$

That is, *the matrix for a linear operator on R^n with respect to the standard basis is the same as the standard matrix for T.*

EXAMPLE 1
Matrix of a
Linear Operator
with Respect to
a Basis B

Let $T : R^2 \to R^2$ be the linear operator whose standard matrix is

$$[T] = \begin{bmatrix} 3 & 2 \\ 2 & 3 \end{bmatrix} \tag{8}$$

Find the matrix for T with respect to the basis $B = \{\mathbf{v}_1, \mathbf{v}_2\}$, where

$$\mathbf{v}_1 = \begin{bmatrix} \frac{1}{\sqrt{2}} \\ -\frac{1}{\sqrt{2}} \end{bmatrix} \quad \text{and} \quad \mathbf{v}_2 = \begin{bmatrix} \frac{1}{\sqrt{2}} \\ \frac{1}{\sqrt{2}} \end{bmatrix}$$

Solution The images of the basis vectors under the operator T are

$$T(\mathbf{v}_1) = [T]\mathbf{v}_1 = \begin{bmatrix} 3 & 2 \\ 2 & 3 \end{bmatrix} \begin{bmatrix} \frac{1}{\sqrt{2}} \\ -\frac{1}{\sqrt{2}} \end{bmatrix} = \begin{bmatrix} \frac{1}{\sqrt{2}} \\ -\frac{1}{\sqrt{2}} \end{bmatrix} = \mathbf{v}_1 = \mathbf{v}_1 + 0\mathbf{v}_2$$

$$T(\mathbf{v}_2) = [T]\mathbf{v}_2 = \begin{bmatrix} 3 & 2 \\ 2 & 3 \end{bmatrix} \begin{bmatrix} \frac{1}{\sqrt{2}} \\ \frac{1}{\sqrt{2}} \end{bmatrix} = \begin{bmatrix} \frac{5}{\sqrt{2}} \\ \frac{5}{\sqrt{2}} \end{bmatrix} = 5\mathbf{v}_2 = 0\mathbf{v}_1 + 5\mathbf{v}_2$$

so the coordinate matrices of these vectors with respect to B are

$$[T(\mathbf{v}_1)]_B - \begin{bmatrix} 1 \\ 0 \end{bmatrix} \quad \text{and} \quad [T(\mathbf{v}_2)]_B = \begin{bmatrix} 0 \\ 5 \end{bmatrix}$$

Thus, it follows from (6) that

$$[T]_B = \left[[T(\mathbf{v}_1)]_B \mid [T(\mathbf{v}_2)]_B\right] = \begin{bmatrix} 1 & 0 \\ 0 & 5 \end{bmatrix}$$

This matrix reveals geometric information about the operator T that was not evident from the standard matrix. It tells us that the effect of T is to stretch the $\mathbf{v}_2$-coordinate of a vector by a factor of 5 and to leave the $\mathbf{v}_1$-coordinate unchanged. For example, Figure 8.1.1 shows the stretching effect that this operator has on a square of side 1 that is centered at the origin and whose sides align with $\mathbf{v}_1$ and $\mathbf{v}_2$. ■

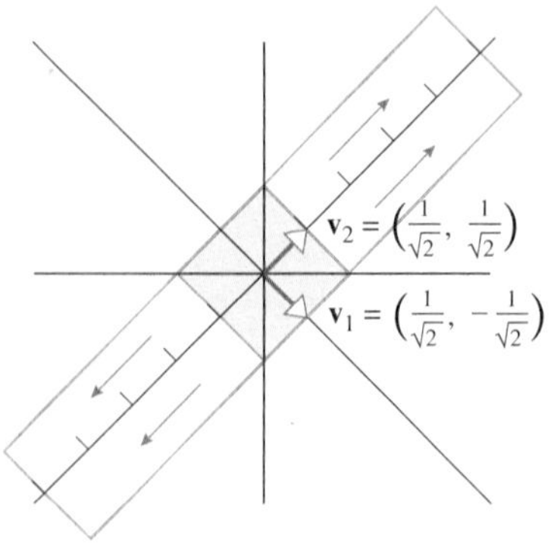

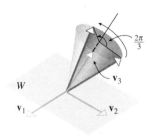

Figure 8.1.1

EXAMPLE 2
Uncovering
Hidden
Geometry

Let $T : R^3 \to R^3$ be the linear operator whose standard matrix is

$$A = \begin{bmatrix} 0 & 0 & 1 \\ 1 & 0 & 0 \\ 0 & 1 & 0 \end{bmatrix} \tag{9}$$

We showed in Example 7 of Section 6.2 that T is a rotation through an angle of $2\pi/3$ about an axis in the direction of the vector $\mathbf{n} = (1, 1, 1)$. Let us now consider how the matrix for T would look with respect to an orthonormal basis $B = \{\mathbf{v}_1, \mathbf{v}_2, \mathbf{v}_3\}$ in which $\mathbf{v}_3 = \mathbf{v}_1 \times \mathbf{v}_2$ is a positive scalar multiple of $\mathbf{n}$ and $\{\mathbf{v}_1, \mathbf{v}_2\}$ is an orthonormal basis for the plane W through the origin that is perpendicular to the axis of rotation (Figure 8.1.2). The rotation leaves the vector $\mathbf{v}_3$ fixed, so

$$T(\mathbf{v}_3) = \mathbf{v}_3 = 0\mathbf{v}_1 + 0\mathbf{v}_2 + 1\mathbf{v}_3$$

and hence

$$[T(\mathbf{v}_3)]_B = \begin{bmatrix} 0 \\ 0 \\ 1 \end{bmatrix}$$

Figure 8.1.2

Also, $T(\mathbf{v}_1)$ and $T(\mathbf{v}_2)$ are linear combinations of $\mathbf{v}_1$ and $\mathbf{v}_2$, since these vectors lie in W. This implies that the third coordinate of both $[T(\mathbf{v}_1)]_B$ and $[T(\mathbf{v}_2)]_B$ must be zero, and the matrix for T with respect to the basis B must be of the form

$$[T]_B = \big[[T(\mathbf{v}_1)]_B \mid [T(\mathbf{v}_2)]_B \mid [T(\mathbf{v}_3)]_B\big] = \begin{bmatrix} \times & \times & 0 \\ \times & \times & 0 \\ 0 & 0 & 1 \end{bmatrix}$$

Since T behaves exactly like a rotation of R^2 in the plane W, the block of missing entries has the form of a rotation matrix in R^2. Thus,

$$[T]_B = \big[[T(\mathbf{v}_1)]_B \mid [T(\mathbf{v}_2)]_B \mid [T(\mathbf{v}_3)]_B\big] = \begin{bmatrix} \cos\frac{2\pi}{3} & -\sin\frac{2\pi}{3} & 0 \\ \sin\frac{2\pi}{3} & \cos\frac{2\pi}{3} & 0 \\ 0 & 0 & 1 \end{bmatrix}$$

This matrix makes it clear that the angle of rotation is $2\pi/3$ and the axis of rotation is in the direction of $\mathbf{v}_3$, facts that are not directly evident from the standard matrix in (9). ∎

CHANGING BASES

It is reasonable to conjecture that two matrices representing the same linear operator with respect to different bases must be related algebraically. To uncover that relationship, suppose that $T : R^n \to R^n$ is a linear operator and that $B = \{\mathbf{v}_1, \mathbf{v}_2, \ldots, \mathbf{v}_n\}$ and $B' = \{\mathbf{v}'_1, \mathbf{v}'_2, \ldots, \mathbf{v}'_n\}$ are bases for R^n. Also, let $P = P_{B \to B'}$ be the transition matrix from B to B' (so $P^{-1} = P_{B' \to B}$ is the transition matrix from B' to B). To find the relationship between $[T]_B$ and $[T]_{B'}$, consider

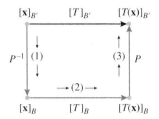

Figure 8.1.3

the diagram in Figure 8.1.3, which links together the following four relationships schematically:

$$[T]_B[\mathbf{x}]_B = [T(\mathbf{x})]_B, \qquad\qquad [T]_{B'}[\mathbf{x}]_{B'} = [T(\mathbf{x})]_{B'}$$

$$P[T(\mathbf{x})]_B = P_{B \to B'}[T(\mathbf{x})]_B = [T(\mathbf{x})]_{B'}, \qquad P[\mathbf{x}]_B = P_{B \to B'}[\mathbf{x}]_B = [\mathbf{x}]_{B'}$$

The diagram shows two different paths from $[\mathbf{x}]_{B'}$ to $[T(\mathbf{x})]_{B'}$, each of which corresponds to a different relationship between these vectors:

1. The direct path from $[\mathbf{x}]_{B'}$ to $[T(\mathbf{x})]_{B'}$ across the top of the diagram corresponds to the relationship

$$[T]_{B'}[\mathbf{x}]_{B'} = [T(\mathbf{x})]_{B'} \qquad\qquad (10)$$

2. The path from $[\mathbf{x}]_{B'}$ to $[T(\mathbf{x})]_{B'}$ that goes down the left side, across the bottom, and up the right side corresponds to computing $[T(\mathbf{x})]_{B'}$ from $[\mathbf{x}]_{B'}$ by three successive matrix multiplications:

   (i) Multiply $[\mathbf{x}]_{B'}$ on the left by P^{-1} to obtain $P^{-1}[\mathbf{x}]_{B'} = [\mathbf{x}]_B$.
   (ii) Multiply $[\mathbf{x}]_B$ on the left by $[T]_B$ to obtain $[T]_B[\mathbf{x}]_B = [T(\mathbf{x})]_B$.
   (iii) Multiply $[T(\mathbf{x})]_B$ on the left by P to obtain $[T(\mathbf{x})]_{B'}$.

This process produces the relationship

$$(P[T]_B P^{-1})[\mathbf{x}]_{B'} = [T(\mathbf{x})]_{B'} \qquad\qquad (11)$$

Thus, (10) and (11) together imply that

$$(P[T]_B P^{-1})[\mathbf{x}]_{B'} = [T]_{B'}[\mathbf{x}]_{B'}$$

Since this holds for all $\mathbf{x}$ in R^n, it follows from Theorem 7.11.6 that

$$P[T]_B P^{-1} = [T]_{B'}$$

Thus, we have established the following theorem that provides the relationship between the matrices for a fixed linear operator with respect to different bases.

> **Theorem 8.1.2** *If $T : R^n \to R^n$ is a linear operator, and if $B = \{\mathbf{v}_1, \mathbf{v}_2, \ldots, \mathbf{v}_n\}$ and $B' = \{\mathbf{v}'_1, \mathbf{v}'_2, \ldots, \mathbf{v}'_n\}$ are bases for R^n, then $[T]_B$ and $[T]_{B'}$ are related by the equation*
>
> $$[T]_{B'} = P[T]_B P^{-1} \qquad\qquad (12)$$
>
> *in which*
>
> $$P = P_{B \to B'} = \big[[\mathbf{v}_1]_{B'} \mid [\mathbf{v}_2]_{B'} \mid \cdots \mid [\mathbf{v}_n]_{B'} \big] \qquad\qquad (13)$$
>
> *is the transition matrix from B to B'. In the special case where B and B' are orthonormal bases the matrix P is orthogonal, so (12) is of the form*
>
> $$[T]_{B'} = P[T]_B P^T \qquad\qquad (14)$$

When convenient, Formula (12) can be rewritten as

$$[T]_B = P^{-1}[T]_{B'} P \qquad\qquad (15)$$

and in the case where the bases are orthonormal this equation can be expressed as

$$[T]_B = P^T[T]_{B'} P \qquad\qquad (16)$$

REMARK When applying all of these formulas it is easy to lose track of whether P is the transition matrix from B to B', or vice versa, particularly if other notations are used for the bases.

A good way to keep everything straight is to draw Figure 8.1.3 with appropriate adjustments in notation. When creating the diagram you can choose either direction for the transition matrix P as long as you adhere to that direction when constructing the associated formula.

Since many linear operators are defined by their standard matrices, it is important to consider the special case of Theorem 8.1.2 in which $B' = S$ is the standard basis for R^n. In this case $[T]_{B'} = [T]_S = [T]$, and the transition matrix P from B to B' has the simplified form

$$P = P_{B \rightarrow B'} = P_{B \rightarrow S} = \big[[\mathbf{v}_1]_S \mid [\mathbf{v}_2]_S \mid \cdots \mid [\mathbf{v}_n]_S \big] = [\mathbf{v}_1 \mid \mathbf{v}_2 \mid \cdots \mid \mathbf{v}_n]$$

Thus, we have the following result.

Theorem 8.1.3   *If $T : R^n \rightarrow R^n$ is a linear operator, and if $B = \{\mathbf{v}_1, \mathbf{v}_2, \ldots, \mathbf{v}_n\}$ is a basis for R^n, then $[T]$ and $[T]_B$ are related by the equation*

$$[T] = P[T]_B P^{-1} \tag{17}$$

in which

$$P = [\mathbf{v}_1 \mid \mathbf{v}_2 \mid \cdots \mid \mathbf{v}_n] \tag{18}$$

is the transition matrix from B to the standard basis. In the special case where B is an orthonormal basis the matrix P is orthogonal, so (17) is of the form

$$[T] = P[T]_B P^T \tag{19}$$

When convenient, Formula (17) can be rewritten as

$$[T]_B = P^{-1}[T]P \tag{20}$$

and in the case where B is an orthonormal basis this equation can be expressed as

$$[T]_B = P^T[T]P \tag{21}$$

Formula (17) [or (19) in the orthogonal case] tells us that the process of changing from the standard basis for R^n to a basis B produces a factorization of the standard matrix for T as

$$[T] = P[T]_B P^{-1} \tag{22}$$

in which P is the transition matrix from the basis B to the standard basis S. To understand the geometric significance of this factorization, let us use it to compute $T(\mathbf{x})$ by writing

$$T(\mathbf{x}) = [T]\mathbf{x} = (P[T]_B P^{-1})\mathbf{x} = P[T]_B (P^{-1}\mathbf{x})$$

Reading from right to left, this equation tells us that $T(\mathbf{x})$ can be obtained by first mapping the standard coordinates of $\mathbf{x}$ to B-coordinates using the matrix P^{-1}, then performing the operation on the B-coordinates using the matrix $[T]_B$, and then using the matrix P to map the resulting vector back to standard coordinates.

EXAMPLE 3
Example 1
Revisited from
the Viewpoint
of Factorization

In Example 1 we considered the linear operator $T : R^2 \rightarrow R^2$ whose standard matrix is

$$A = [T] = \begin{bmatrix} 3 & 2 \\ 2 & 3 \end{bmatrix}$$

and we showed that

$$[T]_B = \begin{bmatrix} 1 & 0 \\ 0 & 5 \end{bmatrix}$$

with respect to the orthonormal basis $B = \{\mathbf{v}_1, \mathbf{v}_2\}$ that is formed from the vectors

$$\mathbf{v}_1 = \begin{bmatrix} \frac{1}{\sqrt{2}} \\ -\frac{1}{\sqrt{2}} \end{bmatrix} \quad \text{and} \quad \mathbf{v}_2 = \begin{bmatrix} \frac{1}{\sqrt{2}} \\ \frac{1}{\sqrt{2}} \end{bmatrix}$$

In this case the transition matrix from B to S is

$$P = [\mathbf{v}_1 \mid \mathbf{v}_2] = \begin{bmatrix} \frac{1}{\sqrt{2}} & \frac{1}{\sqrt{2}} \\ -\frac{1}{\sqrt{2}} & \frac{1}{\sqrt{2}} \end{bmatrix}$$

so it follows from (17) that $[T]$ can be factored as

$$\underbrace{\begin{bmatrix} 3 & 2 \\ 2 & 3 \end{bmatrix}}_{[T]} = \underbrace{\begin{bmatrix} \frac{1}{\sqrt{2}} & \frac{1}{\sqrt{2}} \\ -\frac{1}{\sqrt{2}} & \frac{1}{\sqrt{2}} \end{bmatrix}}_{P} \underbrace{\begin{bmatrix} 1 & 0 \\ 0 & 5 \end{bmatrix}}_{[T]_B} \underbrace{\begin{bmatrix} \frac{1}{\sqrt{2}} & -\frac{1}{\sqrt{2}} \\ \frac{1}{\sqrt{2}} & \frac{1}{\sqrt{2}} \end{bmatrix}}_{P^{-1}}$$

Reading from right to left, this equation tells us that $T(\mathbf{x})$ can be computed by first transforming standard coordinates to B-coordinates, then stretching the $\mathbf{v}_2$-coordinate by a factor of 5 while leaving the $\mathbf{v}_1$-coordinate fixed, and then transforming B-coordinates back to standard coordinates. ∎

EXAMPLE 4
Example 2
Revisited from
the Viewpoint
of Factorization

In Example 2 we considered the rotation $T : R^3 \to R^3$ whose standard matrix is

$$A = [T] = \begin{bmatrix} 0 & 0 & 1 \\ 1 & 0 & 0 \\ 0 & 1 & 0 \end{bmatrix}$$

and we showed that

$$[T]_B = \begin{bmatrix} \cos\frac{2\pi}{3} & -\sin\frac{2\pi}{3} & 0 \\ \sin\frac{2\pi}{3} & \cos\frac{2\pi}{3} & 0 \\ 0 & 0 & 1 \end{bmatrix}$$

with respect to any orthonormal basis $B = \{\mathbf{v}_1, \mathbf{v}_2, \mathbf{v}_3\}$ in which $\mathbf{v}_3 = \mathbf{v}_1 \times \mathbf{v}_2$ is a positive multiple of the vector $\mathbf{n} = (1, 1, 1)$ along the axis of rotation and $\{\mathbf{v}_1, \mathbf{v}_2\}$ is an orthonormal basis for the plane W that passes through the origin and is perpendicular to the axis of rotation. To find a specific basis of this form, recall from Example 7 of Section 6.2 that the equation of the plane W is

$$x + y + z = 0$$

and recall from Example 10 of Section 7.9 that the vectors

$$\mathbf{v}_1 = \begin{bmatrix} -\frac{1}{\sqrt{2}} \\ \frac{1}{\sqrt{2}} \\ 0 \end{bmatrix} \quad \text{and} \quad \mathbf{v}_2 = \begin{bmatrix} -\frac{1}{\sqrt{6}} \\ -\frac{1}{\sqrt{6}} \\ \frac{2}{\sqrt{6}} \end{bmatrix}$$

form an orthonormal basis for W. Since

$$\mathbf{v}_3 = \mathbf{v}_1 \times \mathbf{v}_2 = \begin{vmatrix} \mathbf{i} & \mathbf{j} & \mathbf{k} \\ -\frac{1}{\sqrt{2}} & \frac{1}{\sqrt{2}} & 0 \\ -\frac{1}{\sqrt{6}} & -\frac{1}{\sqrt{6}} & \frac{2}{\sqrt{6}} \end{vmatrix} = \frac{1}{\sqrt{3}}\mathbf{i} + \frac{1}{\sqrt{3}}\mathbf{j} + \frac{1}{\sqrt{3}}\mathbf{k}$$

the transition matrix from $B = \{\mathbf{v}_1, \mathbf{v}_2, \mathbf{v}_3\}$ to the standard basis is

$$P = [\mathbf{v}_1 \mid \mathbf{v}_2 \mid \mathbf{v}_3] = \begin{bmatrix} -\frac{1}{\sqrt{2}} & -\frac{1}{\sqrt{6}} & \frac{1}{\sqrt{3}} \\ \frac{1}{\sqrt{2}} & -\frac{1}{\sqrt{6}} & \frac{1}{\sqrt{3}} \\ 0 & \frac{2}{\sqrt{6}} & \frac{1}{\sqrt{3}} \end{bmatrix}$$

Since this matrix is orthogonal, it follows from (19) that $[T]$ can be factored as

$$\begin{bmatrix} 0 & 0 & 1 \\ 1 & 0 & 0 \\ 0 & 1 & 0 \end{bmatrix} = \begin{bmatrix} -\frac{1}{\sqrt{2}} & -\frac{1}{\sqrt{6}} & \frac{1}{\sqrt{3}} \\ \frac{1}{\sqrt{2}} & -\frac{1}{\sqrt{6}} & \frac{1}{\sqrt{3}} \\ 0 & \frac{2}{\sqrt{6}} & \frac{1}{\sqrt{3}} \end{bmatrix} \begin{bmatrix} \cos\frac{2\pi}{3} & -\sin\frac{2\pi}{3} & 0 \\ \sin\frac{2\pi}{3} & \cos\frac{2\pi}{3} & 0 \\ 0 & 0 & 1 \end{bmatrix} \begin{bmatrix} -\frac{1}{\sqrt{2}} & \frac{1}{\sqrt{2}} & 0 \\ -\frac{1}{\sqrt{6}} & -\frac{1}{\sqrt{6}} & \frac{2}{\sqrt{6}} \\ \frac{1}{\sqrt{3}} & \frac{1}{\sqrt{3}} & \frac{1}{\sqrt{3}} \end{bmatrix}$$

$$\quad [T] \qquad = \qquad\qquad P \qquad\qquad\qquad\qquad [T]_B \qquad\qquad\qquad\qquad P^T$$

Reading from right to left, this tells us that $T(\mathbf{x})$ can be computed by first transforming standard coordinates to B-coordinates, then rotating through an angle of $2\pi/3$ about an axis in the direction of $\mathbf{v}_3$, and then transforming B-coordinates back to standard coordinates. ∎

EXAMPLE 5
Factoring the
Standard Matrix
for a Reflection

Recall from Formula (2) of Section 6.2 that the standard matrix for the reflection T of R^2 about the line L through the origin making an angle θ with the positive x-axis of a rectangular xy-coordinate system is

$$[T] = H_\theta = \begin{bmatrix} \cos 2\theta & \sin 2\theta \\ \sin 2\theta & -\cos 2\theta \end{bmatrix}$$

The fact that this matrix represents a reflection is not immediately evident because the standard unit vectors along the positive x- and y-axes have no special relationship to the line L. Suppose, however, that we rotate the coordinate axes through the angle θ to align the new x'-axis with L, and we let $\mathbf{v}_1$ and $\mathbf{v}_2$ be unit vectors along the x'- and y'-axes, respectively (Figure 8.1.4). Since

$$T(\mathbf{v}_1) = \mathbf{v}_1 = \mathbf{v}_1 + 0\mathbf{v}_2 \quad \text{and} \quad T(\mathbf{v}_2) = -\mathbf{v}_2 = 0\mathbf{v}_1 + (-1)\mathbf{v}_2$$

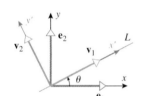

Figure 8.1.4

it follows that the matrix for T with respect to the basis $B = \{\mathbf{v}_1, \mathbf{v}_2\}$ is

$$[T]_B = \big[[T(\mathbf{v}_1)]_B \mid [T(\mathbf{v}_2)]_B\big] = \begin{bmatrix} 1 & 0 \\ 0 & -1 \end{bmatrix}$$

Also, it follows from Example 8 of Section 7.11 that the transition matrices between the standard basis S and the basis B are

$$P = P_{B \to S} = \begin{bmatrix} \cos\theta & -\sin\theta \\ \sin\theta & \cos\theta \end{bmatrix} \quad \text{and} \quad P^T = P_{S \to B} = \begin{bmatrix} \cos\theta & \sin\theta \\ -\sin\theta & \cos\theta \end{bmatrix}$$

Thus, Formula (19) implies that

$$\begin{bmatrix} \cos 2\theta & \sin 2\theta \\ \sin 2\theta & -\cos 2\theta \end{bmatrix} = \begin{bmatrix} \cos\theta & -\sin\theta \\ \sin\theta & \cos\theta \end{bmatrix} \begin{bmatrix} 1 & 0 \\ 0 & -1 \end{bmatrix} \begin{bmatrix} \cos\theta & \sin\theta \\ -\sin\theta & \cos\theta \end{bmatrix}$$

$$\quad [T] \qquad\qquad = \qquad\qquad P \qquad\qquad [T]_B \qquad\qquad P^T$$

Reading from right to left, this equation tells us that $T(\mathbf{x})$ can be computed by first rotating the xy-axes through the angle θ to convert standard coordinates to B-coordinates, then reflecting about the x'-axis, and then rotating through the angle $-\theta$ to convert back to standard coordinates. ∎

**MATRIX OF A LINEAR
TRANSFORMATION WITH
RESPECT TO A PAIR OF
BASES**

Up to now we have focused on matrix representations of linear operators. We will now consider the corresponding idea for linear transformations. Recall that every linear transformation $T: R^n \to R^m$ has an associated $m \times n$ standard matrix

$$[T] = [T(\mathbf{e}_1) \mid T(\mathbf{e}_2) \mid \cdots \mid T(\mathbf{e}_n)]$$

with the property that

$$T(\mathbf{x}) = [T]\mathbf{x}$$

If B and B' are bases for R^n and R^m, respectively, then the transformation

$$\mathbf{x} \xrightarrow{\;T\;} T(\mathbf{x})$$

creates an associated transformation

$$[\mathbf{x}]_B \to [T(\mathbf{x})]_{B'}$$

that maps the coordinate matrix $[\mathbf{x}]_B$ into the coordinate matrix $[T(\mathbf{x})]_{B'}$. As in the operator case, this associated transformation is linear and hence must be a matrix transformation; that is, there must be a matrix A such that

$$A[\mathbf{x}]_B = [T(\mathbf{x})]_{B'}$$

The following generalization of Theorem 8.1.1 shows how to find A. The proof is similar to that of Theorem 8.1.1 and will be omitted.

Theorem 8.1.4 *Let $T : R^n \to R^m$ be a linear transformation, let $B = \{\mathbf{v}_1, \mathbf{v}_2, \dots, \mathbf{v}_n\}$ and $B' = \{\mathbf{u}_1, \mathbf{u}_2, \dots, \mathbf{u}_m\}$ be bases for R^n and R^m, respectively, and let*

$$A = \left[[T(\mathbf{v}_1)]_{B'} \mid [T(\mathbf{v}_2)]_{B'} \mid \cdots \mid [T(\mathbf{v}_n)]_{B'} \right] \tag{23}$$

Then

$$[T(\mathbf{x})]_{B'} = A[\mathbf{x}]_B \tag{24}$$

for every vector $\mathbf{x}$ in R^n. Moreover, the matrix A given by Formula (23) is the only matrix with property (24).

The matrix A in (23) is called the ***matrix for T with respect to the bases B and B'*** and is denoted by the symbol $[T]_{B',B}$. With this notation Formulas (23) and (24) can be expressed as

$$[T]_{B',B} = \left[[T(\mathbf{v}_1)]_{B'} \mid [T(\mathbf{v}_2)]_{B'} \mid \cdots \mid [T(\mathbf{v}_n)]_{B'} \right] \tag{25}$$

and

$$[T(\mathbf{x})]_{B'} = [T]_{B',B}[\mathbf{x}]_B \tag{26}$$

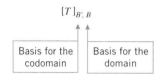

$[T]_{B',B}$

| Basis for the codomain | Basis for the domain |

Figure 8.1.5

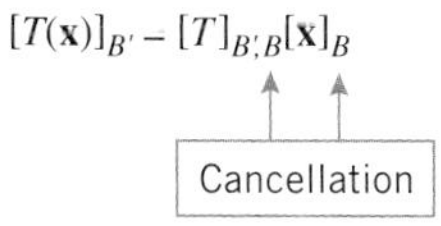

$[T(\mathbf{x})]_{B'} = [T]_{B',B}[\mathbf{x}]_B$

Cancellation

Figure 8.1.6

REMARK Observe the order of the subscripts in the notation $[T]_{B',B}$—the right subscript denotes the basis for the domain and the left subscript denotes the basis for the codomain (Figure 8.1.5). Also, note how the basis for the domain seems to "cancel" in Formula (26) (Figure 8.1.6).

Recalling that the components of a vector in R^n or R^m are the same as its coordinates with respect to the standard basis for that space, it follows from (25) that if $S = \{\mathbf{e}_1, \mathbf{e}_2, \dots, \mathbf{e}_n\}$ is the standard basis for R^n and S' is the standard basis for R^m, then

$$[T]_{S',S} = \left[[T(\mathbf{e}_1)]_{S'} \mid [T(\mathbf{e}_2)]_{S'} \mid \cdots \mid [T(\mathbf{e}_n)]_{S'} \right] = [T(\mathbf{e}_1) \mid T(\mathbf{e}_2) \mid \cdots \mid T(\mathbf{e}_n)] = [T]$$

That is, the matrix for a linear transformation from R^n to R^m with respect to the standard bases for those spaces is the same as the standard matrix for T.

EXAMPLE 6
Matrix of a
Linear
Transformation

Let $T : R^2 \to R^3$ be the linear transformation defined by

$$T\left(\begin{bmatrix} x_1 \\ x_2 \end{bmatrix} \right) = \begin{bmatrix} x_2 \\ -5x_1 + 13x_2 \\ -7x_1 + 16x_2 \end{bmatrix}$$

Find the matrix for T with respect to the bases $B = \{\mathbf{v}_1, \mathbf{v}_2\}$ for R^2 and $B' = \{\mathbf{v}_1', \mathbf{v}_2', \mathbf{v}_3'\}$ for R^3, where

$$\mathbf{v}_1 = \begin{bmatrix} 3 \\ 1 \end{bmatrix}, \quad \mathbf{v}_2 = \begin{bmatrix} 5 \\ 2 \end{bmatrix}; \quad \mathbf{v}_1' = \begin{bmatrix} 1 \\ 0 \\ -1 \end{bmatrix}, \quad \mathbf{v}_2' = \begin{bmatrix} -1 \\ 2 \\ 2 \end{bmatrix}, \quad \mathbf{v}_3' = \begin{bmatrix} 0 \\ 0 \\ 2 \end{bmatrix}$$

Solution Using the given formula for T we obtain

$$T(\mathbf{v}_1) = \begin{bmatrix} 1 \\ -2 \\ -5 \end{bmatrix}, \quad T(\mathbf{v}_2) = \begin{bmatrix} 2 \\ 1 \\ -3 \end{bmatrix}$$

(verify), and expressing these vectors as linear combinations of $\mathbf{v}_1'$, $\mathbf{v}_2'$, and $\mathbf{v}_3'$ we obtain (verify)

$$T(\mathbf{v}_1) = -\mathbf{v}_2' - \tfrac{3}{2}\mathbf{v}_3' \quad \text{and} \quad T(\mathbf{v}_2) = \tfrac{5}{2}\mathbf{v}_1' + \tfrac{1}{2}\mathbf{v}_2' - \tfrac{3}{4}\mathbf{v}_3'$$

Thus,

$$[T]_{B',B} = \left[[T(\mathbf{v}_1)]_{B'} \mid [T(\mathbf{v}_2)]_{B'}\right] = \begin{bmatrix} 0 & \tfrac{5}{2} \\ -1 & \tfrac{1}{2} \\ -\tfrac{3}{2} & -\tfrac{3}{4} \end{bmatrix} \quad\blacksquare$$

EFFECT OF CHANGING BASES ON MATRICES OF LINEAR TRANSFORMATIONS

Theorems 8.1.2 and 8.1.3 and the related factorizations all have analogs for linear transformations. For example, suppose that B_1 and B_2 are bases for R^n, that B_1' and B_2' are bases for R^m, that U is the transition matrix from B_2 to B_1, and that V is the transition matrix from B_2' to B_1'. Then the diagram in Figure 8.1.7 suggests that

$$[T]_{B_1', B_1} = V[T]_{B_2', B_2} U^{-1} \tag{27}$$

In particular, if B_1 and B_1' are the standard bases for R^n and R^m, respectively, and if B and B' are any bases for R^n and R^m, respectively, then it follows from (27) that

$$[T] = V[T]_{B',B}U^{-1} \tag{28}$$

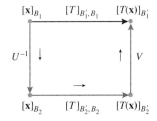

Figure 8.1.7

where U is the transition matrix from B to the standard basis for R^n and V is the transition matrix from B' to the standard basis for R^m.

REPRESENTING LINEAR OPERATORS WITH TWO BASES

A linear operator $T : R^n \to R^n$ can be viewed as a linear transformation in which the domain and codomain are the same. Thus, instead of choosing a single basis B and representing T by the matrix $[T]_B$, we can choose two different bases for R^n, say B and B', and represent T by the matrix $[T]_{B',B}$. Indeed, we will ask you to show in the exercises that

$$[T]_B = [T]_{B,B}$$

That is, the single-basis representation of T with respect to B can be viewed as the two-basis representation in which both bases are B.

EXAMPLE 7
Matrices of Identity Operators

Recall from Example 5 of Section 6.1 that the operator $T_I(\mathbf{x}) = \mathbf{x}$ that maps each vector in R^n into itself is called the *identity operator* on R^n.

(a) Find the standard matrix for T_I.

(b) Find the matrix for T_I with respect to an arbitrary basis B.

(c) Find the matrix for T_I with respect to a pair of arbitrary bases B and B'.

Solution (a) The standard matrix for T_I is the $n \times n$ identity matrix, since

$$[T_I] = [T_I(\mathbf{e}_1) \mid T_I(\mathbf{e}_2) \mid \cdots \mid T_I(\mathbf{e}_n)] = [\mathbf{e}_1 \mid \mathbf{e}_2 \mid \cdots \mid \mathbf{e}_n] = \begin{bmatrix} 1 & 0 & \cdots & 0 \\ 0 & 1 & \cdots & 0 \\ \vdots & \vdots & \ddots & \vdots \\ 0 & 0 & \cdots & 1 \end{bmatrix}$$

Solution (b) If $B = \{\mathbf{v}_1, \mathbf{v}_2, \ldots, \mathbf{v}_n\}$ is any basis for R^n, then

$$[T_I]_B = \left[[T_I(\mathbf{v}_1)]_B \mid [T_I(\mathbf{v}_2)]_B \mid \cdots \mid [T_I(\mathbf{v}_n)]_B \right] = \left[[\mathbf{v}_1]_B \mid [\mathbf{v}_2]_B \mid \cdots \mid [\mathbf{v}_n]_B \right]$$

But for each of these column vectors we have $[\mathbf{v}_i]_B = \mathbf{e}_i$ (why?), so

$$[T_I]_B = [\mathbf{e}_1 \mid \mathbf{e}_2 \mid \cdots \mid \mathbf{e}_n]$$

That is, $[T_I]_B$ is the $n \times n$ identity matrix.

Solution (c) If $B = \{\mathbf{v}_1, \mathbf{v}_2, \ldots, \mathbf{v}_n\}$ and $B' = \{\mathbf{v}'_1, \mathbf{v}'_2, \ldots, \mathbf{v}'_n\}$ are any bases for R^n, then

$$[T_I]_{B', B} = \left[[T_I(\mathbf{v}_1)]_{B'} \mid [T_I(\mathbf{v}_2)]_{B'} \mid \cdots \mid [T_I(\mathbf{v}_n)]_{B'} \right] = \left[[\mathbf{v}_1]_{B'} \mid [\mathbf{v}_2]_{B'} \mid \cdots \mid [\mathbf{v}_n]_{B'} \right]$$

which is the transition matrix $P_{B \to B'}$ [see Formula (11) of Section 7.11]. ■

Exercise Set 8.1

In Exercises 1 and 2, let $T : R^2 \to R^2$ be the linear operator whose standard matrix $[T]$ is given. Find the matrix $[T]_B$ with respect to the basis $B = \{\mathbf{v}_1, \mathbf{v}_2\}$, and verify that Formula (7) holds for every vector $\mathbf{x}$ in R^2.

1. $[T] = \begin{bmatrix} 1 & -1 \\ 1 & 1 \end{bmatrix}$; $\mathbf{v}_1 = \begin{bmatrix} 1 \\ 1 \end{bmatrix}$, $\mathbf{v}_2 = \begin{bmatrix} -1 \\ 0 \end{bmatrix}$

2. $[T] = \begin{bmatrix} 1 & 0 \\ 3 & -2 \end{bmatrix}$; $\mathbf{v}_1 = \begin{bmatrix} 1 \\ 2 \end{bmatrix}$, $\mathbf{v}_2 = \begin{bmatrix} -2 \\ 1 \end{bmatrix}$

3. Factor the matrix in Exercise 1 as $[T] = P[T]_B P^{-1}$, where P is an appropriate transition matrix.

4. Factor the matrix in Exercise 2 as $[T] = P[T]_B P^{-1}$, where P is an appropriate transition matrix.

In Exercises 5 and 6, let $T : R^3 \to R^3$ be the linear operator whose standard matrix $[T]$ is given. Find the matrix $[T]_B$ with respect to the basis $B = \{\mathbf{v}_1, \mathbf{v}_2, \mathbf{v}_3\}$, and verify that Formula (7) holds for every vector $\mathbf{x}$ in R^3.

5. $[T] = \begin{bmatrix} 1 & -1 & 0 \\ -1 & 1 & 0 \\ 1 & 0 & -1 \end{bmatrix}$; $\mathbf{v}_1 = \begin{bmatrix} 1 \\ 0 \\ 1 \end{bmatrix}$, $\mathbf{v}_2 = \begin{bmatrix} 0 \\ 1 \\ 1 \end{bmatrix}$, $\mathbf{v}_3 = \begin{bmatrix} 1 \\ 1 \\ 0 \end{bmatrix}$

6. $[T] = \begin{bmatrix} 0 & -2 & 3 \\ 0 & 0 & 0 \\ 1 & 0 & -2 \end{bmatrix}$; $\mathbf{v}_1 = \begin{bmatrix} -1 \\ 0 \\ 2 \end{bmatrix}$, $\mathbf{v}_2 = \begin{bmatrix} 3 \\ 3 \\ 3 \end{bmatrix}$,

$\mathbf{v}_3 = \begin{bmatrix} -1 \\ 2 \\ 0 \end{bmatrix}$

7. Factor the matrix in Exercise 5 as $[T] = P[T]_B P^{-1}$, where P is an appropriate transition matrix.

8. Factor the matrix in Exercise 6 as $[T] = P[T]_B P^{-1}$, where P is an appropriate transition matrix.

In Exercises 9 and 10, a formula for a linear operator $T : R^2 \to R^2$ is given. Find the matrices $[T]_B$ and $[T]_{B'}$ with respect to the bases $B = \{\mathbf{v}_1, \mathbf{v}_2\}$ and $B' = \{\mathbf{v}'_1, \mathbf{v}'_2\}$, respectively, and confirm that these matrices satisfy Formula (14) of Theorem 8.1.2.

9. $T\left(\begin{bmatrix} x_1 \\ x_2 \end{bmatrix} \right) = \begin{bmatrix} x_1 - 2x_2 \\ x_2 \end{bmatrix}$; $\mathbf{v}_1 = \begin{bmatrix} -1 \\ 5 \end{bmatrix}$, $\mathbf{v}_2 = \begin{bmatrix} 5 \\ -3 \end{bmatrix}$,

$\mathbf{v}'_1 = \begin{bmatrix} 2 \\ 1 \end{bmatrix}$, $\mathbf{v}'_2 = \begin{bmatrix} -3 \\ 4 \end{bmatrix}$

10. $T\left(\begin{bmatrix} x_1 \\ x_2 \end{bmatrix} \right) = \begin{bmatrix} x_1 + 7x_2 \\ 3x_1 - 4x_2 \end{bmatrix}$; $\mathbf{v}_1 = \begin{bmatrix} 2 \\ 22 \end{bmatrix}$, $\mathbf{v}_2 = \begin{bmatrix} 4 \\ -1 \end{bmatrix}$,

$\mathbf{v}'_1 = \begin{bmatrix} 1 \\ 3 \end{bmatrix}$, $\mathbf{v}'_2 = \begin{bmatrix} -1 \\ -1 \end{bmatrix}$

11. Factor the matrix $[T]_B$ in Exercise 9 as $[T]_B = P^{-1}[T]_{B'} P$, where P is an appropriate transition matrix.

12. Factor the matrix $[T]_B$ in Exercise 10 as $[T]_B = P^{-1}[T]_{B'} P$, where P is an appropriate transition matrix.

In Exercises 13 and 14, a formula for a linear operator $T : R^2 \to R^2$ is given. Find the standard matrix $[T]$ and the matrix $[T]_B$ with respect to the basis $B = \{\mathbf{v}_1, \mathbf{v}_2\}$, and confirm that these matrices satisfy Formula (17) of Theorem 8.1.3.

13. $T\left(\begin{bmatrix} x_1 \\ x_2 \end{bmatrix}\right) = \begin{bmatrix} x_1 - 2x_2 \\ x_2 \end{bmatrix}$; $\mathbf{v}_1 = \begin{bmatrix} -1 \\ 5 \end{bmatrix}$, $\mathbf{v}_2 = \begin{bmatrix} 5 \\ -3 \end{bmatrix}$

14. $T\left(\begin{bmatrix} x_1 \\ x_2 \end{bmatrix}\right) = \begin{bmatrix} x_1 + 7x_2 \\ 3x_1 - 4x_2 \end{bmatrix}$; $\mathbf{v}_1 = \begin{bmatrix} 2 \\ 22 \end{bmatrix}$, $\mathbf{v}_2 = \begin{bmatrix} 4 \\ -1 \end{bmatrix}$

15. Let $T : R^2 \to R^2$ be the linear operator that is defined by $T(x, y) = (2x + 3y, x - 2y)$, let $B = \{\mathbf{v}_1, \mathbf{v}_2\}$ be the basis for R^2 in which $\mathbf{v}_1 = (-1, 2)$ and $\mathbf{v}_2 = (2, 1)$, and let $\mathbf{x} = (5, -3)$.
  (a) Find $[T(\mathbf{x})]_B$, $[T]_B$, and $[\mathbf{x}]_B$.
  (b) Confirm that $[T(\mathbf{x})]_B = [T]_B[\mathbf{x}]_B$, as guaranteed by Formula (7).

16. Let $T : R^3 \to R^3$ be the linear operator that is defined by $T(x, y, z) = (x + y + z, 2y + 4z, 4z)$, let $B = \{\mathbf{v}_1, \mathbf{v}_2, \mathbf{v}_3\}$ be the basis for R^3 in which $\mathbf{v}_1 = (1, 1, 0)$, $\mathbf{v}_2 = (-1, 1, 0)$, $\mathbf{v}_3 = (0, 0, 1)$, and let $\mathbf{x} = (2, -3, 4)$.
  (a) Find $[T(\mathbf{x})]_B$, $[T]_B$, and $[\mathbf{x}]_B$.
  (b) Confirm that $[T(\mathbf{x})]_B = [T]_B[\mathbf{x}]_B$, as guaranteed by Formula (7).

In Exercises 17 and 18, a formula for a linear transformation $T : R^2 \to R^3$ is given. Find the matrix $[T]_{B', B}$ with respect to the bases $B = \{\mathbf{v}_1, \mathbf{v}_2\}$ and $B' = \{\mathbf{v}'_1, \mathbf{v}'_2, \mathbf{v}'_3\}$, and verify that Formula (26) holds for all $\mathbf{x}$ in R^2.

17. $T\left(\begin{bmatrix} x_1 \\ x_2 \end{bmatrix}\right) = \begin{bmatrix} x_1 + 2x_2 \\ -x_1 \\ 0 \end{bmatrix}$; $\mathbf{v}_1 = \begin{bmatrix} 1 \\ 3 \end{bmatrix}$, $\mathbf{v}_2 = \begin{bmatrix} -2 \\ 4 \end{bmatrix}$,

$\mathbf{v}'_1 = \begin{bmatrix} 1 \\ 1 \\ 1 \end{bmatrix}$, $\mathbf{v}'_2 = \begin{bmatrix} 2 \\ 2 \\ 0 \end{bmatrix}$, $\mathbf{v}'_3 = \begin{bmatrix} 3 \\ 0 \\ 0 \end{bmatrix}$

18. $T\left(\begin{bmatrix} x_1 \\ x_2 \end{bmatrix}\right) = \begin{bmatrix} x_2 + x_1 \\ x_2 - x_1 \\ 3x_1 + 2x_2 \end{bmatrix}$; $\mathbf{v}_1 = \begin{bmatrix} 2 \\ -1 \end{bmatrix}$, $\mathbf{v}_2 = \begin{bmatrix} 2 \\ 1 \end{bmatrix}$,

$\mathbf{v}'_1 = \begin{bmatrix} 1 \\ 2 \\ 0 \end{bmatrix}$, $\mathbf{v}'_2 = \begin{bmatrix} 0 \\ 1 \\ 2 \end{bmatrix}$, $\mathbf{v}'_3 = \begin{bmatrix} 1 \\ 0 \\ 2 \end{bmatrix}$

In Exercises 19 and 20, a formula for a linear transformation $T : R^3 \to R^2$ is given. Find the matrix $[T]_{B', B}$ with respect to the bases $B = \{\mathbf{v}_1, \mathbf{v}_2, \mathbf{v}_3\}$ and $B' = \{\mathbf{v}'_1, \mathbf{v}'_2\}$, and verify that Formula (26) holds for all $\mathbf{x}$ in R^3.

19. $T\left(\begin{bmatrix} x_1 \\ x_2 \\ x_3 \end{bmatrix}\right) = \begin{bmatrix} 3x_1 + 2x_3 \\ 2x_1 - x_2 \end{bmatrix}$; $\mathbf{v}_1 = \begin{bmatrix} 1 \\ 1 \\ 0 \end{bmatrix}$, $\mathbf{v}_2 = \begin{bmatrix} 1 \\ 0 \\ 1 \end{bmatrix}$,

$\mathbf{v}_3 = \begin{bmatrix} 0 \\ 1 \\ 1 \end{bmatrix}$, $\mathbf{v}'_1 = \begin{bmatrix} -3 \\ 2 \end{bmatrix}$, $\mathbf{v}'_2 = \begin{bmatrix} 1 \\ 4 \end{bmatrix}$

20. $T\left(\begin{bmatrix} x_1 \\ x_2 \\ x_3 \end{bmatrix}\right) = \begin{bmatrix} 2x_1 - x_2 + 4x_3 \\ x_1 + x_2 + x_3 \end{bmatrix}$; $\mathbf{v}_1 = \begin{bmatrix} 0 \\ 1 \\ 1 \end{bmatrix}$, $\mathbf{v}_2 = \begin{bmatrix} 1 \\ 0 \\ 1 \end{bmatrix}$,

$\mathbf{v}_3 = \begin{bmatrix} 1 \\ 1 \\ 0 \end{bmatrix}$, $\mathbf{v}'_1 = \begin{bmatrix} 2 \\ -3 \end{bmatrix}$, $\mathbf{v}'_2 = \begin{bmatrix} 4 \\ 1 \end{bmatrix}$

21. Consider the basis $B = \{\mathbf{v}_1, \mathbf{v}_2\}$ for R^2 in which $\mathbf{v}_1 = (-1, 2)$ and $\mathbf{v}_2 = (2, 1)$, and let $T : R^2 \to R^2$ be the linear operator whose matrix with respect to B is

$$[T]_B = \begin{bmatrix} 1 & 3 \\ -2 & 5 \end{bmatrix}$$

  (a) Find $[T(\mathbf{v}_1)]_B$ and $[T(\mathbf{v}_2)]_B$.
  (b) Find $T(\mathbf{v}_1)$ and $T(\mathbf{v}_2)$.
  (c) Find a formula for $T(x_1, x_2)$.
  (d) Use the formula obtained in part (c) to compute $T(1, 1)$.

22. Consider the bases

$$B = \{\mathbf{v}_1, \mathbf{v}_2, \mathbf{v}_3, \mathbf{v}_4\} \quad \text{and} \quad B' = \{\mathbf{v}'_1, \mathbf{v}'_2, \mathbf{v}'_3\}$$

for R^4 and R^3, respectively, in which $\mathbf{v}_1 = (0, 1, 1, 1)$, $\mathbf{v}_2 = (2, 1, -1, -1)$, $\mathbf{v}_3 = (1, 4, -1, 2)$, $\mathbf{v}_4 = (6, 9, 4, 2)$, $\mathbf{v}'_1 = (0, 8, 8)$, $\mathbf{v}'_2 = (-7, 8, 1)$, $\mathbf{v}'_3 = (-6, 9, 1)$, and let $T : R^4 \to R^3$ be the linear transformation whose matrix with respect to B and B' is

$$[T]_{B', B} = \begin{bmatrix} 3 & -2 & 1 & 0 \\ 1 & 6 & 2 & 1 \\ -3 & 0 & 7 & 1 \end{bmatrix}$$

  (a) Find $[T(\mathbf{v}_1)]_{B'}$, $[T(\mathbf{v}_2)]_{B'}$, $[T(\mathbf{v}_3)]_{B'}$, $[T(\mathbf{v}_4)]_{B'}$.
  (b) Find $T(\mathbf{v}_1)$, $T(\mathbf{v}_2)$, $T(\mathbf{v}_3)$, and $T(\mathbf{v}_4)$.
  (c) Find a formula for $T(x_1, x_2, x_3, x_4)$.
  (d) Use the formula obtained in part (c) to compute $T(2, 2, 0, 0)$.

23. Let $T_I : R^2 \to R^2$ be the identity operator, and consider the bases $B = \{\mathbf{v}_1, \mathbf{v}_2\}$ and $B' = \{\mathbf{v}'_1, \mathbf{v}'_2\}$ for R^2 in which $\mathbf{v}_1 = (2, 3)$, $\mathbf{v}_2 = (-1, 4)$, $\mathbf{v}'_1 = (1, 7)$, $\mathbf{v}'_2 = (6, -9)$. Find $[T]$, $[T]_B$, $[T]_{B'}$, and $[T]_{B', B}$.

24. Let $T_I : R^3 \to R^3$ be the identity operator, and consider the bases $B = \{\mathbf{v}_1, \mathbf{v}_2, \mathbf{v}_3\}$ and $B' = \{\mathbf{v}'_1, \mathbf{v}'_2, \mathbf{v}'_3\}$ for R^3 in which $\mathbf{v}_1 = (3, 4, 1)$, $\mathbf{v}_2 = (0, 5, 2)$, $\mathbf{v}_3 = (3, 9, 4)$, $\mathbf{v}'_1 = (2, 8, -1)$, $\mathbf{v}'_2 = (7, -8, 5)$, $\mathbf{v}'_3 = (9, 0, 2)$. Find $[T]$, $[T]_B$, $[T]_{B'}$, and $[T]_{B', B}$.

25. Show that if $T : R^n \to R^m$ is the zero transformation, then the matrix for T with respect to any bases for R^n and R^m is a zero matrix.

26. Show that if $T : R^n \to R^n$ is a contraction or a dilation of R^n (see Section 6.2), then the matrix for T with respect to any basis for R^n is a positive scalar multiple of the identity matrix.

27. Let $B = \{\mathbf{v}_1, \mathbf{v}_2, \mathbf{v}_3\}$ be a basis for R^3. Find the matrix with respect to B for the linear operator $T : R^3 \to R^3$ defined by $T(\mathbf{v}_1) = \mathbf{v}_3$, $T(\mathbf{v}_2) = \mathbf{v}_1$, $T(\mathbf{v}_3) = \mathbf{v}_2$.

28. Let $B = \{\mathbf{v}_1, \mathbf{v}_2, \mathbf{v}_3, \mathbf{v}_4\}$ be a basis for R^4. Find the matrix with respect to B for the linear operator $T : R^4 \to R^4$ defined by $T(\mathbf{v}_1) = \mathbf{v}_2$, $T(\mathbf{v}_2) = \mathbf{v}_3$, $T(\mathbf{v}_3) = \mathbf{v}_4$, $T(\mathbf{v}_4) = \mathbf{v}_1$.

29. Let $T : R^2 \to R^2$ be the linear operator whose standard matrix is

$$[T] = \begin{bmatrix} 1 & 5 \\ 5 & 1 \end{bmatrix}$$

and let $B = \{\mathbf{v}_1, \mathbf{v}_2\}$ be the orthonormal basis for R^2 in which $\mathbf{v}_1 = \left(\frac{1}{\sqrt{2}}, -\frac{1}{\sqrt{2}}\right)$ and $\mathbf{v}_2 = \left(\frac{1}{\sqrt{2}}, \frac{1}{\sqrt{2}}\right)$. Find $[T]_B$, and use that matrix to describe the geometric effect of the operator T.

30. Let $T : R^3 \to R^3$ be the linear operator whose standard matrix is

$$[T] = \begin{bmatrix} 0 & -2 & 0 \\ 0 & 0 & -2 \\ 2 & 0 & 0 \end{bmatrix}$$

and let $B = \{\mathbf{v}_1, \mathbf{v}_2, \mathbf{v}_3\}$ be the orthonormal basis for R^3 in which

$$\mathbf{v}_1 = \left(\tfrac{1}{\sqrt{3}}, -\tfrac{1}{\sqrt{3}}, \tfrac{1}{\sqrt{3}}\right), \quad \mathbf{v}_2 = \left(\tfrac{1}{\sqrt{6}}, \tfrac{2}{\sqrt{6}}, \tfrac{1}{\sqrt{6}}\right)$$
$$\mathbf{v}_3 = \left(-\tfrac{1}{\sqrt{2}}, 0, \tfrac{1}{\sqrt{2}}\right)$$

Find $[T]_B$, and use that matrix to describe the geometric effect of the operator T.

Discussion and Discovery

D1. If $\mathbf{v}_1 = (2, 0)$ and $\mathbf{v}_2 = (0, 4)$, and if $T : R^2 \to R^2$ is the linear operator for which $T(\mathbf{v}_1) = \mathbf{v}_2$ and $T(\mathbf{v}_2) = \mathbf{v}_1$, then the standard matrix for T is _____, and the matrix for T with respect to the basis $\{\mathbf{v}_1, \mathbf{v}_2\}$ is _____.

D2. Suppose that T is a linear operator on R^n, that B_1 and B_2 are bases for R^n, and that C is the transition matrix from B_2 to B_1. Construct a diagram like that in Figure 8.1.3, and use it to find a relationship between $[T]_{B_1}$ and $[T]_{B_2}$.

D3. Suppose that T is a linear transformation from R^n to R^m, that B_1 and B_2 are bases for R^n, that B_3 and B_4 are bases for R^m, that C is the transition matrix from B_2 to B_1, and that D is the transition matrix from B_4 to B_3. Construct a diagram like that in Figure 8.1.7, and use it to find a relationship between $[T]_{B_3, B_1}$ and $[T]_{B_4, B_2}$.

D4. Indicate whether the statement is true (T) or false (F). Justify your answer.

(a) If $T_1 : R^n \to R^n$ and $T_2 : R^n \to R^n$ are linear operators, and if $[T_1]_{B', B} = [T_2]_{B', B}$ with respect to two bases B and B' for R^n, then $T_1(\mathbf{x}) = T_2(\mathbf{x})$ for every vector $\mathbf{x}$ in R^n.

(b) If $T_1 : R^n \to R^n$ is a linear operator, and if $[T_1]_B = [T_1]_{B'}$ with respect to two bases B and B' for R^n, then $B = B'$.

(c) If $T : R^n \to R^n$ is a linear operator, and if $[T]_B = I_n$ with respect to some basis B for R^n, then T is the identity operator on R^n.

(d) If $T : R^n \to R^n$ is a linear operator, and if $[T]_{B', B} = I_n$ with respect to two bases B and B' for R^n, then T is the identity operator on R^n.

D5. Since the standard basis for R^n is so easy to work with, why would one want to represent a linear operator on R^n with respect to another basis?

Working with Proofs

P1. Prove that Formula (3) defines a linear transformation.

P2. Suppose that $T_1 : R^n \to R^k$ and $T_2 : R^k \to R^m$ are linear transformations, and suppose that B, B', and B'' are bases for R^n, R^k, and R^m, respectively. Prove that

$$[T_2 \circ T_1]_{B'', B} = [T_2]_{B'', B'}[T_1]_{B', B}$$

[*Note:* This generalizes Formula (2) of Section 6.4.]

P3. Suppose that $T : R^n \to R^n$ is a one-to-one linear operator. Prove that $[T]_B$ is invertible for every basis B for R^n and that $[T^{-1}]_B = [T]_B^{-1}$. [*Note:* This generalizes Formula (14) of Section 6.4.]

P4. Prove that if $T : R^n \to R^n$ is a linear operator and B is a basis for R^n, then $[T]_B = [T]_{B, B}$.

Technology Exercises

T1. Let $T : R^5 \to R^3$ be the linear operator given by the formula

$T(x_1, x_2, x_3, x_4, x_5) =$

$(7x_1 + 12x_2 - 5x_3, \ 3x_1 + 10x_2 + 13x_4 + x_5, \ -9x_1 - x_3 - 3x_5)$

and let $B = \{\mathbf{v}_1, \mathbf{v}_2, \mathbf{v}_3, \mathbf{v}_4, \mathbf{v}_5\}$ and $B' = \{\mathbf{v}'_1, \mathbf{v}'_2, \mathbf{v}'_3\}$ be the bases for R^5 and R^3 in which $\mathbf{v}_1 = (1, 1, 0, 0, 0)$, $\mathbf{v}_2 = (0, 1, 1, 0, 0)$, $\mathbf{v}_3 = (0, 0, 1, 1, 0)$, $\mathbf{v}_4 = (0, 0, 0, 1, 1)$,

$\mathbf{v}_5 = (1, 0, 0, 0, 1)$, $\mathbf{v}'_1 = (1, 2, -1)$, $\mathbf{v}'_2 = (2, 1, 3)$, and $\mathbf{v}'_3 = (1, 1, 1)$.

(a) Find the matrix $[T]_{B', B}$.

(b) For the vector $\mathbf{x} = (3, 7, -4, 5, 1)$, find $[\mathbf{x}]_B$ and use the matrix obtained in part (a) to compute $[T(\mathbf{x})]_{B'}$.

T2. Let $[T]$ be the standard matrix for the linear transformation in Exercise T1 and let B and B' be the bases in that exercise. Find the factorization of $[T]$ stated in Formula (28).

Section 8.2 Similarity and Diagonalizability

Diagonal matrices are, for many purposes, the simplest kinds of matrices to work with. In this section we will determine conditions under which a linear operator can be represented by a diagonal matrix with respect to some basis. A knowledge of such conditions is fundamental in the mathematical analysis of linear operators and has important implications in science, engineering, and economics.

SIMILAR MATRICES In our study of how a change of basis affects the matrix representation of a linear operator we encountered matrix equations of the form

$$C = P^{-1}AP$$

[see Formula (15) of Section 8.1, for example]. Such equations are so important that they have some terminology associated with them.

> **Definition 8.2.1** If A and C are square matrices with the same size, then we say that C is *similar to A* if there is an invertible matrix P such that $C = P^{-1}AP$.

REMARK If C is similar to A, then it is also true that A is similar to C. You can see this by letting $Q = P^{-1}$ and rewriting the equation $C = P^{-1}AP$ as

$$A = PCP^{-1} = (P^{-1})^{-1}C(P^{-1}) = Q^{-1}CQ$$

When we want to emphasize that similarity goes both ways, we will say that *A and C are similar*.

The following theorem gives an interpretation of similarity from an operator point of view.

> **Theorem 8.2.2** *Two matrices are similar if and only if there exist bases with respect to which the matrices represent the same linear operator.*

Proof We will show first that if A and C are similar $n \times n$ matrices, then there exist bases with respect to which they represent the same linear operator on R^n. For this purpose, let $T : R^n \to R^n$ denote multiplication by A; that is,

$$A = [T] \tag{1}$$

Since A and C are similar, there exists an invertible matrix P such that $C = P^{-1}AP$, so it follows from (1) that

$$C = P^{-1}[T]P \tag{2}$$

If we assume that the column-vector form of P is

$$P = [\mathbf{v}_1 \mid \mathbf{v}_2 \mid \cdots \mid \mathbf{v}_n]$$

then the invertibility of P and Theorem 7.4.4 imply that $B = \{\mathbf{v}_1, \mathbf{v}_2, \ldots, \mathbf{v}_n\}$ is a basis for R^n. It now follows from Formula (2) above and Formula (20) of Section 8.1 that

$$C = P^{-1}[T]P = [T]_B$$

Thus we have shown that A is the matrix for T with respect to the standard basis, and C is the matrix for T with respect to the basis B, so this part of the proof is complete.

Conversely, assume that C represents the linear operator $T : R^n \to R^n$ with respect to some basis B, and A represents the same operator with respect to a basis B'; that is,

$$C = [T]_B \quad \text{and} \quad A = [T]_{B'}$$

If we let $P = P_{B \to B'}$, then it follows from Formula (12) in Theorem 8.1.2 that

$$[T]_{B'} = P[T]_B P^{-1} \quad \text{or, equivalently,} \quad A = PCP^{-1}$$

Rewriting the last equation as $C = P^{-1}AP$ shows that A and C are similar. ∎

SIMILARITY INVARIANTS

There are a number of basic properties of matrices that are shared by similar matrices. For example, if $C = P^{-1}AP$, then

$$\det(C) = \det(P^{-1}AP) = \det(P^{-1})\det(A)\det(P) = \frac{1}{\det(P)}\det(A)\det(P) = \det(A)$$

which shows that similar matrices have the same determinant. In general, any property that is shared by similar matrices is said to be a *similarity invariant*. The following theorem lists some of the most important similarity invariants.

> **Theorem 8.2.3**
>
> (a) *Similar matrices have the same determinant.*
> (b) *Similar matrices have the same rank.*
> (c) *Similar matrices have the same nullity.*
> (d) *Similar matrices have the same trace.*
> (e) *Similar matrices have the same characteristic polynomial and hence have the same eigenvalues with the same algebraic multiplicities.*

We have already proved part (a). We will prove part (e) and leave the proofs of the other three parts as exercises.

Proof (e) We want to prove that if A and C are similar matrices, then

$$\det(\lambda I - C) = \det(\lambda I - A) \tag{3}$$

As a first step we will show that if A and C are similar matrices, then so are $\lambda I - A$ and $\lambda I - C$ for any scalar λ. To see this, suppose that $C = P^{-1}AP$ and write

$$\lambda I - C = \lambda I - P^{-1}AP = \lambda P^{-1}P - P^{-1}AP = P^{-1}(\lambda P - AP)$$
$$= P^{-1}(\lambda IP - AP) = P^{-1}(\lambda I - A)P$$

This shows that $\lambda I - A$ and $\lambda I - C$ are similar, so (3) now follows from part (a). ∎

EXAMPLE 1
Similarity

Show that there do not exist bases for R^2 with respect to which the matrices

$$A = \begin{bmatrix} 1 & 2 \\ 2 & 6 \end{bmatrix} \quad \text{and} \quad C = \begin{bmatrix} 2 & 1 \\ 1 & 3 \end{bmatrix}$$

represent the same linear operator.

Solution For A and C to represent the same linear operator, the two matrices would have to be similar by Theorem 8.2.2. But this cannot be, since $\text{tr}(A) = 7$ and $\text{tr}(C) = 5$, contradicting the fact that the trace is a similarity invariant. ∎

CONCEPT PROBLEM Do you think that two $n \times n$ matrices with the same trace must be similar? Explain your reasoning.

EIGENVECTORS AND EIGENVALUES OF SIMILAR MATRICES

Recall that the solution space of

$$(\lambda_0 I - A)\mathbf{x} = \mathbf{0}$$

is called the *eigenspace* of A corresponding to λ_0. We call the dimension of this solution space the **geometric multiplicity** of λ_0. Do not confuse this with the *algebraic multiplicity* of λ_0, which, as you may recall, is the number of repetitions of the factor $\lambda - \lambda_0$ in the complete factorization of the characteristic polynomial of A.

EXAMPLE 2
Algebraic and
Geometric
Multiplicities

Find the algebraic and geometric multiplicities of the eigenvalues of

$$A = \begin{bmatrix} 2 & 0 & 0 \\ 1 & 3 & 0 \\ -3 & 5 & 3 \end{bmatrix}$$

Solution Since A is triangular its characteristic polynomial is

$$p(\lambda) = (\lambda - 2)(\lambda - 3)(\lambda - 3) = (\lambda - 2)(\lambda - 3)^2$$

This implies that the distinct eigenvalues are $\lambda = 2$ and $\lambda = 3$ and that

$\lambda = 2$ has algebraic multiplicity 1

$\lambda = 3$ has algebraic multiplicity 2

One way to find the geometric multiplicities of the eigenvalues is to find bases for the eigenspaces and then determine the dimensions of those spaces from the number of basis vectors. Let us do this. By definition, the eigenspace corresponding to an eigenvalue λ is the solution space of $(\lambda I - A)\mathbf{x} = \mathbf{0}$, which in this case is

$$\begin{bmatrix} \lambda - 2 & 0 & 0 \\ -1 & \lambda - 3 & 0 \\ 3 & -5 & \lambda - 3 \end{bmatrix} \begin{bmatrix} x_1 \\ x_2 \\ x_3 \end{bmatrix} = \begin{bmatrix} 0 \\ 0 \\ 0 \end{bmatrix} \tag{4}$$

If $\lambda = 2$, this system becomes

$$\begin{bmatrix} 0 & 0 & 0 \\ -1 & -1 & 0 \\ 3 & -5 & -1 \end{bmatrix} \begin{bmatrix} x_1 \\ x_2 \\ x_3 \end{bmatrix} = \begin{bmatrix} 0 \\ 0 \\ 0 \end{bmatrix} \tag{5}$$

We leave it for you to show that a general solution of this system is

$$\mathbf{x} = \begin{bmatrix} x_1 \\ x_2 \\ x_3 \end{bmatrix} = \begin{bmatrix} \frac{1}{8}t \\ -\frac{1}{8}t \\ t \end{bmatrix} = t \begin{bmatrix} \frac{1}{8} \\ -\frac{1}{8} \\ 1 \end{bmatrix} \tag{6}$$

which shows that the eigenspace corresponding to $\lambda = 2$ has dimension 1 and that the column vector on the right side of (6) is a basis for this eigenspace. Similarly, it follows from (4) that the eigenspace corresponding to $\lambda = 3$ is the solution space of

$$\begin{bmatrix} 1 & 0 & 0 \\ -1 & 0 & 0 \\ 3 & -5 & 0 \end{bmatrix} \begin{bmatrix} x_1 \\ x_2 \\ x_3 \end{bmatrix} = \begin{bmatrix} 0 \\ 0 \\ 0 \end{bmatrix} \tag{7}$$

We leave it for you to show that a general solution of this system is

$$\mathbf{x} = \begin{bmatrix} x_1 \\ x_2 \\ x_3 \end{bmatrix} = \begin{bmatrix} 0 \\ 0 \\ t \end{bmatrix} = t \begin{bmatrix} 0 \\ 0 \\ 1 \end{bmatrix} \tag{8}$$

which shows that the eigenspace corresponding to $\lambda = 3$ has dimension 1 and that the column vector on the right side of (8) is a basis for this eigenspace. Since both eigenspaces have dimension 1, we have shown that

$\lambda = 2$ has geometric multiplicity 1
$\lambda = 3$ has geometric multiplicity 1   ■

EXAMPLE 3
Algebraic and
Geometric
Multiplicities

Find the algebraic and geometric multiplicities of the eigenvalues of

$$A = \begin{bmatrix} 0 & 0 & -2 \\ 1 & 2 & 1 \\ 1 & 0 & 3 \end{bmatrix}$$

Solution We leave it for you to confirm that the characteristic polynomial of A is

$$p(\lambda) = \det(\lambda I - A) = \lambda^3 - 5\lambda^2 + 8\lambda - 4 = (\lambda - 1)(\lambda - 2)^2$$

This implies that the eigenvalues of A are $\lambda = 1$ and $\lambda = 2$ and that

$\lambda = 1$ has algebraic multiplicity 1
$\lambda = 2$ has algebraic multiplicity 2

By definition, the eigenspace corresponding to an eigenvalue λ is the solution space of the system $(\lambda I - A)\mathbf{x} = \mathbf{0}$, which in this case is

$$\begin{bmatrix} \lambda & 0 & 2 \\ -1 & \lambda - 2 & -1 \\ -1 & 0 & \lambda - 3 \end{bmatrix} \begin{bmatrix} x_1 \\ x_2 \\ x_3 \end{bmatrix} = \begin{bmatrix} 0 \\ 0 \\ 0 \end{bmatrix} \tag{9}$$

We leave it for you to show that a general solution of this system for $\lambda = 1$ is

$$\mathbf{x} = \begin{bmatrix} x_1 \\ x_2 \\ x_3 \end{bmatrix} = \begin{bmatrix} -2t \\ t \\ t \end{bmatrix} = t \begin{bmatrix} -2 \\ 1 \\ 1 \end{bmatrix} \tag{10}$$

and that a general solution for $\lambda = 2$ is

$$\mathbf{x} = \begin{bmatrix} x_1 \\ x_2 \\ x_3 \end{bmatrix} = \begin{bmatrix} -s \\ t \\ s \end{bmatrix} = \begin{bmatrix} -s \\ 0 \\ s \end{bmatrix} + \begin{bmatrix} 0 \\ t \\ 0 \end{bmatrix} = s \begin{bmatrix} -1 \\ 0 \\ 1 \end{bmatrix} + t \begin{bmatrix} 0 \\ 1 \\ 0 \end{bmatrix} \tag{11}$$

This shows that the eigenspace corresponding to $\lambda = 1$ has dimension 1 and that the column vector on the right side of (10) is a basis for this eigenspace, and it shows that the eigenspace corresponding to $\lambda = 2$ has dimension 2 and that the column vectors on the right side of (11) are a basis for this eigenspace. Thus,

$\lambda = 1$ has geometric multiplicity 1
$\lambda = 2$ has geometric multiplicity 2   ■

REMARK It is not essential to find bases for the eigenspaces to determine the geometric multiplicities of the eigenvalues. For example, to find the dimensions of the eigenspaces in Example 2 we could have calculated the ranks of the coefficient matrices in (5) and (7) by row reduction and then used the relationship rank + nullity = 3 to determine the nullities.

The next theorem shows that eigenvalues and their multiplicities are similarity invariants.

Theorem 8.2.4 *Similar matrices have the same eigenvalues and those eigenvalues have the same algebraic and geometric multiplicities for both matrices.*

Proof Let us assume first that A and C are similar matrices. Since similar matrices have the same characteristic polynomial, it follows that A and C have the same eigenvalues with the same algebraic multiplicities. To show that an eigenvalue λ has the same geometric multiplicity for both matrices, we must show that the solution spaces of

$$(\lambda I - A)\mathbf{x} = \mathbf{0} \quad \text{and} \quad (\lambda I - C)\mathbf{x} = \mathbf{0}$$

have the same dimension, or equivalently, that the matrices

$$\lambda I - A \quad \text{and} \quad \lambda I - C \tag{12}$$

have the same nullity. But we showed in the proof of Theorem 8.2.3 that the similarity of A and C implies the similarity of the matrices in (12). Thus, these matrices have the same nullity by part (c) of Theorem 8.2.3. ∎

Do not read more into Theorem 8.2.4 than it actually says; the theorem states that similar matrices have the same eigenvalues with the same algebraic and geometric multiplicities, but it does not say that similar matrices have the same eigenspaces. The following theorem establishes the relationship between the eigenspaces of similar matrices.

Theorem 8.2.5 *Suppose that $C = P^{-1}AP$ and that λ is an eigenvalue of A and C.*

(a) *If $\mathbf{x}$ is an eigenvector of C corresponding to λ, then $P\mathbf{x}$ is an eigenvector of A corresponding to λ.*

(b) *If $\mathbf{x}$ is an eigenvector of A corresponding to λ, then $P^{-1}\mathbf{x}$ is an eigenvector of C corresponding to λ.*

We will prove part (a) and leave the proof of part (b) as an exercise.

Proof (a) Assume that $\mathbf{x}$ is an eigenvector of C corresponding to λ, so $\mathbf{x} \neq \mathbf{0}$ and $C\mathbf{x} = \lambda\mathbf{x}$. If we substitute $P^{-1}AP$ for C, we obtain

$$P^{-1}AP\mathbf{x} = \lambda\mathbf{x}$$

which we can rewrite as

$$AP\mathbf{x} = P\lambda\mathbf{x} \quad \text{or equivalently,} \quad A(P\mathbf{x}) = \lambda(P\mathbf{x}) \tag{13}$$

Since P is invertible and $\mathbf{x} \neq \mathbf{0}$, it follows that $P\mathbf{x} \neq \mathbf{0}$. Thus, the second equation in (13) implies that $P\mathbf{x}$ is an eigenvector of A corresponding to λ. ∎

DIAGONALIZATION   Diagonal matrices play an important role in many applications because, in many respects, they represent the simplest kinds of linear operators. For example, suppose that $T : R^n \to R^n$ is a linear operator whose matrix with respect to a basis $B = \{\mathbf{v}_1, \mathbf{v}_2, \ldots, \mathbf{v}_n\}$ is

$$D = \begin{bmatrix} d_1 & 0 & \cdots & 0 \\ 0 & d_2 & \cdots & 0 \\ \vdots & \vdots & \ddots & \vdots \\ 0 & 0 & \cdots & d_n \end{bmatrix}$$

If $\mathbf{w}$ is a vector in R^n, and if $\mathbf{x} = [\mathbf{w}]_B$ is the coordinate matrix for $\mathbf{w}$ with respect to B, then

$$D\mathbf{x} = \begin{bmatrix} d_1 & 0 & \cdots & 0 \\ 0 & d_2 & \cdots & 0 \\ \vdots & \vdots & \ddots & \vdots \\ 0 & 0 & \cdots & d_n \end{bmatrix} \begin{bmatrix} x_1 \\ x_2 \\ \vdots \\ x_n \end{bmatrix} = \begin{bmatrix} d_1 x_1 \\ d_2 x_2 \\ \vdots \\ d_n x_n \end{bmatrix}$$

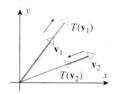

If T is represented by a diagonal matrix with respect to the basis $B = \{\mathbf{v}_1, \mathbf{v}_2\}$, then T contracts or dilates vectors that are parallel to $\mathbf{v}_1$ or $\mathbf{v}_2$ (with possible reversals of direction).

Figure 8.2.1

Thus, multiplying $\mathbf{x}$ by D has the effect of "scaling" each coordinate of $\mathbf{w}$ (with a sign reversal for negative d's). In particular, the effect of T on a vector that is parallel to one of the basis vectors $\mathbf{v}_1, \mathbf{v}_2, \ldots, \mathbf{v}_n$ is to contract or dilate that vector (with a possible reversal of direction) (Figure 8.2.1).

We will now consider the problem of determining conditions under which a linear operator can be represented by a diagonal matrix with respect to some basis. Since we will generally know the standard matrix for a linear operator, we will consider the following form of this problem.

The Diagonalization Problem Given a square matrix A, does there exist an invertible matrix P for which $P^{-1}AP$ is a diagonal matrix, and if so, how does one find such a P? If such a matrix P exists, then A is said to be ***diagonalizable***, and P is said to ***diagonalize*** A.

The following theorem will lead to a solution of the diagonalization problem.

Theorem 8.2.6 *An $n \times n$ matrix A is diagonalizable if and only if A has n linearly independent eigenvectors.*

Proof We will show first that if the matrix A is diagonalizable, then it has n linearly independent eigenvectors. The diagonalizability of A implies that there is an invertible matrix P and a diagonal matrix D, say

$$P = \begin{bmatrix} p_{11} & p_{12} & \cdots & p_{1n} \\ p_{21} & p_{22} & \cdots & p_{2n} \\ \vdots & \vdots & & \vdots \\ p_{n1} & p_{n2} & \cdots & p_{nn} \end{bmatrix} \quad \text{and} \quad D = \begin{bmatrix} \lambda_1 & 0 & \cdots & 0 \\ 0 & \lambda_2 & \cdots & 0 \\ \vdots & \vdots & \ddots & \vdots \\ 0 & 0 & \cdots & \lambda_n \end{bmatrix} \tag{14}$$

such that $P^{-1}AP = D$. If we rewrite this as $AP = PD$ and substitute (14), we obtain

$$AP = PD = \begin{bmatrix} p_{11} & p_{12} & \cdots & p_{1n} \\ p_{21} & p_{22} & \cdots & p_{2n} \\ \vdots & \vdots & & \vdots \\ p_{n1} & p_{n2} & \cdots & p_{nn} \end{bmatrix} \begin{bmatrix} \lambda_1 & 0 & \cdots & 0 \\ 0 & \lambda_2 & \cdots & 0 \\ \vdots & \vdots & \ddots & \vdots \\ 0 & 0 & \cdots & \lambda_n \end{bmatrix} = \begin{bmatrix} \lambda_1 p_{11} & \lambda_2 p_{12} & \cdots & \lambda_n p_{1n} \\ \lambda_1 p_{21} & \lambda_2 p_{22} & \cdots & \lambda_n p_{2n} \\ \vdots & \vdots & & \vdots \\ \lambda_1 p_{n1} & \lambda_2 p_{n2} & \cdots & \lambda_n p_{nn} \end{bmatrix} \tag{15}$$

Thus, if we denote the column vectors of P by $\mathbf{p}_1, \mathbf{p}_2, \ldots, \mathbf{p}_n$, then the left side of (15) can be expressed as

$$AP = A[\mathbf{p}_1 \quad \mathbf{p}_2 \quad \cdots \quad \mathbf{p}_n] = [A\mathbf{p}_1 \quad A\mathbf{p}_2 \quad \cdots \quad A\mathbf{p}_n] \tag{16}$$

and the right side of (15) as

$$[\lambda_1 \mathbf{p}_1 \quad \lambda_2 \mathbf{p}_2 \quad \cdots \quad \lambda_n \mathbf{p}_n] \tag{17}$$

It follows from (16) and (17) that

$$A\mathbf{p}_1 = \lambda_1 \mathbf{p}_1, \quad A\mathbf{p}_2 = \lambda_2 \mathbf{p}_2, \ldots, \quad A\mathbf{p}_n = \lambda_n \mathbf{p}_n$$

and it follows from the invertibility of P that $\mathbf{p}_1, \mathbf{p}_2, \ldots, \mathbf{p}_n$ are nonzero, so we have shown that $\mathbf{p}_1, \mathbf{p}_2, \ldots, \mathbf{p}_n$ are eigenvectors of A corresponding to $\lambda_1, \lambda_2, \ldots, \lambda_n$, respectively. Moreover, the invertibility of P also implies that $\mathbf{p}_1, \mathbf{p}_2, \ldots, \mathbf{p}_n$ are linearly independent (Theorem 7.4.4 applied to P), so the column vectors of P form a set of n linearly independent eigenvectors of A.

Conversely, assume that A has n linearly independent eigenvectors, $\mathbf{p}_1, \mathbf{p}_2, \ldots, \mathbf{p}_n$, and that the corresponding eigenvalues are $\lambda_1, \lambda_2, \ldots, \lambda_n$, so

$$A\mathbf{p}_1 = \lambda_1 \mathbf{p}_1, \quad A\mathbf{p}_2 = \lambda_2 \mathbf{p}_2, \ldots, \quad A\mathbf{p}_n = \lambda_1 \mathbf{p}_n$$

If we now form the matrices

$$P = [\mathbf{p}_1 \quad \mathbf{p}_2 \quad \cdots \quad \mathbf{p}_n] = \begin{bmatrix} p_{11} & p_{12} & \cdots & p_{1n} \\ p_{21} & p_{22} & \cdots & p_{2n} \\ \vdots & \vdots & & \vdots \\ p_{n1} & p_{n2} & \cdots & p_{nn} \end{bmatrix} \quad \text{and} \quad D = \begin{bmatrix} \lambda_1 & 0 & \cdots & 0 \\ 0 & \lambda_2 & \cdots & 0 \\ \vdots & \vdots & \ddots & \vdots \\ 0 & 0 & \cdots & \lambda_n \end{bmatrix}$$

then we obtain

$$\begin{aligned} AP = A[\mathbf{p}_1 \quad \mathbf{p}_2 \quad \cdots \quad \mathbf{p}_n] &= [A\mathbf{p}_1 \quad A\mathbf{p}_2 \quad \cdots \quad A\mathbf{p}_n] \\ &= [\lambda_1\mathbf{p}_1 \quad \lambda_2\mathbf{p}_2 \quad \cdots \quad \lambda_n\mathbf{p}_n] \\ &= \begin{bmatrix} \lambda_1 p_{11} & \lambda_2 p_{12} & \cdots & \lambda_n p_{1n} \\ \lambda_1 p_{21} & \lambda_2 p_{22} & \cdots & \lambda_n p_{2n} \\ \vdots & \vdots & & \vdots \\ \lambda_1 p_{n1} & \lambda_2 p_{n2} & \cdots & \lambda_n p_{nn} \end{bmatrix} \\ &= \begin{bmatrix} p_{11} & p_{12} & \cdots & p_{1n} \\ p_{21} & p_{22} & \cdots & p_{2n} \\ \vdots & \vdots & & \vdots \\ p_{n1} & p_{n2} & \cdots & p_{nn} \end{bmatrix} \begin{bmatrix} \lambda_1 & 0 & \cdots & 0 \\ 0 & \lambda_2 & \cdots & 0 \\ \vdots & \vdots & \ddots & \vdots \\ 0 & 0 & \cdots & \lambda_n \end{bmatrix} = PD \end{aligned}$$

However, the matrix P is invertible, since its column vectors are linearly independent, so it follows from this computation that $D = P^{-1}AP$, which shows that A is diagonalizable. ∎

REMARK Keeping in mind that a set of n linearly independent vectors in R^n must be a basis for R^n, Theorem 8.2.6 is equivalent to saying that an $n \times n$ *matrix A is diagonalizable if and only if there is a basis for R^n consisting of eigenvectors of A.*

A METHOD FOR DIAGONALIZING A MATRIX

Theorem 8.2.6 guarantees that an $n \times n$ matrix A with n linearly independent eigenvectors is diagonalizable, and its proof provides the following method for diagonalizing A in that case.

Diagonalizing an $n \times n$ Matrix with n Linearly Independent Eigenvectors

Step 1. Find n linearly independent eigenvectors of A, say $\mathbf{p}_1, \mathbf{p}_2, \ldots, \mathbf{p}_n$.
Step 2. Form the matrix $P = [\mathbf{p}_1 \quad \mathbf{p}_2 \quad \cdots \quad \mathbf{p}_n]$.
Step 3. The matrix $P^{-1}AP$ will be diagonal and will have the eigenvalues corresponding to $\mathbf{p}_1, \mathbf{p}_2, \ldots, \mathbf{p}_n$, respectively, as its successive diagonal entries.

EXAMPLE 4
Diagonalizing a Matrix

We showed in Example 3 that the matrix

$$A = \begin{bmatrix} 0 & 0 & -2 \\ 1 & 2 & 1 \\ 1 & 0 & 3 \end{bmatrix}$$

has eigenvalues $\lambda = 1$ and $\lambda = 2$ and that basis vectors for these eigenspaces are

$$\mathbf{p}_1 = \begin{bmatrix} -2 \\ 1 \\ 1 \end{bmatrix} \quad \text{and} \quad \mathbf{p}_2 = \begin{bmatrix} -1 \\ 0 \\ 1 \end{bmatrix}, \quad \mathbf{p}_3 = \begin{bmatrix} 0 \\ 1 \\ 0 \end{bmatrix}$$
$$\lambda = 1 \qquad\qquad \lambda = 2$$

It is a straightforward matter to show that these three vectors are linearly independent, so A is diagonalizable and is diagonalized by

$$P = \begin{bmatrix} -2 & -1 & 0 \\ 1 & 0 & 1 \\ 1 & 1 & 0 \end{bmatrix}$$

As a check, we leave it for you to verify that

$$P^{-1}AP = \begin{bmatrix} -1 & 0 & -1 \\ 1 & 0 & 2 \\ 1 & 1 & 1 \end{bmatrix} \begin{bmatrix} 0 & 0 & -2 \\ 1 & 2 & 1 \\ 1 & 0 & 3 \end{bmatrix} \begin{bmatrix} -2 & -1 & 0 \\ 1 & 0 & 1 \\ 1 & 1 & 0 \end{bmatrix} = \begin{bmatrix} 1 & 0 & 0 \\ 0 & 2 & 0 \\ 0 & 0 & 2 \end{bmatrix}$$ ■

REMARK There is no preferred order for the columns of a diagonalizing matrix P—the only effect of changing the order of the columns is to change the order in which the eigenvalues appear along the main diagonal of $D = P^{-1}AP$. For example, had we written the column vectors of P in Example 4 in the order

$$P = [\mathbf{p}_3 \quad \mathbf{p}_1 \quad \mathbf{p}_2] = \begin{bmatrix} 0 & -2 & -1 \\ 1 & 1 & 0 \\ 0 & 1 & 1 \end{bmatrix}$$

then the resulting diagonal matrix would have been

$$P^{-1}AP = \begin{bmatrix} 2 & 0 & 0 \\ 0 & 1 & 0 \\ 0 & 0 & 2 \end{bmatrix}$$

EXAMPLE 5
A Matrix That
Is Not
Diagonalizable

We showed in Example 2 that the matrix

$$A = \begin{bmatrix} 2 & 0 & 0 \\ 1 & 3 & 0 \\ -3 & 5 & 3 \end{bmatrix}$$

has eigenvalues $\lambda = 2$ and $\lambda = 3$ and that bases for the corresponding eigenspaces are

$$\mathbf{p}_1 = \begin{bmatrix} \frac{1}{8} \\ -\frac{1}{8} \\ 1 \end{bmatrix} \quad \text{and} \quad \mathbf{p}_2 = \begin{bmatrix} 0 \\ 0 \\ 1 \end{bmatrix}$$
$$\lambda = 2 \qquad\qquad \lambda = 3$$

These eigenvectors are linearly independent, since they are not scalar multiples of one another, but it is impossible to produce a third linearly independent eigenvector since all other eigenvectors must be scalar multiples of one of these two. Thus, A is not diagonalizable.   ■

**LINEAR INDEPENDENCE
OF EIGENVECTORS**

The following theorem is useful for finding linearly independent sets of eigenvectors.

> **Theorem 8.2.7** *If $\mathbf{v}_1, \mathbf{v}_2, \ldots, \mathbf{v}_k$ are eigenvectors of a matrix A that correspond to distinct eigenvalues $\lambda_1, \lambda_2, \ldots, \lambda_k$, then the set $\{\mathbf{v}_1, \mathbf{v}_2, \ldots, \mathbf{v}_k\}$ is linearly independent.*

Proof We will assume that $\mathbf{v}_1, \mathbf{v}_2, \ldots, \mathbf{v}_k$ are linearly dependent and obtain a contradiction. If $\mathbf{v}_1, \mathbf{v}_2, \ldots, \mathbf{v}_k$ are linearly dependent, then some vector in this sequence must be a linear combination of predecessors (Theorem 7.1.2). If we let $\mathbf{v}_{r+1}$ be the *first* vector in the sequence that is a linear combination of predecessors, then $\mathbf{v}_1, \mathbf{v}_2, \ldots, \mathbf{v}_r$ are linearly independent, and there exist scalars $c_1, c_2, \ldots, c_k$ such that

$$\mathbf{v}_{r+1} = c_1\mathbf{v}_1 + c_2\mathbf{v}_2 + \cdots + c_r\mathbf{v}_r \tag{18}$$

Multiplying both sides of (18) by A and using the fact that $A\mathbf{v}_j = \lambda_j\mathbf{v}_j$ for each j yields

$$\lambda_{r+1}\mathbf{v}_{r+1} = c_1\lambda_1\mathbf{v}_1 + c_2\lambda_2\mathbf{v}_2 + \cdots + c_r\lambda_r\mathbf{v}_r \tag{19}$$

Now multiplying (18) by λ_{r+1} and subtracting from (19) yields

$$\mathbf{0} = c_1(\lambda_1 - \lambda_{r+1})\mathbf{v}_1 + c_2(\lambda_2 - \lambda_{r+1})\mathbf{v}_2 + \cdots + c_r(\lambda_r - \lambda_{r+1})\mathbf{v}_r \qquad (20)$$

Since $\mathbf{v}_1, \mathbf{v}_2, \ldots, \mathbf{v}_r$ are linearly independent, it follows that all of the coefficients on the right side of (20) are zero. However, the eigenvalues are all distinct, so it must be that

$$c_1 = c_2 = \cdots = c_r = 0$$

But this and (18) imply that $\mathbf{v}_{r+1} = \mathbf{0}$, which is impossible since eigenvectors are nonzero. Thus, $\mathbf{v}_1, \mathbf{v}_2, \ldots, \mathbf{v}_k$ must be linearly independent. ∎

REMARK   If $\lambda_1, \lambda_2, \ldots, \lambda_k$ are distinct eigenvalues of a matrix A, then Theorem 8.2.7 tells us that a linearly independent set is produced by choosing one eigenvector from each of the corresponding eigenspaces. More generally, it can be proved that if one chooses linearly independent *sets* of eigenvectors from distinct eigenspaces and combines them into a single set, then that combined set will be linearly independent. For example, for the matrix A in Example 4 we had an eigenvector $\mathbf{p}_1$ from the eigenspace corresponding to $\lambda = 1$ and two linearly independent eigenvectors $\mathbf{p}_2$ and $\mathbf{p}_3$ from the eigenspace corresponding to $\lambda = 2$, so we are guaranteed without any computations that $\{\mathbf{p}_1, \mathbf{p}_2, \mathbf{p}_3\}$ is a linearly independent set.

It follows from Theorems 8.2.6 and 8.2.7 that an $n \times n$ matrix with n distinct real eigenvalues must be diagonalizable, since we can produce a set of n linearly independent eigenvectors by choosing one eigenvector from each eigenspace.

Theorem 8.2.8   *An $n \times n$ matrix with n distinct real eigenvalues is diagonalizable.*

EXAMPLE 6
Diagonalizable
Matrix with
Distinct
Eigenvalues

The 3×3 matrix

$$A = \begin{bmatrix} 2 & 0 & 0 \\ 1 & 3 & 0 \\ -3 & 5 & 4 \end{bmatrix}$$

is diagonalizable, since it has three distinct eigenvalues, $\lambda = 2$, $\lambda = 3$, and $\lambda = 4$. ∎

The converse of Theorem 8.2.8 is *false*; that is, it is possible for an $n \times n$ matrix to be diagonalizable without having n distinct eigenvalues. For example, the matrix A in Example 4 was seen to be diagonalizable, even though it had only two distinct eigenvalues, $\lambda = 1$ and $\lambda = 2$. The diagonalizability was a consequence of the fact that the eigenspaces had dimensions 1 and 2, respectively, thereby allowing us to produce three linearly independent eigenvectors. Thus, we see that the key to diagonalizability rests with the dimensions of the eigenspaces.

Theorem 8.2.9   *An $n \times n$ matrix A is diagonalizable if and only if the sum of the geometric multiplicities of its eigenvalues is n.*

Proof   Let $\lambda_1, \lambda_2, \ldots, \lambda_k$ be the distinct eigenvalues of A, let $E_1, E_2, \ldots, E_k$ denote the corresponding eigenspaces, let $B_1, B_2, \ldots, B_k$ be any bases for these eigenspaces, and let B be the linearly independent set that results when the bases are merged into a single set (i.e., B is the union of the bases). If the sum of the geometric multiplicities is n, then B is a set of n linearly independent eigenvectors, so A is diagonalizable by Theorem 8.2.6. The proof of the converse is left for more advanced courses. ∎

EXAMPLE 7
Diagonalizability
and Geometric
Multiplicity

We showed in Example 2 that the matrix

$$A = \begin{bmatrix} 2 & 0 & 0 \\ 1 & 3 & 0 \\ -3 & 5 & 3 \end{bmatrix}$$

has eigenvalues $\lambda = 2$ and $\lambda = 3$, both with geometric multiplicity 1. Since the sum of the geometric multiplicities is less than 3, the matrix is not diagonalizable. Also, we showed in Example 3 that the matrix

$$A = \begin{bmatrix} 0 & 0 & -2 \\ 1 & 2 & 1 \\ 1 & 0 & 3 \end{bmatrix}$$

has eigenvalues $\lambda = 1$ and $\lambda = 2$ with geometric multiplicities 1 and 2, respectively. Since the sum of the geometric multiplicities is 3, the matrix is diagonalizable (see Example 4). ∎

RELATIONSHIP BETWEEN ALGEBRAIC AND GEOMETRIC MULTIPLICITY

A full excursion into the study of diagonalizability will be left for more advanced courses, but we will mention one result that is important for a full understanding of the diagonalizability question: It can be proved that the geometric multiplicity of an eigenvalue cannot exceed its algebraic multiplicity. For example, if the characteristic polynomial of some 6×6 matrix A is

$$p(\lambda) = (\lambda - 3)(\lambda - 5)^2(\lambda - 6)^3$$

then, depending on the particular matrix A, the eigenspace corresponding to $\lambda = 6$ might have dimension 1, 2, or 3, the eigenspace corresponding to $\lambda = 5$ might have dimension 1 or 2, and the eigenspace corresponding to $\lambda = 3$ must have dimension 1. For the matrix A to be diagonalizable there would have to be six linearly independent eigenvectors, and this will only occur if the geometric and algebraic multiplicities are the same; that is, if the eigenspace corresponding to $\lambda = 6$ has dimension 3, the eigenspace corresponding to $\lambda = 5$ has dimension 2, and the eigenspace corresponding to $\lambda = 3$ has dimension 1. The following theorem, whose proof is outlined in the exercises, summarizes these ideas.

Theorem 8.2.10 *If A is a square matrix, then:*

(a) *The geometric multiplicity of an eigenvalue of A is less than or equal to its algebraic multiplicity.*

(b) *A is diagonalizable if and only if the geometric multiplicity of each eigenvalue of A is the same as its algebraic multiplicity.*

A UNIFYING THEOREM ON DIAGONALIZABILITY

The following unifying theorem ties together some of the results we have considered in this section.

Theorem 8.2.11 *If A is an $n \times n$ matrix, then the following statements are equivalent.*

(a) *A is diagonalizable.*

(b) *A has n linearly independent eigenvectors.*

(c) *R^n has a basis consisting of eigenvectors of A.*

(d) *The sum of the geometric multiplicities of the eigenvalues of A is n.*

(e) *The geometric multiplicity of each eigenvalue of A is the same as the algebraic multiplicity.*

CONCEPT PROBLEM State a relationship between the rank of a diagonalizable matrix and its nonzero eigenvalues.

Exercise Set 8.2

In Exercises 1–4, show that A and B are not similar matrices.

1. $A = \begin{bmatrix} 1 & 1 \\ 3 & 2 \end{bmatrix}$, $B = \begin{bmatrix} 1 & 0 \\ 3 & -2 \end{bmatrix}$

2. $A = \begin{bmatrix} 4 & -1 \\ 2 & 4 \end{bmatrix}$, $B = \begin{bmatrix} 4 & 1 \\ 2 & 4 \end{bmatrix}$

3. $A = \begin{bmatrix} 1 & 2 & 3 \\ 0 & 1 & 2 \\ 0 & 0 & 1 \end{bmatrix}$, $B = \begin{bmatrix} 1 & 2 & 0 \\ \frac{1}{2} & 1 & 0 \\ 0 & 0 & 1 \end{bmatrix}$

4. $A = \begin{bmatrix} 1 & 0 & 1 \\ 2 & 0 & 2 \\ 3 & 0 & 3 \end{bmatrix}$, $B = \begin{bmatrix} 1 & 1 & 0 \\ 2 & 2 & 0 \\ 0 & 1 & 1 \end{bmatrix}$

In Exercises 5 and 6, the characteristic polynomial of a matrix A is given. Find the size of the matrix, list its eigenvalues with their algebraic multiplicities, and discuss the possible dimensions of the eigenspaces.

5. (a) $\lambda(\lambda + 1)^2(\lambda - 1)^2$
  (b) $(\lambda + 3)(\lambda + 1)^3(\lambda - 8)^7$

6. (a) $\lambda(\lambda - 1)(\lambda + 2)(\lambda - 3)^2$
  (b) $\lambda^2(\lambda - 6)(\lambda - 2)^3$

In Exercises 7–10, find the eigenvalues of A and their algebraic and geometric multiplicities.

7. $A = \begin{bmatrix} 1 & 1 & 4 \\ 0 & 1 & 1 \\ 0 & 0 & 2 \end{bmatrix}$

8. $A = \begin{bmatrix} 1 & 0 & 0 \\ 2 & 3 & 0 \\ 7 & 1 & 5 \end{bmatrix}$

9. $A = \begin{bmatrix} 5 & 0 & 0 \\ 1 & 5 & 1 \\ -1 & 0 & 3 \end{bmatrix}$

10. $A = \begin{bmatrix} 1 & -2 & 0 \\ -2 & 1 & 0 \\ 2 & 2 & 3 \end{bmatrix}$

In Exercises 11 and 12, find the geometric multiplicities of the eigenvalues of A by computing the rank of $\lambda I - A$ for each eigenvalue by row reduction and then using the relationship between rank and nullity.

11. $A = \begin{bmatrix} -1 & -1 & 1 \\ -1 & -1 & 1 \\ 1 & 1 & -1 \end{bmatrix}$

12. $A = \begin{bmatrix} 2 & -2 & 4 \\ 2 & -6 & 11 \\ 1 & -4 & 7 \end{bmatrix}$

In Exercises 13 and 14, find the rank of $\lambda I - A$ for each eigenvalue by row reduction, and use the results to show that A is diagonalizable.

13. $A = \begin{bmatrix} 4 & 0 & 1 \\ 2 & 3 & 2 \\ 1 & 0 & 4 \end{bmatrix}$

14. $A = \begin{bmatrix} 0 & -1 & 1 \\ -1 & 0 & 1 \\ 1 & 1 & 0 \end{bmatrix}$

In Exercises 15–18, find a matrix P that diagonalizes the matrix A, and determine $P^{-1}AP$.

15. $A = \begin{bmatrix} -14 & 12 \\ -20 & 17 \end{bmatrix}$

16. $A = \begin{bmatrix} 1 & 0 \\ 6 & -1 \end{bmatrix}$

17. $A = \begin{bmatrix} 1 & 0 & 0 \\ 0 & 1 & 1 \\ 0 & 1 & 1 \end{bmatrix}$

18. $A = \begin{bmatrix} 2 & 0 & -2 \\ 0 & 3 & 0 \\ 0 & 0 & 3 \end{bmatrix}$

In Exercises 19–24, determine whether A is diagonalizable. If so, find a matrix P that diagonalizes the matrix A, and determine $P^{-1}AP$.

19. $A = \begin{bmatrix} -1 & 4 & -2 \\ -3 & 4 & 0 \\ -3 & 1 & 3 \end{bmatrix}$

20. $A = \begin{bmatrix} 19 & -9 & -6 \\ 25 & -11 & -9 \\ 17 & -9 & -4 \end{bmatrix}$

21. $A = \begin{bmatrix} 5 & 0 & 0 \\ 1 & 5 & 0 \\ 0 & 1 & 5 \end{bmatrix}$

22. $A = \begin{bmatrix} 0 & 0 & 0 \\ 0 & 0 & 0 \\ 3 & 0 & 1 \end{bmatrix}$

23. $A = \begin{bmatrix} -2 & 0 & 0 & 0 \\ 0 & -2 & 0 & 0 \\ 0 & 0 & 3 & 0 \\ 0 & 0 & 1 & 3 \end{bmatrix}$

24. $A = \begin{bmatrix} -2 & 0 & 0 & 0 \\ 0 & -2 & 5 & -5 \\ 0 & 0 & 3 & 0 \\ 0 & 0 & 0 & 3 \end{bmatrix}$

25. Show that if an upper triangular matrix with 1's on the main diagonal is diagonalizable, then it is the identity matrix.

26. Show that if a 3×3 matrix has a three-dimensional eigenspace, then it must be diagonal. State a generalization of this result.

27. Show that similar matrices are either both invertible or both singular.

28. Suppose that A, P, and D are $n \times n$ matrices such that D is diagonal, the columns of P are nonzero vectors, and $AP = PD$. Show that the diagonal entries of D are eigenvalues of A and that the kth column vector of P is an eigenvector corresponding to the kth diagonal entry of D. [*Suggestion:* Partition P into column vectors.]

If $T: R^n \to R^n$ is a linear operator and B is any basis for R^n, then we know that $[T]$ and $[T]_B$ are similar matrices, so it follows from Theorem 8.2.4 that $[T]$ and $[T]_B$ have the same eigenvalues with the same algebraic and geometric multiplicities. Since these common properties of $[T]$ and $[T]_B$ are independent of the basis B, we can regard them to be properties of the operator T. Thus, for example, we call the eigenvalues of $[T]$ the *eigenvalues of T*, we call the eigenvectors of $[T]$ the *eigenvectors of T*, and we say that T is a *diagonalizable operator* if and only if $[T]$ is a diagonalizable matrix. These ideas are used in Exercises 29–31.

29. Consider the linear operator $T: R^3 \to R^3$ defined by the formula

$$T(x_1, x_2, x_3) = (-2x_1 + x_2 - x_3, x_1 - 2x_2 - x_3,$$
$$-x_1 - x_2 - 2x_3)$$

Find the eigenvalues of T and show that T is diagonalizable.

30. Consider the linear operator $T: R^3 \to R^3$ defined by the formula

$$T(x_1, x_2, x_3) = (-x_2 + x_3, -x_1 + x_3, x_1 + x_2)$$

Find the eigenvalues of T and show that T is diagonalizable.

31. Suppose that $T: R^n \to R^n$ is a linear operator and λ is an eigenvalue of T. Show that if B is any basis for R^n, then $\mathbf{x}$ is an eigenvector of T corresponding to λ if and only if $[\mathbf{x}]_B$ is an eigenvector of $[T]_B$ corresponding to λ.

32. Let

$$A = \begin{bmatrix} a & b \\ c & d \end{bmatrix}$$

Show that

(a) A is diagonalizable if $(a - d)^2 + 4bc > 0$;

(b) A is not diagonalizable if $(a - d)^2 + 4bc < 0$.

Discussion and Discovery

D1. Devise a method for finding two $n \times n$ matrices that are not similar. Use your method to find two 3×3 matrices that are not similar.

D2. Indicate whether the statement is true (T) or false (F). Justify your answer.

(a) Every square matrix is similar to itself.

(b) If A is similar to B, and B is similar to C, then A is similar to C.

(c) If A and B are similar invertible matrices, then A^{-1} and B^{-1} are similar.

(d) If every eigenvalue of A has algebraic multiplicity 1, then A is diagonalizable.

D3. Indicate whether the statement is true (T) or false (F). Justify your answer.

(a) Singular matrices are not diagonalizable.

(b) If A is diagonalizable, then there is a unique matrix P such that $P^{-1}AP$ is a diagonal matrix.

(c) If $\mathbf{v}_1$, $\mathbf{v}_2$, and $\mathbf{v}_3$ are nonzero vectors that come from different eigenspaces of A, then it is impossible to express $\mathbf{v}_3$ as a linear combination of $\mathbf{v}_1$ and $\mathbf{v}_2$.

(d) If an invertible matrix A is diagonalizable, then A^{-1} is also diagonalizable.

(e) If R^n has a basis of eigenvectors for the matrix A, then A is diagonalizable.

D4. Suppose that the characteristic polynomial of a matrix A is

$$p(\lambda) = (\lambda - 1)(\lambda - 3)^2(\lambda - 4)^3$$

(a) What size is A?

(b) What can you say about the dimensions of the eigenspaces of A?

(c) What can you say about the dimensions of the eigenspaces if you know that A is diagonalizable?

(d) If $\{\mathbf{v}_1, \mathbf{v}_2, \mathbf{v}_3\}$ is a linearly independent set of eigenvectors of A all of which correspond to the same eigenvalue of A, what can you say about that eigenvalue?

D5. Suppose that A is a 6×6 matrix with three distinct eigenvalues, λ_1, λ_2, and λ_3.

(a) What can you say about the diagonalizability of A if λ_1 has geometric multiplicity 2 and λ_2 has geometric multiplicity 3?

(b) What can you say about the diagonalizability of A if λ_1 has geometric multiplicity 2, λ_2 has geometric multiplicity 1, and λ_3 has geometric multiplicity 2?

(c) What can you say about the diagonalizability of A if λ_1 and λ_2 have geometric multiplicity 2?

Working with Proofs

P1. Prove parts (b) and (c) of Theorem 8.2.3. [*Hint:* Use the results in Exercise P7 of Section 7.5.]

P2. Prove part (d) of Theorem 8.2.3.

P3. Prove part (b) of Theorem 8.2.5.

P4. Prove that if A and B are similar, then so are A^k and B^k for every positive integer k.

P5. Prove that if A is diagonalizable, then so is A^k for every positive integer k.

P6. This problem will lead you through a proof of the fact that the algebraic multiplicity of an eigenvalue of an $n \times n$ matrix A is greater than or equal to the geometric multiplicity. For this purpose, assume that λ_0 is an eigenvalue with geometric multiplicity k.

(a) Prove that there is a basis $B = \{\mathbf{u}_1, \mathbf{u}_2, \ldots, \mathbf{u}_n\}$ for R^n in which the first k vectors of B form a basis for the eigenspace corresponding to λ_0.

(b) Let P be the matrix having the vectors in B as columns. Prove that the product AP can be expressed as

$$AP = P \begin{bmatrix} \lambda_0 I_k & X \\ 0 & Y \end{bmatrix}$$

[*Hint:* Compare the first k column vectors on both sides.]

(c) Use the result in part (b) to prove that A is similar to

$$C = \begin{bmatrix} \lambda_0 I_k & X \\ 0 & Y \end{bmatrix}$$

and hence that A and C have the same characteristic polynomial.

(d) By considering $\det(\lambda I - C)$, prove that the characteristic polynomial of C (and hence A) contains the factor $(\lambda - \lambda_0)$ at least k times, thereby proving that the algebraic multiplicity of λ_0 is greater than or equal to the geometric multiplicity k. [*Hint:* See the instructions preceding Exercises 38 and 39 of Section 4.2.]

Technology Exercises

T1. Most linear algebra technology utilities have specific commands for diagonalizing a matrix. If your utility has this capability, then you may find it described as a "Jordan decomposition" or some similar name involving the word "Jordan." Use this command to diagonalize the matrix in Example 4.

T2. (a) Show that the matrix

$$A = \begin{bmatrix} -13 & -60 & -60 \\ 10 & 42 & 40 \\ -5 & -20 & -18 \end{bmatrix}$$

is diagonalizable by finding the nullity of $\lambda I - A$ for each eigenvalue λ and calling on an appropriate theorem.

(b) Find a basis for R^3 consisting of eigenvectors of A.

T3. Construct a 4×4 diagonalizable matrix A whose entries are nonzero and whose characteristic equation is

$$p(\lambda) = (\lambda - 2)^2(\lambda + 3)^2$$

and check your result by diagonalizing A. [*Hint:* See the instructions for Exercises 38 and 39 of Section 4.2.]

Section 8.3 Orthogonal Diagonalizability; Functions of a Matrix

Symmetric matrices arise more frequently in applications than any other class of matrices, so in this section we will consider the diagonalization properties of such matrices. We will also discuss methods for defining functions of matrices.

ORTHOGONAL SIMILARITY

Recall from the last section that two $n \times n$ matrices A and C are said to be *similar* if there exists an invertible matrix P such that $C = P^{-1}AP$. The special case in which there is an orthogonal matrix P such that $C = P^{-1}AP = P^TAP$ is of special importance and has some terminology associated with it.

Definition 8.3.1 If A and C are square matrices with the same size, then we say that *C is orthogonally similar to A* if there exists an orthogonal matrix P such that $C = P^TAP$.

Using the remark following Definition 8.2.1 as a guide, you should be able to show that if C is orthogonally similar to A, then A is orthogonally similar to C, so we will usually say that *A and C are orthogonally similar* to emphasize that the relationship goes both ways.

The following analog of Theorem 8.2.2, whose proof is left as an exercise, gives an interpretation of orthogonal similarity from an operator point of view.

Theorem 8.3.2 *Two matrices are orthogonally similar if and only if there exist orthonormal bases with respect to which the matrices represent the same linear operator.*

Our main concern in this section is determining conditions under which a matrix will be orthogonally similar to a diagonal matrix.

The Orthogonal Diagonalization Problem   Given a square matrix A, does there exist an orthogonal matrix P for which $P^T A P$ is a diagonal matrix, and if so, how does one find such a P? If such a matrix P exists, then A is said to be ***orthogonally diagonalizable***, and P is said to ***orthogonally diagonalize*** A.

REMARK If you think of A as the standard matrix for a linear operator, then the orthogonal diagonalization problem is equivalent to asking whether this operator can be represented by a diagonal matrix with respect to some orthonormal basis.

The first observation we should make about orthogonal diagonalization is that there is no hope of orthogonally diagonalizing a *nonsymmetric* matrix. To see why this is so, suppose that

$$D = P^T A P \tag{1}$$

where P is orthogonal and D is diagonal. Since $P^T P = P P^T = I$, we can rewrite (1) as

$$A = P D P^T$$

Transposing both sides of this equation and using the fact that $D^T = D$ yields

$$A^T = (P D P^T)^T = (P^T)^T D^T P^T = P D P^T = A$$

which shows that an orthogonally diagonalizable matrix must be symmetric. Of course, this still leaves open the question of which symmetric matrices, if any, are orthogonally diagonalizable. The following analog of Theorem 8.2.6 will help us to answer this question.

Theorem 8.3.3 *An $n \times n$ matrix A is orthogonally diagonalizable if and only if there exists an orthonormal set of n eigenvectors of A.*

Proof We will show first that if A is orthogonally diagonalizable, then there exists an orthonormal set of n eigenvectors of A. The orthogonal diagonalizability of A implies that there exists an orthogonal matrix P and a diagonal matrix D such that $P^T A P = D$. However, since the column vectors of an orthogonal matrix are orthonormal, and since the column vectors of P are eigenvectors of A (see the proof of Theorem 8.2.6), we have established that the column vectors of P form an orthonormal set of n eigenvectors of A.

Conversely, assume that there exists an orthonormal set $\{\mathbf{p}_1, \mathbf{p}_2, \ldots, \mathbf{p}_n\}$ of n eigenvectors of A. We showed in the proof of Theorem 8.2.6 that the matrix

$$P = [\mathbf{p}_1 \quad \mathbf{p}_2 \quad \cdots \quad \mathbf{p}_n]$$

diagonalizes A. However, in this case P is an orthogonal matrix, since its column vectors are orthonormal. Thus, P orthogonally diagonalizes A. ∎

REMARK Recalling that an orthonormal set of n vectors in R^n is an orthonormal basis for R^n, Theorem 8.3.3 is equivalent to saying that *an $n \times n$ matrix A is orthogonally diagonalizable if and only if there is an orthonormal basis for R^n consisting of eigenvectors of A.*

We saw above that an orthogonally diagonalizable matrix must be symmetric. The following two-part theorem states that *all* symmetric matrices are orthogonally diagonalizable and gives a property of symmetric matrices that will lead to a method for orthogonally diagonalizing them.

Theorem 8.3.4

    (*a*) *A matrix is orthogonally diagonalizable if and only if it is symmetric.*

    (*b*) *If A is a symmetric matrix, then eigenvectors from different eigenspaces are orthogonal.*

We will prove part (*b*); the proof of part (*a*) is outlined in the exercises.

Proof (*b*) Let $\mathbf{v}_1$ and $\mathbf{v}_2$ be eigenvectors corresponding to distinct eigenvalues λ_1 and λ_2, respectively. The proof that $\mathbf{v}_1 \cdot \mathbf{v}_2 = 0$ will be facilitated by using Formula (26) of Section 3.1 to write $\lambda_1(\mathbf{v}_1 \cdot \mathbf{v}_2) = (\lambda_1\mathbf{v}_1) \cdot \mathbf{v}_2$ as the matrix product $(\lambda_1\mathbf{v}_1)^T\mathbf{v}_2$. The rest of the proof now consists of manipulating this expression in the right way:

$$
\begin{aligned}
\lambda_1(\mathbf{v}_1 \cdot \mathbf{v}_2) = (\lambda_1\mathbf{v}_1)^T\mathbf{v}_2 = (A\mathbf{v}_1)^T\mathbf{v}_2 &= (\mathbf{v}_1^T A^T)\mathbf{v}_2 \quad &&[\text{$\mathbf{v}_1$ is an eigenvector corresponding to λ_1.}]\\
&= (\mathbf{v}_1^T A)\mathbf{v}_2 \quad &&[\text{Symmetry of } A]\\
&= \mathbf{v}_1^T(A\mathbf{v}_2)\\
&= \mathbf{v}_1^T(\lambda_2\mathbf{v}_2) \quad &&[\text{$\mathbf{v}_2$ is an eigenvector corresponding to λ_2.}]\\
&= \lambda_2\mathbf{v}_1^T\mathbf{v}_2\\
&= \lambda_2(\mathbf{v}_1 \cdot \mathbf{v}_2) \quad &&[\text{Formula (26) of Section 3.1}]
\end{aligned}
$$

This implies that $(\lambda_1 - \lambda_2)(\mathbf{v}_1 \cdot \mathbf{v}_2) = 0$, so $\mathbf{v}_1 \cdot \mathbf{v}_2 = 0$ as a result of the fact that $\lambda_1 \neq \lambda_2$. ∎

A METHOD FOR ORTHOGONALLY DIAGONALIZING A SYMMETRIC MATRIX

To orthogonally diagonalize an $n \times n$ symmetric matrix A we have to construct an orthogonal matrix P whose column vectors are eigenvectors of A. One way to do this is to find a basis for each eigenspace, then use the Gram–Schmidt process to produce an orthonormal basis for each eigenspace, and then combine those vectors into a single set. We know from Theorem 8.2.9 that this combined set will have n vectors, and we know from part (*b*) of Theorem 8.3.4 that the eigenvectors in the set that come from distinct eigenspaces will be orthogonal. This implies that the entire set of n vectors will be orthonormal, so any matrix P that has these vectors as columns will be orthogonal and will diagonalize A. In summary, we have the following procedure for orthogonally diagonalizing symmetric matrices:

> **Orthogonally Diagonalizing an $n \times n$ Symmetric Matrix**
>
> **Step 1.** Find a basis for each eigenspace of A.
>
> **Step 2.** Apply the Gram–Schmidt process to each of these bases to produce orthonormal bases for the eigenspaces.
>
> **Step 3.** Form the matrix $P = [\mathbf{p}_1 \ \ \mathbf{p}_2 \ \ \cdots \ \ \mathbf{p}_n]$ whose columns are the vectors constructed in Step 2. The matrix P will orthogonally diagonalize A, and the eigenvalues on the diagonal of $D = P^T A P$ will be in the same order as their corresponding eigenvectors in P.

EXAMPLE 1

Orthogonally Diagonalizing a Symmetric Matrix

Find a matrix P that orthogonally diagonalizes the symmetric matrix

$$
A = \begin{bmatrix} 4 & 2 & 2 \\ 2 & 4 & 2 \\ 2 & 2 & 4 \end{bmatrix}
$$

Solution The characteristic equation of A is

$$
\det(\lambda I - A) = \det\begin{bmatrix} \lambda - 4 & -2 & -2 \\ -2 & \lambda - 4 & -2 \\ -2 & -2 & \lambda - 4 \end{bmatrix} = (\lambda - 2)^2(\lambda - 8) = 0 \tag{2}
$$

Thus, the eigenvalues of A are $\lambda = 2$ and $\lambda = 8$. Using the method given in Example 3 of

Section 8.2, it can be shown that the vectors

$$\mathbf{v}_1 = \begin{bmatrix} -1 \\ 1 \\ 0 \end{bmatrix} \quad \text{and} \quad \mathbf{v}_2 = \begin{bmatrix} -1 \\ 0 \\ 1 \end{bmatrix} \tag{3}$$

form a basis for the eigenspace corresponding to $\lambda = 2$ and that

$$\mathbf{v}_3 = \begin{bmatrix} 1 \\ 1 \\ 1 \end{bmatrix} \tag{4}$$

is a basis for the eigenspace corresponding to $\lambda = 8$. Applying the Gram–Schmidt process to the bases $\{\mathbf{v}_1, \mathbf{v}_2\}$ and $\{\mathbf{v}_3\}$ yields the orthonormal bases $\{\mathbf{u}_1, \mathbf{u}_2\}$ and $\{\mathbf{u}_3\}$, where

$$\mathbf{u}_1 = \begin{bmatrix} -\frac{1}{\sqrt{2}} \\ \frac{1}{\sqrt{2}} \\ 0 \end{bmatrix}, \quad \mathbf{u}_2 = \begin{bmatrix} -\frac{1}{\sqrt{6}} \\ -\frac{1}{\sqrt{6}} \\ \frac{2}{\sqrt{6}} \end{bmatrix} \quad \text{and} \quad \mathbf{u}_3 = \begin{bmatrix} \frac{1}{\sqrt{3}} \\ \frac{1}{\sqrt{3}} \\ \frac{1}{\sqrt{3}} \end{bmatrix}$$

Thus, A is orthogonally diagonalized by the matrix

$$P = \begin{bmatrix} -\frac{1}{\sqrt{2}} & -\frac{1}{\sqrt{6}} & \frac{1}{\sqrt{3}} \\ \frac{1}{\sqrt{2}} & -\frac{1}{\sqrt{6}} & \frac{1}{\sqrt{3}} \\ 0 & \frac{2}{\sqrt{6}} & \frac{1}{\sqrt{3}} \end{bmatrix}$$

As a check, we leave it for you to confirm that

$$P^T A P = \begin{bmatrix} -\frac{1}{\sqrt{2}} & \frac{1}{\sqrt{2}} & 0 \\ -\frac{1}{\sqrt{6}} & -\frac{1}{\sqrt{6}} & \frac{2}{\sqrt{6}} \\ \frac{1}{\sqrt{3}} & \frac{1}{\sqrt{3}} & \frac{1}{\sqrt{3}} \end{bmatrix} \begin{bmatrix} 4 & 2 & 2 \\ 2 & 4 & 2 \\ 2 & 2 & 4 \end{bmatrix} \begin{bmatrix} -\frac{1}{\sqrt{2}} & -\frac{1}{\sqrt{6}} & \frac{1}{\sqrt{3}} \\ \frac{1}{\sqrt{2}} & -\frac{1}{\sqrt{6}} & \frac{1}{\sqrt{3}} \\ 0 & \frac{2}{\sqrt{6}} & \frac{1}{\sqrt{3}} \end{bmatrix} = \begin{bmatrix} 2 & 0 & 0 \\ 0 & 2 & 0 \\ 0 & 0 & 8 \end{bmatrix} \quad \blacksquare$$

SPECTRAL DECOMPOSITION

If A is a symmetric matrix that is orthogonally diagonalized by

$$P = [\mathbf{u}_1 \quad \mathbf{u}_2 \quad \cdots \quad \mathbf{u}_n]$$

and if $\lambda_1, \lambda_2, \ldots, \lambda_n$ are the eigenvalues of A corresponding to $\mathbf{u}_1, \mathbf{u}_2, \ldots, \mathbf{u}_n$, then we know that $D = P^T A P$, where D is a diagonal matrix with the eigenvalues in the diagonal positions. It follows from this that the matrix A can be expressed as

$$A = PDP^T = [\mathbf{u}_1 \quad \mathbf{u}_2 \quad \cdots \quad \mathbf{u}_n] \begin{bmatrix} \lambda_1 & 0 & \cdots & 0 \\ 0 & \lambda_2 & \cdots & 0 \\ \vdots & \vdots & \ddots & \vdots \\ 0 & 0 & \cdots & \lambda_n \end{bmatrix} \begin{bmatrix} \mathbf{u}_1^T \\ \mathbf{u}_2^T \\ \vdots \\ \mathbf{u}_n^T \end{bmatrix} = [\lambda_1 \mathbf{u}_1 \quad \lambda_2 \mathbf{u}_2 \quad \cdots \quad \lambda_n \mathbf{u}_n] \begin{bmatrix} \mathbf{u}_1^T \\ \mathbf{u}_2^T \\ \vdots \\ \mathbf{u}_n^T \end{bmatrix}$$

Multiplying out using the column-row rule (Theorem 3.8.1), we obtain the formula

$$A = \lambda_1 \mathbf{u}_1 \mathbf{u}_1^T + \lambda_2 \mathbf{u}_2 \mathbf{u}_2^T + \cdots + \lambda_n \mathbf{u}_n \mathbf{u}_n^T \tag{5}$$

which is called a *spectral decomposition of A* or an *eigenvalue decomposition of A* (sometimes abbreviated as the EVD of A).[*]

---

[*]The terminology *spectral decomposition* is derived from the fact that the set of all eigenvalues of a matrix A is sometimes called the ***spectrum*** of A. The terminology *eigenvalue decomposition* is due to Professor Dan Kalman, who introduced it in an award-winning paper entitled "A Singularly Valuable Decomposition: The SVD of a Matrix," *The College Mathematics Journal*, Vol. 27, No. 1, January 1996.

To explain the geometric significance of this result, recall that if $\mathbf{u}$ is a unit vector in R^n that is expressed in column form, then the outer product $\mathbf{u}\mathbf{u}^T$ is the standard matrix for the orthogonal projection of R^n onto the line through the origin that is spanned by $\mathbf{u}$ [Theorem 7.7.3 and Formula (17) of Section 7.7]. Thus, the spectral decomposition of A tells us that $A\mathbf{x}$ can be computed by projecting $\mathbf{x}$ onto the lines determined by the eigenvectors of A, then scaling those projections by the eigenvalues, and then adding the scaled projections. Here is an example.

EXAMPLE 2

A Geometric Interpretation of the Spectral Decomposition

The matrix

$$A = \begin{bmatrix} 1 & 2 \\ 2 & -2 \end{bmatrix}$$

has eigenvalues $\lambda_1 = -3$ and $\lambda_2 = 2$ with corresponding eigenvectors

$$\mathbf{x}_1 = \begin{bmatrix} 1 \\ -2 \end{bmatrix} \quad \text{and} \quad \mathbf{x}_2 = \begin{bmatrix} 2 \\ 1 \end{bmatrix}$$

(verify). Normalizing these basis vectors yields

$$\mathbf{u}_1 = \frac{\mathbf{x}_1}{\|\mathbf{x}_1\|} = \begin{bmatrix} \frac{1}{\sqrt{5}} \\ -\frac{2}{\sqrt{5}} \end{bmatrix} \quad \text{and} \quad \mathbf{u}_2 = \frac{\mathbf{x}_2}{\|\mathbf{x}_2\|} = \begin{bmatrix} \frac{2}{\sqrt{5}} \\ \frac{1}{\sqrt{5}} \end{bmatrix}$$

so a spectral decomposition of A is

$$\begin{bmatrix} 1 & 2 \\ 2 & -2 \end{bmatrix} = \lambda_1\mathbf{u}_1\mathbf{u}_1^T + \lambda_2\mathbf{u}_2\mathbf{u}_2^T = (-3)\begin{bmatrix} \frac{1}{\sqrt{5}} \\ -\frac{2}{\sqrt{5}} \end{bmatrix}\begin{bmatrix} \frac{1}{\sqrt{5}} & -\frac{2}{\sqrt{5}} \end{bmatrix} + (2)\begin{bmatrix} \frac{2}{\sqrt{5}} \\ \frac{1}{\sqrt{5}} \end{bmatrix}\begin{bmatrix} \frac{2}{\sqrt{5}} & \frac{1}{\sqrt{5}} \end{bmatrix}$$

$$= (-3)\begin{bmatrix} \frac{1}{5} & -\frac{2}{5} \\ -\frac{2}{5} & \frac{4}{5} \end{bmatrix} + (2)\begin{bmatrix} \frac{4}{5} & \frac{2}{5} \\ \frac{2}{5} & \frac{1}{5} \end{bmatrix} \tag{6}$$

where the 2×2 matrices on the right are the standard matrices for the orthogonal projections onto the eigenspaces.

Now let us see what this decomposition tells us about the image of the vector $\mathbf{x} = (1, 1)$ under multiplication by A. Writing $\mathbf{x}$ in column form, it follows that

$$A\mathbf{x} = \begin{bmatrix} 1 & 2 \\ 2 & -2 \end{bmatrix}\begin{bmatrix} 1 \\ 1 \end{bmatrix} = \begin{bmatrix} 3 \\ 0 \end{bmatrix} \tag{7}$$

and from (6) that

$$A\mathbf{x} = \begin{bmatrix} 1 & 2 \\ 2 & -2 \end{bmatrix}\begin{bmatrix} 1 \\ 1 \end{bmatrix} = (-3)\begin{bmatrix} \frac{1}{5} & -\frac{2}{5} \\ -\frac{2}{5} & \frac{4}{5} \end{bmatrix}\begin{bmatrix} 1 \\ 1 \end{bmatrix} + (2)\begin{bmatrix} \frac{4}{5} & \frac{2}{5} \\ \frac{2}{5} & \frac{1}{5} \end{bmatrix}\begin{bmatrix} 1 \\ 1 \end{bmatrix}$$

$$= (-3)\begin{bmatrix} -\frac{1}{5} \\ \frac{2}{5} \end{bmatrix} + (2)\begin{bmatrix} \frac{6}{5} \\ \frac{3}{5} \end{bmatrix} = \begin{bmatrix} \frac{3}{5} \\ -\frac{6}{5} \end{bmatrix} + \begin{bmatrix} \frac{12}{5} \\ \frac{6}{5} \end{bmatrix} \tag{8}$$

It follows from (7) that the image of $(1, 1)$ under multiplication by A is $(3, 0)$, and it follows from (8) that this image can also be obtained by projecting $(1, 1)$ onto the eigenspaces corresponding to $\lambda_1 = -3$ and $\lambda_2 = 2$ to obtain the vectors $\left(-\frac{1}{5}, \frac{2}{5}\right)$ and $\left(\frac{6}{5}, \frac{3}{5}\right)$, then scaling by the eigenvalues to obtain $\left(\frac{3}{5}, -\frac{6}{5}\right)$ and $\left(\frac{12}{5}, \frac{6}{5}\right)$, and then adding these vectors (see Figure 8.3.1). ■

REMARK The spectral decomposition (5) expresses a symmetric matrix A as a linear combination of rank 1 matrices in which the coefficients of the matrices are the eigenvalues of A. In the exercises we will ask to you to show a kind of converse; namely, if $\{\mathbf{u}_1, \mathbf{u}_2, \ldots, \mathbf{u}_n\}$ is an

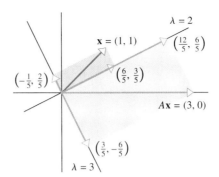

Figure 8.3.1

orthonormal basis for R^n, and if A can be expressed as

$$A = c_1\mathbf{u}_1\mathbf{u}_1^T + c_2\mathbf{u}_2\mathbf{u}_2^T + \cdots + c_n\mathbf{u}_n\mathbf{u}_n^T$$

then A is symmetric and has eigenvalues $c_1, c_2, \ldots, c_n$.

POWERS OF A DIAGONALIZABLE MATRIX

There are many applications that require the computation of high powers of square matrices. Since such computations can be time consuming and subject to roundoff error, there is considerable interest in techniques that can reduce the amount of computation involved. We will now consider an important method for computing high powers of diagonalizable matrices (symmetric matrices, for example). To explain the idea, suppose that A is an $n \times n$ matrix and P is an invertible $n \times n$ matrix. Then

$$(P^{-1}AP)^2 = (P^{-1}AP)(P^{-1}AP) = P^{-1}APP^{-1}AP = P^{-1}AIAP = P^{-1}A^2P$$

and more generally, if k is any positive integer, then

$$(P^{-1}AP)^k = P^{-1}A^k P \tag{9}$$

In particular, if A is diagonalizable and $P^{-1}AP = D$ is a diagonal matrix, then it follows from (9) that

$$P^{-1}A^k P = D^k \tag{10}$$

which we can rewrite as

$$A^k = PD^k P^{-1} \tag{11}$$

This equation, which is valid for any diagonalizable matrix, expresses the kth power of A in terms of the kth power of D, thereby taking advantage of the fact that powers of diagonal matrices are easy to compute [see Formula (3) of Section 3.6].

EXAMPLE 3
Powers of a
Diagonalizable
Matrix

Use Formula (11) to find A^{13} for the diagonalizable matrix

$$A = \begin{bmatrix} 0 & 0 & -2 \\ 1 & 2 & 1 \\ 1 & 0 & 3 \end{bmatrix}$$

Solution We showed in Example 4 of Section 8.2 that

$$P^{-1}AP = \begin{bmatrix} -1 & 0 & -1 \\ 1 & 0 & 2 \\ 1 & 1 & 1 \end{bmatrix} \begin{bmatrix} 0 & 0 & -2 \\ 1 & 2 & 1 \\ 1 & 0 & 3 \end{bmatrix} \begin{bmatrix} -2 & -1 & 0 \\ 1 & 0 & 1 \\ 1 & 1 & 0 \end{bmatrix} = \begin{bmatrix} 1 & 0 & 0 \\ 0 & 2 & 0 \\ 0 & 0 & 2 \end{bmatrix}$$

Thus,

$$A^{13} = \begin{bmatrix} -2 & -1 & 0 \\ 1 & 0 & 1 \\ 1 & 1 & 0 \end{bmatrix} \begin{bmatrix} 1^{13} & 0 & 0 \\ 0 & 2^{13} & 0 \\ 0 & 0 & 2^{13} \end{bmatrix} \begin{bmatrix} -1 & 0 & -1 \\ 1 & 0 & 2 \\ 1 & 1 & 1 \end{bmatrix} = \begin{bmatrix} -8190 & 0 & -16{,}382 \\ 8191 & 8192 & 8191 \\ 8191 & 0 & 16{,}383 \end{bmatrix} \quad (12)$$

$$\underbrace{}_{P} \qquad \underbrace{}_{D^{13}} \qquad \underbrace{}_{P^{-1}}$$

With this method most of the work is diagonalizing A. Once that work is done, it need not be repeated to compute other powers of A. For example, to compute A^{1000} we need only change the exponents from 13 to 1000 in (12). ∎

In the special case where A is a symmetric matrix with a spectral decomposition

$$A = \lambda_1 \mathbf{u}_1 \mathbf{u}_1^T + \lambda_2 \mathbf{u}_2 \mathbf{u}_2^T + \cdots + \lambda_n \mathbf{u}_n \mathbf{u}_n^T$$

the matrix

$$P = [\mathbf{u}_1 \quad \mathbf{u}_2 \quad \cdots \quad \mathbf{u}_n]$$

orthogonally diagonalizes A, so (11) can be expressed as

$$A^k = PD^k P^T$$

We leave it for you to show that this equation can be written as

$$A^k = \lambda_1^k \mathbf{u}_1 \mathbf{u}_1^T + \lambda_2^k \mathbf{u}_2 \mathbf{u}_2^T + \cdots + \lambda_n^k \mathbf{u}_n \mathbf{u}_n^T \quad (13)$$

from which it follows that A^k is a symmetric matrix whose eigenvalues are the kth powers of the eigenvalues of A.

CAYLEY–HAMILTON THEOREM

No discussion of powers of a matrix would be complete without mention of the following result, called the *Cayley–Hamilton theorem*, the general proof of which is omitted (see Exercise P1 of Section 4.4, however).

> **Theorem 8.3.5** (*Cayley–Hamilton Theorem*) *Every square matrix satisfies its characteristic equation; that is, if A is an $n \times n$ matrix whose characteristic equation is*
>
> $$\lambda^n + c_1 \lambda^{n-1} + \cdots + c_n = 0$$
>
> *then*
>
> $$A^n + c_1 A^{n-1} + \cdots + c_n I = 0 \quad (14)$$

The Cayley–Hamilton theorem makes it possible to express all positive integer powers of an $n \times n$ matrix A in terms of $I, A, \ldots, A^{n-1}$ by solving (14) for A^n. In the case where A is invertible, it also makes it possible to express A^{-1} (and hence all negative powers of A) in terms of $I, A, \ldots, A^{n-1}$ by rewriting (14) as

$$A\left(-\frac{1}{c_n} A^{n-1} - \frac{c_1}{c_n} A^{n-2} - \cdots - \frac{c_{n-1}}{c_n} I\right) = I \quad (15)$$

(verify), from which it follows that A^{-1} is the parenthetical expression on the left.

REMARK We are guaranteed that $c_n \neq 0$ in (15), for otherwise $\lambda = 0$ would be a root of the characteristic equation, contradicting the invertibility of A [see parts (c) and (h) of Theorem 7.4.4].

EXAMPLE 4   We showed in Example 2 of Section 8.2 that the characteristic polynomial of

$$A = \begin{bmatrix} 2 & 0 & 0 \\ 1 & 3 & 0 \\ -3 & 5 & 3 \end{bmatrix}$$

is

$$p(\lambda) = (\lambda - 2)(\lambda - 3)^2 = \lambda^3 - 8\lambda^2 + 21\lambda - 18$$

so the Cayley–Hamilton theorem implies that

$$A^3 - 8A^2 + 21A - 18I = 0 \tag{16}$$

This equation can be used to express A^3 and all higher powers of A in terms of I, A, and A^2. For example,

$$A^3 = 8A^2 - 21A + 18I$$

and using this equation we can write

$$A^4 = AA^3 = 8A^3 - 21A^2 + 18A = 8(8A^2 - 21A + 18I) - 21A^2 + 18A$$
$$= 43A^2 - 150A + 144I$$

Equation (16) can also be used to express A^{-1} as a polynomial in A by rewriting it as

$$A(A^2 - 8A + 21I) = 18I$$

from which it follows that (verify)

$$A^{-1} = \tfrac{1}{18}(A^2 - 8A + 21I) = \begin{bmatrix} \frac{1}{2} & 0 & 0 \\ -\frac{1}{6} & \frac{1}{3} & 0 \\ \frac{7}{9} & -\frac{5}{9} & \frac{1}{3} \end{bmatrix}$$ ∎

EXPONENTIAL OF A MATRIX
(Calculus Required)

In Section 3.2 we defined polynomial functions of square matrices. Recall from that discussion that if A is an $n \times n$ matrix and

$$p(x) = a_0 + a_1 x + a_2 x^2 + \cdots + a_m x^m$$

then the matrix $p(A)$ is defined as

$$p(A) = a_0 I + a_1 A + a_2 A^2 + \cdots + a_m A^m$$

Other functions of square matrices can be defined using power series. For example, if the function f is represented by its Maclaurin series

$$f(x) = f(0) + f'(0)x + \frac{f''(0)}{2!}x^2 + \cdots + \frac{f^m(0)}{m!}x^m + \cdots \tag{17}$$

on some interval, then we define $f(A)$ to be

$$f(A) = f(0)I + f'(0)A + \frac{f''(0)}{2!}A^2 + \cdots + \frac{f^m(0)}{m!}A^m + \cdots \tag{18}$$

where we interpret this to mean that the ijth entry of $f(A)$ is the sum of the series of the ijth entries of the terms on the right.[*] In the special case where A is a diagonal matrix, say

$$A = \begin{bmatrix} d_1 & 0 & \cdots & 0 \\ 0 & d_2 & \cdots & 0 \\ \vdots & \vdots & \ddots & \vdots \\ 0 & 0 & \cdots & d_n \end{bmatrix}$$

and f is defined at the points $d_1, d_2, \ldots, d_k$, each matrix on the right side of (18) is diagonal, and hence so is $f(A)$. In this case, equating corresponding diagonal entries on the two sides of (18) yields

$$(f(A))_{kk} = f(0) + f'(0)d_k + \frac{f''(0)}{2!}d_k^2 + \cdots + \frac{f^m(0)}{m!}d_k^m + \cdots = f(d_k)$$

Thus, we can avoid the series altogether in the diagonal case and compute $f(A)$ directly as

$$f(A) = \begin{bmatrix} f(d_1) & 0 & \cdots & 0 \\ 0 & f(d_2) & \cdots & 0 \\ \vdots & \vdots & \ddots & \vdots \\ 0 & 0 & \cdots & f(d_n) \end{bmatrix} \qquad (19)$$

For example, if

$$A = \begin{bmatrix} 1 & 0 & 0 \\ 0 & 3 & 0 \\ 0 & 0 & -2 \end{bmatrix}, \quad \text{then} \quad e^A = \begin{bmatrix} e & 0 & 0 \\ 0 & e^3 & 0 \\ 0 & 0 & e^{-2} \end{bmatrix}$$

Now let us consider how we might use these ideas to find functions of diagonalizable matrices without summing infinite series. If A is an $n \times n$ diagonalizable matrix and $P^{-1}AP = D$, where

$$D = \begin{bmatrix} \lambda_1 & 0 & \cdots & 0 \\ 0 & \lambda_2 & \cdots & 0 \\ \vdots & \vdots & \ddots & \vdots \\ 0 & 0 & \cdots & \lambda_n \end{bmatrix}$$

then (10) and (18) suggest that

$$P^{-1}f(A)P = f(0)I + f'(0)(P^{-1}AP) + \frac{f''(0)}{2!}(P^{-1}A^2 P) + \cdots + \frac{f^m(0)}{m!}(P^{-1}A^m P) + \cdots$$

$$= f(0)I + f'(0)D + \frac{f''(0)}{2!}D^2 + \cdots + \frac{f^m(0)}{m!}D^m + \cdots$$

$$= f(D)$$

This tells us that $f(A)$ can be expressed as

$$f(A) = Pf(D)P^{-1} \qquad (20)$$

which suggests the following theorem.

Theorem 8.3.6 *Suppose that A is an $n \times n$ diagonalizable matrix that is diagonalized by P and that $\lambda_1, \lambda_2, \ldots, \lambda_n$ are the eigenvalues of A corresponding to the successive column vectors of P. If f is a real-valued function whose Maclaurin series converges on some interval containing the eigenvalues of A, then*

$$f(A) = P \begin{bmatrix} f(\lambda_1) & 0 & \cdots & 0 \\ 0 & f(\lambda_2) & \cdots & 0 \\ \vdots & \vdots & \ddots & \vdots \\ 0 & 0 & \cdots & f(\lambda_n) \end{bmatrix} P^{-1} \qquad (21)$$

---

[*]Conditions under which this series converges are discussed in the book *Calculus*, Vol. II, by Tom M. Apostol, John Wiley & Sons, New York, 1969.

Here is an example.

EXAMPLE 5
Exponentials of
Diagonalizable
Matrices

Find e^{tA} for the diagonalizable matrix

$$A = \begin{bmatrix} 0 & 0 & -2 \\ 1 & 2 & 1 \\ 1 & 0 & 3 \end{bmatrix}$$

Solution We showed in Example 8.3 of Section 8.2 that

$$P^{-1}AP = \begin{bmatrix} -1 & 0 & -1 \\ 1 & 0 & 2 \\ 1 & 1 & 1 \end{bmatrix} \begin{bmatrix} 0 & 0 & -2 \\ 1 & 2 & 1 \\ 1 & 0 & 3 \end{bmatrix} \begin{bmatrix} -2 & -1 & 0 \\ 1 & 0 & 1 \\ 1 & 1 & 0 \end{bmatrix} = \begin{bmatrix} 1 & 0 & 0 \\ 0 & 2 & 0 \\ 0 & 0 & 2 \end{bmatrix}$$

so applying Formula (21) with $f(A) = e^{tA}$ implies that

$$e^{tA} = P \begin{bmatrix} e^t & 0 & 0 \\ 0 & e^{2t} & 0 \\ 0 & 0 & e^{2t} \end{bmatrix} P^{-1} = \begin{bmatrix} -2 & -1 & 0 \\ 1 & 0 & 1 \\ 1 & 1 & 0 \end{bmatrix} \begin{bmatrix} e^t & 0 & 0 \\ 0 & e^{2t} & 0 \\ 0 & 0 & e^{2t} \end{bmatrix} \begin{bmatrix} -1 & 0 & -1 \\ 1 & 0 & 2 \\ 1 & 1 & 1 \end{bmatrix}$$

$$= \begin{bmatrix} 2e^t - e^{2t} & 0 & 2e^t - 2e^{2t} \\ e^{2t} - e^t & e^{2t} & e^{2t} - e^t \\ e^{2t} - e^t & 0 & 2e^{2t} - e^t \end{bmatrix} \qquad \blacksquare$$

In the special case where A is a symmetric matrix with a spectral decomposition

$$A = \lambda_1 \mathbf{u}_1 \mathbf{u}_1^T + \lambda_2 \mathbf{u}_2 \mathbf{u}_2^T + \cdots + \lambda_n \mathbf{u}_n \mathbf{u}_n^T$$

the matrix

$$P = [\mathbf{u}_1 \quad \mathbf{u}_2 \quad \cdots \quad \mathbf{u}_n]$$

orthogonally diagonalizes A, so (20) can be expressed as

$$f(A) = Pf(D)P^T$$

We will ask you to show in the exercises that this equation can be written as

$$f(A) = f(\lambda_1)\mathbf{u}_1\mathbf{u}_1^T + f(\lambda_2)\mathbf{u}_2\mathbf{u}_2^T + \cdots + f(\lambda_n)\mathbf{u}_n\mathbf{u}_n^T \tag{22}$$

(Exercise P3), which tells us that $f(A)$ is a symmetric matrix whose eigenvalues can be obtained by evaluating f at the eigenvalues of A.

DIAGONALIZATION AND LINEAR SYSTEMS

The problem of diagonalizing a square matrix A is closely related to the problem of solving the linear system $A\mathbf{x} = \mathbf{b}$. For example, suppose that A is diagonalizable and $P^{-1}AP = D$. If we define a new vector $\mathbf{y} = P^{-1}\mathbf{x}$, and if we substitute

$$\mathbf{x} = P\mathbf{y} \tag{23}$$

in $A\mathbf{x} = \mathbf{b}$, then we obtain a new linear system $AP\mathbf{y} = \mathbf{b}$ in the unknown $\mathbf{y}$. Multiplying both sides of this equation by P^{-1} and using the fact that $P^{-1}AP = D$ yields

$$D\mathbf{y} = P^{-1}\mathbf{b}$$

Since this system has a diagonal coefficient matrix, the solution for $\mathbf{y}$ can be read off immediately, and the vector $\mathbf{x}$ can then be computed using (23).

Many algorithms for solving large-scale linear systems are based on this idea. Such algorithms are particularly effective in cases in which the coefficient matrix can be orthogonally diagonalized since multiplication by orthogonal matrices does not magnify roundoff error.

THE NONDIAGONALIZABLE CASE

In cases where A is not diagonalizable it is still possible to achieve considerable simplification in the form of $P^{-1}AP$ by choosing the matrix P appropriately. We will consider two such theorems for matrices with real entries that involve orthogonal similarity. The proofs will be omitted. The first theorem, due to the German mathematician Issai Schur (1875–1941), states that every square matrix A with real eigenvalues is orthogonally similar to an *upper triangular matrix* that has the eigenvalues of A on the main diagonal.

Theorem 8.3.7 (*Schur's Theorem*) *If A is an $n \times n$ matrix with real entries and real eigenvalues, then there is an orthogonal matrix P such that P^TAP is an upper triangular matrix of the form*

$$P^TAP = \begin{bmatrix} \lambda_1 & \times & \times & \cdots & \times \\ 0 & \lambda_2 & \times & \cdots & \times \\ 0 & 0 & \lambda_3 & \cdots & \times \\ \vdots & \vdots & \vdots & \ddots & \vdots \\ 0 & 0 & 0 & \cdots & \lambda_n \end{bmatrix} \tag{24}$$

in which $\lambda_1, \lambda_2, \ldots, \lambda_n$ are the eigenvalues of the matrix A repeated according to multiplicity.

It is common to denote the upper triangular matrix in (24) by S (for Schur), in which case that equation can be rewritten as

$$A = PSP^T \tag{25}$$

which is called a ***Schur decomposition*** of A.

REMARK Recall from Theorem 4.4.12 that if A is a square matrix, then the trace of A is the sum of the eigenvalues of A and the determinant of A is the product of the eigenvalues of A. Since we know that the determinant and trace are similarity invariants, these facts become obvious by inspection of (24) for the cases where A has real eigenvalues.

The next theorem, due to the German mathematician Gerhard Hessenberg (1894–1925), states that every square matrix with real entries is orthogonally similar to a matrix in which each entry below the first ***subdiagonal*** is zero (Figure 8.3.2). Such a matrix is said to be in ***upper Hessenberg form***.

Figure 8.3.2

First subdiagonal

Theorem 8.3.8 (*Hessenberg's Theorem*) *Every square matrix with real entries is orthogonally similar to a matrix in upper Hessenberg form; that is, if A is an $n \times n$ matrix, then there is an orthogonal matrix P such that P^TAP is a matrix of the form*

$$P^TAP = \begin{bmatrix} \times & \times & \cdots & \times & \times & \times \\ \times & \times & \cdots & \times & \times & \times \\ 0 & \times & \ddots & \times & \times & \times \\ \vdots & \vdots & \ddots & \vdots & \vdots & \vdots \\ 0 & 0 & \cdots & \times & \times & \times \\ 0 & 0 & \cdots & 0 & \times & \times \end{bmatrix} \tag{26}$$

Linear Algebra in History

The life of the German mathematician Issai Schur is a sad reminder of the effect that Nazi policies had on Jewish intellectuals during the 1930s. Schur was a brilliant mathematician and a popular lecturer who attracted many students and researchers to the University of Berlin, where he worked and taught. His lectures sometimes attracted so many students that opera glasses were needed to see him from the back row. Schur's life became increasingly difficult under Nazi rule, and in April of 1933 he was forced to "retire" from the university under a law that prohibited non-Aryans from holding "civil service" positions. There was an outcry from many of his students and colleagues who respected and liked him, but it did not stave off his complete dismissal in 1935. Schur, who thought of himself as a German, rather than a Jew, never understood the persecution and humiliation he received at Nazi hands. He left Germany for Palestine in 1939, a broken man. Lacking in financial resources, he had to sell his beloved mathematics books and lived in poverty until his death in 1941.

Issai Schur
(1875–1941)

REMARK The diagonal entries in (26) will usually not be the eigenvalues of A.

It is common to denote the upper Hessenberg matrix in (26) by H (for Hessenberg), in which case that equation can be rewritten as

$$A = PHP^T \tag{27}$$

which is called an ***upper Hessenberg decomposition*** of A.

In many numerical LU- and QR-algorithms the initial matrix is first converted to upper Hessenberg form, thereby reducing the amount of computation in the algorithm itself. Some computer programs have built-in commands for finding Schur or Hessenberg decompositions.

Exercise Set 8.3

In Exercises 1–4, a symmetric matrix A is given. Find the dimensions of the eigenspaces of A by inspection of the characteristic polynomial.

1. $A = \begin{bmatrix} 1 & 2 \\ 2 & 4 \end{bmatrix}$

2. $A = \begin{bmatrix} 1 & -4 & 2 \\ -4 & 1 & -2 \\ 2 & -2 & -2 \end{bmatrix}$

3. $A = \begin{bmatrix} 1 & 1 & 1 \\ 1 & 1 & 1 \\ 1 & 1 & 1 \end{bmatrix}$

4. $A = \begin{bmatrix} 3 & 2 & 2 \\ 2 & 3 & 2 \\ 2 & 2 & 3 \end{bmatrix}$

In Exercises 5 and 6, verify that eigenvectors from distinct eigenspaces of the symmetric matrix A are orthogonal, as guaranteed by Theorem 8.3.4.

5. The matrix A in Exercise 3.

6. The matrix A in Exercise 4.

In Exercises 7–14, find a matrix P that orthogonally diagonalizes A, and determine the diagonal matrix $D = P^TAP$.

7. $A = \begin{bmatrix} 3 & 1 \\ 1 & 3 \end{bmatrix}$

8. $A = \begin{bmatrix} 6 & -2 \\ -2 & 3 \end{bmatrix}$

9. $A = \begin{bmatrix} -3 & 1 & 2 \\ 1 & -3 & 2 \\ 2 & 2 & 0 \end{bmatrix}$

10. $A = \begin{bmatrix} -2 & 0 & -36 \\ 0 & -3 & 0 \\ -36 & 0 & -23 \end{bmatrix}$

11. $A = \begin{bmatrix} 1 & 1 & 0 \\ 1 & 1 & 0 \\ 0 & 0 & 0 \end{bmatrix}$

12. $A = \begin{bmatrix} 2 & -1 & -1 \\ -1 & 2 & -1 \\ -1 & -1 & 2 \end{bmatrix}$

13. $A = \begin{bmatrix} 3 & 1 & 0 & 0 \\ 1 & 3 & 0 & 0 \\ 0 & 0 & 0 & 0 \\ 0 & 0 & 0 & 0 \end{bmatrix}$

14. $A = \begin{bmatrix} -7 & 24 & 0 & 0 \\ 24 & 7 & 0 & 0 \\ 0 & 0 & -7 & 24 \\ 0 & 0 & 24 & 7 \end{bmatrix}$

In Exercises 15–18, find the spectral decomposition of the matrix A.

15. The matrix A in Exercise 7.

16. The matrix A in Exercise 8.

17. The matrix A in Exercise 9.

18. The matrix A in Exercise 10.

In Exercises 19–22, use the method of Example 3 to compute the stated power of A.

19. $A = \begin{bmatrix} 8 & 9 \\ -6 & -7 \end{bmatrix}$; A^{10}

20. $A = \begin{bmatrix} -30 & 16 \\ -56 & 30 \end{bmatrix}$; A^{10}

21. $A = \begin{bmatrix} -5 & 0 & 6 \\ -3 & 1 & 3 \\ -4 & 0 & 5 \end{bmatrix}$; A^{1000}

22. $A = \begin{bmatrix} -4 & -2 & 6 \\ -2 & -1 & 3 \\ -3 & -2 & 5 \end{bmatrix}$; A^{1000}

23. Consider the matrix $A = \begin{bmatrix} 3 & -2 & 1 \\ 2 & -2 & 2 \\ 3 & -6 & 5 \end{bmatrix}$.

(a) Verify that A satisfies its characteristic equation, as guaranteed by the Cayley–Hamilton theorem.

(b) Find an expression for A^4 in terms of A^2, A, and I, and use that expression to evaluate A^4.

(c) Find an expression for A^{-1} in terms of A^2, A, and I.

24. Follow the directions of Exercise 23 for the matrix A given in Exercise 21.

In Exercises 25–28, compute e^{tA} for the given diagonalizable matrix A.

25. A is the matrix in Exercise 7.

26. A is the matrix in Exercise 8.

27. A is the matrix in Exercise 9.

28. A is the matrix in Exercise 10.

29. Compute $\sin(\pi A)$ for the matrix in Exercise 9.

30. Compute $\cos(\pi A)$ for the matrix in Exercise 10.

31. Consider the matrix $A = \begin{bmatrix} 0 & 0 & 0 \\ 1 & 0 & 0 \\ 2 & 1 & 0 \end{bmatrix}$.

(a) Show that A is nilpotent.

(b) Compute e^A by substituting the nonzero powers of A into Formula (17).

32. For the matrix A in Exercise 31, compute $\sin(\pi A)$ and $\cos(\pi A)$.

33. Show that if A is a symmetric orthogonal matrix, then 1 and -1 are the only possible eigenvalues.

Discussion and Discovery

D1. Indicate whether the statement is true (T) or false (F). Justify your answer.

(a) If A is a square matrix, then AA^T is orthogonally diagonalizable.

(b) If A is an $n \times n$ diagonalizable matrix, then there exists an orthonormal basis for R^n consisting of eigenvectors of A.

(c) An orthogonal matrix is orthogonally diagonalizable.

(d) If A is an invertible orthogonally diagonalizable matrix, then A^{-1} is orthogonally diagonalizable.

(e) If A is orthogonally diagonalizable, then A has real eigenvalues.

D2. (a) Find a 3×3 symmetric matrix with eigenvalues $\lambda_1 = -1, \lambda_2 = 3, \lambda_3 = 7$ and corresponding eigenvectors $\mathbf{v}_1 = (0, 1, -1)$, $\mathbf{v}_2 = (1, 0, 0)$, $\mathbf{v}_3 = (0, 1, 1)$.

(b) Is there a 3×3 symmetric matrix with eigenvalues $\lambda_1 = -1, \lambda_2 = 3, \lambda_3 = 7$ and corresponding eigenvectors $\mathbf{v}_1 = (0, 1, -1)$, $\mathbf{v}_2 = (1, 0, 0)$, $\mathbf{v}_3 = (1, 1, 1)$? Explain your reasoning.

D3. Let A be a diagonalizable matrix with the property that eigenvectors from distinct eigenvalues are orthogonal. Is A necessarily symmetric? Why or why not?

Working with Proofs

P1. Prove: Two matrices are orthogonally similar if and only if there exist orthonormal bases with respect to which the matrices represent the same linear operator. [*Hint:* See Theorem 8.2.2.]

P2. Prove: If $\{\mathbf{u}_1, \mathbf{u}_2, \ldots, \mathbf{u}_n\}$ is an orthonormal basis for R^n, and if A can be expressed as

$$A = c\mathbf{u}_1\mathbf{u}_1^T + c_2\mathbf{u}_2\mathbf{u}_2^T + \cdots + c_n\mathbf{u}_n\mathbf{u}_n^T$$

then A is symmetric and has eigenvalues $c_1, c_2, \ldots, c_n$.

P3. Prove that if A is a symmetric matrix whose spectral decomposition is

$$A = \lambda_1\mathbf{u}_1\mathbf{u}_1^T + \lambda_2\mathbf{u}_2\mathbf{u}_2^T + \cdots + \lambda_n\mathbf{u}_n\mathbf{u}_n^T$$

then

$$f(A) = f(\lambda_1)\mathbf{u}_1\mathbf{u}_1^T + f(\lambda_2)\mathbf{u}_2\mathbf{u}_2^T + \cdots + f(\lambda_n)\mathbf{u}_n\mathbf{u}_n^T$$

P4. In this exercise we will help you to prove that a matrix A is orthogonally diagonalizable if and only if it is symmetric. We have shown that an orthogonally diagonalizable matrix is symmetric. The harder part is to prove that a symmetric matrix A is orthogonally diagonalizable. We will proceed in two steps: first we will show that A is diagonalizable, and then we will build on that result to show that A is orthogonally diagonalizable.

(a) Assume that A is a symmetric $n \times n$ matrix. One way to prove that A is diagonalizable is to show that for each eigenvalue λ_0 the geometric multiplicity is equal to the algebraic multiplicity. For this purpose, assume that the geometric multiplicity of λ_0 is k, let $B_0 = \{\mathbf{u}_1, \mathbf{u}_2, \ldots, \mathbf{u}_k\}$ be an orthonormal basis for the eigenspace corresponding to λ_0, extend this to an orthonormal basis $B = \{\mathbf{u}_1, \mathbf{u}_2, \ldots, \mathbf{u}_n\}$ for R^n, and let P be the matrix having the vectors of B as columns. As shown in Exercise P6(b) of Section 8.2, the product AP can be written as

$$AP = P\begin{bmatrix} \lambda_0 I_k & X \\ 0 & Y \end{bmatrix}$$

Use the fact that B is an orthonormal basis to prove that $X = 0$ [a zero matrix of size $n \times (n - k)$].

(b) It follows from part (a) and Exercise P6(c) of Section 8.2 that A has the same characteristic polynomial as

$$C = \begin{bmatrix} \lambda_0 I_k & 0 \\ 0 & Y \end{bmatrix}$$

Use this fact and Exercise P6(d) of Section 8.2 to prove that the algebraic multiplicity of λ_0 is the same as the geometric multiplicity of λ_0. This establishes that A is diagonalizable.

(c) Use part (*b*) of Theorem 8.3.4 and the fact that A is diagonalizable to prove that A is orthogonally diagonalizable.

Technology Exercises

T1. Most linear algebra technology utilities do not have a specific command for orthogonally diagonalizing a symmetric matrix, so other commands must usually be pieced together for that purpose. Use the commands for finding eigenvectors and performing the Gram–Schmidt process to find a matrix P that orthogonally diagonalizes the following matrix A. Use your result to factor A as $A = PDP^T$, where D is diagonal.

$$A = \begin{bmatrix} \frac{1}{2} & 0 & \frac{3}{2} & 0 \\ 0 & \frac{1}{2} & 0 & \frac{3}{2} \\ \frac{3}{2} & 0 & \frac{1}{2} & 0 \\ 0 & \frac{3}{2} & 0 & \frac{1}{2} \end{bmatrix}$$

T2. Confirm that the matrix A in Exercise T1 satisfies its characteristic equation, in accordance with the Cayley–Hamilton theorem.

T3. Compute e^A for the matrix A in Exercise T1.

T4. Find the spectral decomposition of the matrix A given in Exercise T1.

Section 8.4   Quadratic Forms

In this section we will use matrix methods to study real-valued functions of several variables in which each term is either the square of a variable or the product of two variables. Such functions arise in a variety of applications, including geometry, vibrations of mechanical systems, statistics, and electrical engineering.

DEFINITION OF A QUADRATIC FORM

Expressions of the form

$$a_1 x_1 + a_2 x_2 + \cdots + a_n x_n$$

occurred in our study of linear equations and linear systems. If $a_1, a_2, \ldots, a_n$ are treated as fixed constants, then this expression is a real-valued function of the n variables $x_1, x_2, \ldots, x_n$ and is called a ***linear form*** on R^n. All variables in a linear form occur to the first power and there are no products of variables. Here we will be concerned with ***quadratic forms*** on R^n, which are functions of the form

$$a_1 x_1^2 + a_2 x_2^2 + \cdots + a_n x_n^2 + \text{(all possible terms } a_k x_i x_j \text{ in which } x_i \text{ and } x_j \text{ are distinct)}$$

The terms of the form $a_k x_i x_j$ are called ***cross product terms***. It is common to combine the cross product terms involving $x_i x_j$ with those involving $x_j x_i$ to avoid duplication. Thus, a general quadratic form on R^2 would typically be expressed as

$$a_1 x_1^2 + a_2 x_2^2 + 2a_3 x_1 x_2 \tag{1}$$

and a general quadratic form on R^3 as

$$a_1 x_1^2 + a_2 x_2^2 + a_3 x_3^2 + 2a_4 x_1 x_2 + 2a_5 x_1 x_3 + 2a_6 x_2 x_3 \tag{2}$$

If, as usual, we do not distinguish between the number a and the 1×1 matrix $[a]$, and if we let $\mathbf{x}$ be the column vector of variables, then (1) and (2) can be expressed in matrix form as

$$a_1 x_1^2 + a_2 x_2^2 + 2a_3 x_1 x_2 = \begin{bmatrix} x_1 & x_2 \end{bmatrix} \begin{bmatrix} a_1 & a_3 \\ a_3 & a_2 \end{bmatrix} \begin{bmatrix} x_1 \\ x_2 \end{bmatrix} = \mathbf{x}^T A \mathbf{x}$$

$$a_1 x_1^2 + a_2 x_2^2 + a_3 x_3^2 + 2a_4 x_1 x_2 + 2a_5 x_1 x_3 + 2a_6 x_2 x_3 = \begin{bmatrix} x_1 & x_2 & x_3 \end{bmatrix} \begin{bmatrix} a_1 & a_4 & a_5 \\ a_4 & a_2 & a_6 \\ a_5 & a_6 & a_3 \end{bmatrix} \begin{bmatrix} x_1 \\ x_2 \\ x_3 \end{bmatrix} = \mathbf{x}^T A \mathbf{x}$$

(verify). Note that the matrix A in these formulas is symmetric and that its diagonal entries are the coefficients of the squared terms and its off-diagonal entries are half the coefficients of the cross product terms. In general, if A is a symmetric $n \times n$ matrix and $\mathbf{x}$ is an $n \times 1$ column vector of variables, then we call the function

$$Q_A(\mathbf{x}) = \mathbf{x}^T A \mathbf{x} \tag{3}$$

the *quadratic form associated with A*. When convenient, (3) can be expressed in dot product notation as

$$Q_A(\mathbf{x}) = \mathbf{x}^T A \mathbf{x} = \mathbf{x}^T (A\mathbf{x}) = \mathbf{x} \cdot A\mathbf{x} = A\mathbf{x} \cdot \mathbf{x} \tag{4}$$

In the case where A is a diagonal matrix, the quadratic form Q_A has no cross product terms; for example, if A is the $n \times n$ identity matrix, then

$$Q_A(\mathbf{x}) = \mathbf{x}^T I \mathbf{x} = \mathbf{x}^T \mathbf{x} = \mathbf{x} \cdot \mathbf{x} = \|\mathbf{x}\|^2 = x_1^2 + x_2^2 + \cdots + x_n^2$$

and if A has diagonal entries $\lambda_1, \lambda_2, \ldots, \lambda_n$, then

$$Q_A(\mathbf{x}) = \mathbf{x}^T A \mathbf{x} = [x_1 \ \ x_2 \ \ \cdots \ \ x_n] \begin{bmatrix} \lambda_1 & 0 & \cdots & 0 \\ 0 & \lambda_2 & \cdots & 0 \\ \vdots & \vdots & \ddots & \vdots \\ 0 & 0 & \cdots & \lambda_n \end{bmatrix} \begin{bmatrix} x_1 \\ x_2 \\ \vdots \\ x_n \end{bmatrix} = \lambda_1 x_1^2 + \lambda_2 x_2^2 + \cdots + \lambda_n x_n^2$$

EXAMPLE 1
Expressing
Quadratic
Forms in
Matrix Notation

In each part, express the quadratic form in the matrix notation $\mathbf{x}^T A \mathbf{x}$, where A is symmetric.

(a) $2x^2 + 6xy - 5y^2$ (b) $x_1^2 + 7x_2^2 - 3x_3^2 + 4x_1 x_2 - 2x_1 x_3 + 8x_2 x_2$

Solution The diagonal entries of A are the coefficients of the squared terms, and the off-diagonal entries are half the coefficients of the cross product terms, so we obtain

$$2x^2 + 6xy - 5y^2 = [x \ \ y] \begin{bmatrix} 2 & 3 \\ 3 & -5 \end{bmatrix} \begin{bmatrix} x \\ y \end{bmatrix}$$

$$x_1^2 + 7x_2^2 - 3x_3^2 + 4x_1 x_2 - 2x_1 x_3 + 8x_2 x_3 = [x_1 \ \ x_2 \ \ x_3] \begin{bmatrix} 1 & 2 & -1 \\ 2 & 7 & 4 \\ -1 & 4 & -3 \end{bmatrix} \begin{bmatrix} x_1 \\ x_2 \\ x_3 \end{bmatrix} \quad \blacksquare$$

CHANGE OF VARIABLE IN A QUADRATIC FORM

There are three important kinds of problems that occur in applications of quadratic forms:

1. If $\mathbf{x}^T A \mathbf{x}$ is a quadratic form on R^2 or R^3, what kind of curve or surface is represented by the equation $\mathbf{x}^T A \mathbf{x} = k$?

2. If $\mathbf{x}^T A \mathbf{x}$ is a quadratic form on R^n, what conditions must A satisfy for $\mathbf{x}^T A \mathbf{x}$ to have positive values for $\mathbf{x} \neq \mathbf{0}$?

3. If $\mathbf{x}^T A \mathbf{x}$ is a quadratic form on R^n, what are its maximum and minimum values if $\mathbf{x}$ is constrained to satisfy $\|\mathbf{x}\| = 1$?

We will consider the first two problems in this section and the third problem in the next section.
Many of the techniques for solving these problems are based on simplifying the quadratic form $\mathbf{x}^T A \mathbf{x}$ by making a substitution

$$\mathbf{x} = P\mathbf{y} \tag{5}$$

that expresses the variables $x_1, x_2, \ldots, x_n$ in terms of new variables $y_1, y_2, \ldots, y_n$. If P is invertible, then we call (5) a *change of variable*, and if P is orthogonal, we call (5) an *orthogonal change of variable*.

If we make the change of variable $\mathbf{x} = P\mathbf{y}$ in the quadratic form $\mathbf{x}^T A \mathbf{x}$, then we obtain

$$\mathbf{x}^T A \mathbf{x} = (P\mathbf{y})^T A (P\mathbf{y}) = \mathbf{y}^T P^T A P \mathbf{y} = \mathbf{y}^T (P^T A P) \mathbf{y} \tag{6}$$

The matrix $B = P^T A P$ is symmetric (verify), so the effect of the change of variable is to produce a new quadratic form $\mathbf{y}^T B \mathbf{y}$ in the variables $y_1, y_2, \ldots, y_n$. In particular, if we choose P to orthogonally diagonalize A, then the new quadratic form will be $\mathbf{y}^T D \mathbf{y}$, where D is a diagonal matrix with the eigenvalues of A on the main diagonal; that is,

$$\mathbf{x}^T A \mathbf{x} = \mathbf{y}^T D \mathbf{y} = [y_1 \quad y_2 \quad \cdots \quad y_n] \begin{bmatrix} \lambda_1 & 0 & \cdots & 0 \\ 0 & \lambda_2 & \cdots & 0 \\ \vdots & \vdots & \ddots & \vdots \\ 0 & 0 & \cdots & \lambda_n \end{bmatrix} \begin{bmatrix} y_1 \\ y_2 \\ \vdots \\ y_n \end{bmatrix} = \lambda_1 y_1^2 + \lambda_2 y_2^2 + \cdots + \lambda_n y_n^2$$

Thus, we have the following result, called the ***principal axes theorem***, for reasons that we will explain shortly.

> **Theorem 8.4.1 (The Principal Axes Theorem)**   *If A is a symmetric $n \times n$ matrix, then there is an orthogonal change of variable that transforms the quadratic form $\mathbf{x}^T A \mathbf{x}$ into a quadratic form $\mathbf{y}^T D \mathbf{y}$ with no cross product terms. Specifically, if P orthogonally diagonalizes A, then making the change of variable $\mathbf{x} = P\mathbf{y}$ in the quadratic form $\mathbf{x}^T A \mathbf{x}$ yields the quadratic form*
>
> $$\mathbf{x}^T A \mathbf{x} = \mathbf{y}^T D \mathbf{y} = \lambda_1 y_1^2 + \lambda_2 y_2^2 + \cdots + \lambda_n y_n^2$$
>
> *in which $\lambda_1, \lambda_2, \ldots, \lambda_n$ are the eigenvalues of A corresponding to the eigenvectors that form the successive columns of P.*

EXAMPLE 2
An Illustration of the Principal Axes Theorem

Find an orthogonal change of variable that eliminates the cross product terms in the quadratic form $Q = x_1^2 - x_3^2 - 4x_1 x_2 + 4x_2 x_3$, and express Q in terms of the new variables.

Solution   The quadratic form can be expressed in matrix notation as

$$Q = \mathbf{x}^T A \mathbf{x} = [x_1 \quad x_2 \quad x_3] \begin{bmatrix} 1 & -2 & 0 \\ -2 & 0 & 2 \\ 0 & 2 & -1 \end{bmatrix} \begin{bmatrix} x_1 \\ x_2 \\ x_3 \end{bmatrix}$$

The characteristic equation of the matrix A is

$$\begin{vmatrix} \lambda - 1 & 2 & 0 \\ 2 & \lambda & -2 \\ 0 & -2 & \lambda + 1 \end{vmatrix} = \lambda^3 - 9\lambda = \lambda(\lambda + 3)(\lambda - 3) = 0$$

so the eigenvalues are $\lambda = 0, -3, 3$. We leave it for you to show that orthonormal bases for the three eigenspaces are

$$\lambda = 0: \begin{bmatrix} \frac{2}{3} \\ \frac{1}{3} \\ \frac{2}{3} \end{bmatrix}, \quad \lambda = -3: \begin{bmatrix} -\frac{1}{3} \\ -\frac{2}{3} \\ \frac{2}{3} \end{bmatrix}, \quad \lambda = 3: \begin{bmatrix} -\frac{2}{3} \\ \frac{2}{3} \\ \frac{1}{3} \end{bmatrix}$$

Thus, a substitution $\mathbf{x} = P\mathbf{y}$ that eliminates the cross product terms is

$$\begin{bmatrix} x_1 \\ x_2 \\ x_3 \end{bmatrix} = \begin{bmatrix} \frac{2}{3} & -\frac{1}{3} & -\frac{2}{3} \\ \frac{1}{3} & -\frac{2}{3} & \frac{2}{3} \\ \frac{2}{3} & \frac{2}{3} & \frac{1}{3} \end{bmatrix} \begin{bmatrix} y_1 \\ y_2 \\ y_3 \end{bmatrix}$$

This produces the new quadratic form

$$Q = \mathbf{y}^T(P^TAP)\mathbf{y} = \begin{bmatrix} y_1 & y_2 & y_3 \end{bmatrix} \begin{bmatrix} 0 & 0 & 0 \\ 0 & -3 & 0 \\ 0 & 0 & 3 \end{bmatrix} \begin{bmatrix} y_1 \\ y_2 \\ y_3 \end{bmatrix} = -3y_2^2 + 3y_3^2$$

in which there are no cross product terms. ∎

There are other methods for eliminating cross product terms from a quadratic form, which we will not discuss here. Two such methods, *Lagrange's reduction* and *Kronecker's reduction*, are discussed in more advanced texts.

REMARK If A is a symmetric $n \times n$ matrix, then the quadratic form $\mathbf{x}^TA\mathbf{x}$ is a real-valued function whose range is the set of all possible values for $\mathbf{x}^TA\mathbf{x}$ as $\mathbf{x}$ varies over R^n. It can be shown that a change of variable $\mathbf{x} = P\mathbf{y}$ does not alter the range of a quadratic form; that is, the set of all values for $\mathbf{x}^TA\mathbf{x}$ as $\mathbf{x}$ varies over R^n is the same as the set of all values for $\mathbf{y}^T(P^TAP)\mathbf{y}$ as $\mathbf{y}$ varies over R^n.

QUADRATIC FORMS IN GEOMETRY

Recall that a *conic section* or *conic* is a curve that results by cutting a double-napped cone with a plane (Figure 8.4.1). The most important conic sections are ellipses, hyperbolas, and parabolas, which occur when the cutting plane does not pass through the vertex. Circles are special cases of ellipses that result when the cutting plane is perpendicular to the axis of symmetry of the cone. If the cutting plane passes through the vertex, then the resulting intersection is called a *degenerate conic*. The possibilities are a point, a pair of intersecting lines, or a single line.

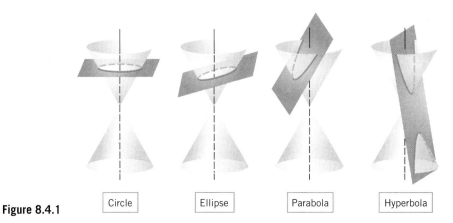

| Circle | Ellipse | Parabola | Hyperbola |

Figure 8.4.1

Quadratic forms on R^2 arise naturally in the study of conic sections. For example, it is shown in analytic geometry that an equation of the form

$$ax^2 + 2bxy + cy^2 + dx + ey + f = 0 \tag{7}$$

in which a, b, and c are not all zero, represents a conic section.[*] If $d = e = 0$ in (7), then there are no linear terms, and the equation becomes

$$ax^2 + 2bxy + cy^2 + f = 0 \tag{8}$$

and is said to represent a *central conic*. These include circles, ellipses, and hyperbolas, but not parabolas. Furthermore, if $b = 0$ in (8), then there is no cross product term, and the equation

$$ax^2 + cy^2 + f = 0 \tag{9}$$

is said to represent a central conic in *standard position*.

_____

[*]We must also allow for the possibility that there are no real values of x and y that satisfy the equation, as with $x^2 + y^2 + 1 = 0$. In such cases we say that the equation has *no graph* or has *an empty graph*.

If $f \neq 0$ in (9), then we can divide through by $-f$ and rewrite this equation in the form

$$a'x^2 + b'y^2 = 1 \tag{10}$$

Furthermore, if the coefficients a' and b' are both positive or if one is positive and one is negative, then this equation represents a nondegenerate conic and can be rewritten in one of the four forms shown in Table 8.4.1 by putting the coefficients in the denominator. These are called the ***standard forms*** of the nondegenerate central conics. In the case where $\alpha = \beta$ the ellipses shown in the table are circles.

Table 8.4.1

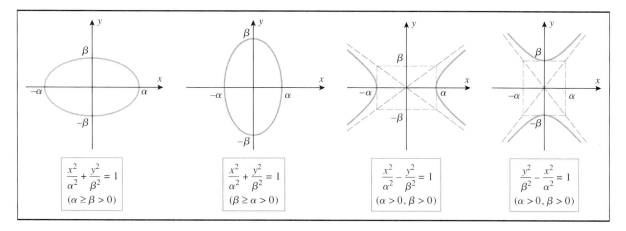

$\dfrac{x^2}{\alpha^2} + \dfrac{y^2}{\beta^2} = 1$	$\dfrac{x^2}{\alpha^2} + \dfrac{y^2}{\beta^2} = 1$	$\dfrac{x^2}{\alpha^2} - \dfrac{y^2}{\beta^2} = 1$	$\dfrac{y^2}{\beta^2} - \dfrac{x^2}{\alpha^2} = 1$
$(\alpha \geq \beta > 0)$	$(\beta \geq \alpha > 0)$	$(\alpha > 0, \beta > 0)$	$(\alpha > 0, \beta > 0)$

We assume that you are familiar with the basic properties of conic sections, so we will not discuss such matters in this text. However, you will need to understand the geometric significance of the constants α and β that appear in the standard forms of the central conics, so let us review their interpretations. In the case of an ellipse, 2α is its length in the x-direction and 2β its length in the y-direction (Table 8.4.1). For a noncircular ellipse, the larger of these numbers is the length of the ***major axis*** and the smaller the length of the ***minor axis***. In the case of a hyperbola, the numbers 2α and 2β are the lengths of the sides of a box whose diagonals are along the asymptotes of the hyperbola (Table 8.4.1). Central conics in standard position are symmetric about both coordinate axes and have no cross product terms. A central conic whose equation has a cross product term results by rotating a conic in standard position about the origin and hence is said to be ***rotated out of standard position*** (Figure 8.4.2).

Quadratic forms on R^3 arise in the study of geometric objects called ***quadric surfaces*** (or ***quadrics***). The most important surfaces of this type have equations of the form

$$ax^2 + by^2 + cz^2 + 2dxy + 2exz + 2fyz + g = 0$$

in which a, b, and c are not all zero. These are called ***central quadrics***. A problem involving quadric surfaces appears in the exercises.

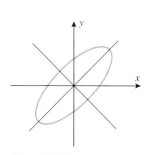

Figure 8.4.2

IDENTIFYING CONIC SECTIONS

We are now ready to consider the first of the three problems posed earlier, identifying the curve or surface represented by an equation $\mathbf{x}^T A \mathbf{x} = k$ in two or three variables. We will focus on the two-variable case. We noted above that an equation of the form

$$ax^2 + 2bxy + cy^2 + f = 0 \tag{11}$$

represents a central conic. If $b = 0$, then the conic is in standard position, and if $b \neq 0$, it is rotated. It is an easy matter to identify central conics in standard position by matching the equation with one of the standard forms. For example, the equation

$$9x^2 + 16y^2 - 144 = 0$$

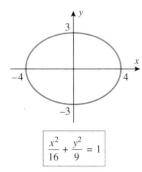

$$\frac{x^2}{16} + \frac{y^2}{9} = 1$$

Figure 8.4.3

can be rewritten as

$$\frac{x^2}{16} + \frac{y^2}{9} = 1$$

which, by comparison with Table 8.4.1, is the ellipse shown in Figure 8.4.3.

If a central conic is rotated out of standard position, then it can be identified by first rotating the coordinate axes to put it in standard position and then matching the resulting equation with one of the standard forms in Table 8.4.1. To find a rotation that eliminates the cross product term in the equation $ax^2 + 2bxy + cy^2 + f = 0$, it will be convenient to take the constant term to the right side and express the equation in the form

$$ax^2 + 2bxy + cy^2 = k$$

or in matrix notation as

$$\mathbf{x}^T A\mathbf{x} = [x \quad y]\begin{bmatrix} a & b \\ b & c \end{bmatrix}\begin{bmatrix} x \\ y \end{bmatrix} = k \tag{12}$$

To rotate the coordinate axes, we need to make an orthogonal change of variable

$$\mathbf{x} = P\mathbf{x}'$$

in which $\det(P) = 1$, and if we want this rotation to eliminate the cross product term, we must choose P to orthogonally diagonalize A. If we make a change of variable with these two properties, then in the rotated coordinate system Equation (12) will become

$$\mathbf{x}'^T D\mathbf{x}' = [x' \quad y']\begin{bmatrix} \lambda_1 & 0 \\ 0 & \lambda_2 \end{bmatrix}\begin{bmatrix} x' \\ y' \end{bmatrix} = k \tag{13}$$

where λ_1 and λ_2 are the eigenvalues of A. The conic can now be identified by writing (13) in the form

$$\lambda_1 x'^2 + \lambda_2 y'^2 = k \tag{14}$$

and performing the necessary algebra to match it with one of the standard forms in Table 8.4.1. For example, if λ_1, λ_2, and k are positive, then (14) represents an ellipse with an axis of length $2\sqrt{k/\lambda_1}$ in the x'-direction and $2\sqrt{k/\lambda_2}$ in the y'-direction. The first column vector of P, which is a unit eigenvector corresponding to λ_1, is along the positive x'-axis; and the second column vector of P, which is a unit eigenvector corresponding to λ_2, is a unit vector along the y'-axis. These are called the ***principal axes*** of the ellipse, which explains why Theorem 8.4.1 is called "the principal axes theorem." Also, since P is the transition matrix from $x'y'$-coordinates to xy-coordinates, it follows from Formula (29) of Section 7.11 that the matrix P can be expressed in terms of the rotation angle θ as

$$P = \begin{bmatrix} \cos\theta & -\sin\theta \\ \sin\theta & \cos\theta \end{bmatrix} \tag{15}$$

(Figure 8.4.4).

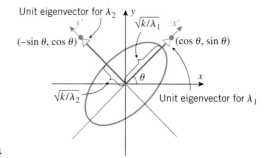

Figure 8.4.4

EXAMPLE 3
Identifying a Conic by Eliminating the Cross Product Term

(a) Identify the conic whose equation is $5x^2 - 4xy + 8y^2 - 36 = 0$ by rotating the xy-axes to put the conic in standard position.

(b) Find the angle θ through which you rotated the xy-axes in part (a).

Solution (a) The given equation can be written in the matrix form

$$\mathbf{x}^T A \mathbf{x} = 36$$

where

$$A = \begin{bmatrix} 5 & -2 \\ -2 & 8 \end{bmatrix}$$

The characteristic polynomial of A is

$$\begin{vmatrix} \lambda - 5 & 2 \\ 2 & \lambda - 8 \end{vmatrix} = (\lambda - 4)(\lambda - 9)$$

so the eigenvalues are $\lambda = 4$ and $\lambda = 9$. We leave it for you to show that orthonormal bases for the eigenspaces are

$$\lambda = 4: \quad \begin{bmatrix} \frac{2}{\sqrt{5}} \\ \frac{1}{\sqrt{5}} \end{bmatrix}, \quad \lambda = 9: \quad \begin{bmatrix} -\frac{1}{\sqrt{5}} \\ \frac{2}{\sqrt{5}} \end{bmatrix}$$

Thus, A is orthogonally diagonalized by

$$P = \begin{bmatrix} \frac{2}{\sqrt{5}} & -\frac{1}{\sqrt{5}} \\ \frac{1}{\sqrt{5}} & \frac{2}{\sqrt{5}} \end{bmatrix} \tag{16}$$

Moreover, it happens by chance that $\det(P) = 1$, so we are assured that the substitution $\mathbf{x} = P\mathbf{x}'$ performs a rotation of axes. Had it been the case that $\det(P) = -1$, then we would have interchanged the columns to reverse the sign. It follows from (13) that the equation of the conic in the $x'y'$-coordinate system is

$$[x' \ \ y'] \begin{bmatrix} 4 & 0 \\ 0 & 9 \end{bmatrix} \begin{bmatrix} x' \\ y' \end{bmatrix} = 36$$

which we can write as

$$4x'^2 + 9y'^2 = 36 \quad \text{or} \quad \frac{x'^2}{9} + \frac{y'^2}{4} = 1$$

We can now see from Table 8.4.1 that the conic is an ellipse whose axis has length $2\alpha = 6$ in the x'-direction and length $2\beta = 4$ in the y'-direction.

Solution (b) It follows from (15) that

$$P = \begin{bmatrix} \frac{2}{\sqrt{5}} & -\frac{1}{\sqrt{5}} \\ \frac{1}{\sqrt{5}} & \frac{2}{\sqrt{5}} \end{bmatrix} = \begin{bmatrix} \cos\theta & -\sin\theta \\ \sin\theta & \cos\theta \end{bmatrix}$$

which implies that

$$\cos\theta = \frac{2}{\sqrt{5}}, \quad \sin\theta = \frac{1}{\sqrt{5}}, \quad \tan\theta = \frac{\sin\theta}{\cos\theta} = \frac{1}{2}$$

Thus, $\theta = \tan^{-1} \frac{1}{2} \approx 26.6°$ (Figure 8.4.5). ∎

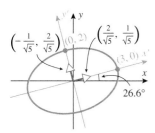

Figure 8.4.5

REMARK In the exercises we will ask you to show that if $b \neq 0$, then the cross product term in the equation

$$ax^2 + 2bxy + cy^2 = k$$

can be eliminated by a rotation through an angle θ that satisfies

$$\cot 2\theta = \frac{a - c}{2b} \tag{17}$$

We leave it for you to confirm that this is consistent with part (b) of the last example.

POSITIVE DEFINITE QUADRATIC FORMS

We will now consider the second of the two problems posed earlier, determining conditions under which $\mathbf{x}^T A\mathbf{x} > 0$ for all nonzero values of $\mathbf{x}$. We will explain why this is important shortly, but first we introduce some terminology.

> **Definition 8.4.2** A quadratic form $\mathbf{x}^T A\mathbf{x}$ is said to be
>
> *positive definite* if $\mathbf{x}^T A\mathbf{x} > 0$ for $\mathbf{x} \neq \mathbf{0}$
> *negative definite* if $\mathbf{x}^T A\mathbf{x} < 0$ for $\mathbf{x} \neq \mathbf{0}$
> *indefinite* if $\mathbf{x}^T A\mathbf{x}$ has both positive and negative values

The terminology in this definition is also applied to the matrix A; that is, we say that *a symmetric matrix A is positive definite, negative definite, or indefinite in accordance with whether the associated quadratic form $\mathbf{x}^T A\mathbf{x}$ has that property.*

The following theorem provides a way of using eigenvalues to determine whether a matrix A and its associated quadratic form $\mathbf{x}^T A\mathbf{x}$ are positive definite, negative definite, or indefinite.

> **Theorem 8.4.3** *If A is a symmetric matrix, then:*
> (a) $\mathbf{x}^T A\mathbf{x}$ *is positive definite if and only if all eigenvalues of A are positive.*
> (b) $\mathbf{x}^T A\mathbf{x}$ *is negative definite if and only if all eigenvalues of A are negative.*
> (c) $\mathbf{x}^T A\mathbf{x}$ *is indefinite if and only if A has at least one positive eigenvalue and at least one negative eigenvalue.*

Proofs (a) and (b) It follows from the principal axes theorem (Theorem 8.4.1) that there is an orthogonal change of variable $\mathbf{x} = P\mathbf{y}$ for which

$$\mathbf{x}^T A\mathbf{x} = \mathbf{y}^T D\mathbf{y} = \lambda_1 y_1^2 + \lambda_2 y_2^2 + \cdots + \lambda_n y_n^2 \tag{18}$$

Moreover, it follows from the invertibility of P that $\mathbf{y} \neq \mathbf{0}$ if and only if $\mathbf{x} \neq \mathbf{0}$, so the values of $\mathbf{x}^T A\mathbf{x}$ for $\mathbf{x} \neq \mathbf{0}$ are the same as the values of $\mathbf{y}^T D\mathbf{y}$ for $\mathbf{y} \neq \mathbf{0}$. Thus, it follows from (18) that $\mathbf{x}^T A\mathbf{x} > 0$ for $\mathbf{x} \neq \mathbf{0}$ if and only if all of the λ's in that equation (which are the eigenvalues of A) are positive, and that $\mathbf{x}^T A\mathbf{x} < 0$ for $\mathbf{x} \neq \mathbf{0}$ if and only if all of the eigenvalues are negative. This proves parts (a) and (b).

Proof (c) Assume that A has at least one positive eigenvalue and at least one negative eigenvalue, and to be specific, suppose that $\lambda_1 > 0$ and $\lambda_2 < 0$ in (18). Then

$$\mathbf{x}^T A\mathbf{x} > 0 \quad \text{if} \quad y_1 = 1 \text{ and all other } y\text{'s are 0}$$

and

$$\mathbf{x}^T A\mathbf{x} < 0 \quad \text{if} \quad y_2 = 1 \text{ and all other } y\text{'s are 0}$$

which proves that $\mathbf{x}^T A \mathbf{x}$ is indefinite. Conversely, if $\mathbf{x}^T A \mathbf{x} > 0$ for some $\mathbf{x}$, then $\mathbf{y}^T D \mathbf{y} > 0$ for some $\mathbf{y}$, so at least one of the λ's in (18) must be positive. Similarly, if $\mathbf{x}^T A \mathbf{x} < 0$ for some $\mathbf{x}$, then $\mathbf{y}^T D \mathbf{y} < 0$ for some $\mathbf{y}$, so at least one of the λ's in (18) must be negative, which completes the proof. ∎

REMARK   The three classifications in Definition 8.4.2 do not exhaust all of the possibilities. For example, a quadratic form for which $\mathbf{x}^T A \mathbf{x} \geq 0$ if $\mathbf{x} \neq \mathbf{0}$ is called *positive semidefinite*, and one for which $\mathbf{x}^T A \mathbf{x} \leq 0$ if $\mathbf{x} \neq \mathbf{0}$ is called *negative semidefinite*. Every positive definite form is positive semidefinite, but not conversely, and every negative definite form is negative semidefinite, but not conversely (why?). By adjusting the proof of Theorem 8.4.3 appropriately, one can prove that $\mathbf{x}^T A \mathbf{x}$ is positive semidefinite if and only if all eigenvalues of A are nonnegative and is negative semidefinite if and only if all eigenvalues of A are nonpositive.

EXAMPLE 4
Positive Definite Quadratic Forms

One cannot usually tell from the signs of the entries in a symmetric matrix A whether that matrix is positive definite, negative definite, or indefinite. For example, the entries of the matrix

$$A = \begin{bmatrix} 3 & 1 & 1 \\ 1 & 0 & 2 \\ 1 & 2 & 0 \end{bmatrix}$$

are nonnegative, but the matrix is indefinite, since its eigenvalues are $\lambda = 1,\ 4,\ -2$ (verify). To see this another way, let us write out the quadratic form as

$$Q_A(\mathbf{x}) = \mathbf{x}^T A \mathbf{x} = \begin{bmatrix} x_1 & x_2 & x_3 \end{bmatrix} \begin{bmatrix} 3 & 1 & 1 \\ 1 & 0 & 2 \\ 1 & 2 & 0 \end{bmatrix} \begin{bmatrix} x_1 \\ x_2 \\ x_3 \end{bmatrix} = 3x_1^2 + 2x_1 x_2 + 2x_1 x_3 + 4x_2 x_3$$

We can now see, for example, that $Q_A = 4$ for $x_1 = 0,\ x_2 = 1,\ x_3 = 1$ and $Q_A = -4$ for $x_1 = 0, x_2 = 1, x_3 = -1$. ∎

CONCEPT PROBLEM   Positive definite and negative definite matrices are invertible. Why?

CLASSIFYING CONIC SECTIONS USING EIGENVALUES

If $\mathbf{x}^T B \mathbf{x} = k$ is the equation of a conic, and if $k \neq 0$, then we can divide through by k and rewrite the equation in the form

$$\mathbf{x}^T A \mathbf{x} = 1 \tag{19}$$

where $A = (1/k)B$. If we now rotate the coordinate axes to eliminate the cross product term (if any) in this equation, then the equation of the conic in the new coordinate system will be of the form

$$\lambda_1 x'^2 + \lambda_2 y'^2 = 1 \tag{20}$$

in which λ_1 and λ_2 are the eigenvalues of A. The kind of conic represented by this equation will depend on the signs of the eigenvalues λ_1 and λ_2. For example, you should be able to see from (20) that:

- $\mathbf{x}^T A \mathbf{x} = 1$ represents an ellipse if $\lambda_1 > 0$ and $\lambda_2 > 0$.
- $\mathbf{x}^T A \mathbf{x} = 1$ has no graph if $\lambda_1 < 0$ and $\lambda_2 < 0$.
- $\mathbf{x}^T A \mathbf{x} = 1$ represents a hyperbola if λ_1 and λ_2 have opposite signs.

In the case of the ellipse, Equation (20) can be rewritten as

$$\frac{x'^2}{(1/\sqrt{\lambda_1})^2} + \frac{y'^2}{(1/\sqrt{\lambda_2})^2} = 1 \tag{21}$$

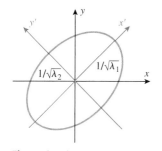

Figure 8.4.6

so the axes of the ellipse have lengths $2/\sqrt{\lambda_1}$ and $2/\sqrt{\lambda_2}$ (Figure 8.4.6).

The following theorem is an immediate consequence of this discussion and Theorem 8.4.3.

Theorem 8.4.4   *If A is a symmetric* 2×2 *matrix, then:*

(a) $\mathbf{x}^T A \mathbf{x} = 1$ *represents an ellipse if A is positive definite.*

(b) $\mathbf{x}^T A \mathbf{x} = 1$ *has no graph if A is negative definite.*

(c) $\mathbf{x}^T A \mathbf{x} = 1$ *represents a hyperbola if A is indefinite.*

In Example 3 we performed a rotation that showed that the equation

$$5x^2 - 4xy + 8y^2 - 36 = 0$$

represents an ellipse with a major axis of length 6 and a minor axis of length 4. This conclusion can also be obtained by rewriting the equation in the form

$$\tfrac{5}{36}x^2 - \tfrac{1}{9}xy + \tfrac{2}{9}y^2 = 1$$

and showing that the associated matrix

$$A = \begin{bmatrix} \frac{5}{36} & -\frac{1}{18} \\ -\frac{1}{18} & \frac{2}{9} \end{bmatrix}$$

has eigenvalues $\lambda_1 = \tfrac{1}{9}$ and $\lambda_2 = \tfrac{1}{4}$. These eigenvalues are positive, so the matrix A is positive definite and the equation represents an ellipse. Moreover, it follows from (20) that the axes of the ellipse have lengths $2/\sqrt{\lambda_1} = 6$ and $2/\sqrt{\lambda_2} = 4$, which is consistent with Example 3.

REMARK   Motivated by part (a) of Theorem 8.4.4, a set of points in R^n that satisfy an equation of the form $\mathbf{x}^T A \mathbf{x} = 1$ in which A is a positive definite symmetric matrix is called a central **ellipsoid**. In the case where A is the identity matrix the ellipsoid is called the **unit sphere** because its equation $\mathbf{x}^T \mathbf{x} = 1$ is equivalent to the condition $\|\mathbf{x}\| = 1$.

IDENTIFYING POSITIVE DEFINITE MATRICES

Positive definite matrices are the most important symmetric matrices in applications, so it will be useful to learn a little more about them. We already know that a symmetric matrix is positive definite if and only if its eigenvalues are all positive; now we will give a criterion that can be used to determine whether a symmetric matrix is positive definite without finding the eigenvalues. For this purpose we define the **kth principal submatrix** of an $n \times n$ matrix A to be the $k \times k$ submatrix consisting of the first k rows and columns of A. For example, here are the principal submatrices of a general 4×4 matrix:

$$\begin{bmatrix} a_{11} & a_{12} & a_{13} & a_{14} \\ a_{21} & a_{22} & a_{23} & a_{24} \\ a_{31} & a_{32} & a_{33} & a_{34} \\ a_{41} & a_{42} & a_{43} & a_{44} \end{bmatrix} \quad \begin{bmatrix} a_{11} & a_{12} & a_{13} & a_{14} \\ a_{21} & a_{22} & a_{23} & a_{24} \\ a_{31} & a_{32} & a_{33} & a_{34} \\ a_{41} & a_{42} & a_{43} & a_{44} \end{bmatrix} \quad \begin{bmatrix} a_{11} & a_{12} & a_{13} & a_{14} \\ a_{21} & a_{22} & a_{23} & a_{24} \\ a_{31} & a_{32} & a_{33} & a_{34} \\ a_{41} & a_{42} & a_{43} & a_{44} \end{bmatrix} \quad \begin{bmatrix} a_{11} & a_{12} & a_{13} & a_{14} \\ a_{21} & a_{22} & a_{23} & a_{24} \\ a_{31} & a_{32} & a_{33} & a_{34} \\ a_{41} & a_{42} & a_{43} & a_{44} \end{bmatrix}$$

First principal submatrix   Second principal submatrix   Third principal submatrix   Fourth principal submatrix $= A$

The following theorem, which we state without proof, provides a determinant test for determining whether a symmetric matrix is positive definite.

Theorem 8.4.5   *A symmetric matrix A is positive definite if and only if the determinant of every principal submatrix is positive.*

EXAMPLE 5
Working with
Principal
Submatrices

The matrix

$$A = \begin{bmatrix} 2 & -1 & -3 \\ -1 & 2 & 4 \\ -3 & 4 & 9 \end{bmatrix}$$

is positive definite since the determinants

$$|2| = 2, \quad \begin{vmatrix} 2 & -1 \\ -1 & 2 \end{vmatrix} = 3, \quad \begin{vmatrix} 2 & -1 & -3 \\ -1 & 2 & 4 \\ -3 & 4 & 9 \end{vmatrix} = 1$$

are all positive. Thus, we are guaranteed that all eigenvalues of A are positive and $\mathbf{x}^T A \mathbf{x} > 0$ for $\mathbf{x} \neq \mathbf{0}$. ∎

CONCEPT PROBLEM If the determinants of all principal submatrices of a symmetric 3×3 matrix are negative, is the matrix negative definite? Explain your reasoning.

The following theorem provides two important factorization properties of positive definite matrices.

Theorem 8.4.6 *If A is a symmetric matrix, then the following statements are equivalent.*

(a) *A is positive definite.*

(b) *There is a symmetric positive definite matrix B such that $A = B^2$.*

(c) *There is an invertible matrix C such that $A = C^T C$.*

Proof $(a) \Rightarrow (b)$ Since A is symmetric, it is orthogonally diagonalizable. This means that there is an orthogonal matrix P such that $P^T A P = D$, where D is a diagonal matrix whose entries are the eigenvalues of A. Moreover, since A is positive definite, its eigenvalues are positive, so we can write D as $D = D_1^2$, where D_1 is the diagonal matrix whose entries are the square roots of the eigenvalues of A. Thus, we have $P^T A P = D_1^2$, which we can rewrite as

$$A = P D_1^2 P^T = P D_1 D_1 P^T = P D_1 P^T P D_1 P^T = (P D_1 P^T)(P D_1 P^T) = B^2 \tag{22}$$

where $B = P D_1 P^T$. We leave it for you to confirm that B is symmetric. We will show that B is positive definite by proving that it has positive eigenvalues. The eigenvalues of B are the same as the eigenvalues of D_1, since eigenvalues are a similarity invariant and B is similar to D_1. Thus, the eigenvalues of B are positive, since they are the square roots of the eigenvalues of A.

Proof $(b) \Rightarrow (c)$ Assume that $A = B^2$, where B is symmetric and positive definite. Then $A = B^2 = BB = B^T B$, so take $C = B$.

Proof $(c) \Rightarrow (a)$ Assume that $A = C^T C$, where C is invertible. We will show that A is positive definite by showing that $\mathbf{x}^T A \mathbf{x} > 0$ for $\mathbf{x} \neq \mathbf{0}$. To do this we use Formula (26) of Section 3.1 and part (e) of Theorem 3.2.10 to write

$$\mathbf{x}^T A \mathbf{x} = \mathbf{x}^T C^T C \mathbf{x} = (C\mathbf{x})^T (C\mathbf{x}) = C\mathbf{x} \cdot C\mathbf{x} = \|C\mathbf{x}\|^2 \geq 0$$

But the invertibility of C implies that $C\mathbf{x} \neq \mathbf{0}$ if $\mathbf{x} \neq \mathbf{0}$, so $\mathbf{x}^T A \mathbf{x} > 0$ for $\mathbf{x} \neq \mathbf{0}$. ∎

EXAMPLE 6
The
Factorization
$A = B^2$

In Example 1 of Section 8.3 we showed that the matrix

$$A = \begin{bmatrix} 4 & 2 & 2 \\ 2 & 4 & 2 \\ 2 & 2 & 4 \end{bmatrix}$$

Linear Algebra in History

The history of the Cholesky factorization is somewhat murky, but it is generally credited to André-Louis Cholesky, a French military officer, who, in addition to being a battery commander, was involved with geodesy and surveying in Crete and North Africa before World War I. Cholesky was killed in action during the war, and his work was published posthumously on his behalf by a fellow officer.

André-Louis Cholesky
(1875–1918)

has eigenvalues $\lambda = 2$ and $\lambda = 8$ and that

$$P^TAP = \begin{bmatrix} -\frac{1}{\sqrt{2}} & \frac{1}{\sqrt{2}} & 0 \\ -\frac{1}{\sqrt{6}} & -\frac{1}{\sqrt{6}} & \frac{2}{\sqrt{6}} \\ \frac{1}{\sqrt{3}} & \frac{1}{\sqrt{3}} & \frac{1}{\sqrt{3}} \end{bmatrix} \begin{bmatrix} 4 & 2 & 2 \\ 2 & 4 & 2 \\ 2 & 2 & 4 \end{bmatrix} \begin{bmatrix} -\frac{1}{\sqrt{2}} & -\frac{1}{\sqrt{6}} & \frac{1}{\sqrt{3}} \\ \frac{1}{\sqrt{2}} & -\frac{1}{\sqrt{6}} & \frac{1}{\sqrt{3}} \\ 0 & \frac{2}{\sqrt{6}} & \frac{1}{\sqrt{3}} \end{bmatrix} = \begin{bmatrix} 2 & 0 & 0 \\ 0 & 2 & 0 \\ 0 & 0 & 8 \end{bmatrix} = D$$

Since the eigenvalues are positive, Theorem 8.4.6 implies that A can be factored as $A = B^2$ for some symmetric positive definite matrix B. One way to obtain such a B is to use Formula (22) and take $B = PD_1P^T$, where D_1 is the diagonal matrix whose diagonal entries are the square roots of the diagonal entries of D. This yields

$$B = PD_1 P^T = \begin{bmatrix} -\frac{1}{\sqrt{2}} & -\frac{1}{\sqrt{6}} & \frac{1}{\sqrt{3}} \\ \frac{1}{\sqrt{2}} & -\frac{1}{\sqrt{6}} & \frac{1}{\sqrt{3}} \\ 0 & \frac{2}{\sqrt{6}} & \frac{1}{\sqrt{3}} \end{bmatrix} \begin{bmatrix} \sqrt{2} & 0 & 0 \\ 0 & \sqrt{2} & 0 \\ 0 & 0 & \sqrt{8} \end{bmatrix} \begin{bmatrix} -\frac{1}{\sqrt{2}} & \frac{1}{\sqrt{2}} & 0 \\ -\frac{1}{\sqrt{6}} & -\frac{1}{\sqrt{6}} & \frac{2}{\sqrt{6}} \\ \frac{1}{\sqrt{3}} & \frac{1}{\sqrt{3}} & \frac{1}{\sqrt{3}} \end{bmatrix}$$

$$= \begin{bmatrix} \frac{4\sqrt{2}}{3} & \frac{\sqrt{2}}{3} & \frac{\sqrt{2}}{3} \\ \frac{\sqrt{2}}{3} & \frac{4\sqrt{2}}{3} & \frac{\sqrt{2}}{3} \\ \frac{\sqrt{2}}{3} & \frac{\sqrt{2}}{3} & \frac{4\sqrt{2}}{3} \end{bmatrix}$$

We leave it for you to confirm that $A = B^2$. ∎

CHOLESKY FACTORIZATION   We know from Theorem 8.4.6 that if A is a symmetric positive definite matrix, then it can be factored as $A = C^TC$, where C is invertible. More specifically, one can prove that if A is symmetric and positive definite, then it can be factored as $A = R^TR$, where R is upper triangular and has positive entries on the main diagonal. This is called a **Cholesky factorization** of A. Cholesky factorizations are important in many kinds of numerical algorithms, and programs such as MATLAB, *Maple*, and *Mathematica* have built-in commands for computing them.

Exercise Set 8.4

In Exercises 1 and 2, express the quadratic form in the matrix notation $\mathbf{x}^TA\mathbf{x}$, where A is a symmetric matrix.

1. (a) $3x_1^2 + 7x_2^2$ (b) $4x_1^2 - 9x_2^2 - 6x_1x_2$

(c) $9x_1^2 - x_2^2 + 4x_3^2 + 6x_1x_2 - 8x_1x_3 + x_2x_3$

2. (a) $5x_1^2 + 5x_1x_2$ (b) $-7x_1x_2$

(c) $x_1^2 + x_2^2 - 3x_3^2 - 5x_1x_2 + 9x_1x_3$

In Exercises 3 and 4, find a formula for the quadratic form that does not use matrices.

3. $\begin{bmatrix} x & y \end{bmatrix} \begin{bmatrix} 2 & -3 \\ -3 & 5 \end{bmatrix} \begin{bmatrix} x \\ y \end{bmatrix}$

4. $\begin{bmatrix} x_1 & x_2 & x_3 \end{bmatrix} \begin{bmatrix} -2 & \frac{7}{2} & 1 \\ \frac{7}{2} & 0 & 6 \\ 1 & 6 & 3 \end{bmatrix} \begin{bmatrix} x_1 \\ x_2 \\ x_3 \end{bmatrix}$

In Exercises 5–8, find an orthogonal change of variables that eliminates the cross product terms in the quadratic form Q, and express Q in terms of the new variables.

5. $Q = 2x_1^2 + 2x_2^2 - 2x_1x_2$

6. $Q = 5x_1^2 + 2x_2^2 + 4x_3^2 + 4x_1x_2$

7. $Q = 3x_1^2 + 4x_2^2 + 5x_3^2 + 4x_1x_2 - 4x_2x_3$

8. $Q = 2x_1^2 + 5x_2^2 + 5x_3^2 + 4x_1x_2 - 4x_1x_3 - 8x_2x_3$

In Exercises 9 and 10, write the quadratic equation in the matrix form $\mathbf{x}^T A\mathbf{x} + K\mathbf{x} + f = 0$, where $\mathbf{x}^T A\mathbf{x}$ is the associated quadratic form and K is an appropriate matrix.

9. (a) $2x^2 + xy + x - 6y + 2 = 0$
  (b) $y^2 + 7x - 8y - 5 = 0$
10. (a) $x^2 - xy + 5x + 8y - 3 = 0$
  (b) $5xy = 8$

In Exercises 11 and 12, identify the type of conic represented by the equation.

11. (a) $2x^2 + 5y^2 = 20$ (b) $x^2 - y^2 - 8 = 0$
  (c) $7y^2 - 2x = 0$ (d) $x^2 + y^2 - 25 = 0$
12. (a) $4x^2 + 9y^2 = 1$ (b) $4x^2 - 5y^2 = 20$
  (c) $-x^2 = 2y$ (d) $x^2 - 3 = -y^2$

In Exercises 13–16, identify the type of conic represented by the equation by rotating the xy-axes to put the conic in standard position. Find the equation of the conic in the rotated coordinate system, and state the angle θ through which you rotated the axes.

13. $2x^2 - 4xy - y^2 + 8 = 0$ **14.** $5x^2 + 4xy + 5y^2 = 9$
15. $11x^2 + 24xy + 4y^2 - 15 = 0$
16. $x^2 + xy + y^2 = \frac{1}{2}$

In Exercises 17 and 18, determine by inspection whether the matrix is positive definite, negative definite, indefinite, positive semidefinite, or negative semidefinite.

17. (a) $\begin{bmatrix} 1 & 0 \\ 0 & 2 \end{bmatrix}$ (b) $\begin{bmatrix} -1 & 0 \\ 0 & -2 \end{bmatrix}$ (c) $\begin{bmatrix} -1 & 0 \\ 0 & 2 \end{bmatrix}$

   (d) $\begin{bmatrix} 1 & 0 \\ 0 & 0 \end{bmatrix}$ (e) $\begin{bmatrix} 0 & 0 \\ 0 & -2 \end{bmatrix}$

18. (a) $\begin{bmatrix} 2 & 0 \\ 0 & -5 \end{bmatrix}$ (b) $\begin{bmatrix} -2 & 0 \\ 0 & -5 \end{bmatrix}$ (c) $\begin{bmatrix} 2 & 0 \\ 0 & 5 \end{bmatrix}$

   (d) $\begin{bmatrix} 0 & 0 \\ 0 & -5 \end{bmatrix}$ (e) $\begin{bmatrix} 2 & 0 \\ 0 & 0 \end{bmatrix}$

In Exercises 19–24, classify the quadratic form as positive definite, negative definite, indefinite, positive semidefinite, or negative semidefinite.

19. $x_1^2 + x_2^2$ **20.** $-x_1^2 - 3x_2^2$
21. $(x_1 - x_2)^2$ **22.** $-(x_1 - x_2)^2$
23. $x_1^2 - x_2^2$ **24.** $x_1 x_2$

In Exercises 25 and 26, show that the matrix A is positive definite by first using Theorem 8.4.3 and then using Theorem 8.4.5.

25. (a) $A = \begin{bmatrix} 5 & -2 \\ -2 & 5 \end{bmatrix}$ (b) $A = \begin{bmatrix} 2 & -1 & 0 \\ -1 & 2 & 0 \\ 0 & 0 & 5 \end{bmatrix}$

26. (a) $A = \begin{bmatrix} 2 & 1 \\ 1 & 2 \end{bmatrix}$ (b) $A = \begin{bmatrix} 3 & -1 & 0 \\ -1 & 2 & -1 \\ 0 & -1 & 3 \end{bmatrix}$

27. Express the symmetric positive definite matrices in Exercise 25 in the form $A = B^2$, where B is symmetric and positive definite.

28. Express the symmetric positive definite matrices in Exercise 26 in the form $A = B^2$, where B is symmetric and positive definite.

In Exercises 29 and 30, find all values of k for which the given quadratic form is positive definite.

29. $5x_1^2 + x_2^2 + kx_3^2 + 4x_1x_2 - 2x_1x_3 - 2x_2x_3$
30. $3x_1^2 + x_2^2 + 2x_3^2 - 2x_1x_3 + 2kx_2x_3$

31. Consider the matrix $A = \begin{bmatrix} 9 & 6 \\ 6 & 9 \end{bmatrix}$.

  (a) Show that the matrix A is positive definite and find two different symmetric positive definite matrices B such that $A = B^2$. [*Hint:* The matrices P and D_1 in Formula (22) are not unique.]
  (b) Find an invertible upper triangular matrix C such that $A = C^T C$. [*Hint:* First find the eigenvalues of C.]

32. Let $\mathbf{x}^T A\mathbf{x}$ be a quadratic form in the variables $x_1, x_2, \ldots, x_n$, and define $T : R^n \to R$ by $T(\mathbf{x}) = \mathbf{x}^T A\mathbf{x}$.
  (a) Show that $T(\mathbf{x} + \mathbf{y}) = T(\mathbf{x}) + 2\mathbf{x}^T A\mathbf{y} + T(\mathbf{y})$.
  (b) Show that $T(c\mathbf{x}) = c^2 T(\mathbf{x})$.

33. Express the quadratic form $(c_1 x_1 + c_2 x_2 + \cdots + c_n x_n)^2$ in the matrix notation $\mathbf{x}^T A\mathbf{x}$, where A is symmetric.

34. In statistics the quantities

$$\overline{x} = \frac{1}{n}(x_1 + x_2 + \cdots + x_n)$$

and

$$s_x^2 = \frac{1}{n-1}\left[(x_1 - \overline{x})^2 + (x_2 - \overline{x})^2 + \cdots + (x_n - \overline{x})^2\right]$$

are called, respectively, the **sample mean** and **sample variance** of $\mathbf{x} = (x_1, x_2, \ldots, x_n)$.
  (a) Express the quadratic form s_x^2 in the matrix notation $\mathbf{x}^T A\mathbf{x}$, where A is symmetric.
  (b) Is s_x^2 a positive definite quadratic form? Explain.

35. The graph in an xyz-coordinate system of an equation of form $ax^2 + by^2 + cz^2 = 1$ in which a, b, and c are positive is a surface called a **central ellipsoid in standard position** (see the accompanying figure). This is the three-dimensional generalization of the ellipse $ax^2 + by^2 = 1$ in the xy-plane.

The intersections of the ellipsoid $ax^2 + by^2 + cz^2 = 1$ with the coordinate axes determine three line segments called the **axes** of the ellipsoid. If a central ellipsoid is rotated about the origin so two or more of its axes do not coincide with any of the coordinate axes, then the resulting equation will have one or more cross product terms.

(a) Show that the equation

$$\tfrac{4}{3}x^2 + \tfrac{4}{3}y^2 + \tfrac{4}{3}z^2 + \tfrac{4}{3}xy + \tfrac{4}{3}xz + \tfrac{4}{3}yz = 1$$

represents an ellipsoid, and find the lengths of its axes. [*Suggestion:* Write the equation in the form $\mathbf{x}^T A \mathbf{x} = 1$ and make an orthogonal change of variable to eliminate the cross product terms.]

(b) What property must a symmetric 3×3 matrix have in order for the equation $\mathbf{x}^T A \mathbf{x} = 1$ to represent an ellipsoid?

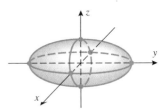

Figure Ex-35

Discussion and Discovery

D1. Indicate whether the statement is true (T) or false (F). Justify your answer.

(a) A symmetric matrix with positive entries is positive definite.

(b) $x_1^2 - x_2^2 + x_3^2 + 4x_1x_2x_3$ is a quadratic form.

(c) $(x_1 - 3x_2)^2$ is a quadratic form.

(d) A positive definite matrix is invertible.

(e) A symmetric matrix is either positive definite, negative definite, or indefinite.

(f) If A is positive definite, then $-A$ is negative definite.

D2. Indicate whether the statement is true (T) or false (F). Justify your answer.

(a) If $\mathbf{x}$ is a vector in R^n, then $\mathbf{x} \cdot \mathbf{x}$ is a quadratic form.

(b) If $\mathbf{x}^T A \mathbf{x}$ is a positive definite quadratic form, then so is $\mathbf{x}^T A^{-1} \mathbf{x}$.

(c) If A is a matrix with positive eigenvalues, then $\mathbf{x}^T A \mathbf{x}$ is a positive definite quadratic form.

(d) If A is a symmetric 2×2 matrix with positive entries and a positive determinant, then A is positive definite.

(e) If $\mathbf{x}^T A \mathbf{x}$ is a quadratic form with no cross product terms, then A is a diagonal matrix.

(f) If $\mathbf{x}^T A \mathbf{x}$ is a positive definite quadratic form in x and y, and if $c \neq 0$, then the graph of the equation $\mathbf{x}^T A \mathbf{x} = c$ is an ellipse.

D3. What property must a symmetric 2×2 matrix A have for $\mathbf{x}^T A \mathbf{x} = 1$ to represent a circle?

Working with Proofs

P1. Prove: If $b \neq 0$, then the cross product term can be eliminated from the quadratic form $ax^2 + 2bxy + cy^2$ by rotating the coordinate axes through an angle θ that satisfies the equation

$$\cot 2\theta = \frac{a - c}{2b}$$

P2. We know from Definition 8.4.2 and Theorem 8.4.3 that if A is a symmetric matrix with positive eigenvalues, then $\mathbf{x}^T A \mathbf{x} > 0$ for $\mathbf{x} \neq \mathbf{0}$. Prove that if A is a symmetric matrix with nonnegative eigenvalues, then $\mathbf{x}^T A \mathbf{x} \geq 0$ for $\mathbf{x} \neq \mathbf{0}$.

Technology Exercises

T1. Find an orthogonal change of variable that eliminates the cross product terms from the quadratic form

$$Q = 2x_1^2 - x_2^2 + 3x_3^2 + 4x_4^2 - x_1x_2 - 3x_1x_4 + 2x_2x_3 - x_3x_4$$

and express Q in terms of the new variables.

T2. Many linear algebra technology utilities have a command for finding a Cholesky factorization of a positive definite symmetric matrix. The Hilbert matrix

$$A = \begin{bmatrix} 1 & \tfrac{1}{2} & \tfrac{1}{3} & \tfrac{1}{4} \\ \tfrac{1}{2} & \tfrac{1}{3} & \tfrac{1}{4} & \tfrac{1}{5} \\ \tfrac{1}{3} & \tfrac{1}{4} & \tfrac{1}{5} & \tfrac{1}{6} \\ \tfrac{1}{4} & \tfrac{1}{5} & \tfrac{1}{6} & \tfrac{1}{7} \end{bmatrix}$$

is obviously symmetric. Show that it is positive definite by finding its eigenvalues, and then find a Cholesky factorization $A = R^T R$, where R is upper triangular. Check your result by computing $R^T R$.

Section 8.5   Application of Quadratic Forms to Optimization

Quadratic forms arise in various problems in which the maximum or minimum value of some quantity is required. In this section we will discuss some problems of this type. Calculus is required in this section.

RELATIVE EXTREMA OF FUNCTIONS OF TWO VARIABLES

Recall that if a function $f(x, y)$ has first partial derivatives, then its relative maxima and minima, if any, occur at points where

$$f_x(x, y) = 0 \quad \text{and} \quad f_y(x, y) = 0$$

These are called ***critical points*** of f. The specific behavior of f at a critical point (x_0, y_0) is determined by the sign of

$$D(x, y) = f(x, y) - f(x_0, y_0) \tag{1}$$

at points (x, y) that are close to, but different from, (x_0, y_0):

Relative minimum at $(0, 0, 0)$

(a)

- If $D(x, y) > 0$ at points (x, y) that are sufficiently close to, but different from, (x_0, y_0), then $f(x_0, y_0) < f(x, y)$ at such points and f is said to have a ***relative minimum*** at (x_0, y_0) (Figure 8.5.1*a*).

- If $D(x, y) < 0$ at points (x, y) that are sufficiently close to, but different from, (x_0, y_0), then $f(x_0, y_0) > f(x, y)$ at such points and f is said to have a ***relative maximum*** at (x_0, y_0) (Figure 8.5.1*b*).

- If $D(x, y)$ has both positive and negative values inside *every* circle centered at (x_0, y_0), then there are points (x, y) that are arbitrarily close to (x_0, y_0) at which $f(x_0, y_0) < f(x, y)$ and points (x, y) that are arbitrarily close to (x_0, y_0) at which $f(x_0, y_0) > f(x, y)$. In this case we say that f has a ***saddle point*** at (x_0, y_0) (Figure 8.5.1*c*).

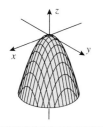

Relative maximum at $(0, 0, 0)$

(b)

In general, it can be difficult to determine the sign of (1) directly. However, the following theorem, which is proved in calculus, makes it possible to analyze critical points using derivatives.

Theorem 8.5.1 (*Second Derivative Test*) *Suppose that (x_0, y_0) is a critical point of $f(x, y)$ and that f has continuous second-order partial derivatives in some circular region centered at (x_0, y_0). Then:*

(*a*) *f has a relative minimum at (x_0, y_0) if*

$$f_{xx}(x_0, y_0)f_{yy}(x_0, y_0) - f_{xy}^2(x_0, y_0) > 0 \quad \text{and} \quad f_{xx}(x_0, y_0) > 0$$

(*b*) *f has a relative maximum at (x_0, y_0) if*

$$f_{xx}(x_0, y_0)f_{yy}(x_0, y_0) - f_{xy}^2(x_0, y_0) > 0 \quad \text{and} \quad f_{xx}(x_0, y_0) < 0$$

(*c*) *f has a saddle point at (x_0, y_0) if*

$$f_{xx}(x_0, y_0)f_{yy}(x_0, y_0) - f_{xy}^2(x_0, y_0) < 0$$

(*d*) *The test is inconclusive if*

$$f_{xx}(x_0, y_0)f_{yy}(x_0, y_0) - f_{xy}^2(x_0, y_0) = 0$$

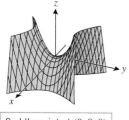

Saddle point at $(0, 0, 0)$

(c)

Figure 8.5.1

Our interest here is in showing how to express this theorem in terms of quadratic forms. For this purpose we consider the matrix

$$H(x, y) = \begin{bmatrix} f_{xx}(x, y) & f_{xy}(x, y) \\ f_{xy}(x, y) & f_{yy}(x, y) \end{bmatrix}$$

which is called the **Hessian** or **Hessian matrix** of f in honor of the German mathematician and scientist Ludwig Otto Hesse (1811–1874). The notation $H(x, y)$ emphasizes that the entries in the matrix depend on x and y. The Hessian is of interest because

$$\det[H(x_0, y_0)] = \begin{vmatrix} f_{xx}(x_0, y_0) & f_{xy}(x_0, y_0) \\ f_{xy}(x_0, y_0) & f_{yy}(x_0, y_0) \end{vmatrix} = f_{xx}(x_0, y_0) f_{yy}(x_0, y_0) - f_{xy}^2(x_0, y_0)$$

is the expression that appears in Theorem 8.5.1. We can now reformulate the second derivative test as follows.

> **Theorem 8.5.2 (Hessian Form of the Second Derivative Test)**   *Suppose that (x_0, y_0) is a critical point of $f(x, y)$ and that f has continuous second-order partial derivatives in some circular region centered at (x_0, y_0). If $H = H(x_0, y_0)$ is the Hessian of f at (x_0, y_0), then:*
>
> (a) *f has a relative minimum at (x_0, y_0) if H is positive definite.*
> (b) *f has a relative maximum at (x_0, y_0) if H is negative definite.*
> (c) *f has a saddle point at (x_0, y_0) if H is indefinite.*
> (d) *The test is inconclusive otherwise.*

Proof (a) If H is positive definite, then Theorem 8.4.5 implies that the principal submatrices of H have positive determinants. Thus,

$$\det[H] = \begin{vmatrix} f_{xx}(x_0, y_0) & f_{xy}(x_0, y_0) \\ f_{xy}(x_0, y_0) & f_{yy}(x_0, y_0) \end{vmatrix} = f_{xx}(x_0, y_0) f_{yy}(x_0, y_0) - f_{xy}^2(x_0, y_0) > 0 \tag{2}$$

and

$$\det[f_{xx}(x_0, y_0)] = f_{xx}(x_0, y_0) > 0 \tag{3}$$

so f has a relative minimum at (x_0, y_0) by part (a) of Theorem 8.5.1.

Proof (b) If H is negative definite, then the matrix $-H$ is positive definite, so the principal submatrices of $-H$ have positive determinants. Thus,

$$\det[-H] = \begin{vmatrix} -f_{xx}(x_0, y_0) & -f_{xy}(x_0, y_0) \\ -f_{xy}(x_0, y_0) & -f_{yy}(x_0, y_0) \end{vmatrix} = f_{xx}(x_0, y_0) f_{yy}(x_0, y_0) - f_{xy}^2(x_0, y_0) > 0$$

and

$$\det[-f_{xx}(x_0, y_0)] = -f_{xx}(x_0, y_0) > 0$$

so f has a relative maximum at (x_0, y_0) by part (b) of Theorem 8.5.1.

Proof (c) If H is indefinite, then it has one positive eigenvalue and one negative eigenvalue. Since $\det[H]$ is the product of the eigenvalues of H, it follows that

$$\det[H] = \begin{vmatrix} f_{xx}(x_0, y_0) & f_{xy}(x_0, y_0) \\ f_{xy}(x_0, y_0) & f_{yy}(x_0, y_0) \end{vmatrix} = f_{xx}(x_0, y_0) f_{yy}(x_0, y_0) - f_{xy}^2(x_0, y_0) < 0$$

so f has a saddle point at (x_0, y_0) by part (c) of Theorem 8.5.1.

Proof (d) If H is not positive definite, negative definite, or indefinite, then one or both of its eigenvalues must be zero; and since $\det[H]$ is the product of the eigenvalues, we must have

$$\det[H] = \begin{vmatrix} f_{xx}(x_0, y_0) & f_{xy}(x_0, y_0) \\ f_{xy}(x_0, y_0) & f_{yy}(x_0, y_0) \end{vmatrix} = f_{xx}(x_0, y_0) f_{yy}(x_0, y_0) - f_{xy}^2(x_0, y_0) = 0$$

so the test is inconclusive by part (d) of Theorem 8.5.1.   ∎

EXAMPLE 1
Using the
Hessian to
Classify
Relative
Extrema

Find the critical points of the function

$$f(x, y) = \tfrac{1}{3}x^3 + xy^2 - 8xy + 3$$

and use the eigenvalues of the Hessian matrix at those points to determine which of them, if any, are relative maxima, relative minima, or saddle points.

Solution To find both the critical points and the Hessian matrix we will need to calculate the first and second partial derivatives of f. These derivatives are

$$f_x(x, y) = x^2 + y^2 - 8y, \qquad f_y(x, y) = 2xy - 8x, \qquad f_{xy}(x, y) = 2y - 8$$
$$f_{xx}(x, y) = 2x, \qquad\qquad f_{yy}(x, y) = 2x$$

Thus, the Hessian matrix is

$$H(x, y) = \begin{bmatrix} f_{xx}(x, y) & f_{xy}(x, y) \\ f_{xy}(x, y) & f_{yy}(x, y) \end{bmatrix} = \begin{bmatrix} 2x & 2y - 8 \\ 2y - 8 & 2x \end{bmatrix}$$

To find the critical points we set f_x and f_y equal to zero. This yields the equations

$$f_x(x, y) = x^2 + y^2 - 8y = 0 \quad \text{and} \quad f_y(x, y) = 2xy - 8x = 2x(y - 4) = 0$$

Solving the second equation yields $x = 0$ or $y = 4$. Substituting $x = 0$ in the first equation and solving for y yields $y = 0$ or $y = 8$; and substituting $y = 4$ into the first equation and solving for x yields $x = 4$ or $x = -4$. Thus, we have four critical points:

$$(0, 0), \quad (0, 8), \quad (4, 4), \quad (-4, 4)$$

Evaluating the Hessian matrix at these points yields

$$H(0, 0) = \begin{bmatrix} 0 & -8 \\ -8 & 0 \end{bmatrix}, \qquad H(0, 8) = \begin{bmatrix} 0 & 8 \\ 8 & 0 \end{bmatrix}$$

$$H(4, 4) = \begin{bmatrix} 8 & 0 \\ 0 & 8 \end{bmatrix}, \qquad H(-4, 4) = \begin{bmatrix} -8 & 0 \\ 0 & -8 \end{bmatrix}$$

We leave it for you to find the eigenvalues of these matrices and deduce the following classifications of the stationary points:

Stationary Point (x_0, y_0)	λ_1	λ_2	Classification
$(0, 0)$	8	-8	Saddle point
$(0, 8)$	8	-8	Saddle point
$(4, 4)$	8	8	Relative minimum
$(-4, 4)$	-8	-8	Relative maximum

■

CONSTRAINED EXTREMUM PROBLEMS

We now turn to the third problem posed in the last section, finding the maximum and minimum values of a quadratic form $\mathbf{x}^T A \mathbf{x}$ subject to the constraint $\|\mathbf{x}\| = 1$.

To visualize this problem geometrically in the case where $\mathbf{x}^T A \mathbf{x}$ is a quadratic form on R^2, view $z = \mathbf{x}^T A \mathbf{x}$ as the equation of some surface in a rectangular xyz-coordinate system and $\|\mathbf{x}\| = 1$ as the unit circle centered at the origin of the xy-plane. Geometrically, the problem of finding the maximum and minimum values of $\mathbf{x}^T A \mathbf{x}$ subject to the constraint $\|\mathbf{x}\| = 1$ amounts to finding the highest and lowest points on the intersection of the surface with the right circular cylinder determined by the circle (Figure 8.5.2).

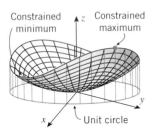

Constrained minimum

Constrained maximum

Unit circle

Figure 8.5.2

The following theorem, whose proof is deferred to the end of the section, is the key result for solving problems of this type.

Theorem 8.5.3 (*Constrained Extremum Theorem*) *Let A be a symmetric n × n matrix whose eigenvalues in order of decreasing size are $\lambda_1 \geq \lambda_2 \geq \cdots \geq \lambda_n$. Then:*

(a) *There is a maximum value and a minimum value for $\mathbf{x}^T A \mathbf{x}$ on the unit sphere $\|\mathbf{x}\| = 1$.*

(b) *The maximum value is λ_1 (the largest eigenvalue), and this maximum occurs if $\mathbf{x}$ is a unit eigenvector of A corresponding to λ_1.*

(c) *The minimum value is λ_n (the smallest eigenvalue), and this minimum occurs if $\mathbf{x}$ is a unit eigenvector of A corresponding to λ_n.*

REMARK The condition $\|\mathbf{x}\| = 1$ in this theorem is called a *constraint*, and the maximum or minimum value of $\mathbf{x}^T A \mathbf{x}$ subject to the constraint is called a *constrained extremum*. This constraint can also be expressed as $\mathbf{x}^T \mathbf{x} = 1$ or as $x_1^2 + x_2^2 + \cdots + x_n^2 = 1$, when convenient.

EXAMPLE 2
Finding Constrained Extrema

Find the maximum and minimum values of the quadratic form

$$z = 5x^2 + 5y^2 + 4xy$$

subject to the constraint $x^2 + y^2 = 1$.

Solution The quadratic form can be expressed in matrix notation as

$$z = 5x^2 + 5y^2 + 4xy = \mathbf{x}^T A \mathbf{x} = [x \quad y] \begin{bmatrix} 5 & 2 \\ 2 & 5 \end{bmatrix} \begin{bmatrix} x \\ y \end{bmatrix}$$

We leave it for you to show that the eigenvalues of A are $\lambda_1 = 7$ and $\lambda_2 = 3$ and that corresponding eigenvectors are

$$\lambda_1 = 7: \quad \begin{bmatrix} 1 \\ 1 \end{bmatrix}, \quad \lambda_2 = 3: \quad \begin{bmatrix} -1 \\ 1 \end{bmatrix}$$

Normalizing these eigenvectors yields

$$\lambda_1 = 7: \quad \begin{bmatrix} \frac{1}{\sqrt{2}} \\ \frac{1}{\sqrt{2}} \end{bmatrix}, \quad \lambda_2 = 3: \quad \begin{bmatrix} -\frac{1}{\sqrt{2}} \\ \frac{1}{\sqrt{2}} \end{bmatrix} \tag{4}$$

Thus, the constrained extrema are

constrained maximum: $z = 7$ at $(x, y) = \left(\frac{1}{\sqrt{2}}, \frac{1}{\sqrt{2}} \right)$

constrained minimum: $z = 3$ at $(x, y) = \left(-\frac{1}{\sqrt{2}}, \frac{1}{\sqrt{2}} \right)$ ∎

REMARK Since the negatives of the eigenvectors in (4) are also unit eigenvectors, they too produce the maximum and minimum values of z; that is, the constrained maximum $z = 7$ also occurs at the point $(x, y) = (-1/\sqrt{2}, -1/\sqrt{2})$ and the constrained minimum $z = 3$ at $(x, y) = (1/\sqrt{2}, -1/\sqrt{2})$.

EXAMPLE 3
A Constrained Extremum Problem

A rectangle is to be inscribed in the ellipse $4x^2 + 9y^2 = 36$, as shown in Figure 8.5.3. Use eigenvalue methods to find nonnegative values of x and y that produce the inscribed rectangle with maximum area.

Solution The area z of the inscribed rectangle is given by $z = 4xy$, so the problem is to maximize the quadratic form $z = 4xy$ subject to the constraint $4x^2 + 9y^2 = 36$. In this problem, the graph of the constraint equation is an ellipse rather than the unit circle, but we can remedy

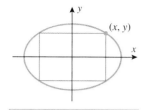

A rectangle inscribed in the ellipse $4x^2 + 9y^2 = 36$

Figure 8.5.3

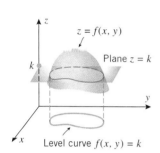

Figure 8.5.4

this problem by rewriting the constraint as

$$\left(\frac{x}{3}\right)^2 + \left(\frac{y}{2}\right)^2 = 1$$

and defining new variables, x_1 and y_1, by the equations

$$x = 3x_1 \quad \text{and} \quad y = 2y_1$$

Thus, the problem can be reformulated as follows:

maximize $z = 4xy = 24x_1y_1$

subject to the constraint

$$x_1^2 + y_1^2 = 1$$

To solve this problem, we will write the quadratic form $z = 24x_1y_1$ as

$$z = \mathbf{x}^T A \mathbf{x} = [x_1 \quad y_1] \begin{bmatrix} 0 & 12 \\ 12 & 0 \end{bmatrix} \begin{bmatrix} x_1 \\ y_1 \end{bmatrix}$$

We now leave it for you to show that the largest eigenvalue of A is $\lambda = 12$ and that the only corresponding unit eigenvector with nonnegative entries is

$$\mathbf{x} = \begin{bmatrix} x_1 \\ y_1 \end{bmatrix} = \begin{bmatrix} \frac{1}{\sqrt{2}} \\ \frac{1}{\sqrt{2}} \end{bmatrix}$$

Thus, the maximum area is $z = 12$, and this occurs when

$$x = 3x_1 = \frac{3}{\sqrt{2}} \quad \text{and} \quad y = 2y_1 = \frac{2}{\sqrt{2}} \qquad \blacksquare$$

CONSTRAINED EXTREMA AND LEVEL CURVES

A useful way of visualizing the behavior of a function $f(x, y)$ of two variables is to consider the curves in the xy-plane along which $f(x, y)$ is constant. These curves have equations of the form

$$f(x, y) = k$$

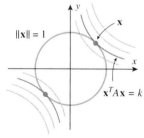

Figure 8.5.5

and are called the ***level curves*** of f (Figure 8.5.4). In particular, the level curves of a quadratic form $\mathbf{x}^T A \mathbf{x}$ on R^2 have equations of the form

$$\mathbf{x}^T A \mathbf{x} = k \tag{5}$$

so the maximum and minimum values of $\mathbf{x}^T A \mathbf{x}$ subject to the constraint $\|\mathbf{x}\| = 1$ are the largest and smallest values of k for which the graph of (5) intersects the unit circle. Typically, such values of k produce level curves that just touch the unit circle (Figure 8.5.5). Moreover, the points at which these level curves just touch the circle produce the components of the vectors that maximize or minimize $\mathbf{x}^T A \mathbf{x}$ subject to the constraint $\|\mathbf{x}\| = 1$.

EXAMPLE 4
Example 2 Revisited Using Level Curves

In Example 2 (and its following remark) we found the maximum and minimum values of the quadratic form

$$z = 5x^2 + 5y^2 + 4xy$$

subject to the constraint $x^2 + y^2 = 1$. We showed that the constrained maximum $z = 7$ occurs at the points $(x, y) = (1/\sqrt{2}, 1/\sqrt{2})$ and $(-1/\sqrt{2}, -1/\sqrt{2})$, and that the constrained minimum $z = 3$ occurs at the points $(x, y) = (-1/\sqrt{2}, 1/\sqrt{2})$ and $(1/\sqrt{2}, -1/\sqrt{2})$. Geometrically, this means that the level curve

$$5x^2 + 5y^2 + 4xy = 7$$

should just touch the unit circle at the points $(1/\sqrt{2},\, 1/\sqrt{2})$ and $(-1/\sqrt{2},\, -1/\sqrt{2})$, and the level curve

$$5x^2 + 5y^2 + 4xy = 3$$

should just touch it at the points $(-1/\sqrt{2},\, 1/\sqrt{2})$ and $(1/\sqrt{2},\, -1/\sqrt{2})$. It also means that there can be no level curve

$$5x^2 + 5y^2 + 4xy = k$$

with $k > 7$ or $k < 3$ that intersects the unit circle (Figure 8.5.6). ■

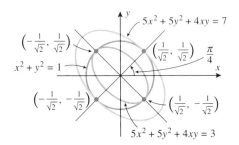

Figure 8.5.6

Proof of Theorem 8.5.3 The first step in the proof is to show that $Q = \mathbf{x}^T A \mathbf{x}$ has constrained maximum and minimum values for $\|\mathbf{x}\| = 1$. Since A is symmetric, the principal axes theorem (Theorem 8.4.1) implies that there is an orthogonal change of variable $\mathbf{x} = P\mathbf{y}$ such that

$$\mathbf{x}^T A \mathbf{x} = \lambda_1 y_1^2 + \lambda_2 y_2^2 + \cdots + \lambda_n y_n^2 \tag{6}$$

in which $\lambda_1, \lambda_2, \ldots, \lambda_n$ are the eigenvalues of A. Let us assume that $\|\mathbf{x}\| = 1$ and that the column vectors of P (which are unit eigenvectors of A) have been ordered so that

$$\lambda_1 \geq \lambda_2 \geq \cdots \geq \lambda_n \tag{7}$$

Since P is an orthogonal matrix, multiplication by P is length preserving, so $\|\mathbf{y}\| = \|\mathbf{x}\| = 1$; that is,

$$y_1^2 + y_2^2 + \cdots + y_n^2 = 1$$

It follows from this equation and (7) that

$$\lambda_n = \lambda_n(y_1^2 + y_2^2 + \cdots + y_n^2) \leq \lambda_1 y_1^2 + \lambda_2 y_2^2 + \cdots + \lambda_n y_n^2 \leq \lambda_1(y_1^2 + y_2^2 + \cdots + y_n^2) = \lambda_1$$

and hence from (6) that

$$\lambda_n \leq \mathbf{x}^T A \mathbf{x} \leq \lambda_1$$

This shows that all values of $\mathbf{x}^T A \mathbf{x}$ for which $\|\mathbf{x}\| = 1$ lie between the largest and smallest eigenvalues of A. Now let $\mathbf{x}$ be a unit eigenvector corresponding to λ_1. Then

$$\mathbf{x}^T A \mathbf{x} = \mathbf{x}^T(\lambda_1 \mathbf{x}) = \lambda_1 \mathbf{x}^T \mathbf{x} = \lambda_1 \|\mathbf{x}\|^2 = \lambda_1$$

which shows that $\mathbf{x}^T A \mathbf{x}$ has λ_1 as a constrained maximum and this maximum occurs if $\mathbf{x}$ is a unit eigenvector of A corresponding to λ_1. Similarly, if $\mathbf{x}$ is a unit eigenvector corresponding to λ_n, then

$$\mathbf{x}^T A \mathbf{x} = \mathbf{x}^T(\lambda_n \mathbf{x}) = \lambda_n \mathbf{x}^T \mathbf{x} = \lambda_n \|\mathbf{x}\|^2 = \lambda_n$$

so $\mathbf{x}^T A \mathbf{x}$ has λ_n as a constrained minimum and this minimum occurs if $\mathbf{x}$ is a unit eigenvector of A corresponding to λ_n. This completes the proof. ■

Exercise Set 8.5

1. (a) Show that the function $f(x, y) = 4xy - x^4 - y^4$ has critical points at $(0, 0)$, $(1, 1)$, and $(-1, -1)$.

   (b) Use the Hessian form of the second derivative test to show f has relative maxima at $(1, 1)$ and $(-1, -1)$ and a saddle point at $(0, 0)$.

2. (a) Show that the function $f(x, y) = x^3 - 6xy - y^3$ has critical points at $(0, 0)$ and $(-2, 2)$.

   (b) Use the Hessian form of the second derivative test to show f has a relative maximum at $(-2, 2)$ and a saddle point at $(0, 0)$.

In Exercises 3–6, find the critical points of f, if any, and classify them as relative maxima, relative minima, or saddle points.

3. $f(x, y) = x^3 - 3xy - y^3$

4. $f(x, y) = x^3 - 3xy + y^3$

5. $f(x, y) = x^2 + 2y^2 - x^2y$

6. $f(x, y) = x^3 + y^3 - 3x - 3y$

In Exercises 7–10, find the maximum and minimum values of the quadratic form $z = f(x, y)$ subject to the constraint $x^2 + y^2 = 1$, and determine the values of x and y at which those extreme values occur.

7. $z = 5x^2 - y^2$ **8.** $z = xy$

9. $z = 3x^2 + 7y^2$ **10.** $z = 5x^2 + 5xy$

In Exercises 11 and 12, find the maximum and minimum values of the quadratic form $w = f(x, y, z)$ subject to the constraint $x^2 + y^2 + z^2 = 1$, and determine the values of x, y, and z at which those extreme values occur.

11. $w = 9x^2 + 4y^2 + 3z^2$

12. $w = 2x^2 + y^2 + z^2 + 2xy + 2xz$

In Exercises 13 and 14, find the maximum and minimum values of the quadratic form $z = f(x, y)$ subject to the given constraint. [*Suggestion:* As in Example 3, rewrite the constraint and make appropriate substitutions to reduce the problem to standard form.]

13. $z = xy$; $4x^2 + 8y^2 = 16$

14. $z = x^2 + xy + 2y^2$; $x^2 + 3y^2 = 16$

In Exercises 15 and 16, draw the unit circle and the level curves corresponding to the constrained extrema of the quadratic form. Show that the unit circle touches each of these curves in exactly two places, label the points of intersection, and confirm that the constrained extrema occur at those points.

15. The quadratic form and constraint in Exercise 7.

16. The quadratic form and constraint in Exercise 8.

17. A rectangle whose center is at the origin and whose sides are parallel to the coordinate axes is to be inscribed in the ellipse $x^2 + 25y^2 = 25$. Use the method of Example 3 to find nonnegative values of x and y that produce the inscribed rectangle with maximum area.

18. Suppose that the temperature at a point (x, y) on a metal plate is $T(x, y) = 4x^2 - 4xy + y^2$. An ant, walking on the plate, traverses a circle of radius 5 centered at the origin. What are the highest and lowest temperatures encountered by the ant?

Discussion and Discovery

D1. (a) Show that the functions $f(x, y) = x^4 + y^4$ and $g(x, y) = x^4 - y^4$ have critical points at $(0, 0)$ but the second derivative test is inconclusive at that point.

   (b) Give a reasonable argument to show that f has a relative minimum at $(0, 0)$ and g has a saddle point at $(0, 0)$.

D2. Suppose that the Hessian matrix of a certain quadratic form $f(x, y)$ is

$$H = \begin{bmatrix} 2 & 4 \\ 4 & 2 \end{bmatrix}$$

What can you say about the location and classification of the critical points of f?

D3. Suppose that A is an $n \times n$ symmetric matrix and

$$q(\mathbf{x}) = \mathbf{x}^T A \mathbf{x}$$

where $\mathbf{x}$ is a vector in R^n that is expressed in column form. What can you say about the value of q if $\mathbf{x}$ is a unit eigenvector corresponding to an eigenvalue λ of A?

Working with Proofs

P1. Prove: If $\mathbf{x}^T A \mathbf{x}$ is a quadratic form whose minimum and maximum values subject to the constraint $\|\mathbf{x}\| = 1$ are m and M, respectively, then for each number c in the interval $m \leq c \leq M$, there is a unit vector $\mathbf{x}_c$ such that $\mathbf{x}_c^T A \mathbf{x}_c = c$. [*Hint:* In the case where $m < M$, let $\mathbf{u}_m$ and $\mathbf{u}_M$ be unit eigenvectors of A such that $\mathbf{u}_m^T A \mathbf{u}_m = m$ and $\mathbf{u}_M^T A \mathbf{u}_M = M$, and let

$$\mathbf{x}_c = \sqrt{\frac{M - c}{M - m}}\, \mathbf{u}_m + \sqrt{\frac{c - m}{M - m}}\, \mathbf{u}_M$$

Show that $\mathbf{x}_c^T A \mathbf{x}_c = c$.]

Technology Exercises

T1. Find the maximum and minimum values of the quadratic form

$$q(\mathbf{x}) = x_1^2 + x_2^2 - x_3^2 - x_4^2 + 2x_1x_2 - 10x_1x_4 + 4x_3x_4$$

subject to the constraint $\|\mathbf{x}\| = 1$.

Section 8.6   Singular Value Decomposition

In this section we will discuss an extension of the diagonalization theory for $n \times n$ symmetric matrices to general $m \times n$ matrices. The results that we will develop in this section have applications to the compression, storage, and transmission of digitized information and are the basis for many of the best computational algorithms that are currently available for solving linear systems.

SINGULAR VALUE DECOMPOSITION OF SQUARE MATRICES

We know from our work in Section 8.3 that symmetric matrices are orthogonally diagonalizable and are the only matrices with this property (Theorem 8.3.4). The orthogonal diagonalizability of an $n \times n$ symmetric matrix A means it can be factored as

$$A = PDP^T \tag{1}$$

where P is an $n \times n$ orthogonal matrix of eigenvectors of A, and D is the diagonal matrix whose diagonal entries are the eigenvalues corresponding to the column vectors of P. In this section we will call (1) an ***eigenvalue decomposition*** of A (abbreviated EVD of A).

If an $n \times n$ matrix A is not symmetric, then it does not have an eigenvalue decomposition, but it does have a Hessenberg decomposition

$$A = PHP^T$$

in which P is an orthogonal matrix and H is in upper Hessenberg form (Theorem 8.3.8). Moreover, if A has real eigenvalues, then it has a Schur decomposition

$$A = PSP^T$$

in which P is an orthogonal matrix and S is upper triangular (Theorem 8.3.7).

The eigenvalue, Hessenberg, and Schur decompositions are important in numerical algorithms not only because H and S have simpler forms than A, but also because the orthogonal matrices that appear in these factorizations do not magnify roundoff error. To see why this is so, suppose that $\hat{\mathbf{x}}$ is a column vector whose entries are known exactly and that

$$\mathbf{x} = \hat{\mathbf{x}} + \mathbf{e}$$

is the vector that results when roundoff error is present in the entries of $\hat{\mathbf{x}}$. If P is an orthogonal matrix, then the length-preserving property of orthogonal transformations implies that

$$\|P\mathbf{x} - P\hat{\mathbf{x}}\| = \|\mathbf{x} - \hat{\mathbf{x}}\| = \|\mathbf{e}\|$$

which tells us that the error in approximating $P\hat{\mathbf{x}}$ by $P\mathbf{x}$ has the same magnitude as the error in approximating $\hat{\mathbf{x}}$ by $\mathbf{x}$.

There are two main paths that one might follow in looking for other kinds of factorizations of a general square matrix A: One might look for factorizations of the form

$$A = PJP^{-1}$$

in which P is invertible but not necessarily orthogonal, or one might look for factorizations of the form

$$A = U\Sigma V^T$$

in which U and V are orthogonal but not necessarily the same. The first path leads to factorizations in which J is either diagonal (Theorem 8.2.6) or a certain kind of block diagonal matrix, called a *Jordan canonical form* in honor of the French mathematician Camille Jordan (1838–1922). Jordan canonical forms, which we will not consider in this text, are important theoretically and in certain applications, but they are of lesser importance numerically because of the roundoff difficulties that result from the lack of orthogonality in P. Our discussion in this section will focus on the second path, starting with the following diagonalization theorem.

Theorem 8.6.1 *If A is an $n \times n$ matrix of rank k, then A can be factored as*

$$A = U\Sigma V^T$$

where U and V are $n \times n$ orthogonal matrices and Σ is an $n \times n$ diagonal matrix whose main diagonal has k positive entries and $n - k$ zeros.

Proof The matrix A^TA is symmetric, so it has an eigenvalue decomposition

$$A^TA = VDV^T$$

where the column vectors of V are unit eigenvectors of A^TA and D is the diagonal matrix whose diagonal entries are the corresponding eigenvalues of A^TA. These eigenvalues are nonnegative, for if λ is an eigenvalue of A^TA and $\mathbf{x}$ is a corresponding eigenvector, then Formula (12) of Section 3.2 implies that

$$\|A\mathbf{x}\|^2 = A\mathbf{x} \cdot A\mathbf{x} = \mathbf{x} \cdot A^TA\mathbf{x} = \mathbf{x} \cdot \lambda\mathbf{x} = \lambda(\mathbf{x} \cdot \mathbf{x}) = \lambda\|\mathbf{x}\|^2$$

from which it follows that $\lambda \geq 0$.

Since Theorems 7.5.8 and 8.2.3 imply that

$$\text{rank}(A) = \text{rank}(A^TA) = \text{rank}(D)$$

and since A has rank k, it follows that there are k positive entries and $n - k$ zeros on the main diagonal of D. For convenience, suppose that the column vectors $\mathbf{v}_1, \mathbf{v}_2, \ldots, \mathbf{v}_n$ of V have been ordered so that the corresponding eigenvalues of A^TA are in nonincreasing order

$$\lambda_1 \geq \lambda_2 \geq \cdots \geq \lambda_n \geq 0$$

Thus,

$$\lambda_1 \geq \lambda_2 \geq \cdots \geq \lambda_k > 0 \quad \text{and} \quad \lambda_{k+1} = \lambda_{k+2} = \cdots = \lambda_n = 0 \tag{2}$$

Now consider the set of image vectors

$$\{A\mathbf{v}_1, A\mathbf{v}_2, \ldots, A\mathbf{v}_n\}$$

This is an orthogonal set, for if $i \neq j$, then the orthogonality of $\mathbf{v}_i$ and $\mathbf{v}_j$ implies that

$$A\mathbf{v}_i \cdot A\mathbf{v}_j = \mathbf{v}_i \cdot A^T A\mathbf{v}_j = \mathbf{v}_i \cdot \lambda_j \mathbf{v}_j = \lambda_j(\mathbf{v}_i \cdot \mathbf{v}_j) = 0 \qquad (3)$$

Moreover,

$$\|A\mathbf{v}_i\|^2 = A\mathbf{v}_i \cdot A\mathbf{v}_i = \mathbf{v}_i \cdot A^T A\mathbf{v}_i = \mathbf{v}_i \cdot \lambda_i \mathbf{v}_i = \lambda_i(\mathbf{v}_i \cdot \mathbf{v}_i) = \lambda_i \|\mathbf{v}_i\|^2 = \lambda_i$$

from which it follows that

$$\|A\mathbf{v}_i\| = \sqrt{\lambda_i} \qquad (i = 1, 2, \ldots, n) \qquad (4)$$

Since $\lambda_1 > 0$ for $i = 1, 2, \ldots, k$, it follows from (3) and (4) that

$$\{A\mathbf{v}_1, A\mathbf{v}_2, \ldots, A\mathbf{v}_k\} \qquad (5)$$

is an orthogonal set of k nonzero vectors in the column space of A; and since we know that the column space of A has dimension k (since A has rank k), it follows that (5) is an orthogonal basis for the column space of A. If we now normalize these vectors to obtain an orthonormal basis $\{\mathbf{u}_1, \mathbf{u}_2, \ldots, \mathbf{u}_k\}$ for the column space, then Theorem 7.9.7 guarantees that we can extend this to an orthonormal basis

$$\{\mathbf{u}_1, \mathbf{u}_2, \ldots, \mathbf{u}_k, \mathbf{u}_{k+1}, \ldots, \mathbf{u}_n\}$$

for R^n. Since the first k vectors in this set result from normalizing the vectors in (5), we have

$$\mathbf{u}_j = \frac{A\mathbf{v}_j}{\|A\mathbf{v}_j\|} = \frac{1}{\sqrt{\lambda_j}} A\mathbf{v}_j \qquad (1 \leq j \leq k)$$

which implies that

$$A\mathbf{v}_1 = \sqrt{\lambda_1}\mathbf{u}_1, \quad A\mathbf{v}_2 = \sqrt{\lambda_2}\mathbf{u}_2, \ldots, \quad A\mathbf{v}_k = \sqrt{\lambda_k}\mathbf{u}_k \qquad (6)$$

Now let U be the orthogonal matrix

$$U = \begin{bmatrix} \mathbf{u}_1 & \mathbf{u}_2 & \cdots & \mathbf{u}_k & \mathbf{u}_{k+1} & \cdots & \mathbf{u}_n \end{bmatrix}$$

and let Σ be the diagonal matrix

$$\Sigma = \begin{bmatrix} \sqrt{\lambda_1} & & & & & & & 0 \\ & \sqrt{\lambda_2} & & & & & & \\ & & \ddots & & & & & \\ & & & \sqrt{\lambda_k} & & & & \\ & & & & 0 & & & \\ & & & & & \ddots & & \\ 0 & & & & & & & 0 \end{bmatrix} \qquad (7)$$

It follows from (2) and (4) that $A\mathbf{v}_j = \mathbf{0}$ for $j > k$, so

$$U\Sigma = [\sqrt{\lambda_1}\mathbf{u}_1 \quad \sqrt{\lambda_2}\mathbf{u}_2 \quad \cdots \quad \sqrt{\lambda_k}\mathbf{u}_k \quad \mathbf{0} \quad \cdots \quad \mathbf{0}]$$
$$= [A\mathbf{v}_1 \quad A\mathbf{v}_2 \quad \cdots \quad A\mathbf{v}_k \quad A\mathbf{v}_{k+1} \quad \cdots \quad A\mathbf{v}_n] = AV$$

which we can rewrite as $A = U\Sigma V^T$ using the orthogonality of V. ∎

It is important to keep in mind that the positive entries on the main diagonal of Σ are not eigenvalues of A, but rather square roots of the nonzero eigenvalues of $A^T A$. These numbers are called the ***singular values*** of A and are denoted by

$$\sigma_1 = \sqrt{\lambda_1}, \quad \sigma_2 = \sqrt{\lambda_2}, \dots, \quad \sigma_k = \sqrt{\lambda_k}$$

With this notation, the factorization obtained in the proof of Theorem 8.6.1 has the form

$$A = U\Sigma V^T = [\mathbf{u}_1 \quad \mathbf{u}_2 \quad \cdots \quad \mathbf{u}_k \quad \mathbf{u}_{k+1} \quad \cdots \quad \mathbf{u}_n] \begin{bmatrix} \sigma_1 & & & & & & & 0 \\ & \sigma_2 & & & & & & \\ & & \ddots & & & & & \\ & & & \sigma_k & & & & \\ & & & & 0 & & & \\ & & & & & \ddots & & \\ 0 & & & & & & & 0 \end{bmatrix} \begin{bmatrix} \mathbf{v}_1^T \\ \mathbf{v}_2^T \\ \vdots \\ \mathbf{v}_k^T \\ \mathbf{v}_{k+1}^T \\ \vdots \\ \mathbf{v}_n^T \end{bmatrix} \quad (8)$$

Linear Algebra in History

The term *singular value* is apparently due to the British-born mathematician Harry Bateman, who used it in a research paper published in 1908. Bateman emigrated to the United States in 1910, teaching at Bryn Mawr College, Johns Hopkins University, and finally at the California Institute of Technology. Interestingly, he was awarded his Ph.D. in 1913 by Johns Hopkins at which point in time he was already an eminent mathematician with 60 publications to his name.

Harry Bateman
(1882–1946)

which is called the ***singular value decomposition*** of A (abbreviated SVD of A) [*]
The vectors $\mathbf{u}_1, \mathbf{u}_2, \dots, \mathbf{u}_k$ are called ***left singular vectors*** of A and $\mathbf{v}_1, \mathbf{v}_2, \dots, \mathbf{v}_k$ are called ***right singular vectors*** of A.

The following theorem is a restatement of Theorem 8.6.1 and spells out some of the results that were established in the course of proving that theorem.

Theorem 8.6.2 (***Singular Value Decomposition of a Square Matrix***) *If A is an $n \times n$ matrix of rank k, then A has a singular value decomposition $A = U\Sigma V^T$ in which:*

(a) $V = [\mathbf{v}_1 \quad \mathbf{v}_2 \quad \cdots \quad \mathbf{v}_n]$ *orthogonally diagonalizes $A^T A$.*

(b) *The nonzero diagonal entries of Σ are*

$$\sigma_1 = \sqrt{\lambda_1}, \sigma_2 = \sqrt{\lambda_2}, \dots, \sigma_k = \sqrt{\lambda_k}$$

where $\lambda_1, \lambda_2, \dots, \lambda_k$ are the nonzero eigenvalues of $A^T A$ corresponding to the column vectors of V.

(c) *The column vectors of V are ordered so that $\sigma_1 \geq \sigma_2 \geq \cdots \geq \sigma_k > 0$.*

(d) $\mathbf{u}_i = \dfrac{A\mathbf{v}_i}{\|A\mathbf{v}_i\|} = \dfrac{1}{\sigma_i} A\mathbf{v}_i \quad (i = 1, 2, \dots, k)$

(e) $\{\mathbf{u}_1, \mathbf{u}_2, \dots, \mathbf{u}_k\}$ *is an orthonormal basis for $\mathrm{col}(A)$.*

(f) $\{\mathbf{u}_1, \mathbf{u}_2, \dots, \mathbf{u}_k, \mathbf{u}_{k+1}, \dots, \mathbf{u}_n\}$ *is an extension of $\{\mathbf{u}_1, \mathbf{u}_2, \dots, \mathbf{u}_k\}$ to an orthonormal basis for R^n.*

REMARK In the special case where the matrix A is invertible, it follows that $k = \mathrm{rank}(A) = n$, so there are no zeros on the diagonal of Σ. Also, there is no extension to be made in part (f) in this case, since the n vectors in part (d) themselves form an orthonormal basis for R^n.

EXAMPLE 1
Singular Value
Decomposition
of a Square
Matrix

Find the singular value decomposition of the matrix

$$A = \begin{bmatrix} \sqrt{3} & 2 \\ 0 & \sqrt{3} \end{bmatrix}$$

---

[*] Strictly speaking we should refer to (8) as "a" singular value decomposition of A and to (1) as "an" eigenvalue decomposition of A, since U, V, and P are not unique. However, we will usually refer to these factorizations as "the" singular value decomposition and "the" eigenvalue decomposition to avoid awkward phrasing that would otherwise occur.

Solution   The first step is to find the eigenvalues of the matrix

$$A^TA = \begin{bmatrix} \sqrt{3} & 0 \\ 2 & \sqrt{3} \end{bmatrix} \begin{bmatrix} \sqrt{3} & 2 \\ 0 & \sqrt{3} \end{bmatrix} = \begin{bmatrix} 3 & 2\sqrt{3} \\ 2\sqrt{3} & 7 \end{bmatrix}$$

The characteristic polynomial of A^TA is

$$\lambda^2 - 10\lambda + 9 = (\lambda - 9)(\lambda - 1)$$

so the eigenvalues of A^TA are $\lambda_1 = 9$ and $\lambda_2 = 1$, and the singular values of A are

$$\sigma_1 = \sqrt{\lambda_1} = \sqrt{9} = 3, \quad \sigma_2 = \sqrt{\lambda_2} = \sqrt{1} = 1$$

We leave it for you to show that unit eigenvectors of A^TA corresponding to the eigenvalues $\lambda_1 = 9$ and $\lambda_2 = 1$ are

$$\mathbf{v}_1 = \begin{bmatrix} \frac{1}{2} \\ \frac{\sqrt{3}}{2} \end{bmatrix} \quad \text{and} \quad \mathbf{v}_2 = \begin{bmatrix} -\frac{\sqrt{3}}{2} \\ \frac{1}{2} \end{bmatrix}$$

respectively. Thus,

$$\mathbf{u}_1 = \frac{1}{\sigma_1}A\mathbf{v}_1 = \frac{1}{3}\begin{bmatrix} \sqrt{3} & 2 \\ 0 & \sqrt{3} \end{bmatrix}\begin{bmatrix} \frac{1}{2} \\ \frac{\sqrt{3}}{2} \end{bmatrix} = \begin{bmatrix} \frac{\sqrt{3}}{2} \\ \frac{1}{2} \end{bmatrix}, \quad \mathbf{u}_2 = \frac{1}{\sigma_2}A\mathbf{v}_2 = (1)\begin{bmatrix} \sqrt{3} & 2 \\ 0 & \sqrt{3} \end{bmatrix}\begin{bmatrix} -\frac{\sqrt{3}}{2} \\ \frac{1}{2} \end{bmatrix} = \begin{bmatrix} -\frac{1}{2} \\ \frac{\sqrt{3}}{2} \end{bmatrix}$$

so

$$U = [\mathbf{u}_1 \quad \mathbf{u}_2] = \begin{bmatrix} \frac{\sqrt{3}}{2} & -\frac{1}{2} \\ \frac{1}{2} & \frac{\sqrt{3}}{2} \end{bmatrix} \quad \text{and} \quad V = [\mathbf{v}_1 \quad \mathbf{v}_2] = \begin{bmatrix} \frac{1}{2} & -\frac{\sqrt{3}}{2} \\ \frac{\sqrt{3}}{2} & \frac{1}{2} \end{bmatrix}$$

It now follows that the singular value decomposition of A is

$$\begin{matrix} \begin{bmatrix} \sqrt{3} & 2 \\ 0 & \sqrt{3} \end{bmatrix} & = & \begin{bmatrix} \frac{\sqrt{3}}{2} & -\frac{1}{2} \\ \frac{1}{2} & \frac{\sqrt{3}}{2} \end{bmatrix} & \begin{bmatrix} 3 & 0 \\ 0 & 1 \end{bmatrix} & \begin{bmatrix} \frac{1}{2} & \frac{\sqrt{3}}{2} \\ -\frac{\sqrt{3}}{2} & \frac{1}{2} \end{bmatrix} \\ A & = & U & \Sigma & V^T \end{matrix}$$

You may want to confirm the validity of this equation by multiplying out the matrices on the right side.   ■

SINGULAR VALUE DECOMPOSITION OF SYMMETRIC MATRICES

A symmetric matrix A has both an eigenvalue decomposition $A = PDP^T$ and a singular value decomposition $A = U\Sigma V^T$, so it is reasonable to ask what relationship, if any, might exist between the two. To answer this question, suppose that A has rank k and that the nonzero eigenvalues of A are ordered so that

$$|\lambda_1| \geq |\lambda_2| \geq \cdots \geq |\lambda_k| > 0$$

In the case where A is symmetric we have $A^TA = A^2$, so the eigenvalues of A^TA are the squares of the eigenvalues of A. Thus, the nonzero eigenvalues of A^TA in nonincreasing order are

$$\lambda_1^2 \geq \lambda_2^2 \geq \cdots \geq \lambda_k^2 > 0$$

and the singular values of A in nonincreasing order are

$$\sigma_1 = \sqrt{\lambda_1^2} = |\lambda_1|, \quad \sigma_2 = \sqrt{\lambda_2^2} = |\lambda_2|, \ldots, \quad \sigma_k = \sqrt{\lambda_k^2} = |\lambda_k|$$

This shows that the *singular values of a symmetric matrix A are the absolute values of the nonzero eigenvalues of A*; and it also shows that if A is a symmetric matrix with nonnegative eigenvalues, then the singular values of A are the same as its nonzero eigenvalues.

EXAMPLE 2
Obtaining a
Singular Value
Decomposition
from an
Eigenvalue
Decomposition

It follows from the computations in Example 2 of Section 8.3 that the symmetric matrix

$$A = \begin{bmatrix} 1 & 2 \\ 2 & -2 \end{bmatrix}$$

has the eigenvalue decomposition

$$A = PDP^T = \begin{bmatrix} \frac{1}{\sqrt{5}} & \frac{2}{\sqrt{5}} \\ -\frac{2}{\sqrt{5}} & \frac{1}{\sqrt{5}} \end{bmatrix} \begin{bmatrix} -3 & 0 \\ 0 & 2 \end{bmatrix} \begin{bmatrix} \frac{1}{\sqrt{5}} & -\frac{2}{\sqrt{5}} \\ \frac{2}{\sqrt{5}} & \frac{1}{\sqrt{5}} \end{bmatrix}$$

We can find a singular value decomposition of A using the following procedure to "shift" the negative sign from the diagonal factor to the second orthogonal factor:

$$A = \begin{bmatrix} \frac{1}{\sqrt{5}} & \frac{2}{\sqrt{5}} \\ -\frac{2}{\sqrt{5}} & \frac{1}{\sqrt{5}} \end{bmatrix} \begin{bmatrix} 3 & 0 \\ 0 & 2 \end{bmatrix} \begin{bmatrix} -1 & 0 \\ 0 & 1 \end{bmatrix} \begin{bmatrix} \frac{1}{\sqrt{5}} & -\frac{2}{\sqrt{5}} \\ \frac{2}{\sqrt{5}} & \frac{1}{\sqrt{5}} \end{bmatrix} = \begin{bmatrix} \frac{1}{\sqrt{5}} & \frac{2}{\sqrt{5}} \\ -\frac{2}{\sqrt{5}} & \frac{1}{\sqrt{5}} \end{bmatrix} \begin{bmatrix} 3 & 0 \\ 0 & 2 \end{bmatrix} \begin{bmatrix} -\frac{1}{\sqrt{5}} & \frac{2}{\sqrt{5}} \\ \frac{2}{\sqrt{5}} & \frac{1}{\sqrt{5}} \end{bmatrix} = U\Sigma V^T$$

Alternatively, we could have shifted the negative sign to the first orthogonal factor (verify). This technique works for any *symmetric* matrix. ∎

POLAR DECOMPOSITION   The following theorem provides another kind of factorization that has many theoretical and practical applications.

> **Theorem 8.6.3 (*Polar Decomposition*)**   *If A is an $n \times n$ matrix of rank k, then A can be factored as*
>
> $$A = PQ \tag{9}$$
>
> *where P is an $n \times n$ positive semidefinite matrix of rank k, and Q is an $n \times n$ orthogonal matrix. Moreover, if A is invertible (rank n), then there is a factorization of form (9) in which P is positive definite.*

Proof Rewrite the singular value decomposition of A as

$$A = U\Sigma V^T = U\Sigma U^T U V^T = (U\Sigma U^T)(U V^T) = PQ \tag{10}$$

The matrix $Q = U V^T$ is orthogonal because it is a product of orthogonal matrices (Theorem 6.2.3), and the matrix $P = U\Sigma U^T$ is symmetric (verify). Also, the matrices Σ and $P = U\Sigma U^T$ are orthogonally similar, so they have the same rank and same eigenvalues. This implies that P has rank k and that its eigenvalues are nonnegative (since this is true of Σ). Thus, P is a positive semidefinite matrix of rank k (see the remark following Theorem 8.4.3). Furthermore, if A is invertible, then there are no zeros on the diagonal of Σ (see the remark preceding Example 1), so the eigenvalues of P are positive, which means that P is positive definite. ∎

REMARK A factorization $A = PQ$ in which Q is orthogonal and P is positive semidefinite is called a ***polar decomposition***[*] of A. Such decompositions play an important role in engineering problems that involve deformation of material—the matrix P describes the stretching and compressing effects of the deformation, and the matrix Q describes the twisting (with a possible reflection).

---

[*]This terminology has its origin in complex number theory. As discussed in Appendix B, every nonzero complex number $z = x + iy$ can be expressed as $z = re^{i\theta}$, where (r, θ) are polar coordinates of z. In this representation, $r > 0$ (analogous to the positive definite P) and multiplying a complex number by $e^{i\theta}$ causes the vector representing that complex number to be rotated through the angle θ [analogous to the multiplicative effect of the orthogonal matrix Q when $\det(Q) = 1$].

EXAMPLE 3
Polar
Decomposition

Find a polar decomposition of the matrix

$$A = \begin{bmatrix} \sqrt{3} & 2 \\ 0 & \sqrt{3} \end{bmatrix}$$

and interpret it geometrically.

Solution We found a singular value decomposition of A in Example 1. Using the matrices U, V, and Σ in that example and the expressions for P and Q in Formula (10) we obtain

$$P = U\Sigma U^T = \begin{bmatrix} \frac{\sqrt{3}}{2} & -\frac{1}{2} \\ \frac{1}{2} & \frac{\sqrt{3}}{2} \end{bmatrix} \begin{bmatrix} 3 & 0 \\ 0 & 1 \end{bmatrix} \begin{bmatrix} \frac{\sqrt{3}}{2} & \frac{1}{2} \\ -\frac{1}{2} & \frac{\sqrt{3}}{2} \end{bmatrix} = \begin{bmatrix} \frac{5}{2} & \frac{\sqrt{3}}{2} \\ \frac{\sqrt{3}}{2} & \frac{3}{2} \end{bmatrix}$$

and

$$Q = UV^T = \begin{bmatrix} \frac{\sqrt{3}}{2} & -\frac{1}{2} \\ \frac{1}{2} & \frac{\sqrt{3}}{2} \end{bmatrix} \begin{bmatrix} \frac{1}{2} & \frac{\sqrt{3}}{2} \\ -\frac{\sqrt{3}}{2} & \frac{1}{2} \end{bmatrix} = \begin{bmatrix} \frac{\sqrt{3}}{2} & \frac{1}{2} \\ -\frac{1}{2} & \frac{\sqrt{3}}{2} \end{bmatrix}$$

Thus, a polar decomposition of A is

$$\begin{bmatrix} \sqrt{3} & 2 \\ 0 & \sqrt{3} \end{bmatrix} = \begin{bmatrix} \frac{5}{2} & \frac{\sqrt{3}}{2} \\ \frac{\sqrt{3}}{2} & \frac{3}{2} \end{bmatrix} \begin{bmatrix} \frac{\sqrt{3}}{2} & \frac{1}{2} \\ -\frac{1}{2} & \frac{\sqrt{3}}{2} \end{bmatrix}$$

$$A \qquad = \qquad P \qquad\qquad Q$$

To understand what this equation says geometrically, let us rewrite it as

$$\begin{bmatrix} 1 & \frac{2}{\sqrt{3}} \\ 0 & 1 \end{bmatrix} \begin{bmatrix} \sqrt{3} & 0 \\ 0 & \sqrt{3} \end{bmatrix} = \begin{bmatrix} \frac{5}{2} & \frac{\sqrt{3}}{2} \\ \frac{\sqrt{3}}{2} & \frac{3}{2} \end{bmatrix} \begin{bmatrix} \frac{\sqrt{3}}{2} & \frac{1}{2} \\ -\frac{1}{2} & \frac{\sqrt{3}}{2} \end{bmatrix} \qquad (11)$$

$$A \text{ (factored)} \qquad = \qquad P \qquad\qquad Q$$

The right side of this equation tells us that multiplication by A is the same as multiplication by Q followed by multiplication by P. In the exercises we will ask you to show that the multiplication by the orthogonal matrix Q produces a rotation about the origin through an angle of $-30°$ (or $330°$) and that the multiplication by the symmetric matrix P stretches R^2 by a factor of $\lambda_1 = 3$ in the direction of its unit eigenvector $\mathbf{u}_1 = (\sqrt{3}/2, 1/2)$ and by a factor of $\lambda_2 = 1$ in the direction of its unit eigenvector $\mathbf{u}_2 = (-1/2, \sqrt{3}/2)$ (i.e., no stretching). On the other hand, the left side of (11) tells us that multiplication by A produces a dilation of factor $\sqrt{3}$ followed by a shear of factor $2/\sqrt{3}$ in the x-direction. Thus, the dilation followed by the shear must have the same effect as the rotation followed by the expansions along the eigenvectors (Figure 8.6.1). ■

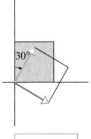

Unit square
rotated $-30°$

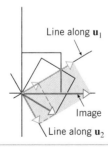

Image of purple edge obtained
by projecting onto the lines
along $\mathbf{u}_1$ and $\mathbf{u}_2$, scaling by
factor 3 in the direction of $\mathbf{u}_1$,
by factor 1 in the direction of
$\mathbf{u}_2$, then adding.

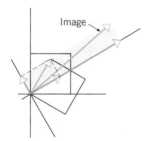

Image of green edge obtained
by projecting onto the lines
along $\mathbf{u}_1$ and $\mathbf{u}_2$, scaling by
factor 3 in the direction of $\mathbf{u}_1$,
by factor 1 in the direction of
$\mathbf{u}_2$, then adding.

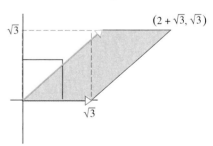

The rotation and scalings produce
the same image of the unit square
as dilating by a factor of $\sqrt{3}$ and
then shearing by a factor of $2/\sqrt{3}$
in the x-direction.

Figure 8.6.1

**SINGULAR VALUE
DECOMPOSITION OF
NONSQUARE MATRICES**

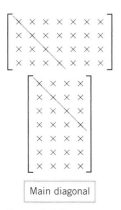

Main diagonal

Figure 8.6.2

Thus far we have focused on singular value decompositions of square matrices. However, the real power of singular value decomposition rests with the fact that it can be extended to general $m \times n$ matrices. To make this extension we define the ***main diagonal*** of an $m \times n$ matrix $A = [a_{ij}]$ to be the line of entries for which $i = j$. In the case of a square matrix, this line runs from the upper left corner to the lower right corner, but if $n > m$ or $m > n$, then the main diagonal is as pictured in Figure 8.6.2.

If A is an $m \times n$ matrix, then $A^T A$ is an $n \times n$ symmetric matrix and hence has an eigenvalue decomposition, just as in the case where A is square. Except for appropriate size adjustments to account for the possibility that $n > m$ or $m < n$, the proof of Theorem 8.6.1 carries over without change and yields the following generalization of Theorem 8.6.2.

Theorem 8.6.4 (*Singular Value Decomposition of a General Matrix*)   *If A is an $m \times n$ matrix of rank k, then A can be factored as*

$$A = U\Sigma V^T = [\mathbf{u}_1 \ \ \mathbf{u}_2 \ \ \cdots \ \ \mathbf{u}_k \mid \mathbf{u}_{k+1} \ \ \cdots \ \ \mathbf{u}_m] \begin{bmatrix} \sigma_1 & 0 & \cdots & 0 & \\ 0 & \sigma_2 & \cdots & 0 & \\ \vdots & \vdots & \ddots & \vdots & 0_{k \times (n-k)} \\ 0 & 0 & \cdots & \sigma_k & \\ \hline & 0_{(m-k)\times k} & & & 0_{(m-k)\times(n-k)} \end{bmatrix} \begin{bmatrix} \mathbf{v}_1^T \\ \mathbf{v}_2^T \\ \vdots \\ \mathbf{v}_k^T \\ \mathbf{v}_{k+1}^T \\ \vdots \\ \mathbf{v}_n^T \end{bmatrix}$$

$$(12)$$

in which U, Σ, and V have sizes $m \times m$, $m \times n$, and $n \times n$, respectively, and in which:

(a) *$V = [\mathbf{v}_1 \ \ \mathbf{v}_2 \ \ \cdots \ \ \mathbf{v}_n]$ orthogonally diagonalizes $A^T A$.*

(b) *The nonzero diagonal entries of Σ are $\sigma_1 = \sqrt{\lambda_1}, \sigma_2 = \sqrt{\lambda_2}, \ldots, \sigma_k = \sqrt{\lambda_k}$, where $\lambda_1, \lambda_2, \ldots, \lambda_k$ are the nonzero eigenvalues of $A^T A$ corresponding to the column vectors of V.*

(c) *The column vectors of V are ordered so that $\sigma_1 \geq \sigma_2 \geq \cdots \geq \sigma_k > 0$.*

(d) *$\mathbf{u}_i = \dfrac{A\mathbf{v}_i}{\|A\mathbf{v}_i\|} = \dfrac{1}{\sigma_i} A\mathbf{v}_i \qquad (i = 1, 2, \ldots, k)$*

(e) *$\{\mathbf{u}_1, \mathbf{u}_2, \ldots, \mathbf{u}_k\}$ is an orthonormal basis for $\mathrm{col}(A)$.*

(f) *$\{\mathbf{u}_1, \mathbf{u}_2, \ldots, \mathbf{u}_k, \mathbf{u}_{k+1}, \ldots, \mathbf{u}_m\}$ is an extension of $\{\mathbf{u}_1, \mathbf{u}_2, \ldots, \mathbf{u}_k\}$ to an orthonormal basis for R^m.*

As in the square case, the numbers $\sigma_1, \sigma_2, \ldots, \sigma_k$ are called the ***singular values*** of A, the vectors $\mathbf{u}_1, \mathbf{u}_2, \ldots, \mathbf{u}_k$ are called the ***left singular vectors*** of A, the vectors $\mathbf{v}_1, \mathbf{v}_2, \ldots, \mathbf{v}_k$ are called the ***right singular vectors*** of A, and $A = U\Sigma V^T$ is called the ***singular value decomposition*** of the matrix A.

EXAMPLE 4
Singular Value
Decomposition
of a Matrix That
Is Not Square

Find the singular value decomposition of the matrix

$$A = \begin{bmatrix} 1 & 1 \\ 0 & 1 \\ 1 & 0 \end{bmatrix}$$

Solution The first step is to find the eigenvalues of the matrix

$$A^T A = \begin{bmatrix} 1 & 0 & 1 \\ 1 & 1 & 0 \end{bmatrix} \begin{bmatrix} 1 & 1 \\ 0 & 1 \\ 1 & 0 \end{bmatrix} = \begin{bmatrix} 2 & 1 \\ 1 & 2 \end{bmatrix}$$

The characteristic polynomial of $A^T A$ is

$$\lambda^2 - 4\lambda + 3 = (\lambda - 3)(\lambda - 1)$$

so the eigenvalues of $A^T A$ are $\lambda_1 = 3$ and $\lambda_2 = 1$ and the singular values of A in order of decreasing size are

$$\sigma_1 = \sqrt{\lambda_1} = \sqrt{3}, \quad \sigma_2 = \sqrt{\lambda_2} = \sqrt{1} = 1$$

We leave it for you to show that unit eigenvectors of $A^T A$ corresponding to the eigenvalues $\lambda_1 = 3$ and $\lambda_2 = 1$ are

$$\mathbf{v}_1 = \begin{bmatrix} \frac{\sqrt{2}}{2} \\ \frac{\sqrt{2}}{2} \end{bmatrix} \quad \text{and} \quad \mathbf{v}_2 = \begin{bmatrix} \frac{\sqrt{2}}{2} \\ -\frac{\sqrt{2}}{2} \end{bmatrix}$$

respectively. These are the column vectors of V, and

$$\mathbf{u}_1 = \frac{1}{\sigma_1} A\mathbf{v}_1 = \frac{\sqrt{3}}{3} \begin{bmatrix} 1 & 1 \\ 0 & 1 \\ 1 & 0 \end{bmatrix} \begin{bmatrix} \frac{\sqrt{2}}{2} \\ \frac{\sqrt{2}}{2} \end{bmatrix} = \begin{bmatrix} \frac{\sqrt{6}}{3} \\ \frac{\sqrt{6}}{6} \\ \frac{\sqrt{6}}{6} \end{bmatrix}, \quad \mathbf{u}_2 = \frac{1}{\sigma_2} A\mathbf{v}_2 = (1) \begin{bmatrix} 1 & 1 \\ 0 & 1 \\ 1 & 0 \end{bmatrix} \begin{bmatrix} \frac{\sqrt{2}}{2} \\ -\frac{\sqrt{2}}{2} \end{bmatrix} = \begin{bmatrix} 0 \\ -\frac{\sqrt{2}}{2} \\ \frac{\sqrt{2}}{2} \end{bmatrix}$$

are two of the three column vectors of U. Note that $\mathbf{u}_1$ and $\mathbf{u}_2$ are orthonormal, as expected. We could extend the set $\{\mathbf{u}_1, \mathbf{u}_2\}$ to an orthonormal basis for R^3 using the method of Example 2 of Section 7.4 and the Gram–Schmidt process directly. However, the computations will be easier if we first remove the messy radicals by multiplying $\mathbf{u}_1$ and $\mathbf{u}_2$ by appropriate scalars. Thus, we will look for a unit vector $\mathbf{u}_3$ that is orthogonal to

$$\sqrt{6}\,\mathbf{u}_1 = \begin{bmatrix} 2 \\ 1 \\ 1 \end{bmatrix} \quad \text{and} \quad \sqrt{2}\,\mathbf{u}_2 = \begin{bmatrix} 0 \\ -1 \\ 1 \end{bmatrix}$$

To satisfy these two orthogonality conditions, the vector $\mathbf{u}_3$ must be a solution of the homogeneous linear system

$$\begin{bmatrix} 2 & 1 & 1 \\ 0 & -1 & 1 \end{bmatrix} \begin{bmatrix} x_1 \\ x_2 \\ x_3 \end{bmatrix} = \begin{bmatrix} 0 \\ 0 \\ 0 \end{bmatrix}$$

We leave it for you to show that a general solution of this system is

$$\begin{bmatrix} x_1 \\ x_2 \\ x_3 \end{bmatrix} = t \begin{bmatrix} -1 \\ 1 \\ 1 \end{bmatrix}$$

Normalizing the vector on the right yields

$$\mathbf{u}_3 = \begin{bmatrix} -\frac{1}{\sqrt{3}} \\ \frac{1}{\sqrt{3}} \\ \frac{1}{\sqrt{3}} \end{bmatrix}$$

Thus, the singular value decomposition of A is

$$\underbrace{\begin{bmatrix} 1 & 1 \\ 0 & 1 \\ 1 & 0 \end{bmatrix}}_{A} = \underbrace{\begin{bmatrix} \frac{\sqrt{6}}{3} & 0 & -\frac{1}{\sqrt{3}} \\ \frac{\sqrt{6}}{6} & -\frac{\sqrt{2}}{2} & \frac{1}{\sqrt{3}} \\ \frac{\sqrt{6}}{6} & \frac{\sqrt{2}}{2} & \frac{1}{\sqrt{3}} \end{bmatrix}}_{U} \underbrace{\begin{bmatrix} \sqrt{3} & 0 \\ 0 & 1 \\ 0 & 0 \end{bmatrix}}_{\Sigma} \underbrace{\begin{bmatrix} \frac{\sqrt{2}}{2} & \frac{\sqrt{2}}{2} \\ \frac{\sqrt{2}}{2} & -\frac{\sqrt{2}}{2} \end{bmatrix}}_{V^T}$$

You may want to confirm the validity of this equation by multiplying out the matrices on the right side. ∎

SINGULAR VALUE DECOMPOSITION AND THE FUNDAMENTAL SPACES OF A MATRIX

The following theorem shows that the singular value decomposition of a matrix A links together the four fundamental spaces of A in a beautiful and natural way.

Theorem 8.6.5 *If A is an $m \times n$ matrix with rank k, and if $A = U\Sigma V^T$ is the singular value decomposition given in Formula (12), then:*

(a) $\{\mathbf{u}_1, \mathbf{u}_2, \ldots, \mathbf{u}_k\}$ *is an orthonormal basis for* $\text{col}(A)$.

(b) $\{\mathbf{u}_{k+1}, \ldots, \mathbf{u}_m\}$ *is an orthonormal basis for* $\text{col}(A)^\perp = \text{null}(A^T)$.

(c) $\{\mathbf{v}_1, \mathbf{v}_2, \ldots, \mathbf{v}_k\}$ *is an orthonormal basis for* $\text{row}(A)$.

(d) $\{\mathbf{v}_{k+1}, \ldots, \mathbf{v}_n\}$ *is an orthonormal basis for* $\text{row}(A)^\perp = \text{null}(A)$.

Proofs (a) and (b) We already know from Theorem 8.6.4 that $\{\mathbf{u}_1, \mathbf{u}_2, \ldots, \mathbf{u}_k\}$ is a basis for $\text{col}(A)$ and that $\{\mathbf{u}_1, \mathbf{u}_2, \ldots, \mathbf{u}_m\}$ is an extension of that basis to an orthonormal basis for R^m. Since each of the vectors in the set $\{\mathbf{u}_{k+1}, \ldots, \mathbf{u}_m\}$ is orthogonal to each of the vectors in the set $\{\mathbf{u}_1, \mathbf{u}_2, \ldots, \mathbf{u}_k\}$, it follows that each of the vectors in $\{\mathbf{u}_{k+1}, \ldots, \mathbf{u}_m\}$ is orthogonal to $\text{span}\{\mathbf{u}_1, \mathbf{u}_2, \ldots, \mathbf{u}_k\} = \text{col}(A)$. Thus, $\{\mathbf{u}_{k+1}, \ldots, \mathbf{u}_m\}$ is an orthonormal set of $m - k$ vectors in $\text{col}(A)^\perp = \text{null}(A^T)$. But the dimension of $\text{null}(A^T)$ is $m - k$ [see Formula (5) of Section 7.5], so $\{\mathbf{u}_{k+1}, \ldots, \mathbf{u}_m\}$ must be an orthonormal basis for $\text{null}(A^T)$.

Proofs (c) and (d) The vectors $\mathbf{v}_1, \mathbf{v}_2, \ldots, \mathbf{v}_n$ form an orthonormal set of eigenvectors of $A^T A$ and are ordered so that the corresponding eigenvalues of $A^T A$ (all of which are nonnegative) are in the nonincreasing order

$$\lambda_1 \geq \lambda_2 \geq \cdots \geq \lambda_n \geq 0$$

We know from Theorem 8.6.4 that the first k of these eigenvalues are positive and the subsequent $n - k$ are zero. Thus, $\{\mathbf{v}_{k+1}, \ldots, \mathbf{v}_n\}$ is an orthonormal set of $n - k$ vectors in the null space of $A^T A$, which is the same as the null space of A (Theorem 7.5.8). Since the dimension of $\text{null}(A)$ is $n - k$ [see Formula (5) of Section 7.5], it follows that $\{\mathbf{v}_{k+1}, \ldots, \mathbf{v}_n\}$ is an orthonormal basis for $\text{null}(A)$. Moreover, since each of the vectors in the set $\{\mathbf{v}_{k+1}, \ldots, \mathbf{v}_n\}$ is orthogonal to each of the vectors in the set $\{\mathbf{v}_1, \mathbf{v}_2, \ldots, \mathbf{v}_k\}$, it follows that each of the vectors in the set $\{\mathbf{v}_1, \mathbf{v}_2, \ldots, \mathbf{v}_k\}$ is orthogonal to $\text{span}\{\mathbf{v}_{k+1}, \ldots, \mathbf{v}_n\} = \text{null}(A)$. Thus, $\{\mathbf{v}_1, \mathbf{v}_2, \ldots, \mathbf{v}_k\}$ is an orthonormal set of k vectors in $\text{null}(A)^\perp = \text{row}(A)$. But $\text{row}(A)$ has dimension k, so $\{\mathbf{v}_1, \mathbf{v}_2, \ldots, \mathbf{v}_k\}$ must be an orthonormal basis for $\text{row}(A)$. ∎

REDUCED SINGULAR VALUE DECOMPOSITION

Algebraically, the zero rows and columns of the matrix Σ in Formula (12) are superfluous and can be eliminated by multiplying out the expression $U \Sigma V^T$ using block multiplication and the partitioning shown in that formula. The products that involve zero blocks as factors drop out, leaving

$$A = [\mathbf{u}_1 \quad \mathbf{u}_2 \quad \cdots \quad \mathbf{u}_k] \begin{bmatrix} \sigma_1 & 0 & \cdots & 0 \\ 0 & \sigma_2 & \cdots & 0 \\ \vdots & \vdots & \ddots & \vdots \\ 0 & 0 & \cdots & \sigma_k \end{bmatrix} \begin{bmatrix} \mathbf{v}_1^T \\ \mathbf{v}_2^T \\ \vdots \\ \mathbf{v}_k^T \end{bmatrix} \tag{13}$$

which is called a ***reduced singular value decomposition*** of A. In this text we will denote the matrices on the right side of (13) by U_1, Σ_1, and V_1^T, respectively, and we will write this equation as

$$A = U_1 \Sigma_1 V_1^T \tag{14}$$

Note that the sizes of U_1, Σ_1, and V_1^T are $m \times k$, $k \times k$, and $k \times n$, respectively, and that the matrix Σ_1 is invertible, since its diagonal entries are positive.

CONCEPT PROBLEM Write out Σ_1^{-1}.

If we multiply out on the right side of (13) using the column-row rule of Theorem 3.8.1, then we obtain

$$A = \sigma_1 \mathbf{u}_1 \mathbf{v}_1^T + \sigma_2 \mathbf{u}_2 \mathbf{v}_2^T + \cdots + \sigma_k \mathbf{u}_k \mathbf{v}_k^T \tag{15}$$

which is called a ***reduced singular value expansion*** of A. This result applies to *all* matrices, whereas the spectral decomposition [Formula (5) of Section 8.3] applies only to symmetric matrices. You should also compare (15) to the column-row expansion of a general matrix A given in Theorem 7.6.5. In the singular value expansion the **u**'s and **v**'s are orthonormal, whereas the **c**'s and **r**'s in Theorem 7.6.5 need not be so.

EXAMPLE 5
Reduced Singular Value Decomposition

Find a reduced singular value decomposition and a reduced singular value expansion of the matrix

$$A = \begin{bmatrix} 1 & 1 \\ 0 & 1 \\ 1 & 0 \end{bmatrix}$$

Solution In Example 4 we found the singular value decomposition

$$\begin{bmatrix} 1 & 1 \\ 0 & 1 \\ 1 & 0 \end{bmatrix} = \begin{bmatrix} \frac{\sqrt{6}}{3} & 0 & -\frac{1}{\sqrt{3}} \\ \frac{\sqrt{6}}{6} & -\frac{\sqrt{2}}{2} & \frac{1}{\sqrt{3}} \\ \frac{\sqrt{6}}{6} & \frac{\sqrt{2}}{2} & \frac{1}{\sqrt{3}} \end{bmatrix} \begin{bmatrix} \sqrt{3} & 0 \\ 0 & 1 \\ 0 & 0 \end{bmatrix} \begin{bmatrix} \frac{\sqrt{2}}{2} & \frac{\sqrt{2}}{2} \\ \frac{\sqrt{2}}{2} & -\frac{\sqrt{2}}{2} \end{bmatrix} \tag{16}$$

$$\quad A \quad\quad = \quad\quad\quad\quad U \quad\quad\quad\quad\quad\quad \Sigma \quad\quad\quad V^T$$

Since A has rank 2 (verify), it follows from (13) with $k = 2$ that the reduced singular value decomposition of A corresponding to (16) is

$$\begin{bmatrix} 1 & 1 \\ 0 & 1 \\ 1 & 0 \end{bmatrix} = \begin{bmatrix} \frac{\sqrt{6}}{3} & 0 \\ \frac{\sqrt{6}}{6} & -\frac{\sqrt{2}}{2} \\ \frac{\sqrt{6}}{6} & \frac{\sqrt{2}}{2} \end{bmatrix} \begin{bmatrix} \sqrt{3} & 0 \\ 0 & 1 \end{bmatrix} \begin{bmatrix} \frac{\sqrt{2}}{2} & \frac{\sqrt{2}}{2} \\ \frac{\sqrt{2}}{2} & -\frac{\sqrt{2}}{2} \end{bmatrix}$$

This yields the reduced singular value expansion

$$
\begin{bmatrix} 1 & 1 \\ 0 & 1 \\ 1 & 0 \end{bmatrix} = \sigma_1 \mathbf{u}_1 \mathbf{v}_1^T + \sigma_2 \mathbf{u}_2 \mathbf{v}_2^T = \sqrt{3} \begin{bmatrix} \frac{\sqrt{6}}{3} \\ \frac{\sqrt{6}}{6} \\ \frac{\sqrt{6}}{6} \end{bmatrix} \begin{bmatrix} \frac{\sqrt{2}}{2} & \frac{\sqrt{2}}{2} \end{bmatrix} + (1) \begin{bmatrix} 0 \\ -\frac{\sqrt{2}}{2} \\ \frac{\sqrt{2}}{2} \end{bmatrix} \begin{bmatrix} \frac{\sqrt{2}}{2} & -\frac{\sqrt{2}}{2} \end{bmatrix}
$$

$$
= \sqrt{3} \begin{bmatrix} \frac{\sqrt{3}}{3} & \frac{\sqrt{3}}{3} \\ \frac{\sqrt{3}}{6} & \frac{\sqrt{3}}{6} \\ \frac{\sqrt{3}}{6} & \frac{\sqrt{3}}{6} \end{bmatrix} + (1) \begin{bmatrix} 0 & 0 \\ -\frac{1}{2} & \frac{1}{2} \\ \frac{1}{2} & -\frac{1}{2} \end{bmatrix}
$$

Note that the matrices in the expansion have rank 1, as expected. ■

DATA COMPRESSION AND IMAGE PROCESSING

Singular value decompositions can be used to "compress" visual information for the purpose of reducing its required storage space and speeding up its electronic transmission. The first step in compressing a visual image is to represent it as a numerical matrix from which the visual image can be recovered when needed.

For example, a black and white photograph might be scanned as a rectangular array of pixels (points) and then stored as a matrix A by assigning each pixel a numerical value in accordance with its gray level. If 256 different gray levels are used (0 = white to 255 = black), then the entries in the matrix would be integers between 0 and 255. The image can be recovered from the matrix A by printing or displaying the pixels with their assigned gray levels.

Linear Algebra in History

The theory of singular value decompositions can be traced back to the work of five people: the Italian mathematician Eugenio Beltrami, the French mathematician Camille Jordan, the English mathematician James Sylvester (see p. 81), and the German mathematicians Erhard Schmidt (see p. 412) and Herman Weyl. More recently, the pioneering efforts of the American mathematician Gene Golub produced a stable and efficient algorithm for computing it. Beltrami and Jordan were the progenitors of the decomposition—Beltrami gave a proof of the result for real, invertible matrices with distinct singular values in 1873 (which appeared, interestingly enough, in the *Journal of Mathematics for the Use of the Students of the Italian Universities*). Subsequently, Jordan refined the theory and eliminated the unnecessary restrictions imposed by Beltrami. Sylvester, apparently unfamiliar with the work of Beltrami and Jordan, rediscovered the result in 1889 and suggested its importance. Schmidt was the first person to show that the singular value decomposition could be used to approximate a matrix by another matrix with lower rank, and, in so doing, he transformed it from a mathematical curiosity to an important practical tool. Weyl showed how to find the lower rank approximations in the presence of error.

Eugenio Beltrami
(1835–1900)

Marie Ennemond Camille Jordan
(1838–1922)

Herman Klaus Hugo Weyl
(1885–1955)

Gene H. Golub
(1932–   )

If the matrix A has size $m \times n$, then one might store each of its mn entries individually. An alternative procedure is to compute the reduced singular value decomposition

$$A = \sigma_1 \mathbf{u}_1 \mathbf{v}_1^T + \sigma_2 \mathbf{u}_2 \mathbf{v}_2^T + \cdots + \sigma_k \mathbf{u}_k \mathbf{v}_k^T \tag{17}$$

in which $\sigma_1 \geq \sigma_2 \geq \cdots \geq \sigma_k$, and store the σ's, the $\mathbf{u}$'s, and the $\mathbf{v}$'s. When needed, the matrix A (and hence the image it represents) can be reconstructed from (17). Since each $\mathbf{u}_j$ has m entries and each $\mathbf{v}_j$ has n entries, this method requires storage space for

$$km + kn + k = k(m + n + 1)$$

numbers. Suppose, however, that the singular values $\sigma_{r+1}, \ldots, \sigma_k$ are sufficiently small that dropping the corresponding terms in (17) produces an acceptable approximation

$$A_r = \sigma_1 \mathbf{u}_1 \mathbf{v}_1^T + \sigma_2 \mathbf{u}_2 \mathbf{v}_2^T + \cdots + \sigma_r \mathbf{u}_r \mathbf{v}_r^T \tag{18}$$

to A and the image that it represents. We call (18) the ***rank r approximation of A***. This matrix requires storage space for only

$$rm + rn + r = r(m + n + 1)$$

numbers, compared to mn numbers required for entry-by-entry storage of A. For example, the rank 100 approximation of a 1000×1000 matrix A requires storage for only

$$100(1000 + 1000 + 1) = 200,100$$

numbers, compared to the 1,000,000 numbers required for entry-by-entry storage of A—a compression of almost 80%.

Figure 8.6.3 shows some approximations of a digitized mandrill image obtained using (18).

 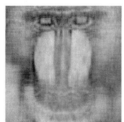   

Rank 4   Rank 10   Rank 20   Rank 50   Rank 128

Figure 8.6.3

REMARK It can be proved that A_r has rank r, that A_r does not depend on the basis vectors used in Formula (18), and that A_r is the best possible approximation to A by $m \times n$ matrices of rank r in the sense that the sum of the squares of the differences between the entries of A and A_r is as small as possible.

SINGULAR VALUE DECOMPOSITION FROM THE TRANSFORMATION POINT OF VIEW

If A is an $m \times n$ matrix and $T_A : R^n \to R^m$ is multiplication by A, then the matrix

$$\Sigma = \begin{bmatrix} \sigma_1 & 0 & \cdots & 0 & \\ 0 & \sigma_2 & \cdots & 0 & \\ \vdots & \vdots & \ddots & \vdots & 0_{k \times (n-k)} \\ 0 & 0 & \cdots & \sigma_k & \\ \hline & 0_{(m-k) \times k} & & & 0_{(m-k) \times (n-k)} \end{bmatrix}$$

in (12) is the matrix for T_A with respect to the bases $\{\mathbf{v}_1, \mathbf{v}_2, \ldots, \mathbf{v}_n\}$ and $\{\mathbf{u}_1, \mathbf{u}_2, \ldots, \mathbf{u}_m\}$ for

R^n and R^m, respectively (verify). Thus, when vectors are expressed in terms of these bases, we see that the effect of multiplying a vector by A is to scale the first k coordinates of the vector by the factors $\sigma_1, \sigma_2, \ldots, \sigma_k$, map the rest of the coordinates to zero, and possibly to discard coordinates or append zeros, if needed, to account for a decrease or increase in dimension. This idea is illustrated in Figure 8.6.4 for a 2×3 matrix A of rank 2. The effect of multiplication by A on the unit sphere in R^3 is to collapse the three dimensions of the domain into the two dimensions of the range and then stretch or compress components in the directions of the left singular vectors $\mathbf{u}_1$ and $\mathbf{u}_2$ in accordance with the magnitudes of the factors σ_1 and σ_2 to produce an ellipse in R^2.

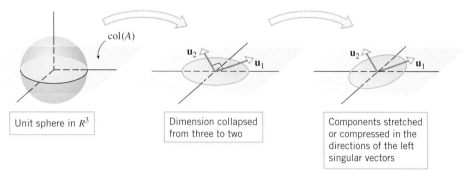

Figure 8.6.4

Some further insight into the singular value decomposition and reduced singular value decomposition of a matrix A can be obtained by focusing on the algebraic properties of the linear transformation $T_A(\mathbf{x}) = A\mathbf{x}$. Since $\text{row}(A)^\perp = \text{null}(A)$, it follows from Theorem 7.7.4 that every vector $\mathbf{x}$ in R^n can be expressed uniquely as

$$\mathbf{x} = \mathbf{x}_{\text{row}(A)} + \mathbf{x}_{\text{null}(A)}$$

where $\mathbf{x}_{\text{row}(A)}$ is the orthogonal projection of $\mathbf{x}$ on the row space of A and $\mathbf{x}_{\text{null}(A)}$ is its orthogonal projection on the null space of A. Since $A\mathbf{x}_{\text{null}(A)} = \mathbf{0}$, it follows that

$$T_A(\mathbf{x}) = A\mathbf{x} = A\mathbf{x}_{\text{row}(A)} + A\mathbf{x}_{\text{null}(A)} = A\mathbf{x}_{\text{row}(A)}$$

This tells us three things:

1. The image of any vector in R^n under multiplication by A is the same as the image of the orthogonal projection of that vector on $\text{row}(A)$.

2. The range of the transformation T_A, namely $\text{col}(A)$, is the image of $\text{row}(A)$.

3. T_A maps distinct vectors in $\text{row}(A)$ into distinct vectors in R^m (why?). Thus, even though T_A may not be one-to-one when considered as a transformation with domain R^n, it is one-to-one if its domain is restricted to $\text{row}(A)$.

Since the behavior of a matrix transformation T_A is completely determined by its action on $\text{row}(A)$, it makes sense, in the interest of efficiency, to eliminate the superfluous part of the domain and consider T_A as a transformation with domain $\text{row}(A)$. The matrix for this restricted transformation with respect to the bases $\{\mathbf{v}_1, \mathbf{v}_2, \ldots, \mathbf{v}_k\}$ for $\text{row}(A)$ and $\{\mathbf{u}_1, \mathbf{u}_2, \ldots, \mathbf{u}_k\}$ for $\text{col}(A)$ is the matrix

$$\Sigma_1 = \begin{bmatrix} \sigma_1 & & & & 0 \\ & \sigma_2 & & & \\ & & \ddots & & \\ 0 & & & & \sigma_k \end{bmatrix}$$

that occurs in the reduced singular value decomposition of A.

REMARK Loosely phrased, the preceding discussion tells us that *"hiding" inside of every nonzero matrix transformation T_A there is a one-to-one matrix transformation that maps the row space of A onto the column space of A. Moreover, that hidden transformation is represented by the reduced singular value decomposition of A with respect to appropriate bases.*

Exercise Set 8.6

In Exercises 1–4, find the distinct singular values of A.

1. $A = \begin{bmatrix} 1 & 2 & 0 \end{bmatrix}$

2. $A = \begin{bmatrix} 3 & 0 \\ 0 & 4 \end{bmatrix}$

3. $A = \begin{bmatrix} 1 & -2 \\ 2 & 1 \end{bmatrix}$

4. $A = \begin{bmatrix} \sqrt{2} & 0 \\ 1 & \sqrt{2} \end{bmatrix}$

In Exercises 5–12, find a singular value decomposition of A.

5. $A = \begin{bmatrix} 1 & -1 \\ 1 & 1 \end{bmatrix}$

6. $A = \begin{bmatrix} -3 & 0 \\ 0 & -4 \end{bmatrix}$

7. $A = \begin{bmatrix} 4 & 6 \\ 0 & 4 \end{bmatrix}$

8. $A = \begin{bmatrix} 3 & 3 \\ 3 & 3 \end{bmatrix}$

9. $A = \begin{bmatrix} -2 & 2 \\ -1 & 1 \\ 2 & -2 \end{bmatrix}$

10. $A = \begin{bmatrix} -2 & -1 & 2 \\ 2 & 1 & -2 \end{bmatrix}$

11. $A = \begin{bmatrix} 1 & 0 \\ 1 & 1 \\ -1 & 1 \end{bmatrix}$

12. $A = \begin{bmatrix} 6 & 4 \\ 0 & 0 \\ 4 & 0 \end{bmatrix}$

In Exercises 13 and 14, use the singular value decomposition of A and the method of Example 3 to find a polar decomposition of A.

13. The matrix A in Exercise 7.

14. The matrix A in Exercise 8.

In Exercises 15 and 16, use the singular value decomposition of A and the method of Example 5 to find a reduced singular value decomposition of A and a reduced singular value expansion of A.

15. The matrix A in Exercise 11.

16. The matrix A in Exercise 10.

In Exercises 17 and 18, find an eigenvalue decomposition of the given symmetric matrix A, and then use the method of Example 2 to find a singular value decomposition of A.

17. $A = \begin{bmatrix} 1 & 2 & 0 \\ 2 & 1 & 0 \\ 0 & 0 & 3 \end{bmatrix}$

18. $A = \begin{bmatrix} 0 & 0 & -2 \\ 0 & -2 & 0 \\ -2 & 0 & 3 \end{bmatrix}$

19. Suppose that A has the singular value decomposition

$$A = \begin{bmatrix} \frac{1}{2} & \frac{1}{2} & \frac{1}{2} & \frac{1}{2} \\ \frac{1}{2} & -\frac{1}{2} & -\frac{1}{2} & \frac{1}{2} \\ \frac{1}{2} & -\frac{1}{2} & \frac{1}{2} & -\frac{1}{2} \\ \frac{1}{2} & \frac{1}{2} & -\frac{1}{2} & -\frac{1}{2} \end{bmatrix} \begin{bmatrix} 24 & 0 & 0 \\ 0 & 12 & 0 \\ 0 & 0 & 0 \\ 0 & 0 & 0 \end{bmatrix} \begin{bmatrix} \frac{2}{3} & -\frac{1}{3} & \frac{2}{3} \\ \frac{2}{3} & \frac{2}{3} & -\frac{1}{3} \\ -\frac{1}{3} & \frac{2}{3} & \frac{2}{3} \end{bmatrix}$$

(a) Find orthonormal bases for the four fundamental spaces of A.

(b) Find the reduced singular value decomposition of A.

20. Let $T : R^n \to R^m$ be a linear transformation whose standard matrix A has the singular value decomposition $A = U\Sigma V^T$, and let $B = \{\mathbf{v}_1, \mathbf{v}_2, \ldots, \mathbf{v}_n\}$ and $B' = \{\mathbf{u}_1, \mathbf{u}_2, \ldots, \mathbf{u}_m\}$ be the column vectors of V and U, respectively. Show that $\Sigma = [T]_{B',B}$.

21. Show that the singular values of A^TA are the squares of the singular values of A.

22. Show that if $A = U\Sigma V^T$ is a singular value decomposition of A, then U orthogonally diagonalizes AA^T.

23. Let $A = PQ$ be the polar decomposition in Example 3. Show that multiplication by Q is a rotation about the origin through an angle of $330°$ and that multiplication by P stretches R^2 by a factor of 3 in the direction of the vector $\mathbf{u}_1 = (\sqrt{3}/2, 1/2)$ and by a factor of 1 in the direction of $\mathbf{u}_2 = (-1/2, \sqrt{3}/2)$.

Discussion and Discovery

D1. (a) If $A = U\Sigma V^T$ is a singular value decomposition of an $m \times n$ matrix of rank k, then U has size _____, Σ has size _____, and V has size _____.

(b) If $A = U_1 \Sigma_1 V_1^T$ is a reduced singular value decomposition of an $m \times n$ matrix of rank k, then U_1 has size _____, Σ_1 has size _____, and V_1 has size _____.

D2. If $A = U\Sigma V^T$ is the singular value decomposition of an invertible matrix A, then $V\Sigma^{-1}U^T$ is the singular value decomposition of _____. Justify your answer.

D3. Do orthogonally similar matrices have the same singular values? Justify your answer.

D4. If P is the standard matrix for the orthogonal projection of R^n onto a subspace W, what can you say about the singular values of P?

D5. (a) The accompanying figure suggests that multiplication by an invertible 2×2 matrix A transforms the unit circle into an ellipse. Write a paragraph that explains more precisely what the figure indicates.

(b) Draw a picture for the matrix in Example 1.

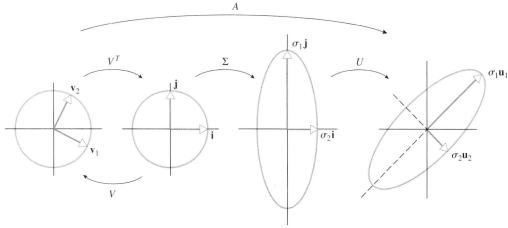

Figure Ex-D5

Technology Exercises

T1. (*Singular value decomposition*) Most linear algebra technology utilities have a command for finding the singular value decompositions, but they vary considerably. Some utilities produce the reduced singular value decomposition, some require that entries be in decimal form, and some produce U in transposed form, so you will need to be alert to this. Find the reduced singular value decomposition of the matrix

$$A = \begin{bmatrix} 3 & 0 & \frac{3}{2} \\ 1 & -2 & \frac{5}{2} \\ 1 & -2 & \frac{5}{2} \\ 3 & 0 & \frac{3}{2} \end{bmatrix}$$

and check your answer by multiplying out the factors and comparing the product to A.

T2. Find the singular values of the matrix

$$A = \begin{bmatrix} 3 & 2 & -5 \\ -6 & -8 & -6 \\ -5 & -5 & 8 \end{bmatrix}$$

by finding the square roots of the eigenvalues of A^TA. Compare your results to those produced by the command for computing the singular value decomposition of A.

T3. Construct a 3×6 matrix of rank 1, and confirm that the rank of the matrix is the same as the number of nonzero singular values. Do the same for 3×6 matrices of rank 2 and rank 3.

T4. (**MATLAB** *and Internet access required*) This problem, which is specific to MATLAB, will enable you to use singular value decompositions to recreate the mandrill pictures in Figure 8.6.3 from a scanned image that we have stored on our Web site for you. The following steps will enable you to produce a rank r approximation of the mandrill image.

> **Step 1.** Download the uncompressed scanned image mandrill.bmp from our Web site
>
> http://www.wiley.com/college/anton
>
> by following the directions posted on the site. As a check, view this image using any program or utility for viewing bitmap images (.bmp files). This image should look like the rank 128 picture in Figure 8.6.3.

Step 2. Use the MATLAB command

```
graymandrill = imread('mandrill.bmp')
```

to assign the pixels in the bitmap image integer values representing their gray levels. This produces a matrix of gray level integers named "graymandrill."

Step 3. Use the MATLAB command

```
A = double(graymandrill)
```

to convert the matrix of gray level integers to a matrix A with floating-point entries.

Step 4. Use the MATLAB command

```
[u,s,v] = svd(A)
```

to compute the matrices in the singular value decomposition of A.

Step 5. To create a rank r approximation of the mandrill image, use appropriate MATLAB commands to form the matrices ur, sr, and vr, where ur consists of the first r columns of u, sr is the matrix formed from the first r rows and the first r columns of s, and vr consists of the first r columns of v. Use appropriate MATLAB commands to compute the product $(\text{ur})(\text{sr})(\text{vr})^T$ and name it Ar.

Step 6. Use the MATLAB command

```
graylevelr = uint8(Ar)
```

to convert the entries of Ar to gray level integer values; the matrix of gray level values is named "graylevelr."

Step 7. Use the MATLAB command

```
imwrite(graylevelr, 'mandrillr.bmp')
```

to create a bitmap file of the rank r picture named "mandrillr.bmp" that can be viewed using any program or utility for viewing bitmap images.

Use the steps outlined to create and view the rank 50, rank 20, rank 10, and rank 4 approximations in Figure 8.6.3.

Section 8.7   The Pseudoinverse

The notion of an inverse applies to square matrices. In this section we will generalize this idea and consider the concept of a "pseudoinverse," which is applicable to matrices that are not square. We will see that the pseudoinverse has important applications to the study of least squares solutions of linear systems.

THE PSEUDOINVERSE   If A is an invertible $n \times n$ matrix with reduced singular value decomposition

$$A = U_1 \Sigma_1 V_1^T$$

then U_1, Σ_1, and V_1 are all $n \times n$ invertible matrices (why?), so the orthogonality of U_1 and V_1 implies that

$$A^{-1} = V_1 \Sigma_1^{-1} U_1^T \tag{1}$$

If A is not square or if it is square but not invertible, then this formula does not apply. However, we noted earlier that the matrix Σ_1 is always invertible, so the product on the right side of (1) is defined for every matrix A, though it is only for invertible A that it represents A^{-1}. If A is a nonzero $m \times n$ matrix, then we call the $n \times m$ matrix

$$A^+ = V_1 \Sigma_1^{-1} U_1^T \tag{2}$$

the *pseudoinverse*[*] of A. If $A = 0$, then we define $A^+ = 0$. The pseudoinverse is the same as the ordinary inverse for invertible matrices, but it is more general in that it applies to *all* matrices.

---

[*] It can be shown that the pseudoinverse does not depend on the bases used to form U_1 and V_1, so the terminology "the" pseudoinverse is appropriate. The pseudoinverse is also called the ***Moore–Penrose inverse*** in honor of the American mathematician E. H. Moore (1862–1932) and the British mathematician and physicist Roger Penrose (1931–   ) who developed the basic concept independently.

EXAMPLE 1
Finding the
Pseudoinverse
from the
Reduced SVD

Find the pseudoinverse of the matrix

$$A = \begin{bmatrix} 1 & 1 \\ 0 & 1 \\ 1 & 0 \end{bmatrix}$$

using the reduced singular value decomposition that was obtained in Example 5 of Section 8.6.

Solution In Example 5 of Section 8.6 we obtained the reduced singular value decomposition

$$A = [\mathbf{u}_1 \quad \mathbf{u}_2] \begin{bmatrix} \sigma_1 & 0 \\ 0 & \sigma_2 \end{bmatrix} \begin{bmatrix} \mathbf{v}_1^T \\ \mathbf{v}_2^T \end{bmatrix} = \begin{bmatrix} \frac{\sqrt{6}}{3} & 0 \\ \frac{\sqrt{6}}{6} & -\frac{\sqrt{2}}{2} \\ \frac{\sqrt{6}}{6} & \frac{\sqrt{2}}{2} \end{bmatrix} \begin{bmatrix} \sqrt{3} & 0 \\ 0 & 1 \end{bmatrix} \begin{bmatrix} \frac{\sqrt{2}}{2} & \frac{\sqrt{2}}{2} \\ \frac{\sqrt{2}}{2} & -\frac{\sqrt{2}}{2} \end{bmatrix}$$

Thus, it follows from (2) that

$$A^+ = [\mathbf{v}_1 \quad \mathbf{v}_2] \begin{bmatrix} \frac{1}{\sigma_1} & 0 \\ 0 & \frac{1}{\sigma_2} \end{bmatrix} \begin{bmatrix} \mathbf{u}_1^T \\ \mathbf{u}_2^T \end{bmatrix}$$

$$= \begin{bmatrix} \frac{\sqrt{2}}{2} & \frac{\sqrt{2}}{2} \\ \frac{\sqrt{2}}{2} & -\frac{\sqrt{2}}{2} \end{bmatrix} \begin{bmatrix} \frac{1}{\sqrt{3}} & 0 \\ 0 & 1 \end{bmatrix} \begin{bmatrix} \frac{\sqrt{6}}{3} & \frac{\sqrt{6}}{6} & \frac{\sqrt{6}}{6} \\ 0 & -\frac{\sqrt{2}}{2} & \frac{\sqrt{2}}{2} \end{bmatrix} = \begin{bmatrix} \frac{1}{3} & -\frac{1}{3} & \frac{2}{3} \\ \frac{1}{3} & \frac{2}{3} & -\frac{1}{3} \end{bmatrix} \quad ∎$$

The following theorem provides an alternative way of computing A^+ when A has full column rank.

> **Theorem 8.7.1** *If A is an $m \times n$ matrix with full column rank, then*
>
> $$A^+ = (A^T A)^{-1} A^T \tag{3}$$

Proof Let $A = U_1 \Sigma_1 V_1^T$ be a reduced singular value decomposition of A. Then

$$A^T A = (V_1 \Sigma_1^T U_1^T)(U_1 \Sigma_1 V_1^T) = V_1 \Sigma_1^2 V_1^T$$

Since A has full column rank, the matrix $A^T A$ is invertible (Theorem 7.5.10) and V_1 is an $n \times n$ orthogonal matrix. Thus,

$$(A^T A)^{-1} = V_1 \Sigma_1^{-2} V_1^T$$

from which it follows that

$$(A^T A)^{-1} A^T = (V_1 \Sigma_1^{-2} V_1^T)(U_1 \Sigma_1 V_1^T)^T = (V_1 \Sigma_1^{-2} V_1^T)(V_1 \Sigma_1 U_1^T) = V_1 \Sigma_1^{-1} U_1^T = A^+ \quad ∎$$

EXAMPLE 2
Pseudoinverse
in the Case of
Full Column
Rank

We computed the pseudoinverse of

$$A = \begin{bmatrix} 1 & 1 \\ 0 & 1 \\ 1 & 0 \end{bmatrix}$$

in Example 1 using singular value decomposition. However, A has full column rank so its pseudoinverse can also be computed from Formula (3). To do this we first compute

$$A^T A = \begin{bmatrix} 1 & 0 & 1 \\ 1 & 1 & 0 \end{bmatrix} \begin{bmatrix} 1 & 1 \\ 0 & 1 \\ 1 & 0 \end{bmatrix} = \begin{bmatrix} 2 & 1 \\ 1 & 2 \end{bmatrix}$$

from which it follows that

$$A^+ = (A^TA)^{-1}A^T = \begin{bmatrix} \frac{2}{3} & -\frac{1}{3} \\ -\frac{1}{3} & \frac{2}{3} \end{bmatrix} \begin{bmatrix} 1 & 0 & 1 \\ 1 & 1 & 0 \end{bmatrix} = \begin{bmatrix} \frac{1}{3} & -\frac{1}{3} & \frac{2}{3} \\ \frac{1}{3} & \frac{2}{3} & -\frac{1}{3} \end{bmatrix}$$

This agrees with the result obtained in Example 1. ∎

PROPERTIES OF THE PSEUDOINVERSE

The following theorem states some algebraic facts about the pseudoinverse, the proofs of which are left as exercises.

Theorem 8.7.2 *If A^+ is the pseudoinverse of an $m \times n$ matrix A, then:*

(a) $AA^+A = A$

(b) $A^+AA^+ = A^+$

(c) $(AA^+)^T = AA^+$

(d) $(A^+A)^T = A^+A$

(e) $(A^T)^+ = (A^+)^T$

(f) $A^{++} = A$

The next theorem states some properties of the pseudoinverse from the transformation point of view. We will prove the first three parts, and leave the last two as exercises.

Theorem 8.7.3 *If $A^+ = V_1\Sigma_1^{-1}U_1^T$ is the pseudoinverse of an $m \times n$ matrix A of rank k, and if the column vectors of U_1 and V_1 are $\mathbf{u}_1, \mathbf{u}_2, \ldots, \mathbf{u}_k$ and $\mathbf{v}_1, \mathbf{v}_2, \ldots, \mathbf{v}_k$, respectively, then:*

(a) $A^+\mathbf{y}$ *is in* row(A) *for every vector* $\mathbf{y}$ *in* R^m.

(b) $A^+\mathbf{u}_i = \dfrac{1}{\sigma_i}\mathbf{v}_i \quad (i = 1, 2, \ldots, k)$

(c) $A^+\mathbf{y} = \mathbf{0}$ *for every vector* $\mathbf{y}$ *in* null(A^T).

(d) AA^+ *is the orthogonal projection of* R^m *onto* col(A).

(e) A^+A *is the orthogonal projection of* R^n *onto* row(A).

Proof (a) If $\mathbf{y}$ is a vector in R^m, then it follows from (2) that

$$A^+\mathbf{y} = V_1\Sigma_1^{-1}U_1^T\mathbf{y} = V_1(\Sigma_1^{-1}U_1^T\mathbf{y})$$

so $A^+\mathbf{y}$ must be a linear combination of the column vectors of V_1. Since Theorem 8.6.5 states that these vectors are in row(A), it follows that $A^+\mathbf{y}$ is in row(A).

Proof (b) Multiplying A^+ on the right by U_1 yields

$$A^+U_1 = V_1\Sigma_1^{-1}U_1^TU_1 = V_1\Sigma_1^{-1}$$

The result now follows by comparing corresponding column vectors on the two sides of this equation.

Proof (c) If $\mathbf{y}$ is a vector in null(A^T), then $\mathbf{y}$ is orthogonal to each vector in col(A), and, in particular, it is orthogonal to each column vector of $U_1 = [\mathbf{u}_1 \quad \mathbf{u}_2 \quad \cdots \quad \mathbf{u}_k]$. This implies that $U_1^T\mathbf{y} = \mathbf{0}$ (why?), and hence that

$$A^+\mathbf{y} = V_1\Sigma_1^{-1}U_1^T\mathbf{y} = (V_1\Sigma_1^{-1})U_1^T\mathbf{y} = \mathbf{0}$$ ∎

EXAMPLE 3
Orthogonal
Projection
Using the
Pseudoinverse

Use the pseudoinverse of

$$A = \begin{bmatrix} 1 & 1 \\ 0 & 1 \\ 1 & 0 \end{bmatrix}$$

to find the standard matrix for the orthogonal projection of R^3 onto the column space of A.

Solution The pseudoinverse of A was computed in Example 2. Using that result we see that the orthogonal projection of R^3 onto col(A) is

$$AA^+ = \begin{bmatrix} 1 & 1 \\ 0 & 1 \\ 1 & 0 \end{bmatrix} \begin{bmatrix} \frac{1}{3} & -\frac{1}{3} & \frac{2}{3} \\ \frac{1}{3} & \frac{2}{3} & -\frac{1}{3} \end{bmatrix} = \begin{bmatrix} \frac{2}{3} & \frac{1}{3} & \frac{1}{3} \\ \frac{1}{3} & \frac{2}{3} & -\frac{1}{3} \\ \frac{1}{3} & -\frac{1}{3} & \frac{2}{3} \end{bmatrix}$$ ∎

CONCEPT PROBLEM Without performing any computations, make a conjecture about the orthogonal projection of R^2 onto the row space of the matrix A in Example 3, and confirm your conjecture by computing A^+A

**PSEUDOINVERSE AND
LEAST SQUARES**

The pseudoinverse is important because it provides a way of using singular value decompositions to solve least squares problems. Recall that the least squares solutions of a linear system $A\mathbf{x} = \mathbf{b}$ are the exact solutions of the normal equation $A^TA\mathbf{x} = A^T\mathbf{b}$. In the case where A has full column rank the matrix A^TA is invertible and there is a unique least squares solution

$$\mathbf{x} = (A^TA)^{-1}A^T\mathbf{b} = A^+\mathbf{b} \tag{4}$$

Thus, in the case of full column rank the least squares solution can be obtained by multiplying $\mathbf{b}$ by the pseudoinverse of A. In the case where A does not have full column rank the matrix A^TA is not invertible and there are infinitely many solutions of the normal equation, each of which is a least squares solution of $A\mathbf{x} = \mathbf{b}$. However, we know that among these least squares solutions there is a unique least squares solution in the row space of A (Theorem 7.8.3), and we also know that it is the least squares solution of minimum norm. The following theorem generalizes (4).

> **Theorem 8.7.4** *If A is an $m \times n$ matrix, and $\mathbf{b}$ is any vector in R^m, then*
>
> $$\mathbf{x} = A^+\mathbf{b}$$
>
> *is the least squares solution of $A\mathbf{x} = \mathbf{b}$ that has minimum norm.*

Proof We will show first that $\mathbf{x} = A^+\mathbf{b}$ satisfies the normal equation $A^TA\mathbf{x} = A^T\mathbf{b}$ and hence is a least squares solution. For this purpose, let $A = U_1\Sigma_1 V_1^T$ be a reduced singular value decomposition of A, so

$$A^+\mathbf{b} = V_1\Sigma_1^{-1}U_1^T\mathbf{b}$$

Thus,

$$(A^TA)A^+\mathbf{b} = V_1\Sigma_1^2 V_1^T V_1\Sigma_1^{-1}U_1^T\mathbf{b} = V_1\Sigma_1^2\Sigma_1^{-1}U_1^T\mathbf{b} = V_1\Sigma_1 U_1^T\mathbf{b} = A^T\mathbf{b}$$

which shows that $\mathbf{x} = A^+\mathbf{b}$ satisfies the normal equation $A^TA\mathbf{x} = A^T\mathbf{b}$.

To show that $\mathbf{x} = A^+\mathbf{b}$ is the least squares solution of minimum norm, it suffices to show that this vector lies in the row space of A (Theorem 7.8.3). But we know this to be true by part (*a*) of Theorem 8.7.3.   ∎

Some of the ideas we have been discussing are illustrated by the Strang diagram in Figure 8.7.1. The linear system $A\mathbf{x} = \mathbf{b}$ represented in that diagram is inconsistent, since $\mathbf{b}$ is not in

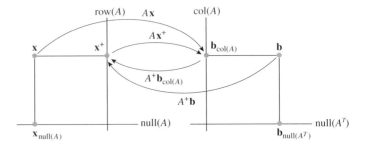

Figure 8.7.1

col(A). We have split $\mathbf{x}$ and $\mathbf{b}$ into orthogonal terms as

$$\mathbf{x} = \mathbf{x}_{\text{row}(A)} + \mathbf{x}_{\text{null}(A)} \quad \text{and} \quad \mathbf{b} = \mathbf{b}_{\text{col}(A)} + \mathbf{b}_{\text{null}(A^T)}$$

and have denoted $\mathbf{x}_{\text{row}(A)}$ by $\mathbf{x}^+$ for brevity. This vector is the least squares solution of minimum norm and is an exact solution of the equation $A\mathbf{x} = \mathbf{b}_{\text{col}(A)}$; that is,

$$A\mathbf{x}^+ = \mathbf{b}_{\text{col}(A)}$$

To solve this equation for $\mathbf{x}^+$, we can first multiply through by the pseudoinverse A^+ to obtain

$$A^+A\mathbf{x}^+ = A^+\mathbf{b}_{\text{col}(A)}$$

and then use Theorem 8.7.3(e) and the fact that $\mathbf{x}^+ = \mathbf{x}_{\text{row}(A)}$ is in the row space of A to obtain

$$\mathbf{x}^+ = A^+\mathbf{b}_{\text{col}(A)}$$

Thus, A maps $\mathbf{x}^+$ into $\mathbf{b}_{\text{col}(A)}$ and A^+ recovers $\mathbf{x}^+$ from $\mathbf{b}_{\text{col}(A)}$.

CONDITION NUMBER AND NUMERICAL CONSIDERATIONS

Singular value decomposition plays an important role in the analysis and solution of linear systems that are difficult to solve accurately because of their sensitivity to roundoff error. In the case of a consistent linear system $A\mathbf{x} = \mathbf{b}$ this typically occurs when the coefficient matrix is "nearly singular" in the sense that one or more of its singular values is close to zero. Such linear systems are said to be ***ill conditioned***. A good measure of how roundoff error will affect the accuracy of a computed solution is given by the ratio of the largest singular value of A to the smallest singular values of A. This ratio, called the ***condition number*** of A, is denoted by

$$\text{cond}(A) = \frac{\sigma_1}{\sigma_k} \tag{5}$$

The larger the condition number, the more sensitive the system to small roundoff errors. In fact, it is shown in books on numerical methods of linear algebra that if the entries of A and $\mathbf{b}$ are accurate to r significant digits, and if the condition number of A exceeds 10^c (for positive integer c), then a computed solution is unlikely to be accurate to more than $r - c$ significant digits. Thus, for example, if $\text{cond}(A) = 10^2$ and the entries of A and $\mathbf{b}$ are accurate to five significant digits, then one should expect an accuracy of at most $5 - 2 = 3$ significant digits in any computed solution.

The basic method for finding least squares solutions of a linear system $A\mathbf{x} = \mathbf{b}$ is to solve the normal equations $A^TA\mathbf{x} = A^T\mathbf{b}$ exactly. However, the singular values of A^TA are the squares of the singular values of A (Exercise 21 of Section 8.6), so $\text{cond}(A^TA)$ is the square of the condition number of A. Thus, if $A\mathbf{x} = \mathbf{b}$ is ill conditioned, then the normal equations are even worse! In theory, one could determine the condition number of A by finding the singular value decomposition and then use that decomposition to compute the pseudoinverse and the least squares solution $\mathbf{x} = A^+\mathbf{b}$ if the system is not ill conditioned.

While all of this sounds reasonable, the difficulty is that the singular values of A are the square roots of the eigenvalues of A^TA, and calculating those singular values directly from the problematical A^TA may produce an inaccurate estimate of the condition number as well as an inaccurate least squares solution. Fortunately, there are methods for finding singular value decompositions that do not involve computing with A^TA. These produce some of the best algorithms known for finding least squares solutions of linear systems and are discussed in books on numerical methods of linear algebra. Two standard books on the subject are *Matrix Computations,* by G. H. Golub and C. F. Van Loan, Johns Hopkins University Press, Baltimore, 1996; and *Numerical Recipes in C, The Art of Scientific Computing,* by William H. Press, Saul A. Teukolsky, William T. Vetterling, and Brian P. Flannery, Cambridge University Press, New York, 1999.

Exercise Set 8.7

In Exercises 1–4, a matrix A with full column rank is given. Use Theorem 8.7.1 to find the pseudoinverse of A.

1. $A = \begin{bmatrix} 3 \\ 4 \end{bmatrix}$

2. $A = \begin{bmatrix} 1 & 1 \\ 2 & 3 \\ 2 & 1 \end{bmatrix}$

3. $A = \begin{bmatrix} 7 & 1 \\ 0 & 0 \\ 5 & 5 \end{bmatrix}$

4. $A = \begin{bmatrix} 4 \\ 5 \end{bmatrix}$

5. Confirm that the six properties in Theorem 8.7.2 hold for the matrices A and A^+ in Exercise 1.

6. Confirm that the six properties in Theorem 8.7.2 hold for the matrices A and A^+ in Exercise 2.

In Exercises 7–10, use the reduced singular value decomposition of A to compute the pseudoinverse of A.

7. The matrix in Exercise 1.

8. The matrix in Exercise 2.

9. The matrix in Exercise 3.

10. The matrix in Exercise 4.

In Exercises 11 and 12, an invertible matrix A is given. Confirm that $A^{-1} = A^+$.

11. $A = \begin{bmatrix} 2 & 2 \\ -1 & 1 \end{bmatrix}$

12. $A = \begin{bmatrix} 2 & 2 \\ 0 & 2 \end{bmatrix}$

In Exercises 13 and 14, show that Formula (3) is not applicable, and then use any appropriate method to find A^+.

13. $A = \begin{bmatrix} 1 & 1 \\ 1 & 1 \end{bmatrix}$

14. $A = \begin{bmatrix} 1 & 0 \\ 0 & 0 \\ 0 & 0 \end{bmatrix}$

In Exercises 15 and 16, use the pseudoinverse of A to find the standard matrix for the orthogonal projection of R^3 onto the column space of A.

15. The matrix A in Exercise 2.

16. The matrix A in Exercise 3.

In Exercises 17 and 18, use an appropriate pseudoinverse to find the least squares solution of minimum norm for the linear system.

17. $\begin{aligned} x_1 + x_2 &= 1 \\ 2x_1 + 2x_2 &= 0 \\ 2x_1 + 2x_2 &= -1 \end{aligned}$

18. $\begin{aligned} x_1 + x_2 &= 1 \\ 2x_1 + 3x_2 &= 1 \\ 2x_1 + x_2 &= 1 \end{aligned}$

19. The matrix $A = \begin{bmatrix} 1 & 2 & 3 \end{bmatrix}$ does not have full column rank, but its transpose does. Thus, $(A^T)^+$ can be computed using Theorem 8.7.1, even though A^+ cannot. Use that theorem to compute $(A^T)^+$ and then use your result to find A^+.

20. Use the idea in Exercise 19 to find the pseudoinverse of the matrix

$$A = \begin{bmatrix} 1 & 2 & 2 \\ 1 & 3 & 1 \end{bmatrix}$$

without finding its reduced singular value decomposition.

21. Show that Formula (3) simplifies to $A^+ = A^{-1}$ in the case where A is invertible.

Discussion and Discovery

D1. What can you say about the pseudoinverse of a matrix A with orthogonal column vectors?

D2. If A is a matrix of rank 1, then Formula (15) of Section 8.6 implies that the reduced singular value expansion of A has the form $A = \sigma \mathbf{u}\mathbf{v}^T$, where $\mathbf{u}$ and $\mathbf{v}$ are unit vectors.

(a) What is the reduced singular value expansion of A^+ in this case?

(b) Use the result in part (a) to compute A^+A and AA^+ in this case, and explain why the resulting expressions could have been anticipated geometrically.

D3. If c is a nonzero scalar, how are $(cA)^+$ and A^+ related?

D4. Find A^+, A^+A, and AA^+ for $A = \begin{bmatrix} 2 & 1 & -2 \end{bmatrix}$ given that its reduced singular value decomposition $A = U\Sigma V^T$ is

$$\begin{bmatrix} 2 & 1 & -2 \end{bmatrix} = [-1][3]\begin{bmatrix} -\frac{2}{3} & -\frac{1}{3} & \frac{2}{3} \end{bmatrix}$$

D5. (a) What properties of AA^+ and A^+A tell you that these matrices must be idempotent?

(b) Use Theorem 8.7.2 to show that AA^+ and A^+A are idempotent.

Working with Proofs

P1. Use Formula (2) to prove that if A is an $m \times n$ matrix, then $AA^+A = A$.

P2. Use Formula (2) to prove that if A is an $m \times n$ matrix, then $(AA^+)^T = AA^+$.

P3. Use Formula (2) to prove that if A is an $m \times n$ matrix, then $(A^T)^+ = (A^+)^T$.

P4. Use Formula (2) to prove that if A is an $m \times n$ matrix, then $A^{++} = A$.

P5. Use the results in Exercises P4 and P1 to prove that if A is an $m \times n$ matrix, then $A^+AA^+ = A^+$.

P6. Use the results in Exercises P4 and P2 to prove that if A is an $m \times n$ matrix, then $(A^+A)^T = A^+A$.

P7. Use Formula (2) to prove that if A is an $m \times n$ matrix, then AA^+ is the orthogonal projection of R^n onto the column space of A. [*Hint:* First show that $AA^+ = U_1 U_1^T$.]

P8. Apply the result in Exercise P7 to A^T, and use appropriate parts of Theorem 8.7.2 to prove that if A is an $m \times n$ matrix, then A^+A is the orthogonal projection of R^n onto the row space of A.

Technology Exercises

T1. (*Pseudoinverse*) Some linear algebra technology utilities provide a command for finding the pseudoinverse of a matrix. Determine whether your utility has this capability; if so, use that command to find the pseudoinverse of the matrix in Example 1.

T2. Use a reduced singular value decomposition to find the pseudoinverse of the matrix

$$A = \begin{bmatrix} 1 & 2 & 3 \\ -2 & 1 & -1 \\ 3 & 4 & 7 \end{bmatrix}$$

If your technology utility has a command for finding pseudoinverses, use it to check the result you have obtained.

T3. The rank of a matrix is the number of linearly independent rows (or columns). In practice, it is difficult to find the rank of a matrix exactly because of roundoff error, particularly for matrices of large size. A common procedure for estimating the rank of a matrix A is to set some error tolerance ϵ (depending on the accuracy of the data), compute A^TA (or AA^T), which is symmetric and has the same rank as A, and estimate the rank of A to be the number of singular values of A^TA (or AA^T) that are greater than ϵ. This is called the ***effective rank*** of A for the given tolerance.

(a) Using a tolerance of $\epsilon = 1$, find the effective rank of the matrix

$$A = \begin{bmatrix} 20.1 & -12.3 & -21.1 & 28.9 & -53.5 \\ 14.3 & 13.2 & 12.7 & 14.8 & 11.6 \\ -11.4 & -9.9 & 16.9 & -38.2 & 18.4 \\ 14.3 & 18.1 & -3.4 & 35.8 & 0.4 \\ -22.1 & -16.7 & 10.6 & -49.4 & 16.0 \\ 9.8 & 5.2 & 7.8 & 7.2 & 3.2 \end{bmatrix}$$

(b) Compare the effective rank of A to the rank of A that results by reducing A to row echelon form or reduced row echelon form.

T4. Consider the inconsistent linear system $A\mathbf{x} = \mathbf{b}$ in which

$$A = \begin{bmatrix} 1 & 2 & 3 \\ -2 & -3 & -5 \\ 1 & 3 & 4 \end{bmatrix}; \quad \mathbf{b} = \begin{bmatrix} 1 \\ 1 \\ 2 \end{bmatrix}$$

Show that the system has infinitely many least squares solutions, and use the pseudoinverse of A to find the least squares solution that has minimum norm.

Section 8.8 Complex Eigenvalues and Eigenvectors

Up to now we have focused primarily on real eigenvalues and eigenvectors. However, complex eigenvalues and eigenvectors have important applications and geometric interpretations, so it will be desirable for us to give them the same status as their real counterparts. That is the primary goal of this section.

VECTORS IN C^n

To establish the foundation for our study of complex eigenvalues and eigenvectors we make the following definition.

> **Definition 8.8.1** If n is a positive integer, then a ***complex n-tuple*** is a sequence of n complex numbers $(v_1, v_2, \ldots, v_n)$. The set of all complex n-tuples is called ***complex n-space*** and is denoted by C^n.

The terminology used for n-tuples of real numbers applies to complex n-tuples without change. Thus, if $v_1, v_2, \ldots, v_n$ are complex numbers, then we call $\mathbf{v} = (v_1, v_2, \ldots, v_n)$ a ***vector*** in C^n and $v_1, v_2, \ldots, v_n$ its ***components***. Some examples of vectors in C^3 are

$$\mathbf{u} = (1 + i, -4i, 3 + 2i), \quad \mathbf{v} = (0, i, 5), \quad \mathbf{w} = \left(6 - \sqrt{2}i, 9 + \tfrac{1}{2}i, \pi i\right)$$

Scalars for C^n are complex numbers, and addition, subtraction, and scalar multiplication are performed componentwise, just as in R^n. It can be proved that the eight properties in Theorem 1.1.5 hold for vectors in C^n, from which it follows that vectors in C^n have the same algebraic properties as vectors in R^n. (See Looking Ahead following Theorem 1.1.6.)

Recall that if $z = a + bi$ is a complex number, then $\bar{z} = a - bi$ is called the ***complex conjugate*** of z, the real numbers $\text{Re}(z) = a$ and $\text{Im}(z) = b$ are called the ***real part*** and ***imaginary part*** of z, respectively, and $|z| = \sqrt{a^2 + b^2}$ is called the ***modulus*** (or ***absolute value***) of z. As illustrated in Figure 8.8.1, a complex number $z = a + bi$ can be represented geometrically as a point or vector (a, b) in a rectangular coordinate system called the ***complex plane***. The angle ϕ shown in the figure is called an ***argument*** of z, and the real and imaginary parts of z can be expressed in terms of this angle as

$$\text{Re}(z) = |z| \cos \phi \quad \text{and} \quad \text{Im}(z) = |z| \sin \phi \tag{1}$$

Thus, z itself can be written as

$$z = |z|(\cos \phi + i \sin \phi) \tag{2}$$

which is called the ***polar form*** of z.

A vector

$$\mathbf{v} = (v_1, v_2, \ldots, v_n) = (a_1 + b_1 i, a_2 + b_2 i, \ldots, a_n + b_n i) \tag{3}$$

in C^n can be expressed as

$$\mathbf{v} = (v_1, v_2, \ldots, v_n) = (a_1, a_2, \ldots, a_n) + i(b_1, b_2, \ldots, b_n) = \text{Re}(\mathbf{v}) + i\,\text{Im}(\mathbf{v}) \tag{4}$$

where the vectors

$$\text{Re}(\mathbf{v}) = (a_1, a_2, \ldots, a_n) \quad \text{and} \quad \text{Im}(\mathbf{v}) = (b_1, b_2, \ldots, b_n)$$

are called the ***real*** and ***imaginary parts*** of $\mathbf{v}$, respectively. The vector

$$\bar{\mathbf{v}} = (\bar{v}_1, \bar{v}_2, \ldots, \bar{v}_n) = (a_1 - b_1 i, a_2 - b_2 i, \ldots, a_n - b_n i) \tag{5}$$

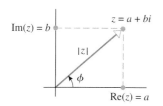

Figure 8.8.1

is called the ***complex conjugate*** of $\mathbf{v}$ and can be expressed in terms of $\text{Re}(\mathbf{v})$ and $\text{Im}(\mathbf{v})$ as

$$\bar{\mathbf{v}} = (a_1, a_2, \ldots, a_n) - i(b_1, b_2, \ldots, b_n) = \text{Re}(\mathbf{v}) - i\,\text{Im}(\mathbf{v}) \tag{6}$$

It follows from (4) that the vectors in R^n can be viewed as the vectors in C^n whose imaginary parts are $\mathbf{0}$, and from (6) that a vector $\mathbf{v}$ in C^n is in R^n if and only if $\bar{\mathbf{v}} = \mathbf{v}$.

In this section we will also need to consider matrices with complex entries, and henceforth we will call a matrix A a ***real matrix*** if its entries are real numbers and a ***complex matrix*** if its entries are complex numbers. Note that a real matrix *must* have real entries, whereas a complex matrix may or may not have real entries. The standard operations on real matrices carry over to complex matrices without change, and all of the familiar properties of matrices continue to hold. If A is a complex matrix, then $\text{Re}(A)$ and $\text{Im}(A)$ are the matrices formed from the real and imaginary parts of A, and $\bar{A}$ is the matrix formed by taking the complex conjugate of each entry in A.

EXAMPLE 1
Real and
Imaginary Parts
of Vectors and
Matrices

Let

$$\mathbf{v} = (3 + i, -2i, 5) \quad \text{and} \quad A = \begin{bmatrix} 1 + i & -i \\ 4 & 6 - 2i \end{bmatrix}$$

Then

$$\bar{\mathbf{v}} = (3 - i, 2i, 5), \quad \text{Re}(\mathbf{v}) = (3, 0, 5), \quad \text{Im}(\mathbf{v}) = (1, -2, 0)$$

$$\bar{A} = \begin{bmatrix} 1 - i & i \\ 4 & 6 + 2i \end{bmatrix}, \quad \text{Re}(A) = \begin{bmatrix} 1 & 0 \\ 4 & 6 \end{bmatrix}, \quad \text{Im}(A) = \begin{bmatrix} 1 & -1 \\ 0 & -2 \end{bmatrix}$$

$$\det(A) = \begin{vmatrix} 1 + i & -i \\ 4 & 6 - 2i \end{vmatrix} = (1 + i)(6 - 2i) - (-i)(4) = 8 + 8i$$ ∎

ALGEBRAIC PROPERTIES OF THE COMPLEX CONJUGATE

The next two theorems list some properties of complex vectors and matrices that we will need in this section; we leave some of the proofs as exercises.

Theorem 8.8.2   *If $\mathbf{u}$ and $\mathbf{v}$ are vectors in C^n, and if k is a scalar, then:*

(a) $\bar{\bar{\mathbf{u}}} = \mathbf{u}$

(b) $\overline{k\mathbf{u}} = \bar{k}\bar{\mathbf{u}}$

(c) $\overline{\mathbf{u} + \mathbf{v}} = \bar{\mathbf{u}} + \bar{\mathbf{v}}$

(d) $\overline{\mathbf{u} - \mathbf{v}} = \bar{\mathbf{u}} - \bar{\mathbf{v}}$

Theorem 8.8.3   *If A is an $m \times k$ complex matrix and B is a $k \times n$ complex matrix, then:*

(a) $\bar{\bar{A}} = A$

(b) $\overline{(A^T)} = (\bar{A})^T$

(c) $\overline{AB} = \bar{A}\,\bar{B}$

THE COMPLEX EUCLIDEAN INNER PRODUCT

The following definition extends the notions of dot product and norm to C^n.

Definition 8.8.4   If $\mathbf{u} = (u_1, u_2, \ldots, u_n)$ and $\mathbf{v} = (v_1, v_2, \ldots, v_n)$ are vectors in C^n, then the ***complex Euclidean inner product*** of $\mathbf{u}$ and $\mathbf{v}$ (also called the ***complex dot product***) is denoted by $\mathbf{u} \cdot \mathbf{v}$ and is defined as

$$\mathbf{u} \cdot \mathbf{v} = u_1\bar{v}_1 + u_2\bar{v}_2 + \cdots + u_n\bar{v}_n \tag{7}$$

We also define the ***Euclidean norm*** on C^n to be

$$\|\mathbf{v}\| = \sqrt{\mathbf{v} \cdot \mathbf{v}} = \sqrt{|v_1|^2 + |v_2|^2 + \cdots + |v_n|^2} \tag{8}$$

As in the real case, we call $\mathbf{v}$ a ***unit vector*** in C^n if $\|\mathbf{v}\| = 1$, and we say two vectors $\mathbf{u}$ and $\mathbf{v}$ are ***orthogonal*** if and only if $\mathbf{u} \cdot \mathbf{v} = 0$.

REMARK The complex conjugates in (7) ensure that $\|\mathbf{v}\|$ is a real number, for without them the quantity $\mathbf{v} \cdot \mathbf{v}$ in (8) might be imaginary. Also, note that (7) becomes the dot product on R^n if $\mathbf{u}$ and $\mathbf{v}$ have real components.

Recall from Formula (26) of Section 3.1 that if $\mathbf{u}$ and $\mathbf{v}$ are *column vectors* in R^n, then their dot product can be expressed as

$$\mathbf{u} \cdot \mathbf{v} = \mathbf{u}^T \mathbf{v} = \mathbf{v}^T \mathbf{u}$$

The analogous formulas in C^n are (verify)

$$\mathbf{u} \cdot \mathbf{v} = \mathbf{u}^T \overline{\mathbf{v}} = \overline{\mathbf{v}}^T \mathbf{u} \tag{9}$$

EXAMPLE 2
Complex Euclidean Inner Product and Norm

Find $\mathbf{u} \cdot \mathbf{v}$, $\mathbf{v} \cdot \mathbf{u}$, $\|\mathbf{u}\|$, and $\|\mathbf{v}\|$ for the vectors

$$\mathbf{u} = (1 + i, i, 3 - i) \quad \text{and} \quad \mathbf{v} = (1 + i, 2, 4i)$$

Solution

$\mathbf{u} \cdot \mathbf{v} = (1 + i)(\overline{1 + i}) + i(\overline{2}) + (3 - i)(\overline{4i}) = (1 + i)(1 - i) + 2i + (3 - i)(-4i) = -2 - 10i$

$\mathbf{v} \cdot \mathbf{u} = (1 + i)(\overline{1 + i}) + 2(\overline{i}) + (4i)(\overline{3 - i}) = (1 + i)(1 - i) - 2i + 4i(3 + i) = -2 + 10i$

$\|\mathbf{u}\| = \sqrt{|1 + i|^2 + |i|^2 + |3 - i|^2} = \sqrt{2 + 1 + 10} = \sqrt{13}$

$\|\mathbf{v}\| = \sqrt{|1 + i|^2 + |2|^2 + |4i|^2} = \sqrt{2 + 4 + 16} = \sqrt{22}$ ∎

Example 2 reveals a major difference between the dot product on R^n and the complex Euclidean inner product on C^n. For the dot product we always have $\mathbf{v} \cdot \mathbf{u} = \mathbf{u} \cdot \mathbf{v}$ (the *symmetry property* of the dot product), but for the complex Euclidean inner product the corresponding relationship is $\mathbf{u} \cdot \mathbf{v} = \overline{\mathbf{v} \cdot \mathbf{u}}$, which is called its ***antisymmetry*** property. The following theorem is an analog of Theorem 1.2.6. We omit the proof.

Theorem 8.8.5 *If $\mathbf{u}$, $\mathbf{v}$, and $\mathbf{w}$ are vectors in C^n, and if k is a scalar, then the complex Euclidean inner product has the following properties:*

(a) $\mathbf{u} \cdot \mathbf{v} = \overline{\mathbf{v} \cdot \mathbf{u}}$ **[Antisymmetry property]**

(b) $\mathbf{u} \cdot (\mathbf{v} + \mathbf{w}) = \mathbf{u} \cdot \mathbf{v} + \mathbf{u} \cdot \mathbf{w}$ **[Distributive property]**

(c) $k(\mathbf{u} \cdot \mathbf{v}) = (k\mathbf{u}) \cdot \mathbf{v}$ **[Homogeneity property]**

(d) $\mathbf{v} \cdot \mathbf{v} \geq 0$ *and* $\mathbf{v} \cdot \mathbf{v} = 0$ *if and only if* $\mathbf{v} = \mathbf{0}$ **[Positivity property]**

Part (*c*) of this theorem states that a scalar multiplying a complex Euclidean inner product can be regrouped with the first vector, but to regroup it with second vector you must first take its complex conjugate (see if you can justify the steps):

$$k(\mathbf{u} \cdot \mathbf{v}) = k(\overline{\mathbf{v} \cdot \mathbf{u}}) = \overline{\overline{k}}\,(\overline{\mathbf{v} \cdot \mathbf{u}}) = \overline{\overline{k}\,(\mathbf{v} \cdot \mathbf{u})} = \overline{(\overline{k}\mathbf{v}) \cdot \mathbf{u}} = \mathbf{u} \cdot (\overline{k}\mathbf{v}) \tag{10}$$

By substituting $\overline{k}$ for k and using the fact that $\overline{\overline{k}} = k$, you can also write this relationship as

$$\mathbf{u} \cdot k\mathbf{v} = \overline{k}(\mathbf{u} \cdot \mathbf{v}) \tag{11}$$

VECTOR SPACE CONCEPTS IN C^n

Except for the use of complex scalars, notions of linear combination, linear independence, subspace, spanning, basis, and dimension carry over without change to C^n, as do most of the theorems we have given in this text about them.

CONCEPT PROBLEM Is R^n a subspace of C^n? Explain.

Eigenvalues and eigenvectors are defined for complex matrices exactly as for real matrices. If A is an $n \times n$ matrix with complex (or real) entries, then the complex roots of the characteristic equation $\det(\lambda I - A) = 0$ are called *complex eigenvalues* of A. As in the real case, λ is a complex eigenvalue of A if and only if there exists a nonzero vector $\mathbf{x}$ in C^n such that $A\mathbf{x} = \lambda\mathbf{x}$. Each such $\mathbf{x}$ is called a *complex eigenvector* of A corresponding to λ. The complex eigenvectors of A corresponding to λ are the nonzero solutions of the linear system $(\lambda I - A)\mathbf{x} = \mathbf{0}$, and the set of all such solutions is a subspace of C^n, called the *eigenspace* of A corresponding to λ.

The following theorem states that if a real matrix has complex eigenvalues, then those eigenvalues and their corresponding eigenvectors occur in conjugate pairs.

Theorem 8.8.6 *If λ is an eigenvalue of a real $n \times n$ matrix A, and if $\mathbf{x}$ is a corresponding eigenvector, then $\bar{\lambda}$ is also an eigenvalue of A, and $\bar{\mathbf{x}}$ is a corresponding eigenvector.*

Proof Since λ is an eigenvalue of A and $\mathbf{x}$ is a corresponding eigenvector, we have

$$\overline{A\mathbf{x}} = \overline{\lambda\mathbf{x}} = \bar{\lambda}\bar{\mathbf{x}} \tag{12}$$

However, $\bar{A} = A$, since A has real entries, so it follows from part (c) of Theorem 8.8.3 that

$$\overline{A\mathbf{x}} = \bar{A}\bar{\mathbf{x}} = A\bar{\mathbf{x}} \tag{13}$$

Equations (12) and (13) together imply that

$$A\bar{\mathbf{x}} = \overline{A\mathbf{x}} = \bar{\lambda}\bar{\mathbf{x}}$$

in which $\bar{\mathbf{x}} \neq \mathbf{0}$ (why?); this tells us that $\bar{\lambda}$ is an eigenvalue of A and $\bar{\mathbf{x}}$ is a corresponding eigenvector. ■

EXAMPLE 3
Complex
Eigenvalues
and
Eigenvectors

Find the eigenvalues and bases for the eigenspaces of

$$A = \begin{bmatrix} -2 & -1 \\ 5 & 2 \end{bmatrix}$$

Solution The characteristic polynomial of A is

$$\begin{vmatrix} \lambda + 2 & 1 \\ -5 & \lambda - 2 \end{vmatrix} = \lambda^2 + 1 = (\lambda - i)(\lambda + i)$$

so the eigenvalues of A are $\lambda = i$ and $\lambda = -i$. Note that these eigenvalues are complex conjugates, as guaranteed by Theorem 8.8.6. To find the eigenvectors we must solve the system

$$\begin{bmatrix} \lambda + 2 & 1 \\ -5 & \lambda - 2 \end{bmatrix} \begin{bmatrix} x_1 \\ x_2 \end{bmatrix} = \begin{bmatrix} 0 \\ 0 \end{bmatrix}$$

with $\lambda = i$ and then with $\lambda = -i$. With $\lambda = i$, this system becomes

$$\begin{bmatrix} i + 2 & 1 \\ -5 & i - 2 \end{bmatrix} \begin{bmatrix} x_1 \\ x_2 \end{bmatrix} = \begin{bmatrix} 0 \\ 0 \end{bmatrix} \tag{14}$$

We could solve this system by reducing the augmented matrix

$$\begin{bmatrix} i + 2 & 1 & 0 \\ -5 & i - 2 & 0 \end{bmatrix} \tag{15}$$

to reduced row echelon form by Gauss–Jordan elimination, though the complex arithmetic is somewhat tedious. A simpler procedure here is to first observe that the reduced row echelon form of (15) must have a row of zeros because (14) has nontrivial solutions. This being the case, each row of (15) must be a scalar multiple of the other, and hence the first row can be made into a row of zeros by adding a suitable multiple of the second row to it. Accordingly, we can simply set the entries in the first row to zero, then interchange the rows, and then multiply the new first row by $-\frac{1}{5}$ to obtain the reduced row echelon form

$$\begin{bmatrix} 1 & \frac{2}{5} - \frac{1}{5}i & 0 \\ 0 & 0 & 0 \end{bmatrix}$$

Thus, a general solution of the system is

$$x_1 = \left(-\tfrac{2}{5} + \tfrac{1}{5}i\right)t, \quad x_2 = t$$

This tells us that the eigenspace corresponding to $\lambda = i$ is one-dimensional and consists of all complex scalar multiples of the basis vector

$$\mathbf{x} = \begin{bmatrix} -\frac{2}{5} + \frac{1}{5}i \\ 1 \end{bmatrix} \tag{16}$$

As a check, let us confirm that $A\mathbf{x} = i\mathbf{x}$. We obtain

$$A\mathbf{x} = \begin{bmatrix} -2 & -1 \\ 5 & 2 \end{bmatrix} \begin{bmatrix} -\frac{2}{5} + \frac{1}{5}i \\ 1 \end{bmatrix} = \begin{bmatrix} -2\left(-\frac{2}{5} + \frac{1}{5}i\right) - 1 \\ 5\left(-\frac{2}{5} + \frac{1}{5}i\right) + 2 \end{bmatrix} = \begin{bmatrix} -\frac{1}{5} - \frac{2}{5}i \\ i \end{bmatrix} = i\mathbf{x}$$

We could find a basis for the eigenspace corresponding to $\lambda = -i$ in a similar way, but the work is unnecessary, since Theorem 8.8.6 implies that

$$\overline{\mathbf{x}} = \begin{bmatrix} -\frac{2}{5} - \frac{1}{5}i \\ 1 \end{bmatrix} \tag{17}$$

must be a basis for this eigenspace. The following computations confirm that $\overline{\mathbf{x}}$ is an eigenvector of A corresponding to $\lambda = -i$:

$$A\overline{\mathbf{x}} = \begin{bmatrix} -2 & -1 \\ 5 & 2 \end{bmatrix} \begin{bmatrix} -\frac{2}{5} - \frac{1}{5}i \\ 1 \end{bmatrix}$$
$$= \begin{bmatrix} -2\left(-\frac{2}{5} - \frac{1}{5}i\right) - 1 \\ 5\left(-\frac{2}{5} - \frac{1}{5}i\right) + 2 \end{bmatrix} = \begin{bmatrix} -\frac{1}{5} + \frac{2}{5}i \\ -i \end{bmatrix} = -i\overline{\mathbf{x}} \qquad ■$$

A PROOF THAT REAL SYMMETRIC MATRICES HAVE REAL EIGENVALUES In Theorem 4.4.10 we proved that 2×2 real symmetric matrices have real eigenvalues, and we stated without proof that the result is true for all real symmetric matrices. We now have all of the mathematical machinery required to prove the general result. The key to the proof is to regard a real symmetric matrix as a complex matrix whose entries have an imaginary part of zero.

Theorem 8.8.7 *If A is a real symmetric matrix, then A has real eigenvalues.*

Proof Suppose that λ is an eigenvalue of A and $\mathbf{x}$ is a corresponding eigenvector, where we allow for the possibility that λ is complex and $\mathbf{x}$ is in C^n. Thus,

$$A\mathbf{x} = \lambda\mathbf{x}$$

where $\mathbf{x} \neq \mathbf{0}$. If we multiply both sides of this equation by $\overline{\mathbf{x}}^T$ and use the fact that

$$\overline{\mathbf{x}}^T A\mathbf{x} = \overline{\mathbf{x}}^T (\lambda \mathbf{x}) = \lambda(\overline{\mathbf{x}}^T \mathbf{x}) = \lambda(\mathbf{x} \cdot \mathbf{x}) = \lambda \|\mathbf{x}\|^2$$

then we obtain

$$\lambda = \frac{\overline{\mathbf{x}}^T A\mathbf{x}}{\|\mathbf{x}\|^2}$$

Since the denominator in this expression is real, we can prove that λ is real by showing that

$$\overline{\overline{\mathbf{x}}^T A\mathbf{x}} = \overline{\mathbf{x}}^T A\mathbf{x} \tag{18}$$

But, A is symmetric and has real entries, so it follows from the second equality in (9) and properties of the conjugate that

$$\overline{\overline{\mathbf{x}}^T A\mathbf{x}} = \overline{\overline{\mathbf{x}}}^T \, \overline{A\mathbf{x}} = \mathbf{x}^T \, \overline{A\mathbf{x}} = (\overline{A\mathbf{x}})^T \mathbf{x} = (A\overline{\mathbf{x}})^T \mathbf{x} = (A\overline{\mathbf{x}})^T \mathbf{x} = \overline{\mathbf{x}}^T A^T \mathbf{x} = \overline{\mathbf{x}}^T A\mathbf{x} \qquad \blacksquare$$

A GEOMETRIC INTERPRETATION OF COMPLEX EIGENVALUES OF REAL MATRICES

The following theorem is the key to understanding the geometric significance of complex eigenvalues of real 2×2 matrices.

Theorem 8.8.8   *The eigenvalues of the real matrix*

$$C = \begin{bmatrix} a & -b \\ b & a \end{bmatrix} \tag{19}$$

are $\lambda = a \pm bi$. If a and b are not both zero, then this matrix can be factored as

$$\begin{bmatrix} a & -b \\ b & a \end{bmatrix} = \begin{bmatrix} |\lambda| & 0 \\ 0 & |\lambda| \end{bmatrix} \begin{bmatrix} \cos\phi & -\sin\phi \\ \sin\phi & \cos\phi \end{bmatrix} \tag{20}$$

where ϕ is the angle from the positive x-axis to the ray from the origin through the point (a, b) (Figure 8.8.2).

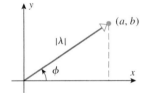

Figure 8.8.2

Geometrically, this theorem states that multiplication by a matrix of form (19) can be viewed as a rotation through the angle ϕ followed by a scaling with factor $|\lambda|$ (Figure 8.8.3).

Proof The characteristic equation of C is $(\lambda - a)^2 + b^2 = 0$ (verify), from which it follows that the eigenvalues of C are $\lambda = a \pm bi$. Assuming that a and b are not both zero, let ϕ be the angle from the positive x-axis to the ray through the origin and the point (a, b). The angle ϕ is an argument of the eigenvalue $\lambda = a + bi$, so we use (1) to express the real and imaginary parts of λ as

$$a = |\lambda| \cos\phi \quad \text{and} \quad b = |\lambda| \sin\phi$$

It follows from this that (19) can be written as

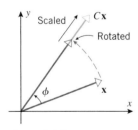

Figure 8.8.3

$$\begin{bmatrix} a & -b \\ b & a \end{bmatrix} = \begin{bmatrix} |\lambda| & 0 \\ 0 & |\lambda| \end{bmatrix} \begin{bmatrix} \frac{a}{|\lambda|} & -\frac{b}{|\lambda|} \\ \frac{b}{|\lambda|} & \frac{a}{|\lambda|} \end{bmatrix} = \begin{bmatrix} |\lambda| & 0 \\ 0 & |\lambda| \end{bmatrix} \begin{bmatrix} \cos\phi & -\sin\phi \\ \sin\phi & \cos\phi \end{bmatrix} \qquad \blacksquare$$

The following theorem, whose proof is considered in the exercises, shows that every real 2×2 matrix with complex eigenvalues is similar to a matrix of form (19).

Theorem 8.8.9   *Let A be a real 2×2 matrix with complex eigenvalues $\lambda = a \pm bi$ (where $b \neq 0$). If $\mathbf{x}$ is an eigenvector of A corresponding to $\lambda = a - bi$, then the matrix $P = [\text{Re}(\mathbf{x}) \quad \text{Im}(\mathbf{x})]$ is invertible and*

$$A = P \begin{bmatrix} a & -b \\ b & a \end{bmatrix} P^{-1} \tag{21}$$

EXAMPLE 4
A Factorization
Using Complex
Eigenvalues

Factor the matrix in Example 3 into form (21) using the eigenvalue $\lambda = -i$ and the corresponding eigenvector that was given in (17).

Solution For consistency with the notation in Theorem 8.8.9, let us denote the eigenvector in (17) that corresponds to $\lambda = -i$ by $\mathbf{x}$ (rather than $\bar{\mathbf{x}}$ as before). For this λ and $\mathbf{x}$ we have

$$a = 0, \quad b = 1, \quad \mathrm{Re}(\mathbf{x}) = \begin{bmatrix} -\frac{2}{5} \\ 1 \end{bmatrix}, \quad \mathrm{Im}(\mathbf{x}) = \begin{bmatrix} -\frac{1}{5} \\ 0 \end{bmatrix}, \quad P = [\mathrm{Re}(\mathbf{x}) \quad \mathrm{Im}(\mathbf{x})] = \begin{bmatrix} -\frac{2}{5} & -\frac{1}{5} \\ 1 & 0 \end{bmatrix}$$

so A can be factored in form (21) as

$$\begin{bmatrix} -2 & -1 \\ 5 & 2 \end{bmatrix} = \begin{bmatrix} -\frac{2}{5} & -\frac{1}{5} \\ 1 & 0 \end{bmatrix} \begin{bmatrix} 0 & -1 \\ 1 & 0 \end{bmatrix} \begin{bmatrix} 0 & 1 \\ -5 & -2 \end{bmatrix}$$

You may want to confirm this by multiplying out the right side.   ∎

To understand what Theorem 8.8.9 says geometrically, let us denote the matrices on the right side of (20) by S and R_ϕ, respectively, and then use (20) to rewrite (21) as

$$A = P S R_\phi P^{-1} = P \begin{bmatrix} |\lambda| & 0 \\ 0 & |\lambda| \end{bmatrix} \begin{bmatrix} \cos\phi & -\sin\phi \\ \sin\phi & \cos\phi \end{bmatrix} P^{-1} \tag{22}$$

If we now view P as the transition matrix from the basis $B = \{\mathrm{Re}(\mathbf{x}), \mathrm{Im}(\mathbf{x})\}$ to the standard basis, then (22) tells us that computing a product $A\mathbf{x}_0$ can be broken down into a three-step process:

1. Map $\mathbf{x}_0$ from standard coordinates into B-coordinates by forming the product $P^{-1}\mathbf{x}_0$.
2. Rotate and scale the vector $P^{-1}\mathbf{x}_0$ by forming the product $S R_\phi P^{-1}\mathbf{x}_0$.
3. Map the rotated and scaled vector back to standard coordinates to obtain
$A\mathbf{x}_0 = P S R_\phi P^{-1}\mathbf{x}_0$.

EXAMPLE 5
An Elliptical
Orbit Explained

At the end of Section 6.1 we showed that if

$$A = \begin{bmatrix} \frac{1}{2} & \frac{3}{4} \\ -\frac{3}{5} & \frac{11}{10} \end{bmatrix} \quad \text{and} \quad \mathbf{x}_0 = \begin{bmatrix} 1 \\ 1 \end{bmatrix}$$

then repeated multiplication by A produces a sequence of points

$$\mathbf{x}_0, \quad A\mathbf{x}_0, \quad A^2\mathbf{x}_0, \dots, \quad A^n\mathbf{x}_0, \dots$$

that follow the elliptical orbit about the origin shown in Figure 6.1.15. We are now in a position to explain that behavior. As the first step, we leave it for you to show that A has eigenvalues $\lambda = \frac{4}{5} \pm \frac{3}{5}i$ and that corresponding eigenvectors are

$$\lambda_1 = \tfrac{4}{5} - \tfrac{3}{5}i: \quad \mathbf{v}_1 = \left(\tfrac{1}{2} + i, 1\right) \quad \text{and} \quad \lambda_2 = \tfrac{4}{5} + \tfrac{3}{5}i: \quad \mathbf{v}_2 = \left(\tfrac{1}{2} - i, 1\right)$$

If we take $\lambda = \lambda_1 = \frac{4}{5} - \frac{3}{5}i$ and $\mathbf{x} = \mathbf{v}_1 = \left(\frac{1}{2} + i, 1\right)$ in (21) and use the fact that $|\lambda| = 1$, then we obtain the factorization

$$\underset{A}{\begin{bmatrix} \frac{1}{2} & \frac{3}{4} \\ -\frac{3}{5} & \frac{11}{10} \end{bmatrix}} = \underset{P}{\begin{bmatrix} \frac{1}{2} & 1 \\ 1 & 0 \end{bmatrix}} \underset{R_\phi}{\begin{bmatrix} \frac{4}{5} & -\frac{3}{5} \\ \frac{3}{5} & \frac{4}{5} \end{bmatrix}} \underset{P^{-1}}{\begin{bmatrix} 0 & 1 \\ 1 & -\frac{1}{2} \end{bmatrix}} \tag{23}$$

where R_ϕ is a rotation about the origin through the angle ϕ whose tangent is

$$\tan\phi = \frac{\sin\phi}{\cos\phi} = \frac{3/5}{4/5} = \frac{3}{4} \qquad \left(\phi = \tan^{-1}\tfrac{3}{4} \approx 36.9°\right)$$

The matrix P in (23) is the transition matrix from the basis

$$B = \{\mathrm{Re}(\mathbf{x}), \mathrm{Im}(\mathbf{x})\} = \left\{\left(\tfrac{1}{2}, 1\right), (1, 0)\right\}$$

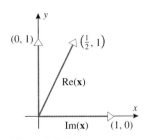

Figure 8.8.4

to the standard basis, and P^{-1} is the transition matrix from the standard basis to the basis B (Figure 8.8.4). Next, observe that if n is a positive integer, then (23) implies that

$$A^n \mathbf{x}_0 = (PR_\phi P^{-1})^n \mathbf{x}_0 = PR_\phi^n P^{-1} \mathbf{x}_0$$

so the product $A^n \mathbf{x}_0$ can be computed by first mapping $\mathbf{x}_0$ into the point $P^{-1}\mathbf{x}_0$ in B-coordinates, then multiplying by R_ϕ^n to rotate this point about the origin through the angle $n\phi$, and then multiplying $R_\phi^n P^{-1}\mathbf{x}_0$ by P to map the resulting point back to standard coordinates. We can now see what is happening geometrically. In B-coordinates each successive multiplication by A causes the point $P^{-1}\mathbf{x}_0$ to advance through an angle ϕ, thereby tracing a circular orbit about the origin. However, the basis B is *skewed* (not orthogonal), so when the points on the circular orbit are transformed back to standard coordinates, the effect is to distort the circular orbit into the elliptical orbit traced by $A^n \mathbf{x}_0$ (Figure 8.8.5a). Here are the computations for the first step (successive steps are illustrated in Figure 8.8.5b):

$$\begin{bmatrix} \frac{1}{2} & \frac{3}{4} \\ -\frac{3}{5} & \frac{11}{10} \end{bmatrix} \begin{bmatrix} 1 \\ 1 \end{bmatrix} = \begin{bmatrix} \frac{1}{2} & 1 \\ 1 & 0 \end{bmatrix} \begin{bmatrix} \frac{4}{5} & -\frac{3}{5} \\ \frac{3}{5} & \frac{4}{5} \end{bmatrix} \begin{bmatrix} 0 & 1 \\ 1 & -\frac{1}{2} \end{bmatrix} \begin{bmatrix} 1 \\ 1 \end{bmatrix}$$

$$= \begin{bmatrix} \frac{1}{2} & 1 \\ 1 & 0 \end{bmatrix} \begin{bmatrix} \frac{4}{5} & -\frac{3}{5} \\ \frac{3}{5} & \frac{4}{5} \end{bmatrix} \begin{bmatrix} 1 \\ \frac{1}{2} \end{bmatrix} \qquad \text{[$\mathbf{x}_0$ is mapped to B-coordinates.]}$$

$$= \begin{bmatrix} \frac{1}{2} & 1 \\ 1 & 0 \end{bmatrix} \begin{bmatrix} \frac{1}{2} \\ 1 \end{bmatrix} \qquad \text{[The point $\left(1, \frac{1}{2}\right)$ is rotated through the angle ϕ.]}$$

$$= \begin{bmatrix} \frac{5}{4} \\ \frac{1}{2} \end{bmatrix} \qquad \text{[The point $\left(\frac{1}{2}, 1\right)$ is mapped to standard coordinates.]} \quad \blacksquare$$

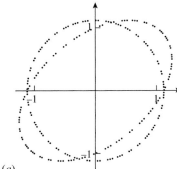

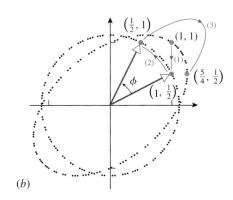

Figure 8.8.5   (a)   (b)

Exercise Set 8.8

In Exercises 1 and 2, find $\bar{\mathbf{u}}$, Re($\mathbf{u}$), Im($\mathbf{u}$), and $\|\mathbf{u}\|$.

1. $\mathbf{u} = (2 - i, 4i, 1 + i)$ **2.** $\mathbf{u} = (6, 1 + 4i, 6 - 2i)$

In Exercises 3 and 4, show that $\mathbf{u}$, $\mathbf{v}$, and k satisfy the relationships stated in Theorem 8.8.2.

3. $\mathbf{u} = (3 - 4i, 2 + i, -6i)$, $\mathbf{v} = (1 + i, 2 - i, 4)$, $k = i$

4. $\mathbf{u} = (6, 1 + 4i, 6 - 2i)$, $\mathbf{v} = (4, 3 + 2i, i - 3)$, $k = -i$

5. Solve the equation $i\mathbf{x} - 3\mathbf{v} = \bar{\mathbf{u}}$ for $\mathbf{x}$, where $\mathbf{u}$ and $\mathbf{v}$ are the vectors in Exercise 3.

6. Solve the equation $(1 + i)\mathbf{x} + 2\mathbf{u} = \bar{\mathbf{v}}$ for $\mathbf{x}$, where $\mathbf{u}$ and $\mathbf{v}$ are the vectors in Exercise 4.

In Exercises 7 and 8, find $\bar{A}$, Re(A), Im(A), det(A), and tr(A).

7. $A = \begin{bmatrix} -5i & 4 \\ 2-i & 1+5i \end{bmatrix}$ **8.** $A = \begin{bmatrix} 4i & 2-3i \\ 2+3i & 1 \end{bmatrix}$

9. Let A be the matrix given in Exercise 7. Confirm that if $B = (1-i, 2i)$ is written in column form, then A and B have the properties stated in Theorem 8.8.3.

10. Let A be the matrix given in Exercise 8. Confirm that if $B = (5i, 1-4i)$ in column form, then A and B have the properties stated in Theorem 8.8.3.

> In Exercises 11 and 12, compute $\mathbf{u} \cdot \mathbf{v}$, $\mathbf{u} \cdot \mathbf{w}$, and $\mathbf{v} \cdot \mathbf{w}$, and show that the properties stated in parts (a), (b), and (c) of Theorem 8.8.5 as well as Formula (9) are satisfied.

11. $\mathbf{u} = (i, 2i, 3)$, $\mathbf{v} = (4, -2i, 1+i)$, $\mathbf{w} = (2-i, 2i, 5+3i)$, $k = 2i$

12. $\mathbf{u} = (1+i, 4, 3i)$, $\mathbf{v} = (3, -4i, 2+3i)$, $\mathbf{w} = (1-i, 4i, 4-5i)$, $k = 1+i$

13. Compute $\overline{(\mathbf{u} \cdot \overline{\mathbf{v}}) - \overline{\mathbf{w} \cdot \mathbf{u}}}$ for the vectors $\mathbf{u}$, $\mathbf{v}$, and $\mathbf{w}$ in Exercise 11.

14. Compute $\overline{(i\mathbf{u} \cdot \mathbf{w})} + (\|\mathbf{u}\|\mathbf{v}) \cdot \mathbf{u}$ for the vectors $\mathbf{u}$, $\mathbf{v}$, and $\mathbf{w}$ in Exercise 12.

> In Exercises 15–18, find the eigenvalues and bases for the eigenspaces of A.

15. $A = \begin{bmatrix} 4 & -5 \\ 1 & 0 \end{bmatrix}$ **16.** $A = \begin{bmatrix} -1 & -5 \\ 4 & 7 \end{bmatrix}$

17. $A = \begin{bmatrix} 5 & -2 \\ 1 & 3 \end{bmatrix}$ **18.** $A = \begin{bmatrix} 8 & 6 \\ -3 & 2 \end{bmatrix}$

> In Exercises 19–22, a matrix C of form (19) is given, so Theorem 8.8.8 implies that C can be factored as the product of a scaling matrix with factor $|\lambda|$ and a rotation matrix with angle ϕ. Find $|\lambda|$ and the angle ϕ such that $-\pi < \phi \le \pi$.

19. $C = \begin{bmatrix} 1 & -1 \\ 1 & 1 \end{bmatrix}$ **20.** $C = \begin{bmatrix} 0 & 5 \\ -5 & 0 \end{bmatrix}$

21. $C = \begin{bmatrix} 1 & \sqrt{3} \\ -\sqrt{3} & 1 \end{bmatrix}$ **22.** $C = \begin{bmatrix} \sqrt{2} & \sqrt{2} \\ -\sqrt{2} & \sqrt{2} \end{bmatrix}$

> In Exercises 23–26, a matrix A with complex eigenvalues is given. Find an invertible matrix P and a matrix C of form (19) such that $A = PCP^{-1}$.

23. $A = \begin{bmatrix} -1 & -5 \\ 4 & 7 \end{bmatrix}$ **24.** $A = \begin{bmatrix} 4 & -5 \\ 1 & 0 \end{bmatrix}$

25. $A = \begin{bmatrix} 8 & 6 \\ -3 & 2 \end{bmatrix}$ **26.** $A = \begin{bmatrix} 5 & -2 \\ 1 & 3 \end{bmatrix}$

27. Find all complex scalars k, if any, for which $\mathbf{u}$ and $\mathbf{v}$ are orthogonal in C^3.
 (a) $\mathbf{u} = (2i, i, 3i)$, $\mathbf{v} = (i, 6i, k)$
 (b) $\mathbf{u} = (k, k, 1+i)$, $\mathbf{v} = (1, -1, 1-i)$

28. Show that if A is a real $n \times n$ matrix and $\mathbf{x}$ is a vector C^n, then $\mathrm{Re}(A\mathbf{x}) = A(\mathrm{Re}(\mathbf{x}))$ and $\mathrm{Im}(A\mathbf{x}) = A(\mathrm{Im}(\mathbf{x}))$.

29. The matrices

$$\sigma_1 = \begin{bmatrix} 0 & 1 \\ 1 & 0 \end{bmatrix}, \quad \sigma_2 = \begin{bmatrix} 0 & -i \\ i & 0 \end{bmatrix}, \quad \sigma_3 = \begin{bmatrix} 1 & 0 \\ 0 & -1 \end{bmatrix}$$

called **Pauli spin matrices**, are used in quantum mechanics to study particles with *half-integral spin*, and the **Dirac matrices**, which are also used in quantum mechanics, are expressed in terms of the Pauli spin matrices and the 2×2 identity matrix I_2 as

$$\beta = \begin{bmatrix} I_2 & 0 \\ 0 & -I_2 \end{bmatrix}, \quad \alpha_x = \begin{bmatrix} 0 & \sigma_1 \\ \sigma_1 & 0 \end{bmatrix}$$

$$\alpha_y = \begin{bmatrix} 0 & \sigma_2 \\ \sigma_2 & 0 \end{bmatrix}, \quad \alpha_z = \begin{bmatrix} 0 & \sigma_3 \\ \sigma_3 & 0 \end{bmatrix}$$

 (a) Show that $\beta^2 = \alpha_x^2 = \alpha_y^2 = \alpha_z^2$.
 (b) Matrices A and B for which $AB = -BA$ are said to be **anticommutative**. Show that the Dirac matrices are anticommutative.

Discussion and Discovery

D1. If $\mathbf{u} \cdot \mathbf{v} = a + bi$, then $(i\mathbf{u}) \cdot \mathbf{v} = $ _____, $\mathbf{u} \cdot (i\mathbf{v}) = $ _____, and $\mathbf{v} \cdot (i\mathbf{u}) = $ _____.

D2. If k is a real scalar and $\mathbf{v}$ is a vector in R^n, then Theorem 1.2.2 states that $\|k\mathbf{v}\| = |k|\|\mathbf{v}\|$. Is this relationship also true if k is a complex scalar and $\mathbf{v}$ is a vector in C^n? Justify your answer.

D3. If $\lambda = a + bi$ is an eigenvalue of a real 2×2 matrix A and $\mathbf{x} = (u_1 + v_1i, u_2 + v_2i)$ is a corresponding eigenvector in which u_1, u_2, v_1, and v_2 are real numbers, then _____ is also an eigenvalue of A and _____ is a corresponding eigenvector.

D4. Show that the eigenvalues of the symmetric matrix

$$A = \begin{bmatrix} 1 & 4i \\ 4i & 3 \end{bmatrix}$$

are not real. Does this contradict Theorem 8.8.7?

Working with Proofs

P1. Prove part (c) of Theorem 8.8.2.

P2. Prove Theorem 8.8.3.

P3. Prove that if $\mathbf{u}$ and $\mathbf{v}$ are vectors in C^n, then

$$\mathbf{u} \cdot \mathbf{v} = \tfrac{1}{4} \|\mathbf{u} + \mathbf{v}\|^2 - \tfrac{1}{4} \|\mathbf{u} - \mathbf{v}\|^2$$
$$+ \tfrac{i}{4} \|\mathbf{u} + i\mathbf{v}\|^2 - \tfrac{i}{4} \|\mathbf{u} - i\mathbf{v}\|^2$$

P4. It follows from Theorem 8.8.8 that the eigenvalues of the rotation matrix

$$R_\phi = \begin{bmatrix} \cos\phi & -\sin\phi \\ \sin\phi & \cos\phi \end{bmatrix}$$

are $\lambda = \cos\phi \pm i\sin\phi$. Prove that if $\mathbf{x}$ is an eigenvector corresponding to either eigenvalue, then $\mathrm{Re}(\mathbf{x})$ and $\mathrm{Im}(\mathbf{x})$ are orthogonal and have the same length. [*Note:* This implies that $P = [\mathrm{Re}(\mathbf{x}) \mid \mathrm{Im}(\mathbf{x})]$ is a real scalar multiple of an orthogonal matrix.]

P5. Prove Theorem 8.8.9 as follows:

(a) For notational simplicity, let

$$M = \begin{bmatrix} a & -b \\ b & a \end{bmatrix}$$

and let $\mathbf{u} = \mathrm{Re}(\mathbf{x})$ and $\mathbf{v} = \mathrm{Im}(\mathbf{x})$, so $P = [\mathbf{u} \mid \mathbf{v}]$. Show that the relationship $A\mathbf{x} = \lambda\mathbf{x}$ implies that $A\mathbf{x} = (a\mathbf{u} + b\mathbf{v}) + i(-b\mathbf{u} + a\mathbf{v})$, and then equate real

and imaginary parts in this equation to show that

$$AP = [A\mathbf{u} \mid A\mathbf{v}] = [a\mathbf{u} + b\mathbf{v} \mid -b\mathbf{u} + a\mathbf{v}] = PM$$

(b) Show that P is invertible, thereby completing the proof, since the result in part (a) implies that $A = PMP^{-1}$. [*Hint:* If P is not invertible, then one of its column vectors is a real scalar multiple of the other, say $\mathbf{v} = c\mathbf{u}$. Substitute this into the equations $A\mathbf{u} = a\mathbf{u} + b\mathbf{v}$ and $A\mathbf{v} = -b\mathbf{u} + a\mathbf{v}$ obtained in part (a), and show that $(1 + c^2)b\mathbf{u} = \mathbf{0}$. Finally, show that this leads to a contradiction, thereby proving that P is invertible.]

P6. In this problem you will prove the complex analog of the Cauchy–Schwarz inequality.

(a) Prove: If k is a complex number, and $\mathbf{u}$ and $\mathbf{v}$ are vectors in C^n, then

$$(\mathbf{u} - k\mathbf{v}) \cdot (\mathbf{u} - k\mathbf{v}) = \mathbf{u} \cdot \mathbf{u} - \overline{k}(\mathbf{u} \cdot \mathbf{v})$$
$$- k\overline{(\mathbf{u} \cdot \mathbf{v})} + k\overline{k}(\mathbf{v} \cdot \mathbf{v})$$

(b) Use the result in part (a) to prove that

$$0 \le \mathbf{u} \cdot \mathbf{u} - \overline{k}(\mathbf{u} \cdot \mathbf{v}) - k\overline{(\mathbf{u} \cdot \mathbf{v})} + k\overline{k}(\mathbf{v} \cdot \mathbf{v})$$

(c) Take $k = (\mathbf{u} \cdot \mathbf{v})/(\mathbf{v} \cdot \mathbf{v})$ in part (b) to prove that

$$|\mathbf{u} \cdot \mathbf{v}| \le \|\mathbf{u}\| \, \|\mathbf{v}\|$$

Technology Exercises

T1. (*Arithmetic operations on complex numbers*) Most linear algebra technology programs have a syntax for entering complex numbers and can perform additions, subtractions, multiplications, divisions, conjugations, modulus and argument determinations, and extractions of real and imaginary parts on them. Enter some complex numbers and perform various computations with them until you feel you have mastered the operations.

T2. (*Vectors and matrices with complex entries*) For most linear algebra technology utilities, operations on vectors and matrices with complex entries are the same as for vectors and matrices with real entries. Enter some complex numbers and perform various computations with them until you feel you have mastered the operations.

T3. Perform the computations in Examples 1 and 2.

T4. (a) Show that the vectors

$$\mathbf{u}_1 = (i, i, i), \quad \mathbf{u}_2 = (0, i, i), \quad \mathbf{u}_3 = (i, 2i, i)$$

are linearly independent.

(b) Use the Gram–Schmidt process to transform $\{\mathbf{u}_1, \mathbf{u}_2, \mathbf{u}_3\}$ into an orthonormal set.

T5. Determine whether there exist scalars c_1, c_2, and c_3 such that

$$c_1(i, 2 - i, 2 + i) + c_2(1 + i, -2i, 2) + c_3(3, i, 6 + i) = (i, i, i)$$

T6. Find the eigenvalues and bases for the eigenspaces of

$$A = \begin{bmatrix} 1 & 2 & 1 \\ -2 & 1 & 0 \\ 1 & 0 & 1 \end{bmatrix}$$

T7. Factor $A = \begin{bmatrix} -1 - \sqrt{3} & 2\sqrt{3} \\ -\sqrt{3} & -1 + \sqrt{3} \end{bmatrix}$ as $A = PCP^{-1}$, where C is of form (19).

Section 8.9   Hermitian, Unitary, and Normal Matrices

We know that every real symmetric matrix is orthogonally diagonalizable and that the symmetric matrices are the only orthogonally diagonalizable matrices. In this section we will consider the diagonalization problem for complex matrices.

HERMITIAN AND UNITARY MATRICES

The transpose operation is less important for complex matrices than for real matrices. A more useful operation for complex matrices is given in the following definition.

Definition 8.9.1 If A is a complex matrix, then the ***conjugate transpose*** of A, denoted by A^*, is defined by

$$A^* = \overline{A}^T \tag{1}$$

REMARK Since part (*b*) of Theorem 8.8.3 states that $(\overline{A^T}) = (\overline{A})^T$, the order in which the transpose and conjugation operations are performed in computing $A^* = \overline{A}^T$ does not matter. Also, in the case where A has real entries we have $A^* = (\overline{A})^T = A^T$, so A^* is the same as A^T for real matrices.

EXAMPLE 1
Conjugate
Transpose

Find the conjugate transpose A^* of the matrix

$$A = \begin{bmatrix} 1+i & -i & 0 \\ 2 & 3-2i & i \end{bmatrix}$$

Solution We have

$$\overline{A} = \begin{bmatrix} 1-i & i & 0 \\ 2 & 3+2i & -i \end{bmatrix} \quad \text{and hence} \quad A^* = \overline{A}^T = \begin{bmatrix} 1-i & 2 \\ i & 3+2i \\ 0 & -i \end{bmatrix} \qquad ∎$$

The following theorem, parts of which are proved in the exercises, shows that the basic algebraic properties of the conjugate transpose operation are similar to those of the transpose (compare to Theorem 3.2.10).

Theorem 8.9.2 *If k is a complex scalar, and if A, B, and C are complex matrices whose sizes are such that the stated operations can be performed, then:*

(*a*) $(A^*)^* = A$

(*b*) $(A + B)^* = A^* + B^*$

(*c*) $(A - B)^* = A^* - B^*$

(*d*) $(kA)^* = \overline{k}A^*$

(*e*) $(AB)^* = B^*A^*$

REMARK Note that the relationship $\mathbf{u} \cdot \mathbf{v} = \overline{\mathbf{v}}^T\mathbf{u}$ in Formula (9) of Section 8.8 can be expressed in terms of the conjugate transpose as

$$\mathbf{u} \cdot \mathbf{v} = \mathbf{v}^*\mathbf{u} \tag{2}$$

We are now ready to define two new classes of matrices that will be important in our study of diagonalization in C^n.

Definition 8.9.3 A square complex matrix A is said to be **unitary** if

$$A^{-1} = A^* \tag{3}$$

and is said to be **Hermitian**[*] if

$$A^* = A \tag{4}$$

If A is a real matrix, then $A^* = A^T$, in which case (3) becomes $A^{-1} = A^T$ and (4) becomes $A^T = A$. Thus, the unitary matrices are complex generalizations of the real orthogonal matrices and Hermitian matrices are complex generalizations of the real symmetric matrices.

EXAMPLE 2
Recognizing
Hermitian
Matrices

Hermitian matrices are easy to recognize because their diagonal entries are real (why?) and the entries that are symmetrically positioned across the main diagonal are complex conjugates. Thus, for example, we can tell by inspection that

$$A = \begin{bmatrix} 1 & i & 1+i \\ -i & -5 & 2-i \\ 1-i & 2+i & 3 \end{bmatrix}$$

is Hermitian. To see this algebraically, observe that

$$\overline{A} = \begin{bmatrix} 1 & -i & 1-i \\ i & -5 & 2+i \\ 1+i & 2-i & 3 \end{bmatrix}, \quad \text{so} \quad A^* = \overline{A}^T = \begin{bmatrix} 1 & i & 1+i \\ -i & -5 & 2-i \\ 1-i & 2+i & 3 \end{bmatrix} = A \quad ■$$

The fact that real symmetric matrices have real eigenvalues is a special case of the following more general result about Hermitian matrices.

Theorem 8.9.4 *The eigenvalues of a Hermitian matrix are real numbers.*

The proof is left for the exercises.

The fact that eigenvectors from different eigenspaces of a real symmetric matrix are orthogonal is a special case of the following more general result about Hermitian matrices.

Theorem 8.9.5 *If A is a Hermitian matrix, then eigenvectors from different eigenspaces are orthogonal.*

Proof Let $\mathbf{v}_1$ and $\mathbf{v}_2$ be eigenvectors corresponding to distinct eigenvalues λ_1 and λ_2. Using the facts that $\lambda_1 = \overline{\lambda}_1$, $\lambda_2 = \overline{\lambda}_2$, and $A = A^*$, we can write

$$\lambda_1(\mathbf{v}_2 \cdot \mathbf{v}_1) = (\lambda_1 \mathbf{v}_1)^* \mathbf{v}_2 = (A\mathbf{v}_1)^* \mathbf{v}_2 = (\mathbf{v}_1^* A^*)\mathbf{v}_2$$
$$= (\mathbf{v}_1^* A)\mathbf{v}_2 = \mathbf{v}_1^*(A\mathbf{v}_2)$$
$$= \mathbf{v}_1^*(\lambda_2 \mathbf{v}_2) = \lambda_2(\mathbf{v}_1^* \mathbf{v}_2) = \lambda_2(\mathbf{v}_2 \cdot \mathbf{v}_1)$$

This implies that $(\lambda_1 - \lambda_2)(\mathbf{v}_2 \cdot \mathbf{v}_1) = 0$ and hence that $\mathbf{v}_2 \cdot \mathbf{v}_1 = 0$ (since $\lambda_1 \neq \lambda_2$). ■

EXAMPLE 3
Eigenvalues
and
Eigenvectors of
a Hermitian
Matrix

Confirm that the Hermitian matrix

$$A = \begin{bmatrix} 2 & 1+i \\ 1-i & 3 \end{bmatrix}$$

has real eigenvalues and that eigenvectors from different eigenspaces are orthogonal.

Solution The characteristic polynomial of A is

$$\det(\lambda I - A) = \begin{vmatrix} \lambda - 2 & -1-i \\ -1+i & \lambda - 3 \end{vmatrix} = (\lambda - 2)(\lambda - 3) - (-1-i)(-1+i) = (\lambda - 1)(\lambda - 4)$$

---

[*]In honor of the French mathematician Charles Hermite (1822–1901).

so the eigenvalues of A are $\lambda = 1$ and $\lambda = 4$, which are real. Bases for the eigenspaces of A can be obtained by solving the linear system

$$\begin{bmatrix} \lambda - 2 & -1 - i \\ -1 + i & \lambda - 3 \end{bmatrix} \begin{bmatrix} x_1 \\ x_2 \end{bmatrix} = \begin{bmatrix} 0 \\ 0 \end{bmatrix}$$

with $\lambda = 1$ and with $\lambda = 4$. We leave it for you to do this and to show that the general solutions of these systems are

$$\lambda = 1: \quad \begin{bmatrix} x_1 \\ x_2 \end{bmatrix} = t \begin{bmatrix} -1 - i \\ 1 \end{bmatrix} \quad \text{and} \quad \lambda = 4: \quad \begin{bmatrix} x_1 \\ x_2 \end{bmatrix} = t \begin{bmatrix} \frac{1}{2}(1 + i) \\ 1 \end{bmatrix}$$

Thus, bases for these eigenspaces are

$$\lambda = 1: \quad \mathbf{v}_1 = \begin{bmatrix} -1 - i \\ 1 \end{bmatrix} \quad \text{and} \quad \lambda = 4: \quad \mathbf{v}_2 = \begin{bmatrix} \frac{1}{2}(1 + i) \\ 1 \end{bmatrix}$$

The vectors $\mathbf{v}_1$ and $\mathbf{v}_2$ are orthogonal since

$$\mathbf{v}_1 \cdot \mathbf{v}_2 = (-1 - i)\left(\tfrac{1}{2}(1 + i)\right) + (1)(1) = \tfrac{1}{2}(-1 - i)(1 - i) + 1 = 0$$

and hence all scalar multiples of them are also orthogonal.   ∎

Unitary matrices are not usually easy to recognize by inspection. However, the following analog of Theorem 6.2.5, part of which is proved in the exercises, provides a way of ascertaining whether a matrix is unitary without computing its inverse.

Theorem 8.9.6   *If A is an $n \times n$ matrix with complex entries, then the following are equivalent.*

    (a) *A is unitary.*

    (b) *$\|A\mathbf{x}\| = \|\mathbf{x}\|$ for all $\mathbf{x}$ in C^n.*

    (c) *$A\mathbf{x} \cdot A\mathbf{y} = \mathbf{x} \cdot \mathbf{y}$ for all $\mathbf{x}$ and $\mathbf{y}$ in C^n.*

    (d) *The column vectors of A form an orthonormal set in C^n with respect to the complex Euclidean inner product.*

    (e) *The row vectors of A form an orthonormal set in C^n with respect to the complex Euclidean inner product.*

EXAMPLE 4
A Unitary
Matrix

Use Theorem 8.9.6 to show that

$$A = \begin{bmatrix} \frac{1}{2}(1 + i) & \frac{1}{2}(1 + i) \\ \frac{1}{2}(1 - i) & \frac{1}{2}(-1 + i) \end{bmatrix}$$

is unitary, and then find A^{-1}.

Solution   We will show that the row vectors

$$\mathbf{r}_1 = \begin{bmatrix} \tfrac{1}{2}(1 + i) & \tfrac{1}{2}(1 + i) \end{bmatrix} \quad \text{and} \quad \mathbf{r}_2 = \begin{bmatrix} \tfrac{1}{2}(1 - i) & \tfrac{1}{2}(-1 + i) \end{bmatrix}$$

are orthonormal. The relevant computations are

$$\|\mathbf{r}_1\| = \sqrt{\left|\tfrac{1}{2}(1 + i)\right|^2 + \left|\tfrac{1}{2}(1 + i)\right|^2} = \sqrt{\tfrac{1}{2} + \tfrac{1}{2}} = 1$$

$$\|\mathbf{r}_2\| = \sqrt{\left|\tfrac{1}{2}(1 - i)\right|^2 + \left|\tfrac{1}{2}(-1 + i)\right|^2} = \sqrt{\tfrac{1}{2} + \tfrac{1}{2}} = 1$$

$$\mathbf{r}_1 \cdot \mathbf{r}_2 = \left(\tfrac{1}{2}(1 + i)\right)\left(\overline{\tfrac{1}{2}(1 - i)}\right) + \left(\tfrac{1}{2}(1 + i)\right)\left(\overline{\tfrac{1}{2}(-1 + i)}\right)$$

$$= \left(\tfrac{1}{2}(1 + i)\right)\left(\tfrac{1}{2}(1 + i)\right) + \left(\tfrac{1}{2}(1 + i)\right)\left(\tfrac{1}{2}(-1 - i)\right) = \tfrac{1}{2}i - \tfrac{1}{2}i = 0$$

Since we now know that A is unitary, it follows that

$$A^{-1} = A^* = \begin{bmatrix} \frac{1}{2}(1-i) & \frac{1}{2}(1+i) \\ \frac{1}{2}(1-i) & \frac{1}{2}(-1-i) \end{bmatrix}$$

We leave it for you to confirm the validity of this result by showing that $AA^* = A^*A = I$. ∎

UNITARY DIAGONALIZABILITY   Since unitary matrices are the complex analogs of the real orthogonal matrices, the following definition is a natural generalization of the idea of orthogonal diagonalizability for real matrices.

> **Definition 8.9.7**   A square complex matrix is said to be ***unitarily diagonalizable*** if there is a unitary matrix P such that $P^*AP = D$ is a complex diagonal matrix. Any such matrix P is said to ***unitarily diagonalize*** A.

Recall that a real symmetric $n \times n$ matrix A has an orthonormal set of n eigenvectors and is orthogonally diagonalized by any $n \times n$ matrix whose column vectors are an orthonormal set of eigenvectors of A. Here is the complex analog of that result.

> **Theorem 8.9.8**   *Every $n \times n$ Hermitian matrix A has an orthonormal set of n eigenvectors and is unitarily diagonalized by any $n \times n$ matrix P whose column vectors are an orthonormal set of eigenvectors of A.*

The procedure for unitarily diagonalizing a Hermitian matrix A is exactly the same as that for orthogonally diagonalizing a symmetric matrix:

Step 1. Find a basis for each eigenspace of A.

Step 2. Apply the Gram–Schmidt process to each of these bases to obtain orthonormal bases for the eigenspaces.

Step 3. Form the matrix P whose column vectors are the basis vectors obtained in the last step. This will be a unitary matrix (Theorem 8.9.6) and will unitarily diagonalize A.

EXAMPLE 5
Unitary
Diagonalization
of a Hermitian
Matrix

Find a matrix P that unitarily diagonalizes the Hermitian matrix

$$A = \begin{bmatrix} 2 & 1+i \\ 1-i & 3 \end{bmatrix}$$

Solution We showed in Example 3 that the eigenvalues of A are $\lambda = 1$ and $\lambda = 4$ and that bases for the corresponding eigenspaces are

$$\lambda = 1: \quad \mathbf{v}_1 = \begin{bmatrix} -1-i \\ 1 \end{bmatrix} \quad \text{and} \quad \lambda = 4: \quad \mathbf{v}_2 = \begin{bmatrix} \frac{1}{2}(1+i) \\ 1 \end{bmatrix}$$

Since each eigenspace has only one basis vector, the Gram–Schmidt process is simply a matter of normalizing these basis vectors. We leave it for you to show that

$$\mathbf{p}_1 = \frac{\mathbf{v}_1}{\|\mathbf{v}_1\|} = \begin{bmatrix} \frac{-1-i}{\sqrt{3}} \\ \frac{1}{\sqrt{3}} \end{bmatrix} \quad \text{and} \quad \mathbf{p}_2 = \frac{\mathbf{v}_2}{\|\mathbf{v}_2\|} = \begin{bmatrix} \frac{1+i}{\sqrt{6}} \\ \frac{2}{\sqrt{6}} \end{bmatrix}$$

Thus, A is unitarily diagonalized by the matrix

$$P = [\mathbf{p}_1 \quad \mathbf{p}_2] = \begin{bmatrix} \frac{-1-i}{\sqrt{3}} & \frac{1+i}{\sqrt{6}} \\ \frac{1}{\sqrt{3}} & \frac{2}{\sqrt{6}} \end{bmatrix}$$

Although it is a little tedious, you may want to check this result by showing that

$$P^*AP = \begin{bmatrix} \frac{-1+i}{\sqrt{3}} & \frac{1}{\sqrt{3}} \\ \frac{1-i}{\sqrt{6}} & \frac{2}{\sqrt{6}} \end{bmatrix} \begin{bmatrix} 2 & 1+i \\ 1-i & 3 \end{bmatrix} \begin{bmatrix} \frac{-1-i}{\sqrt{3}} & \frac{1+i}{\sqrt{6}} \\ \frac{1}{\sqrt{3}} & \frac{2}{\sqrt{6}} \end{bmatrix} = \begin{bmatrix} 1 & 0 \\ 0 & 4 \end{bmatrix}$$ ∎

31. Show that if A

32. Show that the ⟨
either zero or p

33. Show that the ∈
modulus 1.

SKEW-HERMITIAN MATRICES

Recall from Section 3.6 that a square matrix with real entries is said to be skew-symmetric if $A^T = -A$. Analogously, we say that a square matrix with complex entries is *skew-Hermitian* if

$$A^* = -A$$

Discussion a

D1. What can you
both Hermitia

D2. Find a 2×2 r
whose entries

D3. Under what c⟨

$$A = \begin{bmatrix} a \\ 0 \\ 0 \end{bmatrix}$$

A skew-Hermitian matrix must have zeros or pure imaginary numbers on the main diagonal (Exercise 28), and the complex conjugates of entries that are symmetrically positioned about the main diagonal must be negatives of one another. An example of a skew-Hermitian matrix is

$$A = \begin{bmatrix} i & 1-i & 5 \\ -1-i & 2i & i \\ -5 & i & 0 \end{bmatrix}$$

NORMAL MATRICES

Hermitian matrices enjoy many, but not all, of the properties of real symmetric matrices. For example, we know that real symmetric matrices are orthogonally diagonalizable and Hermitian matrices are unitarily diagonalizable. However, whereas the real symmetric matrices are the only orthogonally diagonalizable matrices, the Hermitian matrices do not constitute the entire class of unitarily diagonalizable complex matrices; that is, there exist unitarily diagonalizable matrices that are not Hermitian. Specifically, it can be proved that a square complex matrix A is unitarily diagonalizable if and only if

$$AA^* = A^*A \tag{5}$$

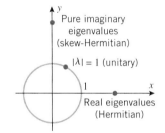

Figure 8.9.1

Matrices with this property are said to be *normal*. Normal matrices include the Hermitian, skew-Hermitian, and unitary matrices in the complex case and the symmetric, skew-symmetric, and orthogonal matrices in the real case. The nonzero skew-symmetric matrices are particularly interesting because they are examples of real matrices that are not orthogonally diagonalizable but are unitarily diagonalizable.

Working wi

P1. Prove: If A i
$(A^*)^{-1} = (A^-$

P2. (a) Use Form
$\det(\overline{A}) =$
(b) Use the r⟨
matrix an
prove tha

P3. Use part (b) ⟨
(a) If A is H⟨
(b) If A is u⟨

A COMPARISON OF EIGENVALUES

We have seen that Hermitian matrices have real eigenvalues. In the exercises we will ask you to show that the eigenvalues of a skew-Hermitian matrix are either zero or purely imaginary (have real part of zero) and that the eigenvalues of unitary matrices have modulus 1. These results are illustrated schematically in Figure 8.9.1.

Technology

T1. Find a matrix

$$A = \begin{bmatrix} \vdots \\ -\vdots \\ 1 - \end{bmatrix}$$

T2. Find the eige
Hermitian m⟨

$$A = \begin{bmatrix} \vdots \\ 3 - \end{bmatrix}$$

and confirm t
8.9.4 and 8.9

Exercise Set 8.9

In Exercises 1 and 2, find A^*.

1. $A = \begin{bmatrix} 2i & 1-i \\ 4 & 3+i \\ 5+i & 0 \end{bmatrix}$

2. $A = \begin{bmatrix} 2i & 1-i & -1+i \\ 4 & 5-7i & -i \end{bmatrix}$

In Exercises 3 and 4, substitute numbers for the ×'s to make A Hermitian.

3. $A = \begin{bmatrix} 1 & i & 2-3i \\ \times & -3 & 1 \\ \times & \times & 2 \end{bmatrix}$

4. $A = \begin{bmatrix} 2 & 0 & 3+5i \\ \times & -4 & -i \\ \times & \times & 6 \end{bmatrix}$

In Exercises 5 and 6, show that A is not Hermitian for any choice of the ×'s.

5. (a) $A = \begin{bmatrix} 1 & i & 2-3i \\ -i & -3 & \times \\ 2-3i & \times & \times \end{bmatrix}$

(b) $A = \begin{bmatrix} \times & \times & 3+5i \\ 0 & i & -i \\ 3-5i & i & \times \end{bmatrix}$

6. (a) $A = \begin{bmatrix} 1 & 1+i & \times \\ 1+i & 7 & \times \\ 6-2i & \times & 0 \end{bmatrix}$

(b) $A = \begin{bmatrix} \\ 3 \end{bmatrix}$

In Exercises 7 ar
firm that the eig
tors from differer
with Theorem 8.

7. $A = \begin{bmatrix} 3 \\ 2+3i \end{bmatrix}$

In Exercises 9–1

9. $A = \begin{bmatrix} \frac{3}{5} & \frac{4}{5} \\ -\frac{4}{5} & \frac{3}{5} \end{bmatrix}$

10. $A = \begin{bmatrix} \frac{1}{\sqrt{2}} \\ -\frac{1}{2}(1 \end{bmatrix}$

11. $A = \begin{bmatrix} \frac{1}{2\sqrt{2}} (\\ \frac{1}{2\sqrt{2}} (1 \end{bmatrix}$

12. $A = \begin{bmatrix} \frac{1}{\sqrt{3}}(-1 \\ \frac{1}{\sqrt{3}} \end{bmatrix}$

In Exercises 13-
unitary matrix P

13. $A = \begin{bmatrix} 4 \\ 1+i \end{bmatrix}$

15. $A = \begin{bmatrix} 6 \\ 2-2i \end{bmatrix}$

17. $A = \begin{bmatrix} 5 \\ 0 \\ 0 & -1 \end{bmatrix}$

18. $A = \begin{bmatrix} 2 \\ -\frac{1}{\sqrt{2}}i \\ \frac{1}{\sqrt{2}}i \end{bmatrix}$

In Exercises 19
make A skew-H

19. $A = \begin{bmatrix} 0 & i \\ \times & 0 \\ \times & \times \end{bmatrix}$

In Exercises 21
for any choice

Section 8.10   Systems of Differential Equations

Many principles of physics, engineering, chemistry, and other sciences are described in terms of "differential equations," that is, equations that involve functions and their derivatives. The study of differential equations is a course in itself, so our goal in this section is simply to illustrate some of the ways in which linear algebra applies to this subject. Calculus is required for this section.

TERMINOLOGY   One of the most basic differential equations is

$$y' = ay \tag{1}$$

where a is a constant, $y = y(t)$ is an unknown function to be determined, and $y' = dy/dt$ is the derivative of y with respect to the independent variable t. This is called a **first-order equation** because it involves only the first derivative of the unknown function. We have used the letter t as the independent variable because this kind of equation commonly arises in problems where $y(t)$ is a function of time.

A **solution** of (1) is any differentiable function $y = y(t)$ for which the equation is satisfied when y and its derivative are substituted. For example,

$$y = ce^{at} \tag{2}$$

is a solution of (1) for any constant c since

$$y' = cae^{at} = a(ce^{at}) = ay$$

for all values of t. Conversely, one can show that every solution of (1) must be of form (2) (Exercise P1), so we call (2) the **general solution** of (1).

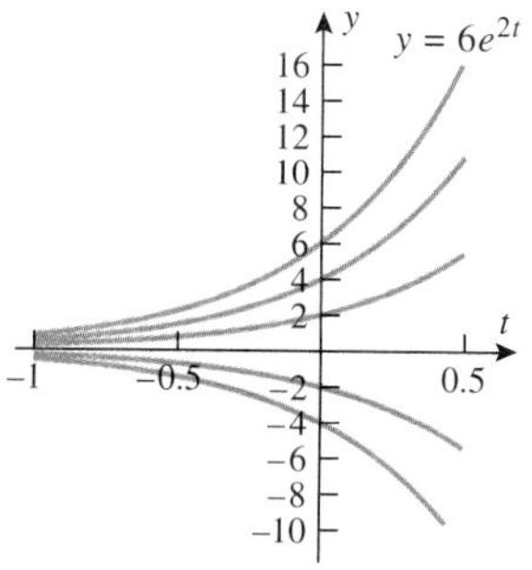

Sometimes we will be interested in a solution of (1) that has a specific value y_0 at a specific time t_0. The requirement $y(t_0) = y_0$ is called an **initial condition**, and we write

$$y' = ay, \quad y(t_0) = y_0 \tag{3}$$

to indicate that we want a solution of the equation $y' = ay$ that satisfies the initial condition. We call (3) an **initial value problem** for (1).

EXAMPLE 1
An Initial Value
Problem

Solve the initial value problem

$$y' = 2y, \quad y(0) = 6$$

Solution   It follows from (2) with $a = 2$ that the general solution of the differential equation is

$$y = ce^{2t} \tag{4}$$

The initial condition requires that $y = 6$ if $t = 0$, and substituting these values in (4) yields $c = 6$ (verify). Thus, the solution of the initial value problem is

$$y = 6e^{2t}$$

Geometrically, the general solution produces a family of curves in a ty-plane that depend on the value of c, and the initial condition isolates the particular solution in the family that passes through the point $(t_0, y_0) = (0, 6)$ (see Figure 8.10.1).   ∎

Figure 8.10.1

LINEAR SYSTEMS OF DIFFERENTIAL EQUATIONS

In this section we will be concerned with systems of differential equations of the form

$$\begin{aligned}
y_1' &= a_{11}y_1 + a_{12}y_2 + \cdots + a_{1n}y_n \\
y_2' &= a_{21}y_1 + a_{22}y_2 + \cdots + a_{2n}y_n \\
&\ \vdots \qquad \vdots \qquad \vdots \qquad\qquad \vdots \\
y_n' &= a_{n1}y_1 + a_{n2}y_2 + \cdots + a_{nn}y_n
\end{aligned} \tag{5}$$

where the a's are real constants and $y_1 = y_1(t)$, $y_2 = y_2(t), \ldots, y_n = y_n(t)$ are differentiable functions to be determined. This is called a ***homogeneous first-order linear system***. When convenient, (5) can be written in matrix notation as

$$\begin{bmatrix} y_1' \\ y_2' \\ \vdots \\ y_n' \end{bmatrix} = \begin{bmatrix} a_{11} & a_{12} & \cdots & a_{1n} \\ a_{21} & a_{22} & \cdots & a_{2n} \\ \vdots & \vdots & & \vdots \\ a_{n1} & a_{n2} & \cdots & a_{nn} \end{bmatrix} \begin{bmatrix} y_1 \\ y_2 \\ \vdots \\ y_n \end{bmatrix} \tag{6}$$

or more briefly as

$$\mathbf{y}' = A\mathbf{y} \tag{7}$$

where $\mathbf{y}$ is the column vector of unknown functions, and $\mathbf{y}'$ is the column vector of their derivatives.[*] Since we will be considering only the case where the number of equations is the same as the number of unknown functions, the matrix A, which is called the ***coefficient matrix*** for the system, will always be square in this section. A ***solution*** of the system can be viewed either as a sequence of differentiable functions

$$y_1 = y_1(t), \quad y_2 = y_2(t), \ldots, \quad y_n = y_n(t)$$

that satisfy all of the equations in (5) or as a column vector

$$\mathbf{y} = \mathbf{y}(t) = \begin{bmatrix} y_1(t) \\ y_2(t) \\ \vdots \\ y_n(t) \end{bmatrix}$$

that satisfies (7). Observe that $\mathbf{y} = \mathbf{0}$ is always a solution of (7). This is called the ***zero solution*** or the ***trivial solution***.

Sometimes we will be interested in finding a solution of (7) that has a specific *value* $\mathbf{y}_0$ at a certain time t_0; that is, we will want $\mathbf{y}(t)$ to satisfy $\mathbf{y}(t_0) = \mathbf{y}_0$. We call this an ***initial condition*** for (7). We denote the problem of solving (7) subject to the initial condition as

$$\mathbf{y}' = A\mathbf{y}, \quad \mathbf{y}(t_0) = \mathbf{y}_0 \tag{8}$$

and call it an ***initial value problem*** for (7). It can be proved that (8) has a unique solution for every $\mathbf{y}_0$.

EXAMPLE 2
An Initial Value Problem

(a) Solve the system

$$\begin{bmatrix} y_1' \\ y_2' \\ y_3' \end{bmatrix} = \begin{bmatrix} 3 & 0 & 0 \\ 0 & -2 & 0 \\ 0 & 0 & 5 \end{bmatrix} \begin{bmatrix} y_1 \\ y_2 \\ y_3 \end{bmatrix}$$

(b) Find the solution of the system that satisfies the initial conditions

$$y_1(0) = 1, \quad y_2(0) = 4, \quad y_3(0) = -2 \tag{9}$$

Solution (a) The fact that the matrix A is diagonal means that each equation in the system involves only one of the three unknowns and hence can be solved individually. These equations

---

[*] In this section we will extend the meaning of the term *vector* to allow for components that are continuous functions. Also, if the components of $\mathbf{y}$ are differentiable functions, then $\mathbf{y}'$ will denote the vector that results by differentiating the components of $\mathbf{y}$. In the next chapter we will consider these ideas in more detail.

are

$$y_1' = 3y_1$$
$$y_2' = -2y_2$$
$$y_3' = 5y_3$$

It follows from (2) that the solutions of these equations are

$$y_1 = c_1 e^{3t}, \quad y_2 = c_2 e^{-2t}, \quad y_3 = c_3 e^{5t}$$

When convenient, the solution of the system can also be expressed as

$$\mathbf{y} = \begin{bmatrix} y_1 \\ y_2 \\ y_3 \end{bmatrix} = \begin{bmatrix} c_1 e^{3t} \\ c_2 e^{-2t} \\ c_3 e^{5t} \end{bmatrix} \tag{10}$$

Observe that (10) involves three arbitrary constants, one from each equation.

Solution (b) Writing the three initial conditions in (9) in column form and using (10), we obtain

$$\mathbf{y}(0) = \begin{bmatrix} y_1(0) \\ y_2(0) \\ y_3(0) \end{bmatrix} = \begin{bmatrix} c_1 e^0 \\ c_2 e^0 \\ c_3 e^0 \end{bmatrix} = \begin{bmatrix} c_1 \\ c_2 \\ c_3 \end{bmatrix} = \begin{bmatrix} 1 \\ 4 \\ -2 \end{bmatrix}$$

Substituting the values $c_1 = 1$, $c_2 = 4$, and $c_3 = -2$ determined by this equation in (10) yields

$$\mathbf{y} = \begin{bmatrix} y_1 \\ y_2 \\ y_3 \end{bmatrix} = \begin{bmatrix} e^{3t} \\ 4e^{-2t} \\ -2e^{5t} \end{bmatrix}$$

which is the solution of the initial value problem. Alternatively, this solution can be expressed as

$$y_1 = e^{3t}, \quad y_2 = 4e^{-2t}, \quad y_3 = -2e^{5t} \qquad \blacksquare$$

FUNDAMENTAL SOLUTIONS

The following theorem provides a basic result about solutions of $\mathbf{y}' = A\mathbf{y}$.

Theorem 8.10.1 *If* $\mathbf{y}_1, \mathbf{y}_2, \ldots, \mathbf{y}_k$ *are solutions of* $\mathbf{y}' = A\mathbf{y}$, *then*

$$\mathbf{y} = c_1\mathbf{y}_1 + c_2\mathbf{y}_2 + \cdots + c_k\mathbf{y}_k$$

is also a solution for every choice of the scalar constants $c_1, c_2, \ldots, c_k$.

Proof Differentiating corresponding components on both sides of the equation

$$\mathbf{y} = c_1\mathbf{y}_1 + c_2\mathbf{y}_2 + \cdots + c_k\mathbf{y}_k \tag{11}$$

yields

$$\mathbf{y}' = c_1\mathbf{y}_1' + c_2\mathbf{y}_2' + \cdots + c_k\mathbf{y}_k' \tag{12}$$

Moreover, since $\mathbf{y}_1, \mathbf{y}_2, \ldots, \mathbf{y}_k$ are solutions of the system $\mathbf{y}' = A\mathbf{y}$, we have

$$\mathbf{y}_1' = A\mathbf{y}_1, \quad \mathbf{y}_2' = A\mathbf{y}_2, \ldots, \quad \mathbf{y}_k' = A\mathbf{y}_k \tag{13}$$

Thus, it follows from (11), (12), and (13) that

$$\mathbf{y}' = c_1 A\mathbf{y}_1 + c_2 A\mathbf{y}_2 + \cdots + c_k A\mathbf{y}_k = A(c_1\mathbf{y}_1 + c_2\mathbf{y}_2 + \cdots + c_k\mathbf{y}_k) = A\mathbf{y}$$

which proves that (11) is a solution of $\mathbf{y}' = A\mathbf{y}$. $\qquad \blacksquare$

If the components of $\mathbf{y}_1, \mathbf{y}_2, \ldots, \mathbf{y}_k$ are continuous functions of t, and if $c_1, c_2, \ldots, c_k$ are scalar constants, then we call

$$\mathbf{y} = c_1\mathbf{y}_1 + c_2\mathbf{y}_2 + \cdots + c_k\mathbf{y}_k$$

a *linear combination* of $\mathbf{y}_1, \mathbf{y}_2, \ldots, \mathbf{y}_k$ with *coefficients* $c_1, c_2, \ldots, c_k$. To take the vector terminology still further, we will say that the set $S = \{\mathbf{y}_1, \mathbf{y}_2, \ldots, \mathbf{y}_k\}$ is *linearly independent* if no vector in S is a linear combination of other vectors in S.

Using this terminology, Theorem 8.10.1 states that every linear combination of solutions of $\mathbf{y}' = A\mathbf{y}$ is also a solution. In particular, this implies that the sum of two solutions is also a solution (closure under addition) and a constant scalar multiple of a solution is also a solution (closure under scalar multiplication). Thus, we will call the solution set of $\mathbf{y}' = A\mathbf{y}$ the *solution space* of the system because it has the properties of a subspace.

The proof of the following theorem can be found in most standard textbooks on differential equations.

> **Theorem 8.10.2**   *If A is an $n \times n$ matrix, then:*
>
> (a) *The equation $\mathbf{y}' = A\mathbf{y}$ has a set of n linearly independent solutions.*
>
> (b) *If $S = \{\mathbf{y}_1, \mathbf{y}_2, \ldots, \mathbf{y}_n\}$ is any set of n linearly independent solutions, then every solution can be expressed as a unique linear combination of the solutions in S.*

If A is an $n \times n$ matrix, and if $S = \{y_1, y_2, \ldots, y_n\}$ is any set of n linearly independent solutions of the system $\mathbf{y}' = A\mathbf{y}$, then we call S a *fundamental set of solutions* of the system, and we call

$$\mathbf{y} = c_1\mathbf{y}_1 + c_2\mathbf{y}_2 + \cdots + c_n\mathbf{y}_n$$

a *general solution* of the system.

EXAMPLE 3
A Fundamental
Set of Solutions

The system

$$\begin{bmatrix} y_1' \\ y_2' \\ y_3' \end{bmatrix} = \begin{bmatrix} 3 & 0 & 0 \\ 0 & -2 & 0 \\ 0 & 0 & 5 \end{bmatrix} \begin{bmatrix} y_1 \\ y_2 \\ y_3 \end{bmatrix}$$

was solved in Example 2. Find a fundamental set of solutions.

Solution   It follows from (10) that every solution $\mathbf{y}$ can be expressed as

$$\mathbf{y} = \begin{bmatrix} c_1 e^{3t} \\ c_2 e^{-2t} \\ c_3 e^{5t} \end{bmatrix} = c_1 \begin{bmatrix} e^{3t} \\ 0 \\ 0 \end{bmatrix} + c_2 \begin{bmatrix} e^{-2t} \\ 0 \\ 0 \end{bmatrix} + c_3 \begin{bmatrix} 0 \\ 0 \\ e^{5t} \end{bmatrix} \tag{14}$$

which is a linear combination of

$$\mathbf{y}_1 = \begin{bmatrix} e^{3t} \\ 0 \\ 0 \end{bmatrix}, \quad \mathbf{y}_2 = \begin{bmatrix} 0 \\ e^{-2t} \\ 0 \end{bmatrix}, \quad \mathbf{y}_3 = \begin{bmatrix} 0 \\ 0 \\ e^{5t} \end{bmatrix} \tag{15}$$

We leave it for you to verify that these vectors are linearly independent by showing that none of them can be expressed as a linear combination of the other two. Thus, (15) is a fundamental set of solutions and (14) is a general solution.   ■

**SOLUTIONS USING
EIGENVALUES AND
EIGENVECTORS**

In light of Theorem 8.10.2, we see that if A is an $n \times n$ matrix, then the general strategy for solving the system $\mathbf{y}' = A\mathbf{y}$ is to find n linearly independent solutions. Let us consider how we might look for solutions.

Since we know that the single equation $y' = ay$ has solutions of the form $y = ce^{at}$, where c is a constant, it seems plausible that there may exist nonzero solutions of

$$\mathbf{y}' = A\mathbf{y} \tag{16}$$

that have the form

$$\mathbf{y} = e^{\lambda t}\mathbf{x} \tag{17}$$

where λ is a constant scalar and $\mathbf{x}$ is a nonzero constant vector. (The exponent λ corresponds to the a in $y = ce^{at}$, and the vector $\mathbf{x}$ corresponds to coefficient c.) Of course, there may or may not be any such solutions, but let us at least explore that possibility by looking for conditions that λ and $\mathbf{x}$ would have to satisfy for $\mathbf{y}$ to be a solution. Differentiating both sides of (17) with respect to t, and keeping in mind that λ and $\mathbf{x}$ do not depend on t, yields

$$\mathbf{y}' = \lambda e^{\lambda t}\mathbf{x}$$

Now substituting this expression and (17) into (16) yields

$$\lambda e^{\lambda t}\mathbf{x} = A e^{\lambda t}\mathbf{x}$$

Canceling the nonzero factor $e^{\lambda t}$ from both sides yields

$$A\mathbf{x} = \lambda\mathbf{x}$$

which shows that if there is a solution of the form (17), then λ must be an eigenvalue of A and $\mathbf{x}$ must be a corresponding eigenvector. In the exercises we will ask you to confirm that every vector of form (17) is, in fact, a solution of (16). This will establish the following result.

Theorem 8.10.3 *If λ is an eigenvalue of A and $\mathbf{x}$ is a corresponding eigenvector, then $\mathbf{y} = e^{\lambda t}\mathbf{x}$ is a solution of the system $\mathbf{y}' = A\mathbf{y}$.*

EXAMPLE 4
Finding a
General
Solution

Use Theorem 8.10.3 to find a general solution of the system

$$\begin{aligned} y_1' &= y_1 + y_2 \\ y_2' &= 4y_1 - 2y_2 \end{aligned}$$

Solution The system can be written in matrix form as $\mathbf{y}' = A\mathbf{y}$, where

$$A = \begin{bmatrix} 1 & 1 \\ 4 & -2 \end{bmatrix}$$

The characteristic polynomial of A is

$$\det(\lambda I - A) = \begin{vmatrix} \lambda - 1 & -1 \\ -4 & \lambda + 2 \end{vmatrix} = \lambda^2 + \lambda - 6 = (\lambda - 2)(\lambda + 3)$$

so the eigenvalues of A are $\lambda = 2$ and $\lambda = -3$. We leave it for you to show that corresponding eigenvectors are

$$\lambda_1 = 2: \quad \mathbf{x}_1 = \begin{bmatrix} 1 \\ 1 \end{bmatrix} \quad \text{and} \quad \lambda_2 = -3: \quad \mathbf{x}_2 = \begin{bmatrix} -\frac{1}{4} \\ 1 \end{bmatrix}$$

so, from Theorem 8.10.3, this produces two solutions of the system; namely,

$$\mathbf{y}_1 = e^{2t}\begin{bmatrix} 1 \\ 1 \end{bmatrix} \quad \text{and} \quad \mathbf{y}_2 = e^{-3t}\begin{bmatrix} -\frac{1}{4} \\ 1 \end{bmatrix}$$

These solutions are linearly independent (since neither vector is a constant scalar multiple of the other), so a general solution of the system is

$$\mathbf{y} = c_1 e^{2t} \begin{bmatrix} 1 \\ 1 \end{bmatrix} + c_2 e^{-3t} \begin{bmatrix} -\frac{1}{4} \\ 1 \end{bmatrix}$$

As a check, you may want to confirm that the components

$$y_1 = c_1 e^{2t} - \tfrac{1}{4} c_2 e^{-3t} \quad \text{and} \quad y_2 = c_1 e^{2t} + c_2 e^{-3t}$$

satisfy the two given equations. ■

In the last example we were able to deduce that y_1 and y_2 were linearly independent by arguing that neither was a constant scalar multiple of the other. The following result eliminates the need to check linear independence.

> **Theorem 8.10.4** *If* $\mathbf{x}_1, \mathbf{x}_2, \dots, \mathbf{x}_k$ *are linearly independent eigenvectors of A, and if* $\lambda_1, \lambda_2, \dots, \lambda_k$ *are corresponding eigenvalues (not necessarily distinct), then*
>
> $$\mathbf{y}_1 = e^{\lambda_1 t} \mathbf{x}_1, \quad \mathbf{y}_2 = e^{\lambda_2 t} \mathbf{x}_2, \dots, \quad \mathbf{y}_k = e^{\lambda_k t} \mathbf{x}_k$$
>
> *are linearly independent solutions of* $\mathbf{y}' = A\mathbf{y}$

Proof We already know from Theorem 8.10.3 that $\mathbf{y}_1, \mathbf{y}_2, \dots, \mathbf{y}_k$ are solutions, so we need only establish the linear independence.

If we assume that

$$c_1 \mathbf{y}_1 + c_2 \mathbf{y}_2 + \cdots + c_k \mathbf{y}_k = \mathbf{0}$$

then it follows that

$$c_1 e^{\lambda_1 t} \mathbf{x}_1 + c_2 e^{\lambda_2 t} \mathbf{x}_2 + \cdots + c_k e^{\lambda_k t} \mathbf{x}_k = \mathbf{0}$$

Setting $t = 0$ in this equation, we obtain

$$c_1 \mathbf{x}_1 + c_2 \mathbf{x}_2 + \cdots + c_k \mathbf{x}_k = \mathbf{0}$$

But $\mathbf{x}_1, \mathbf{x}_2, \dots, \mathbf{x}_k$ are linearly independent, so we must have

$$c_1 = c_2 = \cdots = c_k = 0$$

which proves that $\mathbf{y}_1, \mathbf{y}_2, \dots, \mathbf{y}_k$ are linearly independent. ■

Since we know that an $n \times n$ matrix is diagonalizable if and only if it has n linearly independent eigenvectors (Theorem 8.2.6), the following result follows from Theorem 8.10.4.

> **Theorem 8.10.5** *If A is a diagonalizable $n \times n$ matrix, then a general solution of the system* $\mathbf{y}' = A\mathbf{y}$ *is*
>
> $$\mathbf{y} = c_1 e^{\lambda_1 t} \mathbf{x}_1 + c_2 e^{\lambda_2 t} \mathbf{x}_2 + \cdots + c_n e^{\lambda_n t} \mathbf{x}_n \tag{18}$$
>
> *where $\mathbf{x}_1, \mathbf{x}_2, \dots, \mathbf{x}_n$ are any n linearly independent vectors and $\lambda_1, \lambda_2, \dots, \lambda_n$ are the corresponding eigenvalues.*

Proof Use part (*b*) of Theorem 8.10.2 and Theorem 8.10.4. ■

EXAMPLE 5
A General
Solution Using
Eigenvalues
and
Eigenvectors

Find a general solution of the system

$$y_1' = -2y_3$$
$$y_2' = y_1 + 2y_2 + y_3$$
$$y_3' = y_1 + 3y_3$$

Solution The system can be written in matrix form as

$$\mathbf{y}' = \begin{bmatrix} 0 & 0 & -2 \\ 1 & 2 & 1 \\ 1 & 0 & 3 \end{bmatrix} \mathbf{y}$$

We showed in Examples 3 and 4 of Section 8.2 that the coefficient matrix for this system is diagonalizable by showing that it has an eigenvalue $\lambda = 1$ with geometric multiplicity 1 and an eigenvalue $\lambda = 2$ with geometric multiplicity 2. We also showed that bases for the corresponding eigenspaces are

$$\lambda = 1: \quad \begin{bmatrix} -2 \\ 1 \\ 1 \end{bmatrix} \quad \text{and} \quad \lambda = 2: \quad \begin{bmatrix} -1 \\ 0 \\ 1 \end{bmatrix}, \quad \begin{bmatrix} 0 \\ 1 \\ 0 \end{bmatrix}$$

Thus, it follows from Theorem 8.10.5 that a general solution of the system is

$$\mathbf{y} = c_1 \begin{bmatrix} -2 \\ 1 \\ 1 \end{bmatrix} e^t + c_2 \begin{bmatrix} -1 \\ 0 \\ 1 \end{bmatrix} e^{2t} + c_3 \begin{bmatrix} 0 \\ 1 \\ 0 \end{bmatrix} e^{2t}$$

(Here we have followed a common practice of placing the exponential functions after the column vectors rather than in front as is more common with scalars.) As a check, you may want to verify that the components y_1, y_2, and y_3 of $\mathbf{y}$ satisfy the three given equations. ∎

EXAMPLE 6
A Mixing Problem

Suppose that two tanks are connected as shown in Figure 8.10.2. At time $t = 0$, tank 1 contains 80 liters of water in which 7 kg of salt has been dissolved, and tank 2 contains 80 liters of water in which 10 kg of salt has been dissolved. As indicated in the figure, pure water is pumped into tank 1 at the rate of 30 L/min, the saline mixtures are exchanged between the two tanks at the rates shown, and the mixture in tank 2 drains out at the rate of 30 L/min. Find the amount of salt in each tank at time t.

Solution Observe first that the amount of liquid in each tank remains constant because the rate at which liquid enters each tank is the same as the rate at which it leaves. Thus each tank always contains 80 L of liquid.

Now let

$$y_1(t) = \text{amount of salt in tank 1 at time } t$$
$$y_2(t) = \text{amount of salt in tank 2 at time } t$$

and let us consider the rates of change y_1' and y_2'. Each rate of change can be viewed as the rate at which salt enters the tank minus the rate at which it leaves the tank. For example,

$$\text{rate at which salt enters tank 1} = \left(10 \frac{\text{L}}{\text{min}} \right) \left(\frac{y_2(t)}{80} \frac{\text{kg}}{\text{L}} \right) = \frac{y_2(t)}{8} \frac{\text{kg}}{\text{min}}$$

$$\text{rate at which salt leaves tank 1} = \left(40 \frac{\text{L}}{\text{min}} \right) \left(\frac{y_1(t)}{80} \frac{\text{kg}}{\text{L}} \right) = \frac{y_1(t)}{2} \frac{\text{kg}}{\text{min}}$$

so the rate of change of $y_1(t)$ is

$$y_1'(t) = \text{rate in} - \text{rate out} = \frac{y_2(t)}{8} - \frac{y_1(t)}{2} \tag{19}$$

Similarly,

$$\text{rate at which salt enters tank 2} = \left(40 \frac{\text{L}}{\text{min}} \right) \left(\frac{y_1(t)}{80} \frac{\text{kg}}{\text{L}} \right) = \frac{y_1(t)}{2} \frac{\text{kg}}{\text{min}}$$

$$\text{rate at which salt leaves tank 2} = \left((10 + 30) \frac{\text{L}}{\text{min}} \right) \left(\frac{y_2(t)}{80} \frac{\text{kg}}{\text{L}} \right) = \frac{y_2(t)}{2} \frac{\text{kg}}{\text{min}}$$

Water   Mixture
30 L/min   10 L/min

Tank 1   Tank 2

80 L   80 L

Mixture   Mixture
40 L/min   30 L/min

Figure 8.10.2

so the rate of change of $y_2(t)$ is

$$y_2'(t) = \text{rate in} - \text{rate out} = \frac{y_1(t)}{2} - \frac{y_2(t)}{2} \tag{20}$$

Since the statement of the problem provides the initial conditions

$$y_1(0) = 7 \quad \text{and} \quad y_2(0) = 10 \tag{21}$$

it follows from (19), (20), and (21) that the amount of salt in each tank at time t can be obtained by solving the initial value problem

$$\begin{aligned} y_1' &= -\tfrac{1}{2}y_1 + \tfrac{1}{8}y_2 \\ y_2' &= \tfrac{1}{2}y_1 - \tfrac{1}{2}y_2 \end{aligned} \qquad y_1(0) = 7, \quad y_2(0) = 10$$

The coefficient matrix

$$A = \begin{bmatrix} -\tfrac{1}{2} & \tfrac{1}{8} \\ \tfrac{1}{2} & -\tfrac{1}{2} \end{bmatrix}$$

has eigenvalues $\lambda_1 = -\tfrac{3}{4}$ and $\lambda_2 = -\tfrac{1}{4}$ with corresponding eigenvectors

$$\mathbf{x}_1 = \begin{bmatrix} -1 \\ 2 \end{bmatrix} \quad \text{and} \quad \mathbf{x}_2 = \begin{bmatrix} 1 \\ 2 \end{bmatrix}$$

(verify), so it follows from Theorem 8.10.5 that the general solution of system is

$$\mathbf{y} = c_1 e^{-3t/4}\mathbf{x}_1 + c_2 e^{-t/4}\mathbf{x}_2$$

or equivalently,

$$\begin{aligned} y_1 &= -c_1 e^{-3t/4} + c_2 e^{-t/4} \\ y_2 &= 2c_1 e^{-3t/4} + 2c_2 e^{-t/4} \end{aligned} \tag{22}$$

Substituting the initial conditions $y_1(0) = 7$ and $y_2(0) = 10$ yields the linear system

$$\begin{aligned} -c_1 + c_2 &= 7 \\ 2c_1 + 2c_2 &= 10 \end{aligned}$$

whose solution is $c_1 = -1$, $c_2 = 6$. Now substituting these values into (22) yields the solution of the initial value problem

$$\begin{aligned} y_1 &= e^{-3t/4} + 6e^{-t/4} \\ y_2 &= -2e^{-3t/4} + 12e^{-t/4} \end{aligned}$$

Figure 8.10.3 shows graphs of the two solution curves. ■

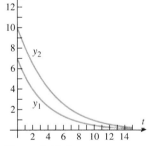

Figure 8.10.3

CONCEPT PROBLEM Give a physical explanation of why the curves in Figure 8.10.3 approach the t-axis as $t \to +\infty$.

EXPONENTIAL FORM OF A SOLUTION

If A is an $n \times n$ matrix, then it follows from (18) that the solution of the initial value problem

$$\mathbf{y}' = A\mathbf{y}, \quad \mathbf{y}(0) = \mathbf{y}_0 \tag{23}$$

can be expressed as

$$\mathbf{y} = c_1 e^{\lambda_1 t}\mathbf{x}_1 + c_2 e^{\lambda_2 t}\mathbf{x}_2 + \cdots + c_n e^{\lambda_n t}\mathbf{x}_n = [\mathbf{x}_1 \ \ \mathbf{x}_2 \ \ \cdots \ \ \mathbf{x}_n] \begin{bmatrix} e^{\lambda_1 t} & 0 & \cdots & 0 \\ 0 & e^{\lambda_2 t} & \cdots & 0 \\ \vdots & \vdots & \ddots & \vdots \\ 0 & 0 & \cdots & e^{\lambda_n t} \end{bmatrix} \begin{bmatrix} c_1 \\ c_2 \\ \vdots \\ c_n \end{bmatrix} \tag{24}$$

where the constants $c_1, c_2, \ldots, c_n$ are chosen to satisfy the initial condition. Since the column vectors of the matrix

$$P = [\mathbf{x}_1 \quad \mathbf{x}_2 \quad \cdots \quad \mathbf{x}_n] \tag{25}$$

form a set of linearly independent eigenvectors of A, this matrix is invertible and diagonalizes A. Moreover, it follows on setting $t = 0$ in (24) that the initial condition is satisfied if

$$\mathbf{y}_0 = [\mathbf{x}_1 \quad \mathbf{x}_2 \quad \cdots \quad \mathbf{x}_n] \begin{bmatrix} c_1 \\ c_2 \\ \vdots \\ c_n \end{bmatrix} = P \begin{bmatrix} c_1 \\ c_2 \\ \vdots \\ c_n \end{bmatrix} \quad \text{or equivalently,} \quad \begin{bmatrix} c_1 \\ c_2 \\ \vdots \\ c_n \end{bmatrix} = P^{-1} \mathbf{y}_0 \tag{26}$$

If we now substitute (26) into (24), we obtain the following expression for the solution of the initial value problem:

$$\mathbf{y} = P \begin{bmatrix} e^{\lambda_1 t} & 0 & \cdots & 0 \\ 0 & e^{\lambda_2 t} & \cdots & 0 \\ \vdots & \vdots & \ddots & \vdots \\ 0 & 0 & \cdots & e^{\lambda_n t} \end{bmatrix} P^{-1} \mathbf{y}_0$$

But Formula (21) of Section 8.3 with $f(x) = e^{tx}$ allows us to rewrite this as $\mathbf{y} = e^{tA}\mathbf{y}_0$, so we have established the following result.

Theorem 8.10.6 *If A is a diagonalizable matrix, then the solution of the initial value problem*

$$\mathbf{y}' = A\mathbf{y}, \quad \mathbf{y}(0) = \mathbf{y}_0$$

can be expressed as

$$\mathbf{y} = e^{tA}\mathbf{y}_0 \tag{27}$$

EXAMPLE 7
Solution Using
the Matrix e^{tA}

Use Theorem 8.10.6 to solve the initial value problem

$$\begin{aligned} y_1' &= -2y_3 \\ y_2' &= y_1 + 2y_2 + y_3 \qquad y_1(0) = 2, \quad y_2(0) = -1, \quad y_3(0) = 0 \\ y_3' &= y_1 + 3y_3 \end{aligned}$$

Solution The coefficient matrix and the vector form of the initial conditions are

$$A = \begin{bmatrix} 0 & 0 & -2 \\ 1 & 2 & 1 \\ 1 & 0 & 3 \end{bmatrix}, \quad \mathbf{y}(0) = \begin{bmatrix} 2 \\ -1 \\ 0 \end{bmatrix}$$

The matrix e^{tA} was computed in Example 5 of Section 8.3. Using that result and (27) yields the solution

$$\mathbf{y} = e^{tA}\mathbf{y}_0 = \begin{bmatrix} 2e^t - e^{2t} & 0 & 2e^t - 2e^{2t} \\ e^{2t} - e^t & e^{2t} & e^{2t} - e^t \\ e^{2t} - e^t & 0 & 2e^{2t} - e^t \end{bmatrix} \begin{bmatrix} 2 \\ -1 \\ 0 \end{bmatrix} = \begin{bmatrix} 4e^t - 2e^{2t} \\ e^{2t} - 2e^t \\ 2e^{2t} - 2e^t \end{bmatrix}$$

or equivalently,

$$y_1 = 4e^t - 2e^{2t}, \quad y_2 = e^{2t} - 2e^t, \quad y_3 = 2e^{2t} - 2e^t \qquad \blacksquare$$

REMARK Formula (27) is an important analytical tool because it provides a formula for the solution of the initial value problem. However, this formula is rarely used in numerical computations because there are better methods available.

THE CASE WHERE *A* IS NOT DIAGONALIZABLE

Although we established Theorem 8.10.6 for diagonalizable matrices only, it can be proved that the formula

$$\mathbf{y} = e^{tA}\mathbf{y}_0 \tag{28}$$

provides the solution of the initial value problem $\mathbf{y}' = A\mathbf{y}$, $\mathbf{y}(0) = \mathbf{y}_0$ in all cases. However, if A is not diagonalizable, then the matrix e^{tA} in this formula cannot be obtained from Formula (21) of Section 8.3; rather, it must somehow be obtained or approximated using the infinite series

$$e^{tA} = I + tA + \frac{t^2 A^2}{2!} + \frac{t^3 A^3}{3!} + \cdots + \frac{t^k A^k}{k!} + \cdots \tag{29}$$

Although this can be difficult, there are certain kinds of engineering problems in which A has properties that simplify the work. Here are some examples.

EXAMPLE 8
A Nondiagonal-izable Initial Value Problem

Solve the initial value problem

$$y_1' = 0$$
$$y_2' = y_1 \qquad y_1(0) = 2, \quad y_2(0) = 1, \quad y_3(0) = 3$$
$$y_3' = y_1 + y_2$$

Solution The coefficient matrix and the vector form of the initial conditions are

$$A = \begin{bmatrix} 0 & 0 & 0 \\ 1 & 0 & 0 \\ 1 & 1 & 0 \end{bmatrix}, \quad \mathbf{y}(0) = \begin{bmatrix} 2 \\ 1 \\ 3 \end{bmatrix}$$

We leave it for you to show that $A^3 = 0$, from which it follows that (29) reduces to the finite sum

$$e^{tA} = I + tA + \frac{t^2 A^2}{2} = \begin{bmatrix} 1 & 0 & 0 \\ t & 1 & 0 \\ t + \frac{1}{2}t^2 & t & 1 \end{bmatrix}$$

(verify). Thus, it follows from (28) that the solution of the initial value problem is

$$\mathbf{y} = \begin{bmatrix} 1 & 0 & 0 \\ t & 1 & 0 \\ t + \frac{1}{2}t^2 & t & 1 \end{bmatrix} \begin{bmatrix} 2 \\ 1 \\ 3 \end{bmatrix} = \begin{bmatrix} 2 \\ 1 + 2t \\ 3 + 3t + t^2 \end{bmatrix}$$

or equivalently,

$$y_1 = 2, \quad y_2 = 1 + 2t, \quad y_3 = 3 + 3t + t^2 \qquad \blacksquare$$

EXAMPLE 9
A Nondiagonal-izable Initial Value Problem

Solve the initial value problem

$$y_1' = y_2$$
$$y_2' = -y_1 \qquad y_1(0) = 2, \quad y_2(0) = 1$$

Solution The coefficient matrix and the vector form of the initial conditions are

$$A = \begin{bmatrix} 0 & 1 \\ -1 & 0 \end{bmatrix}, \quad \mathbf{y}(0) = \begin{bmatrix} 2 \\ 1 \end{bmatrix} \tag{30}$$

We leave it for you to show that $A^2 = -I$, from which it follows that

$$A^4 = I, \ A^6 = -I, \ A^8 = I, \ A^{10} = -I, \ldots$$
$$A^3 = -A, \ A^5 = A, \ A^7 = -A, \ A^9 = A, \ldots$$

Thus, if we make the substitutions $A^{2k} = (-1)^k I$ and $A^{2k+1} = (-1)^k A$ (for $k = 1, 2, \ldots$) in

(29) and use the Maclaurin series for $\sin t$ and $\cos t$, we obtain

$$e^{tA} = I\left(1 - \frac{t^2}{2!} + \frac{t^4}{4!} - \frac{t^6}{6!} + \cdots\right) + A\left(t - \frac{t^3}{3!} + \frac{t^5}{5!} - \frac{t^7}{7!} + \cdots\right)$$

$$= I\cos t + A\sin t = \begin{bmatrix} \cos t & 0 \\ 0 & \cos t \end{bmatrix} + \begin{bmatrix} 0 & \sin t \\ -\sin t & 0 \end{bmatrix} = \begin{bmatrix} \cos t & \sin t \\ -\sin t & \cos t \end{bmatrix}$$

Thus, it follows from (28) that the solution of the initial value problem is

$$\mathbf{y} = \begin{bmatrix} \cos t & \sin t \\ -\sin t & \cos t \end{bmatrix}\begin{bmatrix} 2 \\ 1 \end{bmatrix} = \begin{bmatrix} 2\cos t + \sin t \\ \cos t - 2\sin t \end{bmatrix}$$

or equivalently,

$$y_1 = 2\cos t + \sin t, \quad y_2 = \cos t - 2\sin t \qquad \blacksquare$$

Exercise Set 8.10

In Exercises 1 and 2, solve the initial value problem.

1. $y' = -3y, \; y(0) = 5$ **2.** $y' = 5y, \; y(0) = -3$

In Exercises 3 and 4, solve the system, list a fundamental set of solutions, and find the solution that satisfies the initial conditions.

3. $\begin{bmatrix} y_1' \\ y_2' \\ y_3' \end{bmatrix} = \begin{bmatrix} 1 & 0 & 0 \\ 0 & 4 & 0 \\ 0 & 0 & -2 \end{bmatrix}\begin{bmatrix} y_1 \\ y_2 \\ y_3 \end{bmatrix};\; \begin{bmatrix} y_1(0) \\ y_2(0) \\ y_3(0) \end{bmatrix} = \begin{bmatrix} 1 \\ 1 \\ 1 \end{bmatrix}$

4. $\begin{bmatrix} y_1' \\ y_2' \\ y_3' \end{bmatrix} = \begin{bmatrix} 2 & 0 & 0 \\ 0 & 2 & 0 \\ 0 & 0 & 5 \end{bmatrix}\begin{bmatrix} y_1 \\ y_2 \\ y_3 \end{bmatrix};\; \begin{bmatrix} y_1(0) \\ y_2(0) \\ y_3(0) \end{bmatrix} = \begin{bmatrix} 1 \\ 2 \\ 0 \end{bmatrix}$

In Exercises 5–8, use eigenvalues and eigenvectors to find a general solution of the system, and then find the solution that satisfies the initial conditions.

5. $\begin{aligned} y_1' &= y_1 + 4y_2 \\ y_2' &= 2y_1 + 3y_2 \end{aligned}$ $y_1(0) = 0, \quad y_2(0) = 0$

6. $\begin{aligned} y_1' &= y_1 + 3y_2 \\ y_2' &= 4y_1 + 5y_2 \end{aligned}$ $y_1(0) = 2, \quad y_2(0) = 1$

7. $\begin{bmatrix} y_1' \\ y_2' \\ y_3' \end{bmatrix} = \begin{bmatrix} 4 & 0 & 1 \\ -2 & 1 & 0 \\ -2 & 0 & 1 \end{bmatrix}\begin{bmatrix} y_1 \\ y_2 \\ y_3 \end{bmatrix};\; \begin{bmatrix} y_1(0) \\ y_2(0) \\ y_3(0) \end{bmatrix} = \begin{bmatrix} -1 \\ 1 \\ 0 \end{bmatrix}$

8. $\begin{bmatrix} y_1' \\ y_2' \\ y_3' \end{bmatrix} = \begin{bmatrix} 4 & 2 & 2 \\ 2 & 4 & 2 \\ 2 & 2 & 4 \end{bmatrix}\begin{bmatrix} y_1 \\ y_2 \\ y_3 \end{bmatrix};\; \begin{bmatrix} y_1(0) \\ y_2(0) \\ y_3(0) \end{bmatrix} = \begin{bmatrix} 0 \\ 2 \\ -2 \end{bmatrix}$

9. Suppose that two tanks are connected as shown in the accompanying figure. At time $t = 0$, tank 1 contains 120 L of water in which 30 kg of salt has been dissolved, and tank 2 contains 120 L of water in which 40 kg of salt has been

dissolved. As indicated in the figure, pure water is pumped into tank 1 at the rate of 80 L/min, the saline mixtures are exchanged between the two tanks at the rates shown, and the mixture in tank 2 drains out at the rate of 80 L/min. Find the amount of salt in the two tanks at time t.

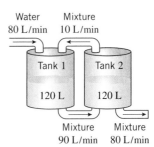

Water   Mixture
80 L/min   10 L/min

Tank 1   Tank 2

120 L   120 L

Mixture   Mixture
90 L/min   80 L/min

Figure Ex-9

10. Suppose that two tanks are connected as shown in the accompanying figure. At time $t = 0$, tank 1 contains 30 L of water in which 2 kg of salt has been dissolved, and tank 2 contains 30 L of water in which 5 kg of salt has been dissolved. Pure water is pumped into tank 1 at the rate of 20 L/min, the saline mixtures are exchanged between the two tanks at the rates shown, the mixture in tank 2 drains out at the rate of 15 L/min, and the mixture in tank 1 drains out at the rate of 5 L/min. Find the amount of salt in the two tanks at time t.

Water   Mixture
20 L/min   6 L/min

Tank 1   Tank 2

30 L   30 L

Mixture   Mixture   Mixture
5 L/min   21 L/min   15 L/min

Figure Ex-10

In Exercises 11–14, use Theorem 8.10.6 and the method of Example 7 to solve the initial value problem.

11. $\begin{bmatrix} y_1' \\ y_2' \end{bmatrix} = \begin{bmatrix} 4 & -2 \\ 1 & 1 \end{bmatrix} \begin{bmatrix} y_1 \\ y_2 \end{bmatrix}$; $\begin{bmatrix} y_1(0) \\ y_2(0) \end{bmatrix} = \begin{bmatrix} 3 \\ -4 \end{bmatrix}$

12. $\begin{bmatrix} y_1' \\ y_2' \end{bmatrix} = \begin{bmatrix} 1 & 2 \\ -1 & 4 \end{bmatrix} \begin{bmatrix} y_1 \\ y_2 \end{bmatrix}$; $\begin{bmatrix} y_1(0) \\ y_2(0) \end{bmatrix} = \begin{bmatrix} 0 \\ 1 \end{bmatrix}$

13. $\begin{bmatrix} y_1' \\ y_2' \\ y_3' \end{bmatrix} = \begin{bmatrix} 1 & -1 & -1 \\ 1 & 3 & 1 \\ -3 & 1 & -1 \end{bmatrix} \begin{bmatrix} y_1 \\ y_2 \\ y_3 \end{bmatrix}$; $\begin{bmatrix} y_1(0) \\ y_2(0) \\ y_3(0) \end{bmatrix} = \begin{bmatrix} 2 \\ 0 \\ -1 \end{bmatrix}$

14. $\begin{bmatrix} y_1' \\ y_2' \\ y_3' \end{bmatrix} = \begin{bmatrix} 0 & -1 & 1 \\ -1 & 0 & 1 \\ 1 & 1 & 1 \end{bmatrix} \begin{bmatrix} y_1 \\ y_2 \\ y_3 \end{bmatrix}$; $\begin{bmatrix} y_1(0) \\ y_2(0) \\ y_3(0) \end{bmatrix} = \begin{bmatrix} 1 \\ 2 \\ -2 \end{bmatrix}$

In Exercises 15 and 16, the system has a nilpotent coefficient matrix. Use the method of Example 8 to solve the initial value problem.

15. $\begin{bmatrix} y_1' \\ y_2' \\ y_3' \end{bmatrix} = \begin{bmatrix} 0 & 1 & 2 \\ 0 & 0 & -1 \\ 0 & 0 & 0 \end{bmatrix} \begin{bmatrix} y_1 \\ y_2 \\ y_3 \end{bmatrix}$; $\begin{bmatrix} y_1(0) \\ y_2(0) \\ y_3(0) \end{bmatrix} = \begin{bmatrix} -1 \\ 4 \\ 2 \end{bmatrix}$

16. $\begin{bmatrix} y_1' \\ y_2' \\ y_3' \end{bmatrix} = \begin{bmatrix} 1 & 5 & -2 \\ 1 & 2 & -1 \\ 3 & 6 & -3 \end{bmatrix} \begin{bmatrix} y_1 \\ y_2 \\ y_3 \end{bmatrix}$; $\begin{bmatrix} y_1(0) \\ y_2(0) \\ y_3(0) \end{bmatrix} = \begin{bmatrix} 1 \\ 2 \\ -1 \end{bmatrix}$

17. Use the method of Example 9 to solve the initial value problem

$$y_1' = -y_2$$
$$y_2' = y_1$$
$$y_1(0) = -1, \quad y_2(0) = 2$$

18. Use the method of Example 9 and the Maclaurin series for $\sinh x$ and $\cosh x$ to solve the initial value problem

$$y_1' = y_2$$
$$y_2' = y_1$$
$$y_1(0) = 3, \quad y_2(0) = -1$$

Working with Proofs

P1. Prove that every solution of $y' = ay$ has the form $y = ce^{at}$. [*Hint:* Let $y = f(t)$ be a solution of the equation and show that $f(t)e^{-at}$ is constant.]

Technology Exercises

T1. (a) Find a general solution of the system

$$y_1' = 3y_1 + 2y_2 + 2y_3$$
$$y_2' = y_1 + 4y_2 + y_3$$
$$y_3' = -2y_1 - 4y_2 - y_3$$

19. This problem illustrates that it is sometimes possible to solve a single differential equation by expressing it as a system.
  (a) Make the substitutions $y_1 = y$ and $y_2 = y'$ in the differential equation $y'' - 6y' - 6y = 0$ to show that the solutions of the equation satisfy the system

$$y_1' = y_2$$
$$y_2' = 6y_1 + 6y_2$$

  (b) Solve the system, and show that its solutions satisfy the original differential equation.

20. Use the idea in Exercise 19 to solve the differential equation $y''' - 6y'' + 11y' - 6y = 0$ by expressing it as a system with three equations.

21. Suppose that two tanks are connected as shown in the accompanying figure. At time $t = 0$, tank 1 contains 60 L of water in which 10 kg of salt has been dissolved, and tank 2 contains 60 L of water in which 7 kg of salt has been dissolved. Pure water is pumped into tank 1 at the rate of 30 L/min, the saline mixture flows from tank 1 to tank 2 at the rate of 10 L/min, the mixture in tank 2 drains out at the rate of 10 L/min, and the mixture in tank 1 drains out at the rate of 20 L/min. Find the amount of salt in the two tanks at time t. [*Hint:* Use exponential methods to solve the resulting system.]

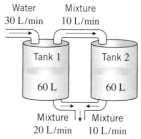

Figure Ex-21

22. Show that if λ is an eigenvalue of the square matrix A, and if $\mathbf{x}$ is a corresponding eigenvector, then $\mathbf{y} = e^{\lambda t}\mathbf{x}$ is a solution of $\mathbf{y}' = A\mathbf{y}$.

by computing appropriate eigenvalues and eigenvectors.
  (b) Find the solution that satisfies the initial conditions $y_1(0) = 0$, $y_2(0) = 1$, $y_3(0) = -3$.

T2. The electrical circuit in the accompanying figure, called a *parallel LRC circuit*, contains a resistor with resistance R ohms (Ω), an inductor with inductance L henries (H), and a capacitor with capacitance C farads (F). It is shown in electrical circuit theory that the current I in amperes (A) through the inductor and the voltage drop V in volts (V) across the capacitor satisfy the system of differential equations

$$\frac{dI}{dt} = \frac{V}{L}$$

$$\frac{dV}{dt} = -\frac{I}{C} - \frac{V}{RC}$$

where the derivatives are with respect to the time t. Find I and V as functions of t if $L = 0.5$ H, $C = 0.2$ F, $R = 2\,\Omega$, and the initial values of V and I are $V(0) = 1$ V and $I(0) = 2$ A.

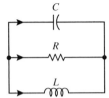

Figure Ex-T2

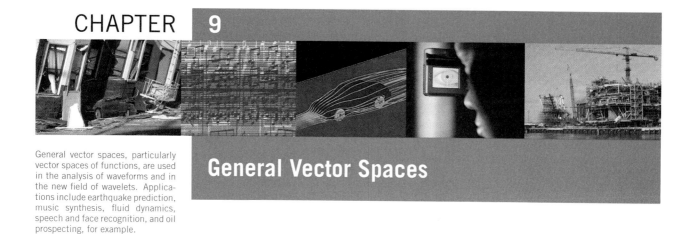

General vector spaces, particularly vector spaces of functions, are used in the analysis of waveforms and in the new field of wavelets. Applications include earthquake prediction, music synthesis, fluid dynamics, speech and face recognition, and oil prospecting, for example.

General Vector Spaces

Section 9.1 Vector Space Axioms

The concept of a vector has been generalized several times in this text. We began by viewing vectors as quantities with length and direction, then as arrows in two or three dimensions, then as ordered pairs or ordered triples of real numbers, and then as n-tuples of real numbers. We also hinted at various times that other useful generalizations of the vector concept exist. In this section we discuss such generalizations. Calculus will be needed in some of the examples.

VECTOR SPACE AXIOMS

Our primary goal in this section is to specify requirements that will allow a general set of objects with two operations to be viewed as a vector space. The basic idea is to impose suitable restrictions on the operations to ensure that the familiar theorems about vectors continue to hold. The idea is not complicated—if you trace the ancestry of the algebraic theorems about R^n and its subspaces (excluding those that involve dot product), you will find that most follow from a small number of core properties:

- All subspaces are closed under scalar multiplication and addition.
- All subspaces contain the vector **0** and the vectors in the subspace satisfy the algebraic properties of Theorem 1.1.5.

Thus, if we have a set of objects with two operations (addition and multiplication by scalars), and if those objects with their operations have these core properties, then those objects with their operations will of necessity also satisfy the familiar algebraic theorems about vectors and hence can reasonably be called vectors. Accordingly, we make the following definition.

Definition 9.1.1 Let V be a nonempty set of objects on which operations of addition and scalar multiplication are defined. By **addition** we mean a rule for associating with each pair of objects **u** and **v** in V a unique object **u** + **v** that we regard to be the sum of **u** and **v**, and by **scalar multiplication** we mean a rule for associating with each scalar k and each object **u** in V a unique object k**u** that we regard to be the scalar product of **u** by k. The set V with these operations will be called a **vector space** and the objects in V will be called **vectors** if the following properties hold for all **u**, **v**, and **w** in V and all scalars k and l.

1. V is **closed under addition**; that is, if **u** and **v** are in V, then **u** + **v** is in V.
2. $\mathbf{u} + \mathbf{v} = \mathbf{v} + \mathbf{u}$
3. $(\mathbf{u} + \mathbf{v}) + \mathbf{w} = \mathbf{u} + (\mathbf{v} + \mathbf{w})$

4. **u** contains an object **0** (which we call a *zero vector*) that behaves like an additive zero in the sense that $\mathbf{u} + \mathbf{0} = \mathbf{u}$ for every **u** in V.

5. For each object **u** in V, there is an object $-\mathbf{u}$ in V (which we call a *negative* of **u**) such that $\mathbf{u} + (-\mathbf{u}) = \mathbf{0}$.

6. V is *closed under scalar multiplication*; that is, if **u** is in V and k is a scalar, then $k\mathbf{u}$ is in V.

7. $k(\mathbf{u} + \mathbf{v}) = k\mathbf{u} + k\mathbf{v}$

8. $(k + l)\mathbf{u} = k\mathbf{u} + l\mathbf{u}$

9. $k(l\mathbf{u}) = (kl)\mathbf{u}$

10. $1\mathbf{u} = \mathbf{u}$

The 10 properties in this definition are called the *vector space axioms*, and a set V with two operations that satisfy these 10 axioms is called a *vector space*.

In the exercises we will help you to use these axioms to prove that the zero vector in a vector space V is unique and that each vector in V has a unique negative. Thus, we will be entitled to talk about *the* zero vector and *the* negative of a vector. If the scalars for a vector space V are required to be real numbers, then we call V a *real vector space*, and if they are allowed to be complex, then we call V a *complex vector space*. In this section we will consider only real vector spaces.

EXAMPLE 1
R^n Is a Vector Space

Since the vector space axioms are based on the closure properties and Theorem 1.1.5 for vectors in R^n, these properties automatically hold for vectors in R^n with the standard operations of addition and scalar multiplication. Thus, R^n is an example of a vector space in the axiomatic sense. ∎

EXAMPLE 2
The Sequence Space R^∞

One natural way to generalize R^n is to allow vectors to have infinitely many components by considering objects of the form

$$\mathbf{v} = (v_1, v_2, \ldots, v_n, \ldots)$$

in which $v_1, v_2, \ldots, v_n, \ldots$ is an infinite sequence of real numbers. We denote this set of infinite sequences by the symbol R^∞. As with vectors in R^n, we regard two infinite sequences to be equal if their corresponding components are equal, and we define operations of scalar multiplication and addition componentwise; that is,

$$k\mathbf{v} = (kv_1, kv_2, \ldots, kv_n, \ldots)$$
$$\mathbf{v} + \mathbf{w} = (v_1, v_2, \ldots, v_n, \ldots) + (w_1, w_2, \ldots, w_n, \ldots)$$
$$= (v_1 + w_1, v_2 + w_2, \ldots, v_n + w_n, \ldots)$$

We leave it as an exercise to confirm that R^∞ with these operations satisfies the 10 vector space axioms. ∎

The proof of the following generalization of Theorem 1.1.6 illustrates how the vector space axioms can be used to extend results from R^n to general vector spaces.

Linear Algebra in History

The Italian mathematician Giuseppe Peano (p. 332) was the first person to state formally the axioms for a vector space. Those axioms, which appeared in a book entitled *Geometrical Calculus*, published in 1888, were not appreciated by many of his contemporaries, but in the end they proved to be a remarkable landmark achievement. Peano's book also formally defined the concept of dimension for general vector spaces.

Theorem 9.1.2   *If* **v** *is a vector in a vector space V, and if k is a scalar, then:*

(a) $0\mathbf{v} = \mathbf{0}$

(b) $k\mathbf{0} = \mathbf{0}$

(c) $(-1)\mathbf{v} = -\mathbf{v}$

Proof (*a*) The scalar product $0\mathbf{v}$ is a vector in V by Axiom 6 and hence has a negative $-0\mathbf{v}$ in V by Axiom 5. Thus,

$$
\begin{aligned}
0\mathbf{v} &= 0\mathbf{v} + \mathbf{0} && \textbf{[Axiom 4]} \\
&= 0\mathbf{v} + [0\mathbf{v} + (-0\mathbf{v})] && \textbf{[Axiom 5]} \\
&= [0\mathbf{v} + 0\mathbf{v}] + (-0\mathbf{v}) && \textbf{[Axiom 3]} \\
&= [0 + 0]\mathbf{v} + (-0\mathbf{v}) && \textbf{[Axiom 8]} \\
&= 0\mathbf{v} + (-0\mathbf{v}) && \textbf{[Property of real numbers]} \\
&= \mathbf{0} && \textbf{[Axiom 5]}
\end{aligned}
$$

Proof (*b*) The scalar product $k\mathbf{0}$ is a vector in V by Axiom 6 and hence has a negative $-k\mathbf{0}$ in V by Axiom 5. Thus,

$$
\begin{aligned}
k\mathbf{0} &= k\mathbf{0} + \mathbf{0} && \textbf{[Axiom 4]} \\
&= k\mathbf{0} + [k\mathbf{0} + (-k\mathbf{0})] && \textbf{[Axiom 5]} \\
&= [k\mathbf{0} + k\mathbf{0}] + (-k\mathbf{0}) && \textbf{[Axiom 3]} \\
&= k[\mathbf{0} + \mathbf{0}] + (-k\mathbf{0}) && \textbf{[Axiom 7]} \\
&= k\mathbf{0} + (-k\mathbf{0}) && \textbf{[Axiom 4]} \\
&= \mathbf{0} && \textbf{[Axiom 5]}
\end{aligned}
$$

Proof (*c*) To prove that $(-1)\mathbf{v}$ is the negative of $\mathbf{v}$, we must show that $\mathbf{v} + (-1)\mathbf{v} = \mathbf{0}$. The argument is as follows:

$$
\begin{aligned}
\mathbf{v} + (-1)\mathbf{v} &= 1\mathbf{v} + (-1)\mathbf{v} && \textbf{[Axiom 10]} \\
&= [1 + (-1)]\mathbf{v} && \textbf{[Axiom 8]} \\
&= 0\mathbf{v} && \textbf{[Property of real numbers]} \\
&= \mathbf{0} && \textbf{[Part (\textit{a}) of this theorem]}
\end{aligned}
$$
■

REMARK Although this theorem can be proved in R^n much more easily by working with components, the resulting theorem would be applicable *only* in R^n. By proving it using the vector space axioms, as we have done here, we have succeeded in creating a theorem that is valid for *all* vector spaces. This is, in fact, the power of the axiomatic approach—one proof serves to establish theorems in many vector spaces.

FUNCTION SPACES Many of the most important vector spaces involve real-valued functions, so we will now consider how such vector spaces arise as a natural generalization of R^n. As a first step, we will need to look at vectors in R^n from a new point of view: If $\mathbf{u} = (u_1, u_2, \ldots, u_n)$ is a vector in R^n, then we can regard the components of $\mathbf{u}$ to be the values of a function f whose independent variable varies over the integers from 1 to n; that is,

$$
f(1) = u_1, \quad f(2) = u_2, \ldots, \quad f(n) = u_n
$$

Thus, for example, the function

$$
f(k) = \tfrac{1}{4}k^2 \qquad (k = 1, 2, 3, 4) \tag{1}
$$

can be regarded as a description of the vector

$$
\mathbf{u} = (f(1), f(2), f(3), f(4)) = \left(\tfrac{1}{4}, 1, \tfrac{9}{4}, 4\right)
$$

If desired, we can graph this function to obtain a visual representation of the vector $\mathbf{u}$ (Figure 9.1.1*a*). This graph consists of isolated points because k assumes only integer values, but suppose that we replace k by a new variable x and allow x to vary unrestricted from $-\infty$ to ∞. Thus, instead of a function whose graph consists of isolated points, we now have a function

$$
f(x) = \tfrac{1}{4}x^2 \qquad (-\infty < x < \infty) \tag{2}
$$

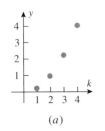

(a)

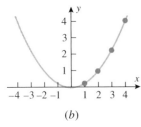

(b)

Figure 9.1.1

whose graph is a continuous curve (Figure 9.1.1b). Moreover, if we compare the forms of Formulas (1) and (2), then it is not an unreasonable jump to regard (2) as the description of a vector with infinitely many components, one for each value of x, and for which the xth component is $f(x)$. Thus, for example, the component at $x = \sqrt{3}$ of the vector represented by Formula (2) is $f(\sqrt{3}) = \frac{3}{4}$.

To continue the analogy between n-tuples and functions, let us denote the set of real-valued functions that are defined for all real values of x by $F(-\infty, \infty)$, and let us agree to consider two such functions f and g to be **equal**, written $f = g$, if and only if their corresponding components are equal; that is,

$$f = g \quad \text{if and only if} \quad f(x) = g(x) \quad \text{for all real values of } x$$

Geometrically, two functions are equal if and only if their graphs are identical. Furthermore, let us agree to perform addition and scalar multiplication on functions componentwise, just as for vectors in R^n; that is, if f and g are functions in $F(-\infty, \infty)$ and c is a scalar, then we define the **scalar multiple** cf and the **sum** $f + g$ by the formulas

$$(cf)(x) = cf(x)$$
$$(f + g)(x) = f(x) + g(x) \tag{3}$$

Geometrically, the graph of $f + g$ is obtained by adding corresponding y-coordinates on the graphs of f and g, and the graph of cf is obtained by multiplying each y-coordinate on the graph of f by c (Figure 9.1.2).

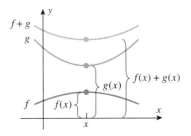

Figure 9.1.2

EXAMPLE 3
$F(-\infty, \infty)$ Is a Vector Space

Show that $F(-\infty, \infty)$ with the operations in (3) is a vector space by confirming that the 10 vector space axioms in Definition 9.1.1 are satisfied.

Solution

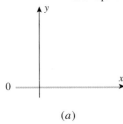

(a)

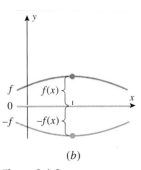

(b)

Figure 9.1.3

- **Axioms 1 and 6**—If f and g are functions in $F(-\infty, \infty)$ and c is a scalar, then the formulas in (3) define $(cf)(x)$ and $(f + g)(x)$ for all real values of x, so cf and $f + g$ are in $F(-\infty, \infty)$.

- **Axiom 4**—Consider the function 0 whose value is zero for all real values of x. Geometrically, the graph of this function coincides with the x-axis (Figure 9.1.3a). If f is any function in $F(-\infty, \infty)$, then

$$f(x) + 0 = f(x)$$

for all real values of x, which implies that $f + 0 = f$. Thus, the zero function is the zero vector in $F(-\infty, \infty)$.

- **Axiom 5**—For each function f in $F(-\infty, \infty)$, define $-f$ to be the function whose values are the negatives of the values of f; that is,

$$(-f)(x) = (-1)f(x)$$

for all real values of x (Figure 9.1.3b). It follows that

$$f(x) + (-f)(x) = f(x) + (-1)f(x) = 0$$

for all real values of x, which implies that $f + (-f) = 0$. Thus, $-f$ is the negative of f.

- **Axioms 2, 3, 7, 8, 9, and 10**—These properties of functions in $F(-\infty, \infty)$ follow from the corresponding properties of real numbers applied to each component. For example, if f and g are functions in $F(-\infty, \infty)$, then the commutative law for real numbers implies that

$$(f + g)(x) = f(x) + g(x) = g(x) + f(x) = (g + f)(x)$$

for all real values of x, and this confirms Axiom 2. The confirmations of the rest of the axioms are similar and are left for the exercises.   ∎

MATRIX SPACES

An $m \times n$ matrix can be viewed as a sequence of m row vectors or n column vectors. However, a matrix can also be viewed as a vector in its own right if we think about it in the right way. For this purpose, let M_{mn} be the set of all $m \times n$ matrices with real entries, and consider the standard operations of matrix addition and multiplication of a matrix by a scalar. These operations satisfy the 10 vector space axioms by previously established theorems, so we can regard M_{mn} as a vector space and the matrices within it as vectors. Thus, for example, the zero vector in this space is the $m \times n$ zero matrix. The vector space of real $m \times 1$ matrices can be viewed as R^m with vectors expressed in column form, and the vector space of real $1 \times n$ matrices can be viewed as R^n with vectors expressed in row form.

UNUSUAL VECTOR SPACES

The definition of a vector space allows any kinds of objects to be vectors and any kinds of operations to serve as addition and scalar multiplication—the only requirement is that the objects and their operations satisfy the 10 vector space axioms. This can lead to some unusual kinds of vector spaces. Here is an example.

EXAMPLE 4
An Unusual Vector Space

Let V be the set of positive real numbers, and for any real number k and any numbers u and v in V define the operations $\oplus$ and $\otimes$ on V to be

$u \oplus v = uv$ [Vector addition is numerical multiplication.]
$k \otimes u = u^k$ [Scalar multiplication is numerical exponentiation.]

Thus, for example, $1 \oplus 1 = 1$ and $2 \otimes 1 = 1^2 = 1$—strange indeed, but nevertheless the set V with these operations satisfies the 10 vector space axioms and hence is a vector space. We will confirm Axioms 4, 5, and 7, and leave some of the others as exercises:

- **Axiom 4**—The zero vector in this space is the number 1 (i.e., $\mathbf{0} = 1$), since

$$u \oplus 1 = u \cdot 1 = u$$

- **Axiom 5**—The negative of a vector u is its reciprocal (i.e. $-u = 1/u$) since

$$u \oplus \frac{1}{u} = u\left(\frac{1}{u}\right) = 1 \;(= \mathbf{0})$$

- **Axiom 7**—$k \otimes (u \oplus v) = (uv)^k = u^k v^k = (k \otimes u) \oplus (k \otimes v)$.   ∎

SUBSPACES

A subset of vectors in a vector space V may itself be a vector space with the operations on V, in which case we have one vector space inside of another.

> **Definition 9.1.3**   If W is a nonempty subset of vectors in a vector space V that is itself a vector space under the scalar multiplication and addition of V, then we call W a **_subspace_** of V.

In general, to show that a set W with a scalar multiplication and an addition is a vector space, it is necessary to verify the 10 vector space axioms. However, if W is part of a larger set V that is already known to be a vector space under the operations on W, and if W is closed under scalar multiplication and addition, then certain axioms need not be verified because they are "inherited" from V. For example, Axiom 2 is an inherited property because if $\mathbf{u} + \mathbf{v} = \mathbf{v} + \mathbf{u}$ holds for all vectors in V, then it holds, in particular, for vectors in W. The other inherited axioms are Axioms 3, 7, 8, 9, and 10. Thus, once the closure axioms (Axioms 1 and 6) are verified for W, we need only confirm the existence of a zero vector in W (Axiom 4) and the existence of a negative in W for each vector in W (Axiom 5) to establish that W is a subspace of V. However, the proof of the following theorem will show that Axioms 4 and 5 for a subspace W follow from the closure axioms for W, so there is no need to check those axioms once the closure axioms for W are established.

> **Theorem 9.1.4** *If W is a nonempty set of vectors in a vector space V, then W is a subspace of V if and only if W is closed under scalar multiplication and addition.*

Proof If W is a subspace of V, then the 10 vector space axioms are satisfied by all vectors in W, so this is true, in particular, for the two closure axioms. Conversely, assume that W is closed under scalar multiplication and addition. As indicated in the discussion preceding this theorem, we need only show that Axioms 4 and 5 hold for W. But the closure axioms imply that $k\mathbf{u}$ is in W for every $\mathbf{u}$ in W and every scalar k. In particular, the vectors $0\mathbf{u} = \mathbf{0}$ and $(-1)\mathbf{u} = -\mathbf{u}$ are in W, which confirms Axioms 4 and 5 for vectors in W. ∎

EXAMPLE 5
Zero Subspace

If $\mathbf{0}$ is the zero vector in any vector space V, then the one-vector set $W = \{\mathbf{0}\}$ is a subspace of V, since $c\mathbf{0} = \mathbf{0}$ for any scalar c and $\mathbf{0} + \mathbf{0} = \mathbf{0}$. We call this the ***zero subspace*** of V. ∎

EXAMPLE 6
Polynomial
Subspaces of
$F(-\infty, \infty)$

Let n be a nonnegative integer, and let P_n be the set of all real-valued functions of the form

$$p(x) = a_0 + a_1 x + \cdots + a_n x^n$$

where $a_0, a_1, \ldots, a_n$ are real numbers; that is, P_n is the set of all polynomials of degree n or less.[*] Show that P_n is a subspace of $F(-\infty, \infty)$.

Solution Polynomials are defined for all real values of x, so P_n is a subset of $F(-\infty, \infty)$. To show that it is a subspace of $F(-\infty, \infty)$ we must show that it is closed under scalar multiplication and addition; that is, we must show that if p and q are polynomials of degree n or less, and if c is a scalar, then cp and $p + q$ are also polynomials of degree n or less. To see that this is so, let

$$p(x) = a_0 + a_1 x + \cdots + a_n x^n \quad \text{and} \quad q(x) = b_0 + b_1 x + \cdots + b_n x^n$$

Thus,

$$cp(x) = ca_0 + (ca_1)x + \cdots + (ca_n)x^n$$

and

$$p(x) + q(x) = (a_0 + b_0) + (a_1 + b_1)x + \cdots + (a_n + b_n)x^n$$

which shows that cp and $p + q$ are polynomials of degree n or less. ∎

CONCEPT PROBLEM The set of all polynomials (no restriction on degree) is denoted by P_∞. In words, explain why P_∞ is a subspace of $F(-\infty, \infty)$.

EXAMPLE 7
Continuous
Functions on
$(-\infty, \infty)$

There are theorems in calculus which state that a constant times a continuous function is continuous and that a sum of continuous functions is continuous. This implies that the real-valued continuous functions on the interval $(-\infty, \infty)$ are closed under scalar multiplication and addition and hence form a subspace of $F(-\infty, \infty)$; we denote this subspace by $C(-\infty, \infty)$. Moreover, since polynomials are continuous functions, it follows that P_∞ and P_n are subspaces of $C(-\infty, \infty)$ for every nonnegative integer n. ∎

EXAMPLE 8
Differentiable
Functions on
$(-\infty, \infty)$

There are theorems in calculus which state that a constant times a differentiable function is differentiable and that a sum of differentiable functions is differentiable. This fact and the closure properties of continuous functions imply that the set of all functions with continuous

---

[*]Some authors regard only the *nonzero* constants to be polynomials of degree zero and do not assign a degree to the constant 0. The reasons for doing this will not be relevant to our work in this text, so we will treat 0 as a polynomial of degree zero.

first derivatives on the interval $(-\infty, \infty)$ are closed under scalar multiplication and addition and hence form a subspace of $F(-\infty, \infty)$; we denote this subspace by $C^1(-\infty, \infty)$. Similarly, the real-valued functions with continuous mth derivatives and the real-valued functions with derivatives of all orders form subspaces of $F(-\infty, \infty)$, which we denote by $C^m(-\infty, \infty)$ and $C^\infty(-\infty, \infty)$, respectively. ■

CONCEPT PROBLEM   Convince yourself that the subspaces of $F(-\infty, \infty)$ that we have discussed so far are "nested" one inside the other as illustrated in Figure 9.1.4.

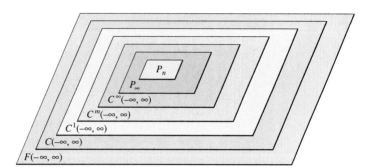

Figure 9.1.4

REMARK   Although we have focused on functions that are defined everywhere on the interval $(-\infty, \infty)$, there are many kinds of problems in which it is preferable to require only that the functions be defined on a specified finite closed interval. Thus, $F[a, b]$ will denote the vector space of functions that are defined for $a \leq x \leq b$, and $C[a, b]$, $C^m[a, b]$, and $C^\infty[a, b]$ denote the obvious subspaces.

EXAMPLE 9
The Invertible
Matrices Are
Not a Subspace
of M_{nn}

Show that the invertible $n \times n$ matrices do not form a subspace of M_{nn}.

Solution   Let W be the set of invertible matrices in M_{nn}. This set fails to be a subspace on both counts—it is closed under neither scalar multiplication nor addition. For example, consider the invertible matrices

$$U = \begin{bmatrix} 1 & 2 \\ 2 & 5 \end{bmatrix} \quad \text{and} \quad V = \begin{bmatrix} -1 & 2 \\ -2 & 5 \end{bmatrix}$$

in M_{22}. The matrix $0U$ is the 2×2 zero matrix, hence is not invertible; and the matrix $U + V$ has a column of zeros, hence is not invertible. You should have no trouble adapting this example in M_{22} to M_{nn}. ■

CONCEPT PROBLEM   Do you think that the symmetric matrices form a subspace of M_{nn}? Explain.

LINEAR INDEPENDENCE, SPANNING, BASIS

The definitions of linear combination, linear independence, spanning, and basis carry over to general vector spaces.

EXAMPLE 10
A Linearly
Dependent Set in
$F(-\infty, \infty)$

Show that the functions $f_1(x) = 1$, $f_2(x) = \sin^2 x$, and $f_3(x) = \cos^2 x$ are linearly dependent in $F(-\infty, \infty)$.

Solution   We can either show that one of the functions is a linear combination of the other two, or we can show that there are scalars c_1, c_2, and c_3 that are not all zero such that

$$c_1(1) + c_2(\sin^2 x) + c_3(\cos^2 x) = 0 \tag{4}$$

for all real values of x; we will do both. We know that the identity $\sin^2 x + \cos^2 x = 1$ is valid for all real values of x. Thus, $f_1(x) = 1$ can be expressed as a linear combination of $f_2(x) = \sin^2 x$

and $f_3(x) = \cos^2 x$ as

$$f_1 = f_2 + f_3$$

Alternatively, we can rewrite this equation as $f_1 - f_2 - f_3 = 0$, which shows that (4) holds with $c_1 = 1, c_2 = -1,$ and $c_3 = -1.$ ∎

EXAMPLE 11
The Span of a Set of Vectors

Describe the subspace of $F(-\infty, \infty)$ that is spanned by the functions $1, x, x^2, \ldots, x^n$.

Solution The subspace of $F(-\infty, \infty)$ that is spanned by $1, x, x^2, \ldots, x^n$ consists of all linear combinations of these functions and hence consists of all functions of the form

$$p(x) = c_0 + c_1 x + \cdots + c_n x^n$$

These are the polynomials of degree n or less, so $\text{span}\{1, x, x^2, \ldots, x^n\} = P_n$. ∎

EXAMPLE 12
A Basis for P_n

We saw in Example 11 that the functions $1, x, x^2, \ldots, x^n$ span P_n. Show that these functions are linearly independent and hence form a basis for P_n. This is called the ***standard basis*** for P_n.

Solution We must show that if

$$c_0 + c_1 x + \cdots + c_n x^n = 0 \tag{5}$$

for all real values of x, then $c_0 = c_1 = \cdots = c_n = 0$. For this purpose, recall from algebra that a *nonzero* polynomial of degree n or less has at most n distinct roots. This implies that all of the coefficients in (5) are zero, for otherwise (5) would be a nonzero polynomial for which every real number x is a root—a contradiction. ∎

EXAMPLE 13
A Basis for M_{22}

The matrices

$$E_1 = \begin{bmatrix} 1 & 0 \\ 0 & 0 \end{bmatrix}, \quad E_2 = \begin{bmatrix} 0 & 1 \\ 0 & 0 \end{bmatrix}, \quad E_3 = \begin{bmatrix} 0 & 0 \\ 1 & 0 \end{bmatrix}, \quad E_4 = \begin{bmatrix} 0 & 0 \\ 0 & 1 \end{bmatrix}$$

form a basis for the vector space M_{22} of 2×2 matrices. To see why this is so, observe that the vectors span M_{22}, since we can write a general 2×2 matrix as

$$\begin{bmatrix} a & b \\ c & d \end{bmatrix} = a \begin{bmatrix} 1 & 0 \\ 0 & 0 \end{bmatrix} + b \begin{bmatrix} 0 & 1 \\ 0 & 0 \end{bmatrix} + c \begin{bmatrix} 0 & 0 \\ 1 & 0 \end{bmatrix} + d \begin{bmatrix} 0 & 0 \\ 0 & 1 \end{bmatrix} = aE_1 + bE_2 + cE_3 + dE_4$$

Moreover, the matrices are linearly independent since the only scalars that satisfy the equation

$$aE_1 + bE_2 + cE_3 + dE_4 = \begin{bmatrix} a & b \\ c & d \end{bmatrix} = \begin{bmatrix} 0 & 0 \\ 0 & 0 \end{bmatrix}$$

are $a = 0, b = 0, c = 0, d = 0.$ ∎

WROŃSKI'S TEST FOR LINEAR INDEPENDENCE OF FUNCTIONS

Although linear independence or dependence of functions can sometimes be established from known identities, as in Example 10, there are no general methods for determining whether or not a set of functions in $F(-\infty, \infty)$ is linearly independent. There is, however, a method, due to the Polish-French mathematician Józef Wroński that is sometimes useful for this purpose. To explain the idea, suppose that $f_1(x), f_2(x), \ldots, f_n(x)$ are functions in $C^{(n-1)}(-\infty, \infty)$. If these functions are linearly dependent, then there exist scalars $k_1, k_2, \ldots, k_n$ that are not all zero and such that

$$k_1 f_1(x) + k_2 f_2(x) + \cdots + k_n f_n(x) = 0$$

for all x in the interval $(-\infty, \infty)$. Combining this equation with those that can be derived from

Linear Algebra in History

Józef Hoëné de Wronski
(1778–1853)

it by $n-1$ differentiations, we see that the following equations hold for all x in the interval $(-\infty, \infty)$:

$$
\begin{aligned}
k_1 f_1(x) &+ k_2 f_2(x) &+ \cdots + k_n f_n(x) &= 0 \\
k_1 f_1'(x) &+ k_2 f_2'(x) &+ \cdots + k_n f_n'(x) &= 0 \\
&\vdots & \vdots \qquad \vdots \\
k_1 f_1^{(n-1)}(x) &+ k_2 f_2^{(n-1)}(x) &+ \cdots + k_n f_n^{(n-1)}(x) &= 0
\end{aligned}
$$

This implies that the linear system

$$
\begin{bmatrix}
f_1(x) & f_2(x) & \cdots & f_n(x) \\
f_1'(x) & f_2'(x) & \cdots & f_n'(x) \\
\vdots & \vdots & & \vdots \\
f_1^{(n-1)}(x) & f_2^{(n-1)}(x) & \cdots & f_n^{(n-1)}(x)
\end{bmatrix}
\begin{bmatrix} k_1 \\ k_2 \\ \vdots \\ k_n \end{bmatrix}
=
\begin{bmatrix} 0 \\ 0 \\ \vdots \\ 0 \end{bmatrix}
$$

has a nontrivial solution for every x in the interval $(-\infty, \infty)$; and this, in turn, implies that the determinant

$$
W(x) =
\begin{vmatrix}
f_1(x) & f_2(x) & \cdots & f_n(x) \\
f_1'(x) & f_2'(x) & \cdots & f_n'(x) \\
\vdots & \vdots & & \vdots \\
f_1^{(n-1)}(x) & f_2^{(n-1)}(x) & \cdots & f_n^{(n-1)}(x)
\end{vmatrix}
\qquad (6)
$$

which is called the ***Wronskian*** of $f_1(x), f_2(x), \ldots, f_n(x)$, is zero for every x in the interval $(-\infty, \infty)$. Stated in contrapositive form, we have shown that if the Wronskian of the functions $f_1(x), f_2(x), \ldots, f_n(x)$ is *not* equal to zero for every x in the interval $(-\infty, \infty)$, then these functions must be linearly independent.

Theorem 9.1.5 (*Wronski's Test*)   *If the functions* $f_1(x), f_2(x), \ldots, f_n(x)$ *have* $n-1$ *continuous derivatives on the interval* $(-\infty, \infty)$, *and if the Wronskian of these functions is not identically zero on* $(-\infty, \infty)$, *then the functions form a linearly independent set in* $F(-\infty, \infty)$.

EXAMPLE 14   Show that the three functions $f_1(x) = 1$, $f_2(x) = e^x$, and $f_3(x) = e^{2x}$ form a linearly independent set in $F(-\infty, \infty)$.

Solution   The Wronskian is

$$
W(x) =
\begin{vmatrix}
1 & e^x & e^{2x} \\
0 & e^x & 2e^{2x} \\
0 & e^x & 4e^{2x}
\end{vmatrix}
= 2e^{3x}
$$

(verify). This function is nonzero for all real values of x and hence is not identically zero on $(-\infty, \infty)$. Thus, the functions are linearly independent.[*] ∎

DIMENSION   We will now consider how to extend the notion of dimension to general vector spaces. For a general vector space V there may or may not exist a finite basis (i.e., a finite set of vectors that is linearly independent and spans V). For example, the set P_∞ of all polynomials in $F(-\infty, \infty)$ is closed under scalar multiplication and addition (verify), and hence is a subspace of $F(-\infty, \infty)$.

---

[*]To prove linear independence we need only show that $W(x) \neq 0$ for *some* value of x in $(-\infty, \infty)$. Thus, our observation in this example that the Wronskian is nonzero for *all* x in $(-\infty, \infty)$ was more than we really needed to say to conclude linear independence—it would have been sufficient to find a single value of x for which $W(x)$ is nonzero.

However, there is no finite set of vectors in P_∞ that spans P_∞ since the polynomials in any finite subset of P_∞ would have some maximum degree, say m, and hence it would not be possible to express any polynomial of degree greater than m as a linear combination of the vectors in the finite subset. Accordingly, we distinguish between two types of vector spaces—those that have a finite basis and those that do not.

> **Definition 9.1.6** A vector space V is said to be *finite-dimensional* if it has a basis with finitely many vectors and *infinite-dimensional* if it does not. In addition, the zero vector space $V = \{\mathbf{0}\}$ is defined to be finite-dimensional.

The proof that we gave of Theorem 7.1.4 carries over without change to prove the following more general result.

> **Theorem 9.1.7** *All bases for a nonzero finite-dimensional vector space have the same number of vectors.*

This theorem allows us to make the following definition.

> **Definition 9.1.8** If V is a nonzero finite-dimensional vector space, then we define the *dimension* of V, denoted by $\dim(V)$, to be the number of vectors in a basis. In addition, we define $V = \{\mathbf{0}\}$ to have dimension zero.

LOOKING AHEAD We will show in Section 9.3 that vectors in an n-dimensional vector space can be matched up with vectors in R^n in such a way that the spaces are algebraically identical except for a difference in notation. The implication of this is that all theorems about vectors in R^n that do not involve notions of inner product, length, distance, angle, or orthogonality are valid in any n-dimensional vector space. This is true, in particular, of Theorem 7.2.4 from which it follows that *a subspace of a finite-dimensional vector space is finite-dimensional*.

EXAMPLE 15
Infinite-
Dimensional
Vector Spaces

We saw in Example 12 that the functions $1, x, x^2, \ldots, x^n$ form a basis for P_n. This implies that all bases for P_n have $n + 1$ vectors and hence that $\dim(P_n) = n + 1$. Also, we showed above that P_∞ is infinite-dimensional, so it follows that $F(-\infty, \infty)$, $C(-\infty, \infty)$, $C^1(-\infty, \infty)$, $C^m(-\infty, \infty)$, and $C^\infty(-\infty, \infty)$ are also infinite-dimensional, since these spaces all contain P_∞ as a subspace (Figure 9.1.4). In the exercises we will ask you to show that R^∞ is infinite-dimensional. ■

**THE LAGRANGE
INTERPOLATING
POLYNOMIALS**

We stated without proof in Theorem 2.3.1 that given any n points in the xy-plane with distinct x-coordinates, there is a unique interpolating polynomial of degree $n - 1$ or less whose graph passes through those points. We eventually proved this theorem in Section 4.3 by showing that the Vandermonde determinant is nonzero [see Formula (9) of Section 4.3 and the related discussion]. We also showed in Section 2.3 that interpolating polynomials can be found by solving linear systems, but we noted that better methods would be forthcoming. We will now discuss such a method.

Suppose, for example, that x_1, x_2, x_3, and x_4 are distinct real numbers, and we want to find the interpolating polynomial of degree 3 or less that passes through the points

$$(x_1, y_1), \quad (x_2, y_2), \quad (x_3, y_3), \quad (x_4, y_4) \tag{7}$$

For this purpose, let us consider the polynomials

$$p_1(x) = \frac{(x - x_2)(x - x_3)(x - x_4)}{(x_1 - x_2)(x_1 - x_3)(x_1 - x_4)}, \quad p_2(x) = \frac{(x - x_1)(x - x_3)(x - x_4)}{(x_2 - x_1)(x_2 - x_3)(x_2 - x_4)}$$

$$p_3(x) = \frac{(x - x_1)(x - x_2)(x - x_4)}{(x_3 - x_1)(x_3 - x_2)(x_3 - x_4)}, \quad p_4(x) = \frac{(x - x_1)(x - x_2)(x - x_3)}{(x_4 - x_1)(x_4 - x_2)(x_4 - x_3)} \tag{8}$$

each of which has degree 3. These polynomials have been constructed so that their values at the points x_1, x_2, x_3, and x_4 are

$$p_1(x_1) = 1, \quad p_1(x_2) = 0, \quad p_1(x_3) = 0, \quad p_1(x_4) = 0$$
$$p_2(x_1) = 0, \quad p_2(x_2) = 1, \quad p_2(x_3) = 0, \quad p_2(x_4) = 0$$
$$p_3(x_1) = 0, \quad p_3(x_2) = 0, \quad p_3(x_3) = 1, \quad p_3(x_4) = 0$$
$$p_4(x_1) = 0, \quad p_4(x_2) = 0, \quad p_4(x_3) = 0, \quad p_4(x_4) = 1$$

(verify). It follows from this that

$$p(x) = y_1 p_1(x) + y_2 p_2(x) + y_3 p_3(x) + y_4 p_4(x) \tag{9}$$

is a polynomial of degree 3 or less whose values at x_1, x_2, x_3, and x_4 are

$$p(x_1) = y_1, \quad p(x_2) = y_2, \quad p(x_3) = y_3, \quad p(x_4) = y_4$$

Thus (9) must be the interpolating polynomial for the points in (7). The four functions in (8) are called the ***Lagrange interpolating polynomials*** at x_1, x_2, x_3, and x_4, and Formula (9) is called the ***Lagrange interpolation formula***, both in honor of the French mathematician Joseph-Louis Lagrange.

What makes the Lagrange interpolating polynomials so useful is that they depend on the x's and not on the y's. Thus, once the Lagrange interpolating polynomials at x_1, x_2, x_3, and x_4 are computed, they can be used to solve all interpolation problems at the points of form (7) simply by substituting the y-values in the Lagrange interpolation formula in (9). This is much more efficient than solving a linear system for each interpolation problem and is less subject to the effect of roundoff error.

EXAMPLE 16
Lagrange
Interpolation

In Example 7 of Section 2.3 we deduced that the interpolating polynomial for the points

$$(1, 3), \quad (2, -2), \quad (3, -5), \quad (4, 0) \tag{10}$$

is

$$p(x) = 4 + 3x - 5x^2 + x^3 \tag{11}$$

by solving a linear system. Use the appropriate Lagrange interpolating polynomials to find this polynomial.

Solution Using the x-coordinates in (10) we obtain the Lagrange interpolating polynomials

$$p_1(x) = \frac{(x - x_2)(x - x_3)(x - x_4)}{(x_1 - x_2)(x_1 - x_3)(x_1 - x_4)} = \frac{(x - 2)(x - 3)(x - 4)}{(1 - 2)(1 - 3)(1 - 4)} = 4 - \tfrac{13}{3}x + \tfrac{3}{2}x^2 - \tfrac{1}{6}x^3$$

$$p_2(x) = \frac{(x - x_1)(x - x_3)(x - x_4)}{(x_2 - x_1)(x_2 - x_3)(x_2 - x_4)} = \frac{(x - 1)(x - 3)(x - 4)}{(2 - 1)(2 - 3)(2 - 4)} = -6 + \tfrac{19}{2}x - 4x^2 + \tfrac{1}{2}x^3$$

$$p_3(x) = \frac{(x - x_1)(x - x_2)(x - x_4)}{(x_3 - x_1)(x_3 - x_2)(x_3 - x_4)} = \frac{(x - 1)(x - 2)(x - 4)}{(3 - 1)(3 - 2)(3 - 4)} = 4 - 7x + \tfrac{7}{2}x^2 - \tfrac{1}{2}x^3$$

$$p_4(x) = \frac{(x - x_1)(x - x_2)(x - x_3)}{(x_4 - x_1)(x_4 - x_2)(x_4 - x_3)} = \frac{(x - 1)(x - 2)(x - 3)}{(4 - 1)(4 - 2)(4 - 3)} = -1 + \tfrac{11}{6}x - x^2 + \tfrac{1}{6}x^3$$

Now using Formula (9) and the y-coordinates in (10) we obtain

$$p(x) = 3\left(4 - \tfrac{13}{3}x + \tfrac{3}{2}x^2 - \tfrac{1}{6}x^3\right) - 2\left(-6 + \tfrac{19}{2}x - 4x^2 + \tfrac{1}{2}x^3\right)$$
$$\qquad - 5\left(4 - 7x + \tfrac{7}{2}x^2 - \tfrac{1}{2}x^3\right) + 0\left(-1 + \tfrac{11}{6}x - x^2 + \tfrac{1}{6}x^3\right)$$
$$= 4 + 3x - 5x^2 + x^3$$

which agrees with (11). ■

LAGRANGE INTERPOLATION FROM A VECTOR POINT OF VIEW

We will now reexamine the Lagrange interpolating polynomials from a vector point of view by considering them to be vectors in P_3. Suppose that x_1, x_2, x_3, and x_4 are distinct real numbers and that $p(x)$ is any polynomial in P_3. Since the graph of $p(x)$ passes through the points

$$(x_1, p(x_1)), \quad (x_2, p(x_2)), \quad (x_3, p(x_3)), \quad (x_4, p(x_4))$$

it must be the interpolating polynomial for these points; and hence it follows from (9) with $y_1 = p(x_1)$, $y_2 = p(x_2)$, $y_3 = p(x_3)$, and $y_4 = p(x_4)$ that $p(x)$ can be expressed as

$$p(x) = p(x_1)p_1(x) + p(x_2)p_2(x) + p(x_3)p_3(x) + p(x_4)p_4(x) \tag{12}$$

Since this is a linear combination of the Lagrange interpolating polynomials, and since $p(x)$ is an arbitrary vector in P_3, we have shown that the Lagrange interpolating polynomials at x_1, x_2, x_3, and x_4 span P_3. However, we also know that P_3 is four-dimensional, so we can conclude further that the Lagrange polynomials form a *basis* for P_3 by the generalization of Theorem 7.2.6 to P_3.

Although we have restricted our discussion of Lagrange interpolation to four points for notational simplicity, the ideas generalize; specifically, if $x_1, x_2, \ldots, x_n$ are distinct real numbers, then the ***Lagrange interpolating polynomials*** $p_1(x)$, $p_2(x), \ldots, p_n(x)$ at these points are defined by

$$p_i(x) = \frac{(x - x_1) \cdots (x - x_{i-1})(x - x_{i+1}) \cdots (x - x_n)}{(x_i - x_1) \cdots (x_i - x_{i-1})(x_i - x_{i+1}) \cdots (x_i - x_n)} \quad (i = 1, 2, \ldots, n) \tag{13}$$

and the interpolating polynomial $p(x)$ of degree $n - 1$ or less whose graph passes through the points

$$(x_1, y_1), (x_2, y_2), \ldots, (x_n, y_n)$$

is given by the ***Lagrange interpolation formula***

$$p(x) = y_1 p_1(x) + y_2 p_2(x) + \cdots + y_n p_n(x) \tag{14}$$

Moreover, it follows, as before, that the Lagrange interpolating polynomials at $x_1, x_2, \ldots, x_n$ span P_{n-1} and hence form a basis for this space, since it is n-dimensional. In summary, we have the following result.

> **Theorem 9.1.9**   *If $x_1, x_2, \ldots, x_n$ are distinct real numbers, then the Lagrange interpolating polynomials at these points form a basis for the vector space of polynomials of degree $n - 1$ or less.*

Linear Algebra in History

The Lagrange interpolation formula was the outgrowth of work on the problem of expressing the roots of a polynomial as functions of its coefficients. The formula was first published by the British mathematician Edward Waring in 1779 and then republished by the French mathematician Joseph-Louis Lagrange in 1795. Interestingly, Waring was a physician who practiced medicine at various hospitals for many years while maintaining a parallel career as a university professor at Cambridge. Lagrange distinguished himself in celestial mechanics as well as mathematics and was regarded as the greatest living mathematician by many of his contemporaries.

Joseph-Louis Lagrange
(1736–1813)   Edward Waring
(1734–1798)

Exercise Set 9.1

1. Let V be the set of all ordered pairs of real numbers, and consider the following addition and scalar multiplication operations on $\mathbf{u} = (u_1, u_2)$ and $\mathbf{v} = (v_1, v_2)$:

$$\mathbf{u} + \mathbf{v} = (u_1 + v_1, u_2 + v_2), \quad k\mathbf{u} = (ku_1, 0)$$

(a) Compute $\mathbf{u} + \mathbf{v}$ and $k\mathbf{u}$ for $\mathbf{u} = (-1, 2)$, $\mathbf{v} = (3, 4)$, and $k = 3$.

(b) In words, explain why V is closed under addition and scalar multiplication.

(c) Since addition on V is the standard addition operation on R^2, certain vector space axioms hold for V because they are known to hold for R^2. Which axioms are they?

(d) Show that Axioms 7, 8, and 9 hold.

(e) Show that Axiom 10 fails and hence that V is not a vector space under the given operations.

2. Let V be the set of all ordered triples of real numbers, and consider the following addition and scalar multiplication operations on $\mathbf{u} = (u_1, u_2, u_3)$ and $\mathbf{v} = (v_1, v_2, v_3)$:

$$\mathbf{u} + \mathbf{v} = (u_1 + v_1, u_2 + v_2, u_3 + v_3), \quad k\mathbf{u} = (ku_1, 0, 0)$$

(a) Compute $\mathbf{u} + \mathbf{v}$ and $k\mathbf{u}$ for $\mathbf{u} = (3, -2, 4)$, $\mathbf{v} = (1, 5, -2)$, and $k = -1$.

(b) In words, explain why V is closed under addition and scalar multiplication.

(c) Since the addition operation on V is the standard addition operation on R^3, certain vector space axioms hold for V because they are known to hold in R^3. Which axioms are they?

(d) Show that Axioms 7, 8, and 9 hold.

(e) Show that Axiom 10 fails and hence that V is not a vector space under the given operations.

3. Let V be the set of all ordered pairs of real numbers, and consider the following addition and scalar multiplication operations on $\mathbf{u} = (u_1, u_2)$ and $\mathbf{v} = (v_1, v_2)$:

$$\mathbf{u} + \mathbf{v} = (u_1 + v_1 + 1, u_2 + v_2 + 1), \quad k\mathbf{u} = (ku_1, ku_2)$$

(a) Compute $\mathbf{u} + \mathbf{v}$ and $k\mathbf{u}$ for $\mathbf{u} = (0, 4)$, $\mathbf{v} = (1, -3)$, and $k = 2$.

(b) Show that $(0, 0) \neq \mathbf{0}$.

(c) Show that $(-1, -1) = \mathbf{0}$.

(d) Show that Axiom 5 holds by producing an ordered pair $-\mathbf{u}$ such that $\mathbf{u} + (-\mathbf{u}) = \mathbf{0}$ for $\mathbf{u} = (u_1, u_2)$.

(e) Find two vector space axioms that fail to hold.

4. Let V be the set of all ordered pairs of real numbers, and consider the following addition and scalar multiplication operations on $\mathbf{u} = (u_1, u_2)$ and $\mathbf{v} = (v_1, v_2)$:

$$\mathbf{u} + \mathbf{v} = (u_1 v_1, u_2 v_2), \quad k\mathbf{u} = (ku_1, ku_2)$$

(a) Compute $\mathbf{u} + \mathbf{v}$ and $k\mathbf{u}$ for $\mathbf{u} = (1, 5)$, $\mathbf{v} = (2, -2)$, and $k = 4$.

(b) Show that $(0, 0) \neq \mathbf{0}$.

(c) Show that $(1, 1) = \mathbf{0}$.

(d) Show that Axiom 5 fails by showing that there is no ordered pair $-\mathbf{u}$ such that $\mathbf{u} + (-\mathbf{u}) = \mathbf{0}$ if $\mathbf{u}$ has a zero component.

(e) Find two other vector space axioms that fail to hold.

5. Which of the following are subspaces of $F(-\infty, \infty)$?

(a) All functions f in $F(-\infty, \infty)$ for which $f(0) = 0$.

(b) All functions f in $F(-\infty, \infty)$ for which $f(0) = 1$.

(c) All functions f in $F(-\infty, \infty)$ for which $f(-x) = f(x)$.

(d) All polynomials of degree 2.

6. Which of the following are subspaces of R^∞?

(a) All sequences $\mathbf{v}$ in R^∞ of the form $\mathbf{v} = (v, 0, v, 0, v, 0, \ldots)$.

(b) All sequences $\mathbf{v}$ in R^∞ of the form $\mathbf{v} = (v, 1, v, 1, v, 1, \ldots)$.

(c) All sequences $\mathbf{v}$ in R^∞ of the form $\mathbf{v} = (v, 2v, 4v, 8v, 16v, \ldots)$.

(d) All sequences in R^∞ whose components are 0 from some point on.

7. Which of the following are subspaces of M_{nn} (the vector space of $n \times n$ matrices)?

(a) The singular matrices.

(b) The diagonal matrices.

(c) The symmetric matrices.

(d) The matrices with trace zero.

8. (*Calculus required*) Which of the following are subspaces of $C(-\infty, \infty)$?

(a) All continuous functions whose Maclaurin series converge on the interval $(-\infty, \infty)$.

(b) All differentiable functions for which $f'(0) = 0$.

(c) All differentiable functions for which $f'(0) = 1$.

(d) All twice differentiable functions $y = f(x)$ for which $y'' + y = 0$.

9. Which of the following are subspaces of P_2?

(a) All polynomials $a_0 + a_1 x + a_2 x^2$ for which $a_0 = 0$.

(b) All polynomials $a_0 + a_1 x + a_2 x^2$ for which $a_0 + a_1 + a_2 = 0$.

(c) All polynomials $a_0 + a_1 x + a_2 x^2$ for which a_0, a_1, and a_2 are integers.

10. (*Calculus required*) Show that the set V of continuous functions $f(x)$ on $[0, 1]$ such that

$$\int_0^1 f(x)\,dx = 0$$

is a subspace of $C[0, 1]$.

11. (a) Show that the set V of all 2×2 upper triangular matrices is a subspace of M_{22}.

(b) Find a basis for V, and state the dimension of V.

12. (a) Show that the set V of all 3×3 skew-symmetric matrices is a subspace of M_{33}.

(b) Find a basis for V, and state the dimension of V.

In Exercises 13 and 14, use appropriate trigonometric identities to show that the stated functions in $F(-\infty, \infty)$ are linearly dependent.

---

13. (a) $f_1(x) = \sin^2 2x$, $f_2(x) = \cos^2 2x$, and $f_3(x) = \cos 4x$

(b) $f_1(x) = \sin^2\left(\frac{1}{2}x\right)$, $f_2(x) = 1$, and $f_3(x) = \cos x$

14. (a) $f_1(x) = \cos^4 x$, $f_2(x) = \sin^4 x$, $f_3(x) = \cos 2x$

(b) $f_1(x) = \sin x$, $f_2(x) = \sin 3x$, $f_3(x) = \sin^3 x$

15. The functions $f_1(x) = x$ and $f_2(x) = \cos x$ are linearly independent in $F(-\infty, \infty)$ because neither function is a scalar multiple of the other. Confirm the linear independence using Wroński's test.

16. The functions $f_1(x) = \sin x$ and $f_2(x) = \cos x$ are linearly independent in $F(-\infty, \infty)$, since neither function is a scalar multiple of the other. Confirm the linear independence using Wroński's test.

17. Use Wroński's test to show that the functions $f_1(x) = e^x$, $f_2(x) = xe^x$, $f_3(x) = x^2 e^x$ span a three-dimensional subspace of $F(-\infty, \infty)$.

18. Use Wroński's test to show that the functions $f_1(x) = \sin x$, $f_2(x) = \cos x$, $f_3(x) = x \cos x$ span a three-dimensional subspace of $F(-\infty, \infty)$.

In Exercises 19 and 20, show that $\{A_1, A_2, A_3, A_4\}$ is a basis for M_{22}, and express A as a linear combination of the basis vectors.

19. $A_1 = \begin{bmatrix} 1 & 0 \\ 1 & 0 \end{bmatrix}$, $A_2 = \begin{bmatrix} 1 & 1 \\ 0 & 0 \end{bmatrix}$, $A_3 = \begin{bmatrix} 1 & 0 \\ 0 & 1 \end{bmatrix}$

$A_4 = \begin{bmatrix} 0 & 0 \\ 1 & 0 \end{bmatrix}$; $A = \begin{bmatrix} 6 & 2 \\ 5 & 3 \end{bmatrix}$

20. $A_1 = \begin{bmatrix} 1 & 1 \\ 1 & 1 \end{bmatrix}$, $A_2 = \begin{bmatrix} 0 & 1 \\ 1 & 1 \end{bmatrix}$, $A_3 = \begin{bmatrix} 0 & 0 \\ 1 & 1 \end{bmatrix}$

$A_4 = \begin{bmatrix} 0 & 0 \\ 0 & 1 \end{bmatrix}$; $A = \begin{bmatrix} 1 & 0 \\ 1 & 0 \end{bmatrix}$

In Exercises 21 and 22, show that $\{\mathbf{p}_1, \mathbf{p}_2, \mathbf{p}_3\}$ is a basis for P_2, and express $\mathbf{p}$ as a linear combination of the basis vectors.

21. $\mathbf{p}_1 = 1 + 2x + x^2$, $\mathbf{p}_2 = 2 + 9x$, $\mathbf{p}_3 = 3 + 3x + 4x^2$; $\mathbf{p} = 2 + 17x - 3x^2$

22. $\mathbf{p}_1 = 1 + x + x^2$, $\mathbf{p}_2 = x + x^2$, $\mathbf{p}_3 = x^2$; $\mathbf{p} = 7 - x + 2x^2$

23. The graph of the polynomial $p(x) = x^2 + x + 1$ passes through the points $(0, 1)$, $(1, 3)$, and $(2, 7)$. Find the Lagrange interpolating polynomials at the points $x = 0, 1, 2$, and express p as a linear combination of those polynomials.

24. The graph of the polynomial $p(x) = x^2 - x + 2$ passes through the points $(1, 2)$, $(2, 4)$, and $(3, 8)$. Find the Lagrange interpolating polynomials for p at $x = 1, 2$, and 3, and express p as a linear combination of those polynomials.

In Exercises 25 and 26, use the appropriate Lagrange interpolating polynomials to find the cubic polynomial whose graph passes through the given points.

25. $(2, 1)$, $(3, 1)$, $(4, -3)$, $(5, 0)$

26. $(1, 2)$, $(2, 1)$, $(3, 3)$, $(6, 1)$

27. Show that the set V of all 2×2 matrices of the form

$$\begin{bmatrix} a & a - b \\ a - b & b \end{bmatrix}$$

is a subspace of M_{22}. Find a basis for the set V, and state its dimension.

28. Verify Axioms 3, 7, 8, 9, and 10 for the vector space given in Example 3.

29. Let V be the set of all real numbers with the two operations $\oplus$ and $\otimes$ defined by

$$u \oplus v = u + v - 1 \quad \text{and} \quad k \otimes u = ku + (1 - k)$$

(a) Compute $1 \oplus 1$.   (b) Compute $0 \otimes 2$.

(c) Show that V is a vector space. [*Suggestion:* You may find it helpful to use the notation $\mathbf{u} = u$, $\mathbf{v} = v$, and $\mathbf{w} = w$ to emphasize the vector interpretation of the numbers in V.]

30. Let V be the vector space in Example 4. Give numerical interpretations of the following relationships whose validity was established in Theorem 9.1.2.

(a) $\mathbf{0} \otimes \mathbf{v} = \mathbf{0}$ (b) $k \otimes \mathbf{0} = \mathbf{0}$

(c) $(-1) \otimes \mathbf{v} = -\mathbf{v}$

31. Let V be the set of real numbers, and consider the following addition and scalar multiplication operations on the numbers u and v:

$$u \oplus v = \text{maximum of } u \text{ and } v, \quad k \otimes u = ku$$

Determine whether the set V is a vector space with these operations.

32. Verify Axioms 3, 8, 9, and 10 for the vector space given in Example 4.

Discussion and Discovery

D1. Consider the set V whose only element is the planet Jupiter. Is V a vector space under the operations $\oplus$ and $\otimes$ defined by the following equations?

Jupiter $\oplus$ Jupiter $=$ Jupiter   and $k \otimes$ Jupiter $=$ Jupiter

Explain your reasoning.

D2. It is impossible to have a vector space V consisting of two distinct vectors. Explain why. [*Hint:* One of the vectors must be $\mathbf{0}$, so assume that $V = \{\mathbf{0}, \mathbf{v}\}$.]

D3. Let B be a fixed 2×2 matrix. Is the set of all 2×2 matrices that commute with B a subspace of M_{22}? Justify your answer.

D4. Draw appropriate pictures to show that a line in R^2 that does not pass through the origin does not satisfy either of the closure axioms for a vector space.

D5. Determine the dimension of the subspace of $F(-\infty, \infty)$ that is spanned by the functions $f_1(x) = \sin x$ and $f_2(x) = 3 \cos\left(\frac{1}{2}\pi - x\right)$. Explain your reasoning.

D6. Let V be the set of positive real numbers, and for any real number k and any numbers u and v in V define the operations $\oplus$ and $\otimes$ on V to be $u \oplus v = uv$ and $k \otimes u = u$. All vector space axioms except one hold. Which one is it?

D7. Show that R^∞ is infinite-dimensional by producing a linearly independent set with infinitely many vectors. [*Note:* See the remark following Definition 3.4.5.]

Working with Proofs

P1. Prove that a vector space V has a unique zero vector. [*Hint:* Assume that $\mathbf{0}_1$ and $\mathbf{0}_2$ are both zero vectors, and compute $\mathbf{0}_1 + \mathbf{0}_2$ two different ways to show that $\mathbf{0}_1 = \mathbf{0}_2$.]

P2. The argument that follows proves that every vector $\mathbf{u}$ in a vector space V has a unique negative. We assume that $\mathbf{u}_1$ and $\mathbf{u}_2$ are both negatives of $\mathbf{u}$ and show that $\mathbf{u}_1 = \mathbf{u}_2$. Justify the steps by filling in the blanks.

$\mathbf{u}_1 + (\mathbf{u} + \mathbf{u}_2) = (\mathbf{u}_1 + \mathbf{u}) + \mathbf{u}_2$ _____

$\mathbf{u}_1 + \mathbf{0} = (\mathbf{u}_1 + \mathbf{u}) + \mathbf{u}_2$ _____

$\mathbf{u}_1 = (\mathbf{u}_1 + \mathbf{u}) + \mathbf{u}_2$ _____

$\mathbf{u}_1 = (\mathbf{u} + \mathbf{u}_1) + \mathbf{u}_2$ _____

$\mathbf{u}_1 = \mathbf{0} + \mathbf{u}_2$ _____

$\mathbf{u}_1 = \mathbf{u}_2 + \mathbf{0}$ _____

$\mathbf{u}_1 = \mathbf{u}_2$ _____

P3. The argument that follows proves that if $\mathbf{u}, \mathbf{v}$, and $\mathbf{w}$ are vectors in a vector space V such that $\mathbf{u} + \mathbf{w} = \mathbf{v} + \mathbf{w}$, then $\mathbf{u} = \mathbf{v}$ (the *cancellation law* for vector addition). Justify the steps by filling in the blanks.

$\mathbf{u} + \mathbf{w} = \mathbf{v} + \mathbf{w}$	Hypothesis
$(\mathbf{u} + \mathbf{w}) + (-\mathbf{w}) = (\mathbf{v} + \mathbf{w}) + (-\mathbf{w})$	Add $-\mathbf{w}$ to both sides.
$\mathbf{u} + [\mathbf{w} + (-\mathbf{w})] = \mathbf{v} + [\mathbf{w} + (-\mathbf{w})]$	_____
$\mathbf{u} + \mathbf{0} = \mathbf{v} + \mathbf{0}$	_____
$\mathbf{u} = \mathbf{v}$	_____

P4. Prove: If $\mathbf{u}$ is a vector in a vector space V and k a scalar such that $k\mathbf{u} = \mathbf{0}$, then either $k = 0$ or $\mathbf{u} = \mathbf{0}$. [*Suggestion:* Show that if $k\mathbf{u} = \mathbf{0}$ and $k \neq 0$, then $\mathbf{u} = \mathbf{0}$. The result then follows as a logical consequence of this.]

P5. Prove: If W_1 and W_2 are subspaces of a vector space V, then $W_1 \cap W_2$ is also a subspace of V.

Section 9.2   Inner Product Spaces; Fourier Series

In this section we will generalize the concept of a dot product, thereby allowing us to introduce notions of length, distance, and orthogonality in general vector spaces. Calculus will be required for some of the examples.

INNER PRODUCT AXIOMS

The definition of a vector space involves addition and scalar multiplication but no analog of the dot product that would allow us to introduce notions of length, distance, angle, and orthogonality. To give a vector space a geometric structure we need to have a third operation that will generalize the concept of a dot product. For this purpose, recall that all of the algebraic properties of the dot product on R^n can be derived from Theorem 1.2.6 (see Looking Ahead at the end of Section 1.2). Thus, if we have an operation that associates real numbers with pairs of vectors in a real vector space V in such a way that the properties in Theorem 1.2.6 hold, then we are guaranteed that the geometric theorems about vectors in R^n will also be valid in V. Accordingly, we make the following definition whose statements duplicate those in Theorem 1.2.6, but in different notation.

Definition 9.2.1 An *inner product* on a real vector space V is a function that associates a unique real number $\langle \mathbf{u}, \mathbf{v} \rangle$ with each pair of vectors $\mathbf{u}$ and $\mathbf{v}$ in V in such a way that the following properties hold for all $\mathbf{u}, \mathbf{v}$, and $\mathbf{w}$ in V and for all scalars k.

1. $\langle \mathbf{u}, \mathbf{v} \rangle = \langle \mathbf{v}, \mathbf{u} \rangle$ [**Symmetry property**]
2. $\langle \mathbf{u} + \mathbf{v}, \mathbf{w} \rangle = \langle \mathbf{u}, \mathbf{w} \rangle + \langle \mathbf{v}, \mathbf{w} \rangle$ [**Additivity property**]
3. $\langle k\mathbf{u}, \mathbf{v} \rangle = k \langle \mathbf{u}, \mathbf{v} \rangle$ [**Homogeneity property**]
4. $\langle \mathbf{v}, \mathbf{v} \rangle \geq 0$ and $\langle \mathbf{v}, \mathbf{v} \rangle = 0$ if and only if $\mathbf{v} = \mathbf{0}$ [**Positivity property**]

A real vector space with an inner product is called a *real inner product space*, and the four properties in this definition are called the *inner product axioms*.

REMARK In this section we will focus exclusively on real inner product spaces and will refer to them simply as *inner product spaces*. Inner products on complex vector spaces are considered in the exercises.

Motivated by Formulas (29) and (31) of Section 1.2, we make the following definition.

Definition 9.2.2 If V is an inner product space, then we define **norm** and **distance** for vectors in V relative to the inner product by the formulas

$$\|\mathbf{v}\| = \sqrt{\langle \mathbf{v}, \mathbf{v} \rangle} \tag{1}$$

$$d(\mathbf{u}, \mathbf{v}) = \|\mathbf{u} - \mathbf{v}\| = \sqrt{\langle \mathbf{u} - \mathbf{v}, \mathbf{u} - \mathbf{v} \rangle} \tag{2}$$

and we define $\mathbf{u}$ and $\mathbf{v}$ to be **orthogonal** if $\langle \mathbf{u}, \mathbf{v} \rangle = 0$.

EXAMPLE 1
The Dot
Product on R^n
Is an Inner
Product

The dot product on R^n (also called the Euclidean inner product) is the most basic example of an inner product. We are guaranteed that $\langle \mathbf{u}, \mathbf{v} \rangle = \mathbf{u} \cdot \mathbf{v}$ satisfies the inner product axioms because those axioms are derived from properties of the dot product. If vectors are expressed in column form, then we can express the dot product as $\langle \mathbf{u}, \mathbf{v} \rangle = \mathbf{u}^T \mathbf{v}$ [Formula (26) of Section 3.1], in which case we can also view it as an inner product on the vector space of real $n \times 1$ matrices. ∎

EXAMPLE 2
Weighted
Euclidean Inner
Products

If $w_1, w_2, \ldots, w_n$ are *positive* real numbers, and if $\mathbf{u} = (u_1, u_2, \ldots, u_n)$ and $\mathbf{v} = (v_1, v_2, \ldots, v_n)$ are vectors in R^n, then the formula

$$\langle \mathbf{u}, \mathbf{v} \rangle = w_1 u_1 v_1 + w_2 u_2 v_2 + \cdots + w_n u_n v_n \tag{3}$$

defines an inner product on R^n that is called the **weighted Euclidean inner product** with **weights** $w_1, w_2, \ldots, w_n$. The proof that (3) satisfies the four inner product axioms is left as an exercise. The norm of a vector $\mathbf{x} = (x_1, x_2, \ldots, x_n)$ relative to (3) is

$$\|\mathbf{x}\| = \sqrt{\langle \mathbf{x}, \mathbf{x} \rangle} = \sqrt{w_1 x_1^2 + w_2 x_2^2 + \cdots + w_n x_n^2}$$

For example, the formula

$$\langle \mathbf{u}, \mathbf{v} \rangle = 2u_1 v_1 + 3u_2 v_2 \tag{4}$$

defines a weighted Euclidean inner product on R^2 with weights $w_1 = 2$ and $w_2 = 3$, and the norm of the vector $\mathbf{x} = (x_1, x_2)$ relative to this inner product is

$$\|\mathbf{x}\| = \sqrt{\langle \mathbf{x}, \mathbf{x} \rangle} = \sqrt{2x_1^2 + 3x_2^2} \tag{5}$$

∎

REMARK Weighted Euclidean inner products are often used in least squares problems in which some data values are known with greater precision than others. The components of the data vectors are then assigned weights in proportion to their precision and one looks for solutions of $A\mathbf{x} = \mathbf{b}$ that minimize $\|\mathbf{b} - A\mathbf{x}\|$ relative to the resulting weighted Euclidean inner product. The minimizing solutions are called **weighted least squares solutions** of $A\mathbf{x} = \mathbf{b}$, and the problem of finding such solutions is called a **weighted least squares problem**. In the exercises we will describe a method for solving weighted least squares problems by transforming them into ordinary least squares problems.

THE EFFECT OF WEIGHTING ON GEOMETRY

It is important to keep in mind that length, distance, angle, and orthogonality depend on which inner product is being used. For example, $\mathbf{x} = (1, 0)$ is a unit vector relative to the Euclidean inner product but is not a unit vector relative to (4), since (5) implies that its length is

$$\|\mathbf{x}\| = \sqrt{2(1)^2 + 3(0)^2} = \sqrt{2}$$

Also, $\mathbf{u} = (1, -1)$ and $\mathbf{v} = (1, 1)$ are orthogonal with respect to the Euclidean inner product on R^2 but not with respect to (4), since with this inner product we have

$$\langle \mathbf{u}, \mathbf{v} \rangle = 2(1)(1) + 3(-1)(1) = -1 \neq 0$$

Since weighting the Euclidean inner product changes lengths, distances, and angles, it should not be surprising that weighting alters orthogonality and shapes of geometric objects. The following example shows that weighting the Euclidean inner product can distort the unit circle in Figure 9.2.1 into an ellipse.

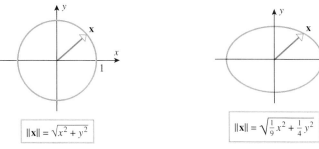

$$\|\mathbf{x}\| = \sqrt{x^2 + y^2}$$

Figure 9.2.1

$$\|\mathbf{x}\| = \sqrt{\tfrac{1}{9}x^2 + \tfrac{1}{4}y^2}$$

Figure 9.2.2

EXAMPLE 3
Unit Circles
Using Weighted
Euclidean Inner
Products on R^2

If V is an inner product space, then we define the **unit circle** (also called the **unit sphere**)[*] to be the set of points in V such that $\|\mathbf{x}\| = 1$. Sketch the unit circle in an xy-coordinate system using the weighted Euclidean inner product

$$\langle \mathbf{u}, \mathbf{v} \rangle = \tfrac{1}{9}u_1 v_1 + \tfrac{1}{4}u_2 v_2 \tag{6}$$

Solution If $\mathbf{x} = (x, y)$, then

$$\|\mathbf{x}\| = \sqrt{\langle \mathbf{x}, \mathbf{x} \rangle} = \sqrt{\tfrac{1}{9}x^2 + \tfrac{1}{4}y^2}$$

so the equation of the unit circle relative to (6) is

$$\sqrt{\tfrac{1}{9}x^2 + \tfrac{1}{4}y^2} = 1 \quad \text{or equivalently,} \quad \frac{x^2}{9} + \frac{y^2}{4} = 1$$

The graph of this equation is the ellipse shown in Figure 9.2.2. ∎

EXAMPLE 4
The Integral
Inner Product
on $C[a, b]$

If f and g are continuous functions on the interval $[a, b]$, then the formula

$$\langle f, g \rangle = \int_a^b f(x)g(x)\, dx \tag{7}$$

defines an inner product on the vector space $C[a, b]$ that we will call the **integral inner product**. The norm of a function f relative to this inner product is

$$\|f\| = \sqrt{\langle f, f \rangle} = \sqrt{\int_a^b [f(x)]^2\, dx} \tag{8}$$

The proof that (7) satisfies the four inner product axioms is as follows:

- **Axiom 1**—If f and g are continuous functions on $[a, b]$, then

$$\langle f, g \rangle = \int_a^b f(x)g(x)\, dx = \int_a^b g(x)f(x)\, dx = \langle g, f \rangle$$

- **Axiom 2**—If f, g, and h are continuous functions on $[a, b]$, then

$$\langle f + g, h \rangle = \int_a^b [f(x) + g(x)]h(x)\, dx = \int_a^b f(x)h(x)\, dx + \int_a^b g(x)h(x)\, dx = \langle f, h \rangle + \langle g, h \rangle$$

---

[*]The term *unit circle* is more appropriate for R^2 and *unit sphere* is more appropriate for R^3. For higher-dimensional spaces either term is reasonable.

- **Axiom 3**—If f and g are continuous functions on $[a, b]$ and k is a scalar, then

$$\langle kf, g \rangle = \int_a^b kf(x)g(x)\, dx = k \int_a^b f(x)g(x)\, dx = k\langle f, g \rangle$$

- **Axiom 4**—If f is a continuous function on $[a, b]$, then

$$\langle f, f \rangle = \int_a^b f(x)f(x)\, dx = \int_a^b [f(x)]^2\, dx \geq 0 \tag{9}$$

Moreover, it follows that $\langle f, f \rangle > 0$ if there is any point x_0 in $[a, b]$ at which $f(x_0) \neq 0$, the reason being that the continuity of f at the point x_0 forces $f(x)$ to be nonzero on some interval of values in $[a, b]$ containing x_0, and this in turn implies that $[f(x)]^2 > 0$ on that interval. Thus, $\langle f, f \rangle = 0$ if and only if $f = 0$, as required by Axiom 4.   ∎

EXAMPLE 5
Orthogonal
Functions in
$C[0, 2\pi]$

Show that if p and q are *distinct* positive integers, then the functions $\cos px$ and $\cos qx$ are orthogonal with respect to the inner product

$$\langle f, g \rangle = \int_0^{2\pi} f(x)g(x)\, dx$$

Solution Let $f(x) = \cos px$ and $g(x) = \cos qx$. We must show that

$$\langle f, g \rangle = \int_0^{2\pi} \cos px \cos qx\, dx = 0 \tag{10}$$

To prove this, we will need the trigonometric identity

$$\cos(A)\cos(B) = \tfrac{1}{2}[\cos(A + B) + \cos(A - B)]$$

Using this identity yields

$$\int_0^{2\pi} \cos px \cos qx\, dx = \frac{1}{2} \int_0^{2\pi} [\cos(px + qx) + \cos(px - qx)]\, dx$$

$$= \frac{1}{2} \int_0^{2\pi} [\cos(p + q)x + \cos(p - q)x]\, dx$$

$$= \frac{\sin[2\pi(p + q)]}{2(p + q)} + \frac{\sin[2\pi(p - q)]}{2(p - q)}$$

$$= 0 + 0 = 0$$

from which (10) follows.   ∎

**ALGEBRAIC PROPERTIES
OF INNER PRODUCTS**

Extending geometric theorems about R^n to general inner product spaces is simply a matter of changing notation appropriately if the theorems do not involve notions of basis or dimension. Extensions of geometric theorems that involve notions of basis or dimension often require some sort of finite-dimensionality condition to be valid. Here are generalized versions of Theorems 1.2.7, 1.2.11, 1.2.12, 1.2.13, and 1.2.15.

> **Theorem 9.2.3** *If $\mathbf{u}$, $\mathbf{v}$, and $\mathbf{w}$ are vectors in an inner product space V, and if k is a scalar, then:*
>
> (a) $\langle \mathbf{0}, \mathbf{v} \rangle = \langle \mathbf{v}, \mathbf{0} \rangle = 0$
>
> (b) $\langle \mathbf{u} + \mathbf{v}, \mathbf{w} \rangle = \langle \mathbf{u}, \mathbf{w} \rangle + \langle \mathbf{v}, \mathbf{w} \rangle$
>
> (c) $\langle \mathbf{u}, \mathbf{v} - \mathbf{w} \rangle = \langle \mathbf{u}, \mathbf{v} \rangle - \langle \mathbf{u}, \mathbf{w} \rangle$
>
> (d) $\langle \mathbf{u} - \mathbf{v}, \mathbf{w} \rangle = \langle \mathbf{u}, \mathbf{w} \rangle - \langle \mathbf{v}, \mathbf{w} \rangle$
>
> (e) $k\langle \mathbf{u}, \mathbf{v} \rangle = \langle \mathbf{u}, k\mathbf{v} \rangle$

Theorem 9.2.4 (*Theorem of Pythagoras*)   *If* **u** *and* **v** *are orthogonal vectors in an inner product space* V, *then*

$$\|\mathbf{u} + \mathbf{v}\|^2 = \|\mathbf{u}\|^2 + \|\mathbf{v}\|^2$$

Theorem 9.2.5 (*Cauchy–Schwarz Inequality*)   *If* **u** *and* **v** *are vectors in an inner product space* V, *then*

$$\langle \mathbf{u}, \mathbf{v} \rangle^2 \leq \|\mathbf{u}\|^2 \|\mathbf{v}\|^2$$

or equivalently (*by taking square roots*),

$$|\langle \mathbf{u}, \mathbf{v} \rangle| \leq \|\mathbf{u}\| \|\mathbf{v}\|$$

Theorem 9.2.6 (*Triangle Inequalities*)   *If* **u**, **v**, *and* **w** *are vectors in an inner product space* V, *then:*

(a) $\|\mathbf{u} + \mathbf{v}\| \leq \|\mathbf{u}\| + \|\mathbf{v}\|$ [Triangle inequality for vectors]

(b) $d(\mathbf{u}, \mathbf{v}) < d(\mathbf{u}, \mathbf{w}) + d(\mathbf{w}, \mathbf{v})$ [Triangle inequality for distances]

REMARK   Formula (30) of Section 1.2 for the angle θ between nonzero vectors **u** and **v** can be extended to general inner product spaces as

$$\theta = \cos^{-1} \left(\frac{\langle \mathbf{u}, \mathbf{v} \rangle}{\|\mathbf{u}\| \|\mathbf{v}\|} \right) \tag{11}$$

since the Cauchy–Schwarz inequality implies that the argument of the inverse cosine is between -1 and 1 (verify).

EXAMPLE 6
The Theorem of Pythagoras

Show that the vectors $\mathbf{u} = (1, -1)$ and $\mathbf{v} = (6, 4)$ are orthogonal with respect to the inner product defined by (4), and then confirm the theorem of Pythagoras for these vectors.

Solution   The vectors are orthogonal since

$$\langle \mathbf{u}, \mathbf{v} \rangle = 2(1)(6) + 3(-1)(4) = 0$$

To confirm the theorem of Pythagoras, we use Formula (5). This yields

$$\|\mathbf{u}\|^2 = 2(1)^2 + 3(-1)^2 = 5$$
$$\|\mathbf{v}\|^2 = 2(6)^2 + 3(4)^2 = 120$$
$$\|\mathbf{u} + \mathbf{v}\|^2 = \|(7, 3)\| = 2(7)^2 + 3(3)^2 = 125$$

Thus, $\|\mathbf{u} + \mathbf{v}\|^2 = \|\mathbf{u}\|^2 + \|\mathbf{v}\|^2$, as guaranteed by the theorem of Pythagoras. Note that the vectors **u** and **v** are not orthogonal with respect to the Euclidean inner product on R^n, so **u** and **v** do not satisfy the theorem of Pythagoras for that inner product.   ■

EXAMPLE 7
Some Famous Integral Inequalities

Suppose that f and g are continuous functions on the interval $[a, b]$ and that $C[a, b]$ has the integral inner product (7) of Example 4. The Cauchy–Schwarz inequality states that

$$|\langle f, g \rangle| \leq \|f\| \|g\|$$

which implies that

$$\left| \int_a^b f(x)g(x)\, dx \right| \leq \sqrt{\int_a^b [f(x)]^2\, dx} \ \sqrt{\int_a^b [g(x)]^2\, dx} \tag{12}$$

and the triangle inequality for vectors states that

$$\|f + g\| \leq \|f\| + \|g\|$$

which implies that

$$\sqrt{\int_a^b [f(x) + g(x)]^2 \, dx} \le \sqrt{\int_a^b [f(x)]^2 \, dx} + \sqrt{\int_a^b [g(x)]^2 \, dx} \tag{13}$$

Formulas (12) and (13) play an important role in many different applications that are beyond the scope of this text. However, the fact that we were able to deduce these inequalities from general results about vector spaces illustrates the power of the methods that we have developed. Formula (13) is sometimes called the ***Minkowski inequality*** for integrals in honor of the German mathematical physicist Hermann Minkowski (1864–1909). ■

ORTHONORMAL BASES   Recall from Theorem 7.9.1 that an orthogonal set of nonzero vectors in R^n is linearly independent. The same is true in a general inner product space V.

EXAMPLE 8
An Orthogonal
Basis for T_n

A function of the form

$$f(x) = (c_0 + c_1 \cos x + c_2 \cos 2x + \cdots + c_n \cos nx)$$
$$+ (d_1 \sin x + d_2 \sin 2x + \cdots + d_n \sin nx)$$

is called a ***trigonometric polynomial***. If c_n and d_n are not both zero, then $f(x)$ is said to have ***order n***. Also, a constant function is regarded to be a trigonometric polynomial of order zero. The set of all trigonometric polynomials of order n or less is the subspace of $C(-\infty, \infty)$ that is spanned by the functions in the set

$$S = \{1, \cos x, \cos 2x, \ldots, \cos nx, \sin x, \sin 2x, \ldots, \sin nx\} \tag{14}$$

We will denote this subspace by T_n. Show that S is an orthogonal basis for T_n with respect to the integral inner product

$$\langle f, g \rangle = \int_0^{2\pi} f(x)g(x) \, dx \tag{15}$$

Solution The set S spans T_n, so it suffices to show that S is an orthogonal set, since the linear independence will then follow from the orthogonality. To prove that S is an orthogonal set we must show that the inner product of any two distinct functions in this set is zero; that is,

$$\langle 1, \cos kx \rangle = \int_0^{2\pi} \cos kx \, dx = 0 \qquad (k = 1, 2, \ldots, n) \tag{16}$$

$$\langle 1, \sin kx \rangle = \int_0^{2\pi} \sin kx \, dx = 0 \qquad (k = 1, 2, \ldots, n) \tag{17}$$

$$\langle \cos px, \cos qx \rangle = \int_0^{2\pi} \cos px \cos qx \, dx = 0 \qquad (p, q = 1, 2, \ldots, n \text{ and } p \ne q) \tag{18}$$

$$\langle \sin px, \sin qx \rangle = \int_0^{2\pi} \sin px \sin qx \, dx = 0 \qquad (p, q = 1, 2, \ldots, n \text{ and } p \ne q) \tag{19}$$

$$\langle \cos px, \sin qx \rangle = \int_0^{2\pi} \cos px \sin qx \, dx = 0 \qquad (p, q = 1, 2, \ldots, n) \tag{20}$$

The first two integrals can be evaluated using basic integration techniques. The integration in (18) was carried out in Example 5, and the integrations in (19) and (20) can be performed similarly. We omit the details. ■

EXAMPLE 9
Orthonormal
Basis for T_n

Find an orthonormal basis for T_n relative to the inner product in (15) by normalizing the functions in the orthogonal basis S given in (14).

Solution It follows from (8) and standard integration procedures that

$$\|1\| = \sqrt{\langle 1, 1 \rangle} = \sqrt{\int_0^{2\pi} [1]^2 \, dx} = \sqrt{2\pi}$$

$$\| \cos px \| = \sqrt{\langle \cos px, \cos px \rangle} = \sqrt{\int_0^{2\pi} \cos^2 px \, dx} = \sqrt{\pi} \qquad (p = 1, 2, \ldots, n)$$

$$\| \sin qx \| = \sqrt{\langle \sin qx, \sin qx \rangle} = \sqrt{\int_0^{2\pi} \sin^2 qx \, dx} = \sqrt{\pi} \qquad (q = 1, 2, \ldots, n)$$

Thus, normalizing the vectors in S yields the orthonormal basis

$$S' = \left\{ \frac{1}{\sqrt{2\pi}}, \frac{\cos x}{\sqrt{\pi}}, \frac{\cos 2x}{\sqrt{\pi}}, \ldots, \frac{\cos nx}{\sqrt{\pi}}, \frac{\sin x}{\sqrt{\pi}}, \frac{\sin 2x}{\sqrt{\pi}}, \ldots, \frac{\sin nx}{\sqrt{\pi}} \right\} \qquad (21)$$

■

BEST APPROXIMATION   We will now consider the following problem.

> **Best Approximation Problem for Functions**   Given a function f that is continuous on an interval $[a, b]$, find the best approximation to f that can be obtained using only functions from a specified finite-dimensional subspace W of $C[a, b]$.

Of course, we will have to clarify what we mean by "best approximation," but here are two typical examples of this type:

1. Find the best approximation to $f(x) = \sin x$ over the interval $[0, 1]$ that can be obtained using a polynomial of degree n or less.

2. Find the best approximation to $f(x) = x$ over the interval $[0, 2\pi]$ that can be obtained using a trigonometric polynomial of order n or less.

Now let us consider how we might interpret the term *best approximation*. Whenever one quantity is approximated by another there needs to be a way of measuring the error to assess the accuracy of the approximation. For example, if we approximate a number x by some other number $\hat{x}$, then it is usual to take the error in the approximation to be

$$E = |x - \hat{x}|$$

where the absolute value serves to eliminate the distinction between positive and negative errors. In the case where we want to approximate a continuous function f by some other continuous function $\hat{f}$ over an interval $[a, b]$, the problem of describing the error is more complicated because we must account for differences between $f(x)$ and $\hat{f}(x)$ over the *entire* interval $[a, b]$, not just at a single point. One such error measure is

$$E = \int_a^b |f(x) - \hat{f}(x)| \, dx \qquad (22)$$

which can be interpreted geometrically as the area between the graphs of f and $\hat{f}$ over the interval $[a, b]$. Thus, the smaller area, the better the approximation (Figure 9.2.3).

Although Formula (22) is appealing geometrically, it is often more convenient to take the error in the approximation of f by $\hat{f}$ to be the distance between f and $\hat{f}$ relative to the integral inner product on $C[a, b]$; that is,

$$E = d(f, \hat{f}) = \| f - \hat{f} \| = \sqrt{\int_a^b [f(x) - \hat{f}(x)]^2 \, dx} \qquad (23)$$

With this measure of error the best approximation for functions now becomes a minimum distance problem that is analogous to that posed at the beginning of Section 7.8.

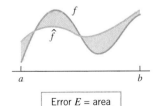

Error E = area between f and $\hat{f}$.

Figure 9.2.3

Minimum Distance Problem for Functions Suppose that $C[a, b]$ has the integral inner product. Given a subspace W of $C[a, b]$ and a function f that is continuous on the interval $[a, b]$, find a function $\hat{f}$ in W that is closest to f in the sense that $\|f - \hat{f}\| < \|f - g\|$ for every function g in W that is distinct from $\hat{f}$. Such a function $\hat{f}$, if it exists, is called a **best mean square approximation** to f from W.

REMARK The terminology *mean square approximation* arises from the fact that any function $\hat{f}$ in W that minimizes (23) also minimizes

$$E_m = \frac{1}{b - a} \int_a^b [f(x) - \hat{f}(x)]^2 \, dx \tag{24}$$

and conversely. Expression (24) is called the **mean square error** in the approximation of f by $\hat{f}$.

Motivated by our experience with minimum distance problems in R^n, it is natural to expect that best mean square approximations are closely related to orthogonal projections. Accordingly, it should not be surprising that the solution of the minimum distance problem is provided by the following analog of Formula (7) in Section 7.9.

Theorem 9.2.7 *If W is a finite-dimensional subspace of $C[a, b]$, and if $\{f_1, f_2, \dots, f_k\}$ is an orthonormal basis for W, then each function f in $C[a, b]$ has a unique best mean square approximation $\hat{f}$ in W, and that approximation is*

$$\hat{f} = \langle f, f_1 \rangle f_1 + \langle f, f_2 \rangle f_2 + \cdots + \langle f, f_k \rangle f_k \tag{25}$$

where

$$\langle f, f_j \rangle = \int_a^b f(x) f_j(x) \, dx \qquad (j = 1, 2, \dots, k)$$

FOURIER SERIES The case in which the basis vectors are trigonometric polynomials is particularly important because the periodicity of these functions makes them a natural choice in the study of periodic phenomena. However, the deeper significance of trigonometric polynomials as basis vectors was first revealed by the French mathematician Joseph Fourier (1768–1830), who, in the course of his studies of heat flow, realized that it is possible to approximate most important kinds of functions to any degree of accuracy by trigonometric polynomials of sufficiently high order. We will now show how to find such approximations.

Let f be a continuous function on the interval $[0, 2\pi]$, and let us consider how we might compute the best mean square approximation to f by a trigonometric polynomial of order n or less. Since we know that this approximation is the orthogonal projection of f on T_n, we can compute it using Formula (25) if we have an orthonormal basis for T_n; we will use the orthonormal basis (21) in Example 9. Let us denote these basis vectors by

$$f_0 = \frac{1}{\sqrt{2\pi}}, \quad f_1 = \frac{\cos x}{\sqrt{\pi}}, \dots, \quad f_n = \frac{\cos nx}{\sqrt{\pi}}, \quad f_{n+1} = \frac{\sin x}{\sqrt{\pi}}, \dots, \quad f_{2n} = \frac{\sin nx}{\sqrt{\pi}}$$

and let the orthogonal projection of f on T_n be

$$\text{proj}_{T_n} f = \frac{a_0}{2} + [a_1 \cos x + \cdots + a_n \cos nx] + [b_1 \sin x + \cdots + b_n \sin nx] \tag{26}$$

where the coefficients are to be determined. It follows from (25) with the appropriate adjustments in notation that these coefficients are

$$a_0 = \frac{2}{\sqrt{2\pi}} \langle f, f_0 \rangle, \quad a_1 = \frac{1}{\sqrt{\pi}} \langle f, f_1 \rangle, \dots, \quad a_n = \frac{1}{\sqrt{\pi}} \langle f, f_n \rangle$$

$$b_1 = \frac{1}{\sqrt{\pi}} \langle f, f_{n+1} \rangle, \dots, \quad b_n = \frac{1}{\sqrt{\pi}} \langle f, f_{2n} \rangle$$

Thus,

$$a_0 = \frac{2}{\sqrt{2\pi}} \langle f, f_0 \rangle = \frac{2}{\sqrt{2\pi}} \int_0^{2\pi} f(x) \frac{1}{\sqrt{2\pi}}\, dx = \frac{1}{\pi} \int_0^{2\pi} f(x)\, dx$$

$$a_1 = \frac{1}{\sqrt{\pi}} \langle f, f_1 \rangle = \frac{1}{\sqrt{\pi}} \int_0^{2\pi} f(x) \frac{\cos x}{\sqrt{\pi}}\, dx = \frac{1}{\pi} \int_0^{2\pi} f(x) \cos x\, dx$$

$$\vdots$$

$$a_n = \frac{1}{\sqrt{\pi}} \langle f, f_n \rangle = \frac{1}{\sqrt{\pi}} \int_0^{2\pi} f(x) \frac{\cos nx}{\sqrt{\pi}}\, dx = \frac{1}{\pi} \int_0^{2\pi} f(x) \cos nx\, dx$$

$$b_1 = \frac{1}{\sqrt{\pi}} \langle f, f_{n+1} \rangle = \frac{1}{\sqrt{\pi}} \int_0^{2\pi} f(x) \frac{\sin x}{\sqrt{\pi}}\, dx = \frac{1}{\pi} \int_0^{2\pi} f(x) \sin x\, dx$$

$$\vdots$$

$$b_n = \frac{1}{\sqrt{\pi}} \langle f, f_{2n} \rangle = \frac{1}{\sqrt{\pi}} \int_0^{2\pi} f(x) \frac{\sin nx}{\sqrt{\pi}}\, dx = \frac{1}{\pi} \int_0^{2\pi} f(x) \sin nx\, dx$$

or more briefly,

$$a_k = \frac{1}{\pi} \int_0^{2\pi} f(x) \cos kx\, dx, \quad b_k = \frac{1}{\pi} \int_0^{2\pi} f(x) \sin kx\, dx \qquad (27\text{–}28)$$

The numbers $a_0, a_1, \ldots, a_n, b_1, \ldots, b_n$ are called the ***Fourier coefficients*** of f, and (26) is called the nth-order Fourier approximation of f. Here is a numerical example.

EXAMPLE 10

(a) Find the second-order Fourier approximation of $f(x) = x$.

(b) Find the nth-order Fourier approximation of $f(x) = x$.

Solution (*a*) The second-order Fourier approximation to x is

$$x \approx \frac{a_0}{2} + [a_1 \cos x + a_2 \cos 2x] + [b_1 \sin x + b_2 \sin 2x] \qquad (29)$$

where a_0, a_1, a_2, b_1, and b_2 are the Fourier coefficients of x. It follows from (27) with $k = 0$ that

$$a_0 = \frac{1}{\pi} \int_0^{2\pi} f(x)\, dx = \frac{1}{\pi} \int_0^{2\pi} x\, dx = 2\pi$$

All other Fourier coefficients can be obtained from (27) and (28) using integration by parts:

$$a_k = \frac{1}{\pi} \int_0^{2\pi} f(x) \cos kx\, dx = \frac{1}{\pi} \int_0^{2\pi} x \cos kx\, dx = \frac{1}{\pi} \left[\frac{1}{k^2} \cos kx + \frac{x}{k} \sin kx \right]_0^{2\pi} = 0 \qquad (30)$$

$$b_k = \frac{1}{\pi} \int_0^{2\pi} f(x) \sin kx\, dx = \frac{1}{\pi} \int_0^{2\pi} x \sin kx\, dx = \frac{1}{\pi} \left[\frac{1}{k^2} \sin kx - \frac{x}{k} \cos kx \right]_0^{2\pi} = -\frac{2}{k} \qquad (31)$$

Thus, $a_1 = 0, a_2 = 0, b_1 = -2$, and $b_2 = -1$. Substituting these values in (29) yields the second-order Fourier approximation

$$x \approx \pi - 2 \sin x - \sin 2x$$

Solution (*b*) The nth-order Fourier approximation to x is

$$x \approx \frac{a_0}{2} + [a_1 \cos x + \cdots + a_n \cos nx] + [b_1 \sin x + \cdots + b_n \sin nx]$$

Linear Algebra in History

The 1807 memoir *On the Propogation of Heat in Solid Bodies* by the French mathematician Joseph Fourier developed the basic work on Fourier series and is now regarded as one of the great milestones in applied mathematics. However, it was quite controversial at the time and heavily criticized for its lack of mathematical precision. Actually, it is only by a stroke of luck that this work even exists, since Fourier was arrested in 1794 during the French Revolution and was in serious danger of going to the guillotine. Fortunately for the world of mathematics the political climate changed and Fourier was freed after Robespierre was beheaded. Fourier subsequently became a scientific advisor to Napoleon during the invasion of Egypt and was embroiled in the political problems of Napoleon's rise and fall for many years.

Jean Baptiste Joseph Fourier
(1768–1830)

so from (30) and (31) this approximation is

$$x \approx \pi - 2\left(\sin x + \frac{\sin 2x}{2} + \frac{\sin 3x}{3} + \cdots + \frac{\sin nx}{n}\right)$$

The graphs of $f(x) = x$ and some of its Fourier approximations are shown in Figure 9.2.4. ∎

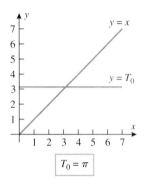

$$T_0 = \pi$$

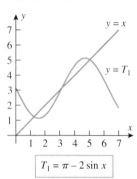

$$T_1 = \pi - 2\sin x$$

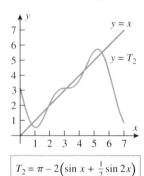

$$T_2 = \pi - 2\left(\sin x + \frac{1}{2}\sin 2x\right)$$

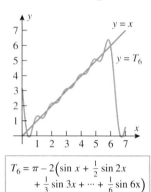

$$T_6 = \pi - 2\left(\sin x + \frac{1}{2}\sin 2x + \frac{1}{3}\sin 3x + \cdots + \frac{1}{6}\sin 6x\right)$$

Figure 9.2.4

It can be proved that if f is continuous on the interval $[0, 2\pi]$, then the mean square error in the nth-order Fourier approximation of f approaches zero as $n \to \infty$. This is denoted by writing

$$f(x) \doteq \frac{a_0}{2} + \sum_{k=1}^{\infty}(a_k \cos kx + b_k \sin kx)$$

which is called the ***Fourier series*** for f.

GENERAL INNER PRODUCTS ON R^n

Inner products on R^n arise naturally in the study of functions of the form

$$\mathbf{x}^T A \mathbf{y} \tag{32}$$

in which A is an $n \times n$ matrix with real entries and $\mathbf{x}$ and $\mathbf{y}$ are vectors in R^n that are expressed in column form. For each $\mathbf{x}$ and $\mathbf{y}$ in R^n the quantity in (32) is a 1×1 matrix and hence can be treated as a scalar. Moreover, if $\mathbf{y}$ is fixed, then the mapping $\mathbf{x} \to \mathbf{x}^T A \mathbf{y}$ is a linear transformation from R^n to R, and if $\mathbf{x}$ is fixed, then the mapping $\mathbf{y} \to \mathbf{x}^T A \mathbf{y}$ is also a linear transformation from R^n to R (Exercise P3); thus, (32) is called a ***bilinear form*** or, more precisely, the ***bilinear form associated with*** A. The following theorem shows that *all* inner products on R^n arise from certain types of bilinear forms.

Theorem 9.2.8 *If A is a positive definite symmetric matrix, and if vectors in R^n are in column form, then*

$$\langle \mathbf{u}, \mathbf{v} \rangle = \mathbf{u}^T A \mathbf{v} \tag{33}$$

is an inner product on R^n, and conversely, if $\langle \mathbf{u}, \mathbf{v} \rangle$ is any inner product on R^n, then there exists a unique positive definite symmetric matrix A for which (33) holds for all column vectors $\mathbf{u}$ and $\mathbf{v}$ in R^n.

The matrix A in this theorem is called the ***matrix for the inner product***.

Proof To prove the first statement in the theorem we must confirm that (33) satisfies the four inner product axioms in Definition 9.2.1:

- **Axiom 1**—Since $\mathbf{u}^T A \mathbf{v}$ is a 1×1 matrix, it is symmetric and hence

$$\langle \mathbf{u}, \mathbf{v} \rangle = \mathbf{u}^T A \mathbf{v} = (\mathbf{u}^T A \mathbf{v})^T = \mathbf{v}^T A^T \mathbf{u} = \mathbf{v}^T A \mathbf{u} = \langle \mathbf{v}, \mathbf{u} \rangle$$

- **Axiom 2**—Using properties of the transpose we obtain

$$\langle \mathbf{u} + \mathbf{v}, \mathbf{w} \rangle = (\mathbf{u} + \mathbf{v})^T A \mathbf{w} = (\mathbf{u}^T + \mathbf{v}^T) A \mathbf{w} = \mathbf{u}^T A \mathbf{w} + \mathbf{v}^T A \mathbf{w} = \langle \mathbf{u}, \mathbf{w} \rangle + \langle \mathbf{v}, \mathbf{w} \rangle$$

- **Axiom 3**—Again using properties of the transpose we obtain

$$\langle k\mathbf{u}, \mathbf{v} \rangle = (k\mathbf{u})^T A \mathbf{v} = (k\mathbf{u}^T) A \mathbf{v} = k(\mathbf{u}^T A \mathbf{v}) = k\langle \mathbf{u}, \mathbf{v} \rangle$$

- **Axiom 4**—Since A is positive definite and symmetric, the expression $\langle \mathbf{v}, \mathbf{v} \rangle = \mathbf{v}^T A \mathbf{v}$ is a positive definite quadratic form, and hence Definition 8.4.2 implies that

$$\langle \mathbf{v}, \mathbf{v} \rangle = \mathbf{v}^T A \mathbf{v} \geq 0 \quad \text{and} \quad \langle \mathbf{v}, \mathbf{v} \rangle = 0 \text{ if and only if } \mathbf{v} = \mathbf{0}$$

To prove the converse, suppose that $\langle \mathbf{u}, \mathbf{v} \rangle$ is an inner product on R^n and that $\mathbf{e}_1, \mathbf{e}_2, \ldots, \mathbf{e}_n$ are the standard unit vectors. It can be shown that

$$A = \begin{bmatrix} \langle \mathbf{e}_1, \mathbf{e}_1 \rangle & \langle \mathbf{e}_1, \mathbf{e}_2 \rangle & \cdots & \langle \mathbf{e}_1, \mathbf{e}_n \rangle \\ \langle \mathbf{e}_2, \mathbf{e}_1 \rangle & \langle \mathbf{e}_2, \mathbf{e}_2 \rangle & \cdots & \langle \mathbf{e}_2, \mathbf{e}_n \rangle \\ \vdots & \vdots & & \vdots \\ \langle \mathbf{e}_n, \mathbf{e}_1 \rangle & \langle \mathbf{e}_n, \mathbf{e}_2 \rangle & \cdots & \langle \mathbf{e}_n, \mathbf{e}_n \rangle \end{bmatrix} \tag{34}$$

is positive definite and symmetric and that $\langle \mathbf{u}, \mathbf{v} \rangle = \mathbf{u}^T A \mathbf{v}$ for all $\mathbf{u}$ and $\mathbf{v}$ in R^n. Moreover, this is the only matrix with this property. ∎

CONCEPT PROBLEM Show that Formula (33) in Theorem 9.2.8 can be written in dot product notation as $\langle \mathbf{u}, \mathbf{v} \rangle = A\mathbf{u} \cdot \mathbf{v}$.

EXAMPLE 11
Weighted
Euclidean Inner
Products
Revisited

If $w_1, w_2, \ldots, w_n$ are positive real numbers, then

$$A = \begin{bmatrix} w_1 & 0 & \cdots & 0 \\ 0 & w_2 & \cdots & 0 \\ \vdots & \vdots & \ddots & \vdots \\ 0 & 0 & \cdots & w_n \end{bmatrix}$$

is positive definite and symmetric (verify), so we know that $\langle \mathbf{u}, \mathbf{v} \rangle = \mathbf{u}^T A \mathbf{v}$ is an inner product. It is, in fact, the weighted Euclidean inner product with weights $w_1, w_2, \ldots, w_n$, since

$$\langle \mathbf{u}, \mathbf{v} \rangle = \mathbf{u}^T A \mathbf{v} = \begin{bmatrix} u_1 & u_2 & \cdots & u_n \end{bmatrix} \begin{bmatrix} w_1 & 0 & \cdots & 0 \\ 0 & w_2 & \cdots & 0 \\ \vdots & \vdots & \ddots & \vdots \\ 0 & 0 & \cdots & w_n \end{bmatrix} \begin{bmatrix} v_1 \\ v_2 \\ \vdots \\ v_n \end{bmatrix} = w_1 u_1 v_1 + w_2 u_2 v_2 + \cdots + w_n u_n v_n$$

∎

EXAMPLE 12
Deriving an Inner
Product from a
Positive Definite
Matrix

Show that $\langle \mathbf{u}, \mathbf{v} \rangle = 6u_1 v_1 - 2u_1 v_2 - 2u_2 v_1 + 3u_2 v_2$ defines an inner product on R^2.

Solution The given expression can be written in matrix form as $\langle \mathbf{u}, \mathbf{v} \rangle = \mathbf{u}^T A \mathbf{v}$, where

$$A = \begin{bmatrix} 6 & -2 \\ -2 & 3 \end{bmatrix}$$

The matrix A is positive definite by Theorem 8.4.5, so $\langle \mathbf{u}, \mathbf{v} \rangle$ is an inner product by Theorem 9.2.8. ∎

Exercise Set 9.2

1. Let $\langle \mathbf{u}, \mathbf{v} \rangle$ be the weighted Euclidean inner product on R^2 defined by

$$\langle \mathbf{u}, \mathbf{v} \rangle = 3u_1v_1 + 2u_2v_2$$

and consider the vectors $\mathbf{u} = (2, -3)$ and $\mathbf{v} = (1, 4)$.
(a) Compute $\langle \mathbf{u}, \mathbf{v} \rangle$.
(b) Compute $\|\mathbf{u}\|$ and $\|\mathbf{v}\|$.
(c) Compute the cosine of the angle between $\mathbf{u}$ and $\mathbf{v}$.
(d) Compute the distance between $\mathbf{u}$ and $\mathbf{v}$.

2. Follow the directions given in Exercise 1 with
$\langle \mathbf{u}, \mathbf{v} \rangle = u_1v_1 + 4u_2v_2$, $\mathbf{u} = (-1, 6)$, and $\mathbf{v} = (2, -5)$.

3. Show that the vectors $\mathbf{u} = (1, 1)$ and $\mathbf{v} = (2, -3)$ are orthogonal with respect to the inner product of Exercise 1, and confirm the theorem of Pythagoras for these vectors.

4. Show that the vectors $\mathbf{u} = (3, 1)$ and $\mathbf{v} = (4, -3)$ are orthogonal with respect to the inner product of Exercise 2, and confirm the theorem of Pythagoras for these vectors.

In Exercises 5 and 6, let $\mathbf{u} = (u_1, u_2, u_3)$ and $\mathbf{v} = (v_1, v_2, v_3)$. The stated expression $\langle \mathbf{u}, \mathbf{v} \rangle$ does not define an inner product on R^3 because one or more of the inner product axioms do not hold. List all axioms in Definition 9.2.1 that fail.

5. (a) $\langle \mathbf{u}, \mathbf{v} \rangle = u_2v_2 + u_3v_3$
(b) $\langle \mathbf{u}, \mathbf{v} \rangle = u_1v_1 + u_2v_2^2 + u_3v_3^3$
(c) $\langle \mathbf{u}, \mathbf{v} \rangle = u_1v_1 - u_2v_2 + u_3v_3$

6. (a) $\langle \mathbf{u}, \mathbf{v} \rangle = u_1v_1 + u_3v_3$
(b) $\langle \mathbf{u}, \mathbf{v} \rangle = \sqrt{u_1v_1 + u_2v_2 + u_3v_3}$
(c) $\langle \mathbf{u}, \mathbf{v} \rangle = u_1v_1 + u_2v_2 - u_3v_3$

7. Verify the Cauchy–Schwarz inequality and the triangle inequality for vectors for the inner product and vectors in Exercise 1.

8. Verify the Cauchy–Schwarz inequality and the triangle inequality for vectors for the inner product and vectors in Exercise 2.

In Exercises 9 and 10, confirm that $S = \{\mathbf{v}_1, \mathbf{v}_2, \mathbf{v}_3\}$ is an orthogonal set relative to the given weighted Euclidean inner product on R^3, and convert S to an orthonormal set by normalizing the vectors.

9. $\langle \mathbf{u}, \mathbf{v} \rangle = u_1v_1 + 3u_2v_2 + 2u_3v_3$; $\mathbf{v}_1 = (1, 1, 1)$, $\mathbf{v}_2 = (1, -1, 1)$, $\mathbf{v}_3 = (2, 0, -1)$

10. $\langle \mathbf{u}, \mathbf{v} \rangle = 2u_1v_1 + u_2v_2 + 3u_3v_3$; $\mathbf{v}_1 = (3, 3, 3)$, $\mathbf{v}_2 = (2, 2, -2)$, $\mathbf{v}_3 = (-4, 8, 0)$

In Exercises 11 and 12, sketch the unit circle for the inner product.

11. $\langle \mathbf{u}, \mathbf{v} \rangle = \frac{1}{16}u_1v_1 + \frac{1}{4}u_2v_2$ **12.** $\langle \mathbf{u}, \mathbf{v} \rangle = \frac{1}{9}u_1v_1 + u_2v_2$

In Exercises 13 and 14, find a weighted Euclidean inner product on R^2 whose unit circle is the given ellipse.

13. **14.**

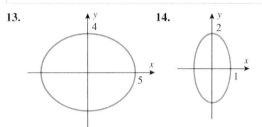

15. Let $C[0, 1]$ have the integral inner product

$$\langle f, g \rangle = \int_0^1 f(x)g(x)\, dx$$

and suppose that $f(x) = x$ and $g(x) = x^2$.
(a) Compute $\langle f, g \rangle$ for the given functions.
(b) Compute $\|f\|$ and $\|g\|$.
(c) Compute the cosine of the angle between f and g.
(d) Compute the distance between f and g.

16. Let $C[-1, 1]$ have the integral inner product

$$\langle f, g \rangle = \int_{-1}^1 f(x)g(x)\, dx$$

and suppose that $f(x) = x^2 - x$ and $g(x) = x + 1$.
(a) Compute $\langle f, g \rangle$ for the given functions.
(b) Compute $\|f\|$ and $\|g\|$.
(c) Compute the cosine of the angle between f and g.
(d) Compute the distance between f and g.

17. Let $C[0, 1]$ have the integral inner product of Exercise 15.
(a) Show that $f_1(x) = 1$ and $f_2(x) = \frac{1}{2} - x$ are orthogonal.
(b) Confirm that f_1 and f_2 satisfy the theorem of Pythagoras.
(c) Find the best mean square approximation of the function $f(x) = x^2$ by functions in span$\{f_1, f_2\}$.

18. Let $C[-1, 1]$ have the integral inner product of Exercise 16.
(a) Show that $f_1(x) = x$ and $f_2(x) = x^2 - 1$ are orthogonal.
(b) Confirm that f_1 and f_2 satisfy the theorem of Pythagoras.
(c) Find the best mean square approximation of the constant function $f(x) = 1$ by functions in span$\{f_1, f_2\}$.

19. Show that if p and q are distinct positive integers, then the functions $f(x) = \sin px$ and $g(x) = \sin qx$ are orthogonal with respect to the inner product in Example 8.

20. Show that if p and q are positive integers, then the functions $f(x) = \cos px$ and $g(x) = \sin qx$ are orthogonal with respect to the inner product in Example 8.

21. Find the second-order Fourier approximation of $f(x) = e^x$.

22. Find the second-order Fourier approximation of $f(x) = e^{-x}$.

23. Find the nth-order Fourier approximation of $f(x) = 3x$.

24. Find the nth-order Fourier approximation of $f(x) = 1+2x$.

> In Exercises 25 and 26, show that the expression for $\langle \mathbf{u}, \mathbf{v} \rangle$ defines an inner product on R^2.

25. $\langle \mathbf{u}, \mathbf{v} \rangle = 2u_1v_1 - u_2v_1 - u_1v_2 + 4u_2v_2$

26. $\langle \mathbf{u}, \mathbf{v} \rangle = 3u_1v_1 - 2u_2v_1 - 2u_1v_2 + 3u_2v_2$

27. Theorem 1.2.14 established the parallelogram law

$$\|\mathbf{u} + \mathbf{v}\|^2 + \|\mathbf{u} - \mathbf{v}\|^2 = 2 \left(\|\mathbf{u}\|^2 + \|\mathbf{v}\|^2 \right)$$

for vectors in R^n. Show that this equality holds in all inner product spaces.

28. Show that the following equality holds in all inner product spaces:

$$\langle \mathbf{u}, \mathbf{v} \rangle = \tfrac{1}{4} \left(\|\mathbf{u} + \mathbf{v}\|^2 - \|\mathbf{u} - \mathbf{v}\|^2 \right)$$

29. Just as the inner product axioms on a real vector space are motivated by the properties of the dot product in Theorem 1.2.6, so the inner product axioms on a complex vector space are motivated by the properties of the complex dot product in Theorem 8.8.5. Thus, if V is a complex vector space, then a **complex inner product** on V is a function that associates a unique complex number $\langle \mathbf{u}, \mathbf{v} \rangle$ with each pair of vectors $\mathbf{u}$ and $\mathbf{v}$ in V in such a way that Axioms 2, 3, and 4 of Definition 9.2.1 hold but Axiom 1 is replaced by $\langle \mathbf{u}, \mathbf{v} \rangle = \overline{\langle \mathbf{v}, \mathbf{u} \rangle}$. A complex vector space with a complex inner product is called a **complex inner product space**. Notions of length, distance, and orthogonality in complex vector spaces are defined by the formulas in Definition 9.2.2, just as in real vector spaces.

(a) Show that the formula $\langle \mathbf{u}, \mathbf{v} \rangle = 3u_1\bar{v}_1 + 2u_2\bar{v}_2$ defines a complex inner product on C^2.

(b) Let $\mathbf{u} = (i, -2i)$ and $\mathbf{v} = (3i, -i)$ and compute $\langle \mathbf{u}, \mathbf{v} \rangle$, $\|\mathbf{u}\|$, $\|\mathbf{v}\|$, and the distance between $\mathbf{u}$ and $\mathbf{v}$ with respect to the inner product in part (a).

Discussion and Discovery

D1. If $\mathbf{u}$ and $\mathbf{v}$ are orthogonal unit vectors in an inner product space, then $\|\mathbf{u} - \mathbf{v}\| = $ _____. Justify your answer with appropriate computations.

D2. Find a value of c for which

$$\langle \mathbf{u}, \mathbf{v} \rangle = 5u_1v_1 - 3u_2v_1 - 3u_1v_2 + cu_2v_2$$

defines an inner product on R^2.

D3. If $U = [u_{ij}]$ and $V = [v_{ij}]$ are 2×2 matrices, then the value of the expression

$$\mathrm{tr}(U^TV) = u_{11}v_{11} + u_{12}v_{12} + u_{21}v_{21} + u_{22}v_{22}$$

is the same as the value of $\mathbf{u} \cdot \mathbf{v}$ if $\mathbf{u} = $ _____ and if $\mathbf{v} = $ _____. It follows that $\langle U, V \rangle = \mathrm{tr}(U^TV)$ satisfies the inner product axioms and hence defines an inner product on the vector space _____.

D4. If p and q are the polynomials $p(x) = c_0 + c_1x + \cdots + c_nx^n$ and $q(x) = d_0 + d_1x + \cdots + d_nx^n$, then the value of the expression

$$\langle p, q \rangle = c_0d_0 + c_1d_1 + \cdots + c_nd_n$$

is the same as the value of $\mathbf{u} \cdot \mathbf{v}$ for the vectors $\mathbf{u} = $ _____ and $\mathbf{v} = $ _____ in the vector space _____. This implies that $\langle p, q \rangle$ satisfies the inner product space axioms and hence defines an inner product on the vector space _____.

D5. Under what conditions on the scalars r and s will the expression $\langle \mathbf{u}, \mathbf{v} \rangle = (r\mathbf{u}) \cdot (s\mathbf{v})$ define an inner product on R^n?

Working with Proofs

P1. Prove that if A is an invertible $n \times n$ matrix and $\mathbf{u}$ and $\mathbf{v}$ are vectors in R^n, then the formula $\langle \mathbf{u}, \mathbf{v} \rangle = A\mathbf{u} \cdot A\mathbf{v}$ defines an inner product on R^n.

P2. Prove that the weighted Euclidean inner product defined by Formula (3) satisfies the inner product axioms.

P3. Prove: If $\mathbf{y}$ is a fixed vector in R^n, then the mapping $\mathbf{x} \to \mathbf{x}^TA\mathbf{y}$ defines a linear transformation from R^n to R.

P4. Prove that the matrix A in Formula (34) is symmetric and positive definite.

P5. Recall that if A is an $m \times n$ matrix, then $\hat{\mathbf{x}}$ is a *least squares solution* of the linear system $A\mathbf{x} = \mathbf{b}$ if and only if $\mathbf{x} = \hat{\mathbf{x}}$ minimizes $\|\mathbf{b} - A\mathbf{x}\|$ with respect to the Euclidean norm on R^m. More generally, if the norm is computed relative to an inner product $\langle \mathbf{u}, \mathbf{v} \rangle$, then $\hat{\mathbf{x}}$ is called a *least squares solution with respect to* $\langle \mathbf{u}, \mathbf{v} \rangle$. Prove that if M is a positive definite symmetric matrix, then $\mathbf{x} = \hat{\mathbf{x}}$ is a least squares solution of $A\mathbf{x} = \mathbf{b}$ with respect to the inner product $\langle \mathbf{u}, \mathbf{v} \rangle = \mathbf{u}^TM\mathbf{v}$ if and only if it is an exact solution of $A^TMA\mathbf{x} = A^TM\mathbf{b}$.

Technology Exercises

T1. Find the Fourier approximations of orders 1, 2, 3, and 4 to the function $f(x) = x^2$, and compare the graphs of those approximations to the graph of f over the interval $[0, 2\pi]$.

T2. In Example 3 of Section 7.8 we found the least squares solution of the linear system

$$x_1 - x_2 = 4$$
$$3x_1 + 2x_2 = 1$$
$$-2x_1 + 4x_2 = 3$$

Use the result in Exercise P5 to find the least squares solution of this system with respect to the weighted Euclidean inner product $\langle \mathbf{u}, \mathbf{v} \rangle = 3u_1v_1 + 2u_2v_2$.

T3. (CAS) Let W be the subspace of $C[-1, 1]$ spanned by the linearly independent polynomials $1, x, x^2$, and x^3. Show

that if the Gram–Schmidt process is applied to these polynomials using the integral inner product

$$\langle f, g \rangle = \int_{-1}^{1} f(x)g(x)\, dx$$

then the resulting orthonormal basis vectors for W are

$$\tfrac{1}{\sqrt{2}}, \quad \sqrt{\tfrac{3}{2}}x, \quad \tfrac{1}{2}\sqrt{\tfrac{5}{2}}(3x^2 - 1), \quad \tfrac{1}{2}\sqrt{\tfrac{7}{2}}(5x^3 - 3x)$$

These are called **_normalized Legendre polynomials_** in honor of the French mathematician Adrien-Marie Legendre (1752–1833) who first recognized their importance in the study of gravitational attraction. [*Note:* To solve this problem you will probably have to piece together integration commands and commands for applying the Gram–Schmidt process to functions.]

Section 9.3   General Linear Transformations; Isomorphism

Up to now our study of linear transformations has focused on transformations from R^n to R^m. In this section we will turn our attention to linear transformations involving general vector spaces, and we will illustrate various ways in which such transformations occur. We will also use our work on general linear transformations to establish a fundamental relationship between general finite-dimensional vector spaces and R^n. Calculus will be needed in some of the examples.

GENERAL LINEAR TRANSFORMATIONS   The definition of a linear transformation from a general vector space V to a general vector space W is similar to Definition 6.1.2.

> **Definition 9.3.1**   If $T : V \to W$ is a function from a vector space V to a vector space W, then T is called a **_linear transformation_** from V to W if the following two properties hold for all vectors $\mathbf{u}$ and $\mathbf{v}$ in V and for all scalars c:
>
> (i) $T(c\mathbf{u}) = cT(\mathbf{u})$ [Homogeneity property]
> (ii) $T(\mathbf{u} + \mathbf{v}) = T(\mathbf{u}) + T(\mathbf{v})$ [Additivity property]
>
> In the special case where $V = W$, the linear transformation T is called a **_linear operator_** on vector space V.

The homogeneity and additivity properties of a linear transformation $T : V \to W$ can be used in combination to show that if $\mathbf{v}_1$ and $\mathbf{v}_2$ are vectors in V and c_1 and c_2 are any scalars, then

$$T(c_1\mathbf{v}_1 + c_2\mathbf{v}_2) = c_1 T(\mathbf{v}_1) + c_2 T(\mathbf{v}_2)$$

More generally, if $\mathbf{v}_1, \mathbf{v}_2, \ldots, \mathbf{v}_k$ are vectors in V and $c_1, c_2, \ldots, c_k$ are any scalars, then

$$T(c_1\mathbf{v}_1 + c_2\mathbf{v}_2 + \cdots + c_k\mathbf{v}_k) = c_1 T(\mathbf{v}_1) + c_2 T(\mathbf{v}_2) + \cdots + c_k T(\mathbf{v}_k) \tag{1}$$

We leave it for you to modify the proof of Theorem 6.1.3 appropriately to prove the following generalization of that theorem.

Theorem 9.3.2 *If $T : V \to W$ is a linear transformation, then:*

(a) $T(\mathbf{0}) = \mathbf{0}$

(b) $T(-\mathbf{u}) = -T(\mathbf{u})$

(c) $T(\mathbf{u} - \mathbf{v}) = T(\mathbf{u}) - T(\mathbf{v})$

EXAMPLE 1
Zero Transformations

If V and W are any two vector spaces, then the mapping $T : V \to W$ for which $T(\mathbf{v}) = \mathbf{0}$ for every vector $\mathbf{v}$ in V is called the ***zero transformation*** from V to W. This transformation is linear, for if $\mathbf{u}$ and $\mathbf{v}$ are any vectors in V and c is any scalar, then $T(c\mathbf{u}) = \mathbf{0}$ and $cT(\mathbf{u}) = c\mathbf{0} = \mathbf{0}$, so

$$T(c\mathbf{u}) = cT(\mathbf{u})$$

Also, $T(\mathbf{u} + \mathbf{v}) = T(\mathbf{u}) = T(\mathbf{v}) = \mathbf{0}$, so

$$T(\mathbf{u} + \mathbf{v}) = T(\mathbf{u}) + T(\mathbf{v})$$

∎

EXAMPLE 2
The Identity
Operator

If V is any vector space, then the mapping $T : V \to V$ for which $T(\mathbf{v}) = \mathbf{v}$ for every vector $\mathbf{v}$ in V is called the ***identity operator*** on V. We leave it for you to verify that T is linear. ∎

EXAMPLE 3
Dilation and
Contraction
Operators

If V is a vector space and k is a scalar, then the mapping $T : V \to V$ given by $T(\mathbf{v}) = k\mathbf{v}$ is a linear operator on V, for if c is any scalar, and if $\mathbf{u}$ and $\mathbf{v}$ are any vectors in V, then

$$T(c\mathbf{u}) = k(c\mathbf{u}) = c(k\mathbf{u}) = cT(\mathbf{u})$$
$$T(\mathbf{u} + \mathbf{v}) = k(\mathbf{u} + \mathbf{v}) = k\mathbf{u} + k\mathbf{v} = T(\mathbf{u}) + T(\mathbf{v})$$

If $0 < k < 1$, then T is called the ***contraction*** of V with factor k, and if $k > 1$, it is called the ***dilation*** of V with factor k (Figure 9.3.1). ∎

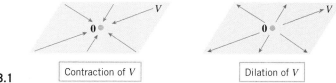

Figure 9.3.1

Contraction of V Dilation of V

EXAMPLE 4
A Linear
Transformation
on an Inner
Product Space

Let V be an inner product space, let $\mathbf{v}_0$ be any fixed vector in V, and let $T : V \to R$ be the transformation

$$T(\mathbf{x}) = \langle \mathbf{x}, \mathbf{v}_0 \rangle$$

that maps a vector $\mathbf{x}$ into its inner product with $\mathbf{v}_0$. This transformation is linear, for if c is any scalar, and if $\mathbf{u}$ and $\mathbf{v}$ are any vectors in V, then it follows from properties of inner products that

$$T(c\mathbf{u}) = \langle c\mathbf{u}, \mathbf{v}_0 \rangle = c\langle \mathbf{u}, \mathbf{v}_0 \rangle = cT(\mathbf{u})$$
$$T(\mathbf{u} + \mathbf{v}) = \langle \mathbf{u} + \mathbf{v}, \mathbf{v}_0 \rangle = \langle \mathbf{u}, \mathbf{v}_0 \rangle + \langle \mathbf{v}, \mathbf{v}_0 \rangle = T(\mathbf{u}) + T(\mathbf{v})$$

∎

EXAMPLE 5
An Evaluation
Transformation

Let V be a subspace of $F(-\infty, \infty)$, let

$$x_1, x_2, \ldots, x_n$$

be distinct real numbers, and let $T : V \to R^n$ be the mapping

$$T(f) = (f(x_1), f(x_2), \ldots, f(x_n)) \tag{2}$$

that associates with f its n-tuple of function values at $x_1, x_2, \ldots, x_n$. We call this the ***evaluation transformation*** on V at $x_1, x_2, \ldots, x_n$. Thus, for example, if

$$x_1 = -1, \quad x_2 = 2, \quad x_3 = 4$$

and if $f(x) = x^2 - 1$, then

$$T(f) = (f(x_1), f(x_2), f(x_3)) = (0, 3, 15)$$

The evaluation transformation in (2) is linear, for if c is any scalar, and if f and g are any functions in V, then

$$\begin{aligned}
T(cf) &= ((cf)(x_1), (cf)(x_2), \ldots, (cf)(x_n)) \\
&= (cf(x_1), cf(x_2), \ldots, cf(x_n)) \\
&= c(f(x_1), f(x_2), \ldots, f(x_n)) = cT(f)
\end{aligned}$$

and

$$\begin{aligned}
T(f + g) &= ((f + g)(x_1), (f + g)(x_2), \ldots, (f + g)(x_n)) \\
&= (f(x_1) + g(x_1), f(x_2) + g(x_2), \ldots, f(x_n) + g(x_n)) \\
&= (f(x_1), f(x_2), \ldots, f(x_n)) + (g(x_1), g(x_2), \ldots, g(x_n)) \\
&= T(f) + T(g)
\end{aligned}$$

EXAMPLE 6
Differentiation
Transformations

Let $V = C^1(-\infty, \infty)$ be the vector space of real-valued functions with continuous first derivatives on $(-\infty, \infty)$, let $W = C(-\infty, \infty)$ be the vector space of continuous real-valued functions on $(-\infty, \infty)$, and let

$$D(f) = f'$$

be the transformation that maps f into its first derivative. The transformation $D: V \to W$ is linear, for if c is any constant, and if f and g are any functions in V, then properties of the derivative imply that

$$D(cf) = cD(f) \quad \text{and} \quad D(f + g) = D(f) + D(g)$$

More generally, if

$$D^k(f) = f^{(k)}$$

denotes the kth derivative of f, then $D^k: V \to W$ is a linear transformation from $V = C^k(-\infty, \infty)$ to $W = C(-\infty, \infty)$. Also, if

$$a_1, a_2, \ldots, a_n$$

are any scalars, then

$$(a_n D^n + a_{n-1} D^{n-1} + \cdots + a_1 D + a_0 I)(f) = a_n f^{(n)} + a_{n-1} f^{(n-1)} + \cdots + a_1 f' + a_0 f$$

is a linear transformation from $V = C^n(-\infty, \infty)$ to $W = C(-\infty, \infty)$.

EXAMPLE 7
An Integral
Transformation

Let $V = C(-\infty, \infty)$ be the vector space of continuous functions on $(-\infty, \infty)$, let $W = C^1(-\infty, \infty)$ be the vector space of functions with continuous first derivatives on $(-\infty, \infty)$, and let $J: V \to W$ be the transformation that maps a function $f(x)$ in V into

$$J(f) = \int_0^x f(t)\, dt$$

For example, if $f(x) = x^2$, then

$$J(f) = \int_0^x t^2\, dt = \frac{t^3}{3}\Big]_0^x = \frac{x^3}{3}$$

The transformation $J: V \to W$ is linear, for if c is any constant, and if f and g are any functions in V, then properties of the integral imply that

$$J(cf) = \int_0^x cf(t)\, dt = c\int_0^x f(t)\, dt = cJ(f)$$

$$J(f + g) = \int_0^x (f(t) + g(t))\, dt = \int_0^x f(t)\, dt + \int_0^x g(t)\, dt = J(f) + J(g)$$

EXAMPLE 8
Some Transformations on Matrix Spaces

Let M_{nn} be the vector space of real $n \times n$ matrices. In each part determine whether the transformation $T : M_{nn} \to R$ is linear for $n > 1$.

(a) $T(A) = \text{tr}(A)$ (b) $T(A) = \det(A)$

Solution (*a*) It follows from parts (*b*) and (*c*) of Theorem 3.2.12 that

$$T(cA) = \text{tr}(cA) = c \, \text{tr}(A) = cT(A)$$
$$T(A + B) = \text{tr}(A) + \text{tr}(B) = T(A) + T(B)$$

so T is a linear transformation.

Solution (*b*) We know from part (*c*) of Theorem 4.2.3 that

$$\det(cA) = c^n \det(A)$$

Thus, the homogeneity property $T(A) = cT(A)$ does *not* hold for all A in M_{nn}. This by itself establishes that T is not linear. However, we also know that $\det(A + B) \neq \det(A) + \det(B)$, in general (Example 8 of Section 4.2), so the additivity property also fails. ∎

CONCEPT PROBLEM Is the transformation $A \to A^T$ a linear operator on M_{nn}? Explain.

KERNEL AND RANGE

The notions of kernel and range are similar to those for transformations from R^n to R^m (compare to Definitions 6.3.1 and 6.3.6).

> **Definition 9.3.3**   If $T : V \to W$ is a linear transformation, then the set of vectors in V that T maps into $\mathbf{0}$ is called the *kernel* of T and is denoted by $\ker(T)$.

> **Definition 9.3.4**   If $T : V \to W$ is a linear transformation, then the *range* of T, denoted by $\text{ran}(T)$, is the set of all vectors in W that are images of at least one vector in V; that is, $\text{ran}(T)$ is the image of the domain V under the transformation T.

EXAMPLE 9
Kernel and Range of Zero

If $T : V \to W$ is the zero transformation, then T maps every vector in V into the vector $\mathbf{0}$ in W. Thus, $\text{ran}(T) = \{\mathbf{0}\}$ and $\ker(T) = V$. ∎

EXAMPLE 10
Kernel and Range of the Identity

Let $T : V \to V$ be the identity operator on V. Since $T(\mathbf{v}) = \mathbf{v}$, every vector in V is the image of a vector in V (namely, itself), so $\text{ran}(T) = V$. Also, $T(\mathbf{v}) = \mathbf{0}$ if and only if $\mathbf{v} = \mathbf{0}$, so $\ker(T) = \{\mathbf{0}\}$. ∎

EXAMPLE 11
Kernel and Range of an Inner Product Transformation

Let V be a nonzero inner product space, let $\mathbf{v}_0$ be a fixed nonzero vector in V, and let $T : V \to R$ be the transformation

$$T(\mathbf{x}) = \langle \mathbf{x}, \mathbf{v}_0 \rangle$$

The kernel of T consists of all vectors $\mathbf{x}$ in V such that $\langle \mathbf{x}, \mathbf{v}_0 \rangle = 0$, so geometrically $\ker(T)$ is the orthogonal complement $\mathbf{v}_0^\perp$ of $\mathbf{v}_0$. Also, $\text{ran}(T) = R$, since every real number k is the image of some vector in V. Specifically,

$$T\left(k \frac{\mathbf{v}_0}{\|\mathbf{v}_0\|^2}\right) = \left\langle k \frac{\mathbf{v}_0}{\|\mathbf{v}_0\|^2}, \mathbf{v}_0 \right\rangle = \frac{k}{\|\mathbf{v}_0\|^2} \langle \mathbf{v}_0, \mathbf{v}_0 \rangle = k$$ ∎

EXAMPLE 12
Kernel and Range of an Evaluation Transformation

Let $x_1, x_2, \ldots, x_n$ be distinct real numbers, and suppose that $T : P_{n-1} \to R^n$ is the evaluation transformation

$$T(p) = (p(x_1), p(x_2), \ldots, p(x_n))$$

defined in Example 5. To find the kernel of T, suppose that $T(p) = \mathbf{0} = (0, 0, \ldots, 0)$. This implies that

$$p(x_1) = 0, \quad p(x_2) = 0, \ldots, \quad p(x_n) = 0$$

which means that $x_1, x_2, \ldots, x_n$ are n distinct roots of the polynomial p. However, a nonzero polynomial of degree $n - 1$ or less can have at most $n - 1$ distinct roots, so it must be the case that $p = 0$ [i.e., $p(x) = 0$ for all x in $(-\infty, \infty)$]. Thus,

$$\ker(T) = \{0\}$$

To find the range of T, let $\mathbf{y} = (y_1, y_2, \ldots, y_n)$ be any vector in R^n. We know from Theorem 2.3.1 that there is a unique polynomial p of degree $n - 1$ or less whose graph passes through the points

$$(x_1, y_1), (x_2, y_2), \ldots, (x_n, y_n)$$

and for this polynomial we have

$$p(x_1) = y_1, \quad p(x_2) = y_2, \ldots, \quad p(x_n) = y_n$$

This means that $T(p) = (y_1, y_2, \ldots, y_n) = \mathbf{y}$, which shows that every vector in R^n is the image of some polynomial in P_{n-1}. Thus,

$$\text{ran}(T) = R^n \qquad\blacksquare$$

EXAMPLE 13
Kernel and Range
of a
Differentiation
Transformation

Let $D: C^1(-\infty, \infty) \to C(-\infty, \infty)$ be the differentiation transformation

$$D(f) = f'$$

To find the kernel of D, suppose that $D(f) = 0$; that is, $f'(x) = 0$ for all x in the interval $(-\infty, \infty)$. We know from calculus that if $f'(x) = 0$ on an interval, then f is constant on that interval. Thus,

$$\ker(D) = \text{the set of constant functions on } (-\infty, \infty)$$

To find the range of D, let $g(x)$ be any function of x that is continuous on $(-\infty, \infty)$, and define the function $f(x)$ to be

$$f(x) = \int_0^x g(t)\, dt$$

It follows from the Fundamental Theorem of Calculus that

$$D(f) = \frac{d}{dx} \int_0^x g(t)\, dt = g(x)$$

which shows that every function g in $C(-\infty, \infty)$ is the image under D of some function in $C^1(-\infty, \infty)$. Thus,

$$\text{ran}(D) = C(-\infty, \infty) \qquad\blacksquare$$

PROPERTIES OF THE KERNEL AND RANGE

In Section 6.3 we proved that the kernel and range of a linear transformation from R^n to R^m are subspaces of R^n and R^m, respectively, and that a linear transformation from R^n to R^m maps subspaces of R^n into subspaces of R^m (see Theorems 6.3.2, 6.3.5, and 6.3.7). If you examine the proofs of those theorems, you will see that they use only the closure properties of subspaces and the homogeneity and additivity properties of linear transformations. Thus, those theorems also hold for general linear transformations.

Theorem 9.3.5   *If $T: V \to W$ is a linear transformation, then T maps subspaces of V into subspaces of W.*

Theorem 9.3.6   *If $T: V \to W$ is a linear transformation, then $\ker(T)$ is a subspace of V and $\text{ran}(T)$ is a subspace of W.*

EXAMPLE 14
Application to
Differential
Equations

Differential equations of the form

$$y'' + \omega^2 y = 0 \qquad (\omega \text{ a positive constant}) \tag{3}$$

arise in the study of vibrations. The set of all solutions of this equation on the interval $(-\infty, \infty)$ is the kernel of the linear transformation $D : C^2(-\infty, \infty) \to C(-\infty, \infty)$, given by

$$D(y) = y'' + \omega^2 y$$

It is proved in standard textbooks on differential equations that the kernel is a two-dimensional subspace of $C^2(-\infty, \infty)$, so that if we can find two linearly independent solutions of (3), then all other solutions can be expressed as linear combinations of those two. We leave it for you to confirm by differentiating that

$$y_1 = \cos \omega x \quad \text{and} \quad y_2 = \sin \omega x$$

are solutions of (3). These functions are linearly independent, since neither is a scalar multiple of the other, and thus

$$y = c_1 \cos \omega x + c_2 \sin \omega x \tag{4}$$

is a "general solution" of (3) in the sense that every choice of c_1 and c_2 produces a solution, and every solution is of this form. ∎

CONCEPT PROBLEM Confirm the linear independence of y_1 and y_2 using Wroński's test (Theorem 9.1.5).

In keeping with the terminology of Definitions 6.3.9 and 6.3.10, we say that a transformation $T : V \to W$ is **one-to-one** if it maps distinct vectors in V into distinct vectors in W, and we say that T is **onto** if its range is all of W. We leave it for you to modify the proof of Theorem 6.3.11 appropriately to prove the following generalization of that theorem.

> **Theorem 9.3.7** *If $T : V \to W$ is a linear transformation, then the following are equivalent.*
>
> (a) *T is one-to-one.*
> (b) $\ker(T) = \{\mathbf{0}\}$.

EXAMPLE 15
Polynomial
Evaluation Is
One-to-One and
Onto

Let $x_1, x_2, \ldots, x_n$ be distinct real numbers and $T : P_{n-1} \to R^n$ be the evaluation transformation

$$T(p) = (p(x_1), p(x_2), \ldots, p(x_n))$$

defined in Example 5. We showed in Example 12 that $\ker(T) = \{\mathbf{0}\}$, so it follows from Theorem 9.3.7 that T is one-to-one. This implies that if p and q are polynomials of degree $n - 1$ or less, and if

$$(p(x_1), p(x_2), \ldots, p(x_n)) = (q(x_1), q(x_2), \ldots, q(x_n))$$

then $p = q$. You should be able to recognize that this is just a restatement of the uniqueness part of Theorem 2.3.1 on polynomial interpolation. ∎

We know from Theorem 6.3.14 that a linear operator on R^n is one-to-one if and only if it is onto. However, if you examine the proof of that theorem, you will see that the finite-dimensionality plays an essential role. The following example shows that a linear operator on an infinite-dimensional vector space can be one-to-one and not onto or onto and not one-to-one.

EXAMPLE 16
One-to-One but
Not Onto and
Onto but Not
One-to-One

Let $V = R^\infty$ be the sequence space discussed in Example 2 of Section 9.1, and consider the "shifting operators" on V defined by

$$T_1(v_1, v_2, \ldots, v_n, \ldots) = (0, v_1, v_2, \ldots, v_{n-1}, \ldots)$$

and

$$T_2(v_1, v_2, \ldots, v_n, \ldots) = (v_2, v_3, v_4, \ldots, v_{n-1}, \ldots)$$

It can be shown that these operators are linear. The operator T_1 is one-to-one because distinct vectors in R^∞ have distinct images, but it is not onto because, for example, no vector in R^∞ maps into $(1, 0, 0, \ldots, 0, \ldots)$. The operator T_2 is not one-to-one because, for example, the vectors $(1, 0, 0, \ldots, 0, \ldots)$ and $(0, 0, 0, \ldots, 0, \ldots)$ both map into $(0, 0, 0, \ldots, 0, \ldots)$, but it is onto (why?).   ∎

REMARK Shifting transformations arise in communications problems in which a sequence of signals $v_1, v_2, \ldots, v_n, \ldots$ is transmitted at clock times $t_1, t_2, \ldots, t_n, \ldots$. Operator T_1 describes a situation in which a transmission delay causes the receiver to record signal v_1 at time t_2, and operator T_2 describes a situation in which the transmitter and receiver clock times are mismatched, so the receiver records signal v_2 when its clock shows the time to be t_1.

ISOMORPHISM Although most of the theorems in this text have been concerned with the vector space R^n, this is not as limiting as it might seem, for we will show that R^n is "the mother of all real finite-dimensional vector spaces" in the sense that every real n-dimensional vector space differs from R^n only in the notation used to represent vectors. To clarify what we mean by this, consider the n-dimensional vector space P_{n-1} (polynomials of degree n or less). Each polynomial in this space can be expressed uniquely in the form

$$p(x) = a_0 + a_1 x + \cdots + a_{n-1} x^{n-1}$$

and hence is uniquely determined by its n-tuple of coefficients

$$(a_0, a_1, \ldots, a_{n-1})$$

Thus, the transformation

$$T(a_0 + a_1 x + \cdots + a_{n-1} x^{n-1}) = (a_0, a_1, \ldots, a_{n-1})$$

which we can also denote as

$$a_0 + a_1 x + \cdots + a_{n-1} x^{n-1} \xrightarrow{\;T\;} (a_0, a_1, \ldots, a_{n-1})$$

is a one-to-one and onto mapping from P_{n-1} to R^n. Moreover, T is linear, for if c is a scalar, and if

$$p(x) = a_0 + a_1 x + \cdots + a_{n-1} x^{n-1} \quad \text{and} \quad q(x) = b_0 + b_1 x + \cdots + b_{n-1} x^{n-1}$$

are polynomials in P_{n-1}, then

$$
\begin{aligned}
T(cp(x)) &= T(ca_0 + ca_1 x + \cdots + ca_{n-1} x^{n-1}) \\
&= (ca_0, ca_1, \ldots, ca_{n-1}) \\
&= c(a_0, a_1, \ldots, a_{n-1}) \\
&= cT(p(x))
\end{aligned}
$$

and

$$
\begin{aligned}
T(p(x) + q(x)) &= T\big((a_0 + b_0) + (a_1 + b_1)x + \cdots + (a_{n-1} + b_{n-1})x^{n-1}\big) \\
&= (a_0 + b_0, a_1 + b_1, \ldots, a_{n-1} + b_{n-1}) \\
&= (a_0, a_1, \ldots, a_{n-1}) + (b_0, b_1, \ldots, b_{n-1}) \\
&= T(p(x)) + T(q(x))
\end{aligned}
$$

The fact that the transformation T is one-to-one, onto, and linear means that it matches up polynomials in P_{n-1} with n-tuples in R^n in such a way that operations on vectors in either space can be performed using their counterparts in the other space. Here is an example.

EXAMPLE 17
Matching P_2 with R^3

The following table shows how the transformation

$$a_0 + a_1 x + a_2 x^2 \xrightarrow{\ T\ } (a_0, a_1, a_2)$$

matches up vector operations in P_2 and R^3.

Operation in P_2	Operation in R^3
$3(1 - 2x + 3x^2) = 3 - 6x + 9x^2$	$3(1, -2, 3) = (3, -6, 9)$
$(2 + x - x^2) + (1 - x + 5x^2) = 3 + 4x^2$	$(2, 1, -1) + (1, -1, 5) = (3, 0, 4)$
$(4 + 2x + 3x^2) - (2 - 4x + 3x^2) = 2 + 6x$	$(4, 2, 3) - (2, -4, 3) = (2, 6, 0)$

Thus, although a polynomial $a_0 + a_1 x + a_2 x^2$ is obviously a different mathematical object from an ordered triple (a_0, a_1, a_2), the *vector spaces* formed by these objects have the same algebraic structure. ∎

In general, if V and W are vector spaces, and if there exists a one-to-one and onto linear transformation from V to W, then the two vector spaces have the same algebraic structure. We describe this by saying that V and W are *isomorphic* (a word that has been pieced together from the Greek words *iso*, meaning "identical," and *morphe*, meaning "form").

Definition 9.3.8 A linear transformation $T : V \to W$ is called an ***isomorphism*** if it is one-to-one and onto, and we say that a vector space V is ***isomorphic*** to a vector space W if there exists an isomorphism from V onto W.

The following theorem, which is one of the most important results in linear algebra, reveals the fundamental importance of the vector space R^n.

Theorem 9.3.9 *Every real n-dimensional vector space is isomorphic to R^n.*

Proof Let V be a real n-dimensional vector space. To prove that V is isomorphic to R^n we must find a linear transformation $T : V \to R^n$ that is one-to-one and onto. For this purpose, let

$$\mathbf{v}_1, \mathbf{v}_2, \ldots, \mathbf{v}_n$$

be any basis for V, let

$$\mathbf{u} = k_1 \mathbf{v}_1 + k_2 \mathbf{v}_2 + \cdots + k_n \mathbf{v}_n$$

be the representation of a vector $\mathbf{u}$ in V as a linear combination of the basis vectors, and define the transformation $T : V \to R^n$ by

$$T(\mathbf{u}) = (k_1, k_2, \ldots, k_n)$$

We will show that T is an isomorphism (linear, one-to-one, and onto). To prove the linearity, let $\mathbf{u}$ and $\mathbf{v}$ be vectors in V, let c be a scalar, and let

$$\mathbf{u} = k_1 \mathbf{v}_1 + k_2 \mathbf{v}_2 + \cdots + k_n \mathbf{v}_n \quad \text{and} \quad \mathbf{v} = d_1 \mathbf{v}_1 + d_2 \mathbf{v}_2 + \cdots + d_n \mathbf{v}_n \quad (5)$$

be the representations of $\mathbf{u}$ and $\mathbf{v}$ as linear combinations of the basis vectors. Then

$$T(c\mathbf{u}) = T(ck_1 \mathbf{v}_1 + ck_2 \mathbf{v}_2 + \cdots + ck_n \mathbf{v}_n)$$
$$= (ck_1, ck_2, \ldots, ck_n) = c(k_1, k_2, \ldots, k_n) = cT(\mathbf{u})$$

Linear Algebra in History

Methods of linear algebra are used in the emerging field of computerized face recognition. Researchers are working with the idea that every human face in a racial group is a combination of a few dozen primary shapes. For example, by analyzing three-dimensional scans of many faces, researchers at Rockefeller University have produced both an average head shape in the Caucasian group—dubbed the ***meanhead*** (top row left in the figure below)—and a set of standardized variations from that shape, called ***eigenheads*** (15 of which are shown in the picture). These are so named because they are eigenvectors of a certain matrix that stores digitized facial information. Face shapes are represented mathematically as linear combinations of the eigenheads.

—Adapted from an article in *Scientific American*, December 1995.

and

$$T(\mathbf{u} + \mathbf{v}) = T\big((k_1 + d_1)\mathbf{v}_1 + (k_2 + d_2)\mathbf{v}_2 + \cdots + (k_n + d_n)\mathbf{v}_n\big)$$
$$= (k_1 + d_1, k_2 + d_2, \ldots, k_n + d_n)$$
$$= (k_1, k_2, \ldots, k_n) + (d_1, d_2, \ldots, d_n)$$
$$= T(\mathbf{u}) + T(\mathbf{v})$$

To show that T is one-to-one, we must show that if $\mathbf{u}$ and $\mathbf{v}$ are distinct vectors, then so are their images under T. But if $\mathbf{u} \neq \mathbf{v}$, and if the representations of these vectors in terms of the basis vectors are as in (5), then we must have $k_i \neq d_i$ for at least one i. Thus,

$$T(\mathbf{u}) = (k_1, k_2, \ldots, k_n) \neq (d_1, d_2, \ldots, d_n) = T(\mathbf{v})$$

which shows that $\mathbf{u}$ and $\mathbf{v}$ have distinct images under T. Finally, the transformation T is also onto, for if

$$\mathbf{y} = (k_1, k_2, \ldots, k_n)$$

is any vector in R^n, then $\mathbf{y}$ is the image under T of the vector

$$\mathbf{u} = k_1\mathbf{v}_1 + k_2\mathbf{v}_2 + \cdots + k_n\mathbf{v}_n \qquad \blacksquare$$

EXAMPLE 18
The Natural Isomorphism from P_{n-1} to R^n

We showed earlier in this section that the mapping

$$a_0 + a_1 x + \cdots + a_{n-1}x^{n-1} \xrightarrow{\ T\ } (a_0, a_1, \ldots, a_{n-1})$$

from P_{n-1} to R^n is one-to-one, onto, and linear. This is called the ***natural isomorphism*** from P_{n-1} to R^n because, as the following computations show, it maps the natural basis $\{1, x, x^2, \ldots, x^{n-1}\}$ for P_{n-1} into the standard basis for R^n:

$$1 = 1 + 0x + 0x^2 + \cdots + 0x^{n-1} \qquad \xrightarrow{\ T\ } \quad (1, 0, 0, \ldots, 0)$$
$$x = 0 + x + 0x^2 + \cdots + 0x^{n-1} \qquad \xrightarrow{\ T\ } \quad (0, 1, 0, \ldots, 0)$$
$$\vdots \qquad\qquad\qquad \vdots \qquad\qquad\qquad \vdots \qquad \vdots$$
$$x^{n-1} = 0 + 0x + 0x^2 + \cdots + x^{n-1} \qquad \xrightarrow{\ T\ } \quad (0, 0, 0, \ldots, 1) \qquad \blacksquare$$

EXAMPLE 19
The Natural Isomorphism from M_{22} to R^4

The matrices

$$E_1 = \begin{bmatrix} 1 & 0 \\ 0 & 0 \end{bmatrix}, \quad E_2 = \begin{bmatrix} 0 & 1 \\ 0 & 0 \end{bmatrix}, \quad E_3 = \begin{bmatrix} 0 & 0 \\ 1 & 0 \end{bmatrix}, \quad E_4 = \begin{bmatrix} 0 & 0 \\ 0 & 1 \end{bmatrix}$$

form a basis for the vector space M_{22} of 2×2 matrices (Example 13 of Section 9.1). Thus, as shown in the proof of Theorem 9.3.9, an isomorphism $T : M_{22} \to R^4$ can be constructed by first writing a matrix A in M_{22} in terms of the basis vectors as

$$A = \begin{bmatrix} a_1 & a_2 \\ a_3 & a_4 \end{bmatrix} = a_1 \begin{bmatrix} 1 & 0 \\ 0 & 0 \end{bmatrix} + a_2 \begin{bmatrix} 0 & 1 \\ 0 & 0 \end{bmatrix} + a_3 \begin{bmatrix} 0 & 0 \\ 1 & 0 \end{bmatrix} + a_4 \begin{bmatrix} 0 & 0 \\ 0 & 1 \end{bmatrix}$$

and then defining T as

$$T(A) = (a_1, a_2, a_3, a_4)$$

Thus, for example,

$$\begin{bmatrix} 1 & -3 \\ 4 & 6 \end{bmatrix} \xrightarrow{\ T\ } (1, -3, 4, 6)$$

More generally, this idea can be used to show that the vector space M_{mn} of $m \times n$ matrices with real entries is isomorphic to R^{mn}. $\qquad \blacksquare$

The fact that every real n-dimensional vector space is isomorphic to R^n makes it possible to apply theorems about vectors in R^n to general finite-dimensional vector spaces. For example, we know that linear transformations from R^n to R^m can be represented by $m \times n$ matrices relative to bases for R^n and R^m. The following example illustrates how the concept of an isomorphism can be used to represent linear transformations from one finite-dimensional vector space to another by matrices.

EXAMPLE 20
Differentiation by Matrix Multiplication

Consider the differentiation operator

$$D : P_3 \to P_2$$

on the vector space of polynomials of degree 3 or less. If we map P_3 and P_2 into R^4 and R^3, respectively, by the natural isomorphisms, then the transformation D produces a corresponding transformation

$$T : R^4 \to R^3$$

between the image spaces. For example, the differentiation transformation produces the polynomial relationship

$$2 + x + 4x^2 - x^3 \xrightarrow{D} 1 + 8x - 3x^2$$

and this corresponds to

$$(2, 1, 4, -1) \xrightarrow{T} (1, 8, -3)$$

in the image spaces.

Since we know from Example 18 that the natural isomorphisms for P_3 and P_2 map their natural bases into the standard bases for R^4 and R^3, let us consider how the standard matrix $[T]$ for T might relate to the transformation D. We know that the column vectors of the matrix $[T]$ are the images under T of the standard basis vectors for R^4, but to find these images we have to first see what D does to the natural basis vectors for P_3. Since

$$1 \xrightarrow{D} 0$$
$$x \xrightarrow{D} 1$$
$$x^2 \xrightarrow{D} 2x$$
$$x^3 \xrightarrow{D} 3x^2$$

the corresponding relationships under the isomorphisms are

$$(1, 0, 0, 0) \xrightarrow{T} (0, 0, 0)$$
$$(0, 1, 0, 0) \xrightarrow{T} (1, 0, 0)$$
$$(0, 0, 1, 0) \xrightarrow{T} (0, 2, 0)$$
$$(0, 0, 0, 1) \xrightarrow{T} (0, 0, 3)$$

and hence the standard matrix for T is

$$[T] = \begin{bmatrix} 0 & 1 & 0 & 0 \\ 0 & 0 & 2 & 0 \\ 0 & 0 & 0 & 3 \end{bmatrix}$$

This matrix performs the differentiation

$$D(a_0 + a_1 x + a_2 x^2 + a_3 x^3) = a_1 + 2a_2 x + 3a_3 x^2$$

by operating on the images of the polynomials under the natural isomorphisms, as confirmed by the computation

$$\begin{bmatrix} 0 & 1 & 0 & 0 \\ 0 & 0 & 2 & 0 \\ 0 & 0 & 0 & 3 \end{bmatrix} \begin{bmatrix} a_0 \\ a_1 \\ a_2 \\ a_3 \end{bmatrix} = \begin{bmatrix} a_1 \\ 2a_2 \\ 3a_3 \end{bmatrix}$$

■

**INNER PRODUCT SPACE
ISOMORPHISMS**

We now know that every real n-dimensional vector space V is isomorphic to R^n and hence has the same algebraic structure as R^n. However, if V is an inner product space, then it has a geometric structure as well, and, of course, R^n has a geometric structure that arises from the Euclidean inner product (the dot product). Thus, it is reasonable to inquire if there exists an isomorphism from V to R^n that preserves the geometric structure as well as the algebraic structure. For example, we would want orthogonal vectors in V to have orthogonal counterparts in R^n, and we would want orthonormal sets in V to correspond to orthonormal sets in R^n.

In order for an isomorphism to preserve geometric structure, it obviously has to preserve inner products, since notions of length, angle, and orthogonality are all based on the inner product. Thus, if V and W are inner product spaces, then we call an isomorphism $T : V \to W$ an **inner product space isomorphism** if

$$\langle T(\mathbf{u}), T(\mathbf{v}) \rangle = \langle \mathbf{u}, \mathbf{v} \rangle$$

It can be proved that if V is any real n-dimensional inner product space and R^n has the Euclidean inner product (the dot product), then there exists an inner product space isomorphism from V to R^n. Under such an isomorphism, the inner product space V has the same algebraic and geometric structure as R^n. In this sense, every n-dimensional inner product space is a carbon copy of R^n with the Euclidean inner product.

EXAMPLE 21
An Inner Product
Space
Isomorphism

Let R^n be the vector space of real n-tuples in comma-delimited form, let M_n be the vector space of real $n \times 1$ matrices, let R^n have the Euclidean inner product $\langle \mathbf{u}, \mathbf{v} \rangle = \mathbf{u} \cdot \mathbf{v}$, and let M_n have the inner product $\langle \mathbf{u}, \mathbf{v} \rangle = \mathbf{u}^T \mathbf{v}$ in which $\mathbf{u}$ and $\mathbf{v}$ are expressed in column form. The mapping $T : R^n \to M_n$ defined by

$$(v_1, v_2, \ldots, v_n) \xrightarrow{T} \begin{bmatrix} v_1 \\ v_2 \\ \vdots \\ v_n \end{bmatrix}$$

is an inner product space isomorphism [see Formula (26) of Section 3.1], so the distinction between the inner product space R^n and the inner product space M_n is essentially notational, a fact that we have used many times in this text. ∎

Exercise Set 9.3

1. (a) If $T : V \to R^2$ is a linear transformation for which $T(\mathbf{u}) = (1, 2)$ and $T(\mathbf{v}) = (-1, 3)$, then $T(2\mathbf{u} + 4\mathbf{v}) =$ _____.

(b) If $T : V \to R^2$ is a linear transformation for which $T(\mathbf{u} + \mathbf{v}) = (2, 4)$ and $T(\mathbf{u} - \mathbf{v}) = (-3, 5)$, then $T(\mathbf{u}) =$ _____ and $T(\mathbf{v}) =$ _____.

2. (a) If $T : V \to W$ is a linear transformation for which $T(\mathbf{u}) = \mathbf{w}_1$ and $T(\mathbf{v}) = \mathbf{w}_2$, then $T(\mathbf{u} - 5\mathbf{v}) =$ _____.

(b) If $T : V \to W$ is a linear transformation for which $T(\mathbf{u} + 2\mathbf{v}) = \mathbf{w}_1$ and $T(\mathbf{u} - 2\mathbf{v}) = \mathbf{w}_2$, then $T(\mathbf{u}) =$ _____ and $T(\mathbf{v}) =$ _____.

In Exercises 3–10, show that T is a linear transformation.

3. $T : P_2 \to P_3$, where $T(p) = x p(x)$.

4. $T : P_2 \to P_2$, where $T(p) = p(x + 1)$.

5. $T : M_{nn} \to M_{nn}$, where $T(A) = A^T$.

6. $T : R^3 \to R^3$, where $\mathbf{x}_0$ is a fixed vector in R^3, and $T(\mathbf{x}) = \mathbf{x}_0 \times \mathbf{x}$. [*Hint:* See Theorem 4.3.8.]

7. $T : C[a, b] \to R$, where $T(f) = \int_a^b f(x)\, dx$.

8. $T : C^1(-\infty, \infty) \to R$, where $T(f) = f'(0)$.

9. $T : M_{nn} \to M_{nn}$, where A_0 is a fixed $n \times n$ matrix and $T(X) = A_0 X$.

10. $T : R^\infty \to R^\infty$, where $T(v_1, v_2, v_3, \ldots, v_n, \ldots) = (0, v_1, v_2, v_3, \ldots, v_{n-1}, \ldots)$.

In Exercises 11–14, show that T is not linear.

11. $T : F(-\infty, \infty) \to F(-\infty, \infty)$, where $T(f) = x^2 f^2(x)$.

12. $T : V \to R$, where $T(\mathbf{x}) = \|\mathbf{x}\|$ and V is an inner product space.

13. $T : V \to V$, where $\mathbf{x}_0$ is a fixed nonzero vector in a vector space V and $T(\mathbf{x}) = \mathbf{x} + \mathbf{x}_0$.

14. $T: C[0, 1] \to R$, where $T(f) = \int_0^1 |f(t)| \, dt$.

In Exercises 15–18, determine whether T is linear.

15. $T: V \to V$, where c_0 is a fixed scalar and $T(\mathbf{x}) = c_0 \mathbf{x} + \mathbf{x}$.

16. $T: R^\infty \to R^\infty$, where
$$T(v_1, v_2, v_3, v_4, \ldots) = (v_1, -v_2, v_3, -v_4, \ldots)$$

17. $T: F(-\infty, \infty) \to R$, where $T(f) = f(0)f(1)$.

18. $T: V \to V$, where $T(\mathbf{x}) = -\mathbf{x}$ for all $\mathbf{x}$ in V.

19. (a) Let $T: P_1 \to P_2$ be the linear transformation defined by $T(p) = xp(x)$. Which of the following, if any, are in the range of T?
$$q_1(x) = 1 + x + x^2, \quad q_2(x) = x + 5x^2, \quad q_3(x) = 0$$

(b) Let $T: P_2 \to R^2$ be the linear transformation defined by $T(p) = (p(-1), p(1))$. Which of the following, if any, are in the kernel of T?
$$q_1(x) = x^2 - 1, \quad q_2(x) = x^2 + 1, \quad q_3(x) = 0$$

20. (a) Let $T: P_2 \to P_2$ be the linear operator defined by $T(a_0 + a_1 x + a_2 x^2) = a_0 + a_1 x$. Which of the following, if any, are in the range of T?
$$q_1(x) = 1 + x + x^2, \quad q_2(x) = 1 + x, \quad q_3(x) = 0$$

(b) Let $T: P_3 \to P_1$ be the linear transformation defined by $T(p) = p''(x)$. Which of the following, if any, are in the kernel of T?
$$q_1(x) = 1 + x + x^2, \quad q_2(x) = 4 + 5x, \quad q_3(x) = 0$$

21. Show that the mapping
$$T\left(\begin{bmatrix} a & b \\ c & d \end{bmatrix}\right) = \begin{bmatrix} a & 0 \\ 0 & a \end{bmatrix}$$
is a linear operator on M_{22}, and find bases for its kernel and range.

22. Show that the mapping
$$T\left(\begin{bmatrix} a & b \\ c & d \end{bmatrix}\right) = \begin{bmatrix} 0 & -b \\ c & a \end{bmatrix}$$
is a linear operator on M_{22}, and find bases for its kernel and range.

In Exercises 23–26, show that the linear transformation is one-to-one, and determine whether it is onto.

23. The linear transformation of Exercise 9 in the case where A_0 is invertible.

24. The linear transformation of Exercise 5.

25. The linear transformation of Exercise 3.

26. The linear transformation of Exercise 4.

27. (a) Find a basis for the kernel of the linear transformation $D: C^2(-\infty, \infty) \to C(-\infty, \infty)$ given by $D(y) = y''(x)$.

(b) Find a general solution of the differential equation $y'' = 0$.

28. (a) Use the results in Example 14 to find a basis for the kernel of the linear transformation
$$D: C^2(-\infty, \infty) \to C(-\infty, \infty)$$
given by $D(y) = y'' + 4y$.

(b) Find a general solution of the differential equation $y'' + 4y = 0$.

29. Let $D: C^2(-\infty, \infty) \to C(-\infty, \infty)$ be given by
$$D(y) = y'' - \omega^2 y$$
It is proved in standard textbooks on differential equations that the kernel of this linear transformation is two-dimensional.

(a) Show that if $w \neq 0$, then the functions $y_1 = e^{-\omega x}$ and $y_2 = e^{\omega x}$ form a basis for $\ker(D)$.

(b) Find a general solution of the differential equation $y'' - \omega^2 y = 0$.

30. Let $D: C^1(-\infty, \infty) \to C(-\infty, \infty)$ be the derivative transformation of Example 6, let $J: C(-\infty, \infty) \to C^1(-\infty, \infty)$ be the integration transformation of Example 7, and let $J \circ D$ be the composition of J with D. Compute $(J \circ D)(f)$ for

(a) $f(x) = x^2 + 2x + 3$ (b) $f(x) = \cos x$

(c) $f(x) = 2e^x + 1$

In Exercises 31 and 32, let $T: P_2 \to R^3$ be the evaluation transformation $T(p) = (p(0), p(1), p(2))$. We know from Example 15 that T is one-to-one and onto, so for each vector $\mathbf{v} = (v_1, v_2, v_3)$ in R^3 there is a unique polynomial in P_2 for which $T(p) = \mathbf{v}$. Find that polynomial.

31. $\mathbf{v} = (1, 3, 7)$ **32.** $\mathbf{v} = (-1, 1, 5)$

33. Let $T: R^\infty \to R^\infty$ be the mapping defined by $T(v_1, v_2, v_3, \ldots) = (v_2, v_3, v_4, \ldots)$. Show that T is linear and find its kernel and range.

34. Consider the weighted Euclidean inner product on R^2 defined by $\langle \mathbf{u}, \mathbf{v} \rangle = 2u_1 v_1 + 3u_2 v_2$, and let $\mathbf{v}_0 = (4, -2)$. We know from Example 4 that the mapping $T(\mathbf{x}) = \langle \mathbf{x}, \mathbf{v}_0 \rangle$ defines a linear transformation from R^2 to R. Let $\mathbf{x} = (x, y)$, and sketch the kernel of T in the xy-plane.

35. Show that if $\{\mathbf{v}_1, \mathbf{v}_2, \ldots, \mathbf{v}_n\}$ is a basis for a vector space V, and if $T: V \to W$ is a linear transformation for which $T(\mathbf{v}_1) = T(\mathbf{v}_2) = \cdots = T(\mathbf{v}_n) = \mathbf{0}$, then T is the zero transformation.

36. Show that if $\{\mathbf{v}_1, \mathbf{v}_2, \ldots, \mathbf{v}_n\}$ is a basis for a vector space V, and if $T: V \to V$ is a linear operator for which
$$T(\mathbf{v}_1) = \mathbf{v}_1, T(\mathbf{v}_2) = \mathbf{v}_2, \ldots, T(\mathbf{v}_n) = \mathbf{v}_n$$
then T is the identity operator.

37. Consider the natural isomorphism from M_{22} to R^4.

(a) What matrix in M_{22} corresponds to the vector $\mathbf{v} = (-1, 2, 0, 3)$ under this isomorphism?

(b) Find a basis for the subspace of R^4 that corresponds to the subspace of M_{22} consisting of symmetric matrices with trace zero.

(c) Find the standard matrix for the linear operator on R^4 corresponding to the linear operator $A \to A^T$ on M_{22}.

38. Consider the natural isomorphism from P_2 to R^3.
  (a) What polynomial corresponds to the vector
      $\mathbf{v} = (2, 3, -1)$ under this isomorphism?
  (b) Find a basis for the subspace of R^3 that corresponds to
      the subspace P_1 of P_2.
  (c) Find the standard matrix for the linear operator on R^3
      that corresponds to the linear operator
      $p(x) \rightarrow p(x + 1)$ on P_2.

39. Consider the natural isomorphisms of P_2 to R^3 and P_3 to
R^4, and let $J : P_2 \rightarrow P_3$ be the integration transformation

$$J(p) = \int_0^x p(t)\, dt$$

Find the standard matrix for the linear transformation from
R^3 to R^4 that corresponds to J, and use that matrix to integrate $p(x) = x^2 + x + 1$ by matrix multiplication. [*Hint:*
See Example 20.]

40. Consider the natural isomorphisms of P_1 to R^2 and P_3 to
R^4, and let $D : P_3 \rightarrow P_1$ be the differentiation transformation $D(p) = p''$. Find the standard matrix for the linear
transformation from R^4 to R^2 that corresponds to D.

Discussion and Discovery

D1. If $T : P_2 \rightarrow P_1$ is a linear transformation for which
$T(1) = 1 + x$, $T(2x) = 1 - 2x$, and $T(3x^2) = 1 + 3x$,
then $T(2 + 4x - x^2) = $ _____.

D2. Indicate whether the statement is true (T) or false (F).
Justify your answer.
  (a) If T is a mapping from a vector space V to a vector
      space W such that $T(c\mathbf{u} + \mathbf{v}) = cT(\mathbf{u}) + T(\mathbf{v})$ for all
      vectors $\mathbf{u}$ and $\mathbf{v}$ in V and all scalars c, then T is a
      linear transformation.
  (b) $T(a, b, c) = ax^2 + bx + c$ defines a one-to-one linear
      transformation from R^3 onto P_2.
  (c) The mapping $T : C^1[a, b] \rightarrow C[a, b]$ defined by

$$T(f) = f'(x) + \int_a^x f(t)\, dt$$

  is a linear transformation.
  (d) The vector space M_{mn} has dimension mn.

  (e) The subspace of all polynomials in $C(-\infty, \infty)$ is
      finite-dimensional.

D3. Let $T : M_{22} \rightarrow M_{22}$ be the mapping that is defined by
$T(A) = A - A^T$.
  (a) Show that T is linear.
  (b) Describe the kernel of T.
  (c) Show that the range of T consists of all 2×2
      skew-symmetric matrices.
  (d) What can you say about the kernel and range of the
      mapping $T : M_{22} \rightarrow M_{22}$ defined by $T(A) = A + A^T$?

D4. Find a linear transformation $T : C^\infty(-\infty, \infty) \rightarrow F(-\infty, \infty)$
whose kernel is P_3.

D5. Describe the kernel of the integration transformation
$T : P_1 \rightarrow R$ defined by

$$T(p) = \int_{-1}^1 p(x)\, dx$$

Working with Proofs

P1. Let V and W be vector spaces, let T, T_1, and T_2 be linear
transformations from V to W, and let k be any scalar. Define
new transformations $(T_1 + T_2) : V \rightarrow W$ and $(kT) : V \rightarrow W$
by

$$(kT)(\mathbf{x}) = kT(\mathbf{x}) \quad \text{and} \quad (T_1 + T_2)(\mathbf{x}) = T_1(\mathbf{x}) + T_2(\mathbf{x})$$

Prove that kT and $T_1 + T_2$ are linear transformations.

P2. Prove Theorem 9.3.2 by modifying the proof of Theorem
6.1.3.

P3. If $T : V \rightarrow W$ is a linear transformation that is one-to-one

and onto, then for each vector $\mathbf{x}$ in W there is a unique
vector $\mathbf{v}$ in V such that $T(\mathbf{v}) = \mathbf{x}$. Prove that the ***inverse
transformation*** $T^{-1} : W \rightarrow V$ defined by $T^{-1}(\mathbf{x}) = \mathbf{v}$ is
linear.

P4. Prove: If V is an n-dimensional vector space and if the transformation $T : V \rightarrow R^n$ is an isomorphism, then there exists
a unique inner product $\langle \mathbf{u}, \mathbf{v} \rangle$ on V such that $T(\mathbf{u}) \cdot T(\mathbf{v}) =
\langle \mathbf{u}, \mathbf{v} \rangle$. [*Hint:* Show that $\langle \mathbf{u}, \mathbf{v} \rangle = T(\mathbf{u}) \cdot T(\mathbf{v})$ defines an
inner product on V.]

Since many of the most important concepts in linear algebra occur as theorem statements, it is important to be familiar with the various ways in which theorems can be structured. This appendix will help you to do that.

CONTRAPOSITIVE FORM OF A THEOREM

The simplest theorems are of the form

$$\text{If } H \text{ is true, then } C \text{ is true.} \tag{1}$$

where H is a statement, called the **hypothesis**, and C is a statement, called the **conclusion**. *The theorem is true if the conclusion is true whenever the hypothesis is true, and the theorem is false if there is some case where the hypothesis is true but the conclusion is false.* It is common to denote a theorem of form (1) as

$$H \Rightarrow C \tag{2}$$

(read, "H implies C"). As an example, the theorem

$$\text{If } a \text{ and } b \text{ are both positive numbers, then } ab \text{ is a positive number.} \tag{3}$$

is of form (2), where

$$H = a \text{ and } b \text{ are both positive numbers} \tag{4}$$

$$C = ab \text{ is a positive number} \tag{5}$$

Sometimes it is desirable to phrase theorems in a *negative* way. For example, the theorem in (3) can be rephrased equivalently as

$$\text{If } ab \text{ is not a positive number, then } a \text{ and } b \text{ are not both positive numbers.} \tag{6}$$

If we write $\sim H$ to mean that (4) is false and $\sim C$ to mean that (5) is false, then the structure of the theorem in (6) is

$$\sim C \Rightarrow \sim H \tag{7}$$

In general, any theorem of form (2) can be rephrased in form (7), which is called the **contrapositive** of (2). If a theorem is true, then so is its contrapositive, and vice versa.

CONVERSE OF A THEOREM

The **converse** of a theorem is the statement that results when the hypothesis and conclusion are interchanged. Thus, the converse of the theorem $H \Rightarrow C$ is the statement $C \Rightarrow H$. Whereas the contrapositive of a true theorem must itself be a true theorem, the converse of a true theorem may or may not be true. For example, the converse of (3) is the *false* statement

$$\text{If } ab \text{ is a positive number, then } a \text{ and } b \text{ are both positive numbers.}$$

but the converse of the true theorem

$$\text{If } a > b, \text{ then } 2a > 2b. \tag{8}$$

is the *true* theorem

$$\text{If } 2a > 2b, \text{ then } a > b. \tag{9}$$

EQUIVALENT STATEMENTS

If a theorem $H \Rightarrow C$ and its converse $C \Rightarrow H$ are both true, then we say that H and C are *equivalent* statements, which we denote by writing

$$H \Leftrightarrow C$$

(read, "H and C are equivalent"). There are various ways of phrasing equivalent statements

A1

as a single theorem. Here are three ways in which (8) and (9) can be combined into a single theorem.

Form 1   If $a > b$, then $2a > 2b$, and conversely, if $2a > 2b$, then $a > b$.

Form 2 $a > b$ if and only if $2a > 2b$.

Form 3   The following statements are equivalent.

  (i) $a > b$
 (ii) $2a > 2b$

THEOREMS INVOLVING THREE OR MORE STATEMENTS

Sometimes two true theorems will give you a third true theorem for free. Specifically, if $H \Rightarrow C$ is a true theorem, and $C \Rightarrow D$ is a true theorem, then $H \Rightarrow D$ must also be a true theorem. For example, the theorems

If opposite sides of a quadrilateral are parallel, then the quadrilateral is a parallelogram.

and

Opposite sides of a parallelogram have equal lengths.

imply the third theorem

If opposite sides of a quadrilateral are parallel, then they have equal lengths.

Sometimes three theorems yield equivalent statements for free. For example, if

$$H \Rightarrow C, \quad C \Rightarrow D, \quad D \Rightarrow H \tag{10}$$

then we have the *implication loop* in Figure A.1 from which we can conclude that

$$C \Rightarrow H, \quad D \Rightarrow C, \quad H \Rightarrow D \tag{11}$$

Combining this with (10) we obtain

$$H \Leftrightarrow C, \quad C \Leftrightarrow D, \quad D \Leftrightarrow H \tag{12}$$

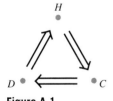

Figure A.1

In summary, if you want to prove the three equivalences in (12), you need only prove the three implications in (10).

APPENDIX B   COMPLEX NUMBERS

Complex numbers arise naturally in the course of solving polynomial equations. For example, the solutions of the quadratic equation $ax^2 + bx + c = 0$, which are given by the quadratic formula

$$x = \frac{-b \pm \sqrt{b^2 - 4ac}}{2a}$$

are complex numbers if the expression inside the radical is negative. In this appendix we will review some of the basic ideas about complex numbers that are used in this text.

COMPLEX NUMBERS

To deal with the problem that the equation $x^2 = -1$ has no real solutions, mathematicians of the eighteenth century invented the "imaginary" number

$$i = \sqrt{-1}$$

which is assumed to have the property

$$i^2 = (\sqrt{-1})^2 = -1$$

but which otherwise has the algebraic properties of a real number. An expression of the form

$$a + bi \quad \text{or} \quad a + ib$$

in which a and b are *real* numbers is called a ***complex number***. Sometimes it will be convenient to use a single letter, typically z, to denote a complex number, in which case we write

$$z = a + bi \quad \text{or} \quad z = a + ib$$

The number a is called the ***real part*** of z and is denoted by $\mathrm{Re}(z)$, and the number b is called the ***imaginary part*** of z and is denoted by $\mathrm{Im}(z)$. Thus,

$$\mathrm{Re}(3 + 2i) = 3, \quad \mathrm{Im}(3 + 2i) = 2$$
$$\mathrm{Re}(1 - 5i) = 1, \quad \mathrm{Im}(1 - 5i) = \mathrm{Im}(1 + (-5)i) = -5$$
$$\mathrm{Re}(7i) = \mathrm{Re}(0 + 7i) = 0, \quad \mathrm{Im}(7i) = 7$$
$$\mathrm{Re}(4) = 4, \quad \mathrm{Im}(4) = \mathrm{Im}(4 + 0i) = 0$$

Two complex numbers are considered ***equal*** if and only if their real parts are equal and their imaginary parts are equal; that is,

$$a + bi = c + di \quad \text{if and only if} \quad a = c \text{ and } b = d$$

A complex number $z = bi$ whose real part is zero is said to be ***pure imaginary***. A complex number $z = a$ whose imaginary part is zero is a real number, so the real numbers can be viewed as a subset of the complex numbers.

Complex numbers are added, subtracted, and multiplied in accordance with the standard rules of algebra but with $i^2 = -1$:

$$(a + bi) + (c + di) = (a + c) + (b + d)i \tag{1}$$
$$(a + bi) - (c + di) = (a - c) + (b - d)i \tag{2}$$
$$(a + bi)(c + di) = (ac - bd) + (ad + bc)i \tag{3}$$

The multiplication formula is obtained by expanding the left side and using the fact that $i^2 = -1$ (verify). Also note that if $b = 0$, then the multiplication formula simplifies to

$$a(c + di) = ac + adi \tag{4}$$

A3

The set of complex numbers with these operations is commonly denoted by the symbol $\mathbb{C}$ and is called the ***complex number system***.

EXAMPLE 1
Multiplying Complex Numbers

As a practical matter, it is usually more convenient to compute products of complex numbers by expansion, rather than substituting in (3). For example,

$$(3 - 2i)(4 + 5i) = 12 + 15i - 8i - 10i^2 = (12 + 10) + 7i = 22 + 7i \qquad \blacksquare$$

THE COMPLEX PLANE

A complex number $z = a + bi$ can be associated with the ordered pair (a, b) of real numbers and represented geometrically by a point or a vector in the xy-plane (Figure B.1). We call this the ***complex plane***. Points on the x-axis have an imaginary part of zero and hence correspond to real numbers, whereas points on the y-axis have a real part of zero and correspond to pure imaginary numbers. Accordingly, we call the x-axis the ***real axis*** and the y-axis the ***imaginary axis*** (Figure B.2).

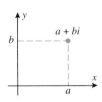

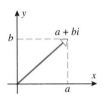

Figure B.1

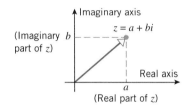

Figure B.2

Complex numbers can be added, subtracted, or multiplied by real numbers geometrically by performing these operations on their associated vectors (Figure B.3, for example). In this sense the complex number system $\mathbb{C}$ is closely related to R^2, the main difference being that complex numbers can be multiplied to produce other complex numbers, whereas there is no multiplication operation on R^2 that produces other vectors in R^2 (the dot product produces a scalar, not a vector in R^2).

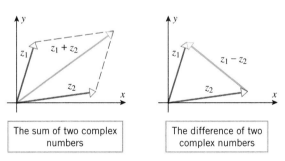

The sum of two complex numbers

The difference of two complex numbers

Figure B.3

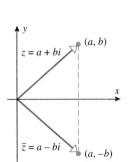

Figure B.4

If $z = a + bi$ is a complex number, then the ***complex conjugate*** of z, or more simply, the ***conjugate*** of z, is denoted by $\bar{z}$ (read, "z bar") and is defined by

$$\bar{z} = a - bi \qquad (5)$$

Numerically, $\bar{z}$ is obtained from z by reversing the sign of the imaginary part, and geometrically it is obtained by reflecting the vector for z about the real axis (Figure B.4).

EXAMPLE 2
Some Complex Conjugates

$$
\begin{aligned}
z &= 3 + 4i & \bar{z} &= 3 - 4i \\
z &= -2 - 5i & \bar{z} &= -2 + 5i \\
z &= i & \bar{z} &= -i \\
z &= 7 & \bar{z} &= 7
\end{aligned}
$$
$\blacksquare$

REMARK The last computation in this example illustrates the fact that a real number is equal to its complex conjugate. More generally, $z = \bar{z}$ if and only if z is a real number.

The following computation shows that the product of any complex number $z = a + bi$ and its conjugate $z = a - bi$ is a nonnegative real number:

$$z\bar{z} = (a + bi)(a - bi) = a^2 - abi + bai - b^2i^2 = a^2 + b^2 \tag{6}$$

You will recognize that

$$\sqrt{z\bar{z}} = \sqrt{a^2 + b^2}$$

is the length of the vector corresponding to z (Figure B.5); we call this length the **modulus** (or **absolute value** of z) and denote it by $|z|$. Thus,

$$|z| = \sqrt{z\bar{z}} = \sqrt{a^2 + b^2} \tag{7}$$

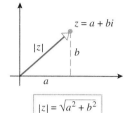

$$|z| = \sqrt{a^2 + b^2}$$

Figure B.5

Note that if $b = 0$, then $z = a$ is a real number and $|z| = \sqrt{a^2} = |a|$, which tells us that the modulus of a real number is the same as its absolute value.

EXAMPLE 3
Some Modulus
Computations

$$z = 3 + 4i \qquad |z| = \sqrt{3^2 + 4^2} = 5$$
$$z = -4 - 5i \qquad |z| = \sqrt{(-4)^2 + (-5)^2} = \sqrt{41}$$
$$z = i \qquad |z| = \sqrt{0^2 + 1^2} = 1$$ ∎

RECIPROCALS AND DIVISION

If $z \neq 0$, then the **reciprocal** (or **multiplicative inverse**) of z is denoted by $1/z$ (or z^{-1}) and is defined to be that complex number, if any, such that

$$\left(\frac{1}{z}\right) z = 1$$

This equation has a unique solution for $1/z$, which we can obtain by multiplying both sides by $\bar{z}$ and using the fact that $z\bar{z} = |z|^2$ [see (7)]. This yields

$$\frac{1}{z} = \frac{\bar{z}}{|z|^2} \tag{8}$$

If $z_2 \neq 0$, then the **quotient** z_1/z_2 is defined to be the product of z_1 and $1/z_2$. This yields the formula

$$\frac{z_1}{z_2} = \frac{\bar{z}_2}{|z_2|^2} z_1 = \frac{z_1 \bar{z}_2}{|z_2|^2} \tag{9}$$

Observe that the expression on the right side of (9) results if the numerator and denominator of z_1/z_2 are multiplied by $\bar{z}_2$. As a practical matter, this is often the best way to perform divisions of complex numbers.

EXAMPLE 4
Division of
Complex
Numbers

Let $z_1 = 3 + 4i$ and $z_2 = 1 - 2i$. Express z_1/z_2 in the form $a + bi$.

Solution We will multiply the numerator and denominator of z_1/z_2 by $\bar{z}_2$. This yields

$$\frac{z_1}{z_2} = \frac{z_1 \bar{z}_2}{z_2 \bar{z}_2} = \frac{3 + 4i}{1 - 2i} \cdot \frac{1 + 2i}{1 + 2i}$$
$$= \frac{3 + 6i + 4i + 8i^2}{1 - 4i^2}$$
$$= \frac{-5 + 10i}{5}$$
$$= -1 + 2i$$ ∎

The following theorems list some useful properties of the modulus and conjugate operations.

Theorem B.1 *The following results hold for any complex numbers z, z_1, and z_2.*

(a) $\overline{z_1 + z_2} = \bar{z}_1 + \bar{z}_2$

(b) $\overline{z_1 - z_2} = \bar{z}_1 - \bar{z}_2$

(c) $\overline{z_1 z_2} = \bar{z}_1 \bar{z}_2$

(d) $\overline{z_1/z_2} = \bar{z}_1/\bar{z}_2$

(e) $\bar{\bar{z}} = z$

Theorem B.2 *The following results hold for any complex numbers z, z_1, and z_2.*

(a) $|\bar{z}| = |z|$

(b) $|z_1 z_2| = |z_1||z_2|$

(c) $|z_1/z_2| = |z_1|/|z_2|$

(d) $|z_1 + z_2| \le |z_1| + |z_2|$

POLAR FORM OF A COMPLEX NUMBER

If $z = a + bi$ is a nonzero complex number, and if ϕ is an angle from the real axis to the vector z, then, as suggested in Figure B.6, the real and imaginary parts of z can be expressed as

$$a = |z| \cos \phi \quad \text{and} \quad b = |z| \sin \phi \tag{10}$$

Thus, the complex number $z = a + bi$ can be expressed as

$$z = |z|(\cos \phi + i \sin \phi) \tag{11}$$

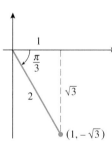

Figure B.6

which is called a ***polar form*** of z. The angle ϕ in this formula is called an ***argument*** of z. The argument of z is not unique because we can add or subtract any multiple of 2π to it to obtain a different argument of z. However, there is only one argument whose radian measure satisfies

$$-\pi < \phi \le \pi \tag{12}$$

This is called the ***principal argument*** of z.

EXAMPLE 5

Polar Form of a Complex Number

Express $z = 1 - \sqrt{3}i$ in polar form using the principal argument.

Solution The modulus of z is

$$|z| = \sqrt{1^2 + (-\sqrt{3})^2} = \sqrt{4} = 2$$

Thus, it follows from (10) with $a = 1$ and $b = -\sqrt{3}$ that

$$1 = 2 \cos \phi \quad \text{and} \quad -\sqrt{3} = 2 \sin \phi$$

and this implies that

$$\cos \phi = \frac{1}{2} \quad \text{and} \quad \sin \phi = -\frac{\sqrt{3}}{2}$$

Figure B.7

The unique angle ϕ that satisfies these equations and whose radian measure is in the interval $-\pi < \phi \le \pi$ is $\phi = -\pi/3$ (Figure B.7). Thus, a polar form of z is

$$z = 2\left(\cos\left(-\frac{\pi}{3}\right) + i \sin\left(-\frac{\pi}{3}\right)\right) = 2\left(\cos\frac{\pi}{3} - i \sin\frac{\pi}{3}\right) \qquad \blacksquare$$

GEOMETRIC INTERPRETATION OF MULTIPLICATION AND DIVISION OF COMPLEX NUMBERS

We now show how polar forms of complex numbers provide geometric interpretations of multiplication and division. Let

$$z_1 = |z_1|(\cos \phi_1 + i \sin \phi_1) \quad \text{and} \quad z_2 = |z_2|(\cos \phi_2 + i \sin \phi_2)$$

be polar forms of the nonzero complex numbers z_1 and z_2. Multiplying, we obtain

$$z_1 z_2 = |z_1||z_2|[(\cos \phi_1 \cos \phi_2 - \sin \phi_1 \sin \phi_2) + i(\sin \phi_1 \cos \phi_2 + \cos \phi_1 \sin \phi_2)]$$

Now applying the trigonometric identities

$$\cos(\phi_1 + \phi_2) = \cos\phi_1 \cos\phi_2 - \sin\phi_1 \sin\phi_2$$
$$\sin(\phi_1 + \phi_2) = \sin\phi_1 \cos\phi_2 + \cos\phi_1 \sin\phi_2$$

yields

$$z_1 z_2 = |z_1||z_2|[\cos(\phi_1 + \phi_2) + i\sin(\phi_1 + \phi_2)] \tag{13}$$

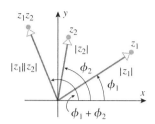

Figure B.8

which is a polar form of the complex number with modulus $|z_1||z_2|$ and argument $\phi_1 + \phi_2$. Thus, we have shown that *multiplying two complex numbers has the geometric effect of multiplying their moduli and adding their arguments* (Figure B.8).

Similar kinds of computations show that

$$\frac{z_1}{z_2} = \frac{|z_1|}{|z_2|}[\cos(\phi_1 - \phi_2) + i\sin(\phi_1 - \phi_2)] \tag{14}$$

which tells us that *dividing complex numbers has the geometric effect of dividing their moduli and subtracting their arguments* (both in the appropriate order).

EXAMPLE 6
Multiplying and
Dividing in
Polar Form

Use polar forms of the complex numbers $z_1 = 1 + \sqrt{3}i$ and $z_2 = \sqrt{3} + i$ to compute $z_1 z_2$ and z_1/z_2.

Solution Polar forms of these complex numbers are

$$z_1 = 2\left(\cos\frac{\pi}{3} + i\sin\frac{\pi}{3}\right) \quad \text{and} \quad z_2 = 2\left(\cos\frac{\pi}{6} + i\sin\frac{\pi}{6}\right)$$

(verify). Thus, it follows from (13) that

$$z_1 z_2 = 4\left[\cos\left(\frac{\pi}{3} + \frac{\pi}{6}\right) + i\sin\left(\frac{\pi}{3} + \frac{\pi}{6}\right)\right] = 4\left[\cos\left(\frac{\pi}{2}\right) + i\sin\left(\frac{\pi}{2}\right)\right] = 4i$$

and from (14) that

$$\frac{z_1}{z_2} = 1 \cdot \left[\cos\left(\frac{\pi}{3} - \frac{\pi}{6}\right) + i\sin\left(\frac{\pi}{3} - \frac{\pi}{6}\right)\right] = \cos\left(\frac{\pi}{6}\right) + i\sin\left(\frac{\pi}{6}\right) = \frac{\sqrt{3}}{2} + \frac{1}{2}i$$

As a check, let us calculate $z_1 z_2$ and z_1/z_2 directly:

$$z_1 z_2 = (1 + \sqrt{3}i)(\sqrt{3} + i) = \sqrt{3} + i + 3i + \sqrt{3}i^2 = 4i$$

$$\frac{z_1}{z_2} = \frac{1 + \sqrt{3}i}{\sqrt{3} + i} = \frac{1 + \sqrt{3}i}{\sqrt{3} + i} \cdot \frac{\sqrt{3} - i}{\sqrt{3} - i} = \frac{\sqrt{3} - i + 3i - \sqrt{3}i^2}{3 - i^2} = \frac{2\sqrt{3} + 2i}{4} = \frac{\sqrt{3}}{2} + \frac{1}{2}i$$

which agrees with the results obtained using polar forms.   ∎

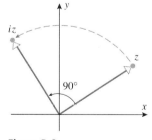

Figure B.9

REMARK   The complex number i has a modulus of 1 and a principal argument of $\pi/2$. Thus, if z is a complex number, then iz has the same modulus as z but its argument is greater by $\pi/2$ ($= 90°$); that is, multiplication by i has the geometric effect of rotating the vector z counterclockwise by $90°$ (Figure B.9).

DEMOIVRE'S FORMULA   If n is a positive integer, and if z is a nonzero complex number with polar form

$$z = |z|(\cos\phi + i\sin\phi)$$

then raising z to the nth power yields

$$z^n = \underbrace{z \cdot z \cdots \cdot z}_{n \text{ factors}} = |z|^n[\cos(\underbrace{\phi + \phi + \cdots + \phi}_{n \text{ terms}})] + i[\sin(\underbrace{\phi + \phi + \cdots + \phi}_{n \text{ terms}})]$$

which we can write more succinctly as

$$z^n = |z|^n(\cos n\phi + i\sin n\phi) \tag{15}$$

In the special case where $|z| = 1$ this formula simplifies to

$$z^n = \cos n\phi + i \sin n\phi$$

which, using the polar form for z, becomes

$$(\cos \phi + i \sin \phi)^n = \cos n\phi + i \sin n\phi \tag{16}$$

This result is called **DeMoivre's formula** in honor of the French mathematician Abraham DeMoivre (1667–1754).

EULER'S FORMULA   If θ is a real number, say the radian measure of some angle, then the **complex exponential** function $e^{i\theta}$ is defined to be

$$e^{i\theta} = \cos \theta + i \sin \theta \tag{17}$$

which is sometimes called **Euler's formula**. One motivation for this formula comes from the Maclaurin series in calculus. Readers who have studied infinite series in calculus can deduce (17) by formally substituting $i\theta$ for x in the Maclaurin series for e^x and writing

$$e^{i\theta} = 1 + i\theta + \frac{(i\theta)^2}{2!} + \frac{(i\theta)^3}{3!} + \frac{(i\theta)^4}{4!} + \frac{(i\theta)^5}{5!} + \frac{(i\theta)^6}{6!} + \cdots$$

$$= 1 + i\theta - \frac{\theta^2}{2!} - i\frac{\theta^3}{3!} + \frac{\theta^4}{4!} + i\frac{\theta^5}{5!} - \frac{\theta^6}{6!} + \cdots$$

$$= \left(1 - \frac{\theta^2}{2!} + \frac{\theta^4}{4!} - \frac{\theta^6}{6!} + \cdots \right) + i \left(\theta - \frac{\theta^3}{3!} + \frac{\theta^5}{5!} - \cdots \right)$$

$$= \cos \theta + i \sin \theta$$

where the last step follows from the Maclaurin series for $\cos \theta$ and $\sin \theta$.

If $z = a + bi$ is any complex number, then the **complex exponential e^z** is defined to be

$$e^z = e^{a+bi} = e^a e^{ib} = e^a (\cos b + i \sin b) \tag{18}$$

It can be proved that complex exponentials satisfy the standard laws of exponents. Thus, for example,

$$e^{z_1} e^{z_2} = e^{z_1 + z_2}, \quad \frac{e^{z_1}}{e^{z_2}} = e^{z_1 - z_2}, \quad \frac{1}{e^z} = e^{-z}$$

▶ **Exercise Set 1.1** (Page 13) ───────────

1. (a) **(b)**

(c) **(d)**

3. (a) 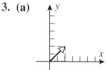 **(b)**

(c) **(d)**

(e)

5. (a) 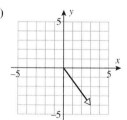 **(b)**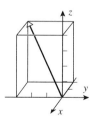

7. (a) $\overrightarrow{P_1P_2} = (-1, 3)$ **(b)** $\overrightarrow{P_1P_2} = (-3, 6, 1)$

9. (a) The terminal point is $B(1, 3)$.
(b) The initial point is $A(-2, -2, -1)$.

11. (a) $\mathbf{v} - \mathbf{w} = (-2, 1, -4, -2, 7)$
(b) $6\mathbf{u} + 2\mathbf{v} = (-10, 6, -4, 26, 28)$
(c) $(2\mathbf{u} - 7\mathbf{w}) - (8\mathbf{v} + \mathbf{u}) = (-77, 8, 94, -25, 23)$

13. $x = \left(-\frac{8}{3}, \frac{1}{2}, \frac{8}{3}, \frac{2}{3}, \frac{11}{6}\right)$

15. (a) not parallel **(b)** parallel **(c)** parallel

17. (a) $\mathbf{u} + \mathbf{v} + \mathbf{w} = (-2, 5)$ **(b)** $\mathbf{u} + \mathbf{v} + \mathbf{w} = (3, -8)$

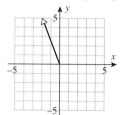 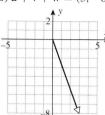

19. $a = 3, b = -1$

21.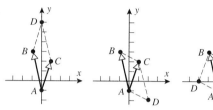

23. $\mathbf{F} = (1, -3)$

25. (a) $\begin{bmatrix} 0 & 1 & 0 & 0 & 0 \\ 1 & 0 & 1 & 0 & 0 \\ 0 & 0 & 0 & 1 & 0 \\ 0 & 0 & 0 & 0 & 1 \\ 1 & 0 & 0 & 0 & 0 \end{bmatrix}$ **(b)** $\begin{bmatrix} 0 & 0 & 0 & 0 & 0 \\ 1 & 0 & 0 & 1 & 1 \\ 1 & 0 & 0 & 1 & 1 \\ 0 & 0 & 0 & 0 & 0 \\ 0 & 0 & 0 & 0 & 0 \end{bmatrix}$

27.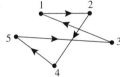

▶ **Exercise Set 1.2** (Page 25) ───────────

1. (a) $\|\mathbf{v}\| = 5$, $\frac{\mathbf{v}}{\|\mathbf{v}\|} = \left(\frac{4}{5}, -\frac{3}{5}\right)$, $-\frac{\mathbf{v}}{\|\mathbf{v}\|} = \left(-\frac{4}{5}, \frac{3}{5}\right)$
(b) $\|\mathbf{v}\| = 2\sqrt{3}$,
$\frac{\mathbf{v}}{\|\mathbf{v}\|} = \left(\frac{1}{\sqrt{3}}, \frac{1}{\sqrt{3}}, \frac{1}{\sqrt{3}}\right)$, $-\frac{\mathbf{v}}{\|\mathbf{v}\|} = \left(-\frac{1}{\sqrt{3}}, -\frac{1}{\sqrt{3}}, -\frac{1}{\sqrt{3}}\right)$
(c) $\|\mathbf{v}\| = \sqrt{15}$,
$\frac{\mathbf{v}}{\|\mathbf{v}\|} = \frac{1}{\sqrt{15}}(1, 0, 2, 1, 3)$, $-\frac{\mathbf{v}}{\|\mathbf{v}\|} = -\frac{1}{\sqrt{15}}(1, 0, 2, 1, 3)$

3. (a) $\|\mathbf{u} + \mathbf{v}\| = \sqrt{83}$ **(b)** $\|\mathbf{u}\| + \|\mathbf{v}\| = \sqrt{17} + \sqrt{26}$
(c) $\|-2\mathbf{u} + 2\mathbf{v}\| = 2\sqrt{3}$ **(d)** $\|3\mathbf{u} - 5\mathbf{v} + \mathbf{w}\| = \sqrt{466}$

5. (a) $\|3\mathbf{u} - 5\mathbf{v} + \mathbf{w}\| = \sqrt{2570}$
(b) $\|3\mathbf{u}\| - 5\|\mathbf{v}\| + \|\mathbf{u}\| = 4\sqrt{46} - 5\sqrt{84}$
(c) $\|-\|\mathbf{u}\|\mathbf{v}\| = 2\sqrt{966}$

7. $k = \frac{5}{7}, k = -\frac{5}{7}$

9. (a) $\mathbf{u} \cdot \mathbf{v} = -8, \mathbf{u} \cdot \mathbf{u} = 26, \mathbf{v} \cdot \mathbf{v} = 24$
(b) $\mathbf{u} \cdot \mathbf{v} = 0, \mathbf{u} \cdot \mathbf{u} = 54, \mathbf{v} \cdot \mathbf{v} = 21$

11. (a) $\|\mathbf{u} - \mathbf{v}\| = \sqrt{14}$ **(b)** $\|\mathbf{u} - \mathbf{v}\| = \sqrt{59}$
(c) $\|\mathbf{u} - \mathbf{v}\| = \sqrt{677}$

13. (a) $\cos\theta = \dfrac{15}{\sqrt{27}\sqrt{17}}$; θ is acute.

(b) $\cos\theta = -\dfrac{4}{\sqrt{6}\sqrt{45}}$; θ is obtuse.

(c) $\cos\theta = -\dfrac{136}{\sqrt{225}\sqrt{180}}$; θ is obtuse.

15. $\mathbf{a}\cdot\mathbf{b} = 45\frac{\sqrt{3}}{2}$ **17.** $\mathbf{x} = \left(-\frac{4}{15}, \frac{14}{15}, -\frac{2}{5}, \frac{2}{15}\right)$

19. (a) $\mathbf{u}\cdot(\mathbf{v}\cdot\mathbf{w})$ does not make sense because $\mathbf{v}\cdot\mathbf{w}$ is a scalar.
(b) $\mathbf{u}\cdot(\mathbf{v}+\mathbf{w})$ makes sense.   (c) $\|\mathbf{u}\cdot\mathbf{v}\|$ does not make sense because the quantity inside the norm is a scalar.
(d) $(\mathbf{u}\cdot\mathbf{v}) - \|\mathbf{u}\|$ makes sense since the terms are both scalars.

21. (a) $|\mathbf{u}\cdot\mathbf{v}| = 10$, $\|\mathbf{u}\|\|\mathbf{v}\| = \sqrt{13}\sqrt{17} \approx 20.025$
(b) $|\mathbf{u}\cdot\mathbf{v}| = 7$, $\|\mathbf{u}\|\|\mathbf{v}\| = \sqrt{10}\sqrt{14} \approx 11.832$
(c) $|\mathbf{u}\cdot\mathbf{v}| = 5$, $\|\mathbf{u}\|\|\mathbf{v}\| = (3)(2) = 6$

25. $\left(\dfrac{b}{\sqrt{a^2+b^2}}, -\dfrac{a}{\sqrt{a^2+b^2}}\right), \left(-\dfrac{b}{\sqrt{a^2+b^2}}, \dfrac{a}{\sqrt{a^2+b^2}}\right)$

27. (a) $k = -\frac{7}{4}$ (b) $k = 0$ or $k = -3$

29. Since $\overrightarrow{BA}\cdot\overrightarrow{BC} = 0$, there is a right angle at the vertex B.

31. (a) valid ISBN   (b) not a valid ISBN

33. $\|\mathbf{v}-\mathbf{w}\| = \left(\displaystyle\sum_{k=1}^{n}(v_k - w_k)^2\right)^{1/2}$

35. (a) $\displaystyle\sum_{k=1}^{n}(a_k + b_k) = (a_1 + b_1) + \cdots + (a_n + b_n) =$
$(a_1 + \cdots + a_n) + (b_1 + \cdots + b_n) = \displaystyle\sum_{k=1}^{n}a_k + \sum_{k=1}^{n}b_k$

(b) $\displaystyle\sum_{k=1}^{n}(a_k - b_k) = (a_1 - b_1) + \cdots + (a_n - b_n) =$
$(a_1 + \cdots + a_n) - (b_1 + \cdots + b_n) = \displaystyle\sum_{k=1}^{n}a_k - \sum_{k=1}^{n}b_k$

(c) $\displaystyle\sum_{k=1}^{n}ca_k = ca_1 + ca_2 + \cdots + ca_n =$
$c(a_1 + a_2 + \cdots + a_n) = c\displaystyle\sum_{k=1}^{n}a_k$

▶ **Exercise Set 1.3** (Page 35)

1. possible answers:
(a) $L_1: x = 1, y = t$ $L_2: x = t, y = 1$
$L_3: x = t, y = t$ $L_4: x = 1-t, y = t$
$(-\infty < t < \infty)$
(b) $L_1: x = 1, y = 1, z = t$ $L_2: x = t, y = 1, z = 1$
$L_3: x = 1, y = t, z = t$ $L_4: x = t, y = t, z = t$
$(-\infty < t < \infty)$

3. (a)   (b)

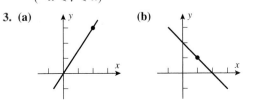

5. possible answers:
(a) vector equation: $(x, y) = t(3, 5)$
parametric equations: $x = 3t, y = 5t$
(b) vector equation: $(x, y, z) = (1, 1, 1) - t(1, 1, 1)$
parametric equations: $x = 1-t, y = 1-t, z = 1-t$
(c) vector equation: $(x, y, z) = (1, -1, 1) + t(1, 2, 0)$
parametric equations: $x = 1+t, y = -1+2t, z = 1$

7. possible answers:
(a) vector equation: $(x, y) = (1, 1) + t(1, 2)$
parametric equations: $x = 1+t, y = 1+2t$
other points: $Q(2, 3)$ corresponds to $t = 1$
$R(3, 5)$ corresponds to $t = 2$
(b) vector equation: $(x, y, z) = (2, 0, 3) + t(1, -1, 1)$
parametric equations: $x = 2+t, y = -t, z = 3+t$
other points: $Q(3, -1, 4)$ corresponds to $t = 1$
$R(1, 1, 2)$ corresponds to $t = -1$
(c) vector equation:
$(x, y, z) = (0, 0, 0) + t(3, 2, -3) = t(3, 2, -3)$
parametric equations: $x = 3t, y = 2t, z = -3t$
other points: $Q(-3, -2, 3)$ corresponds to $t = -1$
$R(6, 4, -6)$ corresponds to $t = 2$

9. $3(x + 1) + 2(y + 1) + (z + 1) = 0$

11. Vector equation: $(x, y, z) = (1, 1, 4) + t_1(1, -4, -3) + t_2(2, 4, -6)$ $(-\infty < t_1, t_2 < \infty)$.
Parametric equations: $x = 1 + t_1 + 2t_2$, $y = 1 - 4t_1 + 4t_2$, $z = 4 - 3t_1 - 6t_2$.
Possible other points on the plane: $S_1(4, 1, -5)$, $(t_1 = 1, t_2 = 1)$; $S_2(0, -7, 7)$, $(t_1 = 1, t_2 = -1)$; $S_3(2, 9, 1)$, $(t_1 = -1, t_2 = 1)$.

13. (a) $(x, y, z) = (2, -1, 0) + t(4, 1, 1)$
(b) $(x, y, z) = (1, -2, 0) + t_1(2, -1, 4) + t_2(1, 5, -1)$
(c) one possible answer: $x = t_1, y = t_2, z = -2 + \frac{3}{2}t_1 + 2t_2$

15. general equation: $x = 1$
vector equation: $(x, y, z) = (1, 0, 0) + t_1(0, 1, 0) + t_2(0, 0, 1)$

17. (a) The line passing through the origin and the point $P(1, -2, 5, 7)$;
parametric equations: $x_1 = t, x_2 = -2t, x_3 = 5t, x_4 = 7t$.
(b) The line passing through $P(4, 5, -6, 1)$ and parallel to $\mathbf{v} = (1, 1, 1, 1)$;
parametric equations: $x_1 = 4+t, x_2 = 5+t, x_3 = -6+t, x_4 = 1+t$.
(c) The plane passing through $P(-1, 0, 4, 2)$ and parallel to $\mathbf{v}_1 = (-3, 5, -7, 4)$ and $\mathbf{v}_2 = (6, 3, -1, 2)$;
parametric equations: $x_1 = -1 - 3t_1 + 6t_2, x_2 = 5t_1 + 3t_2, x_3 = 4 - 7t_1 - t_2, x_4 = 2 + 4t_1 + 2t_2$.

19. (a) These are parametric equations for the line in R^5 that passes through the origin and is parallel to the vector $\mathbf{u} = (3, 4, 7, 1, 9)$.
(b) These are parametric equations for the plane in R^4 that passes through the point $P(3, 4, -2, 1)$ and is parallel to the vectors $\mathbf{v}_1 = (-2, -3, -2, -2)$ and $\mathbf{v}_2 = (5, 6, 7, -1)$.

21. $x = t_1,\ y = t_2,\ z = 3t_1 + 2t_2 - 4$ **23.** (b), (c)

25. $x = 2 + t,\ y = t,\ z = 1 + t$

27. $(x, y, z) = t_1(5, 4, 3) + t_2(1, -1, -2)$

29. possible answer: $x = t_1,\ y = t_2,\ z = \frac{2}{5}t_1 + \frac{3}{5}t_2 + \frac{36}{5}$

31. (a) yes **(b)** no **33. (a)** no **(b)** yes

35. (a) $6(0) + 4t - 4t = 0$ for all t **(b)** The line is parallel to $\mathbf{v} = (0, 1, 1)$, the plane is perpendicular to $\mathbf{n} = (5, -3, 3)$, and $\mathbf{v} \cdot \mathbf{n} = 0$, so the line is parallel to the plane. The line goes through the origin; the plane intersects the z-axis at the point $\left(0, 0, \frac{1}{3}\right)$. Thus the line is below the plane. **(c)** The plane is perpendicular to the vector $\mathbf{n} = (6, 2, -2)$, $\mathbf{v} \cdot \mathbf{n} = 0$, so the line is parallel to the plane. The line goes through the origin, whereas the plane intersects the z-axis at the point $\left(0, 0, \frac{3}{2}\right)$. Thus the line is below the plane.

37. (a) one solution: $x = -12 - 7t,\ y = -41 - 23t,\ z = t$
(b) The planes do not intersect.

39. (a) $\left(-\frac{173}{3}, -\frac{43}{3}, \frac{49}{3}\right)$ **(b)** There is no point of intersection.

41. (a) **(b)**

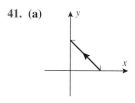
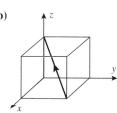

43. $(x, y, z) = (1 - t)(-2, 4, 1) + t(0, 4, 7)$
45. (a) $\left(\frac{9}{2}, -\frac{1}{2}, -\frac{1}{2}\right)$ **(b)** $\left(\frac{23}{4}, -\frac{9}{4}, \frac{1}{4}\right)$

▶ **Exercise Set 2.1** (Page 45) ─────────────

1. (a) and **(c)** are linear; **(b)** and **(d)** are not linear.

3. (a) is linear; **(b)** is linear if $k \neq 0$; **(c)** is linear only if $k = 1$.

5. (a), (d), and **(e)** are solutions; **(b)** and **(c)** are not solutions.

7. The three lines intersect at the point $(1, 0)$. The solution is $x = 1,\ y = 0$.

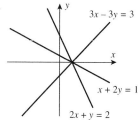

9. possible descriptions of solution sets:
(a) $x = t,\ y = \dfrac{7t - 3}{5}$ $(-\infty < t < \infty)$
(b) $x_1 = s,\ x_2 = t,\ x_3 = \dfrac{7 - 3s + 5t}{4}$ $(-\infty < s, t < \infty)$
(c) $x_1 = r,\ x_2 = \dfrac{1 + 8r + 5s - 6t}{2},\ x_3 = s,\ x_4 = t$
$(-\infty < r, s, t < \infty)$
(d) $v = t_1,\ w = t_2,\ x = t_3,\ y = 3t_1 - 8t_2 + 2t_3 + 4t_4,$
$z = t_4$ $(-\infty < t_1, t_2, t_3, t_4 < \infty)$

11. (a) possible answer: $x - 2y = 5$
(b) Letting $x = t$, we get $y = \frac{1}{2}t - \frac{5}{2}$.

13. possible answer: parametric equations: $x = 1 - 4t,$
$y = 2 + 5t,\ z = t;$
vector equation: $(x, y, z) = (1, 2, 0) + t(-4, 5, 1)$

15. If $k \neq 6$, there is no solution. If $k = 6$, there are infinitely many solutions.

17. $\begin{bmatrix} 3 & -2 & -1 \\ 4 & 5 & 3 \\ 7 & 3 & 2 \end{bmatrix}$ **19.** $\begin{bmatrix} 1 & 2 & 0 & -1 & 1 & 1 \\ 0 & 3 & 2 & 0 & -1 & 2 \\ 0 & 3 & 1 & 7 & 0 & 1 \end{bmatrix}$

21. $\begin{aligned} 2x_1 &= 0 \\ 3x_1 - 4x_2 &= 0 \\ x_2 &= 1 \end{aligned}$ **23.** $\begin{aligned} 7x_1 + 2x_2 + x_3 - 3x_4 &= 5 \\ x_1 + 2x_2 + 4x_3 &= 1 \end{aligned}$

25. (a) B is obtained from A by adding 2 times the first row to the second row. A is obtained from B by adding -2 times the first row to the second row. **(b)** B is obtained from A by multiplying the first row by $\frac{1}{2}$. A is obtained from B by multiplying the first row by 2.

27. $\begin{aligned} 2x + 3y + z &= 7 \\ 2x + y + 3z &= 9 \\ 4x + 2y + 5z &= 16 \end{aligned}$ **29.** $\begin{aligned} x + y + z &= 12 \\ 2x + y + 2z &= 5 \\ -x + z &= 1 \end{aligned}$

31. (a) $\begin{aligned} 3c_1 + c_2 + 2c_3 - c_4 &= 5 \\ c_2 + 3c_3 + 2c_4 &= 6 \\ -c_1 + c_2 + 5c_4 &= 5 \\ 2c_1 + c_2 + 2c_3 &= 5 \end{aligned}$
(b) $\begin{aligned} 3c_1 + c_2 + 2c_3 - c_4 &= 8 \\ c_2 + 3c_3 + 2c_4 &= 3 \\ -c_1 + c_2 + 5c_4 &= -2 \\ 2c_1 + c_2 + 2c_3 &= 6 \end{aligned}$
(c) $\begin{aligned} 3c_1 + c_2 + 2c_3 - c_4 &= 4 \\ c_2 + 3c_3 + 2c_4 &= 4 \\ -c_1 + c_2 + 5c_4 &= 6 \\ 2c_1 + c_2 + 2c_3 &= 2 \end{aligned}$

▶ **Exercise Set 2.2** (Page 59) ─────────────

1. (a), (c), and **(d)** **3. (a)** and **(b)**

5. (a) and **(c)** are in reduced row echelon form; **(b)** is in neither.

7. $\begin{bmatrix} 0 & 0 \\ 0 & 0 \end{bmatrix}, \begin{bmatrix} 0 & 1 \\ 0 & 0 \end{bmatrix}, \begin{bmatrix} 1 & 0 \\ 0 & 1 \end{bmatrix}, \begin{bmatrix} 1 & a \\ 0 & 0 \end{bmatrix}$ (*a* any real number)

9. $x_1 = -3, x_2 = 0, x_3 = 7$

11. $x_1 = -2 + 6s - 3t, x_2 = s, x_3 = 7 - 4t, x_4 = 8 - 5t,$
$x_5 = t \ (-\infty < s, t < \infty)$

13. $x_1 = 8 + 7t, x_2 = 2 - 3t, x_3 = -5 - t, x_4 = t \ (-\infty < t < \infty)$

15. $x_1 = -37, x_2 = -8, x_3 = 5$

17. $x_1 = -11 - 7s + 2t, x_2 = s, x_3 = -4 - 3t,$
$x_4 = 9 - 3t, x_5 = t$

19. $x_1 = -6 + 7t, x_2 = 3 - 4t, x_3 = t, x_4 = 2$

21. $x_1 = 2, x_2 = 1, x_3 = \frac{4}{3}$ **23.** $x_1 = 3, x_2 = 1, x_3 = 2$

25. $x = -1 + t, y = 2s, z = s, w = t$

27. $x_1 = 3, x_2 = 1, x_3 = 2$ **29.** $x_1 = 3, x_2 = 1, x_3 = 2$

31. $x = -1 + t, y = 2s, z = s, w = t$

33. **(a)** has nontrivial solutions   **(b)** only the trivial solution

35. only the trivial solution

37. $w = t, x = -t, y = t, z = 0$

39. $u = \frac{7}{2}s - \frac{5}{2}t, v = -3s + 2t, w = s, x = t$

41. $I_3 = I_2 = I_1 = 0$

43. exactly one solution for every *a*

45. If $a = 3$, there are infinitely many solutions; if $a = -3$, there are no solutions; if $a \neq \pm 3$, there is exactly one solution.

47. **(a)** If $x + y + z = 1$, then $2x + 2y + 2z = 2 \neq 4$; thus there is no solution. The planes are parallel.
(b) If $x + y + z = 0$, then $2x + 2y + 2z = 0$ also; thus the system is redundant and has infinitely many solutions. The planes coincide.

49. $\alpha = \pi/2, \beta = -\pi, \gamma = 0$

51. If $\lambda = 1$, then $x = y = z = 0$. If $\lambda = 2$, then $x = -(t/2)$, $y = 0, z = t$, where $-\infty < t < \infty$.

53. **(c)** $x = \frac{100,000}{49,999}, \ y = \frac{49,997}{49,999}$

► **Exercise Set 2.3** (Page 75)

1.

3. **(a)** $x_3 - x_4 = -500, -x_1 + x_4 = 100, x_1 - x_2 = 300,$
$x_2 - x_3 = 100$ **(b)** $x_1 = -100 + t, x_2 = -400 + t,$
$x_3 = -500 + t, x_4 = t \ (0 \leq t \leq 150)$ **(c)** For all rates to be nonnegative we need $t = 500$ cars per hour, so $x_1 = 400$, $x_2 = 100, x_3 = 0, x_4 = 500$.

5. $I_1 = \frac{13}{5}, I_2 = -\frac{2}{5}, I_3 = \frac{11}{5}$

7. $I_1 = I_4 = I_5 = I_6 = \frac{1}{2}A, I_2 = I_3 = 0A$

9. $x_1 = 1, x_2 = 5, x_3 = 3$, and $x_4 = 4$; the balanced equation is $C_3H_8 + 5O_2 \rightarrow 3CO_2 + 4H_2O$.

11. $x_1 = x_2 = x_3 = x_4 = t$; the balanced equation is $CH_3COF + H_2O \rightarrow CH_3COOH + HF$.

13. $p(x) = x^2 - 2x + 2$ **15.** $p(x) = 1 + \frac{13}{6}x - \frac{1}{6}x^3$

► **Exercise Set 3.1** (Page 90)

1. $a = \frac{9}{2}, b = -\frac{7}{2}, c = -\frac{4}{5}, d = \frac{13}{5}$

3. **(a)** $3 \times 4, 4 \times 2$ **(b)** 3, 11   **(c)** (1, 1), (3, 1), or (3, 2)

(d) $\begin{bmatrix} 3 \\ 3 \\ 9 \\ -4 \end{bmatrix}$ **(e)** $[1 \quad 2]$

5. **(a)** $\begin{bmatrix} 7 & 2 \\ -7 & 4 \\ 9 & 1 \end{bmatrix}$ **(b)** not defined   **(c)** $\begin{bmatrix} 1 & -5 \\ -12 & 15 \end{bmatrix}$

(d) $\begin{bmatrix} 0 & 4 \\ -4 & 0 \end{bmatrix}$ **(e)** $\begin{bmatrix} 8 & -1 & 9 \\ 9 & 1 & 6 \\ 8 & 3 & 11 \end{bmatrix}$ **(f)** not defined

7. **(a)** $\begin{bmatrix} 1 & 1 \\ 6 & 0 \end{bmatrix}$ **(b)** $\begin{bmatrix} 3 & 12 & 6 \\ 5 & -2 & 8 \\ 4 & 5 & 7 \end{bmatrix}$ **(c)** $\begin{bmatrix} 9 & 8 & 19 \\ -2 & 0 & 0 \\ 32 & 9 & 25 \end{bmatrix}$

(d) $\begin{bmatrix} 17 & 18 & 17 \\ 0 & 5 & 3 \end{bmatrix}$ **(e)** $\begin{bmatrix} 5 & -5 & 8 \\ -5 & 10 & -12 \\ 8 & -12 & 16 \end{bmatrix}$

(f) not defined

9. $\begin{bmatrix} 3 \\ -14 \\ 13 \end{bmatrix}$

11. **(a)** $\begin{bmatrix} 2 & -3 & 5 \\ 9 & -1 & 1 \end{bmatrix} \begin{bmatrix} x_1 \\ x_2 \\ x_3 \end{bmatrix} = \begin{bmatrix} 7 \\ -1 \end{bmatrix}$

(b) $\begin{bmatrix} 1 & 1 & 1 \\ 2 & -3 & 4 \\ 1 & 5 & -2 \end{bmatrix} \begin{bmatrix} x_1 \\ x_2 \\ x_3 \end{bmatrix} = \begin{bmatrix} 4 \\ 3 \\ -2 \end{bmatrix}$

13. $5x_1 + 6x_2 - 7x_3 = 2$ **15.** 59
$\quad -x_1 - 2x_2 + 3x_3 = 0$
$\quad\quad\quad 4x_2 - \ x_3 = 3$

17. **(a)** $[67 \quad 41 \quad 41]$ **(b)** $[63 \quad 67 \quad 57]$ **(c)** $\begin{bmatrix} 41 \\ 21 \\ 67 \end{bmatrix}$

19. **(a)** 17   **(b)** 17   **(c)** -59

21. **(a)** 7   **(b)** $\mathbf{uv}^T = \begin{bmatrix} -8 & -10 \\ 12 & 15 \end{bmatrix}$ **23.** -1

31. (a) A has 0's off the main diagonal.   **(b)** A has 0's below the main diagonal.   **(c)** A has 0's above the main diagonal.

(d)
$$\begin{bmatrix} a_{11} & a_{12} & 0 & 0 & 0 & 0 \\ a_{21} & a_{22} & a_{23} & 0 & 0 & 0 \\ 0 & a_{32} & a_{33} & a_{34} & 0 & 0 \\ 0 & 0 & a_{43} & a_{44} & a_{45} & 0 \\ 0 & 0 & 0 & a_{54} & a_{55} & a_{56} \\ 0 & 0 & 0 & 0 & a_{65} & a_{66} \end{bmatrix}$$

33. The total expenditures for purchases during each of the first four months of the year.

▶ **Exercise Set 3.2** (Page 106) ─────────

7. (a) $\dfrac{1}{3}\begin{bmatrix} -48 & -47 & -78 \\ 7 & -47 & -53 \\ 41 & -42 & -22 \end{bmatrix}$ **(b)** $\begin{bmatrix} -10 & 2 & -4 \\ 1 & 2 & 7 \\ 10 & 1 & -1 \end{bmatrix}$

9. (a) $\begin{bmatrix} 2 & -1 \\ -5 & 3 \end{bmatrix}$ **13.** $\begin{bmatrix} 20 & 20 \\ -44 & -46 \end{bmatrix}$

15. (a) $\begin{bmatrix} 9 & -5 \\ -25 & 14 \end{bmatrix}$ **(b)** $\begin{bmatrix} 5 & 1 \\ 5 & 4 \end{bmatrix}$ **(c)** $\begin{bmatrix} 9 & 3 \\ 15 & 6 \end{bmatrix}$

17. $\dfrac{1}{20}\begin{bmatrix} 4 & 3 \\ -4 & 2 \end{bmatrix}$

19. (a) $A = \dfrac{1}{13}\begin{bmatrix} 5 & 1 \\ -3 & 2 \end{bmatrix}$ **(b)** $A = \dfrac{1}{7}\begin{bmatrix} 2 & 7 \\ 1 & 3 \end{bmatrix}$

23. possible answer: $\begin{bmatrix} 1 & 0 & -1 \\ 0 & 2 & -2 \\ -1 & -2 & 3 \end{bmatrix}$

25. $\dfrac{1}{2}\begin{bmatrix} 1 & 1 & -1 \\ -1 & 1 & 1 \\ 1 & -1 & 1 \end{bmatrix}$

31. (a) A is invertible for all θ, and $A^{-1} = \begin{bmatrix} \cos\theta & -\sin\theta \\ \sin\theta & \cos\theta \end{bmatrix}$.
  (b) $x' = x\cos\theta - y\sin\theta,\ y' = x\sin\theta + y\cos\theta$

33. (b) $(A^{-1} + B^{-1})^{-1} = B(A + B)^{-1}A$

37. $A^2 = \begin{bmatrix} 0 & 0 & 0 & 0 & 1 & 3 & 1 \\ 0 & 0 & 0 & 0 & 0 & 0 & 2 \\ 0 & 0 & 0 & 0 & 0 & 0 & 1 \\ 0 & 0 & 0 & 0 & 0 & 0 & 1 \\ 0 & 0 & 0 & 0 & 0 & 0 & 0 \\ 0 & 0 & 0 & 0 & 0 & 0 & 0 \\ 0 & 0 & 0 & 0 & 0 & 0 & 0 \end{bmatrix}$

▶ **Exercise Set 3.3** (Page 119) ─────────

1. (a) elementary   **(b)** not elementary   **(c)** elementary
  (d) not elementary

3. (a) Add 3 times the first row to the second row.
  (b) Multiply the third row by $\frac{1}{3}$.
  (c) Interchange the first and fourth rows.
  (d) Add $\frac{1}{7}$ times the third row to the first row.

5. (a) $\begin{bmatrix} 1 & 0 \\ 3 & 1 \end{bmatrix}$ **(b)** $\begin{bmatrix} 1 & 0 & 0 \\ 0 & 1 & 0 \\ 0 & 0 & \frac{1}{3} \end{bmatrix}$ **(c)** $\begin{bmatrix} 0 & 0 & 0 & 1 \\ 0 & 1 & 0 & 0 \\ 0 & 0 & 1 & 0 \\ 1 & 0 & 0 & 0 \end{bmatrix}$

(d) $\begin{bmatrix} 1 & 0 & \frac{1}{7} & 0 \\ 0 & 1 & 0 & 0 \\ 0 & 0 & 1 & 0 \\ 0 & 0 & 0 & 1 \end{bmatrix}$

7. (a) $\begin{bmatrix} 0 & 0 & 1 \\ 0 & 1 & 0 \\ 1 & 0 & 0 \end{bmatrix}$ **(b)** $\begin{bmatrix} 0 & 0 & 1 \\ 0 & 1 & 0 \\ 1 & 0 & 0 \end{bmatrix}$ **(c)** $\begin{bmatrix} 1 & 0 & 0 \\ 0 & 1 & 0 \\ -2 & 0 & 1 \end{bmatrix}$

(d) $\begin{bmatrix} 1 & 0 & 0 \\ 0 & 1 & 0 \\ 2 & 0 & 1 \end{bmatrix}$

9. $A^{-1} = \begin{bmatrix} 2 & -\frac{1}{2} \\ -\frac{1}{5} & \frac{1}{10} \end{bmatrix}$

11. (a) $\begin{bmatrix} \frac{3}{2} & -\frac{11}{10} & -\frac{6}{5} \\ -1 & 1 & 1 \\ -\frac{1}{2} & \frac{7}{10} & \frac{2}{5} \end{bmatrix}$ **(b)** A is not invertible.

(c) $\dfrac{1}{2}\begin{bmatrix} 1 & -1 & 1 \\ -1 & 1 & 1 \\ 1 & 1 & -1 \end{bmatrix}$

13. $R = \begin{bmatrix} 1 & 2 & 0 \\ 0 & 0 & 1 \\ 0 & 0 & 0 \end{bmatrix}; B = \begin{bmatrix} 1 & -3 & 0 \\ 0 & 1 & 0 \\ -1 & -1 & 1 \end{bmatrix}$ **15.** $c \neq 0, 1$

17. $B = \begin{bmatrix} 0 & 1 & 0 \\ 1 & 0 & 0 \\ 0 & 0 & 6 \end{bmatrix}$ **19.** $\begin{bmatrix} 0 & 0 & 0 & \frac{1}{k_4} \\ 0 & 0 & \frac{1}{k_3} & 0 \\ 0 & \frac{1}{k_2} & 0 & 0 \\ \frac{1}{k_1} & 0 & 0 & 0 \end{bmatrix}$

21. $E_1 = \begin{bmatrix} 1 & 0 \\ 5 & 1 \end{bmatrix}; E_2 = \begin{bmatrix} 1 & 0 \\ 0 & \frac{1}{2} \end{bmatrix}; A^{-1} = E_2 E_1;$

$A = \begin{bmatrix} 1 & 0 \\ -5 & 1 \end{bmatrix}\begin{bmatrix} 1 & 0 \\ 0 & 2 \end{bmatrix}$

23. $A = \begin{bmatrix} 0 & 0 & 1 \\ 0 & 1 & 0 \\ 1 & 0 & 0 \end{bmatrix} \begin{bmatrix} 1 & 0 & 0 \\ 1 & 1 & 0 \\ 0 & 0 & 1 \end{bmatrix} \begin{bmatrix} 1 & 0 & 0 \\ 0 & 1 & 0 \\ 2 & 0 & 1 \end{bmatrix} \begin{bmatrix} 1 & 0 & 0 \\ 0 & 1 & 0 \\ 0 & -1 & 1 \end{bmatrix}$

$\times \begin{bmatrix} 1 & 0 & 0 \\ 0 & 1 & 0 \\ 0 & 0 & -4 \end{bmatrix} \begin{bmatrix} 1 & 0 & 0 \\ 0 & 1 & -1 \\ 0 & 0 & 1 \end{bmatrix} \begin{bmatrix} 1 & 0 & 2 \\ 0 & 1 & 0 \\ 0 & 0 & 1 \end{bmatrix} \begin{bmatrix} 1 & 1 & 0 \\ 0 & 1 & 0 \\ 0 & 0 & 1 \end{bmatrix}$

$A^{-1} = \begin{bmatrix} 1 & -1 & 0 \\ 0 & 1 & 0 \\ 0 & 0 & 1 \end{bmatrix} \begin{bmatrix} 1 & 0 & -2 \\ 0 & 1 & 0 \\ 0 & 0 & 1 \end{bmatrix} \begin{bmatrix} 1 & 0 & 0 \\ 0 & 1 & 1 \\ 0 & 0 & 1 \end{bmatrix}$

$\times \begin{bmatrix} 1 & 0 & 0 \\ 0 & 1 & 0 \\ 0 & 0 & -\frac{1}{4} \end{bmatrix} \begin{bmatrix} 1 & 0 & 0 \\ 0 & 1 & 0 \\ 0 & 1 & 1 \end{bmatrix} \begin{bmatrix} 1 & 0 & 0 \\ 0 & 1 & 0 \\ -2 & 0 & 1 \end{bmatrix}$

$\times \begin{bmatrix} 1 & 0 & 0 \\ -1 & 1 & 0 \\ 0 & 0 & 1 \end{bmatrix} \begin{bmatrix} 0 & 0 & 1 \\ 0 & 1 & 0 \\ 1 & 0 & 0 \end{bmatrix}$

25. first system: $x_1 = -9, x_2 = 4, x_3 = 0$
second system: $x_1 = 4, x_2 = -4, x_3 = 4$

29. $b_1 = 2b_2$ **31.** $2b_1 - b_2 + b_3 = 0$

33. $E = \begin{bmatrix} 0 & 1 & 0 \\ 1 & 0 & 0 \\ 0 & 0 & 1 \end{bmatrix}, F = \begin{bmatrix} 1 & 0 & 0 \\ 0 & 1 & 0 \\ -2 & 0 & 1 \end{bmatrix}, G = \begin{bmatrix} 1 & 0 & 0 \\ 0 & 1 & 0 \\ 0 & 1 & 1 \end{bmatrix},$

$R = \begin{bmatrix} 1 & 3 & 3 & 8 \\ 0 & 1 & 7 & 8 \\ 0 & 0 & 0 & 0 \end{bmatrix}$

▶ **Exercise Set 3.4** (Page 132) ─────────

1. (a) $(x_1, x_2) = t(1, -1)$; $x_1 = t, x_2 = -t$
 (b) $(x_1, x_2, x_3) = t(2, 1, -4)$; $x_1 = 2t, x_2 = t, x_3 = -4t$
 (c) $(x_1, x_2, x_3, x_4) = t(1, 1, -2, 3)$; $x_1 = t, x_2 = t$,
 $x_3 = -2t, x_4 = 3t$

3. (a) $(x_1, x_2, x_3) = s(4, -4, 2) + t(-3, 5, 7)$;
 $x_1 = 4s - 3t, x_2 = -4s + 5t, x_3 = 2s + 7t$
 (b) $(x_1, x_2, x_3, x_4) = s(1, 2, 1, -3) + t(3, 4, 5, 0)$;
 $x_1 = s + 3t, x_2 = 2s + 4t, x_3 = s + 5t, x_4 = -3s$

5. (a) yes   **(b)** no

7. (a) linearly dependent   **(b)** linearly independent

9. $\mathbf{u} = 2\mathbf{v} - \mathbf{w}$

11. (a) a line in R^4 passing through the origin and parallel to the
 vector $(2, -3, 1, 4)$ **(b)** a plane in R^4 passing through the
 origin and parallel to $(3, -2, 2, 5)$ and $(6, -4, 4, 0)$

13. general solution: $x_1 = -6s - 11t, x_2 = s, x_3 = 8t$,
 $x_4 = t$ $(-\infty < s, t < \infty)$
 spanning set: $\{(-6, 1, 0, 0), (-11, 0, 8, 1)\}$

15. general solution: $x_1 = -2r - \frac{1}{3}s - \frac{2}{3}t, x_2 = r$,
 $x_3 = -\frac{2}{3}s - \frac{1}{3}t, x_4 = s, x_5 = t$ $(-\infty < r, s, t < \infty)$
 spanning set: $\{(-2, 1, 0, 0, 0), (-\frac{1}{3}, 0, -\frac{2}{3}, 1, 0),$
 $(-\frac{2}{3}, 0, -\frac{1}{3}, 0, 1)\}$

17. (a) $\mathbf{v}_2 = -5\mathbf{v}_1$ **(b)** Theorem 3.4.8

19. (a) linearly independent   **(b)** linearly dependent
 (c) linearly independent   **(d)** linearly dependent

21. (a) The vectors do not lie in a plane.   **(b)** The vectors lie
 in a plane but not on a line.   **(c)** These vectors lie on a line.

23. (a) is a subspace   **(b)** not a subspace; not closed under
 scalar multiplication   **(c)** is a subspace   **(d)** not a subspace; not closed under addition or scalar multiplication

25. A line through the origin in R^4.

27. (a) $7\mathbf{v}_1 - 2\mathbf{v}_2 + 3\mathbf{v}_3 = \mathbf{0}$ **(b)** $\mathbf{v}_1 = \frac{2}{7}\mathbf{v}_2 - \frac{3}{7}\mathbf{v}_3,$
 $\mathbf{v}_2 = \frac{7}{2}\mathbf{v}_1 + \frac{3}{2}\mathbf{v}_3, \mathbf{v}_3 = -\frac{7}{3}\mathbf{v}_1 + \frac{2}{3}\mathbf{v}_2$

33. (a) no; not closed under addition or scalar multiplication
 (b) $\mathbf{p}_{876} = 0.38\mathbf{c} + 0.59\mathbf{m} + 0.73\mathbf{y} + 0.07\mathbf{k}$
 $\mathbf{p}_{216} = 0.83\mathbf{m} + 0.34\mathbf{y} + 0.47\mathbf{k}$
 $\mathbf{p}_{328} = \mathbf{c} + 0.47\mathbf{y} + 0.3\mathbf{k}$
 (c) $(0.19, 0.71, 0.54, 0.27)$

35. (a) $k_1 = k_2 = k_3 = k_4 = \frac{1}{4}$
 (b) $k_1 = k_2 = k_3 + k_4 + k_5 = \frac{1}{5}$
 (c) The components of $\frac{1}{3}\mathbf{r}_1 + \frac{1}{3}\mathbf{r}_2 + \frac{1}{3}\mathbf{r}_3$ represent the average total population of Philadelphia, Bucks, and Delaware
 counties in each of the sampled years.

▶ **Exercise Set 3.5** (Page 141) ─────────

1. (a) $x_1 = -\frac{2}{3}s + \frac{1}{3}t, x_2 = s, x_3 = t$
 (c) $x_1 = 1 - \frac{2}{3}s + \frac{1}{3}t, x_2 = s, x_3 = 1 + t$

3. (a) $x_1 = \frac{1}{3} - \frac{4}{3}s - \frac{1}{3}t, x_2 = s, x_3 = t, x_4 = 1$
 (b) The general solution of the associated homogeneous
 system is $x_1 = -\frac{4}{3}s - \frac{1}{3}t, x_2 = s, x_3 = t, x_4 = 0$.
 A particular solution of the given nonhomogeneous system
 is $x_1 = \frac{1}{3}, x_2 = 0, x_3 = 0, x_4 = 1$.

5. $\mathbf{w} = -2\mathbf{v}_1 + 3\mathbf{v}_2 + \mathbf{v}_3$ **7.** $\mathbf{w}$ is in span$\{\mathbf{v}_1, \mathbf{v}_2, \mathbf{v}_3, \mathbf{v}_4\}$.

9. (a) $x = \frac{3}{2}t, y = t$ $(-\infty < t < \infty)$
 (b) $x = \frac{5}{4}t, y = s, z = t$ $(-\infty < s, t < \infty)$
 (c) $x_1 = -2r + 3s - 7t, x_2 = r, x_3 = s,$
 $x_4 = t$ $(-\infty < s, t, r < \infty)$

11. general solution: $x_1 = -s - t, x_2 = s, x_3 = t$;
 two-dimensional

13. general solution: $x_1 = \frac{3}{7}r - \frac{19}{7}s - \frac{8}{7}t$,
 $x_2 = -\frac{2}{7}r + \frac{1}{7}s + \frac{3}{7}t, x_3 = r, x_4 = s, x_5 = t$;
 three-dimensional

15. (a) general solution: $(1, 0, 0) + s(-1, 1, 0) + t(-1, 0, 1)$
 (b) a plane in R_4 passing through $P(1, 0, 0)$ and parallel to
 $(-1, 1, 0)$ and $(-1, 0, 1)$

17. (a) $\begin{aligned} x + y + z &= 0 \\ -2x + 3y &= 0 \end{aligned}$ **(b)** a line through the origin in R^3
 (c) $x = -\frac{3}{5}t, y = -\frac{2}{5}t, z = t$

19. (a) $\begin{aligned} x_1 + x_2 + 2x_3 + 2x_4 &= 0 \\ 5x_1 + 4x_2 + 3x_3 + 4x_4 &= 0 \end{aligned}$
 (b) a plane in R^4 passing through the origin
 (c) $x_1 = 5s + 4t, x_2 = -7s - 6t, x_3 = s, x_4 = t$

▶ **Exercise Set 3.6** (Page 151) ───────────

1. (a) not invertible   (b) $\begin{bmatrix} -1 & 0 & 0 \\ 0 & \frac{1}{2} & 0 \\ 0 & 0 & 3 \end{bmatrix}$

3. (a) $\begin{bmatrix} 6 & 3 \\ 4 & -1 \\ 4 & 10 \end{bmatrix}$ (b) $\begin{bmatrix} 10 & -2 \\ -20 & -2 \\ 10 & -10 \end{bmatrix}$ (c) $\begin{bmatrix} 30 & -6 \\ 20 & 2 \\ 20 & -20 \end{bmatrix}$

5. (a) $A^2 = \begin{bmatrix} 1 & 0 & 0 \\ 0 & 4 & 0 \\ 0 & 0 & 16 \end{bmatrix}$ (b) $A^{-2} = \begin{bmatrix} 1 & 0 & 0 \\ 0 & \frac{1}{4} & 0 \\ 0 & 0 & \frac{1}{16} \end{bmatrix}$

(c) $A^{-k} = \begin{bmatrix} 1 & 0 & 0 \\ 0 & \left(-\frac{1}{2}\right)^k & 0 \\ 0 & 0 & \left(\frac{1}{4}\right)^k \end{bmatrix}$ **7.** $x \neq 1, -2, 4$

11. $A = \begin{bmatrix} 0 & 0 & 4 \\ 0 & 0 & 1 \\ -4 & -1 & 0 \end{bmatrix}$

13. $a = 11, b = -9$, and $c = -13$

19. $A = \begin{bmatrix} 1 & 0 & 0 \\ 0 & -1 & 0 \\ 0 & 0 & -1 \end{bmatrix}$

23. all vectors of the form $(a, -a)$ (a real)

25. (a) nilpotency index: 2; $(I - A)^{-1} = \begin{bmatrix} 1 & 1 \\ 0 & 1 \end{bmatrix}$

(b) nilpotency index: 3; $(I - A)^{-1} = \begin{bmatrix} 1 & 0 & 0 \\ 1 & 1 & 0 \\ 9 & 1 & 1 \end{bmatrix}$

31. $\mathrm{tr}(A^T A) = n$

▶ **Exercise Set 3.7** (Page 164) ───────────

1. $x_1 = 2, x_2 = 1$ **3.** $x_1 = -2, x_2 = 1, x_3 = -3$

5. $A = LU = \begin{bmatrix} 2 & 0 \\ -1 & 3 \end{bmatrix} \begin{bmatrix} 1 & 4 \\ 0 & 1 \end{bmatrix}; x_1 = 3, x_2 = -1$

7. $A = LU = \begin{bmatrix} 2 & 0 & 0 \\ 0 & -2 & 0 \\ -1 & 4 & 5 \end{bmatrix} \begin{bmatrix} 1 & -1 & -1 \\ 0 & 1 & -1 \\ 0 & 0 & 1 \end{bmatrix};$
$x_1 = -1, x_2 = 1, x_3 = 0$

9. $A = LU = \begin{bmatrix} -1 & 0 & 0 & 0 \\ 2 & 3 & 0 & 0 \\ 0 & -1 & 2 & 0 \\ 0 & 0 & 1 & 4 \end{bmatrix} \begin{bmatrix} 1 & 0 & -1 & 0 \\ 0 & 1 & 0 & 2 \\ 0 & 0 & 1 & 1 \\ 0 & 0 & 0 & 1 \end{bmatrix};$
$x_1 = -3, x_2 = 1, x_3 = 2, x_4 = 1$

13. $A = L'DU = \begin{bmatrix} 1 & 0 & 0 \\ -1 & 1 & 0 \\ 1 & 1 & 1 \end{bmatrix} \begin{bmatrix} 2 & 0 & 0 \\ 0 & 1 & 0 \\ 0 & 0 & 1 \end{bmatrix} \begin{bmatrix} 1 & \frac{1}{2} & -\frac{1}{2} \\ 0 & 1 & 1 \\ 0 & 0 & 1 \end{bmatrix}$

15. (a) permutation matrix   (b) not a permutation matrix
(c) permutation matrix

17. $x_1 = \frac{21}{17}, x_2 = -\frac{14}{17}, x_3 = \frac{12}{17}$

19. $A = \begin{bmatrix} 1 & 0 & 0 \\ 0 & 0 & 1 \\ 0 & 1 & 0 \end{bmatrix} \begin{bmatrix} 3 & 0 & 0 \\ 0 & 2 & 0 \\ 3 & 0 & 1 \end{bmatrix} \begin{bmatrix} 1 & -\frac{1}{3} & 0 \\ 0 & 1 & \frac{1}{2} \\ 0 & 0 & 1 \end{bmatrix};$
$x_1 = -\frac{1}{2}, x_2 = \frac{1}{2}, x_3 = 3$

21. (a) $66.67 \times 10^4 s$ for forward phase, $10s$ for backward phase
(b) 1334

▶ **Exercise Set 3.8** (Page 170) ───────────

1. (a) yes; $\begin{bmatrix} 6 & -7 & -6 \\ -2 & 9 & 7 \\ 9 & -5 & -6 \end{bmatrix}$ (b) Block sizes do not conform.
(c) Block sizes conform.   (d) Block sizes conform.

3. (a) $\left[\begin{array}{cc|c} 3 & 23 & -10 \\ 37 & -13 & 8 \\ 25 & 23 & 41 \end{array}\right]$ (b) $\left[\begin{array}{cc|c} 3 & 23 & -10 \\ 37 & -13 & 8 \\ 25 & 23 & 41 \end{array}\right]$

5. (a) $\begin{bmatrix} -3 & -15 & -11 \\ 21 & -15 & 44 \end{bmatrix}$ (b) $\left[\begin{array}{ccc|c} 4 & -7 & -19 & -43 \\ 2 & 2 & 18 & 17 \\ 0 & 5 & 25 & 35 \\ \hline 2 & 3 & 23 & 24 \end{array}\right]$

7. $\begin{bmatrix} -9 & 12 \\ 17 & -19 \end{bmatrix}$

9. (a) $\left[\begin{array}{cc|cc} 2 & -1 & 0 & 0 \\ -3 & 2 & 0 & 0 \\ \hline 0 & 0 & \frac{1}{7} & \frac{4}{7} \\ 0 & 0 & \frac{1}{7} & -\frac{3}{7} \end{array}\right]$ (b) $\left[\begin{array}{cc|c|cc} -1 & 2 & 0 & 0 & 0 \\ 3 & -5 & 0 & 0 & 0 \\ \hline 0 & 0 & \frac{1}{5} & 0 & 0 \\ \hline 0 & 0 & 0 & 4 & -7 \\ 0 & 0 & 0 & -1 & 2 \end{array}\right]$

11. $\left[\begin{array}{cc|cc} 1 & -1 & 8 & -10 \\ -1 & 2 & 5 & -13 \\ \hline 0 & 0 & -3 & 5 \\ 0 & 0 & 2 & -3 \end{array}\right]$

13. $MN = \begin{bmatrix} I & AB + BA \\ 0 & A^2 \end{bmatrix}$ **15.** Let $B_1 = \begin{bmatrix} 0 & -\frac{1}{2} \\ 0 & \frac{1}{2} \end{bmatrix}$

17. $x_1 = \frac{17}{6}, x_2 = -\frac{43}{3}, x_3 = -\frac{5}{2}, x_4 = \frac{43}{6}$

▶ **Exercise Set 4.1** (Page 182) ───────────

1. 22   **3.** 59   **5.** $a^2 - 5a + 21$ **7.** -65 **9.** -123

11. (a) $\{4, 1, 3, 5, 2\}$; $-a_{14}a_{21}a_{33}a_{45}a_{52}$
(b) $\{5, 3, 4, 2, 1\}$; $-a_{15}a_{23}a_{34}a_{42}a_{51}$
(c) $\{4, 2, 5, 3, 1\}$; $-a_{14}a_{22}a_{35}a_{43}a_{51}$
(d) $\{5, 4, 3, 2, 1\}$; $+a_{15}a_{24}a_{33}a_{42}a_{51}$
(e) $\{1, 2, 3, 4, 5\}$; $+a_{11}a_{22}a_{33}a_{44}a_{55}$
(f) $\{1, 4, 2, 3, 5\}$; $+a_{11}a_{24}a_{32}a_{43}a_{55}$

13. $\lambda = 1$ or -3 **15.** $\lambda = 1$ or -1 **17.** $x = \dfrac{3 \pm \sqrt{33}}{4}$

19. (a) -1 (b) 0 (c) 6

21. $M_{11} = 29, C_{11} = 29$ $\quad M_{12} = 21, C_{12} = -21$
$M_{13} = 27, C_{13} = 27$ $\quad M_{21} = -11, C_{21} = 11$
$M_{22} = 13, C_{22} = 13$ $\quad M_{23} = -5, C_{23} = 5$
$M_{31} = -19, C_{31} = -19$ $\quad M_{32} = -19, C_{32} = 19$
$M_{33} = 19, C_{33} = 19$

23. (a) $M_{13} = 0, C_{13} = 0$ (b) $M_{23} = -96, C_{23} = 96$
(c) $M_{22} = -48, C_{22} = -48$ (d) $M_{21} = 72, C_{21} = -72$

25. (all parts) 152 **27.** -40 **29.** 0 **31.** -240

33. The determinant is $\sin^2 \theta + \cos^2 \theta = 1$. **35.** $d_2 = d_1 + \lambda$

▶ **Exercise Set 4.2** (Page 192) ───────────────

3. (a) -4 (b) 0 (c) -30

5. (a) -6 (b) 72 (c) -6 (d) 18

9. Each value makes two rows of the matrix proportional.

11. $\det(A) = 30$ **13.** $\det(A) = -17$ **15.** $\det(A) = 39$

17. $\det(A) = -\frac{1}{6}$ **25.** (a) $k \neq \dfrac{5 \pm \sqrt{17}}{2}$ (b) $k \neq -1$

27. (a) 189 (b) $\frac{1}{7}$ (c) $\frac{8}{7}$ (d) $\frac{1}{56}$
31. 39 **39.** (a) -3 (b) -24

▶ **Exercise Set 4.3** (Page 207) ───────────────

1. $\mathrm{adj}(A) = \begin{bmatrix} -3 & 5 & 5 \\ 3 & -4 & -5 \\ -2 & 2 & 3 \end{bmatrix}$; $A^{-1} = \begin{bmatrix} 3 & -5 & -5 \\ -3 & 4 & 5 \\ 2 & -2 & -3 \end{bmatrix}$

3. $\mathrm{adj}(A) = \begin{bmatrix} 2 & 6 & 4 \\ 0 & 4 & 6 \\ 0 & 0 & 2 \end{bmatrix}$; $A^{-1} = \begin{bmatrix} \frac{1}{2} & \frac{3}{2} & 1 \\ 0 & 1 & \frac{3}{2} \\ 0 & 0 & \frac{1}{2} \end{bmatrix}$

5. $x_1 = \frac{9}{8}; x_2 = \frac{13}{8}$ **7.** $x = -\frac{144}{55}; y = -\frac{61}{55}; z = \frac{46}{11}$

9. $x_1 = 5; x_2 = 8; x_3 = 3; x_4 = -1$ **11.** $x = \frac{3}{2}$

13. $\det(A) = 1; A^{-1} = \begin{bmatrix} \cos\theta & -\sin\theta & 0 \\ \sin\theta & \cos\theta & 0 \\ 0 & 0 & 1 \end{bmatrix}$

15. $k \neq 0$ and $k \neq 4$ **17.** $y \neq 0$ and $y \neq \pm x$ **19.** 3

21. 4 **23.** 3 **25.** 7 **27.** 16

29. The vectors do not lie in the same plane.

31. $\pm\dfrac{1}{\sqrt{5}}(0, 2, 1)$ **33.** $\dfrac{12\sqrt{13}}{49}$

35. (a) $(32, -6, -4)$ (b) $(-14, -20, -82)$
(c) $(27, 40, -42)$

37. (a) $(18, 36, -18)$ (b) $(-3, 9, -3)$

43. (a) $\sqrt{59}$ (b) $\sqrt{101}$ **45.** $\dfrac{\sqrt{374}}{2}$ **49.** $(8, 4, 4)$

51. $A^{-1} = \begin{bmatrix} -\frac{34}{7} & 3 & -\frac{1}{7} \\ 2 & -1 & 0 \\ -\frac{1}{7} & 0 & \frac{1}{7} \end{bmatrix}$ **57.** $p(x) = 1 - x - 2x^2 + x^3$

▶ **Exercise Set 4.4** (Page 221) ───────────────

1. (a) $\begin{bmatrix} t \\ t \end{bmatrix}$ $(-\infty < t < \infty)$ (b) $\begin{bmatrix} t \\ t \end{bmatrix}$ $(-\infty < t < \infty)$ **3.** 5

5. (a) $\lambda^2 - 2\lambda - 3 = 0$;
$\lambda = 3$, multiplicity 1; $\lambda = -1$, multiplicity 1
(b) $\lambda^2 - 8\lambda + 16 = 0$; $\lambda = 4$, multiplicity 2
(c) $\lambda^2 - 4\lambda + 4 = 0$; $\lambda = 2$, multiplicity 2

7. (a) $\lambda^3 - 6\lambda^2 + 11\lambda - 6 = 0$; $\lambda = 1$, multiplicity 1;
$\lambda = 2$, multiplicity 1; $\lambda = 3$, multiplicity 1
(b) $\lambda^3 - 4\lambda^2 + 4\lambda = 0$; $\lambda = 0$, multiplicity 1;
$\lambda = 2$, multiplicity 2
(c) $\lambda^3 - \lambda^2 - 8\lambda + 12 = 0$; $\lambda = 2$, multiplicity 1; $\lambda = -3$,
multiplicity 1

9. (a) For $\lambda = 3$, $\mathbf{x} = t \begin{bmatrix} 1 \\ 2 \end{bmatrix}$; the line $y = 2x$ in the xy-plane.

For $\lambda = -1$, $\mathbf{x} = t \begin{bmatrix} 0 \\ 1 \end{bmatrix}$; the line $x = 0$.

(b) For $\lambda = 4$, $\mathbf{x} = t \begin{bmatrix} 3 \\ 2 \end{bmatrix}$; the line $y = \frac{2}{3}x$.

(c) For $\lambda = 2$, $\mathbf{x} = t \begin{bmatrix} 0 \\ 1 \end{bmatrix}$; the line $x = 0$.

11. (a) For $\lambda = 1$, $\mathbf{x} = t \begin{bmatrix} 0 \\ 1 \\ 0 \end{bmatrix}$; a line through the origin in R^3.

For $\lambda = 2$, $\mathbf{x} = t \begin{bmatrix} 1 \\ -2 \\ -2 \end{bmatrix}$; a line through the origin in R^3.

For $\lambda = 3$, $\mathbf{x} = t \begin{bmatrix} -1 \\ 1 \\ 1 \end{bmatrix}$; a line through the origin in R^3.

(b) For $\lambda = 0$, $\mathbf{x} = t \begin{bmatrix} 5 \\ 1 \\ 3 \end{bmatrix}$; a line through the origin in R^3.

For $\lambda = 2$, $\mathbf{x} = s \begin{bmatrix} 5 \\ 0 \\ 2 \end{bmatrix} + t \begin{bmatrix} 5 \\ 2 \\ 0 \end{bmatrix}$; a plane through the

origin in R^3.

(c) For $\lambda = 2$, $\mathbf{x} = t \begin{bmatrix} -1 \\ 1 \\ 3 \end{bmatrix}$; a line through the origin in R^3.

For $\lambda = -3$, $\mathbf{x} = t \begin{bmatrix} -1 \\ 1 \\ -2 \end{bmatrix}$; a line through the origin in R^3.

13. (a) $p(\lambda) = \lambda^2 - 4\lambda - 5$; $-1, 5$
 (b) $p(\lambda) = \lambda^3 - 11\lambda^2 + 31\lambda - 21$; $3, 7, 2$
 (c) $p(\lambda) = \lambda^4 - \frac{5}{6}\lambda^3 - \frac{7}{18}\lambda^2 + \frac{1}{6}\lambda + \frac{1}{18}$; $-\frac{1}{3}, -1, \frac{1}{2}$

15. $(\lambda^2 - 8\lambda + 9)(\lambda^2 - 9)$; $3, -3, 9, -1$

17. $\lambda = -1$ with eigenvectors $\mathbf{x} = t \begin{bmatrix} 2 \\ -1 \\ 1 \end{bmatrix}$;

$\lambda = 1$ with eigenvectors $\mathbf{x} = s \begin{bmatrix} 0 \\ -1 \\ 1 \end{bmatrix} + t \begin{bmatrix} 1 \\ -1 \\ 0 \end{bmatrix}$.

The eigenvalues of A^{25} are $\lambda = (-1)^{25} = -1$ and $\lambda = (1)^{25} = 1$. The corresponding eigenvectors are the same as above.

19. $3, -1$ **21.**

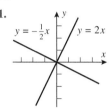

$y = -\frac{1}{2}x$ $y = 2x$

23. (a) $y = 2x$ and $y = x$ **(b)** no invariant lines **(c)** $y = 0$

25. $a = 1, b = 4$ **27.** $\lambda = 1, \lambda = -1$

31. (a) $1, -\frac{1}{4}$ **(b)** $1, -\frac{1}{64}$ **(c)** $-8, -3$ **(d)** $-20, 5$
 (e) $-14, 6$

► **Exercise Set 5.1** (Page 232) ──────────

1. (a) stochastic **(b)** not stochastic **(c)** stochastic
 (d) not stochastic

3. $\begin{bmatrix} 0.54545 \\ 0.45455 \end{bmatrix}$

5. (a) regular **(b)** not regular **(c)** regular

7. $\begin{bmatrix} \frac{8}{17} \\ \frac{9}{17} \end{bmatrix}$ **9.** $\begin{bmatrix} \frac{4}{11} \\ \frac{4}{11} \\ \frac{3}{11} \end{bmatrix}$

11. (a) probability that something in state 1 stays in state 1
 (b) probability that something in state 2 moves to state 1
 (c) 0.8 **(d)** 0.85

13. (a) $\begin{bmatrix} 0.95 & 0.55 \\ 0.05 & 0.45 \end{bmatrix}$ **(b)** 0.93 **(c)** 0.142 **(d)** 0.63

15. (a)

Year	1	2	3	4	5
City	95,750	91,840	88,243	84,933	81,889
Suburbs	29,250	33,160	36,757	40,067	43,111

(b)

City	46,875
Suburbs	78,125

17. (a) $\frac{23}{100}$ **(b)** $\begin{bmatrix} \frac{46}{159} \\ \frac{22}{53} \\ \frac{47}{159} \end{bmatrix}$ **(c)** 35, 50, 35

► **Exercise Set 5.2** (Page 239) ──────────

1. (a) $\begin{bmatrix} 0.50 & 0.25 \\ 0.25 & 0.10 \end{bmatrix}$ **(b)** $\begin{bmatrix} \$25{,}290 \\ \$22{,}581 \end{bmatrix}$

3. (a) $\begin{bmatrix} 0.1 & 0.6 & 0.4 \\ 0.3 & 0.2 & 0.3 \\ 0.4 & 0.1 & 0.2 \end{bmatrix}$ **(b)** $\begin{bmatrix} \$31{,}500 \\ \$26{,}500 \\ \$26{,}300 \end{bmatrix}$ **5.** $\begin{bmatrix} 123.08 \\ 202.56 \end{bmatrix}$

7. (a) productive by Theorem 5.2.1
 (b) productive by Exercise P2

9. $(I - C)^{-1} = \begin{bmatrix} 16 & 10 & 5 \\ 30 & 25 & 10 \\ 28 & 20 & 10 \end{bmatrix}$

► **Exercise Set 5.3** (Page 247) ──────────

1. Jacobi: $\begin{bmatrix} 2.875 \\ 1.375 \end{bmatrix}$; Gauss–Seidel: $\begin{bmatrix} 3.031 \\ 1.016 \end{bmatrix}$; exact: $\begin{bmatrix} 3 \\ 1 \end{bmatrix}$

3. Jacobi: $\begin{bmatrix} 0.492 \\ 0.006 \\ -0.996 \end{bmatrix}$; Gauss–Seidel: $\begin{bmatrix} 0.4989 \\ 0.0004 \\ -0.9998 \end{bmatrix}$;
 exact: $\begin{bmatrix} 0.5 \\ 0 \\ -1 \end{bmatrix}$

5. (a) no **(b)** yes **(c)** yes

7. The matrix is not diagonally dominant, since
$|-1| < |4| + |-6|$; $\begin{bmatrix} 5 & 2 & 2 \\ 3 & -7 & -3 \\ -1 & 4 & -6 \end{bmatrix}$ is strictly diagonally
dominant.

9. The matrix is not diagonally dominant, since
$|1| < |-2| + |3|$. Interchanging rows will not make it diagonally dominant.

▶ **Exercise Set 5.4** (Page 261) ─────────────

1. **(a)** λ_3 dominant   **(b)** no dominant eigenvalue

3. dominant eigenvalue: $\lambda = 5$
   sequence (2): $6.310, 5.210, 5.040, 5.008 \ldots$
   dominant eigenvector: $\begin{bmatrix} -\frac{1}{\sqrt{2}} \\ \frac{1}{\sqrt{2}} \end{bmatrix}$

5. dominant eigenvalue: $\lambda = 2 + \sqrt{10} \approx 5.16228$
   sequence (9): $1.0, 4.6, 5.131, 5.161 \ldots$
   dominant eigenvector: $\begin{bmatrix} 1 \\ 3 - \sqrt{10} \end{bmatrix}$

7. $2.99993; \begin{bmatrix} 0.99180 \\ 1.00000 \end{bmatrix}$

▶ **Exercise Set 6.1** (Page 277) ─────────────

1. **(a)** domain: R^2; codomain: R^3
   (b) domain: R^3; codomain: R^2
   (c) domain: R^3; codomain: R^3

3. $R^2, R^3, (-1, 2, 3)$ 5. **(a)** $\begin{bmatrix} -1 \\ 1 \end{bmatrix}$ **(b)** $\begin{bmatrix} 3 \\ 13 \end{bmatrix}$

7. **(a)** $\begin{bmatrix} -1 \\ 1 \\ 0 \end{bmatrix} + t \begin{bmatrix} -6 \\ 3 \\ 1 \end{bmatrix}$ (t any real number)   **(b)** no solution

9. In (a), (c), and (d), T is linear. In (b), T is not linear; both additivity and homogeneity fail.

11. In (a) and (c), T is linear. In (b) T is not linear; both additivity and homogeneity fail.

13. The domain is R^3, the codomain is R^2, and T is linear.

15. The domain is R^3, the codomain is R^3, and T is linear.

17. $T = \begin{bmatrix} 3 & -1 \\ 2 & 3 \\ 4 & 0 \end{bmatrix}$

19. **(a)** $\begin{bmatrix} -1 & 1 \\ 0 & 1 \end{bmatrix} \begin{bmatrix} -1 \\ 4 \end{bmatrix} = \begin{bmatrix} 5 \\ 4 \end{bmatrix}$

   (b) $\begin{bmatrix} 2 & -1 & 1 \\ 0 & 1 & 1 \\ 0 & 0 & 0 \end{bmatrix} \begin{bmatrix} 2 \\ 1 \\ -3 \end{bmatrix} = \begin{bmatrix} 0 \\ -2 \\ 0 \end{bmatrix}$

21. **(a)** $\begin{bmatrix} 3 & 5 & -1 \\ 4 & -1 & 1 \\ 3 & 2 & -1 \end{bmatrix}$ **(b)** $\begin{bmatrix} 3 \\ -2 \\ -3 \end{bmatrix}$

23. **(a)** $(-2, -1)$ **(b)** $(-1, 2)$ **(c)** $(-2, 0)$ **(d)** $(0, 1)$

25. **(a)** $\left(-\frac{\sqrt{2}}{2}, \frac{7\sqrt{2}}{2} \right)$ **(b)** $(-4, 3)$ **(c)** $(-3, -4)$
   (d) $\left(2 + \frac{3\sqrt{3}}{2}, -\frac{3}{2} + 2\sqrt{3} \right)$

27. **(a)** $\left(-\frac{3}{2} + 2\sqrt{3}, 2 + \frac{3\sqrt{3}}{2} \right)$ **(b)** $\left(\frac{3}{4} + \sqrt{3}, \frac{3}{4}(4 + \sqrt{3}) \right)$

29. $\theta = 3\pi/4$ 33. **(a)** $\left(-\frac{7}{5}, \frac{24}{5} \right)$ **(b)** $\frac{13}{10}(1, 3)$

35. **(a)** $\begin{bmatrix} -1 \\ 2 \\ 4 \end{bmatrix}, \begin{bmatrix} 3 \\ 1 \\ 5 \end{bmatrix}, \begin{bmatrix} 0 \\ 2 \\ -3 \end{bmatrix}$ **(b)** $\begin{bmatrix} 2 \\ 5 \\ 6 \end{bmatrix}$ **(c)** $\begin{bmatrix} 0 \\ 14 \\ -21 \end{bmatrix}$

37. **(a)** $\begin{bmatrix} 1 & 0 & 0 \\ 0 & 0 & 1 \\ 0 & 1 & 0 \end{bmatrix}$ **(b)** $\begin{bmatrix} 0 & 1 & 0 \\ 1 & 0 & 0 \\ 0 & 0 & 1 \end{bmatrix}$ **(c)** $\begin{bmatrix} 0 & 0 & 1 \\ 0 & 1 & 0 \\ 1 & 0 & 0 \end{bmatrix}$

39. $\begin{bmatrix} -1 & 0 \\ 0 & 0 \end{bmatrix}$

▶ **Exercise Set 6.2** (Page 293) ─────────────

1. $A^{-1} = A^T = \begin{bmatrix} \frac{3}{5} & \frac{4}{5} \\ -\frac{4}{5} & \frac{3}{5} \end{bmatrix}$

3. $A^{-1} = A^T = \begin{bmatrix} \frac{4}{5} & -\frac{9}{25} & \frac{12}{25} \\ 0 & \frac{4}{5} & \frac{3}{5} \\ -\frac{3}{5} & -\frac{12}{25} & \frac{16}{25} \end{bmatrix}$

5. **(a)** rotation about the origin through $\theta = 3\pi/4$
   (b) reflection about a line making an angle of $\theta = \pi/3$ with the x-axis

7. **(a)** $A = \begin{bmatrix} \frac{1}{5} & 0 \\ 0 & \frac{1}{5} \end{bmatrix}$ **(b)** $A = \begin{bmatrix} \frac{1}{3} & 0 \\ 0 & 1 \end{bmatrix}$ **(c)** $A = \begin{bmatrix} 1 & 0 \\ 0 & 6 \end{bmatrix}$
   (d) $A = \begin{bmatrix} 1 & 2 \\ 0 & 1 \end{bmatrix}$

9. **(a)** expansion in the x-direction with factor 3
   (b) contraction with factor $\frac{1}{4}$
   (c) shear in x-direction with factor 4
   (d) shear in y-direction with factor -4

11. $A = \begin{bmatrix} 0 & 0 \\ 3 & 0 \end{bmatrix}$ 13. $A = \begin{bmatrix} 1 & 0 & 0 \\ 0 & 0 & 0 \\ 0 & 0 & 0 \end{bmatrix}$

15. **(a)** **(b)** **(c)**

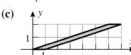

17. (a)   **(b)**

(c)

19. (a) $(2, 5, -3)$ **(b)** $(2, -5, 4)$ **(c)** $(-2, 5, 3)$

21. (a) $A = \begin{bmatrix} 1 & 0 & 0 \\ 0 & 0 & -1 \\ 0 & 1 & 0 \end{bmatrix}$ **(b)** $A = \begin{bmatrix} 0 & 0 & 1 \\ 0 & 1 & 0 \\ -1 & 0 & 0 \end{bmatrix}$

(c) $A = \begin{bmatrix} 0 & 1 & 0 \\ -1 & 0 & 0 \\ 0 & 0 & 1 \end{bmatrix}$

23. (a) $\left(-2, -1 + \frac{\sqrt{3}}{2}, \frac{1}{2} + \sqrt{3} \right)$ **(b)** $(0, 1, 0)$

(c) $\left(-1 + \frac{\sqrt{3}}{2}, \frac{1}{2} + \sqrt{3}, 2 \right)$

25. axis of rotation: x-axis; angle of rotation: $\pi/2$

27. Trace $= 1$ and $\mathbf{v} = (2x, 0, 0)$, so the result is the same as Exercise 25.

29. (a) $M_1 = \begin{bmatrix} 1 & 0 & 0 \\ 0 & 0 & 0 \\ 0 & 0 & 0 \end{bmatrix}$, $M_2 = \begin{bmatrix} 0 & 0 & 0 \\ 0 & 1 & 0 \\ 0 & 0 & 0 \end{bmatrix}$,

$M_3 = \begin{bmatrix} 0 & 0 & 0 \\ 0 & 0 & 0 \\ 0 & 0 & 1 \end{bmatrix}$ **(b)**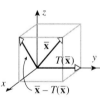

▶ **Exercise Set 6.3** (Page 303) ─────────

1. (a) $\ker(T)$ is the y-axis; $\operatorname{ran}(T)$ is the x-axis; T is neither one-to-one nor onto.   **(b)** $\ker(T)$ is the x-axis; $\operatorname{ran}(T)$ is the yz-plane; T is neither one-to-one nor onto.

(c) $\ker(T) = \left\{ \begin{bmatrix} 0 \\ 0 \end{bmatrix} \right\}$; $\operatorname{ran}(T) = R^2$; T is both one-to-one

and onto.   **(d)** $\ker(T) = \left\{ \begin{bmatrix} 0 \\ 0 \\ 0 \end{bmatrix} \right\}$; $\operatorname{ran}(T) = R^3$; T is

both one-to-one and onto.

3. $\ker(T) = \operatorname{span} \left\{ \begin{bmatrix} -1 \\ 1 \end{bmatrix} \right\}$ **5.** $\ker(T) = \operatorname{span} \left\{ \begin{bmatrix} -5 \\ 4 \\ 1 \end{bmatrix} \right\}$

7. $\begin{bmatrix} 1 & 0 & -1 \\ 0 & 1 & -1 \\ 1 & -1 & 0 \\ 1 & 1 & 1 \end{bmatrix} \begin{bmatrix} x \\ y \\ z \end{bmatrix} = \begin{bmatrix} 0 \\ 0 \\ 0 \\ 0 \end{bmatrix}$; solution is $x = 0$, $y = 0$, $z = 0$; $\ker(T) = \operatorname{span}\{\mathbf{0}\}$.

9. (a) $\mathbf{b}$ is not in the column space of A.

(b) $\begin{bmatrix} 3 \\ -1 \\ 8 \end{bmatrix} = \frac{4}{3} \begin{bmatrix} 1 \\ -2 \\ 1 \end{bmatrix} + \frac{5}{6} \begin{bmatrix} 2 \\ 2 \\ 8 \end{bmatrix}$

11. yes   **13.** $\begin{bmatrix} 2 & -3 \\ 5 & 1 \end{bmatrix}$; both one-to-one and onto

15. $\begin{bmatrix} -1 & 3 & 2 \\ 2 & 0 & 4 \\ 1 & 3 & 6 \end{bmatrix}$; neither one-to-one nor onto

17. possible answer: $(1, 1)$

19. (a) one-to-one   **(b)** not one-to-one

21. (a) $6b_1 + 3b_2 - 4b_3 = 0$ **(b)** $t_1 \begin{bmatrix} 1 \\ 0 \\ \frac{3}{2} \end{bmatrix} + t_2 \begin{bmatrix} 0 \\ 1 \\ \frac{3}{4} \end{bmatrix}$

(c) $t_1 \begin{bmatrix} -1 \\ -1 \\ 1 \\ 0 \end{bmatrix} + t_2 \begin{bmatrix} -1 \\ 1 \\ 0 \\ 1 \end{bmatrix}$

▶ **Exercise Set 6.4** (Page 315) ─────────

1. $T_B \circ T_A = \begin{bmatrix} 5 & -1 & 21 \\ 10 & -8 & 4 \\ 45 & 3 & 25 \end{bmatrix}$, $T_A \circ T_B = \begin{bmatrix} -8 & -3 & 1 \\ -5 & -15 & -8 \\ 44 & -11 & 45 \end{bmatrix}$

3. (a) $T_1 = \begin{bmatrix} 1 & 1 \\ 1 & -1 \end{bmatrix}$, $T_2 = \begin{bmatrix} 3 & 0 \\ 2 & 4 \end{bmatrix}$

(b) $T_2 \circ T_1 = \begin{bmatrix} 3 & 3 \\ 6 & -2 \end{bmatrix}$, $T_1 \circ T_2 = \begin{bmatrix} 5 & 4 \\ 1 & -4 \end{bmatrix}$

(c) $T_2(T_1(x_1, x_2)) = (3x_1 + 3x_2, 6x_1 - 2x_2)$
$T_1(T_2(x_1, x_2)) = (5x_1 + 4x_2, x_1 - 4x_2)$

5. (a) $\begin{bmatrix} 1 & 0 \\ 0 & -1 \end{bmatrix}$ **(b)** $\begin{bmatrix} 0 & 0 \\ 0 & \frac{1}{3} \end{bmatrix}$ **(c)** $\begin{bmatrix} -3 & 0 \\ 0 & 3 \end{bmatrix}$

7. (a) $\begin{bmatrix} -1 & 0 & 0 \\ 0 & 0 & 0 \\ 0 & 0 & 1 \end{bmatrix}$ **(b)** $\begin{bmatrix} 1 & 0 & 1 \\ 0 & \sqrt{2} & 0 \\ -1 & 0 & 1 \end{bmatrix}$

(c) $\begin{bmatrix} -1 & 0 & 0 \\ 0 & 1 & 0 \\ 0 & 0 & 0 \end{bmatrix}$

9. (a) $\begin{bmatrix} \frac{\sqrt{3}}{8} & -\frac{\sqrt{3}}{16} & \frac{1}{16} \\ \frac{1}{8} & \frac{3}{16} & -\frac{\sqrt{3}}{16} \\ 0 & \frac{1}{8} & \frac{\sqrt{3}}{8} \end{bmatrix}$ **(b)** $\begin{bmatrix} 0 & 0 & 0 \\ 0 & -1 & 0 \\ 0 & 0 & -1 \end{bmatrix}$

11. **(a)** $\begin{bmatrix} 2 & 0 \\ 0 & 1 \end{bmatrix}\begin{bmatrix} 1 & 0 \\ 0 & 3 \end{bmatrix}$; expansion in y-direction with
factor 3 followed by expansion in x-direction with factor 2.

(b) $\begin{bmatrix} 0 & 1 \\ 1 & 0 \end{bmatrix}\begin{bmatrix} 1 & 0 \\ 2 & 1 \end{bmatrix}$; shear in y-direction with factor 2
followed by reflection about the line $y = x$.

(c) $\begin{bmatrix} 0 & 1 \\ 1 & 0 \end{bmatrix}\begin{bmatrix} 1 & 0 \\ 0 & -1 \end{bmatrix}\begin{bmatrix} 1 & 0 \\ 0 & 2 \end{bmatrix}\begin{bmatrix} 4 & 0 \\ 0 & 1 \end{bmatrix}$; expansion in
x-direction with factor 4, followed by expansion in
y-direction with factor 2, followed by reflection about the
x-axis, followed by reflection about the line $y = x$.

(d) $\begin{bmatrix} 1 & 0 \\ 4 & 1 \end{bmatrix}\begin{bmatrix} 1 & 0 \\ 0 & 18 \end{bmatrix}\begin{bmatrix} 1 & -3 \\ 0 & 1 \end{bmatrix}$; shear in x-direction with
factor of -3, followed by expansion in y-direction with
factor of 18, followed by shear in y-direction with
factor of 4.

13. **(a)** reflection of R^2 about the x-axis   **(b)** rotation of R^2
through $-\pi/4$ **(c)** contraction of R^2, $k = \frac{1}{3}$
(d) expansion of R^2 in y-direction with $k = 2$

15. $T^{-1} = \begin{bmatrix} -1 & 2 \\ 1 & -1 \end{bmatrix}$, $T^{-1}(w_1, w_2) = (-w_1 + 2w_2, w_1 - w_2)$

17. $T^{-1} = \begin{bmatrix} 0 & 1 \\ -1 & 0 \end{bmatrix}$, $T^{-1}(w_1, w_2) = (w_2, -w_1)$

19. $T^{-1} = \begin{bmatrix} 1 & -2 & 4 \\ -1 & 2 & -3 \\ -1 & 3 & -5 \end{bmatrix}$, $T^{-1}(w_1, w_2, w_3) =$
$(w_1 - 2w_2 + 4w_3, -w_1 + 2w_2 - 3w_3, -w_1 + 3w_2 - 5w_3)$

21. $T^{-1} = \begin{bmatrix} -\frac{3}{2} & -\frac{3}{2} & \frac{11}{2} \\ \frac{1}{2} & \frac{1}{2} & -\frac{3}{2} \\ -\frac{1}{2} & \frac{1}{2} & -\frac{1}{2} \end{bmatrix}$, $T^{-1}(w_1, w_2, w_3)$
$= \left(-\frac{3}{2}w_1 - \frac{3}{2}w_2 + \frac{11}{2}w_3, \frac{1}{2}w_1 + \frac{1}{2}w_2 - \frac{3}{2}w_3, -\frac{1}{2}w_1 + \frac{1}{2}w_2 - \frac{1}{2}w_3\right)$

23. **(a)** yes   **(b)** yes   **(c)** no   **25.** $\begin{bmatrix} \frac{1}{2} & \frac{\sqrt{3}}{2} \\ -\frac{\sqrt{3}}{2} & \frac{1}{2} \end{bmatrix}$

27. area $= 3$

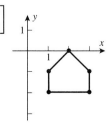

▶ **Exercise Set 6.5** (Page 325) ────

1.    **3.**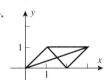

5. $V = \begin{bmatrix} 1 & -1 & -1 & 1 \\ 1 & 1 & -1 & -1 \end{bmatrix}$; $C = \begin{bmatrix} 1 & 1 & 1 & 0 \\ 1 & 1 & 0 & 1 \\ 1 & 0 & 1 & 1 \\ 0 & 1 & 1 & 1 \end{bmatrix}$

7. $V = \begin{bmatrix} 1 & 0 & -1 & -1 & 1 \\ 0 & 1 & 0 & -1 & -1 \end{bmatrix}$; $C = \begin{bmatrix} 1 & 1 & 0 & 0 & 1 \\ 1 & 1 & 1 & 0 & 0 \\ 0 & 1 & 1 & 1 & 0 \\ 0 & 0 & 1 & 1 & 1 \\ 1 & 0 & 0 & 1 & 1 \end{bmatrix}$

9. $\begin{bmatrix} 2 & 0 & -2 & -2 & 2 \\ 0 & 3 & 0 & -3 & -3 \end{bmatrix}$

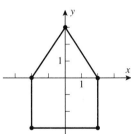

11. $\begin{bmatrix} 1 & 2 & -1 & -3 & -1 \\ 0 & 1 & 0 & -1 & -1 \end{bmatrix}$

13. $\begin{bmatrix} 2 & 1 & 0 & 0 & 2 \\ 1 & 2 & 1 & 0 & 0 \end{bmatrix}$

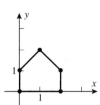

15. $\begin{bmatrix} 3 & 2 & 1 & 1 & 3 \\ -1 & 0 & -1 & -2 & -2 \end{bmatrix}$

17. (a) $\begin{bmatrix} \frac{\sqrt{3}}{2} & -\frac{1}{2} & 0 \\ \frac{1}{2} & \frac{\sqrt{3}}{2} & 0 \\ 0 & 0 & 1 \end{bmatrix}$ **(b)** $\begin{bmatrix} 1 & 0 & 0 \\ 0 & -1 & 0 \\ 0 & 0 & 1 \end{bmatrix}$

19. (a) $\begin{bmatrix} \frac{1}{2} & -\frac{\sqrt{3}}{2} & 0 & 0 \\ \frac{\sqrt{3}}{2} & \frac{1}{2} & 0 & 0 \\ 0 & 0 & 1 & 0 \\ 0 & 0 & 0 & 1 \end{bmatrix}$ **(b)** $\begin{bmatrix} -1 & 0 & 0 & 0 \\ 0 & 1 & 0 & 0 \\ 0 & 0 & 1 & 0 \\ 0 & 0 & 0 & 1 \end{bmatrix}$

21. (a) $\begin{bmatrix} 1 & 0 & 1 \\ 0 & 1 & 2 \\ 0 & 0 & 1 \end{bmatrix}$ **(b)** $(4, 6)$

23. (a) $\begin{bmatrix} 1 & 0 & 0 & 5 \\ 0 & 1 & 0 & 3 \\ 0 & 0 & 1 & -1 \\ 0 & 0 & 0 & 1 \end{bmatrix}$ **(b)** $(10, -1, 0)$

25. $\begin{bmatrix} \frac{1}{2} & -\frac{\sqrt{3}}{2} & 3 \\ \frac{\sqrt{3}}{2} & \frac{1}{2} & -1 \\ 0 & 0 & 1 \end{bmatrix}$ **27.** $\begin{bmatrix} -1 & 0 & 0 & 2 \\ 0 & 2 & 1 & -3 \\ 1 & 1 & 0 & 5 \\ 0 & 0 & 0 & 1 \end{bmatrix}$

▶ **Exercise Set 7.1** (Page 334) ―――――――

1. (a) $\mathbf{v}_2 = 2\mathbf{v}_1$ **(b)** $\mathbf{v}_3 = \mathbf{v}_1 - \mathbf{v}_2$ **(c)** $\mathbf{v}_4 = -\mathbf{v}_1 + 2\mathbf{v}_2 + 6\mathbf{v}_3$
3. $\mathbf{v}_3 = -\frac{7}{3}\mathbf{v}_1 + \frac{2}{3}\mathbf{v}_2$
5. possible answers: **(a)** $\{(1, 2)\}; \{(2, 4)\}; \{(-1, -2)\}$
  (b) $\{(1, -1, 0), (2, 0, -1)\}; \{(1, 1, -1), (0, 2, -1)\};$
  $\{(2, 0, -1), (3, -1, -1)\}$
7. $\left\{\left(-\frac{1}{4}, -\frac{1}{4}, 1, 0\right), (0, -1, 0, 1)\right\};$ dimension $= 2$
9. $\{(-1, 1, 0, 0, 0), (-1, 0, -1, 0, 1)\};$ dimension $= 2$
11. (a) $\{(-2, 1, 0), (3, 0, 1)\}$
  (b) $\{(0, 1, 0, 1), (0, 0, 1, -4), (0, 4, 1, 0)\}$

▶ **Exercise Set 7.2** (Page 340) ―――――――

1. (a) dependent **(b)** do not span **(c)** dependent
5. (a) basis **(b)** not a basis
7. (b) $\mathbf{v} = \mathbf{v}_1 + \mathbf{v}_2 + \mathbf{v}_3 = \frac{1}{2}\mathbf{v}_2 + \frac{2}{3}\mathbf{v}_3 + \mathbf{v}_4 = 7\mathbf{v}_1 + 4\mathbf{v}_2 + 3\mathbf{v}_3 - 6\mathbf{v}_4$
9. possible answer: $\{\mathbf{v}_1, \mathbf{v}_2, (1, 0, 0)\}$
11. possible answer: $\{\mathbf{v}_1, \mathbf{v}_2, \mathbf{v}_4\}$
13. $\mathbf{v} = -3\mathbf{v}_1 + 4\mathbf{v}_2 + \mathbf{v}_3$
15. (a) not a subspace **(b)** subspace
17. (a) $\left(\frac{26}{5}, \frac{13}{5}, -4\right)$
  (b) $\left(\frac{7}{5}a + \frac{3}{5}b + \frac{1}{5}c, \frac{1}{5}a + \frac{4}{5}b - \frac{2}{5}c, -a + c\right)$
  (c) $\begin{bmatrix} \frac{7}{5} & \frac{3}{5} & \frac{1}{5} \\ \frac{1}{5} & \frac{4}{5} & -\frac{2}{5} \\ -1 & 0 & 1 \end{bmatrix}$

19. $(x, y, z) = \left(-\frac{7}{50}x - \frac{17}{100}y + \frac{1}{100}z\right)\mathbf{v}_1 +$
  $\left(-\frac{1}{2}x - \frac{1}{4}y + \frac{1}{4}z\right)\mathbf{v}_2 + \left(\frac{33}{50}x + \frac{23}{100}y - \frac{19}{100}z\right)\mathbf{v}_3$

▶ **Exercise Set 7.3** (Page 349) ―――――――

1. All multiples of $(-7, 1, 2)$ **5.** $y = -\frac{1}{2}x$
7. possible answer: $x = \frac{5}{2}s - 2t, y = s, z = t$ (s and t any real numbers)
9. possible answer: W: $\{(1, 0, -16), (0, 1, -19)\}; W^{\perp}$: $\{(16, 19, 1)\}$
11. possible answer: W: $\{(1, 0, 0, 0), (0, 1, 0, 0), (0, 0, 1, 1)\};$
  $W^{\perp}$: $\{(0, 0, -1, 1)\}$
13. possible answer:
  W: $\{(1, 0, 0, 2, 0), (0, 1, 0, 3, 0), (0, 0, 1, 4, 0),$
  $(0, 0, 0, 0, 1)\};$
  $W^{\perp}$: $\{(-2, -3, -4, 1, 0)\}$
15. $16x_1 + 19x_2 + x_3 = 0$ **17.** $-x_3 + x_4 = 0$
19. $16b_1 + 19b_2 + b_3 = 0$ **21.** $-b_3 + b_4 = 0$ **23.** $\mathbf{b}_1, \mathbf{b}_2$
27. $\{(1, 3, 0, 4, 0, 0), (0, 0, 1, 2, 0, 0), (0, 0, 0, 0, 1, 0),$
  $(0, 0, 0, 0, 0, 1)\}$
31. $\begin{bmatrix} 1 & 4 & 1 & 0 \\ -2 & 0 & 0 & 1 \end{bmatrix}$

▶ **Exercise Set 7.4** (Page 357) ―――――――

1. $\text{rank}(A) = 2, \text{nullity}(A) = 1$
3. $\text{rank}(A) = 2, \text{nullity}(A) = 2$
5. $\text{rank}(A) = 3, \text{nullity}(A) = 2$
7. (a) pivots $= 3$, parameters $= 5$
  (b) pivots $= 2$, parameters $= 2$
  (c) pivots $= 2$, parameters $= 4$
9. (a) maximum rank $= 3$, minimum nullity $= 0$
  (b) maximum rank $= 3$, minimum nullity $= 2$
  (c) maximum rank $= 4$, minimum nullity $= 0$
11. $\left\{\mathbf{v}_1, \mathbf{v}_2, \left(\frac{4}{3}, -\frac{4}{3}, 1, 0\right), \left(\frac{5}{3}, -\frac{5}{3}, 0, 1\right)\right\}$ is a basis.

13. (a) $A = \begin{bmatrix} 1 \\ -2 \end{bmatrix}\begin{bmatrix} 1 \\ -7 \end{bmatrix}^T$ **(c)** $A = \begin{bmatrix} 1 \\ -2 \\ 3 \end{bmatrix}\begin{bmatrix} 1 \\ 1 \\ 3 \\ 3 \\ -9 \end{bmatrix}^T$

15. $\mathbf{uu}^T = \begin{bmatrix} 4 & 6 & 2 & 2 \\ 9 & 6 & 3 & 3 \\ 2 & 3 & 1 & 1 \\ 2 & 3 & 1 & 1 \end{bmatrix}$

17. rank 1 if $t = 1$, rank 2 if $t = -2$, rank 3 otherwise
21. Possible answer: $\{(1, 2, 0), (3, 0, 2)\}; W^{\perp}$ is a hyperplane.

▶ **Exercise Set 7.5 (Page 368)**

1. $\dim(\text{row}(A)) = \dim(\text{col}(A)) = 2, \dim(\text{null}(A)) = 3$,
$\dim(\text{null}(A^\perp)) = 2$, parameters $= 3$

3. $\text{rank}(A) = \text{rank}(A^T) = 3$

5.

	(a)	(b)	(c)	(d)	(e)	(f)	(g)
$\dim(\text{row}(A))$	3	2	1	2	2	0	2
$\dim(\text{col}(A))$	3	2	1	2	2	0	2
$\dim(\text{null}(A))$	0	1	2	7	3	4	0
$\dim(\text{null}(A^\perp))$	0	1	2	3	7	4	4

7. (a) consistent; parameters $= 0$ (b) inconsistent
(c) consistent; parameters $= 2$
(d) consistent; parameters $= 7$

9. (a) full column rank   (b) neither   (c) full row rank
(d) both

11. (a) $\det(A^TA) = 149, \det(AA^T) = 0$ (b) $\det(A^TA) = 0$,
$\det(AA^T) = 0$ (c) $\det(A^TA) = 0, \det(AA^T) = 66$
(d) $\det(A^TA) = 64, \det(AA^T) = 64$

17. Inconsistent unless $(b_1, b_2, b_3, b_4, b_5)$ satisfies the equations
$b_3 = -3b_1 + 4b_2, b_4 = 2b_1 - b_2, b_5 = -7b_1 + 8b_2$

▶ **Exercise Set 7.6 (Page 377)**

1. column space: $\{(1, 5, 7), (-1, -4, -6)\}$
row space: $\{(1, -1, 3), (5, -4, -4)\}$

3. column space: $\{(1, 2, -1), (4, 1, 3)\}$
row space: $\{(1, 4, 5, 2), (2, 1, 3, 0)\}$

5. column space:
$\{(1, 0, 2, 3, -2), (-3, 3, -3, -6, 9), (2, 6, -2, 0, 2)\}$
row space:
$\{(1, -3, 2, 2, 1), (0, 3, 6, 0, -3), (2, -3, -2, 4, 4)\}$

7. $\{\mathbf{v}_1, \mathbf{v}_2, \mathbf{v}_3\}$ **9.** $\{\mathbf{v}_1, \mathbf{v}_3\}, \mathbf{v}_2 = 2\mathbf{v}_1, \mathbf{v}_4 = \mathbf{v}_1 + \mathbf{v}_3$

11. $\left\{\left(1, -\frac{1}{2}, 0\right), (0, 0, 1)\right\}$

13. $\text{row}(A): \{(1, 0, -1, 4), (0, 1, -3, -7)\}$
$\text{col}(A): \{(1, -2, 4, 0), (-1, 1, -3, -1)\}$
$\text{null}(A): \{(1, 3, 1, 0), (4, 7, 0, 1)\}$
$\text{null}(A^T): \left\{\left(1, 0, -\frac{1}{4}, -\frac{1}{4}\right), \left(0, 1, \frac{1}{2}, -\frac{1}{2}\right)\right\}$

15. (a) $\{(4, 0, 0, 0, 1), (8, 5, 0, 3, 2), (8, 15, 4, 9, 2)\}$
(b) $\{(4, 16, 8, 8), (0, 0, 5, 15), (0, 0, 0, 4)\}$
(c) $\{(-4, 1, 0, 0)\}$
(d) $\left\{\left(0, -\frac{3}{5}, 0, 1, 0\right), \left(-\frac{1}{4}, 0, 0, 0, 1\right)\right\}$

17. $A = \begin{bmatrix} 2 & -4 \\ 1 & 1 \\ 3 & -9 \\ 0 & 6 \end{bmatrix} \begin{bmatrix} 1 & 0 & -\frac{7}{3} & -\frac{1}{3} \\ 0 & 1 & -\frac{5}{3} & -\frac{5}{3} \end{bmatrix}$

$= \begin{bmatrix} 2 \\ 1 \\ 3 \\ 0 \end{bmatrix} \begin{bmatrix} 1 & 0 & -\frac{7}{3} & -\frac{1}{3} \end{bmatrix} + \begin{bmatrix} -4 \\ 1 \\ -9 \\ 6 \end{bmatrix} \begin{bmatrix} 0 & 1 & -\frac{5}{3} & -\frac{5}{3} \end{bmatrix}$

▶ **Exercise Set 7.7 (Page 391)**

1. $\left(\frac{11}{5}, \frac{22}{5}\right)$ **3.** $\left(\frac{3}{5}, \frac{6}{5}\right)$ **5.** $\left(0, \frac{2}{5}, -\frac{1}{5}\right), \left(1, \frac{3}{5}, \frac{6}{5}\right)$

7. $\left(\frac{1}{5}, -\frac{1}{5}, \frac{1}{10}, -\frac{1}{10}\right), \left(\frac{9}{5}, \frac{6}{5}, \frac{9}{10}, \frac{21}{10}\right)$ **9.** $\frac{20}{7}$ **11.** $\frac{\sqrt{33}}{3}$

13. $\frac{1}{30} \begin{bmatrix} 1 & -5 & -2 \\ -5 & 25 & 10 \\ -2 & 10 & 4 \end{bmatrix}$ **15.** $\frac{1}{257} \begin{bmatrix} 113 & -84 & 96 \\ -84 & 208 & 56 \\ 96 & 56 & 193 \end{bmatrix}$

17. $\begin{bmatrix} 1 & 0 & 0 \\ 0 & 0 & 0 \\ 0 & 0 & 1 \end{bmatrix}$ **19.** $\frac{1}{3} \begin{bmatrix} 2 & -1 & -1 \\ -1 & 2 & -1 \\ -1 & -1 & 2 \end{bmatrix}, \left(\frac{1}{3}, \frac{7}{3}, -\frac{8}{3}\right)$

21. $\frac{1}{220} \begin{bmatrix} 89 & -105 & -3 & 25 \\ -105 & 135 & -15 & 15 \\ -3 & -15 & 31 & -75 \\ 25 & 15 & -75 & 185 \end{bmatrix}$

23. $\frac{1}{3} \begin{bmatrix} 1 & 0 & -1 & 1 \\ 0 & 1 & -1 & -1 \\ -1 & -1 & 2 & 0 \\ 1 & -1 & 0 & 2 \end{bmatrix}$

25. $\frac{1}{35} \begin{bmatrix} 21 & 14 & -7 & 7 \\ 14 & 11 & -8 & -2 \\ -7 & -8 & 9 & 11 \\ 7 & -2 & 11 & 29 \end{bmatrix} \cdot \frac{1}{6} \begin{bmatrix} 2 & -2 & 2 \\ -2 & 5 & 1 \\ 2 & 1 & 5 \end{bmatrix}$

27. $\left(-\frac{3}{25}, \frac{4}{25}, -\frac{3}{25}, -\frac{4}{25}\right)$ **29.** $\frac{1}{3} \begin{bmatrix} 1 & 1 & 1 \\ 1 & 1 & 1 \\ 1 & 1 & 1 \end{bmatrix}$

31. $\left(-\frac{38}{89}, \frac{23}{89}, \frac{28}{89}, \frac{25}{89}\right)$

▶ **Exercise Set 7.8 (Page 404)**

1. $\left(\frac{20}{11}, -\frac{8}{11}\right)$ **3.** $\left(\frac{28}{11}, \frac{16}{11}, \frac{40}{11}\right)$

5. Solution: $\hat{\mathbf{x}} = \left(\frac{1}{10}, \frac{6}{5}\right)$; least squares error: $\frac{4}{5}\sqrt{5}$

7. Solutions: $\hat{\mathbf{x}} = \left(\frac{7}{6}, -\frac{7}{6}, 0\right) + t(1, -1, 0)$ (t a real number);
least squares error: $\frac{7}{2}\sqrt{14}$

9. $y = -\frac{3}{10} + \frac{7}{10}x$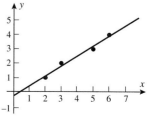

11. $y = 1 - \frac{11}{6}x + \frac{2}{3}x^2$

13. $\begin{bmatrix} 5 & 15 & 55 & 225 \\ 15 & 55 & 225 & 979 \\ 55 & 225 & 979 & 4425 \\ 225 & 979 & 4425 & 20515 \end{bmatrix} \begin{bmatrix} a_0 \\ a_1 \\ a_2 \\ a_3 \end{bmatrix} = \begin{bmatrix} 216.8 \\ 916.0 \\ 4087.4 \\ 18822.4 \end{bmatrix}$

▶ **Exercise Set 7.9** (Page 414) ─────────

1. (a) not orthogonal
  (b) orthogonal; $\left(-\frac{\sqrt{2}}{2}, \frac{\sqrt{2}}{2}\right), \left(\frac{\sqrt{2}}{2}, \frac{\sqrt{2}}{2}\right)$
  (c) orthogonal; $\left(-\frac{\sqrt{6}}{3}, \frac{\sqrt{6}}{6}, \frac{\sqrt{6}}{6}\right), \left(\frac{\sqrt{5}}{5}, 0, \frac{2\sqrt{5}}{5}\right),$
    $\left(-\frac{\sqrt{30}}{15}, -\frac{\sqrt{30}}{6}, \frac{\sqrt{30}}{30}\right)$ (d) not orthogonal

3. (a) orthonormal   (b) not orthonormal; $\mathbf{v}_2 \cdot \mathbf{v}_3 \neq 0$
  (c) not orthonormal; $\|\mathbf{v}_3\| \neq 1$

5. (a) $\left(1, \frac{11}{6}, -\frac{1}{3}, \frac{1}{6}\right)$ (b) $\left(\frac{23}{18}, \frac{11}{6}, -\frac{1}{18}, -\frac{17}{18}\right)$

7. (a) $\left(\frac{3}{2}, \frac{3}{2}, -1, -1\right)$ (b) $\left(\frac{7}{4}, \frac{5}{4}, -\frac{3}{4}, -\frac{5}{4}\right)$

9. $\mathbf{w} = \frac{5}{\sqrt{6}}\mathbf{v}_2 + \frac{11}{\sqrt{66}}\mathbf{v}_3$ **11.** $\mathbf{w} = -\frac{1}{9}\mathbf{v}_1 + \frac{5}{33}\mathbf{v}_2 + \frac{5}{9}\mathbf{v}_3 - \frac{2}{33}\mathbf{v}_4$

13. $\frac{1}{9}\begin{bmatrix} 5 & 4 & 2 \\ 4 & 5 & -2 \\ 2 & -2 & 8 \end{bmatrix}$ **15.** $\left(\frac{2}{9}, \frac{16}{9}, -\frac{28}{9}\right)$

17. $\begin{bmatrix} \frac{1}{2} & \frac{1}{2} & 0 \\ \frac{1}{2} & \frac{1}{2} & 0 \\ 0 & 0 & 1 \end{bmatrix}$ **19.** $\left(-\frac{1}{2}, -\frac{1}{2}, 0\right)$

21. $\text{tr}(P) = 2$ **23.** 2   **25.** 1

27. $\left\{ \left(\frac{1}{\sqrt{10}}, -\frac{3}{\sqrt{10}}\right), \left(\frac{3}{\sqrt{10}}, \frac{1}{\sqrt{10}}\right) \right\}$

29. $\left\{ \left(\frac{1}{\sqrt{3}}, \frac{1}{\sqrt{3}}, \frac{1}{\sqrt{3}}\right), \left(-\frac{1}{\sqrt{2}}, \frac{1}{\sqrt{2}}, 0\right), \left(\frac{1}{\sqrt{6}}, \frac{1}{\sqrt{6}}, -\frac{2}{\sqrt{6}}\right) \right\}$

31. $\left\{ \left(0, \frac{2}{\sqrt{5}}, \frac{1}{\sqrt{5}}, 0\right), \left(\frac{5}{\sqrt{30}}, -\frac{1}{\sqrt{30}}, \frac{2}{\sqrt{30}}, 0\right), \right.$
    $\left. \left(\frac{1}{\sqrt{10}}, \frac{1}{\sqrt{10}}, -\frac{2}{\sqrt{10}}, -\frac{2}{\sqrt{10}}\right), \left(\frac{1}{\sqrt{15}}, \frac{1}{\sqrt{15}}, -\frac{2}{\sqrt{15}}, \frac{3}{\sqrt{15}}\right) \right\}$

33. $\left\{ \left(\frac{1}{\sqrt{2}}, \frac{1}{\sqrt{2}}, 0\right), \left(\frac{1}{\sqrt{2}}, -\frac{1}{\sqrt{2}}, 0\right), (0, 0, 1) \right\}$

35. $\left\{ \left(0, \frac{1}{\sqrt{5}}, \frac{2}{\sqrt{5}}\right), \left(-\frac{5}{\sqrt{30}}, -\frac{2}{\sqrt{30}}, \frac{1}{\sqrt{30}}\right) \right\}$

37. $\mathbf{w}_1 = \left(-\frac{4}{5}, 2, \frac{3}{5}\right), \mathbf{w}_2 = \left(\frac{9}{5}, 0, \frac{12}{5}\right)$

▶ **Exercise Set 7.10** (Page 426) ─────────

1. possible answer: $Q = \begin{bmatrix} \frac{1}{\sqrt{5}} & -\frac{2}{\sqrt{5}} \\ \frac{2}{\sqrt{5}} & \frac{1}{\sqrt{5}} \end{bmatrix}, R = \begin{bmatrix} \sqrt{5} & \sqrt{5} \\ 0 & \sqrt{5} \end{bmatrix}$

3. possible answer: $Q = \begin{bmatrix} \frac{1}{3} & \frac{8}{3\sqrt{26}} \\ -\frac{2}{3} & \frac{11}{3\sqrt{26}} \\ \frac{2}{3} & \frac{7}{3\sqrt{26}} \end{bmatrix}, R = \begin{bmatrix} 3 & \frac{1}{3} \\ 0 & \frac{\sqrt{26}}{3} \end{bmatrix}$

5. possible answer: $Q = \begin{bmatrix} \frac{1}{\sqrt{2}} & \frac{1}{\sqrt{38}} & -\frac{3}{\sqrt{19}} \\ \frac{1}{\sqrt{2}} & -\frac{1}{\sqrt{38}} & \frac{3}{\sqrt{19}} \\ 0 & \frac{6}{\sqrt{38}} & \frac{1}{\sqrt{19}} \end{bmatrix},$
$R = \begin{bmatrix} \sqrt{2} & \frac{3\sqrt{2}}{2} & \sqrt{2} \\ 0 & \frac{\sqrt{38}}{2} & \frac{3\sqrt{38}}{19} \\ 0 & 0 & \frac{\sqrt{19}}{19} \end{bmatrix}$

7. $\left(-\frac{5}{26}, \frac{19}{26}\right)$ **9.** $(-8, -5, 16)$

11. $H = \frac{1}{7}\begin{bmatrix} 3 & 2 & 6 \\ 2 & 6 & 3 \\ -6 & 3 & 2 \end{bmatrix}; H\mathbf{b} = \left(-\frac{5}{7}, \frac{20}{7}, -\frac{4}{7}\right)$

13. $H = \frac{1}{3}\begin{bmatrix} 1 & 2 & -2 \\ 2 & 1 & 2 \\ -2 & 2 & 1 \end{bmatrix}$

15. $H = \frac{1}{11}\begin{bmatrix} 11 & 0 & 0 & 0 \\ 0 & 9 & 2 & -6 \\ 0 & 2 & 9 & 6 \\ 0 & -6 & 6 & -7 \end{bmatrix}$

17. (a) $H = \begin{bmatrix} \frac{3}{5} & \frac{4}{5} \\ \frac{4}{5} & -\frac{3}{5} \end{bmatrix}$ (b) $H = \begin{bmatrix} -\frac{4}{5} & \frac{3}{5} \\ \frac{3}{5} & \frac{4}{5} \end{bmatrix}$
  (c) $H = \begin{bmatrix} \frac{1}{\sqrt{2}} & \frac{1}{\sqrt{2}} \\ \frac{1}{\sqrt{2}} & -\frac{1}{\sqrt{2}} \end{bmatrix}$

19. $H = \frac{1}{15}\begin{bmatrix} -10 & 5 & 10 \\ 5 & 14 & -2 \\ 10 & -2 & 11 \end{bmatrix}$

21. $Q = \begin{bmatrix} \frac{1}{\sqrt{2}} & -\frac{1}{\sqrt{2}} \\ -\frac{1}{\sqrt{2}} & -\frac{1}{\sqrt{2}} \end{bmatrix}, R = \begin{bmatrix} \sqrt{2} & -\frac{1}{\sqrt{2}} \\ 0 & -\frac{5}{\sqrt{2}} \end{bmatrix}$

23. $Q = \begin{bmatrix} \frac{1}{\sqrt{2}} & 0 & -\frac{1}{\sqrt{2}} \\ -\frac{1}{\sqrt{2}} & 0 & -\frac{1}{\sqrt{2}} \\ 0 & 1 & 0 \end{bmatrix}, R = \begin{bmatrix} \sqrt{2} & 2\sqrt{2} & -\sqrt{2} \\ 0 & 4 & 5 \\ 0 & 0 & -2\sqrt{2} \end{bmatrix}$

25. $(1, 1, 1)$

▶ **Exercise Set 7.11** (Page 438)

1. (a) $(\mathbf{w})_B = (3, -7)$, $[\mathbf{w}]_B = \begin{bmatrix} 3 \\ -7 \end{bmatrix}$

(b) $(\mathbf{w})_B = \left(\frac{5}{28}, \frac{3}{14} \right)$, $[\mathbf{w}]_B = \begin{bmatrix} \frac{5}{28} \\ \frac{3}{14} \end{bmatrix}$

3. $(\mathbf{w})_B = (3, -2, 1)$, $[\mathbf{w}]_B = \begin{bmatrix} 3 \\ -2 \\ 1 \end{bmatrix}$

5. $(6, -1, 3)$ **7.** $(-2\sqrt{2}, 5\sqrt{2})$ **9.** $\left(\frac{5}{\sqrt{2}}, -\frac{2}{\sqrt{3}}, \frac{1}{\sqrt{6}} \right)$

11. (a) $\mathbf{u} = \left(\frac{7}{5}, -\frac{1}{5} \right)$, $\mathbf{v} = \left(\frac{13}{5}, \frac{16}{5} \right)$ **(b)** $\sqrt{2}, \sqrt{17}, 3$

13. $\sqrt{15}, \sqrt{35}, \sqrt{30}, \sqrt{105}, 5, 20$

15. $(0, \sqrt{2}), (1, 0), (-1, \sqrt{2}), (a - b, b\sqrt{2})$

17. (a) $\begin{bmatrix} 2 & -3 \\ 1 & 4 \end{bmatrix}$ **(b)** $\frac{1}{11} \begin{bmatrix} 4 & 3 \\ -1 & 2 \end{bmatrix}$ **(d)** $[\mathbf{w}]_B = \begin{bmatrix} 1 \\ -1 \end{bmatrix}$,

$[\mathbf{w}]_S = \begin{bmatrix} 5 \\ -3 \end{bmatrix}$ **(e)** $[\mathbf{w}]_S = \begin{bmatrix} 3 \\ -5 \end{bmatrix}$, $[\mathbf{w}]_B = \begin{bmatrix} -\frac{3}{11} \\ -\frac{13}{11} \end{bmatrix}$

19. (a) $\frac{1}{10} \begin{bmatrix} 13 & -5 \\ -4 & 0 \end{bmatrix}$ **(b)** $\frac{1}{2} \begin{bmatrix} 0 & -5 \\ -4 & -13 \end{bmatrix}$

(d) $[\mathbf{w}]_{B_1} = \begin{bmatrix} -\frac{7}{10} \\ \frac{8}{5} \end{bmatrix}$, $[\mathbf{w}]_{B_2} = \begin{bmatrix} -4 \\ -9 \end{bmatrix}$

(e) $[\mathbf{w}]_{B_2} = \begin{bmatrix} -4 \\ -7 \end{bmatrix}$, $[\mathbf{w}]_{B_1} = \begin{bmatrix} -\frac{17}{10} \\ \frac{8}{5} \end{bmatrix}$

21. (a) $\begin{bmatrix} \frac{3}{4} & \frac{3}{4} & \frac{1}{12} \\ -\frac{3}{4} & -\frac{17}{12} & -\frac{17}{12} \\ 0 & \frac{2}{3} & \frac{2}{3} \end{bmatrix}$

(b) $[\mathbf{w}]_{B_1} = \begin{bmatrix} 1 \\ 1 \\ 1 \end{bmatrix}$, $[\mathbf{w}]_{B_2} = \begin{bmatrix} \frac{19}{12} \\ -\frac{43}{12} \\ \frac{4}{3} \end{bmatrix}$

23. $P_{B_1 \to B_2} = \begin{bmatrix} \frac{2}{\sqrt{6}} & \frac{1}{2} & -\frac{1}{2\sqrt{3}} \\ \frac{1}{3} & 0 & \frac{4}{3\sqrt{2}} \\ -\frac{2}{3\sqrt{2}} & \frac{3}{2\sqrt{3}} & \frac{1}{6} \end{bmatrix}$,

$P_{B_2 \to B_1} = \begin{bmatrix} \frac{2}{\sqrt{6}} & \frac{1}{3} & -\frac{2}{3\sqrt{2}} \\ \frac{1}{2} & 0 & \frac{3}{2\sqrt{3}} \\ -\frac{1}{2\sqrt{3}} & \frac{4}{3\sqrt{2}} & \frac{1}{6} \end{bmatrix}$

25. (a) $P_{B \to S} = \begin{bmatrix} 0 & 1 \\ 1 & 0 \end{bmatrix}$

(b) $P^T = P_{B \to S}^T = \begin{bmatrix} 0 & 1 \\ 1 & 0 \end{bmatrix} = P_{S \to B}$

27. (a) $x' = -\dfrac{x}{\sqrt{2}} + \dfrac{y}{\sqrt{2}}$, $y' = -\dfrac{x}{\sqrt{2}} - \dfrac{y}{\sqrt{2}}$

xy-coordinates of $(-2, 6) \Rightarrow x'y'$-coordinates of $(4\sqrt{2}, -2\sqrt{2})$

(b) $x = -\dfrac{x'}{\sqrt{2}} - \dfrac{y'}{\sqrt{2}}$, $y = \dfrac{x'}{\sqrt{2}} - \dfrac{y'}{\sqrt{2}}$

$x'y'$-coordinates of $(5, 2) \Rightarrow xy$-coordinates of $\left(-\frac{7}{2}\sqrt{2}, \frac{3}{2}\sqrt{2} \right)$

29. (a) $\left(\frac{1}{\sqrt{2}}, \frac{3}{\sqrt{2}}, 5 \right)$ **(b)** $\left(-\frac{5}{\sqrt{2}}, \frac{7}{\sqrt{2}}, -3 \right)$

31. $\begin{bmatrix} \frac{\sqrt{2}}{4} & \frac{\sqrt{6}}{4} & \frac{\sqrt{2}}{2} \\ -\frac{\sqrt{3}}{2} & \frac{1}{2} & 0 \\ -\frac{\sqrt{2}}{4} & -\frac{\sqrt{6}}{4} & \frac{\sqrt{2}}{2} \end{bmatrix}$

▶ **Exercise Set 8.1** (Page 453)

1. $[T]_B = \begin{bmatrix} 2 & -1 \\ 2 & 0 \end{bmatrix}$

3. $\begin{bmatrix} 1 & -1 \\ 1 & 1 \end{bmatrix} = \begin{bmatrix} 1 & -1 \\ 1 & 0 \end{bmatrix} \overset{T_B}{\begin{bmatrix} 2 & -1 \\ 2 & 0 \end{bmatrix}} \begin{bmatrix} 0 & 1 \\ -1 & 1 \end{bmatrix}$

5. $[T]_B = \begin{bmatrix} 1 & -\frac{3}{2} & \frac{1}{2} \\ -1 & \frac{1}{2} & \frac{1}{2} \\ 0 & \frac{1}{2} & -\frac{1}{2} \end{bmatrix}$

7. $\begin{bmatrix} 1 & -1 & 0 \\ -1 & 1 & 0 \\ 1 & 0 & -1 \end{bmatrix} = \begin{bmatrix} 1 & 0 & 1 \\ 0 & 1 & 1 \\ 1 & 1 & 0 \end{bmatrix} \begin{bmatrix} 1 & -\frac{3}{2} & \frac{1}{2} \\ -1 & \frac{1}{2} & \frac{1}{2} \\ 0 & \frac{1}{2} & -\frac{1}{2} \end{bmatrix}$

$\times \begin{bmatrix} \frac{1}{2} & -\frac{1}{2} & \frac{1}{2} \\ -\frac{1}{2} & \frac{1}{2} & \frac{1}{2} \\ \frac{1}{2} & \frac{1}{2} & -\frac{1}{2} \end{bmatrix}$

9. $[T]_B = \begin{bmatrix} -\frac{4}{11} & \frac{9}{11} \\ -\frac{25}{11} & \frac{26}{11} \end{bmatrix}$, $[T]_{B'} = \begin{bmatrix} \frac{3}{11} & -\frac{32}{11} \\ \frac{2}{11} & \frac{19}{11} \end{bmatrix}$

11. $\begin{bmatrix} -\frac{4}{11} & \frac{9}{11} \\ -\frac{25}{11} & \frac{26}{11} \end{bmatrix} = \begin{bmatrix} \frac{1}{2} & \frac{1}{2} \\ \frac{1}{2} & -\frac{1}{2} \end{bmatrix} \begin{bmatrix} \frac{3}{11} & -\frac{32}{11} \\ \frac{2}{11} & \frac{19}{11} \end{bmatrix} \begin{bmatrix} 1 & 1 \\ 1 & -1 \end{bmatrix}$

13. $[T] = \begin{bmatrix} 1 & -2 \\ 0 & 1 \end{bmatrix}$, $[T]_B = \begin{bmatrix} -\frac{4}{11} & \frac{9}{11} \\ -\frac{25}{11} & \frac{26}{11} \end{bmatrix}$

15. (a) $[T(x)]_B = \begin{bmatrix} \frac{21}{5} \\ \frac{13}{5} \end{bmatrix}$, $[T]_B = \begin{bmatrix} -\frac{14}{5} & -\frac{7}{5} \\ \frac{3}{5} & \frac{14}{5} \end{bmatrix}$,

$[x]_B = \begin{bmatrix} -\frac{11}{5} \\ \frac{7}{5} \end{bmatrix}$

17. $[T]_{B'B} = \begin{bmatrix} 0 & 0 \\ -\frac{1}{2} & 1 \\ \frac{8}{3} & \frac{4}{3} \end{bmatrix}$

19. $[T]_{B',B} = \begin{bmatrix} -\frac{11}{14} & -\frac{9}{7} & -\frac{9}{14} \\ \frac{9}{14} & \frac{8}{7} & \frac{1}{14} \end{bmatrix}$

21. **(a)** $[T(\mathbf{v}_1)]_B = \begin{bmatrix} 1 \\ -2 \end{bmatrix}$, $[T(\mathbf{v}_2)]_B = \begin{bmatrix} 3 \\ 5 \end{bmatrix}$

  (b) $T(\mathbf{v}_1) = \begin{bmatrix} -5 \\ 0 \end{bmatrix}$, $T(\mathbf{v}_2) = \begin{bmatrix} 7 \\ 11 \end{bmatrix}$

  (c) $T(x_1, x_2) = \left(\frac{1}{5}(19x_1 - 3x_2), \frac{1}{5}(22x_1 + 11x_2)\right)$

  (d) $T(1, 1) = \left(\frac{16}{5}, \frac{33}{5}\right)$

23. $[T] = \begin{bmatrix} 1 & 0 \\ 0 & 1 \end{bmatrix}$, $[T]_B = \begin{bmatrix} 1 & 0 \\ 0 & 1 \end{bmatrix}$, $[T]_{B'} = \begin{bmatrix} 1 & 0 \\ 0 & 1 \end{bmatrix}$,

  $[T]_{B',B} = \begin{bmatrix} \frac{12}{17} & \frac{5}{17} \\ \frac{11}{51} & -\frac{11}{51} \end{bmatrix}$

27. $T = \begin{bmatrix} 0 & 0 & 1 \\ 1 & 0 & 0 \\ 0 & 1 & 0 \end{bmatrix}$

29. $[T]_B = \begin{bmatrix} -4 & 0 \\ 0 & 6 \end{bmatrix}$. Rotate counterclockwise through angle $\pi/4$ to new $x'y'$-coordinates. Then stretch by 4 units in the x'-direction, reflect about the y'-axis, and stretch by 6 units in the y'-direction. Finally, rotate clockwise through angle $\pi/4$.

▶ **Exercise Set 8.2 (Page 466)** ─────────────

1. Possible reason: Determinants are different.

3. Possible reason: Ranks are different.

5. size: 5×5;

Eigenvalue	Multiplicity	Possible Dimensions of the Eigenspace
0	1	1
1	2	1 or 2
−1	2	1 or 2

7. Eigenvalues are $\lambda = 1$ with algebraic multiplicity 2 and geometric multiplicity 1 and $\lambda = 2$ with algebraic multiplicity 1 and geometric multiplicity 1.

9. Eigenvalues are $\lambda = 3$ with algebraic multiplicity 1 and geometric multiplicity 1 and $\lambda = 5$ with algebraic multiplicity 2 and geometric multiplicity 1.

11. Eigenvalues are $\lambda = 0$ with geometric multiplicity 2 and $\lambda = -3$ with geometric multiplicity 1.

13. $\text{rank}(3I - A) = 1$, $\text{rank}(5I - A) = 2$

15. $P = \begin{bmatrix} 4 & 3 \\ 5 & 4 \end{bmatrix}$; $P^{-1}AP = \begin{bmatrix} 1 & 0 \\ 0 & 2 \end{bmatrix}$

17. $P = \begin{bmatrix} 0 & 1 & 0 \\ -1 & 0 & 1 \\ 1 & 0 & 1 \end{bmatrix}$; $P^{-1}AP = \begin{bmatrix} 0 & 0 & 0 \\ 0 & 1 & 0 \\ 0 & 0 & 2 \end{bmatrix}$

19. A is diagonalizable. $P = \begin{bmatrix} 1 & 1 & 1 \\ 1 & 3 & \frac{3}{2} \\ 1 & 4 & \frac{3}{2} \end{bmatrix}$

21. A is not diagonalizable.

23. A is not diagonalizable.

29. Eigenvalues are $\lambda = 0$ with multiplicity 1 and $\lambda = -3$ with multiplicity 2.

▶ **Exercise Set 8.3 (Page 479)** ─────────────

1. Eigenvalues: 0, 5; both eigenspaces have dimension 1.

3. Eigenvalues: 0, 3; corresponding eigenspaces have dimensions 2 and 1.

7. $P = \begin{bmatrix} -\frac{1}{\sqrt{2}} & \frac{1}{\sqrt{2}} \\ \frac{1}{\sqrt{2}} & \frac{1}{\sqrt{2}} \end{bmatrix}$, $D = \begin{bmatrix} 2 & 0 \\ 0 & 4 \end{bmatrix}$

9. $P = \begin{bmatrix} \frac{1}{\sqrt{6}} & \frac{1}{\sqrt{30}} & -\frac{2}{\sqrt{5}} \\ \frac{1}{\sqrt{6}} & -\frac{5}{\sqrt{30}} & 0 \\ \frac{2}{\sqrt{6}} & \frac{2}{\sqrt{30}} & \frac{1}{\sqrt{5}} \end{bmatrix}$, $D = \begin{bmatrix} 2 & 0 & 0 \\ 0 & -4 & 0 \\ 0 & 0 & -4 \end{bmatrix}$

11. $P = \begin{bmatrix} 0 & -\frac{1}{\sqrt{2}} & \frac{1}{\sqrt{2}} \\ 0 & \frac{1}{\sqrt{2}} & \frac{1}{\sqrt{2}} \\ 1 & 0 & 0 \end{bmatrix}$, $D = \begin{bmatrix} 0 & 0 & 0 \\ 0 & 0 & 0 \\ 0 & 0 & 2 \end{bmatrix}$

13. $P = \begin{bmatrix} 0 & 0 & -\frac{1}{\sqrt{2}} & \frac{1}{\sqrt{2}} \\ 0 & 0 & \frac{1}{\sqrt{2}} & \frac{1}{\sqrt{2}} \\ 0 & 1 & 0 & 0 \\ 1 & 0 & 0 & 0 \end{bmatrix}$, $D = \begin{bmatrix} 0 & 0 & 0 & 0 \\ 0 & 0 & 0 & 0 \\ 0 & 0 & 2 & 0 \\ 0 & 0 & 0 & 4 \end{bmatrix}$

15. $\begin{bmatrix} 3 & 1 \\ 1 & 3 \end{bmatrix} = 2\begin{bmatrix} \frac{1}{2} & -\frac{1}{2} \\ -\frac{1}{2} & \frac{1}{2} \end{bmatrix} + 4\begin{bmatrix} \frac{1}{2} & \frac{1}{2} \\ \frac{1}{2} & \frac{1}{2} \end{bmatrix}$

17. $A = 2\begin{bmatrix} \frac{1}{\sqrt{6}} \\ \frac{1}{\sqrt{6}} \\ \frac{2}{\sqrt{6}} \end{bmatrix}\begin{bmatrix} \frac{1}{\sqrt{6}} & \frac{1}{\sqrt{6}} & \frac{2}{\sqrt{6}} \end{bmatrix} -$

  $4\begin{bmatrix} \frac{1}{\sqrt{30}} \\ -\frac{5}{\sqrt{30}} \\ \frac{2}{\sqrt{30}} \end{bmatrix}\begin{bmatrix} \frac{1}{\sqrt{30}} & -\frac{5}{\sqrt{30}} & \frac{2}{\sqrt{30}} \end{bmatrix} - 4\begin{bmatrix} -\frac{2}{\sqrt{5}} \\ 0 \\ \frac{1}{\sqrt{5}} \end{bmatrix}\begin{bmatrix} -\frac{2}{\sqrt{5}} & 0 & \frac{1}{\sqrt{5}} \end{bmatrix}$

19. $\begin{bmatrix} 3070 & 3069 \\ -2046 & -2045 \end{bmatrix}$ **21.** $\begin{bmatrix} 1 & 0 & 0 \\ 0 & 1 & 0 \\ 0 & 0 & 1 \end{bmatrix}$

23. **(b)** $A^4 = 24A^2 - 64A + 48I$ **(c)** $A^{-1} = \frac{1}{8}A^2 - \frac{3}{4}A + \frac{3}{2}I$

25. $e^{tA} = \frac{1}{2}\begin{bmatrix} e^{2t} + e^{4t} & -e^{2t} + e^{4t} \\ -e^{2t} + e^{4t} & e^{2t} + e^{4t} \end{bmatrix}$

27. $e^{tA} = \frac{1}{6}\begin{bmatrix} 5e^{-4t} + e^{-2t} & -e^{-4t} + e^{-2t} & -2e^{-4t} + 2e^{-2t} \\ -e^{-4t} + e^{-2t} & 5e^{-4t} + e^{-2t} & -2e^{-4t} + 2e^{-2t} \\ -2e^{-4t} + 2e^{-2t} & -2e^{-4t} + 2e^{-2t} & 2e^{-4t} + 4e^{-2t} \end{bmatrix}$

29. $\sin(\pi A) = \begin{bmatrix} 0 & 0 & 0 \\ 0 & 0 & 0 \\ 0 & 0 & 0 \end{bmatrix}$ **31. (b)** $e^A = \begin{bmatrix} 1 & 0 & 0 \\ 1 & 1 & 0 \\ \frac{5}{2} & 1 & 1 \end{bmatrix}$

▶ **Exercise Set 8.4** (Page 492) ─────────

1. (a) $\begin{bmatrix} x_1 & x_2 \end{bmatrix} \begin{bmatrix} 3 & 0 \\ 0 & 7 \end{bmatrix} \begin{bmatrix} x_1 \\ x_2 \end{bmatrix}$ **(b)** $\begin{bmatrix} x_1 & x_2 \end{bmatrix} \begin{bmatrix} 4 & -3 \\ -3 & 9 \end{bmatrix} \begin{bmatrix} x_1 \\ x_2 \end{bmatrix}$

(c) $\begin{bmatrix} x_1 & x_2 & x_3 \end{bmatrix} \begin{bmatrix} 9 & 3 & -4 \\ 3 & -1 & \frac{1}{2} \\ -4 & \frac{1}{2} & 4 \end{bmatrix} \begin{bmatrix} x_1 \\ x_2 \\ x_3 \end{bmatrix}$

3. $2x^2 + 5y^2 - 6xy$

5. $\begin{bmatrix} y_1 \\ y_2 \end{bmatrix} = \begin{bmatrix} \frac{1}{\sqrt{2}} & -\frac{1}{\sqrt{2}} \\ \frac{1}{\sqrt{2}} & \frac{1}{\sqrt{2}} \end{bmatrix} \begin{bmatrix} x_1 \\ x_2 \end{bmatrix}$; $Q = y_1^2 + 3y_2^2$

7. $\begin{bmatrix} y_1 \\ y_2 \\ y_3 \end{bmatrix} = \begin{bmatrix} -\frac{2}{3} & \frac{2}{3} & \frac{1}{3} \\ \frac{2}{3} & \frac{1}{3} & \frac{2}{3} \\ \frac{1}{3} & \frac{2}{3} & -\frac{2}{3} \end{bmatrix}$; $Q = y_1^2 + 4y_2^2 + 7y_3^2$

9. (a) $\begin{bmatrix} x & y \end{bmatrix} \begin{bmatrix} 2 & \frac{1}{2} \\ \frac{1}{2} & 0 \end{bmatrix} \begin{bmatrix} x \\ y \end{bmatrix} + \begin{bmatrix} 1 & -6 \end{bmatrix} \begin{bmatrix} x \\ y \end{bmatrix} + 2 = 0$

(b) $\begin{bmatrix} x & y \end{bmatrix} \begin{bmatrix} 0 & 0 \\ 0 & 1 \end{bmatrix} \begin{bmatrix} x \\ y \end{bmatrix} + \begin{bmatrix} 7 & -8 \end{bmatrix} \begin{bmatrix} x \\ y \end{bmatrix} - 5 = 0$

11. (a) ellipse **(b)** hyperbola **(c)** parabola **(d)** circle

13. hyperbola: $2(y')^2 - 3(x')^2 = 8$; $\theta \approx -63.4°$

15. hyperbola: $4(x')^2 - (y')^2 = 3$; $\theta = -53.13°$

17. (a) positive definite **(b)** negative definite **(c)** indefinite **(d)** positive semidefinite **(e)** negative semidefinite

19. positive definite **21.** positive semidefinite

23. indefinite

27. (a) $B = \begin{bmatrix} \frac{\sqrt{3}}{2} + \frac{\sqrt{7}}{2} & \frac{\sqrt{3}}{2} - \frac{\sqrt{7}}{2} \\ \frac{\sqrt{3}}{2} - \frac{\sqrt{7}}{2} & \frac{\sqrt{3}}{2} + \frac{\sqrt{7}}{2} \end{bmatrix}$

(b) $B = \begin{bmatrix} \frac{1}{2} + \frac{\sqrt{3}}{2} & \frac{1}{2} - \frac{\sqrt{3}}{2} & 0 \\ \frac{1}{2} - \frac{\sqrt{3}}{2} & \frac{1}{2} + \frac{\sqrt{3}}{2} & 0 \\ 0 & 0 & \sqrt{5} \end{bmatrix}$

29. $k > 2$

31. (a) possible answers: $B = \frac{1}{2} \begin{bmatrix} \sqrt{3} + \sqrt{15} & -\sqrt{3} + \sqrt{15} \\ -\sqrt{3} + \sqrt{15} & \sqrt{3} + \sqrt{15} \end{bmatrix}$,

$B = \frac{1}{2} \begin{bmatrix} -\sqrt{3} + \sqrt{15} & \sqrt{3} + \sqrt{15} \\ \sqrt{3} + \sqrt{15} & -\sqrt{3} + \sqrt{15} \end{bmatrix}$ **(b)** $C = \begin{bmatrix} 3 & 2 \\ 0 & \sqrt{5} \end{bmatrix}$

33. The ij entry on A is $c_i c_j$.

35. (b) It must be positive definite.

▶ **Exercise Set 8.5** (Page 501) ─────────

3. critical points: $(-1, 1)$, relative maximum; $(0, 0)$, saddle point

5. critical points: $(0, 0)$, relative minimum; $(2, 1)$ and $(-2, 1)$, saddle points

7. minimum: $z = -1$ at $(0, 1)$ and $(0, -1)$; maximum: $z = 5$ at $(1, 0)$ and $(-1, 0)$

9. minimum: $z = 3$ at $(1, 0)$ and $(-1, 0)$; maximum: $z = 7$ at $(0, 1)$ and $(0, -1)$

11. minimum: $w = 3$ at $(0, 0, 1)$ and $(0, 0, -1)$; maximum: $w = 9$ at $(1, 0, 0)$ and $(-1, 0, 0)$

13. minimum: $z = -\sqrt{2}$; maximum: $z = \sqrt{2}$ **15.**

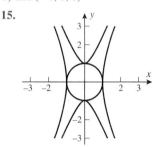

17. corner points: $x = \frac{5}{\sqrt{2}}, y = \frac{1}{\sqrt{2}}$

▶ **Exercise Set 8.6** (Page 516) ─────────

1. $\sqrt{5}$ **3.** $\sqrt{5}$

5. $A = \begin{bmatrix} \frac{1}{\sqrt{2}} & -\frac{1}{\sqrt{2}} \\ \frac{1}{\sqrt{2}} & \frac{1}{\sqrt{2}} \end{bmatrix} \begin{bmatrix} \sqrt{2} & 0 \\ 0 & \sqrt{2} \end{bmatrix} \begin{bmatrix} 1 & 0 \\ 0 & 1 \end{bmatrix}$

7. $A = \begin{bmatrix} \frac{2}{\sqrt{5}} & -\frac{1}{\sqrt{5}} \\ \frac{1}{\sqrt{5}} & \frac{2}{\sqrt{5}} \end{bmatrix} \begin{bmatrix} 8 & 0 \\ 0 & 2 \end{bmatrix} \begin{bmatrix} \frac{1}{\sqrt{5}} & \frac{2}{\sqrt{5}} \\ -\frac{2}{\sqrt{5}} & \frac{1}{\sqrt{5}} \end{bmatrix}$

9. $A = \begin{bmatrix} \frac{2}{3} & \frac{1}{\sqrt{2}} & \frac{\sqrt{2}}{6} \\ \frac{1}{3} & 0 & -\frac{2\sqrt{2}}{3} \\ -\frac{2}{3} & \frac{1}{\sqrt{2}} & -\frac{\sqrt{2}}{6} \end{bmatrix} \begin{bmatrix} 3\sqrt{2} & 0 \\ 0 & 0 \\ 0 & 0 \end{bmatrix} \begin{bmatrix} -\frac{1}{\sqrt{2}} & \frac{1}{\sqrt{2}} \\ \frac{1}{\sqrt{2}} & \frac{1}{\sqrt{2}} \end{bmatrix}$

11. $A = \begin{bmatrix} \frac{1}{\sqrt{3}} & 0 & \frac{2}{\sqrt{6}} \\ \frac{1}{\sqrt{3}} & \frac{1}{\sqrt{2}} & -\frac{1}{\sqrt{6}} \\ -\frac{1}{\sqrt{3}} & \frac{1}{\sqrt{2}} & \frac{1}{\sqrt{6}} \end{bmatrix} \begin{bmatrix} \sqrt{3} & 0 \\ 0 & \sqrt{2} \\ 0 & 0 \end{bmatrix} \begin{bmatrix} 0 & 1 \\ 1 & 0 \end{bmatrix}$

13. $A = \begin{bmatrix} \frac{34}{5} & \frac{12}{5} \\ \frac{12}{5} & \frac{16}{5} \end{bmatrix} \begin{bmatrix} \frac{4}{5} & \frac{3}{5} \\ -\frac{3}{5} & \frac{4}{5} \end{bmatrix}$

15. $A = \begin{bmatrix} \frac{1}{\sqrt{3}} & 0 \\ \frac{1}{\sqrt{3}} & \frac{1}{\sqrt{2}} \\ -\frac{1}{\sqrt{3}} & \frac{1}{\sqrt{2}} \end{bmatrix} \begin{bmatrix} \sqrt{3} & 0 \\ 0 & \sqrt{2} \end{bmatrix} \begin{bmatrix} 1 & 0 \\ 0 & 1 \end{bmatrix}$

$A = \sqrt{3} \begin{bmatrix} \frac{1}{\sqrt{3}} \\ \frac{1}{\sqrt{3}} \\ -\frac{1}{\sqrt{3}} \end{bmatrix} \begin{bmatrix} 1 & 0 \end{bmatrix} + \sqrt{2} \begin{bmatrix} 0 \\ \frac{1}{\sqrt{2}} \\ \frac{1}{\sqrt{2}} \end{bmatrix} \begin{bmatrix} 0 & 1 \end{bmatrix}$

17. $A = \begin{bmatrix} -\frac{1}{\sqrt{2}} & 0 & \frac{1}{\sqrt{2}} \\ \frac{1}{\sqrt{2}} & 0 & \frac{1}{\sqrt{2}} \\ 0 & 1 & 0 \end{bmatrix} \begin{bmatrix} 1 & 0 & 0 \\ 0 & 3 & 0 \\ 0 & 0 & 3 \end{bmatrix} \begin{bmatrix} \frac{1}{\sqrt{2}} & -\frac{1}{\sqrt{2}} & 0 \\ 0 & 0 & 1 \\ \frac{1}{\sqrt{2}} & \frac{1}{\sqrt{2}} & 0 \end{bmatrix}$

19. (a) basis for col(A): $\left\{ \left(\frac{1}{2}, \frac{1}{2}, \frac{1}{2}, \frac{1}{2}\right), \left(\frac{1}{2}, -\frac{1}{2}, -\frac{1}{2}, \frac{1}{2}\right) \right\}$

basis for null(A^T): $\left\{ \left(\frac{1}{2}, -\frac{1}{2}, \frac{1}{2}, -\frac{1}{2}\right), \left(\frac{1}{2}, \frac{1}{2}, -\frac{1}{2}, -\frac{1}{2}\right) \right\}$

basis for row(A): $\left\{ \left(\frac{2}{3}, -\frac{1}{3}, \frac{2}{3}\right), \left(\frac{2}{3}, \frac{2}{3}, -\frac{1}{3}\right) \right\}$

basis for null(A): $\left\{ \left(-\frac{1}{3}, \frac{2}{3}, \frac{2}{3}\right) \right\}$

(b) $\begin{bmatrix} 12 & 0 & 6 \\ 4 & -8 & 10 \\ 4 & -8 & 10 \\ 12 & 0 & 6 \end{bmatrix} = \begin{bmatrix} \frac{1}{2} & \frac{1}{2} \\ \frac{1}{2} & -\frac{1}{2} \\ \frac{1}{2} & -\frac{1}{2} \\ \frac{1}{2} & \frac{1}{2} \end{bmatrix} \begin{bmatrix} 24 & 0 \\ 0 & 12 \end{bmatrix} \begin{bmatrix} \frac{2}{3} & -\frac{1}{3} & \frac{2}{3} \\ \frac{2}{3} & \frac{2}{3} & -\frac{1}{3} \end{bmatrix}$

▶ **Exercise Set 8.7** (Page 523) ───────────

1. $A^+ = \begin{bmatrix} \frac{3}{25} & \frac{4}{25} \end{bmatrix}$ **3.** $A^+ = \begin{bmatrix} \frac{1}{6} & 0 & -\frac{1}{30} \\ -\frac{1}{6} & 0 & \frac{7}{30} \end{bmatrix}$

7. $A^+ = \begin{bmatrix} \frac{3}{25} & \frac{4}{25} \end{bmatrix}$ **9.** $A^+ = \begin{bmatrix} \frac{1}{6} & 0 & -\frac{1}{30} \\ -\frac{1}{6} & 0 & \frac{7}{30} \end{bmatrix}$

13. $A^+ = \begin{bmatrix} \frac{1}{4} & \frac{1}{4} \\ \frac{1}{4} & \frac{1}{4} \end{bmatrix}$ **15.** $\begin{bmatrix} \frac{1}{9} & \frac{2}{9} & \frac{2}{9} \\ \frac{2}{9} & \frac{17}{18} & -\frac{1}{18} \\ \frac{2}{9} & -\frac{1}{18} & \frac{17}{18} \end{bmatrix}$

17. $\begin{bmatrix} -\frac{1}{18} \\ -\frac{1}{18} \end{bmatrix}$ **19.** $\begin{bmatrix} \frac{1}{14} \\ \frac{2}{14} \\ \frac{3}{14} \end{bmatrix}$

▶ **Exercise Set 8.8** (Page 532) ───────────

1. $\bar{\mathbf{u}} = (2+i, -4i, 1-i)$, $\text{Re}(\mathbf{u}) = (2, 0, 1)$,
$\text{Im}(\mathbf{u}) = (-i, 4i, i)$, $\|\mathbf{u}\| = \sqrt{23}$

5. $\mathbf{x} = (7 - 6i, -4 - 8i, 6 - 12i)$

7. $\bar{A} = \begin{bmatrix} 5i & 4 \\ 2+i & 1-5i \end{bmatrix}$, $\text{Re}(A) = \begin{bmatrix} 0 & 4 \\ 2 & 1 \end{bmatrix}$,

$\text{Im}(A) = \begin{bmatrix} -5 & 0 \\ -i & 5 \end{bmatrix}$, $\det(A) = 17 - i$, $\text{tr}(A) = 1$

11. $\mathbf{u} \cdot \mathbf{v} = -1 + i$, $\mathbf{u} \cdot \mathbf{w} = 18 - 7i$, $\mathbf{v} \cdot \mathbf{w} = 12 + 6i$

13. $-11 - 14i$

15. $\lambda_1 = 2 - i, \mathbf{x}_1 = \begin{bmatrix} 2 - i \\ 1 \end{bmatrix}$; $\lambda_2 = 2 + i, \mathbf{x}_1 = \begin{bmatrix} 2 + i \\ 1 \end{bmatrix}$

17. $\lambda_1 = 4 - i, \mathbf{x}_1 = \begin{bmatrix} 1 - i \\ 1 \end{bmatrix}$; $\lambda_2 = 4 + i, \mathbf{x}_1 = \begin{bmatrix} 1 + i \\ 1 \end{bmatrix}$

19. $|\lambda| = \sqrt{2}, \phi = \pi/4$ **21.** $|\lambda| = 2, \phi = -\pi/3$

23. $P = \begin{bmatrix} 5 & 0 \\ -4 & 2 \end{bmatrix}$, $C = \begin{bmatrix} 3 & -2 \\ 2 & 3 \end{bmatrix}$

25. $P = \begin{bmatrix} -1 & -1 \\ 1 & 0 \end{bmatrix}$, $C = \begin{bmatrix} 5 & -3 \\ 3 & 5 \end{bmatrix}$

27. (a) $k = -\frac{8}{3}i$ **(b)** none

▶ **Exercise Set 8.9** (Page 539) ───────────

1. $A^* = \begin{bmatrix} -2i & 4 & 5-i \\ 1+i & 3-i & 0 \end{bmatrix}$

3. $A = \begin{bmatrix} 1 & i & 2-3i \\ -i & -3 & 1 \\ 2+3i & 1 & 2 \end{bmatrix}$

5. (a) $a_{13} \neq \bar{a}_{31}$ **(b)** $a_{22} \neq \bar{a}_{22}$

9. $A^* = A^{-1} = \begin{bmatrix} \frac{3}{5} & -\frac{4}{5} \\ -\frac{4}{5}i & -\frac{3}{5}i \end{bmatrix}$

11. $A^* = A^{-1} = \begin{bmatrix} \frac{-i+\sqrt{3}}{2\sqrt{2}} & \frac{1-i\sqrt{3}}{2\sqrt{2}} \\ \frac{1+i\sqrt{3}}{2\sqrt{2}} & \frac{-i-\sqrt{3}}{2\sqrt{2}} \end{bmatrix}$

13. $P = \begin{bmatrix} \frac{-1+i}{\sqrt{3}} & \frac{1-i}{\sqrt{6}} \\ \frac{1}{\sqrt{3}} & \frac{2}{\sqrt{6}} \end{bmatrix}$, $D = \begin{bmatrix} 3 & 0 \\ 0 & 6 \end{bmatrix}$

15. $P = \begin{bmatrix} \frac{-1-i}{\sqrt{6}} & \frac{1+i}{\sqrt{3}} \\ \frac{2}{\sqrt{6}} & \frac{1}{\sqrt{3}} \end{bmatrix}$, $D = \begin{bmatrix} 2 & 0 \\ 0 & 8 \end{bmatrix}$

17. $P = \begin{bmatrix} 0 & 0 & 1 \\ \frac{2}{\sqrt{6}} & \frac{-1+i}{\sqrt{6}} & 0 \\ \frac{1+i}{\sqrt{6}} & \frac{2}{\sqrt{6}} & 0 \end{bmatrix}$, $D = \begin{bmatrix} -2 & 0 & 0 \\ 0 & 1 & 0 \\ 0 & 0 & 5 \end{bmatrix}$

19. $A = \begin{bmatrix} 0 & i & 2-3i \\ i & 0 & 1 \\ -2-3i & -1 & 4i \end{bmatrix}$

21. (a) $a_{13} \neq -(\overline{a_{31}})$ **(b)** $a_{11} \neq -\overline{a_{11}}$

29. (c) B and C must commute.

▶ **Exercise Set 8.10** (Page 552) ───────────

1. $y = 5e^{-3t}$ **3.** general solution: $\begin{bmatrix} y_1 \\ y_2 \\ y_3 \end{bmatrix} = \begin{bmatrix} c_1 e^t \\ c_2 e^{4t} \\ c_3 e^{-2t} \end{bmatrix}$

fundamental set of solutions: $\left\{ \begin{bmatrix} e^t \\ 0 \\ 0 \end{bmatrix}, \begin{bmatrix} 0 \\ e^{4t} \\ 0 \end{bmatrix}, \begin{bmatrix} 0 \\ 0 \\ e^{-2t} \end{bmatrix} \right\}$

solution of initial value problem: $\begin{bmatrix} y_1 \\ y_2 \\ y_3 \end{bmatrix} = \begin{bmatrix} e^t \\ e^{4t} \\ e^{-2t} \end{bmatrix}$

5. general solution: $\begin{bmatrix} y_1 \\ y_2 \end{bmatrix} = c_1 \begin{bmatrix} -2 \\ 1 \end{bmatrix} e^{-t} + c_2 \begin{bmatrix} 1 \\ 1 \end{bmatrix} e^{5t}$
for initial conditions: $c_1 = c_2 = 0$

7. general solution: $\begin{bmatrix} y_1 \\ y_2 \\ y_3 \end{bmatrix} = c_1 \begin{bmatrix} 0 \\ 1 \\ 0 \end{bmatrix} e^t + c_2 \begin{bmatrix} -1 \\ 2 \\ 2 \end{bmatrix} e^{2t} +$

$c_3 \begin{bmatrix} -1 \\ 1 \\ 1 \end{bmatrix} e^{3t}$; for initial conditions: $c_1 = 1, c_2 = -1, c_3 = 2$

9. $y_1(t) = \frac{25}{3} e^{-t} + \frac{65}{3} e^{-(1/2)t}$ **11.** $y_1 = -11e^{2t} + 14e^{3t}$
$y_2(t) = -25e^{-t} + 65e^{-(1/2)t}$ $y_2 = -11e^{2t} + 7e^{3t}$

13. $\begin{bmatrix} y_1 \\ y_2 \\ y_3 \end{bmatrix} = \frac{1}{5} \begin{bmatrix} e^{-2t} + 10e^{2t} - e^{3t} \\ -e^{-2t} + e^{3t} \\ 4e^{-2t} - 10e^{2t} + e^{3t} \end{bmatrix}$

15. $\begin{bmatrix} y_1 \\ y_2 \\ y_3 \end{bmatrix} = \begin{bmatrix} -1 + 8t - t^2 \\ 4 - 2t \\ 2 \end{bmatrix}$ **17.** $y_1 = -\cos(t) - 2\sin(t)$
$y_2 = 2\cos(t) - \sin(t)$

19. $y(t) = c_1 e^{(3+\sqrt{15})t} + c_2 e^{(3-\sqrt{15})t}$

21. $\begin{bmatrix} y_1 \\ y_2 \end{bmatrix} = \begin{bmatrix} 10e^{-(1/2)t} \\ -5e^{-(1/2)t} + 12e^{-(1/6)t} \end{bmatrix}$

▶ **Exercise Set 9.1** (Page 566) ───────────

1. (a) $\mathbf{u} + \mathbf{v} = (2, 6), 3\mathbf{u} = (-3, 0)$ **(c)** Axioms 1–5

3. (a) $\mathbf{u} + \mathbf{v} = (2, 2), 2\mathbf{u} = (0, 8)$
   (d) $-\mathbf{u} = (-u_1 - 2, -u_2 - 2)$ **(e)** Axioms 7 and 8

5. (a) subspace   **(b)** not a subspace   **(c)** subspace
   (d) not a subspace

7. (a) not a subspace   **(b)** subspace   **(c)** subspace
   (d) subspace

9. (a) subspace   **(b)** subspace   **(e)** not a subspace

11. (b) $\begin{bmatrix} 1 & 0 \\ 0 & 0 \end{bmatrix}, \begin{bmatrix} 0 & 0 \\ 0 & 1 \end{bmatrix}, \begin{bmatrix} 0 & 1 \\ 0 & 0 \end{bmatrix}$; dimension: 3

13. (a) $f_1(x) - f_2(x) + f_3(x) = 0$
   (b) $f_1(x) - \frac{1}{2}f_2(x) + \frac{1}{2}f_3(x) = 0$

15. $W(x) = -x\sin x - \cos x \neq 0$ for some x

17. $W(x) = 2e^{3x} \neq 0$ **19.** $A = A_1 + 2A_2 + 3A_3 + 4A_4$

21. $\mathbf{p} = \mathbf{p}_1 + 2\mathbf{p}_2 - \mathbf{p}_3$

23. $\mathbf{p}_1 = \frac{1}{2}x^2 - \frac{3}{2}x + 1, \mathbf{p}_2 = -x^2 + 2x,$
   $\mathbf{p}_3 = \frac{1}{2}x^2 - \frac{1}{2}x, \mathbf{p} = \mathbf{p}_1 + 3\mathbf{p}_2 + 7\mathbf{p}_3$

25. $\frac{11}{6}x^3 - \frac{37}{2}x^2 + \frac{173}{3}x - 55$

27. basis: $\begin{bmatrix} 1 & 1 \\ 1 & 0 \end{bmatrix}, \begin{bmatrix} 0 & -1 \\ -1 & 1 \end{bmatrix}$; dimension: 2

29. (a) 1   **(b)** 1   **31.** not a vector space

▶ **Exercise Set 9.2** (Page 580) ───────────

1. (a) -18 **(b)** $\|\mathbf{u}\| = \sqrt{30}, \|\mathbf{v}\| = \sqrt{35}$ **(c)** $-\frac{18}{\sqrt{30}\sqrt{35}}$
   (d) $\sqrt{101}$

5. (a) Axiom 4 fails.   **(b)** Axioms 1 and 4 fail.
   (c) Axiom 4 fails.

9. $\frac{1}{\sqrt{6}}(1, 1, 1), \frac{1}{\sqrt{6}}(1, -1, 1), \frac{1}{\sqrt{6}}(2, 0, -1)$

11.

13. $\langle \mathbf{u}, \mathbf{v} \rangle = \frac{1}{25}u_1v_1 + \frac{1}{16}u_2v_2$

15. (a) $\frac{1}{4}$ **(b)** $\frac{\sqrt{3}}{3}, \frac{\sqrt{5}}{5}$ **(c)** $\frac{\sqrt{15}}{4}$ **(d)** $\frac{\sqrt{30}}{30}$

17. (c) $x - \frac{1}{6}$

21. $\frac{e^{2\pi} - 1}{\pi} \left(\frac{1}{2} + \frac{1}{2}\cos x - \frac{1}{2}\sin x + \frac{1}{5}\cos 2x - \frac{2}{5}\sin 2x \right)$

23. $3\pi - \sum_{k=1}^{n} \frac{6\sin kx}{k}$

29. (b) $13, \sqrt{11}, \sqrt{29}, \frac{13}{\sqrt{11}\sqrt{29}}, \sqrt{14}$

▶ **Exercise Set 9.3** (Page 592) ───────────

1. (a) $(-2, 16)$ **(b)** $\left(-\frac{1}{2}, \frac{9}{2}\right), \left(\frac{5}{2}, -\frac{1}{2}\right)$

15. linear   **17.** not linear

19. (a) $q_2(x)$ and $q_3(x)$ **(b)** $q_1(x)$ and $q_3(x)$

21. basis for kernel: $\begin{bmatrix} 0 & 1 \\ 0 & 0 \end{bmatrix}, \begin{bmatrix} 0 & 0 \\ 1 & 0 \end{bmatrix}, \begin{bmatrix} 0 & 0 \\ 0 & 1 \end{bmatrix}$
   basis for range: $\begin{bmatrix} 1 & 0 \\ 0 & 1 \end{bmatrix}$

23. onto   **25.** not onto   **27. (a)** $1, x$ **(b)** $a_0 + a_1x$

29. (b) $a_0 e^{-\omega x} + a_1 e^{\omega x}$ **31.** $1 + x + x^2$

33. kernel: multiples of $(1, 0, 0, 0, \ldots)$; range: R^∞

37. (a) $\begin{bmatrix} -1 & 2 \\ 0 & 3 \end{bmatrix}$ **(b)** $(1, 0, 0, -1), (0, 1, 1, 0)$

   (c) $\begin{bmatrix} 1 & 0 & 0 & 0 \\ 0 & 0 & 1 & 0 \\ 0 & 1 & 0 & 0 \\ 0 & 0 & 0 & 1 \end{bmatrix}$

39. $\begin{bmatrix} 0 & 0 & 0 \\ 1 & 0 & 0 \\ 0 & \frac{1}{2} & 0 \\ 0 & 0 & \frac{1}{3} \end{bmatrix}, x + \frac{x^2}{2} + \frac{x^3}{3}$

Chapter 1 Openers from left to right: ©PhotoDisc, ©PhotoDisc, Courtesy NOAA, ©PhotoDisc, ©Corbis Digital Stock. Pages 3, 7, 8, 12, 19, and 23: Courtesy Electronic Publishing Services, Inc., New York City.

Chapter 2 Opener from left to right: ©Corbis Digital Stock, ©Corbis Digital Stock, Courtesy the National Institute of Health, ©Corbis Digital Stock, ©Corbis Digital Stock. Pages 43, 54, and 70: Courtesy Electronic Publishing Services, Inc., New York City.

Chapter 3 Opener from left to right: ©Corbis Digital Stock, Courtesy Goldwave Inc. (www.goldwave.com). Pages 81, 82, 128, and 155: Courtesy Electronic Publishing Services, Inc., New York City. Page 147: ©Royal Society. Page 169: Carolina Biological Supply Company/Phototake.

Chapter 4 Opener: ©PhotoDisc. Pages 181, 197, 200, and 212: Courtesy Electronic Publishing Services, Inc., New York City.

Chapter 5 Openers from left to right: ©PhotoDisc, ©Corbis Digital Stock, ©PhotoDisc, ©Corbis Digital Stock, ©PhotoDisc. Pages 228, 236, 243, and 253: Courtesy Electronic Publishing Services, Inc., New York City.

Chapter 6 Openers: ©PhotoDisc. Page 288: UPI Bettman.

Chapter 7 Openers from left to right: Courtesy NASA, H. Ford (JHU), G. Illingworth (USCS/LO), M. Clampin (STScl), G. Hartig (STScl), the ACS SCIENCE TEAM, and ESA; Courtesy Tamara Munzner and Paul Burchard, Visualizing the Structure of the World Wide Web in 3D Hyperbolic Space, Proc VRML '95, http://graphics.stanford.edu/papers/webviz. Pages 332, 342, 412, and 422: Courtesy Electronic Publishing Services, Inc., New York City. Page 367: Courtesy Chris Brislawn, Los Alamos National Laboratory. Page 400: Courtesy NASA.

Chapter 8 Openers from left to right: ©Corbis Digital Stock, ©PhotoDisc, ©Eyewire, ©Eyewire. Pages 478, 492, 505, 513, and 529: Courtesy Electronic Publishing Services, Inc., New York City.

Chapter 9 Openers from left to right: J. K. Nakata/U.S. Geological Survey/NASA, ©Eyewire, Image was generated from simulations carried out by the Aerospace Engineering and Mechanics Group at the Army HPC Research Center, Minneapolis, MN, ©PhotoDisc, ©PhotoDisc. Pages 563, 566, 574, and 577: Courtesy Electronic Publishing Services, Inc., New York City. Page 589: Courtesy Dr. Joseph Atick, Dr. Norman Redlich, and Dr. Paul Griffith.

A

absolute value, 525
  of a complex number, A5
addition, *see also* sum
  closed under, 123–124
  of matrices, 80, 94
  in vector spaces, 555
  of vectors, 2–4
additivity, 269
adjacency matrices, 12, 256
adjoint (of a matrix), 196–197
Adobe Systems, 319
affine spaces, 131
algebraic multiplicity, 215–217, 458, 465
algorithms, unstable, 58
alleles, 234
alternation, 250
amperes (amps), 68
angle (between vectors), 20–21
angle-preserving operators, 280
anticommutative matrices, 533
antisymmetry property, 527
Argand, Jean Robert, 8
argument, 525
  of a complex number, A6
Aristotle, 3
arithmetic
  exact, 61
  finite-precision, 61
arithmetic mean, 28
associated homogeneous linear systems, 135–136
associative law
  for addition, 94
  for multiplication, 95
  for vector addition, 10
authorities, 257
authority weights, 257
axis(-es)
  of ellipsoid, 494
  major vs. minor, 485
  principal, 486
  of rotation, 289–290

B

back substitution, 55–56
backward phase (Gauss–Jordan elimination), 53
balanced chemical reactions, 70
basis(-es)
  changing, for R^n, 431–433
  coordinates with respect to, 428–438

existence of, 331
finding, by row reduction, 346–347
for fundamental spaces of a matrix, 374–375
matrix of linear operator with respect to, 443–446
matrix of linear representation with respect to pair of, 450–452
orthonormal, 407–414, 574–575
and pivot theorem, 370–374
properties of, 335–337
representing linear operators with two, 452–453
standard, 330
for subspaces, 329–333
transition matrix between, 446–450
Bateman, Harry, 505
Bellavitis, Giusto, 3
Beltrami, Eugenio, 513
best approximation, 575–578
best mean square approximation, 576
bilinear form, 578
block diagonal matrices, 168
block lower triangular matrices, 169
block multiplication, 166–167
block triangular form, 194
block upper triangular matrices, 168–170
blocks, 166
Bôcher, Maxime, 43, 128
bound vectors, 2, 3
Boyle, James, 156
branches, 68
Brin, Sergey, 256
Bunch, J. R., 156
Bunyakovsky, Viktor Yakovlevich, 23

C

C^n, vectors in, 525–532
  complex eigenvalues of real matrices, 530–532
  complex Euclidean inner product of, 526, 527
  definition of, 525
  properties of, 526
  real eigenvalues of symmetric matrices, 529–530
  vector spaces, 527
cancellation law, 97
Carroll, Lewis, 181
Cauchy, Augustin, 23, 189
Cauchy–Schwarz inequality, 23–24, 573
Cayley, Arthur, 81, 99, 101

Cayley–Hamilton theorem, 223, 474–475
center of scaling, 327
central ellipsoid, 490
  in standard position, 493–494
central quadrics, 485
change of variable, 482–483
characteristic equations, 212
characteristic polynomials, 216
check digits, 19
checkerboard matrices, 224
chemical equations, 70–72
chemical formulas, 70
chess, 203
Cholesky, André-Louis, 492
Cholesky factorization, 492
circle(s), 484
  unit, 571
circuits, electrical, 68–70
closed economies, 235
closed loops, 68
closed networks, 65–66
closed under addition, 123–124, 555
closed under linear combinations, 124
closed under scalar multiplication, 123, 124, 555
CMYK color model, 133
CMYK space, 133
codomain, 267
coefficient matrix(-ces), 82, 115, 543
  linear systems with triangular, 144
  solving multiple linear systems with common, 118
coefficients (in linear combination), 11
cofactor expansions, 180–182, 187
cofactors, 179–180
collinear vectors, 10
column operations, elementary, 185–186
column rule for matrix multiplication, 87
column vectors, 12, 81
column-row expansion, 376–377
column-row factorization, 375
column-row rule, 167
column-vector form (vector notation), 11
comma-delimited form (vector notation), 11
commutative law
  for addition, 94
  for multiplication, 95
companion matrices, 223
complete linear factorization, 216
complete reactions, 70
complex conjugate, 525, 526, A4

complex dot product, 526
complex eigenvalues, 215–217, 528–532
  geometric interpretation of, 530–532
complex eigenvectors, 528
complex Euclidean inner product,
    526–527
complex exponentials, A8
complex inner product, 581
complex inner product space, 581
complex n-spaces, 525
complex n-tuples, 525
complex number(s), 525, A3–A8
  argument of, A6
  division of, A7
  modulus of, A5
  polar form of, A6
  reciprocal, A5
complex number system, A4
complex plane, 525, A4–A5
complex vector spaces, 556
components, 5, 6–7
  of C^n, 525
composition(s)
  factoring linear operators into, 310–312
  of linear transformations, 305–314
  of three or more linear transformations,
      308–310
compression
  in the x-direction with factor k, 285,
      286
  in the y-direction with factor k, 285,
      286
computer graphics, 318–325
  data compression for, 513–514
  homogeneous coordinates, use of,
      320–322
  and matrix representations of
      wireframes, 318–320
  three-dimensional, 323–325
computer roundoff error, 403
condensation, 181
condition number, 522
conforming matrix sizes, 85
conic sections (conics), 484
  classification of, using eigenvalues,
      489–490
  identification of, 485–488
  types of, 484
conjugate, A4
conjugate transpose, 535
connections, one- vs. two-way, 12
connectivity graphs, 12
connectivity matrices, 318
consistency, 118–119, 138
  and full column rank, 387–389
  and rank, 362–364

consistent linear systems, 40
constant of proportionality, 269
constrained extremum problems, 497–500
constrained extremum theorem, 498
constraints, 498
consumption matrices, 236
consumption vectors, 236
contraction(s), 285, 583
contrapositive (of a theorem), A1
convergence, 245–246
    power method and rate of, 255
    of power sequences, 260
converse (of a theorem), A1
coordinate axes, rotation of, 437
coordinate maps, 435–436
coordinate planes, reflections about, 289
coordinate systems, vectors in, 5–6
coordinates, 5
    with respect to basis, 428–438
Cramer, Gabriel, 200
Cramer's rule, 199–201
critical points, 495
cross product terms, 481
cross products, 204–207
current, electrical, 68, 69
curves, level, 499–500

D
data compression, 513–514
decomposition
    eigenvalue, 502
    polar, 507–508
    Schur, 478
    singular value, 502–516
    spectral, 471–473
    upper Hessenberg, 479
degenerate conics, 484
degenerate parallelograms, 201–202
demand vectors
    intermediate, 237
    outside, 236
DeMoivre's formula, A8
Derive (computer algebra system), 61
determinant(s), 99–101
    of $A + B$, 190
    effect of elementary row/column
        operations on, 185–186
    and elementary products, 176–179
    evaluation difficulties for higher-order,
        178
    evaluation of, by LU-decomposition,
        189
    expressions for, in terms of
        eigenvalues, 220–221
    by Gaussian elimination, 187–188
    general, 178

geometric interpretation of, 201–202
    of inverse of a matrix, 189–190
    and invertibility, 188
    of matrices with rows or columns that
        have all zeros, 178–179
    and minors, 179–180
    $n \times n$ (nth-order), 178
    of product of matrices, 188–189
    properties of, 184–192
    of triangular matrices, 179
    of 2×2 and 3×3 matrices, 175–176
    in unifying theorem, 190–192
    Vandermonde, 202–204
diagonal, main, 509
diagonal matrices, 143–144, 460–461
diagonalizability, 461
    and unifying theorem on, 465
    unitary, 538–539
diagonalizable matrices
    exponential of, 475–477
    powers of, 473–474
diagonalizable operators, 467
diagonalization, 460–465
    definition of, 461
    and linear independence of
        eigenvectors, 463–465
    and linear systems, 477
    of a matrix, 462–463
    and nondiagonalization case, 478–479
    orthogonal, 469–473
    of symmetric matrix, 470–473
diagonalization problem, 461
Dickson, L. E., 197
difference (of matrices), 80, 94
differential equations, systems of,
    542–552
    eigenvalues and eigenvectors,
        solutions using, 545–549
    exponential form of solution, 549–550
    fundamental solutions of, 544–545
    linear systems of, 542–544
    with nondiagonalizable matrices,
        551–552
dilations, 285
dimension
    definition of, 140, 332–333
    of hyperplane, 333
    of solution space, 140, 333
    of vector spaces, 563–564
dimension theorem (for homogeneous
    linear systems), 58, 352
dimension theorem (for matrices),
    352–357
    consequences of, 353
    as dimension theorem for subspaces,
        354

and extension of linearly independent sets to basis, 352–353
and hyperplanes, 355
and rank 1 matrices, 355–357
and unifying theorem, 354
dimension theorem (for subspaces), 354, 355
Dirac matrices, 533
direct proportionality, 269
directed graphs, 12
direction, 1
of vectors, 10
discrete averaging model, 248
displacement, 1
distance, 570
distributive law
left, 95
right, 95
division (of complex numbers), A7
Dodgson, Charles, 181
domains, 265
dominant eigenvalue, 249–250
dominant eigenvectors, 249–250
Dongarra, J. J., 156
dot product, 18–20, 85–87
complex, 526
and transpose of a matrix, 106
dot product rule (row-column rule), 86
dot-product-preserving operators, 281
double perp theorem, 389–390
Dritz, Kenneth, 156
dynamical systems, 225–227

E

economy(-ies)
closed, 235
inputs/outputs of, 235–236
Leontief model of open, 236–237
open, 235–239
edges, 12
eigenheads, 589
eigenspaces, 212, 250–251, 528
eigenvalue decomposition (EVD), 471, 502
eigenvalues, 211–221
algebraic and geometric multiplicity of, 465
classification of conic sections using, 489–490
complex, 215–217, 528–532
definition of, 211
dominant, 249–250
expressions for determinant and trace in terms of, 220, 221
of Hermitian matrices, 536, 539
of a linear operator, 467

numerical methods for obtaining, 221
of powers of a matrix, 214
of similar matrices, 458–460
solutions to systems of linear equations using, 545–549
symmetric 2×2 matrices, analysis of, 218–220
of triangular matrices, 214
of 2×2 matrices, analysis of, 217–220
in unifying theorem, 215
eigenvectors, 211–213
complex, 528
and diagonalizability, 461–462
dominant, 249–250
linear independence of, 463–465
of a linear operator, 467
of similar matrices, 460
solutions to systems of linear equations, 545–549
Einstein, Albert, 7, 212, 574
Eisenstein, Gotthold, 82
electrical circuits, 68–70
electrical current, 68–69
electrical potential, 68
elementary matrices, 109–111
elementary products, 176–179
elementary row/column operations, 43–44, 185–186
ellipses, 484
ellipsoid, 490
axes of, 494
central, in standard position, 493–494
empty graphs, 484
entries (of a matrix), 12, 79
equal functions, 558
equal (equivalent) vectors, 2
equation(s)
characteristic, 212
homogeneous, 269
Leontief, 237
with no graph, 484
equilibrium, static, 14
equivalent (equal) vectors, 2
equivalent statements, A1
equivalent vectors, 8
error(s)
computer roundoff, 403
mean square, 576
percentage, 255
relative, 255
roundoff, 58
estimated percentage error, 255
estimated relative error, 255
Euclidean distance, 23
Euclidean inner product, *see* dot product
Euclidean norm, 23, 526

Euclidean n-spaces, 25
Euclidean scaling, 251–252
Euler, Leonhard, 12
Euler's formula, A8
evaluation transformation, 583–584
EVD, *see* eigenvalue decomposition
exact arithmetic, 61
expansion
in the x-direction with factor k, 285, 286
in the y-direction with factor k, 285, 286
exponential(s)
complex, A8
of a matrix, 475–477
extrapolated Gauss–Seidel iteration, 246
extrema
constrained extremum problems, 497–500
relative, of two variables, 495–497

F

factor theorem, 216
factorization
complete linear, 216
solving linear systems by, 154–155
Fibonacci matrices, 184
finite-dimensional vector spaces, 564
finite-precision arithmetic, 61
first-order equations, 542
fixed points, 148
trivial, 210–211
flats, 131
floating-point numbers, 160
flops, 160–163
flow (in networks), 65–66
force vector, 1
forward phase (Gauss–Jordan elimination), 53
forward substitution, 155
Fourier, Joseph, 576, 577
Fourier coefficients, 577
Fourier series, 578
free variables, 50
free vector, 2
function(s), 265–267
best approximation problem for, 575
equal, 558
minimum distance problem for, 576
sum of, 558
function spaces, 557–559
fundamental set of solutions, 545
fundamental solutions (of systems of differential equations), 544–545
fundamental spaces (of a matrix), 511
fundamental theorem of algebra, 216

G

Galileo Galilei, 3
Gauss, Carl Friedrich, 54, 176, 244
Gauss–Jordan elimination, 51–53, 242
  for solving linear systems, 118
  theory vs. implementation of, 58–59
Gauss–Seidel iteration, 244–245
  extrapolated, 246
Gaussian elimination, 53, 54, 187–188
  and LU-decomposition, 156, 158–159
  theory vs. implementation of, 58–59
general equation(s)
  of a line, 29
  of a plane, 31
general linear transformations, 582–588
general relativity theory, 7, 212
general solution
  of consistent linear system, 137
  of differential equations, 542
  of linear system, 51, 126
generalized point, 7
generalized vector, 7
genetics, 234
geometric mean, 28
geometric multiplicity, 458, 465
geometry, quadratic forms in, 484–488
Geršgorin's theorem, 529
Gibbs, J. Willard, 204
Global Positioning System (GPS), 63–65
Golub, Gene, 513
Google search engine, 256–259
GPS, *see* Global Positioning System
graph(s), 12
  directed, 12
  equations with no, 484
Grassmann, Hermann, 8
gravity, 7
Grissom, Gus, 309

H

Hamilton, William R., 8, 475
heat, 248
Hermitian matrices, 536–539
  definition of, 536
  eigenvalues of, 536, 539
  and normal matrices, 539
  skew-Hermitian matrices, 539
  unitary diagonalization of, 538–539
Hessenberg, Gerhard, 478, 479
Hessenberg's theorem, 478–479
Hessians (Hessian matrices), 496–497
higher-dimensional spaces, 7
Hilbert, David, 212
Hilbert matrix, 123
Hill, George William, 128

homogeneous coordinates, translation
  using, 320–322
homogeneous equations, 269
homogeneous first-order linear systems,
  543
homogeneous linear equations, 39
homogeneous linear systems, 56–58
  associated, 135–136
  dimension theorem for, 58
  and linear independence, 130–131
  solution space of, 125–127
Hooke, Robert, 270
Householder reflections, 420–426
  in applications, 426
  definition of, 422
  QR-decomposition using, 423–425
hub weights, 257
hubs, 257
Human Genome Project, 169
hyperbolas, 484
hyperplane(s), 138–139, 333
  distance from a point to a, 394
  reflections of R^n about the, 420, 421

I

identity matrix, 97–98
identity operators, 267, 452, 583
ill-conditioned systems, 61
images, 265
imaginary axis (of complex plane), A4
imaginary part (of complex number), 525,
  A3
inconsistent linear systems, 40
indefinite quadratic forms, 488
index of nilpotency, 109, 149
index of summation, 27
infinite-dimensional vector spaces, 212,
  564
initial approximation, 242
initial authority vectors, 257
initial conditions, 542, 543
initial hub vectors, 257
initial point (of vector), 2
initial value problems, 542, 543
inner product(s)
  algebraic properties of, 572–574
  axioms, 569
  complex, 581
  complex Euclidean, 526–527
  integral, 571–572
  least squares solution with respect to,
    570
  matrix, 89, 90
  matrix for the, 578–579
  on real vector space, 569
  weighted Euclidean, 570
inner product space(s)
  complex, 581

general, 572–574
  isomorphisms, 592
  real, 569–572
input-output analysis, 235–239
inputs, 235, 265
integral inner product, 571–572
intermediate demand vectors, 237
Internet searches, 256–260
interpolating polynomials, 72–74
Introduction to Higher Algebra (Maxime
  Bôcher), 128
inverse, 98–101
  algorithm for finding, 112–114
  determinant of, 188–189
  of elementary matrix, 111
  formula for, 197–199
  of linear transformations, 312–315
  of triangular matrices, 145
inverse operations, 110, 111
inverse power method, 260, 262
  shifted, 263
inverse transformations, 594
inversion, matrix
  algorithm for, 112–114
  solving linear systems by, 114–117
invertibile linear operators, 312–315
invertibility, 111–112
  determinant test for, 188
  of linear operators, 312–315
  of symmetric matrices, 147
invertible matrix, 98–102, 104–105
isomorphism(s), 589–591
  definition of, 589
  inner product space, 592
  natural, 590–591
italic fonts, 319
iterative methods, 241–246
  and convergence, 245–246
  Gauss–Seidel iteration, 244–245
  Jacobi iteration, 242–244

J

Jacobi, Karl Gustav Jacob, 243
Jacobi iteration, 242–244
Jordan, Camille, 503, 513
Jordan, Wilhelm, 54
Jordan canonical forms, 503
junctions, 65

K

Kasner, Edward, 256
kernel
  of linear transformation, 296–297,
    585–587
  of matrix transformation, 297–298

Kirchhoff's current law, 68
Kirchhoff's voltage law, 68
Knight's tour, 203
Kronecker's reduction, 484
kth principal submatrix, 490

L
Lagrange, Joseph-Louis, 566
Lagrange interpolating polynomials,
    564–566
Lagrange interpolation formula, 566
Lagrange's reduction, 484
landscape mode, 319
LDU-decomposition, 159–160
leading numbers, 48
leading variables, 50
least squares problem(s), 394–403
    and fitting a curve to experimental
        data, 399–401
    higher-degree polynomials, least
        squares fits by, 401–403
    linear systems, solution of, 394–396
    orthogonality property of least squares
        error vectors, 397–398
    pseudoinverse and solution of,
        521–522
    QR-decomposition for, 419–420
    Strang diagrams for solving, 398–399
    weighted, 570
least squares solution(s)
    with respect to inner product, 581
    weighted, 570
left distributive law, 95
left singular vectors, 509
left-handed coordinate systems, 5
Legendre, Adrien-Marie, 582
Legendre polynomials, 582
length, 1, 16
length-preserving operators, 280
Leontief equation, 237
Leontief input-output models, 235–239
Leontief matrices, 237
Leontief, Wassily, 235, 236
level curves, 499–500
line(s)
    general equation of a, 29
    parametric equations of, 29–30
    through two points, 30–31
    vector equation of, 29, 30
linear combination(s), 10–11, 128, 545
    matrix products as, 88
linear dependence, 128–131
linear equations, 39
    homogeneous, 39
    systems of, see linear system(s)

linear factorization, complete, 216
linear independence, 127–130, 545,
    561–563
    of eigenvectors, 463–465
    and homogeneous linear systems,
        130–131
    and spanning, 338–339
    of vectors, 330–331
linear isometry, 280
linear manifolds, 131
linear operators, 269–270, 582
    factoring, as a composition, 310–312
    geometry of, 280–293
    invertible, 312–315
    lines through the origin, orthogonal
        projections onto, 275–276
    lines through the origin, reflections
        about, 274–275
    matrix of, with respect to basis,
        443–446
    norm-preserving, 280–281
    origin, rotations about the, 273
    orthogonal, 281–285
    on R^3, 287–293
    representation of, with two bases,
        452–453
linear system(s), 40
    augmented matrix of, 43–44
    choosing an algorithm for solving,
        163–164
    coefficient matrix of, 82
    consistency of, 118–119, 138
    consistent, 40
    cost estimates for solving large, 163
    and diagonalization, 477
    of differential equations, 542–544
    elementary row operations for solution
        of, 43–44
    flops and cost of solving, 160–163
    general solution of, 51
    geometry of, 135–140
    homogeneous, see homogeneous linear
        systems
    homogeneous first-order, 543
    inconsistent, 40
    linear independence and homogeneous,
        130–131
    nonhomogeneous, see
        nonhomogeneous linear systems
    one-to-one and onto from viewpoint of,
        301–302
    overdetermined vs. underdetermined,
        364–365
    solution space of, 125–127

solutions to, 40–42
    solving, by factorization, 154–155
    solving, by matrix inversion, 114–117
    solving, by row reduction, 48–59
    solving, with common coefficient
        matrix, 118
    sparse, 241–246
    with triangular coefficient matrices,
        144
    with two or three unknowns, 40–42
linear transformation(s), 268–277
    compositions of, 305–314
    definition of, 269–270, 582
    effect of changing bases on matrices
        of, 452
    general, 582–588
    inverse of, 312–315
    kernel of, 296–297, 585–587
    lines through the origin, orthogonal
        projections onto, 275–276
    lines through the origin, reflections
        about, 274–275
    matrix of, with respect to pair of bases,
        450–452
    origin, rotations about the, 273
    and power sequences, 276–277
    properties of, 270–271
    from R^n to R^m, 271–273
    range of, 298–299, 585–587
    unifying theorem for, 302–303
    of unit square, 276
    vector spaces, involving, 582–588
link matrices, 256
LINPACK, 156
lower limit of summation, 27
lower triangular matrices, 144, 145, 149
    block, 169
LRC circuits, parallel, 554
LU-decomposition, 154–159, 198, 242
    determinant evalution by, 189
    and Gaussian elimination, 156,
        158–159
    matrix inversion by, 159

M
Maclaurin series, 476
magnitude (of a vector), 2, 16
main diagonal, 509
major axis, 485
mantissa, 160
Maple (computer algebra system), 61, 492
mapping, 265, 298
Markov, A. A., 228

Markov chains, 227–237
  definition of, 228
  long-term behavior of, 230–232
  and powers of transition matrix,
      229–230
  regular, 230–231
  steady-state vectors of, 231
Mathematica (computer algebra system),
      61, 492
MATLAB, 156, 492, 517–518
matrix(-ces), 12, 43, 79–90
  adjacency, 12, 256
  adjoint of, 196–197
  anticommutative, 533
  augmented, 43–44
  block diagonal, 168
  block lower triangular, 169
  block upper triangular, 168–170
  checkerboard, 224
  coefficient, *see* coefficient matrix(-ces)
  companion, 223
  connectivity, 318
  consumption, 236
  determinant of a, 99–101
  diagonal, 143–144, 460–461
  diagonalization of a, 462–463
  difference of, 80, 94
  dimension theorem for, 352–357
  Dirac, 533
  elementary, 109–111
  elementary row operations on, 43, 44
  entries of, 79
  equal, 80
  exponential of, 475–477
  Fibonacci, 184
  fixed points of a, 148
  fundamental spaces of, 342–344, 511
  Hessian, 496–497
  Hilbert, 123
  identity, 97–98
  for the inner product, 578–579
  inverse of a, 98–101
  invertibility of, 111–112
  invertible, 98–102, 104–105
  Leontief, 237
  of linear operator with respect to basis,
      443–446
  of linear transformation with respect to
      pair of bases, 450–452
  link, 256
  negative of, 81
  nilpotent, 109, 148–151
  nondiagonalizable, 551, 552
  nonsingular, *see* invertible matrix
  normal, 539
  operations on, 80–81
  orthogonal, 281–285
  partitioned, 166–170
  Pauli spin, 533
  permutation, 160

  positive definite, 490–492
  powers of a, 102
  product of, 81–88
  product of scalar and, 80–81, 94
  real, 526
  rotation, 289
  row-column rule for, 86
  scaling, 311
  similar, 456–460
  singular, 98
  size of, 79
  skew-Hermitian, 539
  skew-symmetric, 146–147
  square, of order n, 79–80
  square root of, 93
  stochastic, 227–228
  sum of, 80, 94
  symmetric, *see* symmetric matrix(-ces)
  for T with respect to the bases B and
      B', 451
  technology, 236
  trace of a, 89, 105–106
  transition, 228–231
  transpose of a, 88–89, 103–106
  triangular, *see* triangular matrices
  unitary, 536–539
  vertex, 318
  zero, 96–97
matrix inner products, 89, 90
matrix inversion
  algorithm for, 112–114
  LU-decomposition, 159
  solving linear systems by, 114–117
matrix operators, 267
matrix outer products, 89–90
matrix polynomials, 103
matrix spaces, 559
matrix theory, 529
matrix transformations, 267–268
  kernel of, 297–298
  range of, 299–300
maximum, relative, 495
maximum entry scaling, 252–254
mean
  arithmetic, 28
  geometric, 28
  sample, 493
mean square error, 576
meanhead, 589
Memoir on the Theory of Matrices
      (Arthur Cayley), 81, 99
method of simultaneous displacements,
      see Jacobi iteration
method of successive displacements, *see*
      Gauss–Seidel iteration
method of successive overrelaxation, 246
minimum, relative, 495
minimum distance problems, 393–394
Minkowski, Hermann, 574
Minkowski inequality, 574

minor axis, 485
minors, 179–180
Möbius, August Ferdinand, 3
modulus, 525
  of complex number, A5
Moler, C. B., 156
Moore–Penrose inverse, 518
multiplication, *see also* dot product;
          product(s)
  by A, 267
  block, 166–167
  closed under scalar, 123, 124
  column and row rules for matrix, 87
  of complex numbers, A6, A7
  of matrices, 80–88, 94–96
  scalar, 555
  of scalars, 4–5
multiplicative inverse, 98
  of a complex number, A5
multiplicity
  algebraic, 458, 465
  geometric, 458, 465

N
$n \times n$ determinants (nth-order
          determinants), 178
Napoleon, 577
natural isomorphisms, 590–591
n-dimensional Euclidean spaces, 25
negative(s), 556
  of a matrix, 81
negative definite quadratic forms, 488
negative semidefinite quadratic forms,
      489
network analysis, 65–67
networks, 65–66
Newton, Isaac, 270
nilpotent matrices, 109, 148–151
nodes, 65, 68
nondiagonalizable matrices, 551–552
nonhomogeneous linear systems,
      135–138
nonsingular matrix, *see* invertible matrix
nontrivial solutions, 56
norm, 15, 16, 570
normal matrices, 539
normal vectors, 31
normalization (of vectors), 16–17
norm-preserving linear operators,
      280–281
n-tuples, ordered, 7
null space, 297–298, 342, 344, 345
nullity, 342
numerical analysis, 48

O
ohms, 68
one-to-one correspondence, 428
one-to-one transformations, 300–302,
      587–588

one-way connections, 12
onto transformations, 300–302
open economies, 235–239
open networks, 65–66
open sectors, 235
operator(s), 267
    diagonalizable, 467
    identity, 267, 452, 583
    linear, 269–270, 582
    matrix, 267
optimization, application of quadratic
        forms to, 495–500
ordered n-tuples, 7
orientation (of axis of rotation), 290
origin, 7
    hyperplanes passing through, 139
    orthogonal projections onto lines
        through, 275–276
    reflections about lines through,
        274–275
    rotations about, 273
orthogonal bases, 407
orthogonal change of variable, 482
orthogonal complement(s), 139
    properties of, 344–345
orthogonal diagonalization, 469–473
    of symmetric matrix, 470–473
orthogonal diagonalization problem, 469
orthogonal matrices, 281–285
orthogonal operators, 280–285
orthogonal projection(s), 294, 379–390
    computing length of, 382
    finding dimension of the range of, 410
    onto general subspaces, 384–386
    onto lines in R^2, 379–380
    onto lines through the origin, 275–276
    onto lines through the origin of R^n,
        380–382
    matrices as representations of, 386–387
    orthonormal bases using, 408–410
    of R^n onto span{$\mathbf{a}$}, 382
    standard matrix for, 383–384
    onto $W^\perp$, 390
    of $\mathbf{x}$ onto span{$\mathbf{a}$}, 381
orthogonal similarity, 468–470
orthogonal vectors, 22, 406–407, 527, 570
orthonormal bases, 407–414
    coordinates with respect to, 430–431
    definition of, 407
    extending orthonormal sets to, 413–414
    finding, 411–413
    and Gram–Schmidt process, 413
    linear combinations of orthonormal
        basis vectors, 410–411
    orthogonal projections using, 408–410
    of T_n, 574–575
    transition between, 436, 437
orthonormal sets, 22, 23
orthonormal vectors, 22, 406–407
outer product rule, 168

outer products, matrix, 89–90
outputs, 235, 265
outside demand vectors, 236
overdetermined linear systems, 364–365

P
Page, Larry, 256
PageRank algorithm, 256
Pantone Matching System, 133
parallel lines/planes, 34
parallel LRC circuits, 554
parallel processing, 169
parallel vectors, 10
parallelogram, degenerate, 201–202
parallelogram equation for vectors, 24–25
parallelogram rule for vector addition, 3, 4
parameters, 29, 32, 124
parametric equations
    of the line, 29–30
    of the plane, 32–34
partial pivoting, 59, 61–62
particular solution (of consistent linear
        system), 137
partitioned matrices, 166–170
partitioning, 81, 166–170
Pauli spin matrices, 533
Peano, Giuseppe, 332, 556
percentage error, 255
permutation matrices, 160
perpendicular vectors, 21–22
perspective projection, 323
Piazzi, Giuseppe, 54
pivot columns, 53
pivot positions, 53
pivot theorem, 372
pivots, 160
pixels, 7
plane(s)
    complex, 525, A4–A5
    in n-dimensional spaces, 34–35
    parametric equations of, 32–34
    point-normal equations of, 31–32
    translation of, 32
    vector equation of, 32–34
PLU-decomposition, 160
point(s), 6
    critical, 495
    distance between, in vector space,
        17–18
    generalized, 7
    lines through two, 30–31
    saddle, 495
    trivial fixed, 210–211
    vanishing, 323
point-normal equations, 31–32
polar decomposition, 507–508
polar form, 525, A6
polarization identity, 281
polynomial interpolation, 72–74

polynomials
    characteristic, 216
    matrix, 103
    trigonometric, 574
portrait mode, 319
positive definite matrices, 490–492
positive definite quadratic forms, 488–489
positive semidefinite quadratic forms, 489
PostScript, 319
power method, 249–260
    with Euclidean scaling, 251–252
    Internet searches, application to,
        256–260
    inverse, 260, 262
    with maximum entry scaling, 252–254
    and rate of convergence, 255
    shifted inverse, 263
    and stopping procedures, 255–256
    variations of, 260
power sequence(s), 276–277
    generated by A, 249
power series representation, 150
power sources, 68
powers (of a matrix), 102
    diagonalizable matrices, 473–474
    eigenvalues of, 214
    and Markov chains, 229–230
preimage, 312
principal argument (of a complex
        number), A6
principal axes (of an ellipse), 486
principal axes theorem, 483
probability, 227
probability vectors, 227
The Problems of Mathematics (David
        Hilbert), 212
process color method, 133
product(s)
    of coefficient matrix and column
        vector, 81–84
    cross, 204–207
    determinant of, 188–189
    dot, *see* dot product
    elementary, 176–179
    linear combinations, matrix products
        as, 88
    of matrix and scalar, 80–81, 94
    matrix inner, 89, 90
    matrix outer, 89–90
    scalar triple, 208
    signed elementary, 176–179
    of triangular matrices, 145
    of two matrices, 84–88
    undefined, 85
production vectors, 236–237
productive open economies, 238–239
products from chemical reactions, 70
projection theorem for subspaces, 384
projections
    orthogonal, 294, 379–390
    perspective, 323

pseudoinverse, 518–522
defintion of, 518
finding, 519–520
and least squares, 521–522
properties of, 520–521
pure imaginary numbers, A3
Pythagoras, theorem of, 15–16

Q
QR-algorithm, 260
QR-decomposition, 417–420
Householder reflections for, 423–425
for least squares problems, 419–420
theorem of, 418
quadratic form(s), 481–490
application of, to optimization, 495–500
associated with, 482
change of variable in, 482–484
and conic sections, 485–490
definition of, 481–482
in geometry, 484–488
matrix notation for, 482
negative definite, 488
negative semidefinite, 489
positive definite, 488–489
quadratic surfaces (quadrics), 485
quotient (complex numbers), A5

R
range(s), 265
of linear transformation, 298–299, 585–587
of matrix transformations, 299, 300
rank
applications of, 367
and consistency, 362–364
definition of, 342
of matrices of the form A^TA and AA^T, 365
unifying theorems related to, 366–367
rank theorem, 360–362
Rayleigh, John, 253
Rayleigh quotient, 253
reactants, 70
real axis (of complex plane), A4
real inner product spaces, 569–572
real matrix, 526
real part, 525, A3
real vector spaces, 556
reciprocal (of a complex number), A5
rectangular coordinates, 5
recurrence relations, 242
reduced row echelon form, 48–55
and homogeneous linear systems, 57–58
uniqueness of, 53

reduced singular value decomposition, 512–513
reduced singular value expansion, 512–513
reflections
coordinate planes about, 289
about lines through the origin, 274–275
orthogonal linear operators as, 284–285
regular Markov chains, 230–231
relative error, 255
relative maximum, 495
relative minimum, 495
relativity, theories of, 212, 574
resistance, 68
resistors, 68
RGB space (RGB color cube), 11
right distributive law, 95
right singular vectors, 505
right-handed coordinate systems, 5
Robespierre, 577
roll (term), 309
roman fonts, 319
rosettes, 133
rotated out of standard position, 485
rotation(s)
axis of, 289–290
about origin, 273
orthogonal linear operators as, 284–285
in R^3, 289–293
rotation matrices, 289
roundoff error, 58
row echelon form, 49
row equivalence, 112
row operations, elementary, 185–186
row reduction, 48–59
row rule for matrix multiplication, 87
row vectors, 12, 81
row-column rule (dot product rule), 86
row-vector form (vector notation), 11

S
saddle point, 495
sample mean, 493
sample variance, 493
scalar(s), 1
multiplication of, 4–5
multiplication of vectors by, 8–10
product of matrix and, 80–81, 94
scalar multiple, 558
scalar multiplication (in vector spaces), 555
scalar triple products, 208
scaling
center of, 327
Euclidean, 251–252
scaling matrices, 311
scaling operator with factor k, 285, 286

Schmidt, Erhard, 412, 513
Schur, Issai, 478
Schur decomposition, 478
Schur's theorem, 478
Schwarz, Hermann, 23
screen coordinates, 323
search set, 256
second derivative test, Hessian form of, 496
sectors, 235
Seidel, Ludwig Philipp von, 244
shear(s), 286–287
in the x-direction with factor k, 286–287
in the xy-direction with factor k, 294
in the y-direction with factor k, 286–287
shifted inverse power method, 263
shifting, 260
sigma notation, 27
signed elementary products, 176–179
similar matrices, 456–460
eigenvectors/eigenvalues of, 458–460
properties shared by, 457
similarity, orthogonal, 468–470
similarity invariants, 457
simultaneous displacements, method of, *see* Jacobi iteration
singular matrix, 98
singular value decomposition (SVD), 502–516
for data compression, 513–514
and fundamental spaces of a matrix, 511
of nonsquare matrices, 509–511
and polar decomposition, 507–508
reduced, 512–513
of square matrices, 502–506
of symmetric matrices, 506–507
from transformation point of view, 514–516
singular value expansion, reduced, 512, 513
skew-Hermitian matrices, 539
skew-symmetric matrices, 146–147
solution(s)
of differentiable function, 542
general, 51
of linear systems, 40–42
nontrivial, 56
of system of differential equations, 543
trivial, 56
solution set, 40
solution space(s), 125–127, 297, 545
dimension of, 140, 333
geometric interpretation of, 139–140
space-time continuum, 7, 574

spanning, 338–339
spans, 124
sparse linear systems, 241–246
spectral decomposition, 471–473
sphere, unit, 490, 571
spot color method, 133
square matrix(-ces)
  invertibility of, 112
  of order n, 79–80
square root (of a matrix), 93
standard basis, 330, 562
standard forms, 485
standard matrix for T, 271
standard position, 484, 485
standard unit vectors, 17
state (of particle system), 8
state of the dynamical system, 225
state of the variable, 225
static equilibrium, 12
steady state, 248
steady-state vectors (of Markov chain),
  231
Stewart, G. W., 156
stochastic matrices, 227–228
stochastic processes, 227
stopping procedures, 256
Strang diagrams, 387–389
strictly diagonally dominant (square
  matrices), 245–246
strictly lower triangular matrices, 149
strictly triangular matrices, 149
strictly upper triangular matrices, 149
string theory, 7
subdiagonals, 478
submatrix(-ces), 166
  kth principal, 490
subspace(s), 123–125, 559–561, 564
  bases for, 329–333
  determining whether a vector is in a
    given, 347–349
  projection theorem for, 384
  as solution space of linear system,
    125–127
  of subspaces, 337
  translated, 131
  trivial, 124
  zero, 124
substitution, forward, 155
subtraction, *see also* difference
  of matrices, 80, 94
  of vectors, 4
successive displacements, method of, *see*
  Gauss–Seidel iteration
successive overrelaxation, method of, 246
sum
  of functions, 558
  of matrices, 80, 94

summation notation, 27
superposition principle, 270
SVD, *see* singular value decomposition
Sylvester, James, 81, 180
symmetric matrix(-ces), 146–148
  eigenvalue analysis of 2×2, 218–220
  orthogonal diagonalization of, 470–473
  positive definite, 490–491
  singular value decomposition of,
    506–507
symmetric rank 1 matrices, 357
symmetry property, 527
systems, linear systems, 241–246

T
Taussky-Todd, Olga, 529
technology matrices, 236
terminal point (of vector), 2
theorem of Pythagoras, 573
theorems, A1–A2
  contrapositive form of, A1
  converse of, A1–A2
  involving three or more statements, A2
thermodynamics, 248–249
3×3 determinants, 175–176
three-dimensional graphics, 323–325
3-space, rectangular coordinate system in,
  5–6
trace, 89
  expressions for, in terms of
    eigenvalues, 220–221
  properties of, 105–106
transformation(s), 265–266
  corresponding to A, 271
  evaluation, 583–584
  inverse , 594
  matrix, 267–268
  one-to-one, 300–302, 587–588
  onto, 300–302
  and singular value decomposition,
    514–516
  zero, 267, 583
transition matrices, 228–231
  between bases, 446–450
  finding, 434–435
  invertibility of, 434
translated subspaces, 131
translation, 32, 131
  vector addition as, 3
transpose, 88–89, 147
  conjugate, 535
  and dot product, 106
  properties of, 103–105
  of triangular matrices, 145
triangle inequality, 573
  for distances, 25
  for vectors, 24

triangle rule for vector addition, 3
triangular matrices, 144–146, 149
  block lower, 169
  block upper, 168–170
  determinants of, 179
  eigenvalues of, 214
trigonometric polynomials, 574
triple products, scalar, 208
trivial fixed points, 210–211
trivial solutions, 56, 543
trivial subspace(s), 124
Turing, Alan, 155
2×2 matrices
  determinants of, 175–176
  eigenvalue analysis of, 217–220
two-point vector equations, 30–31
2-space, rectangular coordinate system in,
  5–6
two-way connections, 12

U
undefined product, 85
underdetermined linear systems, 364–365
unified field theory, 7
uniqueness, 300
unit circle, 571
unit sphere, 490, 571
unit square, linear transformations of the,
  276
unit vector(s), 16–17
  in C^n, 527
unitarily diagonalizable square complex
  matrices, 538–539
unitary matrices, 536–539
  definition of, 536
  diagonalizability, unitary, 538–539
  properties of, 537–538
unknowns (linear systems), 40
unstable algorithms, 58
upper Hessenberg decomposition, 479
upper Hessenberg form, 478
upper limit of summation, 27
upper triangular matrices, 144, 145, 149
  block, 168–170

V
values (of a function), 265
Vandermonde, Alexandre Théophile, 203
Vandermonde determinants, 202–204
vanishing point, 323
variable(s)
  change of, in quadratic form, 482–484
  free, 50
  leading, 50
  state of the, 225

variance, sample, 493
vector(s), 1–13
    addition of, 2–4, 8
    angle between, 20–21
    bound, 2, 3
    in C^n, 525–532
    collinear, 10
    column, 12, 81
    components of, 5–7, 381
    consumption, 236
    in coordinate systems, 5–6
    direction of, 10
    dot product of, 18–20
    equivalent, 2, 8
    force, 1
    free, 2, 3
    generalized, 7
    intermediate demand, 237
    left singular, 509
    length of, 16
    linear combinations, 10–11
    lines/planes parallel to, 34
    magnitude of, 16
    multiplication of, by scalars, 8–10
    norm of, 15–16
    normal, 31
    normalization of, 16–17
    notation for, 2, 11
    orthogonal, 22, 406–407, 527, 570
    orthonormal, 22, 406–407
    parallel, 10
    perpendicular, 21–22
    probability, 227
    production, 236–237

    right singular, 505
    row, 12, 81
    scalars vs., 1
    standard unit, 17
    subtraction of, 4
    unit, 16–17
    in vector spaces, 555–557
    zero, 2
Vector Analysis (Edwin Wilson), 204
vector component
    of $\mathbf{x}$ along $\mathbf{a}$, 381
    of $\mathbf{x}$ onto span$\{\mathbf{a}\}$, 381
    of $\mathbf{x}$ orthogonal to $\mathbf{a}$, 381
vector equation(s)
    of the line, 29, 30
    of the plane, 32–34
vector space(s)
    axioms, 556
    in C^n, 527
    complex, 556
    definition of, 555–556
    dimension of, 563–564
    finite-dimensional, 564
    function spaces, 557–559
    general inner product spaces, 572–574
    infinite-dimensional, 564
    Lagrange interpolating polynomials as
        basis for, 566
    and linear independence, 561–563
    matrix spaces, 559
    real, 556
    real inner product spaces, 569–572
    subspaces of, 559–561
    unusual types of, 559
    vectors in, 555–557

velocity, 1
vertex matrices, 318
vertices, 12, 318
visible space, 7
voltage, 68–69
    drop, 68–69
    rise, 68–69
volts, 68

W
Waring, Edward, 566
weighted Euclidean inner product, 570
weighted least squares problems, 570
weighted least squares solutions, 570
weights, 570
    authority, 257
    hub, 257
Weyl, Herman, 513
Wilson, Edwin, 204
wireframes, 318–320
wires, 318
Wroński, Józef, 562, 563
Wronskians, 563
Wroński's test, 563

Y
yaw, 309

Z
zero matrix, 96–97
zero solution, 543
zero subspace(s), 124, 560
zero transformation, 267, 583
zero vector, 2, 7